STEEL STRUCTURES
Design and Behavior

The **Intext** *Series in*
Civil Engineering
Series Editor
Russell C. Brinker
New Mexico State University

Bouchard and Moffitt—**Surveying, 5th ed.**
Brinker—**Elementary Surveying, 5th ed.**
Clark, Viessman, and Hammer—**Water Supply and Pollution Control, 2d ed.**
Ghali and Neville—**Structural Analysis: A Unified Classical and Matrix Approach**
Jumikis—**Foundation Engineering**
McCormac—**Structural Analysis, 2d ed.**
McCormac—**Structural Steel Design, 2d Ed.**
Meyer—**Route Surveying and Design, 4th ed.**
Moffitt—**Photogrammetry, 2d ed.**
Salmon and Johnson—**Steel Structures: Design and Behavior**
Ural—**Matrix Operations and Use of Computers in Structural Engineering**
Wang—**Matrix Methods of Structural Analysis, 2d ed.**
Wang and Salmon—**Reinforced Concrete Design**
Winfrey—**Economic Analysis for Highways**

STEEL STRUCTURES
Design and Behavior

CHARLES G. SALMON
Professor of Civil Engineering
The University of Wisconsin
Madison, Wisconsin

JOHN E. JOHNSON
Associate Professor of Civil Engineering
The University of Wisconsin
Madison, Wisconsin

INTEXT EDUCATIONAL PUBLISHERS
College Division of Intext
Scranton San Francisco Toronto London

ISBN 0-7002-2341-X

Library of Congress Catalog Card Number 72-160683

Preface

The design of structural steel members has developed from a simple approach involving a few basic properties of steel and elementary mathematics to a sophisticated treatment demanding a thorough knowledge of structural behavior. Present design utilizes knowledge of mechanics of materials, structural analysis, and, particularly, structural stability, in combination with the use of design specifications to select standard steel sections or components from a list of those available. The 1969 American Institute of Steel Construction (AISC), "Specification for the Design, Fabrication, and Erection of Structural Steel for Buildings," is primarily used throughout this textbook. Steel members are selected from the AISC "Manual of Steel Construction."

Throughout the book effort has been made to present in a logical manner the theoretical background needed for developing and explaining design requirements, particularly those of the 1969 AISC Specification. The authors have striven to provide appropriate coverage of background material, including references to pertinent research, development of specific AISC Specification formulas, and a generous number of design examples explaining in detail the selection of minimum-weight members to satisfy given conditions.

Considerable emphasis has been put on presenting for the beginning, as well as the advanced student, the necessary elastic and inelastic stability concepts, the understanding of which is deemed essential to properly apply most of the AISC Specification formulas. This treatment is incorporated into the chapters in such a way that the reader may either study in detail the stability concepts in logical sequence, or omit or postpone study of articles containing the detailed development, merely accepting qualitative explanation and proceeding directly to the design aspect.

The development of theory and background material for the working-stress and plastic-design methods has generally been integrated throughout the book. This has been done because the majority of provisions of both methods are based on the strength at a load equal to a load factor times the service load. In some cases, plastic strength is achievable; in others, the strength may be less than plastic strength because of factors such as instability. The discussion of specific design provisions of AISC and the illustrative examples are, however, separated within the chapters, so that one may use either the working stress or the plastic design parts.

Depending on the proficiency required of the student, this textbook may provide material for two courses of three or four semester-credit hours each. It is suggested that the beginning course in steel structures for undergraduate students might contain the material of Chapters 1 through 7, 9, 10, 12, and 16, except Secs. 6.4, 6.6, 6.12 to 6.18, 7.9 to 7.11, 9.3, 9.4, 9.9, 9.10, 9.12, and 12.6 to 12.9. The second course would cover the same topics as in the first course, but much more rapidly; emphasizing items omitted in the first course. In addition, the remaining chapters—namely Chapter 8 on torsion, Chapter 11 on plate girders, Chapter 13 on connections, Chapter 14 on braced and unbraced frames, and Chapter 15 on frame design—are suggested for inclusion.

The reader will need to have ready access to the AISC Manual throughout the study of the text, particularly when working with the examples. However, it is not the objective of this text that the reader become proficient in the routine use of tables; the tables serve only as a guide to obtaining experience with variation of design parameters and as an aid in arriving at good designs. The AISC Specification and Commentary are contained in the AISC Manual and are therefore not included in the textbook, except for various individual provisions quoted where they are explained.

Experienced teachers know well that the majority of students must work a sufficient number of numerical exercises in order to grasp and retain the basic concepts. For this purpose many examples are shown in detail so as to give students a reference when solving problems provided at the ends of the chapters.

The 1969 AISC Specification has been used throughout, along with some reference to 1963 requirements where needed to give historical support for 1969 revisions. Some references are also made to the highway bridge specification of the American Association of State Highway Officials (AASHO), but no attempt has been made to give complete coverage to that specification.

The 1970 revised standard nomenclature adopted by the American Iron and Steel Institute has been used throughout the book. The wide-flange sections previously referred to as WF shapes are now W-shapes, such as 12WF50 becomes W12 × 50. The American standard sections, formerly called I-beams, become S-shapes; i.e., the 12I50.8 becomes S12 × 50.8. The standard designation for plates was changed from width × thickness × length to thickness × width × length. The numerous other changes in nomenclature and regrouping of some of the rolled shapes have been incorporated throughout the text. In general, however, the shallowest wide-flange section used in the design examples has been the one of 8-in. depth.

In the examples, the selection of members used design tables, charts, and section properties from the 1963 AISC Manual, since the 1970 Manual was not yet available. It is intended that the reader use the 1970 edition, and topical references to the manual correctly correspond to the 1970 edition; however, some dimensional properties used in examples differ slightly from those of the new edition. The differences occurring in the 1970 edition are the result principally of rounding off or reducing the number of significant figures stated. Computation of sectional properties, such as moment of inertia, torsional constant, and r_T for lateral stability, give slightly different values depending on the method used. Such differences are not considered important, particularly when actual dimensional tolerances are expected and slide-rule accuracy is considered adequate for computations.

Special features to be found in the text are; (a) comprehensive treatment of design of I-shaped members subject to torsion (Chapter 8), including a simplified practical method; (b) detailed treatment of plate girder theory (Chapter 11) and a comprehensive design example of a two-span continuous girder utilizing two different grades of steel; (c) extensive treatment of connections (Chapter 13), including significant discussion and illustration of the design of components, in addition to high-strength bolts (Chapter 4) and welds (Chapter 5) as connectors; and (d) presentation of braced and unbraced frame stability analyses (Chapter 14), using stiffness and flexibility coefficients in order to clarify the concept of effective length for compression members in frames.

The authors are indebted to their students who have been using the chapters as they were completed, thus providing a challenging shakedown to help eliminate errors and improve wording so that the book may better serve the students who follow.

The authors are particularly indebted to their families for their patience and encouragement. Appreciation is due Mrs. John E. Johnson who typed several of the chapters. Special credit is given to Mrs. Charles G. Salmon who typed the majority of the manuscript, without whose constant encouragement the senior author would have been unable to see the project to completion. The authors therefore affectionately dedicate this book to their wives.

Madison, Wisconsin
August, 1971

Charles G. Salmon
John E. Johnson

Contents

1 Introduction . . . 1

2 Structural Steels . . . 29

3 Tension Members . . . 62

7 Beams: Laterally Supported . . . 322

8 Torsion . . . 371

1

Introduction

1.1. STRUCTURAL DESIGN

Structural design may be defined as a mixture of art and science, combining the experienced engineer's intuitive feeling for the behavior of a structure with a sound knowledge of the principles of statics, dynamics, mechanics of materials, and structural analysis, to produce a safe economical structure which will serve its intended purpose.

Until about 1850, structural design was largely an art relying on intuition to determine the size and arrangement of the structural elements. Early man-made structures essentially conformed to those which could also be observed in nature; such as, beams and arches. As the principles governing the behavior of structures and structural materials have become better understood, design procedures have become more scientifically oriented.

Computations involving scientific principles should serve as a *guide* to decision making and not be followed blindly. The art or intuitive ability of the experienced engineer is utilized to make the decisions, guided by the computational results.

1.2. PRINCIPLES OF DESIGN

Design is a process by which an optimum solution is obtained. In this text the concern is with the design of structures—in particular, *steel* structures. Whatever the material, a design for obtaining an optimum solution exists. The more complex the design, the more difficult is the achievement of an optimum design.

In any design, certain criteria must be established to evaluate whether or not an optimum has been achieved. For a structure, typical criteria may be (a) minimum cost; (b) minimum weight; (c) minimum construction time; (d) minimum labor; (e) minimum cost of manufacture of owner's products; (f) maximum efficiency of operation to owner. Usually

Alcoa Building, San Francisco. Courtesy Bethlehem Steel Corporation.

several criteria are involved, each of which may require weighting. Observing the above possible criteria, it may be apparent that setting clearly measurable criteria (such as weight and cost) for establishing an optimum frequently will be difficult, and perhaps impossible. In most practical situations the evaluation must be qualitative.

If a specific objective criterion can be expressed mathematically, then optimization techniques may be employed to obtain a maximum or minimum for the objective function. Optimization procedures and techniques comprise an entire subject which is outside the scope of this text. The criterion of minimum weight is emphasized throughout, under the general assumption that minimum material represents minimum cost. Other subjective criteria must be kept in mind, even though the integration of behavioral principles with design of structural steel elements in this text utilizes only simple objective criteria, such as weight or cost.

Design Procedure. The design procedure may be considered as composed of two parts—functional design and structural framework design. Functional design is the design which insures that the intended results are achieved such as (a) providing adequate working areas and clearances; (b) providing for ventilation and/or air conditioning; (c) adequate transportation facilities, such as elevators, stairways, and cranes or materials handling equipment; (d) adequate lighting; and (e) exhibiting architectural attractiveness.

The structural framework design is the selection of the arrangement and sizes of structural elements so that service loads may be safely carried.

The iterative design procedure may be outlined as follows:

(a) *Planning.* Establishment of the functions for which the structure must serve. Set criteria against which to measure the resulting design for being an optimum.

(b) *Preliminary structural configuration.* Arrangement of the elements to serve the functions in (a).

(c) *Establishment of the loads* to be carried.

(d) *Preliminary member selection.* Based on the decisions of (a), (b), and (c) selection of the member sizes to satisfy an objective criterion, such as least weight or cost.

(e) *Analysis.* Structural analysis to ascertain whether members selected are safe, but not excessively so. This would include checking of all strength and stability factors for members and their connections.

(f) *Evaluation.* Are all requirements satisfied and is the result optimum? Compare the result with the predetermined criteria.

(g) *Redesign.* Repetition of any part of the sequence (a) through (f) found necessary or desirable as a result of evaluation. Steps (a) through (f) represent an iterative process. Usually in this text only steps (c) through (f) will be subject to this iteration since the structural configuration and external loading will be prescribed.

(h) *Final Decision.* The determination of whether or not an optimum design has been achieved.

1.3. HISTORICAL BACKGROUND OF STEEL STRUCTURES

Metal as a structural material began with cast iron, used on a 100-ft arch span which was built in England in 1777–1779.[1] A number of cast-iron bridges were built during the period 1780–1820, mostly arch-shaped with main girders consisting of individual cast-iron pieces forming bars or trusses. Cast iron was also used for chain links on suspension bridges until about 1840.

Wrought iron soon after 1840 began replacing cast iron, the earliest important example being the Brittania Bridge over Menai Straits in

England which was built in 1846–50. This was a tubular girder bridge having spans 230—460—460—230 feet, which was made from wrought-iron plates and angles.

The process of rolling various shapes was developing as cast iron and wrought iron received wider usage. Bars were rolled on an industrial scale beginning about 1780. The rolling of rails began about 1820 and was extended to I-shapes by the 1870's.

The developments of the Bessemer process (1855), the introduction of a basic liner in the Bessemer converter (1870), and the open-hearth furnace brought widespread use of iron ore products in building material. Since 1890 steel has replaced wrought iron as the principal metallic building material. Currently, (1971), steels are available for structural uses which have yield stresses varying from 24,000 to 100,000 pounds per square inch. The various steels, their uses and their properties are discussed in detail in Chapter 2.

1.4. LOADS

The accurate determination of the loads to which a structure or structural element will be subjected is not always predictable and frequently more difficult than the design of the members. Even if the loads are well known at one location in a structure, the distribution of load from element to element throughout the structure usually is obtained by making various assumptions and approximations. Some of the most common kinds of loads are discussed in the following sections.

Dead Load. Dead load is a gravity load, since it acts toward the earth and is always acting. The weight of the structure is considered dead load. Also, dead load includes attachments to the structure such as pipes, electrical conduit, air-conditioning and heating ducts, lighting fixtures, floor covering, roof covering, and suspended ceilings which remain throughout the life of the structure.

Dead loads are usually known accurately but not until the design has been completed. Under steps (c), (d), and (e) of the design procedure discussed in Sec. 1.2, the weight of the structure or structural element must be estimated, preliminary section selected, weight recomputed, and member selection revised if necessary. The dead load of attachments is usually known with reasonable accuracy prior to the design.

Live Load. Ordinary loads, including gravity loads, which vary in magnitude and location are termed *live loads*. Examples of live loads are human occupants, furniture, movable equipment, vehicles and stored goods. Some live loads may be practically permanent, others may be highly transient. Because of the unknown nature of the weight, location, and density of live load items, realistic magnitudes and the positions of such loads are very difficult to determine.

Because of the public concern for adequate safety, minimum values of live loads (to be considered) are usually prescribed by state and local building codes. The live loads prescribed in such codes are generally empirical and conservative, based on previous experience and accepted practice rather than accurately computed values. Wherever local codes do not apply, or do not exist, the provisions from one of several regional and national building codes may be used. One such widely recognized code is The American Standard Building Code (ASBC),[2] sponsored by the National Bureau of Standards, 1955. Typical live loads from this code are given in Table 1.4.1. The table is presented to give the reader the relative magnitudes of representative loads, and is not intended to include all situations.

TABLE 1.4.1
Typical Uniformly Distributed Live Loads
(Adapted from Ref. 2)

Occupancy or Use	Live Load, lb/ft^2
1. Hotel guest rooms	40
School class rooms	
Private apartments	
Hospital private rooms	
2. Assembly halls, fixed seat	60
Library reading rooms	
3. Offices	80
4. School corridors	100
Theater aisles, and lobbies	
Assembly halls, movable seats	
Office building lobbies	
Main floor, retail stores	
Dining rooms and restaurants	
Automobile garages	
5. Storage warehouse, light	125
Manufacturing	
6. Library stack rooms	150
7. Storage warehouse, heavy	250
Sidewalks, driveways, subject to trucking	

Live load when applied to the structure should be positioned to give maximum effect, including requiring partial loading, alternate span loading, or full loading as may be necessary. The simplified assumption of full uniform loading everywhere should be used with the understanding that it is approximate. The probability of having the prescribed loading uniformly over an entire floor, as well as over all other floors of a building simultaneously, is almost nonexistent. Most codes recognize this by allowing for some percentage reduction for full loading. For instance, for live loads of 100 psf or less the ASBC–1955 allows the design live load on any member supporting 150 square feet or more to be reduced at the rate of 0.08 percent per square foot of area supported by the member, except in public assembly areas. The greatest benefit from such a reduction will occur in the columns of a tall building where the total area tributary to columns over ten or twenty stories may be very large. The ASBC limits the reduction to

$$R = 100 \frac{D + L}{4.33L} = 60 \tag{1.4.1}$$

where D = dead load in psf, and L = live load in psf.

Highway Live Loads. Highway vehicle loading in the United States has been standardized by the American Association of State Highway Officials (AASHO)[3] into standard truck loads and lane loads which approximate a series of trucks. There are two systems, designated H and HS, which are identified by the number of axles per truck. The H system has two axles, while the HS system has three axles per truck. Altogether there are five classes of loading: H10, H15, H20, HS15, and HS20. The loading is shown in Fig. 1.4.1.

In designing a given bridge, either one equivalent truck loading is applied to the entire structure, *or* the equivalent lane loading is applied. When the lane loading is used the uniform portion is distributed over as

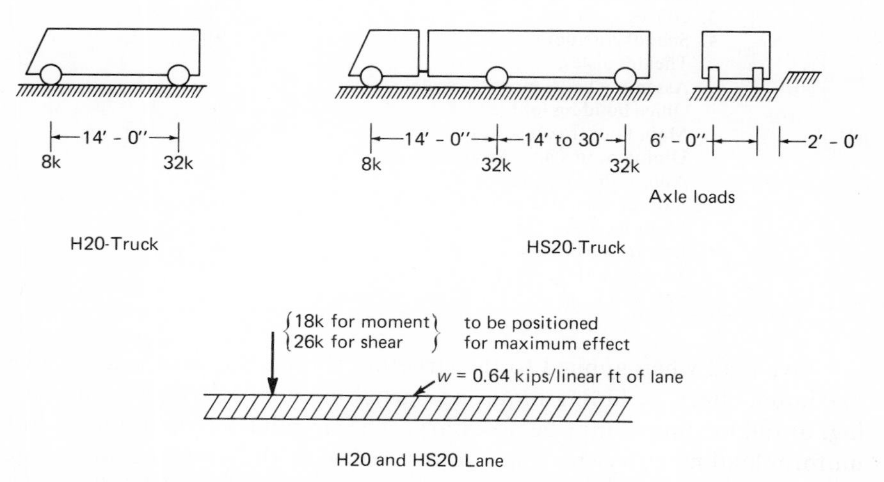

Fig. 1.4.1. AASHO–1969 Highway H20 and HS20 loadings (for H15 and HS15 use 75 percent of H20 and HS20, and for H10 use 50 percent of H20).

much of the span or spans as will cause the maximum effect. In addition, the one concentrated load is positioned for the greatest effect. On continuous structures, in determining maximum negative moment at the support only, an additional concentrated load must be used in a span other than the position of the first one. The load distribution across the width of a bridge to its various supporting members is taken in accordance with semiempirical rules which depend on the type of bridge deck and supporting structure.

The single truck loading provides the effect of a heavy concentrated load and usually governs on relatively short spans. The uniform lane load is to simulate a line of traffic, and the added concentrated load is to account for the possibility of one extra heavy vehicle in the line of traffic. These loads have been used with no apparent difficulty since 1944, before which time a line of trucks was actually used for the loading. On the interstate system of highways a military loading is also used which consists of two 24-kip axle loads spaced 4 ft apart.

Railroad bridges are designed to carry a similar semiempirical loading known as the Cooper E72 train, consisting of a series of concentrated loads a fixed distance apart followed by uniform loading. This loading is prescribed by American Railway Engineering Association (AREA).[4]

Impact. The term impact as ordinarily used in structural design refers to the dynamic effect of a suddenly applied load. In the building of a structure the materials are added slowly; people entering a building are also considered a gradual loading. Dead loads are static loads; i.e., they have no effect other than their weights. Live loads may be either static or they may have a dynamic effect. Persons and furniture would be treated as static live load, but cranes and various types of machinery also have dynamic effects.

Consider the spring-mass system of Fig. 1.4.2a where the spring may be thought of as analogous to an elastic beam. When load is gradually ap-

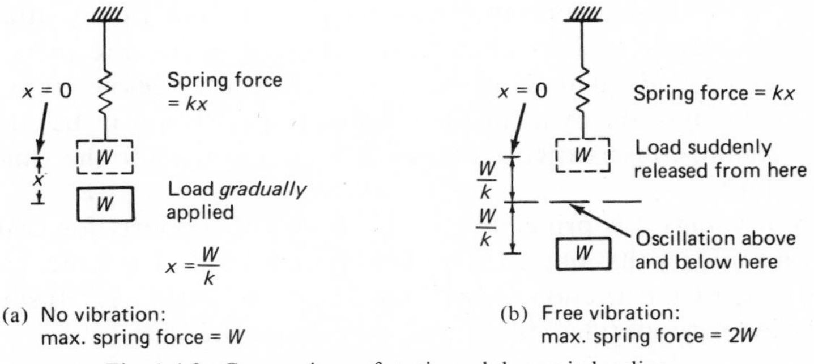

Fig. 1.4.2. Comparison of static and dynamic loading.

plied (i.e., static loading) the mass (weight) deflects an amount x and the load on the spring (beam) is equal to the weight, W. In Fig. 1.4.2b the load is suddenly applied (dynamic loading) and the maximum deflection is $2x$; i.e., the maximum load on the spring (beam) is $2W$. In this case the mass vibrates in simple harmonic motion with its neutral position equal to its static deflected position. In real structures the harmonic (vibratory) motion is damped out (reduced to zero) very rapidly. Once the motion has stopped the force remaining in the spring is the weight, W. To account for the increased force during the time the member is in motion a load equal to twice the static load W should be used—add 100 percent of the static load to represent the dynamic effect. This is called a 100 percent impact factor.

Any live load which can have a dynamic effect should be increased by an impact factor. While a dynamic analysis of a structure could be made to accurately determine these effects, such a procedure is usually too complex or too costly in ordinary design. Thus empirical formulas and impact factors are usually used. In cases where the dynamic effect is small (say where impact would be less than about 20 percent) it is ordinarily accounted for by using a conservative (higher) value for the specified live load. The dynamic effects of persons in buildings and of slow-moving vehicles in parking garages are examples where ordinary design live load is conservative and no explicit impact factor is usually added.

For highway bridge design however, impact is always to be considered. AASHO-1969[3] prescribes empirically that the impact factor expressed as a portion of live load is

$$I = \frac{50}{L + 125} \tag{1.4.2}$$

but not greater than 30 percent (0.30). In Eq. 1.4.2, L expressed in feet is the length of the portion of the span which is loaded to give maximum stress in the member. Since vehicles travel directly on the superstructure, all parts of it are subjected to vibration and must be designed to include impact. The substructure, including all portions not rigidly attached to the superstructure such as abutments, retaining walls, and piers, are assumed to have adequate damping or be sufficiently remote from the application point of the dynamic load so that impact is not to be considered there. Again, conservative static loads may account for the smaller dynamic effects.

In buildings it is principally in the design of supports for cranes and heavy machinery that impact is explicitly considered. The American Institute of Steel Construction (AISC) Specification[5] (AISC–1.3.3) states that if not otherwise specified, the impact percentage shall be:

For supports of elevators...100%

For traveling crane support girders and their connections 25%
For supports of light machinery, shaft or motor driven, not less than 20%
For supports of reciprocating machinery or power driven units, not less than 50%
For hangers supporting floors and balconies 33%

In the design of crane runway beams (see Fig. 1.4.3) and their connections, the horizontal forces caused by moving crane trolleys must be

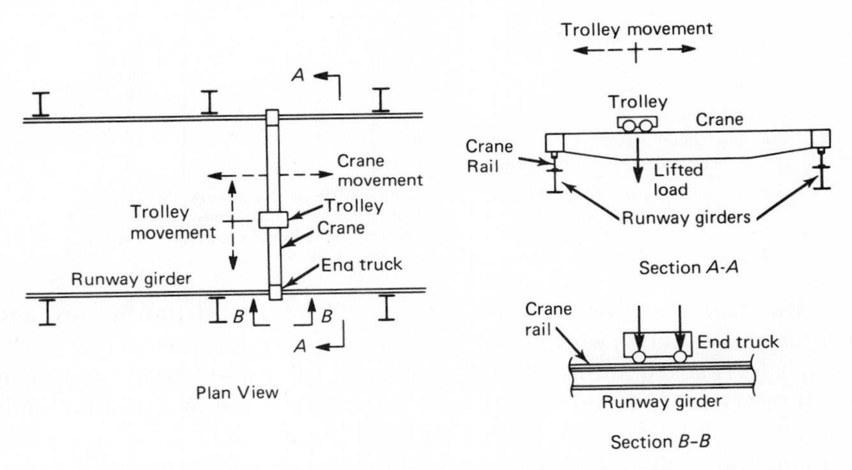

Fig. 1.4.3. Crane arrangement, showing movements which contribute impact loading.

considered. AISC–1.3.4 prescribes using "20 percent of the sum of the weights of the lifted load and of the crane trolley (but exclusive of other parts of the crane)." It also states that "The force shall be assumed to be applied at the top of the rail, one-half on each side of the runway, and shall be considered as acting in either direction normal to the runway rail."

In addition, due to acceleration and deceleration of the entire crane, a longitudinal force is transmitted through friction of the end truck wheels with the crane rail to the runway girder. AISC–1.3.4 says "the longitudinal force shall, if not otherwise specified, be taken as 10 percent of the maximum wheel loads of the crane applied at the top of the rail."

Snow Load. The live loading for which roofs are designed is either totally or primarily a snow load. Since snow has a variable specific gravity, even if one knows the depth of snow for which design is to be made, the weight per square foot of roof is at best only a guess. Snow loads used by various building codes are essentially empirical since little theoretical basis for snow load design has been established.

Probably the most complete study of snow loads is that by the U.S. Housing and Home Finance Agency in 1952.[6] Table 1.4.2 gives suggested minimum uniformly distributed loads based on that study.

TABLE 1.4.2
Minimum Design Snow Loads; Uniformly Distributed on Horizontal Projection (Adapted from Ref. 6)

Region	Loading, psf			
	Slope of Roof			
	3 in 12 or less	6 in 12	9 in 12	12 in 12 or more
Southern states	20	15	12	10
Central states	25	20	15	10
Northern states	30	25	17	10
Great Lakes, New England*	40	30	20	10

*Includes northern Michigan, western and northern New York, Massachusetts, Vermont, New Hampshire, and Maine. These loads also apply to mountain areas; above 2,000 ft in the Appalachians, above 1,000 ft in the mountains near the Pacific Coast, and the Rocky Mountain areas above 4,000 ft.

Reference 6 suggests a specific gravity of 0.1 for newly fallen snow and 0.2 for packed snow, which gives about 6 pcf for fresh dry snow and about 12 pcf for packed snow. The 40 psf from Table 1.4.2 would represent $6\frac{2}{3}$ ft of dry newly fallen snow or over 3 ft of packed snow. Design loads from Table 1.4.2 or from various building codes should serve as a guide for minimum values, but the designer should try to ascertain actual snowfall conditions on which to base the design. In many sections of the country, the minimum values have proven to be insufficient when adding the extra weight of persons attempting to remove the snow.

It is apparent that the steeper the roof the less snow can accumulate. Also partial snow loading should be considered, in addition to full loading, if it is believed such loading can occur and would cause maximum stresses. Wind may also act on a structure which is carrying snow load. It is certainly unlikely, however, that maximum snow and wind loads would act simultaneously. Flat roofs in normally warm climates should be designed for 20 psf even when such accumulation of snow may seem doubtful. This loading may be thought of as due to people gathered on such a roof. Furthermore, though wind is frequently ignored as a vertical force on a roof, nevertheless it may cause such an effect. For these reasons a 20-psf minimum loading, even though it may not always be snow, is reasonable.

Wind Load. All structures are subject to wind load, but it is usually only those more than three or four stories high, other than long bridges, for which special consideration of wind is required. The aerodynamics of wind on various shapes and from various directions has received considerable attention from researchers, but is outside the scope of this text. The reader is referred to the Reports of the Task Committee on Wind Forces[7] where the purpose was to "assemble, correlate, and summarize existing information on the factors that determine wind forces on structures." The Final Report[7] also contains an extended bibliography.

On any typical building of rectangular plan and elevation, wind exerts pressure on the windward side and suction on the leeward side, as well as either uplift or downward pressure on the roof. For most ordinary situations vertical roof loading from wind is neglected on the assumption that snow loading will require a greater strength than wind loading. Furthermore, the total lateral wind load, windward and leeward effect, is assumed to be applied to the windward face of the building.

In accordance with Bernoulli's theorem for an ideal fluid striking an object, the increase in static pressure equals the decrease in dynamic pressure, or

$$q = \frac{1}{2}\rho V^2 \tag{1.4.3}$$

where q = dynamic pressure on the object
ρ = density of air
V = wind velocity

In order to account for the shape of the structure exposed to wind, the dynamic pressure must be multiplied by a coefficient to obtain the equivalent static pressure. Thus

$$p = C_D q \tag{1.4.4}$$

where p = equivalent static pressure

C_D = coefficient to account for shape of structure, nearby topographic features, and gust effects

The Task Committee[7] suggests that for buildings with plane surfaces normal to the wind a coefficient C_D of 1.3 may represent a typical building. Using in Eq. 1.4.3 p = 0.07651 pcf, corresponding to air at sea level and 15°C, and velocity in miles per hour gives

$$p = C_D\left(\frac{1}{2}\right)\left(\frac{0.07651}{32.2}\right)\left(\frac{5280V}{3600}\right)^2 = 0.002558 C_D V^2$$

and applying C_D = 1.3 gives

$$p = 0.00332 V^2 \tag{1.4.5}$$

where p is average pressure on projected elevation of building in psf, and V is design wind velocity in miles per hour.

The commonly used wind pressure of 20 psf, as specified by many building codes, corresponds to a velocity of 77.8 miles per hour from Eq. 1.4.5. Equation 1.4.5 assumes an airtight building, and should not be used when significant openings are present. Figure 1.4.4 shows the relation between horizontal wind pressure and wind velocity for winds of up to 120 miles per hour with C_D taken as unity.

For buildings with nonplanar surfaces, plane surfaces inclined to the wind, or having significant openings, more extensive wind investigation should be made using, for example, the Task Committee Report.[7]

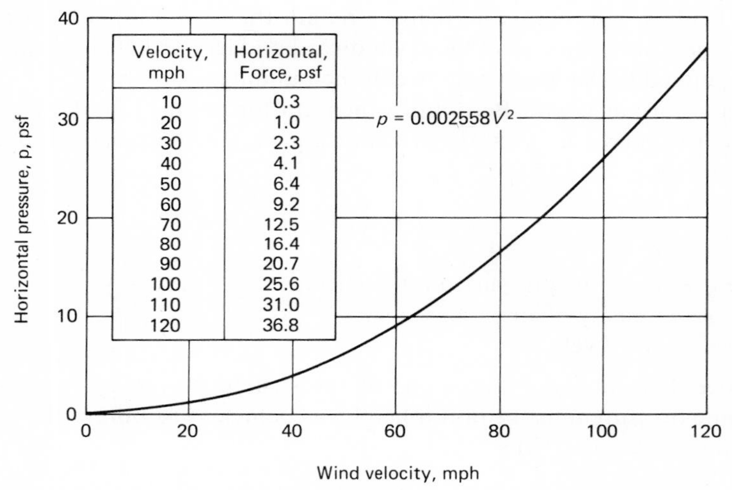

Velocity, mph	Horizontal, Force, psf
10	0.3
20	1.0
30	2.3
40	4.1
50	6.4
60	9.2
70	12.5
80	16.4
90	20.7
100	25.6
110	31.0
120	36.8

Fig. 1.4.4. Horizontal wind pressure as a function of wind velocity (C_D = 1.0).

Earthquake Load. An earthquake consists of horizontal and vertical ground motions, with the vertical motion usually having much the smaller magnitude. Since the horizontal motion of the ground causes the most significant effect it is that effect which is usually thought of as earthquake load. When the ground under an object (structure) having a certain mass suddenly moves, the inertia of the mass tends to resist the movement, as shown in Fig. 1.4.5. A shear force is developed between the ground and the mass. The simple illustration of Fig. 1.4.5 greatly simplifies the behavior; nevertheless, most building codes having earthquake provisions require the designer to consider a lateral force CW which is usually empirically prescribed. The dynamics of earthquake action on structures is outside the scope of this text, but the designer is referred to Refs. 10, 11, and 12.

One of the most commonly used codes is that of the Structural

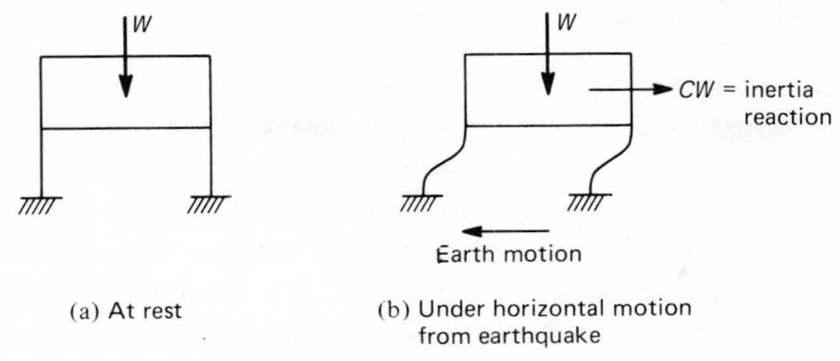

Fig. 1.4.5. Force developed by earthquake.

Engineers Association of California (SEAOC). In its 1959 recommendations the shear force to designed for is

$$V = KCW \tag{1.4.6}$$

where V = base shear to represent the dynamic effect of the inertia force
W = weight of building
C = $0.05/\sqrt[3]{T}$, the seismic coefficient, equivalent to the maximum acceleration in terms of acceleration due to gravity
T = natural period of the structure, i.e., time for one cycle of vibration
K = coefficient varying from 0.67 to 1.50, which indicates capacity of the members to absorb plastic deformation (low values indicate high ductility).

The total lateral load is recommended[13] to be distributed in accordance with the following:

$$F_n = \frac{W_n h_n}{\Sigma Wh} V \tag{1.4.7}$$

where F_n = lateral at the nth-floor level
W_n = weight at the nth-floor level
h_n = height above ground of the nth-floor level
ΣWh = total sum of Wh for all floor levels

If the natural period T cannot be determined by a rational means from technical data, it may be assumed to be

$$T = 0.05H/\sqrt{D} \tag{1.4.8}$$

where H = height of the building above its base
D = dimension of the building in the direction parallel to the applied forces (D to be in the same units as H)

The foregoing discussion based on the SEAOC is presented to show how earthquake loads may be treated empirically by code procedures.

1.5. TYPES OF STRUCTURAL STEEL MEMBERS

As discussed in Sec. 1.2 the function of a structure is the principal factor determining the structural configuration. Using the structural configuration along with the design loads, individual elements, or components, are selected to properly support and transmit loads throughout the structure. Steel members are selected from among the standard rolled shapes adopted by the American Institute of Steel Construction. Of course, welding permits combining plates and/or other rolled shapes to obtain any shape the designer or architect may require.

Typical rolled shapes, the dimensions for which are found in the AISC Manual,[14] are shown in Fig. 1.5.1. The AISC Manual will be fre-

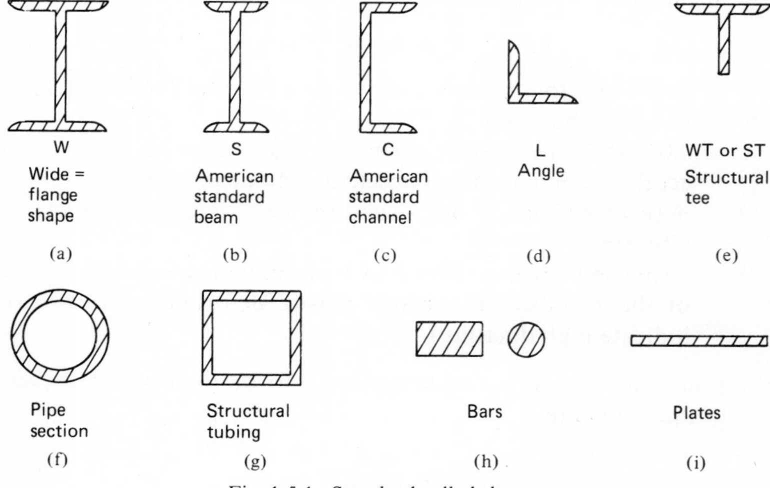

Fig. 1.5.1. Standard rolled shapes.

quently referred to throughout this text and should be available to the reader. The most commonly used section is the wide-flange shape (Fig. 1.5.1a) which is formed by hot rolling in the steel mill. The wide-flange shape is designated by the nominal depth and the weight per foot, such as a W18 × 96 which is nominally 18 in. deep (actual depth = 18.16 in. according to AISC Manual) and weighs 96 pounds per foot. Two sets of dimensions are found in the AISC Manual, one set stated in decimals for the designer to use in computations, and another set expressed in fractions ($\frac{1}{16}$ in. as the smallest increment) for the detailer to use on plans and shop drawings. Rolled W shapes are also designated by ASTM in accordance with web thickness as Groups I through V, with the thinnest web sections in Group I.

The American Standard beam (Fig. 1.5.1b) commonly called the

I-beam, has relatively narrow and sloping flanges and a thick web compared to the wide-flange shape. Use of most I-beams has become relatively uncommon in recent years because of excessive material in the web and relative lack of lateral stiffness due to the narrow flanges.

The channel (Fig. 1.5.1c) and angle (Fig. 1.5.1d) are commonly used either alone or in combination with other sections. The channel is designated, for example, as C12 × 20.7, a nominal 12 inch deep channel having a weight of 20.7 pounds per foot. Angles are designated by their leg length (long leg first) and thickness, such as, L6 × 4 × $\frac{3}{8}$.

The structural tee (Fig. 1.5.1e) is made by cutting wide-flange or I-beams in half and is commonly used for chord members in trusses. The tee is designated, for example, as WT5 × 44.5, where the 5 is the nominal depth and 44.5 is the weight in pounds per foot; this tee being out from a W10 × 89.

Pipe sections (Fig. 1.5.1f) are designated "standard," "extra strong," and "double-extra strong" in accordance with the thickness and are also nominally prescribed by diameter; thus, 10 in. diam. double-extra strong is an example of a particular pipe size.

Structural tubing (Fig. 1.5.1g) is used where pleasing architectural appearance is desired with exposed steel. Tubing is designated by outside dimensions and thickness, such as structural tubing, 8 × 6 × $\frac{1}{4}$.

The sections shown in Fig. 1.5.1 are all hot-rolled; that is, they are formed from hot billet steel (blocks of steel) by passing through rolls numerous times to obtain the final shapes.

Many other shapes are cold-formed from thin plate material (light-gage steel) having a thickness of less than $\frac{3}{16}$-in. as shown in Fig. 1.5.2.

Regarding size and designation of light-gage steel members, there are no truly standard shapes even though the properties of many common shapes are given in the Light-Gage Cold-Formed Steel Design Manual.[15] Various manufacturers produce many varieties of proprietary shapes.

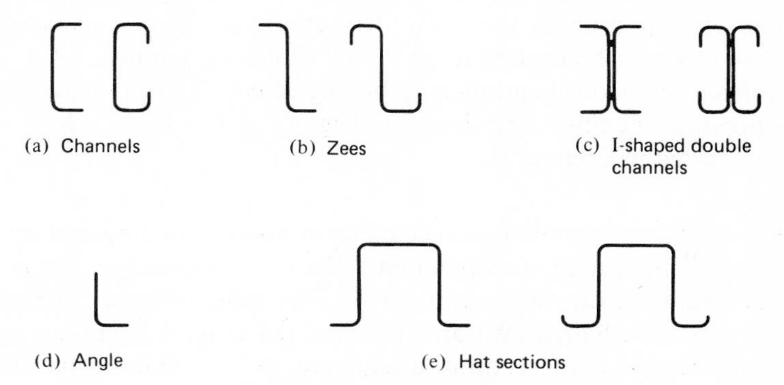

Fig. 1.5.2. Some cold-formed shapes.

Tension Members. The tension member occurs commonly as a chord member in a truss, as diagonal bracing in many types of structures, as direct support for balconies, as cables in suspended roof systems, and as suspension bridge main cables and suspenders which support the roadway. The behavior of tension members is not dependent on the shape or distribution of its area but is particularly related to its connection. Typical cross sections of tension members are shown in Fig. 1.5.3, and their design (except for special factors relating to suspension-type cable supported structures) is treated in Chapter 3.

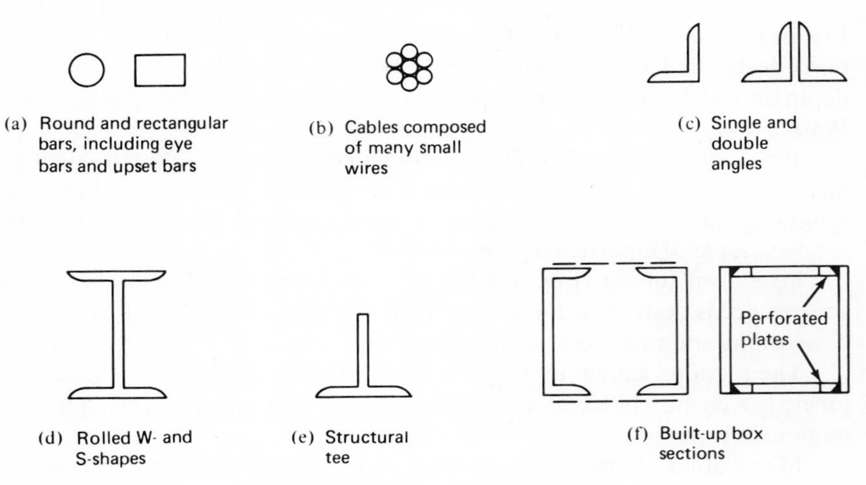

Fig. 1.5.3. Typical tension members.

Compression Members. Since compression member strength is a function of the cross-sectional shape (radius of gyration), the area is generally spread out as much as is practical. Chord members in trusses, and many interior columns in buildings are examples of members subject to axial compression. Even under the most ideal condition, pure axial compression is not attainable; so, design for "axial" loading assumes the effect of any small simultaneous bending may be neglected. Typical cross sections of compression members are shown in Fig. 1.5.4 and their behavior and design is treated in Chapter 6.

Beams. Beams are members subjected to transverse loading and are most efficient when their area is distributed so as to be located at the greatest practical distance from the neutral axis. The most common beams sections are the wide-flange (W) and I-beams (S) (Fig. 1.5.5a), as well as smaller rolled I-shaped sections designated as "miscellaneous shapes" (M).

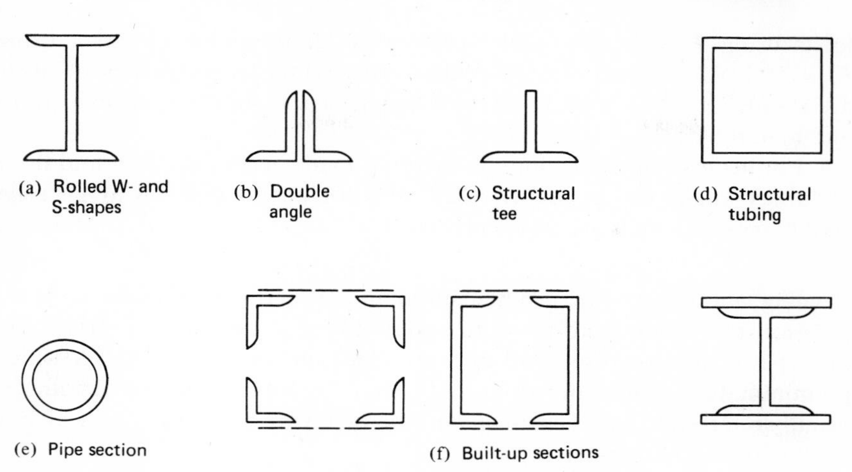

Fig. 1.5.4. Typical compression members.

For deeper and thinner-webbed sections than can economically be rolled, welded I shapes (Fig. 1.5.5b) are used; including stiffened plate girders.

For moderate spans carrying light loads, open-web "joists" are often used (Fig. 1.5.5c). These are designated "J-Series" using material with a minimum yield point stress of 36 ksi, "H-Series" using chord material with a minimum yield point stress of 50 ksi; "LJ-Series" longspan steel

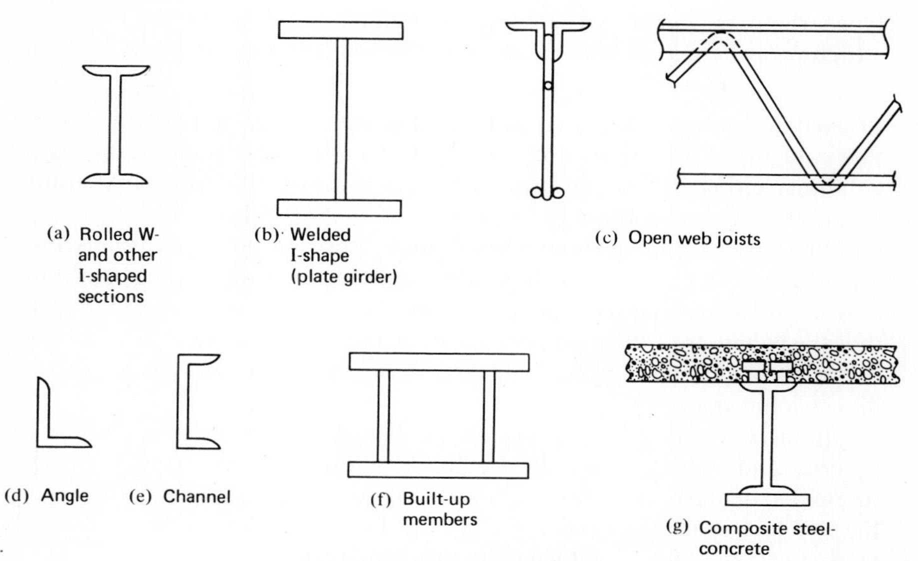

Fig. 1.5.5. Typical beam members.

joists using $F_y = 36$ ksi; and "LH-Series" longspan steel joists using $F_y = 50$ ksi for chord material. The cross-sectional properties of joists are standardized by various manufacturers using standard specifications and load tables.[16]

For beams (known as lintels) carrying loads across window and door openings, angles are frequently used; and for beams (known as girts) in wall panels, channels are frequently used.

Bending and Axial Load. When simultaneous action of tension or compression along with bending occurs, a combined stress problem arises and the type of member used will be dependent on the type of stress which predominates. A member subjected to axial compression and bending is usually referred to as a *beam-column*, the behavior and design of which is dealt with in Chapter 13.

The aforementioned illustrations of types of members to resist various kinds of stress is intended only to show common and representative types of members and not to be all inclusive.

1.6. STEEL STRUCTURES

Structures may be divided into three general categories: (a) framed structures, where elements may consist of tension members, columns, beams, and members under combined bending and axial load; (b) shell-type structures, where axial stresses predominate and, (c) suspension-type structures, where axial tension predominates the principal support system.

Framed Structures. Most typical building and bridge construction is in this category. The multistory building usually consists of beams and columns, either rigidly connected or having simple end connections along with diagonal bracing to provide stability. While the multistory building is three-dimensional, when designed for rigid joints it usually has much greater stiffness in one direction than another so that it may reasonably be treated as a series of plane frames. However, if the framing is such that the behavior of the members in one plane substantially influences the behavior in another plane, the frame must be treated as a three-dimensional space frame.

Industrial buildings and special one-story buildings such as churches, schools, and arenas, generally are either wholly or partly framed structures. Particularly the roof system may be a series of plane trusses (see Fig. 1.6.1), a space truss (see Fig. 1.6.2), a dome (see Fig. 1.6.3), or it may be part of a rigid flat or gabled one-story rigid frame.

Bridges are mostly framed structures, such as continuous beams and

Fig. 1.6.1. Floor joists (plane trusses) and steel decking.

plate girders (see Fig. 1.6.4); or trusses, usually continuous. (See Fig. 1.6.5).

Most of this text is devoted to behavior and design of elements in framed structures.

Shell-type Structures. In this type of structure the shell serves a use function in addition to participation in carrying loads. One common type where the main stress is tension is the containment vessel used to store liquids (for both high and low temperatures), of which the elevated water tank is a notable example. Storage bins, tanks, and the hulls of ships are other examples. On many shell-type structures, a framed structure may be used in conjunction with the shell.

On walls and flat roofs the "skin" elements may be in compression while they act together with a framework. The aircraft body is another such example.

Shell-type structures are usually designed by a specialist and are not within the scope of this text.

Fig. 1.6.2. Space truss roof for Upjohn Company Office Building, Kalamazoo, Michigan. Courtesy Whitehead and Kales Company, Detroit.

Suspension-type Structures. In the suspension-type structure tension cables are major supporting elements. A roof may be cable supported, as shown in Fig. 1.6.6. Probably the most common structure of this type is the suspension bridge, as shown in Fig. 1.6.7. Usually a subsystem of the structure consists of a framed structure, as in the stiffening truss for the suspension bridge. Since the tension element is the most efficient way of carrying load, structures utilizing this concept are coming into increasing use.

Many unusual structures utilizing various combinations of framed, shell-type, and suspension-type structures have been built. However, the typical designer must principally understand the design and behavior of framed structures. When encountering special structures he must seek the help of specialists.

1-7. SPECIFICATIONS AND BUILDING CODES

Structural steel design of buildings in the United States is principally based on the specifications of the American Institute of Steel Construction.[5] The American Institute of Steel Construction (AISC) is composed

Fig. 1.6.3. Dome roof, Brown University auditorium. Courtesy Bethlehem Steel Corporation.

Fig. 1.6.4. Continuous orthotropic plate girder, Poplar Street Bridge, St. Louis, Missouri. Courtesy Bethlehem Steel Corporation.

Fig. 1.6.5. Continuous truss bridge. Outerbridge Crossing, Staten Island, New York.

Fig. 1.6.6. Cable suspended roof for Madison Square Garden Sports and Entertainment Center, New York. Courtesy Bethlehem Steel Corporation.

of steel fabricator and manufacturing companies, as well as individuals interested in steel design and research. The AISC Specification is the result of the combined judgment of researchers and practicing engineers. The research efforts have been synthesized into practical design pro-

Fig. 1.6.7 Newport Bridge, Rhode Island. Courtesy Bethlehem Steel Corporation.

cedures to provide a safe, economical structure. The advent of the digital computer into design practice has generally made feasible more elaborate specification rules, a trend which began with the adoption of the 1961 AISC specification and has continued with the introduction of the 1969 specification. Throughout this text unless otherwise indicated, specification sections referred to by number are from the 1969 edition.

A specification containing a set of rules is intended to insure safety; however, the designer must understand the behavior for which the rule applies, otherwise an absurd, a grossly conservative, and sometimes unsafe design may result. The authors contend that it is virtually impossible to write rules which fully apply to every situation. Behavioral understanding must come first; application of rules then follows. No matter what set of rules is applicable, *the designer has the ultimate responsibility for a safe structure*.

A specification when adopted by AISC is actually a set of recommendations put forth by a highly respected group of experts in the field of steel research and design. Only when governmental bodies, such as city, state, and federal agencies, who have legal responsibility for public safety, adopt or incorporate a specification such as AISC–1969 into their building codes does it become legally official.

The design of steel bridges is generally in accordance with specifications of the American Association of State Highway Officials (AASHO).[3] This becomes a legal set of rules since it has been adopted by the States (usually the state highway departments have this responsibility). The 1969 edition is the most recent one, but yearly supplements with changes are usually approved.

Railroad bridges are designed in accordance with the specifications adopted by the American Railway Engineering Association (AREA),[4] which is approved and used by the various railroads. In this case the railroads have the responsibility for safety and through their own organization adopt the rules to insure safe designs.

The term *building code* is sometimes used synonomously with specifications. More correctly a building code is a broadly based document, either a legal document such as a state or local building code, or a document widely recognized even though not legal which covers the same wide range of topics as the state or local building code. Building codes generally treat all facets relating to safety, such as structural design, architectural details, fire protection, heating and air conditioning, plumbing and sanitation, and lighting. On the other hand, specifications frequently refer to rules set forth by the architect or engineer which pertain to only one particular building while under construction. Building codes also ordinarily prescribe standard loads for which the structure is to be designed, as discussed in Sec. 1.4.

The reader should not be disturbed by the interchangeable use of building code and specification, but should clearly understand that which is legally required for design and that which could be thought of as recommended practice.

1.8. WORKING-STRESS DESIGN AND PLASTIC DESIGN

Two philosophies of design are in current use. One of them, the working-stress method has been the principal one used during the past eighty years. Using this method, one designs a structural element so that stresses computed under the action of working, or service, loads do not exceed some predesignated allowable values. These allowable stresses are prescribed by a building code or specification (such as AISC–1969) to provide a factor of safety against attainment of some limiting stress. Most often the upper limiting stress is the minimum specified yield strength of the steel. Thus, stresses computed under the action of working loads are well within the elastic range; i.e., stresses are proportional to strains.

The other method is the ultimate-strength method, usually referred to as plastic design for steel structures. The plastic-design approach has come into wide use during the past fifteen years. In plastic design, the service loads are first increased by some factor (usually referred to as the load factor) to obtain the desired maximum strength and the structure is then designed to have that maximum strength. Because steel is a ductile material, achievement of a yield condition at one location in a continuous structure does not limit the capacity; but rather, the stresses due to additional load are distributed differently throughout the structure. Moment redistribution is introduced in Chapter 7 and treated in detail in Chapter 10. Once a structure has no further ability to carry an increased load it is said to have reached a "collapse mechanism." The ductility of steel which permits yielding to occur without a decrease in strength is defined as plastic behavior; hence the term *plastic design* is used when the resulting plastic strength is consciously considered in the design process.

It is to be noted that regardless of which of the above philosophies is employed in a design, the actual stresses under service load are in the *elastic range*. Furthermore, criteria other than strength, such as instability, excessive deflection, and fatigue, frequently control the design. Any one or a combination of these factors may provide a criterion for the limit of structural usefulness. For steel structures, where thin elements such as flanges and webs are commonly subject to compressive stress, instability is a major factor in design. Chapter 6 and succeeding chapters place strong emphasis on buckling, or instability, problems as they relate to (a) achievement of a yield condition as is the primary objective in the working-stress method, and (b) achievement of large strains after reaching a yield condition as is the main objective in plastic design.

Historically, the maximum strength was the earliest method for design of most structures because the ultimate load (maximum strength) could be measured by test without a knowledge of the magnitude or distribution of internal stresses. With the interest in and understanding of the elastic methods of analysis in the early 1900's, the elastic working-stress method was adopted almost universally by specification and building codes as the best for design. Since steel is a material with essentially perfect elasticity up to the yield point, it seems ideally suited for such method. As more detailed understanding was gained of the actual behavior of steel structures subjected to loads in excess of the service load, adjustments in the theory and the design procedures were made. Adjustments in the working-stress method have continued until today nearly all of the 1969 AISC Specification is actually based on maximum strength.

The authors believe the term *maximum strength* best defines the underlying concepts in both working-stress and plastic-design methods, since it inherently includes consideration of the major factors of elastic and inelastic stability.

The design of connectors and connections under the working-stress method has always been based on maximum strength, since to do otherwise would have required a highly complex analysis. Column design has been also based on maximum strength since the earliest specifications.

With the adoption of the 1961 AISC specification, working-stress design of beams became based more on maximum strength and the design of plate girders was completely revised to be based on maximum strength.

The plastic design method was first approved by AISC in 1958 and by 1971 the method has become widely used.

It would appear that design procedures based on maximum strength, using load factors applied to service load moments, is the logical approach for the future, whether one expects as the limit for structural usefulness, yielding, instability, or full plastic collapse strength.

1.9. FACTOR OF SAFETY

Structures and structural members must always be designed to carry some reserve load beyond that which is expected under normal use. The reserve capacity is provided to account primarily for the possibility of overload; in addition, however, it also accounts for the possibility of understrength. Deviations in the dimensions of rolled sections, even though within accepted tolerances, can result in understrength. Occasionally a steel section will have a yield-point stress slightly below the minimum specified value; thus, giving a reduced strength. Overloads can arise from changing the use for which a particular structure was designed, from underestimation of the effect of loads by oversimplifications in calculation procedures; and from variations in erection procedures. Normally, the possible drastic change of use is not explicitly considered nor intended to be covered by the load factor; however, erection procedures known to cause particular stress conditions must be considered in design and not left for the load factor, or safety factor, to take into account.

The safety that is required for design is really a combination of economics and statistics. Obviously it will not be economically feasible to design a structure so that it will be impossible for it to fail—that is, it is not feasible to design so that there is zero probability of failure. The load factor, or safety factor, is intended to limit the probability below a certain reasonable level.

The apparently simple task of defining what is meant by the term "factor of safety" was a principal goal of the ASCE Task Committee on Factors of Safety during the period 1956–66. The Final Report[17] of the Task Committee summarizes concepts necessary to a full understanding of structural safety and its relationship to theory of probability. The Final Report[17] indicates "the Committee has not been successful in its efforts to resolve the 'factor of safety' question, it is the belief ... that the probability approach deserves considerable more study than it has received." More recently, safety requirements for structures based on probability have been proposed,[18] which are intended "to enhance realism and improve consistency in the code treatment of uncertainty of both the measurable and the 'professional' kinds." Measurable uncertainty refers to such things as statistical differences in material strengths and strengths related to dimensional tolerances. "Professional" kinds of uncertainty refers to things seldom found in statistical observations, such as incomplete knowledge of structural performance, or situations in which it is not economically feasible to apply an "exact" analysis.

Most building codes have not identified the various factors which are involved in determining the safety requirements. One may state that the minimum resistance must exceed the maximum applied load by some prescribed amount. Suppose the actual load (or design load) exceeds the ordinary design load by an amount ΔS, and the actual resistance is less than the usual expected resistance by an amount ΔR. A structure that is just adequate would have

$$R - \Delta R = S + \Delta S$$

$$R(1 - \Delta R/R) = S(1 + \Delta S/S) \tag{1.9.1}$$

The degree of safety, or safety factor, would be the ratio of the nominal strength to nominal design load, R/S; or

$$\text{FS} = \frac{R}{S} = \frac{1 + \Delta S/S}{1 - \Delta R/R} \tag{1.9.2}$$

Equation 1.9.2 illustrates the effect of overload ($\Delta S/S$) and undercapacity ($\Delta R/R$); however, it does not identify all the factors which may contribute to either. If one assumes that the occasional overload ($\Delta S/S$) may be 40 percent greater than its mean value, and that an occasional understrength ($\Delta R/R$) may be 15 percent less than its mean value; then

$$\text{FS} = \frac{1 + 0.4}{1 - 0.15} = \frac{1.4}{0.85} = 1.65$$

Obviously even if the percentage variations are correct they only relate to a certain probability of the variation occurring. It does *not* mean there is zero probability of having greater variations.

A treatment of statistics and probability is outside the scope of the text, but the reader is referred to Refs. 17 and 18 for extended treatment.

The 1969 AISC uses FS = 1.67 as the basic value for the working stress method, and uses FS = 1.7 as the value for plastic design; i.e., essentially the same value for both. Dividing capacities by 1.67 gives a multiplier of 0.6 (a convenient multiple) in the working stress method. In plastic design, the loads are multiplied by the factor 1.7; a convenient value.

SELECTED REFERENCES

1. Hans Straub, *A History of Civil Engineering*, M.I.T. Press, Cambridge, Mass., 1964, pp. 173–180.
2. "American Standard Building Code Requirements for Minimum Design Loads in Buildings and Other Structures," American Standards Association, ASA A58.1–1955.
3. *Standard Specifications for Highway Bridges*, American Association of State Highway Officials, 10th ed., Washington, D.C., 1969.
4. "Specifications for Steel Railway Bridges," American Railway Engineering Association, Chicago, Ill., 1965.
5. "Specification for the Design Fabrication and Erection of Structural Steel for Buildings," American Institute of Steel Construction, New York, adopted Feb. 12, 1969.
6. "Snow Load Studies," Housing Research Paper No. 19, U.S. Housing and Home Finance Agency, Washington, D.C., May 1952.
7. "Wind Forces on Structures," Task Committee on Wind Forces, Committee on Loads and Stresses, Structural Division, ASCE, Preliminary Reports, *J. Structural Div. ASCE*, Vol. 84, No. ST4 (July 1958); and Final Report, *Trans. ASCE*, Vol. 126, pt. II (1961), pp. 1124–1198.
8. ASCE Subcommittee No. 31, "Wind Bracing in Steel Buildings," *Trans. ASCE*, Vol. 105 (1940), pp. 1713–1738.
9. H. C. S. Thom, "New Distributions of Extreme Winds in the United States," *J. Structural Div. ASCE*, Vol. 94, No. ST7 (July 1968), pp. 1787–1801.
10. "Lateral Forces of Earthquake and Wind," Joint Committee of San Francisco Section, ASCE, and Structural Engineers Association of Northern California, *Trans. ASCE*, Vol. 117 (1952), pp. 716–780. (Includes extensive bibliography.)
11. John M. Biggs, *Introduction to Structural Dynamics*, McGraw-Hill Book Company, New York, 1964, Chap. 6.
12. C. H. Norris et al., *Structural Design for Dynamic Loads*, McGraw-Hill Book Company, New York, 1959, Chaps. 16–18.
13. *Recommended Lateral Force Requirements*, Seismology Committee, Structural Engineers Association of California, San Francisco, July 1959.
14. *Manual of Steel Construction*, 7th ed., American Institute of Steel Construction, Inc., New York, 1970.
15. *Light Gage Cold-Formed Steel Design Manual*, American Iron and Steel Institute, New York, 1968.
16. "Standard Specifications and Load Tables," 1969 ed., Steel Joist Institute, Washington, D.C.
17. Alfred M. Freudenthal, Jewell M. Garrelts, and Masanobu Shinozuka, "The Analysis of Structural Safety," *J. Structural Div. ASCE*, Vol. 92, No. ST1 (February 1966), pp. 267–325.
18. C. Allin Cornell, "A Probability-Based Structural Code," *ACI J. Proc.*, Vol. 66 (December 1969), pp. 974–985.

2

Structural Steels

2.1. INTRODUCTION

During most of the period from the introduction of structural steel as a major building material until about 1960, the steel used was classified as a carbon steel with the ASTM (American Society for Testing and Materials) designation A7, and had a minimum specified yield stress of 33 ksi. Most designers merely referred to "steel" without further identification, and the AISC specification prescribed allowable stresses and procedures only for the A7 type of steel. Other structural steels, such as a special corrosion resistant low alloy (A242) and a more readily weldable steel (A373), were available but they were rarely used in buildings. Bridge design made occasional use of these other steels.

The many steels available to the designer permit him to increase the strength of the material in highly stressed regions rather than greatly increase the size of members. He can decide whether maximum rigidity or least weight is the more desirable attribute. Corrosion resistance, hence elimination of frequent painting, may be a highly important factor. Some steels now oxidize to form a dense protective coating that prevents further oxidation (corrosion), acquiring a pleasing even-textured dark red-brown appearance. Since painting is not required, it may be economical to use these weathering steels even though the initial cost is somewhat higher than traditional carbon steels.

Certain steels provide better weldability than others; some are more suitable than others for pressure vessels, either at temperatures well above or well below room temperatures.

Structural steels are referred to by ASTM designations, and also by many proprietary names. For design purposes the yield stress is the quantity which specifications, such as AISC, use as the material property variable to establish allowable unit stresses under various types of member loading. The term *yield stress* is used to include either "yield point," the well-defined deviation from perfect elasticity exhibited by most of the common structural steels; or "yield strength," the unit stress at a certain

Fig. 2.1. IBM Building, Pittsburgh. Courtesy United States Steel Corporation.

offset strain for steels having no well-defined yield point. In 1971, steels are readily available with yield stresses from 24 ksi to 100 ksi. Figure 2.1.1 shows the various ASTM designated structural steels and their range of yield stresses (many of which vary with thickness) and tensile strengths. Pressure vessel quality steels are not shown.

In the following sections, the various ASTM designated steels are discussed, followed by general consideration of the important mechanical properties.

2.2. CARBON STEELS

Carbon steel is the term applied to steels containing the following maximum percentages of elements other than iron: (a) carbon, 1.7, (b)

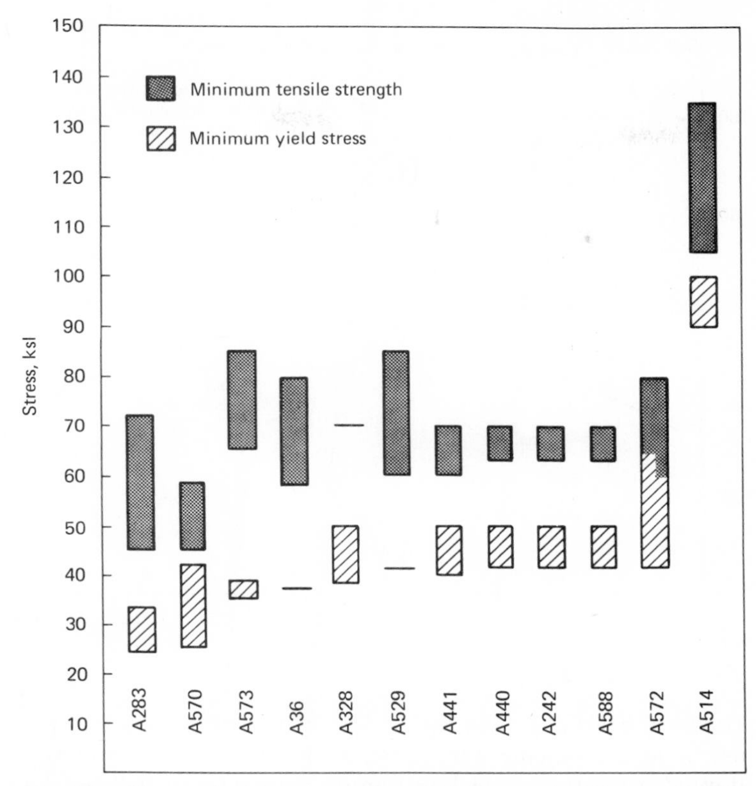

Fig. 2.1.1. Range of minimum specified yield stress and minimum specified tensile strength for various ASTM specified structural steel. (From Ref. 2.)

manganese, 1.65, (c) silicon, 0.60, and (d) copper, 0.60. Carbon and manganese are the main elements to increase strength over that of pure iron. The category includes material from ingot iron containing essentially no carbon to cast iron which has at least 1.7 percent carbon. These steels are divided into four categories: low carbon (less than 0.15 percent); mild carbon (0.15–0.29 percent); medium carbon (0.30–0.59 percent); and high carbon (0.60–1.70 percent). Structural carbon steels are in the mild-carbon category; a steel such as A36 has maximum carbon, varying from 0.25 to 0.29 percent, depending on thickness. These structural carbon steels exhibit marked yield points as shown in curve (a) of Fig. 2.2.1. Increased percentage of carbon raises the yield stress but reduces ductility. Reduction in ductility creates increased problems with welding. Satisfactory economical welding without preheat, postheat, or special welding electrodes can usually be accomplished when carbon content does not exceed 0.30 percent. The various carbon steels have had increased limitations on carbon content as the need for good weldability has increased.

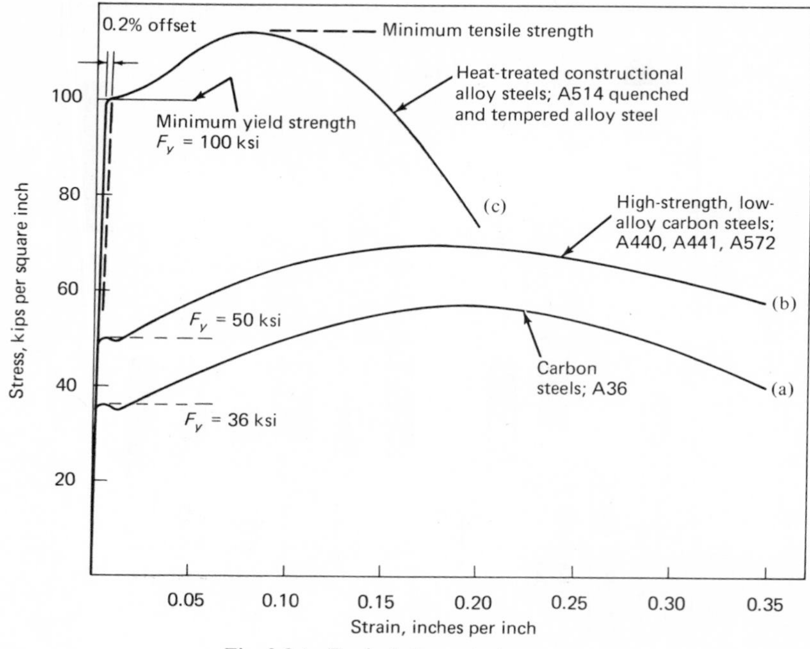

Fig. 2.2.1. Typical stress-strain curves.

A36, Structural Steel (F_y = 36 ksi). This is the primary structural steel for construction, replacing the previous primary steel known as A7. First adopted in 1960 as a steel with more consistent properties than A7, A36 had limitations imposed on carbon, manganese, sulfur, and phosphorus, and on silicon in plates exceeding 1½ in. in thickness. The need for better weldability for bridge construction led to a revision in 1962 to further restrict the carbon and manganese content. Subsequently, A36 steel was approved by the U.S. Bureau of Public Roads for welded bridges. Later revisions removed an earlier limitation of an 8-in. maximum thickness (min. F_y = 32 ksi for these plates thicker than 8 in.).

Where high strength-to-weight ratios are not important, and where bulk for rigidity is desired, A36 is usually the best choice. It is easily welded or bolted, and is available in the large variety of standard shapes, as well as in nearly any plate width and thickness.

A245, Flat-Rolled Carbon Steel Sheets of Structural Quality (F_y = 25 to 33 ksi). This steel has been used for light gage shapes, including steel joists.

A283, Low and Intermediate Tensile Strength Carbon Steel Plate of Structural Quality ($F_y \leqq$ 33 ksi). For use in machine and equipment manufacture, and for oil and water storage vessels, this steel which is usually available only in plates up to 15 in. thick, is cheaper than A36. It is lower strength carbon steel with no specified content for carbon or manganese. It is available in Grades A, B, C, and D, for minimum specified yield stresses of 24, 27, 30 and 33 ksi, respectively. The Grade D is essentially the same as the old A7 steel. This steel is used where rigidity is highly important and lower strength is adequate.

A501, Hot-Formed Welded and Seamless Carbon Steel Structural Tubing (F_y = 36 ksi). Since about 1961 structural tubing sections have become widely used, particularly for exposed columns where appearance is important. This tubing steel has essentially the same properties as A36. The tubular cross-sectional shapes have outstanding strength properties in proportion to their weight. They have good lateral stability as beams and columns, and excellent torsional strength. Connections have typically been complex. Riveted or bolted connections for such sections are not feasible and welded connections have been more expensive than welded connections on other standard shapes because of the extra labor involved. Savings in weight, desire for the improved appearance, and ready availability of standard sections have made tubing a widely used material.

A53, Welded and Seamless Steel Pipe (F_y = 25 to 35 ksi). This material includes seamless and welded black and hot-dipped galvanized steel pipe in three types: (a) Type F, furnace-butt welded; (b) Type E, electric-resistance welded, Grades A and B; and (c) Type S, seamless, Grades A and B. Furnace-butt-welded pipe is produced in continuous lengths, having its longitudinal butt-joint forge welded by the mechanical pressure during the rolling process. Electric-resistance welded pipe is also produced in continuous lengths and has its longitudinal joint formed by a coalescence produced by the combination of heat obtained from resistance of the pipe to the flow of electric current in a circuit of which the pipe is a part, and by applied pressure.

Pipe sections are defined by diameter and as standard-weight, extra-strong-weight, and double-extra-strong-weight, in accordance with the wall thickness.

The increasing use of pipe for structural purposes is the result of the same factors discussed for structural tubing (A501).

A573, Structural Carbon Steel Plates of Improved Toughness (F_y = 35 and 38 ksi). This is a plate material for welded structures, such as vessels serving at ordinary temperatures, where improved toughness (the resistance to fracture in the presence of a notch) is important. A573 is available in two grades and for plates up to 1½ in. thick.

A529, Structural Steel with 42,000 psi Minimum Yield Point. **(F_y = 42 ksi).** This higher strength carbon steel is available in plates only to ½-in. thick and for the smaller rolled shapes (designated Structural Group 1 by ASTM specification A6). It is widely used in structural joists and rigid frames which are manufactured in standard sizes.

A285, A515, and A516, Pressure Vessel Steels (F_y = 30 to 38 ksi). These are plate steels furnished in *pressure-vessel quality* only and are intended for critical applications such as welded pressure vessels. A515 and A516 are furnished in four grades, Grades 55, 60, 65, and 70 (indicating tensile strength), corresponding to minimum yield stresses of 30, 32, 35, and 38 ksi, respectively. A515 is for "intermediate and higher temperature service" and A516 is for "atmospheric and lower temperature service."

A570, Hot-Rolled Carbon Steel Sheets and Strip, Structural Quality (F_y = 25 to 33 ksi). This is material under ½-in. thick, widely used as cold-formed shapes in industrial, residential, and commercial buildings.

A307, Low-Carbon Steel Externally and Internally Threaded Standard Fasteners. This material is used for what are commonly referred to as "machine bolts." These are usually used only for temporary installations. Included are Grade A bolts for general applications, which have a *minimum* tensile strength of 55 ksi; and Grade B bolts for flanged joints in piping systems where one or both flanges are cast iron. The Grade B bolts have a *maximum* tensile strength limitation of 90 ksi. No well-defined yield point is exhibited by these bolts, and no minimum yield strength (for instance, 0.2 percent offset strength) is specified.

A325, High-Strength Bolts for Structural Steel Joints Including Suitable Nuts and Plain Hardened Washers. This quenched and tempered medium carbon steel is used for bolts commonly known as "high-strength structural bolts," or high-strength bolts. This material has maximum carbon of 0.30 percent. It is heat-treated by quenching and then by reheating (tempering) to a temperature of at least 800°F. (The effects of this heat treatment are discussed in more detail later in the chapter). This steel behaves in a tension test more similarly to the heat treated low alloy steels (Sec. 2.4) than to carbon steel. It has an ultimate tensile strength of 105 ksi (1⅛ to 1½ in. diam. bolts) to 120 ksi (½ to 1 in. diam. bolts). Its yield strength, measured at 0.2 percent offset, is prescribed at 81 ksi minimum for 1¼, 1⅜, and 1½ in. dia. bolts. It will be higher for the smaller diameter bolts, but a minimum value is not prescribed.

A233, Mild Steel Covered Arc Welding Electrodes (Also AWS A5.1). This specification covers electrodes classified E60XX and several E70XX categories. The "E" denotes electrodes. The first two digits indicate the tensile strength in kips per square inch; thus the tensile strength is 60 or 70 ksi which is high enough for welding of most carbon steels. The third digit indicates the welding positions for which it can be used; the digit "1" indicates welding in all positions. The last digits refer to whether a-c or d-c current is to be used and whether it is to be single or reverse polarity.

A558, Bare Mild Steel Electrodes and Fluxes for Submerged Arc Welding. (also AWS A5.17). This material, used with granular fluxes designated F6X or F7X, has a minimum tensile strength of 60 or 70 ksi and is suitable for submerged-arc welding (Chapter 5) of mild and low-alloy steels.

A559, Mild Steel Electrodes for Gas Metal-Arc Welding. (also AWS A5.18). This is for classification of mild steel electrodes for gas metal-arc welding (see Chapter 5) of mild and low-alloy steel. The minimum tensile strength of the material is 60 or 70 ksi.

2.3. HIGH-STRENGTH LOW-ALLOY STEELS

This category includes steels having yield stresses from 40 to 70 ksi, with yield stresses actually being well-defined yield points as shown in curve (b) of Fig. 2.2.1 the same as shown by carbon steels. The addition to carbon steels of small amounts of alloy elements such as chromium, columbium, copper, manganese, molybdenum, nickel, phosporus, vanadium or zirconium, improve some of the mechanical properties. Whereas carbon steels gain their strength by increasing carbon content, the alloy elements create increased strength from a fine rather than coarse microstructure obtained during cooling of the steel. High-strength low-alloy steels are used in the as-rolled or normalized condition; i.e., no heat treating is used.

A242, High-Strength Low-Alloy Structural Steel (F_y = 42 to 50 ksi). This steel was developed during the 1940's to provide a higher-strength steel than A7. Since there are limitations only on carbon (0.22 percent max.) and manganese (1.25 percent max.), many combinations of alloying elements are possible. Since the chemistry of this steel is not altered for different thicknesses, the yield point varies with thickness. Hot working by increasing the number of passes through the rolls to produce the thin sections increases the yield point.

This steel is used currently mainly for obtaining high corrosion resistance (about four times that of A36 without copper), and occasionally for weathering steel. Weathering is the self-sealing oxidation which produces the attractive deep-brown color. The weathering grade of A242 must be specified since only a certain combination of alloy elements produces the effect. Other steels described in the following sections have essentially replaced A242, particularly A588 as a weathering steel (see Sec. 2.12).

A374, A375, High-Strength Low-Alloy Cold-Rolled Steel Sheets and Strip, and High-Strength-Low-Alloy Hot-Rolled Steel Sheets and Strip. These steels are used in making light gage steel sections, including steel joists.

A440, High-Strength Structural Steel (F_y = 42 to 50 ksi). This steel was developed in the late 1950's as a high-strength steel cheaper than A242. Based on its chemistry A440 is not really a high-strength low-alloy steel but it has similar mechanical properties and so is included. Its use is intended for bolted construction; it should not be used for welded construction. The higher strength over A36 is obtained mainly by higher carbon content (0.28 percent vs. 0.26 percent for A36 shapes) and higher manganese (1.60 percent vs. 1.20 percent for A36). Its chemistry is more rigorously controlled than in A242; therefore, it is more consistent in its properties. Lower cost by using carbon and manganese as alloys is obtained at the expense of the necessary ductility for economical welding.

As with A242, A440 varies in yield point with thickness: 50 ksi for a thickness not exceeding $\frac{3}{4}$ in. and for Group 1 and 2 shapes; 46 ksi for a thickness greater than $\frac{3}{4}$ in. but not exceeding $1\frac{1}{2}$ in. and for Group 3 shapes; and 42 ksi for a thickness greater than $1\frac{1}{2}$ in. but not exceeding 4 in. and for Groups 4 and 5 shapes. Since the chemistry is constant for all thicknesses the increased hot working to obtain thin sections increases the yield point.

A441, High-Strength Low-Alloy Structural Manganese Vanadium Steel (F_y = 40 to 50 ksi). This steel was introduced with A440 to be cheaper than A242 where high corrosion resistance was not the primary factor. For welded construction, this steel was intended for use when strength is the primary consideration. It has the same carbon and manganese limitation as A242 which assures good weldability, but has its other chemistry more restricted than does A242. Corrosion resistance of this steel is about twice that of carbon steel. Use of A441 is rapidly declining because A572 is better where high strength is important and A242 or A588 are better for corrosion resistance and for weathering steel.

As with A242 and A440, the chemistry is constant for all thicknesses, so that yield point is lowest for the thickest sections, varying in the same way as for A242 and A440. However, for a thickness exceeding 4 in. (not permitted for A242 or A440) and up to 8 inches the minimum yield point is 40 ksi.

A588, High-Strength Low-Alloy Structural Steel With 50,000 psi Minimum Yield Point to 4 in. Thick (F_y = 46 to 50 ksi). This is a new steel approved by ASTM in 1968, which because of constant yield point (up to 4-in. thickness) and its excellent weathering properties is receiving wide usage replacing A441 in welded construction, particularly bridge construction. This steel is more expensive than A441 partly because it requires different percentages of alloying elements for different thicknesses in order to obtain the constant 50 ksi yield point. The higher cost is offset by the increased strength in heavier sections (50 instead of 46 or 42 ksi) and by the greater ease of design using constant yield point.

The weathering grade is usually restricted to low carbon (say 0.12 percent max) and the strength and weathering properties are gained by various combinations of manganese, chromium, copper, nickel, and vanadium. The greater use of these more expensive alloys instead of inexpensive carbon is an additional factor in raising costs. A588 permits a wide range of chemistry (seven categories) but its maximum carbon content is limited well below what is permitted under A242 or A441.

For the heaviest wide-flange shapes (Group 5; 14 × 16; 605 to 730 lb/ft) the minimum specified yield point is 46 ksi, though these are not widely available in A588 steel.

A572, High-Strength Low-Alloy Columbium-Vanadium Steels of Structural Quality (F_y = 42 to 65 ksi). This specification covers a group of non-weathering steels in six separate grades. The steels are designated Grades 42, 45, 50, 55, 60, and 65 with the number representing the minimum yield point. Within any one grade the specified minimum yield point is constant for all thicknesses which are available.

A572 Grades 42, 45, and 50 are intended for use in bolted and welded construction of bridges, buildings, and other structures. Grades 55, 60 and 65 are intended for bolted construction and for welding where dynamic and fatigue loadings are not considered important in design. For welded bridge construction, special requirements are necessary to insure adequate toughness and may only be expected after negotiation with the steel producer.

For these steels carbon content limitations are similar to A441 (0.21 percent for Grade 42 to 0.26 percent for Grade 65). The appropriate strengths are obtained primarily by the use of columbium (0.005–0.05 percent) and/or vanadium (0.01–0.10 percent).

A572 is available in Grades 42 to 55 for all shapes to 426 lb/ft inclusive and for plates and bars not exceeding 1½ in. (Grade 42 up to 4 in. inclusive). Grade 60 covers plates 1 in. or less and shapes in Groups 1 and 2. Grade 65 covers plates ½ in. or less and shapes in Group 1.

The A572 steels are becoming widely used, replacing A529, A242, A440, and A441 when strength is the most important factor.

A316, Low-Alloy Steel Covered Arc-Welding Electrodes (Also AWS A5.5). This specification covers electrodes designated E70XX, E80XX, E90XX, E100XX, E110XX, and E120XX, where the meaning is the same as defined under A233. The numbers refer to various categories which depend on the following[3]:

(a) Mechanical properties of the deposited weld metal
(b) Type of coating
(c) Welding position
(d) Type of welding current
(e) Chemical composition of deposited weld metal

These electrodes are suitable for welding of any of the steels discussed in this chapter.

2.4. HEAT-TREATED LOW-ALLOY STEELS

Low-alloyed steels may be quenched and tempered to obtain yield strengths of 80 to 110 ksi. Yield strength is usually defined as the stress at 0.2 percent offset strain, since these steels do not exhibit a well-defined yield point. A typical stress-strain curve is shown in Fig. 2.2.1, curve (c). These steels are weldable with proper procedures, and ordinarily require no additional heat treatment after they have been welded. For special uses, stress relieving may occasionally be required. Some carbon steels, such as certain pressure vessel steels, may be quenched and tempered to give yield strengths in the 80 ksi range, but most steels of this strength are low-alloy steels. These low-alloy steels generally have a maximum carbon content of about 0.20 percent in order to limit the hardness of any coarse-grain microstructure (martensite) which may form during heat treating or welding, thus reducing the danger of cracking.

The heat treatment consists of quenching (rapid cooling with water from 1625–1675°F to about 300–400°F); then tempering by reheating to about 1100°F and allowing to cool. Tempering, even though reducing the strength and hardness somewhat from the quenched material, greatly improves the toughness and ductility. Reduction in strength and hardness with increasing temperature is somewhat counteracted by occurrence of a secondary hardening, resulting from precipitation of fine columbium, titanium, or vanadium carbides. This precipitation begins at about 950°F and accelerates up to about 1250°F. Tempering at or near 1250°F to get maximum benefit from precipitating carbides may result in entering the transformation zone; thus producing the weaker microstructure that would have been obtained without quenching and tempering.

In summary, the quenching produces martensite, a very hard, strong, and brittle microstructure; reheating reduces the strength and hardness somewhat while increasing the toughness and ductility. For more detailed information concerning the metallurgy of the quenching and tempering process, the reader is referred to the *Welding Handbook*.[3,4] Therein, it is stated that this heat treatment "produces the best combinations of mechanical properties of which a steel is capable."

A514, High-Yield-Strength, Quenched and Tempered Alloy Steel Plate, Suitable for Welding (F_y = 90 or 100 ksi). Steels meeting this specification were first available in the early 1950's; however, it was not covered by ASTM until adoption of A514 in 1964. A514 covers plate steel in thicknesses 4 in. and under; the yield strength for a thickness 2½ in. or less is 100 ksi, while it is 90 ksi for greater thicknesses. Since this steel does not exhibit a well-defined yield point, yield strength is defined as the stress at 0.2 percent offset strain or the stress at 0.5 percent extension under load. These two methods are not greatly different, as shown in curve (c) of the enlarged stress-strain curves in Fig. 2.4.1.

A514 steel is available in eleven grades; A through J and K through M. These are all for structural grade plate but provide differing special features. Grades A through D and J are limited to plates of maximum thickness 1¼ in. Grades G, H, K, and M are limited to a maximum thickness of 2 in., while Grades E and F provide for a maximum thickness of 4 in.

This steel is used for welded bridges, highrise buildings, television towers, missile transporting and erecting equipment, construction machinery, trucks and trailers, and water tanks. These steels are used when high yield strength-to-weight is desired and where good notch toughness is important. Abrasion resistance is also an important feature.

Because this steel has become widely used in building and bridge construction, several rolled shapes have become available in Grades E and F, even though not explicitly covered under A514.

A517 (F_y = 90 ksi to 100 ksi). This steel is similar to A514 but covers pressure vessel quality material which emphasizes maximum toughness.

A490, Quenched and Tempered Alloy Bolts for Structural Steel Joints. This material has carbon content which may range up to 0.53 percent for 1½ in. diam. bolts, and has alloying elements in amounts similar to the A514 steels. After quenching in oil the material is tempered by reheating to at least 900°F. The minimum yield strength, obtained by 0.2 percent offset, ranges from 115 ksi (over 2½ in. to 4 in. diameter) to 130 ksi (for 2½ in. diameters and under).

It is to be noted that many steel producers have other steels identified by brand names which are either heat-treated carbon steels (for which there is not presently a covering ASTM specification) or low-alloy quenched and tempered steels. Such steels may provide additional or different features than those provided for by A514 or A517. Many high-alloy steels are also available which are rarely if ever considered by the structural engineer.

In the foreseeable future, other steels unknown to structural engineers will probably become standard design materials, particularly in buildings.

As an example, stainless steel, once considered only as an architectural material, is being used structurally. Stainless steel achieves strength levels higher than A514. Its disadvantage in addition to high cost is that it is relatively difficult to work or to weld. The accelerating use of stainless steels is indicated by the 1968 American Iron and Steel Institute specification[5] covering cold-formed members made from the several types of stainless steel sheets and strip in the "annealed and strain flattened condition most significant in building construction."

2.5. STRESS-STRAIN BEHAVIOR (TENSION TEST) AT ATMOSPHERIC TEMPERATURES

Typical stress-strain curves for tension are shown in Fig. 2.2.1 for the three categories of steel already discussed; carbon, high-strength low-alloy, and heat-treated high-strength low-alloy. The same behavior occurs in compression when support is provided so as to preclude buckling. The portion of each of the stress-strain curves of Fig. 2.2.1 which is utilized in ordinary design is shown enlarged in Fig. 2.4.1.

The stress-strain curves of Fig. 2.2.1 are determined using a unit stress obtained by dividing the load by the original cross-sectional area of the specimen, and the strain (inches per inch) is obtained as the elongation divided by the original length. Such curves are known as *engineering stress-strain* curves and rise to a maximum stress level (known as the ultimate tensile strength) and then fall off with increasing strain until they terminate as the specimen breaks. Insofar as the material itself is concerned the unit stress continues to rise until failure occurs. The so-called *true-stress/true-strain* curve is obtained by using the actual cross-section even after necking down begins and using the instantaneous incremental

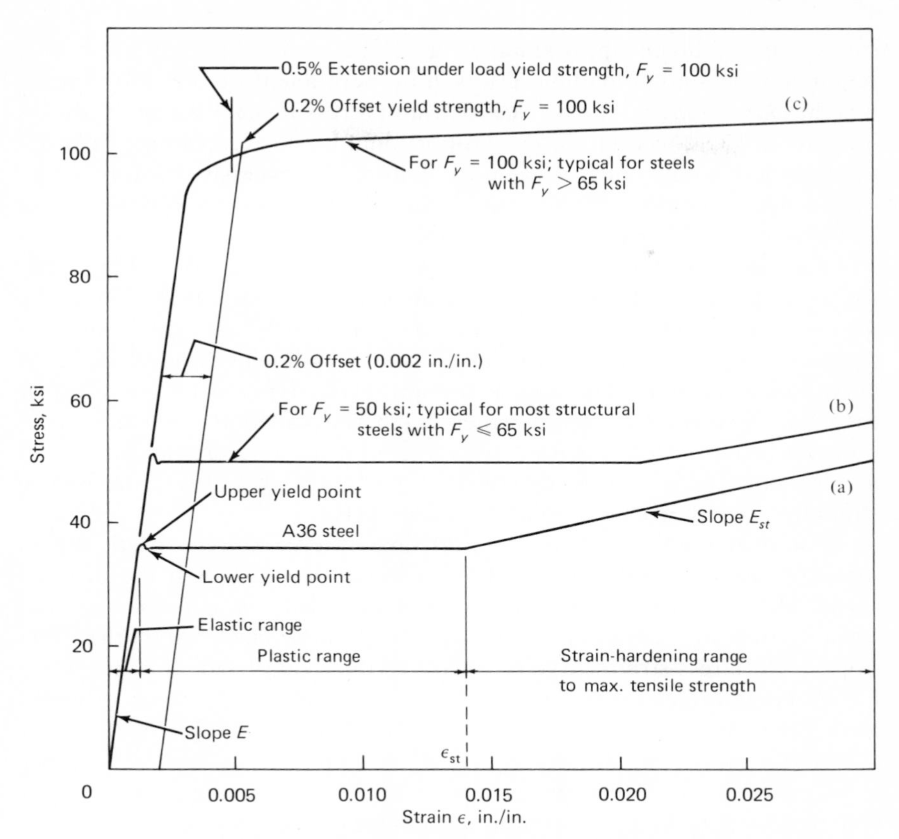

Fig. 2.4.1. Enlarged typical stress-strain curves for different yield stresses.

strain. The engineering strain-strain curve permits the practical use of the curve to determine the maximum *load* which can be carried (ultimate tensile strength).

Stress-strain curves (as per Fig. 2.4.1) show a straight line relationship up to a point known as the *proportional limit*, which essentially coincides with the yield point for most structural steels with yield points not exceeding 65 ksi. For the quenched and tempered low-alloy steels the deviation from a straight line occurs gradually, as in curve (c), Fig. 2.4.1. Since the term yield point is not appropriate to curve (c), *yield strength* is used for the stress at an offset strain of 0.2 percent; or alternatively, a 0.5 percent extension under load, as shown in Fig. 2.4.1. *Yield stress* is the general term to include both the unit stress at a yield point, when such exists; and a unit stress corresponding to a specified strain for material showing a gradual nonlinear stress-strain behavior.

The ratio of stress to strain in the initial straight line region is known as the modulus of elasticity, or Young's modulus, E, which for structural steels may be taken approximately as 29,000 ksi. In the straight-line region loading and unloading results in no permanent deformation; hence it is the *elastic range.* The service load unit stress in steel design is always intended to be safely within the proportional limit, even though in order to ascertain safety factors against failure or excessive deformation, knowledge is required of the stress-strain behavior up to a strain about 15 to 20 times the maximum elastic strain.

For steels exhibiting yield points, as curves (a) and (b) of Fig. 2.4.1, the large strain for which essentially constant stress exists is known as the *plastic range.* The plastic design method consciously uses this range for determining plastic strength (which is usually assumed to be ultimate or maximum strength). The higher strength steels typified by curve (c) of Fig. 2.4.1 also have a region which might be called the plastic range; however, even in this zone the stress is continuously increasing as strain increases instead of remaining constant, so that as yet (1971) the plastic strength methods are not applied to these steels.

For strains greater than 15 to 20 times the maximum elastic strain the stress again increases but with a much flatter slope than the original elastic slope. This increase in strength is called *strain hardening*; the strain-hardening range continuing up to ultimate tensile strength. The slope of the stress-strain curve is known as the strain-hardening modulus, E_{st}. Average values for this modulus and the strain, ϵ_{st} at which it begins have been determined[1] for two steels as: A36 steel, E_{st} = 900 ksi at ϵ_{st} = 0.014 in. per in.; and for A441, E_{st} = 700 ksi at ϵ_{st} = 0.021 in. per in. General use of the strain-hardening range is not made in design, but certain of the buckling limitations are conservatively derived to preclude buckling even at strains well beyond onset of strain hardening.

The stress-strain curve also indicates the *ductility.* Ductility is defined as the amount of permanent strain (i.e., strain exceeding proportional limit) up to the point of fracture. Measurement of ductility is obtained from the tension test by determining the percent elongation (comparing final and original cross-sectional areas) of the specimen. Ductility is important because it permits locally high stresses to be redistributed. Design procedures based on ultimate strength behavior require large inherent ductility, particularly for treatment of stresses near holes or abrupt change in member shape, as well as for design of connections.

2.6. TOUGHNESS AND RESILIENCE

Toughness and resilience are measures of the ability of a metal to absorb mechanical energy. For uniaxial stress these quantities are obtainable from the tension test (engineering stress-strain) curves, such as those of Fig. 2.2.1.

Resilience relates to the elastic energy absorption of the material. Sometimes referred to as *modulus of resilience*, resilience is the amount of elastic energy able to be absorbed by a unit volume of material loaded in tension; i.e., it equals the area under the stress-strain diagram up to the yield stress.

Toughness relates to the total energy, both elastic and inelastic, able to be absorbed by a unit volume of material before it fractures. For uniaxial tension, it is the area under the tension stress-strain curve out to the fracture point where the diagram terminates. This area is sometimes called the *modulus of toughness*. Because all parts of the tensile specimen do not deform the maximum amount, the area only gives an approximate value for the metal's toughness. To illustrate the magnitude of these quantities for some typical steels, Ref. 1 gives the following values:

Steels	Resilience inch pounds/in.^{3}	Toughness, inch pounds/in.^{3}
Carbon (A36 with F_y = 36 ksi)	22	12,000
High-Strength Low-Alloy (A441 with F_y = 50 ksi)	43	15,000
Quenched and Tempered Carbon (F_y = 70 to 80 ksi)	110	18,000
Quenched and Tempered Low Alloy (A514 with F_y = 100 ksi)	170	19,000

The values for the A36, A441, and A514 agree closely with values computed from the curves of Fig. 2.2.1.

Since uniaxial tension rarely exists in real structures, particularly in the region of connections, a more practical index of toughness is used based on the more complex stress condition (probably triaxial) below the root of a notch. *Notch toughness* is the term used to describe the resistance of a metal to the start and propagation of a crack at the base of a standard notch. Notch toughness is most commonly measured by the Charpy V-notch test. This test uses a small rectangular simply supported beam with a V-notch at midlength. The bar is fractured by a blow from a swinging pendulum. The amount of energy absorbed is calculated from the height the pendulum raises after breaking the specimen.

The Charpy V-notch test is widely used to determine the transition temperature from brittle to ductile behavior. For different temperatures the fracture energy absorption is determined and plotted as in Fig. 2.6.1a. The temperature at the point where the slope is steepest (point A of Fig. 2.6.1a) is the transition temperature. Since brittleness and ductility are qualitative terms, the various structural steels have different degrees of ductility which are important only when related to the task they must

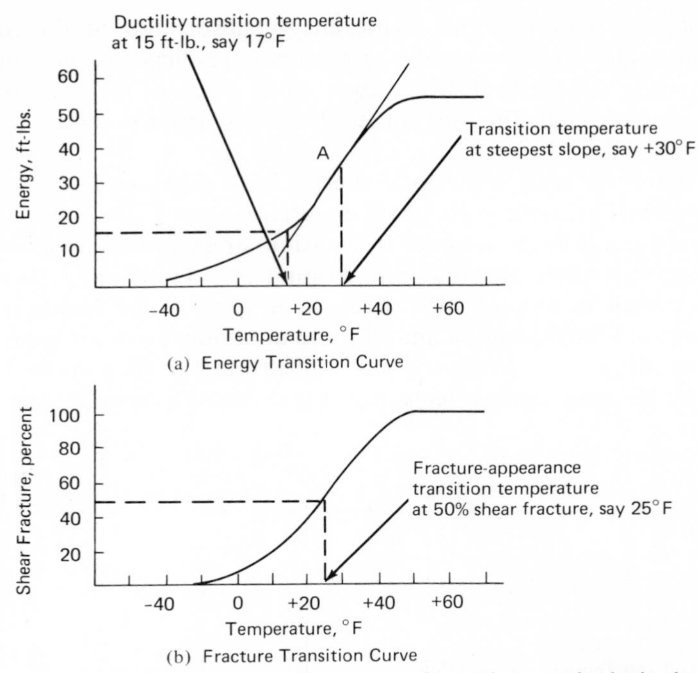

(a) Energy Transition Curve

(b) Fracture Transition Curve

Fig. 2.6.1. Transition-temperature curves for carbon steel obtained from Charpy V-notch impact tests. (Adapted from Ref. 1.)

perform. Thus, for the steels used in ordinary structures an arbitrary amount of energy absorption is often required, such as 15 ft-lb, and the temperature at which the energy equals that value is the *ductility transition temperature*, commonly called just transition temperature. Some typical values of ductility transition temperatures are given in Table 2.6.1 for 1-in.-thick plates.

The various alloying elements used to increase the strength of steel all have the effect of increasing the ductility transition temperature; i.e., adversely affecting notch toughness. The few elements which improve notch toughness either (a) result in high shrinkage (due to removing undissolved oxygen because of a high affinity for it) during the process of finishing a heat of steel (hence, lower quantities yielded); (b) are themselves too expensive, or; (c) decrease weldability. Such elements as carbon, vanadium, and nitrogen increase the transition temperature about 5 to 6°F per 1,000 psi increase in yield point. Certain combinations of vanadium, nitrogen, and columbium, as specified in A572 steel have been found to increase transition temperature only about 4°F per 1,000 psi increase in yield point.

When reduced transition temperatures are necessary, such as for pressure vessels at $-50°F$, heat treatment such as quenching and tempering is required.

TABLE 2.6.1
Typical Ductility Transition Temperatures for 1-in. Thick Plates Based on a Charpy V-Notch Impact Test Value of 15 ft-lb (from Ref. 1)

Carbon Steels	A36	+30°F
	A285 Grade C	+45
	A515 Grade 70	+30
	A516 Grade 70	−20
High-Strength, Low-Alloy Steels	A242	0
	A441	+15
	A440	+40
	A572, Grade 50	+30
Heat-Treated Carbon Steels	F_y = 50 ksi, Normalized	−75
	F_y = 60 ksi, Quenched and Tempered	−90
	F_y = 70 ksi, Quenched and Tempered	−25
Heat-Treated Low-Alloy Steels	A514	−50

An alternative use of the Charpy V-notch test involves examining the appearance of the fracture surface. The percentage of the surface which appears to have fractured by shear is plotted against temperature, as in Fig. 2.6.1b. The shear fracture portion gives a fine fibrous appearance, while the remainder appears brittle or crystalline. The temperature at which the shear fracture portion is 50 percent may be called the *fracture-appearance transition temperature.*[1]

2.7. YIELD STRENGTH FOR MULTIAXIAL STATES OF STRESS

Only when the load-carrying member is also subject to uniaxial stress, can the properties from the tension test be expected to be identical with those of the structural member. In most structural design, one is rarely aware that yielding in a real structure is *not* the well-defined behavior observed in the tension test. Yield is commonly assumed to be achieved when any one component of stress reaches the uniaxial value F_y.

For all states of stress other than uniaxial, a definition of yielding is needed. These definitions, and there are frequently several for a given state of stress, are called *yield conditions* (or theories of failure) and are equations of interaction between the stresses acting.

Energy-of-Distortion (Hencky-vonMises) Yield Criterion. This most commonly accepted theory gives the uniaxial yield stress in terms of the three principal stresses. The yield criterion may be stated

$$\sigma_y^2 = \tfrac{1}{2}[(\sigma_1 - \sigma_2)^2 + (\sigma_2 - \sigma_3)^2 + (\sigma_3 - \sigma_1)^2] \tag{2.7.1}$$

where σ_1, σ_2, σ_3 are the tensile or compressive stresses which act in the three principal directions; i.e., the stresses which act in the three mutually perpendicular planes of zero shear.

For most structural-design situations, one of the principal stresses is either zero or small enough to be neglected; hence Eq. 2.7.1 reduces to the following for the case of plane stress (all stresses considered are acting in a plane)

$$\sigma_y^2 = \sigma_1^2 + \sigma_2^2 - \sigma_1\sigma_2 \tag{2.7.2}$$

When stresses on thin plates are involved, the principal stress acting transverse to the plane of the plate is usually zero (at least to first-order approximation). Flexural stresses on beams assume zero principal stress perpendicular to the plane of bending. Furthermore, structural shapes are all comprised of thin plate components, so that each component is subject to Eq. 2.7.2. The plane stress yield criterion, Eq. 2.7.2, is the one used throughout the remaining chapters where needed, and is illustrated in Fig. 2.7.1.

Shear Yield Stress. The yield point for pure shear could be determined from a stress-strain curve with shear loading, or if the multiaxial yield criterion is known, that relationship can be used. Pure shear occurs on 45° planes to the principal planes when $\sigma_2 = -\sigma_1$, and the shear stress $\tau = \sigma_1$. Substitution of $\sigma_2 = -\sigma_1$ into Eq. 2.7.2 gives

$$\sigma_y^2 = \sigma_1^2 + \sigma_1^2 - \sigma_1(-\sigma_1) = 3\sigma_1^2 \tag{2.7.3}$$

$$\sigma_1 = \sigma_y/\sqrt{3} = \text{shear yield} \tag{2.7.4}$$

which indicates that the yield condition when shear stress alone is acting equals $\sigma_y/\sqrt{3}$.

Poisson's Ratio, μ. When stress is applied in one direction, strains are induced not only in the direction of applied stress but also in the other two mutually perpendicular directions. The usual value of μ used is that obtained from the uniaxial stress condition, where it is the ratio of the transverse strain to longitudinal strain under load. For structural steels, Poisson's ratio is approximately 0.3 in the elastic range where the material is compressible and approaches 0.5 when in the plastic range where the material is essentially incompressible (i.e., constant resistance no matter what the strain).

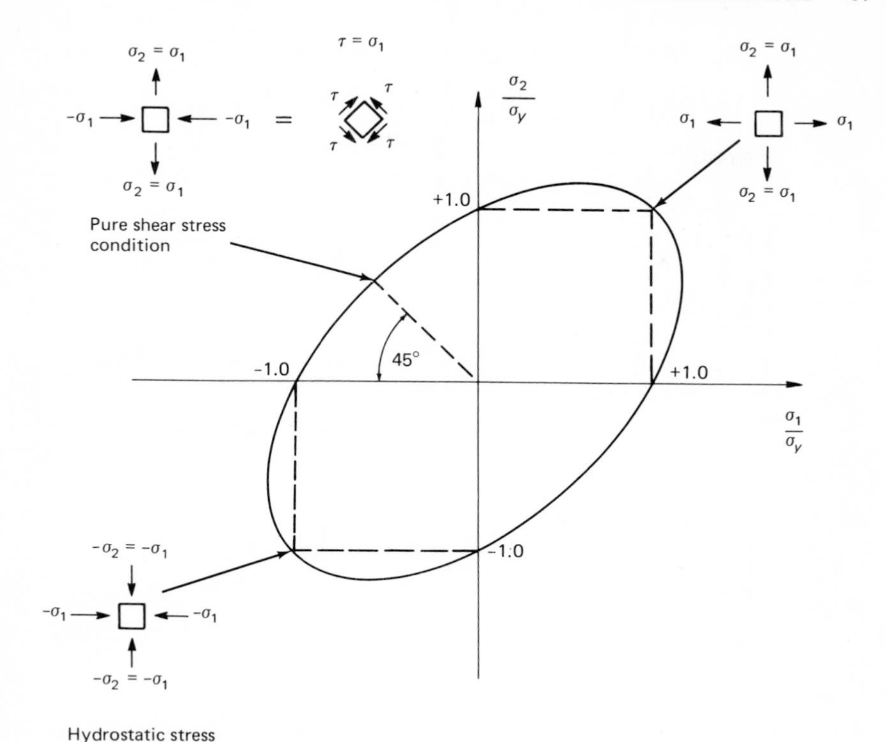

Fig. 2.7.1. Hencky-von Mises, energy-of-distortion yield criterion for plane stress.

Modulus of Elasticity-Shear. Loading in pure shear produces a stress-strain curve with a straight-line portion whose slope represents the shear modulus of elasticity. If Poisson's ratio μ and the tension-compression modulus of elasticity E are known, the shear modulus G is defined by the theory of elasticity as

$$G = \frac{E}{2(1 + \mu)} \tag{2.7.5}$$

which for structural steel is just over 11,000 ksi.

2.8. HIGH-TEMPERATURE BEHAVIOR

The design of structures to serve under atmospheric temperature rarely involves concern about high-temperature behavior. Knowledge of such behavior is desirable when specifying welding procedures; and is necessary when concerned with the effects of fire.

When temperatures exceed about 200°F the stress-strain curve begins to become nonlinear, gradually eliminating the well-defined yield point.

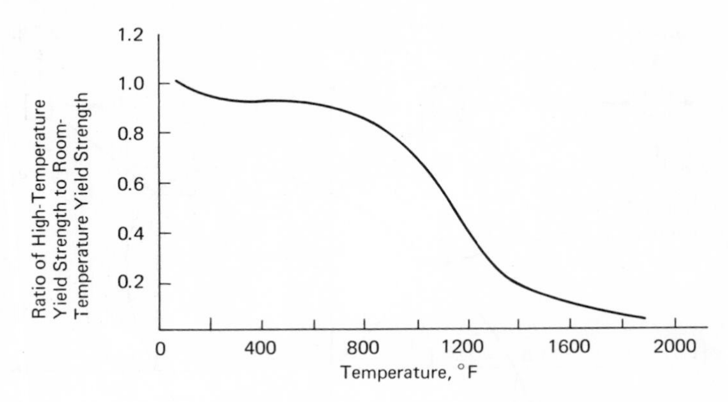

(a) Average Effect of Temperature on Yield Strength

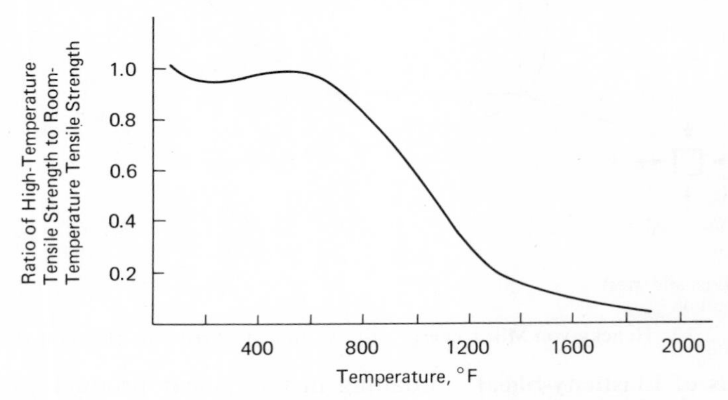

(b) Average Effect of Temperature on Tensile Strength

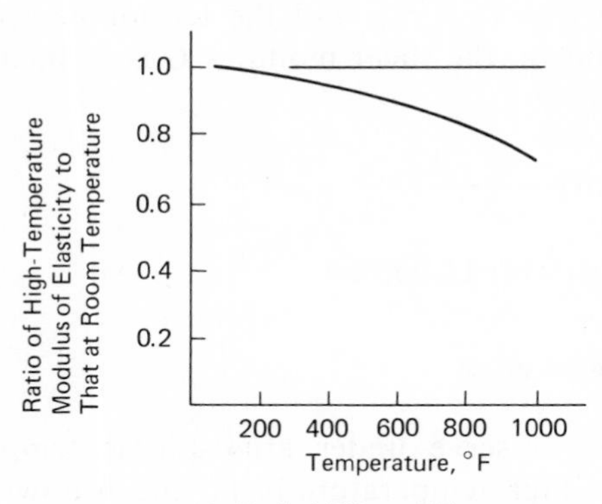

(c) Typical Effect of Temperature on Modules of Elasticity

Fig. 2.8.1. Typical effect of high temperature on stress-strain curve properties of structural steels. (Adapted from Ref. 1.)

The modulus of elasticity, yield strength, and tensile strength all reduce as temperature increases. The range from 800 to 1000°F is where the rate of decrease is maximum. While each steel because of its different chemistry and microstructure behaves somewhat differently, the general relationships are shown in Fig. 2.8.1. Steels having relatively high percentages of carbon, such as A36 and A440, exhibit "strain aging" in the range 300 to 700°F. This is evidenced by a relative rise in yield point and tensile strength in that temperature range over what is shown as average in Fig. 2.8.1a and b. Tensile strength may rise to about 10 percent above that at room temperature and yield point may recover to about its room temperature value when the temperature reaches 500 to 600°F. Strain aging results in decreased ductility.

The modulus of elasticity decrease is moderate up to 1000°F; thereafter it decreases rapidly. More importantly, when temperatures above about 500 to 600°F exist, steels exhibit deformation which increases with increasing time under load, a phenomenon known as *creep*. Creep is well known in concrete structures; and its effect in steel, which does not occur at atmospheric temperatures, increases with increasing temperature.

Other high-temperature effects are (a) improved notch impact resistance up to about 150–200°F, as discussed in Sec. 2.6; (b) increased brittleness due to metallurgical changes, such as carbide precipitation discussed in Sec. 2.4, begins to occur at about 950°F; and (c) corrosion resistance of structural steels increases for temperatures up to about 1000°F. Most steels are used in applications below 1000°F, and some heat treated steels should be kept below about 800°F.

2.9. COLD WORK AND STRAIN HARDENING

Behavior of steel after first yield has been exceeded by a large amount is of interest because when reloaded after unloading the stress-strain properties may differ from those observed during the initial loading. Elastic loading and unloading results in no residual strain; however, initial loading beyond the yield point such as to point A of Fig. 2.9.1 results in unloading to a strain at point B. A permanent set OB has occurred. The ductility capacity has been reduced from a strain OF to the strain BF. Reloading exhibits behavior as if the stress-strain origin were at point B; the plastic zone prior to strain hardening is also reduced.

When loading has occurred until point C is reached, unloading follows the dashed line to point D; i.e., the origin for a new loading is now point D. The length of the line CD is greater, indicating that the yield point has increased. The increased yield point is referred to as a strain hardening effect; the ductility remaining when loading from point D is severely reduced from its original value prior to the initial loading. The

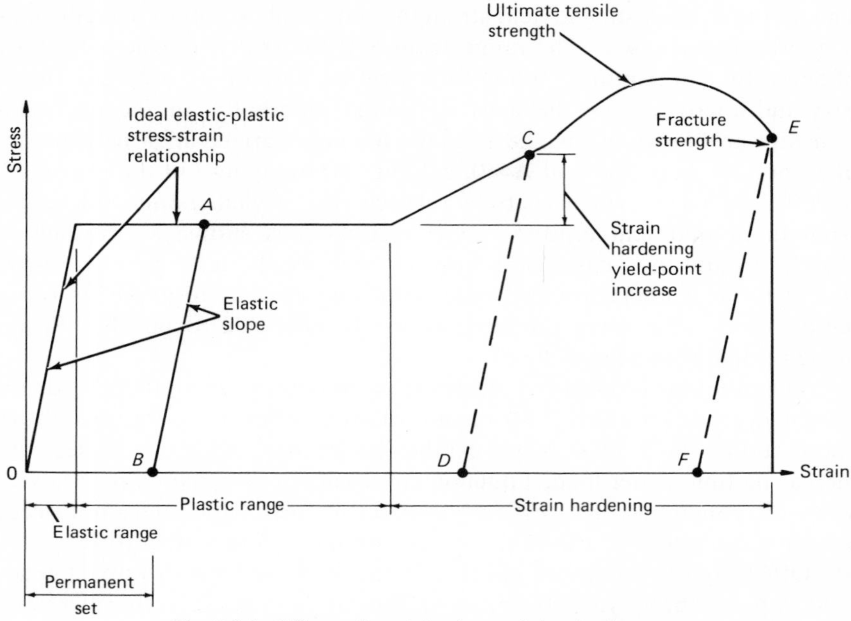

Fig. 2.9.1. Effects of straining beyond the elastic range.

process of loading beyond the elastic range to cause a change in available ductility, when done at atmospheric temperature, is known as *cold work*. Since real structures are not loaded in uniaxial tension-compression the cold work effect is much more complex and any theoretical study of it is outside the scope of the text.

When structural shapes are made by cold-forming from plates at atmospheric temperature, inelastic deformations occur at the bends. Cold working into the strain hardening range at the bend locations increases the yield strength, which design specifications may permit taking into account. The Light-Gage Steel Specification[6] has such provisions.

Based on the previous discussion it would at first appear that the increase in strength is obtained at the expense of ductility, and with the loss of the original well-defined yield point and its associated constant stress plastic range. Upon unloading and after a period of time, the steel will have acquired different properties from those represented by points *D*, *C*, and *E* of Fig. 2.9.1 by a phenomenon known as *strain aging*. Strain aging, as shown in Fig. 2.9.2, produces an additional increase in yield point, restores a plastic zone of constant stress, and gives a new strain-hardening zone at an elevated stress. The original shape of the stress-strain diagram is restored but the ductility is reduced. The new stress-strain diagram may be used as if it were the original for analyzing

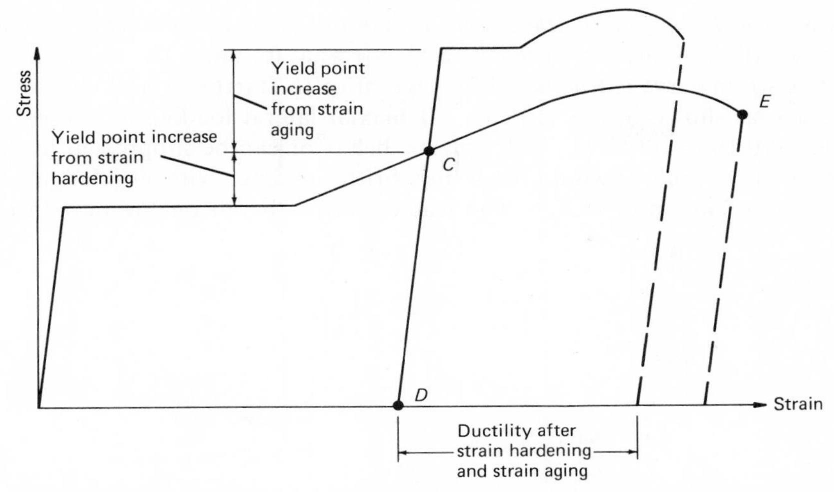

Fig. 2.9.2. Effect of strain aging after straining into strain-hardening range and unloading.

cold-formed sections, as long as the ductility which remains is sufficient. The corner regions of cold-formed shapes generally would not require high ductility for rotational strain about the axis of the bend.

Stress relieving by annealing will eliminate the effects of cold work should that be desired. Annealing involves heating to a temperature above transformation range and allowing to slowly cool; a recrystallization occurs to restore the original properties.

2.10. BRITTLE FRACTURE

As has been discussed in several sections, ordinarily highly ductile steel can become brittle under various conditions. The designer must understand the causes in order to preclude brittle fracture.

Effect of Temperature. Notch toughness, as determined by the Charpy impact transition temperature curves (see Sec. 2.6), is an indication of the susceptibility to brittle fracture. Temperature is a vital factor in several ways: (a) the value below which notch toughness is inadequate; (b) in the 600 to 800°F range causes formation of brittle microstructure; and (c) over 1000°F causes precipitation of carbides of alloying elements to give more brittle microstructure. The other temperature factors have already been discussed in earlier articles.

Effect of Multiaxial Stress. The complex stress condition found in usual structures, particularly at joints, is another major factor affecting brittleness. The *Primer of Brittle Fracture*[7] has provided an excellent rational presentation of this and forms the basis for what follows. The engineering stress-strain curve is for uniaxial stress; prior to fracture a necking down occurs, as shown in Fig. 2.10.1a. If biaxial lateral loading as shown in Fig. 2.10.1b could be applied, "plastic behavior can be suppressed to the point where the bar would break in a brittle manner with no elongation and no reduction in area." The fracture stress based on the unreduced

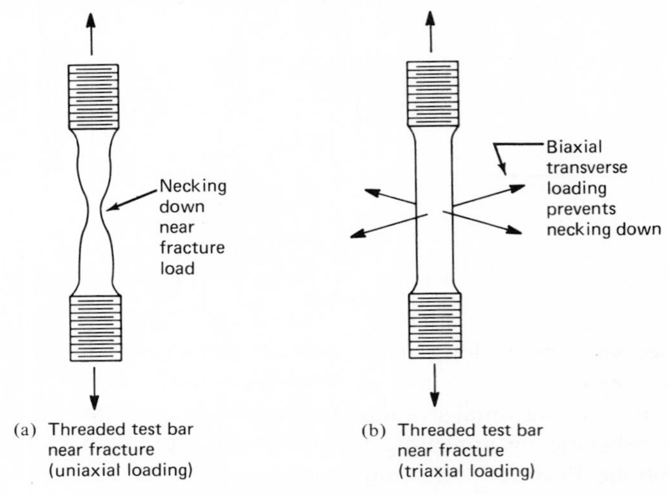

Fig. 2.10.1. Uniaxial and triaxial loading.

cross-sectional area would be the same high value as that based on the necked-down cross section in the uniaxial tension case. The unit stress would be far above the nominal maximum tensile strength of the engineering stress-strain curve, which is always computed on the basis of original cross section. This is a further extension of the yield criterion (or failure criterion) concept discussed in Sec. 2.7.

Also the effects of notches have been alluded to in the discussion of notch toughness in Sec. 2.6. The notch serves somewhat the same effect as the theoretical triaxial loading of Fig. 2.10.1, in that it restrains plastic flow which otherwise would occur and thus at some higher stress may likely fail in a brittle manner. Fig. 2.10.2 shows the effect of a notch in a tensile test specimen. The cross-sectional area at the base of the notch corresponds to the area of the original specimen of Fig. 2.10.1b. The reduced section tries to become narrower as the axial tension increases, but is resisted by the diagonal pull which develops in the corners, as shown in Fig. 2.10.2. The test bar will fail at high stress by brittle fracture.

Notches can occur in real structures by use of unfilleted corners in design or from improperly made welds which may crack. Such occur-

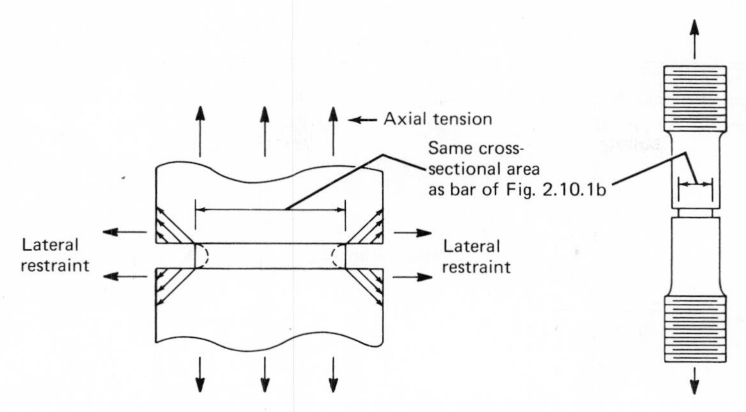

Fig. 2.10.2. Effect of notch on uniaxial tension test.

rences can lead to brittleness. Notches and cracked welds can, however, be minimized by good design and welding procedures.

Unusual configurations and changes in section should be made gradually so the stress flow lines are not required to make abrupt changes. Whenever the complexity is such as to give rise to three-dimensional stresses, the tendency for brittleness increases. Castings, for instance, have the reputation for brittleness. Primarily this is because of the built-in three-dimensional continuity.

Multiaxial Stress Induced by Welding. In general welding creates a built-in continuity which gives rise to biaxial and triaxial stress and strain conditions, which result in brittle behavior. To illustrate, consider the loaded simply supported beam of Fig. 2.10.3, which in turn supports a plate in tension. Due to flexure, the bottom flange of the beam is in tension; therefore, the stress at point A is uniaxial tension (neglecting the small effects of beam width and attachment of flange to web). Application of the tension plate using angles and bolts puts the flange bolts and the angles essentially in uniaxial tension and the bolt which passes through the suspender plate in shear, so that there is no appreciable effect on the stress at point A. In other words, the stress conditions in the connection of Fig. 2.10.3a are approximately uniaxial in nature.

Next, consider the tensile suspender plate welded to the tension flange of the beam, as in Fig. 2.10.3b. The stress at point A is now biaxial because of the direct attachment to the flange at that point. The weld region, therefore, is subject to triaxial stress; biaxial from the directly applied loads, plus the resistance to deformation along the axis of the welds resulting from contiriuous attachment (Poisson's ratio effect). The design of

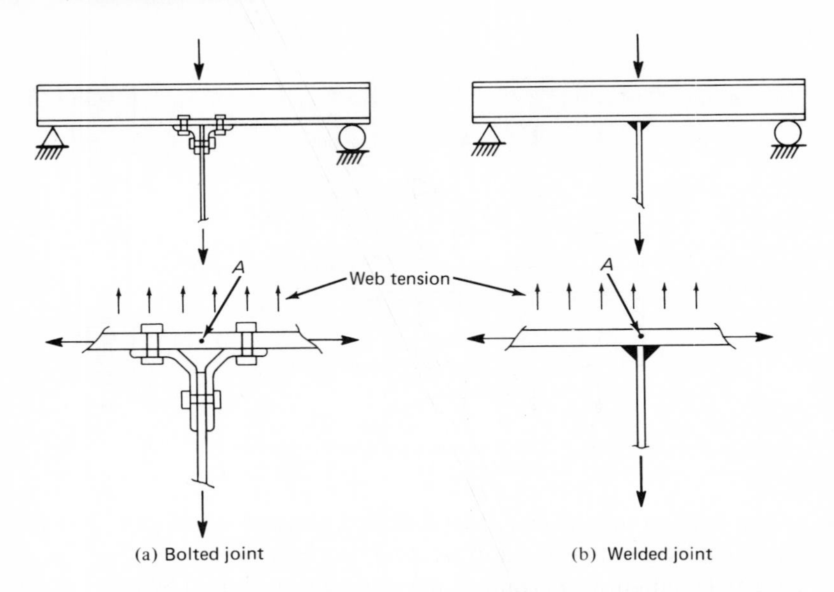

Fig. 2.10.3. Comparison of stress condition in bolted joint with welded joint.

welded joints should consider the possibilities of brittleness due to three-dimensional stressing.

Effect of Thickness. As discussed in Sec. 2.7, if plane stress exists, such as with thin plates where stress in the direction transverse to the plane of the plate may be disregarded, the third dimensional effect is eliminated. For thick plates, because of the three-dimensional effects, the tendency for brittleness increases. From the manufacturing process, thick plates also tend to be more brittle than thin ones; the slower cooling rate gives rise to a coarser microstructure and the higher carbon content, as discussed in Sec. 2.2 through 2.5, which is necessary to obtain the same yield strength as obtained by additional hot working in thin sections, also produces a more brittle material.

Effect of Dynamic Loading. The stress-strain properties referred to so far have been for static loading slowly applied. More rapid loading, such as that of forge drop hammers, earthquake, or nuclear blast changes the stress-strain properties. Ordinarily, the increased strain rate from dynamic loading increases the yield point, tensile strength, and ductility. At temperatures about 600°F there is a moderate decrease in strength. Some increased brittleness has been noted with high strain rate, but it seems principally associated with other factors already discussed, such as notches where stress concentrations exist and the temperature effect on toughness. The more important factor relating to dynamic load application is not that a rapid increasing strain rate occurs, but that it is combined with a rapid *decreasing* strain rate. The effect of stress *variation* is discussed in the next section on fatigue.

Table 2.10.1, from Ref. 7, provides a list of factors "to help determine whether or not the risk of brittle fracture is serious and requires special design considerations."

TABLE 2.10.1
The Element of Risk: Factors to Analyze in Estimating Seriousness of Brittle Fracture (from Ref. 7).

1. What is the minimum anticipated service temperature? The lower the temperature, the greater the susceptibility to brittle fracture.
2. Are tension stresses involved? Brittle fracture can occur only under condition of tensile stress.
3. How thick is the material? The thicker the steel, the greater the susceptibility to brittle fracture.
4. Is there three-dimensional continuity? Three-dimensional continuity tends to restrain the steel from yielding and increases susceptibility to brittle fracture.
5. Are notches present? The presence of sharp notches increases susceptibility to brittle fracture.
6. Are multiaxial stress conditions likely to occur? Multiaxial stresses will tend to restrain yielding and increase susceptibility to brittle fracture.
7. Is loading applied at a high rate? The higher the rate of loading, the greater susceptibility to brittle fracture.
8. Is there a changing rate of stress? Brittle fracture occurs only under conditions of increasing rate of stress.
9. Is welding involved? Weld cracks can act as severe notches.

2.11. FATIGUE STRENGTH

Repeated loading and unloading even if the yield point is never exceeded may result in eventual failure. Such a phenomenon is known as *fatigue*. While fatigue may be observed even if all conditions are ideal; i.e., excellent notch toughness, no stress concentrations from holes or notches, uniaxial stress condition, ductile microstructure, etc., adverse conditions affecting ductility and the existence of multiaxial stress conditions greatly reduce fatigue strength. Actually, the factors discussed throughout this chapter are all interrelated.

Consider the possible stress cycles of Fig. 2.11.1. The most extreme variation is the full reversal; zero to maximum tension, unload to zero, load in compression to same numerical stress as in tension, unload to zero. The ratio of S_{max} (tension) to S_{min} (compression) is said to be $R = -1$, which results in the lowest fatigue strength. The least extreme, of course, is static load without variation; with the stress ratio, $R = +1$. When the maximum stress, S_{max}, is plotted against the number of

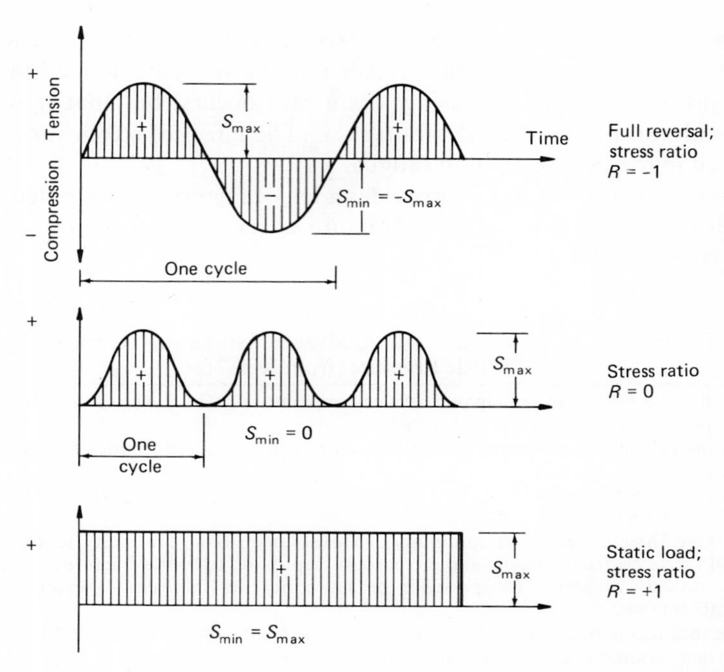

Fig. 2.11.1. Types of stress cycles, showing extreme range of stress ratio from $R = +1$ (nonfatigue condition) to $R = -1$ (stress reversal).

cycles to which it was subjected before failure occurred, a curve such as in Fig. 2.11.2 results. Knowing the maximum number of cycles to which the structure will be subjected, along with the stress ratio, the fatigue strength can be determined. When plotted to log-log scale, as in Fig. 2.11.3, the curves closely approximate straight lines. When a curve reaches a constant stress which is independent of the number of cycles of loading, the corresponding stress is referred to as the *fatigue limit* or *endurance limit*.

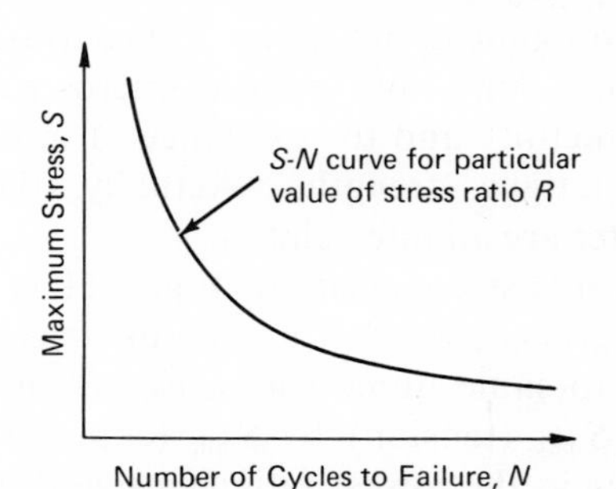

Fig. 2.11.2. Maximum stress S obtainable for various number of cycles N of loading (plotted to linear scale).

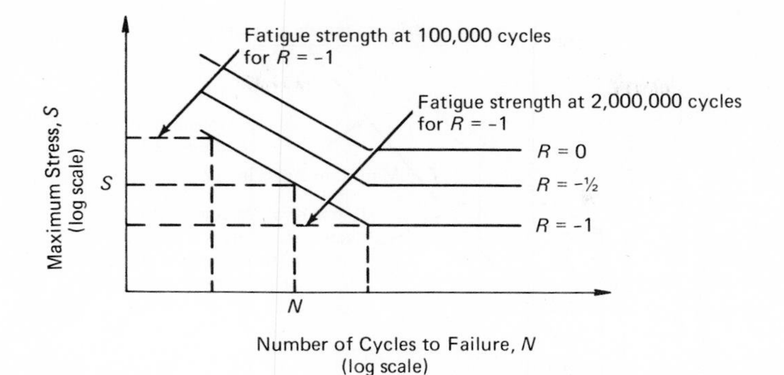

Fig. 2.11.3. Typical S–N curves (approximation by straight lines) for several stress ratios plotted to log-log scale.

This usually occurs at about 2 million cycles of loading. At the lower end, there is usually little strength reduction for fewer than 100,000 cycles. Since most buildings are subject to 100,000 cycles of loading or less, fatigue ordinarily is not considered. Highway bridges usually are expected to have more than 100,000 cycles of loading, so that fatigue is an important consideration in their design.

The sloping portion of Fig. 2.11.3 may be expressed[8] as

$$F_n = S(N/n)^k \tag{2.11.1}$$

where F_n = fatigue strength computed for failure at n cycles
S = stress which produced failure at N cycles
k = slope of the best-fit straight line representing the data

As shown in Fig. 2.11.3 each stress ratio requires a different S–N curve. Usually a diagram known as a *Goodman diagram* is used to summarize the results for the various types of stress cycles. Figure 2.11.4 shows such a diagram from Ref. 8.

Frequently for design purposes a modified Goodman diagram is used on which the effects of several different life cycles may be presented, as shown in Fig. 2.11.5. Stress Category B from 1969 AISC Appendix B has been used for illustration. The maximum tensile stress has been taken as 30 ksi ($0.60F_y$ for $F_y = 50$ ksi). For static tension ($R = +1$) the upper limit represented by the horizontal line at $0.6F_y$ governs. For stress ratios (R) between $+\frac{1}{2}$ and $+1$, fatigue has no effect; i.e., so long as the minimum stress is not less than one-half the maximum and is of the same sign.

When the stress ratio is less than $+1/2$, some reduction for fatigue is required when the number of cycles N exceeds 2 million. The sloping line intersecting the $R = 0$ line at 15 ksi indicates a stress variation of 15 ksi is permitted when $N > 2{,}000{,}000$. Notice that full reversal ($R = -1$) would allow the stress to vary from 7.5 ksi in tension to 7.5 ksi in compression. For any given stress ratio R, one may graphically obtain the

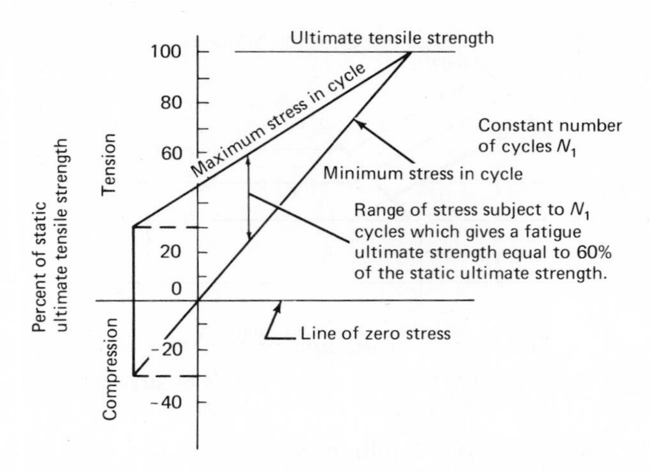

Fig. 2.11.4. Typical Goodman diagram showing effect of various stress ratios on fatigue strength for N_1 cycles of loading. (Adapted from Ref. 8.)

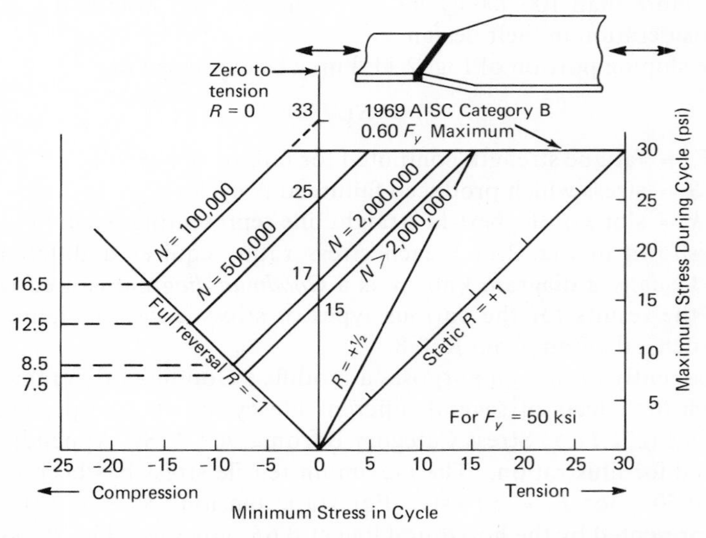

Fig. 2.11.5. Modified Goodman diagram, giving 1969 AISC allowable stresses for base metal and weld metal at full penetration groove welded splices at transitions in width and thickness.

maximum stress permitted during the cycle by means of a modified Goodman diagram.

According to the Appendix B of the 1969 AISC Specification some reduction in strength due to fatigue may be expected under certain stress ratios and types of loading when the number of expected cycles exceeds 20,000 (approximately 2 applications per day for 25 years).

The mechanism of fatigue is still not entirely understood, but it is known to be closely related to the factors relating to ductility. Welding, in particular, may have a dramatic effect on fatigue strength. For a more detailed treatment the reader is referred to a publication of the Welding Research Council.[8]

2.12. CORROSION RESISTANCE AND WEATHERING STEELS

Since the earliest uses of steel, one of the important drawbacks was that painting was required to prevent the deterioration of the metal by corrosion (rusting). The lower-strength carbon steels were inexpensive but very vulnerable to corrosion. Corrosion resistance may be improved by the addition of copper as an alloy element. However, copper-bearing carbon steel is too expensive for general use.

High-strength low-alloy steels have several times[9] the corrosion resistance of structural carbon steel, with or without the addition of copper, as shown in Fig. 2.12.1. The high-strength low-alloy steels do not pit as

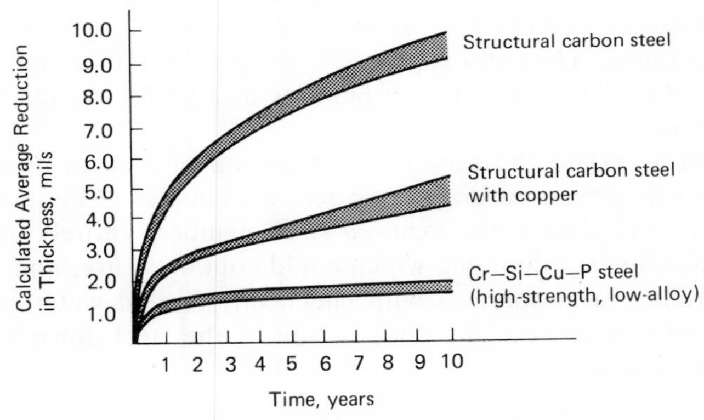

Fig. 2.12.1. Comparative corrosion of steels in an industrial atmosphere. Shaded areas indicate range for individual specimens. (Adapted from Ref. 9.)

severely as carbon steels and the rust that forms becomes a protective coating to prevent further deterioration. With certain alloy elements the high-strength low-alloy steel will develop an oxide protective coating that is pleasing in appearance which has been described as follows:* "It is a very dense corrosion—actually a deeply colored brown, red, purple. . . . It

**Architectural Record*, August 1962.

has a texture and color which cannot be reproduced artificially—a character only nature can give, as with stone, marble, and granite." When steels are to be unpainted and left exposed they are called weathering steels.

As might be expected, the corrosion properties of any steel, including the weathering steels, are dependent on the chemical composition, the degree of pollution in the atmosphere, and the frequency of wetting and drying of the steel.

Since its first major use in 1958, for the Administrative Center for Deere & Company in Moline, Illinois, the use of weathering steel has received considerable attention. At first such steels were specified under ASTM A242 which as previously discussed is very general, allowing a wide variation in chemistry.

With the adoption of A588 steel in 1969, this steel is the high-strength, low-alloy weathering steel. A242 is now essentially obsolete. A588 has been accepted for welded exposed bridges and is increasingly being used for that purpose.

Fabrication and erection of weathering steel requires care. Unsightly gouges, scratches, and dents should be avoided. Painting, even for identification, should be minimized, since all marks must be removed after the erection is completed. Scale and discoloration from welding also must be removed. The extra expense resulting from fabrication and erection is offset by the elimination of painting at intervals during the life of the structure.

To a large extent the need for fireproofing of steel in buildings has slowed the use of exposed steel. Two recent examples of innovative solutions involve (1) keeping the exposed steel members entirely outside the enclosed portion of a building which could contain a fire; and (2) filling exposed hollow steel columns with chemically treated water which will act as a heat sink to keep the temperature of the steel down should the columns be subject to fire.[10]

SELECTED REFERENCES

1. R. L. Brockenbrough and B. G. Johnston, *Steel Design Manual*, U.S. Steel Corporation, Pittsburgh, Pa., 1968, Chap. 1.
2. Highway Structures Design Handbook, Vol. 1, U.S. Steel Corporation, Pittsburgh, Pa., 1965.
3. *Welding Handbook*, 6th ed., Vol. 1, *Fundamentals of Welding*, American Welding Society, New York, 1968, Secs. 2.50–2.72.
4. *Welding Handbook*, 5th ed., Vol. 4, *Metals and Their Weldability*, American Welding Society, New York, 1966, Secs. 61.10, 63.1–63.17.
5. "Design of Light Gage Cold-Formed Stainless Steel Structural Members," American Iron and Steel Institute, New York, 1968.

6. "Specification for the Design of Light Gage Cold-Formed Steel Structural Members," American Iron and Steel Institute, New York, 1968.
7. "A Primer on Brittle Fracture," Booklet 1960-A, Steel Design File, Bethlehem Steel Company, Bethlehem, Pa.
8. W. H. Munse and LaMotte Grover, *Fatigue of Welded Structures*, Welding Research Council, New York, 1964.
9. C. P. Larrabee, "Corrosion Resistance of High-Strength Low-Alloy Steels as Influenced by Composition and Environment," *Corrosion Magazine*, Vol. 9, No. 8 (August 1953), pp. 259-271.
10. "Weathering Steels Become Loadbearing," *Progressive Architecture*, September 1967.

3

Tension Members

3.1. INTRODUCTION

Tension members are encountered in most steel structures. They occur as principal structural members in bridge and roof trusses, in truss structures such as transmission towers and wind bracing systems in multistoried buildings. They frequently appear as secondary members, being used as tie rods to stiffen a trussed floor system or to provide intermediate support for a wall girt system. Tension members may consist of a single structural shape or they may be built up from a number of structural shapes. The cross sections of some typical tension members are shown in Fig. 3.1.1.

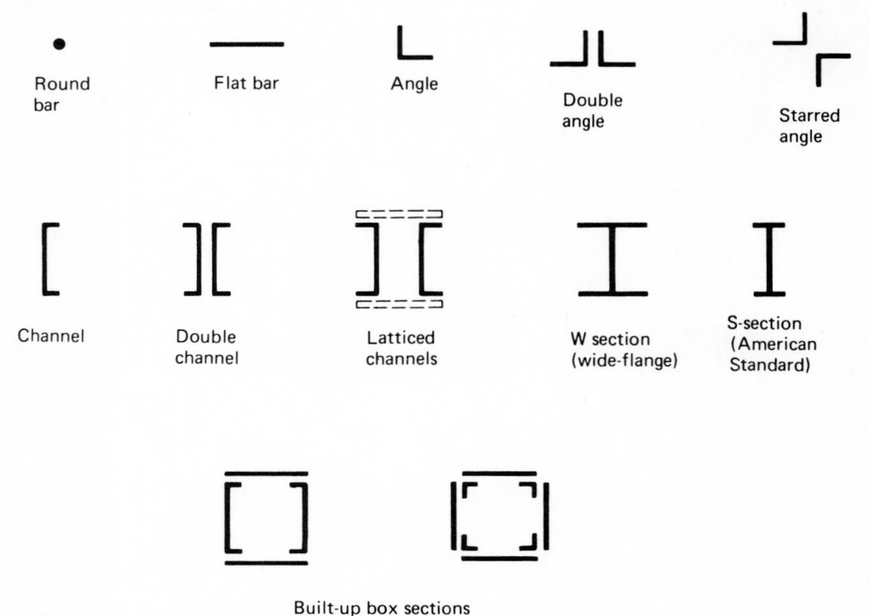

Fig. 3.1.1. Cross section of typical tension members.

In general, the use of single structural shapes is more economical than the built-up sections. However, built-up members may be required when (a) the tensile capacity of a single rolled section is not sufficient, (b) the L/r ratio (the ratio of the unbraced length to the minimum radius of gyration) does not provide sufficient rigidity, (c) the effect of bending combined with the tensile behavior requires a larger lateral stiffness, (d) unusual connection details require a particular cross section, or (e) esthetics control.

3.2. STRENGTH AS A DESIGN CRITERION

The design of a tension member is one of the simplest and most straightforward problems in structural engineering. Since stability is only of secondary concern, the problem of designing a tension member is basically one of providing a member with a sufficient cross-sectional area to resist the applied loads with an adequate factor of safety against failure.

Usual design procedures, though nominally prescribed as working stress method, actually are based on ultimate strength. The ultimate capacity of a tension member is

$$T_u = F_y A_{\text{net}} \tag{3.2.1}$$

Referring to Fig. 3.2.1, once all fibers have reached yield strain, un-

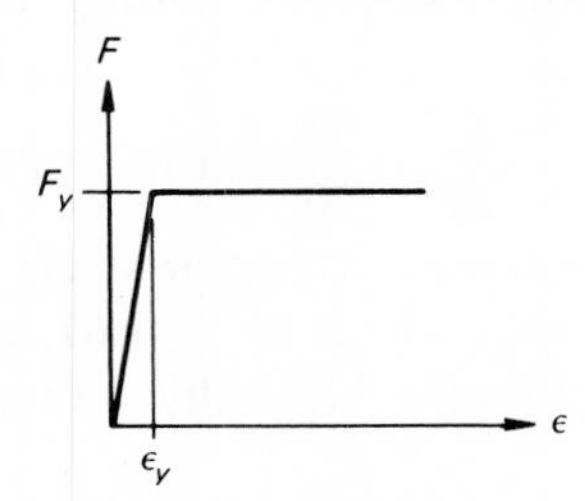

Fig. 3.2.1. Stress-strain curve for steel.

restricted plastic flow takes place, and ultimate capacity has been achieved.

If the ultimate load is divided by a factor of safety the safe service condition (i.e. working) load is obtained. Thus

$$T_w = \frac{F_y A_{\text{net}}}{FS} = F_t A_{\text{net}} \tag{3.2.2}$$

where F_t may be termed the allowable stress for working conditions.

It should be understood that reductions in area, such as holes in members, or threading on rods, cause stress concentrations so that under

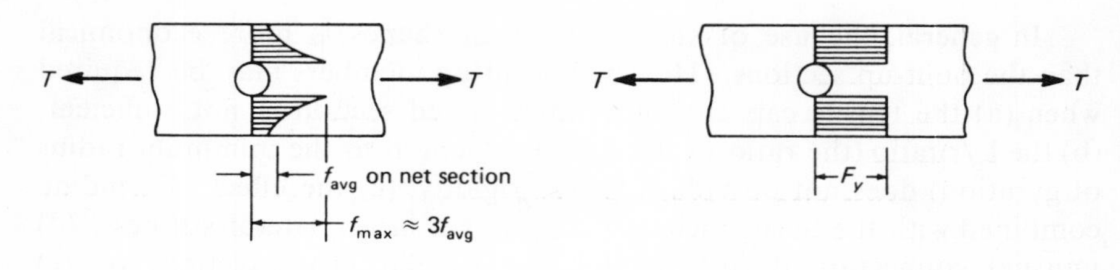

Fig. 3.2.2. Stress distribution with holes present.

service conditions the stress distribution may be as in Fig. 3.2.2a rather than uniform. Theory of elasticity shows that tensile stress adjacent to a hole will be about three times the average stress on the net area. However, as each fiber reaches yield strain its stress then becomes a constant F_y, with deformation continuing with increasing load until finally all fibers have achieved or exceeded yield strain (Fig. 3.2.2b).

According to the nominal working-stress method for design, the area required to resist a given tensile load T in a member is simply

$$\text{Reqd } A_{\text{net}} = \frac{T}{F_t}$$

where F_t is the allowable tensile stress $= F_y/(\text{Factor of safety})$. According to previous practice and AISC specifications, the allowable tensile stress for the now obsolete A7 steel was 20 ksi. Since A7 steel had a specified minimum yield point of 33 ksi, the resulting factor of safety was equal to 1.65: i.e.,

$$\frac{F_y}{F_t} = \frac{33}{20} = 1.65$$

Hence the basic allowable tensile stress for steels with other yield points has been prescribed as $0.6F_y$, which is $F_y/1.67$, approximately the same as the earlier value for A7 steel. Allowable stresses for axial tension are given in Table 3.2.1.

TABLE 3.2.1
Allowable Stresses in Tension on Net Section

AISC–1.5.1.1.–1969	
On net section except at pin holes,	$F_t = 0.60F_y$
On net section at pin holes in eyebars, pin-connected plates or built-up members,	$F_t = 0.45F_y$
Threaded parts of steel meeting requirements of Section 1.4.1	$F_t = 0.60F_y$*
*Applied to tensile stress area equal to $0.7854\left(D - \frac{0.9743}{n}\right)^2$ where D is the major thread diameter and n the number of threads per inch.	
AASHO–1.7.1–1969	
On net section (except where fatigue limitations require lower values)	$F_t = 0.55F_y$

3.3. NET SECTION

Whenever a tension member is to be fastened by means of bolting or riveting, holes must be provided at the location of the connection. As a result, the cross sectional area is reduced and the allowable tensile load in the member must also be reduced. The amount by which the allowable tensile load must be reduced is dependent upon the size and spacing of the holes.

There are a number of methods which can be used to make holes depending upon the special requirements of the connection. The most common method is to punch full-sized holes $\frac{1}{16}$ in. larger than the diameter of the rivet or bolt. During the punching operation the metal at the edge of the holes is damaged. This difficulty is overcome by assuming that the extent of the damage is limited to a distance of $\frac{1}{32}$ in. around the hole. The total diameter to be deducted becomes, therefore, the diameter of the bolt or rivet plus $\frac{1}{16} + 2(\frac{1}{32})$ or $\frac{1}{8}$ in. Full-sized punched holes are the least expensive to form.

A second method consists essentially of subpunching the holes $\frac{3}{16}$ in. diam. undersize and then reaming the holes to the finished size after the pieces being joined are assembled. This method is more expensive than that of punching full-sized holes but does offer the advantages of very accurate alignment of holes.

A third method consists of drilling the holes to a diameter of the bolt or rivet plus $\frac{1}{32}$ in. This method is used to join very thick pieces, and is the most expensive of the common methods.

An additional criterion imposed by the AISC specifications is that the effective net area shall not exceed 85 percent of the gross area whenever there is a hole (or holes) in a tension member. The reason for this is that due to the effect of stress concentrations, the strength of such a tension member is substantially reduced. Experiments have shown that a tension member will have about a 15 percent reduction in ultimate strength even though the area removed by the hole is substantially less.

EXAMPLE 3.3.1

What is the net area of the tension member shown in Fig. 3.3.1?

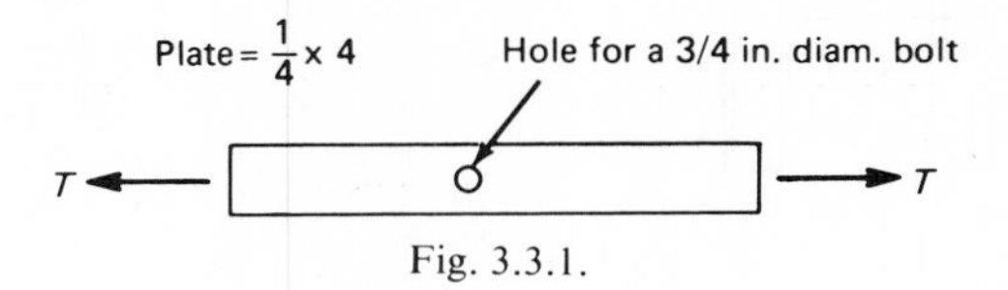

Fig. 3.3.1.

Solution

$$\text{Gross area} = 4(0.250) = 1.000 \text{ sq in.}$$
$$\text{Diameter to be deducted for hole} = 0.750 + 0.125 = 0.875 \text{ in.}$$

$$\begin{aligned}\text{Net area} &= \text{Gross area} - (\text{Dia. for Hole}) \cdot (\text{thickness of plate})\\ &= 1.000 - 0.875(0.250)\\ &= 1.000 - 0.219 = \mathit{0.781\ sq\ in.}\\ \text{Max. net area} &= 0.85(1.0) = 0.85 \text{ sq in.} > 0.78 \text{ sq in.}\end{aligned}$$

3.4 TENSION RODS

A very common and simple tension member is the threaded rod. Such rods are usually secondary members where the design stress is small, such as (a) sag rods to help support purlins in industrial buildings (Fig. 3.4.1a); (b) vertical ties to help support girts in industrial building walls (Fig. 3.4.1b); (c) hangers, such as tie rods supporting a balcony (Fig. 3.4.1c); and (d) tie rods to resist the thrust of an arch.

Tie rods are frequently used with an initial tension as diagonal wind bracing in walls, roofs, and towers. The initial tension effectively adds to

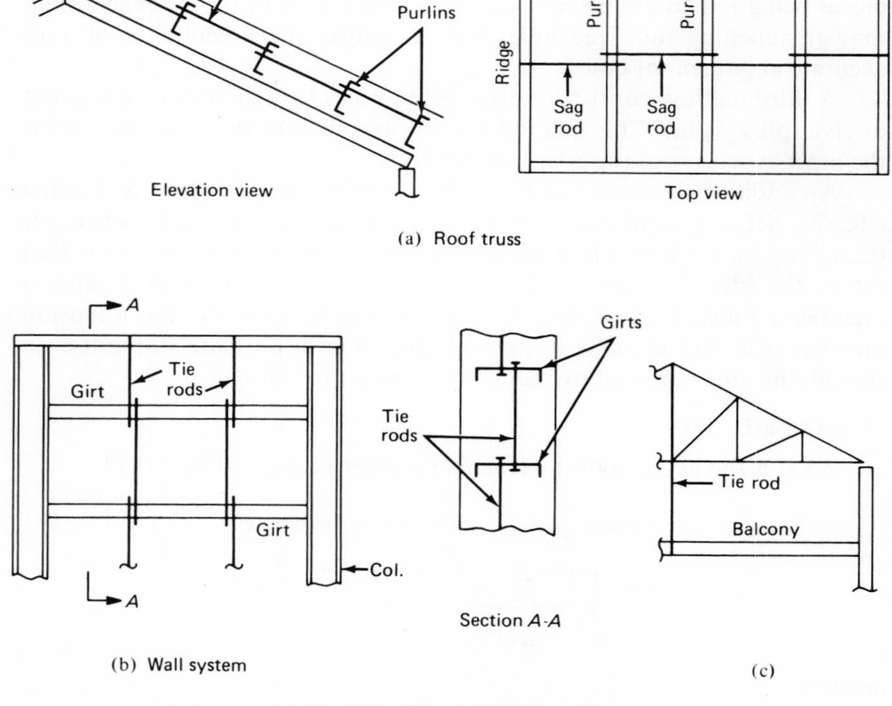

Fig. 3.4.1. Uses of tension rods.

the stiffness and reduces deflection and vibrational motion which tends to cause fatigue failures in the connections. Such initial tension can be obtained by designing the member something on the order of $^1/_{16}$ in. short for a 20-ft length.

EXAMPLE 3.4.1

Determine the size of a threaded round steel tension rod to carry 9 kips using steel having $F_y = 36$ ksi ($F_t = 0.6F_y = 21.6$ ksi. AISC uses $F_t = 22$ ksi).

Solution

The net area required is

$$A_{\text{net}} = \frac{9}{22} = 0.409 \text{ sq in.}$$

Information on standard threaded rods is to be found in the section, "Threaded Fasteners," of the AISC Manual, where the tensile stress area should be used as the basis for design. This is slightly larger than the area through the root of the threads and is an empirical quantity to account for the added strength resulting when loads bear against the threads at an angle to the axis of the member.

Select $\dfrac{^7/_8\text{-in. dia. rod}}{\text{(9 threads per inch)}}$ (A_{net}(tensile stress) = 0.462 sq in.)

Note: AISC–1.15.1 requires "connections carrying calculated stresses, except for lacing, sag bars, and girts, shall be designed to support not less than 6 kips."

EXAMPLE 3.4.2

Design sag rods to support the purlins of the industrial building roof of Fig. 3.4.2. Sag rods are spaced at $^1/_3$ points between roof trusses, which are spaced 24 ft apart. Assume $F_a = 22$ ksi.

Solution

(a) Loads: Roofing—corrugated asbestos = 3 psf
Purlin (assume given) = $3\,^1/_2$ psf

Since purlins must have been designed first, their weight is known:

Snow 20 psf (cos 25°) = 18.1 psf
Total (per sq ft of roof area) = 24.6 psf

Note that American Standard Building Code (excerpt in AISC Manual) requires design for "not less than 20 psf of horizontal projection in addition to the dead load, and in addition to either the wind or earthquake

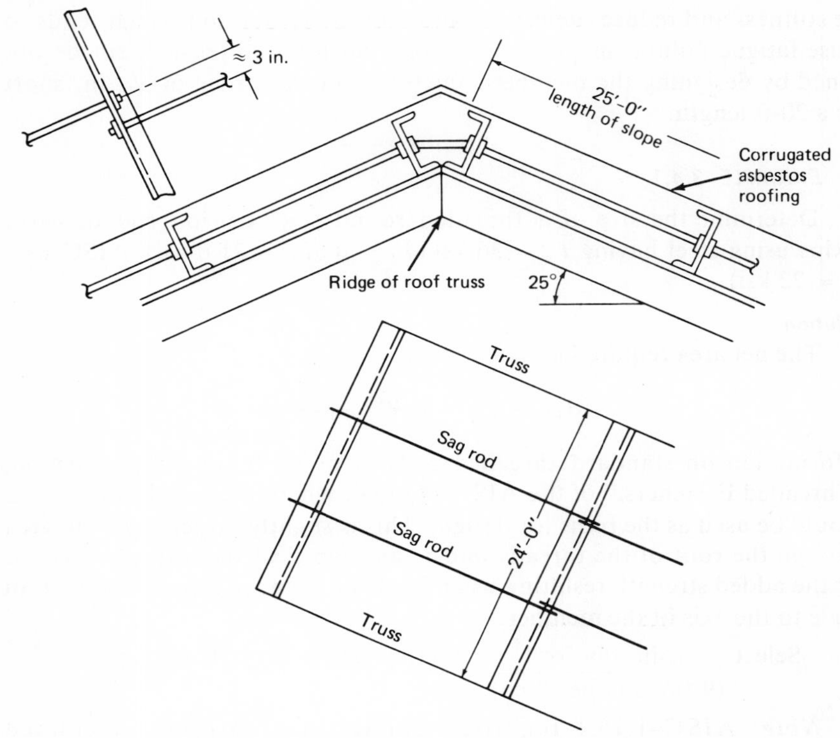

Fig. 3.4.2. Details for Example 3.4.2.

load, whichever produces the greater stresses." This would account properly for snow loading in some areas, but in northern areas where moderate to heavy snowfall is expected 30 to 40 psf should be used.

(b) Sag rods are expected to carry the component of the load acting parallel to the roof.

$$w \text{ (parallel to roof)} = 24.6 \sin 25^\circ = 24.6(0.423) = 10.4 \text{ psf}$$

(c) Choose rod size.

Load carried by one rod:

$$T = w \text{ (tributary area)} = 10.4(25)(8) = 2{,}080 \text{ lb}$$

$$A_{\text{net}} = \frac{2080}{22{,}000} = 0.0945 \text{ sq in.}$$

Select for strength a $\frac{1}{2}$-in. diam. threaded rod. $A_{\text{net}} = 0.142$ sq in. For good practice (to give stiffness; see Sec. 3.7) the smallest standard rod should be $\frac{5}{8}$ in. diam.

Use <u>$\frac{5}{8}$ in. diam. rod.</u>

3.5. EFFECT OF HOLE SPACING ON NET SECTION

Whenever there is more than one row of holes in a tension member, a number of potential transverse failure lines are frequently possible. The designer must then determine the failure line which will yield the minimum net section. The minimum section thus determined is called the "critical" net section and becomes the basis for design, representing the section at which failure is most likely to occur.

In Fig. 3.5.1a the failure plane is seen to be along the section A–B. In Fig. 3.5.1.b the failure plane would not be along A–C, since the length

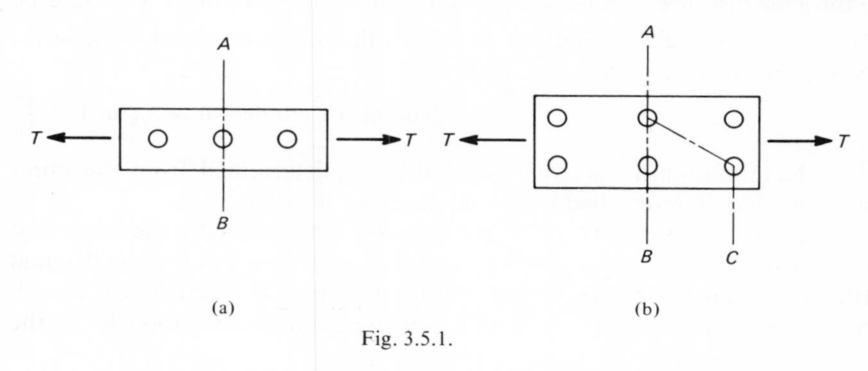

Fig. 3.5.1.

of the line through A–C would be greater than the length of the line through A–B. In case (b), both planes A–B and A–C would require deductions for two holes. The minimum or critical net area would therefore be along the plane A–B.

Figure 3.5.2 shows two rows of staggered holes with a spacing p,

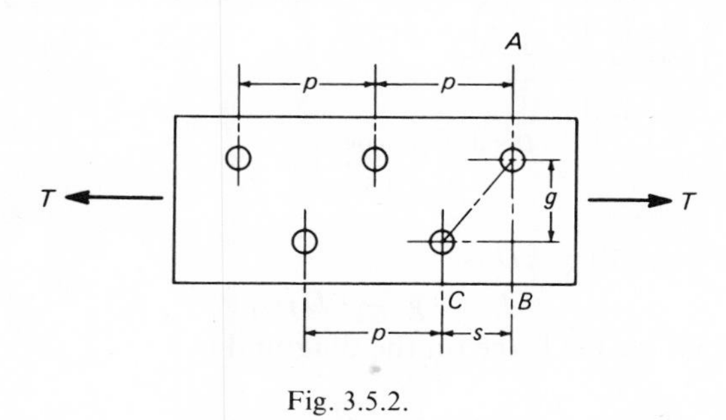

Fig. 3.5.2.

and separated by a gage distance g. In this case it is not immediately evident whether A–B or A–C is the critical section. At first glance one might think section A–B is critical since the path A–B is obviously shorter than path A–C. However, from path A–B, only one hole would be deducted while two holes would have to be deducted from path A–C. In order to determine the actual critical section, both paths A–B and A–C must be investigated. Accurate checking of strength along path A–C is complex. However, a simplified empirical relationship[1] has been adopted by the AISC specifications to account for the difference between the path A–C and the path A–B expressed as a length correction,

$$\frac{s^2}{4g}$$

where s is the stagger, or spacing of adjacent holes (see Fig. 3.5.2), and g the gage distance. Thus the net lengths of paths A–B and A–C would be

$$\text{Net length of } A\text{–}B = \text{length of } (A\text{–}B) - (\text{diam. of connector} + \tfrac{1}{8} \text{ in.})$$
$$\text{Net length of } A\text{–}C = \text{length of } (A\text{–}B) - 2(\text{diam. of connector} + \tfrac{1}{8} \text{ in.}) + \frac{s^2}{4g}$$

The minimum net section would then be determined from the minimum net length multiplied by the thickness of the plate.

The development of the $s^2/4g$ expression began with the idea that the tensile stress across a transverse section of length g–kD should equal the maximum principal stress along a diagonal section of length $\sqrt{g^2 + s^2} - D$ (see Fig. 3.5.3). Thus if the axial tensile stress is f_t, the

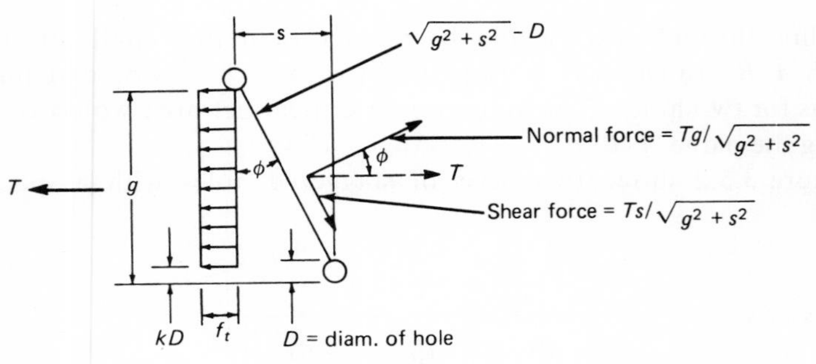

Fig. 3.5.3.

tensile force over length g–kD is

$$T = (g - kD)tf_t \tag{3.5.1}$$

while the principal tensile force on the diagonal is

$$f_t'(\sqrt{g^2 + s^2} - D)t = \frac{T}{2} g/\sqrt{g^2 + s^2} + \sqrt{\left[\frac{T}{2} g/\sqrt{g^2 + s^2}\right]^2 + [Ts/\sqrt{g^2 + s_2}]^2} \tag{3.5.2}$$

which upon solving for T, gives

$$T = \frac{2f_t'[g^2 + s^2 - D\sqrt{g^2 + s^2}]}{g + \sqrt{g^2 + 4s^2}}\,t \tag{3.5.3}$$

The quantity kD becomes then a length correction which would eliminate concern about principal stresses. If the maximum principal stresses cannot exceed the axial stress f_t, then Eqs. 3.5.1 and 3.5.2 can be equated and solved for the length correction kD. Thus

$$kD = g - \frac{2[g^2 + s^2 - D\sqrt{g^2 + s^2}]}{g + \sqrt{g^2 + 4s^2}} \tag{3.5.4}$$

based on equal principal tensile stresses. Cochrane[1] proposed that if k were taken as $\frac{1}{2}$ in Eq. 3.5.4 the transverse net area gt would about equal the diagonal net area, $[\sqrt{g^2 + s^2} - D]t$;
i.e., for equal areas,

$$\begin{aligned} g &= \sqrt{g^2 + s^2} - D \\ (g + D)^2 &= g^2 + s^2 \\ s &= \sqrt{2gD + D^2} \end{aligned} \tag{3.5.5}$$

Since many considered the equal area concept to be the more logical method anyway, this combination proposal of Cochrane's was endorsed. His final proposal was that k vary from zero for no stagger (ordinary deduction for the holes) to $k = \frac{1}{2}$ for $s = \sqrt{2gD + D^2}$. Further, he assumed that k generally varied as s^2. Thus

$$k = 1 - \frac{s^2}{2(2gD + D^2)} \tag{3.5.6}$$

If D is small compared to g, Eq. 3.5.6 becomes

$$kD = D - \frac{s^2}{4g} \tag{3.5.7}$$

It is therefore apparent if one is considering the diagonal path and deducts two holes, one should add on the quantity

$$D - kD = \frac{s^2}{4g} \tag{3.5.8}$$

This would be the same as considering the transverse section with the fraction kD of one hole deducted:

$$\begin{aligned} \text{Net width} &= g - kD \\ &= g - (D - s^2/4g) \\ &= g - D + s^2/4g \end{aligned} \tag{3.5.9}$$

The $s^2/4g$ approach gives good agreement with the maximum stress theory. In the years since Cochrane proposed the simple $s^2/4g$ expression many investigators have proposed other rules[2-5] but none of them give significantly better results and all are more complicated.

Consistent with the general trend toward using design approaches that relate to the ultimate strength, the work of Bijlaard[6] and others[7] has provided limit analysis theories to obtain net area in tension. These theories also do not deviate from the $s^2/4g$ method by more than 10 to 15 percent.

The reader is referred to McGuire[11] for a more complete coverage of this subject of net section through staggered lines of connectors.

Most of the recent studies, both empirical based on tests, and analytical based on plastic theories indicate that the effective net section with holes cannot exceed 85 percent of the gross section.[8-10]

EXAMPLE 3.5.1

Determine the minimum net area of the member shown in Fig. 3.5.4, assuming $\frac{7}{8}$ in. diam. bolts are to be used.

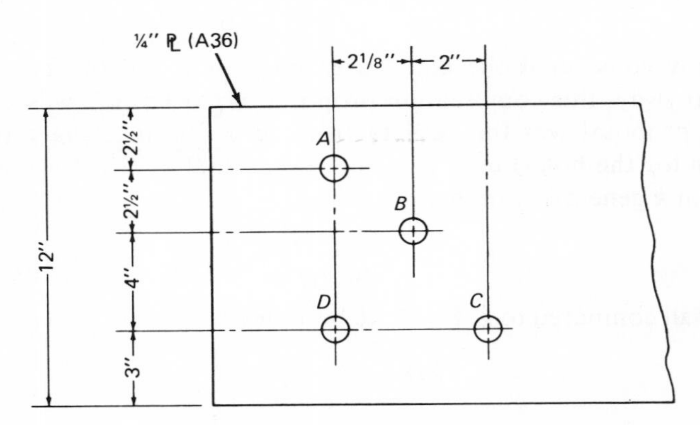

Fig. 3.5.4. Example 3.5.1.

Solution

Path *AD*:

$$[12 - 2(0.875 + 0.125)]0.250 = 2.500 \text{ sq in.}$$

Path *ABD*:

$$\left[12 - 3(0.875 + 0.125) + \frac{(2.125)^2}{4(2.50)} + \frac{(2.125)^2}{4(4.00)}\right]0.25 = 2.434 \text{ sq in.}$$

Path *ABC*:

$$\left[12 - 3(0.875 + 0.125) + \frac{(2.125)^2}{4(2.5)} + \frac{(2.00)^2}{4(4.00)}\right]0.25 = 2.426 \text{ sq in.} \quad \underline{\text{Ans.}}$$

Check 85% of gross area:

$$(0.85)12(0.25) = 2.550 \text{ sq in.}$$

EXAMPLE 3.5.2

Calculate the least net section for the single lap joint in Fig. 3.5.5 and show free-body diagrams of portions of top plate with sections taken through each line of holes.

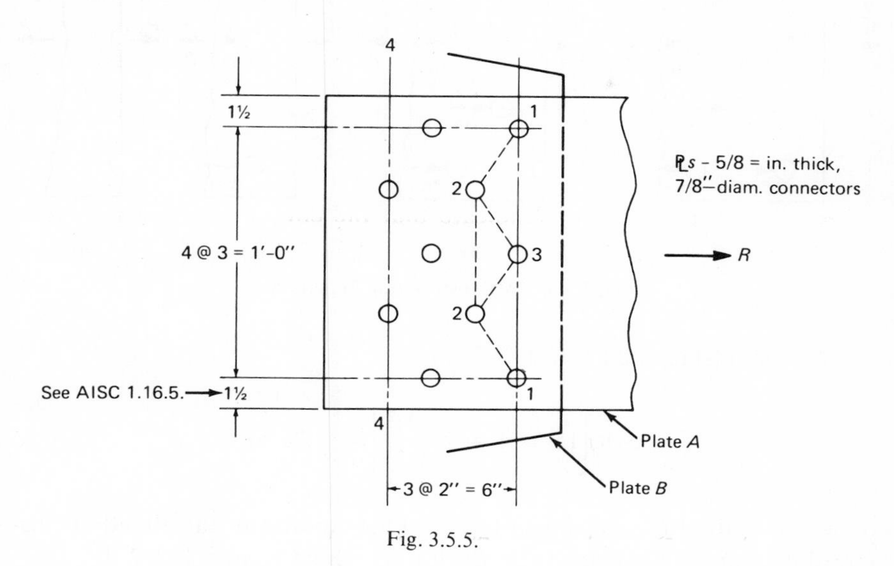

Fig. 3.5.5.

Solution

This problem illustrates use of the basic assumption of connection design that each equal size connector carries an equal share of the load. The merits of such an assumption are treated in Chapter 4. In effect, then, the section through the first line of connector holes (section 1–1) in plate *A* and the section through the last line of connector holes (section 4-4) in plate *B*, must each carry the full load *P*. Sections in between in either plate carry less than *P*.

$$\text{Max. effective net area} = .85A_{\text{gross}} = 0.85(15.0)(5/8) = 7.96 \text{ sq in.}$$

$$\text{Net area (section 1–1)} = 5/8(15 - 3) = 7.50 \text{ sq in.}$$

on which 100 percent of *P* acts (Fig. 3.5.6d).

Net area (staggered path 1–2–3–2–1):

$$= 5/8\left[15 - 5(1) + 4\,\frac{(2)^2}{4(3)}\right] = 7.08 \text{ sq in.} \qquad \underline{\text{Ans.}}$$

$s^2/4g$

on which 100 percent of *P* also acts.

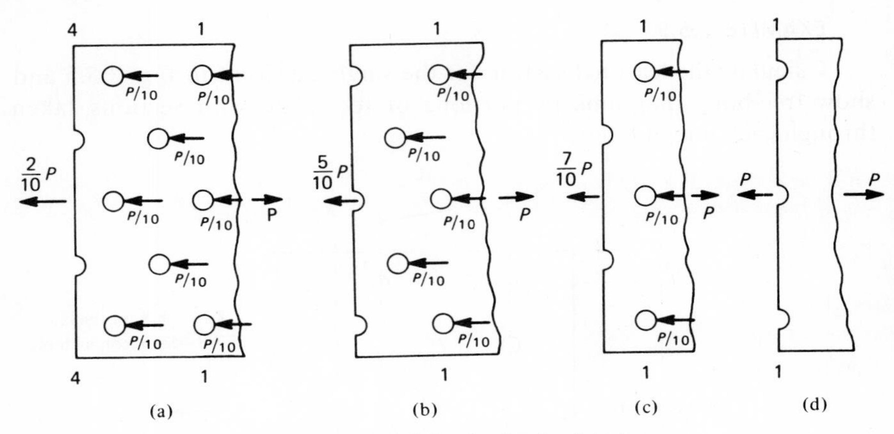

Fig. 3.5.6. Load distribution in plate A.

Net area (staggered path 1–2–2–1):

$$= 5/8 \left[15 - 4 + 2\,\frac{(2)^2}{4(3)}\right] = 7.29 \text{ sq in.}$$

on which 9/10 of P is presumed to act. One connector has already transferred its share of the load (1/10) prior to reaching section 1–2–2–1.

The distribution presumed to act through the plate A is shown in Fig. 3.5.6.

3.6. NET SECTION OF ANGLES WITH STAGGERED HOLES

The net area of an angle is determined on the basis of its thickness and its net length. The net length is assumed to be along the centerline of the angle shown in Fig. 3.6.1. and is equal to:

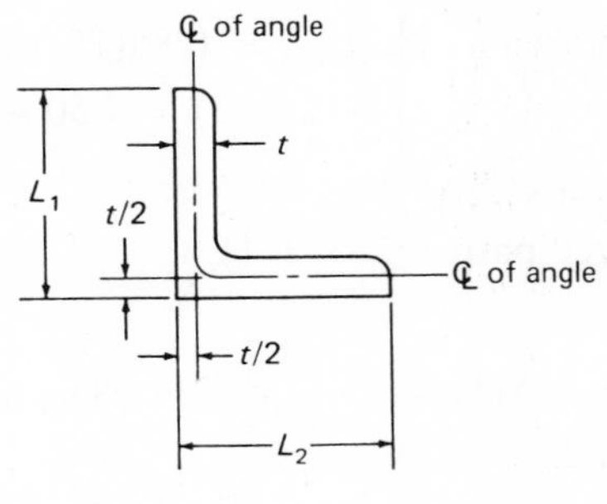

Fig. 3.6.1.

$$\text{Net length} = L_1 + L_2 - t/2 - t/2$$
$$= L_1 + L_2 - t$$

Every rolled angle has a standard value for the location of holes (i.e., gage distances) depending upon the size of the angle. Table 3.6.1. shows the usual gages for angles as listed in the AISC Manual.

Unless a designer specifically indicates a different gage distance, the steel detailer will specify the standard gage distances shown in Table 3.6.1. These gage distances will also be used for the standard beam connections as discussed in Chapter 13. Gage distances other than standard should be avoided whenever possible, because higher fabrication costs will result.

TABLE 3.6.1
Usual Gages for Angles, Inches (from AISC Manual)

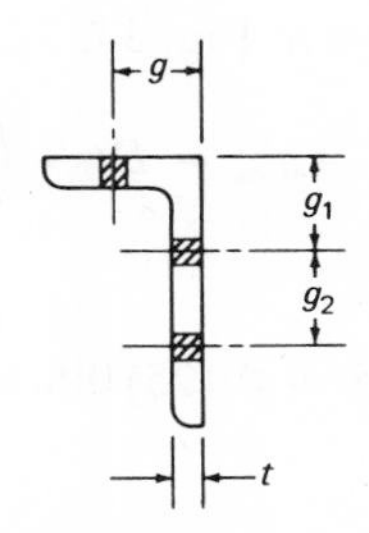

Leg	8	7	6	5	4	3½	3	2½	2	1¾	1½	1⅜	1¼	1
g	4½	4	3½	3	2½	2	1¾	1⅜	1⅛	1	⅞	⅞	¾	⅝
g_1	3	2½	2¼	2										
g_2	3	3	2½	1¾										

EXAMPLE 3.6.1

Determine the net section and allowable capacity of the angle shown in Fig. 3.6.2 if A36 steel and 7/8 in. diam. bolts are used. Assume that the bolt strength does not govern.

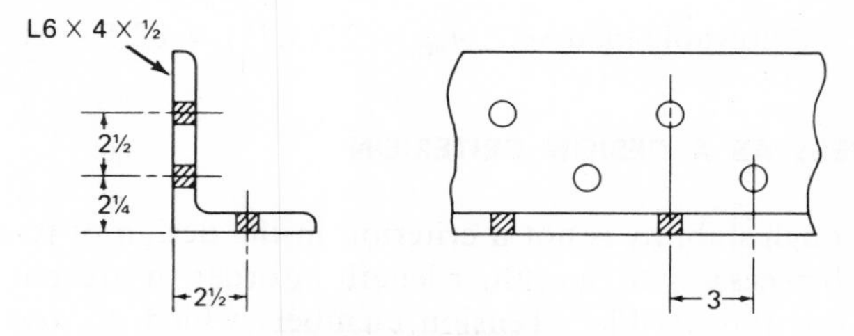

Fig. 3.6.2. Example 3.6.1.

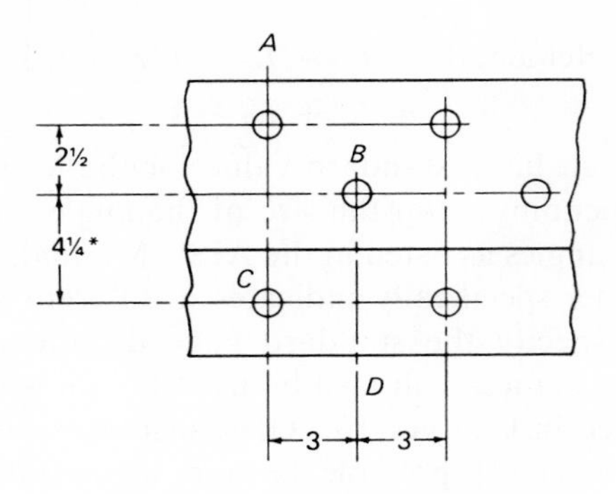

Fig. 3.6.3.

Solution

Assuming that the net length is along the centerline of the angle, the section is "developed" as shown in Fig. 3.6.3.

$$A_{\text{net}} = A_{\text{gross}} - Dt + \left(\frac{s^2}{4g}\right) t$$

Path AC:

$$4.75 - 2(0.875 + 0.125)0.5 = 3.75 \text{ sq in.}$$

Path ABC:

$$4.75 - 3(0.875 + 0.125)0.5 + \left[\frac{(3)^2}{4(2.50)} + \frac{(3)^2}{4(4.25)}\right](0.5) = 3.96 \text{ sq in.}$$

Path ABD:

$$4.75 - 2(0.875 + 0.125)0.5 + \frac{(3)^2}{4(2.50)}(0.5) = 4.20 \text{ sq in.}$$

Check 85% of gross area:

$$0.85A_g = 0.85(4.75) = 4.04 \text{ sq in.}$$

Path AC governs:

$$\text{Allowable load} = F_t A_{\text{net}} = 22(3.75) = \underline{82.5 \text{ kips}}$$

3.7 STIFFNESS AS A DESIGN CRITERION

Even though stability is not a criterion in the design of tension members, it is still necessary to limit their length in order to prevent a member from becoming too flexible. Tension members which are too long may sag excessively due to their own weight. In addition, they may also vibrate when subjected to wind forces as in an open truss or when supporting vibrating equipment such as fans or compressors.

In order to reduce the problems associated with excessive deflections and vibrations, it is necessary to establish a stiffness criterion. This criterion is based on the slenderness ratio, L/r, of a member where L is the length and r the radius of gyration. The accepted maximum slenderness ratios for tension members (see AISC–1.8.4) are:

	AISC	AASHO
For main members	240	200
For bracing and other secondary members	300	240

In applying the stiffness criterion to tension members, the higher slenderness ratio based on the two principal axis must be used. A symmetrical member may have two different radii of gyration, and for nonsymmetrical members one must consider the weakest principal axis. When tension members are built up from a number of sections, the radii of gyration must be computed from the relationship

$$r = \sqrt{\frac{I}{A}}$$

where I is the moment of inertia and A is the cross sectional area. The value for r will be about the same axis as that of the moment of inertia used in the above relationship. If a tension member is not restrained laterally, the lowest value for r will give the maximum slenderness ratio.

EXAMPLE 3.7.1

Determine the maximum length permitted by the AISC specifications for a tension member whose cross section is a flat bar 1 × 6.

Solution

Determine the least radius of gyration, which may be shown to be a function of the lateral dimension of the member only.

$$I = bt^3/12; \qquad A = bt$$

$$r = \sqrt{\frac{I}{A}} = \sqrt{\frac{bt^3/12}{bt}} = t\sqrt{\frac{1}{12}} = 0.288t$$

Thus the least r occurs with respect to the least lateral dimension:

$$r_{min} = 0.288(1) = 0.288 \text{ in.}$$

$$\frac{L}{r} = 240 = \frac{L}{0.288t}; \qquad L = 69.4 \text{ in.}$$

EXAMPLE 3.7.2

Determine the maximum unsupported length permitted by the AISC specifications for the cross sections indicated.

(a) C12 × 20.7
(b) L5 × 5 × ½
(c) 2 – L6 × 4 × ⅜ (long legs back-to-back)

Solution

The values for r are taken from the AISC Manual.

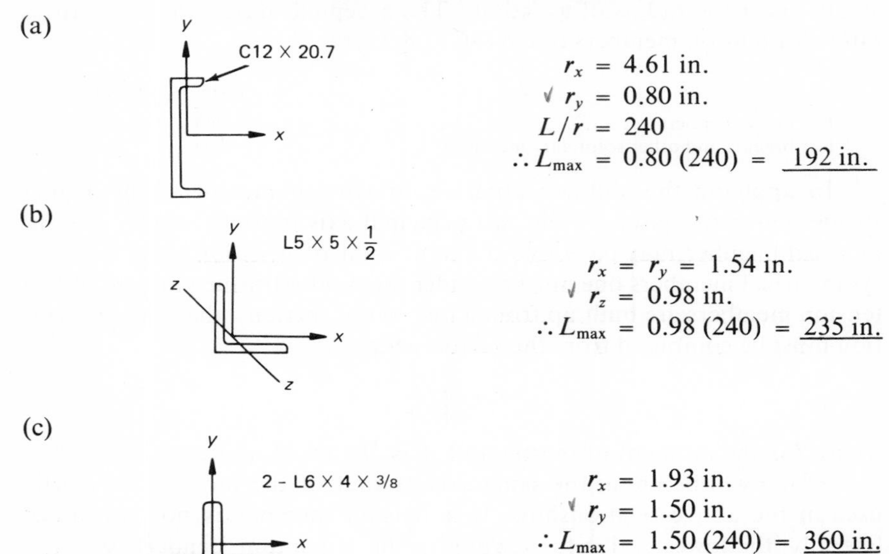

(a)

$$r_x = 4.61 \text{ in.}$$
$$r_y = 0.80 \text{ in.}$$
$$L/r = 240$$
$$\therefore L_{max} = 0.80\,(240) = \underline{192 \text{ in.}}$$

(b)

$$r_x = r_y = 1.54 \text{ in.}$$
$$r_z = 0.98 \text{ in.}$$
$$\therefore L_{max} = 0.98\,(240) = \underline{235 \text{ in.}}$$

(c)

$$r_x = 1.93 \text{ in.}$$
$$r_y = 1.50 \text{ in.}$$
$$\therefore L_{max} = 1.50\,(240) = \underline{360 \text{ in.}}$$

3.8. DESIGN PROCEDURE—AISC 1969

The following example illustrates a rational procedure for combining (a) the strength requirement; and (b) the stiffness requirement, to select standard rolled shapes containing holes.

EXAMPLE 3.8.1

A tension diagonal member for a roof truss is to be selected of steel with F_y = 50 ksi. The axial tension is 60 kips and the member is 12 ft long. Assume ⅞ in. diam. bolts will be located on a single gage line.

(a) Select the lightest single angle member.
(b) Select the lightest double angle member.

Solution

Considering that 85 percent of gross area is the maximum effective net area, the minimum required gross area may be established. Also, the minimum r to satisfy AISC–1.8.4 may be determined.

$$F_t = 0.60\,F_y = 0.60\,(50) = 30 \text{ ksi}$$

$$\text{Reqd } A_n = \frac{60}{30} = 2.00 \text{ sq in.}$$

$$\text{Min. } A_g = \frac{2.00}{0.85} = 2.35 \text{ sq in.}$$

$$\text{Min. } r = \frac{144}{240} = 0.6 \text{ in.}$$

(a) Select single angle member. The required gross area in each case depends on the area deducted for one hole, which in turn depends on the thickness. The following tabular procedure may be found useful in making the selection:

Standard Thickness t	Reduction for One Hole	Reqd Gross Area	Choices from AISC Manual Single-Angle Properties
1/4 in.	0.250*	2.35 min.	
5/16	0.313	2.35 min.	L4 × 4 × 5/16, A = 2.40, r = 0.79†
3/8	0.375	2.38‡	L5 × 3 × 5/16, A = 2.40, r = 0.66
7/16	0.438	2.44	
1/2	0.500	2.50	

*(7/8 + 1/8) 0.25 = 0.250 sq in.
†*Note*: Min. $r = r_z$ for single angles.
‡Reqd A_g = 2.00 + 0.375 = 2.375, say 2.38 > 2.35 min.

In searching for possible choices, the L4 × 4 × 5/16 was the first one found; thereafter no section with an area exceeding 2.40 was written down.

Use L4 × 4 × 5/16 Single-Angle Member.

(b) Select double-angle member. For this type of section two holes must be deducted. Selection should be made from the double angle properties in AISC Manual.

Standard Thickness, t	Reduction for Two Holes	Reqd Gross Area	Choices from AISC Manual Double Angle Properties
1/4	0.500	2.50	L3 × 2½ × 1/4, A = 2.62, r = 0.95
5/16	0.626	2.63	L2½ × 2 × 5/16, A = 2.62, r = 0.78
3/8	0.750	2.75	

Use 2 – L3 × 2½ × 1/4 with long legs back-to-back. This is preferred since for the same area it gives the greater r value.

PROBLEMS

Note: For all problems assume connector strength is adequate and does not control. Use AISC Specifications unless otherwise indicated.

3.1. Select a standard threaded rod to carry a tensile force of 10 kips. Use A572 Grade 50 steel.

3.2. Select a standard threaded rod to carry a tensile force of 6 kips. Use A36 steel.

3.3. Design sag rods to support the purlins of an industrial building roof. Sag rods are placed at $\frac{1}{3}$ points between roof trusses, which are spaced 30 ft apart. Assume roofing and purlin weight is 9 psf of roof surface. Use standard threaded rods and A36 steel.

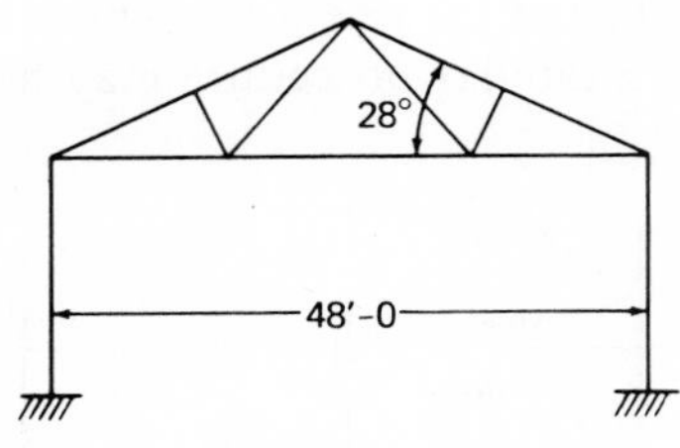

Prob. 3.3.

3.4. Design an eyebar to carry 100 kips, using flame-cut A572 Grade 50 steel plate. (Refer to AISC–1.14.6.).

3.5. Determine the allowable tensile load on an L6 × 4 × $\frac{3}{4}$ using (a) A36 steel, (b) steel with $F_y = 50$ ksi. Assume welded connections.

3.6. Determine the allowable tensile load on the angle in Prob. 3.5 if a single row of 7/8 in. diam. bolts is used in the 4 in. leg and a double row of 7/8 in. diam. bolts is used in the 6-in. leg. Assume no stagger of rows.

3.7. Determine the allowable tensile load on an A36 steel bar with a cross section of 12 in. by $\frac{1}{4}$ in. with a single line of holes for 7/8 in. diam. bolts.

3.8. Select a single angle of A36 steel to support a tensile load of 55 kips assuming a single row of 3/4 in. diam. bolts.

3.9. Repeat Prob. 3.8 using A572 Grade 50 steel.

3.10. Select a pair of angles to support a tensile load of 90 kips using A36 steel. Assume a 3/8-in. gusset plate between the angles and that the connection is to be welded.

3.11. Repeat Prob. 3.10 using A572 Grade 55 steel.

3.12. Repeat Prob. 3.8 if the length of this main member is 20 ft.

3.13. Repeat Prob. 3.10 if the length of this main member is 36 ft.

3.14. What is the maximum length of the angle used in Prob. 3.5 permitted by the AASHO specifications.

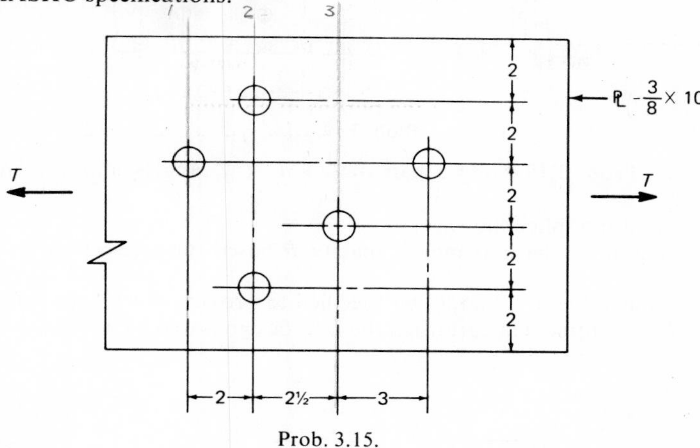

Prob. 3.15.

3.15. Find the net area of the plate shown in the accompanying illustration and determine the maximum value for T if A36 steel is specified and the holes are for 3/4 in. diam. bolts.

3.16. Repeat Prob. 3.15 using A572 Grade 60 steel and 7/8 in. diam. bolts.

3.17. Determine the maximum allowable tensile load for a single C15 × 33.9 fastened to a ½-in. gusset plate. Use A36 steel and AISC specifications. Assume holes are for 7/8 in. diam. bolts.

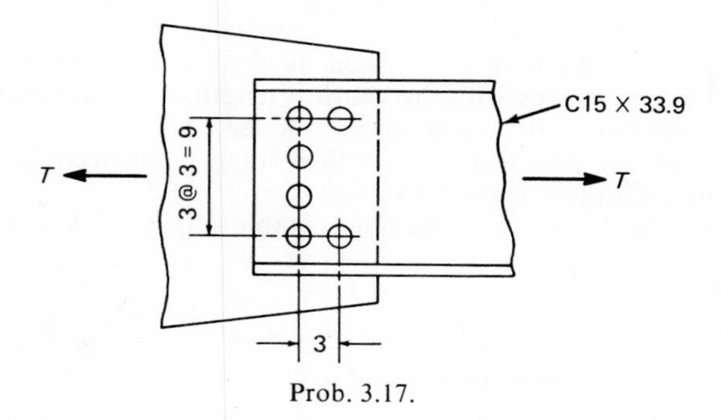

Prob. 3.17.

3.18. Repeat Prob. 3.17 using an MC18 × 42.7.

3.19. Determine the tensile load permitted by AISC for a pair of 6 × 4 × ½ angles with the standard gage distances shown in Table 3.6.1. Assume A36 steel and 3/4 in. diam. bolts.

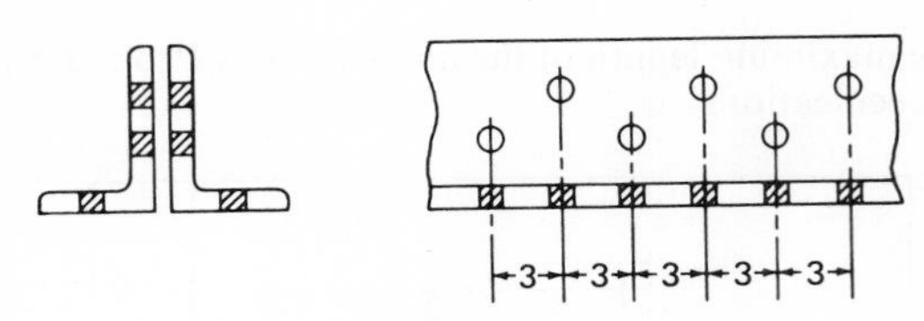

Prob. 3.19.

3.20. Repeat Prob. 3.19 using a pair of 8 × 6 × ¾ angles and 7/8 in. diam. bolts.

3.21. Given the lap splice shown:
(a) Determine the maximum capacity P based on net section, using A36 steel.
(b) What value of s should be specified to provide the maximum capacity P as computed in part (a), if the final design is to have $s_1 = s_2 = s$?

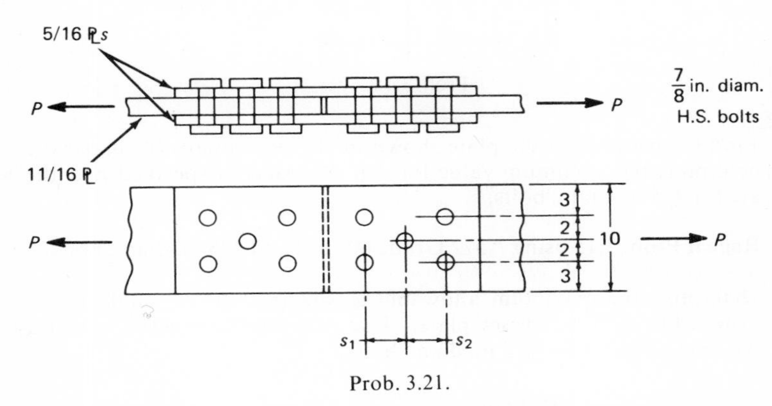

Prob. 3.21.

3.22. A 5 × 3½ × ½ angle is to be used as a tension member. In order to carry a load of 68 kips with the shortest length of connection, two rows of high strength bolts will be used in the 5-in. leg.
(a) What is the minimum acceptable stagger, theoretical and specified (½-in. multiples), using A36 steel?
(b) What would be the minimum acceptable stagger if A572 Grade 50 steel is used?

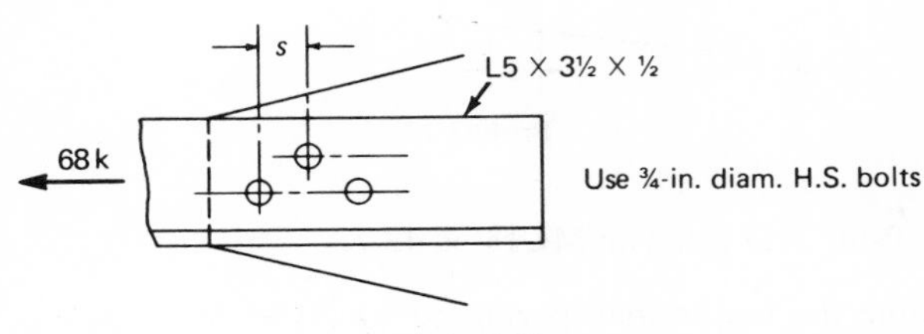

Prob. 3.22.

3.23. A 5 × 3½ × ½ angle is to be used as a tension member carrying a load of 69 kips. Using one gage line of high strength bolts (7/8 in. diam.) in each leg, what would be the minimum stagger between the first connector on each gage line required to accomplish this? Use A36 steel.

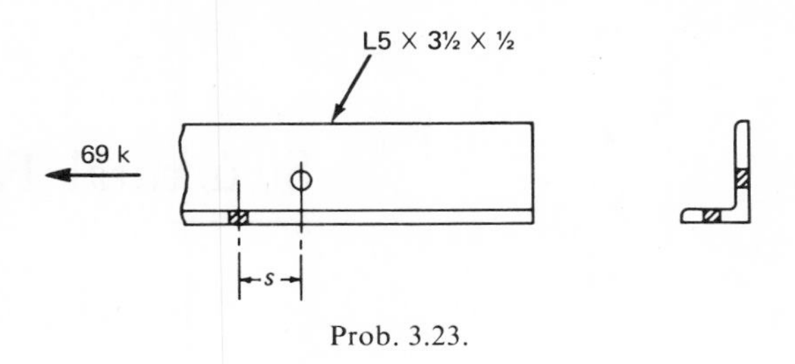

Prob. 3.23.

SELECTED REFERENCES

1. V. H. Cochrane, "Rules for Rivet-Hole Deductions in Tension Members," *Engineering News-Record*, Vol. 89 (1922), pp. 847–848.
2. W. M. Wilson, Discussion of "Tension Tests of Large Rivet Joints," *Trans. ASCE*, Vol. 105 (1942), p. 1268.
3. W. M. Wilson, W. H. Munse, and M. A. Cayci, "A Study of the Practical Efficiency under Static Loading of Riveted Joints Connecting Plates," U. of Illinois Engg. Experiment Station Bulletin 402, 1952.
4. F. W. Schutz, "Effective Net Section of Riveted Joints," Proc. Second Illinois Structural Engg. Conf., November 1952.
5. "Here's a Better Way to Design Splices," *Engineering News-Record*, Vol. 150, Part I (Jan. 8, 1953), p. 41.
6. P. P. Bijlaard, Discussion of "Investigation and Limit Analysis of Net Area in Tension," *Trans. ASCE*, Vol. 120 (1955), pp. 1156–1163.
7. G. W. Brady and D. C. Drucker, "Investigation and Limit Analysis of Net Area in Tension," *Trans. ASCE*, Vol. 120) (1955), pp. 1133–1154.
8. Frank Baron, Edward W. Larson, Jr., and Keith J. Kenworthy, "The Effect of Certain Rivet Patterns on the Fatigue and Static Strengths of Joints," Research Council on Riveted and Bolted Structural Joints, Engg. Foundation, February 1955.
9. W. H. Munse and E. Chesson, Jr., "Riveted and Bolted Joints: Net Section Design," *J. Structural Div. ASCE*, Vol. 89 (February 1963), pp. 107–126.
10. E. Chesson and W. H. Munse, "Behavior of Riveted Connections in Truss-Type Members," *J. Structural Div. ASCE*, Vol. 83, No. ST1 (January 1957).
11. William McGuire, *Steel Structures*, Prentice-Hall, Inc., Englewood Cliffs, N.J., 1968, pp. 310–328.

4

Structural Fasteners

4.1. TYPES OF FASTENERS

Every structure is an assemblage of individual parts or members which must be fastened together, usually at the ends of its members, by some means. One such means is welding which is treated in Chapter 5. The other is bolting and, in a few isolated cases, riveting. This chapter is primarily concerned with bolting; in particular, high-strength bolts. High-strength bolts have for the most part replaced rivets as the principal means of making nonwelded connections. However, for completeness, a brief description of the other fasteners, including rivets and unfinished machine bolts, is given.

High-Strength Bolts. The two basic types of high-strength bolts are designated by ASTM as A325 and A490, the material properties of which are discussed in Chapter 2. These bolts are heavy hexagon-head bolts, used with heavy semifinished hexagon nuts, as shown in Fig. 4.1.1b. The threaded portion is shorter than for bolts in nonstructural applications, and may be cut or rolled. A325 bolts are of heat-treated *medium carbon* steel having an approximate yield strength of 81 to 92 ksi depending on diameter. A490 bolts are also heat-treated but are of *alloy* steel having an approximate yield strength of 115 to 130 ksi depending on diameter.

High-strength bolts range in diameter from ½ to 1½-in. The most common diameters used in building construction are 3/4 in. and 7/8 in., whereas the most common sizes in bridge design are 7/8 in. and 1 in.

High-strength bolts are tightened to develop high tensile stress in them which results in a predictable clamping force on the joint. The actual transfer of service loads through a joint is therefore, due to the friction developed in the pieces being joined. Joints containing high-strength bolts are designed either as friction type, where slip is the basis

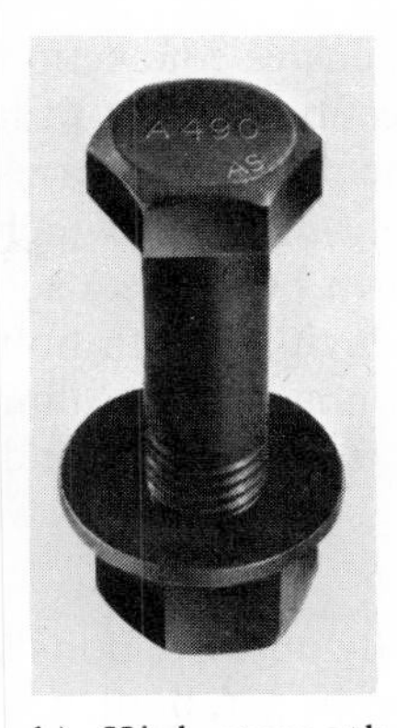

a) Rivet b) High-strength hexagon head bolt c) High-strength interference-body bolt

Fig. 4.1.1. Types of fasteners.

for ultimate strength; or as bearing type, where bearing of the bolt shank against the hole is the basis for ultimate strength.

Installation of these bolts may be either with calibrated torque wrenches, or more commonly with any ordinary wrench using the "turn-of-the-nut" method. The latter method involves making an additional angular turn of the nut starting from the snug position.

Rivets. For many years rivets were the accepted means of connecting members but in recent years have become virtually obsolete. Undriven rivets are formed from bar steel, a cylindrical shaft with a head formed on one end, as shown in Fig. 4.1.1a. Rivet steel is a mild carbon steel designated by ASTM as A502 Grade 1 (F_y = 28 ksi) and Grade 2 (F_y = 38 ksi), with the minimum specified yield strengths based on bar stock as rolled. The forming of undriven rivets and the driving of rivets cause changes in the mechanical properties.

The method of installation is essentially that of heating the rivet to a light cherry-red color, inserting it into a hole and then applying pressure to the preformed head while at the same time squeezing the plain end of the rivet to form a rounded head. During this process the shank of the rivet completely or nearly fills the hole into which it had been inserted. Upon cooling, the rivet shrinks, thereby providing a clamping force. However, the amount of clamping produced by the cooling of the rivet varies from rivet to rivet and therefore cannot be counted on in design calculations. Rivets may also be installed cold but then they do not develop the clamping force since they do not shrink after driving.

Unfinished Bolts. These bolts are made from low carbon steel, designated as ASTM A307, and are the least expensive type of bolt. They may not, however, produce the least expensive connection since many more may be required in a particular connection. Their primary use is in light structures, secondary or bracing members, platforms, catwalks, purlins, girts, small trusses and similar applications in which the loads are primarily small and static in nature. Such bolts are also used as temporary fitting up connectors in cases where high-strength bolts, rivets, or welding may be the permanent means of connection. Unfinished bolts are sometimes called common, machine, or rough bolts and may come with square heads and square nuts.

Turned Bolts. These practically obsolete bolts are machined from hexagonal stock to much closer tolerances (about $^1/_{50}$ in.) than unfinished bolts. This type of bolt was primarily used in connections which required close-fitting bolts in drilled holes, such as in riveted construction where it was not possible to drive satisfactory rivets. They are sometimes useful in aligning mechanical equipment and structural members which require precise positioning. They are now (1971) rarely if ever used in ordinary structural connections, since high-strength bolts are better and cheaper.

Ribbed Bolts. These bolts of ordinary rivet steel which have a rounded head and raised ribs parallel to the shank were used for many years as an alternative to rivets. The actual diameter of a given size of ribbed bolt is slightly larger than the hole into which it is driven. In driving a ribbed bolt, the bolt actually cuts into the edges around the hole producing a relatively tight fit. This type of bolt was particularly useful in bearing connections and in connections which had stress reversals.

A modern variation of the ribbed bolt is the *interference-body bolt* shown in Fig. 4.1.1c which is of A325 bolt steel and instead of longitudinal ribs has serrations around the shank as well as parallel to the shank. Because of the serrations around the shank through the ribs, this bolt is often called an *interrupted-rib* bolt. Ribbed bolts were also difficult to drive when several layers of plates were to be connected. The current high-strength A325 interference-body bolt may also be more difficult to insert through several plates; however, it is used when tight fit of the bolt in the hole is desired and it permits tightening by means of turning the nut without the simultaneous holding of the bolt head as may be required with smooth loose fitting ordinary A325 bolts.

4.2. HISTORICAL BACKGROUND OF HIGH-STRENGTH BOLTS

The first experiments indicating the possibility of using high-strength bolts in steel-framed construction was reported by Batho and Bateman[1] in

1934. Batho and Bateman concluded that bolts with a minimum yield strength of 54 ksi could be relied on to prevent slippage of the connected parts. Follow-up tests by Wilson and Thomas[2] substantiated the earlier work by reporting that high-strength bolts smaller in diameter than the holes in which they were inserted had fatigue strengths equal to that of well-driven rivets provided that the bolts were sufficiently pretensioned.

The next major step occured in 1947 with the formation of the Research Council on Riveted and Bolted Structural Joints. This organization began by using and extrapolating information from studies of riveted joints; in particular, the extensive annotated Bibliography by De Jonge,[3] completed in 1956, was used. From this beginning, the Research Council has continued to organize and sponsor research on high-strength bolted connections, and issue specifications at intervals on the basis of research findings.

The American Railway Engineering Association (AREA) also became interested in 1948 and initiated studies on the use of high-strength bolts in railroad bridge maintenance. In the same year the Association of American Railroads initiated a number of field test installations confirming the adequacy of connections made with high-strength bolts.

By 1950 the concept of using high-strength bolts and a summary of research and behavior was presented[4] to practising engineers and the steel-fabrication industry. As the next step, in 1951 the Research Council published its first specifications, permitting the replacement of rivets with bolts on a one-to-one basis. It was conservatively assumed that friction transfer of the load was necessary in all joints under service load conditions. The factor of safety against slip was established at a high enough level so that good fatigue resistance (i.e., no slip under varying stress or stress reversal consisting of many load cycles) was provided in every joint, similar to or better than that shown by riveted joints.

In 1954 a revision was made in the specifications to include the use of flat washers on 1:20 sloping surfaces and to allow the use of impact wrenches for installing high-strength bolts. Also, the 1954 revision permitted the surfaces in contact to be painted when the bolts were to create a *bearing-type* connection; i.e., when the ultimate strength of the connection was to be based on the bolt in bearing against the side of the hole.

In 1956 W. H. Munse summarized[5] bolt behavior and concluded that if high-strength bolts are to be as efficient and economical as possible, an initial tension as high as practicable must be induced in the bolts. By 1960 much additional research justified increasing the minimum bolt tension, recognized that the *bearing-type* connection was ordinarily an acceptable substitute for a riveted connection, and accepted that the connection with its strength based on slippage, known as a *friction-type* joint, may only be necessary when direct tension acts on the bolts or when fatigue conditions are important. Also, a turn-of-the-nut procedure was introduced as an alternative to the torque wrench method for tightening the high-strength bolts.

The previous requirement of two washers was reduced to require only one under the element (head or nut) being turned, if the turn-of-the-nut method of installation was used.

In 1962 the Research Council again liberalized the Specifications, completely deleting the requirement for washers.[6]

In 1964, the A490 bolt was included in the specification for use with high-strength steels, while in 1966 the required initial tension for the A490 bolts was reduced.[7] Finally, the most recent modification of the specification[7] in 1970 has added provisions relating to the use of high-strength bolts on galvanized steel.

An excellent summary of high-strength bolting, including the 1966 modifications, has been presented by Munse.[8] Furthermore, an extensive bibliography[9] has been developed under the sponsorship of the ASCE Task Committee on Structural Connections.

4.3. CAUSES OF RIVET OBSOLESCENCE

Riveting is a method of connecting a joint by inserting ductile metal pins into holes in the pieces being joined and forming a head at each end to prevent the joint from coming apart. Typical types of rivets are shown in Fig. 4.3.1. The principal causes for rivet obsolescence have been the

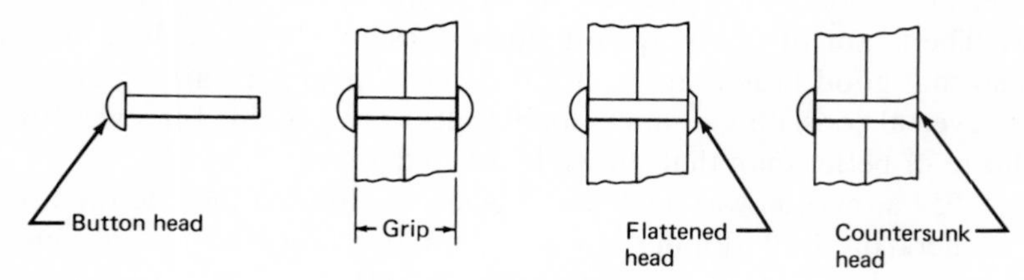

Fig. 4.3.1. Types of rivets.

advent of high-strength bolts and development of welding techniques. The development of welding procedures is discussed in Chapter 5 and the development of high-strength bolting forms the basis for this chapter. However, in addition to the above mentioned causes, a number of inherent disadvantages have hastened the obsolescence of structural riveting, especially field riveting.

Riveting crews require 4 or 5 men each of whom must be experienced. Such experienced crews have not always been available in all sections of the country. On the other hand, the crews required for high-strength bolts do not need to be highly skilled and are readily available. Inspection procedures in riveted connections are difficult and unfortunately, even the most experienced riveting crews require rigid inspection. Whenever bad rivets are found, they must be cut out and replaced. This is an expensive procedure at best and not met with favor among the riveting crews. Even the pre-heating immediately prior to driving is critical in developing the necessary tightness after cooling.

Some additional disadvantages of rivets are the ever present danger of fire and the high level of noise caused by driving the rivets. The matter of noise levels is usually critical whenever existing buildings are nearby. The level of sound produced by riveting can be very disruptive to business and has caused many expensive lawsuits.

Historically, rivets had a higher fatigue strength than standard bolts because the rivets more or less completely filled the holes as a result of the driving and head forming procedure. High fatigue strength is particularly important to structures in which there is a cycling of stresses or stress reversals as in the case of railroad bridges. Therefore, railroad bridge designers were among the last to continue requiring the use of rivets.

The principal factor which delayed immediate acceptance of high-strength bolts was the cost of the material. High-strength A325 bolts are about three times the cost of rivets. In addition, the hardened washers added cost. In the early 1950's the reduced labor cost for installing bolts did not offset the higher bolt material cost. Once the washers could be reduced to one or eliminated, and the greater strength of a bolt over that of a rivet could be utilized in design, high-strength bolts became economical. Now (1971) with ever higher labor costs and connection design generally requiring fewer bolts than would be required for rivets, the economy is clearly with high-strength bolts.

Shop fabrication using rivets persisted long after field erection use of high-strength bolts was standard for nonwelded connections. Since extensive labor was involved in the shop fabrication of the holes little extra labor was needed for riveting. In 1971, welding procedures have advanced so that shop fabrication is nearly always welded; thus, the last use of rivets has become uneconomical. In addition, properly welded joints are simpler and better.

Since the use of high-strength bolts provides all the advantages of rivets while not incurring any of the disadvantages, rivets are obsolete.

4.4. TYPES OF HIGH-STRENGTH BOLTS AND INSTALLATION PROCEDURES

At the present time the two basic types of high strength bolts are the A325 bolt, for most common situations, and the A490, for use on higher strength steels where an excessive number of A325 bolts might otherwise be required. The A325 bolt is identified by three radial lines spaced 120 degrees apart on the head while the A490 bolt has A490 and the manufacturer's symbol on the head. Both types of bolts have heavy hexagon heads and come with heavy semifinished hexagon nuts. Another characteristic of both the A325 and A490 bolts is their shorter thread lengths, the shorter thread lengths making it easier to exclude the threads from the shearing plane. Figure 4.4.1 and Table 4.4.1 show the control dimensions for A325 and A490 bolts and the methods of identification.

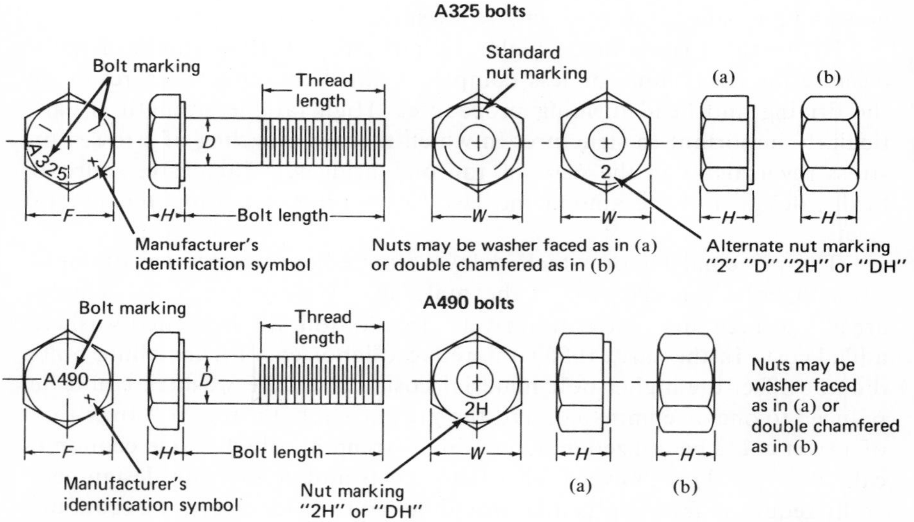

Fig. 4.4.1. Control dimension for high-strength bolts. (From ASTM A325 and A490 specifications.)

TABLE 4.4.1
High-Strength Bolt Dimensions

Nominal bolt size, *D*	Bolt Dimensions, In.			Nut Dimensions, In.	
	Heavy Hex Structural Bolts			Heavy Hex Nuts	
	Width across flats *F*	Height, *H*	Thread length	Width across flats *W*	Height, *H*
1/2	7/8	5/16	1	7/8	31/64
5/8	1 1/16	25/64	1 1/4	1 1/16	39/64
3/4	1 1/4	15/32	1 3/8	1 1/4	47/64
7/8	1 7/16	35/64	1 1/2	1 7/16	55/64
1	1 5/8	39/64	1 3/4	1 5/8	63/64
1 1/8	1 13/16	11/16	2	1 13/16	1 7/64
1 1/4	2	25/32	2	2	1 7/32
1 3/8	2 3/16	27/32	2 1/4	2 3/16	1 11/32
1 1/2	2 3/8	15/16	2 1/4	2 3/8	1 15/32

Occasionally, ASTM A449 bolts are substituted for A325 bolts when a longer thread dimension is required, the A449 bolts having the same hexagon head and thread length as standard A307 bolts but having the same strength as A325 bolts.

A variation of the standard A325 and A490 used for bearing connections is the *interference-body* bolt described in Sec. 4.1. These bolts have a button head and may be used with the standard high-strength nuts or with self-locking nuts. The interference-body type bolts have particular applications where high bearing capacity is desired together with stress reversals or vibratory loads.

Proof Load and Bolt Tension. The primary requirement when installing high-strength bolts is to provide a sufficient *pretension* force. The pretension should be as high as possible without chancing permanent deformation or failure of the bolt. Bolt material exhibits a stress-strain (load-deformation) behavior as shown in Fig. 4.4.2 which has no well-defined yield point. The proportional limit corresponds to the *proof load.* ASTM prescribes this proof load for each diameter connector instead of specifying a unit stress to represent proportional limit. The proof load is 70 percent and 80 percent of the minimum ultimate tensile strength for A325 and 80 percent for A490 bolts respectively.

From the early 1950's the minimum required pretension was equal to the proof load. When the turn-of-the-nut became the recommended method no difficulty was found in obtaining proof load for A325 bolts with $^2/_3$ turn of the nut from the snug position, as shown in Fig. 4.4.2. With the A490 bolt, the $^2/_3$ turn from snug may not achieve the proof load, as seen in Fig. 4.4.2.

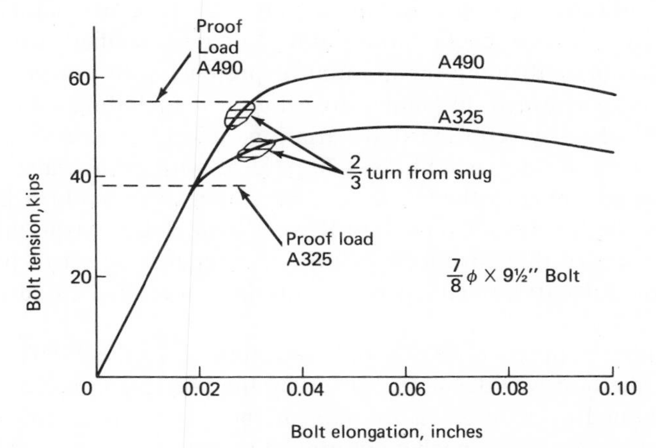

Fig. 4.4.2. Typical load-elongation curves for A325 and A490 bolts. (Adapted from Ref. 7.)

Thus 1969 AISC–1.23.5, in accordance with the 1966 bolt specification, requires a pretension equal to 70 percent of minimum tensile strength, as given in Table 4.4.2. This equals the proof load for A325 bolts and about 85 to 90 percent of proof load for A490 bolts.

The magnitude of pretension that is desirable and necessary has been the subject of considerable study by researchers.[10–12]

TABLE 4.4.2
AISC–1.23.5

Bolt Size, Inches	Minimum Bolt Tension,[1] Kips	
	A325 and A449 Bolts	A490 Bolts
1/2	12	15
5/8	19	24
3/4	28	35
7/8	39	49
1	51	64
1 1/8	56	80
1 1/4	71	102
1 3/8	85	121
1 1/2	103	148
Over 1 1/2		0.7 × T.S.

[1] Equal to 70 percent of specified minimum tensile strengths of bolts, rounded off to the nearest kip.

Installation Techniques. There are two general methods of developing the required pretension indicated in Table 4.4.2. One is called the *calibrated wrench* method and the other the *turn-of-the-nut* method.

The calibrated wrench method includes the use of manual torque wrenches and power wrenches adjusted to stall at a specified torque value. Early studies of controlling the amount of pretension by torque measurements were performed by many investigators including Stewart and Maney.[13–15]. Maney reported that variations in tensile stresses produced by a given torque were as high as ±30 percent with an average variation of ±10 percent. Due to this variation, which was also confirmed by field experience, the Research Council on Riveted and Bolted Structural Joints has recommended that torque or calibrated wrenches be set to produce a bolt tension 5 to 10 percent in excess of the values indicated in Table 4.4.2.

Beginning in the early 1950's and continuing into the 1960's the turn-of-the-nut method was developed whereby the pretensioning force in the bolt is obtained by a *specified rotation* of the nut from an initially *snug* position which causes a specific amount of strain in the bolt. According to the Research Council on Riveted and Bolted Structural Joints, a nut is considered to be snug when impacting first begins as it is being

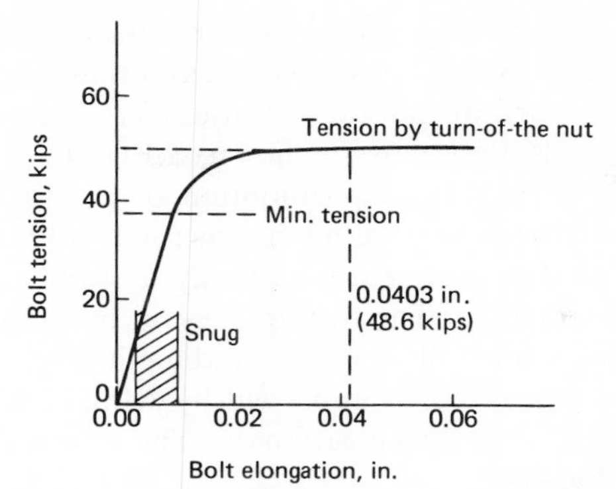

Fig. 4.4.3. Bolt elongations in a typical test joint. (From Ref. 20.)

tightened by an impact wrench. Although snugness or initial tightness can vary due to the condition of the surfaces being tightened, this variation is not significant as can be seen from Fig. 4.4.3 taken from Fisher et al.[20] The clamping force of 48.6 kips corresponding to half a turn-of-the-nut is seen to occur sufficiently far along the flat portion of the tension curve to make any variation in the snug tension insignificant to the overall behavior of the bolt.

One may wonder whether there is danger of having inadequate reserve strength if the pretension exceeds the proof load; i.e., when it approaches 90 percent of ultimate strength. Figure 4.4.4 from Ref. 10 shows the effect of various turns of the nut with the margin of safety indicated. If the calibrated wrench method is used *strength* is the critical factor, with the typical safety margin shown in Fig. 4.4.4. The possibility of over-

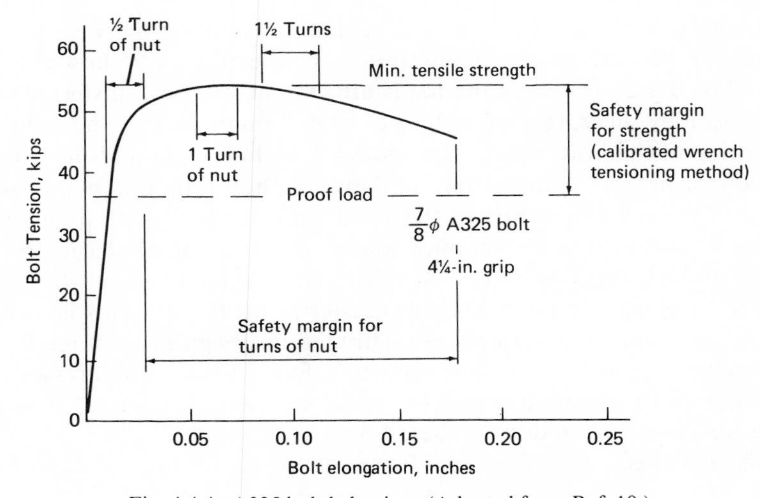

Fig. 4.4.4. A325 bolt behavior. (Adapted from Ref. 10.)

torquing the bolts with power wrenches is not considered a problem since such overtorquing usually fractures the bolts and they are replaced during installation. In the turn-of-the-nut method, *deformation* is the critical factor with the typical safety margin shown in Fig. 4.4.4. For either installation process one can expect a minimum of $2\frac{1}{4}$ turns from snug to fracture. When the turn-of-the-nut method is used and bolts are tensioned using $\frac{1}{8}$ turn increments, frequently as many as four turns may be obtained from snug to fracture. The turn-of-the-nut method is the cheapest, is more reliable, and is generally the preferred method.

The method requirements, as approved by AISC–1.23.5, are given by the Research Council 1966 specifications.[7] The approved nut rotations are indicated in Table 4.4.3.

TABLE 4.4.3
Nut Rotation[a] from Snug Tight Condition (from Ref. 7)

Disposition of Outer Faces of Bolted Parts		
Both faces normal to bolt axis, or one face normal to axis and other face sloped not more than 1 : 20 (bevel washer not used)		Both faces sloped not more than 1 : 20 from normal to bolt axis (bevel washers not used)
Bolt length[b] not exceeding 8 diameters or 8 inches	Bolt length[b] exceeding 8 diameters or 8 inches	For all length of bolts
$\frac{1}{2}$ turn	$\frac{2}{3}$ turn	$\frac{3}{4}$ turn

[a] Nut rotation is rotation relative to bolt regardless of the element (nut or bolt) being turned. Tolerance on rotation: 30° over or under. For coarse thread heavy hex structural bolts of all sizes and length and heavy hex semi-finished nuts.
[b] Bolt length is measured from underside of head to extreme end of point.

4.5. NOMINAL STRESSES

Loads are transferred from one member to another by means of the connection between them. A few typical connections are shown in Fig. 4.5.1. For design purposes the loads are assumed to be transferred without friction by shear, tension or by a combination of both in the bolts and by bearing of the bolts against the sides of the holes. In addition, design procedures neglect deformation in between holes in the plates being joined. Thus, all connectors of the same diameter in a joint are assumed to offer equal resistance to applied loads. Since, as discussed in the last article, high-strength bolts are installed with initial tension which produces a clamping effect, actual load transfer in nearly all joints is through friction; consequently, the stresses computed in design procedures are *not* real stresses but only serve as criteria of safety. Such false stresses used for design purposes are known as *nominal* stresses. The actual frictional forces are dealt with in detail in Sec. 4.6.

The net effect of the frictional forces developed during bolting will be

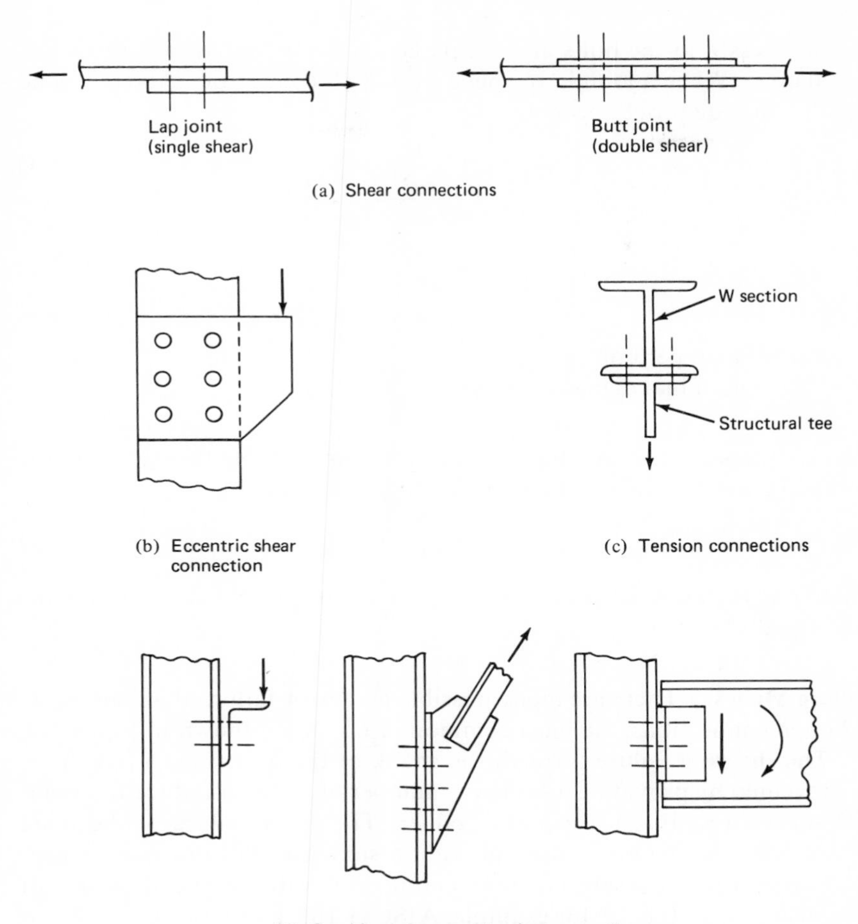

Fig. 4.5.1. Typical bolted connections.

to create a more even distribution of stresses under service loads when rows or groupings of bolts are used. Once the friction is overcome and slip occurs due to overload, the transfer of load begins to approach that assumed when computing nominal stresses. In effect, connection design is based on ultimate strength behavior where the ultimate connector capacity is divided by a factor to reduce it into the service load range. The resulting allowable capacity is then used to design for service loads. As will be observed in later chapters many procedures of working stress design utilize behavior which occurs only under loads approaching ultimate capacity.

Shear and Tensile Stress. For design purposes, the shear stress f_v and the tensile stress f_t in the bolts are computed on the nominal bolt cross-sectional area. When a joint is arranged so that all connectors share the load equally, the equations become

$$f_v = \frac{P}{\frac{\pi D^2}{4}} \tag{4.5.1}$$

$$f_t = \frac{P}{\frac{\pi D^2}{4}} \tag{4.5.2}$$

where P = load per bolt
D = nominal diameter of bolt

Bearing Stress. The bearing stress f_p is computed on the basis of the nominal diameter and the thickness of the plate; thus,

$$f_p = \frac{P}{Dt} \tag{4.5.3}$$

where t = thickness of plate. Again, it is assumed connectors share the load equally.

Failure Modes. There are many possible modes of failure that can occur in bolted connections, the most common of which are shown in Fig. 4.5.2.

The shearing failure through the shank of the bolt (Fig. 4.5.2a) may be prevented by providing a sufficient number of bolts to reduce the nominal shear stress to an acceptable value. The shear failure of the plate (Fig. 4.5.2b) can be avoided by providing sufficient end distance as generally prescribed by specifications empirically without the necessity of computing stress (i.e., as for example, AISC–1.16.5 and 1.16.6).

A bearing failure of the bolt itself (Fig. 4.5.2c) is rare, occurring only when the steel in the plates is harder than that in the bolts. A more common bearing failure is the type in which the plate fails (Fig. 4.5.2d) due to excessive elongation of the hole. This type of failure can be prevented by providing thicker plates. The tensile failure shown in Fig. 4.5.2e may result from overtightening the bolts, or from excessive externally applied tensile forces. This type of failure is not common; in erection, failure by overtightening is corrected by immediate replacement of the bolt and external tension may be reduced by increasing the number of bolts or using larger-diameter bolts. The bending failure indicated in Fig. 4.5.2f is caused by using excessively long bolts in excessively thick connections made up of several layers of plates. This type of failure can be prevented by redesigning the connection to use a larger-diameter connector.

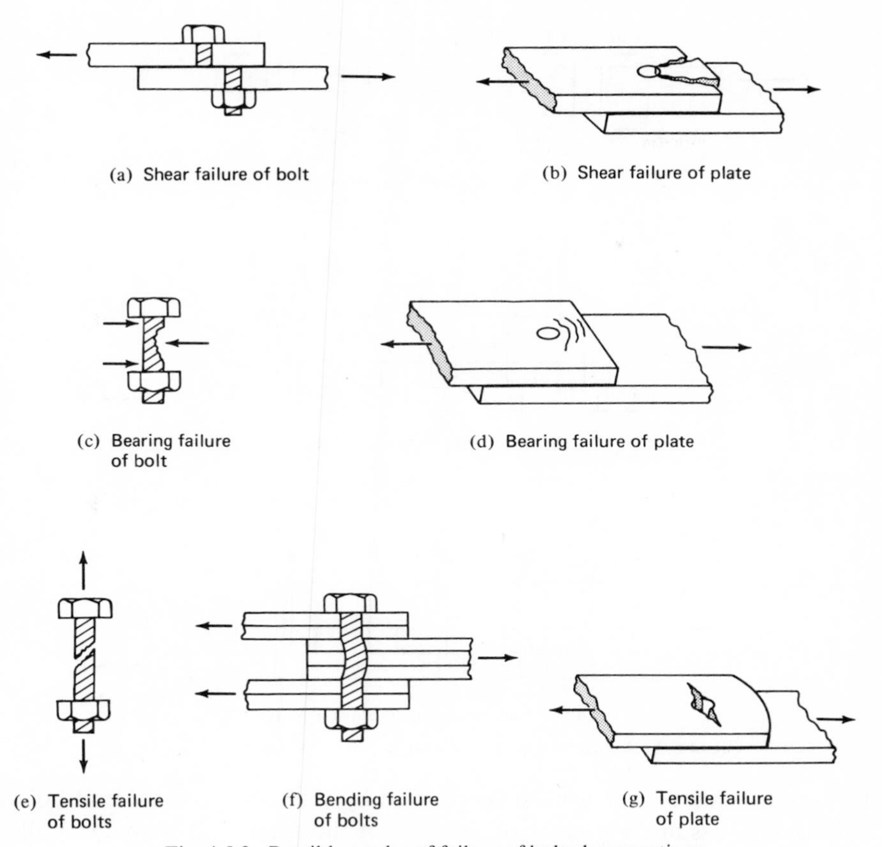

Fig. 4.5.2. Possible modes of failure of bolted connections.

Performance Assumptions Relating to Nominal Stresses. In practice, one nearly always uses one or more lines of bolts in a group such as shown in Fig. 4.5.3 instead of a single bolt. Figure 4.5.3a shows the assumed force distribution acting along each of the plates as discussed in Sec. 3.5, beginning with the full value of T before a bolt is encountered and decreasing linearly towards the end of the plate. This is based on the assumption that each of the bolts take an amount of load in proportion to the total number of bolts; i.e., friction is neglected. Also, this behavior assumption requires that the plates are rigid with all the deformations occurring in the bolts as shown in Fig. 4.5.3b. In reality, however, the plates are not rigid but do elongate in proportion to the forces along the plate. Thus in any given tension connection having a number of bolts, the plate would elongate more between the first two bolts than between the last two. This behavior results in an uneven dis-

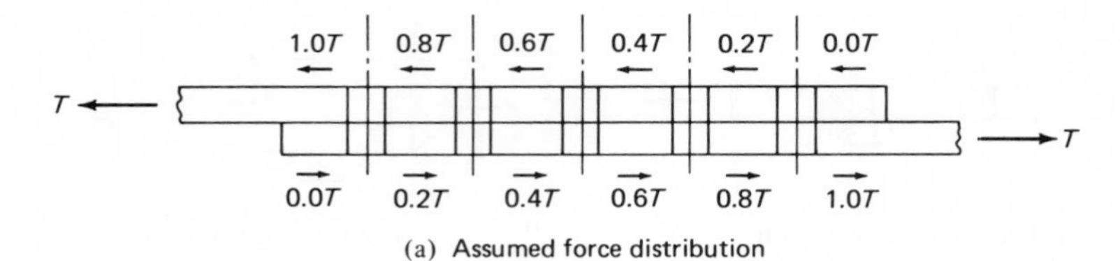

(a) Assumed force distribution

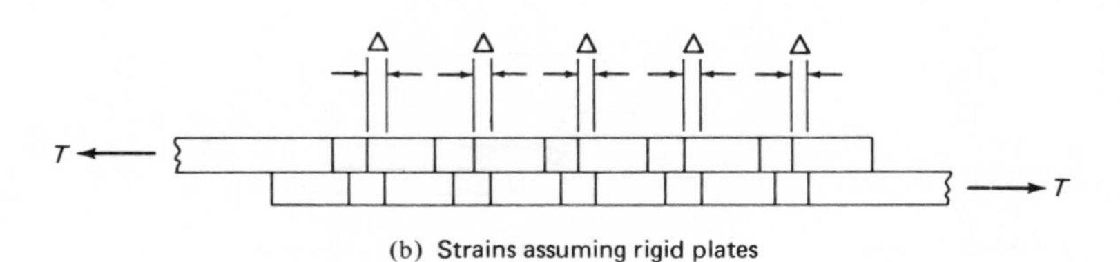

(b) Strains assuming rigid plates

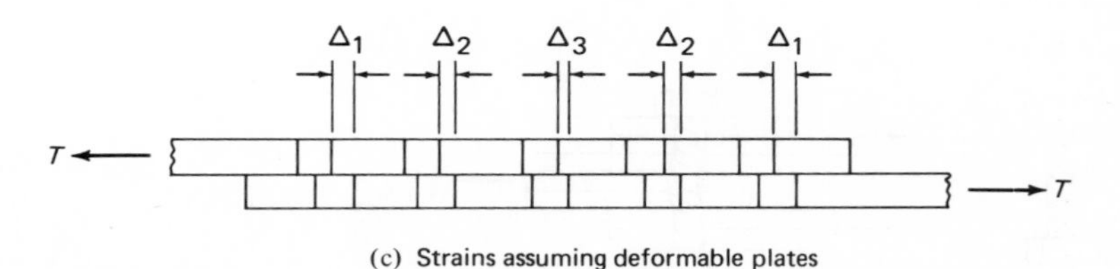

(c) Strains assuming deformable plates

Fig. 4.5.3. Assumed forces and resulting strains in a row of bolts.

tribution of forces which must be resisted by the bolts, i.e., the end bolts will have to assume a greater proportion of the total load than the bolts in the center of the grouping. Fig. 4.5.3c shows the net strains on the bolts if the plates are assumed to deform elastically. A theoretical solution to this mode of behavior would be to provide a scarfed joint as shown in Fig. 4.5.4. However, such a solution is impracticable and is never attempted in actual practice.

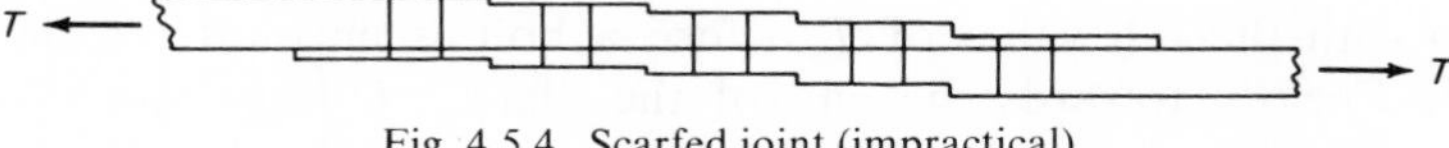

Fig. 4.5.4. Scarfed joint (impractical).

It should be recalled in view of the discussion in this article, the tensile stress computed on net section through holes as discussed in Chapter 3 is also a *nominal* stress.

There are several other reasons, in addition to the most important ones of neglecting friction and considering plates as rigid, why the nominal stresses which are computed for design purposes as caused by service loads differ from actual service load stresses. The following list of reasons *why nominal stresses differ from actual* is presented in summary:

(a) Friction resistance to slip is neglected.
(b) Deformation of plates is neglected.
(c) Tensile stress concentrations at holes is neglected.
(d) Shearing deformation of connectors is assumed proportional to shearing stress.
(e) Shearing stress is assumed uniform over connector cross section.
(f) Bearing stress is assumed uniform over a nominal contact surface equal to connector diameter times plate thickness.
(g) Bending of connectors is neglected.

Thus it may be apparent that even if desired, the computation of actual stresses and the force distribution in a bolted connection is complex. Furthermore, in addition to items (a) through (g) the actual stresses are dependent on many factors often beyond the control of the designer; such as poor alignment of holes, loose or unequal tightening, eccentric loading not anticipated, and poor construction.

In view of the above discussion, it is fortuitious that the excellent ductility inherent to steel eliminates the necessity of complex stress computations for design. Prior to reaching ultimate capacity, the bolts and plates are capable of deforming plastically an amount sufficient to redistribute any unequal forces. Under ultimate conditions, the assumptions used for nominal stresses at service load are reasonably correct. The result is that the designer is able to proportion connections by determining the loads acting on each bolt, using elementary mechanics and nominal stresses, with the knowledge that even though service loads stresses are unknown the connection will have an adequate factor of safety with regard to strength.

4.6. FRICTION-TYPE AND BEARING-TYPE CONNECTIONS

AISC 1969 provides for two categories of high-strength bolted joints, the *friction-type* and the *bearing-type*. Bolts in either type of joint are installed by the same process so that the same pretension is provided. Performance of the bolted joints under service loads is identical; service load transmission is by friction. The difference between the friction-type and bearing-type connection lies entirely in the factor of safety provided against slip under overloads.

The friction-type is so called because it has the higher factor of safety against slip and thus is most suitable when stress reversal or cyclical loading may occur. The higher factor of safety provides good fatigue resistance.

The bearing-type joint is so named for use when it is not deemed critical if slip occurs under an occasional overload to bring the bolt shank into contact with the side of the hole. For any subsequent loading, the stress is transferred by friction in combination with bearing on the plate. As long as the loading is static such slip will occur only once; thereafter the bolt is already bearing against the material at the side of the hole.

Friction-Type Connection. In developing rationale for the AISC–1969 provision for a friction-type connection, it is helpful to study the behavior of high strength bolts under loads. Figure 4.6.1 shows the forces present

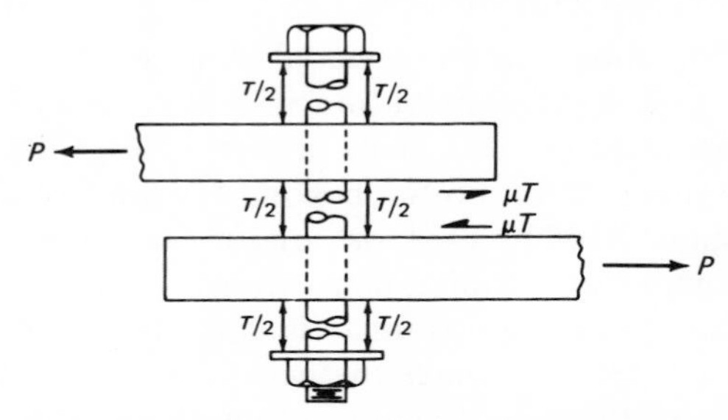

Fig. 4.6.1. Forces assumed acting on a friction connection.

in a typical high-strength bolted connection. The pretension (proof load for A325 bolts) acting on the head and nut of the bolt produce a clamping force T which brings the two plates into contact. The clamping force in turn produces a resistance to shear μT, which is a product of the clamping force T and the coefficient of friction, μ, between two steel surfaces.

The coefficient of friction depends on many factors such as the condition of the plates, including the presence of oil, paint, or mill scale. Determination of the coefficient of friction, or more properly called the "slip coefficient," has been the study of a number of investigators using a variety of steels and joint arrangements.[18,21–25] The two terms are not identical in that the measurements are made using slightly different methods and criteria.

The slip coefficient is determined using the load at which "the friction bond is definitely broken and the two surfaces slip with respect to one another by a relatively large amount."[24] The coefficient of friction is based on the slip load "at which movement of one entire joined element is first detected, e.g., with dial gages."[24]

Since the coefficient applied to design is the result of tests, the procedure for measurement becomes important. The coefficient of slip as reported is about 20 to 30 percent higher than the coefficient of friction, both as defined above. Due to the strain compatibility requirement as discussed in regard to Figs. 4.5.3 and 4.5.4 one would expect movement at the end of a joint prior to having a general slippage. The coefficient of friction, using the slip coefficient definition, is generally the one of interest and has been found to range from 0.30 to 0.40. The value 0.35 has been used by the Research Council.[7]

EXAMPLE 4.6.1

Determine the amount of force P required to cause a $\frac{7}{8}$ in. diam. A325 bolt to slip. Use a slip coefficient (coefficient of friction) of 0.35.

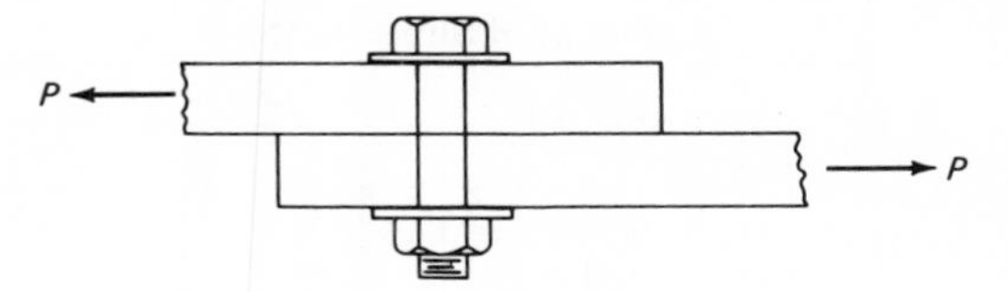

Fig. 4.6.2. Example 4.6.1.

Solution

Using the pretension load from Table 4.4.2,

$$T_i = 39 \text{ kips}$$
$$P = \mu T_i = 0.35(39) = 13.65 \text{ kips}$$

Since the overall action of the joint is a shearing effect the nominal shear stress in the $\frac{7}{8}$-in. high-strength bolt to cause slippage to begin is

$$\frac{13.65}{\text{Area}} = \frac{13.65}{0.6013} = 22.7 \text{ ksi}$$

AISC–1.5.2 limits the nominal shear stress in friction-type connections using A325 bolts to 15 ksi. The factor of safety FS, which this value provides against slippage is

$$\text{FS} = \frac{\text{Frictional shear resistance}}{\text{Applied shear force}} = \frac{\mu\,(\text{clamping force})}{(15 \text{ ksi})(\text{area of bolt})} \tag{4.6.1}$$

The value of 15 ksi for A325 bolts is that which has been used successfully in riveted construction for many years. Originally it was established at 75 percent of the basic allowable stress for A7 steel which was 20 ksi. Many tests on riveted construction had established that shear failures would not occur if the nominal shear stress was kept below 75 percent of the tensile stress on net section.[26]

EXAMPLE 4.6.2

Determine the factor of safety against slippage of a $\frac{7}{8}$-in. dia. A325 bolt.

Solution

Applying Eq. 4.6.1,

$$FS = \frac{0.35(39)}{15(0.6013)} = 1.51$$

For other bolt sizes, the variation in the factor of safety against slip may be observed from Table 4.6.1. The clamping force is the initial tension from Table 4.4.2.

TABLE 4.6.1
Factor of Safety Against Slip, Friction-Type Connection
(Coefficient of Slip = 0.35)

Bolt Diameter	Nominal Area	Factor of Safety	
		A325	A490
5/8	0.3068	1.44	1.37
3/4	0.4418	1.48	1.39
7/8	0.6013	1.51	1.43
1	0.7854	1.52	1.43
1 1/8	0.9940	1.31	1.41
1 1/4	1.2272	1.35	1.45

Under static loading conditions the factor of safety against slippage as determined in Example 4.6.2 represents the margin against a once in the life of the structure movement such as a shock or severe wind loading. This type of a loading is not generally susceptable to reversals in direction and therefore a lower value for the factor of safety such as 1.4 to 1.5 (compared with the basic 1.67) is acceptable. Furthermore, slippage does *not* constitute failure. The bolt will come into bearing against the adjoining material. Subsequent loads will be carried mostly by friction with some help of direct bearing.

For unusual design conditions the factor of safety against slippage may be desired to be increased; which may be accomplished by using the allowable nominal shear stress below the values prescribed by the AISC and AASHO specifications. For normal design cases, including stress reversals and other fatigue loadings where fewer than 20,000 repetitions are expected (less than 2 applications per day for 25 years), the nominal stress allowable values prescribed by AISC and AASHO are entirely suitable. Table 4.6.2 summarizes the allowable nominal shear stresses permitted by AISC.

EXAMPLE 4.6.3

Determine the tensile capacity of the connection shown in Fig. 4.6.3 if 7/8-in. dia. A325 bolts are used. Assume the plates to be A572–Grade 50 steel, and the connection is to be a friction-type according to AISC specifications.

TABLE 4.6.2
Allowable Shear Stress on Nominal Bolt Cross-Sectional Area (from AISC–1.5.2)

Description of Fastener	Shear (F_v) Friction-Type Connections	Shear (F_v) Bearing-Type Connections
A502, Grade 1, hot-driven rivets		15.0
A502, Grade 2, hot-driven rivets		20.0
A307 bolts		10.0
Threaded parts of steel meeting the requirements of Sec. 1.4.1		$0.30\,F_y$
A325 and A449 bolts, when threading is *not* excluded from shear planes	15.0	15.0
A325 and A449 bolts, when threading is excluded from shear planes	15.0	22.0
A490 bolts, when threading is *not* excluded from shear planes	20.0	22.5
A490 bolts, when threading is excluded from shear planes	20.0	32.0

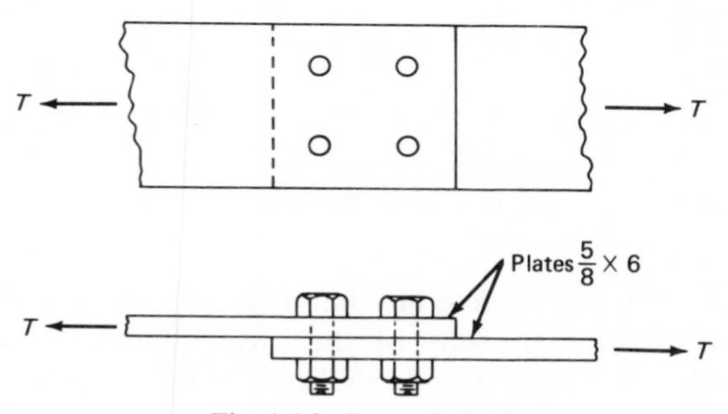

Fig. 4.6.3. Example 4.6.3.

Solution

Consider first the capacity of the plates based on net section.

$$T = [6 - 2(7/8 + 1/8)]0.625(30) = 75 \text{ kips}$$

or

$$T = 0.85(6)\,0.625(30) = 95.6 \text{ kips}$$

The action of the entire connection is a shear transfer of load between the two plates. The plane of contact may be thought of as the shear plane. When there is a single plane of contact involved in the load transfer, it is referred to as "single shear."

The allowable capacity per bolt, R, in single shear is

$$R_{SS} = F_v A = 15(0.6013) = 9.02 \text{ kips}$$

the total capacity for the connection is

$$T = 4R_{SS} = 4(9.02) = 36.08 \text{ kips}$$

Remember, no slip occurs under a load of 36.08 kips; rather, there is a factor of safety against slip of 1.51 (Table 4.6.1). The bolts are actually in tension and the above calculation is prescribed by AISC only as a simple device to ascertain safety.

EXAMPLE 4.6.4

Determine the number of ³/₄-in. diam. A325 bolts required to develop the full tensile strength of the plates if A572 Grade 65 steel is used. Assume a double row of bolts. Use a friction-type connection.

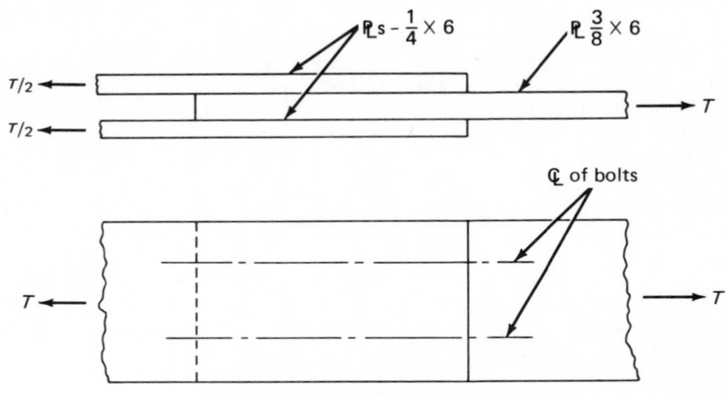

Fig. 4.6.4. Example 4.6.4.

Solution

It is determined by inspection that the cross-sectional area of the center plate is less than the sum of the areas of the two outer plates. Therefore in this case one need only check the capacity of the center plate.

$$\text{Net area} = [6 - 2(3/4 + 1/8)]0.375 = 1.59 \text{ sq in.}$$

$$0.85\, A_{gross} = 0.85(6)\,0.375 = 1.91 \text{ sq in.}$$

$$T = 1.59(39.0) = 62 \text{ kips}$$

In this case the action of the entire connection may be thought of as a shear transfer between plates occurring at the *two* planes of contact. When connectors are positioned so that they cross two shear planes of contact, it is called "double shear." *Double shear* is a symmetrically loaded situation so far as the shear planes and directions of shear transfer are concerned, whereas the single-shear case is unsymmetrical (Fig. 4.6.3).

Allowable capacity per bolt (double shear) is

$$R_{DS} = 2(\text{area})\, F_v = 2(0.4418)(15) = 13.25 \text{ kips}$$

$$\text{Number of bolts} = \frac{62}{13.25} = 4.68$$

Use 6 – ³/₄-in. diam A325 bolts.

Bearing-type Connection. As stated at the beginning of this article, the bearing-type connection is used when static loads are to be carried and it is not considered critical if slip occurs. When slip is not critical a lower factor of safety against slip may be used. This is accomplished by allowing a higher nominal shear stress than for the friction-type connection. AISC–1.5.2 allows 22 ksi for A325 bolts when bolt threads do not extend across the plane of contact between the pieces being joined.

The value arises from tests[22] using A7 steel plates with A325 bolts, wherein it was shown that a tension-shear ratio (tensile stress on net section to nominal shear stress) of 1.0:1.10 provided a so-called balanced design at ultimate strength. Thus since the tensile stress allowed was 20 ksi, the nominal shear value became 1.1(20) = 22 ksi. These stresses imply a factor of safety against ultimate strength of 3.0.

More recent tests[27] have demonstrated that with higher-strength steels the concept of *balance design* (simultaneous attainment of shear failure and tensile ultimate strength) leads to inconsistent allowable bolt stresses. It has also been found that "compact joints" behave differently than long joints.[22,27–29] Compact joints are those where the bolt group has values of g/D (gage to hole diameter ratios) and p/D (pitch to hole diameter ratios) of approximately 4, and a length of joint not greater than 5 pitches. In compact joints high-strength bolts behave similarly in shear regardless of

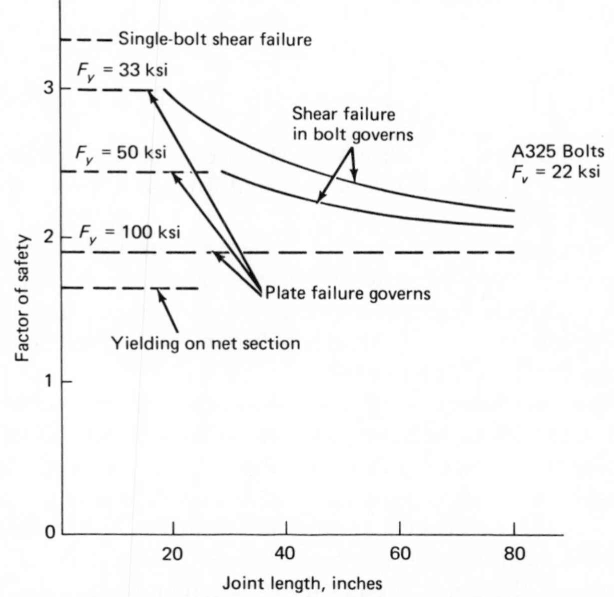

Fig. 4.6.5. Factor of safety against failure when using nominal shear stress, F_v = 22 ksi with A325 bolts.

the type of connected material. Regarding shear failure in the bolts, the factor of safety varies from about 3.3 for a single A325 bolt and for a group of such bolts in a compact joint, to approximately 2.0 for long joints of about 80 in. This is shown in Fig. 4.6.5 from Ref. 27. The conclusion is that shear failure of the bolts is much more likely in long joints than in short ones.

It seems likely that in future specifications the nominal shear stress allowable values will be made a function of joint length; for the present, factor of safety against bolt shear failure is about right for long joints but excessively high for short compact ones.

Regarding the actual slippage under service loads, the bearing-type joint will have factors of safety against slip, using Eq. 4.6.1 with nominal shear stresses of 22 ksi for A325 and 32 ksi for A490, as shown in Table 4.6.3.

TABLE 4.6.3.
Factor of Safety Against Slip, Bearing-Type Connection with Threads Excluded from Shear Plane (Coefficient of Slip = 0.35).

Bolt Diameter	Nominal Area	Factor of Safety Against Slip	
		A325	A490
5/8	0.3068	0.98	0.85
3/4	0.4418	1.01	0.87
7/8	0.6013	1.03	0.89
1	0.7854	1.04	0.89
1 1/8	0.9940	0.89	0.88
1 1/4	1.2271	0.92	0.91

A study of Table 4.6.3 shows the factor of safety against slip to be close to 1.0 for A325 bolts but less than that for A490 bolts. Under actual loading slip may or may not occur.

In the vast majority of situations where connections are designed as bearing-type, the full design service loads rarely occur, the number of connectors used somewhat exceeds the minimum requirement, or the coefficient of friction exceeds 0.35, often becoming as high as 0.40. Thus most bearing-type joints do not slip.

The important feature is that in a bearing-type connection, the joint is designed with the expectation of slip, and as a result, the bolts are expected to bear in direct contact with the sides of the hole. In this case, both the shear force acting across the bolts as well as the bearing stresses in the plates must be checked. Figure 4.6.6 shows the forces assumed to be acting on a bearing-type connection. In determining the allowable value of shear in the bolts, the matter of whether or not the threads are in the shearing plane must be taken into account. This was not done in the friction-type connection because for that type of connection the oc-

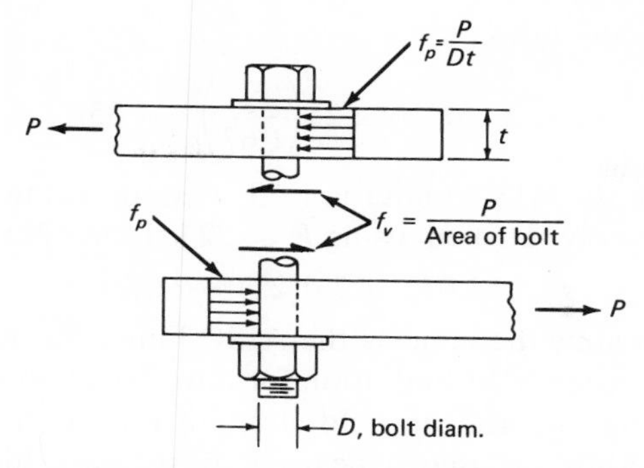

Fig. 4.6.6. Forces assumed acting on a bearing-type connection.

currence of slip was considered as the limit of capacity instead of a true shear failure across the shank of the bolt. However, in the case of the bearing-type connection, the bolts *may* fail in shear and any reduction in the cross-sectional area in the shearing plane due to the presence of the threads must be taken into account. Thus Table 4.6.2 shows for bearing-type connections two values for the allowable shearing stress F_v depending on whether or not the threads are excluded from the shear plane.

The allowable bearing stress F_p on the projected area of bolts in bearing-type connections (also for riveted connections) is dependent on the yield strength of the plates and is given for AISC and for AASHO by Eqs. 4.6.2 and 4.6.3, respectively.

$$F_p = 1.35F_y \quad \text{AISC–1.5.2.2} \tag{4.6.2}$$

$$F_p = 1.22F_y \quad \text{AASHO–1.7.1} \tag{4.6.3}$$

The explanation of the allowable bearing stress lies primarily in the research conducted on riveted joints. The bearing-type connection is considered to be the equal of a riveted connection having the same number of connectors. Studies of the relationship between tension on net section, nominal shear stress, and bearing stress on riveted joints as summarized by Jones[26] form the basis for the present allowable values.

Two important conclusions[26] are as follows:

1. Under static loading, the strength of a joint loaded in tension is *not* reduced by reason of bearing pressure if

$$f_p \leq 2.25f_t$$

where f_p = nominal bearing stress, and f_t = tensile stress on net section.

2. Under static loading the strength of a joint loaded in compression or shear is *not* reduced by reason of bearing pressure if

$$f_p \leq 3.0f_v$$

where f_v = nominal shear stress on bolt cross section.

Using the first limitation

$$F_p = 2.25F_t = 2.25\left(\frac{F_y}{1.67}\right) = 1.35F_y$$

which agrees with the AISC limitation. The second limitation indicated above is automatically satisfied; using $F_v = 22$ ksi for A325 bolts,

$$F_p = 3.0F_v = 3.0(22) = 66 \text{ ksi}$$

which is approximately the same as the AISC limitation for material with $F_y = 50$ ksi; and exceeds that limitation for lower yield point material.

For connecting material of yield strength higher than 50 ksi, A490 bolts should generally be used if bearing stresses are high, so that the AISC allowable bearing stress seems reasonable.

The fact that the allowable bearing stress is higher than the yield stress is seen to be reasonable when one considers the decrease in stresses at the edge of the head or nut as shown in Fig. 4.6.7. The effective length at the edge of the head or nut is given by Eqs. 4.6.4a and b:

$$\text{Effective length} = D + 2\frac{F}{2} = D + F \tag{4.6.4a}$$

or

$$= D + 2\frac{W}{2} = D + W \tag{4.6.4b}$$

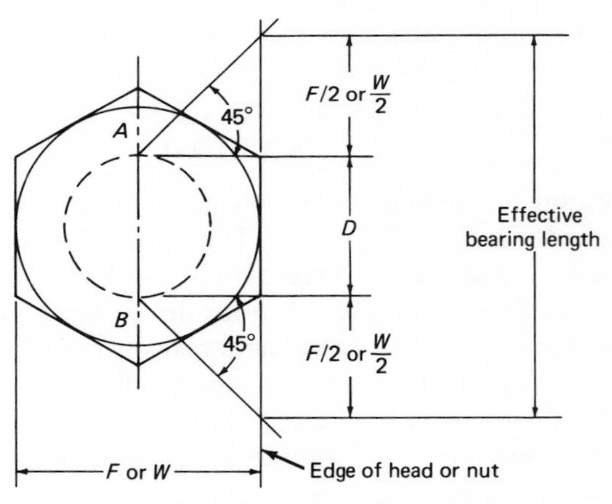

Fig. 4.6.7. Effective bearing length.

As may be seen from Table 4.4.1, F and W are equal for high-strength bolts and, therefore, either of Eqs. (4.6.4) may be used. Taking a $\frac{7}{8}$ in. diam. A325 bolt, the effective length is

$$\text{E.L.} = \tfrac{7}{8} + 1\tfrac{7}{16} = 0.8750 + 1.4376 = 2.3125 \text{ in.}$$

Dividing the allowable bearing stress by the effective length, an apporximation is obtained for the bearing stress at the edge of the head:

$$\frac{1.35F_y}{2.3125} = \frac{F_y}{1.58} \approx 0.63F_y$$

Similarly, a 1 in. diam. A325 bolt would have an allowable bearing stress at the edge of the head equal to approximately $0.52F_y$.

EXAMPLE 4.6.5

Determine the tensile capacity of the connection in Fig. 4.6.3 if (a) the bolt threads are excluded from the shear plane and (b) if the bolt threads are included in the shearing plane. Use AISC specs.

Solution

(a) Threads excluded from shear plane.

$$\text{Plate capacity} = 75 \text{ kips (same as Example 4.6.3)}$$

The tensile capacity based on single shear with $F_v = 22.0$ ksi on $^7/_8$ in. diam. A325 bolts is

$$T = 4(22.0)0.6013 = 52.9 \text{ kips}$$

The tensile capacity based on an allowable bearing stress, $F_p = 1.35F_y$ on $^5/_8$-in. $F_y = 50$ ksi plates is

$$T = 4(1.35)50(0.875)0.625 = 147.8 \text{ kips}$$

<u>Nominal shear stress governs: $T = 52.9$ kips.</u>

(b) Threads included in shear plane. The tensile capacity based on single shear with $F_v = 15.0$ ksi is

$$T = 4(15.0)0.6013 = 36.08 \text{ kips}$$

Nominal shear stress again governs since there is no change in net section tensile capacity or in the capacity based on bearing.

EXAMPLE 4.6.6

Determine the number of $^3/_4$ in. diam. A325 bolts required to develop the full strength of the plates in Example 4.6.4 if it is designed as a bearing-type connection and threads are excluded from the shearing planes. Use AISC specifications.

Solution

Plate capacity = 62 kips (same as Example 4.6.4). Allowable shear force per bolt (double shear) is

$$R_{DS} = 2(0.4418)22 = 19.44 \text{ kips}$$

$$\text{Number of bolts reqd} = \frac{62}{19.44} = 3.19$$

Allowable bearing force per bolt (based on $^3/_8$-in. thick plate):

$$R_B = 1.35(65)0.750(0.375) = 24.7 \text{ kips}$$

$$\text{Number of bolts reqd} = \frac{62}{24.7} = 2.51$$

Shear controls, therefore use 4 bolts.

Shear stresses will govern the design of most bearing-type connections, however, bearing stresses will control in cases where the plates are wide and relatively thin as shown in Example 4.6.7.

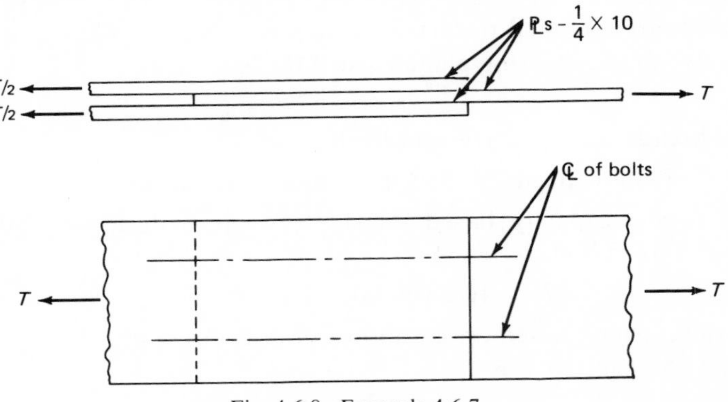

Fig. 4.6.8. Example 4.6.7.

EXAMPLE 4.6.7

Determine the number of $^3/_4$ in. diam. A325 bolts required to develop the full strength of the plates if A36 steel is used. Assume a bearing-type connection with the threads excluded from the shearing plane and a double row of bolts. Use AISC specifications.

Solution

In this example, the center plate will control the tensile capacity based on net section,

$$\text{Capacity} = [10 - 2(^3/_4 + ^1/_8)]0.25(22.0) = 45.4 \text{ kips}$$
$$85\% \ A_{gross} = 0.85(10)0.25(22.0) = 46.8 \text{ kips} > 45.4$$

Design connectors for $T = 45.4$ kips.

For double shear, the allowable force per bolt is $R_{DS} = 19.44$ kips.

$$\text{Number of bolts reqd for shear} = \frac{45.4}{19.44} = 2.33 \text{ bolts}$$

Allowable bearing force per bolt:

$$R_B = 1.35(36)0.75\ (0.250) = 9.11 \text{ kips}$$

$$\text{Number of bolts reqd for bearing} = \frac{45.4}{9.11} = 4.98 \text{ bolts}$$

Since the example called for a double row of bolts, use 6 bolts.

EXAMPLE 4.6.8

Recheck the number of bolts required to develop the allowable tensile force in Example 4.6.7 if AASHO specifications are to be used.

Solution

Assuming that fatigue considerations do not control, the capacity based on net section is

$$T = (\text{Net area})(0.55F_y)$$
$$= [10 - 2(\tfrac{3}{4} + \tfrac{1}{8})]0.25(0.55)36 = 40.8 \text{ kips}$$

Allowable shear per bolt (double shear) is

$$R_{DS} = 2(0.4418)20.0 = 17.67 \text{ kips}$$

$$\text{Number of bolts reqd for shear} = \frac{40.8}{17.67} = 2.31 \text{ bolts}$$

Allowable bearing force per bolt is

$$R_B = 1.22(36)0.750(0.250) = 8.24 \text{ kips}$$

$$\text{Number of bolts reqd for bearing} = \frac{40.8}{8.24} = 4.96 \text{ bolts}$$

Use 6 bolts.

It should be noted that although the number of bolts required was the same as in Example 4.6.7, the allowable tensile force was less.

Summary of Design Considerations for High-Strength Bolted Joints Subjected to Axial Tension or Compression. The following summary is intended to combine factors dealt with in this article with those in Chapter 3 so that strength along with other limitations of design specifications will be considered in the design of axially loaded joints. The elements of a good design are:

1. Adequate *net section* to carry tensile load (see Chapter 3).
2. Adequate *gross area* to carry compressive load.
3. Adequate number of bolts so that allowable nominal *shear stress* is not exceeded (discussed in this article).
4. Adequate number of bolts, or use of thick enough plates, so that allowable *bearing stress* is not exceeded (for *bearing-type* joints only, as discussed in this article).
5. Adequate *edge distance* so that connectors cannot shear out (see Table 4.6.4 for requirements of AISC–1.16.5). When not more

than two connectors are used in a line parallel to the direction of stress, special provisions must be considered according to AISC–1.16.6.

6. Maximum edge distance not exceeded, so that "dishing" or curling of the edge will not occur. (see AISC–1.16.7).
7. Reasonable spacing between connectors, measured center-to-center, so that tearing of plates cannot occur and so that wrenches can easily be applied for bolt tightening. AISC–1.16.4 states that the minimum spacing is $2\frac{2}{3}$ times bolt diameter, but preferably three bolt diameters. Commonly a 3-in. spacing is used for $\frac{3}{4}$ in. and $\frac{7}{8}$ in. diam. bolts.

TABLE 4.6.4
Minimum Edge Distances (AISC–1.16.5)

Rivet or Bolt Diameter (Inches)	Minimum Edge Distance for Punched, Reamed or Drilled Holes (Inches)	
	At Sheared Edges	At Rolled Edges of Plates, Shapes or Bars or Gas Cut Edges**
$\frac{1}{2}$	$\frac{7}{8}$	$\frac{3}{4}$
$\frac{5}{8}$	$1\frac{1}{8}$	$\frac{7}{8}$
$\frac{3}{4}$	$1\frac{1}{4}$	1
$\frac{7}{8}$	$1\frac{1}{2}$*	$1\frac{1}{8}$
1	$1\frac{3}{4}$*	$1\frac{1}{4}$
$1\frac{1}{8}$	2	$1\frac{1}{2}$
$1\frac{1}{4}$	$2\frac{1}{4}$	$1\frac{5}{8}$
Over $1\frac{1}{4}$	$1\frac{3}{4}$ × Diameter	$1\frac{1}{4}$ × Diameter

*These may be $1\frac{1}{4}$-in. at the ends of beam connection angles.
**All edge distances in this column may be reduced $\frac{1}{8}$-in. when the hole is at a point where stress does not exceed 25% of the maximum allowed stress in the element.

8. Long lines of connectors avoided; use compact joints if feasible. Compact joints result when connector spacing both transverse and parallel to the direction of stress is approximately 5 times the diameter of the bolts, and the length of the joint does not exceed about 5 pitches in the direction of stress.
9. Long bolt grips avoided to minimize bending and the resulting nonuniformity of stress. Furthermore, the minimum pretension force causes greater elongation and a reduced clamping force when bolts with long grips are used. The grip, or total thickness of pieces fastened by the bolt, should not exceed 5 diameters; or provisions of AISC–1.16.3 must be applied.

4.7. ECCENTRIC SHEAR

Eccentric shear occurs frequently in structural connections and some typical examples are shown in Fig. 4.7.1. The exact distribution of stresses

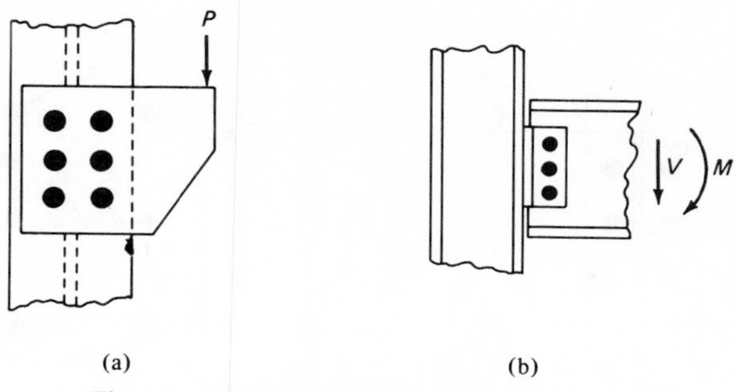

(a) (b)

Fig. 4.7.1. Typical eccentric shear connections.

that exist in such fasteners is complex and depends on many factors such as the tightness of the bolts, the manner in which the bolts fill the holes, the strain in the plates, and the friction between the plates. However, in order to provide a reasonable design approach nominal stresses are used involving essentially the assumptions discussed in Sec. 4.5. The following are the specific assumptions used for developing the nominal stress equations for eccentric shear:

1. The plates making up the connection are perfectly rigid while the fasteners are perfectly elastic.
2. The rotation of the plate produces shearing deformations in the fasteners which are proportional to and normal to the radius from the center of rotation.
3. Stress is proportional to strain in the fasteners.
4. The stress across each fastener is uniform and the force in each fastener acts through its center.

The foregoing assumptions correspond to connection behavior when the load on the connection approaches ultimate strength and plastic deformations have occurred. Under service loads, as discussed in Sec. 4.6, for connections designed as friction-type the transfer of load is always through friction, and for those designed as bearing-type usually the load transfer is through friction.

The method for computing the stresses in an eccentric shear connec-

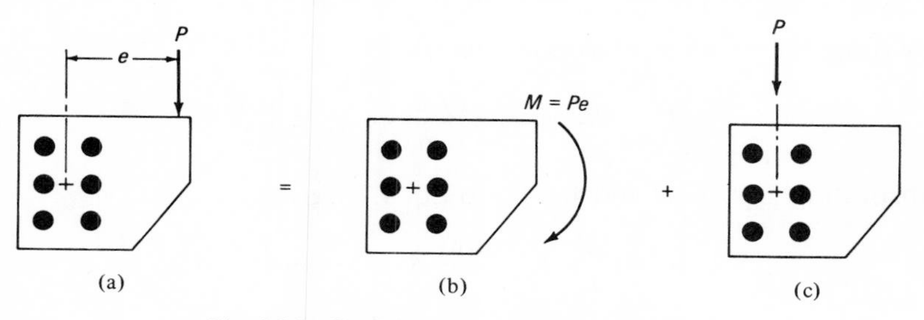

(a) (b) (c)

Fig. 4.7.2. Combined moment and direct shear.

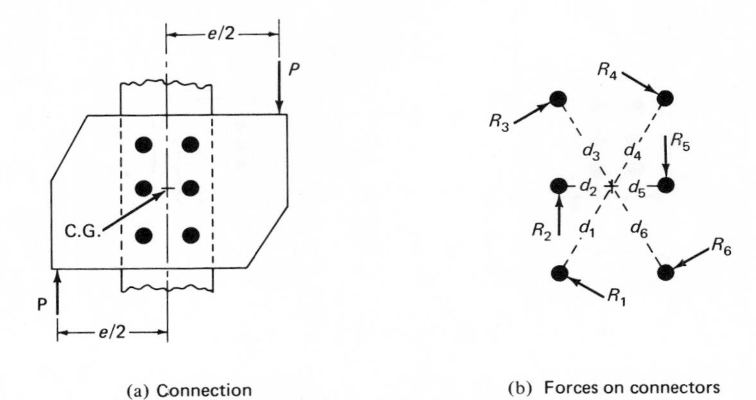

(a) Connection (b) Forces on connectors

Fig. 4.7.3. Pure moment connection.

tion is basically a superposition procedure in which the effect of pure moment is added to the effect of a direct load as shown in Fig. 4.7.2. In developing the procedure for computing the eccentric shear stresses, consider first the pure moment connection shown in Fig. 4.7.3. Neglecting any friction between the plates and assuming the connectors to be perfectly elastic, the force on each connector is proportional to its distance from the centroid of the group. The moment produced by the force system in Fig. 4.7.3a equals the sum of the forces shown in Fig. 4.7.3b times their distances to the centroid, as given in Eq. 4.7.1.

$$M = 2P\left(\frac{e}{2}\right) = Pe = R_1 d_1 + R_2 d_2 + \cdots + R_6 d_6 \qquad (4.7.1a)$$

$$= \Sigma R d \qquad (4.7.1b)$$

Assuming all the connectors to be of the same cross-sectional area A the unit stresses are

$$f_1 = \frac{F_1}{A};\ f_2 = \frac{R_2}{A} \cdots f_6 = \frac{R_6}{A} \qquad (4.7.2)$$

Also, since the stress in each connector is proportional to its distance d,

$$\frac{f_1}{d_1} = \frac{f_2}{d_2} = \cdots \frac{f_6}{d_6} \qquad (4.7.3)$$

Rewriting the stresses in terms of f_1 and d_1,

$$f_1 = \frac{f_1 d_1}{d_1};\quad f_2 = \frac{f_1 d_2}{d_1} \cdots f_6 = \frac{f_1 d_6}{d_1} \qquad (4.7.4)$$

Substituting Eqs. 4.7.2 and 4.7.4 into Eq. 4.7.1 gives

$$M = \frac{f_1 d_1^2}{d_1} A + \frac{f_1 d_2^2}{d_1} A + \cdots + \frac{f_1 d_6^2}{d_1} A$$

$$= \frac{f_1}{d_1} A\,[d_1^2 + d_2^2 + d_3^2 + \cdots + d_6^2] \tag{4.7.5a}$$

$$= \frac{f_1}{d_1} \Sigma A d^2 \tag{4.7.5b}$$

The stress in connector 1 is therefore

$$f_1 = \frac{M d_1}{\Sigma A d^2} \tag{4.7.6a}$$

and by similar reasoning, stresses on the other connectors are

$$f_2 = \frac{M d_2}{\Sigma A d^2};\quad f_3 = \frac{M d_3}{\Sigma A d^2}; \cdots, f_6 = \frac{M d_6}{\Sigma A d^2} \tag{4.7.6b}$$

or in general,

$$f = \frac{M d}{\Sigma A d^2} \tag{4.7.7}$$

which is the same as the familiar mechanics of materials formula for torsion on a circular shaft, Tr/J, which is also discussed in Sec. 8.2. The twisting moment, $T = M$; the radius from the center of rotation to the point at which stress is computed, $r = d$; and the polar moment of inertia, $J = \Sigma A d^2$.

In computing the stresses on the connectors it is usually more convenient to resolve the distances and forces into their horizontal and vertical components as shown in Fig. 4.7.4. The horizontal and vertical com-

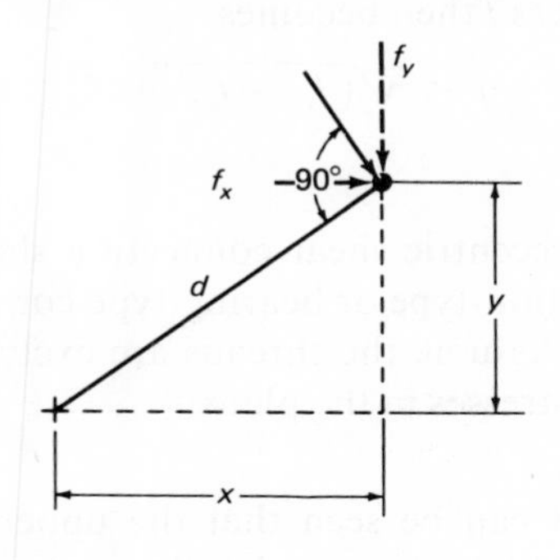

Fig. 4.7.4.

ponents of the stress, f, are f_x and f_y, respectively, and similarly the components of the distance d are x and y. From Fig. 4.7.4,

$$f_x = f\frac{y}{d} \quad \text{and} \quad f_y = f\frac{x}{d} \tag{4.7.8}$$

Substituting Eq. 4.7.8 into Eq. 4.7.7 gives

$$f_x = \frac{My}{\Sigma Ad^2} \quad \text{and} \quad f_y = \frac{Mx}{\Sigma Ad^2} \tag{4.7.9}$$

Assuming that all the connectors have the same area, and noting that

$$d^2 = x^2 + y^2$$

the vertical and horizontal components of the stresses may be written

$$f_x = \frac{My}{A(\Sigma x^2 + \Sigma y^2)} \tag{4.7.10a}$$

$$f_y = \frac{Mx}{A(\Sigma x^2 + \Sigma y^2)} \tag{4.7.10b}$$

By taking the vector sum of f_x and f_y, the total stress f on the connector becomes

$$f = \sqrt{f_x^2 + f_y^2} \tag{4.7.11}$$

In order to compute the total shear stresses in an eccentric connection such as shown in Fig. 4.7.2a, the direct shear stresses f_s of Fig. 4.7.2c must be added to the pure moment condition in Fig. 4.7.2b. Assuming the vertical force is equally shared by all the connectors, the direct shear stress f_s is

$$f_s = \frac{P}{\Sigma A} \tag{4.7.12}$$

The total resultant stress f then becomes

$$f = \sqrt{(f_y + f_s)^2 + f_x^2} \tag{4.7.13}$$

EXAMPLE 4.7.1

Determine if the eccentric shear connection shown in Fig. 4.7.5a can be used as either a friction-type or bearing-type connection if 7/8 in. diam. A325 bolts are used. Assume the threads are excluded from the shearing plane and neglect the stresses in the plate.

Solution

From Fig. 4.7.5 it can be seen that the upper and lower right connectors are the most highly stressed. Since these two connectors are equally stressed only one need be investigated; check upper-right connector. The eccentricity e, as measured from the centroid (center of rotation), is

$$e = 3 + 2 = 5 \text{ in.}$$
$$M = 24(5) = 120 \text{ in.-kips}$$

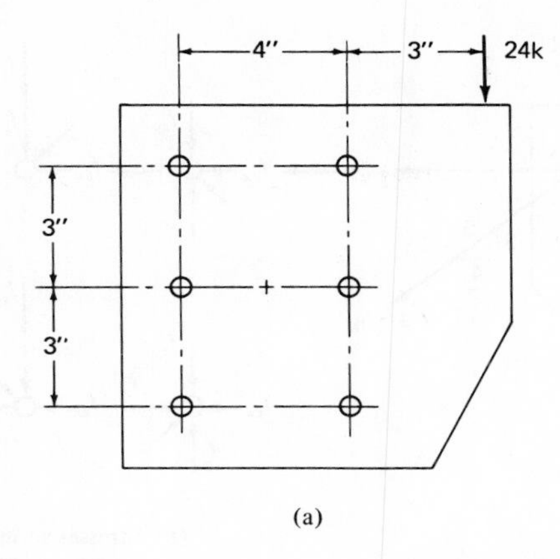

(a)

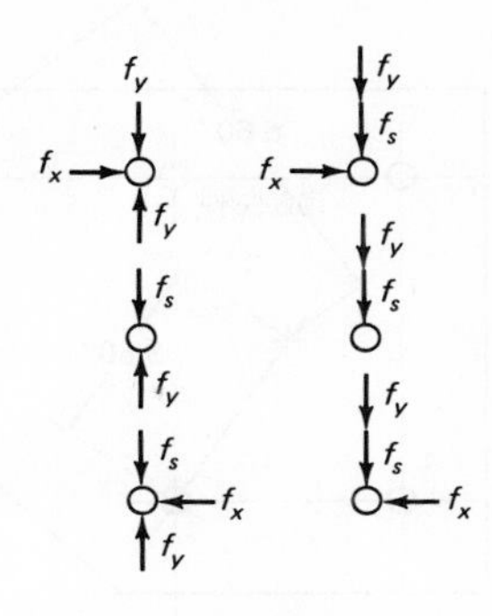

(b) Stress acting on connector

Fig. 4.7.5. Example 4.7.1.

$$\Sigma x^2 + \Sigma y^2 = 6(2)^2 + 4(3)^2 = 60 \text{ sq in.}$$

$$f_x = \frac{My}{A(\Sigma x^2 + \Sigma y^2)} = \frac{120(3)}{0.601(60)} = 9.99 \text{ ksi}$$

$$f_y = \frac{Mx}{A(\Sigma x^2 + \Sigma y^2)} = \frac{120(2)}{0.601(60)} = 6.65 \text{ ksi}$$

$$f_s = \frac{P}{\Sigma A} = \frac{24.0}{6(0.601)} = 6.65 \text{ ksi}$$

$$f = \sqrt{(6.65 + 6.65)^2 + (9.99)^2} = 16.65 \text{ ksi}$$

Since the maximum stress f exceeds $F_v = 15$ ksi but is less than $F_v = 22$ ksi, the connection is inadequate as a friction-type, but is satisfactory as a bearing-type connection.

EXAMPLE 4.7.2

Determine the maximum shear stress in the bottom right connector of the connection shown in Fig. 4.7.6 if 1 in. diam. connectors are used. Solve by (a) considering P only and (b) considering P_x and P_y.

Solution

(a) considering P only,

$$e = 6.60 \text{ in.}$$

$$M = 10(6.60) = 66.0 \text{ in.-kips}$$

$$\Sigma Ad^2 = 4(0.785)(3.60)^2 = 40.7 \text{ sq in.}$$

$$f_x = \frac{My}{\Sigma Ad^2} = \frac{66.0(3)}{40.7} = 4.86 \text{ ksi}$$

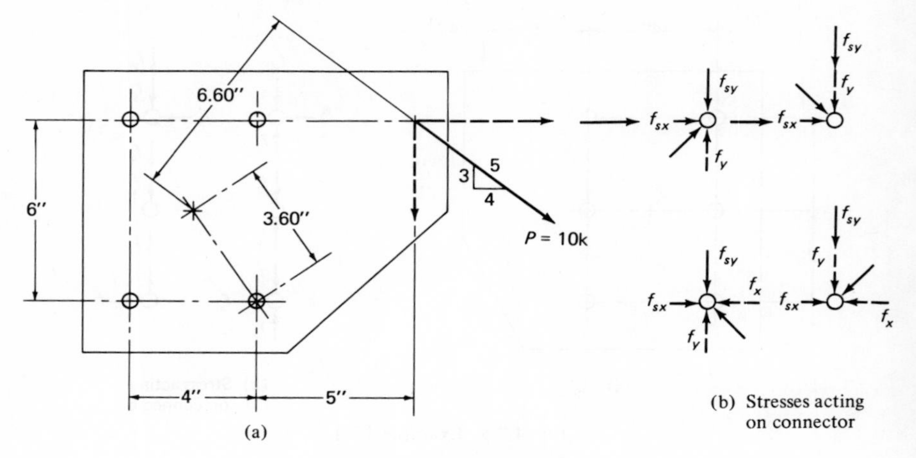

Fig. 4.7.6. Example 4.7.2.

$$f_y = \frac{Mx\text{-}}{\Sigma Ad^2} = \frac{66.0(2)}{40.7} = 3.24 \text{ ksi}$$

$$f_{sx} = \frac{P_x}{\Sigma A} = \frac{0.8(10)}{4(0.785)} = 2.55 \text{ ksi}$$

$$f_{sy} = \frac{P_x}{\Sigma A} = \frac{0.6(10)}{4(0.785)} = 1.91 \text{ ksi}$$

$$f = \sqrt{(f_y + f_{sy})^2 + (f_x + f_{sx})^2}$$

$$= \sqrt{(3.24 + 1.91)^2 + (4.86 + 2.55)^2} = 9.04 \text{ ksi}$$

(Maximum nominal shear stress)

(b) Considering the components P_x and P_y,

$$e_x = 5.0 + 2.0 = 7.0 \text{ in.} \qquad e_y = 3.0 \text{ in.}$$

$$M = M_x + M_y = P_x(e_y) + P_y(e_x)$$

$$0.8\ (10)3.0 + 0.6(10)7.0 - 24.0 + 42.0 = 66.0 \text{ in.-kips}$$

$$\Sigma x^2 + \Sigma y^2 = 4(2)^2 + 4(3)^2 = 52.0 \text{ in.}^2$$

$$f_x = \frac{My}{A(\Sigma x^2 + \Sigma y^2)} = \frac{66.0(3)}{0.785(52)} = 4.86 \text{ ksi}$$

$$f_y = \frac{Mx}{A(\Sigma x^2 + \Sigma y^2)} = \frac{66.0(2)}{0.785(52)} = 3.24 \text{ ksi}$$

The direct shear components f_{sx} and f_{sy}, along with the result are the same as in part (a).

AISC Suggestion for Using Reduced Effective Eccentricity. There have been few reported tests of the behavior of eccentric shear connections. For many years, designers have used the nominal stress procedure developed earlier in this section without any apparent difficulty. The conservatism of the procedure was verified in 1964 by tests sponsored by AISC.[30] As a result the AISC recommendation given in the AISC Manual (though *not* a specification requirement) is to use a reduced effective eccentricity of load in accordance with the following:

(a) For fasteners equally spaced on a single gage line (Fig. 4.7.7a)

$$e_{\text{eff}} = e - \frac{1 + 2n}{4} \tag{4.7.14}$$

where n = number of fasteners in one line.

(b) For fasteners equally spaced on two or more gage lines,

$$e_{\text{eff}} = e - \frac{1 + n}{2} \tag{4.7.15}$$

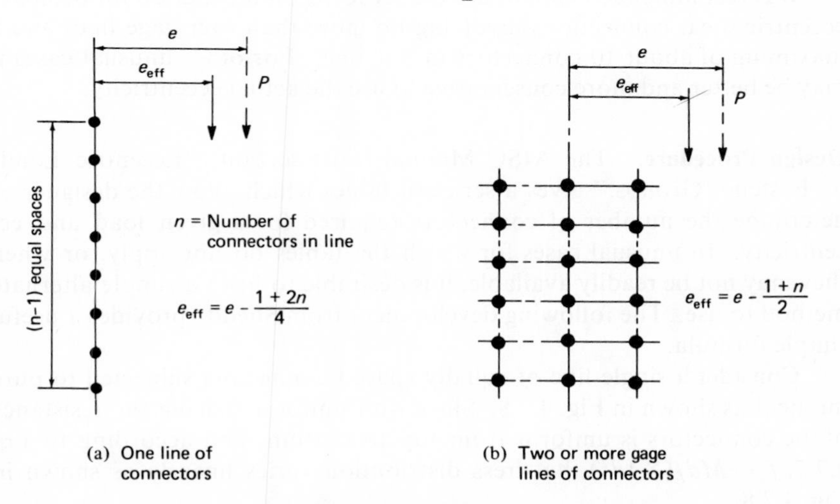

Fig. 4.7.7. AISC recommendation for effective eccentricity.

EXAMPLE 4.7.3

Repeat Example 4.7.1 using the reduced effective eccentricity of Eq. 4.7.15 and determine the percentage reduction in the computed nominal stress on the most highly stressed connector (see Fig. 4.7.5).

Solution

The reduced effective eccentricity is

$$e_{\text{eff}} = 5 - \frac{1 + 3}{2} = 3 \text{ inches}$$

$$M = 24(3) = 72 \text{ in.-kips}$$

From the stress equations it is apparent the stress is directly proportional to M; then, the maximum stress is

$$f = 16.65\frac{72}{120} = 10.0 \text{ ksi}$$

a 40 percent reduction, which would permit the connection to be used as a friction-type connection.

It should be noted that on many simple connections which have small eccentricity of load, it has long been the practice to ignore such eccentricity. This concept of a reduced effective eccentricity is in agreement with that practice.

A more recent study[31] has developed procedures for determining ultimate strength of a variety of bolt-group configurations and presented interaction relationships in graphical form relating the direct load effect and the moment (torsion) effect.

It is recommended here to use the AISC recommendation for ordinary eccentric shear connections involving no more than four gage lines and a maximum of about 10 connectors in one line. For other unusual cases it may be better and more conservative to use the actual eccentricity.

Design Procedure. The AISC Manual in its section, "Eccentric Loads on Fastener Groups," gives a series of tables which allow the designer to determine the number of connectors required for a given load and eccentricity. In unusual cases for which the tables do not apply, or when they may not be readily available, it is desirable to have a simple alternate method to use. The following development from Shedd* provides a useful simple formula.

Consider a single line of equally spaced connectors subjected to pure moment as shown in Fig. 4.7.8. Since with uniform spacing the resistance of the connectors is uniform from top to bottom, and according to Eq. 4.7.7, $f = Md/(\Sigma Ad^2)$ the stress distribution varies linearly as shown in Fig. 4.7.8.

Assuming R is the force, fA, in the outermost connector, and that it represents the accumulation of stress which would occur on a rectangular resisting section over the height p, one may designate the average load per inch of height at the outermost connector as R/p.

Using similar triangles the load per inch at the extreme fiber may be determined,

*Thomas C. Shedd, *Structural Design in Steel* (New York: John Wiley & Sons, Inc., 1934), p. 287.

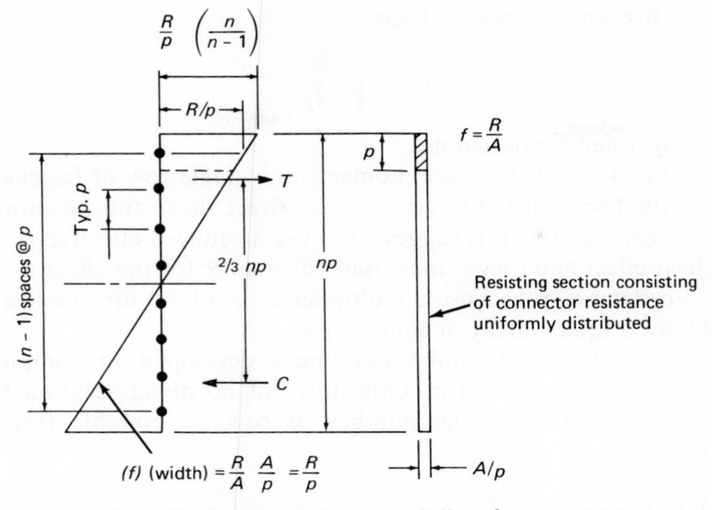

Fig. 4.7.8. Pure moment on a single line of connectors.

$$\frac{\text{Extreme fiber value}}{\dfrac{np}{2}} = \frac{B/p}{\dfrac{(n-1)p}{2}} \tag{4.7.16}$$

$$\text{Extreme fiber value} = \frac{R}{p}\left(\frac{n}{n-1}\right) \tag{4.7.17}$$

The tensile force is the area of the triangle represented by the force per inch diagram,

$$T = \frac{1}{2}\left(\frac{np}{2}\right)\left(\frac{R}{p}\right)\left(\frac{n}{n-1}\right) = \frac{Rn^2}{4(n-1)} \tag{4.7.18}$$

The internal resisting moment is

$$M = T(\tfrac{2}{3}np) \tag{4.7.19}$$

Substitution of Eq. 4.7.18 into Eq. 4.7.19 gives

$$M = \frac{Rn^2}{4(n-1)}\left(\frac{2}{3}np\right) = \frac{Rn^3p}{6(n-1)} \tag{4.7.20}$$

Solving Eq. 4.7.20 for n^2, one obtains

$$n = \sqrt{\frac{6M}{Rp}\left(\frac{n-1}{n}\right)} \tag{4.7.21}$$

which as a first approximation becomes

$$n = \sqrt{\frac{6M}{Rp}} \tag{4.7.22}$$

which is suggested for design use.

Since Eq. 4.7.22 is for pure moment on a single row of fasteners, the R value should be adjusted to account for direct shear and for more than one row of connectors. It is suggested to use a reduced effective R for the direct-shear effect and use an increased effective R for the effect of lateral spread. For lateral spread use a multiplier on R of 1.0 for one line up to about 2.0 for a square array of connectors.

More complicated formulas have been developed to compute the maximum stress or force on a connector, but no direct solution for the number of connectors or the number of rows is possible from such equations.

EXAMPLE 4.7.4

Determine the required number of ⅞ in. diam. A325 bolts for one vertical line of connectors Ⓐ–Ⓐ in the bracket shown in Fig. 4.7.9. Assume it to be a bearing-type connection.

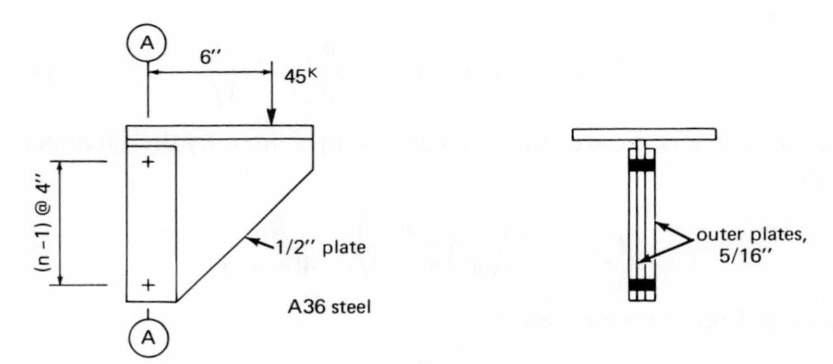

Fig. 4.7.9. Example 4.7.4.

Solution

(a) Using allowable nominal stresses determine the maximum allowable load per connector.

$$F_v = 22 \text{ ksi}$$
$$R_{DS} = 22(0.6013)(2) = 26.4 \text{ kips}$$

It is a case of double shear, since two shear planes are resisting failure of each connector.

$$F_p = 1.35F_y = 1.35(36) = 48.5 \text{ ksi}$$
$$R_B = 48.5(7/8)(0.5) = 21.3 \text{ kips controls!}$$

The plates are assumed to be adequate.

(b) Estimate the number of bolts required using Eq. 4.7.22. Reduce R to be used in the formula to estimate the horizontal component, say use 17 kips instead of the full 21.3 kips. Also, the effective eccentricity is estimated knowing that three connectors are needed for direct shear alone.

$$n = \sqrt{\frac{6M}{Rp}} = \sqrt{\frac{6(45)(4.5)}{17(4)}} = 4.2$$

(c) Check adequacy of 5 bolts.

$$f_s = \frac{P}{\Sigma A} = \frac{P}{An}$$

or multiplying by A, the force per connector is obtained:

$$f_s A = \frac{P}{n} = \frac{45}{5} = 9 \text{ kips} \downarrow$$

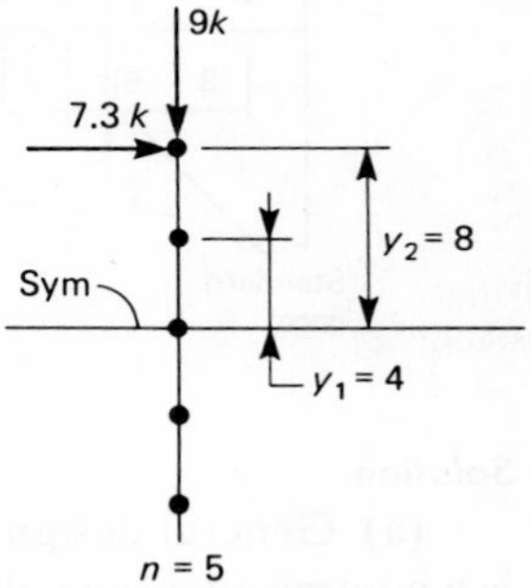

the moment component is

$$f_x = \frac{My}{A(\Sigma x^2 + \Sigma y^2)}$$

$$f_x A = \frac{My}{\Sigma x^2 + \Sigma y^2}$$

9k
7.3 k
Sym
$y_2 = 8$
$y_1 = 4$
$n = 5$

$$\Sigma x^2 + \Sigma y^2 = 0 + 2[(4)^2 + (8)^2] = 160$$

$$e_{\text{eff}} = e - \frac{1 + 2n}{4} = 6 - \frac{1 + 10}{4} = 3.25 \text{ in.}$$

$$f_x A = \frac{45(3.25)8}{160} = 7.3 \text{ kips} \rightarrow$$

$$\text{Actual } R = \sqrt{(9)^2 + (7.3)^2} = 11.6 \text{ kips} < 21.3$$

Too low.

(d) Try 4 bolts in one line.

$$f_s A = \frac{P}{\Sigma n} = \frac{45}{4} = 11.25 \text{ kips} \downarrow$$

$$e_{\text{eff}} = 6 - \frac{1 + 8}{4} = 3.75 \text{ in.}$$

$$\Sigma x^2 + \Sigma y^2 = 0 + 2[(2)^2 + (6)^2] = 80$$

$$f_x A = \frac{45(3.75)6}{80} = 11.25 \text{ kips} \rightarrow$$

$$\text{Actual } R = \sqrt{(11.25)^2 + (11.25)^2} = 15.9 \text{ kips} < 21.3$$

Use 4 – A325 bolts @ 4 in. pitch.

EXAMPLE 4.7.5

Determine the required number of ¾ in. diam. A325 bolts for the bracket plate of Fig. 4.7.10, assuming 4 vertical rows. Use a friction-type connection.

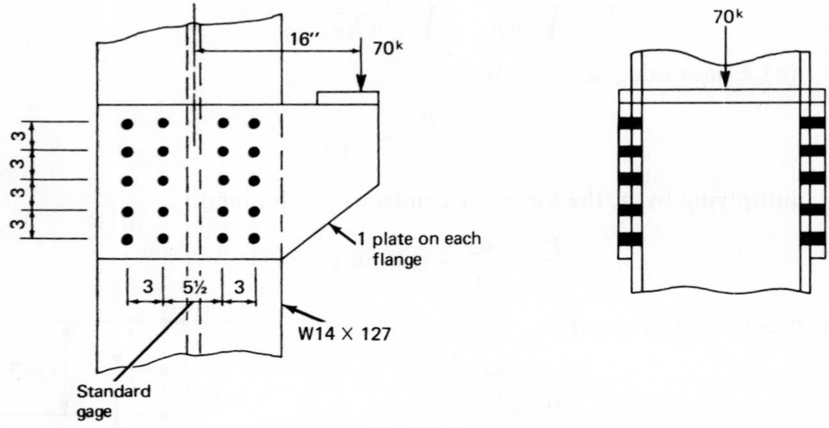

Fig. 4.7.10.

Solution

(a) General design method. Choose a vertical pitch of 3 in., which is a commonly used value. The maximum allowable load per connector is

$$R_{SS} \text{ (single shear)} = 15(0.4418) = 6.63 \text{ kips}$$

Bearing is not to be considered for friction-type connections.

Estimate the adjustment factor on R as about 1.3 to account for the four rows.

Estimate the adjustment on eccentricity to be such that $e_{\text{eff}} \approx 13$ to 13.5 in., say 13 in.

$$n = \sqrt{\frac{6M}{Rp}} = \sqrt{\frac{6(35/4)(13)}{1.3(6.63)(3)}} = 5.1$$

In the above equation, the load per plate is 35 kips, and the load per line of connectors is 35/4, which must be used since Eq. 4.7.22 applies to one line of connectors.

Check 5 bolts per row.

$$f_s A = \frac{P}{n} = \frac{35}{20} = 1.75 \text{ kips} \downarrow$$

$$e_{\text{eff}} = e - \frac{1+n}{2} = 16 - \frac{1+5}{2} = 13 \text{ in.}$$

$$\Sigma x^2 + \Sigma y^2 = 10[(2.75)^2 + (5.75)^2] + 8[(3)^2 + (6)^2]$$

$$= 405.6 + 360 = 765.6 \text{ in.}^2$$

$$f_x A = \frac{My}{\Sigma x^2 + \Sigma y^2} = \frac{35(13)6}{765.6} = 3.57 \text{ kips} \rightarrow$$

$$f_y A = \frac{Mx}{\Sigma x^2 + \Sigma y^2} = \frac{35(13)5.75}{765.6} = 3.42 \text{ kips} \downarrow$$

$$\text{Actual } R = \sqrt{(1.75 + 3.42)^2 + (3.57)^2} = 6.29 \text{ kips} < 6.63 \text{ o.k.}$$

Use–5¾ in. diam. A325 bolts per row.

(b) Design using AISC Tables, "Eccentric Loads on Fasteners Groups."

$$r_v = \text{allowable fastener load} = 6.63 \text{ kips}$$

$$P = 35 \text{ kips for one plate}$$

$$\text{Coefficient } C = \frac{P}{r_v} = \frac{35}{6.63} = 5.29$$

The connector spacing of Fig. 4.7.10 fits AISC Table XIII, $D = 11\frac{1}{2}$ (total width of group).

	$n = 4$	$n = 5$	
$d_{eff} = 12$	4.42	5.93	
$= 13$	4.23	5.60	$\leftarrow n = 4.8$
$= 14$	3.93	5.27	for $C = 5.29$

Use 5 bolts per row.

General Formula for Eccentric Loads on Fastener Groups. It is possible to develop complex formulas for computing capacity of eccentric loads on fastener groups, and the AISC Manual Tables, "Eccentric Loads on Fastener Groups," present the formulas and tabulate values for a number of cases. The formula for the case shown in Fig. 4.7.11 is developed as typical of the procedure.

Eq. 4.7.13 applies for this case:

$$f = \sqrt{(f_y + f_s)^2 + f_x^2} \qquad [4.7.13]$$

where

$$f_s = \frac{P}{\Sigma A}$$

$$f_x = \frac{My}{\Sigma A d^2} \quad \text{and} \quad f_y = \frac{Mx}{\Sigma A d^2}$$

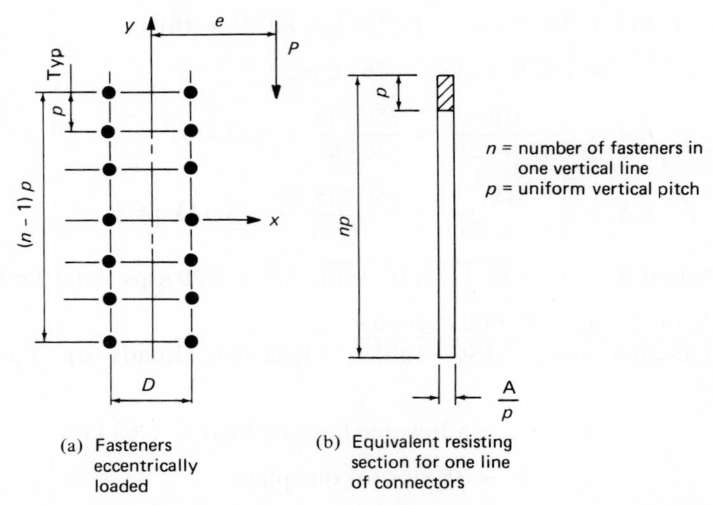

(a) Fasteners eccentrically loaded
(b) Equivalent resisting section for one line of connectors

Fig. 4.7.11. Development of general formula for eccentric loads on fastener group.

It is noted that Ad^2 is the polar moment of inertia J about the centroid of the fastener group, and may also be expressed as

$$J = I_x + I_y = \Sigma Ay^2 + \Sigma Ax^2$$

For a line of fasteners the effective section may be thought of as distributed into a rectangle of width A/p and height np, as in Fig. 4.7.11b. The cross-hatched area representing one fastener has an area $p(A/p) = A$.

The moment of inertia for the rectangle is $(A/p)(np)^3/12$. Since ordinarily the moment of inertia of the connector area about its own centroid is neglected (i.e., as in ΣAy^2), the effect of the moment of inertia of the cross-hatched part about its own centroid should be subtracted for each fastener in the group; thus

$$I_x = \left(\frac{A}{p}\right)(np)^3\frac{1}{12} - n\left(\frac{A}{p}\right)p^3\frac{1}{12} = \frac{Ap^2}{12}n(n^2 - 1) \qquad (4.7.23)$$

for *one* vertical line. For m vertical lines, I_x becomes

$$I_x = \frac{mAp^2n(n^2 - 1)}{12} \qquad (4.7.24)$$

Similarly for moment of inertia about the y axis, if q is the horizontal pitch,

$$I_y = \frac{nAq^2m(m^2 - 1)}{12} \qquad (4.7.25)$$

and

$$J = I_x + I_y = \frac{Amn}{12}[p^2(n^2 - 1) + q^2(m^2 - 1)] \qquad (4.7.26)$$

For this case with $m = 2$, $q = D$, and $M = Pe$,

$$f_s A = \frac{P}{2n}$$

$$f_x A = \frac{Pe\,(n-1)p/2}{\frac{2n}{12}[\,p^2(n^2-1) + 3D^2]} \tag{4.7.27a}$$

$$f_y A = \frac{PeD/2}{\frac{2n}{12}[\,p^2(n^2-1) + 3D^2]} \tag{4.7.27b}$$

Applying Eq. 4.7.13 with all terms multiplied by the area gives the resultant force R on the most highly stressed connector.

$$R = \sqrt{\frac{P}{2n} + \left(\frac{PeD}{n\left[\frac{p^2}{3}(n^2-1) + D^2\right]}\right)^2 + \left(\frac{Pe(n-1)p}{n\left[\frac{p^2}{3}(n^2-1) + D^2\right]}\right)^2} \tag{4.7.28}$$

Solving for P gives

$$P = \frac{nR}{\sqrt{\left[\frac{eD}{D^2 + \frac{1}{3}(n^2-1)p^2} + \frac{1}{2}\right]^2 + \left[\frac{e(n-1)p}{D^2 + \frac{1}{3}(n^2-1)p^2}\right]^2}} \tag{4.7.29}$$

which agrees with the AISC Manual formula for this case. It is apparent that Eq. 4.7.29 is only suitable for analysis; however, for given values of p (AISC tables use 3 in.) and D (AISC tables use 3 in. and $5\frac{1}{2}$ in.), the multiplier of R is called the coefficient C and is tabulated as the interaction between e (AISC uses l_{eff}) and n.

4.8. FASTENERS ACTING IN AXIAL TENSION

Fasteners acting in tension without simultaneous shear occur in tension members such as hangers, (see Fig. 4.5.1c) or other members whose line of action is perpendicular to the member to which it is fastened. When such tension members are not perpendicular to their connecting members, the fasteners are subjected to both axial tension and shear. The latter, more typical case, is discussed in Sec. 4.9.

Table 4.8.1 shows values for the allowable tensile forces according to AISC and AASHO specifications. The allowable tensile stress, F_t, for both high-strength bolts and rivets is based on their nominal cross-sectional area. In the case of A307 bolts, the allowable tensile stress is

TABLE 4.8.1
Allowable Tensile Stress on Nominal Cross Sectional Area of Fasteners (from AISC–1.5.2)

Description of Fastener	Tension, F_t	
	AISC	AASHO
A502, Grade 1, hot-driven rivets	20.0	Not Permitted on Rivets
A502, Grade 2, hot-driven rivets	27.0	
A307 bolts	20.0[1]	13.5
Threaded parts[3] of steel meeting the requirements of Sect. 1.4.1	$0.60F_y^1$	
A325 and A449 bolts, when threading is *not* excluded from shear planes	40.0[2]	36.0[2]
A325 and A449 bolts, when threading is excluded from shear planes	40.0[2]	36.0[2]
A490 bolts, when threading is *not* excluded from shear planes	54.0[2,4]	48.0[2,4]
A490 bolts, when threading is excluded from shear planes	54.0[2,4]	48.0[2,4]

[1]Applied to tensile stress area equal to $0.7854 \left(D - \frac{0.9743}{n}\right)^2$ where D is the major thread diameter and n is the number of threads per inch.
[2]Applied to the nominal bolt area.
[3]Since the nominal area of an upset rod is less than the stress area, the former area will govern.
[4]Static loading only.

based on a modified cross-sectional area, known as tensile-stress area, given by Eq. 4.8.1 and discussed in Chapter 3.

$$\text{Modified area} = 0.7854\left(D - \frac{0.9743}{n}\right)^2 \qquad (4.8.1)$$

where D = major or nominal thread diameter, in.
n = number of threads per in.

Equation 4.8.1 approximates the minimum cross-sectional area that results from passing a plane through the threaded part of the bolt.

No such reduction is necessary in the case of high strength bolts since the allowable F_t is substantially less than the pretension stress. Any permissible tensile force applied to a pretensioned high-strength bolt will only decrease the clamping force produced by the bolt and not increase the stress in the bolt.

Prestress Effect of High-Strength Bolts Under External Tension. In order to understand the effect of an externally applied load on a pretension high-strength bolt consider a single bolt and the tributary portion of the connected pieces as shown in Fig. 4.8.1a. The pieces being joined are of

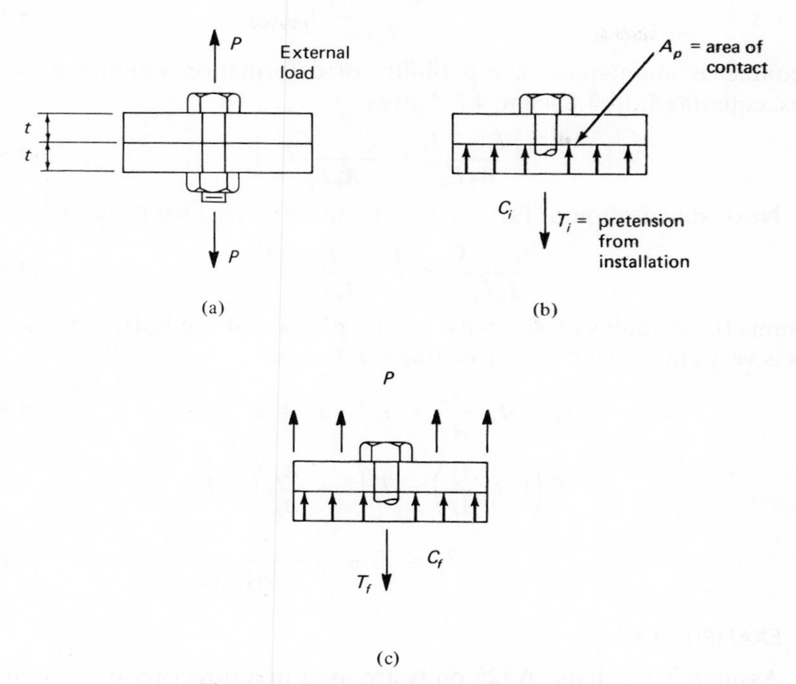

Fig. 4.8.1. Prestress effect on bolted joint.

thickness t and the area of contact between the pieces is A_p. Prior to applying external load the situation is as shown in Fig. 4.8.1b, where the bolt has been installed to have a pretension force T_i (values as in Table 4.4.2). The pieces being joined are compressed an amount C_i. For equilibrium,

$$C_i = T_i \tag{4.8.2}$$

The external load P is then applied and the forces acting are shown in Fig. 4.8.1c. This time equilibrium requires

$$P + C_f = T_f \tag{4.8.3}$$

where the subscript f refers to final conditions after application of the load P.

The increased bolt tension lengthens the bolt an amount δ_b.

$$\delta_b = \frac{T_f - T_i}{A_b E_b} t \tag{4.8.4}$$

The external tension reduces the compression which increases the plate thickness an amount δ_p.

$$\delta_p = \frac{C_i - C_f}{A_p E_p} t \tag{4.8.5}$$

If contact is maintained, compatibility of deformation requires $\delta_b = \delta_p$; thus, equating Eqs. 4.8.4 and 4.8.5 gives

$$\frac{T_f - T_i}{A_b E_b} = \frac{C_i - C_f}{A_p E_p} \tag{4.8.6}$$

Next, substitution of Eq. 4.8.2 for C_i and Eq. 4.8.3 for C_f gives

$$\frac{T_f - T_i}{A_b E_b} = \frac{T_i - T_f + P}{A_p E_p} \tag{4.8.7}$$

Assume the modulus of elasticity for the plates and the bolts is the same; this is very closely true. Then solving for T_f gives

$$(T_f - T_i)\frac{A_p}{A_b} = T_i - T_f + P \tag{4.8.8}$$

$$T_f\left(1 + \frac{A_p}{A_b}\right) = T_i\left(1 + \frac{A_p}{A_b}\right) + P$$

$$T_f = T_i + \frac{P}{1 + A_p/A_b} \tag{4.8.9}$$

EXAMPLE 4.8.1

Assume ⁷⁄₈ in. diam. A325 bolts are used in a direct-tension situation such as in Fig. 4.8.2. With bolts spaced 3 in. apart and having 1½-in.

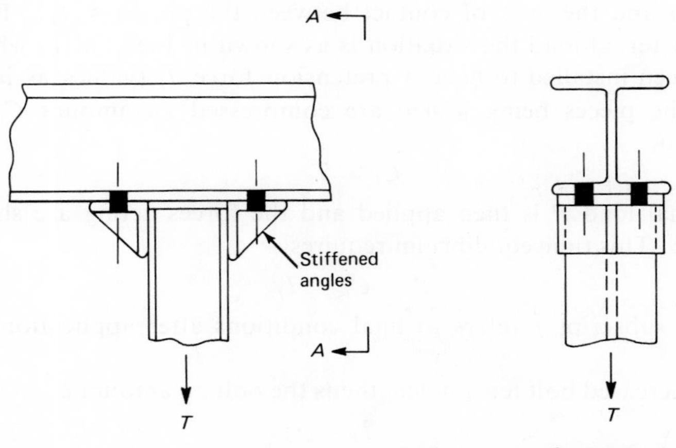

Fig. 4.8.2. Example 4.8.1.

edge distances, the tributary area of contact may reasonably be about 9 sq in. If the maximum external tensile load permitted by AISC is applied, how much does the bolt tension increase?

Solution

(a) Maximum applied tensile force P. From Table 4.8.1,

$$P = F_t A_b = 40.0(0.6013) = 24.05 \text{ kips}$$

(b) Initial tensile force in $^7/_8$ in. diam. A325 bolt. From Table 4.4.2,

$$T_i = 39 \text{ kips}$$

(c) Final tensile force in bolt:

$$\frac{A_p}{A_b} = \frac{9.0}{0.6013} = 15$$

This neglects subtracting the bolt area from the total tributary area, but little difference results. Using Eq. 4.8.9 then gives

$$T_f = 39 + \frac{24.05}{1 + 15} = 39 + 1.5 = 40.5 \text{ kips}$$

The increase in tension is 3.8 percent. The variation in actual pretension from installation may be expected to exceed this amount, so that this increase is not of concern. Furthermore, the tributary area used for the example (9.0 sq in.) is probably the minimum one might encounter in practice, since less than a 3-in. pitch and gage is rarely used.

A conclusion from this example is that no significant increase in bolt tension arises until the external load equals or exceeds the pretension, in which case the pieces do not remain in contact and the applied force equals the bolt tension.

If the connection can distort and give rise to "prying forces" these must also be considered. (See AISC Commentary–1.5.2.1 and the treatment in the "Split Beam Tee Connections" part of Sec. 13.6.)

The use of the 40 ksi for nominal stress in direct tension provides a factor of safety compatible with the other factors in design. In the situation of Example 4.8.1 the approximate factor of safety is

$$\text{FS} = \frac{T_i}{P} = \frac{39}{24.05} = 1.62$$

In general, in the case of A325 bolts under AISC–1.5.2 and 1.23.5, the margin against F_t exceeding the proof load is approximately 1.6 for diameters up to 1 in. and approximately 1.4 for diameters over 1 in. Similarly, use of the AASHO Specification would result in values of 1.8 and 1.5, respectively.

EXAMPLE 4.8.2

Compute the allowable load T as determined by the tensile capacity of fasteners if $^7/_8$ in. diam. A307 bolts with 9 threads per inch are used.

Solution

The modified cross-sectional area as given by Eq. 4.8.1 is

$$0.7854\left(0.8750 - \frac{0.9743}{9}\right)^2 = 0.462 \text{ sq in.}$$

$$\begin{aligned} T &= (0.462)(F_t)(\text{number of bolts}) \\ &= 0.462(20.0)4 = 36.96 \text{ kips} \end{aligned}$$

EXAMPLE 4.8.3

Compute the allowable load T in Example 4.8.2 if $^7/_8$-in. A325 bolts are used.

Solution

$$\begin{aligned} T &= (\text{area of bolt})(F_t)(\text{number of bolts}) \\ &= 0.601(40.0)4 = 96.2 \text{ kips} \end{aligned}$$

EXAMPLE 4.8.4

Determine the required number of $^3/_4$-in. A490 bolts required for the connection shown in Fig. 4.8.3. Assume that the pieces making up the

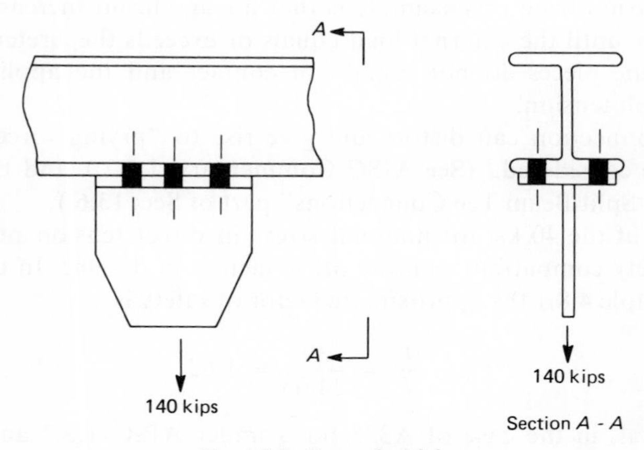

Fig. 4.8.3. Example 4.8.3.

connection are adequate and that the nominal tensile stresses on the bolts govern.

Solution

The number of bolts reqd is

$$N = \frac{T}{(\text{Area of bolt})(F_t)}$$

$$= \frac{140}{0.4418(54)} = 5.88$$

Use–$6\frac{3}{4}$ in. diam. A490 bolts.

4.9. COMBINED SHEAR AND TENSION

In a large number of commonly used connections, both shear and tension occur and must be considered in their design. Figure 4.9.1 shows

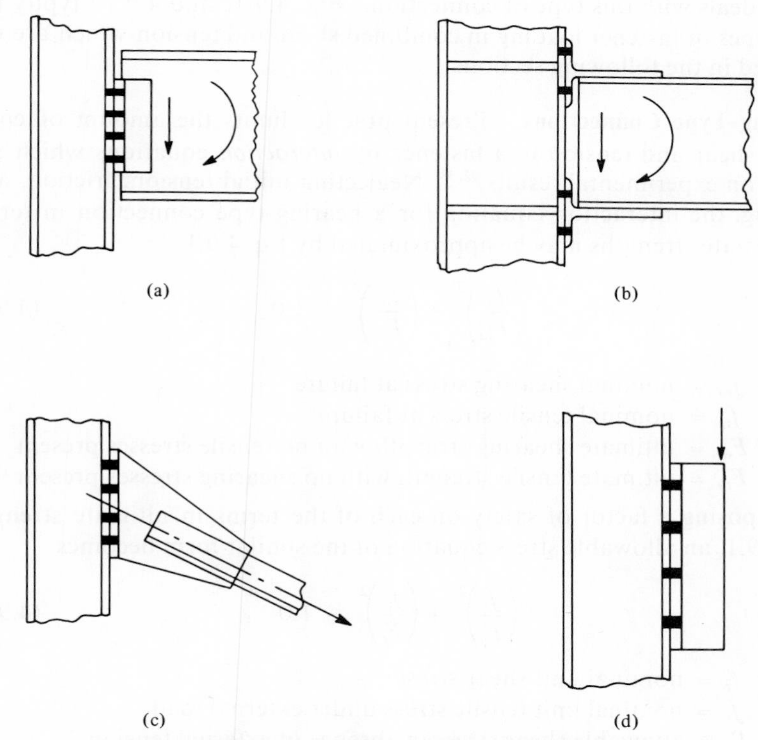

Fig. 4.9.1. Typical combined shear and tension connections.

a few typical connections in which the connectors are simultaneously subjected to both shear and tension. The connection shown in Fig. 4.9.1a is the most commonly used connection in which the beam is bolted to the column by means of a pair of angles. From the moment force indicated in the figure, the upper fasteners are subjected to tension in proportion to the magnitude of the applied moment. However, one may recall from structural analysis that only a small amount of end rotation on a beam is necessary to change from a fixed end to a hinged end condition. In addition the web carries only a very small moment. Thus, one may intuitively sense that the moment shown will be relieved before a significant tension force can be developed in the connectors. Such connections are used when little end moment is desired to be transmitted. An exception to this occurs in the case of a very deep beam such as a plate girder.

Referring next to Fig. 4.9.1b in which the applied moment is transmitted through the flanges of the beam the situation is different. In this case a large applied moment is intended to be transmitted so the connection is made at the flanges which carry most of the moment. Chapter 13 deals with this type of connection. Fig. 4.9.1c and 4.9.1d typify the two types of fastener loading in combined shear and tension which are developed in the following sections.

Bearing-Type Connections. Present practice limits the amount of combined shear and tension in a fastener by *interaction* equations which are based on experimental results.[32,33] Neglecting initial tension, friction, and bearing, the interaction equation for a bearing-type connection in terms of ultimate strengths may be approximated by Eq. 4.9.1:

$$\left(\frac{f_{vu}}{F_{vu}}\right)^2 + \left(\frac{f_{tu}}{F_{tu}}\right)^2 \leqq 1.0 \tag{4.9.1}$$

where f_{vu} = nominal shearing stress at failure
f_{tu} = nominal tensile stress at failure
F_{vu} = ultimate shearing strength with no tensile stresses present
F_{tu} = ultimate tensile strength with no shearing stresses present

By imposing a factor of safety on each of the terms in ultimate strength Eq. 4.9.1, an allowable stress equation of the similar form becomes

$$\left(\frac{f_v}{F_v}\right)^2 + \left(\frac{f_t}{F_t}\right)^2 \leqq 1.0 \tag{4.9.2}$$

where f_v = nominal unit shear stress
f_t = nominal unit tensile stress under external load
F_v = allowable shear stress in absence of external tension
F_t = allowable tensile stress in absence of shear

AISC has chosen to use a simpler straight-line interaction which requires a reduction for combined effects only in the most severe loading cases. The straight-line expression

$$\frac{f_v}{F_v} + \frac{f_t}{F_t} \leqq \text{const.} \tag{4.9.3}$$

is shown in Fig. 4.9.2 compared with the more exact relationship. Multi-

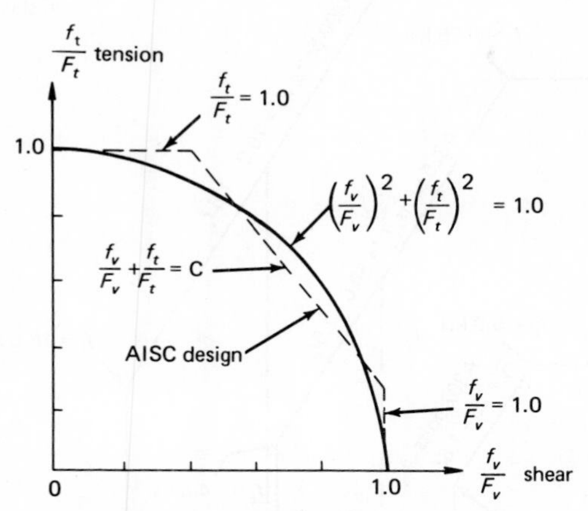

Fig. 4.9.2. Nondimensionalized shear-tension interaction curves: bearing-type connections.

plying Eq. 4.9.3 by F_t and solving for f_t gives

$$f_t \leqq F_t(\text{const.}) - \frac{F_t}{F_v} f_v \tag{4.9.4}$$

Since AISC–1.5.2 gives $F_t/F_v = 40/22 = 1.82$ for A325 bolts and $F_t/F_v = 54/32 = 1.69$ for A490 bolts, it will be conservative to take $F_t/F_v = 1.6$ and the constant as approximately 1.25. Using the term F_t' instead of f_t to signify an allowable stress in tension which cannot exceed F_t. Thus AISC–1.6.3 prescribes

$$F_t' \leq C' - 1.6 f_v \leq F_t \tag{4.9.5}$$

which is summarized in Table 4.9.1, and shown by diagrams in Fig. 4.9.3.

TABLE 4.9.1
Allowable Tensile Stress, F_t' (ksi), AISC–1.6.3

Connector	F_t'
A307 bolts	$F_t' = 28.0 - 1.6 f_v \leq 20.0$
A325 and A449 (in bearing-type connections)	$F_t' = 50.0 - 1.6 f_v \leq 40.0$
A490 (in bearing-type connections)	$F_t' = 70.0 - 1.6 f_v \leq 54.0$

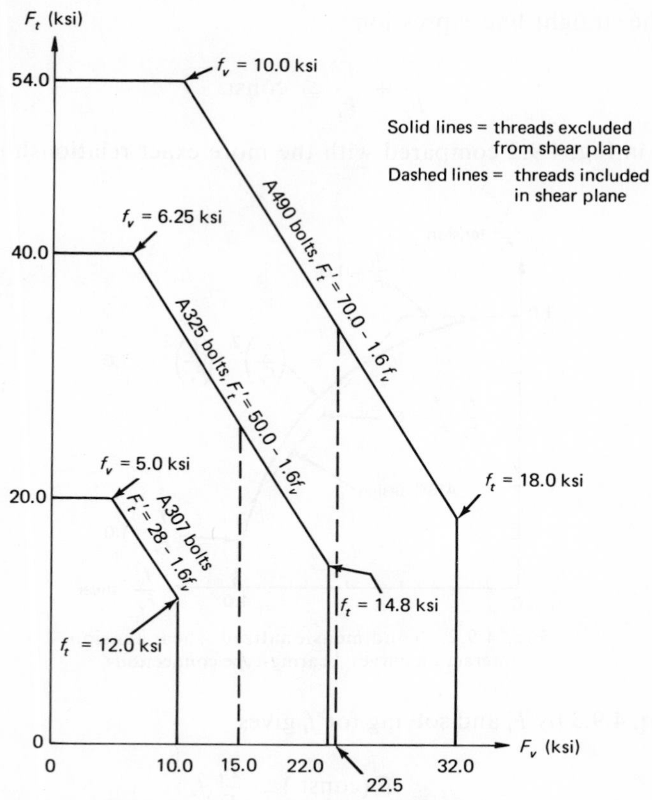

Fig. 4.9.3. Interaction relationship for combined shear and tension in bearing-type connections, AISC–1969.

Under AASHO–1.7.5, the approach is to use the circle of Eq. 4.9.2 for bearing-type connections, except multiply through by F_v^2, thus:

$$f_v^2 + \left(f_t \frac{F_v}{F_t}\right)^2 \leq F_v^2 \tag{4.9.6}$$

where the allowable values for either shear F_v or tension F_t in the absence of the other (see Table 4.9.1) give a ratio for

$$\text{A325 bolts of } \frac{F_v}{F_t} = \frac{20}{36} = 0.555$$

Under AASHO, bolts in bearing-type joints *must* have threads excluded from the shear plane. Thus the interaction equation, Eq. 4.9.6, becomes

$$f_v^2 + (0.555F_t)^2 \leq F_v^2 \tag{4.9.7}$$

Friction-Type Connections. Since a higher factor of safety is necessary for friction-type connections than for bearing-type, any reduction in the clamping forces (developed during the pretensioning of the bolts) due to an externally applied tensile stress f_t will reduce the clamping force, and hence reduce the friction force.

Again, a straight-line interaction relationship is used; but one which is more conservative than Eq. 4.9.5 for bearing-type connections. The constant is reduced from about 1.25 for Eq. 4.9.5 to 1.0 for friction-type connections. In addition, the allowable nominal superimposed tensile stress F_t is replaced in the denominator of the tensile stress ratio by the unit tensile stress F_{tp} based on the pretensioning load. Thus the interaction equation is

$$\frac{f_v}{F_v} + \frac{f_t}{F_{tp}} \leq 1.0 \tag{4.9.8}$$

where

$$F_{tp} = \frac{T_i}{A_b} \tag{4.9.9}$$

and T_i = pretension force in the bolt
A_b = nominal cross-sectional area of bolt

Substituting Eq. 4.9.9 into Eq. 4.9.8 gives

$$\frac{f_v}{F_v} + \frac{f_t A_b}{T_i} \leq 1.0 \tag{4.9.10}$$

or, solving for f_v gives

$$f_v \leq F_v\left(1.0 - \frac{f_t A_b}{T_i}\right) \tag{4.9.11}$$

Letting f_v be called F'_v, the allowable shear stress when external tension is also present, and substituting the values for F_v, the allowable shear stress in the absence of tension, one obtains from Eq. 4.9.11 the limitations of AISC–1.6.3:

$$F'_v \geq 15.0\left(1 - \frac{f_t A_b}{T_i}\right) \quad \text{A325 and A449 bolts} \tag{4.9.11}$$

$$F'_v \leq 20.0\left(1 - \frac{f_t A_b}{T_i}\right) \quad \text{A490 bolts} \tag{4.9.12}$$

The 1969 AASHO–1.7.5 also uses the more conservative straight line interaction relationship for friction-type connections. Substitution of the AASHO value of 13.5 ksi for F_v with A325 bolts, and letting f_v be called F'_v as above, Eq. 4.9.11 becomes

$$F'_v \leq 13.5\left(1 - f_t \frac{A_b}{T_i}\right) \tag{4.9.13}$$

Examining several values of A_b/T_i; for $^3/_4$ in. diam. A325 bolts,

$$\frac{A_b}{T_i} = \frac{0.44}{28} = 0.017$$

and for $^7/_8$ in. diam. bolts,

$$\frac{A_b}{T_i} = \frac{0.601}{39} = 0.0154$$

Using the larger value (0.017) in Eq. 4.9.13, one obtains the AASHO limitation

$$F'_v \leq 13.5 - 0.22f_t \tag{4.9.14}$$

EXAMPLE 4.9.1

Determine the adequacy of the fasteners in Fig. 4.9.4 if $^7/_8$ in. diam. A325 bolts and the connection is designed as (a) a friction-type connection

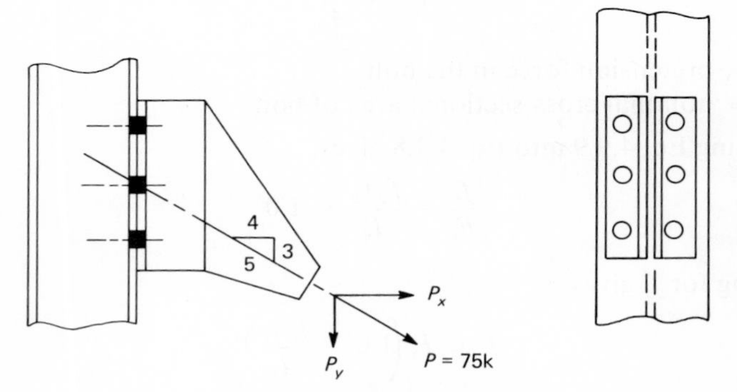

Fig. 4.9.4. Example 4.9.1.

and (b) as a bearing connection. Neglect all stresses in the pieces making up the connection assuming they do not govern and that threads are excluded from the shearing plane.

Solution

(a) Check bolts for use in a friction-type connection.

$$P_x = \frac{4}{5}(75 \text{ k}) = 60 \text{ k}$$

$$P_y = \frac{3}{5}(75 \text{ k}) = 45 \text{ k}$$

$$f_v = \frac{P_y}{\Sigma A} = \frac{45}{6(0.6013)} = 12.48 \text{ ksi (shear)}$$

$$f_t = \frac{P_x}{\Sigma A} = \frac{60}{6(0.6013)} = 16.64 \text{ ksi (tension)}$$

$$F_v' = 15.0\left(1 - \frac{f_t A_b}{T_i}\right) = 15.0\left(1 - \frac{16.64(0.6013)}{39}\right) = 11.17 \text{ ksi}$$

Since $f_v > F_v'$, there is no need to check further. Fasteners *are not adequate* to be used in a friction-type connection.

(b) Check bolts for use in a bearing-type connection.

$$F_t' = 50.0 - 1.6f_v \leq 40.0$$
$$= 50.0 - 1.6(12.48) = 20.04 \text{ ksi}$$

Since $f_t = 16.64 < 20.04$ and $f_v = 12.48 < 22.0$, the fasteners *are* adequate to be used in a bearing-type connection.

EXAMPLE 4.9.2

Determine the maximum value of P in Example 4.9.1 assuming (a) a friction-type connection and (b) a bearing-type connection. Use AISC specs.

Solution

(a) friction-type connection; 7/8 in. diam. A325 bolts.

$$P_x = \frac{4}{5}P = 0.80P$$

$$P_y = \frac{3}{5}P = 0.60P$$

$$f_v = \frac{0.60P}{6(0.6013)} = 0.166P$$

$$f_t = \frac{0.80P}{6(0.6013)} = 0.222P$$

$$F_v' = 15.0\left(1 - \frac{f_t A_b}{T_i}\right)$$

$$F_v' = 15.0\left(1 - \frac{0.222P(0.6013)}{39}\right) = 15.0 - 0.051P$$

Equating f_v to F_v'

$$0.166P = 15.0 - 0.051P$$

$$P = \frac{15.0}{0.217} = 69.2 \text{ kips}$$

$$f_t = 0.222(69.2) = 15.4 \text{ ksi} < 40.0 \text{ ksi} \qquad \text{OK}$$

The maximum capacity as a friction-type connection is 69.2 kips.

(b) bearing-type connection.

$$F_t' = 50.0 - 1.6f_v$$
$$= 40.0 - 1.6(0.166P) = 50.0 - 0.266P$$

Equating f_t to F'_t,

$$0.222P = 50.0 - 0.266P$$

$$P = \frac{50.0}{0.488} = 102.3 \text{ kips}$$

The maximum value of P as limited by $F_v = 22.0$ ksi is

$$P = \frac{F_v(A)}{0.60} = \frac{22.0(6)0.6013}{0.60}$$

$$= 132.2 \text{ kips} > 102.3 \text{ kips}$$

Therefore the maximum value of P determined by the fasteners when designed as a bearing-type connection is 102.3 kips.

EXAMPLE 4.9.3

Determine the number of $^3/_4$ in. diam. A325 bolts required to carry a shear of 70 kips combined with a tension of 120 kips. The connection is to be designed such that the resultant force acts through the centroid of the connection. (a) Use a bearing-type connection with thread excluded from the shear plane; (b) use a friction-type connection.

Solution

In this design problem it is apparent that both the shear and tension forces are of comparable magnitude; thus under such combination neither the full shear allowable F_v nor the full tension allowable F_t may be used, as observed from Fig. 4.9.3. Use of design charts[34] is one possible approach. The following approach[34] may be used when charts are not available.

(a) Bearing-type connection. From AISC–1.6.3, the interaction criterion is

$$F'_t = 50 - 1.6f_v \leq 40$$
$$F_v \leq 22 \text{ ksi}$$

or in general, Eq. 4.9.5,

$$F'_t = C' - 1.6f_v \leq F_t \qquad [4.9.5]$$

Convert Eq. 4.9.5 to a force equation; multiplying by A_b; thus

$$F'_tA_b = C'A_b - 1.6f_vA_b \leq F_tA_b \qquad \text{(a)}$$
$$T = C'A_b - 1.6V \leq F_tA_b \qquad \text{(b)}$$

where T and V are the applied tension and shear forces, respectively. Solving Eq. (b) for A_b gives

$$A_b = \frac{T + 1.6V}{C'} \qquad \text{(c)}$$

the basic *design equation* for bearing-type connections.

For A325 bolts, Eq. (c) becomes

$$\text{Reqd } A_b = \frac{T + 1.6V}{50} \tag{d}$$

For A490 bolts, Eq. (c) becomes

$$\text{Reqd } A_b = \frac{T + 1.6V}{70} \tag{e}$$

For this example, $T = 120$ kips and $V = 70$ kips, and using Eq. (c) which is expected to govern gives

$$\text{Reqd } A_b = \frac{T + 1.6V}{C'} = \frac{120 + 1.6(70)}{50} = 4.64 \text{ sq in.}$$

For this area, the capacities are

Max $T = 40(4.64) = 186$ kips $>$ Actual $T = 120$ kips OK
Max $V = 22(4.64) = 102$ kips $>$ Actual $V = 70$ kips OK

$$\text{Reqd } N = \frac{4.64}{0.44} = 10.5 \text{ bolts}$$

Since usually an even number is used; use 12–¾ in. diam. A325 bolts for a bearing-type connection.

(b) Friction-type connection. From AISC–1.6.3, the interaction expression is

$$F'_v \leq 15\left(1 - f_t \frac{A_b}{T_i}\right) \tag{4.9.11}$$

For ¾ in. diam., $A_b/T_i = 0.44/28 = 0.0157$. Substituting into Eq. 4.9.11 gives

$$F'_v = 15 - 15(0.0157)f_t$$
$$= 15 - 0.236f_t \tag{f}$$

Convert Eq. (f) into a force equation by multiplying by A_b; thus

$$F'_v A_b = 15A_b - 0.236A_b f_t \tag{g}$$

or

$$V = 15A_b - 0.236T \tag{h}$$

$$A_b = \frac{V + 0.236T}{15} \tag{i}$$

Multiplying numerator and denominator by 4, and calling $4(0.236) \approx 1$, gives the interaction design equation, similar to Eq. (c) for bearing-type connections,

$$A_b = \frac{T + 4V}{60} \tag{j}$$

the basic *design equation* for A325 bolts in friction-type connections.

For A490 bolts, the denominator becomes $4(20) = 80$, and

$$A_b = \frac{T + 4V}{80} \tag{k}$$

For this example

$$\text{Reqd } A_b = \frac{T + 4V}{60} = \frac{120 + 4(70)}{60} = 6.67 \text{ sq in.}$$

For this area the capacities are

$$\text{Max } T = 40(6.67) > \text{Actual } T = 120 \text{ kips} \qquad \text{OK}$$

$$\text{Max } V = 15(6.67) = 100 \text{ kips} > \text{Actual } V = 70 \text{ kips} \qquad \text{OK}$$

$$\text{Reqd } N = \frac{6.67}{0.44} = 15.2 \text{ bolts}$$

Use 16–$^3/_4$ in. diam. A325 bolts for a friction-type connection.

Eccentric Shear and Tension. In a typical bracket connection such as shown in Fig. 4.9.1d, the eccentric load produces both shear and tension in the fasteners as shown in Fig. 4.9.5. As in most other connections, the manner in which the pieces making up the bracket behave is complex. However, the stresses in the fasteners may be computed by using one of two approaches; (a) that of neglecting any initial tension in the fasteners or (b) that of considering the initial pretension forces in the fasteners. When fasteners such as A307 bolts are used, the amount of initial tension present is usually small and of an indeterminable amount. Therefore, in this case the neglecting of any initial tension is reasonable and gives conservative results. On the other hand, when high-strength bolts are used the initial pretension forces are of utmost importance and must be taken into account.

Consider the bracket connection in Fig. 4.9.5a in which the eccentric load P with its eccentricity e is shown to be equivalent to the superposition of the vertical load P and an applied moment $M = Pe$. If the pretension forces in the fasteners are neglected, the stresses on bolts can be computed by superimposing the effect of the moment and shear shown in Fig. 4.9.5b. However, if the pretension forces in the fasteners are considered, the stresses developed by these forces must be added to the moment and shear stresses as shown in Fig. 4.9.5c. In either case, for computation purposes, the plates or sections making up the connection are assumed to be rigid and all movement takes place as a result of the elongations of the fasteners.

Neglecting Initial Tension. Considering the pure moment connection in Fig. 4.9.6a the tensile forces are resisted by the fasteners and the compressive forces are resisted by the flange of the bracket pressing against the column. Since initial tension is neglected and the upper part of the bracket is assumed to pull away from the column, the *effective* width, b_e,

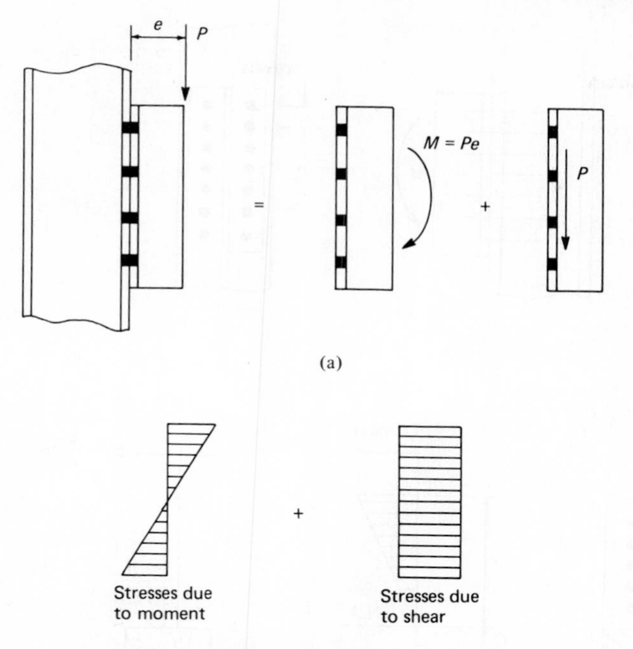

(a)

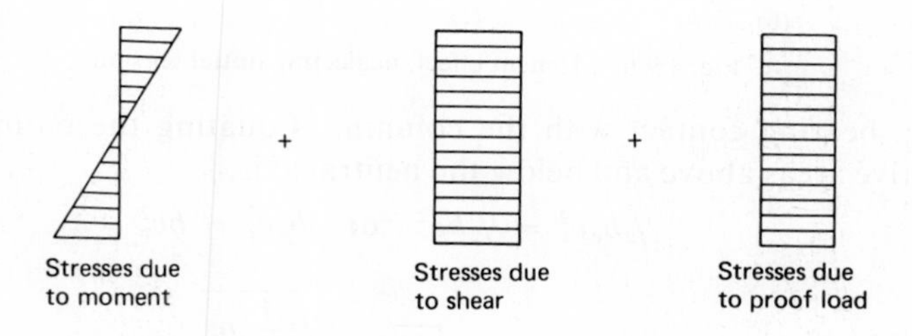

(b) Stresses neglecting pretensioning effect of bolts

Stresses due
to moment

Stresses due
to shear

Stresses due
to proof load

(c) Stresses considering pretensioning effect of bolts

Fig. 4.9.5. Forces and stress distribution on bolts in an eccentric bracket connection.

of the part above the neutral axis is

$$b_e = \frac{A}{p} m$$

where A = cross-sectional area of fastener
p = spacing of fasteners
m = number of rows

The width in compression is assumed to be the width of the bracket in

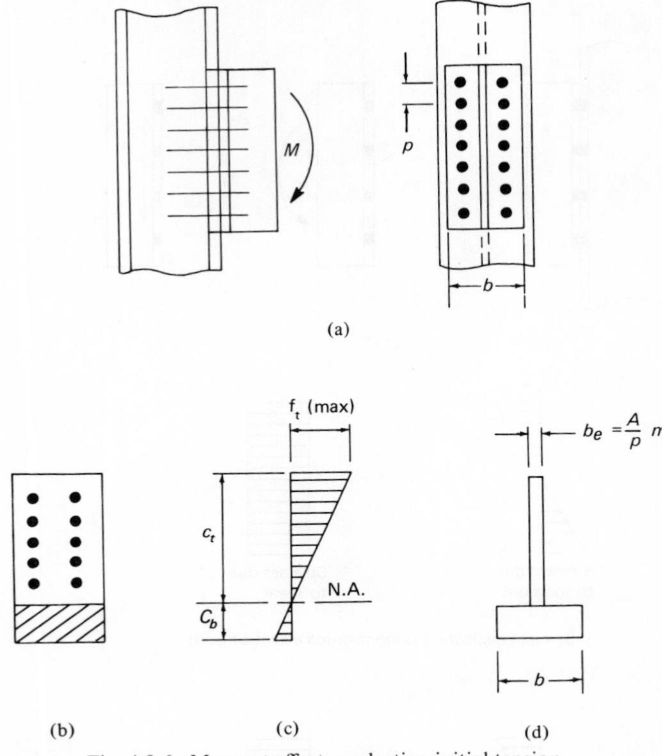

Fig. 4.9.6. Moment effect, neglecting initial tension.

direct bearing contact with the column. Equating the moments of the effective areas above and below the neutral axis,

$$\tfrac{1}{2} b_e c_t^2 = \tfrac{1}{2} b c_b^2 \quad \text{or} \quad b_e c_t^2 = b c_b^2 \tag{4.9.15}$$

or

$$\frac{c_b}{c_t} = \sqrt{\frac{b_e}{b}} = \sqrt{\frac{\frac{A}{p}\, m}{b}} \tag{4.9.16}$$

For a typical problem assuming a double row of ¾-in. fasteners, a pair of 6 × 6 angles (b = 12 in.), and a 3-in. spacing,

$$\frac{c_b}{c_t} = \sqrt{\frac{\frac{0.75}{3}(2)}{12}} \approx \frac{1}{5}$$

In most cases the value of c_b/c_t varies between ⅛ to ¼ in. For computation purposes, the value of c_b/c_t is assumed to be ⅙ for the first approximation to determining the location of the neutral axis. By the repeated application of Eq. 4.9.15, the neutral axis is located when $b_e c_t^2$ equals bc_b^2. The moment of inertia of the equivalent section as indicated in Fig. 4.6.5d is

$$I = \frac{b_e c_t^3}{3} + \frac{bc_b^3}{3} \tag{4.9.17}$$

The maximum tensile stress, $f_{t(\max)}$ is

$$f_t = \frac{Mc_t}{I} \tag{4.9.18}$$

and the maximum tensile force in the top fastener is

$$R_{\max} = f_t A \tag{4.9.19}$$

Equation 4.9.19 does not take into account that the top fastener is located approximate $p/2$ from the top of the connection. However, the error is not significant when there are a large number of fasteners. When there are only two to four fasteners the resulting error may be significant and should be considered. By replacing the value of c_t in Eq. 4.9.18 by $(c_t - p/2)$,

$$f_{t(\text{modified})} = \frac{M(c_t - p/2)}{I} \tag{4.9.20}$$

and

$$R_{(\max)\text{modified}} = f_{t(\text{modified})} A$$

or

$$= f_t A\left[\frac{(c_t - p/2)}{c_t}\right] \tag{4.9.21}$$

In the case of the bracket in Fig. 4.9.5, the effect of the shear stresses must then be added to the effect of the moment stresses as will be shown in the next example.

EXAMPLE 4.9.4

Determine the maximum capacity P which two angles $5 \times 3\frac{1}{2} \times \frac{1}{2}$ can carry based on combined shear and tension when connected by $\frac{7}{8}$ in. diam. A307 bolts, as shown in Fig. 4.9.7. Neglect any small initial tension in the bolts.

Solution

(a) Determine neutral axis assuming no initial tension. The tensile-stress area (Eq. 4.8.1) instead of the nominal area is to be used for tensile-stress calculation.

$$b_e = \frac{A}{p}m = \frac{0.462(2)}{3} = 0.308 \text{ in.}$$

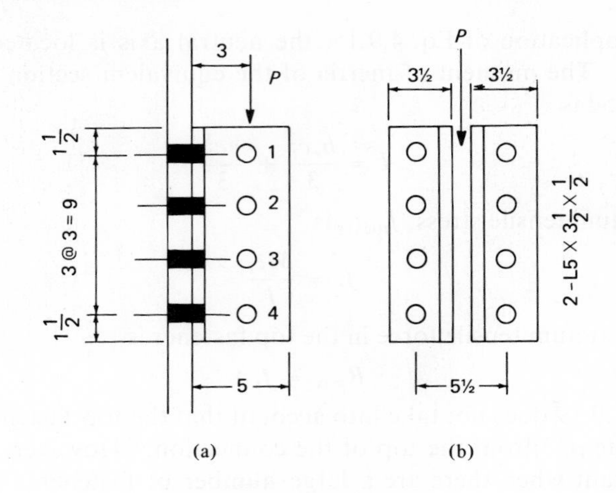

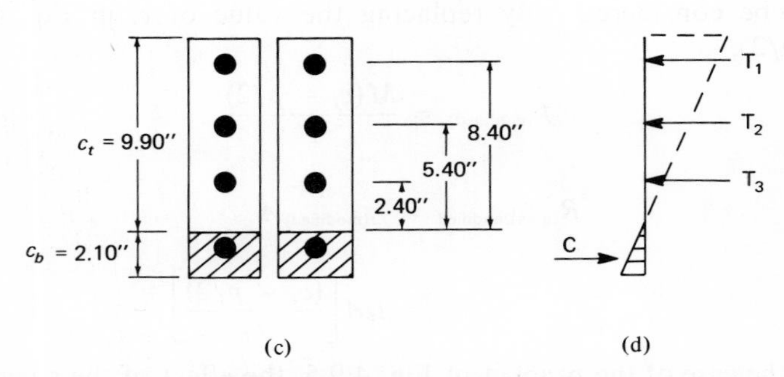

Fig. 4.9.7. Example 4.9.4. No initial tension case.

Using Eq. 4.9.16 as a first approximation,

$$\frac{c_b}{c_t} = \sqrt{\frac{Am}{pb}} = \sqrt{\frac{0.308}{7}} = \frac{1}{4.77}$$

$$\text{Try } c_b = \frac{12 - c_b}{4.77} = \frac{1}{5.77}(12) = 2.08 \text{ in.}$$

		A	y		Ay
Compressive area	$7(2.08) =$	14.56	(-1.04)	=	-15.12
Bolts, 1	$2(0.462) =$	0.924	$(+8.42)$	=	$+\ 7.77$
Bolts, 2	$2(0.462) =$	0.924	$(+5.42)$	=	$+\ 5.00$
Bolts, 3	$2(0.462) =$	0.924	$(+2.42)$	=	$+\ 2.23$
Bolts holes, 4	$\approx 2(0.462) =$	-0.924	(-0.58)	=	$+\ 0.54$
	$\Sigma A =$	16.41;	ΣAy	=	0.42

$$\text{Correction to } c_b = \frac{0.42}{16.41} = 0.025 \text{ in.}$$

Probably close enough! Use $c_b = 2.10$ in. for computing effective moment of inertia.

Compressive area	$7(2.10),\ \frac{1}{3}bc_b^3$	$= 21.6$
Bolts, 1 Ay^2	$0.924(8.40)^2$	$= 65.3$
Bolts, 2	$0.924(5.40)^2$	$= 26.9$
Bolts, 3	$0.924(2.40)^2$	$= 5.3$
Bolt holes, 4	$-0.924(0.60)^2$	$= -0.3$
	I	$= 118.8$ in.^{4}

The tensile stress on bolt No. 1 due to moment is

$$f_t = \frac{Pe(8.40)}{I} = \frac{P(3)(8.40)}{118.8} = 0.212P$$

Note that if Eq. 4.9.17 had been used to compute moment of inertia,

$$I = \frac{0.308(9.90)^3}{3} + \frac{7(2.10)^3}{3} = 99.5 + 21.6 = 121.1 \text{ in.}^4$$

which is little different than that obtained by the more laborious process. Since the assumption of no initial tension is conservative, the simplified process is considered satisfactory for design. Then, in accordance with Eq. 4.9.20,

$$f_t = \frac{P(3)(9.90 - 1.50)}{121.1} = 0.208P$$

For the direct shear component all connectors are assumed to participate, thus:

$$f_v = \frac{P}{\Sigma A} = \frac{P}{8(0.601)} = 0.208P$$

From Table 4.9.1 the AISC criterion is

$$F_t' \leq 28.0 - 1.6f_v \leq 20 \text{ ksi}$$
$$F_t' = 28.0 - 1.6(0.208P) = 0.208P$$

Solving for P gives

$$P(0.208 + 0.333) = 28.0; \quad P = 51.7 \text{ kips}$$

based on combined shear and tension.

$$\text{Check } f_t = 0.208(51.7) = 10.8 \text{ ksi} < 20 \text{ ksi} \qquad \text{OK}$$
$$f_v = 0.208(51.7) = 10.8 \text{ ksi} > 10.0 \text{ ksi} \qquad \text{NG}$$

$$P = \frac{10}{0.208} = 48.0 \text{ kips} \qquad \text{Governs!}$$

based on shear alone.

Considering Initial Tension. When the fasteners used in the bracket in Fig. 4.9.5a are A325 or A490 bolts, an initial tension results from the installation procedure as discussed in Art. 4.4. Assuming that the proof load in the bolts is not exceeded, the plates or sections making up the connection will be in complete contact with the column. Considering the pure-moment connection in Fig. 4.9.8a the neutral axis will occur at the

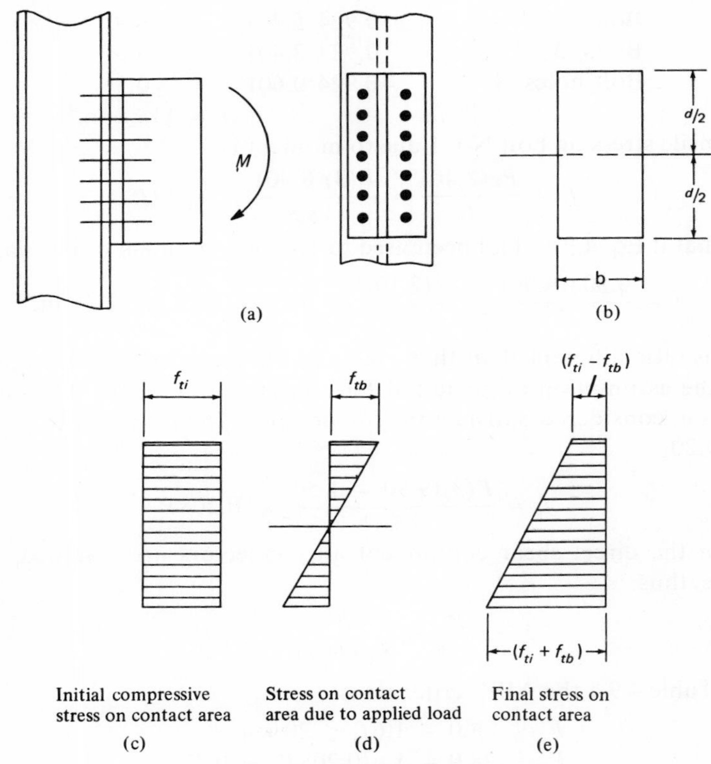

Fig. 4.9.8. Considering initial tension.

midpoint of the connection as indicated in Fig. 4.9.8b. The initial bearing pressure, f_{ti} as shown in Fig. 4.9.8c is assumed to be uniform over the area bd and is equal to

$$f_{ti} = \frac{\Sigma T_i}{bd} \tag{4.9.22}$$

where ΣT_i = the pretension load times the number of bolts. The tensile stress f_{tb} at the top due to the applied moment is

$$f_{tb} = \frac{Md/2}{I} = \frac{6M}{bd^2} \tag{4.9.23}$$

and should not exceed f_{ti} for this method to be valid.

The "net" tension load on the bolt is equal to the product of the tributary area for each fastener times f_{tb}.

$$T_{net} = f_{tb} bp \tag{4.9.24}$$

Substituting Eq. 4.9.23 into Eq. 4.9.24 gives

$$T_{net} = \frac{6M}{bd^2} bp = \frac{6Mp}{d^2} \tag{4.9.25}$$

The nominal tensile stress f_t in the top bolt is

$$f_t = \frac{T_{net}}{A} = \frac{6Mp}{Ad^2} \tag{4.9.26}$$

Assuming the top bolt is approximately $p/2$ from the top, the value of f_t can be modified to be

$$f_{t(\text{modified})} = f_t \frac{(d - p/2)}{d}$$

$$= \frac{6Mp}{Ad^2}\left(\frac{d - p}{d}\right) \tag{4.9.27}$$

EXAMPLE 4.9.5

Determine the capacity P for the connection of Fig. 4.9.7a based on the connectors subject to shear and tension if the connectors are $^3/_4$ in. diam. A325 bolts in a bearing-type connection with no threads in the shear plane. AISC specs.

Solution

Referring to Fig. 4.9.8d, the neutral axis for flexure is at mid-depth of the contact area. Using Eq. 4.9.23,

$$f_{tb} = \frac{6M}{bd^2} = \frac{6P(3)}{7(12)^2} = 0.0179P$$

or if reduced to the stress at the extreme bolt:

$$f_{tb} = 0.179P\,\frac{4.5}{6} = 0.0134P$$

Then, according to Eq. 4.9.24, the tension force on the two most highly stressed bolts is

$$T = f_{tb} bp = 0.0134P(7)(3) = 0.281P$$

and the nominal stress on the two bolts is

$$f_t = \frac{T}{A_b} = \frac{0.281P}{2(0.44)} = 0.320P$$

The direct-shear component is

$$f_v = \frac{P}{\Sigma A} = \frac{P}{8(0.44)} = 0.284P$$

From Table 4.9.1 the AISC criterion is

$$F_t' \le 50 - 1.6 f_v \le 40 \text{ ksi}$$

$$0.320P = 50 - 1.6(0.284P)$$

$$P = \frac{50}{0.320 + 0.455} = 64.5 \text{ kips}$$

$$\text{Check: } f_t = 0.320(64.5) = 20.6 \text{ ksi} < 40 \text{ ksi} \quad \text{OK}$$
$$f_v = 0.284(64.5) = 18.3 \text{ ksi} < 22 \text{ ksi} \quad \text{OK}$$

Therefore the capacity is 64.5 kips.

Considering Initial Tension—Simplified Procedure. As long as initial tension is not exceeded, one may compute tensile stress by the flexure formula

$$f_t = \frac{My}{I} = \frac{My}{\Sigma A y^2} \tag{4.9.28}$$

If $d = np$, where n = number of fasteners in one line, Eq. 4.9.27 becomes

$$f_t = \frac{6Mp}{An^2p^2}\left(\frac{np - p}{np}\right) = \frac{12M}{An^3p^2}\left(\frac{p(n-1)}{2}\right) \tag{4.9.29}$$

Noting that $p(n - 1)/2$ is the distance from mid-depth to the outermost fastener, and that from Eq. 4.7.23

$$\Sigma A y^2 \approx \frac{1}{12} A n^3 p^2$$

Thus Eq. 4.9.29 is the same as Eq. 4.9.28.

EXAMPLE 4.9.6

Show that the tensile stress for Example 4.9.5 may be obtained using the simplified Eq. 4.9.28.

Solution

$$\Sigma A y^2 = 4(0.44)[(1.5)^2 + (4.5)^2] = 39.6 \text{ in.}^4$$

$$f_t = \frac{My}{\Sigma A y^2} = \frac{P(3)(4.5)}{39.6} = 0.34P$$

which is approximately the same as the $0.32P$ using the volume of the stress solid tributary to the extreme fasteners. The simple method is fully as satisfactory as the more complex method and is the one recommended in the "Eccentric Loads on Fastener Groups" section of the AISC Manual.

EXAMPLE 4.9.7

For the connection of the bracket of Fig. 4.9.9 to the column, determine the number of $\frac{7}{8}$ in. diam. A325 bolts required to transmit the

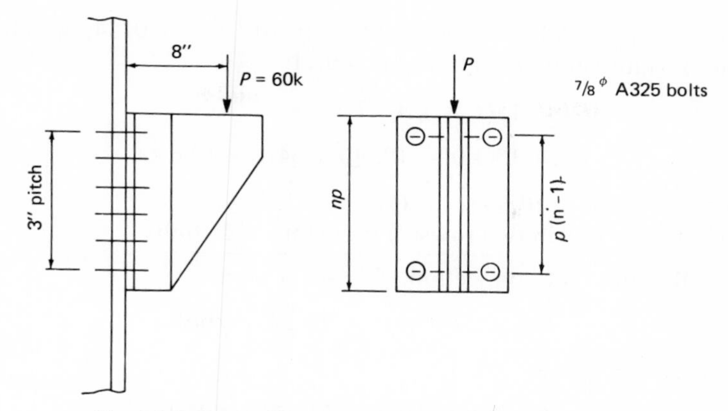

Fig. 4.9.9. Example 4.9.7. Design for shear and tension.

shear and tension forces. Use 3-in. vertical pitch. (a) Use friction-type connection; (b) use bearing-type connection where threads are excluded from the shear plane.

Solution

Since it has been shown that when there is initial tension, the procedure for eccentric shear is also applicable for computing tension; the design equation, Eq. 4.7.22, may also be used here:

$$n = \sqrt{\frac{6M}{Rp}} \qquad [4.7.22]$$

(a) Friction-type connection. For shear alone,

$$R_{ss} = 15(0.601) = 9.02 \text{ kips/bolt}$$

and for tension alone

$$R_T = 40(0.601) = 24.05 \text{ kips/bolt}$$

Assuming a value of R less than R_T, say 18 kips, to be used in Eq. 4.7.22:

$$n = \sqrt{\frac{6(240)}{18(3)}} = 5.16, \text{ say 6 per line where } M = 60(8)/2 = 240$$

in-kips per vertical line of fasteners,

Check stresses:

$$\Sigma Ay^2 = 4(0.601)[(1.5)^2 + (4.5)^2 + (7.5)^2] = 189.5 \text{ in.}^4$$

$$f_t = \frac{My}{\Sigma Ay^2} = \frac{60(8)7.5}{189.5} = 19.0 \text{ ksi} < 40 \text{ ksi} \qquad \text{OK}$$

$$f_v = \frac{P}{\Sigma A} = \frac{60}{12(0.601)} = 8.3 \text{ ksi} < 15 \text{ ksi} \qquad \text{OK}$$

For ⅞ in. diam. A325 bolts, $A_b/T_i = 0.601/39 = 0.0154$; which when used in the interaction equation, Eq. 4.9.11

$$F'_v = 15(1 - f_t A_b/T_i)$$
$$= 15(1.0 - 19.0(0.0154)) = 10.6 \text{ ksi}$$

Since $f_v = 8.3 \text{ ksi} < 10.6 \text{ ksi}$ OK
Use 6 bolts per row in two rows; ⅞ in. diam. A325 bolts.

(b) Bearing-type connection. For shear alone,

$$R_{ss} = 22(0.601) = 13.2 \text{ kips/bolt}$$

and for tension alone,

$$R_T = 40(0.601) = 24.05 \text{ kips/bolt}$$

The estimate using $n = \sqrt{6M/Rp}$ would have been done the same as for part (a); however, instead of using 6 per row, 5 per row would be tried since the basic shear allowable is higher (13.2 kips) than it was (9.02) for the friction-type connection.

Try 5 per row:

$$\Sigma Ay^2 = 4(0.601)[(3.0)^2 + (6.0)^2] = 108.0 \text{ in.}^4$$

$$f_t = \frac{My}{Ay^2} = \frac{60(8)6.0}{108.0} = 26.7 \text{ ksi} < 40 \text{ ksi} \qquad \text{OK}$$

$$f_v = \frac{P}{\Sigma A} = \frac{60}{10(0.601)} = 10.0 \text{ ksi}$$

$$F'_t = 50.0 - 1.6(10.0) = 34.0 \text{ ksi}$$

Since $f_t = 26.7 \text{ ksi} < F'_t = 34.0 \text{ ksi}$ OK
Use 5 bolts per row in two rows; ⅞ in. diam. A325 bolts.

For many commonly encountered design situations tables are available in the AISC Manual section "Eccentric Loads on Fastener Groups." Charts are also available for this type of problem.[35]

4.10. HIGH-STRENGTH BOLTED CONNECTIONS IN PLASTIC DESIGN

In general, the approach for plastic design is the same as for working stress design. It has already been stated that nominal stresses approximately are based on ultimate strength. Allowable nominal stresses under service load provide a factor of safety against connector failure ranging from about 2.2 for long joints to about 3.0 for compact joints.

Under AISC–2.8 when factored loads are used for design, the connection must carry moments and forces which have been factored (1.7 for gravity loads). The allowable stresses are also multiplied by the same factor. The ultimate strength is assumed to be the load factor (say 1.7) times the service load capacity; a safe procedure which still provides a greater safety factor with regard to connectors than for the members joined. Having increased the shear, bearing, and tension capacities of the connectors, the approach to analysis and design is identical with that presented in Secs. 4.6 through 4.9.

PROBLEMS

Note: Use the latest AISC Specification for all problems except where indicated.

4.1. Determine the tensile capacity of the connection if $^3/_4$-in. A325 bolts are used and it is designed (a) as a friction-type connection and (b) as a bearing-type connection. Assume the plates to be rolled from A36 steel and that the threads are excluded from the shearing plane.

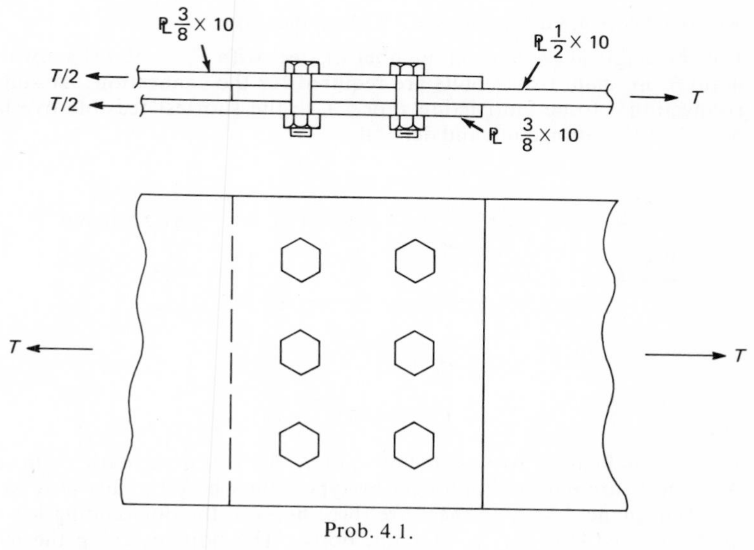

Prob. 4.1.

4.2. Rework Prob. 4.1 using A572 Grade 55 steel.

4.3. Rework Prob. 4.1 using A572 Grade 60 steel and $^7/_8$-in. A325 bolts.

4.4. Rework Prob. 4.1 using A572 Grade 65 steel and $^7/_8$-in. A490 bolts.

4.5. Determine the factor of safety against slip for the connection of Prob. 4.1.

4.6. Determine the factor of safety against slip for the connection of Prob. 4.4.

4.7. Determine the capacity P capable of being carried across the butt splice of Prob. 3.21 if $s_1 = s_2 = 2$ in. and A325 bolts are used with no threads in the shear planes, if (a) a friction-type connection is used; and (b) a bearing-type connection is used.

4.8. Determine the number of ¾-in. A325 bolts required to develop the capacity of the angles when designed (a) as a friction-type and (b) as a bearing-type connection with the threads excluded from the shearing plane. The angles are rolled from A572 Grade 50 steel. Assume a double row of bolts without stagger. Detail the connection.

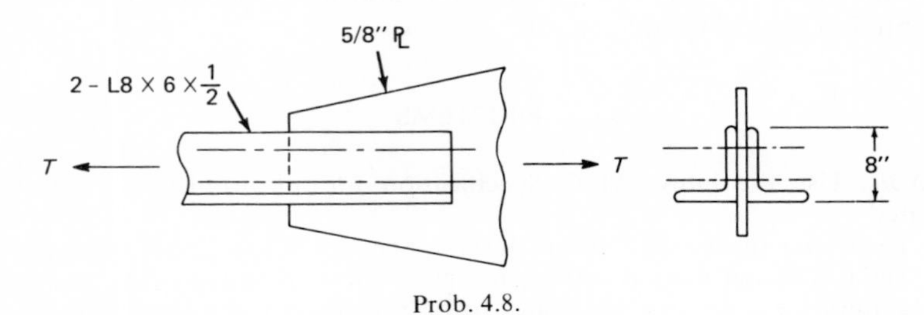

Prob. 4.8.

4.9. Rework Prob. 4.8 using ⅞-in. A325 bolts.

4.10. Rework Prob. 4.8 using the AASHO Specification.

4.11. For the single angle tension member of steel with F_y = 50 ksi shown, how many ¾ in. diam. A325 bolts are required for the connection? Assume the connection is to be the friction type. Use the shortest feasible overlap of pieces for the connection and detail it.

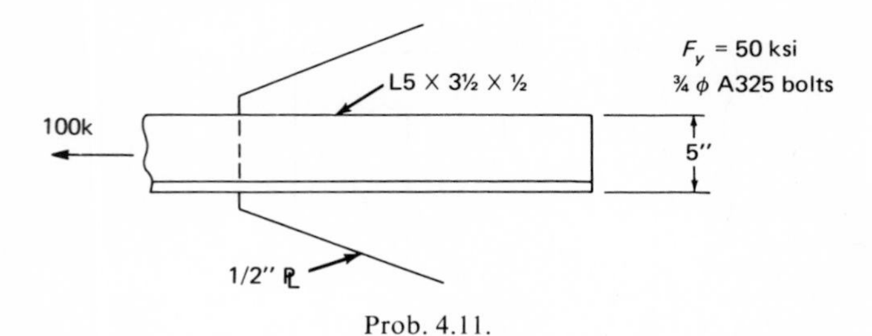

Prob. 4.11.

4.12. For the single-angle tension member of A36 steel, how many ⅞ in. diam. A325 bolts are required in a bearing-type joint where threads may exist in the shear plane? Assume there are three holes in the outstanding leg which are to be used later for ⅞ in. diam. bolts. The bolts carrying the 69 kips are to be located in a single line with the first of the bolts located a distance S ahead of the first empty hole. Detail the connection.

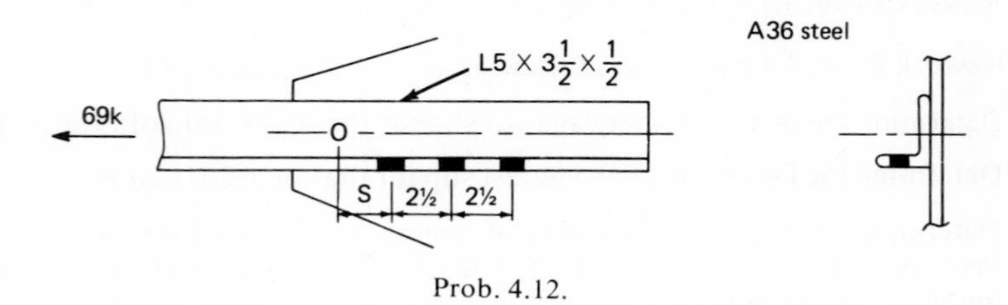

Prob. 4.12.

4.13. Design and detail the double lap splice shown to develop maximum tensile capacity. A36 steel is used and 7/8 in. diam. A325 bolts are to be used in a bearing-type connection with no threads in the shear plane. What is the resulting capacity of the joint?

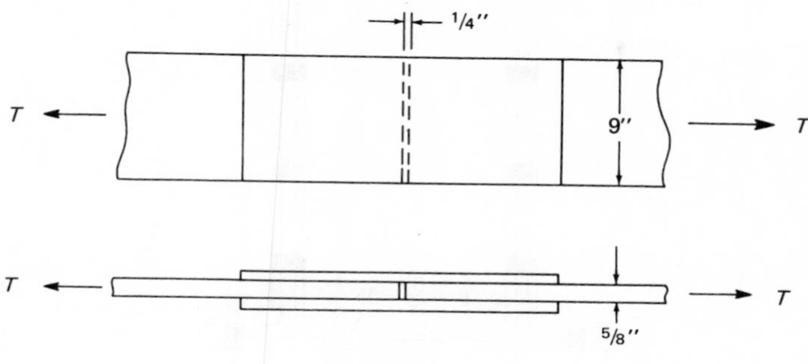

Prob. 4.13.

4.14. Determine the safe capacity of this tension chord splice of two channels. The material is A36 steel, and the 7/8 in. diam. A325 bolts are bearing-type with *no* threads in shear planes.

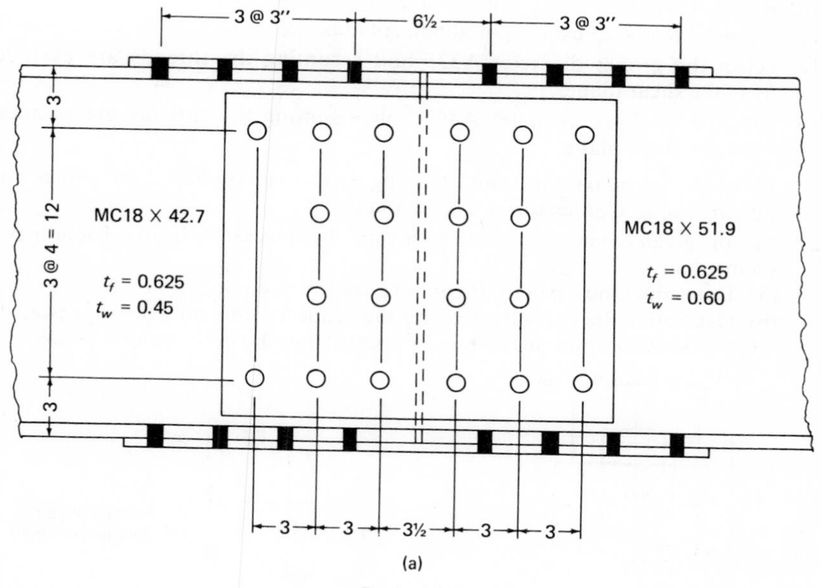

Prob. 4.14.

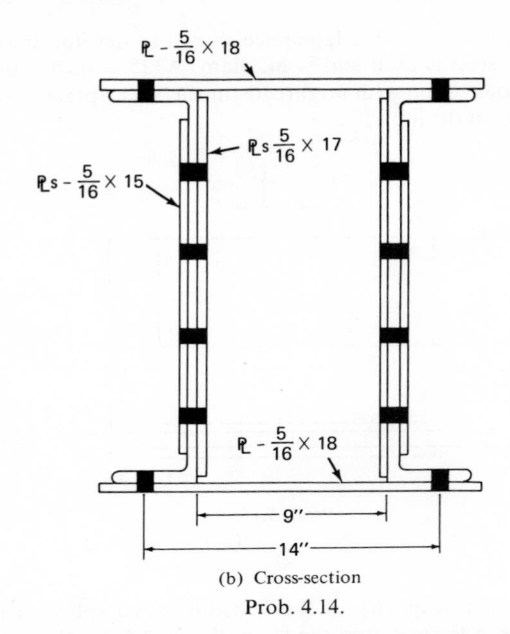

(b) Cross-section

Prob. 4.14.

4.15. Compute the maximum stress on the fastener group if $\frac{7}{8}$ in. diam. A325 bolts are used and the threads are excluded from the shear plane. Assume the bracket plate has adequate strength.

4.16. Repeat Prob. 4.15 using 1 in. diam. A325 bolts.

4.17. Repeat Prob. 4.15 using $\frac{3}{4}$ in. diam. A490 bolts.

4.18. Select the proper diameter A325 bolt assuming the threads are excluded from the shear plane.

4.19. Select the proper diameter A490 bolt assuming the threads are excluded from the shear plane.

4.20. Compute the maximum unit shearing stress and indicate on which connector it acts. Connectors are $\frac{3}{4}$-in. diam.

4.21. For the given connection made with $\frac{3}{4}$ in. diam. A325 bolts in a friction-type connection:
   (a) Determine the capacity P applying basic principles.
   (b) Determine the capacity P using the reduced moment arm suggested by the AISC Manual section on "Eccentric Loads on Fastener Groups."

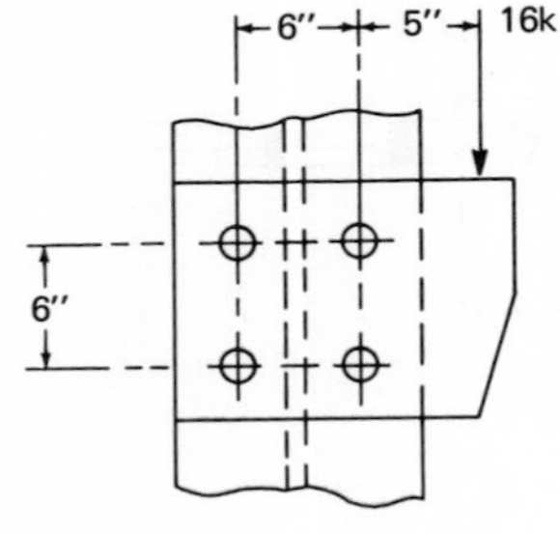

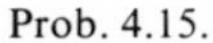
Prob. 4.15.

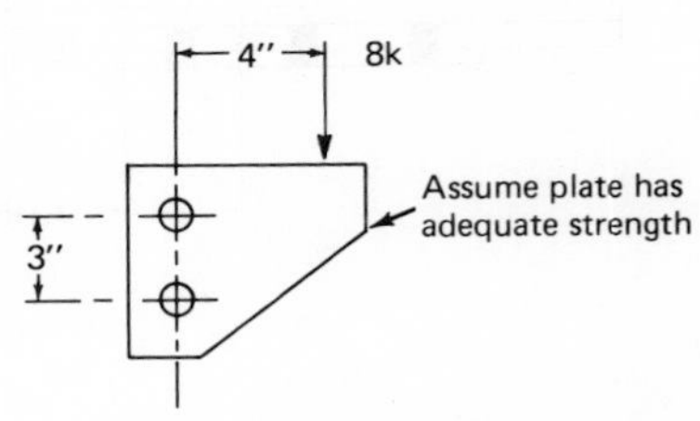

Probs. 4.18 and 4.19.

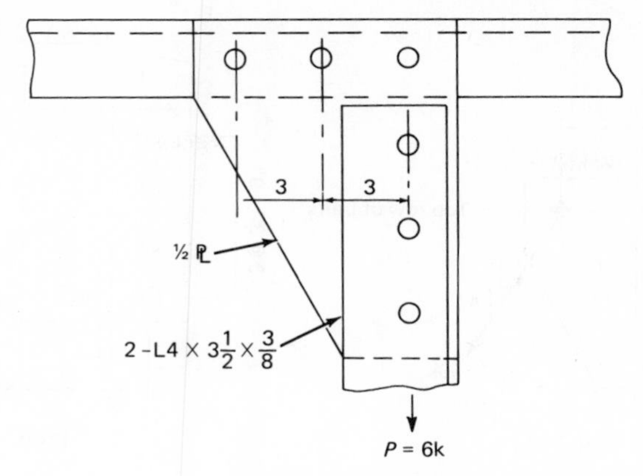

Prob. 4.20.

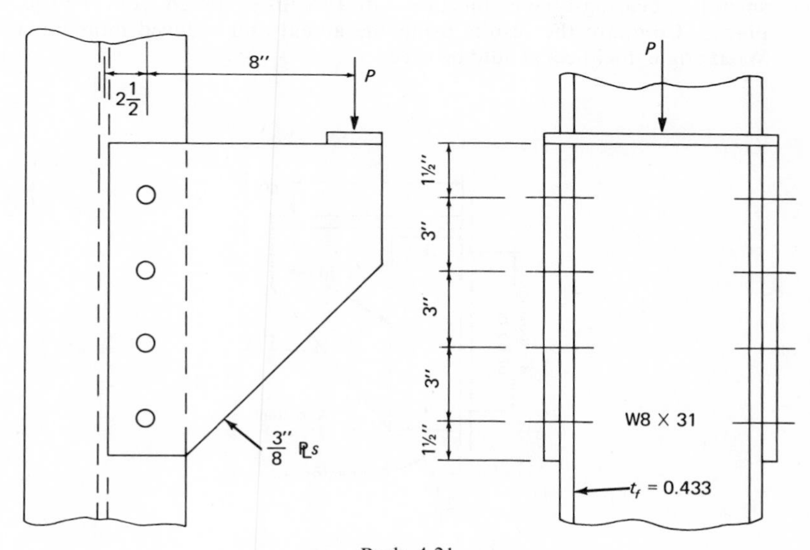

Prob. 4.21.

4.22. Assuming a double row of ⅞-in. A325 bolts, and a 3-in. vertical spacing, select the proper number of bolts. Assume the threads are excluded from shear plane.

4.23. Repeat Prob. 4.22 using ¾ in. diam. A325 bolts.

4.24. For the connection shown using ⅞ in. diam. A325 bolts in a bearing-type connection with threads excluded from the shear plane:
   (a) Determine the load (kips) on each fastener and indicate its direction using basic principles. Is the connection adequate according to AISC specs?
   (b) Repeat part (a) using the reduced moment arm method suggested in the AISC Manual (see also Fig. 4.7.7).

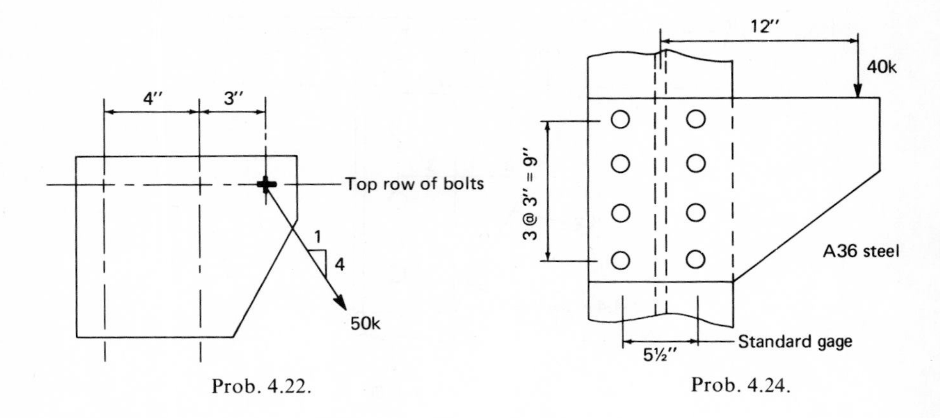

Prob. 4.22.

Prob. 4.24.

4.25. Determine the number of $\frac{7}{8}$ in. diam. A325 bolts required to resist eccentric shear in a bearing-type connection with threads excluded from the shearing plane. Compare the results using the actual and reduced moment arms. What angle thickness should be used?

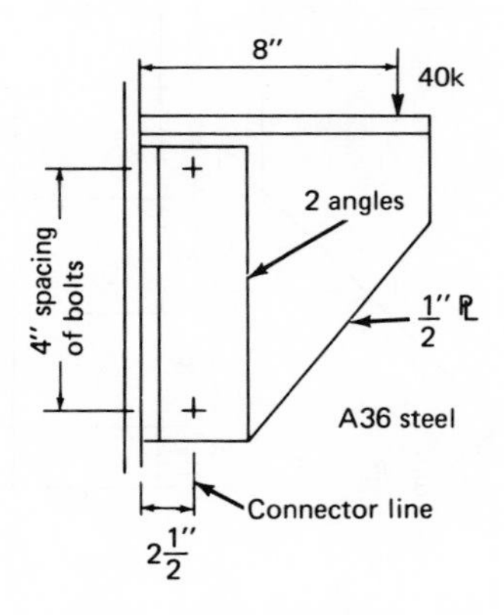

Probs. 4.25 and 4.28.

4.26. Assume two angles, $4 \times 3\frac{1}{2} \times \frac{5}{8}$ have been selected to carry their maximum capacity as a tension member of A36 steel. Assume the connection of the angles to the structural tee web will be along a single gage line as shown. Determine the number and positioning of $\frac{7}{8}$ in. diam. A325 high-strength bolts required in a bearing-type connection to attach the WT to the flange of a W section. The flanges of the WT and the W shapes are each $\frac{3}{4}$-in. thick, and the material is A36 steel.

4.27. Repeat Prob. 4.26 if a friction-type connection is used.

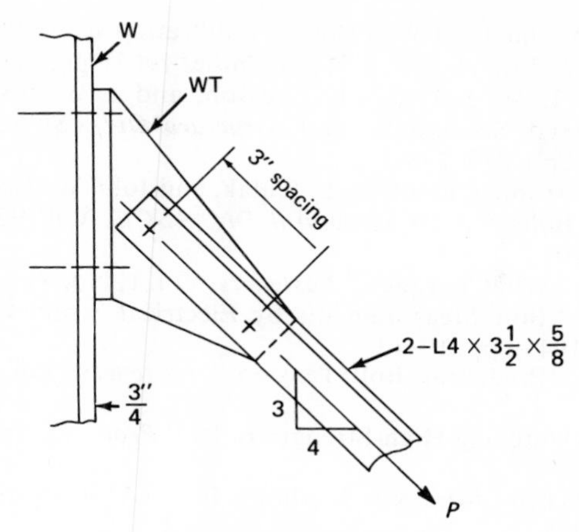

Probs. 4.26 and 4.27.

4.28. Determine the number and placement of ⅞ in. diam. A325 bolts for the combined shear and tension connection of the bracket of Prob. 4.25 to a ¾-in.-thick column flange.
(a) Friction-type connection.
(b) Bearing-type connection, with threads excluded from shear plane.

SELECTED REFERENCES

1. C. Batho and E. H. Bateman, "Investigations on Bolts and Bolted Joints," second Report of the Steel Structures Research Committee, His Majesty's Stationery Office, London, 1934.
2. W. M. Wilson and F. P. Thomas, "Fatigue Tests on Riveted Joints," Bulletin 302, Engg. Experiment Station, U. of Illinois, Urbana, Ill., 1938.
3. A. E. R. De Jonge, "Riveted Joints; a Critical Review of the Literature Covering Their Development," American Society of Mechanical Engineers, New York, 1945.
4. "Symposium on High-Strength Bolts," Proc. AISC National Engineering Conference, 1950, pp. 22–43.
5. William H. Munse, "Research on Bolted Connections," *Trans. ASCE*, Vol. 121 (1956), pp. 1255–1266.
6. "Rivets and High-Strength Bolts, A Symposium," *Trans. ASCE*, Vol. 126, Part II (1961), pp. 693–820.
7. Specification for Structural Joints Using ASTM A325 or A490 Bolts, Research Council on Riveted and Bolted Structural Joints of the Engineering Foundation, Sept. 1, 1966.
8. William H. Munse, "High Strength Bolting," *Engineering Journal, AISC*, Vol. 4, No. 1 (January 1967) pp. 18–27.
9. John W. Fisher and Lynn S. Beedle, *Bibliography of Bolted and Riveted Structural Joints*, ASCE Manual No. 48, Task Committee on Structural Connections, ASCE, 1968.

10. John L. Rumpf and John W. Fisher, "Calibration of A325 Bolts," *J. Structural Div. ASCE*, Vol. 89, No. ST6 (December 1963), pp. 215–234.
11. G. H. Sterling, E. W. J. Troup, E. Chesson, and J. W. Fisher, "Calibration Tests of A490 High-Strength Bolts, *J. Structural Div. ASCE*, Vol. 91, No. ST5 (October 1965), pp. 279–298.
12. Richard J. Christopher, Geoffrey L. Kulak, and John W. Fisher, "Calibration of Alloy Steel Bolts," *J. of Structural Div. ASCE*, Vol. 92, No. ST2 (April 1966), pp. 19–40.
13. W. C. Stewart, "What Torque?," *Fasteners*, Vol. 1, No. 4 (1944), pp. 8–10.
14. G. A. Maney, "Bolt Measurements by Electrical Strain Gages," *Fasteners*, Vol. 2, No. 1 (1945), pp. 10–13.
15. G. A. Maney, "Predicting Bolt Tension," *Fasteners*, Vol. 3, No. 5 (1946), pp. 16–18.
16. F. P. Drew, "Tightening High-Strength Bolts," *Proc. No. 786*, *ASCE*, Vol. 81 (August 1955).
17. E. J. Ruble, "Turn-of-the-Nut Method" Proc. AISC National Engineering Conference, 1955.
18. Desi D. Vasarhelyi, Said Y. Beano, Ronald B. Madison, Zung-An Lu, and Umesh C. Vasishth, "Effects of Fabrication Techniques on Bolted Joints," *Trans. ASCE*, Vol. 126, Part II (1961), pp. 764–796.
19. Ethan F. Ball and J. J. Higgins, "Installation and Tightening of Bolts," *Trans. ASCE*, Vol. 126, Part II (1961), pp. 797–810.
20. J. W. Fisher, P. O. Ramseier, and L. S. Beedle, "Strength of A440 Steel Joints Fastened with A325 Bolts," Publications IABSE, Vol. 23 (1963).
21. R. A. Hechtman, D. R. Young, A. G. Chin, and E. R. Savikko, "Slip of Joints Under Static Loads," *Trans. ASCE*, Vol. 120 (1955), pp. 1335–1352.
22. Robert T. Foreman and John L. Rumpf, "Static Tension Tests of Compact Bolted Joints, "*Trans. ASCE*, vol. 126, Part II (1961), pp. 228–254.
23. Gordon H. Sterling and John W. Fisher, "A440 Steel Joints Connected by A490 Bolts, *J. Structural Div. ASCE*, Vol. 92, No. ST3 (June 1966), pp. 101–118.
24. Desi D. Vasarhelyi and Kah Ching Chiang, "Coefficient of Friction in Joints of Various Steels, *J. Structural Div. ASCE*, Vol. 93, No. ST4 (August 1967), pp. 227–243.
25. George C. Brookhart, I. H. Siddiqi, and Desi D. Vasarhelyi, "Surface Treatment of High-Strength Bolted Joints," *J. Structural Div. ASCE*, Vol. 94, No. ST3 (March 1968), pp. 671–681.
26. Jonathan Jones, "Effect of Bearing Ratio on Static Strength of Riveted Joints," *Trans. ASCE*, Vol. 123 (1958), pp. 964–972.
27. John W. Fisher and Lynn S. Beedle, "Criteria for Designing Bearing-Type Bolted Joints," *J. Structural Div. ASCE*, Vol. 91, No. ST5 (October 1965), pp. 129–154.
28. Robert A. Bendigo, Roger M. Hansen, and John L. Rumpf, "Long Bolted Joints," *J. Structural Div. ASCE*, Vol. 89, No. ST6 (December 1963), pp. 187–213.
29. John W. Fisher and John L. Rumpf, "Analysis of Bolted Butt Joints," *J. Structural Div. ASCE*, Vol. 91, No. ST5 (October 1965), pp. 181–203.
30. T. R. Higgins, "New Formulas for Fasteners Loaded Off Center," *Engineering News-Record*, May 21, 1964.
31. A. L. Abolitz, "Plastic Design of Eccentrically Loaded Fasteners," *Engineering Journal*, AISC, Vol. 3, (July 1966), pp. 122–132.

32. T. R. Higgins and W. H. Munse, "How Much Combined Stress Can a Rivet Take?," *Engineering News-Record*, Dec. 4, 1952, pp. 40–42.
33. Eugene Chesson, Jr., Norberto L. Faustino, and William H. Munse, "High-Strength Bolts Subjected to Tension and Shear," *J. Structural Div. ASCE*, Vol. 91, No. ST5 (October 1965), pp. 155–180.
34. Hans William Hagen and Richard C. Penkul, "Design Charts for Bolts with Combined Shear and Tension," *Engineering Journal*, AISC, Vol. 2, No. 2 (April 1965), pp. 42–45.
35. Alfred Zweig, "Design of Bolts or Rivets Subject to Combined Shear and Tension," *Engineering Journal*, AISC, Vol. 3, No. 2 (April 1966), pp. 72–79.

5

Welding

5.1. INTRODUCTION AND HISTORICAL DEVELOPMENT

The process of welding denotes the joining of metal pieces by heating to a plastic or fluid state, with or without pressure. In its simplest form, "welding" has been known and used for several thousand years. Historians have speculated that the early Egyptians may have first used pressure-welding about 5500 B.C. in making copper pipes from sheets by overlapping the edges and hammering. Winterton[1] has reported that Egyptian art objects dating about 3000 B.C. have been found on which gold foil has been hammered and fused onto the base copper. This type of welding, called *forge welding*, was man's first process to join pieces of metal together. A well known early example of forge welding is the Damascus sword which was made by forging layers of iron with different properties. Interestingly, forge welding was sufficiently well developed and important enough to the early Romans that they named one of their gods, Vulcan (the god of fire and metalworking) to represent that art. In recent times, the word vulcanizing has been used in reference to treating rubber with sulfur but originally was used to mean "to harden." Today, forge welding is practically a forgotten art in which the village blacksmith was the last major practitioner. Welding, as we know it today is much more complex and highly refined and the remainder of this section will trace some of the important events which have contributed to the art. Specific welding processes are discussed in Sec. 5.2.

Very little progress in welding technology had been made until 1877, prior to which time most of the then known processes such as forge welding and brazing had been used for at least 3,000 years. The origin of resistance welding began around 1877 when Professor Elihu Thompson began a set of experiments[2,3] reversing the polarity of transformer coils. He received his first patent[4] in 1885 and the first resistance butt welding machine was demonstrated at the American Institute Fair[2] in 1887. In

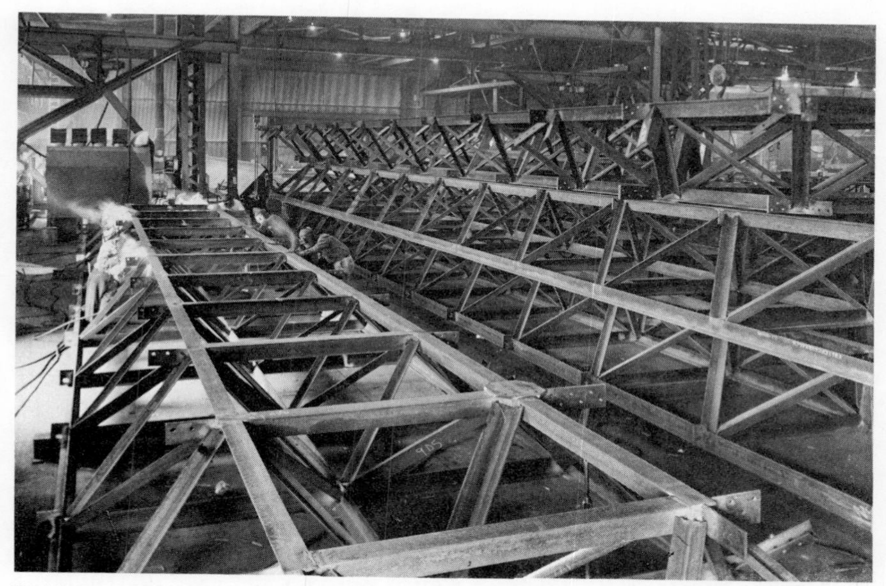

Welding of space trusses—Upjohn Company Office Building—Courtesy Whitehead and Kales Company, Detroit.

1889 Coffin[2] was issued a patent for flash-butt welding which became one of the important butt welding processes.

Zerner, in 1885 introduced the carbon-arc welding process, making use of two carbon electrodes and N. G. Slavinoff[5] in Russia was the first to use the metal-arc process using uncoated, bare electrodes in 1888. Coffin, working independently also investigated the metal-arc process and was issued a U.S. Patent in 1892. In 1889, A. P. Strohmeyer[2] introduced the concept of coated metal electrodes to eliminate many of the problems associated with the use of bare electrodes.

Thomas Fletcher[1] in 1887 used a blowpipe burning hydrogen and oxygen and showed that he could successfully cut or melt metal. In 1901–03 Fouche and Picard developed torches which could be used with acetylene and thus the era of oxyacetylene welding and cutting began.

The period between 1903 and 1918 saw the use of welding primarily as a method of repair, the greatest impetus occurring during World War I (1914–18). Welding techniques proved to be especially adapted to repairing ships which had been damaged. Winterton[6,7] reported that in 1917 there were 103 interned enemy ships alone in the United States which were damaged and the number of persons employed in welding operations rose from 8,000 to 33,000 during the period 1914–18.

After 1919, the use of welding as a construction and fabrication technique began to develop with copper-tungsten alloy electrodes being first used for spot welding techniques[6] in 1920. The period 1930–50 saw many improvements[2,8] in the development of welding machines. The submerged-arc welding process in which the arc is buried under a powdered flux was first used commercially in 1934 and patented in 1935.

Today there are over 50 different welding processes which can be used to join various metals and their alloys. Those of particular interest to the structural engineer are discussed in Sec. 5.2.

5.2. BASIC PROCESSES

As defined by the *Welding Handbook*,[9] "Welding is the process of joining two or more pieces of material, often metallic, by a localized coalescence or union across an interface." Most welding processes involve heating the material to be welded, or at least involve energy input to the material. The heat generation may be categorized according to its source; as electrical, chemical, optical, mechanical, and solid state. Heat is used to melt the base metal and filler material in order that flow of material will occur, i.e., that fusion will take place. In addition, heat is used to increase ductility so that plastic flow can occur even if melting does not take place; further, heating helps to remove contaminating films on the material.

The most common welding processes, particularly for welding structural steel, use electrical energy as the heat source; the most often used is the electric arc. The arc consists of a relatively large current discharge between electrode and base material sustained through a thermally ionized gaseous column, called a plasma.[9] In arc welding, fusion occurs by the flow of material across the arc, without pressure being applied.

Other processes, not ordinarily used for steel structures, involve other energy sources, and some of those processes involve the application of pressure, either in the absence or presence of flow of molten material. Bonding may also occur by diffusion, wherein atomic particles intermix across the interface and melting of the base material does not occur.

There are many welding processes which have special uses for particular metals and for various thicknesses. This section emphasizes those processes which are used in the welding of carbon and low-alloy steel for buildings and bridges. However, in order to present an idea of the broad spectrum of processes, a number of them are listed according to the sources of energy for generating heat and/or inducing the bonding process:

(a) Electrical energy
   1. Arc welding—fusion process; heat from electric arc.

2. Resistance welding—pressure process; heat from resistance to the flow of current.
3. Induction welding—fusion combined with pressure; heat by electromagnetic induction.
4. Electroslag welding—fusion process where molten slag provides heat; electric current is conducted through the slag without an arc.

(b) Chemical energy
1. Oxyacetylene gas welding—fusion process; heat from acetylene burning in the presence of oxygen.
2. Thermit welding—fusion process; heat from chemical reaction between a metal oxide and aluminum.

(c) Optical energy
1. Laser-beam welding—fusion process; heat from application of a concentrated coherent light striking the surfaces to be joined.
2. Electron-beam welding—fusion process; heat from conversion of kinetic energy of high velocity electrons bombarding the surfaces to be joined.

(d) Mechanical energy
1. Friction welding—pressure process without melting of base material; heat generated from friction between a stationary and rotating member subject to high normal force on the interface to be joined.
2. Ultrasonic welding—pressure process without melting of base material; heat generated by high-frequency vibrating energy.
3. Explosion welding—pressure process without melting of base material; heat generated by high velocity impact on the interface resulting from a controlled detonation.

(e) Solid-State Bonding
One might call this category natural welding, wherein pressure is applied and the temperature is raised but no melting occurs. After a period of time diffusion occurs by atoms moving across the original joint interface and intermixing to create a solid state bond. Good clean surface preparation is important in this process.

For design of steel structures, arc welding is the category of processes which are of particular interest. For some situations involving light-gage steel, resistance welding may also be important.

Shielded Metal-Arc Welding. This is the most usual type of welding sometimes called the manual stick electrode method. Heating is accomplished by means of an electric arc between a coated electrode and the materials being joined. The welding circuit is shown in Fig. 5.2.1a.

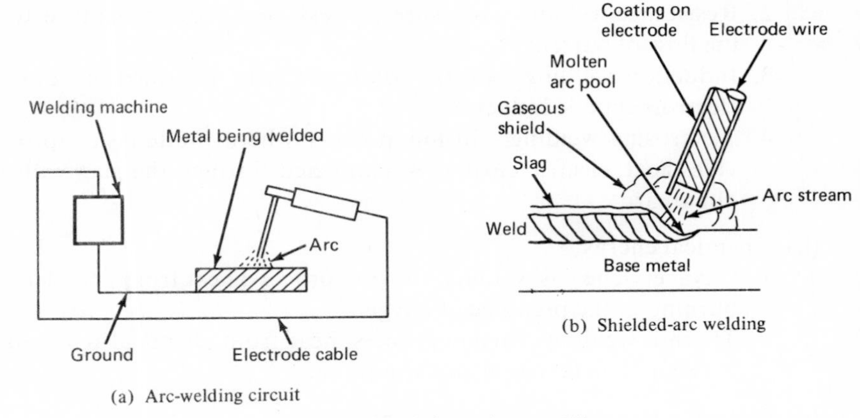

(a) Arc-welding circuit

(b) Shielded-arc welding

Fig. 5.2.1. Shielded metal-arc welding.

The coated electrode is consumed as the metal is transferred from the electrode to the base material during the welding process. The electrode wire becomes filler material and the coating is converted partly into a shielding gas, partly into slag, and some part is absorbed by the weld metal. There is a wide variety of coating materials producing different proportions of gas and slag.

The transfer of metal from electrode to the work being welded is induced by molecular attraction and surface tension, without application of pressure. The shielding of the arc prevents atmospheric contamination of the molten metal in the arc stream and in the arc pool. It prevents nitrogen and oxygen from being picked up and forming nitrides and oxides which may cause embrittlement.

The electrode coating may perform the following functions:

(a) Produces a gaseous shield, as described above.
(b) Introduces other materials, such as deoxidizers, to refine the grain structure of the weld metal.
(c) Produces a blanket of slag over the molten pool and the solidified weld to protect it from oxygen and nitrogen in the air, and also retards cooling.

The electrode material is specified under ASTM A233, (AWS A5.1) for carbon steel welding and ASTM A316 (AWS A5.5) for low alloy steel welding. The strength and designations are referred to in Chapter 2. The designations are E60XX for 60 ksi tensile strength, E70XX for 70 ksi tensile strength, etc. Table 5.12.1 indicates which coated electrodes should be used with each particular structural steel.

For welding high-carbon or low-alloy steels, low hydrogen electrodes with special precautions are recommended, and required by AISC-1.5.3

for use on A242, A441, A572, and A588 steels. The low-hydrogen electrode is a rod with a carbonate of soda, or "lime," coating. The use of this electrode requires a different technique from the conventional electrode in that it requires a short arc and provides globular-type, rather than a spray-type transfer of metal from the electrode to the work. However, it is required by AISC because the as-welded mechanical properties have been found to be superior to properties obtained using other types of electrode coatings.

The process is the principal one used for manual welding of steel structures.

Gas-Shielded Arc Welding. In gas-shielded arc-welding, fusion occurs under a shield of protective gas (inert or reactive) by the heat of an electric arc between a metal electrode (consumable or nonconsumable) and the piece being welded. Essentially two major processes may be considered; those with a nonconsumable electrode and those using a continuously fed consumable electrode.

Gas Tungsten-Arc Welding (TIG). The gas tungsten-arc welding process uses a nonconsumable tungsten electrode which is contained in a holder with a sheath through which the shielding gas passes, as shown in Fig. 5.2.2. This method, commonly called TIG (tungsten inert gas), uses an

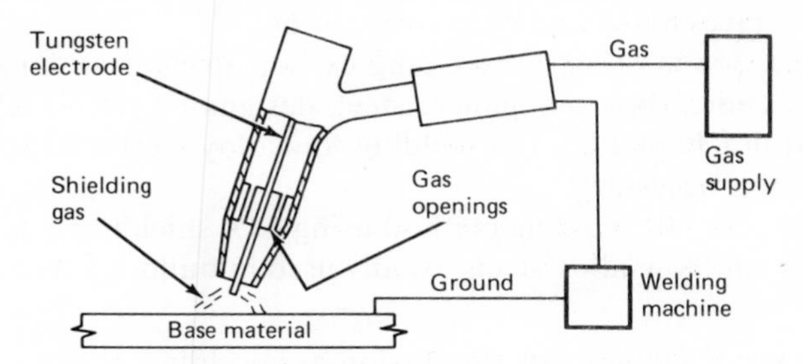

Fig. 5.2.2. Gas tungsten-arc welding.

inert gas, such as argon or helium, for the protection shield. Rarely, if ever, is this method used for welding structural steel, though it is common for welding aluminum and magnesium which are beginning to receive wider structural usage.

Gas Metal-Arc Welding. In this process the electrode is a continuous wire that is fed from a coil through the electrode holder. The shielding, as in the TIG method, is entirely from an externally supplied gas or gas mixture.

Originally, this method was used only with inert gas shielding, hence, the name MIG (metal-inert gas) has been used. Reactive gases alone are generally not practical; the exception is CO_2 (carbon dioxide). The use of CO_2 has become extensively used for welding of steels, either alone or in a mixture with inert gases.

Argon as a shielding gas works for welding virtually all metals; however, it is not recommended for steels because of its expense and the fact that other shielding gases and gas mixtures are acceptable. For welding carbon and some low-alloy steels either (1) 75 percent argon and 25 percent CO_2, or (2) 100 percent CO_2 is recommended.[9] For low alloy steels where toughness is important, it is recommended[9] to use a mixture of 60–70 percent helium, 25–30 percent argon, and 4–5 percent CO_2.

The shielding gas serves the following functions in addition to protecting the molten metal from the atmosphere:

(a) Controls the arc and metal-transfer characteristics.
(b) Affects penetration, width of fusion, and shape of the weld region.
(c) Affects the speed of welding.
(d) Controls undercutting.

By mixing an inert gas with a reactive gas the arc may be made more stable and the spatter during metal transfer may be reduced. The use of CO_2 alone for welding steel is the least expensive procedure, because of its lower cost for shielding gas, higher welding speed, better joint penetration, and sound deposits with good mechanical properties. The only disadvantage is that it gives harsh and excessive spatter.

The electrode material for welding carbon steels is an uncoated mild steel, deoxidized carbon manganese steel, designated A559 (AWS A5.18) as described in Chapter 2. For welding low-alloy steel a deoxidized low-alloy material is necessary.

The gas metal-arc welding method using CO_2 shielding is good for the lower carbon and low-alloy steels usually used in buildings and bridges.

Submerged-Arc Welding. In this fusion-arc welding process the arc is not visible because it is covered by a blanket of granular, fusible material, as shown in Fig. 5.2.3. The bare metal electrode is consumable in that it

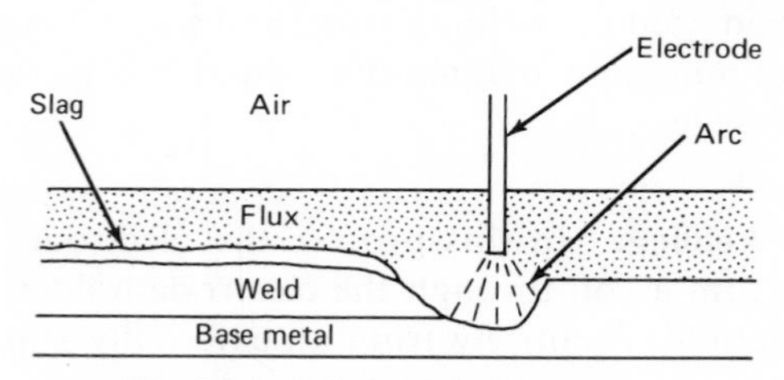

Fig. 5.2.3. Submerged-arc process.

is deposited as filler material. The end of the electrode is kept continuously shielded by the molten flux over which is deposited a layer of unfused flux in its granular condition.

The flux, which is the special feature of this method, provides a cover which allows the weld to be made without spatter, sparks, or smoke. The granular flux is laid usually automatically along the seam ahead of the advancing electrode. It protects the weld pool against the atmosphere, serves to clean the weld metal, and modifies the chemical composition of the weld metal.

Welds made by the submerged-arc process are found to have uniformly high quality; exhibiting good ductility, high impact strength, high density and good corrosion resistance. Mechanical properties of the weld are consistently as good as the base material.

The bare-rod electrodes are of mild steel designated E6XX or E7XX according to ASTM A558 (AWS A5.17), indicating a minimum tensile strength of 60 or 70 ksi, respectively. Fluxes are also classified under A558 (AWS A5.17) and designated by a prefix F followed by a two-digit number indicating tensile strength and impact strength requirements for the resulting welds. The combination of flux and electrode is usually designated together, such as F7X-EXXX, where the last three digits classify the electrode.

The submerged arc method is commonly used to weld steel in shop fabrication operations using automatic or semiautomatic equipment.

Flux-Cored Arc Welding. This is a gas metal-arc process using flux cored filler, and may be with or without external gas shielding. For welding of steel usually a CO_2 shielding gas is used. Instead of a coating on the electrode, which is not feasible for a continuously fed electrode wire; the coating material, or flux, is contained in the core of the electrode.

The electrodes are specified as a mild steel type E60T or E70T under AWS A5.20, and for tensile strengths greater than 70 ksi may be referred to as Grades E80T, E90, E100T, and E110T for tensile strength of 80, 90, 100 and 110 ksi, respectively.

Stud Welding. The most commonly used process of welding a metal stud to a base material is known as arc stud welding, an essentially automatic process but similar in characteristics to manual shielded arc welding. The stud serves as the electrode and an electric arc is created from the end of the stud to the plate. The stud is contained in a gun which controls the timing during the process. Shielding is accomplished by placing a ceramic ferrule around the end of the stud in the gun. The gun is placed in position and the arc is created, during which time the ceramic ferrule contains the molten metal. After a short instant of time, the gun drives the stud into the molten pool and the weld is completed leaving a small fillet around the stud. Full penetration across the shank of the stud is obtained and the weld is completed usually in less than one second.

Oxyacetylene Welding. This early method wherein heat is generated from the combustion of a mixture of oxygen and acetylene has now been essentially replaced by electric arc welding. However the oxyacetylene torch is widely used for cutting steel, as well as for heating to camber and flame-straighten steel.

For most fabrication of steel buildings or bridges, either shielded metal-arc welding using coated electrodes is used for manual operation, or the submerged arc process with bare electrodes and granular flux is used for automatic and semiautomatic operations.

5.3. WELDABILITY OF STRUCTURAL STEEL

Most of the ASTM-specification construction steels can be welded without special precautions or special procedures. Chapter 2 discusses the three basic categories of structural steels:

(a) Carbon steels..(Sec. 2.2)
(b) High-strength low-alloy steels ....(Sec. 2.3)
(c) Heat-treated low-alloy steels............................(Sec. 2.4)

The discussion in Chapter 2 also includes the steels contained in the various commonly used electrodes. Section 5.12 discusses the need to select the proper electrode to join a particular grade of steel and a summary of the "matching" electrodes and the base steel is given in Table 5.12.1.

The *weldability* of a steel is a measure of the ease of producing a crack-free and sound structural joint. Some of the readily available structural steels are more suited to welding than others, and are discussed in Chapter 2. Welding procedures should be based on a steel's chemistry instead of the published maximum alloy content since most mill runs are usually below the maximum alloy limits set by its specification. Table 5.3.1 shows the ideal chemical analysis of the carbon steels. Most mild steels fall well within this range, while higher-strength steels may exceed the ideal analysis shown in Table 5.3.1.

When a mill produces a run of steel it maintains a complete record of its chemical content which follows all sections made from the particular ingot. If the designer is concerned about the chemistry of a particular grade of steel, he may request a Mill Test Report. Any variation in

TABLE 5.3.1
Preferred Analysis of Carbon Steel[10]
for Good Weldability

Element	Normal Range, percent	Percent Requiring Special Care
Carbon	0.06–0.25	0.35
Manganese	0.35–0.80	1.40
Silicon	0.10 max	0.30
Sulfur	0.035 max	0.050
Phosphorus	0.030 max	0.040

chemical content above the ideal values may be evaluated and special welding procedures be set up to insure a properly welded joint.

It should be noted, however, that most structural welding situations do not require the caution implied by the previous paragraph.

5.4. TYPES OF JOINTS

The types of joints used in structural connections depend on many design considerations, including the size and shape of the members coming into the joint, the type of loading, the amount of joint area available for welding, and the relative costs for various types of welds. There are five basic types of welded connections although many variations and combinations are found in practice. The five basic types are the butt, lap, tee, corner, and edge joints, as shown in Fig. 5.4.1.

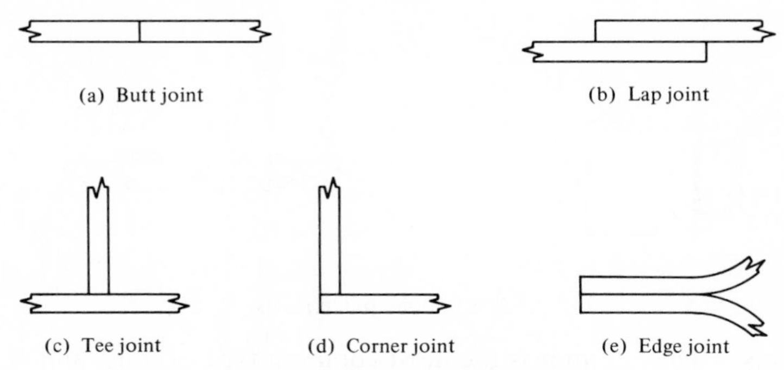

Fig. 5.4.1. Basic types of welded joints.

Butt Joints. The butt joint is used mainly to join the ends of flat plates of the same or nearly the same thicknesses. The principal advantage of this type of joint is to eliminate the eccentricity developed in single lap joints as shown in Fig. 5.4.1b. When used in conjunction with full penetration welds, butt joints minimize the size of a connection and are usually more esthetically pleasing than built-up joints. Their principal disadvantage lies in the fact that the edges to be connected must usually be specially prepared (beveled, or ground flat) and very carefully aligned prior to welding. Little adjustment is possible and the pieces must be carefully detailed and fabricated. As a result, most butt joints are made in the shop where the welding process can be accurately controlled.

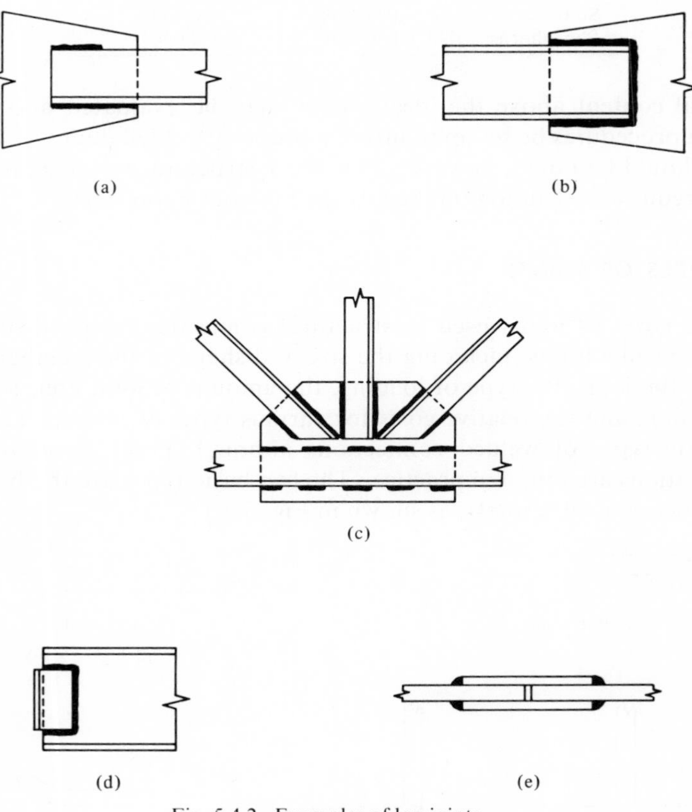

Fig. 5.4.2. Examples of lap joints.

Lap Joints. The lap joint is the most common type of joint and is used in a large variety of connections. Figure 5.4.2 shows a few common applications of the lap joint. There are two principal advantages to using lap joints:

1. *Ease of fitting.* Pieces being joined do not require the preciseness in fabricating as do the other types of joints. The pieces can be slightly shifted to accomodate minor errors in fabrication or to make adjustments in length.

2. *Ease of joining.* The edges of the pieces being joined do not need special preparation and are usually sheared or flame cut. Lap joints are especially adapted to accepting fillet welds and are therefore equally well suited to shop or field welding. The pieces being joined are in most cases simply clamped together without the use of special jigs. Occasionally the pieces are positioned by a small number of erection bolts which may either be left in place or removed after the welding is completed.

A further advantage of the lap joint is the ease in which plates of different thicknesses can be joined, such as in the double lap joint in Fig. 5.4.2e. The reader should especially note the truss connection shown in Fig. 5.4.2c and consider the difficulty in making such a connection by any other type of joint.

Tee Joints. This type of joint is used to fabricate built up sections such as tees, H-shapes, plate girders, bearing stiffeners, hangers, brackets, and in general, pieces framing in at right angles as shown in Fig. 5.4.1c. This type of joint is especially useful in that it permits sections to be built up of flat plates which can be joined by either fillet or groove welds.

Corner Joints. Corner joints are used principally to form built-up rectangular box sections such as used for columns and for beams required to resist high torsional forces.

Edge Joints. Edge joints are generally not structural but are most frequently used to keep two or more plates in a given plane or to maintain initial alignment.

As the reader can infer from the previous discussions, the variations and combinations of the five basic types of welds are virtually infinite. Since there is usually more than one way to connect one structural member to another, the designer is left with the decision for selecting the best joint (or combination of joints) in each given situation.

5.5. TYPES OF WELDS

The four basic types of welds are the groove, fillet, slot, and plug welds as shown in Fig. 5.5.1. Each basic type of weld has specific advantages which determines the extent of its usage. The four basic types of welds and their variations constitute virtually all of the structural welds found in common practice. Broadly speaking, the usage of the basic welds are: groove welds, 15 percent; fillet welds, 80 percent: the remaining 5 percent are made up of the slot, plug, and other special welds.

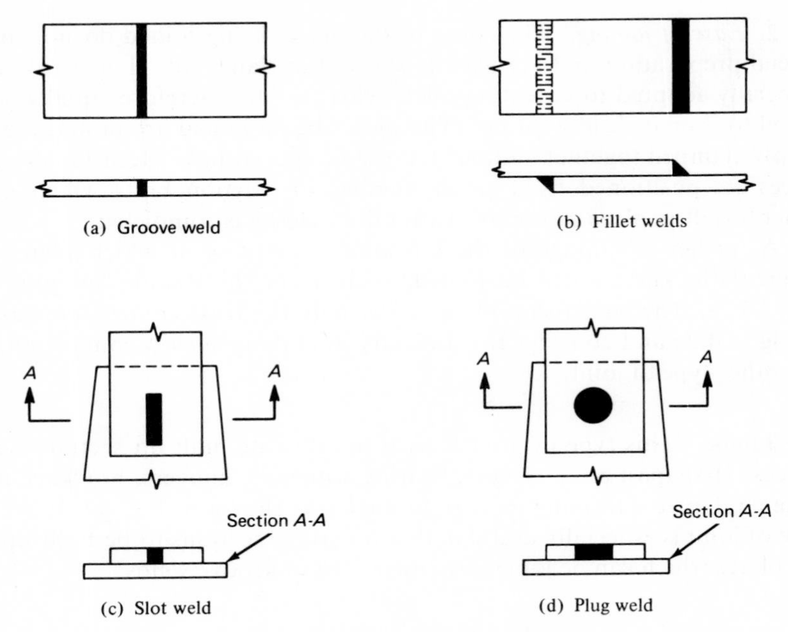

(a) Groove weld

(b) Fillet welds

(c) Slot weld

(d) Plug weld

Fig. 5.5.1. Basic types of welds.

Groove Welds. The principal use of groove welds is to connect structural members which are aligned in the same plane. Since groove welds must transmit the entire load at a particular joint they usually must have at least as good structural properties as the members which they connect; in which case they may be referred to as full penetration welds. There are many variations of groove welds and each is classified according to its particular shape. Each type of groove weld requires a specific edge preparation and is named accordingly.

Figure 5.5.2 shows the common types of groove welds and indicates the end preparations required for each. The selection of the proper groove weld is dependent on the cost of the edge preparations and the cost of making the weld. The final selection as to which type of groove weld to use must also consider the facilities of the fabricator making the welds, his ability to provide the required edge preparations and whether or not the welding can be done on both sides. Groove welds may also be used in connections as shown in Fig. 5.5.3.

Fillet Welds. Fillet welds owing to their overall economy, ease of fabricating, and adaptability are the most widely used of all the basic welds. Fillet welds offer great flexibility to the designer since they are adaptable

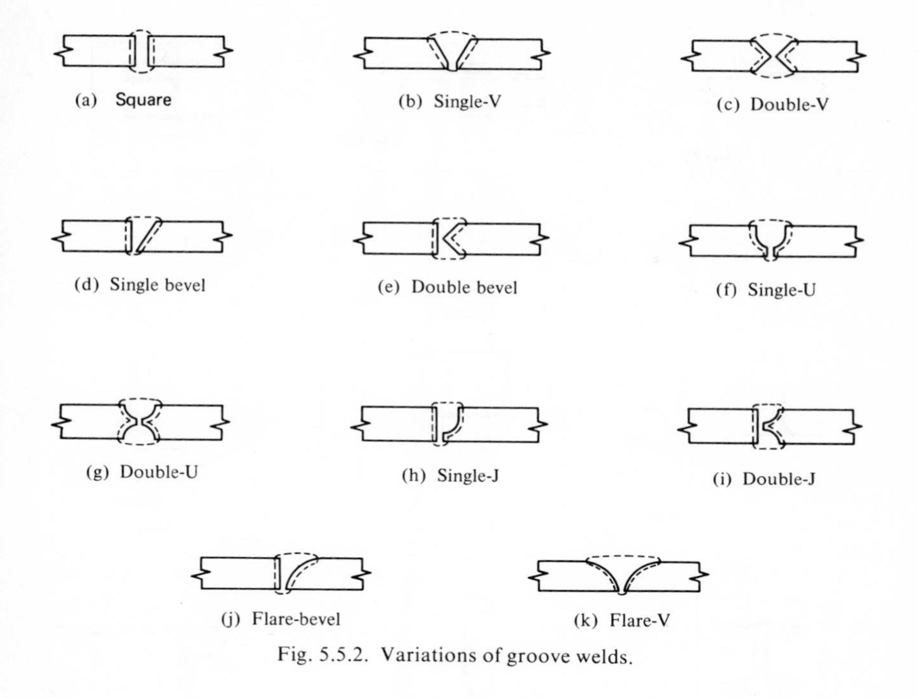

Fig. 5.5.2. Variations of groove welds.

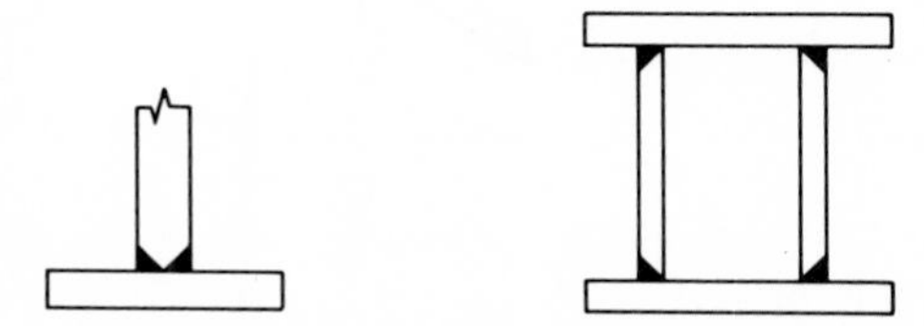

Fig. 5.5.3. Uses of groove welds in tee joints.

to a large variety of connections, a few of which were shown in Fig. 5.5.4. They generally require less precision in the "fitting up" since the plates being joined can be moved about more than groove welds that may require specific gaps or critical alignment. This is particularly advantageous to welding in the field or in realigning members or connections that were fabricated within accepted tolerances but which may not fit as accurately as desired. In addition, the edges of pieces being joined seldom need special preparation such as beveling or squaring since the edge conditions resulting from the usual flame cutting or from shear cutting procedures are generally adequate.

Slot and Plug Welds. Slot and plug welds may be used exclusively in a connection as shown in Figs. 5.5.1c and d, or they may be used in combi-

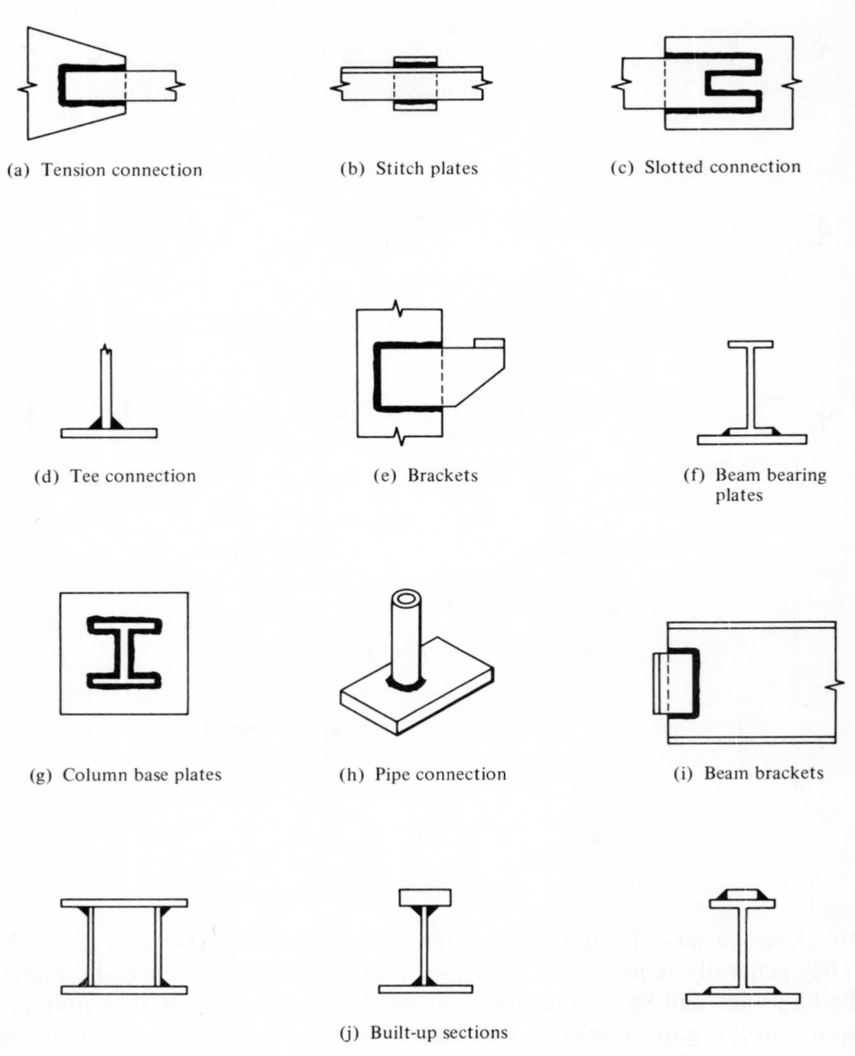

Fig. 5.5.4. Typical uses of fillet welds.

nation with fillet welds as shown in Fig. 5.5.5. A principal use for plug or slot welds is to transmit shear in a lap joint when the size of the connection limits the length available for fillet or other edge welds. Slot and plug welds are also useful in preventing overlapping parts from buckling.

5.6. WELDING SYMBOLS

Before a connection or joint is welded, the designer must in some way be able to instruct the steel detailer and the fabricator as to the type

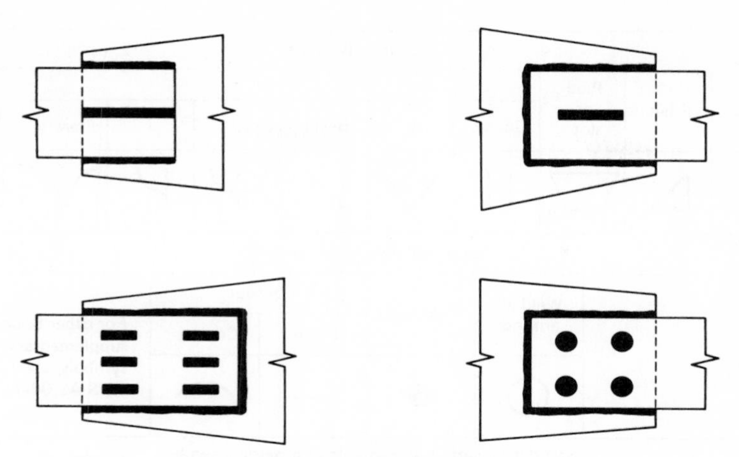

Fig. 5.5.5. Slot and plug welds in combination with fillet welds.

and size of weld required. The basic types of welds and some of their variations are discussed in Sec. 5.5. If individual and detailed instructions were needed each time a connection was made, the task of providing directions for making the joint would indeed be formidable.

The need for a simple and yet accurate method for communicating between the designer and fabricator gave rise to a type of shorthand symbols which characterize the type and size of weld. As a result, the American Welding Society welding symbols, shown in Fig. 5.6.1, indicate the shape of the weld and its size, as well as any special instructions.

Most of the commonly made connections do not require special instructions and are typically specified as shown in Fig. 5.6.2. For a more detailed use of welding symbols the reader is referred to the AISC Manual, various publications of the American Welding Society, and to special publications.[11]

The reader may feel that the number of symbols is burdensome. However, the system of designating welds is broken down into a few basic types which are built up to give a complete set of instructions. Whenever a particular connection is used in many parts of a structure, it may only be necessary to show a typical detail as shown in Fig. 5.6.3a. Whenever special connections are used, they should be detailed sufficiently to leave no doubt as to the designer's intentions, as shown in Fig. 5.6.3b.

In Fig. 5.6.3b the designer specified that the plug weld be made in the shop and ground flush while the double bevel weld connecting the gusset plate to the column be made in the field. Since the designer did not specify where the fillet welds attaching the angle to the gusset plate were to be made, the steel fabricator would be free to make the decision. However, in this particular detail, it would be better to make the fillet welds in the shop since the plug weld might be over stressed during the field erection process. In general, the fabricator will make as many welds in the

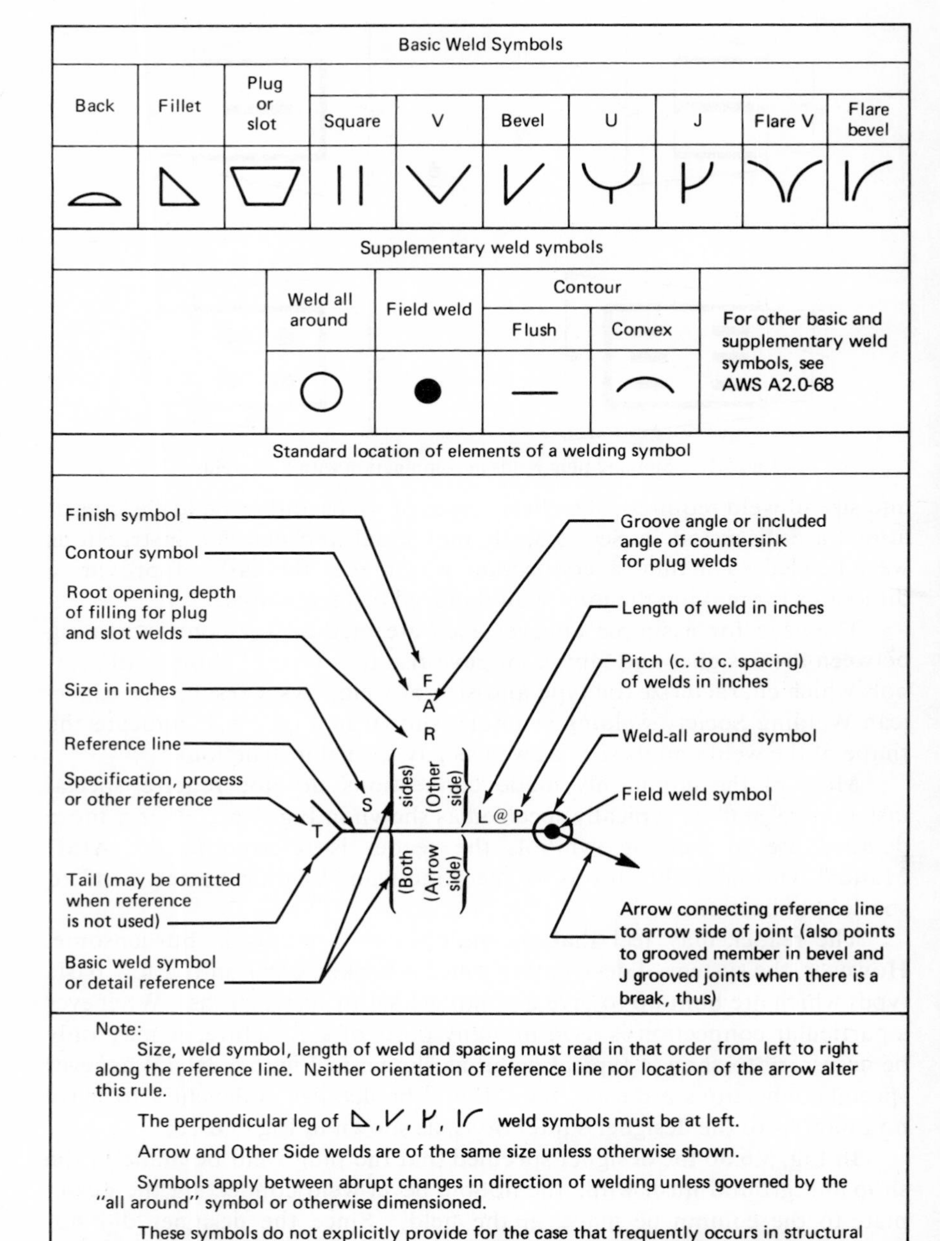

Note:

Size, weld symbol, length of weld and spacing must read in that order from left to right along the reference line. Neither orientation of reference line nor location of the arrow alter this rule.

The perpendicular leg of ◺, V, Ɏ, I⌐ weld symbols must be at left.

Arrow and Other Side welds are of the same size unless otherwise shown.

Symbols apply between abrupt changes in direction of welding unless governed by the "all around" symbol or otherwise dimensioned.

These symbols do not explicitly provide for the case that frequently occurs in structural work, where duplicate material (such as stiffeners) occurs on the far side of a web or gusset plate. The fabricating industry has adopted this convention; that when the billing of the detail material discloses the identity of far side with near side, the welding shown for the near side shall also be duplicated on the far side.

Fig. 5.6.1. Standard welding symbols (from AISC Manual).

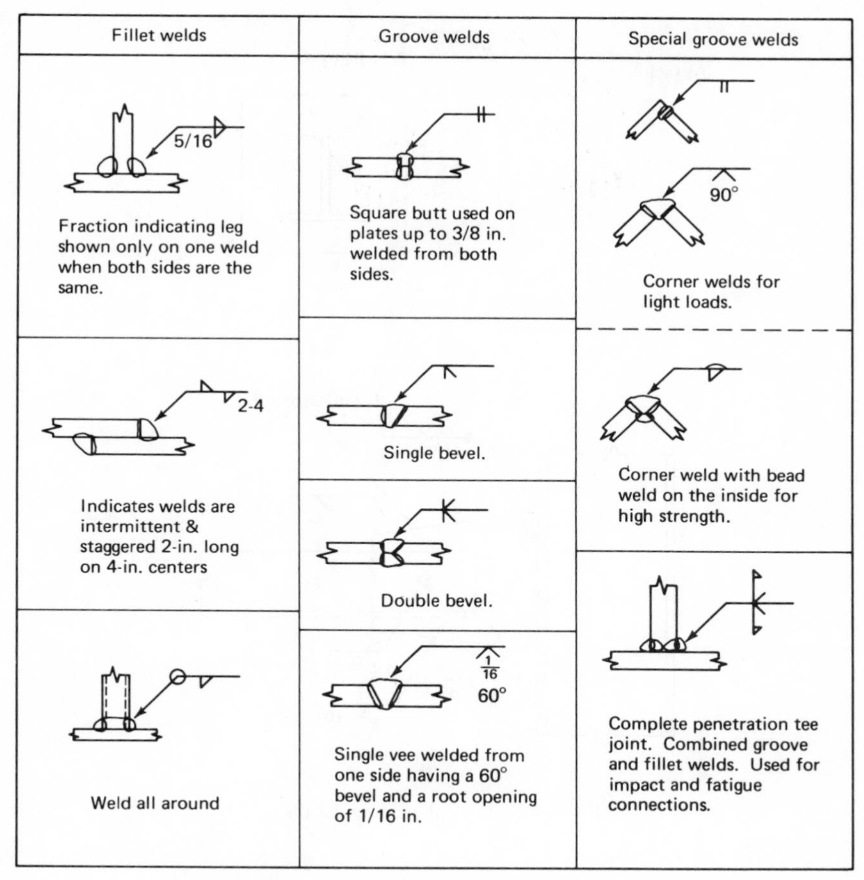

Fillet welds
Groove welds
Special groove welds
5/16
Fraction indicating leg shown only on one weld when both sides are the same.
Square butt used on plates up to 3/8 in. welded from both sides.
90°
Corner welds for light loads.
2-4
Indicates welds are intermittent & staggered 2-in. long on 4-in. centers
Single bevel.
Double bevel.
Corner weld with bead weld on the inside for high strength.
1/16
60°
Weld all around
Single vee welded from one side having a 60° bevel and a root opening of 1/16 in.
Complete penetration tee joint. Combined groove and fillet welds. Used for impact and fatigue connections.

Fig. 5.6.2. Common uses of welding symbols.

shop as he can, due to economic considerations. Therefore, it is most critical that the designer specify those welds he wants to be *field welded.*

5.7. FACTORS AFFECTING THE QUALITY OF WELDED CONNECTIONS

Providing a satisfactory welded connection requires the combination of many individual skills, beginning with the actual design of the weld and ending with the welding operation. A well designed weld does not insure a strong connection unless it is properly made in the shop or field. It is therefore necessary that the structural engineer be aware of the factors that affect the quality of a weld and design his connections accordingly.

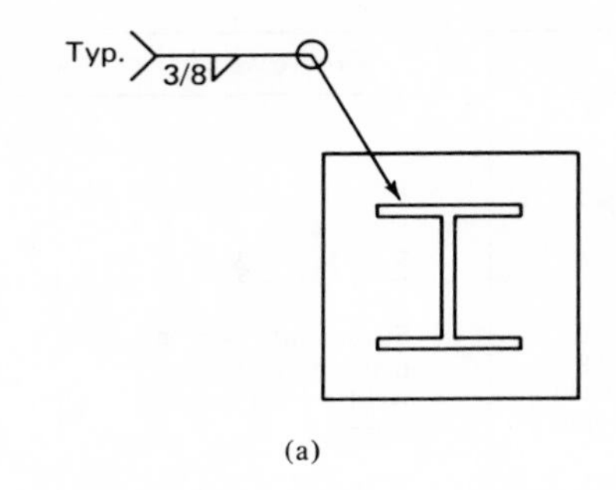

(a)

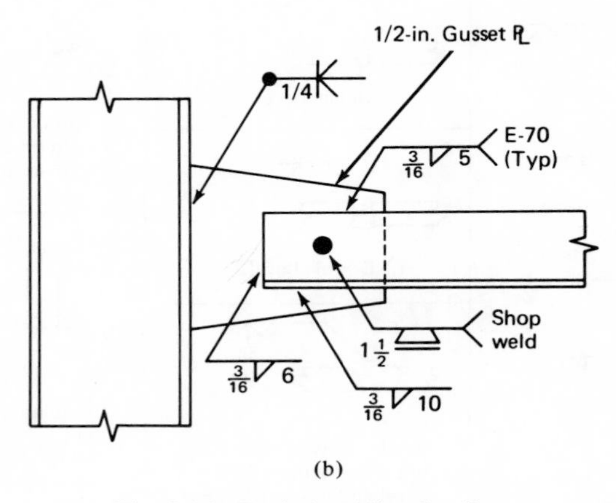

(b)

Fig. 5.6.3. Typical welding details.

Proper Electrodes, Welding Apparatus, and Procedures. After the proper electrode is specified to match the steel in the pieces being joined as indicated in Sec. 5.12 the diameter of the welding electrode must be selected. The particular size of the electrode selected is based on the size of the weld to be made and on the output of the welding apparatus. It is important that the welding apparatus be capable of providing sufficient current for the size of electrode being used. Since most welding machines have controls for reducing the current output, electrodes smaller than the maximum capability can easily be accomodated and should be used.

Since the depositing of weld metal in metal-arc welding is by the electromagnetic field and not by gravity, the welder is not limited to flat or horizontal welding positions. The four basic welding positions are shown in Fig. 5.7.1. The designer should avoid whenever possible the overhead position as shown in Fig. 5.7.1d, since it is the most difficult to make and control. Joints welded in the shop are usually positioned in the

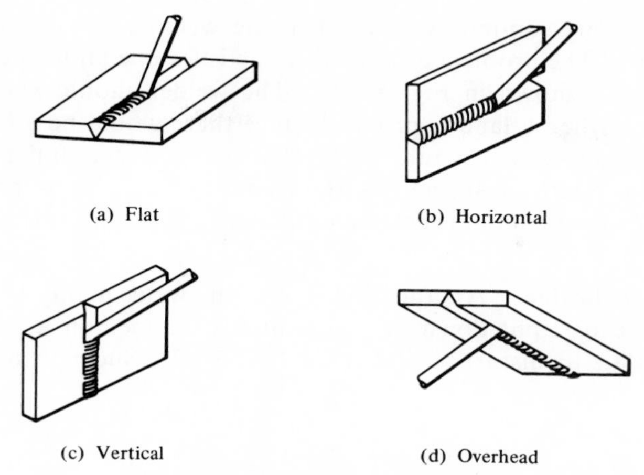

Fig. 5.7.1. Basic welding positions.

flat or horizontal positions but welds made in the field may assume any welding position depending on the orientation of the connection. The designer should therefore use precaution in specifying field welds.

Proper Edge Preparation. Typical edge preparations provided for groove welds are shown in Fig. 5.7.2. The root opening R is the separation of the

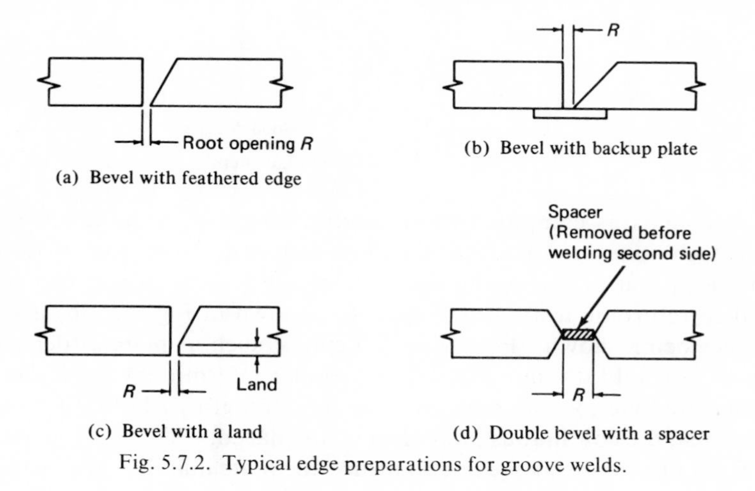

Fig. 5.7.2. Typical edge preparations for groove welds.

pieces being joined and is provided for electrode accessibility to the base of a joint. The smaller the root opening the greater must be the angle of the bevel. The feathered edge as shown in Fig. 5.7.2a is subject to burn-through unless a backup plate is provided as shown in Fig. 5.7.2b.

Backup strips are commonly used when the welding is to be done from one side only. The problem of burn-through is lessened if the bevel is provided a land as shown in Fig. 5.7.2c. The welder should *not* provide a backup plate when a land is provided since there would be a good possibility that a gas pocket would be formed preventing a full penetration weld. Occasionally a spacer, as shown in Fig. 5.7.2d, is provided to prevent burn-through but is gouged out before the second side is welded.

Control of Distortion. Another factor affecting weld quality is shrinkage. If a single bead is put down in a continuous manner on a plate, it will cause the plate to distort as shown in Fig. 5.7.3. Such distortions may

Fig. 5.7.3. Distortion of plate.

easily take place unless care is exercised in both the design of the joint and the welding procedure. Figure 5.7.4 shows the result of using unsym-

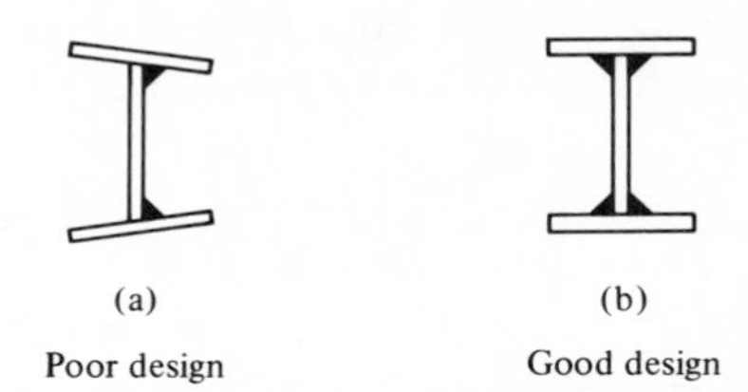

(a) Poor design

(b) Good design

Fig. 5.7.4. Effect of weld placement.

metrical welds as compared to symmetrical welds. Although there are many techniques available for minimizing distortion, the most common one is that of staggering intermittent welds as shown in Fig. 5.7.5a, and then returning to fill in the spaces as shown in Fig. 5.7.5b, a typical sequence being shown. For many structures, such as plate girders, short segments of weld (though not usually regular intermittent welds) may be used at strategic locations to give enough strength to hold all pieces in place; then, the continuous lines of weld are placed.

To minimize shrinkage and to insure adequate ductility, the American Welding Society has established recommendations for minimum preheat and interpass temperatures which are summarized in Table 5.7.1 (see p. 184) from AISC–1.23.6.

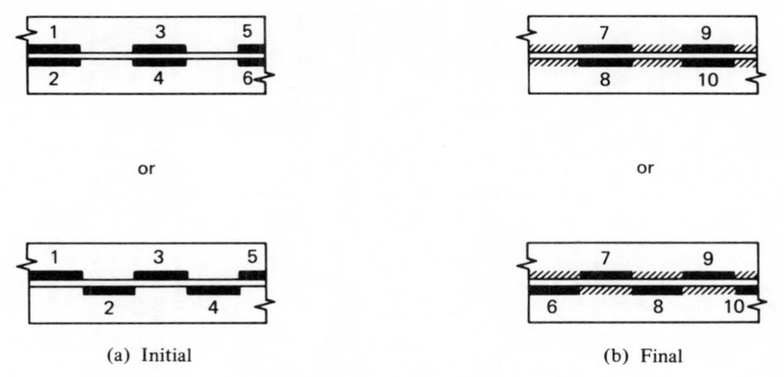

(a) Initial (b) Final

Fig. 5.7.5. Sequences for intermittent welds.

The following summarizes ways of minimizing distortion:

1. Reduce the shrinkage forces by
   (a) Using minimum weld metal; for grooves use no greater root opening than necessary; do not overweld.
   (b) Using as few passes as possible.
   (c) Using proper edge preparation and fit-up.
   (d) Using intermittent weld, at least for preliminary connection.
   (e) Using backstepping; depositing weld segments toward the previously completed weld; i.e., depositing in the direction opposite to the progress of welding the joint.
2. Allow for the shrinkage to occur by
   (a) Tipping the plates so after shrinkage occurs they will be correctly aligned.
   (b) Using prebending of pieces.
3. Balance shrinkage forces by
   (a) Using symmetry in welding; fillets on each side of a piece contribute counteracting effects.
   (b) Using scattered weld segments.
   (c) Using peening; stretching the metal by a series of blows.
   (d) Using of clamps, jigs, etc; this forces weld metal to stretch as it cools.

5.8. POSSIBLE DEFECTS IN WELDS

Unless good welding techniques and procedures are used, a number of possible defects may result relating to discontinuities within the weld. Some of the more common defects are: incomplete fusion, inadequate

TABLE 5.7.1
(AISC–Table 1.23.6)
Minimum Preheat and Interpass Temperature, °F[1]

Thickness of Thickest Part at Point of Welding (inches)	Welding Process				
	Shielded Metal-Arc Welding with other than Low Hydrogen Electrodes	Shielded Metal-Arc Welding with Low Hydrogen Electrodes; Submerged Arc Welding; Gas Metal-Arc Welding; or Flux Cored Arc Welding		Shielded Metal-Arc Welding with Low Hydrogen Electrodes; Submerged Arc Welding with Carbon or Alloy Steel Wire, Neutral Flux; Gas Metal-Arc Welding; or Flux Cored Arc Welding	Submerged Arc Welding with Carbon Steel Wire, Alloy Flux
	ASTM A36; A53 Grade B; A375; A500; A501; A529; A570 Grades D and E	ASTM A36; A242 Weldable Grade; A375; A441; A529; A570 Grades D & E; A572 Grades 42, 45, and 50; A588	ASTM A572 Grades 55, 60, and 65	ASTM A514	ASTM A514
To $^3/_4$, incl.	None[2,3]	None[2]	70	50	50
Over $^3/_4$ to $1^1/_2$, incl.	150	70[4]	150	125	200
Over $1^1/_2$ to $2^1/_2$, incl.	225	150[4]	225	175	300
Over $2^1/_2$	300	225	300	225	400

[1] Welding shall not be done when the ambient temperature is lower than 0°F. When the base metal is below the temperature listed for the welding process being used and the thickness of material being welded, it shall be preheated (except as otherwise provided) in such manner that the surface of the parts on which weld metal is being deposited are at or above the specified minimum temperature for a distance equal to the thickness of the part being welded, but not less than 3 in., both laterally and in advance of the welding. Preheat and interpass temperatures must be sufficient to prevent crack formation. Temperature above the minimum shown may be required for highly restrained weld. For A514 steel the maximum preheat and interpass temperature shall not exceed 400°F for thicknesses up to $1^1/_2$ in., inclusive, and 450°F for greater thicknesses.

[2] When base metal temperature is below 32°F, preheat base metal to at least 70°F and maintain this minimum temperature during welding.

[3] This provision also applies to A36 steel in thicknesses up to 1 in.

[4] Minimum preheat for A36 steel in thicknesses up to 2 in. shall be 50°F.

penetration, porosity, undercutting, inclusion of slag, and cracks. Examples of these defects are shown in Fig. 5.8.1.

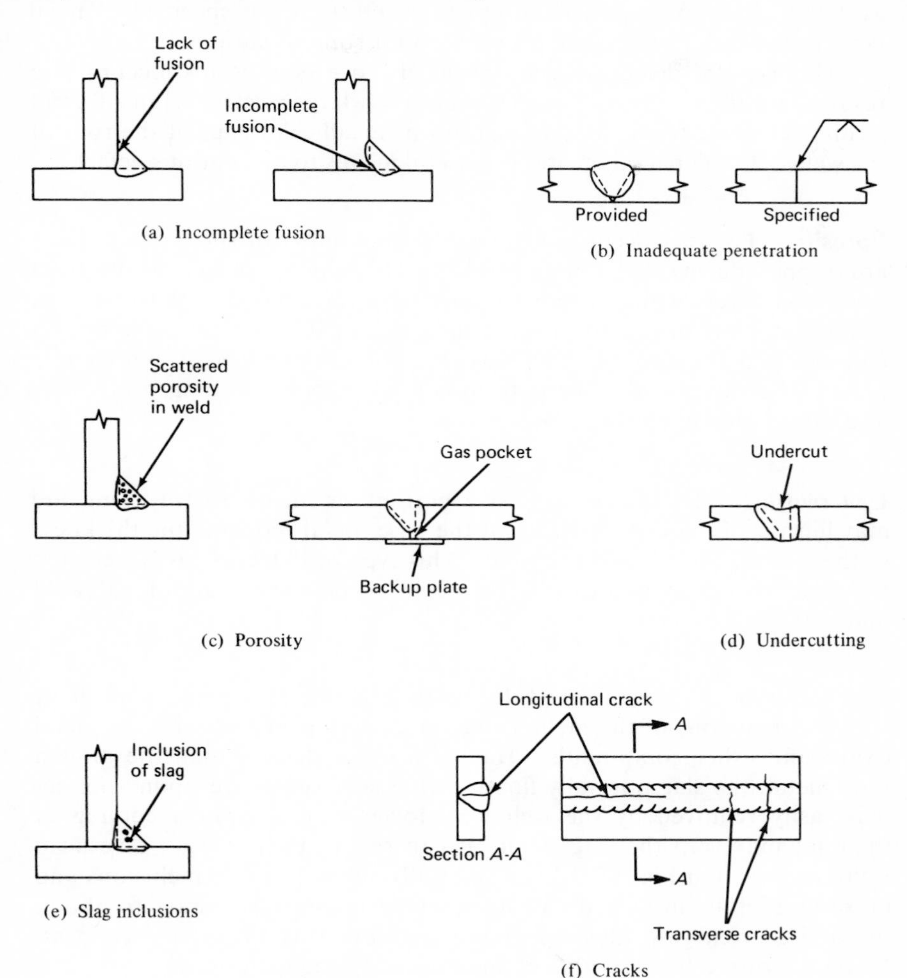

Fig. 5.8.1. Possible weld defects.

Incomplete fusion. Incomplete fusion may occur if the surfaces to be joined have not been properly cleaned and are coated with mill scale, slag, oxides or other foreign materials. Another cause of this type of defect is the use of welding equipment of insufficient current, so that base metal does not reach melting point. Too rapid a rate of welding will also have the same effect.

Inadequate Penetration. Inadequate penetration means the weld extends a shallower distance through the depth of the groove than specified, as shown in Fig. 5.8.1, where complete penetration was specified. Partial penetration may be adequate penetration for some situations.

This type of defect, relating primarily to groove welds, occurs as a result of insufficient groove angles, excessively large electrodes, insufficient welding current, excessive welding rates or insufficient gaps at the root of the welds. Using backup plates is a useful means to prevent this defect.

Porosity. Porosity occurs when voids or a number of small gas pockets are trapped during the cooling process. This type of defect results from using excessively high current or too long an arc length. Porosity may occur uniformly dispersed through the weld, or it may be a large pocket concentrated at the root of a fillet weld or at the root adjacent to a backup plate in a groove weld. The latter is caused by poor welding procedures and careless use of backup plates.

Undercutting. The use of excessive current or an excessively long arc may burn or dig away a portion of the base metal reducing the thickness of the joint at the edge of the weld. This type of defect is easily detected by visual inspection and can be corrected by depositing additional weld material.

Slag Inclusion. Slag is formed during the welding process as a result of chemical reactions of the melted electrode coating and consists of metal oxides and other compounds. Having a lower density than the molten weld metal, the slag normally floats to the surface, where upon cooling, it is easily removed by the welder. However, too rapid a cooling of the joint may trap the slag before it can rise to the surface. Overhead welds as shown in Fig. 5.7.1d are especially subject to slag inclusions and must be carefully inspected. When several passes are necessary to obtain the desired weld size, the welder must remove slag between each pass. Failure to properly do so is a common cause of slag inclusion.

Cracks. Cracks are breaks in the weld metal, either longitudinal or transverse to the line of weld which result from internal stress. Cracks may also extend from the weld metal into the base metal or may be entirely in the base metal in the vicinity of the weld. Cracks are perhaps the most harmful of weld defects; however, tiny cracks called *microfissures* may not have any detrimental effect.

Some cracks form as the weld begins to solidify, generally caused by brittle constituents, either brittle states of iron or alloying elements, forming along the grain boundaries. More uniform heating and slower cooling will prevent the "hot" cracks from forming.

Cracks may also form at room temperature parallel to but under the weld in the base material. These cracks arise in low-alloy steels from the combined effects of hydrogen, a brittle martensite microstructure, and restraint to shrinkage and distortion. Use of low-hydrogen electrodes along with proper preheating and postheating will minimize such "cold" cracking.

5.9. INSPECTION AND CONTROL

The enormous success and growth in recent years in the area of structural welding of buildings and bridges could not have occurred without some means of inspection and control. The welding industry has led in the development of guidelines which, if followed, virtually insure a sound weld. The inspection and control procedure should begin before the first arc is struck, continue throughout the welding procedure, and if necessary, a pretest of the joint should be made to assure its satisfactory performance. Since such close supervision is not possible on every weld made, the following suggestions will serve as a guideline to achieve good structural welds:

(a) Establish good welding procedures.
(b) Use only prequalified welders.
(c) Use qualified inspectors and have them present.
(d) Use special inspection techniques when necessary.

Good welding procedures can be developed from recommendations from the AWS, AISC, and the manufacturers of welding supplies and equipment. The procedure to be followed will depend on the chemical and physical properties of the materials, the types and sizes of weld, and the particular equipment used.

All welders should be required to have passed an American Welding Society Qualification Test before being permitted to make a structural connection. Although this is usually considered adequate, it doesn't prove the ability of the welder to make welds at the actual job site, particularly if the welds are unusual or difficult and were not specified in the Qualification Test. Happily, most welding contractors exercise their own control over their welders in such situations.

The use of qualified welding inspectors at a jobsite generally has the effect of causing welders to perform their best work, feeling that the inspector is able to recognize the quality of their welds. The welding inspector should be a competent welder himself and be able to recognize the possible defects discussed in Sec. 5.8. Any poor or suspicious welds should be cut out and replaced.

The simplest and least expensive method of inspection is *visual* but is dependent on the competence of the observer. A welding gage such as shown in Fig. 5.9.1a offers a rapid means of checking the size of fillet welds.

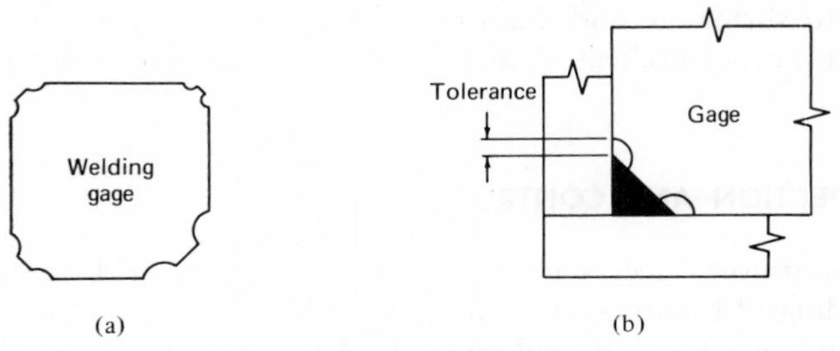

Fig. 5.9.1. Checking size of fillet welds.

For more important structures and for welds whose failure could be catastrophic, more rigid inspection techniques should be used. Some of the useful ones are the ultrasonic, radiographic, and magnetic particle methods. The *ultrasonic* method[12] passes ultrahigh-frequency sound waves through the weldment. Defects in a particular weld will reflect the sound waves while a weld without defects will not impede passage of the waves. *Radiographic* methods include the use of both X-rays and gamma rays. In this method the radiating source is placed on one side of the weld and a photographic plate on the other. This method is expensive and requires special precautions be taken due to the hazards of radiation. However, the method is reliable and furnishes a permanent record. The *magnetic particle* testing method uses iron powder which is spread around the welded area and polarized by passing an electric current through the weld. Small local poles will be formed at the edges of any defects, and this may be interpreted by an experienced observer.

5.10. ECONOMICS OF WELDED MEMBERS AND CONNECTIONS

There are many factors which must be examined when considering the overall economy in welded connections. Some of the factors such as the amount of electrode material used can easily be computed while other factors may be intangible such as the value to be placed on esthetics. The actual economy of welded connections must be viewed from a broad aspect and include the overall design of the structural system.

Welded connections are usually neater in appearance, providing a less cluttered effect in contrast to bolted or riveted connections. Figure 5.10.1 shows a comparison between a section of a riveted or bolted plate girder and a section of a welded plate girder. Besides the neater appearance of

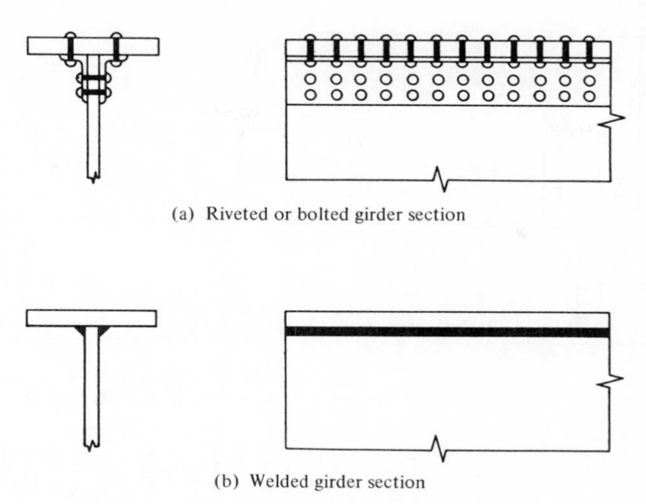

(a) Riveted or bolted girder section

(b) Welded girder section

Fig. 5.10.1. Comparison between plate girders.

the welded joint, welded connections offer the designer more freedom to be innovative in his entire design concept. The designer is not bound to standard sections but may build up any cross section he feels to be most advantageous. Similarly, he can use the best configuration to transmit the loads from one member to another.

Welded connections generally eliminate the need for holes in members except possibly for erection purposes. Therefore, the problem of determining the minimum net section in tension members as discussed in Chapter 3 need not be considered. Since it is usually the holes at the ends that govern the design of a bolted or riveted tension member, a welded connection will generally result in a member with a smaller cross section.

Welded connections can sometime reduce field construction costs by the fact that members may be shifted slightly to accommodate minor errors in fabricating or erection. Also, members may be easily shortened by cutting and rejoining by suitable welding as well as lengthened by splicing a piece of the same cross section.

In addition, there are several direct factors which influence the cost of welding. Generally, welding that is performed in the shop is less expensive than field welding. Some of the reasons for this are availability in the shop of automatic welding machines, a more pleasant and less hostile environment (the weather), and the availability of special jigs for holding the pieces to be welded in a more favorable position. Also, work to be done can be scheduled for a continuous operation whereas field welding must often wait for cranes and special erection equipment. Other operations such as the proper preheating of pieces to be welded can be difficult if not impossible to perform in the field. Other factors which influence

TABLE 5.10.1
Relative Cost Factors for Welded Joints[13]
(Based on a Cost Factor at 1.00 for a 1/4 in. Fillet Weld)

Type / Size (T) in.	Fillet 45°	Fillet 60°–30° (see sketch)	Single V 60°	Single V 90°	Single Bevel 45°	Single Bevel 60°	Double V 60°	Double V 90°	Double Bevel 45°	Double Bevel 60°	Square Groove
1/16	0.55	0.65	—	—	—	—	—	—	—	—	0.55
1/8	0.70	0.80	—	—	—	—	—	—	—	—	0.60
3/16	0.85	0.95	—	—	—	—	—	—	—	—	0.80
1/4	1.00	1.25	1.35	1.75	1.30	1.35	1.20	1.30	—	—	1.00
5/16	1.20	1.75	1.95	2.70	1.80	1.95	1.45	1.75	—	—	1.20
3/8	1.45	2.55	2.70	3.60	2.50	2.70	1.85	2.50	1.75	2.00	1.50
7/16	2.00	3.30	3.45	4.95	3.15	3.45	2.40	2.90	2.20	2.50	4.80
1/2	2.60	4.20	3.80	5.10	3.50	3.90	2.90	3.70	2.70	3.10	5.55
5/8	3.80	6.30	5.55	8.00	5.05	5.55	4.70	5.40	3.80	4.30	—
3/4	5.30	8.90	7.85	11.50	7.10	7.90	5.45	7.55	5.05	5.85	—
7/8	7.10	12.00	10.20	15.30	9.10	10.22	7.05	10.10	6.45	7.55	—
1	9.15	15.50	12.80	20.15	11.50	12.95	9.15	13.25	8.20	9.65	—
1 1/8	11.50	19.50	15.70	25.60	14.15	15.95	11.15	16.45	10.00	11.80	—
1 1/4	14.10	24.00	19.00	30.20	16.90	19.10	13.15	19.85	11.90	14.10	—
1 3/8	17.00	29.10	22.50	35.70	20.15	22.75	15.70	24.00	14.10	16.70	—
1 1/2	20.00	34.50	26.40	42.85	23.60	26.75	18.30	28.10	16.40	19.60	—
1 5/8	—	—	30.90	49.80	27.30	31.20	20.80	32.50	18.70	22.60	—
1 3/4	—	—	35.40	59.00	31.20	35.60	23.40	36.90	21.00	25.40	—
1 7/8	—	—	40.00	69.20	35.50	40.40	26.20	41.15	23.30	28.20	—
2	—	—	45.00	79.00	40.00	45.50	29.00	46.40	25.60	31.10	—
2 1/8	—	—	—	—	—	—	—	—	28.00	—	—
2 1/4	—	—	—	—	—	—	—	—	30.40	—	—
2 3/8	—	—	—	—	—	—	—	—	32.90	—	—
2 1/2	—	—	—	—	—	—	—	—	35.60	—	—
2 5/8	—	—	—	—	—	—	—	—	38.70	—	—
2 3/4	—	—	—	—	—	—	—	—	42.00	—	—
2 7/8	—	—	—	—	—	—	—	—	45.45	—	—
3	—	—	—	—	—	—	—	—	49.00	—	—

Plug Welds (per 100 holes 1 in. deep)	
Dia. of Hole (in.)	Cost Factor
7/16	25.20
9/16	39.50
11/16	57.50
13/16	78.80
15/16	104.00
1 1/16	132.00
1 3/16	165.00

Notes: Factors apply to all weld processes. A single weld position is assumed for entire weld.

welding costs are:

1. Cost of preparing the edges to be welded.
2. The amount of weld material required.
3. The ratio of the actual arc time to overall welding time.
4. The amount of handling required.
5. General overhead costs.

The factors listed above are generally unknown to the designer since the fabricator is usually not selected until after the design has been completed. However the designer must still make decisions—should he specify short large fillet welds or long small fillet welds? Should he specify large fillet welds or groove welds? If he decides to use groove welds he must then select the proper and most economical type.

In most instances the designer is not as concerned with the specific costs of each type of weld as he is with *relative* cost of the various types and sizes. As aid to the designer, Donnelly[13] developed relative cost factors relating the cost of fillet and groove welds of common sizes to the cost of a single-pass $\frac{1}{4}$ in. fillet weld. This work is reproduced in Table 5.10.1. The cost of a specific weld can be determined by Eq. 5.10.1 once the designer can obtain the local costs.

$$\text{Cost of Weld} = (\text{Length of Weld}) \times (\text{Factor from Table 5.10.1}) \times (\text{Local Cost}) \qquad (5.10.1)$$

Currently (1971), welded connections are used for the vast majority of shop connections and a sizable though not a majority of field connections.

5.11. SIZE AND LENGTH LIMITATIONS FOR FILLET WELDS

Since all types of welding involve the heating of the metal pieces, prevention of too rapid a rate of cooling is of fundamental importance to achieving a good weld. It is therefore imperative to consider the effect of the plate thickness on rate of cooling. Consider the two extreme thicknesses of plates in Fig. 5.11.1, each of which has received a bead of fillet weld. Most of the heat energy given off during the welding process is absorbed by plates being joined. The thicker plate shown in Fig. 5.11.1a dissipates the heat vertically as well as horizontally whereas the thinner plate is essentially limited to a horizontal dissipation. In other words, the thicker the plate, the faster heat is removed from the welding area, thereby lowering the temperature in the region of the weld. Since a minimum temperature is required to cause the base metal to become molten, it is therefore necessary to provide as a minimum, a weld of sufficient size (and heat content) to prevent the plate from removing the heat at a faster rate than it is being supplied. Unless a proper temperature is maintained in the area being welded a lack of fusion will result.

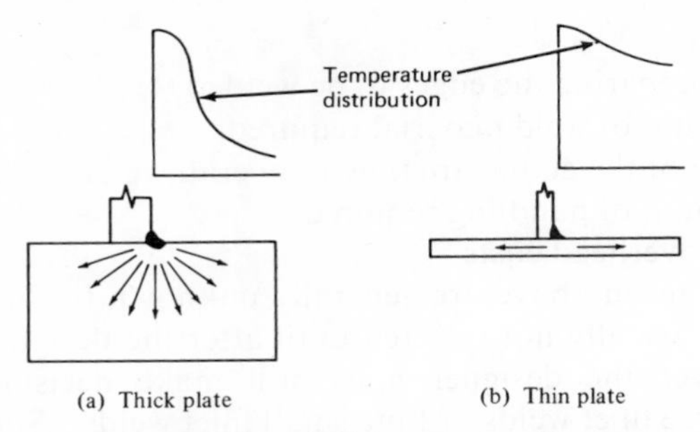

Fig. 5.11.1. Effect of thickness on cooling rate.

Minimum Fillet Weld Size. Recognizing the effect of plate thickness in the cooling rate, *minimum* fillet weld sizes have been established to insure fusion and to minimize distortion. Table 5.11.1 shows the minimum sizes of fillet welds depending on the plate thicknesses.

TABLE 5.11.1
(AISC–Table 1.17.5)
Minimum Fillet Weld Sizes

Material Thickness of Thicker Part Joined, inches	Minimum Size of Fillet Weld, inches
To $\frac{1}{4}$ inclusive	$\frac{1}{8}$
Over $\frac{1}{4}$ to $\frac{1}{2}$	$\frac{3}{16}$
Over $\frac{1}{2}$ to $\frac{3}{4}$	$\frac{1}{4}$
Over $\frac{3}{4}$ to $1\frac{1}{2}$	$\frac{5}{16}$
Over $1\frac{1}{2}$ to $2\frac{1}{4}$	$\frac{3}{8}$
Over $2\frac{1}{4}$ to 6	$\frac{1}{2}$
Over 6	$\frac{5}{8}$

According to the AISC–1.17.5, the above fillet weld sizes are governed by the thicker of the two pieces being joined, except that the weld size need not exceed the thickness of the thinner piece joined unless a larger size is required by calculated stress.

Maximum Fillet Weld Size Along Edges. The *maximum* size of fillet welds used along the edges of pieces being joined is limited by the thickness of the thinner piece. The maximum permitted by AISC–1.17.6 is, as shown in Fig. 5.11.2.

1. Along edges of material less than $\frac{1}{4}$ in. thick, the maximum size may be equal to the thickness of the material.

2. Along edges of material $\frac{1}{4}$ in. or more in thickness, the maximum size shall be $\frac{1}{16}$ in. less than the thickness of the material, unless the weld is especially designated on the drawings to be built out to obtain full throat thickness.

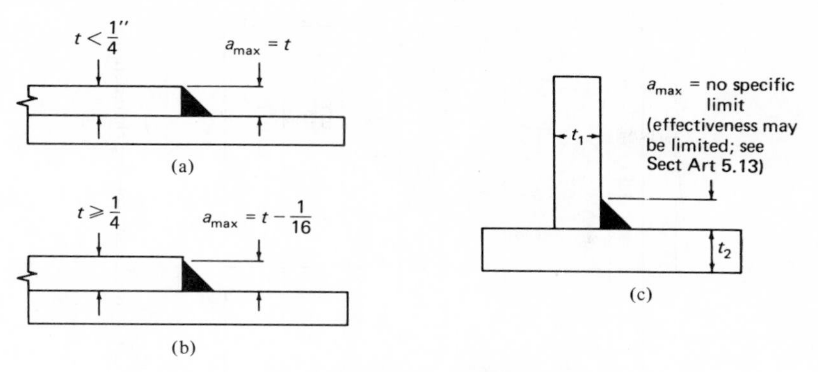

Fig. 5.11.2. Maximum weld size.

Minimum Effective Length of Fillet Welds. When placing a fillet weld, the welder builds up the weld to its full dimension as near the beginning of the weld as he can. However, there is always a slight tapering off in the region where the weld is started and where it ends. AISC–1.17.7 therefore limits the minimum effective length of fillet welds to four times their nominal size. If this criterion is not met, the size of the weld shall be considered to be one-fourth of the effective length.

AISC–1.17.10 also recommends the use of end returns, whenever practicable as shown in Fig. 5.11.3. For other limitations the reader is referred to the AISC Specification.

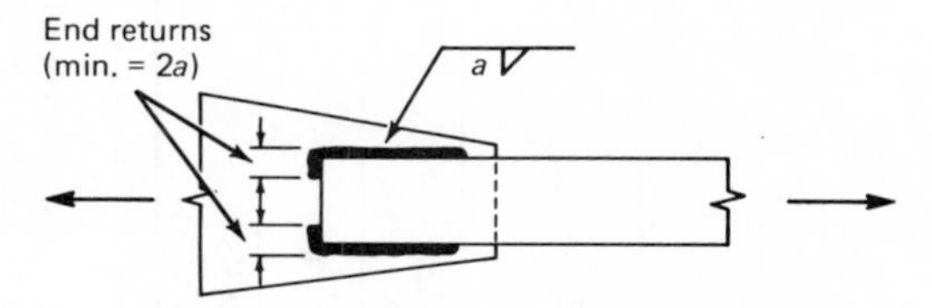

Fig. 5.11.3. Use of end returns.

5.12. ALLOWABLE STRESSES

Since welds must transmit the entire load from one member to another, welds must be sized accordingly and be formed from the correct electrode material.

While distribution of stresses in fillet welds is discussed in Sec. 5.14, it is noted here that fillet welds are assumed for design purposes to transmit loads through *shear stress* on the effective area no matter how the fillets are oriented on the structural connection. Groove welds transmit loads exactly as the pieces they join.

The required electrode material for groove welds depends on the base

TABLE 5.12.1
(AISC–Table 1.17.2)
Electrodes for Use with Various Steels

Base Metal[3]	Welding Process[1,2]			
	Shielded Metal-Arc	Submerged-Arc	Gas Metal-Arc	Flux Cored-Arc
ASTM A36, A53 Gr. B, A375, A500, A501, A529, and A570 Gr. D and E	AWS A5.1 or A5.5, E60XX or E70XX[3]	AWS A5.17 F6X or F7X-EXXX	AWS A5.18 E70S-X or E70U-1	AWS 5.20 E60T-X or E70T-X (except EXXT-2 and EXX-3)
ASTM A242, A441, A572 Grades 42 thru 60 and A588[4]	AWS A5.1 or A5.5, E70XX[5]	AWS A5.17 F7X-EXXX	AWS A5.18 E70S-X or E70U-1	AWS 5.20 E70T-X (except E70T-2 and E70T-3)
ASTM A572 Grade 65	AWS A5.5 E80XX[5]	Grade F80	Grade E80S	Grade E80T
ASTM A514 over 2½″ thick	AWS A5.5 E100XX[5]	Grade F100	Grade E100S	Grade E100T
ASTM A514 2½″ thick and under	AWS A5.5 E110XX[5]	Grade F110	Grade E110S	Grade E110T

Use of the same type filler metal having next higher mechanical properties is permitted.
[1]When welds are to be stress relieved the deposited weld metal shall not exceed 0.05 percent vanadium.
[2]See Article 422 of AWS D1.0-69 for electroslag and electrogas weld metal requuirements.
[3]On joints involving base metals of different yield strengths, filler metals appllicable to lower yield strength may be used.
[4]For architectural exposed bare unpainted applications, the deposited weld metal shall have similar atmospheric corrosion resistance and coloring characteristics as the base metal used. The steel manufacturer's recommendation shall be followed.
[5]Low hydrogen classifications.

metal in the members and is given in Table 5.12.1 for various welding processes. When the electrodes specified in Table 5.12.1 are used, the allowable tension, compression, and shear stresses on complete penetration groove welds are the same as for the base metal, as summarized in Table 5.12.2.

TABLE 5.12.2
(from AISC–Table 1.5.3)
Allowable Stresses in Groove Welds

Kind of Stress	Permissable Stress
Tension and compression parallel to axis of any complete-penetration groove weld	Same as for base metal
Tension normal to effective throat of complete-penetration groove weld	Same as allowable tensile stress for base metal
Compression normal to effective throat of complete or partial-penetration groove weld	Same as allowable compressive stress for base metal
Shear on effective throat of complete- and partial-penetration groove welds	Same as allowable shear stress for base metal

The allowable stresses for (1) shear on the effective area of fillet welds, (2) the tensile stress normal to the axis on the effective area of a partial penetration groove weld; and (3) the shear stress on the effective area of a plug or slot weld are shown in Table 5.12.3 and are equal to 0.3 times the electrode tensile strength.

The allowable stresses on fillet welds (Table 5.12.3) have been significantly increased in the 1969 AISC Specification over those previously used. For the E60 and E70 electrodes, the only ones included in 1963 AISC Specification, the increase is about 33 percent. Recent tests[14] and reevaluation of earlier tests have resulted in this increase. Reference 14 states "in recognition of improvements in the art of welding that have taken place in the past 40 years, the more liberal working stress provision —0.3 times the electrode tensile strength—appears fully justified."

The effective areas are discussed in Sec. 5.13 and the limitations on the maximum and minimum size of fillet welds are discussed in Sec. 5.11.

5.13. ESTABLISHING EFFECTIVE AREAS OF WELDS

The allowable stresses on the various types of welds summarized in Sec. 5.12 are dependent upon their *effective areas*. The effective areas of groove or fillet welds are considered to be the product of the *effective throat* dimension t_e, times the length of the weld.

The effective throat dimension is based on the nominal size and the shape of the weld. The nominal throat dimension may be thought of as the minimum width of the expected failure plane. The following discus-

TABLE 5.12.3
Allowable Stresses in Fillet Welds and Plug or Slot Welds
(from AISC–1.5.3)

Kind of Stress	Permissible Stress	Required Electrode[2]	"Matching" Base Metal[2]
Shear stress on effective throat of fillet weld regardless of direction of application of load; tension normal[1] to the axis on the effective throat of a partial-penetration groove weld; and shear stress on effective area of a plug or slot weld. The given stresses shall also apply to such welds made with the specified electrode on steel having a yield stress greater than that of the "matching" base metal. The permissible stress, regardless of electrode classification used, shall not exceed that given in the table for the weaker "matching" base metal being joined.	18.0 ksi	AWS A5.1, E60XX electrodes AWS A5.17, F6X-EXXX flux-electrode combination AWS A5.20, E60T-X electrodes	A500 Grade A A570 Grade D
	21.0 ksi	AWS A5.1 or A5.5, E70XX electrodes AWS A5.17, F7X-EXXX flux-electrode combination AWS A5.18, E70S-X or E70U-1 electrodes AWS A5.20, E70T-X electrodes	A36 A53 Grade B A242 A375 A441 A500 Grade B A501 A529 A570 Grade E A572 Grades 42 to 60 A588
	24.0 ksi	AWS A5.5, E80XX electrodes Grade 80 Submerged Arc, Gas Metal-Arc or Flux Cored Arc Weld Metal	A572 Grade 65
	27.0 ksi	AWS A5.5, E90XX electrodes Grade 90 Submerged Arc, Gas Metal-Arc or Flux Cored Arc Weld Metal	A514 over 2½ in. thick
	30.0 ksi	AWS A5.5, E100XX electrodes Grade 100 Submerged Arc, Gas Metal-Arc or Flux Cored Arc Weld Metal	A514 over 2½ in. thick
	33.0 ksi	AWS A5.5, E110XX electrodes Grade 110 Submerged Arc, Gas Metal-Arc or Flux Cored Arc Weld Metal	A514 2½ in. and less in thickness

[1]Fillet welds and partial penetration groove welds joining the component elements of built-up members, such as flange-to-web connections, may be designed without regard to the tension or compression stress in these elements parallel to the axis of the welds.

[2]Only low-hydrogen electrodes shall be used on A242, A441, A514, A572 and A588.

sion will consider the effective throat dimensions for each of the basic types of welds.

Groove welds. The effective throat dimension of a full penetration groove weld is the thickness of the thinner part joined as shown in Figs. 5.13.1a and b. The effective throat dimension of single and double partial-penetration groove welds is the depth of the groove except in the case of a bevel joint made by the manual shielded metal-arc process. In the latter case, the effective throat dimension shall be taken (AISC–1.14.7) as the

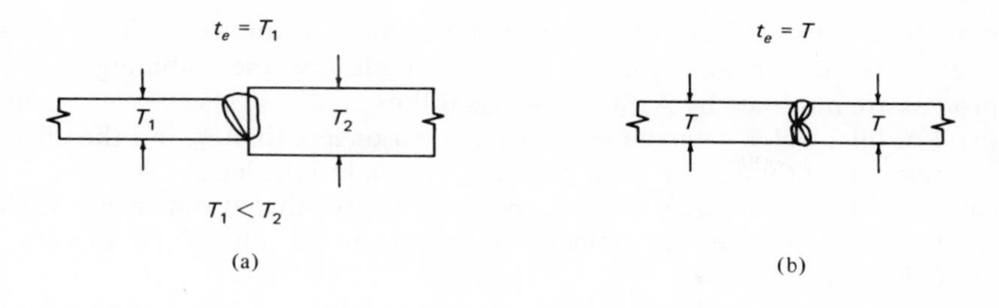

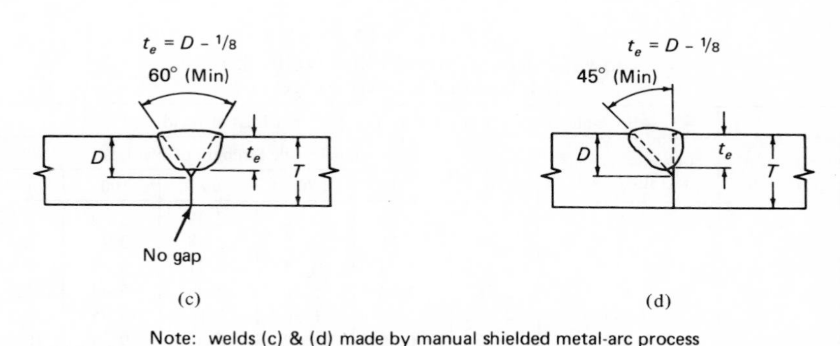

Note: welds (c) & (d) made by manual shielded metal-arc process

Fig. 5.13.1. Effective throat dimensions for groove welds (AISC–1.14.7).

depth of the groove less ⅛ in. but cannot be less than $\sqrt{t_t/6}$ where t_t is the thickness of the thinner piece being joined, as shown in Figs. 5.13.1c and d.

Fillet Welds. The effective throat dimension of fillet welds is the shortest distance from the root to the face of the weld, as shown in Fig. 5.13.2. Assuming the fillet weld to have equal legs of nominal size a, the effective throat, t_e, is $0.707a$. If the fillet weld is designed to be unsymmetrical (a rare situation) with unequal legs, as shown in Fig. 5.13.2b, the value of t_e

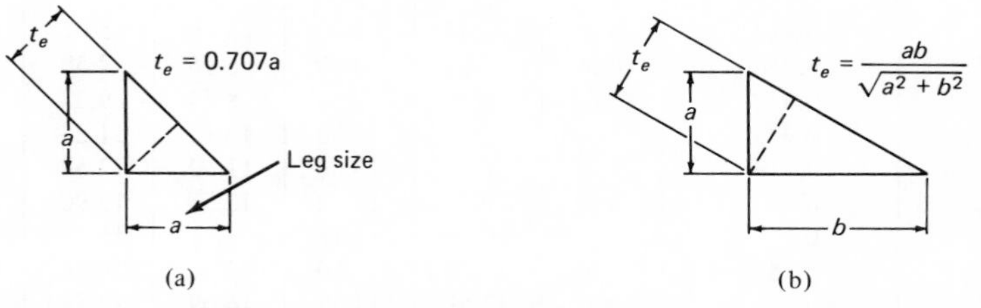

Fig. 5.13.2. Effective throat dimensions for fillet welds (except by submerged-arc process).

must be computed from the diagrammatic shape of the weld. The effective throat dimensions for fillet welds made by the submerged arc process are modified by AISC–1.14.7 as follows:

(a) For fillet welds with the leg size equal to or less than 3/8 in., the effective throat dimension shall be taken as equal to the leg size a.

(b) For fillet welds larger than 3/8 in., the effective throat dimension shall be taken as the theoretical throat dimension plus 0.11 in. (i.e., $0.707a + 0.11$).

Tables 5.13.1 and 5.13.2 summarize the effective throat dimensions and the allowable shear resistance R_w of fillet welds in kips per inch of weld.

TABLE 5.13.1
Allowable Resistance of Fillet Welds, kips/in.
(Metal Shielded-Arc Process)

	Allowable Shear in Fillet Welds, R_w, Kips per Inch of Weld						
Nominal Size, in.	Effective Throat (AISC–1.14.7)	Minimum Tensile Strength of Weld, ksi					
		60	70	80	90	100	110
1/8	0.088	1.59	1.86	2.12	2.39	2.69	2.92
3/16	0.132	2.38	2.78	3.18	3.58	3.97	4.37
1/4	0.177	3.18	3.71	4.24	4.77	5.30	5.83
5/16	0.221	3.98	4.64	5.30	5.96	6.63	7.30
3/8	0.265	4.77	5.57	6.36	7.16	7.95	8.75
7/16	0.309	5.57	6.49	7.42	8.35	9.28	10.21
1/2	0.353	6.36	7.42	8.48	9.54	10.60	11.66
9/16	0.398	7.16	8.35	9.54	10.74	11.93	13.12
5/8	0.442	7.95	9.28	10.61	11.93	13.26	14.58
11/16	0.486	8.75	10.21	11.67	13.12	14.58	16.04
3/4	0.530	9.54	11.13	12.72	14.31	15.91	17.50

TABLE 5.13.2
Allowable Resistance of Fillet Welds, kips/in.
(Submerged-Arc Process)

	Allowable Shear in Fillet Welds, R_w, Kips per Inch of Weld						
Nominal Size, in.	Effective Throat (AISC–1.14.7)	Minimum Tensile Strength of Weld, ksi					
		60	70	80	90	100	110
1/8	0.125	2.25	2.62	3.00	3.37	3.75	4.12
3/16	0.187	3.37	3.94	4.50	5.06	5.62	6.19
1/4	0.250	4.50	5.25	6.00	6.75	7.50	8.25
5/16	0.312	5.62	6.56	7.50	8.44	9.37	10.31
3/8	0.375	6.75	7.87	9.00	10.12	11.25	12.37
7/16	0.419	7.55	8.80	10.06	11.32	12.58	13.84
1/2	0.463	8.34	9.73	11.12	12.51	13.90	15.30
9/16	0.508	9.14	10.66	12.18	13.71	15.23	16.75
5/8	0.552	9.93	11.59	13.25	14.90	16.56	18.21
11/16	0.596	10.73	12.52	14.31	16.09	17.88	19.67
3/4	0.640	11.52	13.44	15.36	17.28	19.21	21.13

Plug and Slot Welds. The effective shearing area of plug or slot welds is their nominal area in the shearing plane. The resistance of plug or slot welds is the product of the nominal cross section times the allowable stress as discussed in Sec. 5.12.

EXAMPLE 5.13.1

Using AISC specs. determine the effective throat dimension of a $^7/_{16}$-in. fillet weld produced by (a) shielded metal-arc process and (b) submerged-arc process.

Solution

(a) $t_e = 0.707a = 0.707(0.4375) = 0.309$ in.
(b) $t_e = 0.707a + 0.11 = 0.707(0.4375) + 0.11 = 0.419$ in.

EXAMPLE 5.13.2

Using AISC specs. determine the allowable shear resistance of a $^3/_8$-in. fillet weld produced by (a) shielded metal-arc process, and (b) submerged-arc process. Assume minimum tensile strength of 70 ksi.

Solution

(a) $t_e = 0.707(0.375) = 0.265$ in.
$R_w = 0.2651(0.3)70 = 0.265(21) = 5.57$ kips/in.
(b) $t_e = 0.375$ in.
$R_w = 0.375(21.0) = 7.87$ kips/in.

The solution to Examples 5.13.1 and 5.13.2 may be checked by referring to Tables 5.13.1 and 5.13.2.

EXAMPLE 5.13.3

Determine the allowable capacity of a $^3/_4$ in. diam. plug weld using E70 electrode material. Use AISC specs.

Solution

Assuming the weld diameter satisfies the limitations of AISC–1.17.12 relating to the dimension of the piece in which the plug weld is made,

$$\text{Capacity} = 0.3f_{ts}(\pi D^2/4)$$
$$= 21(\pi)(0.75)^2/4 = 9.28 \text{ kips}$$

Maximum Effective Fillet Weld Size. In Sec. 5.11 the limitations on maximum and minimum fillet weld size and length relating to practical design considerations were given. Those requirements relate to the size of weld that is actually placed. Regarding strength, however, no fillet weld or welds of whatever size may be permitted in design to carry a greater load than permitted on the adjacent base material.

Consider the two lines of fillet weld transmitting the shear V across

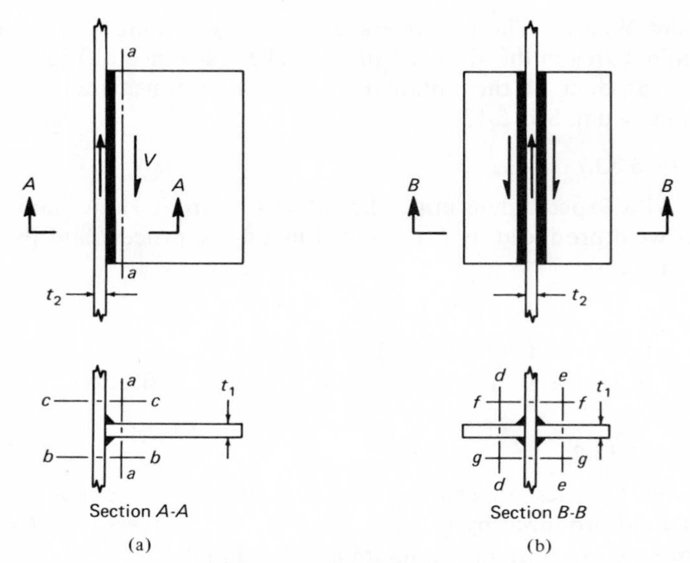

Fig. 5.13.3. Critical sections for possible overstressing of base material.

section *a–a* of Fig. 5.13.3a. Equating the capacity per inch of the weld to capacity per inch in shear (see AISC–1.5.1.2) on the plate gives

$$\overbrace{2a(0.707)(0.3f_{ts})}^{\text{weld}} = \overbrace{0.4F_y t_1}^{\text{plate}} \tag{5.13.1}$$

$$a_{\text{max eff}} = \frac{0.4F_y t_1}{2(0.707)(0.3f_{ts})} = 0.945\,\frac{F_y t_1}{f_{ts}} \tag{5.13.2}$$

where t_1 = thickness of base material
f_{ts} = tensile strength of electrode material (70 for E70 electrodes)
F_y = yield stress of base material

Sections *b–b* and *c–c* will not be critical since two lines of weld transfer load across two sections; maximum effective weld size for the transfer across *b–b* and *c–c* is

$$a(0.707)(0.3f_{ts}) = 0.4F_y t_2 \tag{5.13.3}$$

$$a_{\text{max eff}} = 1.89\,\frac{F_y t_2}{f_{ts}} \tag{5.13.4}$$

Considering the four fillets of Fig. 5.13.3b, sections *d–d* and *e–e* are the same as section *a–a* and Eq. 5.13.2 applies. On sections *f–f* and *g–g* four fillets are transferring load across two sections:

$$4a(0.707)(0.3f_{ts}) = 2(0.4F_y)t_2$$

and the result is again, Eq. 5.13.2.

EXAMPLE 5.13.4

Determine the capacity per inch of weld, R_w, to be used in design of the flange to web connection in Fig. 5.13.4. The plates are A36 steel and E70 electrodes are to be used with (a) the shielded metal-arc process, and (b) the submerged-arc process.

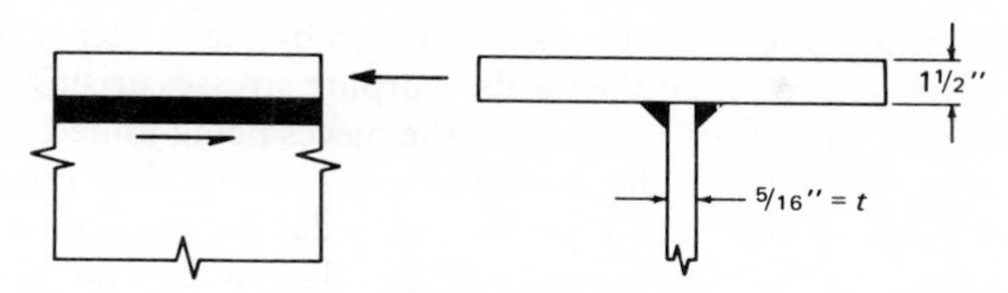

Fig. 5.13.4. Example 5.13.4.

Solution

Minimum weld size = a_{min} = $\frac{5}{16}$ (AISC–1.17.5)

(a) Shielded metal-arc process. Equation 5.13.2 applies

$$a_{\text{max eff}} = 0.945\,\frac{F_y t_1}{f_{ts}} = 0.945\,\frac{36(t)}{70} = 0.485t$$

$$= 0.485(0.3125) = 0.152 \text{ in.}$$

$$R_w = 2(0.152)(0.707)21 = 4.51 \text{ kips/in. for 2 fillets}$$

Even though a $\frac{5}{16}$-in. fillet must be placed, its strength in design may not exceed the strength assuming $a = 0.152$ in.

(b) Submerged arc process. Here the throat dimension may equal the leg size. Equating weld strength to plate strength gives

$$2(a)(21) = 0.4(36)t$$

$$a_{\text{max eff}} = 0.342t = 0.342(\tfrac{5}{16}) = 0.107 \text{ in.}$$

$$R_w = 2(0.107)21 = 4.51 \text{ kips/in.}$$

Again, the same result is obtained as in (a) since 4.51 kips/in. is the allowable capacity of the web plate, and is well below the actual weld capacity.

5.14. STRESS DISTRIBUTION IN FILLET WELDS

The procedure for designing welded connections uses as criteria of safety nominal stresses in the welds. Just as for bolted and riveted connections, nominal stresses for welds assumed the weld is elastic and the plates making up the joint are rigid. If a welded joint is subjected to pure shear, compression, or tension, the stresses in the welds are assumed to be uniform over the length of the welds. If a weld is subjected to pure moment or pure torsion, the stresses are assumed to vary linearly depending on location of the neutral axis. If two or more of the previous design conditions exist simultaneously the stresses are assumed to be their vector sum.

However, the actual stress distribution in a welded connection is complex even in the simplest joint. It must be recognized that the welds *and* the pieces being joined must deform together, otherwise a separation will occur. In addition, the true stresses are altered by the existence of residual stresses due to the cooling of the welds, warping stresses arising from poor weld procedures, and stress relieving in the pieces being joined. A detailed and quantitative study of the actual stresses in welded connection is beyond the scope of this text. However, in order for the reader to obtain an appreciation for the complexity of the problem, a few common types of joints will be shown.

Figure 5.14.1 shows the typical shear distribution in longitudinal fillet

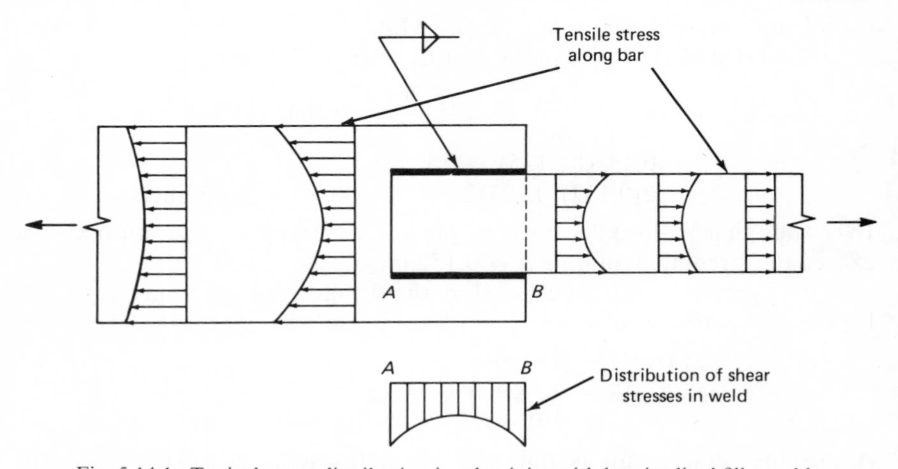

Fig. 5.14.1. Typical stress distribution in a lap joint with longitudinal fillet welds.

welds. The actual magnitude of the variation from points A to B will depend upon length of the weld as well as the ratio of widths of the two plates being joined. Figure 5.14.2 shows the typical shear distribution for transverse fillet welds. Again the magnitude of the shear distribution will depend upon the relative width of the plates and length of the weld.

The stress distributions in fillet welds used to connect tee-joints is somewhat more complex as indicated in Fig. 5.14.3. Due to the tendency of the fillet to rotate about point C, the maximum tensile stress in the y-direction, f_y, is approximately 4 times the nominal or average stress $f_{y\,avg}$.

If an exact analysis were required each time a connection was designed, very few welded connections would be made. In order to eliminate the need for such an exact analysis, a factor of safety is used to

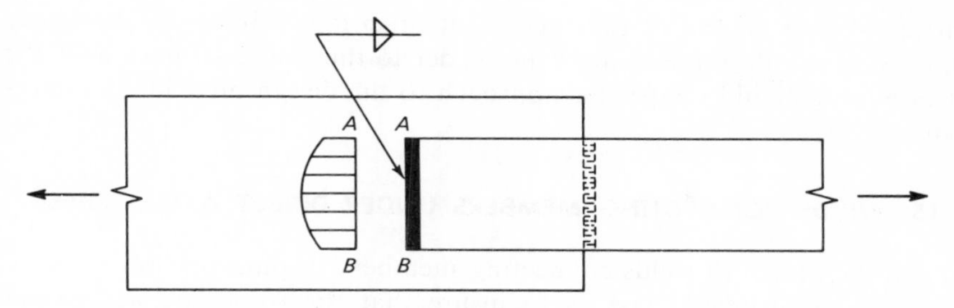

Fig. 5.14.2. Typical stress distribution in a lap joint with transverse fillet welds.

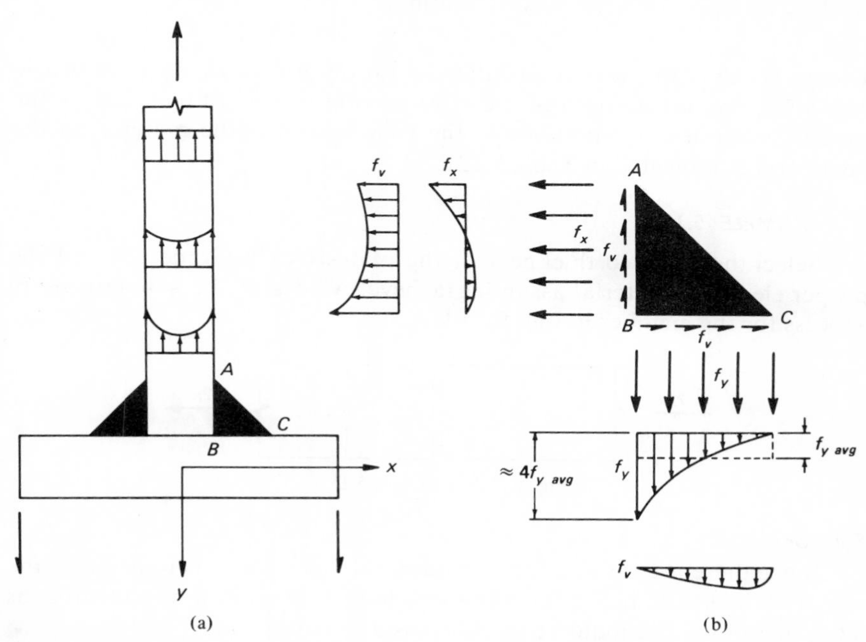

Fig. 5.14.3. Typical stress distribution in a tee joint with fillet welds (shear transverse to fillet).

arrive at the allowable nominal stresses based on the manner of computing nominal stress and the desired relationship between service load and ultimate strength. By this logic, the designer is able to make a reasonable analysis on the basis of nominal stresses and select appropriately sized welds with confidence. The fact that the welds will deform plastically after yielding and thus redistribute unusually high stresses is analogous to the assumptions made for rows of bolts as discussed in Sec. 4.5.

It would serve little purpose to develop an "exact" analysis for welded joints, since in most engineering design problems the loading is virtually never known to that degree of precision. However, the above discussion should serve to alert the reader to the true distribution of the stresses and should temper his approach to the design of welded connections.

5.15. WELDS CONNECTING MEMBERS UNDER DIRECT AXIAL STRESS

In the design of welds connecting members in pure tension or compression, the principal task is to insure that the welds are at least as strong as the members they connect and that the connection does not introduce significant eccentricity of loading.

Groove Welds. In the case of full-penetration groove welds as shown in Fig. 5.5.2, the full strength of the cross section is insured by selecting the proper electrode corresponding to the base material (the material in the members) as indicated in Table 5.12.1.

EXAMPLE 5.15.1

Select the required thickness of the plates (A572 Grade 55) and the proper electrode material assuming a bevel weld and the submerged-arc process for the member in Fig. 5.15.1.

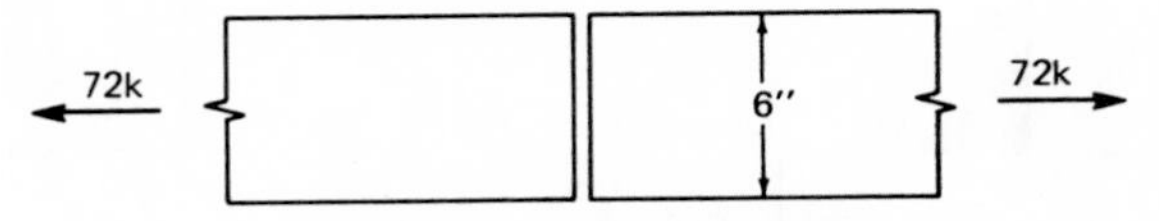

Fig. 5.15.1. Example 5.15.1.

Solution

Since there are no holes to consider, the net section is 6 in. times the required thickness, t_{reqd}. The allowable tensile stress F_t in the member is $0.6F_y$ or 33 ksi. The required thickness is, therefore,

$$t_{reqd} = \frac{T}{6F_t} = \frac{72}{6(33)} = 0.364 \text{ in., use Ps } \tfrac{3}{8} \times 6$$

From Table 5.12.1, use F7X-EXXX flux electrode combination and a full-penetration bevel weld.

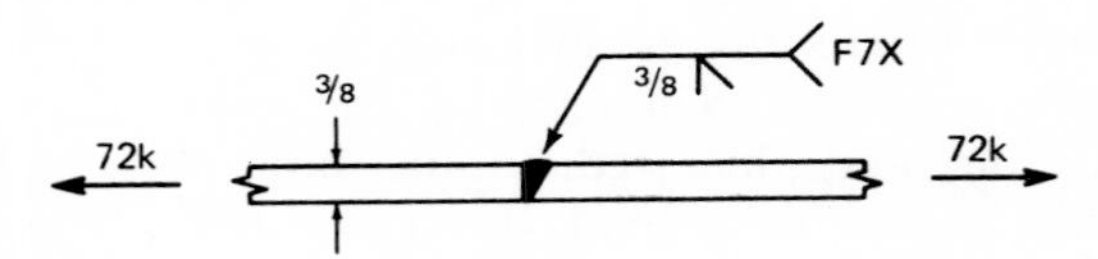

EXAMPLE 5.15.2

Repeat Example 5.15.1 using A572 Grade 65 plates and a square-groove weld with a zero root opening.

Solution

$$t_{\text{reqd}} = \frac{72}{6(0.6)F_y} = \frac{72}{6(0.6)65} = 0.308 \text{ in., use PLs } ^5\!/_{16} \times 6$$

From Table 5.12.1, use Grade 80 submerged-arc process.

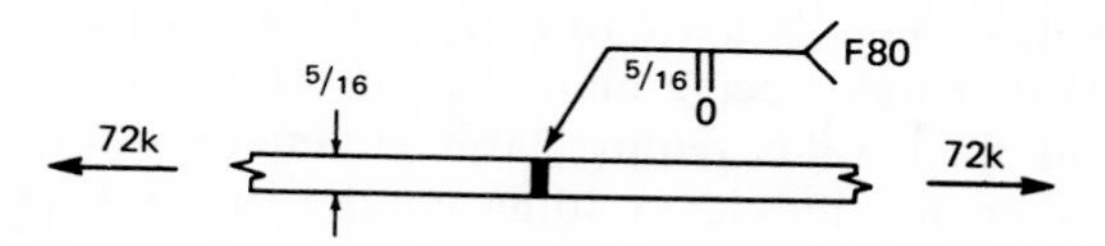

Note: In Examples 5.15.1 and 5.15.2 it was not necessary to include the length of the welds as shown below since they are to be made the full width of the plates unless otherwise specified.

3/8 6 F7X or 5/16 0 6 F80

EXAMPLE 5.15.3

Determine the capacity of the tee connection shown below and detail the proper double-bevel weld, assuming a shielded metal-arc process. Assume the flange of the tee does not control design.

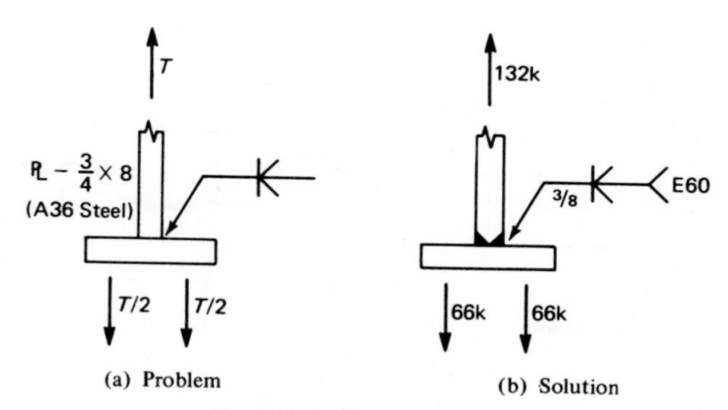

(a) Problem (b) Solution

Fig. 5.15.2. Example 5.15.3.

Solution

The allowable tensile force is

$$T = 8(0.75)22 = 132 \text{ kips}$$

From Table 5.12.1, use E60 electrodes.

Note: On the basis of strength only, a single ¾-in. bevel could have been used instead of the double-bevel weld specified. However, welding the stem of the tee from one side only may cause excessive warping and introduces eccentricity into the connection.

Fillet Welds. The design procedure for fillet welds is based on the nominal shear stress of the fillet weld on the effective area as discussed in Sec. 5.13. The selection of the size of the fillet weld is based on the thickness of the pieces being joined and the available length over which the fillet welds can be made. Other factors such as the type of welding equipment used, whether the welds are to be made in the field or in the shop, and the size of other welds being made will also influence the size of fillet specified. Large fillet welds require larger diameter electrodes which in turn require larger and bulkier welding equipment, not necessarily convenient for field use. The most economical size of fillet weld made in the field is about $\frac{5}{16}$ in. Also, if a certain size of fillet weld is used in adjacent areas to the particular joint in question, it is advisable to use the same size since then the same electrodes and welding equipment could be used and the welder would not have to alter his procedure to accommodate a larger or smaller weld. In addition, inspection of the welds is further simplified.

EXAMPLE 5.15.4

Determine the size and length of the fillet weld for the lap joint shown below assuming the submerged-arc process and the plates are A36 steel.

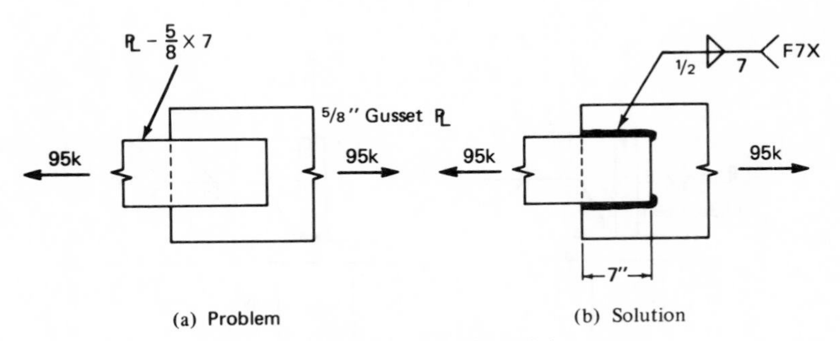

Fig. 5.15.3. Example 5.15.4.

Solution

Referring to Sec. 5.11,

$$\text{Maximum size} = \tfrac{5}{8} - \tfrac{1}{16} = \tfrac{9}{16}$$
$$\text{Minimum size} = \tfrac{1}{4}$$

Assuming that the connection is to be shop-welded, use $\frac{1}{2}$ in. fillet.

From AISC–1.14.7, since the nominal size of fillet weld is over $\frac{3}{8}$ in., the effective throat dimension is equal to the theoretical throat plus 0.11 in.

$$\text{Effective throat} = 0.707(0.50) + 0.11 = 0.4635 \text{ in.}$$

From Table 5.12.3 use F7X-EXXX flux electrode combination.

From Table 5.12.3, the capacity of ½-in. fillet weld per inch of length is

$$R_w = \text{(Effective throat)(allowable stress)}$$
$$= 0.4635(21.0) = 9.73 \text{ kips/in.}$$

Length of weld, L_w, required is

$$L_w = \frac{95}{9.73} = 9.76 \text{ in., use ½-in. fillet, 7 in. on each side.}$$

Fillet length is extended to satisfy AISC–1.17.7 (par. 2).

EXAMPLE 5.15.5

Rework Example 5.15.4 using ¼-in. fillet welds.

Solution

Since the nominal size is less than ⅜ the full leg dimension may be used as the effective throat dimension for submerged-arc welding.

$$R_w = 0.250(21.0) = 5.25 \text{ kips/in.}$$
$$\text{Reqd } L_w = \frac{95}{5.25} = 18.1 \text{ in., say 19 in.}$$

Two possible solutions are shown in Fig. 5.15.4, both of which provide 19 in. of ¼-in. fillet welds.

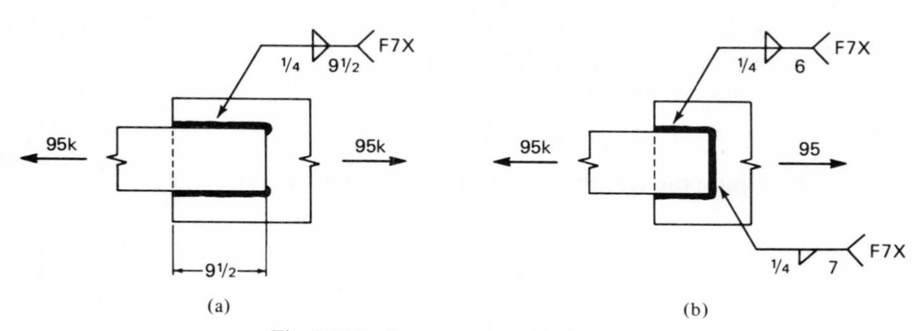

Fig. 5.15.4. Solutions to Example 5.15.5.

The solution in Fig. 5.15.4b is preferred since it is more compact and reduces the overall size of the connection.

In a great number of cases, members subjected to direct axial stress are themselves unsymmetrical and cause eccentricities in welded connections. Consider the angle tension member shown in Fig. 5.15.5 welded as indicated. The force T, applied at some distance from the connection will act along the centroid of the member as shown. The force T will be resisted by the forces F_1, F_2, and F_3 developed by the weld lines. The forces F_1 and F_3 are assumed to act at the top and bottom edges, respectively, of the angle shown. The force F_2 will act at the centroid of the weld which is located at $d/2$. Taking moments about some point A

located on the bottom edge of the member and considering clockwise moments positive,

$$\Sigma M_A = -F_1 d - F_2 d/2 + Ty = 0 \tag{5.15.1}$$

or

$$F_1 = \frac{Ty}{d} - \frac{F_2}{2} \tag{5.15.2}$$

The force F_2 is equal to the resistance of the weld per inch times the length of the weld:

$$F_2 = R_w L_w \tag{5.15.3}$$

Considering horizontal force equilibrium gives

$$\Sigma F_H = T - F_1 - F_2 - F_3 = 0 \tag{5.15.4}$$

Solving Eqs. 5.15.1 and 5.15.4 simultaneously gives

$$F_3 = T\left(1 - \frac{y}{d}\right) - \frac{R_2}{2} \tag{5.15.5}$$

Designing the connection shown in Fig. 5.15.5 to eliminate eccentricity caused by the unsymmetrical weld is called *balancing the welds.* The procedure for balancing the welds may be summarized as follows:

1. After deciding on the proper weld size and electrode, compute the force resisted by the end welds F_2 (if any) on the basis of Eq. 5.15.3.
2. Compute F_1 on the basis of Eq. 5.15.2.
3. Compute F_3 on the basis of Eq. 5.15.5, or

$$F_3 = T - F_1 - F_2 \tag{5.15.6}$$

4. Compute the lengths, L_{w1} and L_{w3}, on the basis of

$$L_{w1} = \frac{F_1}{R_w} \tag{5.15.7a}$$

and

$$L_{w3} = \frac{F_3}{R_w} \tag{5.15.7b}$$

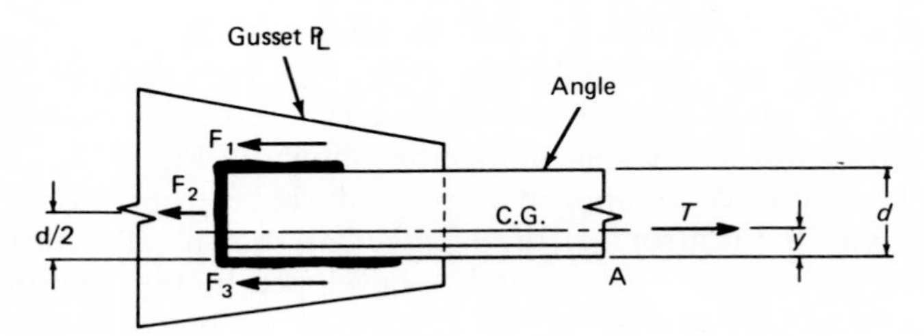

Fig. 5.15.5.

It is noted that approximately balanced welds are desirable; however, unless the member is subject to repeated variations in stress, balanced connections for "single angle, double angle, and similar type members is not required" by AISC–1.15.3.

EXAMPLE 5.15.6.

Design the fillet welds to develop the full strength of the angle shown in Fig. 5.15.6 minimizing the effect of eccentricity. Assume the gusset plate does not govern and the shielded metal-arc process is used.

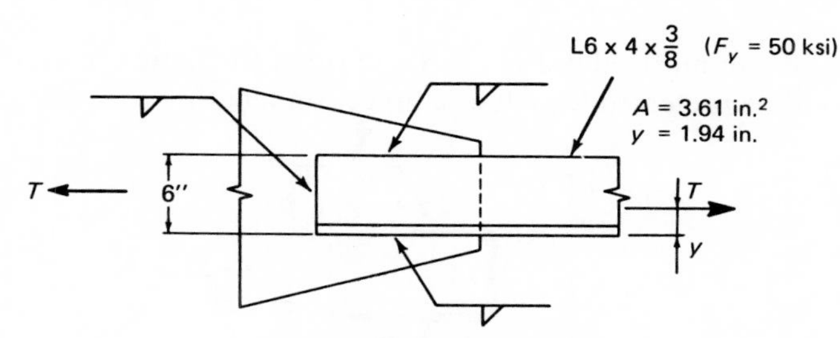

Fig. 5.15.6. Example 5.15.6.

Solution

Using the forces as in Fig. 5.15.7,

$$T = 0.6F_yA = 0.6(50)3.61 = 108.3 \text{ kips}$$

Minimum size fillet weld = $\frac{3}{16}$ in. (Table 5.11.1)

Maximum size fillet weld = $\frac{5}{16}$ in. (Fig. 5.11.2)

Use $\frac{3}{16}$ in. fillet weld with E70 electrodes.

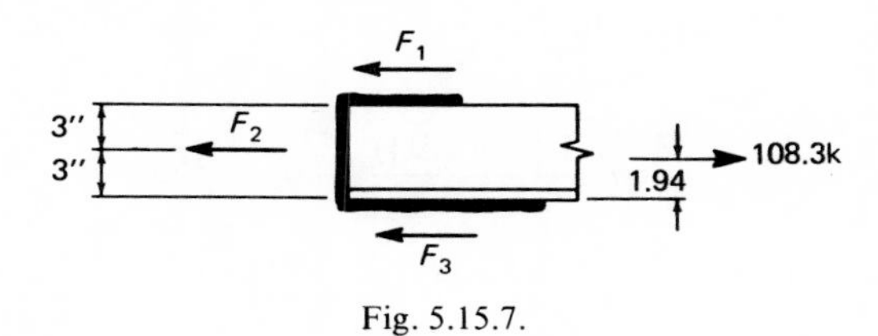

Fig. 5.15.7.

From Eq. 5.15.3,

$$F_2 = R_wL_w = \tfrac{3}{16}(0.707)(21.0)6 = 2.79(6) = 16.7 \text{ kips}$$

From Eq. 5.15.2,

$$F_1 = \frac{Ty}{d} - \frac{F_2}{2} = \frac{108.3(1.94)}{6} - \frac{16.7}{2} = 26.6 \text{ kips}$$

From Eq. 5.15.6

$$F_3 = T - F_1 - F_2 = 108.3 - 26.6 - 16.7 = 65.0 \text{ kips}$$

From Eqs. 5.15.7a and 5.15.7b,

$$L_{w1} = \frac{F_1}{R_w} = \frac{26.6}{2.79} = 9.54 \text{ in., say 10 in.}$$

and

$$L_{w3} = \frac{F_3}{R_w} = \frac{65.0}{2.79} = 23.3 \text{ in., say 24 in.}$$

Use welds as summarized in Fig. 5.15.8, though for better economy larger welds would be preferred to reduce connection length.

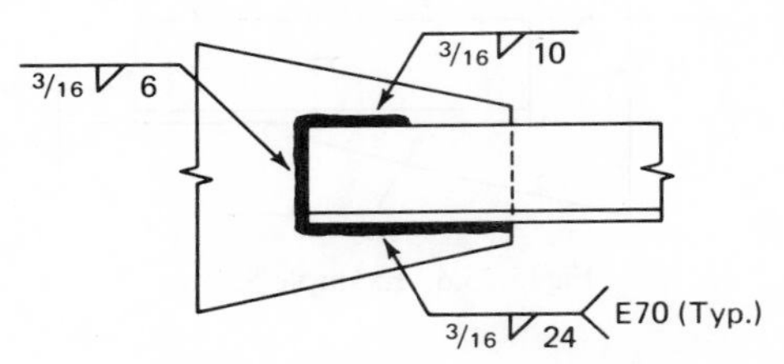

Fig. 5.15.8. Solution for Example 5.15.6.

EXAMPLE 5.15.7.

Rework Example 5.15.6 if the weld at the end of the angle is omitted, and submerged-arc process is used instead of the shielded metal-arc process.

Solution

Again, try $\frac{3}{16}$-in. weld on the $F_y = 50$ ksi base material. Using the forces in Fig. 5.15.9,

$$F_1 = \frac{Ty}{d} - \frac{F_2}{2} = \frac{108.3(1.94)}{6} - 0 = 35.0 \text{ kips}$$

$$F_3 = T - F_1 - F_2 = 108.3 - 35.0 - 0 = 73.3 \text{ kips}$$

$$R_w = (\tfrac{3}{16})(21) = 3.94 \text{ kips/in.}$$

$$L_{w1} = \frac{35.0}{3.94} = 8.9 \text{ in., say 9 in.}$$

$$L_{w3} = \frac{73.3}{3.94} = 18.65 \text{ in., say 19 in.}$$

The length of L_{w3} requires an excessively large joint which probably is uneconomical. Therefore it is advisable to use a larger size fillet weld.

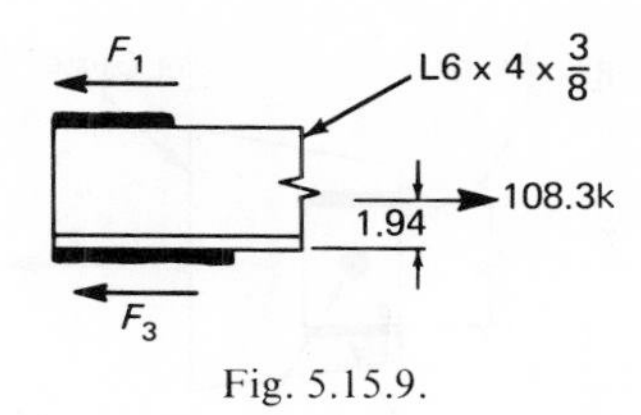

Fig. 5.15.9.

Try a $\frac{5}{16}$-in. weld.

Effective throat dimension $= \frac{5}{16}$ (submerged arc process)

The forces F_1 and F_3 are independent of the weld size and the new required lengths are

$$L_{w1} = \frac{35.0}{\frac{5}{16}(21.0)} = 5.33 \text{ in.,} \quad \text{say } 5\tfrac{1}{2} \text{ in.}$$

$$L_{w3} = \frac{73.3}{\frac{5}{16}(21.0)} = 11.17 \text{ in.,} \quad \text{say } 11\tfrac{1}{2} \text{ in.}$$

Use welds as summarized in Fig. 5.15.10.

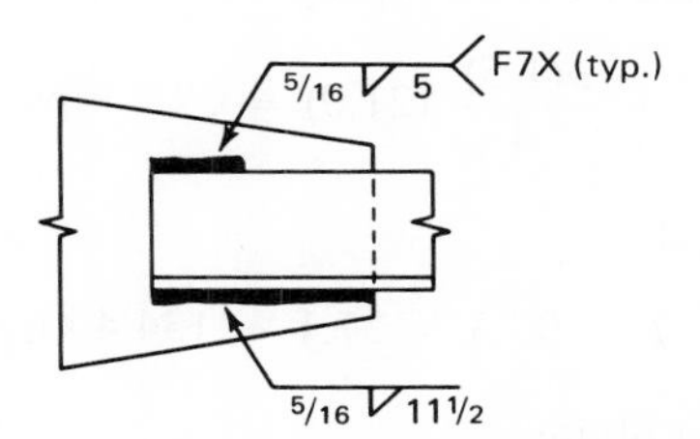

Fig. 5.15.10. Solution for Example 5.15.7.

Slot and Plug Welds. Slot and plug welds depend upon their shear strength in the shearing plane between the plates being joined. As indicated in Sec. 5.7, their principal use is in lap joints. Plug welds are also occasionally used to fill up holes in connections, such as beam to column angles where temporary erection bolts had been placed to align the members prior to welding. This latter usage is generally not considered to be structural although in certain cases, the designer may consider the strength of the plug weld in designing the rest of the welds in a given connection. As a rule, plug and slot welds will be designed to work together with other welds, usually fillet welds, in lap joints as shown in Fig. 5.5.5.

EXAMPLE 5.15.8.

Assuming A36 steel determine the value of T permitted by AISC on the connection in Fig. 5.15.11.

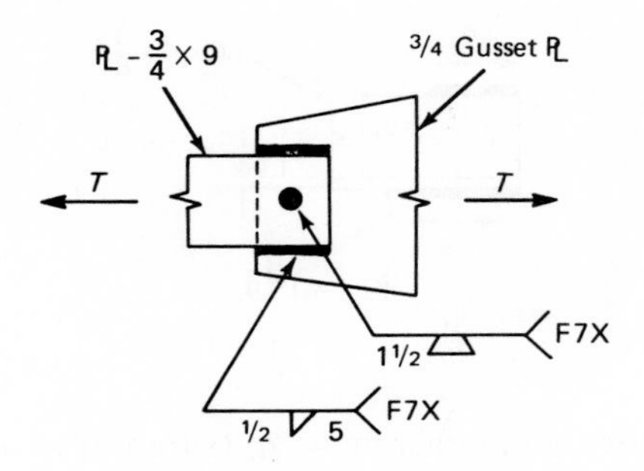

Fig. 5.15.11. Example 5.15.8.

Solution

The resistance per inch supplied by the ½-in. fillet welds, R_w, is

$$R_w = [0.5(0.707) + 0.11](21) = 9.73 \text{ kips/in.}$$

The total resistance provided by the fillet welds

$$= 10(9.73) = 97.3 \text{ kips}$$

The resistance provided by the 1½ plug weld

$$= \frac{\pi(1.5)^2}{4}(21.0) = 37.1 \text{ kips}$$

The value of T is equal to

$$T = 97.3 + 37.1 = 134.4 \text{ kips}$$

Check capacity of plate:

$$T = 9(0.75)22 = 148.5 \text{ kips} > 134.4\text{k} \qquad \text{OK, } \underline{T = 134.4 \text{ kips}}$$

EXAMPLE 5.15.9.

Compute the allowable capacity of the connection shown in Fig. 5.15.12 according to AISC, assuming A572 Grade 50 steel.

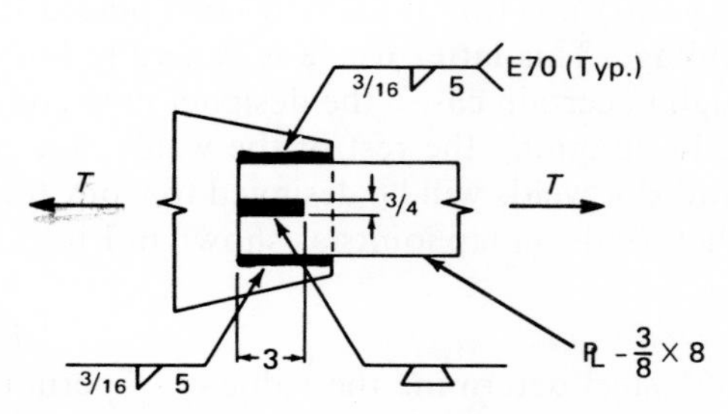

Fig. 5.15.12. Example 5.15.9.

Solution

From Table 5.13.1, a $\frac{3}{16}$ in. E70 fillet weld provides 2.78 k/in. $[(\frac{3}{16})(0.707)(21)]$. The total resistance provided by the fillet welds

$$= 2(5)2.78 = 27.8 \text{ kips}$$

The resistance provided by the slot weld

$$= \tfrac{3}{4}(3)(21) = 47.3 \text{ kips}$$

The value of T is given by

$$T = 27.8 + 47.0 = 74.8 \text{ kips}$$

Check capacity of plate:

$$T = 8(\tfrac{3}{8})30 = 90 \text{ kips} > 74.8\text{k}; \quad \text{thus } \underline{T = 74.8 \text{ kips}}.$$

EXAMPLE 5.15.10.

Design an end connection to develop the full tensile value of an C8 × 13.75 in a lap length of 5 inches. The channel of A572 Grade 50 steel is connected to a $\frac{3}{8}$-in. plate and fillet welds are limited to $\frac{3}{8}$ in. and are to be made by the shielded metal-arc process. Use AISC specifications.

Solution

Channel capacity $= P = F_t A = 0.6(50)4.02 = 120.5$ kips

(a) Fillet weld. Use E70 electrodes.

Minimum $a = \frac{3}{16}$ in. (AISC–1.17.5)

Maximum $a = 0.303 - \frac{1}{16} = 0.24$, say $\frac{1}{4}$ in. (AISC–1.17.6)

While $\frac{1}{4}$-in. weld must be used on one end along the channel web, $\frac{3}{8}$-in. weld could be used along the flanges. It is better not to mix the fillet sizes, so try $\frac{1}{4}$ in. all around.

$$R_w = \tfrac{1}{4}(0.707)(21) = 3.71 \text{ kips/in.}$$

$$\text{Reqd } L_w = \frac{P}{R_w} = \frac{120.5}{3.71} = 32.5 \text{ in.}$$

Since the length all around is only 26 in.; additional capacity from fillet welds in a large slot, slot welds, or plug welds, is necessary.

(b) Slot Weld. Try a slot weld in accordance with AISC–1.17.12.

$$\text{Min width of slot} = \left(t + \frac{5}{16}\right) \text{[rounded to next odd } \tfrac{1}{16} \text{ in.]}$$

$$= 0.303 + 0.3125 = 0.6155$$

$$\text{[rounded to } \tfrac{11}{16} \text{ in.]}$$

$$\text{Max width of slot} = 2\frac{1}{4}\,a = 2\frac{1}{4}\left(\frac{1}{4}\right) = 0.56 = \frac{9}{16}\text{ in.}$$

$$\text{Load to be carried by slot} = 120.5 - (26 - 0.56)3.71 = 26.0\text{ kips}$$

Try $^{9}/_{16}$ in. width of slot:

$$\text{Length reqd} = \frac{26.0}{(^{9}/_{16})(21)} = 2.2\text{ in.}$$

Max length of slot $= 10a = 10(^{1}/_{4}) = 2.25$ in.

Use a slot weld $^{9}/_{16} \times 2^{1}/_{4}$. The final design is shown in Fig. 5.15.13.

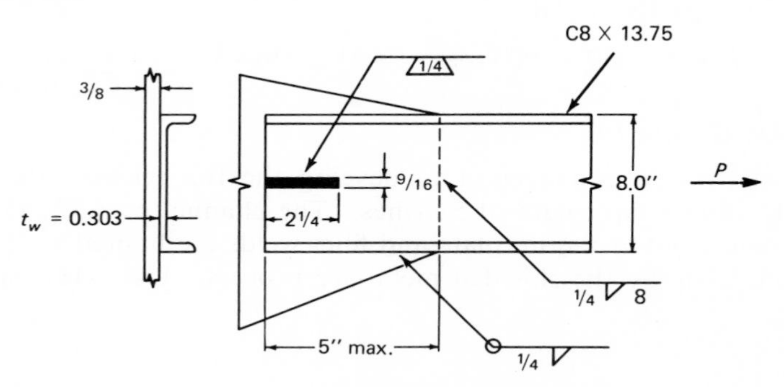

Fig. 5.15.13. Solution for Example 5.15.10.

5.16. ECCENTRIC WELDED CONNECTIONS

The versatility and ease of welding make welded connections desirable for a wide variety of usages. In actual practice, welded connections often use a group of welds rather than a single weld, the result being that the welds may be subjected to more than one type of loading. Examples of various types of loading are shown in Fig. 5.16.1. As in the case of eccentric bolted connections (Sec. 4.7), the exact elastic analysis of the stresses in eccentric welded connections is impractical. In addition to the complex stress distribution discussed in Sec. 5.14, the great variety of possible combinations of loading and grouping of welds makes it necessary to evaluate the adequacy of a connection on the basis of nominal stresses.

The general procedure for investigating the nominal stress in weld groups is based on the general assumptions discussed in Sec. 5.14 and the principles of mechanics. The procedure, in brief, consists of the following steps:

1. Establish the *effective* throat dimension t_e and draw the effective cross section of the weld group.
2. Establish a coordinate system and determine the centroid of the weld group.

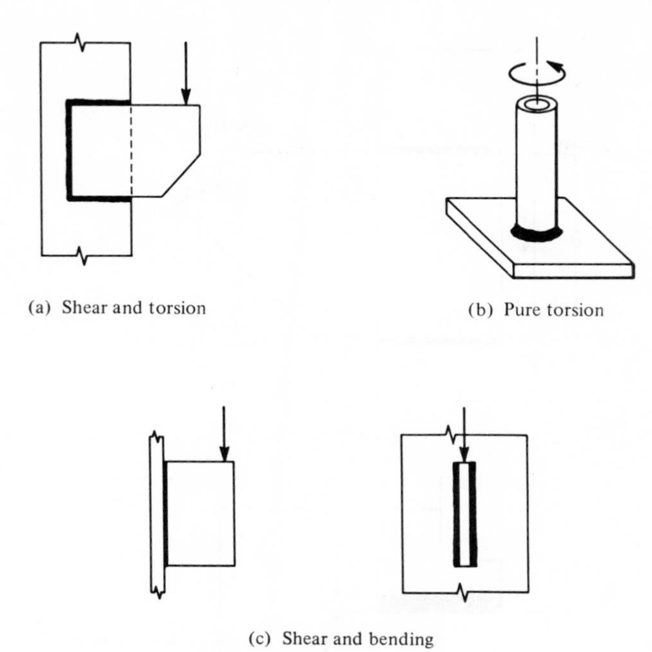

(a) Shear and torsion (b) Pure torsion

(c) Shear and bending

Fig. 5.16.1. Types of eccentric loading.

3. Determine the forces acting on the weld group.
4. Compute the individual stresses in the welds at critical points resulting from direct shear, torsion, and moment.
5. Combine the individual stresses vectorically.

The general procedure outlined above is illustrated in the examples that follow.

Eccentric Shear (Shear and Torsion). In order to expand the general procedure for combined shear and torsion, consider the connection shown in Fig. 5.16.2a. The effective cross section and the applied force system are shown in Fig. 5.16.2b. In developing the procedure the following notation will be used:

$$f' = \frac{P}{A} = \text{stress due to direct shear} \tag{5.16.1}$$

$$f'' = \frac{Tr}{I_p} = \text{stress due to torsional moment} \tag{5.16.2}$$

where r = radial distance from the centroid to point of stress

I_p = polar moment of inertia

For computing nominal stresses the *locations* of the lines of weld are defined by edges along which the fillets are placed, rather than to the center

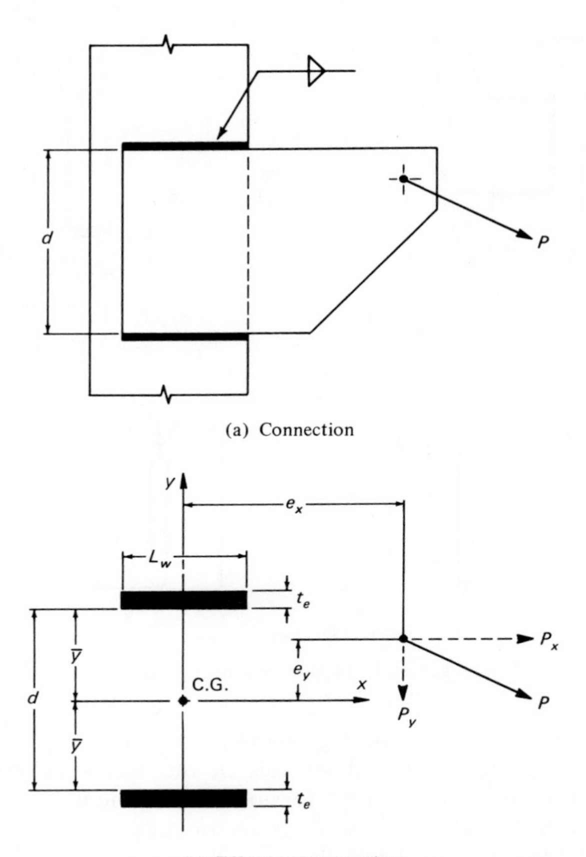

(a) Connection

(b) Effective cross section

Fig. 5.16.2. Eccentric bracket connection.

of the effective throat. This makes little difference, since the throat dimension is usually small. As discussed for eccentrically loaded bolted connections it is convenient to use a cartesian coordinate system and compute x and y components of the forces on welds.

For the general case shown in Fig. 5.16.2, the components of stress due to direct shear and torsion are

$$f'_x = \frac{P_x}{A} \tag{5.16.3a}$$

$$f'_y = \frac{P_y}{A} \tag{5.16.3b}$$

$$f''_x = \frac{Ty}{I_p} = \frac{(P_x e_y + P_y e_x)y}{I_p} \tag{5.16.4a}$$

$$f_y'' = \frac{Tx}{I_p} = \frac{(P_x e_y + P_y e_x)x}{I_p} \tag{5.16.4b}$$

where $I_p = I_x + I_y = \Sigma I_{xx} + \Sigma A\bar{y}^2 + \Sigma I_{yy} + \Sigma A\bar{x}^2$ (5.16.5)

In Eq. 5.16.5, $\bar{x}$ and $\bar{y}$ refer to distances from the center of gravity of the weld group to the center of gravity of the individual weld segments. I_{xx} and I_{yy} refer to the moments of inertia of the individual segments with respect to their own centroidal axes.

Thus, for the situation of Fig. 5.16.2, Eq. 5.16.5 becomes

$$\begin{aligned} I_p &= 2\left[\frac{L_w(t_e)^3}{12}\right] + 2[L_w(t_e)(\bar{y})^2] + 2\left[\frac{t_e(L_w)^3}{12}\right] \\ &= \frac{t_e}{6}[L_w(t_e)^2 + 12L_w(\bar{y})^2 + L_w^3] \end{aligned} \tag{5.16.6}$$

For practical situations, the first term of Eq. 5.16.6 is neglected because, with t_e small, the term is not significant compared to the other terms. Hence

$$I_p \approx \frac{t_e}{6}[12L_w(\bar{y})^2 + L_w^3] \tag{5.16.7}$$

Then, after evaluating the torsional components in accordance with Eqs. 5.16.4, the x and y *components* of the resultant stress are

$$f_x = f_x' + f_x'' \tag{5.16.8a}$$

$$f_y = f_y' + f_y'' \tag{5.16.8b}$$

and the resultant stress f_r is

$$\begin{aligned} f_r &= \sqrt{(f_x)^2 + (f_y)^2} \\ &= \sqrt{(f_x' + f_x'')^2 + (f_y' + f_y'')^2} \end{aligned} \tag{5.16.9}$$

The resultant stress f_r is considered to be a nominal shear stress on the effective throat area of fillet welds. AISC–1.5.3 gives the allowable values as discussed in Sec. 5.12. The safe capacity R_w of a fillet weld per inch of weld is

$$R_w = (t_e)\,(\text{allowable unit stress})$$

or

$$\frac{R_w}{t_e} = \text{allowable unit stress}$$

For satisfactory safety, it is required from analysis that

$$f_r \le \frac{R_w}{t_e} \tag{5.16.10}$$

For *investigating* stresses in weld groups, such as in Fig. 5.16.2, I_p may be computed as illustrated in obtaining Eq. 5.16.7. Note that the throat dimension t_e may be factored out of Eq. 5.16.7; this may always be done. Following through the computation of stress components and obtaining f_r, one will find that t_e appears as a denominator multiplier in the resultant stress f_r. By multiplying f_r by t_e, a resultant load per inch is obtained which may be compared with allowable R_w to establish whether or not safety is adequate. In other words, one may frequently desire to use Eq. 5.16.10 in the form

$$f_r t_e \leq R_w \tag{5.16.11}$$

When *designing* welds, the throat dimension t_e is frequently unknown and is to be solved for. In such cases it is usually convenient to compute I_p assuming $t_e = 1$; thus one may think of having computed $f_r t_e$, the actual load per inch on the weld group. Thus, for design, it is required that

$$t_e \geq \frac{f_r}{\text{allowable stress}} \tag{5.16.12}$$

where f_r is in *ksi units*. Or Eq. 5.16.12 may be written

$$t_e \geq \frac{f_r(1)}{R_w} \tag{5.16.13}$$

where $f_r(1) = f_r t_e$ with $t_e = 1$, and is in *kips per inch units*.

Treating the welds making up the effective cross section in Fig. 5.16.2 as line welds (i.e., as in deriving Eq. 5.16.7 with $t_e = 1$) and using the general terms b and d, as shown in Fig. 5.16.3, Eq. 5.16.7 becomes

$$I_p \approx \frac{1}{6}\left[12b\left(\frac{d}{2}\right)^2 + b^3\right] = \frac{b}{6}[3d^2 + b^2] \tag{5.16.14}$$

Table 5.16.1 gives I_p values treated as properties of lines for other common weld configurations.

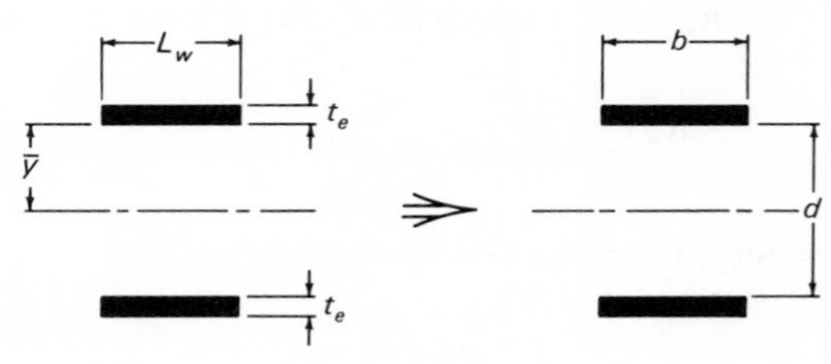

Fig. 5.16.3.

TABLE 5.16.1
Properties of Welds Treated as Lines

Section b = width; d = depth	Section Modulus $I_x/\bar{y}$	Polar Moment of Inertia, I_p about Center of Gravity
1.	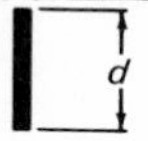 $S = \frac{d^2}{6}$	$I_p = \frac{d^3}{12}$
2.	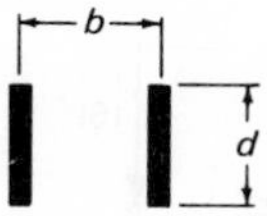 $S = \frac{d^2}{3}$	$I_p = \frac{d(3b^2 + d^2)}{6}$
3.	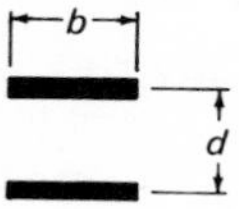 $S = bd$	$I_p = \frac{b(3d^2 + b^2)}{6}$
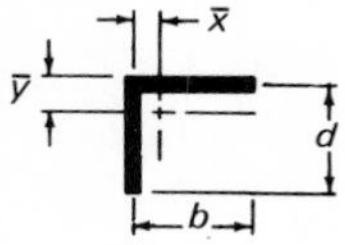 4. $\bar{y} = \frac{d^2}{2(b + d)}$ $\bar{x} = \frac{b^2}{2(b - d)}$	$S = \frac{4bd + d^2}{6}$	$I_p = \frac{(b + d)^4 - 6b^2d^2}{12(b + d)}$
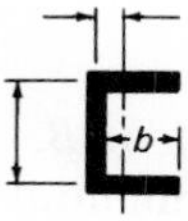 5. $\bar{x} = \frac{b^2}{2b + d}$	$S = bd + \frac{d^2}{6}$	$I_p = \frac{8b^3 + 6bd^2 + d^3}{12} - \frac{b^4}{2b + d}$
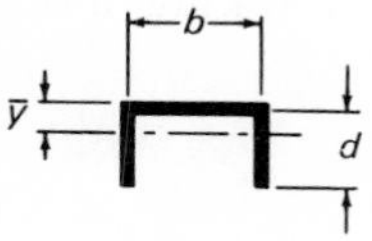 6. $\bar{y} = \frac{d^2}{b + 2d}$	$S = \frac{2bd + d^2}{3}$	$I_p = \frac{b^3 + 6b^2d + 8d^3}{12} - \frac{d^4}{2d + b}$
7.	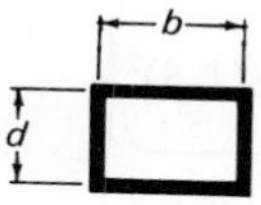 $S = bd + \frac{d^2}{3}$	$I_p = \frac{(b + d)^3}{6}$
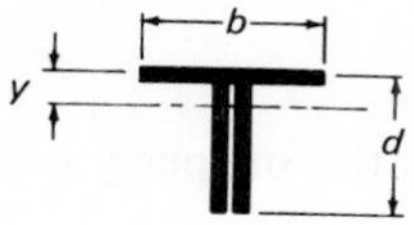 8. $\bar{y} = \frac{d^2}{b + 2d}$	$S = \frac{2bd + d^2}{3}$	$I_p = \frac{b^3 + 8d^3}{12} - \frac{d^4}{b + 2d}$
9.	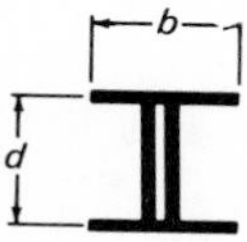 $S = bd + \frac{d^2}{3}$	$I_p = \frac{b^3 + 3b^2 + d^3}{6}$
10.	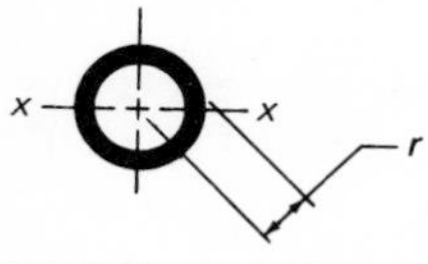 $S = \pi r^2$	$I_p = 2\pi r^3$

EXAMPLE 5.16.1

Compute the size of fillet weld required using E70 electrodes for the connection shown in Fig. 5.16.4. Assume that the plate does not govern or control weld size.

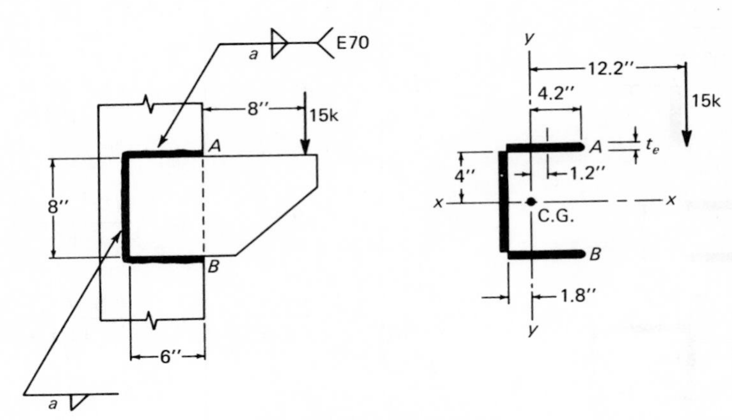

Fig. 5.16.4. Example 5.16.1.

Solution

The highest stress on the weld group will occur at point A or B. The solution will be illustrated by carrying t_e as a factor throughout the computation.

Locate the centroid of the entire configuration by taking moments about the edge of the vertical weld:

$$\bar{x} = \frac{2t_e(6)(3)}{2(6)t_e + 8t_e} = 1.8 \text{ in.}$$

$$I_p = t_e\left\{\underbrace{\frac{(8)^3}{12} + 2[6(4)^2]}_{I_x} + \underbrace{2\left[\frac{(6)^3}{12}\right] + 2[6(1.2)^2] + 8(1.8)^2}_{I_y}\right\}$$

$$= t_e(313.9) \text{ in.}^4$$

The bracketed multiplier of t_e in the I_p expression is the property of lines as found in Table 5.16.1.

$$A = t_e[2(6) + 8] = t_e(20.0) \text{ sq in.}$$

$$f'_y = \frac{P_y}{A} = \frac{15}{t_e(20)} = \frac{0.75}{t_e} \text{ ksi}$$

$$f''_x = \frac{Ty}{I_p} = \frac{15(12.2)4}{t_e(313.9)} = \frac{2.34}{t_e} \text{ ksi}$$

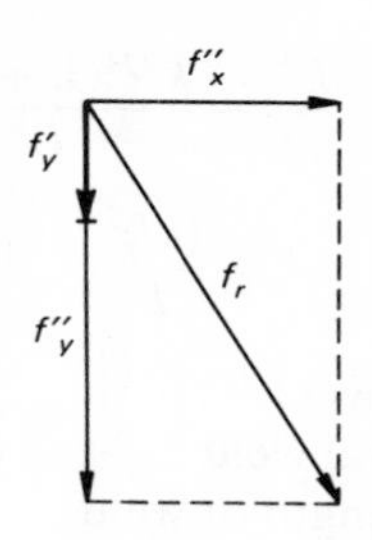

$$f''_y = \frac{Tx}{I_p} = \frac{15(12.2)4.2}{t_e(313.9)} = \frac{2.45}{t_e}\text{ ksi}$$

The resultant stress f_r is

$$f_r = \sqrt{\frac{(2.34)^2 + (2.45 + 0.75)^2}{t_e^2}} = \frac{3.97}{t_e}\text{ksi}$$

It will be apparent from the calculation that one may easily obtain the same numerical result by assuming $t_e = 1$.

Using Eq. 5.16.12, which is merely setting f_r equal to its maximum allowable, value,

$$\text{Reqd } t_e = \frac{f_r}{\text{allowable stress}} = \frac{3.97}{21} = 0.189 \text{ in.}$$

$$\text{Reqd } a = \frac{t_e}{0.707} = \frac{0.189}{0.707} = 0.267 \text{ in., say } 5/16 \text{ in.}$$

Use 5/16-in. E70 fillet welds.

For the general solution of welds under eccentric loading as in Fig. 5.16.4, the AISC Manual has tables, "Eccentric Loads on Weld Groups," using the terms as defined in Fig. 5.16.5. As given by the AISC Manual,

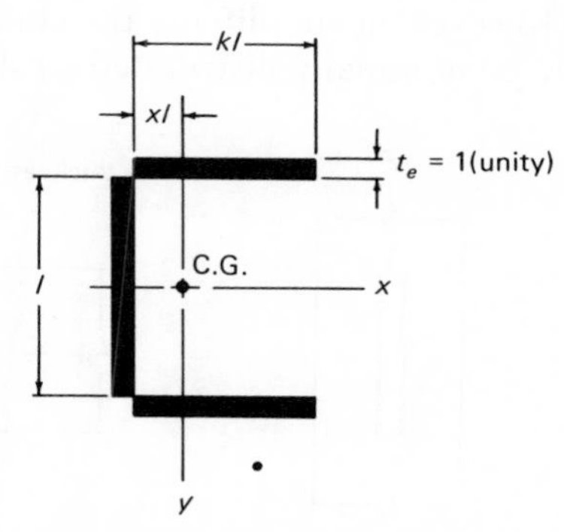

Fig. 5.16.5. Terminology for AISC Manual, "Eccentric Loads on Weld Groups."

$$I_p = \frac{L^3}{12} - \frac{k^2l^4(1 + k)^2}{L} \tag{5.16.15}$$

$$\bar{x} = xl = \frac{(kl)^2}{L} \tag{5.16.16}$$

where l = length of vertical weld
kl = length of horizontal weld
$L = l + 2kl$ = total length of weld

EXAMPLE 5.16.2

Recheck I_p in Example 5.16.1 using the AISC method.

Solution

Using Eq. 5.16.15,

$$kl = 6 \text{ in.}; \quad l = 8 \text{ in.}; \qquad k = \frac{6}{8} = 0.75$$

$$L = 8 + 2(6) = 20 \text{ in.}$$

$$I_p = \frac{(20)^3}{12} - \frac{(0.75)^2(8)^4(1 + 0.75)^2}{20} = 667 - 352 = 315 \text{ in.}^3$$

This agrees within slide-rule error to the previously computed value of 313.9 in.^{3}

Shear and Bending. Combined shear and bending stresses are computed by vectorially adding the *nominal* shear and bending stresses. The procedure is illustrated by considering the bracket shown in Fig. 5.16.6a and the effective cross section of the weld group shown in Fig. 5.16.6b. Figure 5.16.7 shows the variation of the shear and bending stresses. The reader should note that the actual maximum shear and bending stresses occur at different locations. However, in simplifying the computations, the shear stresses are assumed to be nominally distributed as shown in Fig. 5.16.7c.

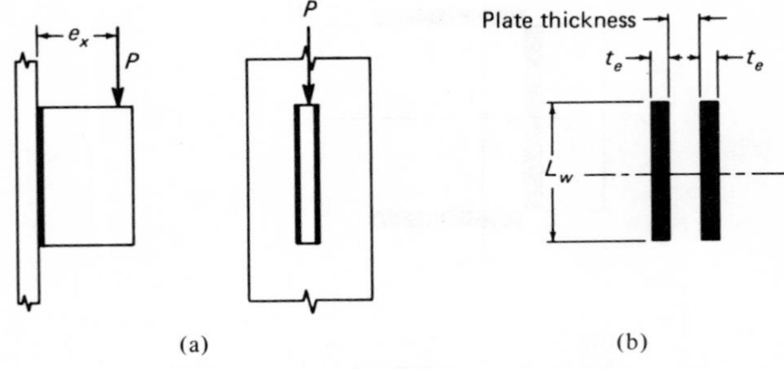

Fig. 5.16.6.

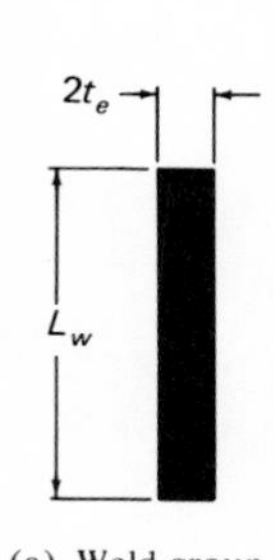

(a) Weld group from Fig. 5.15.6

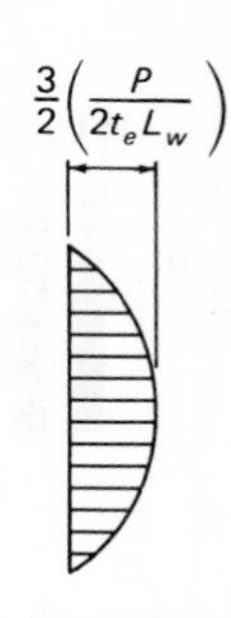

(b) Actual shear stress distribution

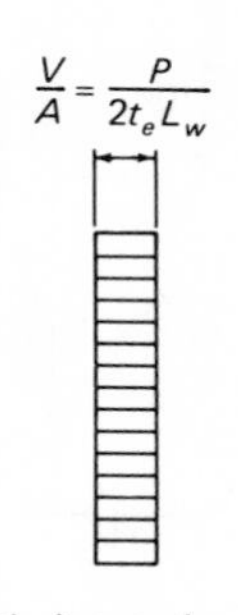

(c) Assumed nominal shear stress distribution

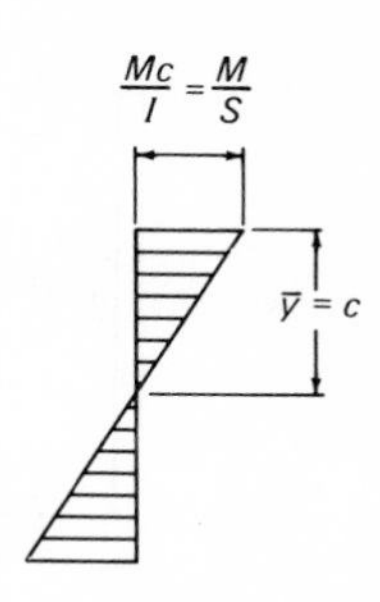

(d) Bending stress distribution

Fig. 5.16.7.

The nominal shear stress is then added *vectorially* to the maximum bending stress.

For this particular case the assumed vertical shear stress from Eq. 5.16.3b is

$$f'_y = \frac{P_y}{A} = \frac{P}{2t_eL_w}$$

and the horizontal stress due to bending is

$$f''_x = \frac{Mc}{I} = \frac{(Pe_x)(L_w/2)}{\left[\frac{2t_e(L_w)^3}{12}\right]} = \frac{3Pe_x}{t_e(L_w)^2}$$

The stress resultant is

$$f_r = \sqrt{(f'_y)^2 + (f''_x)^2}$$

For the flexural stress component, I equals either I_x or I_y, whichever is the axis for bending. The I values may be computed as the properties of line configurations in a manner similar to that used for I_p. For some commonly used configurations, $S = I/\bar{y}$ expressions are given in Table 5.16.1.

EXAMPLE 5.16.3

Compute the size of E70 fillet weld required for the connection shown in Fig. 5.16.8a using shielded metal-arc process. Assume that the column and bracket does not control.

Solution

$$f'_y = \frac{P}{A} = \frac{10}{2(10)1} = 0.50 \text{ ksi}$$

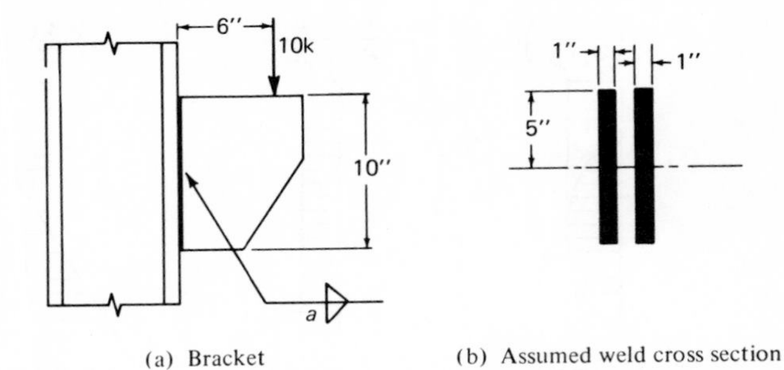

(a) Bracket (b) Assumed weld cross section

Fig. 5.16.8. Example 5.16.3.

$$I_x = \frac{(1)(10)^3}{12} = 83.33 \text{ in.}^4$$

$$f_x'' = \frac{Mc}{I} = \frac{10(6)5}{83.3} = 3.60 \text{ ksi}$$

$$f_r = \sqrt{(0.5)^2 + (3.6)^2} = 3.64 \text{ ksi for 1-in. effective throat.}$$

$$\text{Reqd } t_e = \frac{f_r}{\text{allowable stress}} = \frac{3.64}{21} = 0.173 \text{ in.}$$

$$\text{Reqd } a = \frac{0.173}{0.707} = 0.244 \text{ in.}$$

Use ¼-in. E70 fillet welds.

Design for Lines of Weld Under Moment. Even when there are moderate returns at the top of lines of fillet weld, an estimate of the length required may be obtained by using the same approach as used for determining the number of bolts in a line in Sec. 4.7. In Fig. 4.7.8, R/p has units of kips per inch and corresponds to f_r when t_e is unity for welds.

For pure moment on one line of weld,

$$f_r = \frac{M}{S} = \frac{M}{\left[\frac{1}{6} L_w^2\right]} \text{ kips/in.} \tag{5.16.17}$$

Since the maximum value of f_r is R_w,

$$R_w = \frac{6M}{L_w^2}$$

or

$$\text{Reqd } L_w = \sqrt{\frac{6M}{R_w}} \tag{5.16.18}$$

Equation 5.16.18 for welds corresponds to Eq. 4.7.22 for bolts. Since it is correct only for pure moment, R_w should be taken as a reduced value to account for direct shear.

EXAMPLE 5.16.4

Determine the length L required to carry the load in Fig. 5.16.9, using $\frac{5}{16}$-in. E70 fillet weld.

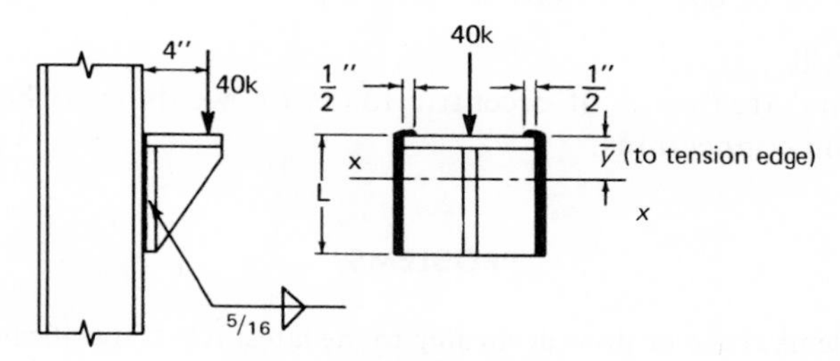

Fig. 5.16.9. Example 5.16.4.

Solution

Estimate L required by using Eq. 5.16.18:

$$R_w = \frac{5}{16}(0.707)21 = 4.65 \text{ kips/in.}$$

$$M = 40(4) = 160 \text{ in.-kips per 2 weld lines}$$

$$\text{Reqd } L \approx \sqrt{\frac{6M}{R_w}} = \sqrt{\frac{6(160/2)}{4.0}} = 11 \text{ in.}$$

where a reduced value of R_w has been used to account for the direct shear effect. The return will add additional strength; try $L = 10$ in.

Investigate configuration using weld properties as lines computed from basic principles,

$$\bar{y} = \frac{2(10)5}{2(10 + 0.5)} = \frac{100}{21} = 4.76 \text{ in.}$$

Direct shear:

$$f'_y = \frac{40}{20} = 2.00 \text{ kips/in.}$$

Since little direct shear is carried by the returns, they are neglected above.

$$I_x = \frac{2L^3}{12} + 2L(5.0 - 4.76)^2 + 2(0.5)(4.76)^2$$

$$= \frac{(10)^3}{6} + 20(0.24)^2 + (4.76)^2 = 190.6 \text{ in.}^3$$

$$S = \frac{I}{\bar{y}} = \frac{190.6}{4.76} = 40.0 \text{ in.}^2$$

Flexure:

$$f''_x = \frac{M}{S} = \frac{160}{40} = 4.0 \text{ kips/in.}$$

Resultant:

$$f_r = \sqrt{(2.00)^2 + (4.00)^2} = 4.48 \text{ kips/in.} < 4.65 \qquad \text{OK}$$

Use $L = 10$ in.

Additional treatment of eccentric loads on welds is to be found in Chapter 13 on connections.

PROBLEMS

All problems are to be done according to the latest AISC specifications, unless otherwise indicated.

5.1. Determine the proper plate thickness and indicate which ones, if any, of the following joints made by the submerged arc process are acceptable according to AISC–1.17.2. Refer to AISC Manual, "Welded Joints."
(a) Single bevel groove, butt joint, welded from one side.
(b) Single-V groove, butt joint, welded one side with no root opening.
(c) Single bevel groove, butt joint, welded from one side with a backup plate.
(d) Double-V groove; butt joint.

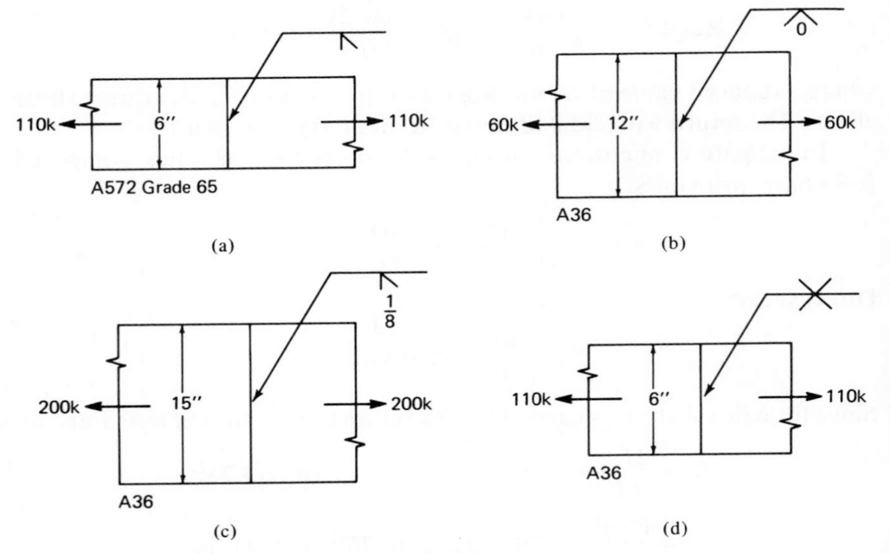

Prob. 5.1.

5.2. Determine either the plate thickness and weld material, or the capacity T of the joint for the given conditions; assuming shielded metal-arc process:
(a) Square groove, butt joint.
(b) Double-bevel groove, butt joint.
(c) Double-bevel groove, butt joint, welded from both sides, partial penetration.
(d) Single-V groove, butt joint, welded from one side, partial penetration.

For cases (a) and (b), is there a maximum thickness for which the type of weld may be used? Show cross section for welds in all four cases. (Refer to AISC Manual, "Welded Joints.")

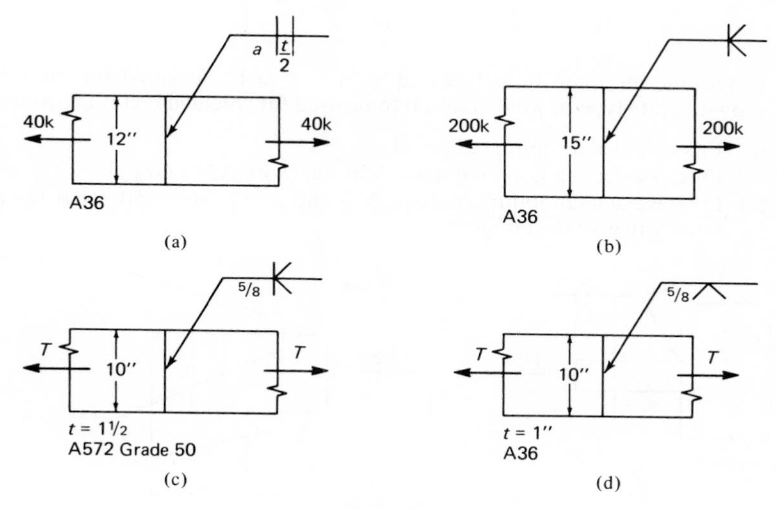

Prob. 5.2.

5.3. Determine the capacity of the connection shown, using the submerged-arc process: (a) use A36 steel, (b) use A572 Grade 55. Assume appropriate electrode material is to be used.

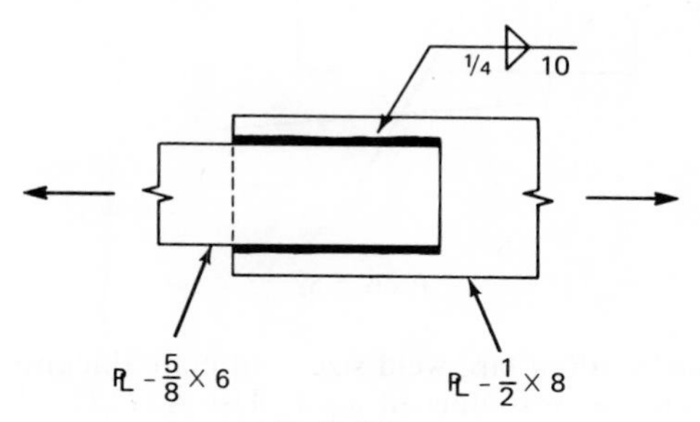

Prob. 5.3.

5.4. Determine the fillet welds required to develop the capacity of the connection shown. Specify the proper electrode for using the submerged-arc process: (a) use A572 Grade 42 steel, (b) use steel with F_y = 100 ksi.

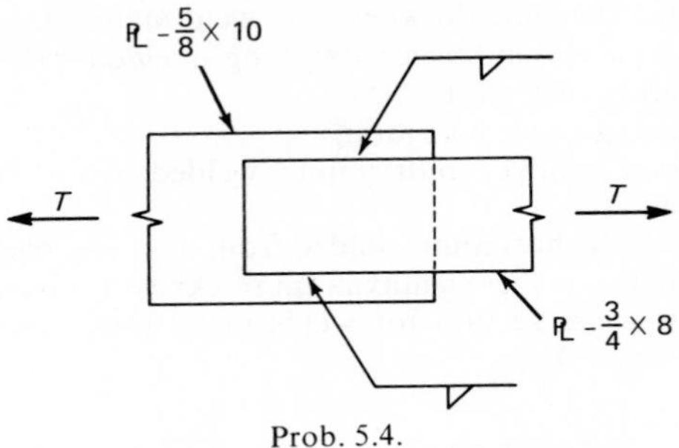

Prob. 5.4.

5.5. Determine the plate thickness and weld size to be specified for the joints shown. State type of weld material to be used for shielded metal-arc process.

(a) Fillet welded tee joint; A36 steel.
(b) Fillet welded lap joint; compare A36 steel with A572 Grade 50.
(c) Fillet welded lap joint; compare A36 and A572 Grade 50; what lap distance will give the best joint?

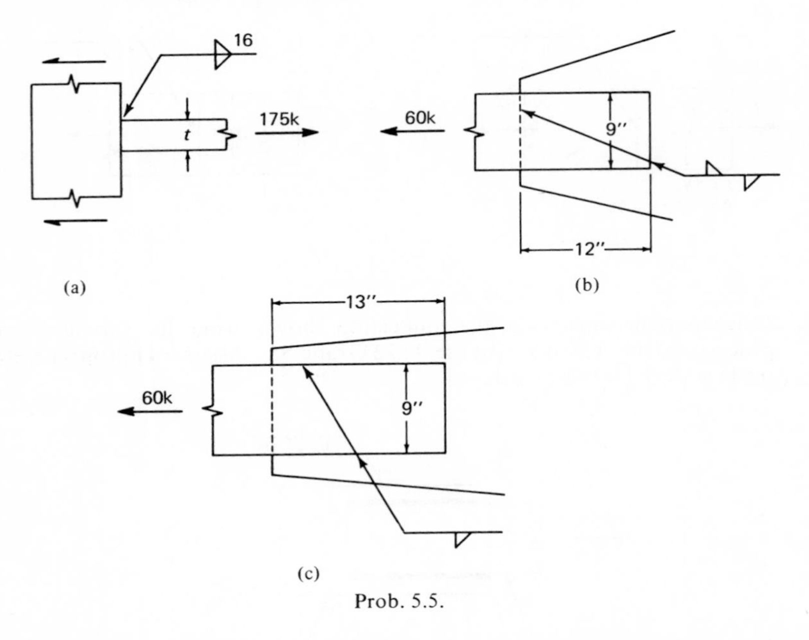

Prob. 5.5.

5.6. Determine the length of lap, weld size, and plate thickness of the 9-in. wide plate, to obtain the most efficient joint. Use A572 Grade 50 steel, and the shielded metal-arc process.

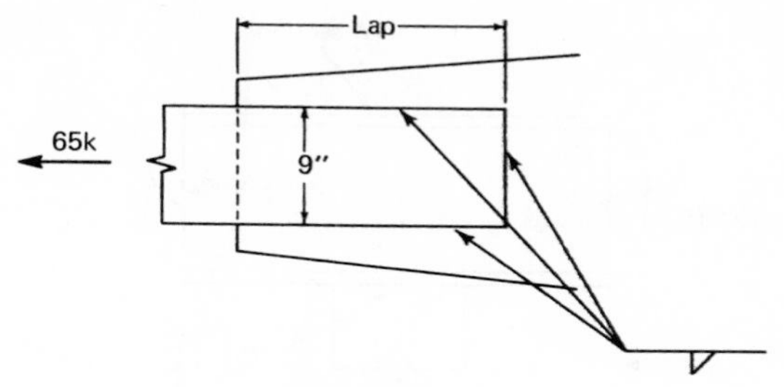

Prob. 5.6.

5.7. Design the reinforced lap joint shown, assuming the plates are 7-in. wide. Use A36 steel and the shielded metal-arc process. (Refer to AISC Manual, "Welded Joints"–BTC p. 4).

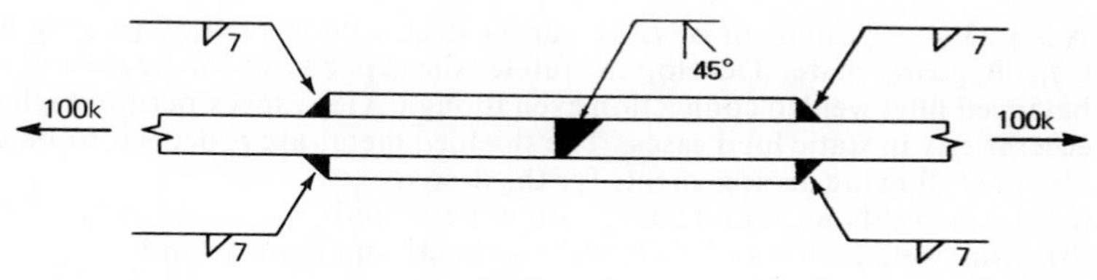

Prob. 5.7.

5.8. Select a pair of channels and design the welds assuming the shielded metal-arc process: (a) use A36 steel, and (b) use A572 Grade 60 steel.

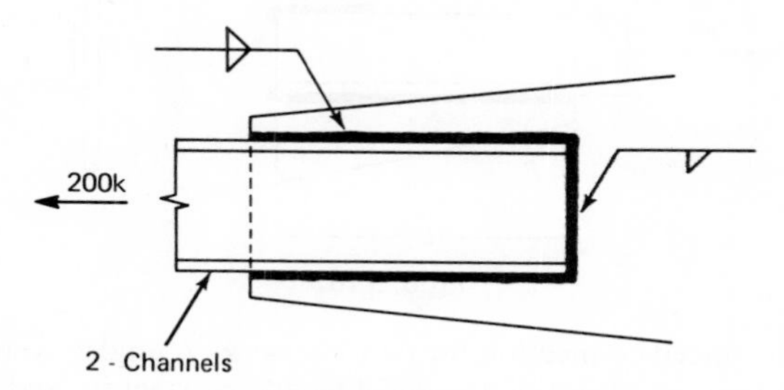

Prob. 5.8.

5.9. Design the plate framing into the W section and the welds, assuming shielded metal-arc process is to be used.

(a) Use A572 Grade 42 steel.

(b) Use A572 Grade 65 steel.

(c) Use A572 Grade 42 steel, with fillet welds instead of groove weld.

(d) Use A572 Grade 65 steel, with fillet welds instead of groove weld.

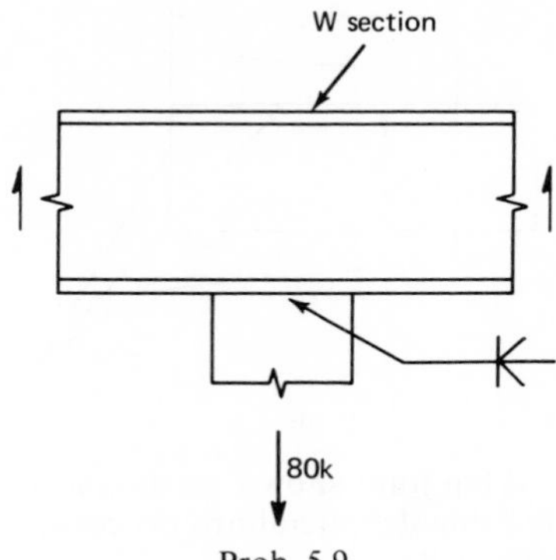

Prob. 5.9.

5.10. A 5 × 3½ × ⅜ angle of A572 Grade 50 steel is connected by its long leg to a 5/16-in. gusset plate. Develop the full tensile capacity of the angle and use a balanced fillet welded connection even though AISC specs permit neglecting eccentricity in static load cases. The shielded metal-arc process is to be used. Use the following arrangements for the design:
(a) 5/16-in. weld on toe and back, with none on end.
(b) ¼-in. weld on toe and ⅜-in. weld on back, and none on end.

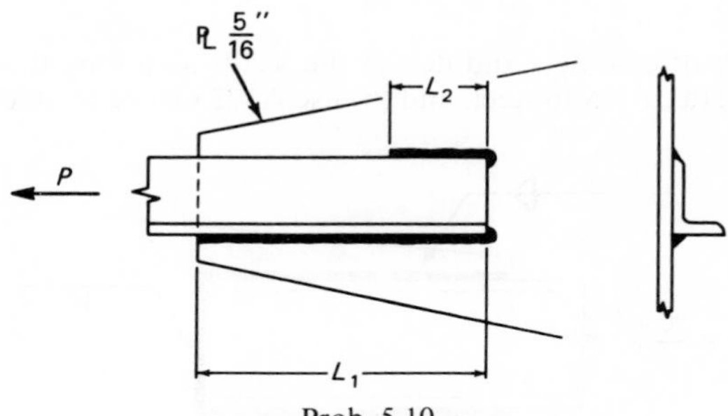

Prob. 5.10.

5.11. Design a balanced connection for two 7 × 4 × ½ angles connected by their long legs to a ⅜-in. gusset plate. A572 Grade 60 steel is used and welding is by the shielded metal-arc process. Detail the joint to balance the loads and still give the shortest possible overlap.

5.12. Design the welds indicated to develop the full strength of the angles and minimize eccentricity. Use the submerged-arc welding process.
(a) Use A36 steel.
(b) Use A572 Grade 50 steel.
(c) Use A572 Grade 65 steel.
(d) Use A36 steel, but omit weld on the end of angles.
(e) Use A572 Grade 65 steel, but omit weld on the ends of angles.

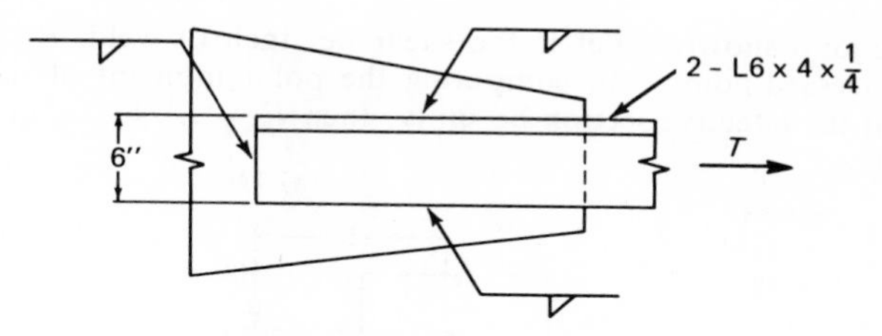

Prob. 5.12.

5.13. Assume the joint of Prob. 5.6 must be redesigned to carry 125 kips, and there exists a possibility of some accidental bending which cannot be computed. Thus, to insure a tighter joint a 1¼ in. diam. plug weld is to be used. Determine the thickness of the smaller plate, the amount of lap, and the weld size for the best joint. Use A572 Grade 50 steel, and assume shielded metal-arc welding process is to be used.

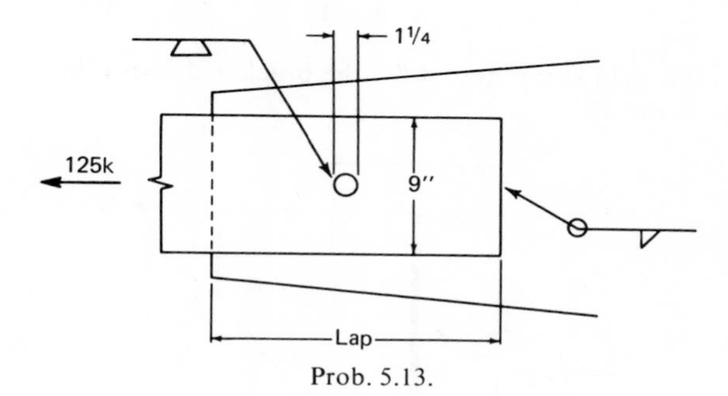

Prob. 5.13.

5.14. Determine the minimum length of slot in order to develop the full strength of a C12 × 20.7 welded to a ⅜-in. plate. Use the same size fillet weld over the entire length, and assume it is to be placed by the shielded metal-arc process.

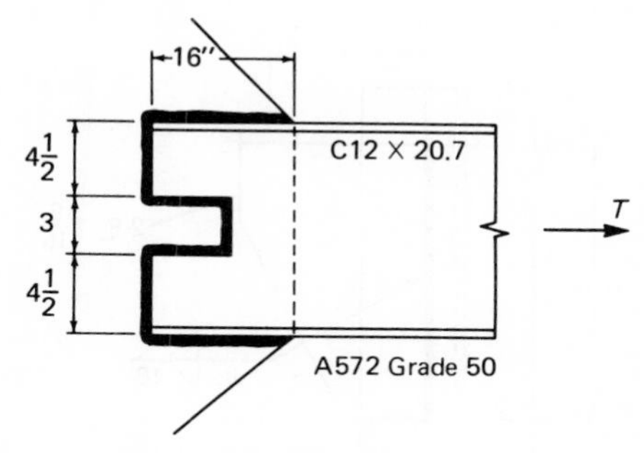

Prob. 5.14.

5.15. For the joint shown, what is the shear per inch of weld, R_w, at the most highly stressed point? (In computing the polar moment of inertia, let the width of the effective section be unity, then $I_p = L^3/12$, as given in Table 5.16.1.)

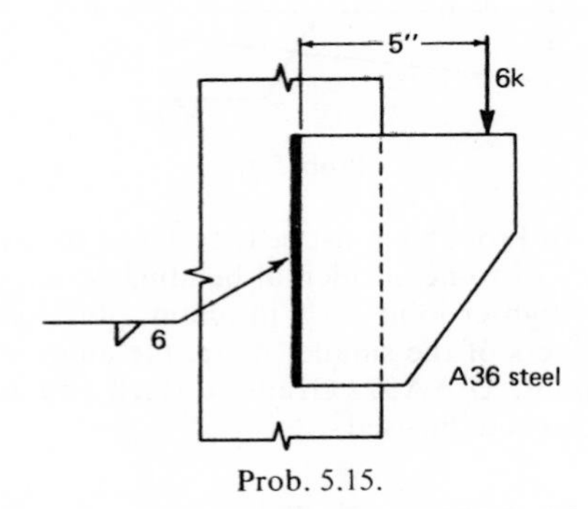

Prob. 5.15.

5.16. What is the weld size required for the bracket shown if the shielded metal-arc process is used?

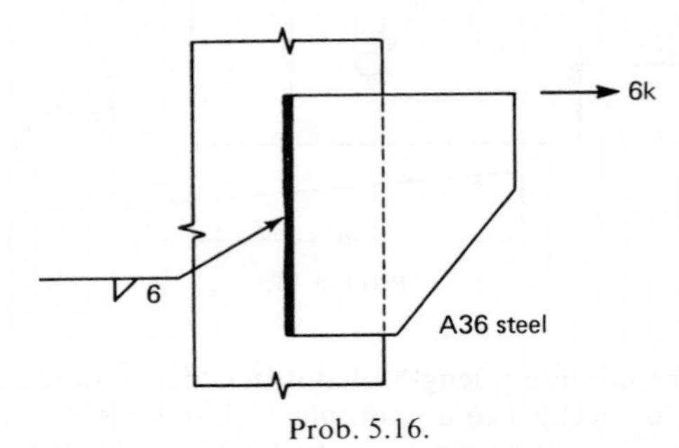

Prob. 5.16.

5.17. For the bracket shown, find the safe capacity P based on the weld. Neglect any returns at ends and assume the shielded metal-arc process is to be used.

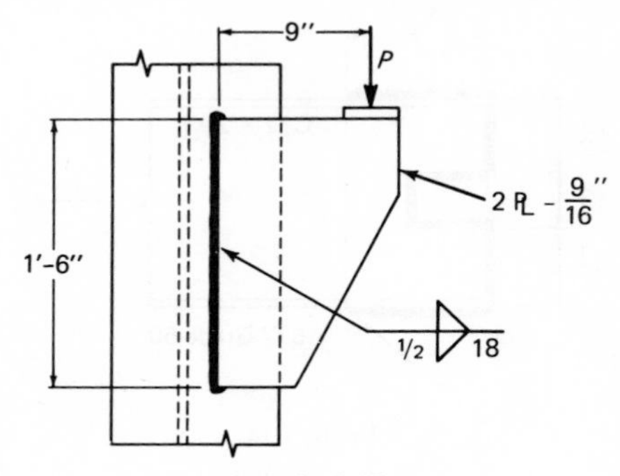

Prob. 5.17.

5.18. Ignoring the effect of returns at the lower end of the connection, and allowing a resultant stress of 2.4 kips per inch of fillet weld, what is the maximum allowable load P for the given connection?

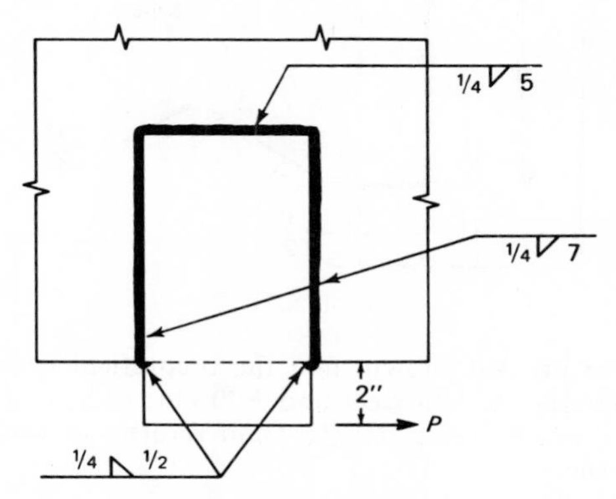

Probs. 5.18 and 5.19.

5.19. Recompute the capacity for the connection of Prob. 5.18, using $\frac{3}{8}$-in. fillet weld on the sides and $\frac{1}{4}$ on the end. Neglect the returns. Assume shielded metal-arc process is to be used and the plates are of A572 Grade 50 steel.

5.20. Derive the general expression for the required weld size on the seat angle (E70 electrodes with shielded metal-arc process) in terms of P, L, and e, using the following assumptions:

(a) Ignoring the returns at the top.

(b) Considering an average return of $L/12$.

(c) Using a return equal to twice the weld size.

If $e = 2\frac{3}{8}$ in. and $L = 6$ in., determine using assumption (a) the weld size needed to carry a load $P = 38.3$ kips. For the weld size selected, check the capacity which may be carried using all three assumptions.

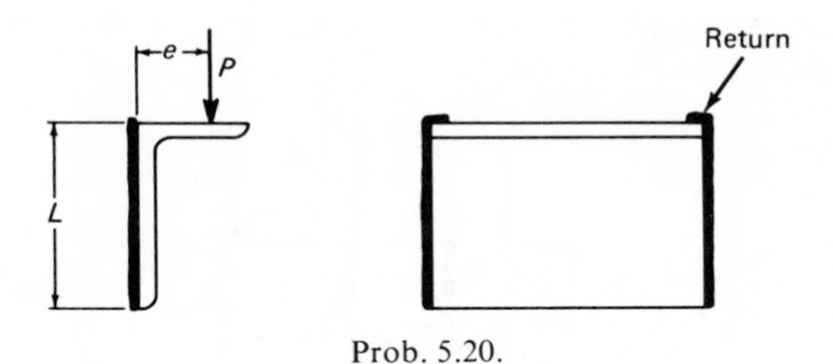

Prob. 5.20.

5.21. Determine the capacity P for the bracket shown. The weld size is $\frac{3}{8}$ in. and E70 electrodes are used with shielded metal-arc process. Compare answer with AISC Manual tables.

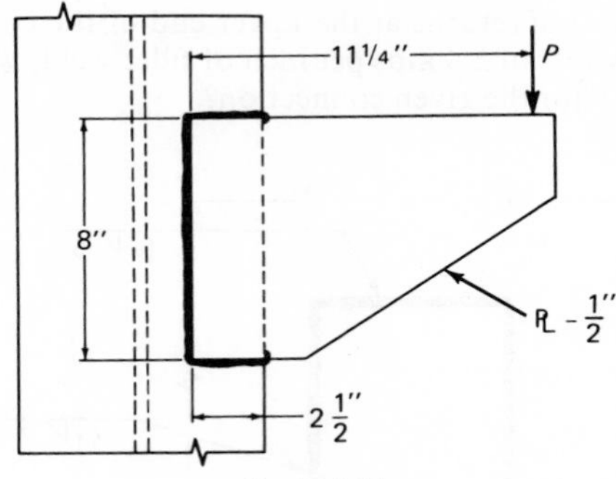

Prob. 5.21.

5.22. For the welded bracket shown, find the theoretical weld size required for strength if material is A36 steel and E70 electrode material is used with shielded metal-arc process. Neglect end returns at the upper and lower right-hand corners.

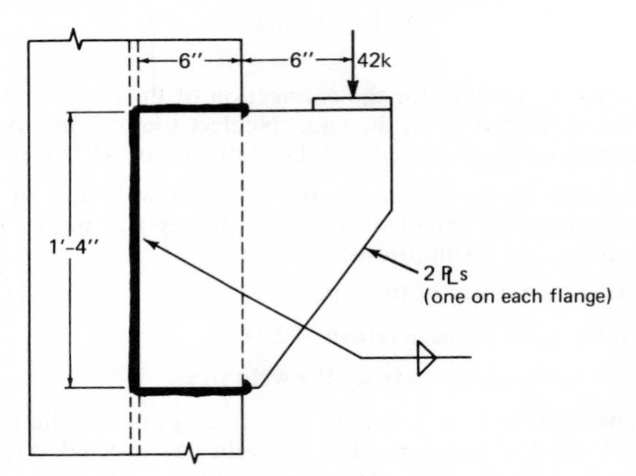

Prob. 5.22.

5.23. Determine the length L required when using $\frac{5}{16}$-in. fillet weld on A36 steel with shielded metal-arc welding.

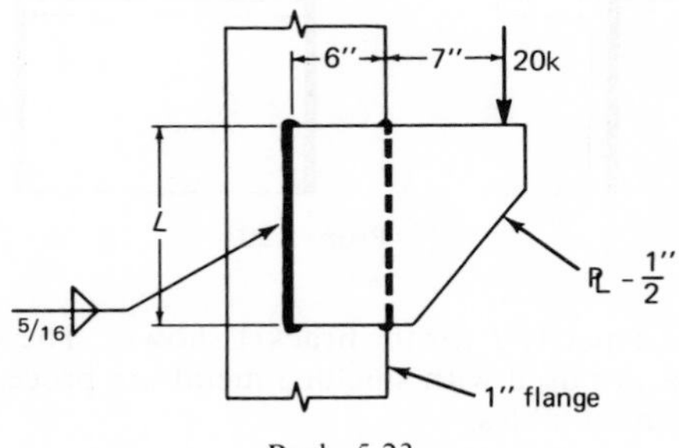

Prob. 5.23.

5.24. Using basic principles, determine the length L required to safely carry the 40-kip load using $^5/_{16}$-in. weld. Assume steel is A572 Grade 50 and shielded metal-arc process is to be used for welding.

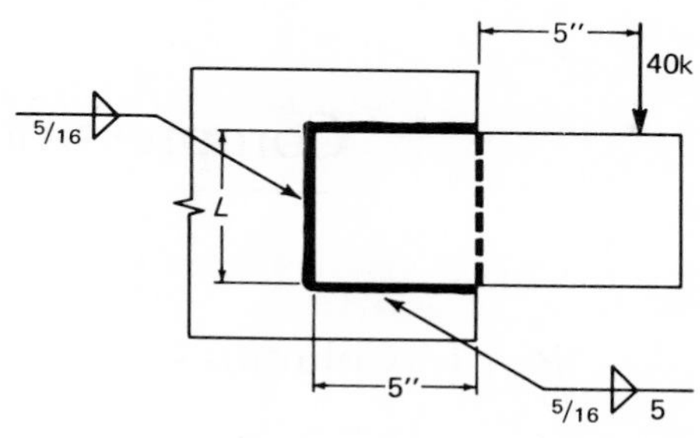

Prob. 5.24.

SELECTED REFERENCES

1. K. Winterton, "A Brief History of Welding Technology," *Welding and Metal Fabrication*, November 1962.
2. "100 Years of Metalworking-Welding, Brazing and Joining," *The Iron Age*, June 1955.
3. H. Carpmael, *Electric Welding and Welding Appliances*, D. Van Nostrand Company, London, 1920.
4. P. M. Hall, "77 Years of Resistance Welding." *Welding Engineer*, February 1954; March 1954; April 1954.
5. G. Herden, *Schweiss und Schneid-Technik,* Carl Marhold Verlag, Halle, East Germany, 1960.
6. K. Winterton, "A Brief History of Welding Technology," *Welding and Metal Fabrication*, December 1962.
7. J. H. Davies, *Modern Methods of Welding*, D. Van Nostrand Company, New York, 1922.
8. E. Viall, *Electric Welding*, McGraw-Hill Book Company, New York, 1921.
9. *Welding Handbook*, 6th ed., Vol. 2, *Welding Processes*; *Gas, Arc, and Resistance*, American Welding Society, New York, 1969.
10. Omer W. Blodgett, *Design of Welded Structures*, James F. Lincoln Arc Welding Foundation, 1966.
11. *Welding Handbook*, 6th Ed., Vol. 1, *Fundamental of Welding*, American Welding Society, New York, 1968.
12. "Weld Defects Sound Off," *The Iron Age*, Mar. 27, 1969.
13. J. A. Donnelly, "Determining the Cost of Welded Joints," *Engineering Journal*, AISC, Vol. 5, No. 4 (October 1968). pp. 146–147.
14. T. R. Higgins and F. R. Preece, "Proposed Working Stresses for Fillet Welds in Building Construction," *Engineering Journal*, AISC, Vol. 6, No. 1 (January 1969), pp. 16–20.

6

Compression Members

Part I: Columns

6.1. GENERAL

In this chapter, members subjected to axial compression stresses are to be treated. Referred to by various terms, such as *column*, *stanchion*, *post*, and *strut*, these members are rarely if ever actually carrying axial compression alone. However, whenever the loading is so arranged that either the end rotational restraint is negligible or the loading is symmetrically applied from members framing in at the column ends, and bending may be considered negligible compared to the direct compression, the member can safely be designed as a concentrically loaded column.

It is well known from basic mechanics of materials that only very short columns can be stressed to their yield point; the usual situation is that buckling, or sudden bending as a result of instability, occurs prior to developing the full material strength of the member. Thus a sound knowledge of compression member stability is necessary for those designing in structural steel.

6.2. EULER ELASTIC BUCKLING AND HISTORICAL BACKGROUND

Column buckling theory originated with Leonhard Euler in 1757. An initially straight concentrically loaded member, in which all fibers remain elastic until buckling occurs, is slightly bent as shown in Fig. 6.2.1. Though Euler dealt with a member built-in at one end and simply supported at the other, the same logic is applied here to the pin-end column, which having zero end rotational restraint represents the members with least buckling strength.

At any location, the bending moment on the slightly bent member is

$$M_x = Py \tag{6.2.1}$$

U. S. Steel Building, Pittsburgh. Courtesy of United States Steel Corporation.

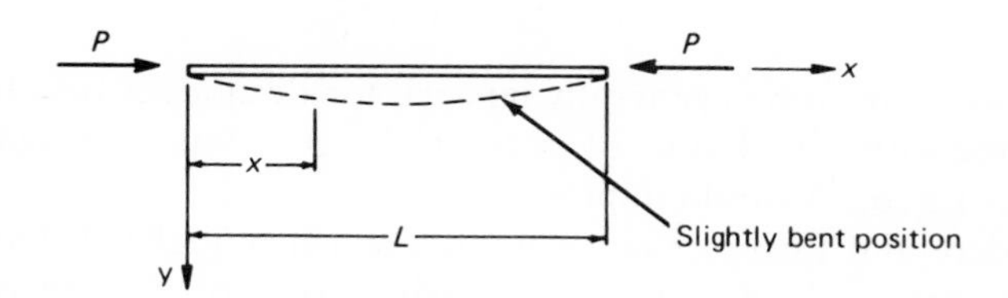

Fig. 6.2.1. Euler column.

and since

$$\frac{d^2y}{dx^2} = -\frac{M_x}{EI} \tag{6.2.2}$$

the differential equation becomes

$$\frac{d^2y}{dx^2} + \frac{P}{EI}\,y = 0 \tag{6.2.3}$$

After letting $k^2 = P/EI$, the solution of this second-order linear differential equation may be expressed

$$y = A \sin kx + B \cos kx \tag{6.2.4}$$

Applying boundary conditions, (a) $y = 0$ at $x = 0$; and (b) $y = 0$ at $x = L$, one obtains for condition (a), $B = 0$; and for condition (b),

$$0 = A \sin kL \tag{6.2.5}$$

Satisfaction of Eq. 6.2.5 may be accomplished in three possible ways; (a) constant $A = 0$, i.e., no deflection; (b) $kL = 0$, i.e., no applied load; and (c) $kL = N\pi$, which is the requirement for buckling to occur. Thus

$$\left(\frac{N\pi}{L}\right)^2 = \frac{P}{EI}$$

$$P = \frac{N^2\pi^2EI}{L^2} \tag{6.2.6}$$

The fundamental buckling mode, a single-curvature deflection ($y = A \sin \pi x/L$ from Eq. 6.2.4.), will occur when $N = 1$; thus, the Euler critical load for a column with both ends pinned is

$$P_{cr} = \frac{\pi^2EI}{L^2} \tag{6.2.7}$$

or in terms of average compressive stress, using $I = A_g^{+2}$

$$F_{cr} = \frac{P_{cr}}{A_g} = \frac{\pi^2E}{(L/r)^2} \tag{6.2.8}$$

Euler's approach was generally ignored for design because test results did not agree with it; columns of ordinary length used in design were not as strong as Eq. 6.2.7 would indicate.

Considère and Engesser in 1889 independently realized that portions of usual length columns become inelastic prior to the occurrence of buckling and that a value of E should be used which could account for some of the compressed fibers being strained beyond the proportional limit. It was thus consciously recognized that in fact *ordinary length* columns fail by inelastic buckling, rather than by elastic buckling.

Complete understanding of the behavior of concentrically loaded columns, however, was not achieved until 1946 when Shanley[6] offered the explanation which now seems obvious. He reasoned that it was actually possible for a column to bend and still have increasing axial compression, but that it *begins* to bend upon reaching what is commonly referred to as the *buckling load*, which includes inelastic effects on some or all fibers of the cross section. These inelastic effects are discussed in detail in Sec. 6.4.

An extensive review of the development of column theory from Euler to Shanley is given by N. J. Hoff,[7] and a brief summary is given by B. G. Johnston.[8]

6.3. BASIC COLUMN STRENGTH

In order to determine a basic column strength certain conditions may be assumed for the ideal column.[3] With regard to material, it may be assumed (a) there are the same compressive stress-strain properties throughout the section; (b) no initial internal stresses exist such as those due to cooling after rolling and those due to welding. Regarding shape and end conditions, it may be assumed (c) the column is perfectly straight and prismatic; (d) the load resultant acts through the centroidal axis of the member until the member begins to bend; and (e) the end conditions must be determinate so that a definite equivalent pinned length may be established. Further assumptions regarding buckling may be made, as (f) the small deflection theory of ordinary bending is applicable and shear may be neglected; and (g) twisting or distortion of the cross section does not occur during bending.

Once the foregoing assumptions have been made, it is now agreed[2] that the strength of a column may be expressed by

$$F_{cr} = \frac{P}{A} = \frac{\pi^2 E_t}{(KL/r)^2} \tag{6.3.1}$$

where P/A = average stress in the member
E_t = tangent modulus at stress P/A
KL/r = equivalent pin-end column slenderness ratio (K value discussed in Sec. 6.9)

It is well known that long compression members fail by elastic buckling and that short stubby compression members may be loaded until the material yields or perhaps even into the strain-hardening range. However, in the vast majority of usual situations failure occurs by buckling after a portion of the cross section has yielded. This is known as *inelastic buckling*.

Actually, pure buckling under axial load occurs only when the assumptions (a) through (g) aforementioned apply. Columns are usually an integral part of a structure and as such *cannot* behave entirely independently. The practical use of the term *buckling* is that it is the boundary between stable and unstable deflections of a compression member, rather than the instantaneous condition which occurs in the isolated slender elastic rod. Many engineers refer to this "practical buckling load" as the "ultimate load."

As previously mentioned, for many years theoretical determinations of "ultimate load" did not agree with test results. Test results included effects of initial crookedness of the member, accidental eccentricity of load, end restraint, local or lateral buckling, and residual stress. A typical curve of observed ultimate loads was as shown in Fig. 6.3.1. Design formulas,

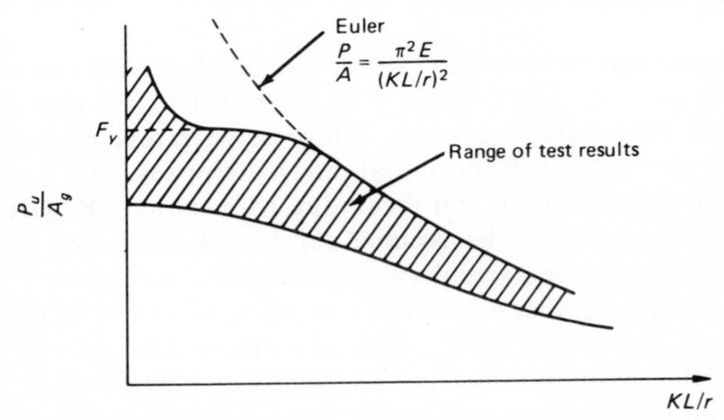

Fig. 6.3.1. Typical range of column strength vs. slenderness ratios.

therefore, were based on such empirical results. Various straight-line and parabolic formulas have been used, as well as other more complex expressions, in order to fit the curve of test results in a reasonably accurate, yet practical manner.

As a general approach, the Euler elastic buckling governs the strength for large slenderness ratios, the yield point is used as a basis for short columns, and a transition curve must be used for inelastic buckling.

6.4.* INELASTIC BUCKLING

Since ordinary length columns buckle when some of their fibers are inelastic, having a modulus of elasticity less than their initial elastic value, the logic of Engesser, Considère, and Shanley will be explained and developed in this article.

*Sections so marked may be omitted without loss of continuity.

Basic Tangent Modulus Theory. Euler's theory pertained only to situations where the compressive stress which acts uniformly over the cross section is below the elastic limit when unstable equilibrium occurs. Engesser[4] and Considère[5] were the first to utilize the possibility of a variable modulus of elasticity. In Engesser's tangent modulus theory the column remains straight up to the moment of failure and the modulus of elasticity at failure is the tangent to the stress-strain curve. The relationships are shown in Fig. 6.4.1. The theory prescribed that at a certain stress, $F_{cr} = P_{cr}/A$,

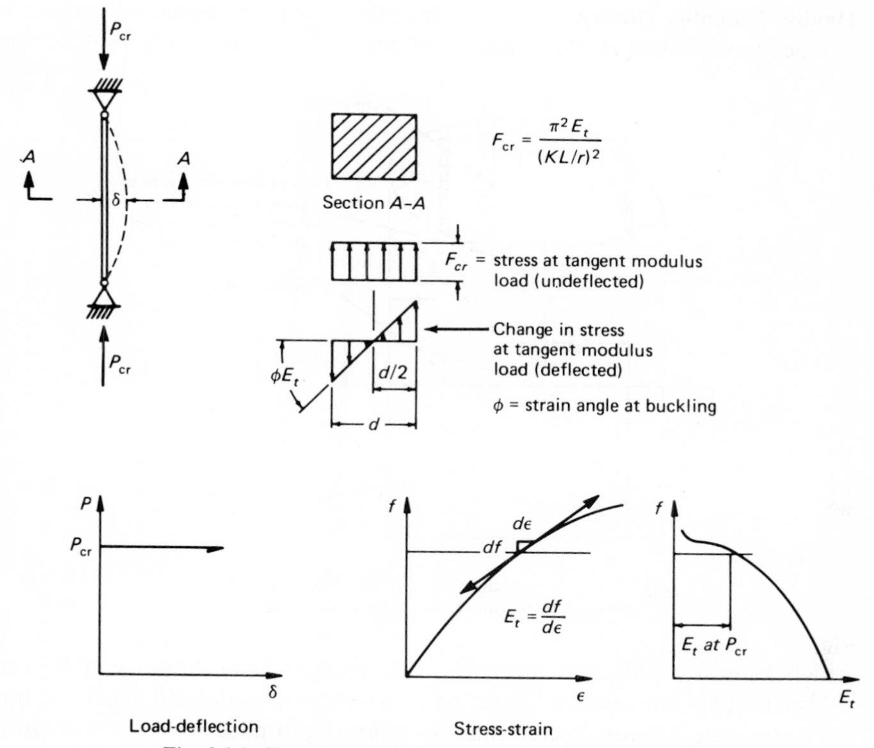

Fig. 6.4.1. Engesser original tangent modulus theory, 1889.

the member could acquire an unstable deflected shape and that the deformation at F_{cr} is governed by $E_t = df/d\epsilon$. Thus Engesser modified Euler's equation to become

$$F_{cr} = \frac{P_t}{A} = \frac{\pi^2 E_t}{(KL/r)^2} \tag{6.4.1}$$

where P_t is the tangent modulus load.

This theory, however, still did not agree with test results, giving computed loads lower than measured ultimate capacities. The principal assumption which caused this tangent modulus theory to be considered erroneous is that as the member changes from a straight to bent form, no strain reversal takes place. In 1895 Engesser changed his theory, reasoning that during bending some fibers are undergoing increased strain (lowered tangent modulus) and some fibers are being unloaded (higher modulus at the reduced strain); therefore, a combined value should be used for the modulus.

Double-Modulus Theory. To examine the process of column bending at stresses beyond the elastic limit, consider the section of Fig. 6.4.2 from

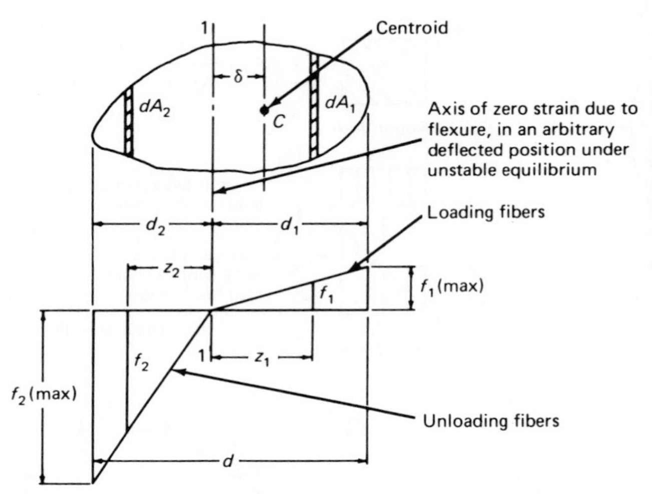

Fig. 6.4.2. Stress distribution in condition of unstable equilibrium (Double Modulus Theory).

which Engesser's double modulus, or "reduced" modulus, will be developed. This concept had logic to it which was generally accepted but gave computed loads higher than measured ultimate loads. Not until Shanley's explanation was the inconsistency resolved.

At unstable equilibrium, the stress at the neutral axis (section 1–1 of Fig. 6.4.2) remains as it was prior to the deflection δ occurring. On the loading fibers where strain is increasing, the stress increase is proportional to $E_t = df/d\epsilon$, whereas on the unloading fibers the decrease in strain relieves the elastic part of the strain; thus, the stress decrease is proportional to the elastic modulus E.

As shown in Fig. 6.4.3, the strain on the cross section will be linear. At the extreme unloaded fiber, applying Hooke's Law the stress becomes

$$f_{2(\max)} = (\text{unit strain})\,E = \frac{\Delta dx}{dx}E \tag{6.4.2}$$

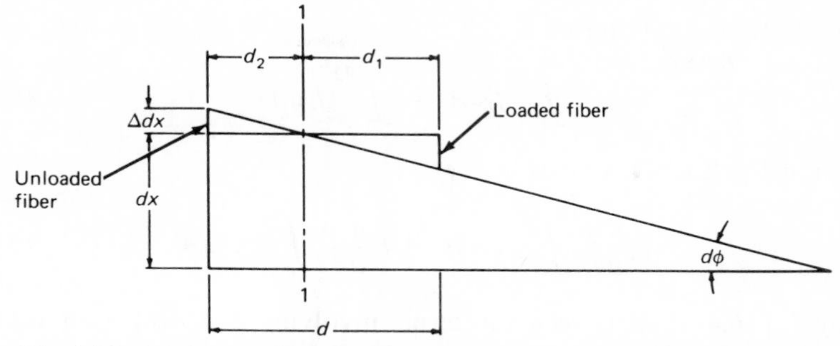

Fig. 6.4.3. An element dx along the axis of the column in the unstable equilibrium position.

and at the loaded fiber,

$$f_{1(\max)} = \frac{\Delta dx\, d_1}{d_2} \frac{E_t}{dx} \tag{6.4.3}$$

$$\frac{\Delta dx}{d_2} = d\phi \tag{6.4.4}$$

Thus

$$f_{2(\max)} = Ed_2 \frac{d\phi}{dx}; \quad f_{1(\max)} = E_t d_1 \frac{d\phi}{dx} \tag{6.4.5}$$

For small curvature,

$$\frac{1}{\text{radius of curvature}} = \frac{M}{E_r I} = \frac{d\phi}{dx} = \frac{d^2y}{dx^2} \tag{6.4.6}$$

where E_r = Engesser's double modulus.

The internal resisting moment for the stress condition of Fig. 6.4.2 gives

$$M = -Py = \int_0^{d_1} f_1(z_1 - \delta)\, dA_1 + \int_0^{d_2} f_2(z_2 + \delta)\, dA_2 \tag{6.4.7}$$

and from linear stress distribution and Eq. 6.4.5,

$$\begin{aligned} f_1 &= f_{1(\max)} \frac{z_1}{d_1} = E_t d_1 \frac{d^2y}{dx^2} \frac{z_1}{d_1} \\ f_2 &= f_{2(\max)} \frac{z_2}{d_2} = Ed_2 \frac{d^2y}{dx^2} \frac{z_2}{d_2} \end{aligned} \tag{6.4.8}$$

Thus Eq. 6.4.7 becomes

$$-Py = E_t \frac{d^2y}{dx^2} \int_0^{d_1} z_1(z_1 - \delta)\, dA_1 + E \frac{d^2y}{dx^2} \int_0^{d_2} z_2(z_2 + \delta)\, dA_2 \tag{6.4.9}$$

Force equilibrium requires

$$\int_0^{d_1} f_1 dA_1 = \int_0^{d_2} f_2 dA_2 \tag{6.4.10}$$

which, using Eq. 6.4.8, gives

$$E_t \frac{d^2y}{dx^2} \int_0^{d_1} z_1 dA_1 = E \frac{d^2y}{dx^2} \int_0^{d_2} z_2 dA_2 \tag{6.4.11}$$

Using Eq. 6.4.11, it is seen the terms involving δ cancel each other in Eq. 6.4.9, thus giving

$$-Py = E_t \frac{d^2y}{dx^2} \int_0^{d_1} z_1^2 \, dA_1 + E \frac{d^2y}{dx^2} \int_0^{d_2} z_2^2 \, dA_2$$

$$\therefore \frac{d^2y}{dx^2} \left[E_t \int_0^{d_1} z_1^2 \, dA_1 + E \int_0^{d_2} z_2^2 \, dA_2 \right] + Py = 0 \tag{6.4.12}$$

Equation 6.4.12 is obviously of the same form as the elastic buckling equation, Eq. 6.2.3. Thus for the double modulus theory,

$$P_{\text{cr}} = \frac{\pi^2}{L^2} \left[E_t \int_0^{d_1} z_1^2 \, dA_1 + E \int_0^{d_2} z_2^2 \, dA_2 \right] \tag{6.4.13}$$

EXAMPLE 6.4.1

(a) Determine the effective stiffness, $(EI)_{\text{eff}}$ for buckling of a wide-flange shape with respect to its weak axis according to the double modulus concept. (b) Construct a column strength curve ($f_{\text{cr}} = P/A_g$ vs. L/r) according to the double modulus concept for a material whose stress-strain properties may be approximated by the straight line segments given in Fig. 6.4.4b. Neglect any residual stress.

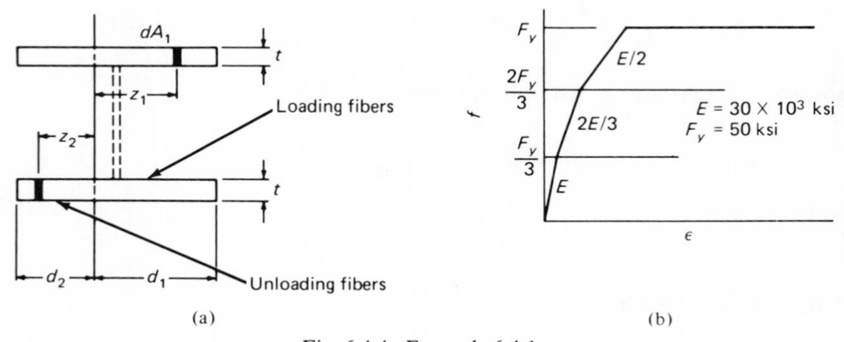

Fig. 6.4.4. Example 6.4.1.

Solution

(a) From Eq. 6.4.13 it can be seen that the effective stiffness is a function of the shape of the cross section. Evaluating Eq. 6.4.13,

$$E_t \int_0^{d_1} z_1^2 \, dA_1 + E \int_0^{d_2} z_2^2 \, dA_2 = 2E_t\left(\frac{1}{3}\right)td_1^3 + 2E\left(\frac{1}{3}\right)td_2^3$$

Letting $E_r I = (EI)_{\text{eff}}$,

$$E_r I = \frac{2t}{3}[E_t d_1^3 + E d_2^3] \tag{6.4.14}$$

where I = moment of inertia for the gross cross section with respect to its weak-axis centroid.

Using Eq. 6.4.11,

$$2E_t t d_1\left(\frac{d_1}{2}\right) = 2E\, t d_2\left(\frac{d_2}{2}\right)$$

$$d_1^2 = \frac{E}{E_t} d_2^2 \tag{6.4.15}$$

Substituting Eq. 6.4.15 into Eq. 6.4.14, taking $I = \frac{2t}{12}(d_1 + d_2)^3$, and solving for E_r gives

$$E_r = \frac{4EE_t}{(E^{1/2} + E_t^{1/2})^2} \tag{6.4.16}$$

for rectangular sections.

(b) Column strength curve:

Stage 1—buckling at $f = P/A_g < F_y/3$,

$$E_t = E$$

$$E_r = \frac{4E(E)}{(E^{1/2} + E^{1/2})^2} = E$$

Using Eq. 6.4.13 expressed in terms of average unit stress,

$$f_{\text{cr}} = \frac{\pi^2 E_r}{(L/r)^2}$$

$$\frac{L}{r} = \sqrt{\frac{\pi^2 E}{F_y/3}} = \sqrt{\frac{\pi^2(30{,}000)(3)}{50}} = 133.3 \quad \text{(point A)}$$

Stage 2—buckling at $f = P/A_g = F_y/3$,

$$E_t = \frac{2E}{3} \text{(loading fibers)}$$

$$E = E \text{ (unloading fibers)}$$

$$E_r = \frac{4E(2E/3)}{\left(E^{1/2} + \left(\frac{2}{3}E\right)^{1/2}\right)^2} = \frac{8E}{3(1 + \sqrt{2/3})^2} = 0.81E$$

$$\frac{L}{r} = \sqrt{\frac{\pi^2 0.81E}{F_y/3}} = \sqrt{\frac{\pi^2(0.81)(30{,}000)(3)}{50}} = 120 \quad \text{(point B)}$$

Stage 3—buckling at $f = P/A_g = 2F_y/3$,

$$E_t = \frac{2E}{3}$$

$$E_r = 0.18E$$

$$\frac{L}{r} = \sqrt{\frac{\pi^2 0.81E}{2F_y/3}} = \sqrt{\frac{\pi^2(0.81)(30{,}000)(3)}{2(50)}} = 84.8 \quad \text{(point } C\text{)}$$

$$E_t = E/2$$

$$E_r = 0.687E$$

$$\frac{L}{r} = \sqrt{\frac{\pi^2 0.687E}{2F_y/3}} = 78 \quad \text{(point } D\text{)}$$

Stage 4—yielding at $f = P/A_g = F_y$

$$E_t = \frac{E}{2}$$

$$E_r = 0.687E$$

$$\frac{L}{r} = \sqrt{\frac{\pi^2 0.687E}{F_y}} = 63.8 \quad \text{(point } E\text{)}$$

The complete strength curve is shown in Fig. 6.4.5.

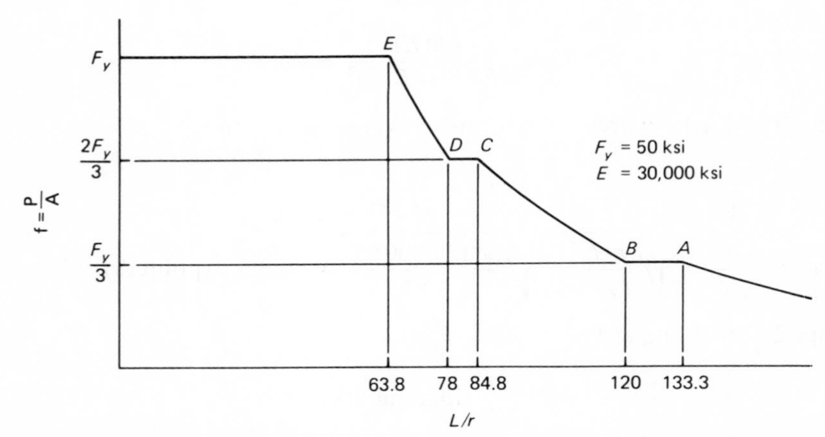

Fig. 6.4.5. Column-strength curve based on double-modulus concept (Example 6.4.1).

Shanley Concept—True Column Behavior. To understand the actual behavior of a column as explained by Shanley[6] in 1946, consider the rectangular section of Fig. 6.4.6 subjected to axial compression. Up until the tangent modulus load P_t is reached, the ideal column remains perfectly straight with zero deflection (point A of Fig. 6.4.6a). The load P_t at

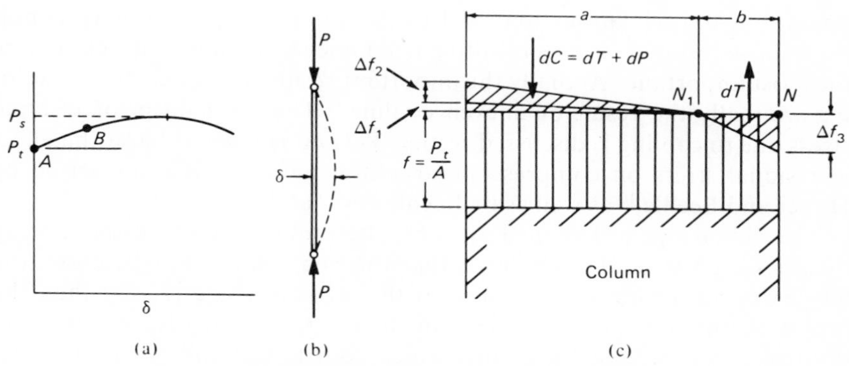

Fig. 6.4.6. Shanley concept—true column behavior.

point A may most correctly be defined as follows:[8] "The tangent modulus load is the smallest value of axial load at which bifurcation of the equilibrium positions can occur regardless of whether or not the transition to the bent position requires an increase of axial load." Consider that at onset of bending (infinitesimal curvature) there will be an infinitesimal increase in axial strain and stress Δf_1. By the time the curvature becomes finite, i.e., the point N moves to N_1, some strain reversal must occur if the column cross section is to develop a resisting moment to maintain equilibrium with the moment due to the external load, $P\delta$. For small but finite values of curvature the increment of load represented by stress on the area of increasing strain exceeds the increment of load represented by stress on the area of decreasing strain; thus, P is increased by an amount dP (point B of Fig. 6.4.6a). As each increment of curvature takes place P will further increase as long as $dC > dT$. The increased compressive force dC is computed using the tangent modulus, E_t, while in the region of strain reversal the elastic modulus, E, is used to compute dT. The double modulus theory, which similarly treated loading and unloading fibers, did not accept $dC > dT$, but rather only considered equilibrium positions near the perfectly straight one.

For practical purposes the increase of capacity from P_t to P_s (Fig. 6.4.6a), can be neglected for design use. Therefore, the tangent modulus load may be treated as the critical load, i.e., the load at which bending begins.

6.5. RESIDUAL STRESS

Residual stresses are stresses that remain in a member after it has been formed into a finished product. Such stresses result from plastic deformations, which in structural steel may result from several sources: (a) uneven cooling which occurs after hot rolling of structural shapes; (b) cold bending or cambering during fabrication; (c) punching of holes and cutting operations during fabrication; and (d) welding. Under ordinary conditions those residual stresses resulting from uneven cooling and welding are the most important. Actually the important residual stresses due to welding are really the result of uneven cooling. The mechanism of residual stress due to cooling is discussed in the *Welding Handbook*[9] and the effect of residual stress on compression structural members is summarized by Huber and Beedle,[10] as well as by Beedle and Tall.[1]

In wide-flange or I-shaped sections, after hot rolling, the flanges, being the thicker parts, cool more slowly than the web region. Furthermore, the flange tips having greater exposure to the air cool more rapidly than the region at the junction of flange to web. Consequently, compressive residual stress exists at flange tips which cool fastest and tensile residual stress exists in the flange at the region where it joins the web and generally throughout the web. There may be compressive residual stress at mid-depth of the web on deeper sections. Figure 6.5.1 shows typical residual

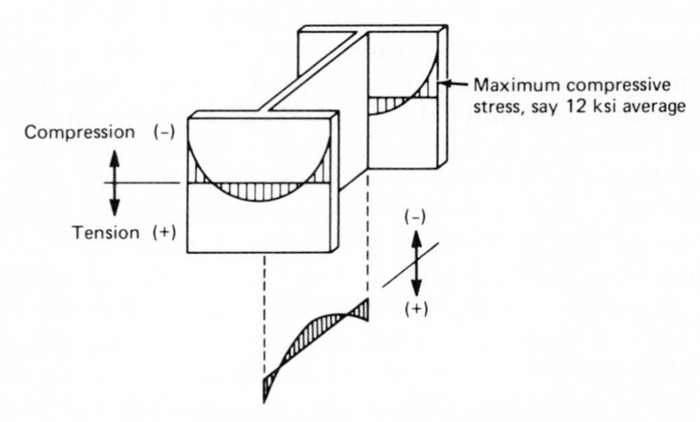

Fig. 6.5.1. Typical residual-stress pattern on rolled shapes.

stress distribution on rolled beams. Considerable variation can be expected as the true pattern will be a function of the dimensions of the beam.

At this point one might wonder whether the general column-strength equation (Eq. 6.3.1) discussed in the preceding section still is applicable. The theory is applicable but all fibers in the cross section cannot be considered as stressed to the same level under the action of the compressive service load. The tangent modulus E_t on one fiber is not the same as that on an adjacent fiber.

In a rolled steel beam the influence of residual stress on the stress-strain curve, using average stress on the gross area as the ordinate is shown in Fig. 6.5.2. It is to be noted that residual stress in an elastic-

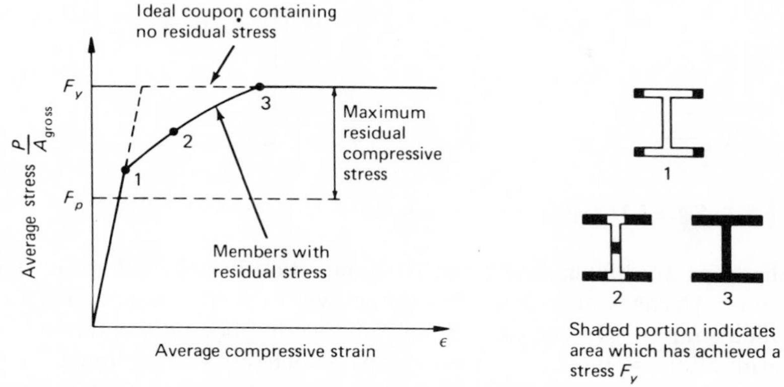

Fig. 6.5.2. Influence of residual stress on average stress-strain curve.

plastic material such as steel gives the same effect as that obtained for a material such as aluminum, which is not perfectly elastic, when it contains no residual stress. Thus, assuming the tangent modulus concept applies, column strength may be said to be based on inelastic buckling because the average stress-strain curve is nonlinear when maximum column strength is reached.

Whereas it was once believed the nonlinear portion of the average stress-strain curve for axially loaded compression members was due entirely to initial curvature and accidental eccentricity, it has been verified[10] that residual stress is the primary cause and the other factors have a relatively minor effect. Residual stresses at flange tips of rolled beams have been measured as high as 20 ksi, a high percentage of the minimum specified yield stress for steels such as A36. Residual stresses, whose magnitudes depend on the cross-sectional shape, are essentially independent of yield point stress.

Welding of built-up shapes is an even greater contributor to residual stress than cooling of hot-rolled shapes.[12] The plates themselves generally have little residual stress initially because of relatively uniform cooling after rolling. However, after the heat is applied to make the welds, the subsequent nonuniform cooling and restraint against distortion cause high residual stresses. Figure 6.5.3 shows typical residual stress patterns on I-shaped and box welded built-up shapes.

One should note that compressive residual stresses typically occurring

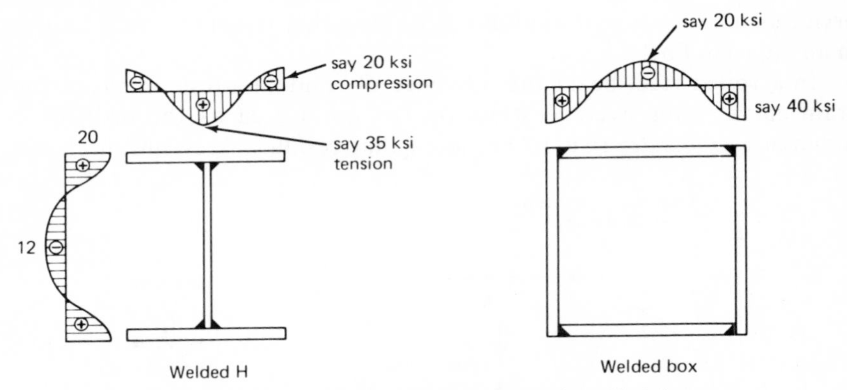

Fig. 6.5.3. Typical residual-stress distribution in welded shapes.

at flange tips are higher in welded H-shapes than in rolled shapes. Thus the column strength of such welded shapes will be lower than corresponding rolled shapes. On the other hand, the welded-box shape, having tensile residual stress in the corner regions which contribute most to the stiffness as a column, will be stronger than a rolled shape having the same slenderness ratio.

Having accepted that residual stresses exist, such information must be used to obtain a column strength curve (average stress vs slenderness ratio) which will form the basis for design. Until the early 1950's, column design was based on many formulas, all of which tried to empirically account for column behavior exhibited by tests. By clearly indicating that the tangent modulus was the proper criterion for strength and by identifying the role of residual stress, the Column Research Council[3] has made a significant contribution.

6.6.* DEVELOPMENT OF COLUMN STRENGTH CURVES INCLUDING RESIDUAL STRESS

The analytical approach which follows and which shows the logic behind the CRC basic column strength equation is essentially as presented by Huber and Beedle.[10] There are two general methods by which column strength can be obtained. Analytically, making use of the residual stress distribution (either measured or assumed) along with the stress-strain diagram for the material (coupon test), the strength can be expressed as a function of the moment of inertia of the unyielded part of the cross section and the slenderness ratio.[11] An alternate method is to experimentally determine an average stress-strain relationship from a short section of rolled shape containing residual stress. Column strength can then be determined using the tangent modulus of this average stress-strain curve along with the slenderness ratio which is proper for strong or weak axis bending. This second approach does not require knowledge about residual stress. Both methods assume symmetrical patterns of residual stress.

The following development is made with the objective of obtaining a relationship between average externally applied stress and the slenderness ratio. Thus the capacity of a member can be obtained by a simple multiplication of safe stress times gross area, without regard to what the actual stress is at any point in the cross section or what is the true residual stress pattern.

As a starting point consider steel which as a material is perfectly elastic until a certain strain ϵ_y is achieved and then is plastic (i.e., constant capacity with increasing strain). A coupon cut from the web of a rolled shape exhibits such behavior which is shown as the dotted lines in Fig. 6.6.1. The solid lines indicate the behavior of a wide-flange shape including residual stress effects.

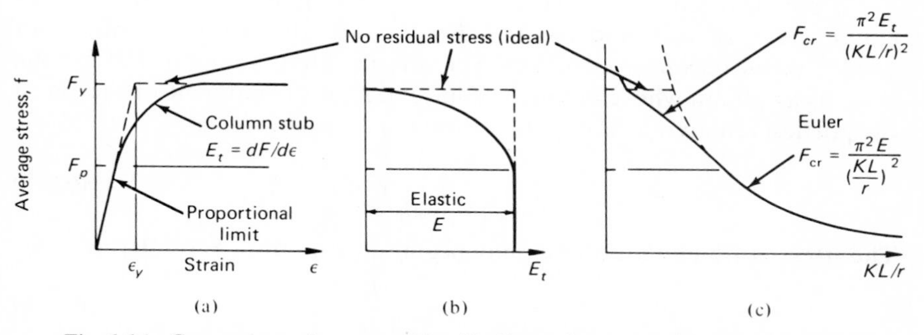

Fig. 6.6.1. Comparison of coupon with wide-flange shape containing residual stress.

To account for the effects of early yielding due to residual stress, consider one fiber at a distance x from the axis of zero strain due to bending (Fig. 6.6.2). The bending is taken as an infinitesimal amount consistant with equilibrium at the tangent modulus load. The bending-moment contribution from stress on the one fiber is

$$dM = (\text{stress})(\text{area})(\text{moment arm}) = (\phi E_t x)(dA)(x) \qquad (6.6.1)$$

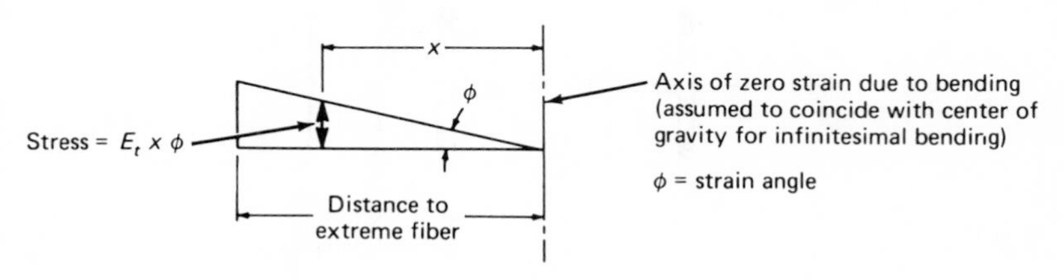

Fig. 6.6.2.

which for the entire cross section becomes

$$M = \int_A \phi E_t x^2\, dA = \phi \int_A E_t x^2\, dA \tag{6.6.2}$$

From elementary bending theory, the radius of curvature is

$$R = \frac{1}{\phi}$$

$$\phi = \frac{1}{R} = \frac{M}{\text{Equivalent } EI} = \frac{M}{E'I} \tag{6.6.3}$$

Thus

$$E'I = \frac{M}{\phi} = \int_A E_t x^2\, dA$$

$$E' = \frac{1}{I} \int_A E_t x^2\, dA \tag{6.6.4}$$

which may be called the "effective modulus."

If the idealized elastic-plastic F-ϵ curve of Fig. 6.6.1a (dotted) is used (for $f < F_y$, $E_t = E$ and for $f = F_y$, $E_t = 0$) the bending stiffness of yielded parts becomes zero; however, the buckling strength will be the same as a column whose moment of inertia, I_e, is the moment of inertia of the portion remaining elastic. Equation 6.6.4 then becomes

$$E' = \frac{E}{I} \int_{A\,(\text{elastic part only})} x^2\, dA = E \frac{I_e}{I} \tag{6.6.5}$$

The stress at which the column may begin to bend, from Eq. 6.3.1, is

$$F_{\text{cr}} = \frac{P}{A} = \frac{\pi^2 E(I_e/I)}{(KL/r)^2} \tag{6.6.6}$$

In order for Eq. 6.6.6 to be useful, the relationship between F_{cr} and I_e must be established.

Case A. Buckling about Weak Axis. A reasonable assumption will be that the flanges become fully plastic before the web yields (see Fig. 6.6.3).

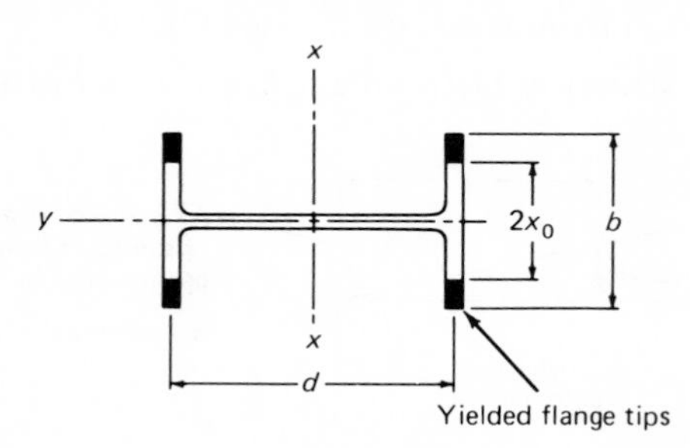

Fig. 6.6.3.

Let k = proportion of the flange remaining elastic
$= 2x_o/b = A_e/A$

Then Eq. 6.6.5 becomes

$$E\,\frac{I_e}{I} = E\,\frac{t_f(2x_o)^3}{12}\left(\frac{12}{t_f b^3}\right) = Ek^3 \tag{6.6.7}$$

if the web is neglected in computing I. Applying the tangent modulus definition,

$$E_t = \frac{\text{Nominal incremental stress}}{\text{Incremental elastic strain}}$$

$$= \frac{dP/A}{\dfrac{dP(L)}{A_e E}\left(\dfrac{1}{L}\right)} = \frac{A_e E}{A} \tag{6.6.8}$$

$$E_t A = A_e E$$

$$= E(A_w + 2kA_f) \tag{6.6.9}$$

where A_w = web area
A_f = gross area of one flange
A = total gross area of section

Solving Eq. 6.6.9 for k and using Eq. 6.6.7 in Eq. 6.6.6 gives

$$k = \frac{E_t A}{2EA_f} - \frac{A_w}{2A_f} \tag{6.6.10}$$

$$F_{cr} = \frac{\pi^2 Ek^3}{(KL/r)^2} = \frac{\pi^2 E}{(KL/r)^2}\left[\frac{AE_t}{2A_f E} - \frac{A_w}{2A_f}\right]^3 \tag{6.6.11}$$

which includes the effect of the elastic web, for buckling with respect to the weak axis (y–y).

Case B. Buckling about Strong Axis. Again, assuming the web is elastic, but neglecting its contribution towards the moment of inertia gives approximately

$$E\,\frac{I_e}{I} \approx E\,\frac{2A_e(d/2)^2}{2A_f(d/2)^2} = Ek \tag{6.6.12}$$

If the elastic web is included,

$$E\,\frac{I_e}{I} = E\left[\frac{2kA_f(d^2/4) + t_w d^3/12}{2A_f(d^2/4) + t_w d^3/12}\right] \tag{6.6.13}$$

$$= E\left[\frac{2kA_f + A_w/3}{2A_f + A_w/3}\right] \tag{6.6.14}$$

Using tangent modulus definition and Eq. 6.6.9,

$$2kA_f = \frac{E_t A}{E} - A_w$$

which, upon eliminating the $2kA_f$ term from Eq. 6.6.14, gives

$$E\frac{I_e}{I} = \left[\frac{E_t A/E - 2A_w/3}{2A_f + A_w/3}\right]E \qquad (6.6.15)$$

Thus

$$F_{cr} = \frac{\pi^2 Ek}{(KL/r)^2} \qquad (6.6.16)$$

is the approximate equation using Eq. 6.6.10 for k, or more exactly using Eq. 6.6.15 in Eq. 6.6.6 gives

$$F_{cr} = \frac{\pi^2 E}{(KL/r)^2}\left[\frac{E_t A/E - 2A_w/3}{2A_f + A_w/3}\right] \qquad (6.6.17)$$

for buckling with respect to the strong axis (x–x).

From the foregoing development it is apparent that two equations are necessary to properly determine column strength, one for strong axis buckling and one for weak axis buckling. Although the value I_e/I is not itself a function of the residual stress distribution provided that it satisfies the general geometric requirement as shown in Fig. 6.6.3; nevertheless, the critical stress, F_{cr}, computed as the buckling load divided by the gross area, has a relationship with KL/r that does depend on residual stress.

EXAMPLE 6.6.1.

Establish the column strength curve (F_{cr} vs. KL/r) for weak axis buckling of an H-shape on high strength steel having a yield stress of 100 ksi, exhibiting perfect elastic-plastic strength in a coupon test (Fig.

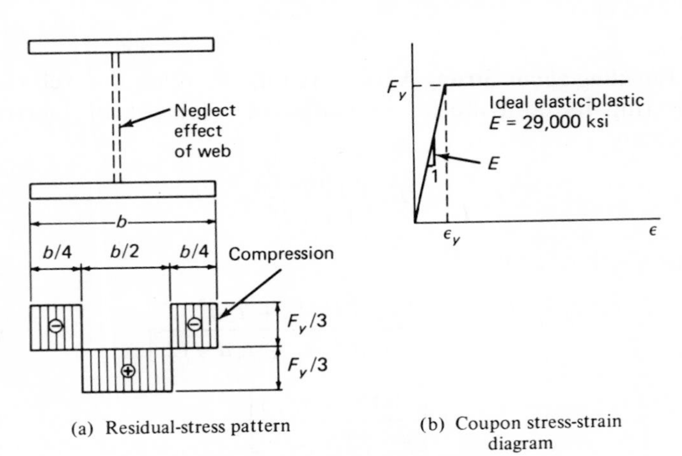

(a) Residual-stress pattern

(b) Coupon stress-strain diagram

Fig. 6.6.4. Data for Example 6.6.1.

6.6.4b), and having the very simplified residual stress pattern shown in Fig. 6.6.4a. Neglect the contribution of the web.

Solution

For any external load the strain on every fiber is the same. Until a fiber reaches the strain ϵ_y at first yield, the applied load is

$$P = \int_A f\,dA = fA$$

After a portion has become plastic, the applied load is

$$P = (A - A_e)F_y + \int_{A_e} f\,dA$$

In this problem for $F_{cr} = P/A \leq 2F_y/3$ the entire section remains elastic, $E_t = E$,

$$F_{cr} = \frac{2F_y}{3} = \frac{\pi^2 E}{(KL/r)^2}$$

$$\frac{KL}{r} = \sqrt{\frac{\pi^2(29{,}000)}{(2/3)(100)}} = 65.4 \text{ (point 1, Fig. 6.6.5)}$$

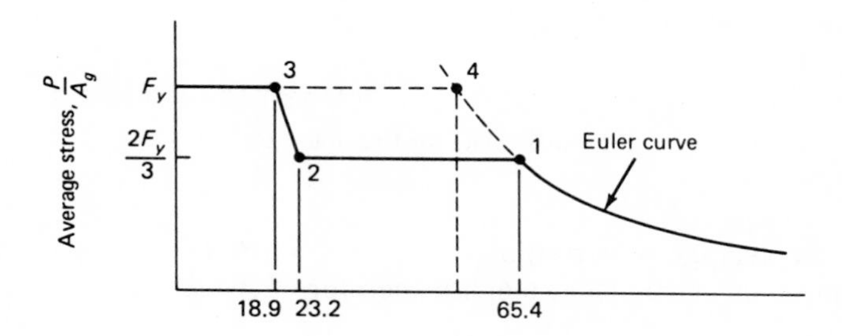

Fig. 6.6.5. Column-strength curve for Example 6.6.1.

When $F_{cr} = P/A > 2F_y/3$, the flange tips have yielded; thus

$$\frac{I_e}{I} = \frac{(b/2)^3}{b^3} = \frac{1}{8}$$

$$F_{cr} = \frac{2F_y}{3} = \frac{\pi^2 E(I_e/I)}{(KL/r)^2} = \frac{\pi^2 E}{8(KL/r)^2}$$

$$\frac{KL}{r} = 23.2 \quad \text{(point 2, Fig. 6.6.5)}$$

for average stress infinitesimally greater than $2F_y/3$. When $F_{cr} = P/A = F_y$,

$$F_{cr} = F_y = \frac{\pi^2 E}{8(KL/r)^2}$$

$$\frac{KL}{r} = 18.9 \quad \text{(point 3, Fig. 6.6.5)}$$

when the total load $P = F_y A$. The results are shown in Fig. 6.6.5. If there had been no residual stress at $F_{cr} = F_y$,

$$\frac{KL}{r} = 53.5 \quad \text{(point 4, Fig. 6.6.5)}$$

EXAMPLE 6.6.2.

Establish the column-strength curve for the more realistic linear distribution of residual stress shown in Fig. 6.6.6. Consider weak-axis buckling of an H-section of steel with F_y = 36 ksi and 100 ksi. Neglect the effect of the web.

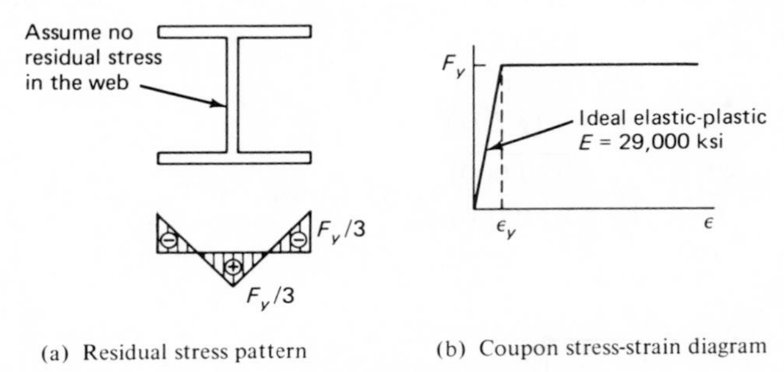

(a) Residual stress pattern (b) Coupon stress-strain diagram

Fig. 6.6.6. Data for Example 6.6.2.

Solution

For an average superimposed stress $f = P/A \leq 2F_y/3$ the entire section remains elastic (Fig. 6.6.7a); therefore $E_t = E$, and

$$F_{cr} = \frac{2F_y}{3} = \frac{\pi^2 E}{(KL/r)^2}$$

For an average stress due to applied load greater than $2F_y/3$, part of the cross-section is plastic and part elastic, as in Fig. 6.6.7b. During this stage, the *change in stress is not the same on all fibers*, because the modulus of elasticity is not the same on all fibers.

$$F_{cr} = \frac{\pi^2 E I_e/I}{(KL/r)^2}$$

$$\frac{I_e}{I} = \frac{2(1/12)(2x_0)^3 t}{2(1/12)b^3 t} = \frac{8x_0^3}{b^3}$$

neglecting the effect of the web,

$$F_{cr} = \frac{8\pi^2 E(x_0/b)^3}{(KL/r)^2} \tag{a}$$

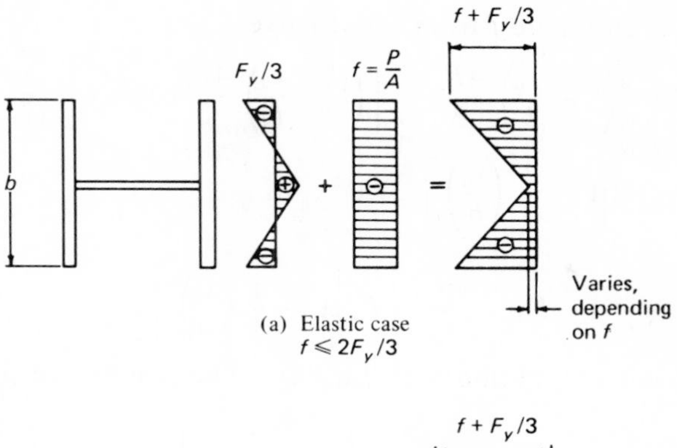

(a) Elastic case
$f \leqslant 2F_y/3$

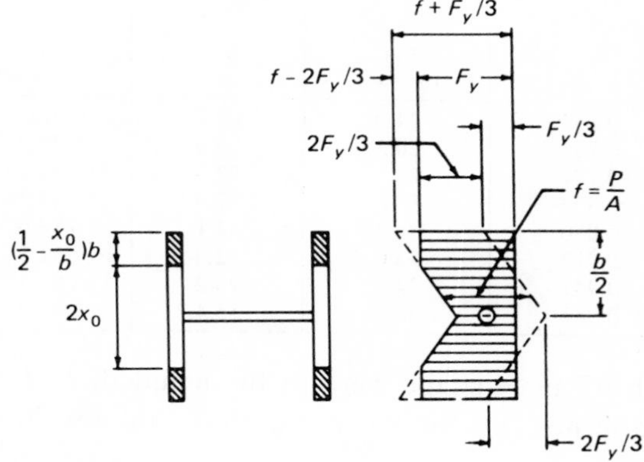

f = superimposed stress on elastic fibers

(b) Elasto-plastic case
$f > 2F_y/3$

Fig. 6.6.7. Stress distribution with linear residual stress variation.

which gives F_{cr} as a function of two variables, x_0/b and KL/r. An additional relationship is required. The total load during the elasto-plastic stage can be expressed

$$P_{cr} = 2\left[fbt - 2\left(\frac{1}{2}\right)\left(f - \frac{2F_y}{3}\right)\left(\frac{1}{2} - \frac{x_0}{b}\right)bt\right] \qquad \text{(b)}$$

which is the shaded area of the stress diagram in Fig. 6.6.7b. From similar triangles on the dotted triangle of Fig. 6.6.7b,

$$\frac{f - 2F_y/3}{\left(\frac{1}{2} - \frac{x_0}{b}\right)b} = \frac{2F_y/3}{b/2}$$

$$f = \left[1 - \frac{x_0}{b}\right]\frac{4F_y}{3} \qquad \text{(c)}$$

Using Eq. (c) to eliminate f from Eq. (b) gives

$$P_{cr} = 2bt\left\{\left(1 - \frac{x_0}{b}\right)\frac{4F_y}{3} - \left[\left(1 - \frac{x_0}{b}\right)\frac{4F_y}{3} - \frac{2F_y}{3}\right]\left(\frac{1}{2} - \frac{x_0}{b}\right)\right\}$$

$$= A_g F_y\left[1 - \frac{4}{3}\left(\frac{x_0}{b}\right)^2\right] \tag{d}$$

Thus

$$F_{cr} = \frac{P}{A_g} = F_y\left[1 - \frac{4}{3}\left(\frac{x_0}{b}\right)^2\right] \tag{e}$$

which is used in combination with Eq. (a). The results are presented in Fig. 6.6.8.

x_0/b	F_{cr}	For $F_y = 36$	$\frac{KL}{r}$	For $F_y = 100$	$\frac{KL}{r}$
0.50	$0.66F_y$	24.0	109.2	66.7	65.4
0.45	0.73	26.3	89.0	73.0	53.4
0.40	0.787	28.3	72.0	78.7	43.1
0.35	0.837	30.2	57.0	83.7	34.2
0.30	0.880	31.7	44.1	88.0	26.5
0.25	0.917	33.0	32.9	91.7	19.7
0.20	0.947	34.1	23.2	94.7	13.9
0.10	0.987	35.5	15.4	98.7	4.8

If the web of an H-shape were to be included, I_e/I could easily include the web terms. Furthermore, Eq. (b) could also have included the

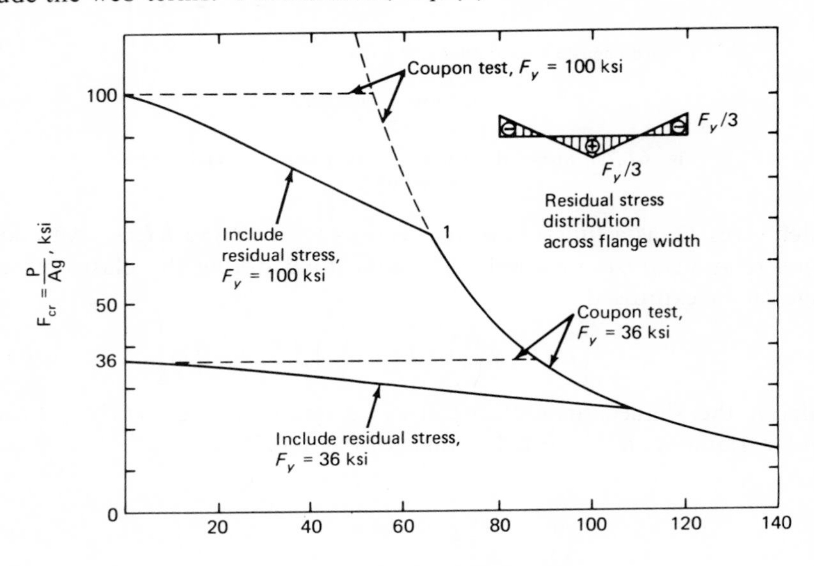

Fig. 6.6.8. Column-strength curves showing effect of residual stress (E = 29,000 ksi). Solution for Example 6.6.2.

web terms. Such inclusion of the effect of the web brings in the variable A_w/A_f and in most cases the effect is small.

Finally, curves similar to those of Fig. 6.6.8 can be obtained by using an average stress-strain curve for a short segment of rolled shape as referred to earlier in this article. In which case Eqs. 6.6.11 and 6.6.17 can be used with the E_t obtained from the "cross section" stress-strain curve.

6.7. COLUMN RESEARCH COUNCIL BASIC STRENGTH CURVE

Based upon the methods discussed in Sec. 6.6, column strength curves can be obtained for weak- or strong-axis buckling with various distributions of residual stress. For most practical situations it has been reported that an assumed linear distribution of residual stress in the flanges results in a fair average column curve.[10] Furthermore, the development in the previous article (Eqs. 6.6.11 and 6.6.16) shows that for the *same* slenderness ratio, H-section columns allowed to bend in the weak direction can carry less load than columns permitted to bend only in the strong direction. Compressive residual stress which is greatest at the flange tips accounts for this strength difference.

Typical column-strength curves for parabolic and linear distribution of residual stress across the flange are shown in Fig. 6.7.1. For structural carbon steels the average value of the maximum compressive residual stress is approximately[2] $0.3F_y$. For the high-strength steels residual stress will generally be a lower fraction of the yield stress.

The basic column-strength curve adopted by the Column Research Council is based on the parabolic equation proposed by Bleich.[13] Since curves (2), (3), and (4) of Fig. 6.7.1 are essentially parabolas, and since test results have generally given essentially a parabolic variation for buckling in the inelastic range, a parabola representing a compromise among the four curves of Fig. 6.7.1 seems reasonable. The curve proposed by Bleich is

$$F_{cr} = F_y - \frac{F_p}{\pi^2 E}(F_y - F_p)\left(\frac{KL}{2}\right)^2 \tag{6.7.1}$$

where F_p = stress at the proportional limit. Since the deviation from elastic behavior on the average stress-strain curve is accounted for by residual stress, the F_p was replaced by

$$F_p = F_y - F_r = \text{yield stress} - \text{residual stress}$$

and Eq. 6.7.1 becomes

$$F_{cr} = F_y - \frac{F_y - F_r}{\pi^2 E}(F_y - F_y + F_r)\left(\frac{KL}{r}\right)^2$$

$$= F_y\left[\frac{F_r}{F_y}(F_y - F_r)\frac{(KL/r)^2}{\pi^2 E}\right] \tag{6.7.2}$$

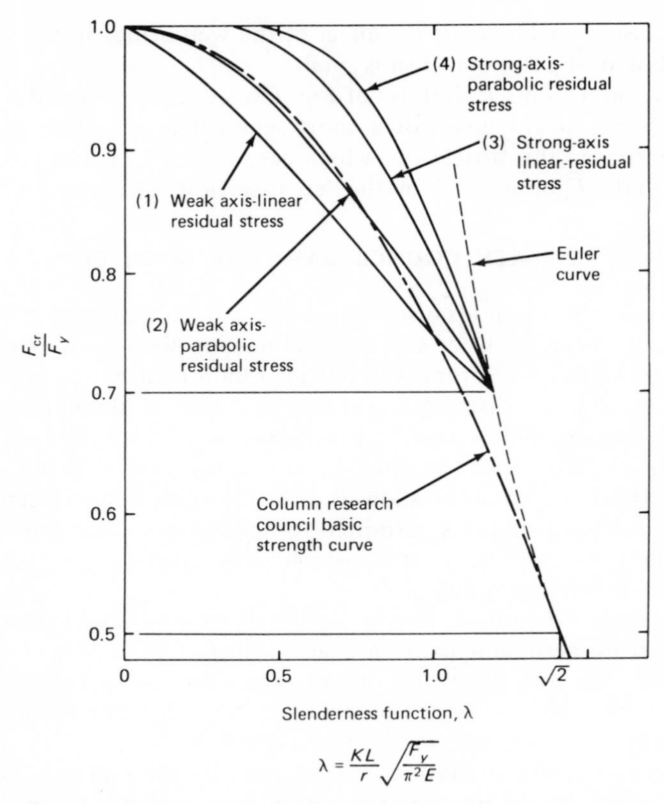

Fig. 6.7.1. Column-strength curves for H-shapes with compressive residual stress at flange tips. (From Ref. 2, p. 23.)

In order for Eq. 6.7.2 to properly represent a compromise between weak and strong axis buckling, the residual stress must be assumed higher than $0.3F_y$. The Column Research Council arbitrarily chose $F_p = 0.5F_y$ so as to give a smooth transition from the Euler curve for elastic buckling to the parabolic curve representing inelastic buckling. The two curves are tangent at $F_{cr}/F_y = 0.5$. The CRC curve then becomes

$$F_{cr} = F_y\left[1 - \frac{F_y}{4\pi^2 E}\left(\frac{KL}{r}\right)^2\right] \tag{6.7.3}$$

In Fig. 6.7.1 where the CRC curve is compared with other more exact curves which distinguish between residual stress patterns and axes of bending, it may appear that the CRC compromise curve lies too far above curve (1) over much of its range. When it is realized that typically the residual stress pattern is somewhere between linear and parabolic the true weak axis strength curve is actually closer to the CRC curve.

While design formulas based on various empirical rules and "curve fitting" still exist, it is nevertheless generally accepted that the CRC curve, which uses the tangent modulus concept and accounts for residual stress, has merit because of its generality and inherent correctness.[2]

6.8. AISC DESIGN EQUATIONS

The basic strength curve of the Column Research Council, Eq. 6.7.3, divided by a factor of safety becomes the allowable stress equation of the AISC specification for slenderness ratios where inelastic buckling controls. Thus, using a critical stress of $0.5F_y$ (assumed as the proportional limit) in Eq. 6.7.3 gives the maximum value of KL/r, defined as C_c,

$$0.5F_y = F_y\left[1 - \frac{F_y}{4\pi^2 E}\left(\frac{KL}{r}\right)^2\right]$$

$$C_c = \frac{KL}{r} = \sqrt{\frac{2\pi^2 E}{F_y}} = \frac{755}{\sqrt{F_y}} \tag{6.8.1}$$

when $E = 29{,}000$ ksi and F_y is in ksi.

Therefore, Eq. 6.7.3 upon substitution of C_c becomes the AISC column formula which governs inelastic buckling for slenderness ratios, $KL/r \leqq C_c$,

$$F_a = \frac{F_y}{\text{FS}}\left[1 - \frac{(KL/r)^2}{2C_c^2}\right] \tag{6.8.2}$$

where F_a = allowable stress on gross area under service, or working, load; KL/r is the slenderness ratio of the *equivalent* pin-end column; and FS = factor of safety.

While Eq. 6.8.2 is ordinarily applicable for design, and is 1969 AISC Formula (1.5–1); the specification also provides for a reduced efficiency of the cross section for thin-walled sections which do not satisfy the basic requirements of AISC–1.9.

1969 AISC, therefore, adds in Appendix C a form factor Q to Eqs. 6.8.1 and 6.8.2. $Q = 1.0$ for all sections satisfying AISC–1.9. Thus

$$F_a = \frac{QF_y}{\text{FS}}\left[1.0 - \frac{(KL/r)^2}{2C_c^2}\right] \tag{6.8.3}$$

which is AISC Formula (C5–1). Also,

$$C_c = \sqrt{\frac{2\pi^2 E}{QF_y}} \tag{6.8.4}$$

Since ordinary designs will use $Q = 1$, the development of the logic behind the use of such a factor is reserved for Sec. 6.18 in Part II on plate buckling.

Since all columns contain some initial curvature and accidental eccentricity, the factor of safety should reflect such conditions. For short members which are negligibly affected by small eccentricity of load or by residual stress, the factor of safety need not be greater than is deemed desirable for tension members, i.e., 1.67 under AISC specifications. For long columns accidental eccentricity definitely affects strength so that the factor of safety should be increased. The AISC prescribes a 15 percent increase over the basic value, i.e., a maximum factor of 1.92. Arbitrarily, to obtain a smooth transition from FS = 1.67 for $KL/r = 0$ to FS = 1.92 for $KL/r = C_c$, a cubic equation approximation to a quarter sine wave was used:

$$\text{FS} = \frac{5}{3} + \frac{3}{8}\frac{KL/r}{C_c} - \frac{1}{8}\left(\frac{KL/r}{C_c}\right)^3 \tag{6.8.5}$$

which is used with Eq. 6.8.3. Equation 6.8.5 is illustrated graphically in Fig. 6.8.1.

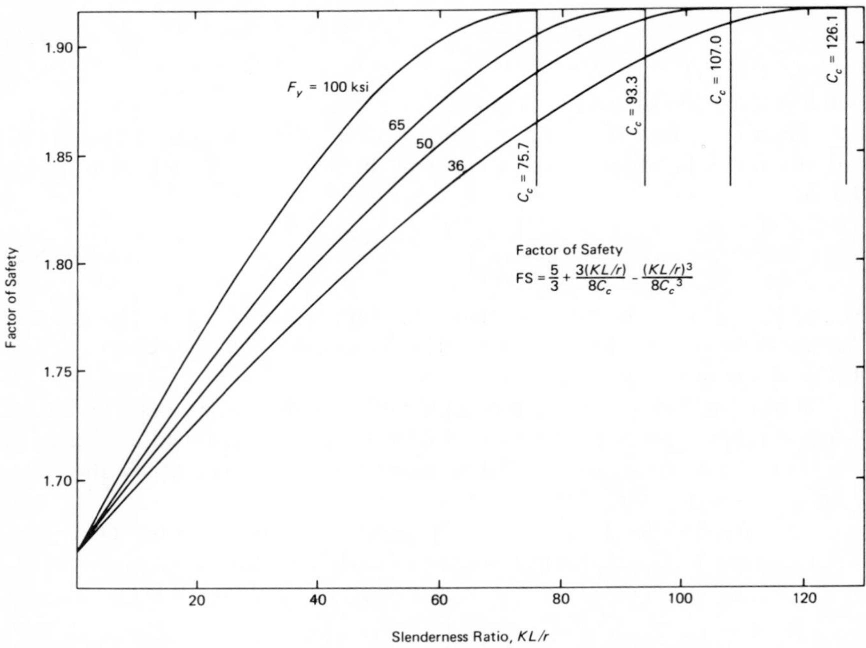

Fig. 6.8.1. Factor of safety for members under axial compression according to AISC specifications.

For columns with slenderness ratios exceeding $KL/r = C_c$, elastic buckling controls as defined by the Euler equation (Eq. 6.3.1 with $E_t = E$),

$$F_a = \frac{\pi^2 E}{\text{FS}\,(KL/r)^2} \tag{6.8.6}$$

which gives allowable stress on gross area for long columns. In this situation the factor of safety is taken constant at 1.92. Using $E = 29{,}000$ ksi and FS = 1.92, Eq. 6.8.6 becomes

$$F_a = \frac{149{,}000}{(KL/r)^2} \text{ ksi} \tag{6.8.7}$$

or, using FS = 23/12 which Eq. 6.8.5 gives for $KL/r = C_c$,

$$F_a = \frac{12\pi^2 E}{23(KL/r)^2} \tag{6.8.8}$$

which is AISC Formula (1.5–2).

The criteria of safety used in establishing the factor to be applied assume the members to be principal load carrying members. It is logical, therefore, that reduced factors of safety are appropriate for bracing and secondary members. The AISC retains from the 1946 specifications the following empirical expression which serves to increase the allowable stress on bracing and secondary members whose slenderness ratio exceeds 120,

$$F_a = \frac{\text{Eq. 6.8.3 or 6.8.8}}{\left(1.6 - \dfrac{L/r}{200}\right)} \tag{6.8.9}$$

It is to be noted that C_c may be either more or less than 120, but for most practical purposes it would be sufficient to assume Eq. 6.8.8 acceptable for use in Eq. 6.8.9.

After one understands the basis for the design formulas, routine designs should utilize the tables of allowable stresses which are in the AISC Appendix which forms an integral part of the specification. For many of the common types of steel the allowable stress curves are shown in Fig. 6.8.2.

6.9. EFFECTIVE LENGTH

Discussion of column strength to this point has assumed hinged ends or ends with no moment restraint. Zero moment restraint at ends constitutes the weakest situation for compression members when translation of one end relative to the other is prevented. For such pin-end columns the effective length equals the actual length, i.e., $K = 1.0$.

For most real situations moment restraint at the ends does exist and the inflection points in the buckled shape curve may occur at locations other than the ends of the member. The distance between the points of inflection, either real or imaginary, is the effective or equivalent pinned length for the column.

It is difficult in most situations to evaluate the degree of moment restraint offered by members which frame into the given column, by the

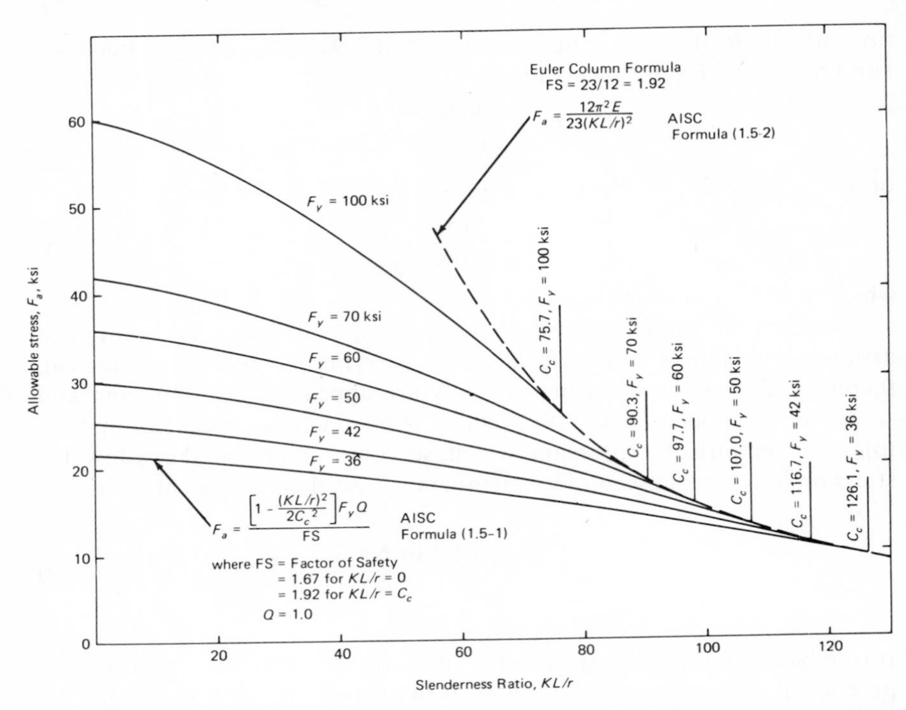

Fig. 6.8.2. Allowable stresses on gross area due to axial compression. (AISC–1.5.1.3)

footing and soil under it, and indeed the full interaction of all the members of a structural steel frame.

Whether or not the designer is able to determine the degree of end restraint accurately, he should recognize that structural frames can be separated into two categories, those whose joints are adequately prevented from translation and those whose joints can translate. A *braced* frame whose joints cannot translate is one in which "lateral stability is provided by adequate attachment to diagonal bracing, shear walls, an adjacent structure having adequate lateral stability, or to floor slabs or roof decks secured horizontally by walls or bracing systems parallel to the plane of the frame." An *unbraced* frame is one "where lateral stability is dependent upon the bending stiffness of rigidly connected beams and columns." The effective length in braced frames is never greater than the actual length, i.e., $K \leq 1.0$. Since this chapter is concerned essentially with compression members containing no intentional bending moment, the theoretical development to justify the aforementioned frame relationships appears in Chapter 14.

Compression members which can properly be designed as axially loaded are usually interior members in braced structures where the compressive loads are nearly symmetrically applied and either little or no bending results from frame action, i.e., situations where any small bending moment may be safely neglected.

The effective lengths for ideal end restraint in situations where columns might be designed as axially loaded are given in Fig. 6.9.1 along with practical design values as recommended by the Column Research Council. If one wishes a more exact approach than just an estimate based on Fig. 6.9.1, he can make use of the alignment charts which appear as

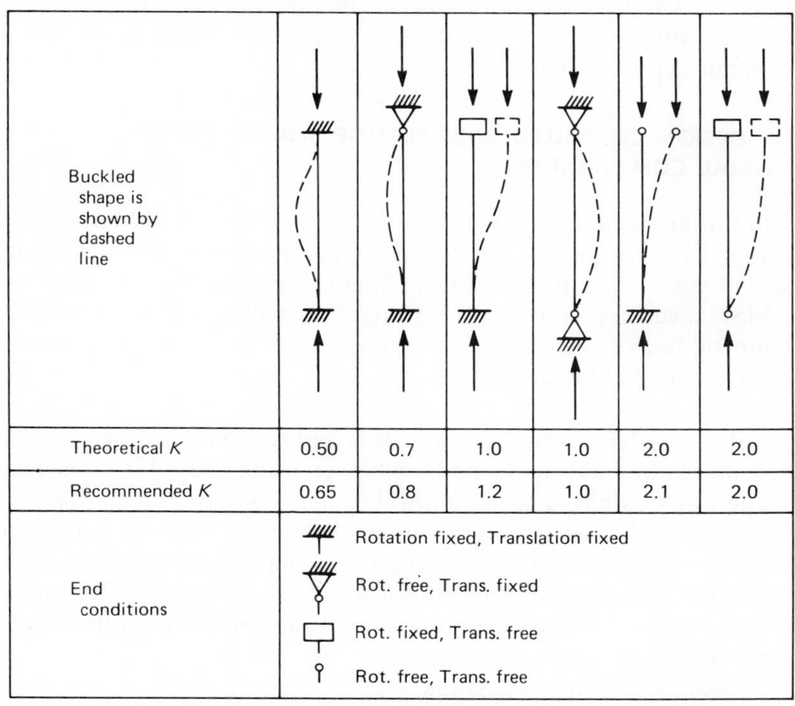

Fig. 6.9.1. Effective-length factors for centrally loaded columns with various idealized end conditions. (Ref. 2, p. 41.)

Fig. 14.3.1 in Chapter 14 and which are based on end restraint factors dependent upon the members which frame together at the ends of the column under consideration.

Safety in column design necessitates knowing the difference between braced and unbraced frames, i.e., whether the effective length is less than or greater than actual length. A thorough study of Chapter 14 is recommended for those involved in the design of compression members in frames.

For truss compression members, end restraint may be present and joint translation is prevented so that K might logically be less than 1.0. Under static loading, stresses in all the members remain in the same proportion to one another for various loads. If all members are designed for minimum weight they will achieve ultimate capacity simultaneously under live load. Thus, restraint offered by members framing at a joint disappears or at least is greatly reduced. The CRC, therefore, recommends using $K = 1.0$ for members of a truss designed for fixed-position loading. When designing for moving load systems on trusses, K can be reduced to 0.85 because conditions causing maximum stress in the member under consideration will not cause maximum stress in the members framing in to provide restraint.[2]

6.10. DESIGN OF ROLLED WIDE-FLANGE SHAPES UNDER AXIAL COMPRESSION

In this article reference will be made to the AISC Steel Manual, Part 1 which gives properties of the rolled shapes and Part 3 which contains column tables. Whether or not the final choice of a section must satisfy the AISC specification, use of this general procedure and use of the Part 3 column tables will enable the designer to get at the least a good preliminary estimate of member size.

General Procedure. The design of compression members, whether rolled shapes or built-up sections, whether using the AISC specification or some other, is based on an allowable stress on gross area. The allowable stress in all cases is a function of slenderness ratio and yield point of the material. Since the allowable stress depends on slenderness ratio KL/r, where r depends on the section selected, the design of compression members is an indirect process unless load tables are available. The general procedure is

1. Assume an allowable stress.
2. Based on computed area requirement, select a section.
3. Based on KL/r for the section selected, compute allowable stress.
4. Compute P/A and compare with allowable stress.
5. If P/A is less than allowable or not more than 2 to 3 percent greater, generally the design would be considered acceptable.

EXAMPLE 6.10.1

Select the lightest W section of A36 steel to serve as a pin-ended main member column 16 ft long to carry an axial compression load of 195 kips in a braced structure (see Fig. 6.10.1). Use 1969 AISC specification, and indicate first three choices.

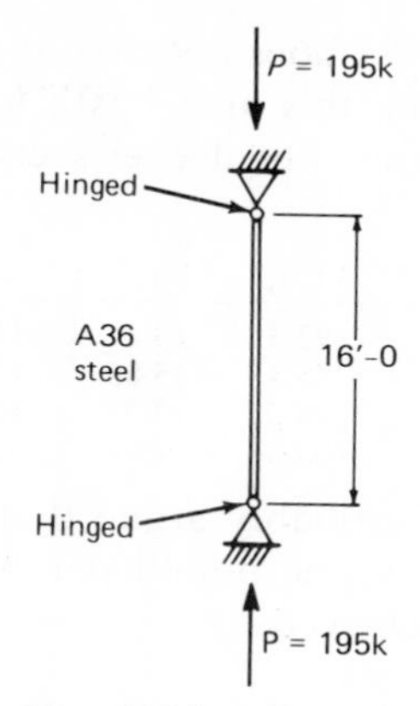

Fig. 6.10.1. Example 6.10.1.

Solution

Since the assumption of hinged ends is made, the effective length equals the actual length, i.e., $K = 1.0$. Considering $KL = 16$ ft as a moderately long length, the slenderness ratio may be estimated at about 70 to 80. For rolled W shapes, $Q = 1.0$.

Estimate allowable stress, using AISC Formula (1.5–1), Eq. 6.8.3 along with Eq. 6.8.5 for factor of safety. In lieu of computing the FS for the KL/r of the as yet unselected member, it may be estimated as about 1.85. Thus,

$$F_a = \frac{36}{1.85}\left[1 - \frac{80^2}{2(126.1)^2}\right] = \frac{36}{1.85}[1 - 0.201] = 15.55 \text{ ksi}$$

or F_a could be estimated directly by referring to Fig. 6.8.2 (for $KL/r = 80$, $F_a \approx 15.4$ ksi):

$$A_{\text{reqd}} = \frac{195}{15.55} = 12.5 \text{ sq in.}$$

Since weak-axis bending will certainly control the strength for W shapes, when KL is the same with respect to both x and y axes, the lightest sections for this problem will be those which have the least r_x/r_y. A high r_x indicates excessive strength with respect to the strong axis, which may only be utilized by providing additional bracing in the weak direction.

Using the column tables for W shapes in Part 3 of the AISC Manual, one might tentatively select a W8 × 48, the lightest W8 which has at least the required area. Furthermore, these W8 sections have the lowest r_x/r_y for a given area.

Try W8 × 48:

$$A = 14.1 \text{ sq in.}, \qquad f = P/A = 13.8 \text{ ksi}$$
$$KL/r_y = 92.3 \qquad F_a = 13.8 \text{ ksi}$$

Since the computed stress P/A does not exceed the allowable value, and since no other section having this area has a lower r_x/r_y, the W8 × 48 is the lightest available section. For deeper sections, heavier sections will be required, as follows:

Section	Area	KL/r_y	F_a	P/A	
W8 × 48	14.1	92.3	13.8	13.8	1st choice
W10 × 49	14.4	75.5	15.85	13.52	2nd choice
W12 × 50	14.7	98.0	13.23	13.25	3rd choice
W14 × 53	15.6	100.0	12.98	12.50	

Some designers might justifiably argue that the W10 × 49 is better than W8 × 48 since for only one pound per foot extra a considerably greater margin of safety is achieved.

EXAMPLE 6.10.2

Select the lightest W section of A36 steel to serve as a main member 30 ft long and carry an axial compression of 160 kips. The member may be assumed pinned at top and bottom and in addition has weak direction support at mid-height (see Fig. 6.10.2). The structure is braced and the 1969 AISC specification is to be used.

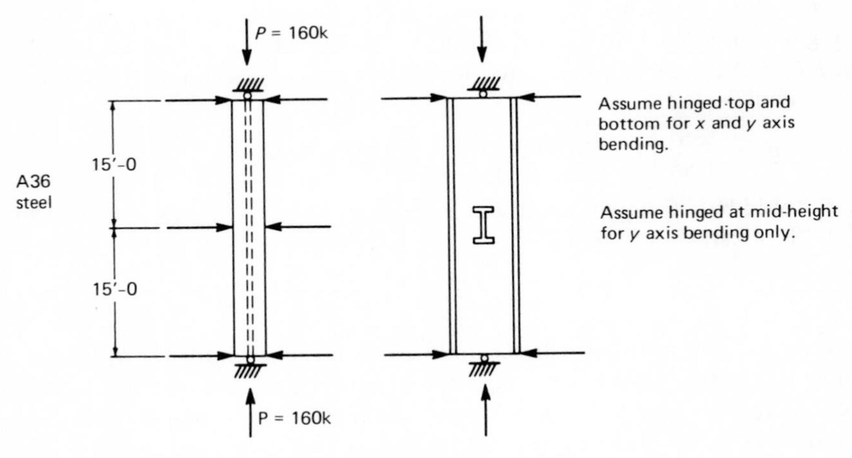

Fig. 6.10.2. Example 6.10.2.

Solution

Use the AISC column tables entering with the length corresponding to that for weak direction bending since tabular loads are based on KL/r_y,

$$P = 160 \text{ kips} \qquad L_y = 15 \text{ ft}$$

Starting with the shallow W8 sections and working toward the deeper sections, find

$$\text{W8} \times 40 \qquad P = 170 \text{ kips} \qquad r_x/r_y = 1.73$$

$$\begin{array}{lll} W10 \times 39 & P = 162 & r_x/r_y = 2.16 \\ W12 \times 40 & P = 163 & r_x/r_y = 2.64 \end{array}$$

Since the actual support conditions are such that $KL_x = 2KL_y$, if $r_x/r_y \geq 2$, weak-axis controls and tabular loads give the correct answer. Thus, W10 × 39 and W12 × 40 are obviously acceptable.

Since r_x/r_y for W8 × 40 is less than 2, strong axis bending controls. The capacity can be checked by determining the maximum unbraced length for weak axis bending which corresponds to $L_x = 30$ ft. For equal strength about x- and y-axis,

$$\frac{L_x}{r_x} = \frac{L_y}{r_y}$$

$$\text{Equiv. } L_y = \frac{L_x}{r_x/r_y} = \frac{30}{1.73} = 17.3 \text{ ft}$$

On an effective span of 17.3, the W8 × 40 can carry only 150 kips; therefore it is not acceptable.

When using tables it is wise to verify any solution thus found:

$$W10 \times 39,\ KL/r_y = 15(12)/1.98 = 91, \qquad F_a = 14.09 \text{ ksi}$$

$$P/A = 160/11.5 = 13.92 \text{ ksi} < 14.09 \text{ ksi} \qquad \text{OK}$$

Use W10 × 39.

EXAMPLE 6.10.3

Select the lightest W section of A572 Grade 60 steel to serve as a main member 22 ft long carrying an axial compression of 300 kips. Assume the member hinged at the top and fixed at the bottom for bending about either principal axis (see Fig. 6.10.3). The member is part of a braced structure and 1969 AISC specifications are to be used.

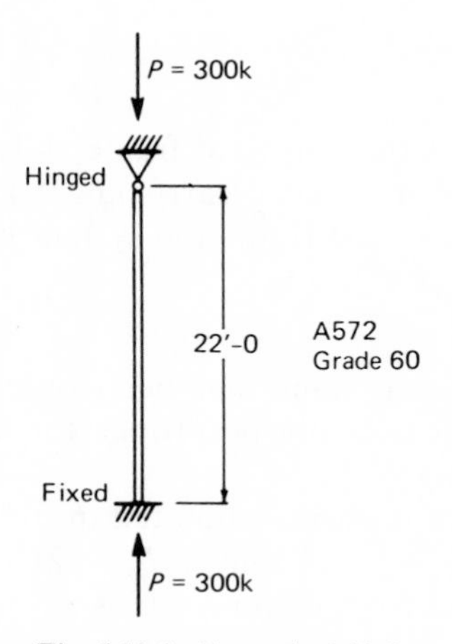

Fig. 6.10.3. Example 6.10.3.

Solution

For this problem there are no AISC column load tables available for direct selection of sections. To get an estimate of the approximate slenderness ratio involved, use AISC column load tables for $F_y = 50$. Since the member is fixed at one end, in accordance with CRC recommendation (Fig. 6.9.1), assume $K = 0.8$. Select for $KL = 0.8(22) = 17.6$ ft and P = 300 kips:

$$\text{W12} \times 58,\ F_y = 50\text{ ksi},\ KL/r = 17.6(12)/2.51 = 84.2$$
$$\text{W10} \times 54,\ F_y = 50\text{ ksi},\ KL/r = 17.6(12)/2.56 = 82.5$$

For $F_y = 60$ ksi,

$$C_c = 755/\sqrt{60} = 97.7$$

$$\text{FS} \approx 1.91 \text{ from Fig. 6.8.1; } Q = 1.0,$$

$$F_a = \frac{F_y}{\text{FS}}\left[1 - \frac{(KL/r)^2}{2C_c^2}\right] \approx \frac{60}{1.91}\left[1 - \frac{1}{2}\left(\frac{82.5}{97.7}\right)^2\right] = 20.2 \text{ ksi}$$

$$\text{Reqd } A = \frac{300}{20.2} = 15 \text{ sq in.}$$

Try W10 × 54, W12 × 53

Check W12 × 53:

$$KL/r = 17.6(12)/2.48 = 85.2$$

$$\text{FS} = \frac{5}{3} + \frac{3}{8}\left(\frac{85.2}{97.7}\right) - \frac{1}{8}\left(\frac{85.2}{97.7}\right)^3 = 1.910$$

$$F_a = \frac{60}{1.91}\left[1 - \frac{1}{2}\left(\frac{85.2}{97.7}\right)^2\right] = 19.45 \text{ ksi}$$

$$f_a = \frac{P}{A} = \frac{300}{15.6} = 19.25 \text{ ksi} < 19.45 \text{ ksi} \qquad \text{OK}$$

Use W12 × 53.

EXAMPLE 6.10.4

Design column A of the unbraced frame shown in Fig. 6.10.4 as an axially loaded compression member carrying 275 kips using A36 steel. In the plane perpendicular to the frame the system is braced, with supports at top and bottom of a 21-ft height.

Solution

While it is rare that a frame member would be designed as axially loaded only, it may occasionally be proper for some interior members with symmetrical loading.

(a) The column size must be estimated in order to determine the K_x factor for the frame. It is given that $K_y L_y = 21$ ft, so that a preliminary size might be determined from column tables.

$$KL = 21 \text{ ft} \qquad \text{Find W12} \times 65, \qquad P = 286 \text{ kips}$$

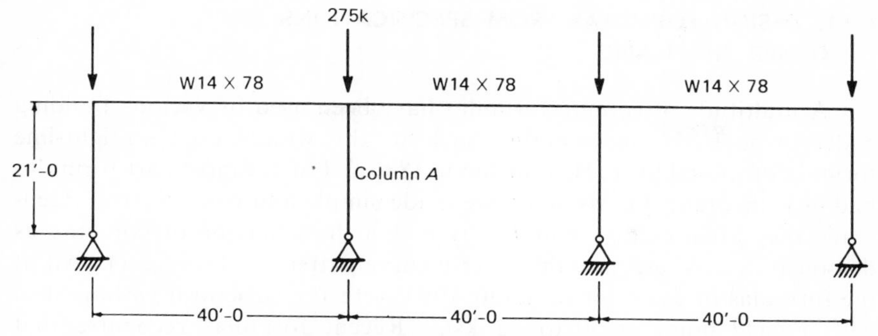

Fig. 6.10.4. Unbraced frame for Example 6.10.4.

(b) Using $I = 533$ in.^{4} for W12 × 65, compute G_{top} in accordance with AISC Commentary Sec. 1.8 (or Fig. 14.3.1).

$$G_{top} = \frac{\Sigma I/L \text{ for columns}}{\Sigma I/L \text{ for beams}} = \frac{533/21}{2(851)/40} = \frac{25.4}{42.5} = 0.60$$

$$G_{bottom} = 10 \text{ (according to AISC–C1.8)}$$

$$K_x = 1.8$$

$$K_x L_x = 1.8(21) = 37.8 \text{ ft}$$

For equal strength about each axis,

$$\frac{r_x}{r_y} = \frac{37.8}{21} = 1.8$$

(c) Select sections:

W12 × 65, $\quad r_x/r_y < 1.8 \quad$ Equiv. $KL_y = 37.8/1.75 = 21.2$ ft
$P = 282$ kips $\quad$ OK

W14 × 61, $\quad r_x/r_y > 2.2 \quad$ $KL_y = 21$ ft
$P = 226$ kips $\quad$ N.G.

(d) Check W12 × 65:

$$I = 533 \text{ in.}^4$$

Since G_{top} and G_{bottom} are not changed from the preliminary calculation, no recheck is needed; thus, $K_x = 1.8$. If weak axis had not been close to controlling so as to give the correct section, as it was in this case, a new computation of G_{top} would have been necessary to obtain a revised K_x.

$$\frac{K_x L_x}{r_x} = \frac{1.8(21)(12)}{5.28} = 86 \qquad \frac{K_y L_y}{r_y} = \frac{1.0(21)(12)}{3.02} = 83.3$$

$$F_a = 14.67 \text{ ksi from AISC Table 1-36.}$$

$$f_a = \frac{275}{19.1} = 14.3 \text{ ksi} < F_a \qquad \text{OK}$$

Use W12 × 65, A36 steel.

6.11. DESIGN FORMULAS FROM SPECIFICATIONS OTHER THAN AISC

A multitude of column formulas have been used by various building codes through the years dating back to the widely used straight-line formula proposed by T. H. Johnson in 1886.[14] For the most part formulas had little theoretical basis and were made simple and conservative. Generally they attempted to empirically match the behavior of columns as demonstrated by tests (see the reverse curve pattern of Fig. 6.3.1). All of the formulas in use prior to about 1960 were for structural carbon steel with a yield point of 30 to 33 ksi. Recent formulas recognize that various strengths of steel are available.

AASHO–1969. The parabolic equations of AASHO[15] for short columns follow the basic logic of the Column Research Council but use higher safety factors than AISC in part to account for the dynamic nature of loads. The equations are, in terms of C_c, as previously defined,
For riveted ends,

$$F_a = \frac{0.55\,F_y}{1.25}\left[1 - \frac{(0.75\,L/r)^2}{2C_c^2}\right] \tag{6.11.1}$$

For pinned ends,

$$F_a = \frac{0.55\,F_y}{1.25}\left[1 - \frac{(0.875\,L/r)^2}{2C_c^2}\right] \tag{6.11.2}$$

which are to be used for $L/r \leq 130$ for A36 steel and $L/r \leq 125$ for A242, A440, and A441 steels.

The L is the actual length of the member since the effective length factor K has been specified in the formula. The factor of safety for columns is taken as 25 percent greater than the basic value of 1.82 used for tension members.

Secant Formula. This long-time rational type of formula could be applied to the entire range of slenderness ratios. In essence such a formula gives the correct relationship if one assumes the deviation from elastic buckling is entirely due to initial curvature and accidental eccentricity. As discussed previously, residual stress is now agreed to be the main cause of inelastic buckling, so that continued use of the unwieldy secant formula has lost much of its appeal.

The equation for maximum stress on a structural member subject to axial compression and uniform bending along its length, as shown in

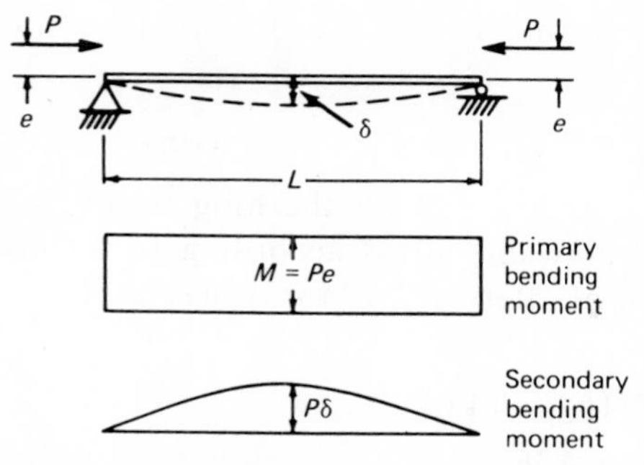

Fig. 6.11.1. Loading and bending moment for secant formula.

Fig. 6.11.1, is

$$f = \frac{P}{A} + \frac{M}{S} \sec \frac{kL}{2} \tag{6.11.3}$$

where $k = \sqrt{P/EI}$. The derivation of the second term of Eq. 6.11.3 occurs in Chapter 12, Sec. 12.2, on beam-columns; however, it is obtained from the differential equation solution which includes the constant primary bending moment, Pe, due to equal end eccentricity, as well as the secondary moment due to axial load times beam deflection.

Placing an upper limit for stress f at the yield point F_y, gives from Eq. 6.11.3

$$F_y = \frac{P}{A}\left(1 + \frac{eA}{S} \sec \frac{L}{2}\sqrt{\frac{P}{EI}}\right) \tag{6.11.4}$$

Assuming an initial value of 0.25 for the eccentricity ratio eA/S even when the loading is presumably concentrically applied and recognizing $I = Ar^2$ gives

$$F_y = \frac{P}{A}\left(1 + 0.25 \sec \frac{L}{2r}\sqrt{\frac{P}{EA}}\right) \tag{6.11.5}$$

which for adequate safety should not exceed F_y even when P is increased by a factor of safety (FS). Thus the allowable average stress on the gross section is

$$F_a = \frac{P}{A} = \frac{F_y/\text{FS}}{1 + 0.25 \sec \dfrac{L}{2r}\sqrt{\dfrac{F_a(\text{FS})}{E}}} \tag{6.11.6}$$

which is similar to the long column equation prescribed by the AASHO and the AREA specifications. In Eq. 6.11.6, L is the effective length, KL, with K taken as 0.75 for riveted ends and 0.875 for pinned ends, and the factor of safety is usually taken about 1.80.

Other column formulas may be found adequately discussed elsewhere.[16,17]

EXAMPLE 6.11.1

For the W8 × 48 selected for the hinged end member in Example 6.10.1, determine allowable stress according to the secant formula, Eq. 6.11.6 with a factor of safety of 1.77 for A36 steel.

Solution

$$L = 0.875(16) = 14 \text{ ft}$$

$$\text{FS} = 1.77 \text{ for A36}$$

$$L/2r = 14(12)/[2(2.08)] = 40.3$$

$$F_a = \frac{36/1.77}{1 + 0.25 \sec (40.3)\sqrt{\dfrac{F_a(1.77)}{29{,}000}}} = \frac{20.3}{1 + 0.25 \sec 0.314\sqrt{F_a}}$$

Solve by trial.

Assume $F_a = 15$ ksi, sec 1.217 = 2.90, $F_a = 11.8$ ksi

$F_a = 14$ ksi, sec 1.172 = 2.58, $F_a = 12.3$

$F_a = 12.9$ ksi, sec 1.128 = 2.33, $F_a = 12.8$ ksi

This shows the allowable stress to be lower than that allowed by the AISC. The example is intended to demonstrate the procedure and difficulty of using the secant formula.

6.12.* SHEAR EFFECT

When built-up members are connected together by means of lacing bars, it is the objective in design to make all of the components act as a unit. As a compression member bends, a shearing component of the axial load is brought in to action and must be considered. The magnitude of the shear effect in reducing column strength is proportional to the amount of deformation which can be attributed to shear. It will be found that solid-webbed sections, such as W shapes, have less shear deformation than do latticed columns (Fig. 6.12.1) where lacing bars and/or batten plates are used.

It further will be shown that shear has an insignificant effect on reducing column strength for solid-webbed shapes and may safely be neglected. The shear effect should not, however, be neglected for latticed columns.

To include the effect of shear, the curvature due to shear should be added to the buckling curvature to obtain the total curvature. It is well known that

$$V = dM_x/dx = P\,dy/dx \tag{6.12.1}$$

after recognizing that $M_x = Py$ from Eq. 6.2.1 (see Fig. 6.2.1).

The slope θ, due to shearing deformation is

$$\theta = \frac{\text{shear stress}}{\text{shear modulus}} = \frac{\beta V}{AG} \tag{6.12.2}$$

where β is a factor to correct for nonuniform stress across sections of various shapes. The shear contribution to curvature then becomes

$$\frac{d\theta}{dx} = \frac{\beta}{AG}\frac{dV}{dx} = \frac{P\beta}{AG}\frac{d^2y}{dx^2} \tag{6.12.3}$$

The total curvature is the sum of Eqs. 6.2.3 and 6.12.3,

$$\frac{d^2y}{dx^2} = -\frac{Py}{EI} + \frac{P\beta}{AG}\frac{d^2y}{dx^2}$$

which gives

$$\frac{d^2y}{dx^2} + \frac{P}{EI}\left[\frac{1}{1 - P\beta/AG}\right]y = 0 \tag{6.12.4}$$

which is of the same form as Eq. 6.2.3; therefore, the modified form of the Euler critical load is

$$P_{\text{cr}} = \frac{\pi^2 EI}{L^2} \underbrace{\frac{1}{\left[1 + \dfrac{\beta}{AG}\dfrac{\pi^2 EI}{L^2}\right]}}_{\substack{\text{modification for} \\ \text{shear effect}}} \tag{6.12.5}$$

In accordance with the previous discussion on basic column strength, G and E can be replaced by the tangent modulus values, G_t and E_t, and $E_t/G_t = 2(1 + \mu)$, and L can be replaced by KL. Further, combining the shear effect with the equivalent pinned length gives

$$F_{\text{cr}} = \frac{P_{\text{cr}}}{A} = \frac{\pi^2 E_t}{(\alpha KL/r)^2} \tag{6.12.6}$$

where $\alpha = \sqrt{1 + 2(1 + \mu)\pi^2\beta/(KL/r)^2}$. It is seen from Eq. 6.12.6 that the shear effect may be accounted for by an adjustment to the equivalent pinned length. For W shapes when bending about the weak axis, β averages about 2. Using $\mu = 0.3$ for steel, typical values for α are

$$\begin{aligned} KL/r &= 50 & \alpha &= 1.01 \\ &= 70 & &= 1.005 \\ &= 100 & &= 1.003 \end{aligned}$$

For slenderness ratios less than about 50, yielding controls, so that the shear effect on solid wide-flange columns is equivalent to an increase in effective length of less than one percent, which can be safely neglected.

Latticed Columns. The following simplified treatment of shear effect on the laced column generally follows that of Timoshenko and Gere.[19] Consider that Eq. 6.12.5 may be written as

$$P_{cr} = P_e \left[\frac{1}{1 + P_e/P_d} \right] \tag{6.12.7}$$

where $P_e = \pi^2 EI/(KL)^2$, the basic Euler load; and $P_d = AG/\beta$. From Eq. 6.12.2, the slope due to shear deformation may be expressed

$$\theta = V/P_d \tag{6.12.8}$$

Thus P_d is the quantity by which the shear force V is divided to obtain the additional slope θ of the deflection curve resulting from shear.

To illustrate the procedure of computing $1/P_d$ for a laced column, use the column of Fig. 6.12.1c. The shear deformation is determined assum-

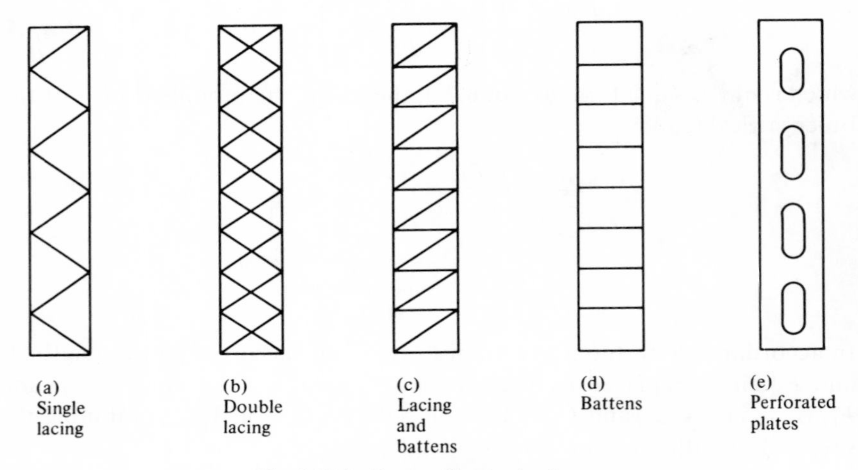

Fig. 6.12.1. Types of latticed columns.

ing the column to be a truss with pinned joints. Thus the shear effect will be overemphasized because in reality the main vertical members are continuous rather than pinned at each joint. It is noted that ordinary bending of latticed members considers only the deformation in the main elements of the column while neglecting entirely deformation in diagonal and horizontal lacing bars.

Referring to Fig. 6.12.2, the slope in a given panel is composed of two parts,

$$\theta = \gamma_1 + \gamma_2 \tag{6.12.9}$$

the sum of the angle change due to change of length in the diagonal plus the change of length in the horizontal member. Thus from Fig. 6.12.2a,

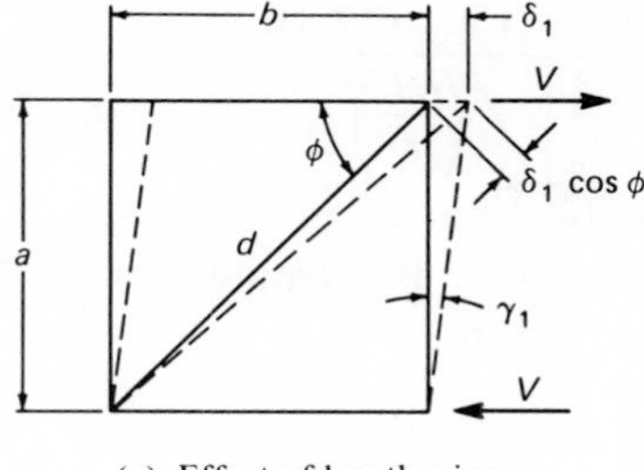

(a) Effect of lengthening of diagonal bars.

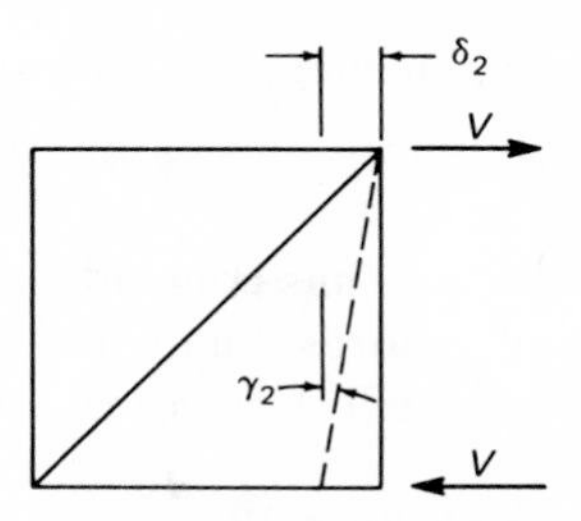

(b) Effect of shortening of horizontal bars.

Fig. 6.12.2. Shear deformation in laced column.

$$\delta_1 \cos\phi = \frac{V}{\cos\phi}\frac{d}{A_d E} \tag{6.12.10}$$

$$\gamma_1 = \frac{\delta_1}{a} = \frac{Vd}{a\cos^2\phi\, A_d E} \tag{6.12.11}$$

The horizontal deformation component from Fig. 6.12.2b is

$$\gamma_2 = \frac{\delta_2}{a} = \frac{Vb}{aA_b E} \tag{6.12.12}$$

and using Eq. 6.12.9 gives

$$\theta = \frac{V}{aE}\left[\frac{d}{\cos^2\phi\, A_d} + \frac{b}{A_b}\right]$$

which upon substitution of $\cos^2\phi = b^2/d^2$ gives

$$\theta = \frac{V}{ab^2E}\left[\frac{d^3}{A_d} + \frac{b^3}{A_b}\right] \tag{6.12.13}$$

From Eq. 6.12.8,

$$\frac{1}{P_d} = \frac{1}{ab^2E}\left[\frac{d^3}{A_d} + \frac{b^3}{A_b}\right] \tag{6.12.14}$$

For design purposes this effect may be incorporated into the equivalent length in accordance with Eq. 6.12.6 where for single lacing and battens (Fig. 6.12.1c)

$$\begin{aligned}\alpha &= \sqrt{1 + P_e/P_d} \\ &= \sqrt{1 + \pi^2\left[\frac{A}{ab^2}\left(\frac{d^3}{A_d} + \frac{b^3}{A_b}\right)\right]\Big/\left(\frac{KL}{r}\right)^2}\end{aligned} \tag{6.12.15}$$

where A = gross compression area of member. For the case without battens (Fig. 6.12.1a) the second term of Eq. 6.12.14 vanishes, thus:

$$\alpha = \sqrt{1 + \pi^2\left[\frac{A}{ab^2}\frac{d^3}{A_d}\right]\Big/\left(\frac{KL}{r}\right)^2} \tag{6.12.16}$$

For double diagonals without horizontal bars (Fig. 6.12.1b),

$$\alpha = \sqrt{1 + \pi^2 \left[\frac{2A}{ab^2} \frac{d^3}{A_d}\right] \Big/ \left(\frac{KL}{r}\right)^2} \tag{6.12.17}$$

Bleich[13] has concluded that any error inherent in not assuming continuous chords (main compression elements) of the frameworks has a negligible effect on conventionally designed columns.

EXAMPLE 6.12.1

Illustrate the shear effect on the column with single diagonal and battens (Fig. 6.12.1c) under typical design conditions.

Solution

Referring to Fig. 6.12.3, and using Eq. 6.12.15, A is the total compressive area of four angles, A_d is the sum of the area of the diagonals on opposite faces of the member at any section, and A_b is the sum of the areas of the horizontal battens on opposite faces of the member at any section:

$$\text{Typical } \frac{A}{A_d} = \frac{\text{area four angles}}{\text{area two diagonal bars}}, \quad \text{say 15 to 25.}$$

While battens might have an area equal to two or three times the diagonal area, a conservative approach could say

$$A_b = A_d$$

overemphasizing the shear effect.

Typical inclination of diagonals might be 45°.

Using $A_b = A_d$, $\phi = 45°$, and $A/A_d = 15$ and 20, Eq. 6.12.15 becomes

$$\alpha = \sqrt{1 + \pi^2(A/A_d)(3.828)/(KL/r)^2}$$

which is shown plotted in Fig. 6.12.3.

The curves of Fig. 6.12.3 are surely high estimates of the increase in effective length due to shear.

The Column Research Council[2] has recommended that "a conservative estimate of the influence of 60° or 45° lacing as generally specified in bridge design practice can be made by modifying the effective length factor" K to a new factor αK, as follows:

$$\text{for } \frac{KL}{r} > 40, \qquad \alpha = \sqrt{1 + 300/(KL/r)^2} \tag{6.12.18}$$

$$\text{for } \frac{KL}{r} \leq 40, \qquad \alpha = 1.1 \tag{6.12.19}$$

This CRC recommendation is shown in Fig. 6.12.3. Such effective-length modification will rarely effect the design of short columns in braced systems.

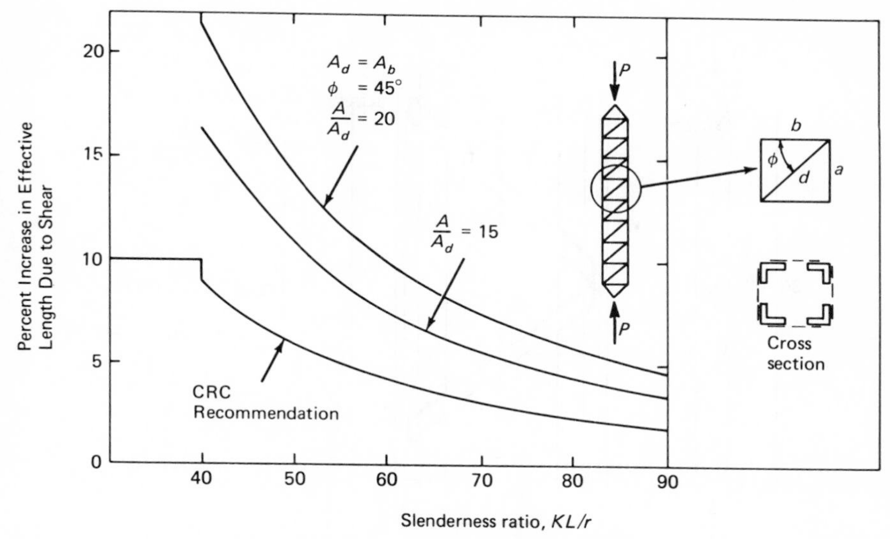

Fig. 6.12.3. Effect of shear on effective length.

One may easily ascertain the effect, if that seems desirable, by using Eqs. 6.12.15 through 6.12.17.

6.13.* DESIGN OF LATTICED MEMBERS

Under most specifications, latticed members are designed according to detailed empirical rules most of which are related to local buckling requirements. AISC–1.18 gives many detailed requirements which will not be enumerated here. Two examples follow which illustrate some of the provisions as well as general procedures for built-up sections. The reader is referred to Blodgett[23] who has summarized the AISC provisions along with other information concerning built-up section design.

EXAMPLE 6.13.1

Design a laced column as shown in Fig. 6.13.1, composed of four angles to carry 600 kips axial compression, with an equivalent pinned length of 30 ft. Assume all connections will be welded and use steel with $F_y = 50$ ksi.

Solution

(a) Establish the depth of section h. It will be found that the radius of gyration of a four angle column is dependent only upon h and essentially is independent of thickness. Thus selecting h establishes the slenderness ratio, and vice versa. Appendix, Table A1 shows the relation-

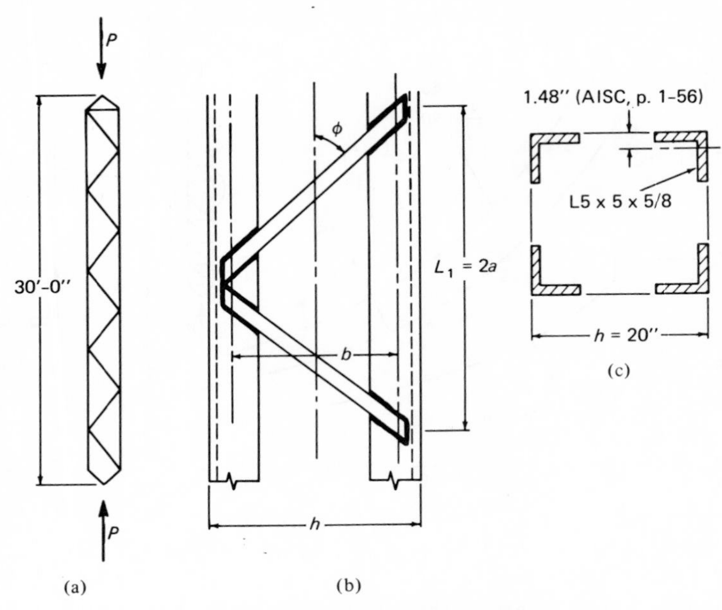

Fig. 6.13.1. Details of Example 6.13.1.

ships between the radius of gyration and the geometry of the cross-section. Thus from Appendix, Table A1,

$$r \approx 0.42h$$

$$\frac{KL}{r} = \frac{360}{0.42h} = \frac{857}{h}$$

h	KL/r	F_a (AISC Table 1-50)	Area Reqd.	Angles	
24	35.7	26.4 ksi	22.7	L6 × 6 × ½	A = 23.0
				L5 × 5 × ⅝	23.44
22	39	25.97	23.1	same as above	
21	40.7	25.79	23.25	L5 × 5 × ⅝	
20	42.8	25.43	23.55	L5 × 5 × ⅝	

It would appear that 20 in. is preferred, since dimension is probably more important than 0.44 sq in. of cross section. Investigate 4 – L5 × 5 × ⅝, assuming $Q = 1.0$,

$$I_x = I_y = 4\,[13.6 + 5.86\,(10.0 - 1.48)^2] = 1{,}755 \text{ in.}^4$$

$$r_x = r_y = \sqrt{\frac{I}{A}} = \sqrt{\frac{1755}{23.44}} = 8.64 \text{ in.} \qquad (\text{Approx. } .42\,(20) = 8.40 \text{ in.})$$

$$\frac{KL}{r} = \frac{360}{8.64} = 41.7, \quad F_a\,(\text{Table 1-50}) = 25.59 \text{ ksi}$$

$$f_a = \frac{P}{A} = \frac{600}{23.44} = 25.6 \text{ ksi} \approx F_a = 25.59 \text{ ksi}$$

Use 4 – L5 × 5 × $\frac{5}{8}$ with $h = 20$ in.

(b) Local buckling. For $Q = 1.0$ to apply, angles must satisfy local buckling requirements of AISC–1.9. In this case,

$$\frac{b}{t} = \frac{5}{5/8} = 8 < \frac{76.0}{\sqrt{50}} = 10.7 \qquad \text{OK}$$

Development of the provisions of AISC–1.9 appears in Chapter 6, Part II on plate strength.

(c) Design single lacing. According to AISC–1.18.2.6, "the inclination of lacing bars to the axis of the member shall preferably be not less than 60 degrees for single lacing . . ."

For $\phi = 60°$ (Fig. 6.13.1b); $b = 20 - 2(3) = 14$ in., assuming use of distance between standard gage lines for bolts. This would be approximately center-to-center of welded connection. Thus

$$L_1 = 2b \tan 30° = 2(14)0.577 = 16.2 \text{ in.}; \qquad \text{use 16 in.}$$

For a single angle,

$$\frac{L_1}{r_z} = \frac{16}{0.97} = 16.5 < 41.7 \text{ (slenderness ratio for entire member)} \qquad \text{OK}$$

According to AISC–1.18.2.6, "Lacing shall be proportioned to resist a shearing stress normal to the axis of the member equal to 2 percent of the total compressive stress in the member."

$$V = 0.02\,(600) = 12 \text{ kips (6 kips per side)}$$

The force in one bar is

$$P = V/\cos 30° = 6/0.866 = 6.93 \text{ kips}$$

$$\frac{L}{r} \leq 140 \text{ for single lacing}$$

$$r = \sqrt{\frac{I}{A}} = \sqrt{\frac{\frac{1}{12}bt^3}{bt}} = 0.288t$$

$$t_{\min} = \frac{66}{0.288\,(140)} = 0.397, \text{ Use } 7/16 \text{ in.}$$

$$\text{For } t = 7/16 \text{ in.}, \frac{L}{r} = \frac{16}{0.288\,(0.4375)} = 127$$

From Table 1–50, $F_a = 9.59$ ksi for secondary members,

$$\text{Reqd area} = \frac{6.93}{9.59} = 0.722 \text{ sq in.}$$

$$\text{Width } b = \frac{0.722}{0.4375} = 1.65 \text{ in.}$$

Since no holes are required for connectors, tension on net section need not be investigated for this design.

Use bars $\frac{7}{16} \times 1\frac{3}{4}$.

(d) Design tie plates at ends (AISC–1.18.2.5). The tie plates should extend along the length of the member a distance equal to the distance b from the end of the member. Use a length of 14 in.

$$t \geq \frac{1}{50}(b) = \frac{1}{50}(14) = 0.28 \text{ in.}$$

Use tie plates $\frac{5}{16} \times 14 \times 1'\text{-}8''$.

(e) Examine the effect of shear on equivalent length, Eq. 6.12.16.

$$\alpha = \sqrt{1 + \pi^2 (A/A_d)(d^3/ab^2)/(KL/r)^2}$$

$$\frac{A}{A_d} = \frac{2(5.86)}{1.75(\frac{7}{16})} = 15.3$$

$$\frac{d^3}{ab^2} = \frac{(16)^3}{8(14)^2} = 2.61$$

$$\alpha = \sqrt{1 + \pi^2(15.3)(2.61)/(41.7)^2} = \sqrt{1.227} = 1.11$$

Thus the equivalent pinned length should have been increased 11 percent due to shear. The neglect of end restraint probably is, in most cases, equal to about the same increase in equivalent length.

EXAMPLE 6.13.2

Redesign the column of Example 6.13.1 using a welded perforated box shape (Fig. 6.13.2).

Solution

It is to be observed that one important advantage of a welded shape is that the number of individual components in the shape is minimized. Four plates can be used for this welded shape, whereas a bolted or riveted section requires four angles in addition to plates.

Using Appendix, Table A1, for a box shape,

$$r \approx 0.40h$$

$$\frac{KL}{r} = \frac{360}{0.4h} = \frac{900}{h}$$

$$\text{Area} \approx 4ht$$

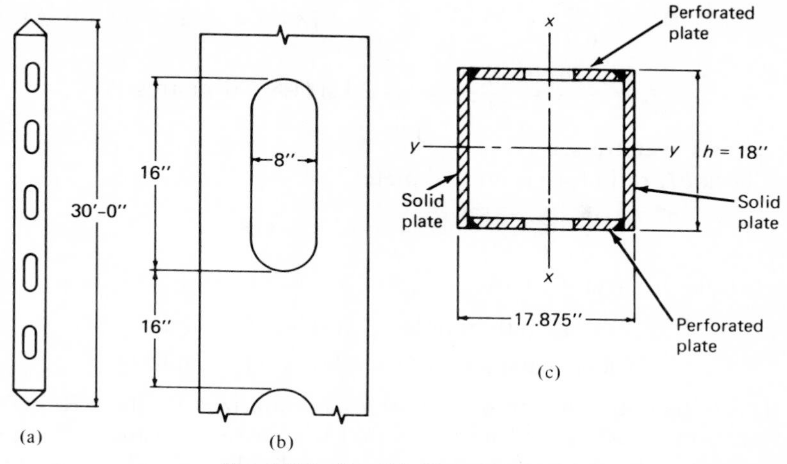

Fig. 6.13.2. Details of Example 6.13.2.

h, in.	KL/r	AISC F_a [Table 1-50 ,] ksi	Area Reqd, sq. in.	Plate Thickness, in.
21	42.8	25.43	23.55	0.28
	$\frac{b}{t}$ (solid plate) = 21/(5/16) = 67 > 33.7 NG			
18	50	24.35	24.6	0.342
16	56.2	23.36	25.7	0.402
15	60	22.72	26.4	0.44

Without perforations, 2PLs – $\frac{7}{16}$ × 16 and 2 PLs – $\frac{7}{16}$ × 15 are probably acceptable, satisfying b/t ratios in accordance with AISC–1.9.2.

Usually, however, such shapes are used on bridges where access is required for maintenance so that perforations may be desirable.

Assume perforations to be 8 in. wide (frequently one-half or less of total width) and $h \approx$ 18 in.

$$\text{Net area available} = 2(18)(7/16) + 2(17 - 8)(1/2)$$
$$= 15.75 + 9.0 = 24.75 \text{ sq. in.}$$

$$I_x = 2(18)\left(\frac{7}{16}\right)(8.719)^2 + \frac{2}{12}[(17)^3 - (8)^3]\frac{1}{2}$$
$$= 1197 + 367 = 1564 \text{ in.}$$

$$I_y = \frac{2}{12}(18)^3\left(\frac{7}{16}\right) + 2(9)\left(\frac{1}{2}\right)(8.75)^2$$
$$= 425 + 689 = 1114 \text{ in.}^4$$

$$r_y = \sqrt{\frac{1114}{24.75}} = 6.71 \text{ in.}$$

$$\frac{KL}{r_y} = \frac{360}{6.71} = 53.6, \quad F_a = 23.78 \text{ ksi} \qquad \text{(Table 1–50, AISC)}$$

$$f_a = \frac{P}{A_{\text{net}}} = \frac{600}{24.75} = 24.2 \text{ ksi } (1.8\% \text{ overstress})$$

Accept this overstress.

Check b/t ratio for perforated plate:

$$\frac{b}{t} = \frac{17}{0.5} = 34 < 45 \qquad \text{OK}$$

Access hole size, (AISC–1.18.2.7):

$$\text{Length of hole} = 2(8) = 16 \text{ in. max}$$

$$\text{Clear spacing between holes} = 17 \text{ in. min}$$

It is to be noted that among the rules summarized by the CRC Guide (p. 79 of Ref. 2), is one which insures that the portions of the column section separated by the perforations can act individually without buckling. The CRC suggestion is that

$$\frac{\text{Length of perforation}}{\text{Radius of gyration of flange}} \leq \frac{1}{3}\left(\frac{KL}{r} \text{ of entire section}\right) \leq 20 \text{ max}$$

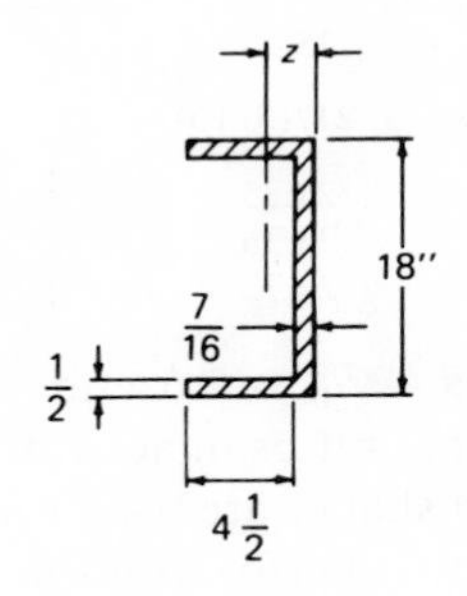

$$z = \frac{4.5(2.6875) + 7.87(7/32)}{4.5 + 7.87}$$

$$= \frac{12.10 + 1.72}{12.37} = 1.118 \text{ in.}$$

$$I = \frac{1}{3}(4.5)^3 - 12.37(1.118)^2$$

$$= 30.35 - 15.45 = 14.90 \text{ in.}^4$$

$$r = \sqrt{\frac{14.90}{12.37}} = 1.098 \text{ in.}$$

$$\frac{L}{r} = \frac{16}{1.098} = 14.55 < \frac{53.6}{3} = 17.9 < 20$$

When this slenderness-ratio restriction is satisfied the CRC suggests that only then may the full net section be considered effective.

Use 2 PLs $\frac{7}{16} \times 18$ (solid) and 2 PLs $\frac{1}{2} \times 17$ (perforated) with holes 16″ by 8″ spaced 2′–8″ center-to-center.

Part II Plates

6.14.* INTRODUCTION TO STABILITY OF PLATES.

All column sections whether rolled WF shapes or built-up sections are composed of plate elements. Up to this point in the chapter consideration has been given only to the possibility of buckling of the member based on the slenderness ratio for the entire cross section. It may be, however, that a local compression failure in one of the plate elements which make up the cross section will occur first. Such local buckling must be considered in all situations where compression stress occurs.

The theory of bending of plates and elastic stability of plates are subjects which should be studied in depth by the advanced student in structural engineering. The brief treatment that follows is intended to give the reader the general idea of plate buckling necessary to properly use and understand current steel specifications. The general approach and terminology follow that of Timoshenko.[18,19]

Before one can treat the stability problem, the differential equation for bending of plates is required, just as the differential equation for the bending of a beam, Eq. 6.2.2, was used in the slender column stability treatment in Sec. 6.2.

Differential Equation for Bending of Homogeneous Plates. First, the strains will be obtained in terms of displacements. Let h = plate thickness and u, v, and w equal the displacements in the x, y, and z directions, respectively. Referring to Fig. 6.14.1, consider an element of a plate $dx\,dy$,

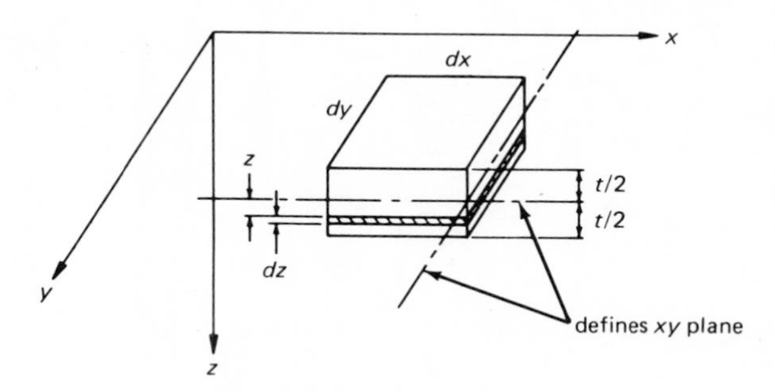

Fig. 6.14.1.

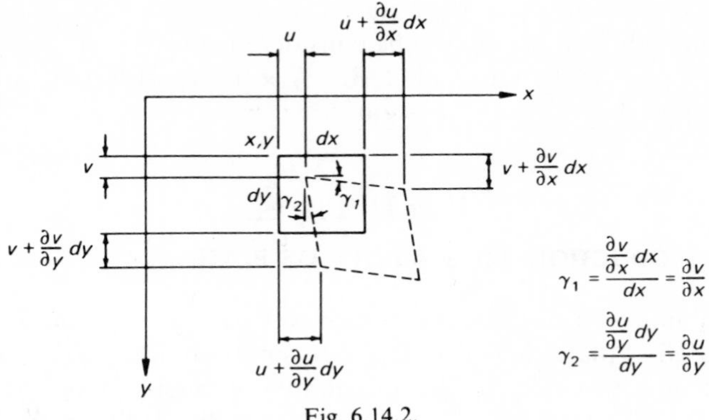

Fig. 6.14.2.

and assume no stretching of the neutral plane at $z = 0$. Examining a slice $dx\,dy\,dz$ of the plate element located at a distance z from the neutral plane shows, in Fig. 6.14.2, the coodinate unit strains ϵ_x, ϵ_y, and the shearing strain γ_{xy}. Thus

$$\epsilon_x = \frac{dx + \left[u + \dfrac{\partial u}{\partial x}\,dx - u\right] - dx}{dx} = \frac{\partial u}{\partial x} \qquad (6.14.1a)$$

$$\epsilon_y = \frac{\delta v}{\partial y} \qquad (6.14.1b)$$

$$\gamma_{xy} = \gamma_1 + \gamma_2 = \frac{\partial y}{\partial x} + \frac{\partial u}{\partial y} \qquad (6.14.1c)$$

Expressing the displacements in the plane of the plate in terms of the lateral deflection w, as shown in Fig. 6.14.3, and recognizing that positive

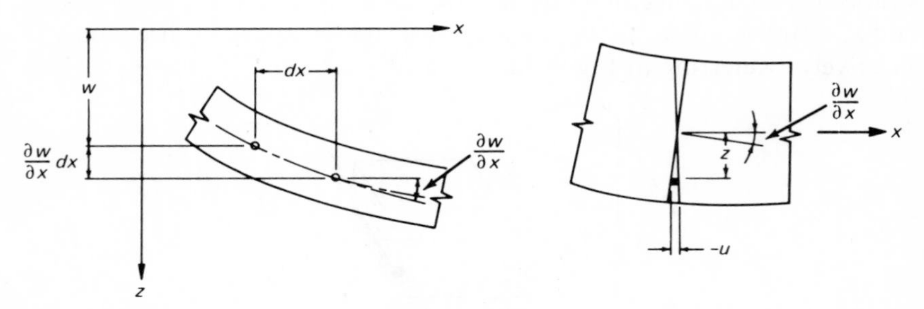

Fig. 6.14.3.

slope gives a negative displacement u or v, one establishes

$$-u = z\,\frac{\partial w}{\partial x}; \qquad -v = z\,\frac{\partial w}{\partial y} \qquad (6.14.2)$$

Substitution of Eqs. 6.14.2 into Eqs. 6.14.1 gives strains in terms of curvatures for x-direction bending, y-direction bending, and twisting

$$\epsilon_x = \frac{\partial u}{\partial x} = -z\frac{\partial^2 w}{\partial x^2} \tag{6.14.3a}$$

$$\epsilon_y = \frac{\partial v}{\partial y} = -z\frac{\partial^2 w}{\partial y^2} \tag{6.14.3b}$$

$$\gamma_{xy} = \frac{\partial v}{\partial x} + \frac{\partial u}{\partial y} = -z\left(\frac{\partial^2 w}{\partial x \partial y} + \frac{\partial^2 w}{\partial x \partial y}\right) = -2z\frac{\partial^2 w}{\partial x \partial y} \tag{6.14.3c}$$

Next, making use of Hooke's Law expressing strains in terms of the stresses σ_x, σ_y, normal stresses in the x and y directions, and τ_{xy}, the shear stress,

$$\epsilon_x = \frac{1}{E}[\sigma_x - \mu\sigma_y] \tag{6.14.4a}$$

$$\epsilon_y = \frac{1}{E}[-\mu\sigma_x + \sigma_y] \tag{6.14.4b}$$

$$\gamma_{xy} = \tau_{xy}/G \tag{6.14.4c}$$

where μ = Poisson's Ratio (see Sec. 2.7), and G = shear modulus of elasticity.

For any stress condition, such as $\sigma_y = -\sigma_x$ which gives pure shear on an element rotated 45° to the x axis, the work done by the equivalent sys-

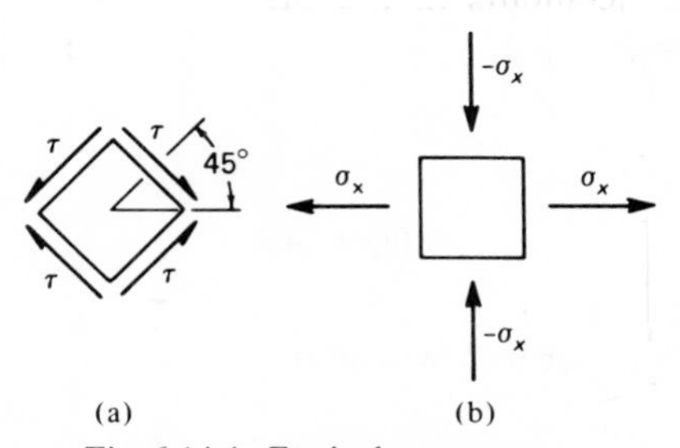

Fig. 6.14.4. Equivalent systems.

tems of Fig. 6.14.4 must be a constant:

$$\tfrac{1}{2}\sigma_x\epsilon_x - \tfrac{1}{2}\sigma_x\epsilon_y = \tfrac{1}{2}\tau\gamma \tag{6.14.5}$$

Substituting Eqs. 6.14.4 into Eq. 6.14.5 gives

$$\frac{\sigma_x}{E}(\sigma_x - \mu\sigma_y + \mu\sigma_x - \sigma_y) = \tau^2/G$$

If $\sigma_y = -\sigma_x$, maximum $\tau = \sigma_x$; thus

$$\frac{1}{E}(1 + \mu + \mu + 1) = \frac{1}{G}$$

$$G = \frac{E}{2(1 + \mu)} \tag{6.14.6}$$

Solving Eqs. 6.14.4 for stresses and substituting Eqs. 6.14.3 give stresses in terms of curvatures,

$$\sigma_x = \frac{-zE}{1 - \mu^2}\left(\frac{\partial^2 w}{\partial x^2} + \mu \frac{\partial^2 w}{\partial y^2}\right) \tag{6.14.7a}$$

$$\sigma_y = \frac{-zE}{1 - \mu^2}\left(\mu \frac{\partial^2 w}{\partial x^2} + \frac{\partial^2 w}{\partial y^2}\right) \tag{6.14.7b}$$

$$\tau_{xy} = -2zG \frac{\partial^2 w}{\partial x \partial y} \tag{6.14.7c}$$

Next it is necessary to relate curvatures and bending moments. Referring to Fig. 6.14.5, and using the right-hand rule for positive twist, it is

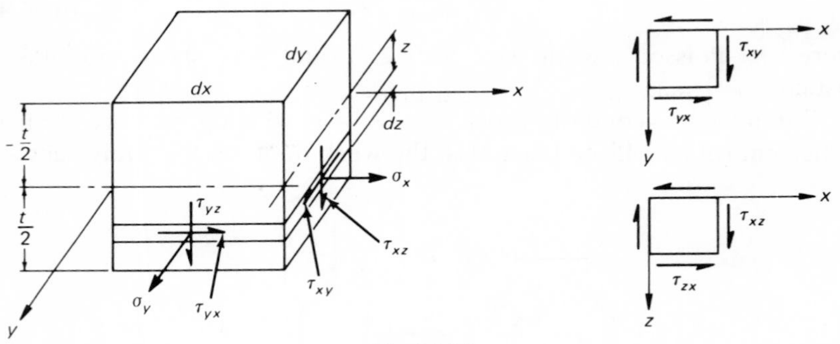

Fig. 6.14.5.

seen that the moments per unit width are

$$M_x = \int_{-t/2}^{t/2} z\sigma_x \, dz = \frac{-Et^3}{12(1 - \mu^2)}\left(\frac{\partial^2 w}{\partial x^2} + \mu \frac{\partial^2 w}{\partial y^2}\right) \tag{6.14.8a}$$

$$M_y = \int_{-t/2}^{t/2} z\sigma_y \, dz = \frac{-Et^3}{12(1 - \mu^2)}\left(\mu \frac{\partial^2 w}{\partial x^2} + \frac{\partial^2 w}{\partial y^2}\right) \tag{6.14.8b}$$

$$M_{xy} = -\int_{-t/2}^{t/2} \tau_{xy} z \, dz = +2G\left(\frac{t^3}{12}\right)\frac{\partial^2 w}{\partial x \partial y} \tag{6.14.8c}$$

Finally, considering all forces and moments acting on the plate element, moments about the x and y axes and force summation in the z direction gives rise to three equations. Figure 6.14.6 is a free body of the

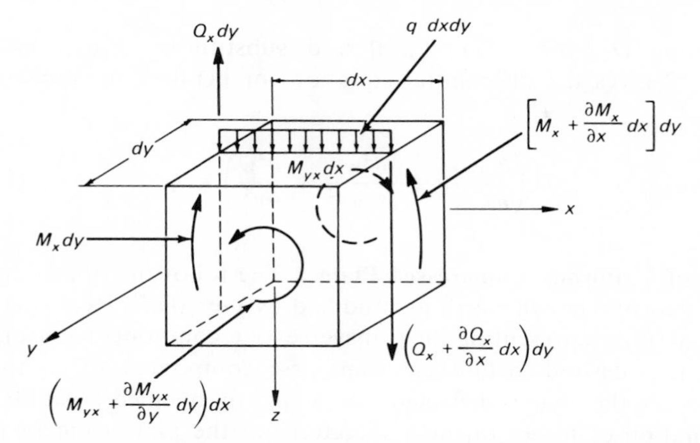

Fig. 6.14.6. Free-body forces involved in rotation about y axis. (Forces involved in x axis rotation not shown.)

plate element showing only the forces involved with moments about the y axis.

Taking moments about the y axis gives

$$\cancel{M_x\, dy} + \frac{\partial M_x}{\partial x}\, dx\, dy - \cancel{M_x\, dy} + \cancel{M_{yx}\, dx} + \frac{\partial M_{yx}}{\partial y}\, dy\, dx - \cancel{M_{yx}\, dx}$$

$$- \left(Q_x\, dy + \underbrace{\cancel{\frac{\partial Q_x}{\partial x}\, dx\, dy}}_{\text{neglect}}\right) \underbrace{dx}_{\text{mom. arm}} - \underbrace{q\, \cancel{dx\, dy}\, \frac{dx}{\cancel{2}}}_{\text{neglect}} = 0$$

Neglecting infinitesimals of higher order and dividing by $dx\ dy$ gives

$$\frac{\partial M_x}{\partial x} + \frac{\partial M_{yx}}{\partial y} - Q_x = 0 \qquad (6.14.9)$$

Similarly for moments about the x axis,

$$\frac{\partial M_y}{\partial y} - \frac{\partial M_{xy}}{\partial x} - Q_y = 0 \qquad (6.14.10)$$

Force equilibrium in the z direction gives

$$\frac{\partial Q_x}{\partial x} + \frac{\partial Q_y}{\partial y} + q = 0 \qquad (6.14.11)$$

Using Eqs. 6.14.9 and 6.14.10 for Q_x and Q_y, and substitution into Eq. 6.14.11 gives

$$\frac{\partial^2 M_x}{\partial x^2} + \frac{\partial^2 M_y}{\partial y^2} - 2\frac{\partial^2 M_{xy}}{\partial x \partial y} = -q \tag{6.14.12}$$

Defining $D = Et^3/[12(1 - \mu^2)]$ and substituting Eq. 6.14.8 into Eq. 6.14.12 gives the differential equation for bending of homogeneous plates,

$$D\left(\frac{\partial^4 w}{\partial x^4} + 2\frac{\partial^4 w}{\partial x^2 \partial y^2} + \frac{\partial^4 w}{\partial y^4}\right) = q \tag{6.14.13}$$

Buckling of Uniformly Compressed Plate. The following approach is essentially that of Timoshenko[19] as modified by Gerstle.[20] Realizing that q is a general term representing the transverse load component causing plate bending, it is desired to find the transverse component of compressive force N_x when the plate is deflected into a slightly buckled position. Taking summation of forces in the z direction on the plate element of Fig. 6.14.7b gives

$$N_x dy \frac{\partial w}{\partial x} - \left(N_x + \frac{\partial N_x}{\partial x} dx\right) dy \left(\frac{\partial w}{\partial x} + \frac{\partial^2 w}{\partial x^2} dx\right) = q\, dx\, dy$$

$$-\left(N_x \frac{\partial^2 w}{\partial x^2} + \frac{\partial N_x}{\partial x}\frac{\partial w}{\partial x} + \frac{\partial N_x}{\partial x} dx \frac{\partial^2 w}{\partial x^2}\right) dy\, dx = q\, dx\, dy \tag{6.14.14}$$

which upon neglecting the higher-order infinitesimal terms gives

$$q = -N_x \frac{\partial^2 w}{\partial x^2} \tag{6.14.15}$$

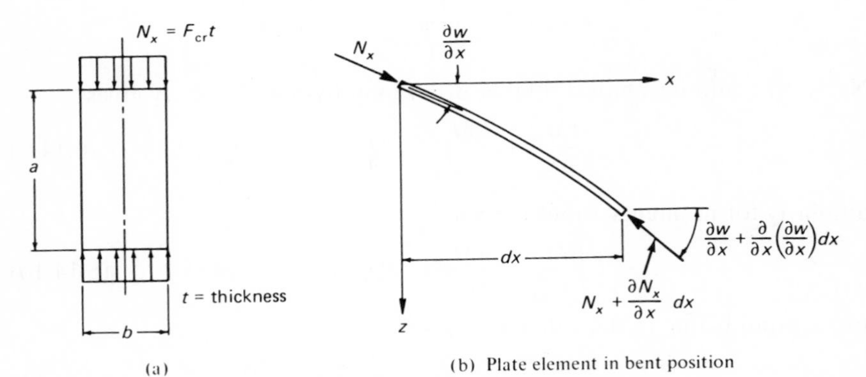

Fig. 6.14.7. Uniformly compressed plate.

The differential equation then becomes

$$\frac{\partial^4 w}{\partial x^4} + 2\frac{\partial^4 w}{\partial x^2 \partial y^2} + \frac{\partial^4 w}{\partial y^4} = -\frac{N_x}{D}\frac{\partial^2 w}{\partial x^2} \qquad (6.14.16)$$

which is a partial differential equation where w is a function of both x and y. It will be reasonable to assume that the deflection w can be expressed as the product of an x function (X) and a y function (Y). Further, buckling may be assumed to give sinusoidal variation in the x direction. Thus

$$w = X(x)Y(y) \qquad (6.14.17)$$

Letting

$$X(x) = \sin\frac{m\pi x}{a}$$

where the X function satisfies the zero deflection and zero moment conditions of simple support at $x = 0$ and $x = a$. Substitution of Eq. 6.14.17 into Eq. 6.14.16 gives, after canceling the common term $\sin m\pi x/a$,

$$\left(\frac{m\pi}{a}\right)^4 Y - 2\left(\frac{m\pi}{a}\right)^2\frac{d^2Y}{dy^2} + \frac{d^4Y}{dy^4} = +\frac{N_x}{D}\left(\frac{m\pi}{a}\right)^2 Y$$

$$\frac{d^4Y}{dy^4} - 2\left(\frac{m\pi}{a}\right)^2\frac{d^2Y}{dy^2} + \left[\left(\frac{m\pi}{a}\right)^4 - \frac{N_x}{D}\left(\frac{m\pi}{a}\right)^2\right]Y = 0 \qquad (6.14.18)$$

an ordinary fourth-order homogeneous differential equation.

The solution may be expressed in the form

$$Y = C_1\sinh\alpha y + C_2\cosh\alpha y + C_3\sin\beta y + C_4\cos\beta y \qquad (6.14.19)$$

where

$$\alpha = \sqrt{\left(\frac{m\pi}{a}\right)^2 + \sqrt{\frac{N_x}{D}\left(\frac{m\pi}{a}\right)^2}} \quad \text{and} \quad \beta = \sqrt{-\left(\frac{m\pi}{a}\right)^2 + \sqrt{\frac{N_x}{D}\left(\frac{m\pi}{a}\right)^2}}$$

Thus the entire plate deflection equation is

$$w = \left(\sin\frac{m\pi x}{a}\right)(C_1\sinh\alpha y + C_2\cosh\alpha y + C_3\sin\beta y + C_4\cos\beta y) \qquad (6.14.20)$$

which must satisfy boundary conditions. Assuming the x axis as an axis of symmetry through the plate, i.e., identical support conditions along the two edges parallel to the direction of loading, the odd function coefficients C_1 and C_3 must be zero. Thus

$$w = (C_2\cosh\alpha y + C_4\cos\beta y)\sin\frac{m\pi x}{a} \qquad (6.14.21)$$

Using simple support conditions at $y = b/2$ and $y = -b/2$, requires that at $y = \pm b/2$,

$$w = 0 = \left(C_2 \cosh \alpha \frac{b}{2} + C_4 \cos \beta \frac{b}{2}\right) \sin \frac{m\pi x}{a}$$

$$(6.14.22)$$

$$\frac{\delta^2 w}{\delta y^2} = 0 = \left(C_2 \alpha^2 \cosh \alpha \frac{b}{2} + C_4 \beta^2 \cos \beta \frac{b}{2}\right) \sin \frac{m\pi x}{a}$$

For a solution other than $C_2 = C_4 = 0$ it is necessary for the determinant of the coefficients to be zero. Thus

$$(\alpha^2 - \beta^2) \cosh \alpha \frac{b}{2} \cos \beta \frac{b}{2} = 0 \qquad (6.14.23)$$

Since $\alpha^2 \neq \beta^2$ unless $N_x = 0$ (a trivial solution), and since $\cosh \alpha \dfrac{b}{2} > 1$, the only way Eq. 6.14.23 can be satisfied in the real problem is for

$$\cos \beta \frac{b}{2} = 0$$

Therefore

$$\beta \frac{b}{2} = \frac{\pi}{2}, \quad \frac{3\pi}{2}, \quad \frac{5\pi}{2}, \text{ etc.}$$

Using the lowest value of $\beta \dfrac{b}{2}$ and substituting into β as defined below Eq. 6.14.19 gives

$$\frac{b}{2} \sqrt{-\left(\frac{m\pi}{a}\right)^2 + \sqrt{\frac{N_x}{D}} \left(\frac{m\pi}{a}\right)^2} = \frac{\pi}{2}$$

$$b^2 \left[-\left(\frac{m\pi}{a}\right)^2 + \sqrt{\frac{N_x}{D}} \left(\frac{m\pi}{a}\right)^2\right] = \pi^2$$

$$\frac{N_x}{D} \left(\frac{m\pi}{a}\right)^2 = \left[\frac{\pi^2}{b^2} + \left(\frac{m\pi}{a}\right)^2\right]^2$$

$$N_x = D \left[\frac{\pi^2 a}{b^2 m\pi} + \frac{m\pi}{a}\right]^2$$

$$N_x = \frac{D\pi^2}{b^2} \left[\frac{1}{m} \frac{a}{b} + m \frac{b}{a}\right]^2 \qquad (6.14.24)$$

Since $N_x = F_{cr}t$ and $D = Et^3/[12(1 - \mu^2)]$, the critical unit stress may be expressed as

$$F_{cr} = k \frac{\pi^2 E}{12(1 - \mu^2)(b/t)^2} \tag{6.14.25}$$

where

$$k = \left[\frac{1}{m}\frac{a}{b} + m\frac{b}{a}\right]^2 \tag{6.14.26}$$

The buckling coefficient k, it should be noted, is a function of the type of stress (in this case uniform compression on two opposite edges) and the edge support conditions (in this case simple support on four edges), in addition to the aspect ratio a/b which appears directly in the equation.

The reader is strongly reminded that Eq. 6.14.25 is based on two assumptions which apparently seem reasonable, but which actually are severely restrictive when actual behavior is considered. The assumptions are: (a) the plate is initially perfectly flat; and (b) transverse displacements to the plate occurring during buckling are small compared to its thickness.

The large reserve strength of a plate girder after web buckling has occurred, as discussed in Sec. 11.4, will illustrate the deficiency of this linear-buckling theory to predict plate strength after buckling has occurred.

The equation for plate buckling, Eq. 6.14.25, may be considered entirely general and the development leading up to it for this one case may be considered illustrative of the procedure. The integer m indicates the number of half-waves which occur in the x direction at buckling. Figure 6.14.8 shows that there is a minimum value of k for any given number of

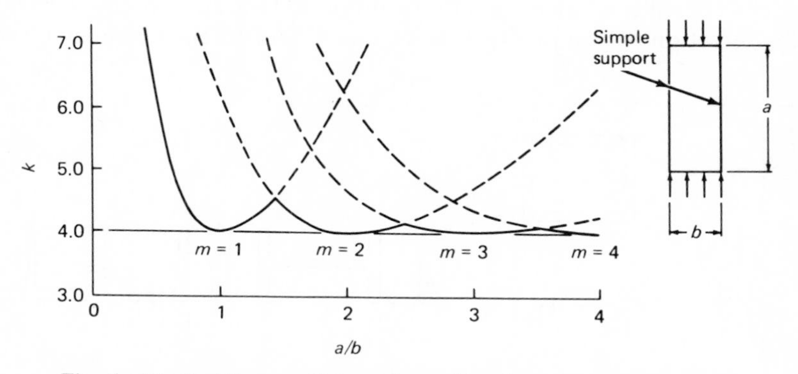

Fig. 6.14.8. Buckling coefficient for uniformly compressed plate—simple support longitudinal edges (Eq. 6.14.26).

half-waves, i.e., the weakest condition. It is noted that this weakest situation occurs when the length is an even multiple of width, and that multiple equals the number of half-waves.

Thus, setting $a/b = m$ gives $k = 4$. Further, as m becomes larger the k equation becomes flatter and approaches a constant value of 4 for large a/b ratio. This gives for the elastic buckling stress equation of plate elements under uniform compression along two edges and simply supported along the two edges parallel to the load,

$$F_{cr} = \frac{4\pi^2 E}{12(1 - \mu^2)(b/t)^2} \tag{6.14.27}$$

6.15.* STRENGTH OF PLATES UNDER UNIFORM EDGE COMPRESSION

Since rolled shapes, as well as built-up shapes, are composed of plate elements, the column strength of the section based on its overall slenderness ratio can only be achieved if the plate elements do not buckle locally. Local buckling of plate elements can cause premature failure of the entire section, or at the least it will cause stresses to become nonuniform and reduce the overall strength.

In Sec. 6.14 the basic approach to elastic stability of plates was developed. The theoretical elastic buckling stress for a plate was shown to be expressible as

$$F_{cr} = k \frac{\pi^2 E}{12(1 - \mu^2)(b/t)^2} \qquad [6.14.25]$$

where k is a constant depending upon type of stress, edge support conditions, and length to width ratio (aspect ratio) of the plate, E the modulus of elasticity, μ Poisson's ratio, and b/t the width-to-thickness ratio.

In general, plate-compression elements can be separated into two categories: (1) stiffened elements; those supported along two edges parallel to the direction of compressive stress; and (2) unstiffened elements; those

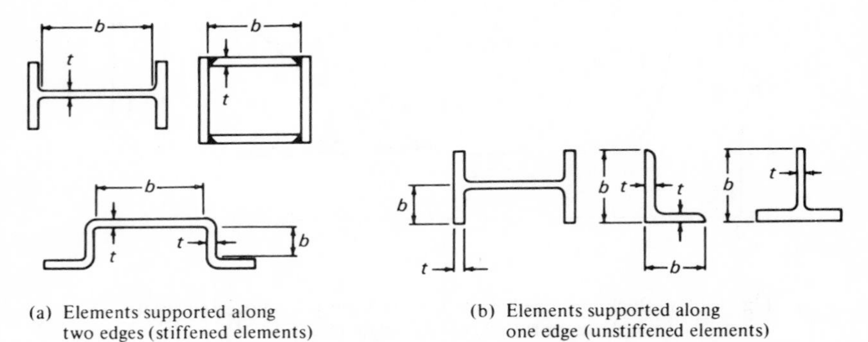

(a) Elements supported along two edges (stiffened elements)

(b) Elements supported along one edge (unstiffened elements)

Fig. 6.15.1. Typical local buckling situations.

supported along one edge and free on the other edge parallel to direction of compressive stress. Refer to Fig. 6.15.1 for typical examples of these two situations.

For the elements shown in Fig. 6.15.1 various degrees of rotation restraint are present. Figure 6.15.2 shows the variation in k with aspect

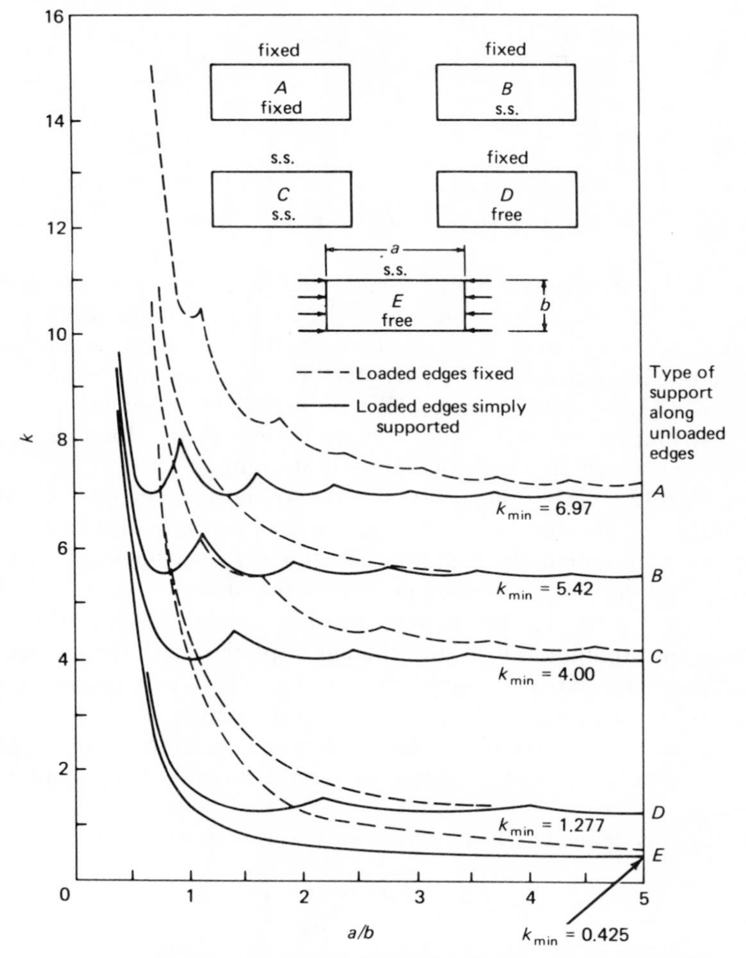

Fig. 6.15.2. Compressive-buckling coefficients for flat rectangular plates. (From Ref. 2, p. 80.)

ratio a/b for most of the idealized edge conditions, i.e., clamped (fixed), simply supported, and free.

Actual plate ultimate strength in compression is dependent upon many of the same factors which affect overall column strength, particularly residual stress. Figure 6.15.3 shows typical behavior of a compressed

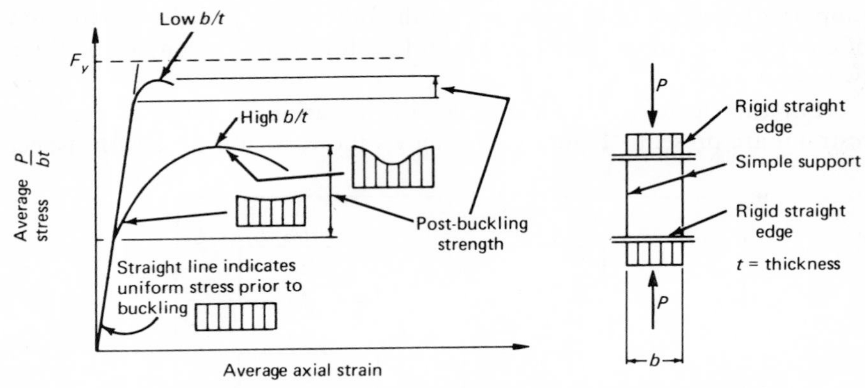

Fig. 6.15.3. Behavior of plate under edge compression. (Adapted from Ref. 21.)

plate loaded to its ultimate load.[21] Assuming ideal elastic-plastic material containing no residual stress the stress distribution remains uniform until the elastic buckling stress F_{cr} is reached. Further increase in load can be achieved but the portion of the plate farthest from its side supports will deflect out of its original plane. This out-of-plane deflection causes the stress distribution to be nonuniform even though the load is applied through ends which are rigid and perfectly straight.

Figure 6.15.3 clearly shows that plate strength under edge compression consists of the sum of two components; (1) elastic or inelastic buckling stress represented by Eq. 6.14.25, and (2) post-buckling strength. Also one should note the higher post-buckling strength as the width-to-thickness ratio (b/t) becomes larger. For low values of b/t, not only will post-buckling strength vanish, but the entire plate may have yielded and reached the strain-hardening condition, so that F_{cr}/F_y may become greater than unity. For plates *without* residual stress (referring to Fig. 6.15.4) three regions must be considered for establishing strength; elastic buckling (Euler hyperbola), yielding (segments AB, $A'B$, and $A''B$), and strain hardening.

If F_{cr}/F_y is defined as $1/\lambda^2$, which for plates then becomes

$$\lambda = \frac{b}{t}\sqrt{\frac{F_y(12)(1-\mu^2)}{\pi^2 Ek}} \tag{6.15.1}$$

it is observed from Fig. 6.15.4 that, when compared with columns (curve a), plates (curves b and c) achieve a strain hardening condition at relatively higher values of λ. In the earlier discussion on columns the value of λ at which strain hardening commences (λ_0) was assumed to be zero because of its relatively small value. The values of λ_0 for columns and plates under uniform edge compression for F_y = 36 ksi are given as

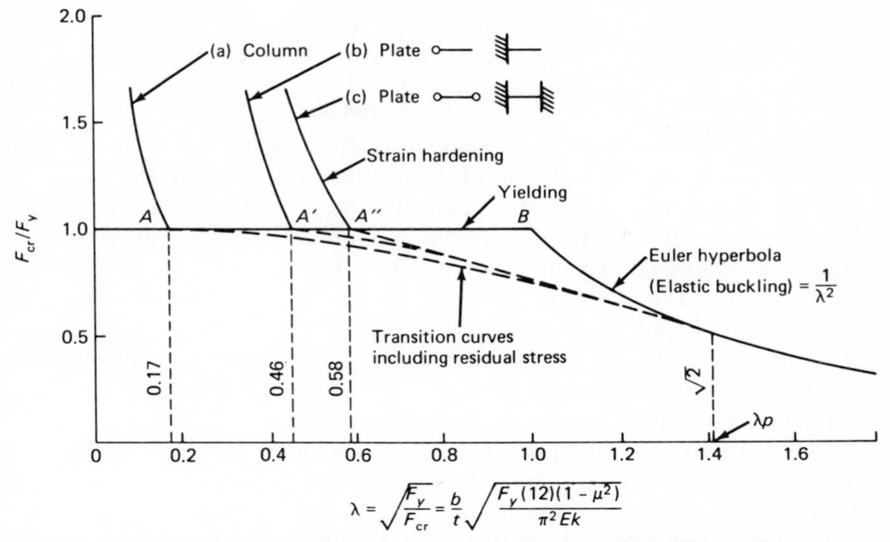

Fig. 6.15.4. Plate buckling compared with column buckling. (Adapted from Ref. 22.)

follows from Ref. 22:

Columns	$\lambda_0 = 0.173$	$(KL/r = 15.7)$
Long hinged flanges..........	$\lambda_0 = 0.455$	$(b/t = 8.15)$
Fixed flanges	$\lambda_0 = 0.461$	$(b/t = 14.3)$
Hinged webs	$\lambda_0 = 0.588$	$(b/t = 32.3)$
Fixed webs	$\lambda_0 = 0.579$	$(b/t = 42.0)$

From the above it should be noted that the important factor determining λ_0 is whether the plate element is supported along one or both edges parallel to loading, while the degree of rotational restraint along the supported edge (simply supported or fixed) has essentially no effect. Thus curves b and c of Fig. 6.15.4 each can represent two cases, where point A' has been taken at $\lambda = 0.46$ and point A'' at $\lambda \doteq 0.58$.

Since plates as well as rolled shapes contain residual stress the true strength should be represented by a transition curve, Fig. 6.15.4, between the Euler curve and the point at which strain hardening commences. It has been suggested[22] that the transition curve for compression elements may be taken as

$$\frac{F_{cr}}{F_y} = 1 - \left(1 - \frac{F_p}{F_y}\right)\left(\frac{\lambda - \lambda_0}{\lambda_p - \lambda_0}\right)^n \tag{6.15.2}$$

where λ_p = slenderness function at F_p = proportional limit.

The CRC parabola for overall column strength, Eq. 6.7.3 is also obtainable from Eq. 6.15.2 when $n = 2$, $\lambda_0 = 0$, and $\lambda_p = \sqrt{2}$ at

$F_{cr}/F_y = 0.5$ on the Euler hyperbola with

$$\lambda = \frac{KL}{r}\sqrt{\frac{F_y}{\pi^2 E}}$$

Since the influence of residual stress on plate buckling is less severe than for columns, a value of n greater than 2 would seem reasonable.

When considering inelastic behavior, the modulus of elasticity used for calculating strain in the direction of maximum stress, σ_x, should be the tangent modulus E_t. Examination of Eq. 6.14.4 shows that for inelastic strains in the x direction but elastic strain in the y direction, E cannot be factored out. Bleich[13] has shown the solution for this case of using different E values, and suggests arbitrarily using $\sqrt{E_t/E}$ as a multiplier for Eq. 6.14.25. This alternative to Eq. 6.15.2 is essentially the same thing, but without specifically defining the variation of E_t.

In summary, the strength of plates under edge compression may be governed by (1) strain hardening, low values of λ; (2) yielding, at λ = say 0.5 to 0.6; (3) inelastic buckling, represented by the transition curve (some fibers elastic and some yielded); (4) elastic buckling represented by the Euler hyperbola, at λ about 1.4; and (5) post-buckling strength with stress redistribution and large deformation, say for λ greater than 1.5.

For design purposes, performance criteria must be established to decide what range of λ values may be acceptable in design and how conservative (and simple) or liberal (and relatively complicated) should be the specification expressions for plate strength.

6.16. AISC BASIC PROVISIONS TO PREVENT PLATE BUCKLING FOR WORKING STRESS DESIGN

For a better understanding of the background for these requirements the reader is invited to delve into the subject of plate stability and strength as introduced in Secs. 6.14 and 6.15. However, it may be sufficient for many purposes merely to understand that components such as flanges, webs, angles, and cover plates, which are combined to form a column section may themselves buckle locally prior to the entire section achieving its maximum capacity. Typical elements are shown in Fig. 6.15.1. The buckled deflection of uniformly compressed plates is shown in Fig. 6.16.1 where two categories are apparent: (1) "unstiffened" plate elements having one free edge parallel to loading; and (2) "stiffened" plate elements supported along both edges parallel to loading.

Plates in compression behave essentially the same as columns and the basic elastic buckling expression corresponding to the Euler equation for columns has been derived as Eq. 6.14.25,

$$F_{cr} = k\frac{\pi^2 E}{12(1-\mu^2)(b/t)^2} \qquad [6.14.25]$$

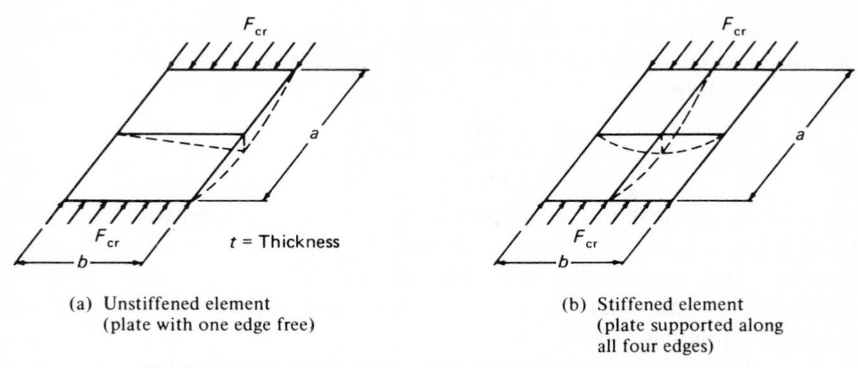

(a) Unstiffened element (plate with one edge free)

(b) Stiffened element (plate supported along all four edges)

Fig. 6.16.1. Buckled deflection of uniformly compressed plates.

where k is a constant depending on type of stress, edge conditions, and length to width ratio; μ is Poisson's ratio, and b/t is the width-to-thickness ratio (see Fig. 6.14.1). Typical k values are given in Fig. 6.15.2.

It is known that for low b/t values, strain hardening is achieved without buckling occurring, for medium values of b/t residual stress and imperfections give rise to inelastic buckling represented by a transition curve, and for large b/t buckling occurs in accordance with Eq. 6.14.25. Actual strength for plates with large b/t ratio exceeds buckling strength, i.e., they exhibit post-buckling strength. Thus strength for plates may be shown in a dimensionless fashion in Fig. 6.16.2.

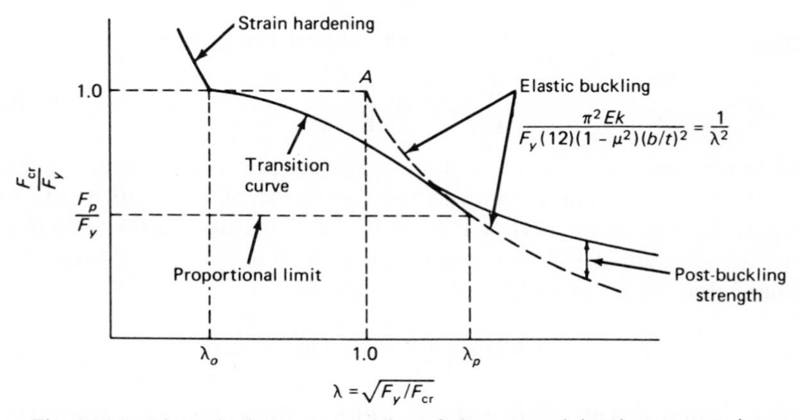

Fig. 6.16.2. Dimensionless representation of plate strength in edge compression.

To establish design requirements, it must be ascertained what performance is desired. Prior to 1969, AISC design procedures provided for designing columns composed of elements capable of reaching yield stress in compression without local buckling occurring. The local buckling of a column component may more logically be restricted so as not to occur prior to achieving full strength of the column based on its overall slenderness ratio. The performance requirement becomes

$$\underset{\substack{\text{Component}\\\text{element,}\\\text{i.e., plate}}}{F_{cr}} \geq \underset{\substack{\text{Overall}\\\text{column}}}{F_{cr}} \tag{6.16.1}$$

which would mean that acceptable b/t ratios would vary depending on the overall slenderness ratio of the column. If post-buckling strength were considered, the relationship would be additionally complicated. Generally, neglect of post-buckling strength for hot rolled plates $^{3}/_{16}$ in. or more in thickness involves little error.

For many years the AISI specification for light gage steel has used the approach of Eq. 6.16.1. The 1969 AISC specification in appendix C has included provisions to consider post-buckling strength, and permits general treatment of plate elements with higher width-to-thickness ratios (slenderness ratios) than the ordinary limitations of AISC–1.9.

To simplify design procedures, the AISC–1.9 basic provisions essentially require the critical buckling stress to be no less than yield stress in plate elements. Buckling, in other words, is prevented prior to achieving an average stress of F_y by application of the basic limits; buckling at lower stress may be considered by resorting to AISC Appendix C. The basic limitation to prevent buckling is

$$F_{cr} = \frac{k\pi^2 E}{12(1 - \mu^2)(b/t)^2} \geq F_y \tag{6.16.2}$$

Using $\mu = 0.3$ for steel, and $E = 29{,}000{,}000$ psi and F_y in psi,

$$\frac{b}{t} \leq 5{,}110 \sqrt{\frac{k}{F_y}} \tag{6.16.3}$$

which is represented by point A at $\lambda = 1.0$ on Fig. 6.16.2, a point lying above the transition curve. Thus a reduced value of λ should be used to minimize the deviation between F_y and the transition curve which accounts for residual stress and imperfections. Thus $\lambda = 0.7$ is taken as a rational value, which gives for b/t

$$\frac{b}{t} \leq 5{,}110\,\lambda \sqrt{\frac{k}{F_y}} = 3{,}570 \sqrt{\frac{k}{F_y}} \tag{6.16.4}$$

where F_y is in psi. Table 6.16.1 shows width to thickness ratios for various situations of uniform compression. The coefficients used by AISC–1969 tend to imply greater accuracy in criteria than is justified. The original coefficients were established using F_y in psi; after some rounding they formed the basis of AISC–1963 specifications. The present values are obtained by dividing by $\sqrt{1000}$ so that F_y can be used in ksi.

TABLE 6.16.1
Width-to-Thickness Ratio Requirements for Plate Elements
Subject to Uniform Compression

Structural Elements (1)	Buckling Coefficients (Fig. 6.15.2) (2)	b/t Eq. 6.16.4 (3)	AISC–1.9, 1969	
			F_y, psi (4)	F_y, ksi (5)
Unstiffened:				
(a) Single angles	0.425	$2{,}330/\sqrt{F_y}$	$2{,}400/\sqrt{F_y}$	$76.0/\sqrt{F_y}$
(b) Flanges	0.70*	$3{,}010/\sqrt{F_y}$	$3{,}000/\sqrt{F_y}$	$95.0/\sqrt{F_y}$
(c) Stems of tees	1.277	$4{,}040/\sqrt{F_y}$	$4{,}000/\sqrt{F_y}$	$127/\sqrt{F_y}$
Stiffened:				
(a) Uniform thickness flanges, such as tubular sections			$7{,}500/\sqrt{F_y}$‡	$238/\sqrt{F_y}$
(b) Perforated cover plates	6.97†	$9{,}450/\sqrt{F_y}$	$10{,}000/\sqrt{F_y}$	$317/\sqrt{F_y}$
(c) Others	5.0**	$7{,}980/\sqrt{F_y}$	$8{,}000/\sqrt{F_y}$	$253/\sqrt{F_y}$

*Arbitrarily selected to fall about mid-way between simply supported and fixed along the supported edge.

**Edge restraint estimated at about 1/3 fixed ($k = 4.0$ for simple support, and $k = 6.97$ for fixed—see Fig. 6.15.2).

†Consider full fixity—use of net plate width will provide adequate reserve.

‡Hollow sections generally receive negligible torsional restraint by the thin supporting edges; thus a coefficient somewhat less than 8,000 is used.

Table 6.16.2 gives AISC–1.9 values for width-to-thickness ratios, as well as AASHO–1969, section 1.7.70 and 1.7.89 values.

TABLE 6.16.2
Basic Width-to-Thickness Ratios for Plate Elements,
AISC—1969 and AASHO–1969.

Structural Elements	AISC–1969–1.9						AASHO–1969
	F_y, ksi						
	36	42	50	60	65	100	36
Unstiffened:							
(a) Single angles	12.7	11.7	10.7	9.8	9.4	7.6	12
(b) Flanges	15.8	14.7	13.4	12.3	11.8	9.5	11½*
(c) Stems of tees	21.2	19.6	18.0	16.4	15.8	12.7	—
Stiffened:							
(a) Uniform thickness flanges, as for tubular sections	39.7	36.7	33.7	30.7	29.5	23.8	—
(b) Cover plates	—	—	—	—	—	—	40
(c) Perforated plates	52.8	48.9	44.8	40.9	39.3	31.7	48
(d) Others	42.2	39.0	35.8	32.7	31.4	25.3	32

*For welded flanges; 12 on riveted girder flange angles and 16 on secondary members.

The AISC and AASHO specifications both permit using larger b/t ratios if the element is not considered fully effective. AISC–1.9 refers to use of Appendix C where provisions for obtaining reduced efficiency are given; i.e., the form factor Q in Eq. 6.8.3 becomes less than one. On the other hand, AASHO–1.7.89 permits an increase in b/t in proportion to the square root of the stress ratio (i.e., using an Euler-type curve). If f_a is the actual stress P/A and F_a is the allowable stress, when $f_a/F_a < 1.0$,

$$\text{AASHO:} \qquad \frac{b}{t} = \frac{\text{const.}}{\sqrt{F_y}} \sqrt{\frac{F_a}{f_a}}$$

The AISC Appendix C approach is the more rational; its development is given in Sec. 6.18, which considers post-buckling strength.

6.17. AISC PLASTIC STRENGTH PLATE BUCKLING PROVISIONS

Since plastic design is not treated until Chapter 7, it is only necessary to state here that the performance requirement of plastic design for plate elements under edge compression is that they be able to undergo strain in excess of that at first yield, ϵ_y. Referring to Fig. 6.17.1, local buckling

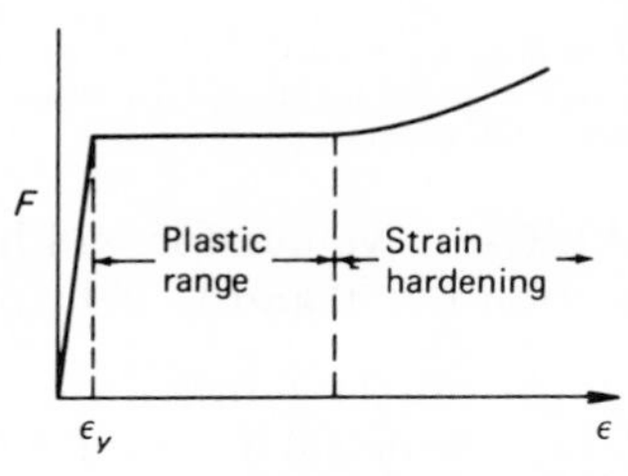

Fig. 6.17.1.

must not occur prior to achieving a compressive strain well into the plastic range and approaching strain hardening. Referring to Fig. 6.16.2, λ must be less than λ_0, i.e., reduced width to thickness ratios must be used compared with those for working stress method wherein it was necessary only to achieve a compressive strain, ϵ_y.

Development of theoretical expressions for local buckling under large plastic strains is outside the scope of this text. It is logical, however, from this brief discussion that λ should be at least reduced to λ_0 (Fig. 6.16.2) which for uniformly compressed plates has been shown[22] to be about 0.46 (Fig. 6.15.4). Thus, using Eq. 6.16.4, which in fact is not strictly applicable, with $\lambda = 0.46$, one obtains

$$\frac{b}{t} \leq 2350 \sqrt{\frac{k}{F_y}} \tag{6.17.1}$$

which if $k = 0.425$ (its least value) for elements supported along one edge (unstiffened), one obtains

$$\frac{b}{t} = \frac{1530}{\sqrt{F_y}}$$

and if $k =$ about 5, assuming typical edge restraint along two edges (stiffened),

$$\frac{b}{t} = \frac{5250}{\sqrt{F_y}}$$

Since residual stresses disappear in the plastic range and material imperfections have less effect, the above b/t limitations are actually overly severe. Thus AISC makes use of:

For unstiffened elements (projecting elements)

$$\frac{b}{t} \leq \frac{1650}{\sqrt{F_y}} \qquad \text{for } F_y \text{ in psi}$$

or

$$\frac{b}{t} \leq \frac{52.2}{\sqrt{F_y}} \qquad \text{for } F_y \text{ in ksi} \tag{6.17.2}$$

Since the coefficients are not precise values, the limitations are specified by a table rather than by a general expression (see Table 6.17.1).

TABLE 6.17.1
Width-to-Thickness Ratio Requirements for Plate Elements to Accommodate Plastic Strain in Axial Compression, AISC–2.7

F_y, ksi	Unstiffened Elements		Stiffened Elements
	Eq. 6.17.2	AISC–2.7	Eq. 6.17.3 and AISC–2.7
36	8.7	8.5	31.7
42	8.05	8.0	29.3
45	7.76	7.4	28.3
50	7.38	7.0	26.9
55	7.04	6.6	25.6
60	6.73	6.3	24.5
65	6.47	6.0	23.6

For stiffened elements,

$$\frac{b}{t} \leq \frac{6000}{\sqrt{F_y}} \qquad \text{for } F_y \text{ in psi}$$

or

$$\leq \frac{190}{\sqrt{F_y}} \quad \text{for } F_y \text{ in ksi} \tag{6.17.3}$$

Typical values from the formulas and AISC–2.7 appear in Table 6.17.1.

6.18.* AISC PROVISIONS TO ACCOUNT FOR THE BUCKLING AND POST-BUCKLING STRENGTHS OF PLATE ELEMENTS

Prior to the 1969 AISC, whenever width-to-thickness ratios exceeded the limits of AISC–1.9, the excess width could be neglected in computing the area and radius of gyration of the cross section. Strength computed on this reduced section is usually lower than actual strength, but is not uniformly so. Actual behavior dictates an alternative procedure, such as that of 1969 AISC.

As discussed in Sec. 6.15 and 6.16, plate elements in compression, either "stiffened" or "unstiffened" (see Fig. 6.16.1), have strength after buckling has occurred, i.e. post-buckling strength. Stiffened elements have a large post-buckling strength while unstiffened elements have only a little. However, since the ultimate strength of such elements can be evaluated, there is good reason to provide for its use in AISC–1969, as has been done in the light gage steel specification[25] since it was first introduced in 1946.

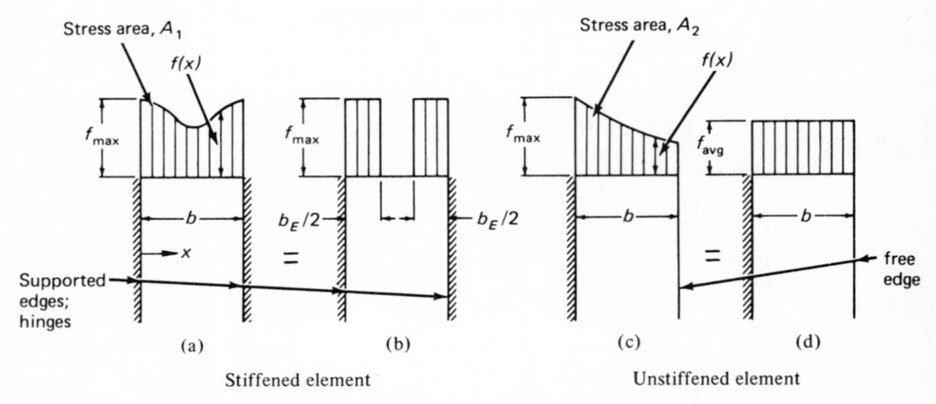

Fig. 6.18.1. Plate elements under axial compression, showing actual stress distribution and an equivalent system.

From Fig. 6.18.1a it is apparent that the strength of a stiffened element might be expressed as

$$P_{ult} = t \int_{A_1} f(x)\, dx \tag{6.18.1}$$

involving an integration of a nonuniform stress distribution; or alternatively, an "effective width" concept (Fig. 6.18.1b) may be used:

$$P_{ult} = tb_E f_{max} \quad \text{(stiffened element)} \tag{6.18.2}$$

where b_E = effective width over which the maximum stress may be considered uniform and give the correct total capacity.

Fig. 6.18.1c shows that Eq. 6.18.1 is equally valid for the unstiffened element with integration over area A_2. If a reduced stress $f_{avg} < f_{max}$ is used, the unstiffened element capacity could be written (Fig. 6.18.1d),

$$P_{ult} = tbf_{avg} \text{ (unstiffened element)} \tag{6.18.3}$$

The 1969 AISC and the AISI[25] have chosen to treat thin elements according to Eqs. 6.18.2 and 6.18.3, although actually either equation could be used for either type of element. Because of the large post-buckling strength of the stiffened element, one can imagine that it *has* buckled and part of the element is no longer active. On the other hand, the unstiffened element, with relatively little post-buckling strength may be thought of as not buckling because of the use of a reduced stress.

Effect on Overall Column Strength. For design stresses, it is desired to use gross section properties; thus for stiffened elements

$$\frac{P_{ult}}{A_{gross}} = \frac{A_{eff}}{A_{gross}} f_{max} = Q_a f_{max} \tag{6.18.4}$$

and for unstiffened elements,

$$\frac{P_{ult}}{A_{gross}} = \frac{f_{avg}}{f_{max}} (f_{max}) \frac{A_{gross}}{A_{gross}} = Q_s f_{max} \tag{6.18.5}$$

where Q_a and Q_s may be thought of as shape, or form, factors.

A compresssion system composed of both stiffened and unstiffened elements would be treated as unstiffened for establishing the stress f_{avg}; then the effective width for the stiffened elements is determined using $f_{max} = f_{avg}$. Thus, the total capacity would be

$$P_{ult} = f_{avg} A_{eff} \tag{6.18.6}$$

which upon dividing by A_{gross} gives

$$\frac{P_{ult}}{A_{gross}} = \frac{f_{avg}}{f_{max}} (f_{max}) \frac{A_{eff}}{A_{gross}} = Q_s Q_a f_{max} \tag{6.18.7}$$

From Eqs. 6.18.4, 6.18.5, and 6.18.7, it is clear that the effect of premature local buckling before the strength of the overall column has been achieved is to multiply the maximum achievable stress by the form factors Q. Neglecting the possibility of strain hardening, the maximum stress is the yield stress, which is therefore to be multiplied by Q. For the basic CRC parabola, Eq. 6.7.3, this will give

$$F_{cr} = QF_y\left[1 - \frac{QF_y}{4\pi^2 E}\left(\frac{KL}{r}\right)^2\right] \tag{6.18.8}$$

where $Q = Q_s Q_a$. From Eq. 6.18.8, when $F_{cr} = QF_y/2$ the slenderness ratio is

$$C_c = \sqrt{\frac{2\pi^2 E}{QF_y}} \tag{6.18.9}$$

Applying the factor of safety to Eq. 6.18.8 gives AISC Formula (C5–1), which is also Eq. 6.8.3. Whenever KL/r exceeds C_c, the effect of local buckling on overall column stability is insignificant,, so that the Euler equation divided by the factor of safety 23/12 is used without any form factor adjustment (see Fig. 6.8.8).

Note that since the form factor Q is used in the allowable stress equations, the radius of gyration r is to be that of the *gross* section.

Form Factor Q_s for Unstiffened Elements. Referring back to Fig. 6.16.2, a Q_s value less than 1.0 means that $\lambda > \lambda_0$. A transition parabola, such as Eq. 6.15.2 could have been used to compute the reduced stress. For simplification, however, a straight line has been used as shown by curve *A* in Fig. 6.18.2. It assumes as did Fig. 6.16.4 that $\lambda = 0.7$ is the maximum value for which $F_{cr} = F_y$, and that the proportional limit occurs at $\lambda_p = \sqrt{2}$, the same as for overall buckling. However, because of some post-buckling strength the Euler-type curve has been raised above the theoretical (curve C) so that on the AISC design curve (curve *B*) when $\lambda_p = \sqrt{2}$, $Q_s = F_{cr}/F_y = 0.65$. Many different expressions could have been used with the same logical results.

AISC Appendix C gives the stress reduction equations for the case of flanges and for the stems of tees. The logic for these reduction equations is the same as described for single angles and shown in Fig. 6.18.2. These other equations will be found to be approximately proportional to $\sqrt{k}$, as seen by reference to Eq. 6.16.4 and Table 6.16.1. The table contains the k values used for the other unstiffened cases.

The limiting proportions for channels and tees given in AISC Appendix Table C1 are to preclude torsional buckling as a failure mode. This concept is discussed in Sec. 8.11.

Form Factor Q_a for Stiffened Elements. The concept of using an effective width over which stress may be considered uniform, even though it is actually non-uniform, was developed by von Karman,[26] and later modi-

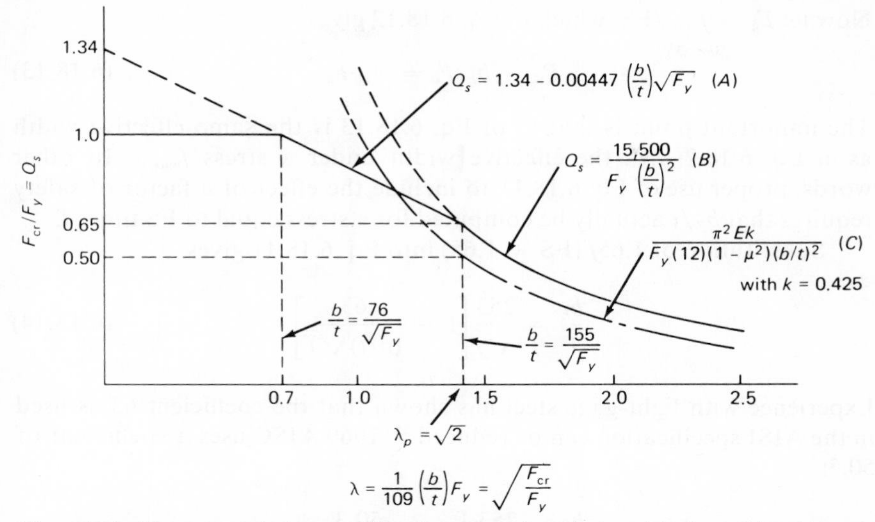

Fig. 6.18.2. Plate strength for unstiffened compression element (single-angle) with one edge hinged and the other free (AISC–1969).

fied by Winter.[27] Winter's equation, which has been used by the light-gage specification[25] since 1946, is

$$\frac{b_E}{t} = 1.9\sqrt{\frac{E}{f}}\left[1.0 - \frac{0.475}{(b/t)}\sqrt{\frac{E}{f}}\right] \tag{6.18.10}$$

where f = stress acting on the element ($f_{\max}$ of Fig. 6.18.1)
b/t = actual width-to-thickness ratio

The form and constants of the equation were essentially determined to agree with experimental results.

Substituting E = 29,000 ksi gives

$$\frac{b_E}{t} = \frac{324}{\sqrt{f}}\left[1.0 - \frac{81}{(b/t)\sqrt{f}}\right] \tag{6.18.11}$$

which would give the correct effective width under the action of a stress f in ksi; i.e., no factor of safety would be involved.

Reexamine Eq. 6.18.2:

$$P_{\text{ult}} = b_E t f_{\max} \qquad [6.18.2]$$

which if divided by a factor of safety, FS, would give the safe service load, P_w; thus

$$P_w = b_E t\,(f_{\max}/\text{FS}) \tag{6.18.12}$$

Now let $F_a = f_{max}/\text{FS}$, which in Eq. 6.18.12 gives

$$P_w = b_E t F_a = A_{eff} F_a \tag{6.18.13}$$

The important point is that b_E of Eq. 6.18.13 is the same effective width as in Eq. 6.18.2; i.e., the effective width under a stress f_{max}. In other words, proper use of Eq. 6.18.11 to include the effect of a factor of safety requires that b_E/t actually be computed for a stress equal to FS times f.

Substitution of $1.65f$ (FS = 1.65) into Eq. 6.18.11 gives

$$\frac{b_E}{t} = \frac{253}{\sqrt{f}}\left[1 - \frac{63}{(b/t)\sqrt{f}}\right] \tag{6.18.14}$$

Experience with light-gage steel has shown that the coefficient 63 as used in the AISI specification can be reduced.[25] 1969 AISC uses a coefficient of 50.3:

$$\frac{b_E}{t} = \frac{253}{\sqrt{f}}\left[1 - \frac{50.3}{(b/t)\sqrt{f}}\right] \tag{6.18.15}$$

which is AISC Formula (C3–1), which applies for "flanges of square and rectangular sections of uniform thickness."

The difference between Eqs. 6.18.14 and 6.18.15 can be partly explained by the difference in the rotational restraint (moment along the supported edges, Fig. 6.18.1b) that is assumed. In sections comprising hot-rolled elements as covered by AISC, the degree of restraint provided by adjacent elements is probably greater than on typical light gage shapes (where t is usually $< 3/16$ in.).

Since a column cross section may include unstiffened elements, the controlling stress permitted on unstiffened elements is used as the applicable maximum stress acting on the stiffened elements. Thus the service load stress is

$$f = 0.6F_{cr} \qquad \text{based on unstiffened elements} \tag{6.18.16}$$

or, since $Q_s = F_{cr}/F_y$

$$f = 0.6Q_s F_y \tag{6.18.17}$$

as prescribed by AISC–C3.

Finally, Q_a as defined by Eq. 6.18.4 is

$$Q_a = \frac{(\text{effective width})t}{\text{actual area}} = \frac{A_{eff}}{A_{gross}} \qquad [6.18.4]$$

where $A_{eff} = A_{gross} - \Sigma(b - b_E)t$.

Design Properties. In computing nominal stress $f_a = P/A$ and $f_b = M/S$ under service loads the following rules apply, in accordance with AISC Appendix–C4.

For axial compression:

(a) Use gross area for P/A.
(b) Use radius of gyration of gross area for KL/r.

This agrees with the philosophy stated earlier in this article when developing Eqs. 6.18.4 through 6.18.7.

For Flexure:

(a) Use reduced section properties for beam with flanges containing stiffened elements.

(b) For sections having symmetrical gross area about the axis of bending; in lieu of computing a neutral axis shift due to reduced effective area on the compression flange, the same reduction may be assumed for the tension flange.

Since the allowable stresses for flexure do not contain the form factors Q, it is entirely appropriate to use section properties based on effective area.

For beam-columns:

(a) Use gross area for P/A.
(b) Use reduced section properties for flexure involving stiffened compression elements for M_x/S_x and M_y/S_y.
(c) Use Q_a and Q_s for determining F_a
(d) For F_b, use formulas based on lateral-torsional buckling as discussed in Chapter 9, but the maximum value is $0.6Q_sF_y$ if unstiffened compression elements are involved.

6.19. DESIGN OF COMPRESSION MEMBERS AS AFFECTED BY LOCAL BUCKLING PROVISIONS

Design of single and double angle struts, structural tees, welded built-up I-shapes, and most other built-up sections, including box-type built-up sections, involves close consideration of local buckling limitations. Rolled W shapes as treated previously are proportioned such that for working stress design the width-to-thickness limitations are satisfied ($Q = 1.0$), and even in plastic design, many sections satisfy local buckling limitations.

The following examples illustrate the treatment of situations where local buckling may occur prior to development of column strength based on the overall slenderness ratio, KL/r.

EXAMPLE 6.19.1

A double-angle compression chord member for the truss of Fig. 6.19.1 is composed of 2-L9 × 4 × ½, with short legs back-to-back, having a length of 28 ft. Bracing is provided in the plane of the truss every 7 ft, but

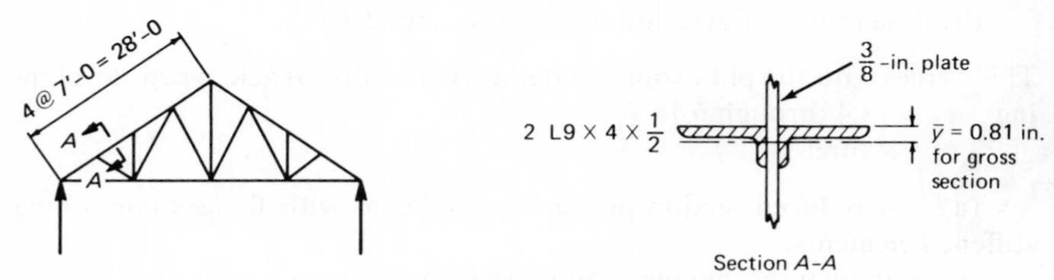

Fig. 6.19.1. Example 6.19.1

bracing between trusses is such that lateral support occurs only at the ends of the 28-ft length. Neglect contribution of roofing to lateral support. What is the maximum axial compression service load which this member can be permitted to carry safely? Use A36 steel and AISC specs.

Solution

The 9-in. legs of this compression member are unstiffened elements according to AISC–1.9.1.

Check width/thickness ratio:

$$\left(\frac{b}{t} = \frac{9}{0.5} = 18 > \frac{b}{t}\right)_{\text{lim}} = \frac{76.0}{\sqrt{F_y}} = 12.7$$

Thus local buckling will control and the section efficiency is reduced. Using AISC Appendix–C2 the reduced stress factor, Q_s is

$$Q_s = 1.340 - 0.00447\left(\frac{b}{t}\right)\sqrt{F_y}$$

$$= 1.340 - 0.00447(18)\sqrt{36} = 0.858$$

For axial compression, properties of the gross section are used. The AISC Manual properties for double angle struts give

$$A = 12.50 \text{ sq in.}$$

$$r_x = 1.05 \text{ in.,} \qquad r_y = 4.55 \text{ in. for } 3/8\text{-in. gusset plate}$$

Assuming $K = 1.0$ for truss members as per Sec. 6.9,

$$\frac{KL_x}{r_x} = \frac{1.0(84)}{1.05} = 80; \qquad \frac{KL_y}{r_y} = \frac{1.0(28)(12)}{4.55} = 74$$

Using AISC Formula (C5–1),

$$C_c = \sqrt{\frac{2\pi^2 E}{Q_s F_y}} = \frac{755}{\sqrt{Q_s F_y}} = \frac{755}{\sqrt{0.858(36)}} = 135.5$$

$$\text{FS} = \frac{5}{3} + \frac{3}{8}\left(\frac{80}{135.5}\right) - \frac{1}{8}\left(\frac{80}{135.5}\right)^3 = 1.863$$

$$F_a = Q_s F_y\left[1 - \frac{(KL/r)^2}{2\,C_c^2}\right]\frac{1}{\text{FS}}$$

$$= 0.858(36)\left[1 - \frac{1}{2}\left(\frac{80}{135.5}\right)^2\right]\frac{1}{1.863} = 13.7 \text{ ksi}$$

Safe permissible load:

$$P = A_g F_a = 12.50(13.7) = 171 \text{ kips}$$

EXAMPLE 6.19.2

Select an economical double-angle compression member for use as a spreader strut used for hoisting large loads, as shown in Fig. 6.19.2. The total lifted load is to be 60 tons. Use $F_y = 60$ ksi with AISC specs.

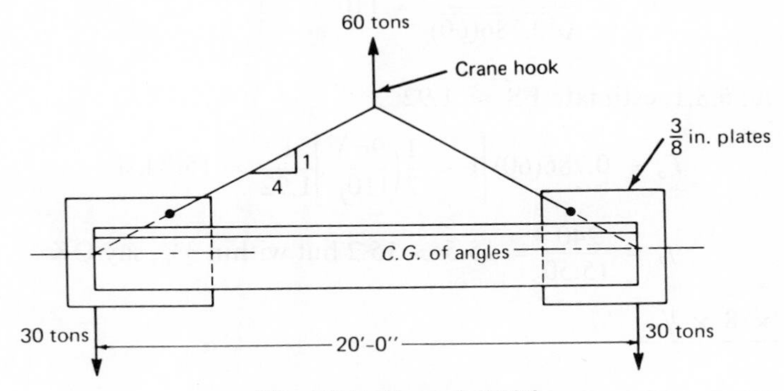

Fig. 6.19.2. Example 6.19.2.

Solution

Assume $K = 1.0$, and use Appendix, Table A1 to estimate r,

$$r_x \approx 0.29h$$
$$r_y \approx 0.24b$$

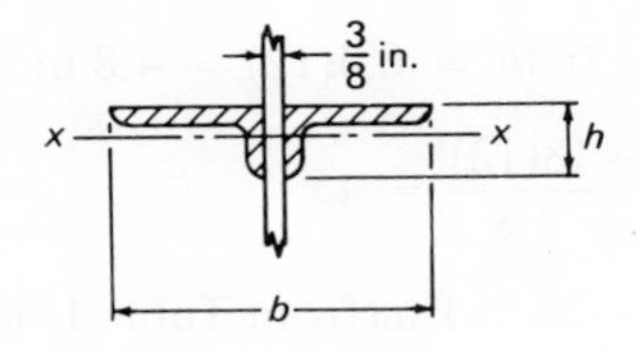

If 8 in. legs are used, $r_y \approx 0.24(8) = 1.92$ in.

$$\text{Estimate} \qquad \frac{KL}{r} = \frac{240}{2} = 120$$

$$F_a = 10.37 \text{ ksi (from AISC Table 1-60)}$$

$$\text{Reqd } A = \frac{240}{10.37} = 23 \text{ sq in.}$$

Try L8 × 8 × ⅝, $r_{\min} = 2.49$ in., $KL/r = 96.5$, $F_a = 16.02$ ksi

$$\text{Revised} \quad \text{Reqd } A = \frac{240}{16.02} = 14.95 \text{ sq in.}$$

Try L8 × 8 × ½, $r_{\min} = 2.50$ in., $A = 15.50$ sq in., $KL/2 = 96$

$$\frac{b}{t} = \frac{8}{0.5} = 16 > \frac{76.0}{\sqrt{F_y}} = \frac{76.0}{\sqrt{60}} = 9.8$$

$$Q_s = 1.340 - 0.00447(16)\sqrt{60} = 0.786$$

$$C_c = \frac{755}{\sqrt{0.786(60)}} = 110$$

Using Fig. 6.8.1, estimate FS ≈ 1.92:

$$F_a = 0.786(60)\left[1 - \frac{1}{2}\left(\frac{96}{110}\right)^2\right]\frac{1}{1.92} = 15.2 \text{ ksi}$$

$$f_a = \frac{240}{15.50} = 15.5 > 15.2 \text{ but within } 3\%, \text{ say OK}$$

Use L8 × 8 × ½.

EXAMPLE 6.19.3

Select the thinnest 12 × 12 structural tube to carry an axial compression of 200 kips over a length $KL = 18$ ft. The member is part of a braced system. Use $F_y = 60$ ksi and AISC specs.

Solution

Use Appendix, Table A1, to estimate radius of gyration,

$$r \approx 0.4h \approx 0.4(12) = 4.8 \text{ in.}$$

$$\frac{KL}{r} = \frac{18(12)}{4.8} = 45$$

$$F_a = 29.35 \text{ ksi (from Table 1-60, AISC)}$$

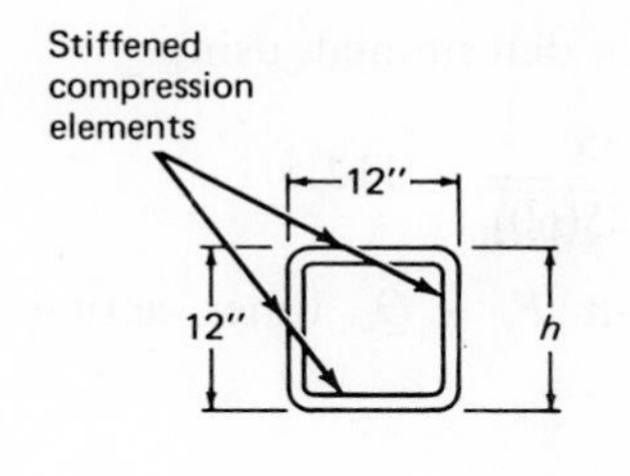

(a) Gross section

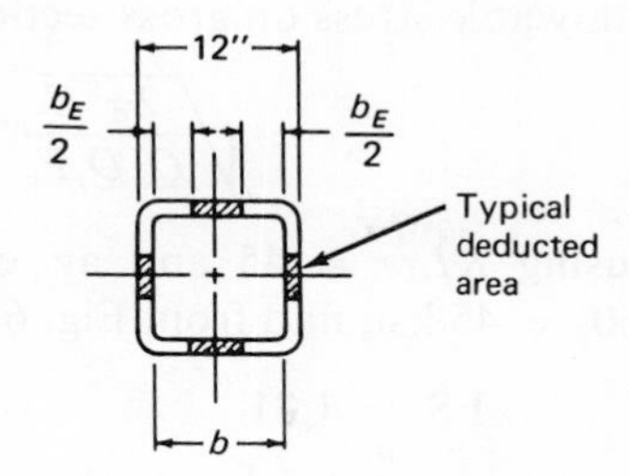

(b) Effective section

Fig. 6.19.3. Example 6.19.3.

$$\text{Reqd } A = \frac{200}{29.35} = 6.82 \text{ sq in.}$$

Try 12 × 12 × ¼ structural tube: $A = 11.48$, $r = 4.77$.
Check b/t on stiffened elements:

$$\frac{b}{t} = \frac{12.0 - 2(0.25)}{0.25} = \frac{11.5}{0.25} = 46$$

$$\left(\frac{b}{t}\right)_{\text{lim}} = \frac{238}{\sqrt{F_y}} = \frac{238}{\sqrt{60}} = 30.8 < 46$$

Local buckling will reduce the efficiency of this section. Compute the form factor Q in accordance with AISC Appendix C. The effective width/thickness ratio is

$$\frac{b_E}{t} = \frac{253}{\sqrt{f}}\left[1 - \frac{50.3}{(b/t)\sqrt{f}}\right] \qquad [6.18.15]$$

where $f = 0.6F_y Q_s$. In this case, a section with no unstiffened elements has $Q_s = 1.0$; thus, $f = 0.6F_y = 36$ ksi.

$$\frac{b_E}{t} = \frac{253}{\sqrt{36}}\left[1 - \frac{50.3}{(46)\sqrt{36}}\right] = 42.2(1.0 - 0.182) = 34.5$$

for all four sides of the tube. The effective area is

$$A_{\text{eff}} = 11.48 - 4\left[\frac{b}{t} - \frac{b_E}{t}\right]t^2$$

$$= 11.48 - 4(46 - 34.5)(0.25)^2 = 8.60 \text{ sq in.}$$

and the form factor is

$$Q_a = \frac{A_{\text{eff}}}{A_{\text{gross}}} = \frac{8.60}{11.48} = 0.75$$

The allowable stress on gross section is then determined, using

$$C_c = \sqrt{\frac{2\pi^2 E}{Q_a Q_s F_y}} = \frac{755}{\sqrt{0.75(60)}} = 112.5$$

then using $KL/r \approx 45$ and an equivalent $F_y = Q_a$ times actual $F_y = 0.75(60) = 45$ ksi, find from Fig. 6.8.1,

$$\text{FS} \approx 1.81$$

$$F_a = \frac{Q}{\text{FS}}\left[1 - \frac{1}{2}\left(\frac{KL/r}{C_c}\right)^2\right]F_y$$

$$= \frac{0.75}{1.81}\left[1 - \frac{1}{2}\left(\frac{45}{112.5}\right)^2\right]60 = \frac{41.4}{1.81} = 22.85 \text{ ksi}$$

or from AISC Table 1-45, $F_a = 22.90$ ksi.

$$f_a = \frac{P}{A_{\text{gross}}} = \frac{200}{11.48} = 17.4 \text{ ksi} < 22.85 \qquad \text{OK}$$

The stress is low but the lightest 12 × 12 section has been checked. Use 12 × 12 × ¼ structural tube, $F_y = 60$ ksi.

EXAMPLE 6.19.4

Determine the axial-compression capacity for the nonstandard shape of Fig. 6.19.4 for an equivalent length, $KL = 15$ ft. Use $F_y = 100$ ksi and AISC specs.

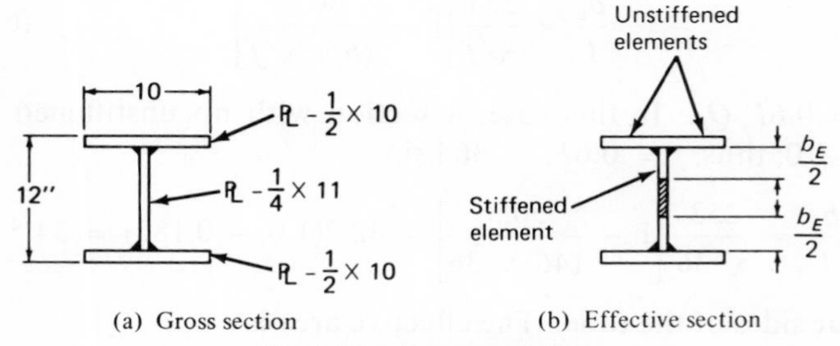

Fig. 6.19.4. Example 6.19.4.

Solution

In axial compression this section contains unstiffened elements (flanges) and a stiffened element (the web). Unstiffened elements must be considered first in order to determine the effective stress level.

(a) Properties of the gross section. Using AISC Manual pp. 2–125 to 2–131.

$$I_y = \text{negligible (web)} + 2(41.7)\text{ (flanges)} = 83.4\text{ in.}^4$$

$$A = 11(0.25) + 2(5.0) = 12.75\text{ in.}^2$$

$$r_y = \sqrt{I/A} = \sqrt{83.4/12.75} = 2.56\text{ in.}$$

(b) Unstiffened elements, AISC–1.9.1.2:

$$\frac{b}{t} = \frac{5.0}{0.5} = 10.0 > \left.\frac{b}{t}\right)_{\text{lim}} = \frac{95.0}{\sqrt{100}} = 9.5$$

Thus local buckling causes reduced efficiency; $Q_s < 1.0$:

$$Q_s = 1.415 - 0.00437\left(\frac{b}{t}\right)\sqrt{F_y} \quad \text{[AISC Formula C2–3]}$$

$$= 1.415 - 0.00437(10)\sqrt{100} = 0.978$$

(c) Stiffened element, AISC–1.9.2.2:

$$\frac{b}{t} = \frac{11.0}{0.25} = 44 > \left.\frac{b}{t}\right)_{\text{lim}} = \frac{253}{\sqrt{100}} = 25.3$$

Thus $Q_a < 1.0$. The stress presumed to act on the stiffened element is that controlled by the unstiffened elements:

$$f = 0.6F_yQ_s = 0.6(100)(0.978) = 58.6\text{ ksi}$$

$$\frac{b_E}{t} = \frac{253}{\sqrt{f}}\left[1 - \frac{44.3}{(b/t)\sqrt{f}}\right] \quad \text{[AISC Formula C3–2]}$$

$$= \frac{253}{\sqrt{58.6}}\left[1 - \frac{44.3}{44\sqrt{58.6}}\right] = 33(1 - 0.131) = 28.7$$

$$A_{\text{eff}} = A_{\text{gross}} - \left(\frac{b}{t} - \frac{b_E}{t}\right)t^2$$

$$= 12.75 - (44 - 28.7)(0.25)^2 = 11.795\text{ sq in.}$$

$$Q_a = \frac{A_{\text{eff}}}{A_{\text{gross}}} = \frac{11.795}{12.75} = 0.925$$

$$Q = Q_sQ_a = 0.978(0.925) = 0.904$$

$$\text{Effective } F_y = QF_y = 0.904(100) = 90.4$$

Use AISC Table 1-90,

$$\frac{KL}{r} = \frac{15(12)}{2.56} = 70.4,\ F_a = 28.7\text{ ksi}$$

or F_a could be computed from formulas as in previous examples.

$$\text{Allowable } P = A_{\text{gross}} F_a = 12.75(28.7) = 366 \text{ kips}$$

PROBLEMS

Note: All problems are to be done according to the latest AISC specifications unless otherwise indicated.

6.1. Select the lightest W section to carry an axial compression load of 100 kips. The member is part of a braced frame and has an equivalent pinned length of 22 ft. (a) Design using A36 steel; (b) Design using steel having a minimum specified yield point, $F_y = 50$ ksi.

6.2. Re-do Prob. 6.1 considering that the member is fixed (built-in) at the bottom of the 22-ft length.

6.3. Select the lightest W section to serve as a main member column 28 ft long, in a braced frame, with additional lateral support in the weak direction at mid-height, to carry an axial load of 210 kips. Assume bottom and top of column are pinned. Use A36 steel and indicate first and second choices for the sections.

6.4. Design the most economical W section to carry an axial compression of 150 kips when the member has an equivalent pinned length of 18 ft in a braced frame, assuming that relative costs of various steels are as follows:

A36	1.00
$F_y = 50$ ksi	1.07
$F_y = 60$ ksi	1.10

Design the section assuming constant yield point for all available sections.

6.5. Redesign the column of Prob. 6.4 assuming additional weak direction support at mid-height.

6.6. Select the lightest W section of A36 steel to carry a concentric load of 308 kips. The member is part of a braced frame and may be assumed pinned at the top and bottom of a 30 ft length, and in addition has support in the weak direction 14 ft from the bottom.

6.7. Select the lightest W section to carry an axial compression load of 400 kips. Assume the member to be part of a braced frame. The idealized support conditions are that the member is hinged in both principal directions at the top of a 30 ft. height; supported in the weak direction at 14 and 22 ft. from the bottom; and fixed in both directions at the bottom. Use A36 steel.

6.8. Design the lightest W section to carry a compressive load of 825 kips on an equivalent pinned length of 25 ft. Use steel with $F_y = 50$ ksi for all sections. If there is a 7% differential in price per pound between this steel and A36, is the section economical when designed using $F_y = 50$ ksi?

6.9. Design the lightest W section for the column shown in the accompanying figure. The member is built into a wall so that it may be considered as continuously braced in the weak direction. Use A36 steel.

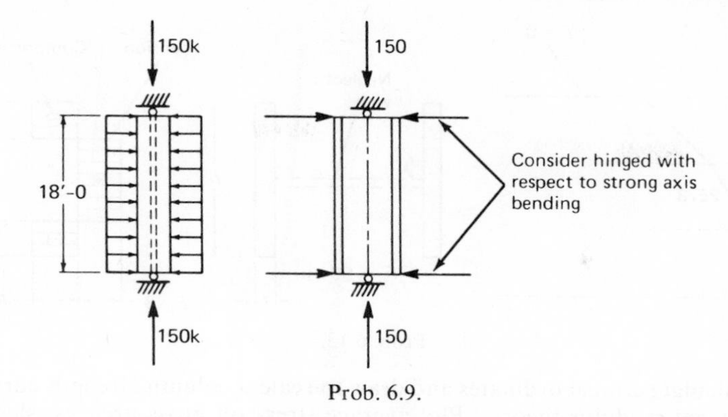

Prob. 6.9.

6.10. Redesign the column of Prob. 6.8 assuming that column ultimate buckling strength may be defined by Euler's Equation, $F_{cr} = \pi^2 E/(KL/r)^2$. Use steel with $F_y = 50$ ksi (Remember F_{cr} cannot exceed F_y), and use the AISC equation for factor of safety only.

6.11. Redesign the column of Prob. 6.8 using the secant formula (Eq. 6.11.6) similar to those used by AASHO and AREA specifications. (Use FS = 1.89)

6.12. Using the tangent modulus theory, draw to scale the column strength (in terms of average unit stress on gross area) vs. slenderness ratio curve for a steel with $F_y = 50$ ksi but having a stress-strain diagram for a column stub as shown in the accompanying figure. Assume no residual stress. Using the strength curve along with the AISC factor of safety, select the lightest W section for the loading and support conditions of Prob. 6.8.

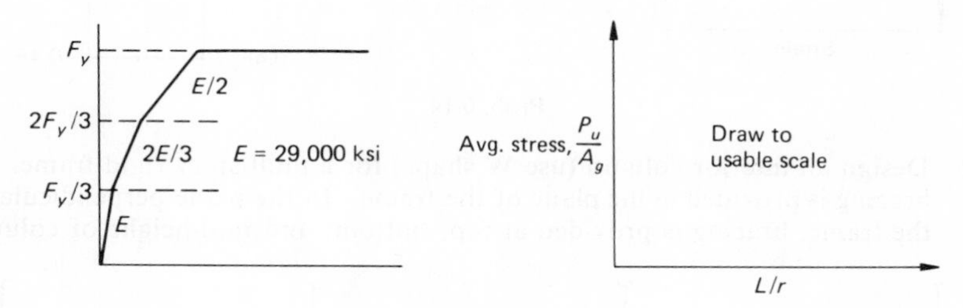

Prob. 6.12.

6.13. Using the tangent modulus theory: (a) construct a column strength curve for a wide-flange shape. Assume weak-axis bending and neglect the effect of the web. Assume an idealized stress-strain relationship for each fiber as shown in the accompanying figure. (b) Using the strength curve ($f_{cr} = P_{cr}/A_g$ vs. KL/r), select the lightest W shape to carry 300 kips on a pin end length of 30 ft. Use AISC factor of safety and $F_y = 50$ ksi. (c) Compare with section obtained using AISC specification.

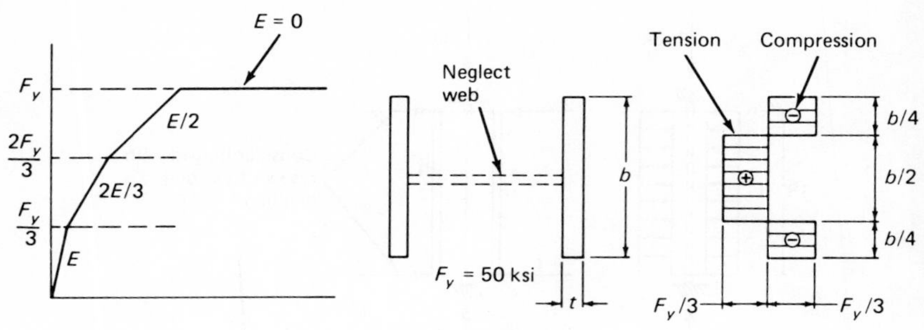

Prob. 6.13.

6.14. Calculate critical ordinates and draw to scale a column strength curve using tangent modulus theory. Plot average stress on gross area vs. slenderness ratio for weak-axis buckling of wide-flange shapes. Assume each fiber has an idealized elastic-plastic stress-strain diagram with F_y = 50 ksi. Neglect the effect of the web and consider each flange to have a residual stress pattern as shown. Using the developed curve along with the AISC factor of safety, select the lightest W section for the loading and support conditions of Prob. 6.8. (Draw conclusions regarding the results of Probs. 6.8, and 6.12, if worked, and discuss briefly the merits of making a comparison.)

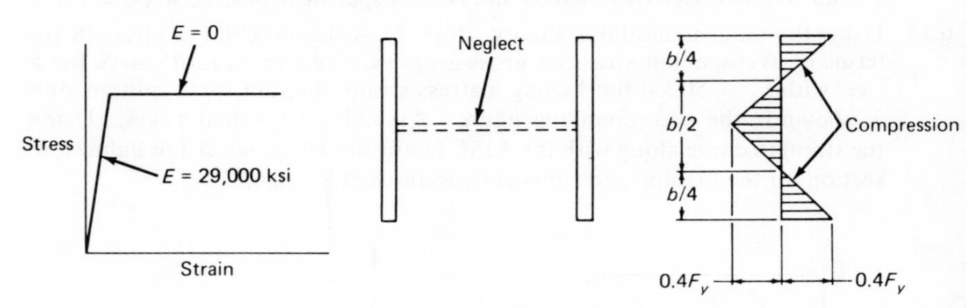

Prob. 6.14.

6.15. Design an interior column (use W shape) for a multistory rigid frame. No bracing is provided in the plane of the frame. In the plane perpendicular to the frame, bracing is provided at top, bottom, and mid-height of columns

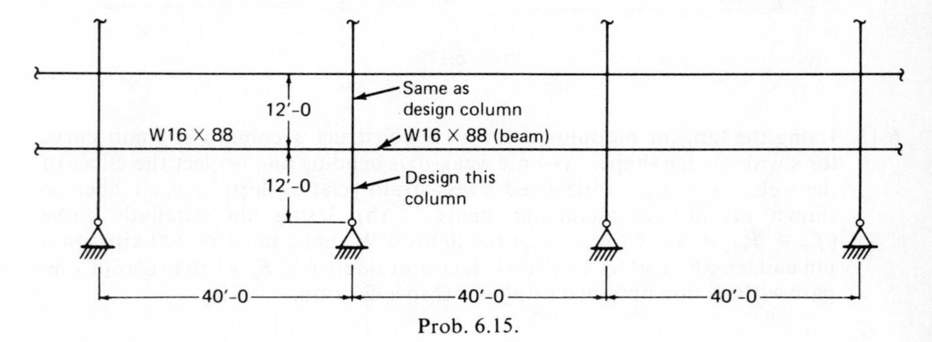

Prob. 6.15.

and simple flexible beam to column connections are used. The axial compressive load is 1,500 kips and bending moments are neglected. Use F_y = 50 ksi.

6.16. Design a welded laced column (use single lacing) composed of four angles to carry an axial compression of 500 kips on an effective length, KL = 28 ft. Use either A36 or F_y = 50 ksi steel, whichever is most economical, if F_y = 50 ksi steel costs 7 percent more per pound of fabricated steel than A36. Satisfy basic requirements of AISC-1.9.

6.17. Repeat Prob. 6.16 if the axial load is 400 kips and the effective length is 32 ft.

6.18. Design a solid welded-box section for the conditions of Prob. 6.16.

6.19. Design a solid welded-box section for the conditions of Prob. 6.17.

6.20. Design a box section with two perforated plates for the conditions of Prob. 6.16.

6.21. Design a box section with two perforated plates for the conditions of Prob. 6.17.

6.22. Determine the safe compressive load on a WT10.5 × 48 (structural tee) when used in a truss location where it is braced in the plane of the truss at 20-ft intervals and braced transverse to the plane of the truss at 10-ft intervals. Material is A36 steel. Apply provisions of AISC-Appendix C if it seems desirable.

6.23. Repeat Prob. 6.22 using F_y = 65 ksi. If this material costs 12% more than A36, is its use justified?

6.24. Select the lightest double-angle compression member of A36 steel to carry 240 kips on an equivalent pinned length of 20 ft. Assume the angles are attached to ⅜-in. gusset plates.

6.25. Select the lightest two-angle main compression member connected to a ⅜-in. gusset plate and to carry a concentric load of 180 kips on a pinned length of 16 ft. Use F_y = 50 ksi steel.

6.26. Select the most economical double angle member to carry a compressive load of 80 kips on an equivalent pinned length of 12 ft. Consider A36, A572 Grade 50, and A572 Grade 60 steels. Assume relative costs are A36(1.0), Grade 50(1.07), and Grade 60(1.10).

6.27. Design a top chord of a roof truss which consists of a double-angle section connected to ½-in. gusset plates, and to carry a load of 160 kips (including wind loading). The member is braced in a vertical plane by adjoining web members connecting in at 5-ft intervals. The chord is braced transverse to the plane of the truss at 10-ft intervals. Neglect bending due to roof loads. Use A36 steel.

6.28. Select the lightest structural tee (WT) for use as a top-chord compression member. Neglect bending. The member has a 9-ft equivalent pinned length for buckling in either the x–x or y–y plane. The axial compressive load is 135 kips. Use A572 Grade 50 steel.

6.29. Select the lightest structural tee (WT) to serve as the compression chord of a truss to carry 85 kips. In the vertical plane the chord is braced by adjoining web members which frame in at 5-ft intervals. Horizontally the chord is braced at 10 ft by a system of lateral purlin supports. Use the most economical material based on relative costs in Prob. 6.23.

6.30. Determine the safe axial compressive load on a 10 × 10 × ¼ structural tube. Effective length $KL = 18$ ft. Use (a) A36 steel. (b) steel with $F_y = 75$ ksi.

6.31. Determine the safe axial compressive load on a 178 × 6 × ¼ structural tube. Effective length, $KL_y = 10$ ft with respect to weak-axis bending, and $KL_x = 15$ ft with respect to strong-axis bending. (a) Use A36 steel; (b) Use steel with $F_y = 60$ ksi.

6.32. Redesign the column of Prob. 6.1, selecting a structural tube instead of a W section.

6.33. Redesign the column of Prob. 6.4, selecting a structural tube instead of a W section.

6.34. Redesign the column of Prob. 6.5, selecting a structural tube instead of a W section.

SELECTED REFERENCES

1. Lynn S. Beedle and Lambert Tall, "Basic Column Strength," *J. Structural Div. ASCE*, No. ST7 (July 1960), pp 139–173.
2. Bruce G. Johnston, *The Column Research Council Guide to Design Criteria for Metal Compression Members*, 2nd ed., John Wiley & Sons, Inc., New York, 1966.
3. Bruce C. Johnston, "A Survey of Progress, 1944–51," Bulletin No. 1, Column Research Council, January 1952.
4. F. Engesser, *Zeitschrift für Architektur und Ingenieurwesen*, 1889, p. 455.
5. A Considère, "Resistance des pièces comprimèes," Congrès International des Procèdés de Construction, Paris, 1891, Vol. 3, p. 371.
6. F. R. Shanley, "The Column Paradox," *J. Aero. Sci.*, Vol. 13, No. 5 (December 1946), p. 678.
7. N. J. Hoff, "Buckling and Stability," *J. Royal Aeronaut. Soc.*, Vol. 58, Aero Reprint No. 123, January 1954.
8. Bruce G. Johnston, "Buckling Behavior Above the Tangent Modulus Load," *J. Eng. Mech. Div.*, ASCE, EM6, December 1961, pp. 79–98.
9. *Welding Handbook*, Sec. 2, American Welding Society, 5th ed., 1963, Sec. 25.22–27.
10. A. W. Huber and L. S. Beedle, "Residual Stress and the Compressive Strength of Steel," *Welding Journal*, December 1954, pp. 589s–614s.
11. C. H. Yang, L. S. Beedle, and B. G. Johnston, "Residual Stress and the Yield Strength of Steel Beams," *Welding Journal*, April 1952, pp. 205s–229s.
12. N. R. Nagaraja Rao, F. R. Estuar, and L. Tall, "Residual Stresses in Welded Shapes," *Welding Journal*, July 1964, pp. 295s–306s.
13. Friedrich Bleich, *Buckling Strength of Metal Structures*, McGraw-Hill Book Company, New York, 1952.
14. T. H. Johnson, "On the Strength of Columns," *Trans. ASCE*, Vol. 15 (July 1886), p. 517.
15. *American Association of State Highway Officials*, Standard Specifications for Highway Bridges, 10th ed., 1969.
16. Jack C. McCormac, *Structural Steel Design*, International Textbook Company, Scranton, Pa., 1965, pp. 62–66.

17. D. H. Young, "Rational Design of Steel Columns," *Trans. ASCE*, Vol. 101 (1936), pp. 422–500.
18. S. Timoshenko and S. Woinowsky-Krieger, *Theory of Plates and Shells*, 2nd ed., McGraw-Hill Book Company, New York, 1959, pp. 79–82.
19. Stephen P. Timoshenko and James M. Gere, *Theory of Elastic Stability*, 2nd ed., McGraw-Hill Book Company, New York, 1961, pp. 319–328, 351–356.
20. Kurt H. Gerstle, *Basic Structural Design*, McGraw-Hill Book Company, New York, 1967, pp. 88–90.
21. L. S. Beedle et al., *Structural Steel Design*, Ronald Press Company New York, 1964, Chaps. 9, 17.
22. Geerhard Haaijer and Bruno Thürlimann, "On Inelastic Buckling in Steel," *Trans. ASCE*, Vol. 125 (1960), pp. 308–344.
23. Omer W. Blodgett, *Design of Welded Structures*, James F. Lincoln Arc Welding Foundation, Cleveland, Ohio, 1966.
24. George Gerard and Herbert Becker, *Handbook of Structural Stability*, Part I —*Buckling of Flat Plates*, Tech. Note 3871, National Advisory Committee for Aeronautics, Washington, D.C., July 1957.
25. Specification for the Design of Light Gage Cold-Formed Steel Structural Members, 1968 ed., American Iron and Steel Institute, New York.
26. Theodore von Kármán, E. E. Sechler, and L. H. Donnell, "The Strength of Thin Plates in Compression," *ASME Trans.*, Vol. 54, APM-54-5 (1932), p. 53.
27. G. Winter, "Strength of Thin Compression Flanges," *ASCE Trans.*, Vol. 112 (1947), pp. 527–576.

7

Beams: Laterally Supported

7.1. INTRODUCTION

A beam is generally considered to be any member subjected principally to transverse gravity loading. The term transverse loading is taken to include end moments. Thus, beams in a structure may also be referred to as *girders* (usually the most important beams which are frequently at wide spacing); *joists* (usually less important beams which are closely spaced, frequently with truss type webs); *purlins* (roof beams spanning between trusses); *stringers* (longitudinal bridge beams spanning between floor beams); *girts* (horizontal wall beams serving principally to resist bending due to wind on the side of an industrial building; frequently supporting corrugated siding); and *lintels* (members supporting a wall over window or door openings). Other terms, such as header, trimmer and rafter, are sometimes used, but beam identification by these terms is not generally applied.

It is obvious, of course, that a beam is a combination of a tension element and a compression element. The concepts of tension members and compression members are now combined in the treatment as a beam. In this chapter, it is assumed that the compression element (one flange) is adequately braced so that its buckling as a column cannot occur prior to its developing the ultimate moment capacity of the section. While it is likely true that most beams used in practical situations are adequately braced so that stability need not be considered, the percentage of stable situations is probably not as high as assumed. The important treatment of lateral stability is found in Chapter 9.

7.2. SIMPLE BENDING OF SYMMETRICAL SHAPES

The most common design situations involve selection of rolled wide-flange shapes from the AISC tables, which often becomes routine and may lead the designer into overconfidence in treatment of beams. It is well

Cobo Arena, Detroit. Courtesy of Whitehead and Kales Co. Detroit

known that the flexure formula ($f = Mc/I$) is applicable to ordinary situations. The stresses on the common sections of Fig. 7.2.1 may be computed by the simple flexure formula when loads are acting in one of

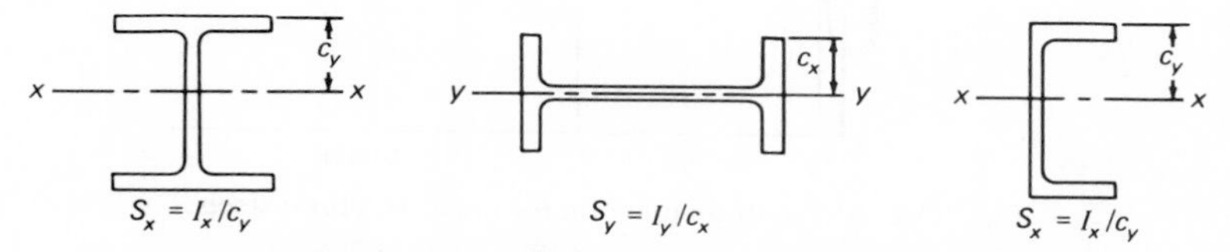

Fig. 7.2.1. Elastic section modulus expressions for symmetrical shapes.

the principal directions. When any section with at least one axis of symmetry and loaded through the shear center is subjected to a bending moment in an arbitrary direction, the components M_{xx} and M_{yy}, in the principal directions, can be obtained and the stress computed as

$$f = \frac{M_{xx}}{S_x} + \frac{M_{yy}}{S_y} \tag{7.2.1}$$

where S_x and S_y are known as section moduli, defined as in Fig. 7.2.1. For members without at least one axis of symmetry the reader is referred to Sec. 7.8.

7.3. DESIGN FOR STRENGTH CRITERION

Assuming adequate lateral stability of the compression flange, beam design is based on development of the maximum bending strength of the section. The stress distribution on a typical wide-flange shape subjected to increasing bending moments is shown in Fig. 7.3.1. This behavior is

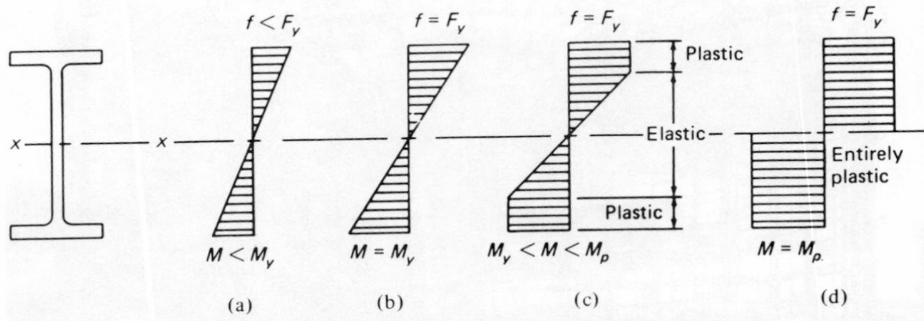

Fig. 7.3.1. Stress distribution at different stages of loading.

based on the material remaining elastic until the yield point is reached; thereafter, additional strain induces no increase in stress. Such stress-strain behavior is shown in Fig. 7.3.2 and is the accepted idealization for structural steels with yield points of about F_y = 65 ksi and less.

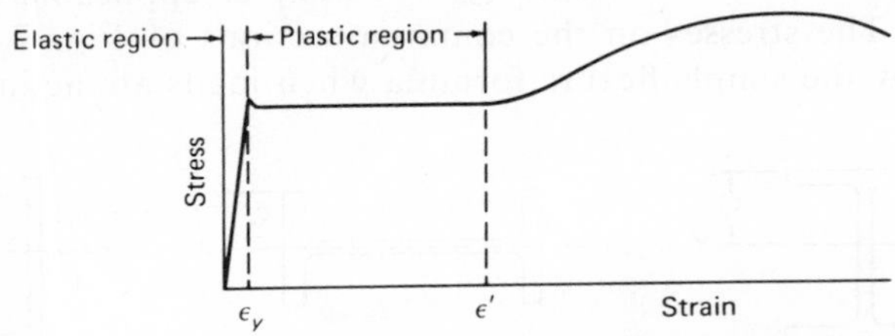

Fig. 7.3.2. Stress-strain diagram for most structural steels.

These steels are referred to as exhibiting elastic-plastic behavior. When the yield stress is achieved at the extreme fiber (Fig. 7.3.1b), the moment capacity is referred to as the yield moment and may be computed as

$$M_y = F_y S_x \tag{7.3.1}$$

When the condition of Fig. 7.3.1d is reached, every fiber has a strain equal to or greater than $\epsilon_y = F_y/E_s$, i.e., it is in the plastic range. The moment capacity is therefore referred to as the plastic moment, and is computed

$$M_p = F_y \int_A y\,dA = F_y Z \tag{7.3.2}$$

where $Z = \int y\,dA$ may be called the *plastic modulus*.

It will be observed that the ratio M_p/M_y is a property of the cross-sectional shape and is independent of the material properties. This ratio is referred to as the shape factor, $\not{f}$,

$$\not{f} = \frac{M_p}{M_y} = \frac{Z}{S} \tag{7.3.3}$$

For wide-flange (W) shapes in flexure about the strong axis (x–x) the shape factor ranges from about 1.09 to about 1.18 with the usual value being about 1.12. One may conservatively say the ultimate bending capacity (plastic moment) of W sections is at least 10 percent greater than the capacity at first yield (M_y).

Design procedures beginning with the 1963 AISC specification have recognized that beams behave in the manner discussed above. During the period since 1946 extensive testing has adequately verified that plastification of the entire cross-section does occur.[1]

EXAMPLE 7.3.1

Determine the shape factor for a rectangular beam of width b and depth d.

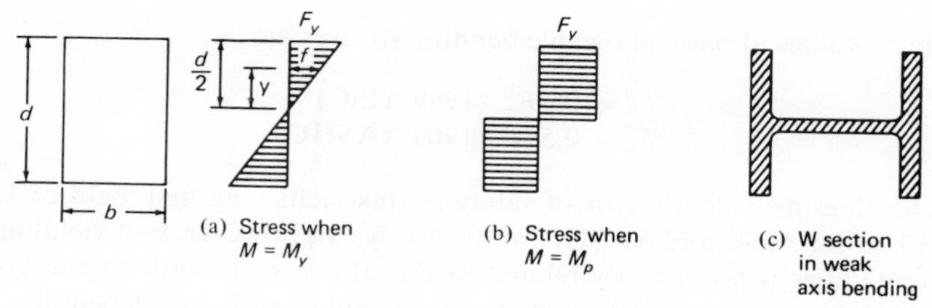

Fig. 7.3.3.

Solution

Referring to Fig. 7.3.3a, the moment at first yield, M_y, is

$$M_y = \int_A f\, y\, dA$$

and

$$f = F_y(y)/(d/2) = F_y \frac{2y}{d}$$

$$M_y = 2\int_0^{d/2} \frac{2F_y}{d} y^2 b\, dy = F_y \frac{bd^2}{6}$$

From Fig. 7.3.3b,

$$M_p = \int_A f\, y\, dA$$

$$= 2\int_0^{d/2} F_y b y\, dy = F_y \frac{bd^2}{4}$$

The shape factor is then

$$\mathcal{f} = \frac{M_p}{M_y} = 1.5$$

which illustrates that there is greater reserve beyond first yield in the bending of a rectangular section than in an I-shaped section bending about its strong axis. The reader is alerted to the fact that the W shape bent about its *weak* axis (y–y) is essentially a rectangular section (two rectangles separated by a distance) (Fig. 7.3.3c).

Working-Stress (or Elastic-Design) Method. Under the working-stress method the service load bending moments at various sections are computed using traditional elastic concepts, including statically indeterminate structural analysis whenever statics alone is not sufficient. Designs are made such that computed stresses under the action of the service load moments do not exceed specified allowable stresses. The allowable bending stress, F_b, is taken as some fraction of the yield point stress. Thus a correctly designed beam would have a service load capacity computed as

$$M_w = F_b S$$

Typical values of basic allowable bending stresses are

$$F_b = 0.6F_y \quad \text{(1969 AISC)}$$
$$F_b = 0.55F_y \quad \text{(1969 AASHO)}$$

Such values provide a factor of safety against achieving first yield of 1.67 and 1.82 for AISC and AASHO, respectively. If, however, first yielding at extreme fiber is not directly related to the ultimate strength of the cross section, then the true safety factor against failure will vary depending on how much the ultimate capacity exceeds the first yield. In other words, the ratio of ultimate moment capacity, M_u, to service load moment, M_w, may more correctly be called the safety factor. When the ultimate moment can reach M_p, the plastic moment, the ratio M_p/M_w is referred to as the load factor (L.F.).

The 1969 AISC specification recognizes the ability of most rolled beams to achieve the plastic moment; such beams are referred to as "compact sections" and the allowable stress is increased 10 percent and 25 percent for bending about the strong and weak axes, respectively. This is because M_p usually exceeds M_y by at least these percentages.

Why are there some beams which cannot achieve the plastic-moment capacity? Local buckling of the compression flanges, as discussed in Sec. 6.16, may occur prior to the achieving of the high compressive strain necessary to achieve M_p. Even though the local buckling provisions of AISC–1.9 are satisfied, only achievement of M_y is assured (i.e., buckling below $f = F_y$ is prevented). The extreme fiber strain is assured only of reaching $\epsilon_y = F_y/E_s$. To achieve greater strain, the values of b/t must be further restricted. To undergo large plastic strain the more severe restrictions discussed in Sec. 6.17 prescribed for 'compact' sections must be satisfied, as given in Table 7.3.1.

TABLE 7.3.1
Local Buckling Requirements for Compact Sections (AISC–1.5.1.4.1)

F_y	Unstiffened Elements (Uniform Compression) $\frac{b_f}{2t_f} = \frac{52.2}{\sqrt{F_y}}$	Stiffened Elements (Uniform Compression) $\frac{b}{t_f} = \frac{190}{\sqrt{F_y}}$	Stiffened Elements (Bending) $\frac{d}{t_w} = \frac{412}{\sqrt{F_y}}$
36	8.7	31.7	68.7
42	8.1	29.3	63.6
45	7.8	28.3	61.4
50	7.4	26.9	58.3
55	7.0	25.6	55.6
60	6.7	24.5	53.2
65	6.5	23.6	51.1

The basic working-stress (or elastic-design) method of AISC–1969 for beams may be summarized (AISC–1.5.1.4) as follows:

(a) For laterally supported compact sections which are symmetrical about, and loaded in, the plane of their minor axis (i.e., sections which have a shape factor of at least 1.10 and which are capable of achieving a plastic moment capacity M_p),

$$F_b = 0.66F_y \tag{7.3.4}$$

(b) For sections in category (a) whose flanges (unstiffened elements) are stiffer than required by AISC–1.9.1 (i.e., $b_f/2t_f < 95.0/\sqrt{F_y}$) but not stiff enough to accommodate the plastic strain required of a 'compact' section (i.e., $b_f/2t_f > 52.2/\sqrt{F_y}$, a linear transition between $0.6F_y$ and $0.66F_y$ is provided in AISC–1.5.1.4.2. For such cases,

$$F_b = F_y\left[0.733 - 0.0014\left(\frac{b_f}{2t_f}\right)\sqrt{F_y}\right] \tag{7.3.5}$$

the variation is shown in Fig. 7.3.4.

(c) For doubly symmetrical I- and H-shaped members, satisfying the unstiffened element local buckling requirement of AISC–1.5.1.4.1b, and bent about the minor axis,

$$F_b = 0.75F_y \tag{7.3.6}$$

in recognition of the higher shape factor for a rectangle in bending, as shown in Example 7.3.1. Lateral bracing is not required for this case.*

(d) For other sections, such as box-types or unsymmetrical, whose

*AISC Specification Addendum (November, 1970) allows a linear transition between 0.60 F_y and 0.75 F_y for weak axis bending under the same conditions applicable to Eq. 7.3.5.

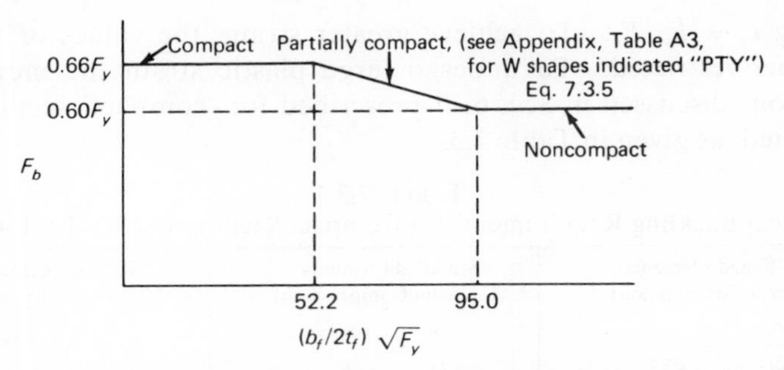

Fig. 7.3.4. Allowable stress on laterally braced I-shaped sections qualifying as "compact" except for an excessive width-to-thickness ratio at the unstiffened compression flange element.

compression flanges are adequately braced, (i.e., sections which can achieve at ultimate moment an extreme fiber strain equal to the yield strain F_y/E_s),

$$F_b = 0.6F_y \tag{7.3.7}$$

Since plastic behavior is not utilized for material with F_y greater than 65 ksi (A572 steel), the higher-yield-point steels are governed by a maximum allowable stress of $0.6F_y$.

It is to be noted that the AASHO–1969 does not recognize the ability of any section to develop a moment capacity exceeding M_y; therefore, the maximum allowable stress is $F_b = 0.55F_y$. Further, because of fatigue considerations involved in highway bridge design, allowable stresses may be significantly less than $0.55F_y$ depending on the maximum and minimum stresses and the number of cycles of loading for which the structure is to be designed.

Plastic-Design Method. In the plastic-design method, the service loads are multiplied by a quantity known as the load factor to obtain the required ultimate capacity which the structure must be capable of carrying without collapse. The ultimate moments are then obtained for the collapse condition. In the statically determinate structure, achieving a plastic moment at one location is sufficient to create a collapse mechanism. The section at which the plastic moment occurs will continue to deform without inducing additional resistance after it reaches M_p. This condition of increasing deformation with a constant resisting moment is called a *plastic hinge*.

Referring to Fig. 7.3.5, it is seen that as the bending moment is increased above the service load value, the strain angle ϕ (rotation in radians/inch) is entirely an elastic (up to M_y) or partially elastic (M_y to M_p) (see Fig. 7.3.1) relationship until M_p is reached. Thereafter, an unstable situation or mechanism occurs such that the deflection increases

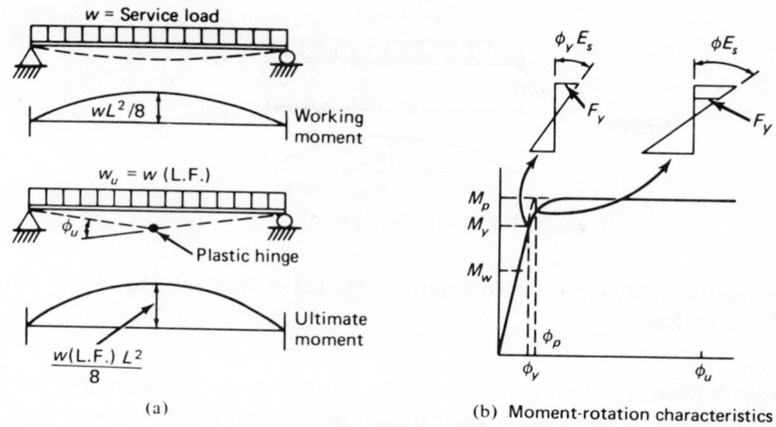

(a) (b) Moment-rotation characteristics

Fig. 7.3.5. Plastic behavior.

without restraint. At the ultimate condition the elastic deformation due to bending on the segments between the ends and midspan is negligible compared to the rotation ϕ_u occurring at the plastic hinge. Thus the analysis may treat the collapse situation as two rigid bodies with an angular discontinuity ϕ_u at midspan. As will be shown later in Art. 10.2, it is only for the statically determinate situations that one can expect all points on the ultimate moment diagram to be equal to a constant times the values of elastic moments. Redistribution of the moments occurs during loading beyond the elastic range in usual statically indeterminate situations.

Under AISC–1969, plastic design requires provision of adequate lateral stability since it is paramount in the method that the plastic moment be achieved under ultimate loading. The load factor prescribed for gravity loading is 1.70.

Just as in the working stress method, where a "compact section" is necessary to use $0.66F_y$, so in plastic design similar local buckling provisions are given in AISC–2.7.

At the present time (1971), plastic behavior of steels up through $F_y = 65$ ksi is recognized by the AISC "compact section" provisions of AISC–1.5.1.4.1 and the plastic design provisions of AISC–Part 2. Steels above $F_y = 65$ ksi do not exhibit the sharp yield point of the lower-strength steels.

EXAMPLE 7.3.2

Using the working stress method, select the lightest W or M section to carry a uniformly distributed load of 1 kip/ft on a 20-ft simply supported span (Fig. 7.3.6). The compression flange of the beam is fully

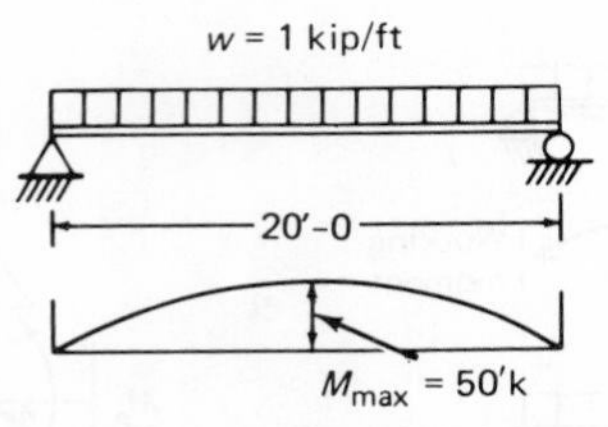

Fig. 7.3.6. Example 7.3.2.

supported against lateral movement. Use (a) A36 steel; (b) $F_y = 50$ ksi; (c) $F_y = 65$ ksi.

Solution

(a) A36 steel. Assume "compact section" since the vast majority of sections satisfy the necessary conditions; thus, the allowable stress is

$$F_b = 0.66F_y = 24 \text{ ksi}$$

Note that rounded (i.e., 0.66 times 36 ksi = 23.7 ksi; use 24 ksi) values are acceptable in accordance with the Appendix A of the 1969 AISC specification.

$$\text{Design } M = \frac{wL^2}{8} = \frac{1(20)^2}{8} = 50 \text{ ft-kips}$$

$$\text{Required } S_x = \frac{M_x}{F_b} = \frac{50(12)}{24} = 25 \text{ in.}^3$$

Select from the AISC Manual "Allowable Stress Design Selection Table," (or from text, Appendix, Table A3), the lightest section having at least $S_x = 25$ in.^{3}:

$$\text{Try W12} \times 22\text{:} \quad S_x = 25.3 \text{ in.}^3$$

Check "compact section" requirements:

$$\frac{b_f}{2t_f} = \frac{4.03}{2(0.424)} = 4.75 < 8.7 \text{ (Table 7.3.1)} \qquad \text{OK}$$

$$\frac{d}{t_w} = \frac{12.31}{0.26} = 47.3 < 68.7 \qquad \text{OK}$$

Use W12 × 22 ($F_y = 36$ ksi).

(b) Steel with $F_y = 50$ ksi

$$F_b = 0.66F_y = 33 \text{ ksi}$$

$$\text{Reqd } S_x = \frac{50(12)}{33} = 18.2 \text{ in.}^3$$

Try M14 × 17.2 with $S_x = 21.0$ as lightest section satisfying section modulus requirement. Check "compact section."

$$\frac{b_f}{2t_f} = \frac{4.0}{2(0.272)} = 7.35 < 7.4 \qquad \text{OK}$$

$$\frac{d}{t_w} = \frac{14.00}{0.21} = 66.7 > 58.3 \qquad \text{N.G.}$$

Therefore, the allowable stress is limited to $0.6F_y$; thus, for noncompact sections,

$$F_b = 0.6F_y = 30 \text{ ksi}$$

$$f = \frac{50(12)}{21.0} = 28.5 \text{ ksi} < 30 \text{ ksi} \qquad \text{OK}$$

Use M14 × 17.2 ($F_y = 50$ ksi).

(c) Steel with $F_y = 65$ ksi

$$F_b = 0.66F_y = 43 \text{ ksi}$$

$$\text{Reqd } S_x = \frac{50(12)}{43} = 14.0 \text{ in.}^3$$

Try W12 × 14: $S_x = 14.8 \text{ in.}^3$

$$\frac{b_f}{2t_f} = \frac{3.97}{2(0.224)} = 8.9 > \frac{52.2}{\sqrt{65}} = 6.5$$

Thus $F_b \neq 0.66F_y$. However,

$$8.9 < \frac{95.0}{\sqrt{65}} = 11.8$$

Therefore

$$F_b = F_y[0.733 - 0.0014(b_f/2t_f)\sqrt{F_y}]$$
$$= F_y[0.733 - 0.0014(8.9)\sqrt{65}] = 0.633F_y = 41.1 \text{ ksi}$$

$$f = \frac{50(12)}{14.8} = 40.5 \text{ ksi} < 41.1 \text{ ksi} \qquad \text{OK}$$

Check:

$$\frac{d}{t_w} = \frac{11.91}{0.20} = 59.5 > 51.1 \qquad \text{N.G.}$$

$$\therefore F_b = 0.6F_y = 39.0 \text{ ksi}$$

Try W12 × 16.5, $S_x = 17.5$, which is noncompact:

$$f = \frac{50(12)}{17.5} = 34.3 \text{ ksi} < 39 \text{ ksi} \qquad \text{OK}$$

Use W12 × 16.5 ($F_y = 65$ ksi).

Note should be made of the fact that moment due to beam weight causes about 2 percent increase in stresses in the above beams and should be allowed for in checking sections. As in this example, this checking may frequently be done by visual inspection.

EXAMPLE 7.3.3

Repeat Example 7.3.2 using the plastic design method.

Solution

The load factor is first applied to the service load and the required plastic moment is computed.

(a) A36 Steel:

$$w_u = 1.0(1.70) = 1.70 \text{ kips/ft}$$
$$\text{Reqd } M_p = 1.70(20)^2/8 = 85 \text{ ft-kips}$$

From Eq. 7.3.2,

$$\text{Reqd } Z = \frac{M_p}{F_y} = \frac{85(12)}{36} = 28.4 \text{ in.}^3$$

Select from AISC Manual "Plastic Design Selection Table," W12 × 22, $Z_x = 29.3$ in.³ The sections tabulated are those which also satisfy the local buckling requirements of Sec. 2.7 for A36 steel, i.e., $b/2t_f \leq 8\frac{1}{2}$ and $d/t_w \leq 68.7$.

<u>Use W12 × 22 (F_y = 36 ksi).</u>

(b) Steel with F_y = 50 ksi:

$$\text{Reqd } Z = \frac{M_p}{F_y} = \frac{85(12)}{50} = 20.4 \text{ in.}^3$$

Select from AISC Manual, W12 × 16.5, Z = 20.6 in.³

$$\frac{b_f}{2t_f} = \frac{4}{2(0.269)} = 7.44 > 7.0 \qquad \text{N.G.}$$

Try W10 × 19, Z = 21.6 in.³ (The M14 × 17.2 is the next lightest possibility but does not satisfy local buckling requirements.)

$$\frac{b_f}{2t_f} = \frac{4.02}{2(0.394)} = 5.1 < 7.0 \qquad \text{OK}$$

$$\frac{d}{t_w} = \frac{10.25}{0.25} = 41 < \frac{412}{\sqrt{50}} = 58.3 \qquad \text{OK}$$

<u>Use W10 × 19 (F_y = 50 ksi).</u>

(c) Steel with F_y = 65 ksi:

$$\text{Reqd } Z = \frac{M_p}{F_y} = \frac{85(12)}{65} = 15.7 \text{ in.}^3$$

Since W10 × 15, W12 × 16.5, and W8 × 17 do not satisfy $b_f/2t_f$ or d/t_w local buckling requirements; try *W*10 × 17, Z = 16.2 in.³

$$\frac{b_f}{2t_f} = \frac{4.01}{2(0.329)} = 6.1 \approx 6.0 \text{ limit, say OK}$$

$$\frac{d}{t_w} = \frac{10.12}{0.24} = 46.4 < 51.1 \qquad \text{OK}$$

Use W10 × 17 (F_y = 65 ksi).

A comparison of results is given below:

	WSD	Plastic Design
F_y = 36 ksi	W12 × 22	W12 × 22
F_y = 50 ksi	M14 × 17.2	W10 × 19
F_y = 65 ksi	W12 × 16.5	W10 × 17

The requirement of plastic design that the section selected must have the capability of developing a plastic hinge resulted in heavier sections for the last two cases under plastic design than for working-stress design. The working-stress method recognized the fact that with the given loading the members used were understressed (i.e., larger than necessary even though no better choices were available).

7.4. DEFLECTIONS

As beams are used on long spans (i.e., as the ratio of the span length to the depth of the section becomes large) or shallower sections of high-strength steels are used, deflection restrictions may control the design. There are few if any difficulties encountered in computing deflections on steel structures under service loads.

There are numerous structural analysis methods available for computing deflections on uniform and variable moment of inertia sections in statically determinate and indeterminate structures.* In general, the maximum deflection in an elastic member may be expressed as

$$y_{\max} = m \frac{WL^3}{48EI} \qquad (7.4.1)$$

where W = total load on the span
L = span length
E = modulus of elasticity (29,000 ksi for steel)
I = moment of inertia
m = coefficient which depends upon the degree of fixity at supports, the variation in moment of inertia along the span, and the distribution of loading (some typical values for the coefficient m may be found in a Portland Cement Association publication).[2]

For continuous beams, the midspan deflection in the common situation of a uniform loading on a prismatic beam with unequal end moments (see Fig. 7.4.1) may be expressed as

$$y_{\text{midspan}} = \frac{5L^2}{48EI}[M_s - 0.1(M_a + M_b)] \qquad (7.4.2)$$

*See, for example, C. K. Wang and C. L. Eckel, *Elementary Theory of Structures* (New York: McGraw-Hill Book Company, 1957), or Jack McCormac, *Structural Analysis*, 2nd ed. (Scranton, Pa.: International Textbook Company, 1966).

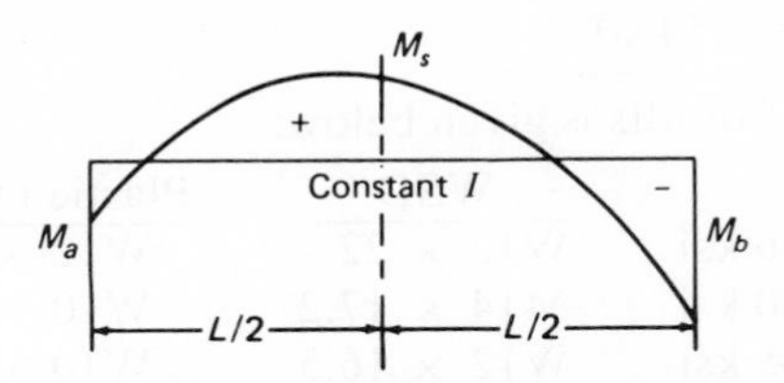

Fig. 7.4.1. Typical bending-moment diagram for uniformly loaded beam.

Equation 7.4.2 will give satisfactory results when considered to be the maximum deflection for nearly all practical uniform section cases. Eq. 7.4.2 may be verified by the use of a method such as conjugate beam:

For uniformly loaded simply supported beams,

$$y_{\max} = \frac{5wL^4}{384EI}$$

which upon substitution of $M = wL^2/8$, $f = Mc/I$, and $c = d/2$, gives

$$y_{\max} = \frac{10fL^2}{48Ed} \tag{7.4.3}$$

Equation 7.4.3 can be used as a good approximation for any simply supported beam as long as the maximum stress occurs near midspan. Refer to Table 7.4.1 for typical values.

TABLE 7.4.1
Deflection Relationships According to Eq. 7.4.3

$y_{\max}$	L/d	$L/d(f = 22)$	$L/d(f = 30)$
$L/360$	$387/f$	17.6	12.9
$L/300$	$464/f$	21.1	15.5
$L/240$	$580/f$	26.3	19.3
$L/200$	$695/f$	31.6	23.2

AISC–1.13.1 states, "Beams and girders supporting floors and roofs shall be proportioned with due regard to the deflection produced by design loads." In addition, live-load deflection where plastered ceilings are supported is limited to $L/360$.

For the $L/360$ limitation, Eq. 7.4.3, using $E = 29{,}000$ ksi, becomes

$$\frac{L}{d} \le \frac{48(29{,}000)}{(10)360f} = \frac{387}{f} \tag{7.4.4}$$

where f is computed in ksi.

When considering deflections, it should be remembered that the dead-load deflections usually can be accounted for during construction by either cambering (negative bending) or thickening the slab or floor topping. It is only the deflection that occurs due to loads applied after construction is completed that may crack ceilings, partitions, or walls.

Specification requirements for limiting deflections are meager because there is no single or standard value for the tolerable deflection. The acceptable amount must of necessity depend on the type and arrangement of materials being supported.

As a guide only, the AISC Commentary suggests the following limitations:

Floor beams and girders, fully stressed, not subject to shock or vibration:

$$\frac{L}{d} \leq \frac{800}{F_y}$$

Floor beams and girders, subject to shock or vibratory loads, supporting large open areas free of partitions or other sources of damping:

$$\frac{L}{d} \leq 20$$

Roof purlins, fully stressed, except sloped roofs steeper than slope of 3 in 12:

$$\frac{L}{d} \leq \frac{1000}{F_y}$$

Based on service load deflections where $F_b = 0.66F_y$ or assuming $f = F_y/1.515$, the AISC Commentary suggestions of 800 and 1000 correspond to L/d values of $528/f$ and $660/f$, respectively. Using Eq. 7.4.3 the simple beam deflection limitations would be approximately $L/260$ and $L/210$.

On continuous spans it is, of course, the actual deflection that is of importance, not the L/d ratio. For continuous beam deflection, a comparison of Eqs. 7.4.2 and 7.4.3 shows that Eq. 7.4.3 can also be used for continuous beams if the stress f is computed using the equivalent bending moment

$$M_e = M_s - 0.1(M_a + M_b) \tag{7.4.5}$$

Ponding of Water on Flat Roofs. When members of a flat roof system deflect, a bowl-shaped volume is created which is capable of retaining water. As water begins to accumulate, deflection increases to provide an increased volumetric capacity. This cyclical process continues until either (1) the succeeding deflection increments become smaller and equilibrium is reached; or (2) succeeding deflection increments are increasing, the system is divergent, and collapse occurs. This retention of water which results solely from the deflection of flat roof framing is what 1969 AISC refers to as ponding.

To prevent ponding or excessive water accumulation on flat roofs, AISC–1969 required supporting members to satisfy the limitation

$$\frac{L}{d} \leq \frac{600}{f_b} \tag{7.4.6}$$

where f_b is the computed bending stress. Using Eq. 7.4.3, this would correspond roughly to a simple span deflection limitation of $L/240$.

The problem of ponding is much more complex than indicated by the above limitation. Marino[3] has provided an extensive treatment which forms the basis for the AISC–1969 provisions. The flat roof is treated as a two-way system of secondary members (say, purlins) elastically supported by primary members (say, girders) which are rigidly supported by walls or columns, as shown in Fig. 7.4.2. Figure 7.4.3a shows the

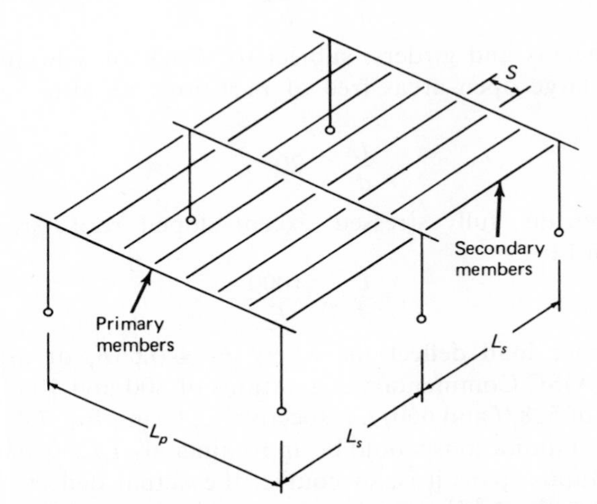

Fig. 7.4.2. Flat-roof arrangement for ponding analysis. (from Ref. 3.)

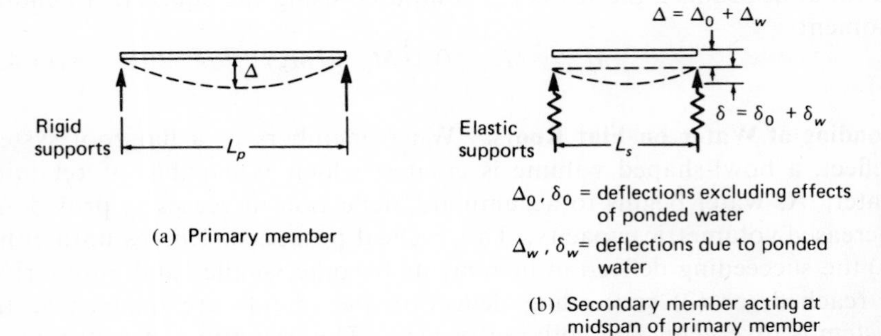

Fig. 7.4.3. Deflection of flat-roof elements.

primary member with midspan deflection equal to Δ. Figure 7.4.3b represents a secondary member intersecting the primary member at midspan; its supports have deflected an amount Δ, with δ representing the secondary member deflection relative to its ends.

The loads assumed for the analysis are shown in Fig. 7.4.4. In each case the deflection curve is assumed as a half sine wave. The loading on the critical midspan secondary member assumes uniform distribution of water of depth equal to $\Delta(\Delta_0$, initial deflection not including that due to

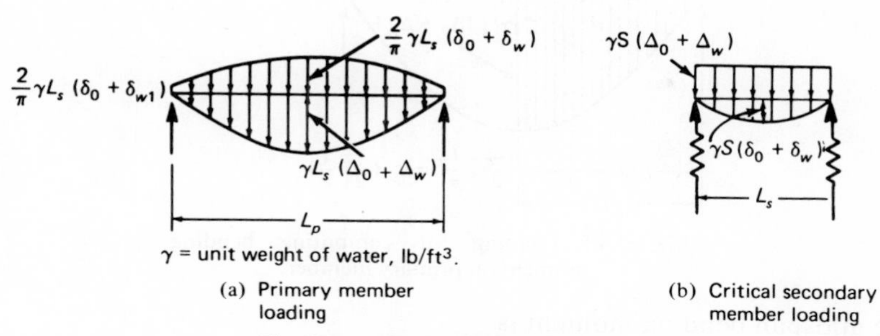

(a) Primary member loading

(b) Critical secondary member loading

Fig. 7.4.4. Assumed ponding loads.

water; plus Δ_w, the additional deflection due to water) due to the deflection of the primary member. There is additional loading which varies as a half sine wave due to the relative deflection of the secondary member.

The primary member is loaded with a half sine curve due to its own deflection plus the reaction from the half sine wave loading due to the deflection of the secondary member. This reaction from the midspan secondary member is

$$\frac{2}{\pi}\gamma L_s(\delta_0 + \delta_w)$$

per foot of width along the primary member. The secondary member framing in near the support of the primary member will have a reduced self-deflection δ_{w1} due to the water; thus, the reaction to the primary member at its end is

$$\frac{2}{\pi}\gamma L_s(\delta_0 + \delta_{w1})$$

per foot of width L_p.

Referring to Fig. 7.4.5, compute the bending moment on the primary member. First determine the beam reaction:

$$R = \frac{\gamma L_s L_p}{\pi}(\Delta_0 + \Delta_w) + \frac{2\gamma L_s L_p}{\pi^2 EI_p}(\delta_w - \delta_{w1}) + \frac{\gamma L_s L_p}{\pi EI_p}(\delta_0 + \delta_{w1})$$

$$= \frac{\gamma L_s L_p}{\pi}\left[(\Delta_0 + \Delta_w) + \frac{2}{\pi}(\delta_w - \delta_{w1}) + (\delta_0 + \delta_{w1})\right] \qquad (7.4.7)$$

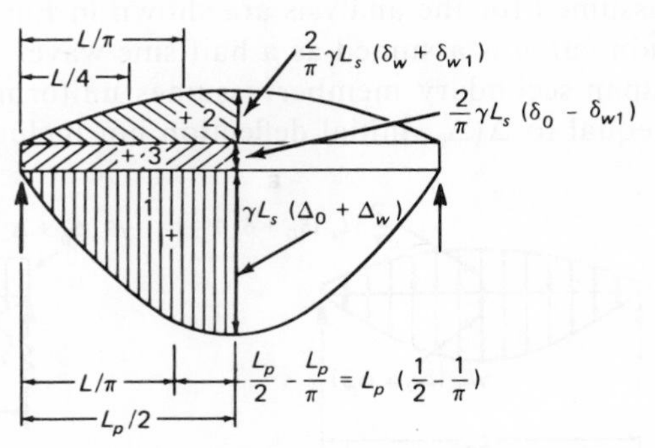

Fig. 7.4.5. Loading for computing bending moment on primary member.

The midspan bending moment is

$$M = \frac{\gamma L_s L_p}{\pi}\left[(\Delta_0 + \Delta_w)\left(\frac{L_p}{2} - \frac{L_p}{2} + \frac{L_p}{\pi}\right) + \frac{2}{\pi}(\delta_w - \delta_{w1})\frac{L_p}{\pi} + (\delta_0 + \delta_{w1})\left(\frac{L_p}{2} - \frac{L_p}{4}\right)\right]$$

$$= \frac{\gamma L_s L_p^2}{\pi}\left[\frac{1}{\pi}(\Delta_0 + \Delta_w) + \frac{2}{\pi^2}(\delta_w - \delta_{w1}) + \frac{1}{4}(\delta_0 + \delta_{w1})\right] \tag{7.4.8}$$

The deflection is found by finding the bending moment on the conjugate beam. Since the original loading was a half sine curve for load terms 1 and 2, the M/EI diagram for those terms is likewise a sine curve. The third term, representing uniform loading gives a parabola for its M/EI loading. Assuming the parabola as a sine curve, the moment at midspan of the conjugate beam is obtained by multiplying by L_p^2/π^2; thus

$$\Delta_w = \frac{\gamma L_s L_p^4}{\pi^3 EI_p}\left[\frac{1}{\pi}(\Delta_0 + \Delta_w) + \frac{2}{\pi^2}(\delta_w - \delta_{w1}) + \frac{1}{4}(\delta_0 + \delta_{w1})\right]$$

$$= \frac{\gamma L_s L_p}{\pi^4 EI_p}\left[(\Delta_0 + \Delta_w) + \frac{2}{\pi}(\delta_w - \delta_{w1}) + \frac{\pi}{4}(\delta_0 + \delta_{w1})\right] \tag{7.4.9}$$

Letting $C_p = \gamma L_s L_p^4/(\pi^4 EI_p)$, and solving for Δ_w gives

$$\Delta_w = \frac{C_p}{1 - C_p}\left(\Delta_0 + \frac{\pi}{4}\delta_0 + \frac{2}{\pi}\delta_w + \frac{\pi}{4}\delta_{w1} - \frac{2}{\pi}\delta_{w1}\right) \tag{7.4.10}$$

Similarly, the midspan deflection on the secondary member may be expressed as

$$\delta_w = \frac{\gamma S L_s^4}{\pi^4 EI_s}\left[\frac{\pi^2}{8}(\Delta_0 + \Delta_w) + \delta_0 + \delta_w\right] \tag{7.4.11}$$

Letting $C_s = \gamma S L_s^4/(\pi^4 EI_s)$, and solving for δ_w gives

$$\delta_w = \frac{C_s}{1 - C_s}\left(\frac{\pi^2}{8}\,\Delta_0 + \delta_0 + \frac{\pi^2}{8}\,\Delta_w\right) \tag{7.4.12}$$

At the support of the primary member, δ_w is δ_{w1} and $\Delta_0 = \Delta_w = 0$. Thus, from Eq. 7.4.12,

$$\delta_{w1} = \frac{C_s}{1 - C_s}\,\delta_0 \tag{7.4.13}$$

Furthermore, the initial deflections, δ_0/Δ_0, are in the same ratio as SL_s^4/EI_s is to $L_s L_p^4/EI_p$; thus

$$\frac{\delta_0}{\Delta_0} = \frac{C_s}{C_p} \tag{7.4.14}$$

Thus, substitution of Eq. 7.4.13 into 7.4.10 and then solving Eqs. 7.4.10 and 7.4.12 for Δ_w and δ_w gives

$$\Delta_w = \frac{\alpha_p \Delta_0 \left[1 + \frac{\pi}{4}\,\alpha_s + \frac{\pi}{4}\,\rho(1 + \alpha_s)\right]}{1 - \frac{\pi}{4}\,\alpha_p \alpha_s} \tag{7.4.15}$$

$$\delta_w = \frac{\alpha_s \delta_0 \left[1 + \frac{\pi^3}{32}\,\alpha_p + \frac{\pi^2}{8\rho}\,(1 + \alpha_p) + 0.185\alpha_s\alpha_p\right]}{1 - \frac{\pi}{4}\,\alpha_p \alpha_s} \tag{7.4.16}$$

where $\rho = \delta_0/\Delta_0 = C_s/C_p$
$\alpha_p = C_p/(1 - C_p)$
$\alpha_s = C_s/(1 - C_s)$

Note that even if infinite elasticity is assumed for the supporting system, collapse occurs as $\alpha_p\alpha_s \rightarrow 4/\pi$. Restrictions are necessary to keep the system safely within the actual elastic range.

Since deflection is proportional to stress,

$$\frac{f_w}{f_0} = \frac{\text{stress under a deflection, } \Delta_w}{\text{stress initially under deflection, } \Delta_0} = \frac{\Delta_w}{\Delta_0} \tag{7.4.17}$$

Using a factor of safety of 1.25, the stress due to ponding may be limited,

$$f_w \le 0.8F_y - f_0 \tag{7.4.18}$$

which gives in terms of deflections, using Eq. 7.4.17,

$$\Delta_w \le \left(\frac{0.8F_y - f_0}{f_0}\right)\Delta_0 \tag{7.4.19}$$

or, substituting Eq. 7.4.15 for Δ_w, gives the final criterion for safety of the primary member,

$$\left[\frac{0.8F_y - f_0}{f_0}\right]_p \geq \frac{\alpha_p\left[1 + \frac{\pi}{4}\alpha_s + \frac{\pi}{4}\rho(1 + \alpha_s)\right]}{1 - \frac{\pi}{4}\alpha_p\alpha_s} \tag{7.4.20}$$

For the secondary member, the criterion becomes

$$\left[\frac{0.8F_y - f_0}{f_0}\right]_s \geq \frac{\alpha_s\left[1 + \frac{\pi^3}{32}\alpha_p + \frac{\pi^2}{8\rho}(1 + \alpha_p) + 0.185\alpha_s\alpha_p\right]}{1 - \frac{\pi}{4}\alpha_p\alpha_s} \tag{7.4.21}$$

Equations 7.4.20 and 7.4.21 are presented with some of the foregoing discussion in the 1969 AISC Commentary, along with charts for evaluating the criteria. Knowing the stress index $U = (0.8F_y - f_0)/f_0$ and C_s, the value of C_p satisfying Eq. 7.4.20 can be determined by a chart (AISC—Fig. C1.13.3.1). Knowing U for the secondary member and C_p, the value of C_s satisfying Eq. 7.4.21 can be determined by chart (AISC—Fig. C1.13.3.2). f_0 is the initial stress in the member under consideration.

AISC–1.13.3 gives a simple, but very conservative criterion

$$C_p + 0.9C_s \leq 0.25 \tag{7.4.22}$$

and

$$I_d \geq 25S^4/10^6 \tag{7.4.23}$$

where $C_p = \dfrac{\gamma L_s L_p^4}{\pi^4 E I_p} = \dfrac{62.4 L_s L_p^4(144)}{\pi^4(29{,}000{,}000)\, I_p} = \dfrac{32 L_s L_p^4}{10^7 I_p}$

$C_s = \dfrac{32 S L_s^4}{10^7 I_s}$

L_p = length of primary member, ft
L_s = length of secondary member, ft
S = spacing of secondary member, ft
I_p = moment of inertia of primary member, in.^4
I_s = moment of inertia of secondary member, in.^4
I_d = moment of inertia of steel deck supported on secondary members, in.^4 per ft

The criterion $C_p + 0.9C_s \leq 0.25$ assumes the members essentially fully stressed before onset of ponding. For instance,

$$\frac{0.8F_y - f_0}{f_0} = \frac{0.8F_y - 0.66F_y}{0.66F_y} = 0.212$$

$$= \frac{0.8F_y - 0.6F_y}{0.6F_y} = 0.33$$

A stress index 0.25 is a reasonable lower limit for that quantity. Examination of AISC Commentary Figs. C1.13.3.1 and C1.13.3.2 will show the above criteria to be conservative, the more so when stresses at onset of ponding are low.

Equation 7.4.23 pertains to roof decking which is supported on secondary members. Since this contributes little to ponding, it may be treated as a one-way system in the manner presented by Chinn.[4]

Consider the decking to span a distance S, be supported on rigid supports, and be loaded with a half sine curve whose maximum ordinate is $\gamma(\Delta_0 + \Delta_w)$ per ft of width. The beam reaction is

$$R = \gamma(\Delta_0 + \Delta_w)\frac{S}{\pi}$$

Computing the midspan bending moment and then using conjugate beam to get the midspan deflection gives

$$\Delta_w = \frac{\gamma S^4}{\pi^4 EI}(\Delta_0 + \Delta_w) \tag{7.4.24}$$

$$\Delta_w = \Delta_0 \frac{\gamma S^4/(\pi^4 EI)}{[1 - \gamma S^4/(\pi^4 EI)]} \tag{7.4.25}$$

which becomes infinite when

$$I = \frac{\gamma S^4}{\pi^4 E} \tag{7.4.26}$$

Applying a factor of 1.25, the member will be safe from ponding failure when

$$I \geq \frac{\gamma S^4}{1.25\pi^4 E} = \frac{62.4(144)\,S^4}{1.25\pi^4(29{,}000{,}000)} = \frac{24.8S^4}{10^6} \tag{7.4.27}$$

Thus AISC–1969 requires for roof deck supported on secondary members,

$$I_d \geq \frac{25S^4}{10^6} \tag{7.4.28}$$

When roof decking *is* the secondary member then it should be treated according to Eq. 7.4.22.

EXAMPLE 7.4.1

Select the lightest W section to support a superimposed load of 1.5 kips per ft, of which 1.0 kips per ft is live load, on a simply supported span of 42 ft. Limit live load deflection to $L/360$. Adequate lateral support is provided. Use steel with F_y = 50 ksi, and working stress method of 1969 AISC specs.

Solution

Estimating beam weight at 70 lb per ft,

$$M = 1.57\,(42)^2/8 = 346 \text{ ft-kips}$$

$$F_b = 0.66\,F_y = 33 \text{ ksi}$$

$$\text{Reqd } S_x = \frac{346\,(12)}{33} = 126 \text{ in.}^3$$

assuming live load stress is $33(1.0)/1.57 = 21$ ksi. From Table 7.4.1 for $L/360$,

$$\frac{L}{d} \leq \frac{387}{21} = 18.4$$

$$\text{Min } d = 42(12)/18.4 = 27.4 \text{ in.}$$

From AISC "Allowable Stress Design Selection Table," try W24 × 68, $S_x = 153$ in.^{3}

$$\text{Live-load stress} = \frac{346(12)}{153}\left(\frac{1.0}{1.57}\right) = 27.1\left(\frac{1.0}{1.57}\right) = 17.25 \text{ ksi}$$

$$\frac{L}{d} \leq \frac{387}{17.25} = 22.4, \ d_{\min} = 22.5 \text{ in.} < 24 \text{ in.} \quad \text{OK}$$

It is noted that since the reqd $S_x = 126$ in.^{3} and the minimum depth of 27 in. are seen to give widely divergent sections from the section modulus table, the trial beam was a compromise due to the fact that the stress was only 17.25 instead of 22 ksi. The W24 × 68 was noncompact; therefore $F_b = 0.6F_y = 30$ ksi > actual $f = 27.1$ ksi.

Use W24 × 68.

EXAMPLE 7.4.2

A flat roof for an industrial building has bays 45 ft by 35 ft. Without regard to ponding, the girders have been selected as W24 × 84 (A36 steel) and the joists spanning 35 ft are 24 J 6 ($F_b = 22$ ksi).

Given:

W24 × 84, $I_p = 2364$ in.^{4}
$f_b = 23.5$ ksi

24 J 6 $I_s = 367(12)/22 = 200$ in.^{4}
(spacing 5′ = 0″) (computed from load-table data)
$f_b = 20$ ksi

Assume that one-fourth of live load is acting at onset of ponding (i.e., assume that this is 60 percent of the total service load) and check adequacy with regard to ponding.

Solution

(a) Check the girder:

$$f_0 = 0.6(23.5) = 14.1 \text{ ksi}$$

$$U = \text{stress index} = \frac{0.8F_y - f_0}{f_0} = \frac{0.8(36) - 14.1}{14.1} = 1.04$$

$$C_p = \frac{32(35)(45)^4}{10^7(2364)} = 0.194$$

$$C_s = \frac{32(5)(35)^4}{10^7(200)} = 0.12$$

$$C_P + 0.9C_s = 0.194 + 0.9(0.12) = 0.30 > 0.25$$

This appears inadequate by this conservative check. Using chart in AISC Commentary, Fig. C1.13.3.1 gives

For $U_p = 1.04$ and $C_s = 0.12$
Find allowable $C_p \approx 0.40 > 0.19$ OK

Thus girder is OK.

(b) Check the joist:

$$f_0 = 0.6(20) = 12.0 \text{ ksi}$$

$$U = \text{stress index} = \frac{0.8F_y - f_0}{f_0} = \frac{0.8(36) - 12.0}{12.0} = 1.40$$

By accurately checking using Fig. C1.13.3.2 of the AISC Commentary,

For $U_s = 1.40$ and $C_p = 0.194$
Find allowable $C_s \approx 0.38 > 0.12$ OK

Thus both members satisfy ponding provisions. When the stress index is above about 0.5, the simple expression of AISC–1.13.3 will be grossly conservative.

7.5. SHEAR ON ROLLED BEAMS

Whereas long beams may be governed by deflection and medium length beams are controlled by flexural stress, short-span beams are frequently governed by shear.

To review the development of the shear stress equation for symmetrical sections, consider the slice dz of the beam of Fig. 7.5.1a, shown as a free body in Fig. 7.5.1b. If the unit shear stress v at a section y_1 from the neutral axis is desired, it is observed from Fig. 7.5.1c that

$$dC' = vt\,dz \tag{7.5.1}$$

The horizontal forces arising from bending moment are

$$C' = \int_{y_1}^{y_2} f\,dA$$

$$C' + dC' = \int_{y_1}^{y_2} (f + df)\,dA$$

Subtracting,

$$dC' = \int_{y_1}^{y_2} df\,dA \tag{7.5.2}$$

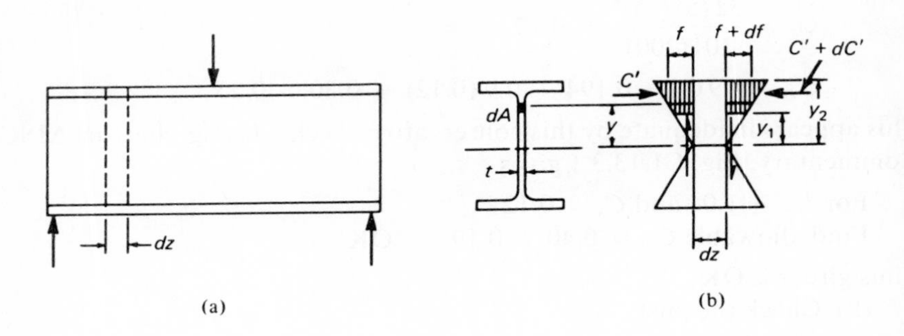

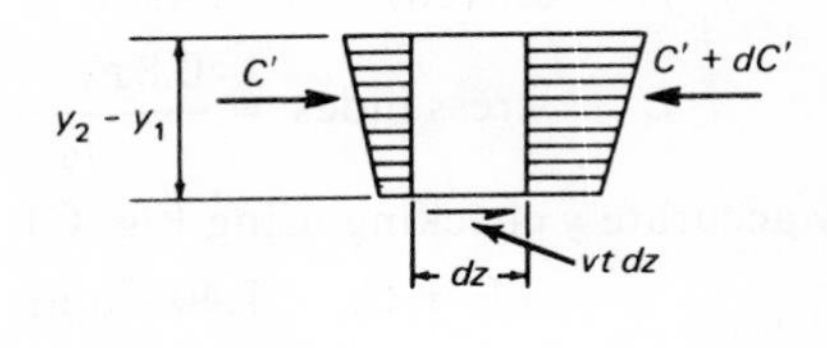

Fig. 7.5.1.

$$df = dM\,y/I \tag{7.5.3}$$

$$dC' = \int_{y_1}^{y_2} \frac{dMy}{I}\,dA = \frac{dM}{I}\int_{y_1}^{y_2} y\,dA \tag{7.5.4}$$

Substituting Eq. 7.5.4 into Eq. 7.5.1 and solving for the shear stress v gives

$$v = \frac{dM}{dz}\left(\frac{1}{tI}\right)\int_{y_1}^{y_2} y\,dA \tag{7.5.5}$$

and upon recognizing that $V = dM/dz$, and letting

$$Q = \int_{y_1}^{y_2} y\,dA$$

the familiar equation

$$v = \frac{VQ}{It} \tag{7.5.6}$$

is obtained where Q is the first moment of area about the x-axis of the area between the extreme fiber and the particular location at which the shearing stress is to be determined, i.e., the shaded area lying between the limits y_1 and y_2 in Fig. 7.5.1b.

Under usual procedures of steel design, the shear stress is computed as the average value over the area of the web, thus:

$$v = \frac{V}{A_w} = \frac{V}{dt_w} \tag{7.5.7}$$

The basis for this is demonstrated by the following example.

EXAMPLE 7.5.1

Determine the shear-stress distribution on a W24 × 94 beam subjected to a shear force of 200 kips. Also compute the portion of the shear carried by the flange and that carried by the web. (See Fig. 7.5.2.)

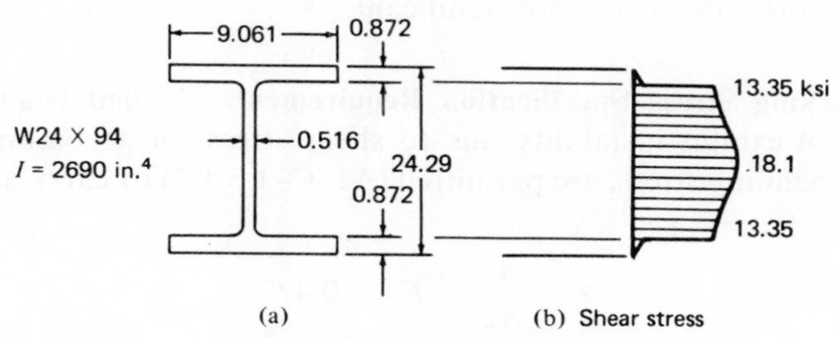

Fig. 7.5.2. Example 7.5.1.

Solution

V = 200 kips

(a) Stress at junction of flange and web:

$$Q = 9.061\,(0.872)\,(12.145 - 0.436) = 92.5 \text{ in.}^3$$

$$v = \frac{200\,(92.5)}{2690(0.516)} = 13.35 \text{ ksi (web)}, \; v = 0.76 \text{ ksi (flange)}$$

(b) Stress at neutral axis:

$$Q = 92.5 + (12.145 - 0.872)^2(0.516)\,(0.5) = 92.5 + 32.8$$
$$= 125.3 \text{ in.}^3$$

$$v = \frac{200\,(125.3)}{2690(0.516)} = 18.1 \text{ ksi}$$

(c) Shear carried by flanges and web: Using an approximate linear variation,

$$V(\text{flanges}) = 2\,(\tfrac{1}{2})\,(0.76)\,(0.872)\,(9.061) = 6 \text{ kips}$$
$$V(\text{web}) = 200 - 6 = 194 \text{ kips}$$

In this case, 97 percent of the shear is carried by the web.

(d) Average shear on web:

$$v\,(\text{average}) = \frac{200}{24.29\,(0.516)} = 15.95 \text{ ksi}$$

which is about 11.5 percent below the maximum value.

Because of this relatively small difference, specifications usually permit and prescribe calculation of average shear stress and correspondingly adjust the allowable values. In computing the average shear stress for rolled beams the overall depth is usually used, while on plate girders the web plate depth is used. The 1969 AISC–1.5.1.2 clearly permits the use of overall depth. Logic would suggest use of the web area lying between the flanges but the difference is not significant.

AISC-Working Stress Specification Requirement. Rolled beams whose webs do not exhibit instability due to shear stress, or a combination of shear and bending stress, are permitted (AISC–1.5.1.2) to carry a shearing stress of

$$v = \frac{V}{A_w} \leq F_v = 0.4F_y \tag{7.5.8}$$

It is the low depth to web thickness ratio which insures achievement of shear yield; thus eliminating the web buckling problem and the need for stiffeners. The use of $F_v = 0.4F_y$ implies h/t ratios which satisfy the AISC Formula (1.10–1) for omission of intermediate stiffeners. The development of Formula (1.10–1) appears in Chapter 11 on plate girders, but from that formula, h/t limits which insure use of $F_v = 0.4F_y$ can be obtained and these are given in Table 7.5.1.

TABLE 7.5.1
Maximum h/t Values for Elimination of Stiffeners
(Based on AISC–1.10.5)

F_y, ksi	h/t
36	62.6
50	53.7
60	49
65	47

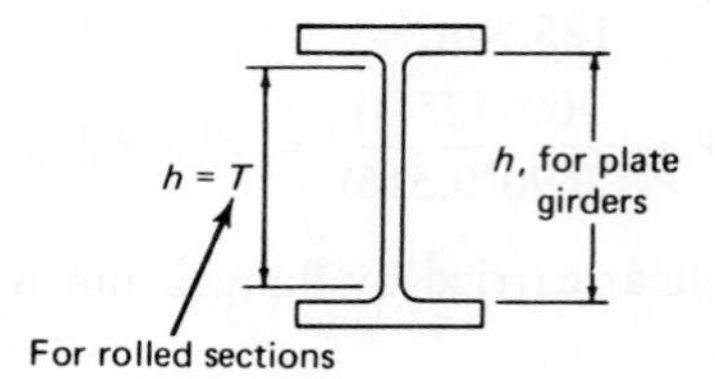

The use of $0.4F_y$ as the maximum allowable shear stress arises by considering theories of failure when shear stresses are simultaneously acting with bending normal stresses. Using the generally accepted Huber-von Mises-Hencky "energy of distortion" theory,* the shear yield point, τ_y, equals the compression-tension yield point, F_y, divided by $\sqrt{3}$, i.e.,

$$\tau_y = \frac{F_y}{\sqrt{3}} \tag{7.5.9}$$

A nominal 1.67 factor of safety would give an allowable shear stress of about $0.35F_y$. There are two reasons why the allowable stress is permitted to exceed this value. First, for low h/t ratios, the shear stresses can actually go into the strain hardening range without the beam exhibiting any distress. Secondly, for about fifty years steel specifications have permitted shear stresses to be two-thirds of the values permitted for bending. Considering the many years of experience with the lower factor of safety and the relatively minor consequences of shear yielding, the value $0.4F_y$ seems reasonable.

EXAMPLE 7.5.2

Select the lightest W section of A36 steel to carry 10 kips per ft on a simply supported span of 6 ft. Lateral bracing is adequate for stability.

Solution

Since the loading is heavy and the span is short, the designer should expect the shear may govern:

$$M = 10(6)^2/8 = 45 \text{ ft-kips}$$

$$V = 10(6)/2 = 30 \text{ kips}$$

Assuming "compact section," $F_b = 0.66F_y = 24$ ksi

$$\text{Reqd } S_x = \frac{45(12)}{24} = 22.5 \text{ in.}^3$$

Select W12 × 22 from AISC "Allowable Stress Design Selection Table" as the lightest beam having $S_x \geq 22.5$ in.^{3}
Check shear.

$$v = \frac{V}{dt_w} = \frac{30}{12.31(0.26)} = 9.35 \text{ ksi}$$

$$\text{Allowable } v = 0.4F_y = 0.4(36) = 14.4 \text{ ksi} > 9.35 \text{ ksi} \qquad \text{OK}$$

This can be verified as acceptable by checking h/t:

$$\frac{h}{t} = \frac{\text{unsupported height of web}}{\text{web thickness}} = \frac{10.75}{0.26} = 41.3$$

which is less than 62.6 as given in Table 7.5.1 and therefore acceptable without stiffeners. The detailed discussion of stiffener requirements and

*See Fred B. Seely and James O. Smith, *Advanced Mechanics of Materials*. 2nd ed. (New York: John Wiley & Sons, Inc., 1952), pp. 76–91.

situations where h/t exceeds the values given in Table 7.5.1 appears in Chapter 11.

AISC-Plastic Design Requirement. The ultimate shear stress v_u, computed as an average on gross web area, when the web is not reinforced by diagonal stiffeners or a doubler plate, is permitted according to AISC–2.5 to be

$$v_u = \frac{V_u}{dt_w} \leq 0.55F_y \tag{7.5.10}$$

which is consistent with the "energy of distortion" theory which says $\tau_y = F_y/\sqrt{3} = 0.58F_y$.

7.6. WEB CRIPPLING AND BEARING PLATES

Web crippling is a localized yielding which arises from high compressive stresses occurring in the vicinity of concentrated loads. Such situations arise when concentrated loads are applied to beams, beam bearing supports, and reactions of beam flanges at connections to columns.

Referring to Fig. 7.6.1, according to AISC–1.10.10.1, a conservative assumption is used that the load is distributed at 45 degrees into the critical section at the toe of the fillet at a distance k from the face of the beam. To establish the reliability of such a procedure two factors should be considered. First, before a crippling failure occurs the distribution of load spreads out over a distance of $N + 5k$ to $N + 7k$, particularly when the bearing length N is small, according to recent tests;[5] and secondly, the yield point in the region between the critical section and the interior face of the flange is likely to be less than the yield point in the web, primarily because the thinner web material has been worked more during forming. The important requirement is not that yielding be prohibited everywhere but that it not occur over a significant length of critical section such that a crippling failure can occur.

AISC-Working Stress Requirement. AISC–1.10.10.1, is based empirically on the 1935 work of Lyse and Godfrey,[6] who realized that this crippling phenomenon is not the same as the overall buckling of the web. Investigators since that time have shown the AISC procedure to be conservative.[5,7]

The compressive stress at the web toes of the fillets resulting from concentrated loads not supported by stiffeners (Fig. 7.6.1) shall not exceed the following values.

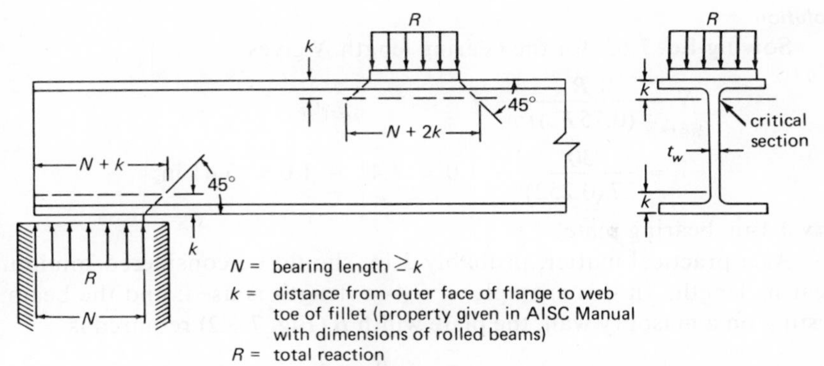

Fig. 7.6.1. Web-crippling considerations for establishing bearing length.

For interior loads,

$$f_b = \frac{R}{t_w(N + 2k)} \leq 0.75F_y \tag{7.6.1}$$

For end reactions,

$$f_b = \frac{R}{t_w(N + k)} \leq 0.75F_y \tag{7.6.2}$$

The 1.33 nominal safety factor against yielding in addition to the conservatively low values of $2k$ and k make these specification requirements reasonably consistent with other procedures.

AISC–Plastic Design Requirement. AISC–2.6 prescribes that at every point of load concentration where a plastic moment is expected to develop under collapse conditions a web stiffener must be used.

No reference is made to other points of load concentration, and it seems proper for those locations to be treated in accordance with the working stress procedure of AISC–1.10.10.1.

Web-crippling problems in beam-column connections are treated in Chapter 13, while the design of bearing stiffeners for cases where stresses exceed the allowables to prevent web crippling is discussed in Sec. 11.10.

EXAMPLE 7.6.1

Determine the size of bearing plate required for an end reaction of 30 kips on a W10 × 25 beam of A36 steel according to 1969 AISC specifications. The beam rests on a wall of concrete with a 28-day compressive strength of 3,000 psi.

Solution

Solving Eq. 7.6.2 for the bearing length N gives

$$N = \frac{R}{(0.75F_y)t} - k$$

$$= \frac{30}{27(0.252)} - 1.0 = 4.41 - 1.0 = 3.41 \text{ in.}$$

Try a 4-in. bearing plate.

As a practical matter, probably 3 in. should be considered minimum bearing length. In this example, if a 4-in. length is used, and the beam is resting on a masonry wall, the plate width B (Fig. 7.6.2) required is

$$B = \frac{R}{NF_P}$$

where F_p = allowable unit stress on the support (say AISC–1.5.5). Assuming a wall with concrete having a 28-day compressive strength, f'_c, of 3,000 psi,

$$F_p = 0.375f'_c = 1125 \text{ psi}$$

$$\text{Reqd } B = \frac{30{,}000}{4(1125)} = 6.67 \text{ in., say 7 in.}$$

To determine the plate thickness, bending must be considered. A rectangular section has a shape factor, f, equal to 1.5 (see Sec. 7.3) which

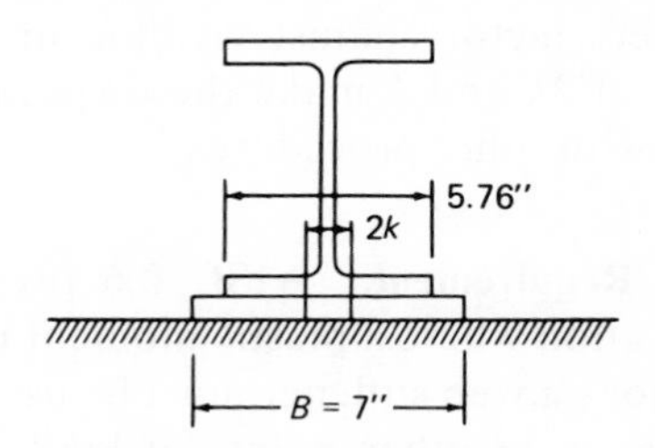

Fig. 7.6.2. Example 7.6.1.

is 30 to 40 percent greater than that of rolled W shapes; therefore, the allowable bending stress on plates (AISC–1.5.1.4.3) is

$$F_b = 0.75F_y$$

which conservatively increases the basic $0.6F_y$ by 25 percent for rectangular plates. A reason for conservatism is that uniform bearing pressure assumes a relatively rigid plate; thus it is really deflection limitations which justify holding the allowable bending stress to $0.75F_y$.

The critical section for bending is taken at the toe of flange to web fillet, a distance of k from center of web, and the beam flange is assumed not to participate. The bending moment and required thickness are then computed:

$$\text{Uniform bearing pressure } p = \frac{30{,}000}{4\,(7)} = 1{,}070 \text{ psi}$$

$$M = \frac{p\,(B/2 - k)^2 N}{2}$$

$$= \frac{1.07\,(3.5 - 1.0)^2\, N}{2} = 3.34\, N$$

$$\text{Section modulus } S = \frac{1}{6}\, Nt^2$$

$$\text{Reqd } S = \frac{M}{F_b} = \frac{3.34\, N}{0.75\,(36)} = 0.124\, N$$

$$\frac{Nt^2}{6} = 0.124\, N, \quad t = \sqrt{6(0.124)} = 0.86 \text{ in.}$$

Use bearing plate, ⅞ × 4 × 0′ − 7″.

Solving for the plate thickness in general,

$$\frac{Nt^2}{6} = \frac{p\,(B/2 - k)^2 N}{2\,F_b}$$

$$t = \sqrt{\frac{3p(B/2 - k)^2}{F_b}}$$

7.7. HOLES IN BEAMS

In modern design practice, holes in beams occur primarily where field connections are made with high-strength bolts. Field splices are rarely made at locations of maximum moment; the extensive concern regarding the effect of holes has largely been eliminated.

For tension members the effect of holes has been discussed in Chapter 3, where holes are deducted and net section is used. For compression members, since the connectors occupy most of the space in the hole, the connectors are assumed in design to completely fill the holes and a deduction for holes has never been the practice in building construction.

It seems clear that the effect of holes must surely be to weaken the beam and not strengthen it. However, unless the section area is significantly reduced by holes the bending capacity at *service loads* seems largely unaffected, primarily because of stress concentrations adjacent to holes.

Within the accuracy of the design methods employed, it does not seem justifiable to calculate a shift in neutral axis as a result of holes in the tension flange. The total *service load* capacity of the tension flange without holes is essentially the same as that with holes; the stresses are merely distributed differently. Thus, the neutral axis does not shift.

Present 1969 AISC specifications (AISC–1.10.1) consider no shift in neutral axis and deduct area for holes only when the area of holes exceeds 15 percent of the gross flange, and then only the excess area over 15 percent of gross flange is deducted.

A reduction in this case seems justifiable because the presence of holes in the tension flange reduces the plastic moment capacity of the section. The plastic moment is computed using yield stress acting over the net area. For cases where ultimate capacity in bending is M_y or less (See Fig. 7.3.1) neglect of the holes gives no major change in safety factor; stress concentrations adjacent to holes provide carrying capacity to replace that lost by making holes. If the ultimate capacity is M_p, the holes will reduce the factor of safety because M_p will reduce almost in proportion to the area of the holes. Thus, AISC–1.10.1 seems to consider that a slight reduction in safety factor is not sufficiently important to warrant deduction for holes, but when the area lost by holes is large some correction should be made to prevent too great a decrease in the safety factor.

Other specifications follow the more conservative procedures which have been used for many years. The American Association of State Highway Officials (AASHO) and the American Railway Engineering Association (AREA) use the net section (holes deducted from both flanges) for computing tension flange stress, and use gross section (no holes from either flange) for computing compression stress.

Examples of procedure for considering holes appear in the Chapter 13 section on beam and girder splices.

7.8. GENERAL FLEXURAL THEORY

Thus far consideration has been given only to symmetrical shapes loaded symmetrically for which $f = Mc/I$ is correct. The development which follows will treat the general bending of arbitrary prismatic beams, i.e., beams which have any cross-sectional shape with no variation along the length of the beam. They are also assumed to be free from twisting.

Consider the straight uniform cross section beam of Fig. 7.8.1 acted upon by pure moment applied in the plane ABCD which makes the angle ϕ with the xz plane. The moments are represented by vectors normal to the plane of action (positive moment defined by using right-hand rule for rotation).

Examine next a portion of the beam of length z as shown in Fig. 7.8.2a. To satisfy equilibrium on the free body of Fig. 7.8.2a requires

$$\Sigma F_z = 0, \qquad \int_A \sigma\, dA = 0 \tag{7.8.1}$$

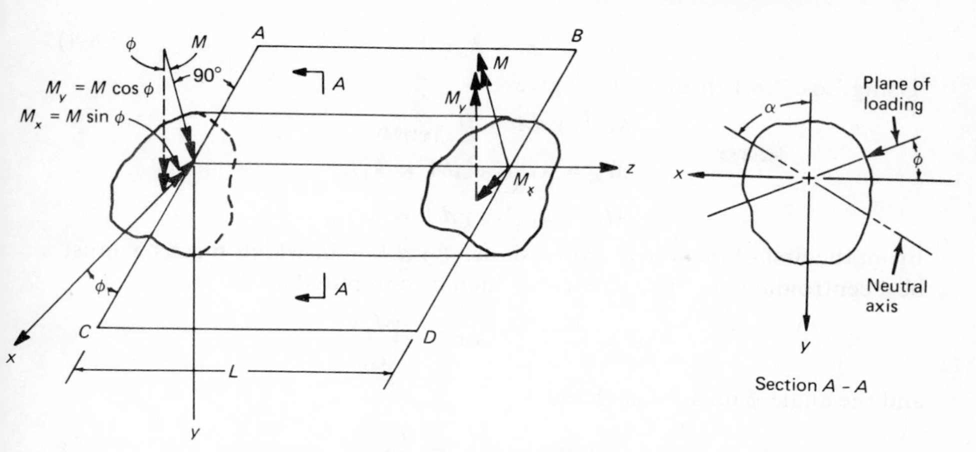

Fig. 7.8.1. Prismatic beam under pure bending.

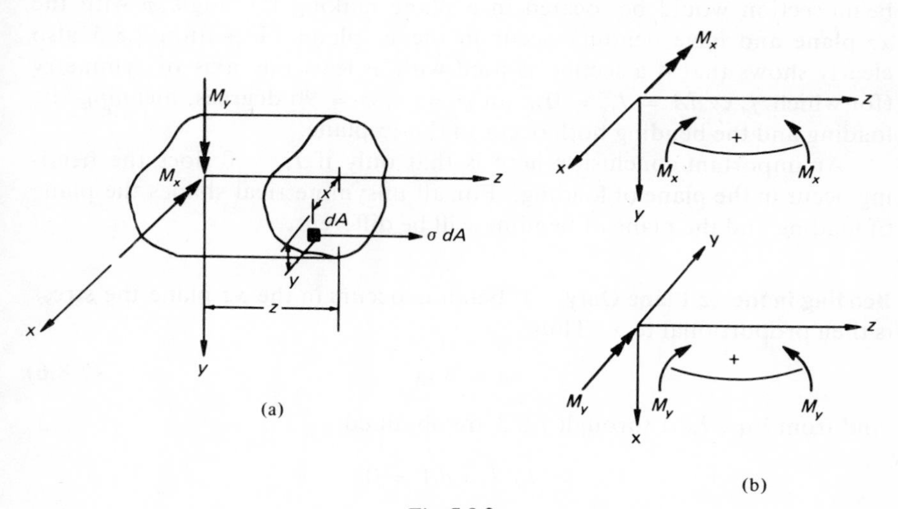

Fig. 7.8.2.

$$\Sigma M_x = 0, \qquad M_x = \int_A y\sigma\, dA \tag{7.8.2}$$

$$\Sigma M_y = 0, \qquad M_y = \int_A x\sigma\, dA \tag{7.8.3}$$

It is to be noted from Fig. 7.8.2b that the moments M_x and M_y are both positive in accordance with the customary convention of calling positive bending that which causes compression in the top portion of the beam.

Bending in *yz* Plane Only. If bending occurs in the *yz* plane, the stress is then proportional to y. Thus

$$\sigma = k_1 y \tag{7.8.4}$$

Using Eqs. 7.8.1 through 7.8.3 gives

$$k_1 \int_A y\,dA = 0$$

$$M_x = k_1 \int_A y^2\,dA = k_1 I_x$$

$$M_y = k_1 \int_A xy\,dA = k_1 I_{xy}$$

From the first of the above expressions, $\int_A y\,dA = 0$, which means x must be a centroidal axis. The stress may then be computed as

$$\sigma = \frac{M_x y}{I_x} \quad \text{or} \quad \frac{M_y y}{I_{xy}}$$

and the angle ϕ must be such that

$$\tan \phi = \frac{M_x}{M_y} = \frac{I_x}{I_{xy}} \tag{7.8.5}$$

As a practical matter it would be unlikely that an unsymmetrical beam section would be located in a plane making the angle ϕ with the xz plane and have bending occur in the yz plane. Equation 7.8.5 also clearly shows that if a section is used with at least one axis of symmetry (for which $\int_A xy\,dA = I_{xy} = 0$), $\tan \phi = \infty$, $\phi = 90$ degrees, meaning the loading and the bending both occur in the yz plane.

An important conclusion here is that only if $I_{xy} = 0$ does the bending occur in the plane of loading. For all unsymmetrical shapes the plane of loading and the plane of bending will be different.

Bending in the *xz* Plane Only. If bending occurs in the xz plane the stress is then proportional to x. Thus

$$\sigma = k_2 x \tag{7.8.6}$$

and from Eqs. 7.8.1 through 7.8.3 are obtained:

$$k_2 \int_A x\,dA = 0$$

which means y must be a centroidal axis. Also,

$$M_x = k_2 \int_A xy\,dA = k_2 I_{xy}$$

$$M_y = k_2 \int_A x^2\,dA = k_2 I_y$$

and

$$\tan \phi = \frac{M_x}{M_y} = \frac{I_{xy}}{I_y} \tag{7.8.7}$$

In the case where $I_{xy} = 0$, $\tan \phi = 0$, i.e., the loading and bending both occur in the xz plane.

Bending in neither *xz* nor *yz* Planes. This is the realistic case when considering unsymmetrical sections in flexure. Since it is assumed all stresses are within the elastic limit, the total stress is the sum of the stresses due to bending in each of the *xz* and *yz* planes. Thus,

$$\sigma = k_1 y + k_2 x \tag{7.8.8}$$

and

$$M_x = k_1 I_x + k_2 I_{xy} \tag{7.8.9}$$

$$M_y = k_1 I_{xy} + k_2 I_y \tag{7.8.10}$$

Solving Eqs. 7.8.9 and 7.8.10 for k_1 and k_2 and substituting into Eq. 7.8.8 gives

$$\sigma = \frac{M_x I_y - M_y I_{xy}}{I_x I_y - I_{xy}^2} y + \frac{M_y I_x - M_x I_{xy}}{I_x I_y - I_{xy}^2} x \tag{7.8.11}$$

which is the general flexure equation. The assumptions inherent in Eq. 7.8.11 are (a) a straight beam; (b) constant cross section; (c) x and y axes are mutually perpendicular centroidal axes; and (d) that stress is proportional to strain and the maximum value is within the proportional limit.

Principal Axes. The principal axes are mutually perpendicular centroidal axes for which the moment of inertia is either a maximum or minimum. Furthermore, these axes are the only mutually perpendicular axes for which the product of inertia I_{xy} is zero. When a section has an axis of symmetry, that axis is a principal axis and Eq. 7.8.11 becomes

$$\sigma = \frac{M_x}{I_x} y + \frac{M_y}{I_y} x \tag{7.8.12}$$

When there is no axis of symmetry, Eq. 7.8.12 can still be used if the principal axes are located and the quantities M_x, M_y, I_x, I_y, x, and y are all corrected so as to refer to the principal axes. Usually such transformations offer no advantage over direct use of Eq. 7.8.11.

Inclination of the Neutral Axis. When the loads acting on a flexural member pass through the centroid of the cross section but are inclined with respect to either of the principal axes, the stresses may be determined by using Eq. 7.8.11 or Eq. 7.8.12. However, note should be made that the neutral axis is not necessarily perpendicular to the plane of loading. As shown from Eqs. 7.8.5 and 7.8.7 and Fig. 7.8.1,

$$\tan \phi = \frac{M_x}{M_y} \tag{7.8.13}$$

Since the stress along the neutral axis is equal to zero, σ may be set equal to zero in Eq. 7.8.11. With $\sigma = 0$, solving for $-x/y$ gives

$$-\frac{x}{y} = \left(\frac{M_x I_y - M_y I_{xy}}{I_x I_y - I_{xy}^2}\right)\left(\frac{I_x I_y - I_{xy}^2}{M_y I_x - M_x I_{xy}}\right) \tag{7.8.14}$$

From Fig. 7.8.1, Section A–A, it is seen that at any point on the neutral axis, $\tan \alpha = -x/y$. Dividing both numerator and denominator of the right side of Eq. 7.8.14 by M_y gives

$$\tan \alpha = \frac{\dfrac{M_x}{M_y} I_y - I_{xy}}{I_x - \dfrac{M_x}{M_y} I_{xy}} \tag{7.8.15}$$

Substitution of Eq. 7.8.13 into Eq. 7.8.15 gives

$$\tan \alpha = \frac{I_y \tan \phi - I_{xy}}{I_x - I_{xy} \tan \phi} \tag{7.8.16}$$

When investigating a section with one axis of symmetry, $I_{xy} = 0$; Eq. 7.8.16 then becomes

$$\tan \alpha = \frac{I_y}{I_x} \tan \phi \tag{7.8.17}$$

EXAMPLE 7.8.1

A W18 × 50 used as a beam is subjected to loads inclined at 5° from the vertical axis as shown in Fig. 7.8.3. Locate the inclination of the neutral axis.

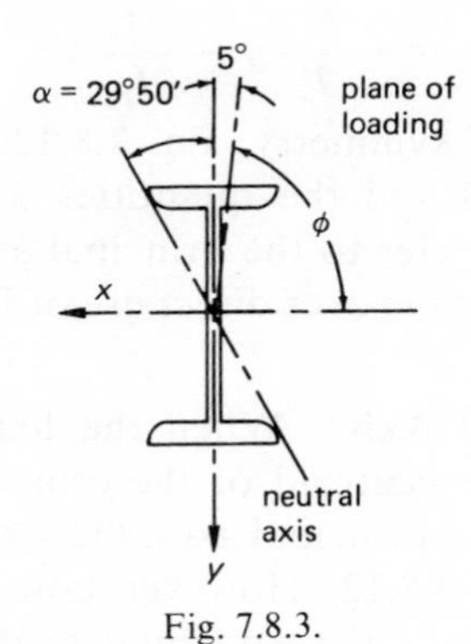

Fig. 7.8.3.

Solution

$I_x = 802 \text{ in.}^4; \qquad I_y = 40.2 \text{ in.}^4$
$\tan 85° = \tan \phi$
$\tan \alpha = 40.2\ (11.430)/802 = 0.573$
$\tan^{-1}(0.573) = 29°\ 50'$

EXAMPLE 7.8.2

Compute the maximum flexural stress in a 6 × 4 × ½ angle with the long leg vertically downward and which carries 0.5 kip/ft over a simply supported span of 10 ft (see Fig. 7.8.4). Compare the value assuming the angle is completely free to bend in any direction with that obtained assuming bending in only the vertical plane.

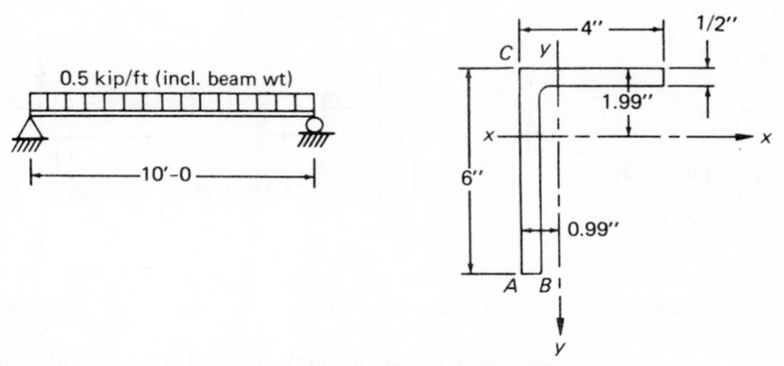

Fig. 7.8.4. Data for Example 7.8.1.

Solution

(a) Angle free to bend in any direction. Using Eq. 7.8.11,

$$I_x = 17.4 \text{ in.}^4, \qquad I_y = 6.3 \text{ in.}^4$$

$$I_{xy} = \tfrac{1}{2}[6(3.00 - 1.99)(-0.99 + 0.25) + 3.5(2.25 - 0.99)(-1.99 + 0.25)]$$
$$= \tfrac{1}{2}[(-4.48) + (-7.67)] = -6.08 \text{ in.}^4$$

$$M_x = 0.5(10)^2/8 = 6.25 \text{ ft-kips} = 75 \text{ in-kips}$$
$$M_y = 0$$

Stress at point A:

$$f_A = \frac{M_x(I_y y - I_{xy} x)}{I_x I_y - I_{xy}^2} = \frac{6.25(12)[6.3(+4.01) - (-6.08)(-0.99)]}{17.4(6.3) - (6.08)^2}$$

$$= \frac{75(19.23)}{72.6} = 75(0.2645) = +19.9 \text{ ksi (tension)}$$

Stress at point B:

$$f_B = \frac{75[6.3(+4.01) - (-6.08)(-0.49)]}{72.6}$$

$$= 75(0.307) = +23.0 \text{ ksi (tension)}$$

Stress at point C:

$$f_C = \frac{75[6.3(-1.99) - (-6.08)(-0.99)]}{72.6}$$

$$= 75(-0.2655) = -19.1 \text{ ksi (comp.)}$$

(b) Angle free to bend in any direction. Use alternate method suggested by Gaylord and Gaylord (pp. 143–147).[8] Compute the stresses assuming first that the beam bends only in the *yz* plane. Conveniently let $M_x = 100$ in.-kips. Then, according to Eq. 7.8.5, $M_y = M_x I_{xy}/I_x$ must also be acting (Fig. 7.8.5a) if bending occurs only in the *yz* plane.

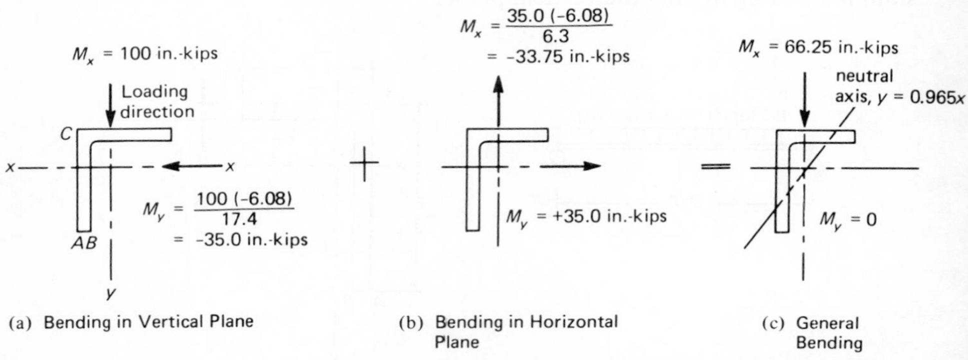

Fig. 7.8.5. Solution by superposition of bending in the vertical and horizontal planes.

Since the real loading has only M_x, M_y must be removed by application of an equal and opposite moment, further considering that bending occurs only in the *xz* plane. This means, according to Eq. 7.8.7, the simultaneous application of $M_x = M_y I_{xy}/I_y$ (see Fig. 7.8.4b).

Bending in vertical plane:

$$M_x = 75.0\left(\frac{100}{66.25}\right) = 113.2 \text{ in.-kips}$$

$$f_{A_1} = f_{B_1} = \frac{113.2(4.01)}{17.4} = 26.1 \text{ ksi (tension)}$$

$$f_{C_1} = \frac{113.2(1.99)}{17.4} = 13.0 \text{ ksi (compression)}$$

Bending in horizontal plane:

$$M_y = -113.2(-6.08)/17.4 = 39.6 \text{ in.-kips}$$

$$f_{A_2} = f_{C_2} = \frac{39.6(0.99)}{6.3} = 6.2 \text{ ksi (compression)}$$

$$f_{B_2} = \frac{39.6(0.49)}{6.3} = 3.1 \text{ ksi (compression)}$$

Total stresses in general bending:

$$f_A = +26.1 - 6.2 = +19.9 \text{ ksi (tension)}$$
$$f_B = +26.1 - 3.1 = +23.0 \text{ ksi (tension)}$$
$$f_C = -13.0 - 6.2 = -19.2 \text{ (compression)}$$

which agree with the values, 19.9, 23.0 and 19.1 ksi, respectively, as computed by the general formula.

The general equation for stress at any point is seen to be

$$f = \frac{113.2y}{17.4} + \frac{39.6x}{6.3}$$

where if $f = 0$, the neutral axis is $y = 0.965x$.

This method permits the designer to visualize what is taking place. If attached construction constrains an unsymmetrical section to bend in the vertical plane, the restraining moment capacity can be computed by using Eq. 7.8.5.

(c) Angle restrained to bend in the vertical plane:

$$f_A = f_B = \frac{75(4.01)}{17.4} = +17.3 \text{ ksi} \quad \text{(tension)}$$

$$f_C = \frac{75(-1.99)}{17.4} = -8.6 \text{ ksi} \quad \text{(compression)}$$

Unless the horizontal constraints actually act, the tensile stress at point B is underestimated by 25 percent, and the compressive stress at C is underestimated by 55 percent.

Frequently designers assume $f = My/I$ is applicable without considering whether or not adequate horizontal restraints are present. Although usually some degree of restraint is present, care should be exercised when investigating unsymmetrical beams. Neglect of the lateral (horizontal) component is always on the unsafe side.

7.9. BIAXIAL BENDING OF SYMMETRICAL SECTIONS

Flexural stresses on sections with at least one axis of symmetry and loaded through the centroid may be computed using Eq. 7.8.12, which when modified to give maximum stress becomes

$$\sigma = \frac{M_x}{S_x} + \frac{M_y}{S_y} \tag{7.9.1}$$

where $S_x = I_x/(d/2)$ and $S_y = I_y/(b/2)$ are the section modulus values.

The investigation of stresses by Eq. 7.9.1 is a simple task; however, the selection of a beam of minimum weight to satisfy a maximum stress limitation involves an indirect process.

Assume the maximum combined stress σ may not exceed a prescribed allowable value, F_b; thus Eq. 7.9.1 becomes

$$\frac{M_x}{S_x} + \frac{M_y}{S_y} \le F_b \tag{7.9.2}$$

and multiplying by S_x and dividing by F_b gives

$$S_x \geq \frac{M_x}{F_b} + \frac{M_y}{F_b}\left(\frac{S_x}{S_y}\right) \tag{7.9.3}$$

For selecting standard rolled shapes one finds that for a given depth of section the ratio S_x/S_y is relatively constant. A typical range of such values appears in Table 7.9.1. The lightest sections in any depth will be the narrowest ones and thus have the higher values of S_x/S_y.

An alternate method is suggested by Gaylord and Gaylord (Ref. 8, p. 168) for use with I-shaped sections (W, M, and S). The quantity S_x/S_y is computed approximately using the beam properties neglecting the web effect. This gives, using b = width and d = depth,

$$\frac{S_x}{S_y} = \frac{I_x(b/2)}{(d/2)I_y} = \frac{2bt(d/2)^2(b/2)}{(d/2)(2)(1/12)tb^3} = \frac{3d}{b} \tag{7.9.4}$$

which is suggested to be increased to $3.5d/b$ to account for a greater error in numerator than in denominator resulting from the neglect of web.

EXAMPLE 7.9.1

Select the lightest W or M section to carry moments M_x = 60 ft-kips and M_y = 25 ft-kips. Use the working-stress method and consider that adequate bracing is provided so that the allowable combined stress is $0.6F_y$. Use steel with F_y = 50 ksi.

TABLE 7.9.1
Typical S_x/S_y Values

Shape	Depth d, in.	S_x/S_y
M	6, 8	5–7
M	10, 12	8–11
M	14, 16	11–12
Light W and M	4–8	3
W	8, 10	3–4
W	12	3–6
W	14 (up to 84 lb/ft)	4–8
W	14 (over 84 lb/ft)	2½–3
W	16, 18, 21	5–9
W	24, 27	6–10
W	30, 33, 36	7–12
S	6–8	d
S	10–18	$.75d$
C	up to 7	$1.5d$
C	8–10	$1.25d$
C	12, 15	d

Solution

In this case $F_b = 0.6F_y = 30$ ksi. Note that under 1969 AISC–1.5.1.4.5 the maximum allowable stress in unsymmetrical bending is $0.6F_y$, which assumes the ultimate capacity is achieved when the extreme fiber reaches F_y. When the section is 'compact' it is probably more logical to use an allowable stress between $0.66F_y$ (F_{bx} for strong axis bending) and $0.75F_y$ (F_{by} for weak axis bending). A straight line interaction would be obtained if F_{bx} and F_{by} were used in the denominators of Eq. 7.9.3. instead of the single value F_b. While AISC–1.5.1.4 seems explicit in allowing only $0.60F_y$ for this case, AISC–1.6.1 implies the straight line interaction for the limiting case of zero axial compression when using the beam-column formulas. Using Eq. 7.9.3,

$$\text{Reqd } S_x \geq \frac{60(12)}{30} + \frac{25(12)}{30}\left(\frac{S_x}{S_y}\right)$$

$$\geq 24 + 10\, S_x/S_y$$

From Table 7.9.1 the ratio S_x/S_y can be expected to be on the order of 3 to 4; thus $S_x \approx 55 - 65$ in.^{3}

Using the AISC "Allowable Stress Design Selection Table," try W16 × 40, $S_x = 64.6$. For this beam, $S_x/S_y = 64.6/8.2$, and it is apparent by inspection the beam is inadequate. Continuing with the "Allowable Stress Design Selection Table," it is found by inspection that the lightest W shapes (boldface type) are inadequate up to about W21 × 68, which has about $S_x/S_y \approx 9$ (Table 7.9.1). Thus 60 lb/ft seems to be the weight range.

Try W10 × 60:

$$S_x/S_y = 67.1/23.1 = 2.9$$

$$\text{Reqd } S_x = 24 + 10(2.9) = 53 \text{ in.}^3 < 67.1 \text{ in.}^3 \qquad \text{OK}$$

For W10 × 49,

$$S_x/S_y = 54.6/18.6 = 2.94$$

$$\text{Reqd } S_x = 24 + 10(2.94) = 53.4 \text{ in.}^3 < 54.6 \text{ in.}^3 \qquad \text{OK}$$

Use W10 × 49.

If one had assumed a 10-in. depth as desirable and used Eq. 7.9.4 with a coefficient of 3.5 instead of 3,

$$\text{Reqd } S_x = 24 + 10(3.5\ d/b)$$

For W10 sections, d/b is either 1 or 1.25.

$$\text{Reqd } S_x \approx 24 + 35 = 59 \text{ in.}^3$$

which would also have required a check of two sections. The more general approach using typical S_x/S_y values seems most useful unless a specific depth is desired.

7.10. SECTION MODULUS POLYGON

Many common biaxial bending situations arise as a result of loading in a plane which makes an angle θ with the plane of the web. If one could merely compute the moment in the plane of bending and use it directly in design, some of the complications illustrated in Sec. 7.9 could be eliminated.

A useful graphical approach was presented a number of years ago by Johnson, Bryan, and Turneaure.[9] The "S polygon" can be developed and constructed for any shape, symmetrical or unsymmetrical; a concise presentation for the unsymmetrical shape is presented by McGuire (Ref. 10, pp. 333–336.).

For the section with $I_{xy} = 0$, Eq. 7.8.11 may be written

$$\sigma = \frac{M_x y}{I_x} + \frac{M_y x}{I_y} \tag{7.10.1}$$

Referring to Fig. 7.10.1a, the components of the applied moment M are

$$M_x = M\cos\theta = \text{bending about } x \text{ axis}$$
$$M_y = M\sin\theta = \text{bending about } y \text{ axis}$$

For maximum stress at $y = d/2$ and $x = b/2$, substitution of $S_x = I_x/(d/2)$ and $S_y = I_y/(b/2)$ into Eq. 7.10.1 gives

$$\sigma = M\left(\frac{\cos\theta}{S_x} + \frac{\sin\theta}{S_y}\right)$$
$$= M\left(\frac{S_y\cos\theta + S_x\sin\theta}{S_x S_y}\right) \tag{7.10.2}$$

Equation 7.10.2 may be stated as

$$\sigma = \frac{M}{S_A} \tag{7.10.3}$$

where

$$S_A = \frac{S_x S_y}{S_y\cos\theta + S_x\sin\theta}$$

which can be thought of as the effective section modulus resisting a moment applied in the plane making the angle θ with the web.

Examination of the expression for S_A will show it to be the equation of a straight line. Let $m = S_A\cos\theta$ and $n = S_A\sin\theta$,

$$S_A = \frac{S_x S_y}{S_y m/S_A + S_x n/S_A} = \frac{S_A S_x S_y}{S_y m + S_x n}$$

Solving for m,

$$m = S_x - \frac{S_x}{S_y} n \tag{7.10.4}$$

which is the equation of a straight line (AB of Fig. 7.10.1b). If one considers that θ may be an angle in any quadrant, the S_A expression defines a polygon $ABCD$ (Fig. 7.10.1b); hence the name "S polygon."

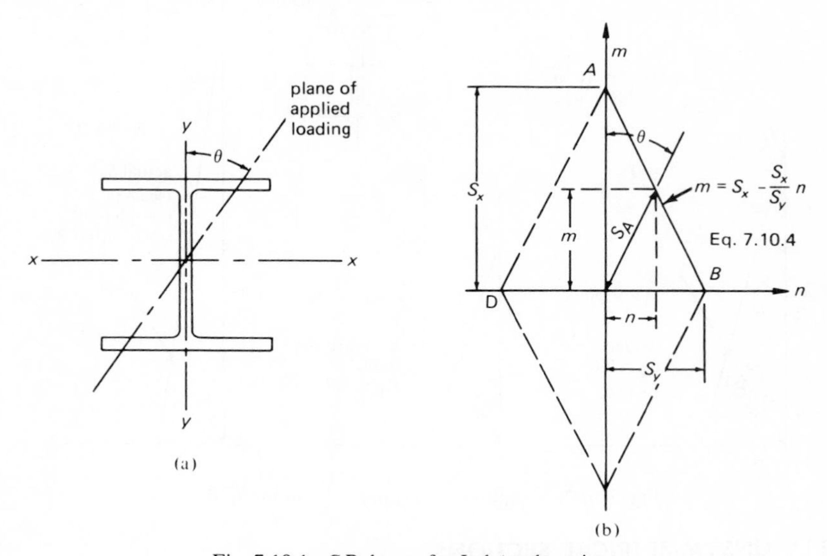

Fig. 7.10.1. S Polygon for I-shaped sections.

EXAMPLE 7.10.1

Design a W section for a simple span of 16 ft to carry a transverse oblique load of 500 lb/ft (including beam weight) at an angle of 30° with the beam web (see Fig. 7.10.1a). Assume the load acts through the center of gravity of the section, i.e., assume the effect of any twisting moment may be carried by the roofing and neglected in the selection of the purlin. Use A36 steel. Assume lateral instability is not possible and that allowable bending stress is $0.6F_y$ (AISC–1.5.1.4.5).

Solution

$$F_b = 0.6F_y = 22 \text{ ksi}$$

$$M = \frac{wL^2}{8} = \frac{0.5(16)^2}{8} = 16 \text{ ft-kips @ } \theta = 30°$$

$$S_A \text{ reqd} = \frac{16(12)}{22} = 8.7 \text{ in.}^3$$

Referring to Fig. 7.10.2 and laying out the required S_A to scale at the proper angle, it is only necessary to find a section whose S_x and S_y give a straight line passing through or closely to the right of the required point.

The graphical solution is very rapid and is shown in Fig. 7.10.2.

Use W8 × 28 in preference to the 4 in. deeper W12 × 27.

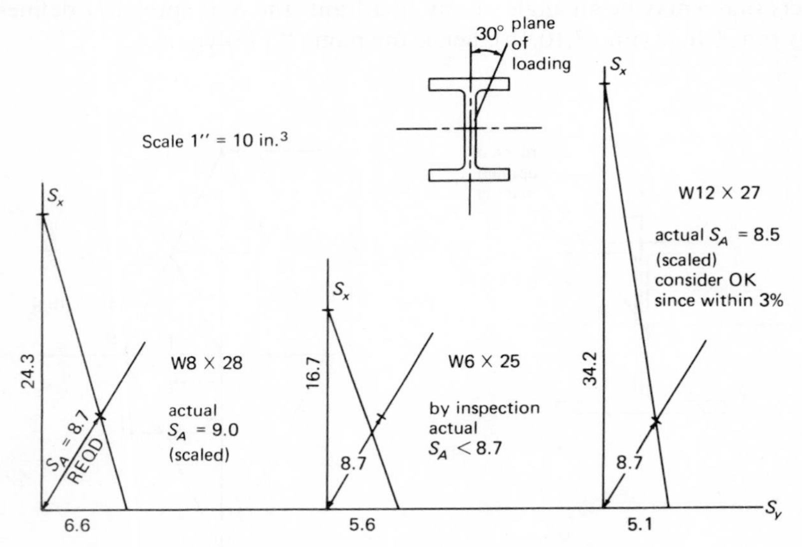

Fig. 7.10.2. Graphical solution for Example 7.10.1.

7.11* UNSYMMETRICAL SECTIONS

With the increasing use of multitypes of steel, and composite construction which combines the compressive strength of concrete with the tensile strength of steel (see Chapter 16), the need to consider unsymmetrical sections increases. While very irregular shapes must be dealt with individually on a trial-and-error basis, sections of the general I-shape which merely have one flange larger than the other can be treated in a systematic manner.

The following development of a useful design equation is essentially that presented by Gaylord and Gaylord[8] (pp. 182–183). The primary shape is assumed to have its centroid at mid-depth, though it need not be I-shaped. Referring to Fig. 7.11.1, the shift in centroid y which results from adding a plate of area A_f to one flange is

$$\bar{y} = \frac{A_f y_1}{A_i + A_f} \tag{7.11.1}$$

which if $p = A_f/A_i$ becomes

$$\bar{y} = \frac{p y_1}{1 + p} \tag{7.11.2}$$

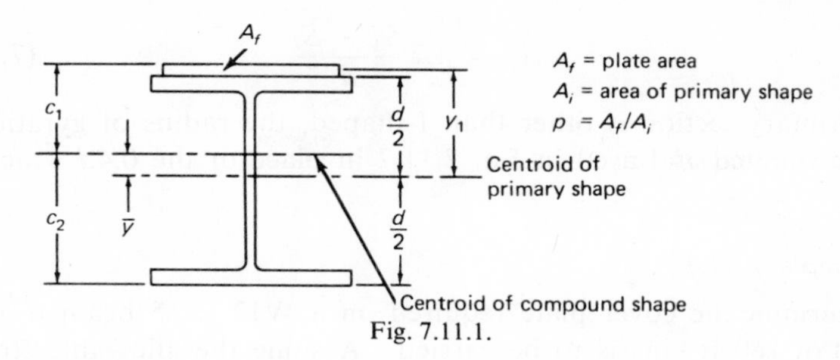

Fig. 7.11.1.

The moment of inertia of the compound section is

$$I = I_i + A_i\bar{y}^2 + A_f(y_1 - \bar{y})^2 \tag{7.11.3}$$

which upon substituting Eq. 7.11.2 becomes

$$I = I_i + A_i\left(\frac{py_1^2}{1 + p}\right) \tag{7.11.4}$$

For the most common situation of an I-shaped section (W, S, or M), the radius of gyration is approximately known (See Appendix, Table A1) to range from $0.38d$ to $0.42d$; thus, using an average value,

$$I_i = A_i r_i^2 = A_i(0.4d)^2 \tag{7.11.5}$$

Next, convert the moment of inertia I into section modulus form, letting $S_1 = I/c_1 \approx I/(y_1 - \bar{y})$; thus, using Eq. 7.11.2 gives

$$S_1 = I\left(\frac{1 + p}{y_1}\right)$$

and putting this into Eq. 7.11.4 gives

$$S_1 = \frac{I_i}{y_1} + \frac{I_i p}{y_1} + A_i p y_1 \tag{7.11.6}$$

and when the second term of Eq. 7.11.6 is modified by substituting Eq. 7.11.5 into it; and further, since $S_i \approx I_i/y_1$,

$$S_1 = S_i + A_i\left[\frac{(0.4d)^2 p}{y_1}\right] + py_1 \tag{7.11.7}$$

and finally, saying $y_1 \approx d/2$,

$$S_1 = S_i + A_f\left(\frac{0.16d^2}{d/2} + \frac{d}{2}\right)$$

$$= S_i + A_f d\ (0.32 + 0.5) \tag{7.11.8}$$

Solving for A_f gives

$$A_f = 1.22\left(\frac{S_1 - S_i}{d}\right) \tag{7.11.9}$$

For design purposes with a primary section of I-shape,

$$A_f \approx 1.2\frac{S_1 - S_i}{d} \tag{7.11.10}$$

If the primary section is other than I-shaped, the radius of gyration, r_i, can be estimated and used in Eq. 7.11.7 in place of the $0.4d$ which was used.

Example 7.11.1

Determine the cover plate required on a W12 × 65 beam if a total moment of 160 ft-kips is to be carried. Assume the allowable stress in tension is 22 ksi while the allowable in compression is 16 ksi. (Such a condition could commonly arise because of lateral stability problems as will be discussed in Chapter 9, or it might occur when designing using a hybrid plate girder having flanges of two different steels.)

Solution

For the W12 × 65, $S_i = 88$ in.³ and $d = 12.12$ in.

The required S_1 to the compression face where the flange plate will be is

$$\text{Reqd } S_1 = \frac{M}{F_b} = \frac{160(12)}{16} = 120 \text{ in.}^3$$

$$\text{Reqd } A_f \approx 1.2\,\frac{120 - 88}{12} = 3.2 \text{ sq in.}$$

One could select a plate and check the stresses. Or, preferably before doing so, examine the section modulus required to satisfy the tension-stress requirement.

Note:

$$S_2 = I/(d/2 + \bar{y}) \approx I/(y_1 + \bar{y})$$

thus

$$S_2 = S_1/(1 + 2p) \tag{7.11.11}$$

For tension stress,

$$\text{Reqd } S_2 = \frac{160\,(12)}{22} = 87.2 \text{ in.}^3$$

Based on reqd. $S_2 = 87.2$ in.³ and $p = A_f/A_i = 3.2/19.1 = 0.167$,

$$\text{Reqd } S_1 = 87.2\,[1 + 2\,(0.167)] = 116 \text{ in.}^3$$

Since this is less than 120 in.³ required based on compression stress, select plate for $A_f = 3.2$ sq in. and check stress.

Try ℞ $^5/_{16}$ × 10: $A_f = 3.125$ sq in.

$$\bar{y} = \frac{3.125\,(6.06 + 0.156)}{19.1 + 3.125} = 0.873 \text{ in.}$$

$$I = 533 + 19.1(0.873)^2 + 3.125(6.216 - 0.873)^2$$

$$I = 533 + 14.6 + 88.8 = 636.4 \text{ in.}^4$$

$$f \text{ (compression)} = \frac{160(12)5.50}{636.4} = 16.6 \text{ ksi} > 16 \text{ ksi} \qquad 3.8\% \text{ high}$$

$$f \text{ (tension)} = \frac{160(12)6.93}{636.4} = 20.9 \text{ ksi} < 22 \text{ ksi} \qquad \text{OK}$$

Use ℞ 3/8 × 10 in preference to 5/16 × 10 which is overstressed by more than three percent.

PROBLEMS

All problems are to be done in accordance with the latest AISC Specifications unless otherwise indicated, and the term "beam section" is intended to include W, S, and M standard shapes.

7.1. Select the lightest beam section to carry a uniformly distributed load of 1 kip per ft on a simply supported span of 35 ft. Assume full lateral support of compression flange and that deflection need not be considered. Use AISC specification and working stress method, with the following steels: (a) A36; (b) F_y = 50 ksi; (c) F_y = 60 ksi; (d) F_y = 100 ksi (A514 steel). Assume all standard sections are available in each of these steels (which they are not).

7.2. Repeat Prob. 7.1, increasing the span to 55 ft.

7.3. Repeat Prob. 7.1, using plastic design for all applicable steels. Does plastic design offer any saving over using working-stress method? By using basic principles, compute the shape factor for the A36 beam selected.

7.4. Repeat Prob. 7.1, assuming that 80 percent of the loading is live load and that live-load deflection may not exceed $L/360$.

7.5. Repeat Problem 7.2, assuming that 60 percent of the loading is live load and that live-load deflection may not exceed $L/300$.

7.6. Determine the allowable superimposed concentrated load at midspan for a special beam whose web is 3/8 × 24 and flange plates are 1/2 × 12. Assume the beam has full lateral support of the compression flange. (a) Use the working-stress method and A36 steel. (b) What is the safe P if computed according to AISC plastic design? Can plastic design be used?

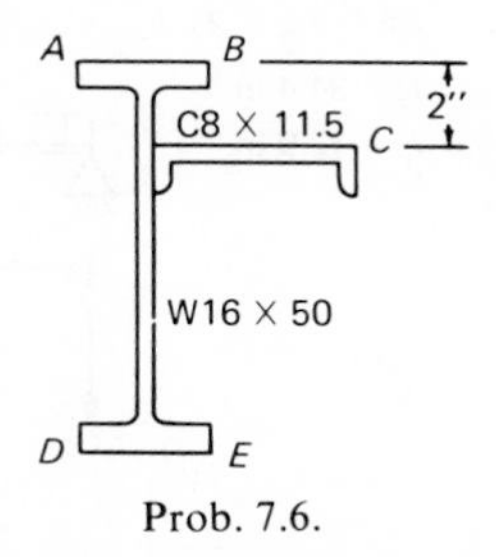

Prob. 7.6.

7.7. Repeat Prob. 7.6 if the flange plates are 3/4 × 12 instead of 1/2 × 12. In addition to parts (a) and (b); use working stress with F_y = 60 ksi.

7.8. Select the lightest beam section to carry 20 kips per ft on a simple span of 7 ft. Compression flange is adequately supported laterally. Use AISC specification with working stress method; and the following steels: (a) A36; (b) A572 Grade 50; (c) $F_y = 60$ ksi; and (d) $F_y = 100$ ksi (A514 steel).

7.9. Repeat Prob. 7.8, using plastic design and only A36 steel.

7.10. A W24 × 94 beam 6 ft long center-to-center of supports underpins a column which brings 280 kips to its top flange 2 ft 6 in. from the left support. The length of the base plate under the column is 12 in. measured along the beam. The length of each of the plates under the ends of the beam is 8 in. There are no end connections. Review this beam for (a) flexure, (b) shear; and (c) web crippling. Specify changes (if any) required if the beam is A36 and AISC working stress method is used.

7.11. A W16 × 78 section of A36 steel is to be used as a simply supported beam on a span of 10 ft with the compression flange laterally supported. Wall bearing is 10 in. (a) what slowly *moving* concentrated load may be safely carried by this beam? (b) If two 1-in. diameter holes are punched in both flanges of the beam at midspan, what would be the safe load for this beam? Use AISC working stress method.

7.12. Determine the size of bearing plate required for an end reaction of 55 kips on a W14 × 53 beam of steel with $F_y = 60$ ksi, according to the AISC specification. The beam rests on a 4″ wall of concrete with a 28-day compressive strength of 3,500 psi.

7.13. Verify the values $V = 90$ kips, $R = 65$ kips, $R_i = 13.4$ kips, and $N_e = 5.3$ in. as given in the AISC Manual Beam Tables for a W12 × 85 of A36 steel.

7.14. For the Zee section shown in the accompanying figure, assume the loading in the plane of the web (yz plane). (a) Determine the safe uniform loading (total) assuming bending occurs in the yz plane and the maximum allowable stress is 30 ksi; (b) Using the loading determined in (a) compute the bending stress at points designated by letters if the beam is free to bend and not restrained to bend in the yz plane. Using the general flexure equation, Eq. 7.8.11, (c) repeat (b) but use the method described in Example 7.8.1b. (d) Locate principal axes and resolve the moment into principal planes, and use $f = M'_x/S'_x + M'_y/S'_y$, (e) State conclusions.

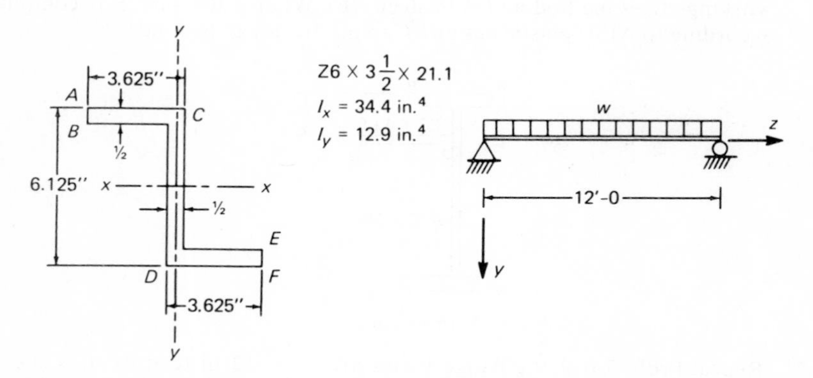

Prob. 7.14.

7.15. Repeat Prob. 7.14, using a 7 × 4 × ½ angle with long leg pointed downward as shown in the accompanying figure. Assume loading in the *yz* plane and neglect any torsional effects.

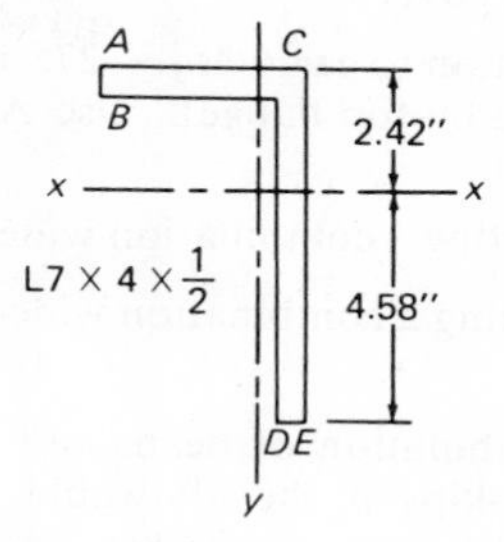

Prob. 7.15.

7.16. Repeat Prob. 7.14, using a combination section, consisting of a W16 × 50 and an C8 × 11.5, as shown in the accompanying figure. Assume loading in the *yz* plane and neglect any torsional effects.

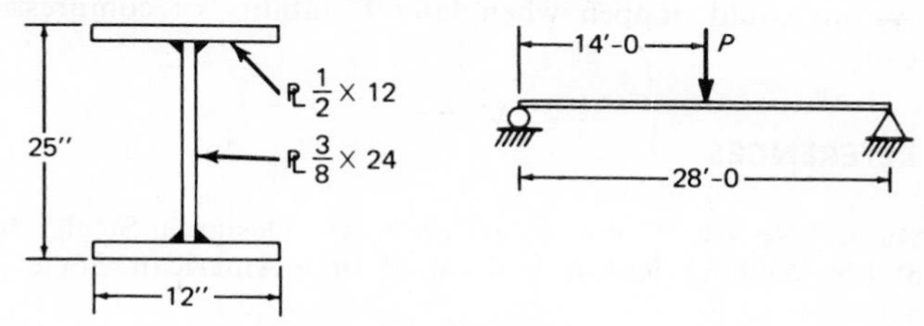

Prob. 7.16.

7.17. Repeat Prob. 7.14, using a combination section, consisting of a C10 × 15.3 and a 4 × 3 × ¼ angle, as shown in the accompanying figure. Assume loading in the *yz* plane and neglect any torsional effects.

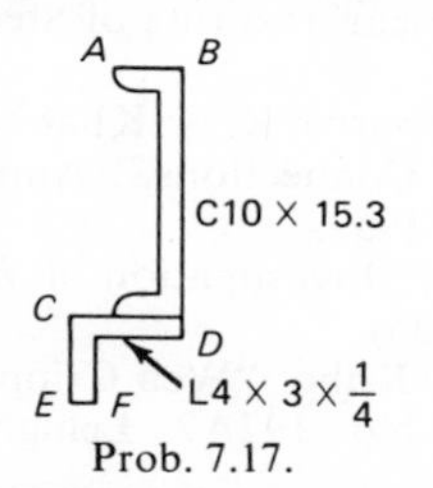

Prob. 7.17.

7.18. Select the most economical W8 section to use as a purlin on a roof sloped 30 degrees to the horizontal. The span is 21 ft, the load is uniform 370 lb/ft (superimposed), and lateral stability is assured by attachment of the roofing to the compression flange. Assume load acts through the beam centroid and that any torsional effect can be resisted by the roofing and therefore neglected in selecting the beam. A36 steel, and working stress method.

7.19. Repeat Prob. 7.18, eliminating size requirement; roof angle is 20 degrees, loading is 500 lb per ft, span is 18 ft, and material has F_y = 50 ksi.

7.20. Select the lightest W section to carry moments, $M_x = 145$ ft-kips and $M_f = 30$ ft-kips (lateral moment resisted by top flange). Assume $M_y \approx 2M_f$ and that torsion is neglected. Use $F_y = 50$ ksi and assume lateral stability does not govern.

7.21. Select the lightest W section to carry $M_x = 275$ ft-kips and $M_f = 100$ ft-kips (lateral moment resisted by top flange). Use A36 steel and assume lateral stability does not govern.

7.22. Repeat Prob. 7.20, selecting a combination wide-flange and channel.

7.23. Repeat Prob. 7.21, selecting a combination wide-flange and channel. AISC Manual.

7.24. Design the lightest combination of beam and one cover plate to carry a moment, $M_x = 500$ ft-kips if the allowable extreme fiber stresses are, $F_{b\,\text{tension}} = 30$ ksi, and $F_{b\,\text{compression}} = 23$ ksi. How much saving in weight is there over using a single W section without plate? Discuss.

7.25. Design a built-up I-shaped welded girder with different size flanges. The moment to be carried is 450 ft-kips and the allowable extreme fiber stresses are $F_{b\,\text{tension}} = 30$ ksi and $F_{b\,\text{compression}} = 15$ ksi. (As will be discussed in Chap. 9, this could happen when lateral stability of compression flange controls.)

SELECTED REFERENCES

1. ASCE Manual No. 41, "Commentary on Plastic Design in Steel," Joint Committee of the Welding Research Council and American Society of Civil Engineers, 1961.
2. "Deflection of Reinforced Concrete Members," No. ST-70, Portland Cement Association, Chicago, 1947.
3. Frank J. Marino, "Ponding of Two-Way Roof Systems," *Engineering Journal*, American Institute of Steel Construction, Vol. 3, No. 3 (July 1966), pp 93–100.
4. James Chinn, "Failure of Simply-Supported Flat Roofs by Ponding of Rain," *Engineering Journal*, American Institute of Steel Construction, Vol. 2, No. 2 (April 1965), pp 38–41.
5. J. D. Graham, A. N. Sherbourne, R. N. Khabbaz, and C. D. Jensen, "Welded Interior Beam-to-Column Connections," American Institute of Steel Construction, Inc., New York, 1959.
6. I. Lyse and H. J. Godfrey, "Investigation of Web Buckling in Steel Beams," *Trans. ASCE*, Vol. 100 (1935).
7. B. G. Johnston and G. G. Kubo, "Web Crippling at Seat Angle Supports," Fritz Laboratory Report No. 192A2, Lehigh University, Bethlehem, Pa., 1941.
8. E. H. Gaylord, Jr., and C. N. Gaylord, *Design of Steel Structures*, McGraw-Hill Book Company, 1957, Chap. 5.
9. J. B. Johnson, C. W. Bryan, and F. E. Turneaure, *Modern Framed Structures*, Vol. 3, 10th ed., John Wiley & Sons, Inc., 1929, pp 458, 479.
10. William McGuire, *Steel Structures*, Prentice-Hall, Inc., Englewood Cliffs, N.J., 1968.

8

Torsion

8.1. INTRODUCTION

In structural design, torsional stress may on occasion be a significant stress for which provision must be made. The most efficient shape for carrying a torque is a hollow circular shaft; extensive treatment of torsion and torsion combined with bending and axial force is to be found in most texts on mechanics of materials.[1]

Frequently torsion is a secondary, though not necessarily a minor effect which must be considered in combination with other types of behavior. The shapes which make good columns and beams, i.e., those which have their material distributed as far from their centroids as practicable, are not equally efficient in resisting torsion. It is found that thin-walled circular and box sections are stronger torsionally than sections with the same area arranged as channel, I, tee, angle, or zee shapes.

When a simple circular solid shaft is twisted, the shearing stress at any point on a transverse cross section varies directly as the distance from the center of the shaft. Thus, during twisting, the cross section which is initially planar remains a plane and rotates only about the axis of the shaft.

In 1853 the French engineer Adhémar Jean Barré de Saint-Venant presented to the French Academy of Sciences the classical torsion theory which forms the basis for present-day analysis.* Saint-Venant showed that when a noncircular bar is twisted, a transverse section that was planar prior to twisting does *not* remain plane after twisting. The original cross-section plane surface becomes a warped surface. In torsion problems it is necessary to recognize the out-of-plane, or warping effect, in addition to the rotation, or pure twisting, effect.

*For a summary of Saint-Venant's work, see Isaac Todhunter and Karl Pearson, *A History of the Theory of Elasticity and of the Strength of Materials*, Vol. II, 1893 (reprinted by Dover Publications, Inc., N.Y., 1960, pp. 17–51).

In this chapter primary emphasis is given to the recognition of torsion on the usual structural members, such as I-shaped, channel, angle, and zee sections; how the torsional stresses may be approximated and how such members may be selected to resist torsional effects.

Also included is a brief treatment of torsional stiffness and the computation of torsional stresses on closed thin-walled sections as well as torsional buckling.

8.2. PURE TORSION OF HOMOGENEOUS SECTIONS

A review of shear stress under pure torsion and of torsional stiffness seems a desirable beginning prior to considering structural shapes in locations where the warping of the cross section is restrained.

Consider a torsional moment T acting on a solid shaft of homogeneous material and uniform cross section, as shown in Fig. 8.2.1. Assume no out-of-plane warping, or at least that out-of-plane warping has negligible effect on the angle of twist ϕ. This assumption will be nearly correct so long as the cross section is small compared to the length of the shaft and also that no significant reentrant corners exist. It is further assumed that no distortion of the cross section occurs during twisting. The rate of twist (twist per unit length) may therefore be expressed as

$$\theta = \text{rate of twist} = \frac{d\phi}{dz} \tag{8.2.1}$$

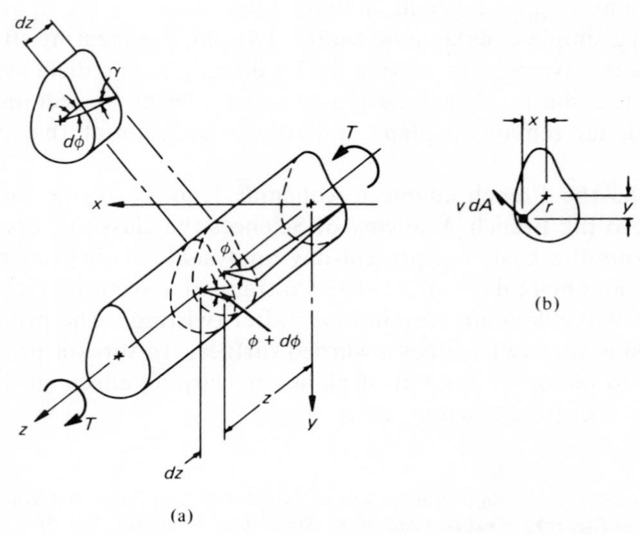

Fig. 8.2.1. Torsion of a prismatic shaft.

which can be thought of as torsional curvature (rate of change of angle). Since it is the relative rotation of the cross sections at z and $z + dz$ that causes strain, the magnitude of displacement at a given point is proportional to the distance r from the center of twist. The strain angle γ, or unit shear strain, at any element r from the center is

$$\gamma \, dz = r \, d\phi$$

$$\gamma = r(d\phi/dz) = r\theta \tag{8.2.2}$$

From Hooke's law, the unit shear stress is

$$v = \gamma G \tag{8.2.3}$$

Thus as shown in Fig. 8.2.1b, the elemental torque is

$$dT = rv \, dA = r\gamma G \, dA = r^2(d\phi/dz)\, G \, dA \tag{8.2.4}$$

The total resisting moment for equilibrium is

$$T = \int_A r^2 \frac{d\phi}{dz} G \, dA$$

and since $d\phi/dz$ and G are constants at any section,

$$T = \frac{d\phi}{dz} G \int_A r^2 \, dA = GK \frac{d\phi}{dz} \tag{8.2.5}$$

where $K = \int_A r^2 \, dA$. Equation 8.2.5 may be thought of as analogous to flexure, i.e., bending moment M equals rigidity EI times curvature, d^2y/dz^2. Here torsional moment T equals torsional rigidity GK times torsional curvature (rate of change of angle).

Shear stress may then be computed using Eqs. 8.2.2 and 8.2.3,

$$v = \gamma G = r \frac{d\phi}{dz} G \tag{8.2.6}$$

and

$$\frac{d\phi}{dz} = \frac{T}{GK}$$

which gives

$$v = \frac{Tr}{K} \tag{8.2.7}$$

Thus as long as the assumptions of this development reasonably apply, torsional shear stress is proportional to the radial distance from the center of twist.

Circular Sections. For the specific case of the circular section of diameter t, no warping of the sections occurs (i.e., no assumption is required) and K = polar moment of inertia = $\pi t^4/32$. Thus, for maximum shear stress at $r = t/2$,

$$v_{max} = \frac{16T}{\pi t^3} \tag{8.2.8}$$

Rectangular Sections. The analysis as applied to rectangles becomes complex since the shear stress is affected by warping, though essentially the angle of twist is unaffected.

As an approximation, consider the element of Fig. 8.2.2 subjected to shear, in which

$$\gamma = t\frac{d\phi}{dz} \tag{8.2.9}$$

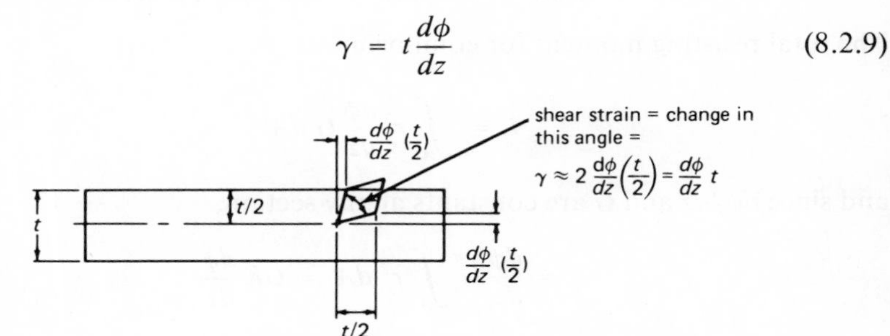

Fig. 8.2.2. Torsion of a rectangular section.

For a thin rectangle, neglecting end effects, the shear stress may be expressed as

$$v = \gamma G = tG\frac{d\phi}{dz} \tag{8.2.10}$$

or using Eq. 8.2.5,

$$v = \frac{Tt}{K} \tag{8.2.11}$$

From the theory of elasticity,[1-3] the maximum shear stress, v_{max}, occurs at the midpoint of the long side of a rectangle and acts parallel to it. The magnitude is a function of the ratio b/t (length/width) and may be expressed as

$$v_{max} = \frac{k_1 T}{bt^2} \tag{8.2.12}$$

and the torsional constant K may be expressed as

$$K = k_2 bt^3 \tag{8.2.13}$$

where the values of k_1 and k_2 may be found in Table 8.2.1.

TABLE 8.2.1
Values of k_1 and k_2 for Eqs. 8.2.12 and 8.2.13

b/t	1.0	1.2	1.5	2.0	2.5	3.0	4.0	5.0	∞
k_1	4.81	4.57	4.33	4.07	3.88	3.75	3.55	3.44	3.00
k_2	0.141	0.166	0.196	0.229	0.249	0.263	0.281	0.291	0.333

I-shaped, Channel, and Tee Sections. As will be observed from a study of Table 8.2.1 the values of k_1 and k_2 become nearly constant for large ratios b/t. Thus, the torsional constants for sections composed of thin rectangles may be computed as the sum of the values for the individual components. Such an approach will give an approximation which neglects the contribution in the fillet region where the components are joined. For most common structural shapes this approximation causes little error, thus:

$$K \approx \Sigma \tfrac{1}{3} bt^3 \tag{8.2.14}$$

where b is the long dimension and t the thin dimension of the rectangular elements.

More accurate expressions for various structural shapes have been developed,[4–7] and many are readily available in useful design aid publications.[8,9]

8.3. SHEAR STRESSES DUE TO BENDING OF THIN-WALLED OPEN CROSS SECTIONS

Before treating the computation of stresses due to torsion of thin-walled open sections restrained from warping, a review of shear stress resulting from general flexure will be developed. Recognition of a torsion situation precedes concern about calculation of resulting stresses. Extensive treatment of thin-walled members of open cross section is given by Timoshenko.[10]

Referring to the general thin-walled section of Fig. 8.3.1, where x and y are centroidal axes, consider equilibrium of the element $t\ ds\ dz$. Force equilibrium in the z direction requires

$$\frac{\partial(\tau t)}{\partial s} ds\, dz + t\frac{\partial \sigma_z}{\partial z} dz\, ds = 0 \tag{8.3.1}$$

or

$$\frac{\partial(\tau t)}{\partial s} = -t\frac{\partial \sigma_z}{\partial z} \tag{8.3.2}$$

(a) Assume moment is applied in the yz plane only, i.e., $M_y = 0$.

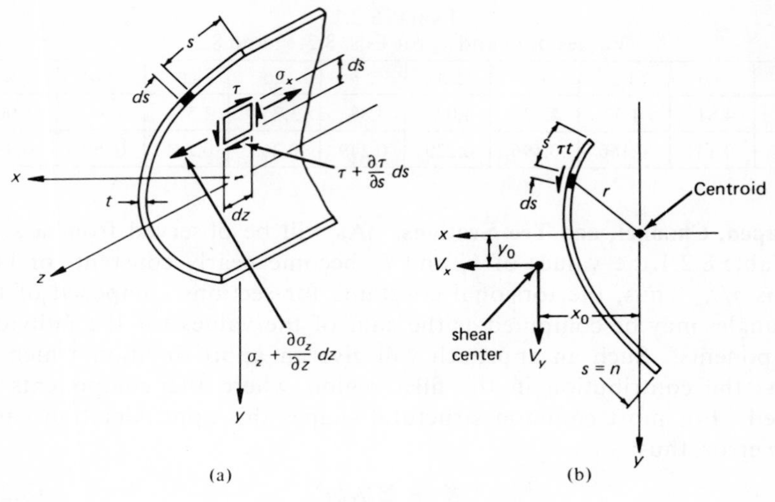

(a) (b)

Fig. 8.3.1.

Using Eq. 7.8.11, the normal stress is

$$\sigma_z = \frac{M_x}{I_x I_y - I_{xy}^2}(I_y y - I_{xy} x)$$

$$\frac{\partial \sigma_z}{\partial z} = \frac{\partial M_x/\partial z}{I_x I_y - I_{xy}^2}(I_y y - I_{xy} x) \tag{8.3.3}$$

Recognizing that $V_y = \partial M_x/\partial z$, and substituting Eq. 8.3.3 into Eq. 8.3.2 gives

$$\frac{\partial \tau t}{\partial s} = \frac{-tV_y}{I_x I_y - I_{xy}^2}(I_y y - I_{xy} x) \tag{8.3.4}$$

Integrating to find τt at a distance s from a free edge, gives the shear flow τt as

$$\tau t = \frac{-V_y}{I_x I_y - I_{xy}^2}\left(I_y \int_0^s yt\, ds - I_{xy} \int_0^s xt\, ds\right) \tag{8.3.5}$$

(b) Assume moment is applied in the xz plane only, i.e., $M_x = 0$. The normal stress from Eq. 7.8.11 becomes

$$\sigma_z = \frac{M_y}{I_x I_y - I_{xy}^2}(-I_{xy} y + I_x x) \tag{8.3.6}$$

taking $\partial\sigma_z | \partial z$, recognizing that $V_x = \partial M_y/\partial z$, and integrating to get τt, gives in a manner similar to Eq. 8.3.5,

$$\tau t = \frac{+V_x}{I_x I_y - I_{xy}^2}\left(I_{xy}\int_0^s yt\,ds - I_x\int_0^s xt\,ds\right) \tag{8.3.7}$$

(c) Moments applied in both yz and xz planes. If shear stresses are desired they can be computed by superimposing the results from Eqs. 8.3.5 and 8.3.7.

It is to be observed from Fig. 8.3.1b that equilibrium requires that the shear in the y direction, V_y, equals the components of τt in the y direction summed over the entire section. Similarly V_x equals the summation of τt components in the x direction. Rotational equilibrium must also be satisfied; the moment about the centroid of the section is (see Fig. 8.3.1b):

$$\int_0^n \tau t r\,ds$$

which will be zero in some cases (such as I-shaped and Z-shaped sections). If such rotational equilibrium is automatically satisfied when the flexural shears act through the centroid then no torsion will occur simultaneously with bending.

8.4. SHEAR CENTER

The shear center is the location in a cross section where no torsion occurs when flexural shears act in planes passing through that location. In other words, loads applied through the shear center will cause no torsional stresses to develop, i.e.,

$$\int_0^n \tau t r\,ds = 0 \tag{8.4.1}$$

Since the shear center does not necessarily coincide with the centroid of the section, it is necessary to be able to locate the shear center in order to evaluate the torsional stress. For I-shaped and Z-shaped sections, the shear center coincides with the centroid, but for such sections as channels and angles it does not.

Referring to Fig. 8.3.1b, consider the shears V_x and V_y acting at distances from the centroid y_0 and x_0 respectively, such that the torsional moment with respect to the centroid is the same as $\int_0^n \tau t r\,ds$; thus,

$$V_y x_0 - V_x y_0 = \int_0^n \tau t r\,ds \tag{8.4.2}$$

In other words, the torsional moment is $(V_y x_0 - V_x y_0)$ when the loads are applied in planes passing through the centroid but is zero if the loads are in planes passing through the shear center, i.e., the point whose coordinates are $x_0 y_0$.

It is observed that the location of shear center is independent of the magnitude or type of loading, but is only dependent upon the cross-sectional configuration.

To determine the shear-center location, first let one of the shears be zero, say $V_y = 0$; then from Eq. 8.4.2

$$y_0 = -\frac{1}{V_x}\int_0^n \tau tr\,ds \tag{8.4.3}$$

where

$$\tau t = \frac{V_x}{I_x I_y - I_{xy}^2}\left[I_{xy}\int_0^s yt\,ds - I_x\int_0^s xt\,ds\right]$$

Alternately, letting $V_x = 0$ gives from Eq. 8.4.2,

$$x_0 = \frac{1}{V_y}\int_0^n \tau tr\,ds \tag{8.4.4}$$

where

$$\tau t = \frac{-V_y}{I_x I_y - I_{xy}^2}\left[I_y\int_0^s yt\,ds - I_{xy}\int_0^s xt\,ds\right]$$

EXAMPLE 8.4.1

Locate the shear center for the channel section of Fig. 8.4.1.

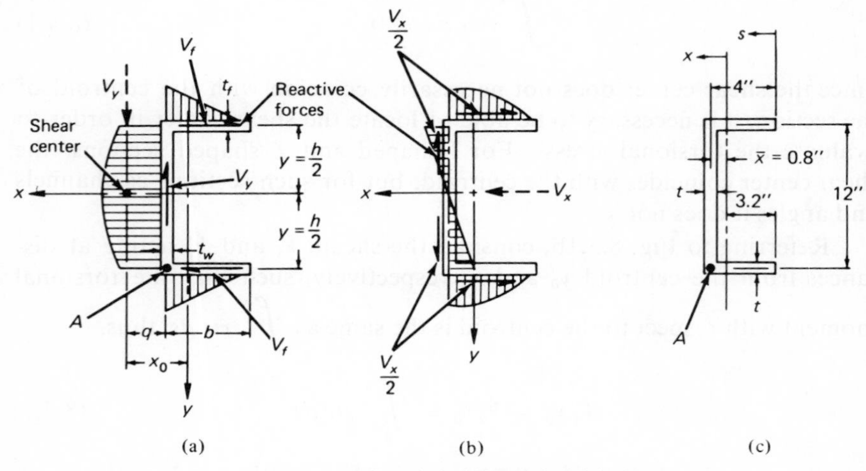

Fig. 8.4.1. Channel of Example 8.4.1.

Solution

It is to be noted that many practical cases can be solved without resort to the general formulas, Eqs. 8.4.3 and 8.4.4. Since the shear center location is a problem in equilibrium, moments may most conveniently be taken through a point which eliminates the greatest number of forces. Thus, letting $V_x = 0$ and taking moments about point A of Fig. 8.4.1a, changes the equilibrium equation, Eq. 8.4.2, to

$$V_y q = V_f h = \int_0^b \tau t h \, ds \tag{a}$$

where

$$\tau t = \frac{-V_y}{I_x} \int_0^s yt \, ds = \frac{-V_y}{I_x} yts \tag{b}$$

Substituting Eq. (b) in Eq. (a), and using $y = -h/2$, give

$$V_y q = \int_0^b \frac{-V_y}{I_x}\left(\frac{-h}{2}\right) ths \, ds \tag{c}$$

$$= \frac{V_y th^2}{2I_x} \int_0^b s \, ds = \frac{V_y th^2 b^2}{4I_x}$$

Thus

$$q = \frac{th^2 b^2}{4I_x} \tag{d}$$

measured in the positive x direction to the left of the channel web.

For the shear-center coordinate measured along the y axis, apply V_x and let $V_y = 0$, and because of symmetry V_x must act at $y = 0$ for equilibrium. To demonstrate, let V_x be applied at the distance y_0 below the x axis and take moments about point A. Satisfying equilibrium,

$$V_x\left(\frac{y}{2} - y_0\right) = \int_0^b \tau t y \, ds \tag{e}$$

where

$$\tau t = \frac{-V_x}{I_y} \int_0^s xt \, ds \tag{f}$$

To illustrate numerically, use $b = 4$ in. and $h = 12$ in., in which case the centroid of the channel (refer to Fig. 8.4.1c) is located

$$\overline{x} = \frac{\Sigma Ax}{\Sigma A} = \frac{b^2}{h + 2b} = \frac{16}{20} = 0.8 \text{ in.}$$

Thus, $s = x + 3.2$ in.

$$I_y = [2(4)^3(\tfrac{1}{3}) - 20(0.8)^2]t = 29.87t$$

$$\tau t = \frac{-V_x t}{29.87t}\int_0^s (s - 3.2)\,ds = \frac{-V_x}{29.87}\left(\frac{s^2}{2} - 3.2s\right)$$

Substitution of τt into Eq. (e) gives

$$V_x\left(\frac{h}{2} - y_0\right) = \int_0^4 \frac{-V_x}{29.87}\left(\frac{s^2}{2} - 3.2s\right)h\,ds$$

$$= \frac{-V_x h}{29.87}\left(\frac{s^3}{6} - 3.2\frac{s^2}{2}\right)\Bigg|_0^4 = \frac{+V_x h}{2}$$

Thus it is proved that the flange shear force equals $V_x/2$ and y_0 is zero, as previously stated.

In computing the shear center location for any unsymmetrical shape, a check should be made by computing the shear force in each element and verifying that statics is satisfied.

8.5. TORSIONAL STRESSES IN I-SHAPED STEEL SECTIONS

While it is certainly true that only occasionally will torsion be severe enough to control the design of a sections, the structural engineer should still recognize a torsion situation and be able to apply approximate design methods and perform a stress analysis when necessary. Rolled-steel sections under uniform and nonuniform torsion have been studied analytically and experimentally by many investigators. The development in this section is similar to that of Timoshenko,[10] and others.[4,6,11] Discussion of some of the practical aspects, along with summary of solutions for various loading and support cases, is given by Hotchkiss[9]; and charts are given in the handbook by Heins and Seaburg.[8]

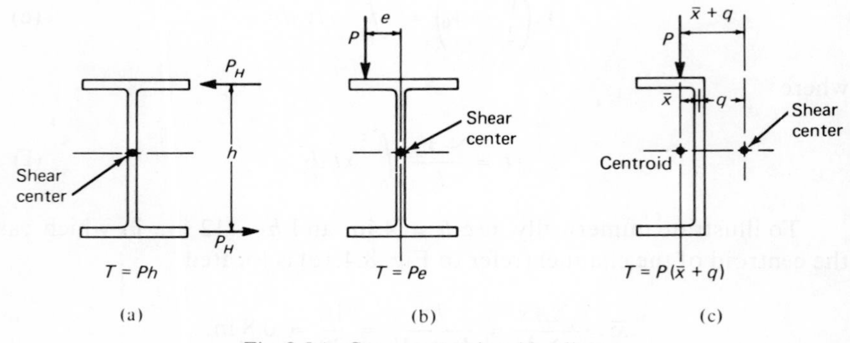

Fig. 8.5.1. Common torsional loadings.

As seen from the preceding article, when any section is loaded in such a manner that the plane of loading does not pass through the shear center, torsion will occur (see Fig. 8.5.1). Shear stresses, or shear stresses plus flexural stresses arising from torsion, must be superimposed on the ordinary shear and normal stresses due to flexure alone.

Torsional stresses are of two types: Pure torsion, or as it is often called, *Saint-Venant's torsion*, and warping torsion. Pure torsion assumes that a cross-sectional plane prior to application of torsion remains a plane and only element rotation occurs during torsion. A circular shaft subjected to torsion is a situation where pure torsion exists as the only type. Warping torsion is the out-of-plane effect which arises when the flanges are laterally displaced during twisting, analogous to bending from laterally applied loads.

(a) *Pure torsion* (*Saint-Venant's torsion*). Just as flexural curvature (change in slope per unit length) can be expressed as $M/EI = d^2y/dz^2$; i.e., moment divided by flexural rigidity equals flexural curvature; in pure torsion the torsional moment divided by the torsional rigidity GK equals the torsional curvature (change in angle of twist per unit length). Recalling previously derived Equation 8.2.5 for T, which now becomes the component, M_s, due to pure torsion,

$$M_s = GK\frac{d\phi}{dz} \tag{8.5.1}$$

where M_s = pure torsional moment (Saint-Venant torsion);
G = shear modulus of elasticity $= E/[2(1 + \mu)]$, in terms of the tension-compression modulus of elasticity, E, and Poisson's ratio, μ
K = torsional constant (see Sec. 8.2)

In accordance with Eq. 8.2.7, stress due to M_s is proportional to the distance from the center of twist.

(b) *Warping torsion.* Consider a beam as in Fig. 8.5.2 subjected to torsion M_z, the compression flange is bent in one direction laterally while the tension flange is bent in the other. Whenever the cross section is such that it would warp (become a nonplanar section) if not restrained, the restrained system has stresses induced. The torsional situation of Fig. 8.5.2 illustrates where a beam is prevented from twisting at each end but the top flange deflects laterally by an amount u_f. This may be called lateral flange bending. This bending causes flexural normal stresses across the flange width, as well as shear stresses.

Thus torsion may be thought of as being composed of two parts; (1) rotation of elements, the pure torsion part, and (2) translation producing lateral bending, the warping part.

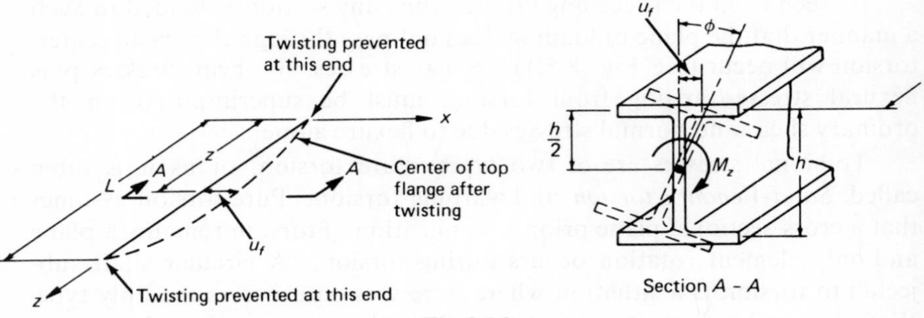

Fig. 8.5.2.

(c) *Differential equation for torsion on* I- *and channel-shaped sections.* Consider the deflected position of a flange centerline, as in Fig. 8.5.2, where u_f is the lateral deflection of one of the flanges at a section a distance z from the end of the member; ϕ is the twist angle at the same section, and V_f (Fig. 8.5.3) is the horizontal shear force developed in the flange at that

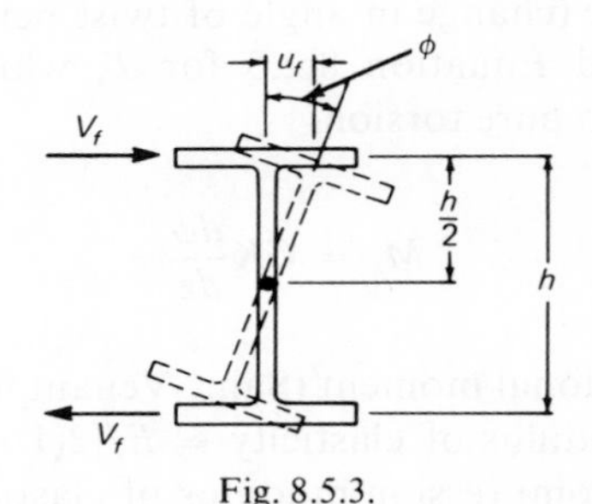

Fig. 8.5.3.

section due to lateral bending. It is to be noted that an important assumption is that the web remains a plane during rotation, so that the flanges deflect laterally an equal amount and remain at 90 degrees to the axis or rotation. Thus it is assumed the web is thick enough compared to the flanges so that it does not bend during twisting as a result of high torsional resistance of the flanges. Except for thin-web plate girders, it has been shown[6,12] that assuming no lateral bending in the web, i.e., no effect on the warping torsion component, is sufficiently correct for practical purposes. Since rarely are thin-web plate girders used without stiffeners, and certainly not when torsional stress exists, such cases are not of practical importance.

From geometry

$$u_f = \phi \frac{h}{2} \tag{8.5.2}$$

for small values of ϕ. For understanding of torsion, Eq. 8.5.2 is the single most important relationship. The twist angle is directly proportional to the lateral deflection. Torsion boundary conditions are analogous to lateral bending boundary conditions.

Differentiating three times with respect to z in Eq. 8.5.2 gives

$$\frac{d^3u_f}{dz^3} = \left(\frac{h}{2}\right)\frac{d^3\phi}{dz^3} \tag{8.5.3}$$

For one flange the curvature relationship is

$$\frac{d^2u_f}{dz^2} = \frac{-M_f}{EI_f} \tag{8.5.4}$$

where M_f is the lateral bending-moment on one flange, I_f is the moment of inertia about the y-axis of the beam for one flange, and the minus sign arises from positive bending as shown in Fig. 8.5.2. Also, since $V = dM/dz$,

$$\frac{d^3u_f}{dz^3} = \frac{-V_f}{EI_f} \tag{8.5.6}$$

Using Eqs. 8.5.3 and 8.5.6 gives

$$V_f = -EI_f\left(\frac{h}{2}\right)\frac{d^3\phi}{dz^3} \tag{8.5.7}$$

Referring to Fig. 8.5.3, the torsional moment component M_w, causing lateral bending of the flanges, equals the flange shear force times the moment arm h. This assumes no shear resistance to warping is contributed by the web,

$$M_w = V_f h = -EI_f\frac{h^2}{2}\frac{d^3\phi}{dz^3} \tag{8.5.8}$$

$$= -EI_w\frac{d^3\phi}{dz^3} \tag{8.5.9}$$

where $I_w = I_f h^2/2$, often referred to as the *warping torsional constant.*

The total torsional moment is composed of the sum of the rotational part, M_s, and the lateral bending part, M_w, which from Eqs. 8.5.1 and 8.5.9 gives

$$M_z = M_s + M_w = GK\frac{d\phi}{dz} - EI_w\frac{d^3\phi}{dz^3} \tag{8.5.10}$$

the differential equation for torsion. The torsional moment M_z, depends upon the loading and in usual situations will be a polynomial in z. Values of the torsion constant K and warping constant I_w are to be found in Table 8.11.1.*

Rewrite Eq. 8.5.10, dividing by EI_w,

$$\frac{d^3\phi}{dz^3} - \frac{GK}{EI_w}\frac{d\phi}{dz} = \frac{-M_z}{EI_w} \tag{8.5.11}$$

Letting $\lambda^2 = GK/EI_w$ ($\lambda = 1/a$ of *Torsion Handbook*[8]), and for the homogeneous solution of Eq. 8.5.11 let $\phi_h = Ae^{mz}$,

$$\frac{d^3\phi}{dz^3} - \lambda^2\frac{d\phi}{dz} = 0 \tag{8.5.12}$$

which upon substitution of the homogeneous solution gives

$$Ae^{mz}(m^3 - \lambda^2 m) = 0 \tag{8.5.13}$$

which requires

$$m(m^2 - \lambda^2) = 0; \qquad \therefore m = 0, m = \pm\lambda$$

Thus

$$\phi_h = A_1 e^{\lambda z} + A_2 e^{-\lambda z} + A_3 \tag{8.5.14}$$

which upon using the hyperbolic function identities and regrouping the constants may be expressed as

$$\phi_h = A \sinh \lambda z + B \cosh \lambda z + C \tag{8.5.15}$$

where

$$\lambda = \frac{1}{a} = \sqrt{\frac{GK}{EI_w}}$$

For the particular solution, since M_z is in general some function of z,

$$M_z = f(z)$$

Let $\phi_p = f_1(z)$, and substitute into Eq. 8.5.11, giving

$$\frac{d^3 f_1(z)}{dz^3} - \lambda^2 \frac{df_1(z)}{dz} = -\frac{1}{EI_w} f(z) \tag{8.5.16}$$

where terms on the left-hand side must be paired with terms on the right side. Rarely will $f_1(z)$ be required to contain higher than second degree terms.

*K and I_w are J and C_w, respectively, in the 1970 AISC Manual.

EXAMPLE 8.5.1

Develop the expressions for the twist angle ϕ, as well as the first, second, and third derivatives, for the case of concentrated torsional moment applied at midspan when the ends are torsionally simply supported, using the differential equation.

Solution

Referring to Fig. 8.5.4, it is apparent that M_z is contant and equal to $T/2$.

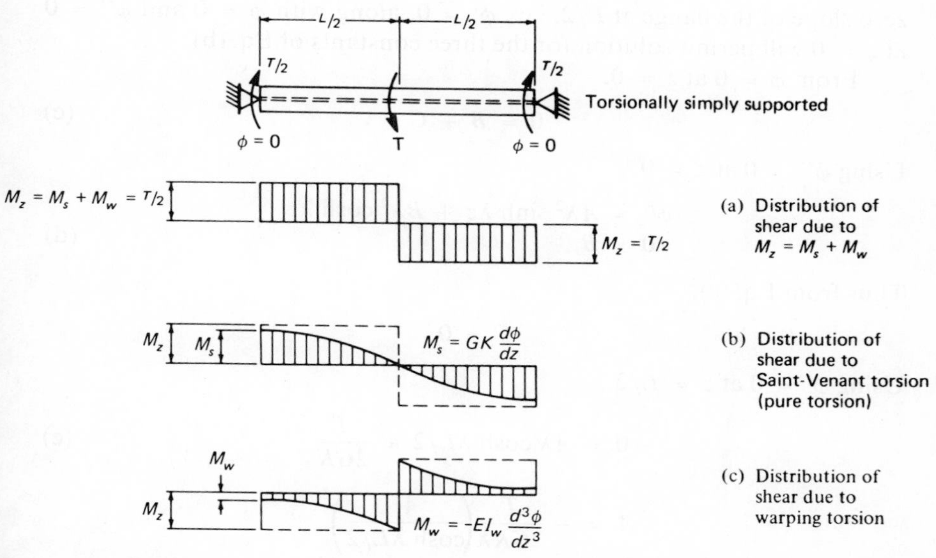

Fig. 8.5.4. Case of Example 8.5.1. Concentrated torsional moment at midspan; torsionally simply supported. (Adapted from Ref. 9.)

Thus, let

$$\phi_p = C_1 + C_2 z \text{ (any polynomial)} \tag{a}$$

Using Eq. 8.5.11 gives

$$-\lambda^2 C_2 = -\frac{1}{EI_w}\left(\frac{T}{2}\right); \qquad \therefore C_2 = \frac{T}{2GK}$$

The other constant C_1 may be combined with C of Eq. 8.5.15. The complete solution for this loading is therefore

$$\phi = A \sinh \lambda z + B \cosh \lambda z + C + \frac{T}{2GK} z \tag{b}$$

Consider the boundary conditions for torsional simple support. Thinking of the lateral bending of the flange (since ϕ is proportional to u_f), simple support conditions mean zero moment and deflection at each end, or for torsion

$$\phi = 0 \quad \text{at} \quad z = 0 \text{ and } z = L$$

$$\frac{d^2\phi}{dz^2} = \phi'' = 0 \quad \text{at} \quad z = 0 \text{ and } z = L$$

In this case the differential equation is discontinuous at $L/2$; thus, using zero slope of the flange at $L/2$, i.e., $\phi' = 0$, along with $\phi = 0$ and $\phi'' = 0$ at $z = 0$ will permit solution for the three constants of Eq. (b).

From $\phi = 0$ at $z = 0$,

$$0 = B + C \tag{c}$$

Using $\phi'' = 0$ at $z = 0$,

$$\phi'' = A\lambda^2 \sinh \lambda z + B\lambda^2 \cosh \lambda z$$
$$0 = B \tag{d}$$

Thus from Eq. (c),

$$C = 0$$

Using $\phi' = 0$ at $z = L/2$,

$$0 = A\lambda \cosh \lambda L/2 + \frac{T}{2GK} \tag{e}$$

$$A = -\frac{T}{2GK\lambda}\left(\frac{1}{\cosh \lambda L/2}\right)$$

Finally, Eq. (b) becomes

$$\phi = \frac{T}{2GK\lambda}\left[\lambda z - \frac{\sinh \lambda z}{\cosh \lambda L/2}\right] \tag{f}$$

also

$$\phi' = \frac{T}{2GK}\left[1 - \frac{\cosh \lambda z}{\cosh \lambda L/2}\right] \tag{g}$$

$$\phi'' = \frac{T\lambda}{2GK}\left[\frac{-\sinh \lambda z}{\cosh \lambda L/2}\right] \tag{h}$$

$$\phi''' = \frac{T\lambda^2}{2GK}\left[\frac{-\cosh \lambda z}{\cosh \lambda L/2}\right] \tag{i}$$

Thus the solution of the differential equation is illustrated. The stress equations making use of the derivatives are developed in the next section.

(d) *Torsional stresses.* The shear stress v_s resulting from the Saint-Venant torsion, M_s, is computed in accordance with Eq. 8.2.11,

$$v_s = \frac{M_s t}{K} \tag{8.2.11}$$

and using Eq. 8.5.1 gives

$$v_s = Gt\, d\phi/dz \tag{8.5.17}$$

whose distribution is shown in Fig. 8.5.5a.

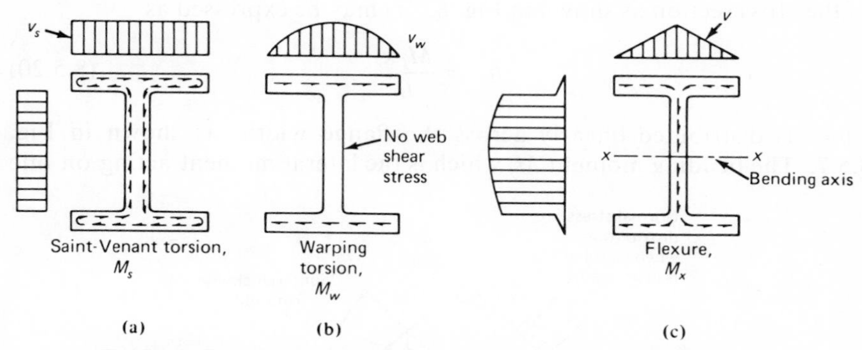

Fig. 8.5.5. Direction and distribution of shear stress in I-shaped sections.

The shear stress v_w which results from warping is distributed parabolically across the flange width as shown in Fig. 8.5.5b and may be computed

$$v_w = \frac{V_f Q_f}{I_f t_f} \tag{8.5.18}$$

where Q_f = statical moment of area about the y axis.

The negligible shear carried by the web is not considered. For maximum shear stress v_w, which actually acts at the face of web but may be approximated as acting at the mid-width of the flange, take Q_f (see Fig. 8.5.6) as

$$Q_f = \frac{bt_f}{2}\left(\frac{b}{4}\right)$$

Substituting Q_f and V_f from Eq. 8.5.7 into Eq. 8.5.18 gives

$$v_w = E\frac{b^2 h}{16}\frac{d^3\phi}{dz^3} \tag{8.5.19}$$

taking the absolute value.

The normal stress due to lateral bending of the flanges (i.e., warping

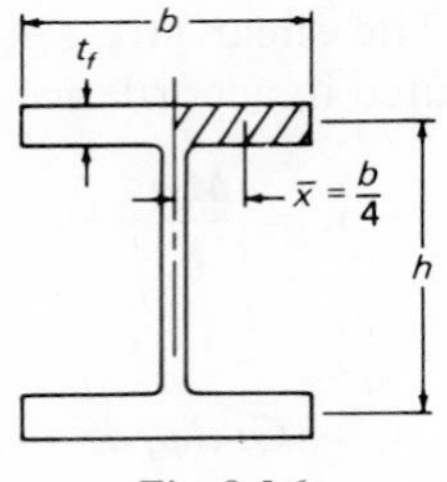

Fig. 8.5.6.

of the cross section as shown in Fig. 8.5.7) may be expressed as

$$f_{bw} = \frac{M_f x}{I_f} \tag{8.5.20}$$

which is distributed linearly across the flange width, as shown in Fig. 8.5.7. The bending moment M_f which is the lateral moment acting on one

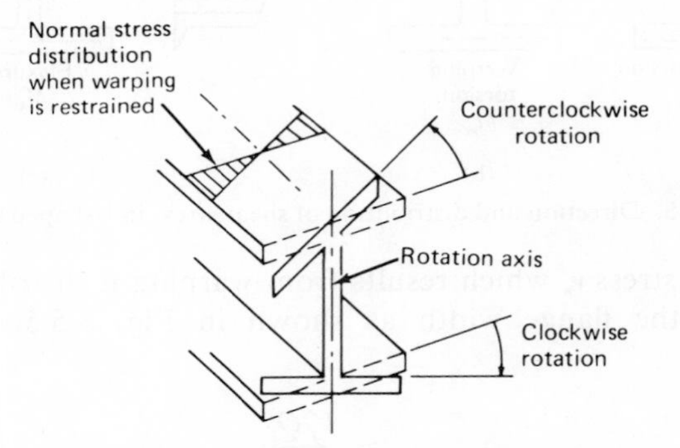

Fig. 8.5.7. Warping of cross section.

flange may be expressed, using Eqs. 8.5.2 and 8.5.4, as

$$M_f = \frac{EI_w}{h}\frac{d^2\phi}{dz^2} \tag{8.5.21}$$

taking the absolute value, since tension occurs on one side while compression occurs on the other.

The maximum stress occurs at $x = b/2$, which when used with Eq. 8.5.21 gives for Eq. 8.5.20,

$$f_{bw} = \frac{EI_w}{h}\left(\frac{b}{2}\right)\frac{1}{I_f}\frac{d^2\phi}{dz^2}$$

$$f_{bw} = \frac{Ebh}{4}\frac{d^2\phi}{dz^2} \tag{8.5.22}$$

In summary, there are three kinds of stresses arising in any I-shaped or channel section arising from torsional loading; (a) shear stresses v_s in web and flanges due to rotation of the elements of the cross section (Saint-Venant torsional moment, M_s); (b) shear stresses v_w in the flanges due to lateral bending (warping torsional moment, M_w); and (c) normal stresses f_{bw} due to lateral bending of the flanges (lateral bending moment on flange, M_f).

EXAMPLE 8.5.2

A W18 × 70 beam on a 24-ft simply supported span is loaded with a concentrated load of 20 kips at midspan. The ends of the member are simply supported with respect to torsional restraint (i.e., $\phi = 0$) and the

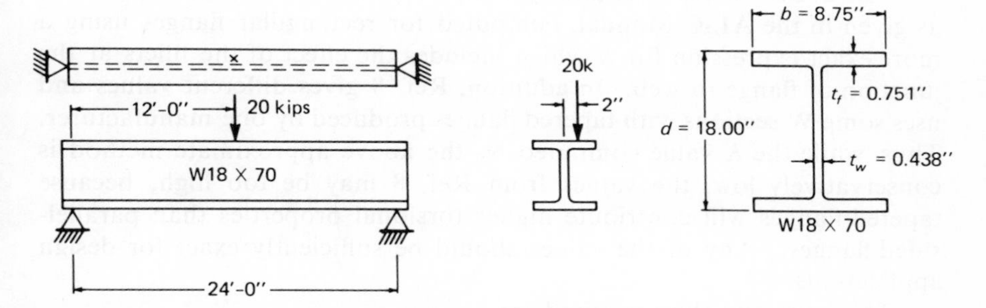

Fig. 8.5.8. Problem of Example 8.5.2.

concentrated load acts with a 2-in. eccentricity from the plane of the web (see Fig. 8.5.8). Compute combined bending and torsional stresses.

Solution

The differential equation solution for this type of loading and end restraint was obtained in Example 8.5.1. The solution as obtained is

$$\phi = \frac{T}{2GK\lambda}\left[\lambda z - \frac{\sinh \lambda z}{\cosh \lambda L/2}\right]$$

In accordance with the derivation (see Fig. 8.5.4), T is the applied torsional moment,

$$T = 20\,(2) = 40 \text{ in.-kips}$$

Recalling from Eq. 8.5.11,

$$\lambda = \sqrt{\frac{GK}{EI_w}} = \sqrt{\frac{2.93}{2.6(6230)}} = \frac{1}{74.3} = 0.01346$$

where

$$\frac{E}{G} = \frac{2E(1 + \mu)}{E} = 2.6 \text{ for } \mu = 0.3$$

$$K \approx \Sigma \frac{bt^3}{3}, \text{ Eq (8.2.14)}$$

$$= \frac{1}{3}[2(8.75)(0.751)^3 + (18 - 1.502)(0.438)^3]$$

$$= \frac{1}{3}(7.41 + 1.39) = 2.93 \text{ in.}^4$$

$$I_w = \frac{I_f h^2}{2} = \frac{1}{12}(8.75)^3(0.751)(18.0 - 0.751)^2/2$$

$$= 6{,}230 \text{ in.}^6$$

The above values of K and I_w compare with $K = 3.13$ and $I_w = 6250$ as given in the AISC Manual, computed for rectangular flanges using a more exact expression for K which includes the effect of the fillets at the junction of flange to web. In addition, Ref. 8 gives different values and uses some W sections with tapered flanges produced by one manufacturer. Thus, while the K value computed by the above approximate method is conservatively low, the values from Ref. 8 may be too high, because tapered flanges will contribute higher torsional properties than parallel-sided flanges. Any of the values should be sufficiently exact for design applications.

The function values required are

$$\lambda L = 23(12)/74.3 = 3.88$$

z	λz	$\sinh \lambda z$	$\cosh \lambda z$
0.1L	0.388	0.398	1.076
0.2L	0.776	0.856	1.316
0.3L	1.164	1.445	1.757
0.4L	1.552	2.255	2.466
0.5L	1.940	3.408	3.551

(a) Pure torsion (Saint-Venant torsion); using Eq. (8.5.17),

$$v_s = Gt\, d\phi/dz$$

$$\frac{d\phi}{dz} = \frac{T}{2GK}\left(1 - \frac{\cosh \lambda z}{\cosh \lambda L/2}\right)$$

$$v_s = \frac{Tt}{2K}\left(1 - \frac{\cosh \lambda z}{3.551}\right) = \frac{40t}{2(2.93)}\left(1 - \frac{\cosh \lambda z}{3.551}\right)$$

The shear stress v_s is a maximum at $z = 0$ and zero at $z = L/2$:

$$v_s \text{ (flange at } z = 0) = \frac{40(0.751)}{2(2.93)}\left(1 - \frac{1}{3.551}\right) = 3.68 \text{ ksi}$$

$$v_s \text{ (web at } z = 0) = 3.68\frac{0.438}{0.751} = 2.15 \text{ ksi}$$

(b) Lateral bending of flanges (warping torsion). Use Eq. 8.5.19 for shear stress in flanges,

$$v_w = E\frac{b^2h}{16}\frac{d^3\phi}{dz^3}$$

$$\frac{d^3\phi}{dz^3} = \frac{T\lambda^2}{2GK}\left(\frac{-\cosh \lambda z}{\cosh \lambda L/2}\right)$$

$$v_w = \frac{T}{2I_w}\frac{b^2h}{16}\left(\frac{-\cosh \lambda z}{\cosh \lambda L/2}\right)$$

This shear stress acts at mid-width of the flange, and the maximum value occurs at $z = L/2$ while the minimum value is at $z = 0$,

$$v_w \text{ (flange at } z = L/2) = \frac{40}{2(6230)}\left(\frac{(8.75)^2 17.249}{16}\right) = 0.27 \text{ ksi}$$

$$v_w \text{ (flange at } z = 0) = 0.27\,\frac{1.0}{3.551} = 0.08 \text{ ksi}$$

For normal stress in flanges due to warping, use Eq. 8.5.22:

$$f_{bw} = \frac{Ebh}{4}\frac{d^2\phi}{dz^2}$$

$$\frac{d^2\phi}{dz^2} = \frac{M\lambda}{2GK}\left[\frac{-\sinh \lambda z}{\cosh \lambda L/2}\right]$$

$$f_{bw} = \frac{M(2.6)\lambda\, bh}{8\,K}\left[\frac{\sinh \lambda z}{\cosh \lambda L/2}\right]$$

which is a maximum at $z = L/2$ and zero at $z = 0$. Thus

$$f_{bw} \text{ (flanges at } z = L/2) = \frac{40(2.6)(8.75)(17.249)}{8(2.93)(74.3)}\left[\frac{3.408}{3.551}\right] = 8.65 \text{ ksi}$$

(c) Ordinary flexure. Maximum normal stress is

$$f_b \text{ (at } z = L/2) = \frac{PL}{4S_x} = \frac{20(24)(12)}{4(128.2)} = 11.22 \text{ ksi}$$

The shear stresses due to flexure are constant from $z = 0$ to $L/2$ and are computed by

$$v = \frac{VQ}{It} = \frac{10Q}{1160t} = \frac{Q}{116t}$$

For maximum flange shear stress,

$$Q = (4.375 - 0.219)(0.751)\left(\frac{17.249}{2}\right) = 26.91 \text{ in.}^3$$

$$v\,(\text{flange at } z = 0) = \frac{26.91}{116(0.751)} = 0.31 \text{ ksi}$$

For maximum web shear stress,

$$Q = 8.75(0.751)\left(\frac{17.249}{2}\right) + \frac{16.498}{2}(0.438)\left(\frac{16.498}{4}\right) = 71.45 \text{ in.}^3$$

$$v\,(\text{web at } z = 0) = \frac{71.45}{116(0.438)} = 1.41 \text{ ksi}$$

A summary of stresses showing combinations is given in Table 8.5.1.

TABLE 8.5.1
Summary of Stresses for Example 8.5.2

Type of Stress	Support ($z = 0$)	Midspan ($z = L/2$)
Normal Stress:		
Vertical bending, f_b	0	11.22
Torsional bending, f_{bw}	0	8.65
→		19.87 ksi
Shear stress, web:		
Saint-Venant torsion, v_s	2.15	0
Vertical bending, v	1.41	1.41
→	3.56 ksi	
Shear stress, flange:		
Saint-Venant torsion, v_s	3.68	0
Warping torsion, v_w	0.08	0.27
Vertical bending, v	0.31	0.31
→	4.07 ksi	

8.6. ANALOGY BETWEEN TORSION AND PLANE BENDING

Because the differential equation solution is time consuming, and really suited only for analysis, design of a beam to include torsion is most conveniently done by making the analogy between torsion and ordinary bending.

Consider that the applied torsional moment T of Fig. 8.6.1 can be converted into a couple P_H times h. The force P_H can then be treated as a lateral load acting on the flange of a beam.

The substitute system will have constant shear over one-half the span, a diagram as given in Fig. 8.5.4a. The true distribution of lateral

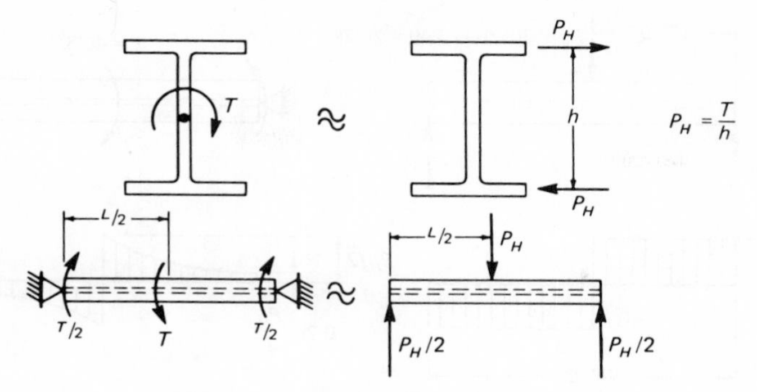

Fig. 8.6.1. Analogy between flexure and torsion.

shear which contributes to lateral deflection is only that part due to warping as shown in Fig. 8.5.4c. Thus the substitute system overestimates the lateral shear force and consequently overestimates the lateral bending moment M_f which causes normal stresses.

In most practical design situations when it is desirable to include the effect of torsion, the normal compressive stress due to the warping component is the quantity of most importance. The shear stress contributions are normally not of significance.

EXAMPLE 8.6.1

Compute the stresses on the W18 × 70 beam of Example 8.5.2 and Fig. 8.5.8 using the flexural analogy rather than the differential equation solution.

Solution

The substitute system is as shown in Fig. 8.6.2a. The lateral bending moment is then

$$M_f = V_f(L/2) = 1.16(12) = 13.9 \text{ ft-kips}$$

acting on one flange. Twice the moment acting on the entire section gives

$$f_{bw} = \frac{2M_f}{S_y} = \frac{2(13.9)(12)}{19.2} = 17.4 \text{ ksi}$$

For torsional shear stress, since $M_z = T/2 = 20$ in-kips,

$$v_s = \frac{M_z t}{K} = \frac{20(0.751)}{2.93} = 5.13 \text{ ksi (flange)}$$

$$v_s = 5.13\left(\frac{0.438}{0.751}\right) = 2.98 \text{ ksi (web)}$$

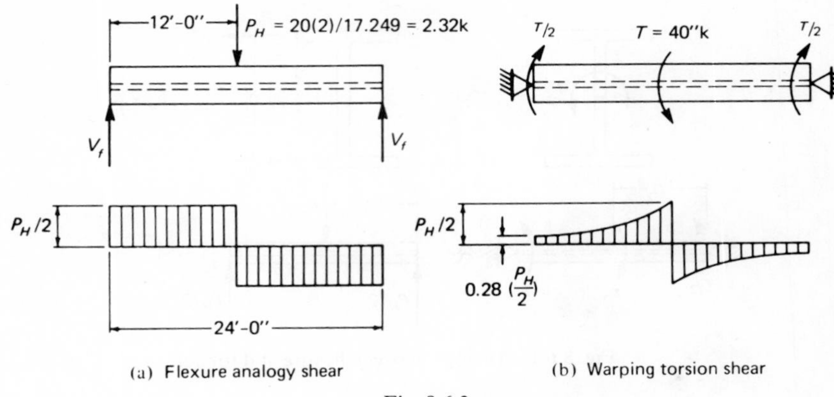

(a) Flexure analogy shear (b) Warping torsion shear

Fig. 8.6.2.

For lateral bending-flange shear stress,

$$v_w = \frac{V_f Q_f}{I_f t_f} = \frac{1.16(7.2)}{(42.0)0.751} = 0.26 \text{ ksi}$$

where $Q_f = (8.75/2)(0.751)(8.75/4) = 7.2 \text{ in.}^3$

The results of the two methods are compared as follows:

Type of stress	Flexural Analogy	Diff. Equation
Normal stress $= f_b + f_{bw} = 11.2 + 17.4 =$	28.6 ksi	19.87 ksi
Web shear stress $= v + v_s = 1.41 + 2.98 =$	4.39 ksi	3.56 ksi
Flange shear stress $= v + v_s + v_w$ $= 0.31 + 5.13 + 0.26 =$	5.70 ksi	4.07 ksi

It is readily apparent that use of the flexure analogy without modification is a very conservative approach. In some situations it is so excessively conservative as to be practically useless. Furthermore, the most important design item, the lateral bending normal stress f_{bw} is overestimated by the greatest amount.

The relationship between the flexural analogy and the true torsion problem is best illustrated by referring to Fig. 8.5.4a which shows that the full torsional shear, from M_s and M_w, is correctly analogous to the lateral flexure problem. Figure 8.5.4b shows the portion of the shear which goes into rotation of elements, while Fig. 8.5.4c shows the portion contributing to lateral flange bending. If one could correctly assess how the shear due to warping torsion compares with the lateral flexure situation, design for torsion could be greatly simplified without being grossly conservative.

Figure 8.6.2b shows the accurate variation of V_f for the problem of Example 8.6.1, computed according to Eq. 8.5.7, whereupon

$$V_f = \frac{T}{2h}\left(\frac{\cosh \lambda z}{\cosh \lambda L/2}\right) \tag{8.6.1}$$

in which the shear from the lateral bending analogy, $T/2h$, is modified by the hyperbolic function.

The lateral bending moment can thus be expressed for this problem as

$$M_f = \beta \frac{T}{2h}\left(\frac{L}{2}\right) \tag{8.6.2}$$

or, in general, the change in lateral moment between the support and location of zero shear is

$$\Delta M_f = \beta \times (\text{area under flexure analogy shear diagram}) \tag{8.6.3}$$

where β is a reduction factor which depends upon λL.

It is to be noted that if Eq. 8.6.2 is multiplied by h, and the concentrated moment T is thought of as a concentrated load, the analogous moment $M_f h$ equals β times the simple beam moment. Thus

$$M_f h = \beta\left(\frac{TL}{4}\right) \tag{8.6.4}$$

for the case of Fig. 8.6.1.

Values of β for most common situations can be estimated for design purposes from Tables 8.6.1 through 8.6.4, and for other unusual cases Table I of Ref. 9 (M_w of Ref. 9 equals $M_f h$ above) or the curves of Ref. 8 may be used.

EXAMPLE 8.6.2

Recompute the stresses due to torsion on the beam of Example 8.6.1, using the flexural analogy along with the β values from Table 8.6.1

TABLE 8.6.1
β Values, Concentrated Load, Torsional Simple Support

T; aL; bL; L

$M_f h = \beta(Tab\ L)$ at $z = aL$

λL	β values				
	$a = 0.5$	$a = 0.4$	$a = 0.3$	$a = 0.2$	$a = 0.1$
0.5	0.98	0.98	0.98	0.99	0.99
1.0	0.92	0.93	0.94	0.95	0.97
2.0	0.76	0.77	0.80	0.84	0.91
3.0	0.60	0.62	0.65	0.72	0.83
4.0	0.48	0.50	0.54	0.62	0.76
5.0	0.39	0.41	0.45	0.54	0.70
6.0	0.33	0.34	0.39	0.47	0.65
8.0	0.25	0.26	0.30	0.37	0.55
10.0	0.20	0.21	0.24	0.31	0.48

TABLE 8.6.2
Values, Concentrated Load, Torsionally Fixed Supports

$M_f h = \beta_1 (T ab^2 L)$ at $z = 0$

$M_f h = \beta_2 (T a^2 bL)$ at $z = L$

λL	$a = 0.5$	$a = 0.4$		$a = 0.3$		$a = 0.2$	
	$\beta_1 = \beta_2$	β_1	β_2	β_1	β_2	β_1	β_2
0.5	0.99	1.00	0.99	1.00	0.99	1.00	0.99
1.0	0.98	0.98	0.98	0.98	0.98	0.99	0.98
2.0	0.92	0.93	0.92	0.94	0.92	0.96	0.92
3.0	0.85	0.86	0.84	0.88	0.84	0.91	0.85
4.0	0.76	0.78	0.75	0.81	0.75	0.86	0.77
5.0	0.68	0.70	0.67	0.74	0.67	0.80	0.69
6.0	0.60	0.63	0.59	0.67	0.60	0.75	0.62
8.0	0.48	0.51	0.47	0.56	0.49	0.65	0.52
10.0	0.39	0.42	0.39	0.47	0.41	0.56	0.44

TABLE 8.6.3
β Values, Uniform Load, Torsional Simple Support

$M_f h = \beta \left(\frac{m}{2} abL^2 \right)$

at $z = aL$

λL	β Values				
	$a = 0.5$	$a = 0.4$	$a = 0.3$	$a = 0.2$	$a = 0.1$
0.5	0.97	0.97	0.98	0.98	0.98
1.0	0.91	0.91	0.91	0.91	0.92
2.0	0.70	0.71	0.71	0.72	0.74
3.0	0.51	0.51	0.52	0.54	0.57
4.0	0.37	0.37	0.38	0.41	0.44
5.0	0.27	0.27	0.29	0.31	0.34
6.0	0.20	0.20	0.22	0.24	0.28
8.0	0.12	0.12	0.13	0.16	0.19
10.0	0.08	0.08	0.09	0.11	0.14

TABLE 8.6.4
β Values, Uniform Load, Torsionally Fixed Supports

$M_f h = \beta \left(\frac{m}{12} L^2 \right)$

at $z = 0$ and $z = L$

λL	0.5	1.0	2.0	3.0	4.0	5.0	6.0	8.0
β	0.99	0.98	0.94	0.88	0.81	0.74	0.67	0.56

TABLE 8.6.5
β Values, Concentrated Load,
Torsionally Fixed Supports

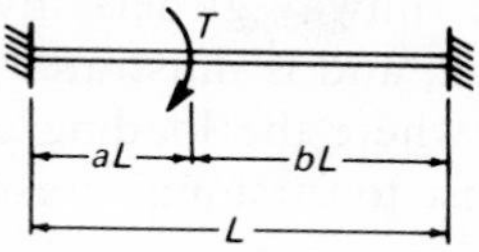

$M_{fh} = \beta$ (positive moment by flexure theory) $= \beta[2Ta^2b^2L]$ at $z = aL$

λL	$a = 0.5$	$a = 0.3$	$a = 0.1$
0.5	0.99	1.00	1.00
1.0	0.98	0.99	1.01
2.0	0.92	0.95	1.05
3.0	0.85	0.91	1.10
4.0	0.76	0.85	1.16
5.0	0.68	0.79	1.21
6.0	0.60	0.73	1.25

Solution.

The flexure analogy gives

$$M_f = 13.9 \text{ ft-kips}$$

as previously computed. Using Ref. 8, λL is found to be

$\lambda L = 24(12)/68.2 = 4.22$ (compares with $\lambda L = 3.88$ as computed in Example 8.5.2).

From Table 8.6.1, $\beta \approx 0.48$, i.e., use about 48 percent of the flexure analogy value. Thus

$$M_f = 13.9(0.48) = 6.67 \text{ ft-kips}$$

$$f_{bw} = \frac{2M_f}{S_y} = \frac{2(6.67)12}{19.2} = 8.35 \text{ ksi}$$

which compares favorably with $f_{bw} = 8.65$ ksi as computed by the differential equation solution using $\lambda L = 3.88$. It is believed that the β values from Tables 8.6.1 through 8.6.4 will give lateral bending moments with an accuracy consistent with the knowledge of torsional end restraint and torsional stiffness in situations normally found in design.

8.7 PRACTICAL SITUATIONS OF TORSIONAL LOADING.

There are relatively few occasions in actual practice where the torsional load can cause significant twisting, and frequently these situations arise during construction. In most building construction the members are laterally restrained by attachments along the length of the member and therefore they are not free to twist. Even though torsional loading exists, it may be self-limiting because the rotation cannot exceed the end slope of the transverse attached members.

Torsional loading exists on spandrel beams, where the torsional loading may be uniformly distributed; it exists where a beam frames into a girder on one side only, or where unequal reactions come to opposite sides of a girder. The design of crane runway girders involves the combination of biaxial bending and torsion, and is illustrated in Sec. 8.8 for laterally stable beams. Any situation where the loading or reaction acts eccentrically to the shear center gives rise to torsional stresses.

Analysis for Torsional Moment. The determination of the torsional moment in a framing system involves a frame analysis where the joints may be rigid or semirigid. While the details of such analysis are outside the scope of this text, some discussion is necessary so that at least the problem is understood. Goldberg[11] has discussed this subject and presented an approximate method suitable for design. Spandrel girders have been treated by Lothers.[13]

Consider Goldberg's example of a floor framing system as shown in Fig. 8.7.1. Spandrel beam AB is subjected to torsion because of the floor beams framing on only one side. Contrary to some common belief, however, the torsional moment is *not* equal to the beam reaction times its eccentricity from the centerline of the girder web. Moment is transmitted across the joint, and the end moment on the beam must equal the torsional moment on the girder. To attack such a problem one must first determine the relationship between the angle of twist ϕ and the applied torsional moment T.

For example, in Fig. 8.7.1 the loading system causes equal torsional moments at the 1/3 points. Assuming the girder torsionally simply supported at ends A and B, using either the differential equation solution

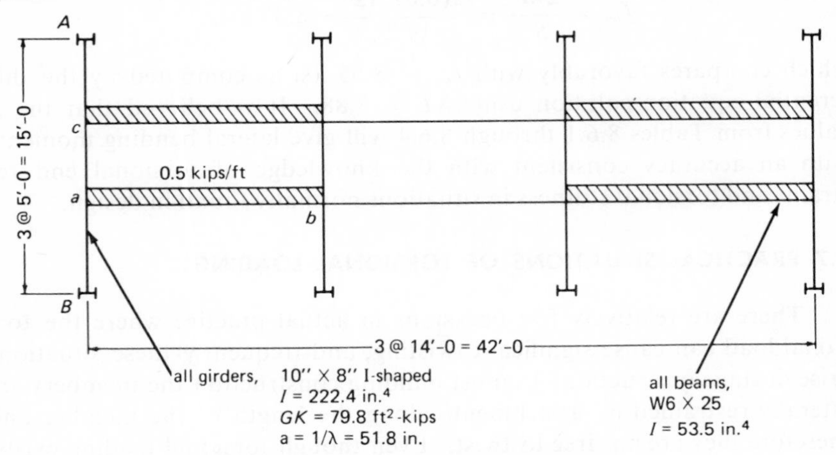

Fig. 8.7.1. Plan view of floor framing.

formulas or the curves of Ref. 8 (case 3), one finds for $\lambda L = 15(12)/51.8 = 3.47$,

$$\phi \frac{GK}{TL} \approx 0.087$$

or

$$\phi_{aa} = 15.7 \frac{T_a}{GK}$$

for T applied at a. In addition the value of ϕ for T applied at c is

$$\phi_{ac} = 0.07\,(180) \frac{T_c}{GK} = 12.6 \frac{T_c}{GK}$$

Finally, for $T_a = T_c = T$,

$$\phi_a = (15.7 + 12.6) \frac{T}{GK} = 28.3 \frac{T}{GK}$$

The twist angle ϕ_a must be compatible with the end slope of the beam; using slope deflection,

$$M_{ab} = M_{Fab} + \frac{2EI}{L_{ab}} \left[-2\phi_a - \phi_b + \frac{3\Delta}{L_{ab}} \right]$$

where M_{Fab} = the fixed-end moment for beam ab at a.
ϕ_a = beam slope at a
ϕ_b = beam slope at b
Δ = relative deflection between a and b

After having established the necessary slope deflection equations for moments, joint equilibrium and shear conditions are necessary, after which simultaneous equations must be solved. One such joint equation is that at joint a,

$$T + M_{ab} = 0$$

After solving for the slopes, then the torsional moments can be found; for the torsional moment at a,

$$T = \phi_a GK/28.3$$

In the example,[11] the value of T was 1.55 in.-kips using an approximate method of satisfying deformation compatibility.

Suppose one had taken the simple beam reaction, 0.5(7) = 3.5 kips, and assume use of an AISC Framed Beam Connection (see Fig. 13.2.1). If the eccentricity had been taken to the bolt line on outstanding leg ($2\frac{1}{4}$ in.), the torsional moment would have been far too great, while if the eccentricity had been taken as one-half the web thickness (0.328/2), the torsional moment would have been far too small.

The proper torsional moment can only be obtained (even approximately) by considering deformation compatibility.

Torsional End Restraint. If a torsional situation is deemed to require design or analysis, the torsional end restraint must be evaluated. Under AISC specifications (AISC–1.2.) three types of construction are permitted; "rigid-frame," "simple" or "conventional," and "semirigid framing." For practical purposes, most construction comes under Type I—rigid frame, i.e., full flexural moment restraint at joints, or Type II—simple connections, i.e., negligible flexural restraint at joints.

The torsional restraint conditions for Type I and Type II construction are given in Fig. 8.7.2. Again, the lateral bending analogy will help in establishing the torsional restraint conditions. Fig. 8.7.2a shows the analogy situation of zero deflection and zero moment which correspond torsionally to $\phi = 0$ and $d^2\phi/dz^2 = 0$. It is to be noted that $\phi = 0$ only if the simple connection extends over a significant portion of the beam depth.

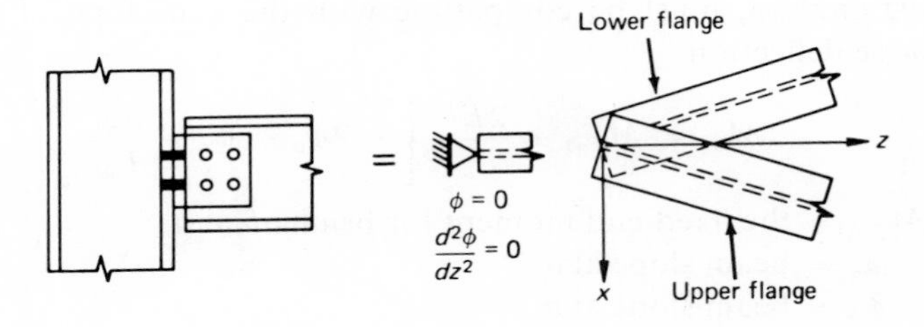

(a) AISC simple framing connection (Type II)
$M_z = M_s + M_w$; $M_f = 0$

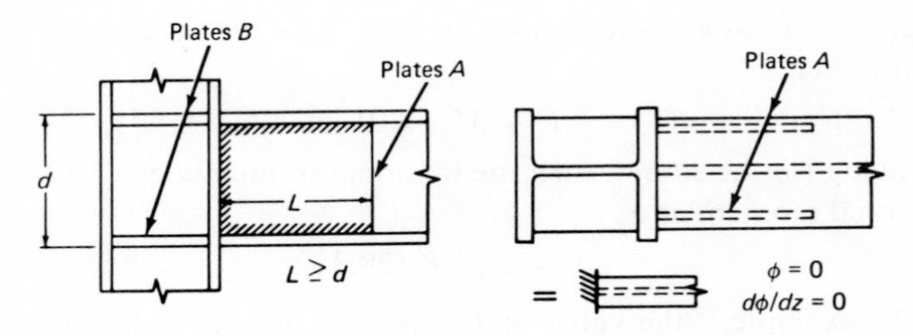

(b) AISC rigid framing connection (Type I) with additional stiffening plates.
$M_z = M_w$; $M_s = 0$

Fig. 8.7.2. Torsional restraint conditions (adapted from Fig. 8 of Ref. 9).

Fig. 8.7.2b shows the analogy situation of zero deflection and zero slope which correspond torsionally to $\phi = 0$ and $d\phi/dz = 0$. Note is made that Hotchkiss[9] states the ends of the beam must be boxed in so as to assure $d\phi/dz = 0$. Stiffener plates (plates A, Fig. 8.7.2b) may be welded between the toes of the flanges and extended along the beam for a length equal at least to the beam depth. Furthermore, if the column has torsionally flexible flanges, column stiffeners (plates B of Fig. 8.7.1b) should be provided opposite the beam flanges.

The structural engineer should remember that in practical situations where no special precautions are taken at the ends, the torsional restraint is neither simple ($d^2\phi/dz^2 = 0$) nor fixed ($d\phi/dz = 0$) but is, however, usually such that the end twist is nearly zero ($\phi = 0$).

8.8. DESIGN FOR COMBINED BENDING AND TORSION—LATERALLY STABLE BEAMS

In order to illustrate the use of some of the concepts developed in the previous articles, several design examples are presented. In all these examples it will be assumed that the combined maximum compression stress is limited to $0.6F_y$ in accordance with AISC–1.5.1.4.6b wherein the unbraced length does not exceed $76.0b_f/\sqrt{F_y}$. Treatment of cases where this length is exceeded is found in Chapter 9.

In essence, present design procedures are based on elastic behavior; therefore, ultimate strength including the effect of residual stress on torsional strength is not considered.

EXAMPLE 8.8.1

Select the lightest W section of A36 steel to carry a superimposed uniform loading of 1.9. kips per ft on a simple supported span of 28 ft, as shown in Fig. 8.8.1. The loading is applied eccentrically 7 in. from the web, and assume that the ends have torsional simple support.

Solution

Estimating the beam weight as 0.13 kips/ft, the flexural bending moment is

$$M_x = 2.03(28)^2/8 = 199 \text{ ft-kips}$$

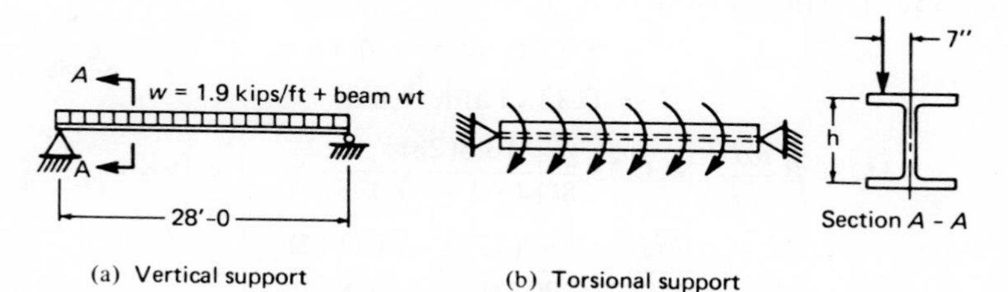

Fig. 8.8.1. Conditions for Example 8.8.1.

The torsional uniformly distributed moment is

$$m = 1.9(7) = 13.3 \text{ in.-kips/ft}$$

Consider m/h as the uniformly distributed lateral load acting on one flange of the beam, and using the flexure analogy, gives

$$m_f = \frac{m}{h}\left(\frac{L^2}{8}\right) = \frac{13.3(28)^2}{h(8)} = \frac{1303}{h} \text{ ft-kips}$$

without regard to the reduction factor β.

As a first approximation, assume $h = 14$ in. and $\beta = 0.5$ (as for example, Table 8.6.3, where $\lambda L = 3$):

$$M_f = \beta \frac{1303}{h} = 0.5 \frac{1303}{14} = 46.5 \text{ ft-kips}$$

The design acceptability criterion is

$$\frac{M_x}{S_x} + \frac{M_y}{S_y} \leq 0.6F_y$$

and using the procedure discussed in Sec. 7.9 gives

$$\text{Reqd } S_x = \frac{M_x}{0.6F_y} + \frac{M_y}{0.6F_y}\left(\frac{S_x}{S_y}\right)$$

$$= \frac{199(12)}{22} + \frac{2(46.5)(12)}{22}(3) = 109.5 + 152$$

$$= 261.5 \text{ in.}^3$$

in which the ratio S_x/S_y is estimated at 3 for medium weight W14 sections, and M_f is doubled to give the moment resisted by two flanges.

This would indicate a W14 × 167 with an $S_x = 267.3$ in.^{3} It will be observed that actual S_x/S_y for sections in this range is about 2.6. Reqd S_x is then reduced to about 241, which would indicate W14 × 150.

From Ref. 8, or by calculation as previously illustrated, for the W14 × 150, $\lambda = 1/71.7$. Thus,

$$\lambda L = 28(12)/71.7 = 4.7$$

Using Fig. 8.6.3, β is reduced to about 0.3, which further reduces required S_x to about 190 in.^{3} Try W14 × 127.

$$\lambda L = 28(12)/75.6 = 4.45$$

$$\beta \approx 0.33 \text{ (Table 8.6.3)}$$

$$M_f = \beta \frac{mL^2}{8h} = 0.33 \frac{13.3(28)^2}{8(14.62 - 0.998)} = 31.6 \text{ ft-kips}$$

$$f_{max} = \frac{M_x}{S_x} + \frac{M_f}{S_y/2} = \frac{199(12)}{202.0} + \frac{31.6(12)}{71.8/2}$$

$$= 11.8 + 10.5 = 22.3 \text{ ksi}$$

Use W14 × 127. It will be found that where high torsional strength is required the wide W14 sections are most suitable. For the *same weight per foot*, deeper sections give a reduced stress from ordinary flexure but an increased stress from torsional warping. The W21 × 127 and the W24 × 130 give about the same maximum stress as the above selected beam.

The differential equation solution gives for the normal stress due to warping torsion 10.6 ksi as compared with 10.5 ksi computed above. The maximum flange shear stress is 11.3 ksi, while that in the web is 9.3 ksi, both computed from the differential equation solution. These are well within allowable values.

EXAMPLE 8.8.2

Design a beam with torsionally fixed ends to carry two concentrated loads eccentric to the plane of the web by 6 in. as shown in Fig. 8.8.2. Assume for conservatism that for ordinary flexure the beam is simple supported. Use A36 steel and assume the maximum combined stress is limited to $0.6F_y$.

Solution

Estimating beam weight as 0.15 kips/ft, the flexural bending moment M_x is

$$M_x = 20(12) + 0.15(36)^2/8 = 240 + 24.3 = 264.3 \text{ ft-kips}$$

The concentrated torsional moment is

$$T = 20(6)/12 = 10 \text{ ft-kips}$$

Considering T/h as the analogous concentrated loads, the fixed end moments are computed; thus

$$M_f h \text{ (at ends)} = \frac{Tab^2}{L^2} + \frac{Ta^2b}{L^2} = \frac{10(12)(24)}{(36)^2}(24 + 12)$$
$$= 53.3 + 26.7 = 80 \text{ ft}^2\text{-kips}$$

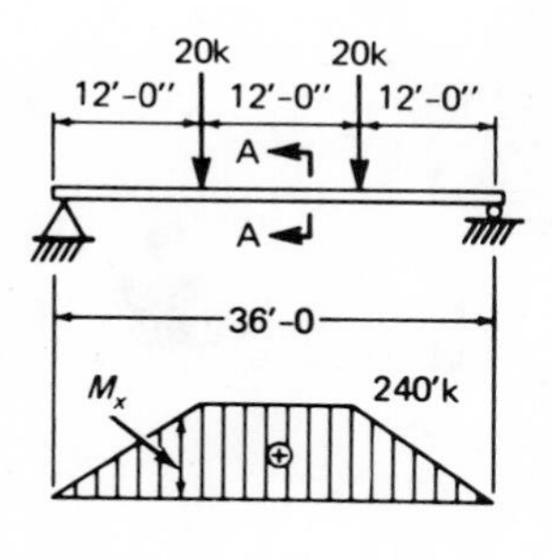

(a) Vertical support and ordinary fleuxural moment.

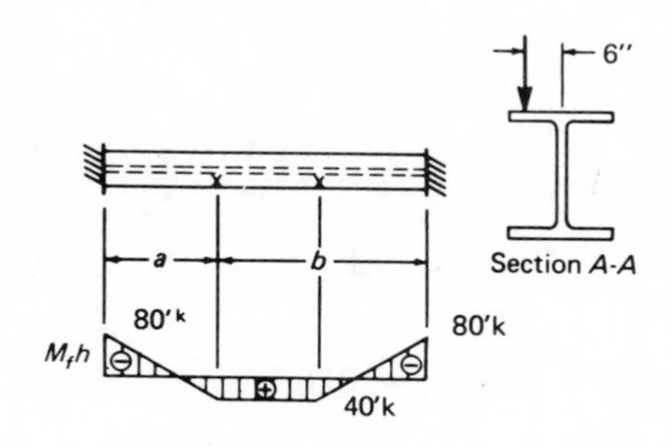

(b) Torsional support and flexure analogy lateral bending moment.

Fig. 8.8.2. Conditions for Example 8.8.2.

and in the positive moment zone,

$$M_f h \text{ (at concentrated loads)} = \frac{TL}{3} - 80.0 = 10(12) - 80.0$$
$$= 40.0 \text{ ft}^2\text{-kips}$$

without regard to the β reduction factor; the flexure analogy gives $M_f h$ values as shown in Fig. 8.8.2b.

Estimating average λL at about 3, and using $a = 0.3$ in Table 8.6.2 for end moments, the analogous fixed-end moments become

$$M_f h \text{ (at ends)} = 0.88(53.3) + 0.84(26.7) = 47 + 22$$
$$= 69 \text{ ft}^2\text{-kips}$$

For the positive moment at 12 ft from the support, refer to Table 8.6.5, estimating β as 0.9, though the exact case being considered is not convered in any of the tables of β values,

$$M_f h \text{ (at } z = 0.3L) \approx 0.9(40) = 36 \text{ ft}^2\text{-kips}$$

which is known to be conservatively high (see Fig. 8.8.3) because the

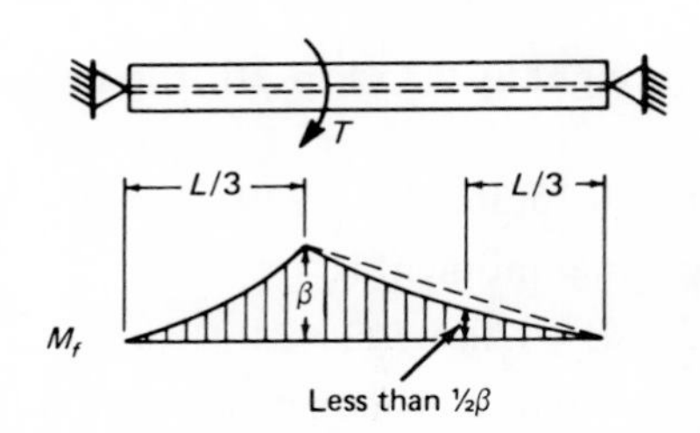

Fig. 8.8.3. M_f variation for concentrated M.

40 value includes the effects of both concentrated torsional moments (Ref. 8, Case 6, indicates that T applied at $z = 0.36$ has negligible effect at $0.7L$.)

Assuming $h = 14$ in.,

$$M_f = 36(12)/14 = 30.9 \text{ ft-kips}$$

$$\text{Reqd } S_x = \frac{M_x}{22} + \frac{2M_f}{22}\left(\frac{S_x}{S_y}\right)$$

$$= \frac{264.3(12)}{22} + \frac{2(30.9)(12)}{22}(2.5) = 144 + 84 = 228 \text{ in.}^3$$

W14 × 136: $S_x = 216 \text{ in.}^3$

$$\lambda L = 36(12)/71.5 = 6.05$$

$$M_f(\text{at } z = 0.3L) = 0.7(40)(12)/14 = 24 \text{ ft-kips}$$

$$\text{Reqd } S_x = 144 + \frac{2(24)(12)}{22}(2.8) = 144 + 73 = 217.0 \text{ in.}^3$$

which seems sufficiently close. Use W14 × 136.

For a more accurate check using Ref. 8, Case 6,

$$\lambda L = 6.05$$

$$M_f h = (\text{Ref. 8 coeff.})\frac{T}{\lambda} = 0.385(10)(71.17)/12 = 22.8 \text{ ft}^2\text{-kips}$$

$$M_f = 22.8(12)/14.0 = 19.55 \text{ ft-kips}$$

$$f = \frac{264.3(12)}{216} + \frac{2(19.55)(12)}{77.0}$$

$$= 14.7 + 6.1 = 20.8 \text{ ksi} < 22 \text{ ksi} \qquad \text{OK}$$

Also the stress at the supports should be checked. Using Table 8.6.2, find $\beta_1 \approx 0.67$ and $\beta_2 \approx 0.6$:

$$M_f h = 0.67(53.3) + 0.6(26.7) = 35.7 + 16 = 51.7 \text{ ft}^2\text{-kips}$$

$$M_f = 51.7(12)/14 = 44.3 \text{ ft-kips}$$

$$f = \frac{2M_f}{S_y} = \frac{2(44.3)12}{77.0} = 13.8 \text{ ksi} < 22 \text{ ksi} \qquad \text{OK}$$

These two examples illustrate that using approximate β values, along with the flexure analogy for lateral bending due to warping torsion, gives sufficiently quick and accurate results for ordinary design. Furthermore, the designer can better visualize what is happening using the flexure analogy rather than working with the hyperbolic functions for ϕ.

For additional treatment of combined torsion and flexure, particularly on channel and zee sections, the reader is referred to the work of Lansing.[14]

Another topic, outside the scope of this text, is the secondary lateral bending moment which arises from the torsional deflection of the compression flange laterally. In the deflected position the compressive force resulting from ordinary flexural moment, M_x, times the lateral flange deflection gives rise to the secondary lateral moment which in turn causes greater lateral deflection. Discussion of this topic appears elsewhere[15] and is similar to the secondary bending moment which occurs in beam-columns, a subject treated in Chapter 12.

8.9. TORSION IN CLOSED THIN-WALLED SECTIONS

In general, it will be found that for situations where a high torsional stiffness is required, a closed box section is preferred over the ordinary open section, such as the I-shape or channel. It is because of this high torsional stiffness that box sections are extensively used in aircraft structural components[17] and curved girders[18] for bridges and buildings. This subject is treated in a number of textbooks,[1-3,16,17] so that only a brief treatment follows.

In the closed section of Fig. 8.9.1, it is assumed that the walls are

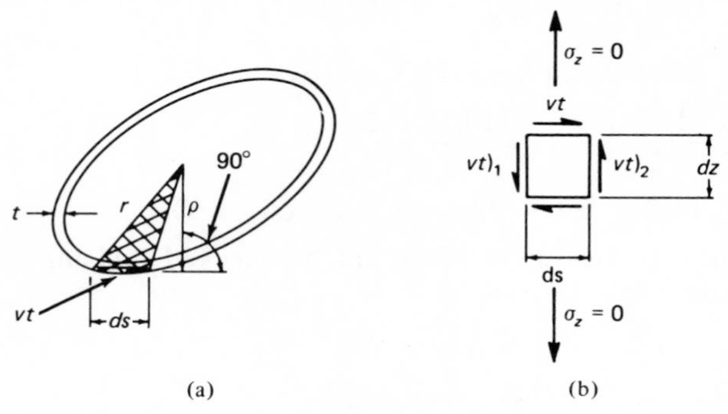

Fig. 8.9.1.

thin so that the shearing stress may be assumed uniformly distributed across the thickness t. If the shear stress is v, then vt is the shear force per unit distance along the wall, usually referred to as shear flow. Since only torsional stress is presently being considered, the normal stresses (σ_z of Fig. 8.9.1b) are zero. Since $\sigma_z = 0$, the shear flow vt cannot vary along the wall; i.e., vt is constant.

Referring to Fig. 8.9.1a, the increment of torsional moment contributed by each element is

$$dT = vt\rho\, ds \tag{8.9.1}$$

Integrating gives the full torsional moment, which is in effect the same as Eq. 8.2.5,

$$T = vt \int_s \rho\, ds \tag{8.9.2}$$

Note that $\frac{1}{2}\rho\, ds$ is the cross-hatched area of the triangular segment in Fig. 8.9.1. Thus the integral

$$\int_s \rho\, ds = 2A \tag{8.9.3}$$

where A = area enclosed by the walls. Finally,

$$T = 2vtA \tag{8.9.4}$$

If a cut is made in the wall of a closed thin-wall section (Fig. 8.9.2), a relative movement (as in Fig. 8.9.2b) will be produced between the two sides in the axial direction of the member. The unit shearing strain along

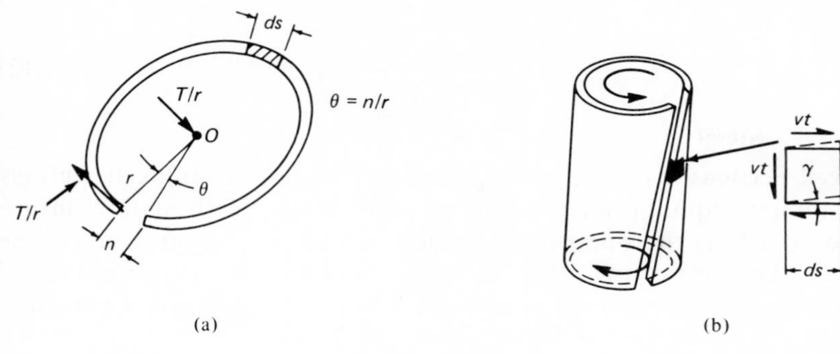

(a)

(b)

Fig. 8.9.2.

the perimeter is

$$\gamma = v/G \tag{8.9.5}$$

The internal strain energy for any elemental length *ds* along the perimeter is

$$dW_i = \tfrac{1}{2} vt\gamma\, ds \tag{8.9.6}$$

$$= \frac{1}{2}\left(\frac{T}{2A}\right)\frac{v}{G}\, ds \tag{8.9.7}$$

The twisting moment, T, about point 0 can now be replaced by a couple, T/r. The external work done by the couple is

$$dW_e = \frac{1}{2}\left(\frac{T}{r}\right)n = \frac{T\theta}{2} \tag{8.9.8}$$

Equating internal and external work per unit length gives

$$\frac{T\theta}{2} = \frac{T}{4AG}\int_s v\, ds \tag{8.9.9}$$

$$\theta = \frac{\int_s v\, ds}{2AG} = \frac{vt\int_s ds/t}{2AG} \tag{8.9.10}$$

since vt is a constant.

In order to obtain more useful forms of the equations, recall from Sec. 8.2, Eq. 8.2.5, that

$$T = GK\theta \qquad [8.2.5]$$

and using Eq. 8.9.10 gives

$$T = GK\,\frac{vt\int ds/t}{2AG} \tag{8.9.11}$$

and eliminating T between Eqs. 8.9.4 and 8.9.11 gives when solving for K

$$K = \frac{4A^2}{\int_s ds/t} \tag{8.9.12}$$

Multicell Sections. In computing the shear stresses in multicell sections under torsion, equilibrium must be satisfied; thus, the shear flow into a junction (such as the one shown shaded in Fig. 8.9.3a) must equal the flow out; the angle of twist θ must be identical for each cell; the sum of the individual torsional resistances must be equal to the applied torsional moment.

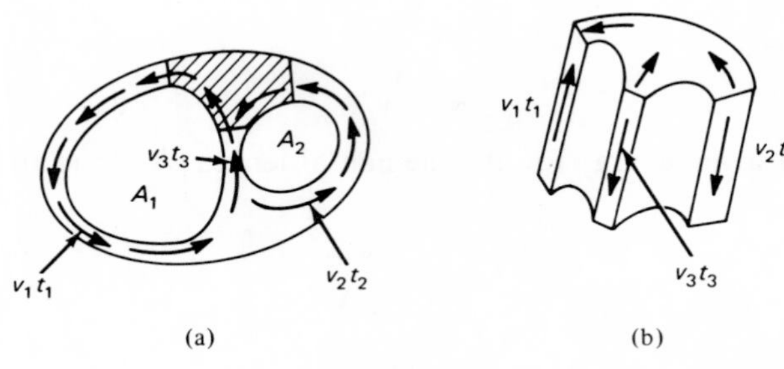

Fig. 8.9.3.

In Fig. 8.9.3 the force equilibrium at joints is satisfied by

$$v_3t_3 = v_1t_1 - v_2t_2 \tag{8.9.11}$$

Secondly, deformation compatibility is obtained when

$$\left.\begin{aligned} \theta &= \frac{\int_1 v\,ds}{2A_1G} \\ \theta &= \frac{\int_2 v\,ds}{2A_2G} \end{aligned}\right\} \tag{8.9.12}$$

Finally, moment equilibrium requires

$$T = 2A_1v_1t_1 + 2A_2v_2t_2 + \cdots + 2A_nv_nt_n \tag{8.9.13}$$

The development of the equations for torsion stiffness and stress in closed thin-walled sections may be alternatively developed using the "membrane analogy" developed by Prandtl, which is discussed in Refs. 1, 2, and 3.

The principal aim in this text is to develop the basic expressions and illustrate the high degree of torsional stiffness which closed sections exhibit as compared to open ones. The following two examples are intended for this purpose.

EXAMPLE 8.9.1

Compare the torsional resisting moment T_t and the torsional stiffness K for the sections of Fig. 8.9.4 if the maximum shear stress is 14 ksi.

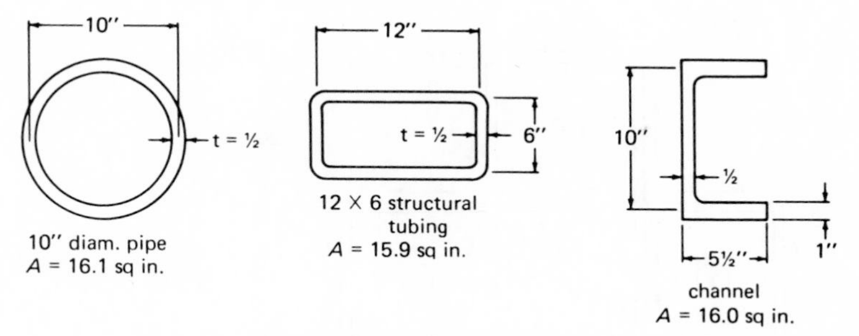

Fig. 8.9.4. Sections for Example 8.9.1.

Solution

(a) Circular thin-wall section. Using Eq. 8.9.4,

$$T = 2vtA = 2(14)(0.5)[\pi(10)^2/4]/12 = 91.5 \text{ ft-kips}$$

$$K = \frac{4A^2}{\int ds/t} = \frac{4(25\pi)^2}{20\pi} = 393 \text{ in.}^4$$

where $\int ds/t = 2\pi(5)/0.5 = 20\pi$.

(b) Rectangular box section:

$$T = 2vtA = 2(14)(0.5)(72) = 84.0 \text{ ft-kips}$$

$$K = \frac{4A^2}{\int ds/t} = \frac{4(72)^2}{(36/0.5)} = 288 \text{ in.}^4$$

(c) Channel section. Since for this open section,

$$v = \frac{Tt}{K}$$

the maximum shear stress will be in the flange. Also,

$$K = \Sigma \frac{bt^3}{3}$$

$$K = \tfrac{1}{3}[10(0.5)^3 + 2(5.5)(1)^3] = 4.08 \text{ in.}^4$$

$$T = \frac{Kv}{t_f} = \frac{4.08(14)}{(1)(12)} = 4.75 \text{ ft-kips}$$

The circular section is best for torsional capacity, the rectangular box is next; these closed sections having torsional stiffness, K, 96 and 69 times that of the channel, respectively. The resisting moments are 19.3 and 17.7, respectively, times that of the channel.

EXAMPLE 8.9.2

Compute the torsional moment capacity of the multicell girder of Fig. 8.9.5 if the maximum shear stress cannot exceed 14 ksi.

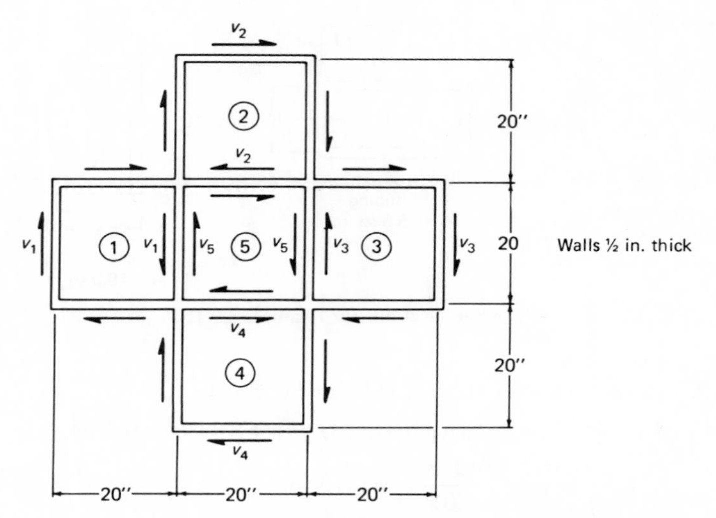

Fig. 8.9.5. Section for Example 8.9.2.

Solution

From symmetry,

$$v_1 = v_2 = v_3 = v_4$$

Also, the net shear stress in the interior walls is $v_1 - v_5$.

From the compatibility of deformation, Eqs. 8.9.12 give

For cell ①,

$$\int v\,ds = 3(20)\,v_1 + 20(v_1 - v_5) = 2G\theta(400)$$

For cell ⑤,

$$\int v\,ds = 4[20(v_5 - v_1)] = 2G\theta(400)$$

The above two equations are then solved for v_1 and v_5:

$$\begin{aligned} 80v_1 - 20v_5 &= 800G\theta \\ -80v_1 + 80v_5 &= 800G\theta \\ \hline 60v_5 &= 1600G\theta \end{aligned}$$

$$v_5 = 26.67G\theta$$
$$v_1 = 16.67G\theta$$

The exterior-wall shear stresses equal $16.67G\theta$, while the interior wall stress is $(26.67 - 16.67)G\theta$. For $v_{max} = 14$ ksi,

$$G\theta = \frac{14}{16.67} = 0.84$$

Using Eq. 8.9.13,

$$\begin{aligned} T &= \Sigma 2Avt \\ &= 4[2(400)(16.67)\,G\theta](0.5) + 2(400)(26.67)\,G\theta(0.5) \\ &= 26{,}667\,G\theta + 10{,}667\,G\theta = 37{,}333\,G\theta \\ &= 37{,}333(0.84)/12 = 2610 \text{ ft-kips} \end{aligned}$$

The stiffness K is

$$K = \frac{T}{G\theta} = 37{,}333 \text{ in.}^4$$

The design of closed sections for bending and torsion is outside the scope of this text, and the reader is referred to the work of Felton and Dobbs.[19]

8.10. TORSION IN SECTIONS WITH OPEN AND CLOSED PARTS

Generally this problem is treated by combining the principles discussed separately for open and closed parts. The procedure to be used for determining resisting moment, stiffness, and shear center location for such sections is presented with examples by Chu and Longinow.[20] The following is a summary of pertinent equations:

Total resisting moment is

$$T = \sum_{i=1}^{n} 2v_i t_i A_i + GK\theta \tag{8.10.1}$$

where $K = \Sigma \frac{1}{3} bt^3$ for open parts only.

In addition, each of the closed cells must satisfy Eq. 8.9.12:

$$\int_s v_i t_i \frac{ds}{t_i} = 2GA_i\theta \tag{8.10.2}$$

for each cell.

8.11. TORSIONAL BUCKLING

Since the differential equation for torsion was developed earlier in this chapter, and buckling of axially loaded columns has previously been treated, a consideration of torsional buckling is desirable. Most centrally loaded columns fail by local buckling or at the tangent-modulus Euler load as discussed in Chapter 6. However, some thin-walled sections such as angles, tees, zees, and channels, with relatively low torsional stiffness may, under axial compression, buckle torsionally while the longitudinal axis remains straight.

The subject of torsional buckling is treated extensively by Timoshenko and Gere[21] and Bleich,[22] and discussed to a lesser extent by Gaylord and Gaylord.[23]

Using concepts previously developed, it is the objective here merely to illustrate how such buckling can occur and the types of situations for which the designer should be cautious.

Consider the doubly symmetrical section in the shape of a cross given in Fig. 8.11.1, whose shear center and centroid coincide. Recalling the

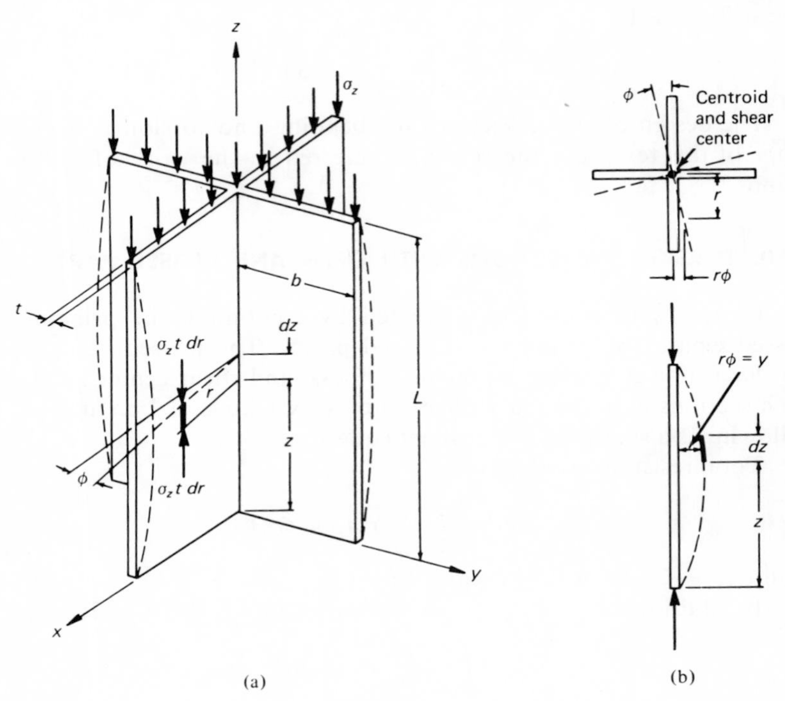

Fig. 8.11.1.

Euler equation,

$$EI \frac{d^2y}{dz^2} + Py = 0$$

which differentiated twice becomes

$$EI \frac{d^4y}{dz^4} = -P \frac{d^2y}{dz^2} \tag{8.11.1}$$

Since $EI\ d^4y/dz^4$ is the loading, the Euler column can be thought of as a beam laterally loaded with the fictitious loading $-P(d^2y/dz^2)$. Thus the compressive force on the element $dr\ dz$, which is $\sigma_z t\ dr$, is statically equivalent to a lateral load whose intensity per unit length is

$$-(\sigma_z t\ dr) \frac{d^2(r\phi)}{dz^2}$$

The increment of torsional moment about the z axis tributary to the length dz equals the load times the moment arm r; thus,

$$dm_z = -(\sigma_z tr\,dr)\frac{rd^2\phi}{dz^2}\,dz \tag{8.11.2}$$

The total torsional moment for the slice dz of the column is

$$m_z = -\sigma_z\frac{d^2\phi}{dz^2}\,dz\int_A r^2\,dr \tag{8.11.3}$$

Equation 8.11.3 represents the contribution to the torsional moment M_z tributary to the element dz:

$$m_z = \frac{dM_z}{dz}$$

The differential equation for torsion on I-shaped sections, Eq. 8.5.10, is

$$M_z = GK\frac{d\phi}{dz} - EI_w\frac{d^3\phi}{dz^3} \qquad [8.5.10]$$

which when differentiated once becomes

$$\frac{dM_z}{dz} = GK\frac{d^2\phi}{dz^2} - EI_w\frac{d^4\phi}{dz^4}$$

For the more general cases, let GK become C, the *torsional rigidity*; and EI_w become C_1, the *warping rigidity*:

$$\frac{dM_z}{dz} = C\frac{d^2\phi}{dz^2} - C_1\frac{d^4\phi}{dz^4} \tag{8.11.4}$$

Referring to Fig. 8.5.2, positive M_z at the section z gives a clockwise rotation; whereas in Fig. 8.11.1 at the section z there is counterclockwise rotation:

$$\frac{dM_z}{dz} = -m_z$$

for use in Eq. 8.11.4. The general differential equation for torsional buckling is then

$$\sigma_z\frac{d^2\phi}{dz^2}\int_A r^2\,dr = C\frac{d^2\phi}{dz^2} - C_1\frac{d^4\phi}{dz^4}$$

or

$$C_1\frac{d^4\phi}{dz^4} - \left(C - \sigma_z\int_A r^2\,dr\right)\frac{d^2\phi}{dz^2} = 0 \tag{8.11.5}$$

in which $\int_A r^2\,dr = J$, the polar moment of inertia about the shear center. When the centroid coincides with the shear center, Eq. 8.11.5 alone determines the buckling condition.

It will be found (Table 8.11.1) that the warping rigidity is zero for shapes consisting of thin rectangular elements intersecting at a common point.

For other cases, Eq. 8.11.5 may be written as

$$\frac{d^4\phi}{dz^4} + p^2 \frac{d^2\phi}{dz^2} = 0 \tag{8.11.6}$$

when

$$p^2 = \frac{\sigma_z J - C}{C_1}$$

for which the general solution is

$$\phi = A_1 \sin pz + A_2 \cos pz + A_3 z + A_4 \tag{8.11.7}$$

Considering the pin-end column, with rotation about z prevented at each end, but with warping not restricted at the ends, gives in a manner similar to the Euler column derivation in Chapter 6, that

$$A_2 = A_3 = A_4 = 0$$

and since A_1 cannot also be zero,

$$\sin pL = 0, \qquad pL = n\pi$$

The elastic buckling stress at which torsional buckling occurs is

$$\frac{\pi^2}{L^2} = \frac{\sigma_z J - C}{C_1} = \frac{\sigma_z J - GK}{EI_w}$$

$$\sigma_{z\ \text{critical}} = \frac{EI_w \pi^2}{JL^2} + \frac{GK}{J} \tag{8.11.8}$$

which is accurate for doubly symmetrical sections whose shear center and centroid coincide, such as I- and Z-shaped sections. For the common single-angle strut, since the distance from centroid to shear center is small, Eq. 8.11.8 will provide a reasonable approximation for the torsional buckling stress. Expressions for the warping constant I_w and the torsion constant K for various shapes are to be found in Table 8.11.1.

The reader should not lose sight of the fact that the most probable buckling mode is still that occurring at the tangent-modulus Euler load because of lateral bending about the x or y axis. Thus, the problem involves three critical values of axial load; bending about either principal axis and twisting about the longitudinal axis. On wide-flange sections, torsional buckling may be important on sections with extra wide flanges and short lengths.[21]

In the general case where the shear center does not coincide with the centroid, the buckling failure is actually a combination of torsion and flexure. For this case, the three differential equations, (1) buckling by lateral bending about the x axis; (2) buckling by lateral bending about the y axis; and (3) twisting about the centroidal axis, are interdependent.

TABLE 8.11.1
Torsional Properties

O = shear center $\quad$ K = torsion constant, $\;I_w$ = warping constant
G = centroid $\quad$ J = polar moment of inertia about shear center

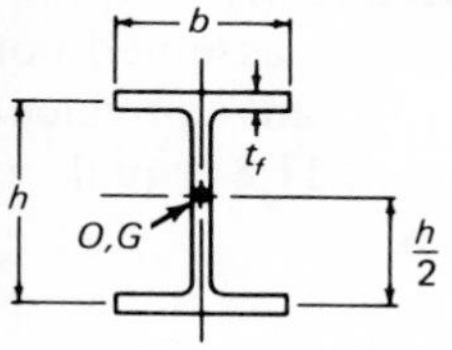

$$K = \tfrac{1}{3}(2bt_f^3 + ht_w^3)$$
$$I_w = \frac{I_f h^2}{2} = \frac{t_f b^3 h^2}{24} = \frac{h^2 I_y}{4}$$
$$J = I_x + I_y$$

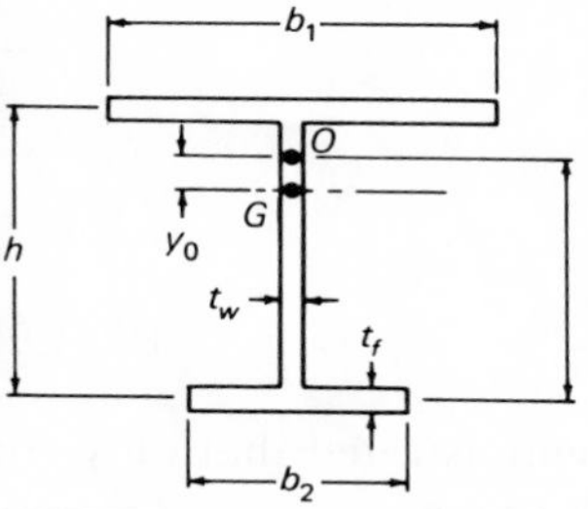

$$e = h\,\frac{b_1^3}{b_1^3 + b_2^3}$$
$$K = \tfrac{1}{3}(b_1 t_f^3 + b_2 t_f^3 + ht_w^3)$$
$$I_w = \frac{t_f h^2}{12}\left(\frac{b_1^3 b_2^3}{b_1^3 + b_2^3}\right)$$
$$J = I_y + I_x + Ay_0^2$$

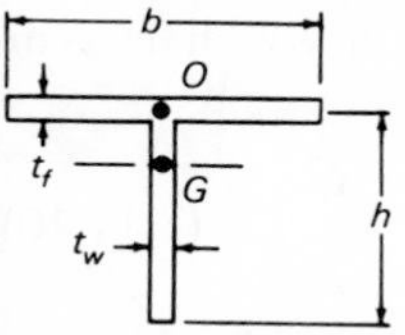

$$K = \tfrac{1}{3}(bt_f^3 + ht_w^3)$$
$$I_w = \frac{1}{36}\left(\frac{b^3 t_f^3}{4} + h^3 t_w^3\right)$$
$$\approx \text{zero for small } t$$

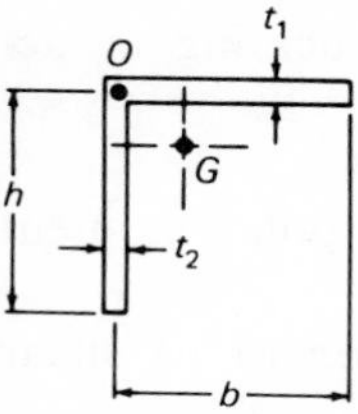

$$K = \tfrac{1}{3}(bt_1^3 + ht_2^3)$$
$$I_w = \frac{1}{36}(b^3 t_1^3 + h^3 t_2^3)$$
$$\approx \text{zero for small } t$$

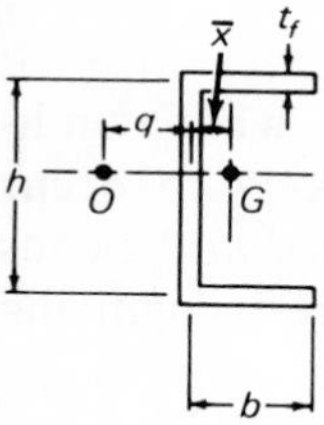

$$K = \tfrac{1}{3}(2bt_f^3 + ht_w^3)$$
$$I_w = \frac{t_f b^3 h^2}{12}\left(\frac{3bt_f + 2ht_w}{6bt_f + ht_w}\right) = \frac{h^2}{4}(I_y + A\bar{x}^2 - q\bar{x}A)$$
$$q = \frac{th^2b^2}{4I_x}$$

Thus, three simultaneous differential equations must be solved to get the buckling loads. The development and solution of these equations is outside the scope of this text and is adequately treated elsewhere.[21,22]

For design purposes, since torsional buckling is frequently not considered by standard design specifications and usually may be assumed not to control, a check may be made to determine whether an equivalent radius of gyration[22] is below r_x and r_y. Setting Eq. 8.11.8 equal to Euler's equation, defining r_E as the equivalent value gives

$$\frac{\pi^2 E}{(L/r_E)^2} = \frac{EI_w\pi^2}{JL^2} + \frac{GK}{J}$$

$$r_E = \sqrt{\frac{I_w}{J} + \frac{GKL^2}{EJ\pi^2}}$$

which for steel with $E/G = 2.6$ gives

$$r_E = \sqrt{\frac{I_w}{J} + 0.04\,\frac{KL^2}{J}} \tag{8.11.9}$$

for doubly symmetrical sections. It has been demonstrated that only for short lengths will r_E be lower than r_x and r_y for W shapes.[22]

The equivalent radius of gyration for a cross section which is symmetrical only about the y-axis is given by

$$\frac{1}{r_e^2} = \frac{1}{2r_E^2} + \frac{1}{2r_y^2} + \sqrt{\left(\frac{1}{2r_E^2} - \frac{1}{2r_y^2}\right)^2 + \left(\frac{y_0}{r_E r_y r_p}\right)^2} \tag{8.11.10}$$

where r_e = equivalent radius of gyration for torsional-flexural buckling
r_E = equivalent radius of gyration for torsional buckling in accordance with Eq. 8.11.9
r_y = radius of gyration for axis of symmetry
r_p = polar radius of gyration = $\sqrt{J/A}$, where J = polar moment of inertia about shear center
y_0 = distance from center of gravity of cross section to its shear center

EXAMPLE 8.11.1

For the sections given in Fig. 8.11.2, determine under what conditions, if any, torsional or flexural-torsional buckling is likely to occur under axial compression loading. Assume the members pinned at the ends of the unbraced lengths, and free to warp at the ends, fully recognizing that these two assumptions minimize buckling strength.

Solution

(a) W8 × 31. Since the centroid and shear center coincide, use Eq. 8.11.9 to get the equivalent r_E:

$$K = \tfrac{1}{3}[2(8)(0.433)^3 + 7.567(0.288)^3] = 0.492 \text{ in.}^4$$

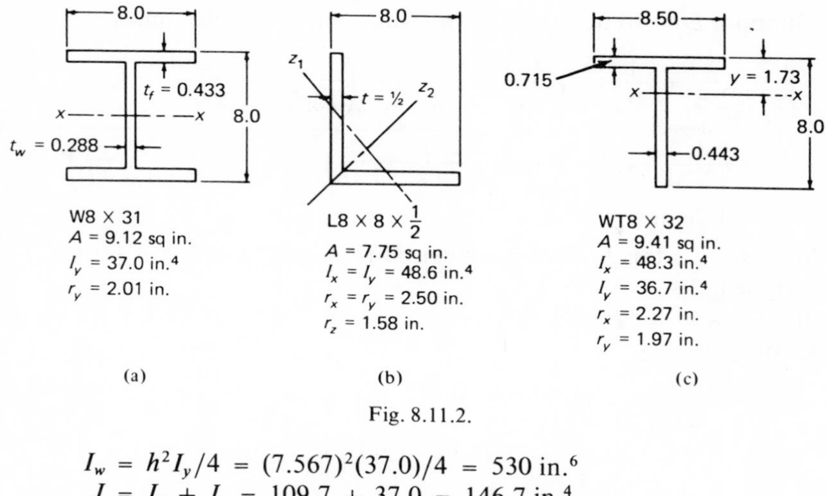

Fig. 8.11.2.

$$I_w = h^2 I_y/4 = (7.567)^2(37.0)/4 = 530 \text{ in.}^6$$
$$J = I_x + I_y = 109.7 + 37.0 = 146.7 \text{ in.}^4$$
$$r_E = \sqrt{\frac{I_w}{J} + 0.04\frac{KL^2}{J}} = \sqrt{\frac{530}{146.7} + 0.04\frac{(0.492)^2 L^2(144)}{146.7}}$$
$$= \sqrt{3.61 + 0.0193L^2}$$

where L is the unsupported length, in feet. Only when L is less than 4.6 ft does r_y exceed r_E; only for a very short column is torsional buckling a possibility, and even when r_E approaches its minimum value (when $L = 0$), which is $r_E(\min) = 1.90$ in., it is only 4 percent less than r_y.

The result using this section is typical of standard W and S shapes and indicates torsional buckling may properly be neglected.

(b) WT8 × 32. The centroid and shear center do not coincide, but the section has one axis of symmetry; using Eq. 8.11.10:

$$K = \tfrac{1}{3}[(8.5)(0.715)^3 + 7.285(0.443)^3] = 1.246 \text{ in.}$$

Referring to Table 8.11.1,

$$I_w = \left[\frac{(8.5)^3(0.715)^3}{4} + (7.285)^3(0.443)^3\right] \Big/ 36 = 2.49 \text{ in.}^6$$

which is actually small enough to be neglected.

$$J = I_x + I_y + Ay_0^2 = 48.3 + 36.7 + 9.41(1.73)^2 = 113.1 \text{ in.}^4$$

For use in Eq. 8.11.10, $y_0 = 1.73$ in., $r_y = 1.97$ in.,

$$r_p = \sqrt{J/A} = \sqrt{113.1/9.41} = 3.47 \text{ in.}$$

$$r_E = \sqrt{\frac{2.49}{113.1} + \frac{0.04(1.246)\,L^2(144)}{113.1}} = \sqrt{0.022 + 0.0635L^2}$$

with L in feet.

Suppose $L = 10$ ft, $r_E = 2.53$ in. Equation 8.11.10 becomes

$$\frac{1}{r_e^2} = \frac{1}{2r_E^2} + \frac{1}{2r_y^2} + \sqrt{\left(\frac{1}{2r_E^2} - \frac{1}{2r_y^2}\right) + \left(\frac{y_0}{r_E r_y r_p}\right)^2}$$

$$= \frac{1}{12.70} + \frac{1}{7.76} + \sqrt{\left(\frac{1}{12.70} - \frac{1}{7.76}\right)^2 + \left(\frac{1.73}{2.53(1.97)(3.47)}\right)^2}$$

$$r_e = 1.76 \text{ in.}$$

In this case r_e is less than r_y and flexural-torsional buckling is critical for ordinary lengths.

(c) L8 × 8 × ½. For this section, Eq. 8.11.10 may be used with the z_2 axis of symmetry.

$$K = \tfrac{1}{3}(2)(7.75)(0.5)^3 = 0.645 \text{ in.}^4$$
$$I_w = \text{neglect}$$
$$J = 2I_x + 2Ay^2 = 2(48.6) + 2(7.75)(2.19)^2 = 171.5 \text{ in.}^4$$
$$r_p = \sqrt{171.5/7.75} = 4.7 \text{ in.}$$
$$x = y = 2.19 \text{ in.}, \; y_0 = y/0.707 = 3.1 \text{ in.}$$
$$r_z = \sqrt{2(2.50)^2 - (1.58)^2} = 3.16 \text{ in.}$$
$$r_E = \sqrt{0.04 \frac{0.645}{171.5} L^2(144)} = \sqrt{0.0217L^2} = 0.147L$$

with L in feet.

Let $L = 10$ ft, $r_E = 1.47$ in. Equation 8.11.10 then becomes

$$\frac{1}{r_e^2} = \frac{1}{2(1.47)^2} + \frac{1}{2(3.16)^2} + \sqrt{\left(\frac{1}{2(1.47)^2} = \frac{1}{2(3.16)^2}\right)^2 + \left(\frac{3.1}{1.47(3.16)(4.7)}\right)^2}$$

$$r_e = 1.40 \text{ in.}$$

Thus, for this angle r_e is less than $r_{z_1} = 1.58$ in., so that flexural-torsional buckling controls for the 10 ft length. In general, the single angle strut is controlled by flexural-torsional buckling; it is for this reason that caution concerning its behavior is given in the AISC Manual.

As a conclusion to this treatment, the designer is cautioned about using open sections in compression having less than two axes of symmetry, particularly when high width to thickness ratios for the elements exist. AISC–1.9.1 on local buckling provides some control over the more critical cases, since local buckling of sections such as angles, flanges, and tees is closely related to torsional buckling. AISC Appendix Table C1 gives limiting proportions for channels and tees to preclude torsional buckling.

PROBLEMS

8.1. For the section shown in the accompanying figure:
(a) Compute and draw to scale the shear flow, vt, distribution.
(b) Determine the total shear force in terms of V_y in each element of the cross

section, and show the direction of each, if the centroidal shear is positive V_y.

(c) Determine the location of the shear center.

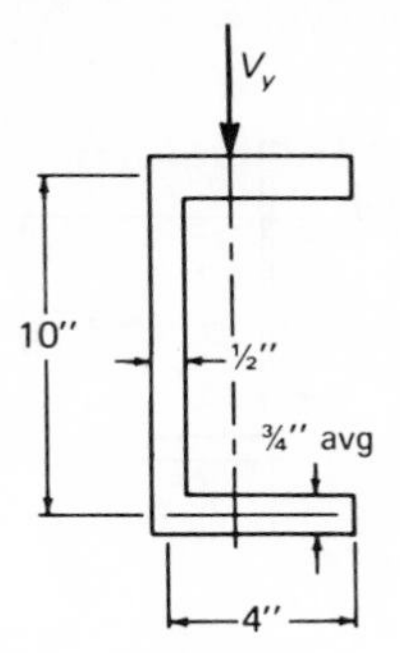

Prob. 8.1.

8.2. Repeat the requirements of Prob. 8.1 for the channel with sloping flanges. Comment on the effect of using average thickness instead of the actual sloping flanges for determining shear center on standard rolled channels.

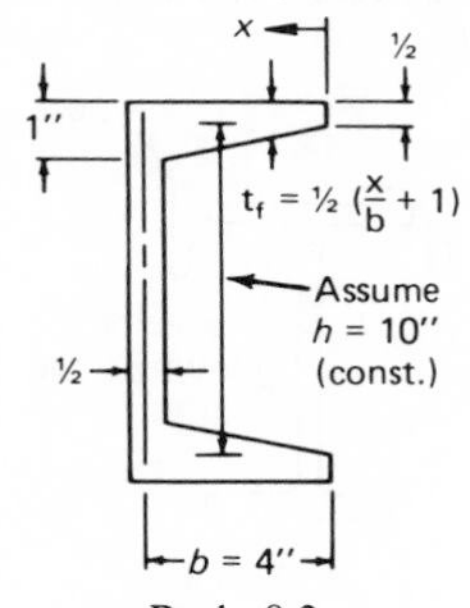

Prob. 8.2.

8.3. Repeat the requirements of Prob. 8.1 for channel with unequal flanges. Parts (a), (b), and (c) are to be done for V_x and V_y applied separately through the centroid. When finding both coordinates of the shear center, verify that statics of the shear forces is satisfied.

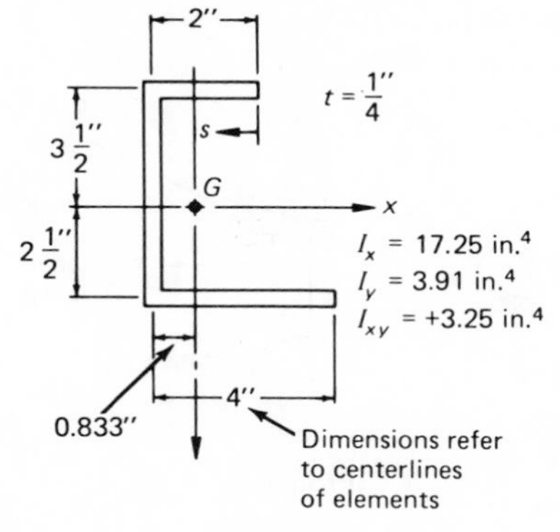

Prob. 8.3.

8.4. Locate the shear center for the angle and zee sections shown, including computing and showing the direction of the total shear force in each element of the cross section under the action separately of V_x and V_y. Show shear flow distribution in each element.

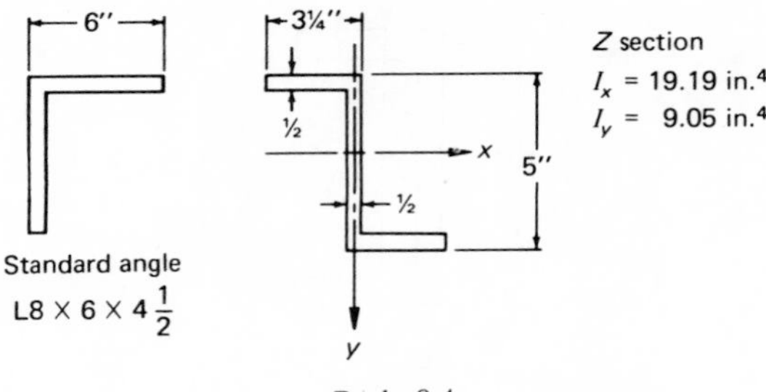

Prob. 8.4.

8.5. Locate the shear center for the combined W and channel crane girder section. Is there significant error in assuming the shear center lies at the centroid? Use average thickness and constant depth for the channel.

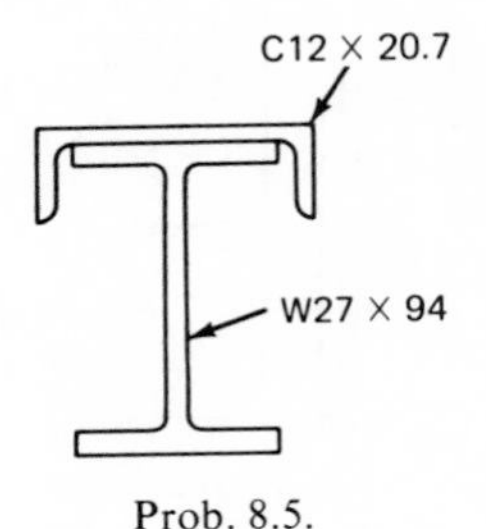

Prob. 8.5.

8.6. Suppose the given 8 × 6 × ½ angle, positioned with its long leg pointing downward, is used as a beam to span 12 ft. The uniform gravity loading (including angle weight) is 0.6 kips/ft, and the horizontal leg is to be restrained laterally so the angle will bend vertically. Assuming the attachment to the horizontal leg is simply supported, what moment capacity against lateral load must it be designed to carry?

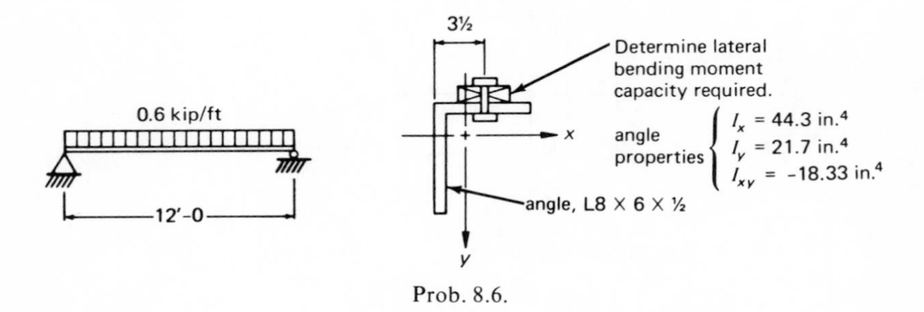

Prob. 8.6.

8.7. An MC 18 × 58, is to be used on a 24 ft simply supported span to carry a total load of 0.8 kips/ft, with the load applied in the plane of the web. Suppose the flanges are to have attachments so that the channel will bend vertically about the x axis. What lateral moment capacity M_y should the attachments be capable of resisting? What percent of M_x does this represent?

8.8. (a) Develop the torsion differential equation solution for the W beam with torsionally fixed ends, having an eccentrically applied concentrated load at midspan.
(b) Compute the torsion constant K, the warping constant I_w, and λ.
(c) Compute the combined bending stress, including warping torsion and ordinary flexure components.
(d) Compute the maximum shear stress in the web, including the Saint-Venant torsion, and flexural shear.
(e) Compute the maximum shear stress in the flange, including Saint-Venant torsion, warping shear, and vertical flexure flange shear. Give a tabular summary of all stresses.

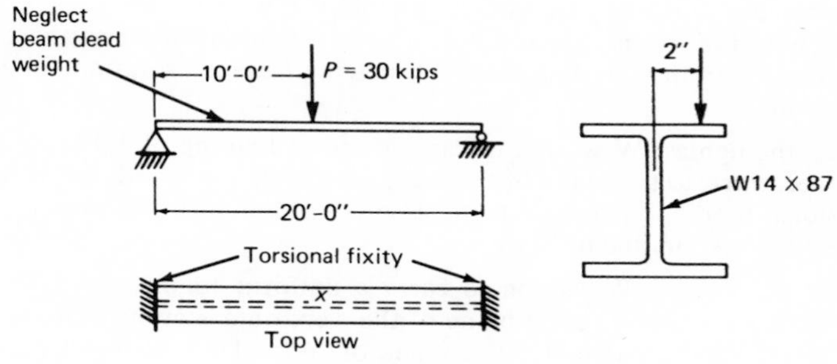

Prob. 8.8.

8.9. Develop the torsion differential equation solution for the cantilever beam with an eccentric concentrated load at its end, and compute constants and stresses as given in items (b) through (e) of Prob. 8.8. Is there any relationship between this problem and Prob. 8.8?

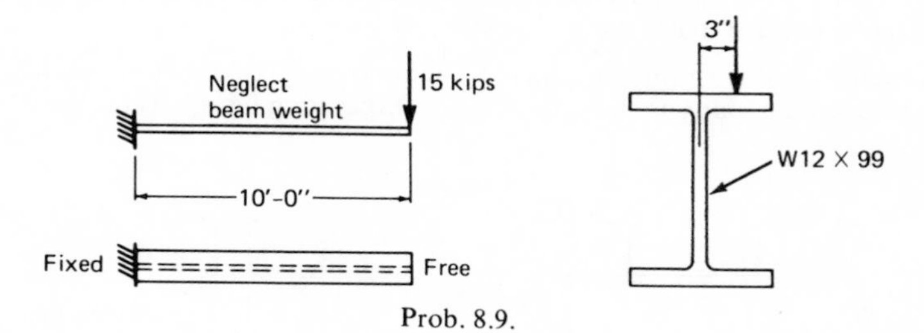

Prob. 8.9.

8.10. Develop the torsion differential equation solution for the uniformly loaded beam with loading applied eccentrically to the web. Consider the ends torsionally simply supported. Compute constants and stresses as given in items (b) through (e) of Prob. 8.8.

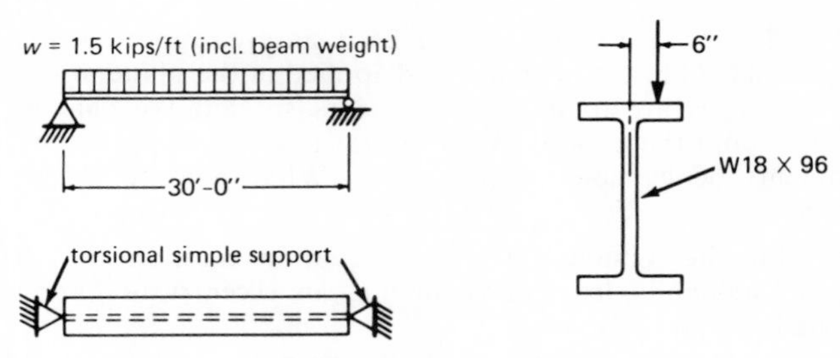

Prob. 8.10.

8.11. Repeat Prob. 8.10, considering the ends torsionally fixed.

8.12. Repeat Prob. 8.8, with the load applied in the plane of the y axis of a channel, C15 × 50.

8.13. Select a W14 section, using the modified flexure analogy approach, to carry a 25-kip concentrated load at midspan of a 20-ft simply supported span. The 25-kip load is eccentric to the web by 4 in. and the allowable combined compressive stress is 22 ksi. Use A36 shear allowables of AISC. The ends of the beam are assumed to have torsional simple support. Check the stresses in the selected section using the "exact" solution for assumed conditions.

8.14. Select the lightest W section to carry uniform loading of 2.0 kips/ft, acting 7-in. eccentric to the plane of the web, on a fixed end (both for vertical and torsional loading) span of 28 ft. Use A36 steel with the maximum compressive stress limited to 22 ksi.

8.15. Select the lightest W section to carry a uniform loading of 1.75 kips/ft, acting 5 in. eccentric to the plane of the web, on a simply supported (both for vertical and torsional loading) span of 26 ft. Use steel with F_y = 50 ksi and assume the maximum compressive stress may be 30 ksi.

8.16. Repeat Prob. 8.15, assuming the ends fixed with respect to torsional restraint only. If Prob. 8.15 has also been solved, discuss the effect of considering torsional restraint as fixed or hinged.

8.17. Repeat Prob. 8.15 using A36 steel with allowable compressive stress of 22 ksi. If F_y = 50 ksi, material costs 7 percent more per pound than A36 steel, is it economical to use the higher strength material?

8.18. Given the 40-ft simply supported span carrying two symmetrically placed concentrated loads as shown in the accompanying figure. If the loads are ec-

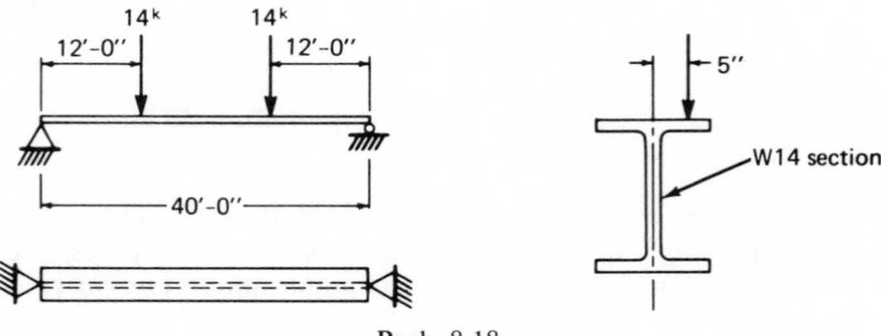

Prob. 8.18.

centric to the web by 5 in., and the member is torsionally simply supported, select the lightest W14 section suitable using the modified flexure analogy method, assuming the maximum combined stress is limited to 22 ksi.

8.19. The 30-ft simply supported (for M_x) span is to carry two symmetrically placed concentrated loads of 22 kips located 10 ft from the supports. The loads are 6-in. eccentric to the web, and full fixity is assumed for torsional restraint. Neglect the dead weight of the beam. Select, using the modified flexure analogy method, the lightest W section such that the maximum combined stress does not exceed 22 ksi.

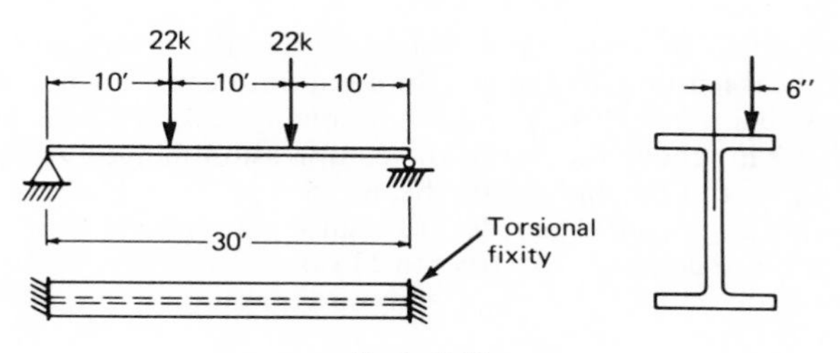

Prob. 8.19.

8.20. Estimate, using the modified flexure analogy design procedure used for W shapes and crane girders, the uniformly distributed load capacity for a C15 × 50 of A36 steel. Assume that the allowable compressive stress is $0.6F_y$. The ends of the beam are restrained against rotation ($\phi = 0$), but the ends are free to warp. Check stresses using the differential equation solution.

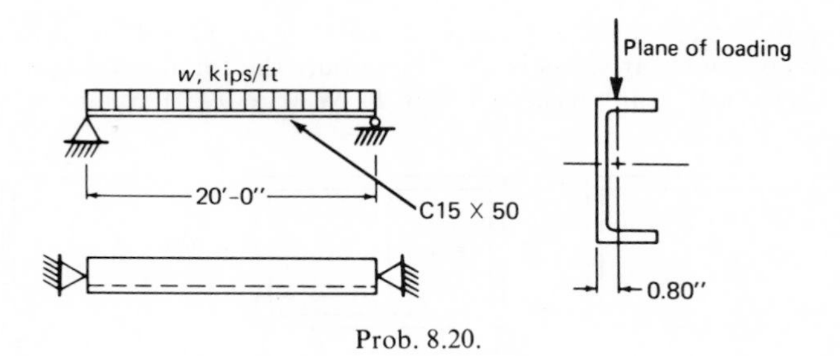

Prob. 8.20.

8.21. A simply supported beam is to carry 0.8 kip/ft on a span of 24 ft. A channel section is to be used, and since it is not laterally restrained, torsion is to be considered in the design. Assume the member is torsionally simply supported. After computing the torsional moment, use the generally accepted approximate flexure analogy to make a selection of an American Standard channel (C-section) from the AISC Manual. Assume the allowable combined stress to be $0.6F_y$. Investigate the stress on the selected section using the exact solution based on the stated loading and support condition. Use A36 steel.

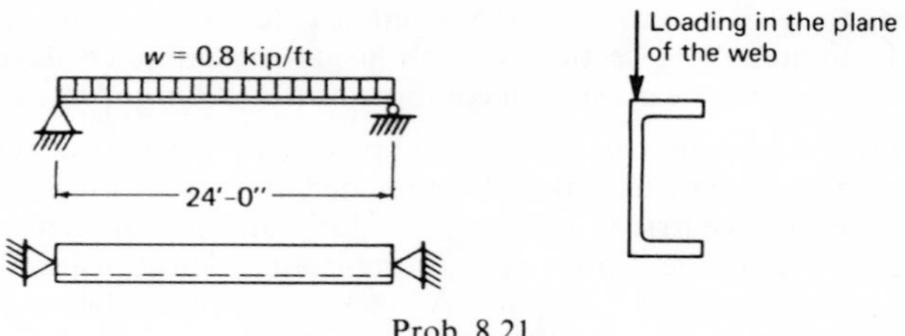

Prob. 8.21.

8.22. Assume a single W section is to serve as a crane runway girder which carries a vertical loading, as shown. In addition, design must include an axial compressive force of 10 kips and a horizontal force of 3 kips on each wheel applied at 3 in. above the top of the compression flange. Assume torsional simple support at the ends of the beam. Select the lightest W14 section of A36 steel using the modified flexure analogy approach, limiting the maximum combined compressive stress to 22 ksi.

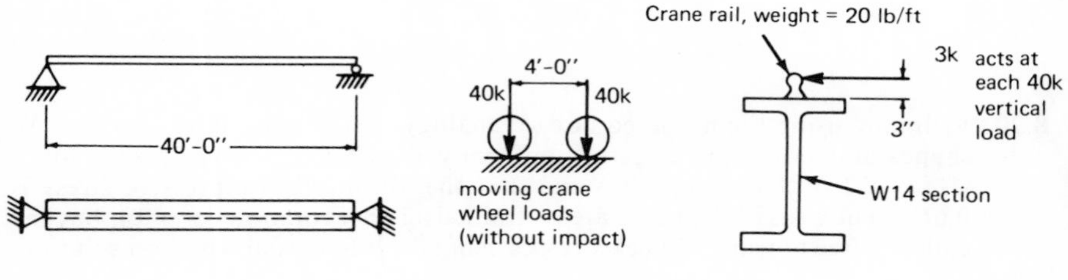

Prob. 8.22.

8.23. Compare the torsional constants K for each of the following sections. If the maximum shear stress is 14 ksi compute the torsional moment capacity of each section. Can this be done for the W30 × 99? Explain.

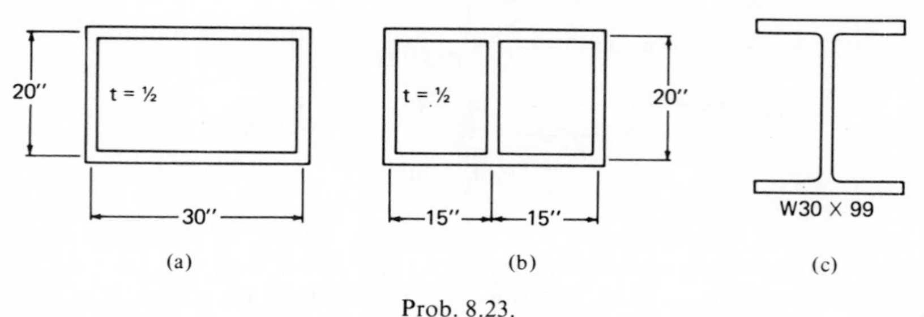

(a) (b) (c)

Prob. 8.23.

8.24. If the maximum shear stress is limited to 14 ksi, compute the torsional moment capacity for the section given. What is the percentage change in capacity if the interior walls are omitted (assuming local buckling does not control)?

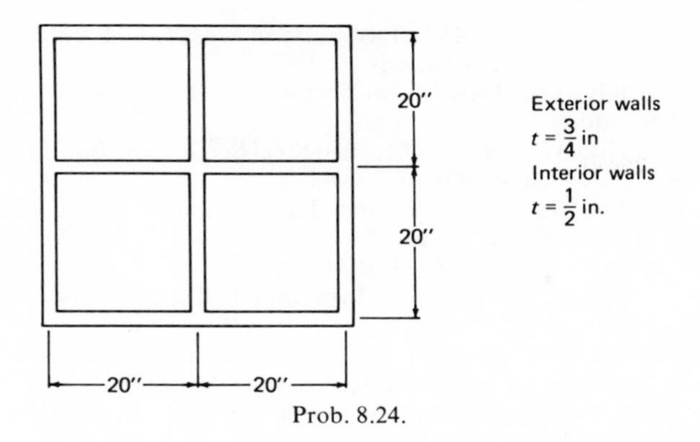

Prob. 8.24.

8.25. Investigate the possibility of torsional buckling occurring on the following beams: W16 × 31, W6 × 15.5, M8 × 6.5. For each, plot equivalent slenderness ratio, r_E, against length of column. Give conclusions.

8.26. Estimate the buckling load, assuming zero residual stress and elastic buckling so that Euler's equation applies, for the following sections on a pin-end length of 10 ft.

(a) MC10 × 6.5
(b) L4 × 4 × ¼
(c) WT7 × 15
(d) Zee section shown.

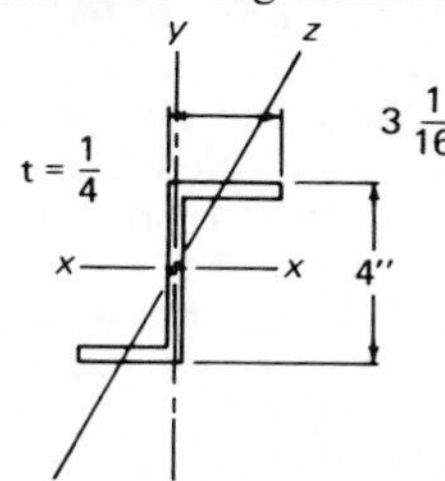

At what length would torsional buckling be likely to control?

SELECTED REFERENCES

1. Fred B. Seely and James O. Smith, *Advanced Mechanics of Materials*, 2nd ed., John Wiley & Sons, Inc., New York, 1952, Chap. 9.
2. S. Timoshenko, *Strength of Materials*, Part II, 2nd ed., D. Van Nostrand Company, Inc., New York, 1941, Chap. 6.
3. William McGuire, *Steel Structures*, Prentice-Hall, Inc., Englewood Cliffs, N.J., 1968, pp. 346–400.
4. I. Lyse and B. G. Johnston, "Structural Beams in Torsion," *Trans. ASCE*, Vol. 101 (1936), pp. 857–926.
5. F. K. Chang and Bruce G. Johnston, "Torsion of Plate Girders," *Trans. ASCE*, Vol. 118 (1953), pp. 337–396.
6. Gerald G. Kubo, Bruce G. Johnston, and William J. Eney, "Nonuniform Torsion of Plate Girders," *Trans. ASCE*, Vol. 121 (1956), pp. 759–785. (Good summary of torsion theory.)
7. I. A. El Darwish and Bruce G. Johnston, "Torsion of Structural Shapes," *J. Structural Div. ASCE*, Vol. 91, ST 1 (February 1965), pp. 203–227. (Er-

rata: ST 1: Feb. 66: 471:4638.) See also *Trans. ASCE*, Vol. 131 (1966), pp. 428–429 for summary of equations.

8. "Torsion Analysis of Rolled Steel Sections," Handbook 1963-B, Bethlehem Steel Corporation.
9. John G. Hotchkiss, "Torsion of Rolled Steel Sections in Building Structures," *AISC Engg. J.*, January 1966, pp. 19–45.
10. S. Timoshenko, "Theory of Bending, Torsion, and Buckling of Thin-Walled Members of Open Cross Section," *J. Franklin Inst.*, Vol. 239, Nos. 3, 4, and 5 (1945), pp. 201–219; 249–268, and 343–361.
11. John E. Goldberg, "Torsion of I-Type and H-Type Beams," *Trans. ASCE*, Vol. 118 (1953), pp. 771–793.
12. J. N. Goodier and M. V. Barton, "The Effects of Web Deformation on the Torsion of I-Beams," *J. Appl. Mech.*, (March 1944), p. A–35.
13. J. E. Lothers, "Torsion in Steel Spandrel Girders," *Trans. ASCE*, Vol. 112 (1947), pp. 345–376.
14. Warner Lansing, "Thin-Walled Members in Combined Torsion and Flexure," *Trans. ASCE*, Vol. 118 (1953), pp. 128–146. (Particular emphasis on channel and zee sections.)
15. Basil Sourochnikoff, "Strength of I-Beams in Combined Bending and Torsion," *Trans. ASCE*, Vol. 116 (1951), pp. 1319–1342.
16. L. C. Maugh, *Statically Indeterminate Structures*, John Wiley & Sons, Inc., New York, 1946, pp. 309–320.
17. Alfred S. Niles and Joseph S. Newell, *Airplane Structures*, 3rd ed., John Wiley & Sons Inc., New York, 1943, Chap. 16.
18. Charles E. Cutts, "Horizontally Curved Box Beams," *Trans. ASCE*, Vol. 118 (1953), pp. 517–544.
19. Lewis P. Felton and M. W. Dobbs, "Optimum Design of Tubes for Bending and Torsion," *J. Structural Division ASCE*, Vol. 93, No. ST4 (August 1967), pp. 185–200.
20. Kuang-Han Chu and Anatole Longinow, "Torsion in Sections with Open and Closed Parts," *J. Structural Division ASCE*, Vol. 93, No. ST6 (December 1967), pp. 213–227.
21. S. P. Timoshenko and J. M. Gere, *Theory of Elastic Stability*, 2nd ed., McGraw-Hill Book Company, 1961, pp. 225–250.
22. Friedrich Bleich, *Buckling Strength of Metal Structures*, McGraw-Hill Book Company, 1952.
23. E. H. Gaylord, Jr., and C. N. Gaylord, *Design of Steel Structures*, McGraw-Hill Book Company, 1957, pp. 132–135.

9

Lateral-Torsional Buckling of Beams

9.1. RATIONAL ANALOGY TO PURE COLUMNS

Emphasis in this chapter is on the stability considerations associated with bending. In beams as in axially-loaded columns it is not possible to achieve perfect loading, i.e., beams are never perfectly straight, not perfectly homogeneous, and are usually not loaded in exactly the plane that is assumed for design and analysis.

Consider the compression zone of the laterally unsupported beam of Fig. 9.1.1. With the loading in the plane of the web, according to ordinary beam theory, points A and B are equally stressed. Imperfections in the beam and accidental eccentricity in loading actually result in different stresses at A and B. Furthermore, residual stresses as discussed in Chapter 6 contribute to unequal stresses across the flange width at any distance from the neutral axis.

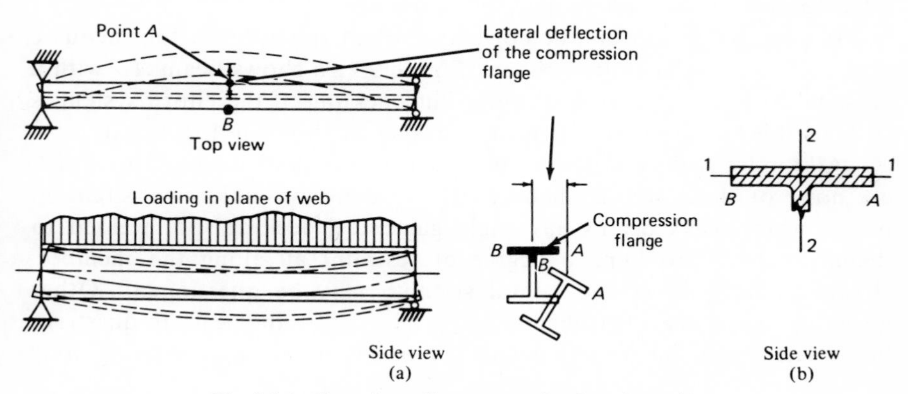

Fig. 9.1.1. Beam laterally supported only at its ends.

In a qualitative way one may look upon the compression flange of a beam as a column, with all the considerations treated in Chapter 6. The rectangular flange as a column would ordinarily buckle in its weak direction, by bending about an axis such as 1–1 of Fig. 9.1.1b, but the web provides continuous support to prevent such buckling. At higher compressive loads the rectangular flange will tend to buckle with respect to axis 2–2 of Fig. 9.1.1b. It is this sudden buckling of the flange about its strong axis in a lateral direction, that is commonly referred to as lateral buckling. The analogy between the compression flange of a beam and a column is intended to present only the general behavior for lateral buckling.

In order to evaluate this behavior more precisely, one must realize that the compression flange is not only braced in its weak direction by its attachment via the web to the stable tension flange, but the web also provides continuous moment and shear restraint against large lateral deflection by means of its stiffness. Thus the entire section is brought into action when lateral motion commences.

9.2. LATERAL SUPPORT

Rarely does the beam exist with its compression flange entirely free of all restraint. Even when it does not have a positive connection to a floor or roof system, there is still friction between the beam flange and whatever it supports. There are two categories of lateral support which are definite and adequate; these are:

(a) Continuous lateral support of high strength by embedment of the compression flange in a concrete floor slab (Fig. 9.2.1a and b).

(b) Lateral support at intervals (Fig. 9.2.1c through g) provided by cross beams, cross frames, ties, or struts, framing in laterally, where the lateral system is itself adequately stiff and braced.

It is necessary to examine not only the individual beam for adequate bracing, but also the entire system. Figure 9.2.2a shows beam *AB* with a cross beam framing in at midlength, but buckling of the entire system is still possible unless the system is braced, such as shown in Fig. 9.2.2b.

All too frequently in design, the engineer encounters situations which are none of these well-defined cases. A common unknown situation occurs when heavy beams have light-gage steel decking spot welded to them; certainly providing a degree of restraint all along the member. However, the stiffness and lateral strength may be questioned. Other questionable cases are (a) where bracing frames into the beam in question, but it is at or near the tension flange; (b) timber or light-gage decking floor

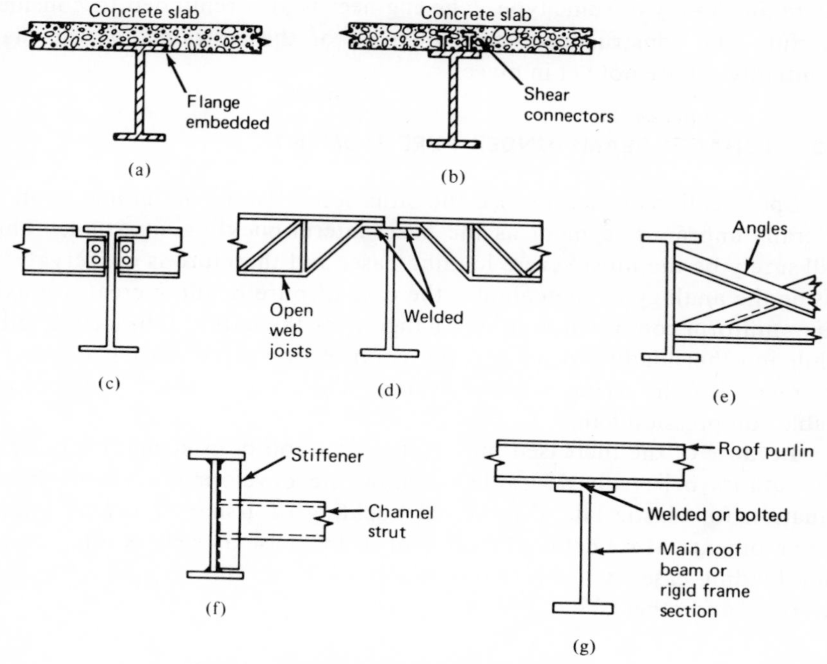

Fig. 9.2.1. Types of definite lateral support.

systems which rest on but are not solidly attached to the beams; and (c) rigid frames which are enclosed in light gage metal sheathing.

However, it is better to assume no lateral support in doubtful situations. Alternatively, it may be possible in some cases to evaluate it as an elastic restraint. Such an analytical approach is discussed in Sec. 9.10.

Lateral-support considerations must not be ignored; probably most failures in steel structures are the result of inadequate bracing against

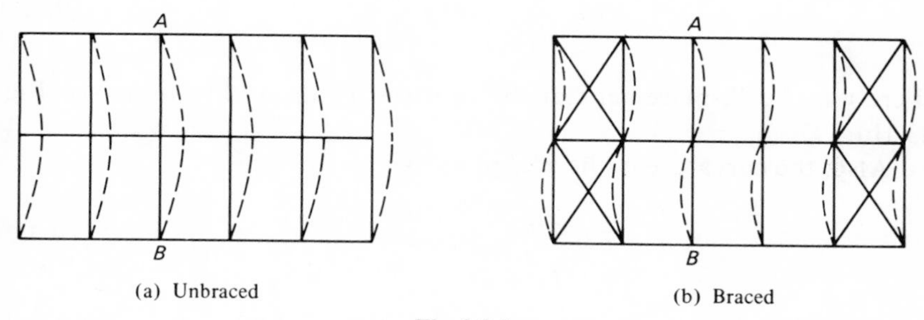

Fig. 9.2.2.

lateral instability of some type. The engineer is also reminded to consider carefully the construction stage when all of the restraints which may eventually act are not yet in place.

9.3.* I-SHAPED BEAMS UNDER PURE MOMENT

Specifications generally use the situation of uniform moment on a laterally unbraced segment as the basic lateral-buckling situation. This will surely be the most severe loading case, and therefore is conservative. Using the analogy to the column, the case of pure bending creates maximum uniform compression in one flange over the entire unbraced length, while in other loading situations the compression stress varies along the unbraced length, giving a lower average compression force and a more stable compression flange.

Because of the increased importance of stability in structural design, the authors believe it desirable to show the development of the basic equation for elastic lateral-torsional buckling of I-shaped beams under pure moment. More detailed treatment of this case as well as other common loading cases is to be found in the work of Timoshenko,[1] Bleich,[2] deVries,[3] and others.[4–7]

Differential Equation for Elastic Lateral Buckling Under Pure Moment. Referring to Fig. 9.3.1., which shows the beam in a buckled position, it is observed that the applied moment M_0 in the yz plane will give rise to moment components $M_{x'}$, $M_{y'}$, and $M_{z'}$, about the x', y', and z' axes respectively. This means there will be bending curvature in both the $x'z'$ and $y'z'$ planes as well as torsional curvature about the z' axis. Assuming small deformation, the bending in the $y'z'$ plane (considering the direction cosine is 1 between y' and y, and z' and z axes) may be written

$$EI_x \frac{d^2v}{dz^2} = M_{x'} = M_0 \tag{9.3.1}$$

where v is the displacement of the centroid in the y direction (see Fig. 9.3.1b).

Also, the curvature in the $x'z'$ plane is

$$EI_y \frac{d^2u}{dz^2} = M_{y'} = M_0\phi \tag{9.3.2}$$

as is seen from Fig. 9.3.1c, where u is the displacement of the centroid in the x direction.

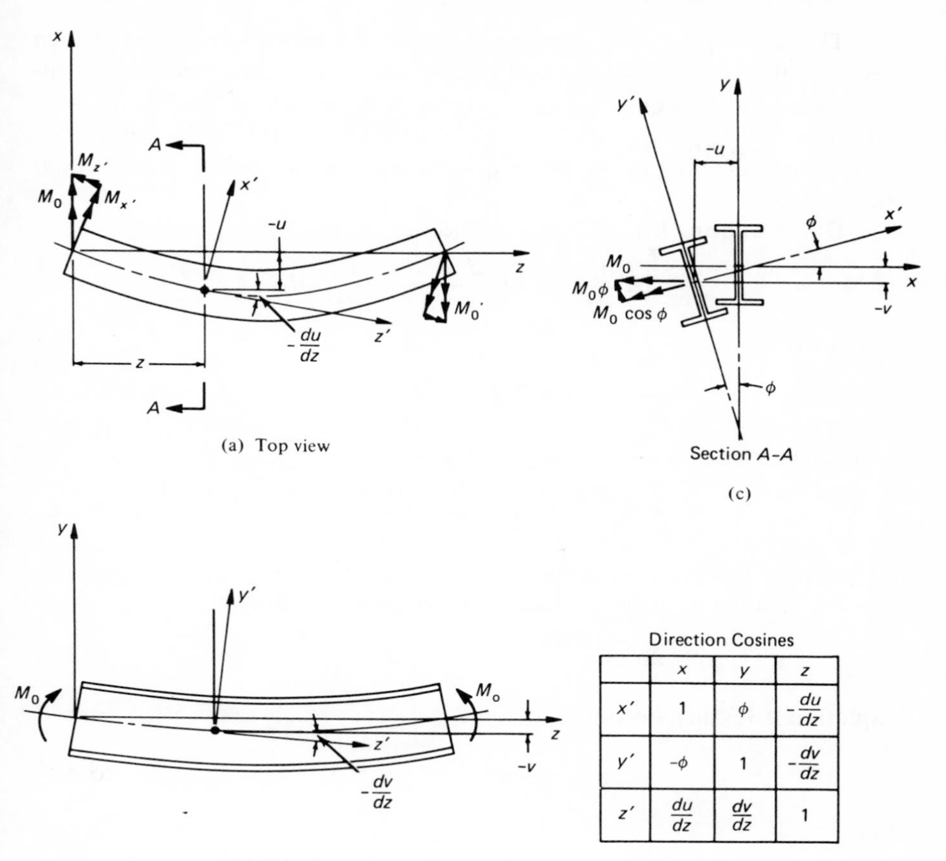

	x	y	z
x′	1	ϕ	$-\frac{du}{dz}$
y′	$-\phi$	1	$-\frac{dv}{dz}$
z′	$\frac{du}{dz}$	$\frac{dv}{dz}$	1

Fig. 9.3.1. I-shaped beam in slightly buckled position.

The differential equation for torsion of I-shaped beams was developed in Chapter 8 as Eq. 8.5.10, as follows:

$$M_{z'} = GK \frac{d\phi}{dz} - EI_w \frac{d^3\phi}{dz^3} \qquad [8.5.10]$$

From Fig. 9.3.1 and the direction cosines, the torsional component of M_0 when the beam is slightly buckled is proportional to the slope of the beam in the xz plane:

$$M_{z'} = -\frac{du}{dz} M_0 \tag{9.3.3}$$

which gives for the torsional differential equation

$$GK \frac{d\phi}{dz} - EI_w \frac{d^3\phi}{dz^3} = -\frac{du}{dz} M_0 \tag{9.3.4}$$

Two assumptions are inherent in Eqs. 9.3.1 and 9.3.2, both of which relate to the assumption of small deformation. It is assumed that properties $I_{x'}$, and $I_{y'}$, equal I_x and I_y, respectively; and also that I_x is large compared to I_y, so that Eq. 9.3.1 is not linked to Eqs. 9.3.2 and 9.3.4. Thus displacement v in the plane of bending does not affect the torsional function ϕ.

Differentiating Eq. 9.3.4 with respect to z gives

$$GK \frac{d^2\phi}{dz^2} - EI_w \frac{d^4\phi}{dz^4} = -\frac{d^2u}{dz^2} M_0 \tag{9.3.5}$$

From Eq. 9.3.2,

$$\frac{d^2u}{dz^2} = \frac{M_0\phi}{EI_y}$$

which when substituted into Eq. 9.3.5 gives

$$EI_w \frac{d^4\phi}{dz^4} - GK \frac{d^2\phi}{dz^2} - \frac{M_0^2}{EI_y} \phi = 0 \tag{9.3.6}$$

which is the differential equation for the angle of twist.

To obtain a solution for Eq. 9.3.6, divide by EI_w and let

$$2\alpha = \frac{GK}{EI_w} \quad \text{and} \quad \beta = \frac{M_0^2}{E^2 I_w I_y} \tag{9.3.7}$$

Equation 9.3.6 then becomes

$$\frac{d^4\phi}{dz^4} - 2\alpha \frac{d^2\phi}{dz^2} - \beta\phi = 0 \tag{9.3.8}$$

Let

$$\left.\begin{aligned} \phi &= Ae^{mz} \\ \frac{d^2\phi}{dz^2} &= Am^2e^{mz} \\ \frac{d^4\phi}{dz^4} &= Am^4e^{mz} \end{aligned}\right\} \tag{9.3.9}$$

Substitution of Eq. 9.3.9 into Eq. 9.3.8 gives

$$Ae^{mz}(m^4 - 2\alpha m^2 - \beta) = 0 \tag{9.3.10}$$

Since e^{mz} cannot be zero and A can be zero only if no buckling has occurred, the bracket expression of Eq. 9.3.10 must be zero:

$$m^4 - 2\alpha m^2 - \beta = 0$$

which gives for the solution

$$m^2 = \alpha \pm \sqrt{\beta + \alpha^2}$$

or

$$m = \pm \sqrt{\alpha \pm \sqrt{\beta + \alpha^2}} \tag{9.3.11}$$

It is apparent from Eq. 9.3.11 that m will consist of two real and two complex roots because

$$\sqrt{\beta + \alpha^2} > \alpha$$

Let

$$n^2 = \alpha + \sqrt{\beta + \alpha^2} \quad \text{(both real roots)} \tag{9.3.12}$$

$$q^2 = -\alpha + \sqrt{\beta + \alpha^2} \quad \text{(real part of complex roots)} \tag{9.3.13}$$

Using the four values for m, the expression for ϕ from Eq. 9.3.9 becomes

$$\phi = A_1 e^{nz} + A_2 e^{-nz} + A_3 e^{iqz} + A_4 e^{-iqz} \tag{9.3.14}$$

Since

$$\left.\begin{aligned} e^{iqz} &= \cos qz + i \sin qz \\ e^{-iqz} &= \cos qz - i \sin qz \end{aligned}\right\} \tag{9.3.15}$$

By using Eq. 9.3.15 and defining new constants A_3 and A_4 which equal $(A_3 + A_4)$ and $(A_3 i - A_4 i)$, respectively, one obtains

$$\phi = A_1 e^{nz} + A_2 e^{-nz} + A_3 \cos qz + A_4 \sin qz \tag{9.3.16}$$

The constants A_1 through A_4 are determined by the end support conditions. For the case of torsional simple support, i.e., beam ends may not twist, but are free to warp, the conditions are

$$\phi = 0, \ \frac{d^2\phi}{dz^2} = 0 \qquad \text{at } z = 0 \text{ and } z = L$$

For $\phi = 0$ at $z = 0$, Eq. 9.3.16 gives

$$0 = A_1 + A_2 + A_3 \tag{9.3.17}$$

For $\dfrac{d^2\phi}{dz^2} = 0$ at $z = 0$,

$$0 = A_1 n^2 + A_2 n^2 - A_3 q^2 \tag{9.3.18}$$

Multiplying Eq. 9.3.17 by n^2 and subtracting Eq. 9.3.18 gives

$$0 = A_3(q^2 + n^2), \qquad \underline{\therefore A_3 = 0}$$

Then, from Eq. 9.3.17,

$$A_1 = -A_2 \tag{9.3.19}$$

Thus Eq. 9.3.16 becomes

$$\phi = A_1(e^{nz} - e^{-nz}) + A_4 \sin qz \tag{9.3.20}$$

which may be written

$$\phi = 2A_1 \sinh nz + A_4 \sin qz \tag{9.3.21}$$

At $z = L$, $\phi = 0$; therefore, from Eq. 9.3.21,

$$0 = 2A_1 \sinh nL + A_4 \sin qL \tag{9.3.22}$$

Also, at $z = L$, $d^2\phi/dz^2 = 0$, which gives

$$0 = 2A_1 n^2 \sinh nL - A_4 q^2 \sin qL \tag{9.3.23}$$

Multiplying Eq. 9.3.22 by q^2 and adding to Eq. 9.3.23 gives

$$2A_1(n^2 + q^2)\sinh nL = 0 \tag{9.3.24}$$

Since $(n^2 + q^2)$ cannot be zero, and sinh nL can be zero only if $n = 0$, therefore A_1 must be zero:

$$\underline{A_1 = -A_2 = 0}$$

Finally, from Eqs. 9.3.21 and 9.3.22,

$$\phi = A_4 \sin qL = 0 \tag{9.3.25}$$

If lateral-torsional buckling occurs, A_4 cannot be zero, so that

$$\sin qL = 0 \tag{9.3.26}$$

$$qL = N\pi$$

where N is any integer.

The elastic-buckling condition is defined by

$$q = \frac{N\pi}{L} \tag{9.3.27}$$

which for the fundamental buckling mode $N = 1$.

The value of M_0 which satisfies Eq. 9.3.27 is said to be the critical moment:

$$q = \sqrt{-\alpha + \sqrt{\beta + \alpha^2}} = \frac{\pi}{L} \tag{9.3.28}$$

Squaring both sides, and substituting the definitions of α and β, from Eqs. 9.3.7,

$$-\frac{GK}{2EI_w} + \sqrt{\frac{M_0^2}{E^2 I_w I_y} + \left(\frac{GK}{2EI_w}\right)^2} = \frac{\pi^2}{L^2} \tag{9.3.29}$$

Solving for $M_0 = M_{cr}$ gives

$$M_{cr}^2 = E^2 I_w I_y \left[\left(\frac{\pi^2}{L^2} + \frac{GK}{2EI_w}\right)^2 - \left(\frac{GK}{2EI_w}\right)^2\right] \tag{9.3.30}$$

$$M_{cr} = \sqrt{\frac{\pi^4 E^2 I_w I_y}{L^4} + \frac{\pi^2 EI_y GK}{L^2}} \tag{9.3.31}$$

Equation 9.3.31 may be converted to unit stress by dividing by the section modulus, S_x,

$$F_{cr} = \frac{M_{cr}}{S_x} = \frac{\pi\sqrt{EI_y GK}}{LS_x}\sqrt{\frac{\pi^2 EI_w}{L^2 GK} + 1} \tag{9.3.32}$$

or alternatively, factoring from the first term of Eq. 9.3.31,

$$F_{cr} = \frac{\pi^2 E\sqrt{I_w I_y}}{L^2 S_x}\sqrt{1 + \frac{L^2 GK}{\pi^2 EI_w}} \tag{9.3.33}$$

Using the torsional property definitions of Chapter 8, $\lambda^2 = 1/a^2 = GK/EI_w$, and $I_w \approx I_y h^2/4$, Eqs. 9.3.32 and 9.3.33 become

$$F_{cr} = \frac{\pi\sqrt{EI_y GK}}{LS_x}\sqrt{\left(\frac{\pi}{\lambda L}\right)^2 + 1} \tag{9.3.34}$$

$$F_{cr} = \frac{\pi^2 EI_y}{L^2 S_x}\left(\frac{h}{2}\right)\sqrt{1 + \left(\frac{\lambda L}{\pi}\right)^2} \tag{9.3.35}$$

Equation 9.3.34 may be a convenient form where torsional stiffness is high (i.e., λL is large), and Eq. 9.3.35 may be a convenient form where lateral bending stiffness controls and torsional stiffness is small (i.e., the Euler-type multiplier, $\pi^2 EI_y/L^2$, and low values of λL.)

Lateral Buckling in the Plastic Range. When design procedures consider moments greater than $M_y = SF_y$ and up to M_p, the plastic moment, lateral buckling is a distinct possibility. Both working-stress and plastic-design procedures inherently require lateral bracing at locations where plastic hinges are expected to occur in the failure mechanism. Upon reaching a plastic hinge at any section, the extreme fibers will be strained near or into the strain hardening region.

If the rigidities EI_y and GK are taken to include the values in the inelastic range as well as the elastic range, the lateral-torsional buckling equilibrium equation for pure moment, Eq. 9.3.31, may also be used for the plastic range. Since for beams where plastic moments are assumed to develop, the distances between lateral support points will be relatively short; it has been determined[10] that the term involving torsional rigidity GK may be neglected. Thus Eq. 9.3.31, neglecting the second term, becomes

$$M_{cr} = \frac{\pi^2 E}{L^2}\sqrt{I_w I_y} \tag{9.3.36}$$

and since M_{cr} must reach $M_p = ZF_y$; including $I_w = I_y h^2/4$ and $I_y = Ar_y^2$, the maximum slenderness ratio is obtained,

$$\frac{L}{r_y} = \sqrt{\frac{\pi^2 E}{2F_y}\left(\frac{hA}{Z}\right)} \tag{9.3.37}$$

for uniform plastic moment, as in Fig. 9.3.2. The extreme fiber strain will

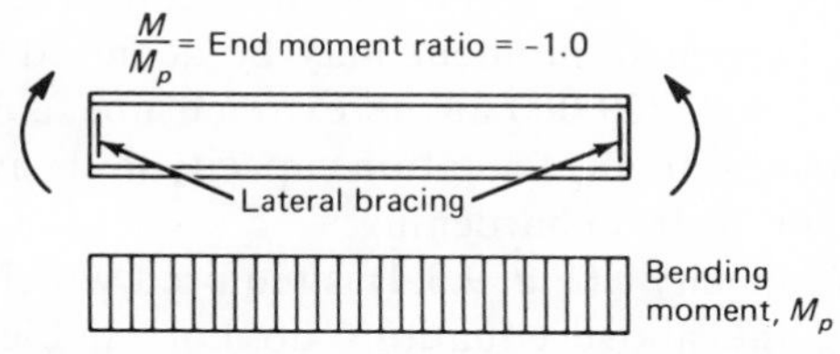

Fig. 9.3.2. Uniform plastic moment over unbraced length.

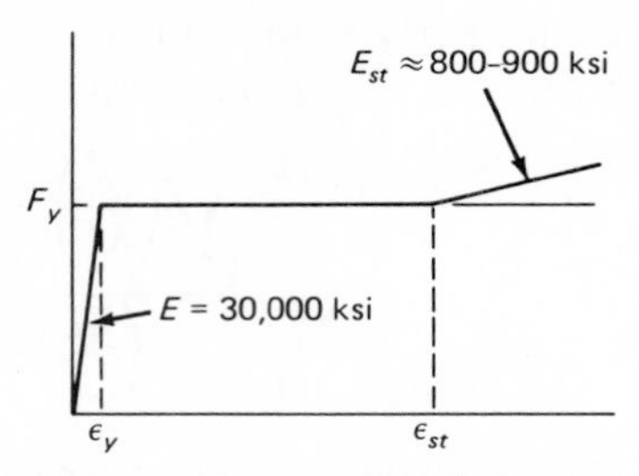

Fig. 9.3.3. Stress-strain relationship.

be near the strain hardening range; thus $E_{st} \approx$ 800–900 ksi (see Fig. 9.3.3) is used in place of the elastic E:

$$\frac{L}{r_y} = \sqrt{\frac{\pi^2(900)}{2F_y}}\sqrt{\frac{h(2A_f + A_w)}{2\left[A_f\frac{h}{2} + \frac{A_w}{2}\left(\frac{h}{4}\right)\right]}}$$

$$= \frac{66.7}{\sqrt{F_y}}\sqrt{\frac{2 + A_w/A_f}{1 + A_w/(4A_f)}}$$

as A_w/A_f varies from about 0.5 to 1.5, $\sqrt{hA/Z}$ varies from 1.49 to 1.60. Considering that uniform plastic moment over an unbraced length is an ideal and very conservative case, use $\sqrt{hA/Z} = 1.6$:

$$\frac{L}{r_y} = \frac{106.7}{\sqrt{F_y}} \tag{9.3.38}$$

which for $F_y = 36$ ksi gives $L/r_y = 18$; a very severe restriction.

Factors not considered in the development but which decrease the likelihood of lateral buckling are: (a) torsional stiffness which was neglected; (b) uniform plastic moment which is an unachievable limiting case—moment gradient should be considered; and (c) end restraint which considerably increases the maximum safe unbraced length.

If the elastic modulus, $E = 29{,}000$ ksi, had been used in Eq. 9.3.37,

$$\frac{L}{r_y} = \frac{605}{\sqrt{F_y}} \tag{9.3.39}$$

which for $F_y = 36$ ksi gives $L/r_y \approx 100$.

From the preceding discussion, one may conclude that if L/r_y is restricted to 100, the plastic moment may be achieved with a rotation θ_p (i.e., a strain of $\epsilon_y = F_y/29{,}000$ at the extreme fiber). If the slenderness ratio, L/r_y is restricted to 18, a rotation capacity θ_{st} is available which will assure achievement of strain hardening.

If a rotation capacity θ (Fig. 9.3.4) is required which is greater than θ_p but less than θ_{st} the above equations do not enable determination of critical L/r_y. The work of Lay and Galambos[10] has provided a procedure

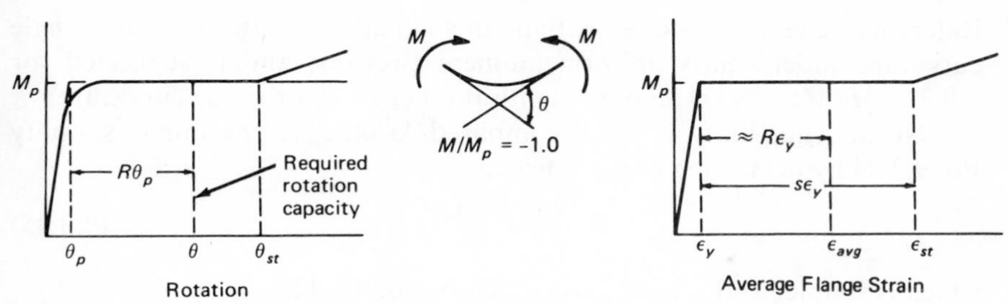

Fig. 9.3.4. Deformation requirements for plastic design.

and an equation for determining the intermediate values of L/r_y. Their equation is

$$\frac{kL}{r_y} = \sqrt{\frac{\pi^2 E}{F_y}} \frac{1}{\sqrt{1 + 0.7 \dfrac{E}{E_{st}} \left(\dfrac{R}{s - 1}\right)}} \tag{9.3.40}$$

where R and s are as defined in Fig. 9.3.4, and k is an effective length factor.

Letting ϕ be the proportion of the unbraced length that has yielded, the average strain ϵ_{avg} for the entire length is

$$\epsilon_{avg} = \epsilon_y(1 - \phi) + \phi s \epsilon_y \tag{9.3.41}$$

$$R = \frac{\epsilon_{avg}}{\epsilon_y} - 1 = \phi(s - 1)$$

$$\phi = \frac{R}{s - 1} \tag{9.3.42}$$

The practical maximum percentage of the span which may become plastified is probably about 80 percent; therefore, $\phi_{max} = 0.80$. The strain hardening modulus of elasticity E_{st} may be taken as

$$E_{st} \approx \frac{E}{F_y} \text{ ksi}$$

Equation 9.3.40 for single curvature with most of the length yielded becomes

$$\frac{kL}{r_y} = \frac{535}{\sqrt{F_y(1 + 0.56F_y)}} \tag{9.3.43}$$

where the effective length factor k should be taken as

$$k = 0.54 \text{ if the adjacent span is elastic}$$
$$= 0.80 \text{ if the adjacent span is yielded}$$

Examining the usual case of an elastic adjacent span, $k = 0.54$ which gives for Eq. 9.3.43,

$$\frac{L}{r_y} = \frac{990}{\sqrt{F_y(1 + 0.56F_y)}} \tag{9.3.44}$$

Reference 22 recommends essentially this equation for design when single curvature under nearly uniform moment prevails, and is suggested for $-0.7 > M/M_p > -1.0$ (end moment ratio negative for single curvature).

Noting that unity is small compared to $0.56F_y$, one might simplify Eq. 9.3.44 by neglecting unity; it becomes

$$\frac{L}{r_y} = \frac{1320}{F_y} \tag{9.3.45}$$

which is simpler and is little different from Eq. 9.3.44.

For situations where there is greater variation of moment along the unbraced length, including double curvature, Eq. 9.3.37 can be used along with multipliers to correct for other variables.

$$\frac{L}{r_y} = \sqrt{\frac{\pi^2 E_{st}}{2F_y}}\sqrt{\frac{hA}{Z}}\;\xi_p\xi_\alpha\xi_s\xi_\gamma \tag{9.3.46}$$

where ξ_p = moment gradient correction, which from Ref. 11 may be approximated as $1.34 + 0.34\,M/M_p$

ξ_α = partial yielding correction; varies from about 1.8 for $M/M_p = 0$ to about 1.3 for $M/M_p = -0.7$

ξ_s = correction for torsional stiffness (pure torsion); approximately[11] $1.08 + 0.04\,M/M_p$

ξ_γ = correction for end fixity, say average is 1.25 for continuous beams with constant EI and equal unbraced segment lengths

Using the approximations indicated, and taking $\sqrt{hA/Z} = 1.60$, Eq. 9.3.46 becomes

$$\frac{L}{r_y} = \sqrt{\frac{\pi^2 E}{2F_y}}\sqrt{\frac{E_{st}}{E}}(1.60)\left(3.3 + 2.2\,\frac{M}{M_p}\right) \tag{9.3.47}$$

Assuming E/E_{st} may be taken as $29{,}000/900 \approx 33$, Eq. 9.3.47 becomes

$$\frac{L}{r_y} = \frac{0.7\pi}{\sqrt{\epsilon_y}}\,\frac{(5.3 + 3.5\,M/M_p)}{5.7} \tag{9.3.48}$$

which for $\epsilon_y = F_y/29{,}000$ gives

$$\frac{L}{r_y} = \frac{375}{\sqrt{F_y}}(0.93 + 0.62\,M/M_p) \tag{9.3.49}$$

which formed the basis of the 1963 AISC limitation for A36 steel, and is considered generally valid for $+1.0 > M/M_p > -0.6$.

As an alternate to the general expression, Eq. 9.3.49, Ref. 22 suggests, as a result of more research, using

$$\frac{L}{r_y} = \frac{375}{\sqrt{F_y}} \tag{9.3.50}$$

which is about what one gets when $M/M_p = 0$, and is recommended[22] when $+1.0 > M/M_p > -0.7$ (large moment gradient, including double curvature).

9.4.* DESIGN PROCEDURES—WORKING STRESS METHOD

The lateral buckling formulas apply to those cases where the ultimate moment capacity is achieved under an applied moment which is less than M_y (see Fig. 7.3.1). Elastic or inelastic buckling controls the maximum safe stress which may be permitted to act.

The Column Research Council (CRC) Guide[8] discusses three methods of considering lateral buckling strength for doubly symmetric I-shaped beams and girders. These are

(a) "Basic" design procedure.
(b) Single formula simplified procedure.
(c) Double formula simplified procedure (AISC–1963 and 1969 method).

For working stress method the critical stress must be divided by the desired factor of safety (1.67 for 1963 and 1969 AISC) to obtain the allowable stress.

"Basic" Design Procedure. This rational method involves using the theoretical elastic stability solution for the actual loading and restraint condition, which as a practical matter can only be done by the use of tables and charts. For the beam under pure moment, torsionally simply supported, this would mean using Eq. 9.3.32 or one of its alternate forms.

For any loading and restraint condition, the critical stress may be expressed[8]

$$F_{cr} = \frac{C_4 \sqrt{EI_y GK}}{S_x L} \tag{9.4.1}$$

where for the pure moment case, comparing with Eq. 9.3.34,

$$C_4 = \pi \sqrt{1 + \left(\frac{\pi}{\lambda L}\right)^2} \tag{9.4.2}$$

but which in general may be written[8]

$$C_4 = C_1 \pi \left[\sqrt{1 + \left(\frac{\pi}{\lambda L}\right)^2 (C_2^2 + 1)} \pm C_2 \frac{\pi}{\lambda L}\right] \tag{9.4.3}$$

where C_1 is a coefficient dependent on the type of load and support condition, and C_2 is a coefficient to account for the position of the load vertically with respect to the centroidal axis; values of C_1 and C_2 are to be found for common cases in Fig. 9.4.1.

It is to be noted that in the above equations the length L is to be taken as the effective length between points of torsional simple support (i.e., rotation about beam axis z is prevented, but rotation about y axis (warping) is not prevented). As a practical matter the full length equals the effective length when at the ends of the beam there is zero rotational restraint about the y axis, and the effective length equals one-half the actual

Case No.	Loading	Bending-Moment Diagram	Condition of Restraint Against Rotation about Vertical Axis at Ends	Equiv. Length Factor, k	Value of Coefficient C_1*	C
Beams Restrained Against Lateral Displacement at Both Ends of Span						
1.	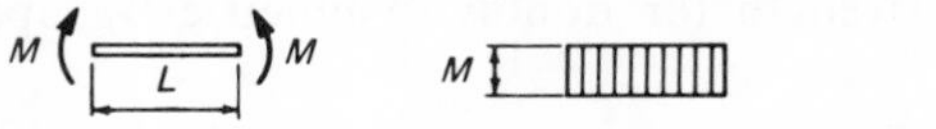 M, L, M	M	Simple support	1.0	1.0	
			Fixed	0.5	1.0	
2.	M, M/2	M, M/2	Simple support	1.0	1.3	
			Fixed	0.5	1.3	
3.	M	M	Simple support	1.0	1.8	
			Fixed	0.5	1.8	
4.	M, M/2	M, M/2	Simple support	1.0	2.4	
			Fixed	0.5	2.3	
5.	M, M	M, M	Simple support	1.0	2.6	
			Fixed	0.5	2.3	
6.	W	$\frac{WL}{8}$	Simple support	1.0	1.1	0.4
			Fixed	0.5	1.0	0.2
7.	W	$\frac{WL}{12}$, $M_{cr} = WL/24$	Simple support	1.0	1.3	1.5
			Fixed	0.5	0.9	0.8
8.	P	$\frac{PL}{4}$	Simple support	1.0	1.4	0.5
			Fixed	0.5	1.1	0.4
9.	P	$\frac{PL}{8}$, $\frac{PL}{8}$	Simple support	1.0	1.7	1.4
			Fixed	0.5	1.0	0.8
10.	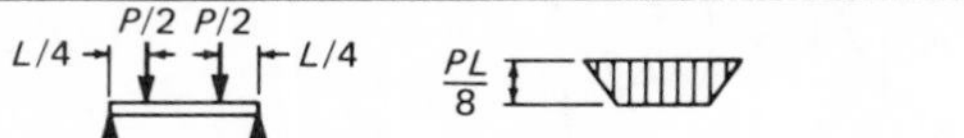 P/2 P/2, L/4, L/4	$\frac{PL}{8}$	Simple support	1.0	1.0	0.4
Cantilever Beams						
11.	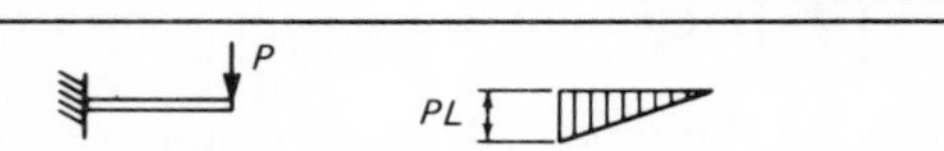 P	PL	Warping restrained at supported end	1.0	1.3	0.6
12.	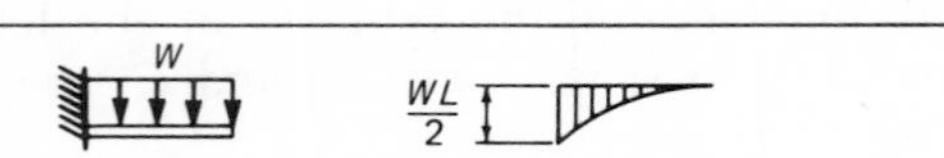 W	$\frac{WL}{2}$	Warping restrained at supported end	1.0	2.1	

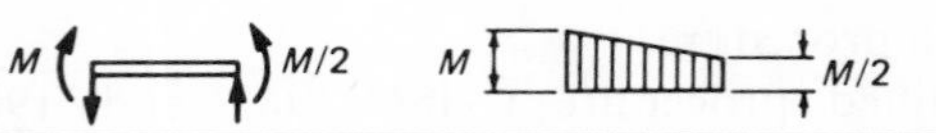

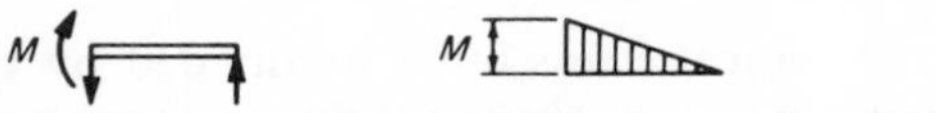

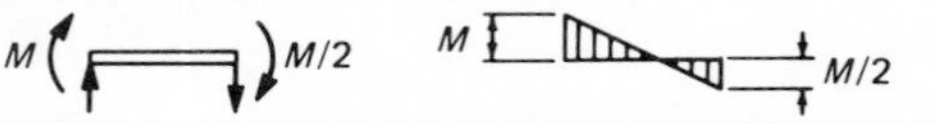

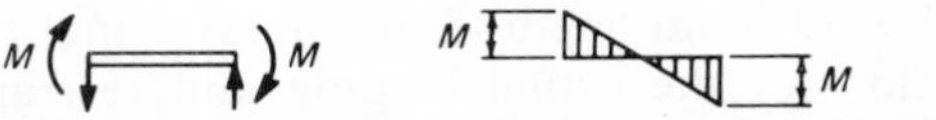

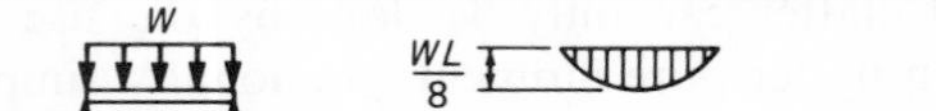

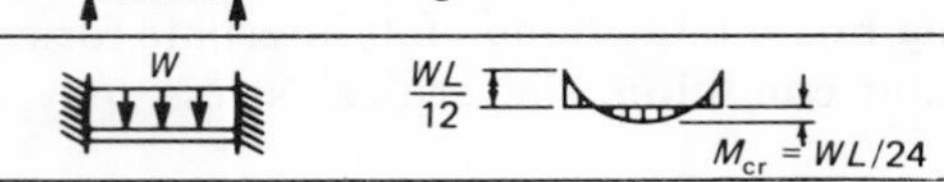

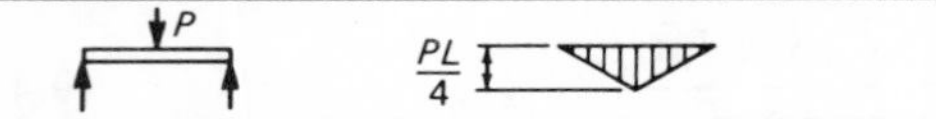

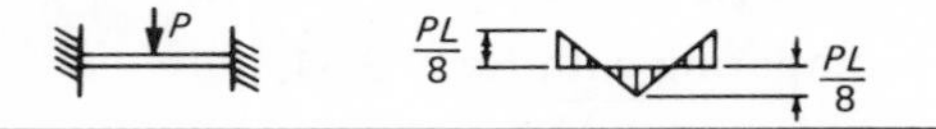

* Approximate minimum values for C_1 are given. Range of values and sources of data are to be found in Ref. 14.

Fig. 9.4.1. Usable values of coefficients in Eq. 9.4.1 for elastic buckling strength of beams (Adapte from Ref. 14).

length when full rotational restraint (torsional fixity) about the y axis exists at the beam ends.

For cases of unequal end moment, the work of Salvadori (p. 1167 of Ref. 9) has shown that the following expression for C_1 is valid:

$$C_1 = 1.75 + 1.05\left(\frac{M_1}{M_2}\right) + 0.3\left(\frac{M_1}{M_2}\right)^2 \leq 2.3 \tag{9.4.4}$$

where M_1 is the smaller of the rotational end moments on the member. A positive ratio indicates double curvature while a negative ratio indicates single curvature. In AISC–1963, bending moments were used rather than end rotational moments so that the sign of the second term was negative.

Because of residual stress, accidental eccentricity of loading, and initial crookedness, truly elastic buckling occurs only when the critical stress computed by Eq. 9.4.1 is less than about $F_y/2$, just as discussed for compression members in Chapter 6. Thus, computed buckling stresses above $F_y/2$ should be adjusted to account for inelastic buckling. The CRC Guide[8] suggests the use of the same parabolic variation between $F_y/2$ and F_y that is used for columns. Use the Euler equation

$$F_e = \frac{\pi^2 E}{\left(\frac{L}{r}\right)^2}$$

but replace F_e by F_{cr} computed from Eq. 9.4.1 for lateral-torsional buckling, and determine an equivalent L/r. In which case

$$\left.\frac{L}{r}\right)_{\text{equiv}} = \pi\sqrt{\frac{E}{F_{cr} \text{ by Eq. 9.4.1}}} \tag{9.4.5}$$

This equivalent slenderness ratio can be used with the basic CRC column formula (Eq. 6.7.3), dividing the result by the desired safety factor.

The ultimate strength of beam-columns as affected by lateral-torsional buckling indicates that for beam-columns acting as contiguous parts of frames the elastic and inelastic theories for the reduced bending strength as treated in this chapter may be overly conservative. (see Sec. 12.8.) End restraint seems to have a significant effect. Only for plastic design of beam-columns has this conservatism been recognized.

Single Formula Simplified Procedure. This procedure amounts to the same as using Eq. 9.3.34 or 9.3.35 but eliminates the need for having the torsional properties by reducing the equation to two parameters. L/r_y and d/t_f.

For typical beam proportions, assume the following reasonable approximations for the variables in those equations. As indicated by Eq. 8.2.14, the torsional constant K may be expressed for an I-shaped section,

$$K = 2\left(\frac{bt_f^3}{3}\right) + \frac{1}{3}\,dt_w^3 = \frac{t_f^2}{3}\left[2bt_f + dt_w\left(\frac{t_w}{t_f}\right)^2\right] \qquad (9.4.6)$$

where t_f = flange thickness, b = flange width, t_w = web thickness, and d = web depth. If $t_w/t_f \approx 0.5$, and if the web area is about 20 percent of the area of the beam ($A_w = 0.2A$), Eq. 9.4.6 becomes

$$K = \frac{t_f^2}{3}\left[2bt_f + dt_w - 0.75dt_w\right]$$

$$A = 2bt_f + dt_w$$

$$A_w = dt_w$$

$$0.75\,A_w = 0.75(.2A) = 0.15A$$

then

$$K \approx \frac{t_f^2}{3}\,[A - 0.15A] \approx .28At_f^2 \qquad (9.4.7)$$

For other variables,

$$G = \text{shear modulus} = \frac{E}{2(1+\mu)} = E/2.6$$

$$I_y = Ar_y^2$$
$$S_x = 2Ar_x^2/d$$
$$r_x \approx 0.41d \text{ (see Appendix, Table A1)}$$
$$h = 0.95d$$

where μ = Poisson's ratio (0.3 for steel)
r_y = radius of gyration with respect to the y axis
r_x = radius of gyration with respect to the x axis
d = overall depth of beam
h = distance between centroids of flanges

Finally, the torsional parameter $\lambda^2 = 1/a^2$ is

$$\lambda^2 = \frac{GK}{EI_w} = \frac{(E/2.6)(0.28At_f^2)}{E(Ar_y^2h^2/4)} = \frac{t_f^2}{2.32r_y^2h^2}$$
$$= \frac{t_f^2}{2.1r_y^2d^2} \qquad (9.4.8)$$

Substitution of these approximations into Eq. 9.3.34 gives

$$F_{cr} = \frac{\pi}{2(0.41)^2}\sqrt{\frac{0.28}{2.6}}\left(\frac{E}{Ld/r_yt_f}\right)\sqrt{1. + 2.1\pi^2\left(\frac{d/t_f}{L/r_y}\right)^2}$$

$$= \frac{3E}{Ld/r_yt_f}\sqrt{1 + 21\left(\frac{d/t_f}{L/r_y}\right)^2} \qquad (9.4.9)$$

or Eq. 9.3.35 gives

$$F_{cr} = \frac{\pi^2(0.95)}{4(0.41)^2} \frac{E}{(L/r_y)^2} \sqrt{1 + \frac{1}{2.1\pi^2}\left(\frac{L/r_y}{d/t_f}\right)^2}$$

$$= \frac{14E}{(L/r_y)^2} \sqrt{1 + \frac{1}{21}\left(\frac{L/r_y}{d/t_f}\right)^2} \qquad (9.4.10)$$

Equations 9.4.9 and 9.4.10 are alternate forms of the same equation which indicate parameters designers are familiar with instead of torsional beam properties. These equations apply correctly to cases of pure moment (C_4 in accordance with Eq. 9.4.2) and underestimate the buckling stress for other loadings with torsional simple support at the ends. Figure 9.4.2 adapted from the CRC Guide permits application of a correction

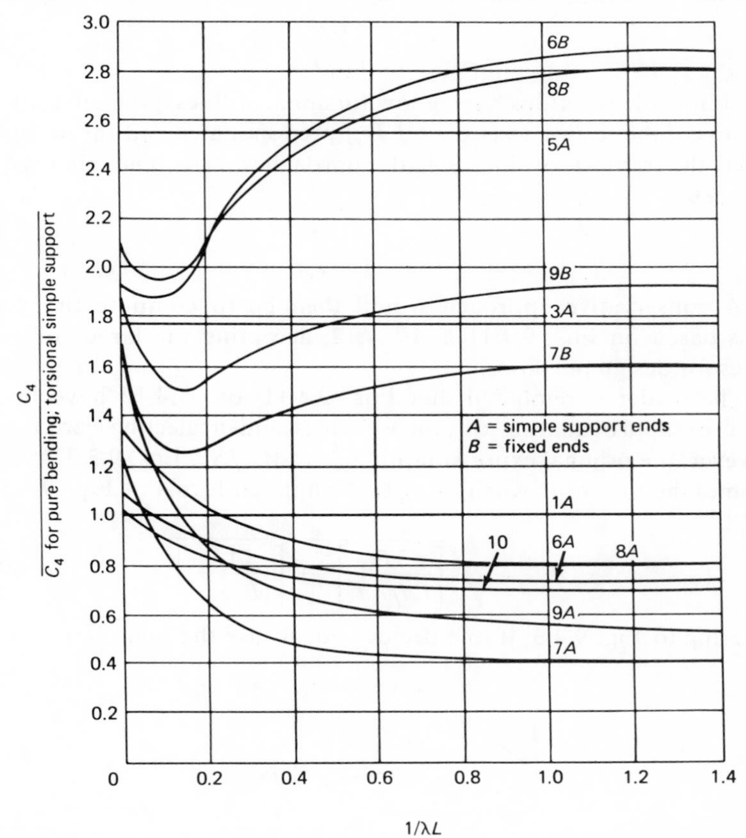

Fig. 9.4.2. Curves of $C_4/(C_4$ for pure bending with torsional simple support at ends) vs. $1/\lambda L$. The curves are numbered in accordance with cases of Fig. 9.4.1. (Adapted from Ref. 8.) (Lateral loading cases assume loads are applied on the top flange.)

multiplier to Eqs. 9.4.9 and 9.4.10 to give more accurate elastic buckling stresses. As in the basic method using torsional properties, any computed critical stress exceeding about $F_y/2$ should be modified to account for residual stress and inelastic buckling by applying Eq. 9.4.5 and using the basic column formula as previously discussed.

Special note should be made that for cases 6 through 10 of Fig. 9.4.1, torsional simple support with loading on the top flange, the critical stress is lower than that for the beam under pure moment (see Fig. 9.4.2).

Double Formula Procedure. In this approach it is recognized that many beams having a high torsional stiffness (shallow, thickwalled sections), Eq. 9.4.9 can be simplified because the second term under the radical may be neglected compared to unity. For such torsionally stiff sections, Eq. 9.4.9 may be written

$$F_{cr} = \frac{3E}{Ld/r_y t_f} \tag{9.4.11}$$

For thin-walled sections having low torsional stiffness, the lateral bending stiffness of the section (based on L/r_y) predominates and the second term under the radical of Eq. 9.4.10 approaches zero and that equation becomes

$$F_{cr} = \frac{14E}{(L/r_y)^2} \tag{9.4.12}$$

A conservative approach would then be to compute the buckling stress based on Eqs. 9.4.11 and 9.4.12, accepting the larger of the two values for design purposes.

The reader is reminded that Eqs. 9.4.11 and 9.4.12 have been obtained by approximating certain variables and neglecting certain terms. However, the original expression for $F_{cr} = M_{cr}/S_x$, Eqs. 9.3.32 or 9.3.33, becomes the following when using the simplified forms of Eqs. 9.4.11 and 9.4.12,

$$F_{cr} = \sqrt{\left[\frac{3E}{Ld/r_y t_f}\right]^2 + \left[\frac{14E}{(L/r_y)^2}\right]^2} \tag{9.4.13}$$

Referring to Fig. 9.4.3, if one decides not to use the complicated expres-

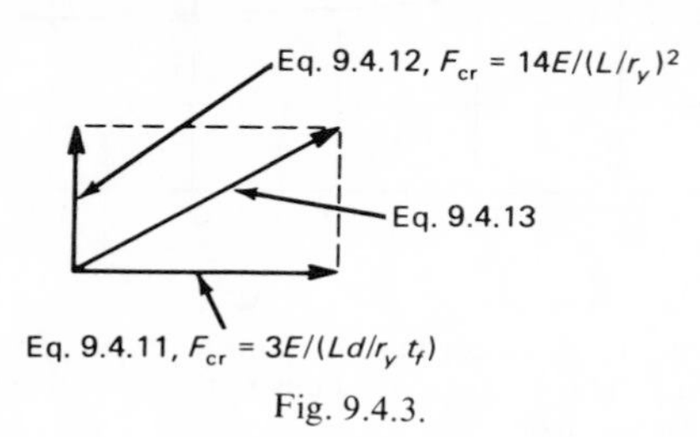

Fig. 9.4.3.

sion for the resultant, the next best is to use the larger of the two legs of the right triangle.

Again, as in the CRC "Basic" method, critical stress values exceeding $F_y/2$ should be adjusted for inelastic buckling, using Eqs. 9.4.5 and 6.7.3. The AISC specification is based on the procedure using the larger of two formulas, as described in the next two articles.

9.5. AISC–1963 WORKING-STRESS DESIGN CRITERIA

The 1963 AISC approach is presented partly because of historical precedent. To permit fuller understanding, this article considers the full range from laterally stable beams to situations where lateral buckling causes considerable strength reduction. Loading in the plane of the web is assumed.

(a) *Case 1: Plastic moment is reached* ($M_u = M_p$) *along with large plastic-rotation capacity*. Assuming local buckling of compression elements (discussed in Chapter 6) does not control, the L/r_y must be restricted so that the necessary plastic strain can occur. If one recognizes that in the working-stress method less plastic strain (rotation capacity, Fig. 9.3.4) is utilized, it is probably satisfactory to use Eq. 9.3.50 as the slenderness requirement for treatment as a "compact" section:

$$\frac{L}{r_y}_{\max} = \frac{375}{\sqrt{F_y}} \tag{9.5.1}$$

which since $r_y \approx 0.2b_f$ (b_f = flange width), and letting $L = L_c$ = maximum unbraced length for treatment as "compact" section, becomes

$$\frac{L_c}{b_f} = \frac{76.0}{\sqrt{F_y}} \quad \left(\text{or } \frac{2400}{\sqrt{F_y}} \text{ for } F_y \text{ in psi}\right) \tag{9.5.2}$$

in accordance with AISC 1.5.1.4.1. L_c values for the W shapes are given in Appendix, Table A3. The L_c/b_f limiting ratios are given in Table 9.5.1. The allowable working stress for cases satisfying the L/b limitation is $0.66F_y$, as discussed in Chapter 7.

TABLE 9.5.1
Maximum Unbraced Length for "Compact" Section

F_y ksi	L_c/b_f
36.0	12.65
42.0	11.7
45.0	11.3
50.0	10.75
55.0	10.2
60.0	9.8
65.0	9.4

(b) *Case 2:* $M_p > M_u > M_y$. Beams which are capable of achieving yield strain at the extreme fiber but are unable to undergo sufficient plastic strain to be in Case 1 are assigned an allowable stress of $0.6F_y$. Essentially, elastic and inelastic stability must be assured until yield strain is reached at the extreme fiber; which is done by the double formula procedure using Eqs. 9.4.11 and 9.4.12.

Since 1946, a formula similar to Eq. 9.4.11 has been used for establishing allowable stresses. Upon substituting $r_y \approx 0.22b$ (see Appendix, Table A1) into Eq. 9.4.11 one obtains

$$F_{cr} = \frac{0.66E}{Ld/bt_f} \tag{9.5.3}$$

which if $A_f = bt$ and a factor of safety of 1.67 is used, an allowable stress equation is obtained

$$F_b = \frac{12{,}000}{Ld/A_f} \qquad \leq 0.6F_y \tag{9.5.4}$$

which is 1963 AISC Formula 5. Under AISC specifications no transition to account for inelastic buckling resulting from residual stress was used between stability controlling ($M_u < M_y$) and yielding controlling ($M_u = M_y$); a logical approach considering residual stress has little effect on torsional stiffness.

From Eq. 9.5.4 it can be determined that the maximum laterally unbraced length, $L = L_u$, for which an allowable stress of $0.6F_y$ may be used is

$$L_u = \frac{12{,}000}{0.6F_y(d/A_f)} = \frac{20{,}000}{F_y(d/A_f)} \tag{9.5.5}$$

Because of the more severe ultimate strain requirements in Case 1, it is proper that

$$L_c \leq L_u \tag{9.5.6}$$

L_c and L_u values for W sections appear in Appendix, Table A3.

(c) *Case 3:* $M_u < M_y$. For this case a reduced allowable stress must be used based on lateral-torsional buckling strength. AISC Formula 5, Eq. 9.5.4, was used as one of the possible controlling equations in the double-formula procedure, and will be found to control for beams with high torsional stiffness.

Since 1961 the AISC specification has also used a second formula, based on Eq. 9.4.12, which applies for deep thin-walled sections, such as plate girders. Since Eq. 9.4.12 is really a column formula (Euler-type equation) representing the elastic lateral bending stiffness, all factors discussed in Chapter 6 apply. Accounting principally for residual stress, the

CRC basic column formula, Eq. 6.7.3 was used instead of the Euler-type equation. The alternate to Eq. 9.4.12 is

$$F_{cr} = \left[1.0 - \frac{(L/r)^2}{2C_c^2 C_b}\right] F_y \tag{9.5.7}$$

which is the CRC formula, with an additional parameter C_b. C_b (C_1 of Eq. 9.4.4) is a term to account for the moment gradient; i.e., it accounts for the fact that the compression force in the beam flange is not a constant over its length, as it is in a pure compression member as discussed in Chapter 6. Note that Eq. 9.5.7 uses a radius of gyration r, which is that of a tee section comprising the flange plus one-sixth of the web area (See Appendix, Table A2, for r values for most rolled W shapes); a logical procedure since certainly some of the web area acts with the compression flange.

Dividing Eq. 9.5.7 by a 1.67 factor of safety, we obtain

$$F_b(4) = \left[1.0 - \frac{(L/r)^2}{2C_c^2 C_b}\right] 0.60 F_y \tag{9.5.8}$$

which is 1963 AISC Formula 4. When L/r is less than 40 the stress reduction may be neglected. Since a parabola replaces an Euler-type equation 9.4.12, it obviously cannot be valid for large L/r. The implication is therefore, that Formula 5, Eq. 9.5.3, governs for large slenderness ratios.

To obtain a better understanding of the significance of using the two formulas, reexamine Eq. 9.4.13 which is Eq. 9.3.31 divided by S_x,

$$F_{cr} = \sqrt{\left(\frac{3E}{Ld/r_y t_f}\right)^2 + \left(\frac{14E}{(L/r_y)^2}\right)^2} \qquad [9.4.13]$$

or, if $r_y \approx 0.22 b_f$ is used in the first term; $r_y \approx \dfrac{r \text{ of Formula 4}}{1.2}$ in the second term,

$$F_{cr} = \sqrt{\left(\frac{0.66E}{Ld/A_f}\right)^2 + \left(\frac{\pi^2 E}{(L/r)^2}\right)^2} \tag{9.5.9}$$

From the form of Eq. 9.5.9 it is evident that it will be conservative to use the *larger* of the squared terms; the first term representing the square of Formula 5 (1.5–7 for AISC–1969) and the second term representing the square of Formula 4 (1.5–6 for AISC–1969). Thus AISC states the allowable stress is to be taken as the *larger* of the values from the two formulas.

A summary of the 1963 AISC working-stress criteria appears in Fig. 9.5.1.

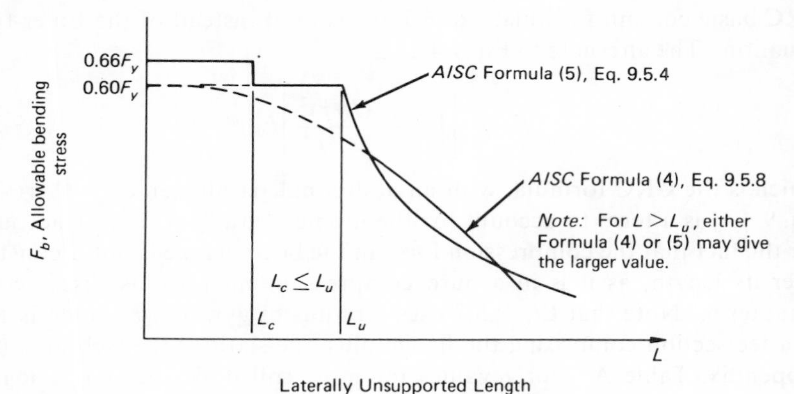

Fig. 9.5.1. Summary of 1963—AISC criteria for working stress design for a typical W section, for bending about the strong axis (x–x).

EXAMPLE 9.5.1

Consider the 50-ft simple span W27 × 84 beam carrying a uniformly distributed load of 1 kip/ft, as shown in Fig. 9.5.2. Lateral support is provided at the ends and at 20 ft from each end. Use A36 steel.

(a) Determine allowable stress according to AISC–1963.

(b) Determine the allowable stress using the CRC Basic Design Procedure, utilizing the differential equation solution as corrected in accordance with Fig. 9.4.1. Use a 1.67 factor of safetv.

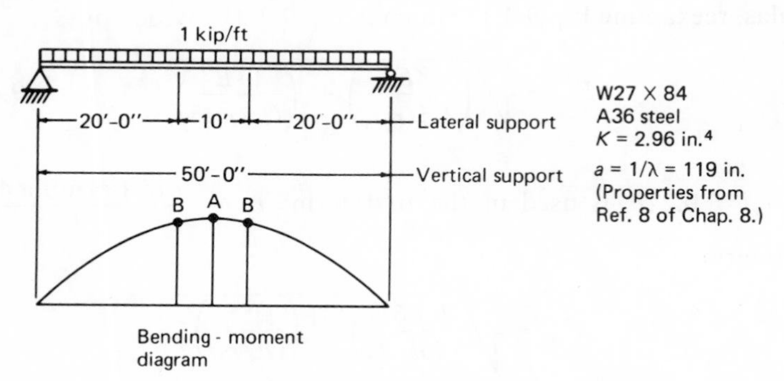

Fig. 9.5.2. Example 9.5.1.

Solution

(a) AISC–1.5.1.4.1.–1963: For Formula 4, compute radius of gyration using the compression flange plus $1/6$ of the web area,

$$r = \sqrt{\frac{\frac{1}{12}(9.96)^3 0.636}{9.96(0.636) + \frac{1}{6}(26.69 - 1.27)0.463}} = 2.51 \text{ in.}$$

In determining the slenderness ratio for Formula 4, it is to be realized that when checking the adequacy of the moment capacity at point A, the 10-ft laterally unsupported length is used for the allowable stress; for adequacy under the smallest moment at B, the longer unbraced length of 20 ft is used. For C_b, Eq. 9.4.4,

$$C_b = 1.75 + 1.05\left(\frac{M_1}{M_2}\right) + 0.3\left(\frac{M_1}{M_2}\right)^2 \leq 2.3$$

where Fig. 9.5.3 defines the two cases to be considered.

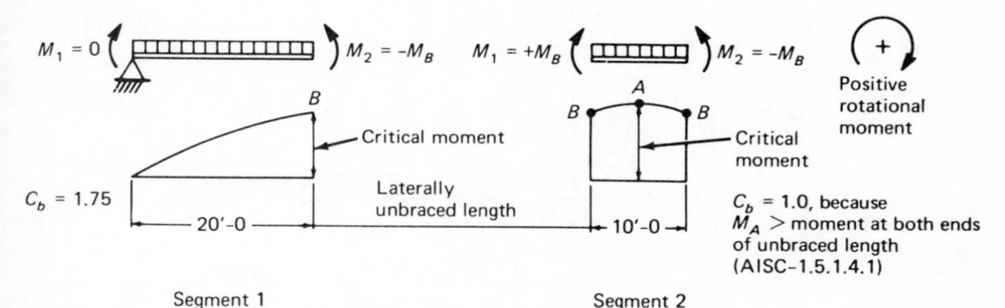

Fig. 9.5.3. C_b values for Example 9.5.1.

Formula 4 for, $F_y = 36$ ksi, as given in the AISC Appendix, is

$$F_b(4) = 22.0 - \frac{0.679}{1000}\left(\frac{L}{r\sqrt{C_b}}\right)^2$$

where

$$\frac{L}{r\sqrt{C_b}} = \frac{20(12)}{2.51\sqrt{1.75}} = 72.1 \quad \text{for Segment 1}$$

$$\frac{L}{r\sqrt{C_b}} = \frac{10(12)}{2.51\sqrt{1.0}} = 48.0 \quad \text{for Segment 2}$$

For Segment 1,

$$F_b(4) = 18.47 \text{ ksi}$$

For Segment 2,

$$F_b(4) = 20.44 \text{ ksi}$$

For Formula (5), using d/A_f from beam properties,

$$L(d/A_f) = 20(12)(4.21) = 1{,}010$$

$$F_b(5) = \frac{12{,}000}{1010} = 11.9 \text{ ksi} \quad \text{(Segment 1)}$$

$$F_b(5) = \frac{12{,}000}{505} = 23.7 \text{ ksi} > 22 \text{ ksi}$$

$$\therefore F_b(5) = 22 \text{ ksi} \quad \text{(Segment 2)}$$

Summary 1963 AISC

	Segment 1	Segment 2
C_b	1.75	1.0
Allowable stress..................	$F_b(4) = 18.47$ ksi	$F_b(5) = 22$ ksi
Critical moment	$M = M_B$	$M = M_A$
L..................................	$L = 20$ ft	$L = 10$ ft

(b) CRC "basic" method using the differential equation solution for pure bending, Eqs. 9.4.1 and 9.4.2:

where
$$F_b = \frac{F_{cr}}{1.67} = \frac{C_4\sqrt{EI_yGK}}{1.67S_xL} \qquad [9.4.1]$$

$$C_4 = \pi\sqrt{1+\left(\frac{\pi}{\lambda L}\right)^2} = \pi\sqrt{1+2.42} = 5.82 \qquad [9.4.2]$$

$$= \pi\sqrt{1+\frac{119\pi}{(20(12))}^{2}} = \pi\sqrt{1+2.42} = 5.82$$

using $G = E/2.6$ in Eq. 9.4.1 above; $I_y = 105$ in.^{4}; $S_x = 212$ in.^{3}

$$F_{cr} = \frac{5.82(29{,}000)\sqrt{105(2.96)/2.6}}{(212)(240)} = 36.3 \text{ ksi} > F_y/2$$

Using Eq. 9.4.5,

$$\left.\frac{KL}{r}\right)_{\text{equiv}} = \pi\sqrt{\frac{29{,}000}{36.3}} = 89$$

Putting this into the CRC basic column formula, Eq. 6.7.3,

$$F_{cr} = F_y\left[1 - \frac{F_y}{4\pi^2E}\left(\frac{L}{r}\right)^2\right]$$

$$= 36\left[1 - \frac{36}{4\pi^2(29{,}000)}\quad(89)^2\right] = 0.751(36) = 27.0 \text{ ksi}$$

$$F_b = \frac{27.0}{1.67} = 16.2 \text{ ksi}$$

which is more conservative than 1963 AISC. Greater accuracy is achieved by correcting for the moment gradient along the beam. Using the general expression for C_4, Eq. 9.4.3, or Fig. 9.4.2,

$$\frac{\text{Actual } C_4}{\text{Pure bending } C_4} \approx 1.78$$

for Case 3 (segment 1 has moment variation approximating a triangle) and $1/\lambda L = 119/240 = 0.495$,

$$F_{cr} \text{ by Eq. 9.4.1} = 36.3(1.78) = 64.5 \text{ ksi} > F_y/2$$

$$L/r)_{equiv} = 67$$

$$F_{cr} \text{ by Eq. 6.7.3} = 0.85\,(36) = 31.0 \text{ ksi}$$

$$F_b = \frac{31.0}{1.67} = 18.5 \text{ ksi}$$

which agrees well with the AISC value for segment 1. (Note is made that AISC–1969–1.5.1.4.6a permits use of a "more precise analysis," which according to the AISC Commentary includes the CRC "basic" design procedure.

9.6. AISC–1969 WORKING-STRESS DESIGN CRITERIA

The 1969 specifications involve little change in concept; the changes are mainly in appearance. One important change is that it permits use of "a more precise analysis" which the Commentary indicates means the CRC "basic" design procedure. The logic and procedure described in Sec. 9.5 for (a) Case 1: $M_u = M_p$, and (b) Case 2: $M_p > M_u > M_y$ have not been changed, insofar as lateral-torsional buckling is concerned. When the possibility of local buckling of flanges is the only factor which prevents a section from being Case 1, an allowable value between $0.60F_y$ and $0.60F_y$ is permitted as discussed in Sec. 7.3. To assist the designer, Appendix Table A3 contains values of L_c, L_u, and the maximum allowable flexural stress, F_b.

AISC-1.5.1.4.3 recognizes the higher shape factors for I-shaped sections bent about their weak axis in allowing $0.75F_y$; properly indicating in an explicit way that lateral-torsional buckling does not occur in such situations.

1963 AISC Formula 5 which is based on the torsional stiffness of I-shaped sections is retained with the addition of C_b to account for moment gradient:

$$F_b = \frac{12{,}000\,C_b}{Ld/A_f} \tag{9.6.1}$$

which is 1969 AISC Formula 1.5–7. Examination of Eq. 9.4.1, the original source of Eq. 9.6.1, shows C_4 as a multiplier, and Eq. 9.4.3 shows C_4 directly proportional to C_1, which is now called C_b. Since $C_b \geq 1.0$, this is a rational liberalizing change in the formula.

Furthermore, AISC–1.5.1.4.6a states that for channels Eq. 9.6.1 only is to be used, and that Eq. 9.6.1 is a valid criterion only when "the compression flange is solid and approximately rectangular in cross section and its area is not less than that of the tension flange." These requirements are consistent with the approximations used to develop the equation in Sec. 9.4.

The principal change involves 1963 AISC Formula (4) which is based on the column strength of the compression flange. Recall that ordinary column design as discussed in Chapter 6 uses two equations; a parabola for the inelastic buckling range, and an Euler-type equation for elastic buckling. Formula (4) now has two equations, essentially the AISC–1963 parabola for inelastic buckling, and an Euler-type expression for large slenderness ratios.

Consider that C_C represented the slenderness ratio for which residual stress was considered to cause inelastic buckling; i.e., the KL/r value when the Euler stress, $F_{cr} = F_y/2$. Thus,

$$C_c^2 = \frac{2\pi^2 E}{F_y}$$

Substitution of C_c^2 into 1963–AISC Formula 4, Eq. 9.5.8, gives

$$F_b = 0.6F_y\left[1.0 - \frac{F_y(L/r)^2}{1144(10^3)C_b}\right] \tag{9.6.2}$$

Then, in order to recognize that using the larger of Formula 4 or 5 is conservative, and also to obtain a continuous stress relationship with unbraced length L, a reduced factor of safety is used so that 0.6 becomes 2/3. Equation 9.6.2 then becomes

$$F_b = \left[\frac{2}{3} - \frac{(2/3)F_y(L/r)^2}{1144(10^3)C_b}\right]F_y$$

$$= \left[\frac{2}{3} - \frac{F_y(L/r)^2}{1720(10^3)C_b}\right]F_y \tag{9.6.3}$$

1969 AISC has arbitrarily adjusted the coefficient in the second term from 1720 to 1530 (i.e., used $^3/_4$ instead of $^2/_3$ on that term) to somewhat offset the reduction in safety factor. Thus, old Formula (4) becomes

$$F_b = \left[\frac{2}{3} - \frac{F_y(L/r_T)^2}{1530\ (10^3)C_b}\right]F_y \tag{9.6.4}$$

which is 1969 AISC Formula 1.5–6a. The symbol r_T is used for the "radius of gyration of a section comprising the compression flange plus one-third of the compression web area, taken about an axis in the plane of the web."* Except for clarification of wording regarding the web contribution, the slenderness ratio used is the same as previously prescribed.

*The r_T values used in this text are computed as $\sqrt{\frac{1}{12}t_f b_f^3/[t_f b_f + (d - 2t_f)t_w/6]}$ and are given for standard wide-flange shapes in Appendix Table A3. The values differ slightly from those given in the 1970 AISC Manual.

Equation 9.6.4 is based on *inelastic* lateral-torsional buckling and usually will give higher values than Eq. 9.6.1 for deep I-shaped sections having thin narrow flanges, such as plate girders.

For *elastic* buckling, the Euler-type expression which involves only the slenderness ratio is used. Equation 9.4.12 was the elastic buckling equation forming the basis for the 1963 AISC Formula 4 parabola, Eq. 9.6.2:

$$F_{cr} = \frac{14E}{(L/r_y)^2} \qquad [9.4.12]$$

in the development of which r_x was approximated as $0.41d$. It has been found that column buckling only controls for large slenderness ratios on thin, or light gage, sections. For such sections, r_x will be better approximated by $0.38d$. In which case, Eq. 9.4.12 would become

$$F_{cr} \approx \frac{16E}{(L/r_y)^2} \qquad (9.6.5)$$

Dividing by 1.92, the same factor used for columns of large slenderness ratio, multiplying by C_b as indicated by Eqs. 9.4.1 and 9.4.3 (where $C_1 = C_b$), and converting r_y to r_T by using $r_y \approx r_T/1.2$ gives

$$F_b = \frac{16(29{,}000)\,C_b}{1.92\,(L/r_T)^2(1.2)^2} \approx \frac{170{,}000\,C_b}{(L/r_T)^2} \qquad (9.6.6)$$

which is 1969 AISC Formula 1.5–6b. Thus the single 1963 Formula (4) is replaced by the two Eqs. 9.6.4 and 9.6.6.

The lowest slenderness ratio for which Formula 1.5–6a is applicable is found by setting Eq. 9.6.4 equal to $0.6F_y$, which gives

$$\frac{2}{3} - \frac{F_y(L/r_T)^2}{1530(10^3)\,C_b} = 0.6$$

$$\frac{L}{r_T} = \sqrt{\frac{0.067(1530)(10^3)\,C_b}{F_y}} = \sqrt{\frac{102(10^3)\,C_b}{F_y}} \qquad (9.6.7)$$

below which slenderness ratio, Formulas 1.5–6 and 1.5–7 will not control. L_u, defined by Eq. 9.5.5, may still exceed the L determined from Eq. 9.6.7, in which case $0.6F_y$ applies to a laterally unbraced length as great as L_u.

To determine the slenderness ratio at which Formula 1.5–6b applies, equate Eqs. 9.6.4 and 9.6.6:

$$\frac{2F_y}{3} - \frac{F_y^2(L/r_T)^2}{1530(10^3)\,C_b} = \frac{170(10^3)\,C_b}{(L/r_T)^2}$$

from which is obtained

$$\frac{L}{r_T} = \sqrt{\frac{510(10^3)\,C_b}{F_y}} \tag{9.6.8}$$

above which slenderness ratio, *elastic* stability controls. Table 9.6.1 gives L/r_T limits and parabolic denominator coefficients in Formula 1.5–6a for various material yield stresses.

The 1969 specification provisions dealing with lateral buckling under the working stress method are realistic and indicate another move toward making design procedures better agree with true behavior.

TABLE 9.6.1
Limits on L/r_T and Coefficients for Formulas (1.5–6)–1969

F_y,	$\sqrt{\frac{102(10^3)C_b}{F_y}}$			$\frac{1530(10^3)C_b}{F_y^2}$, for F_b(1.5–6a)			$\sqrt{\frac{510(10^3)C_b}{F_y}}$		
	C_b			C_b			C_b		
ksi	1.0	1.75	2.3	1.0	1.75	2.3	1.0	1.75	2.3
36	53	70	80	1181	2070	2720	119	158	180*
42	49	65	74	867	1517	1995	110	146	167
50	45	60	68	612	1070	1408	101	134	153
60	41	54	62	425	744	977	92	122	139
65	40	53	61	362	634	832	89	118	135
100	32	42	49	153	268	352	71	94	107

*AISC-1.8.4 limits slenderness ratios for compression members to 200. $L/r_y \approx 200$ when $L/r_T = 200/1.2 = 167$.

EXAMPLE 9.6.1

For the W27 × 84 beam of Example 9.5.1.;

(a) For both 1963 and 1969 AISC Specifications, plot allowable stress vs laterally unbraced length L, using $C_b = 1.75$ as in Segment 1 of Fig. 9.5.3.

(b) Compare 1969 AISC specific values for Example 9.5.1 with those previously obtained.

Solution

(a) For $F_y = 36$ ksi, Case 1, $M_u = M_p$, Eq. 9.5.2

$$L \le L_c = \frac{76.0}{\sqrt{F_y}}b_f = \frac{76.0(9.96)}{\sqrt{36}\,(12)} = 10.5 \text{ ft} \le L_u$$

where

$$L_u = \frac{20{,}000}{F_y(d/A_f)} = \frac{20{,}000}{36(4.21)(12)} = 11.0 \text{ ft} \qquad \text{OK}$$

$$F_b = 0.66F_y \qquad \text{for} \qquad L \leq L_c = 10.5 \text{ ft}$$

for both 1963 and 1969 AISC.

(b) Case 2, $M_p > M_u > M_y$

1963 AISC, Formula (5), Eq. 9.5.4

$$L_u = 11.0 \text{ ft}$$

1969 AISC, Formula 1.5–7, Eq. 9.6.1

$$C_b L_u = 1.75(11.0) = 19.25 \text{ ft}$$

Check minimum L for parabolic formula governing. 1963 AISC, Formula 4, Eq. 9.5.8, no reduction is needed when

$$L < 40\, r_T = \frac{40(2.51)}{12} = 8.37 \text{ ft}$$

Since this is less than $L_u = 11.0$, no reduction below $0.6F_y$ is required for laterally unbraced lengths less than 11.0 ft.
1969 AISC, Formula 1.5–6a, Eq. 9.6.7,

$$\frac{L}{r_T} = \sqrt{\frac{102(10^3)C_b}{F_y}} = \sqrt{\frac{102(10^3)1.75}{36}} = 70.4$$

$$L = 70.4r_T = 70.4(2.51)/12 = 14.7 \text{ ft}$$

where r_T was computed in Example 9.5.1.

(c) For Case 3, $M_u < M_y$:

$$F_b\,(1.5\text{–}6\text{a}) = \frac{2}{3}\,F_y - \frac{F_y^2(L/r_T)^2}{1530(10^3)\,C_b} = 24.0 - \frac{(L/r_T)^2}{2060}$$

$$F_b\,(1.5\text{–}6\text{a}) = 24.0 - L^2/90.0 \text{ ksi}$$

using $r_T = 2.51$ in. and expressing L in feet. For $L >$ Eq. 9.6.8,

$$\frac{L}{r_T} = \sqrt{\frac{510(10^3)\,C_b}{F_y}} = \sqrt{\frac{510(10^3)1.75}{36}} = 157.5$$

$$L = 157.5(2.51)/12 = 32.9 \text{ ft}$$

$$F_b\,(1.5\text{–}6\text{b}) = \frac{170(10^3)\,C_b}{(L/r_T)^2} = \frac{170(10^3)1.75(2.51)^2}{L^2(144)} = \frac{13{,}000}{L^2} \text{ ksi}$$

Also, Formula 1.5–7 is to be investigated:

$$F_b\,(1.5\text{–}7) = \frac{12(10^3)\,C_b}{Ld/A_f} = \frac{12(10^3)1.75}{L(12)(4.21)} = \frac{415}{L} \text{ ksi}$$

To compare with 1963 AISC, using Formulas 4 and 5:

$$F_b(4) = 22.0 - \frac{0.679}{1000C_b}\left(\frac{L}{r}\right)^2$$

$$= 22.0 - \frac{0.679}{1000(1.75)}\left(\frac{L^2(144)}{(2.51)^2}\right)$$

$$= 22.0 - L^2/112 \text{ ksi}$$

$$F_b(5) = \frac{12(10^3)}{Ld/A_f} = \frac{12(10^3)}{L(12)(4.21)} = \frac{237.5}{L} \text{ ksi}$$

The comparison of 1963 and 1969 specifications for this W27 × 84 with $C_b = 1.75$ is shown in Fig. 9.6.1. The shaded areas indicate where

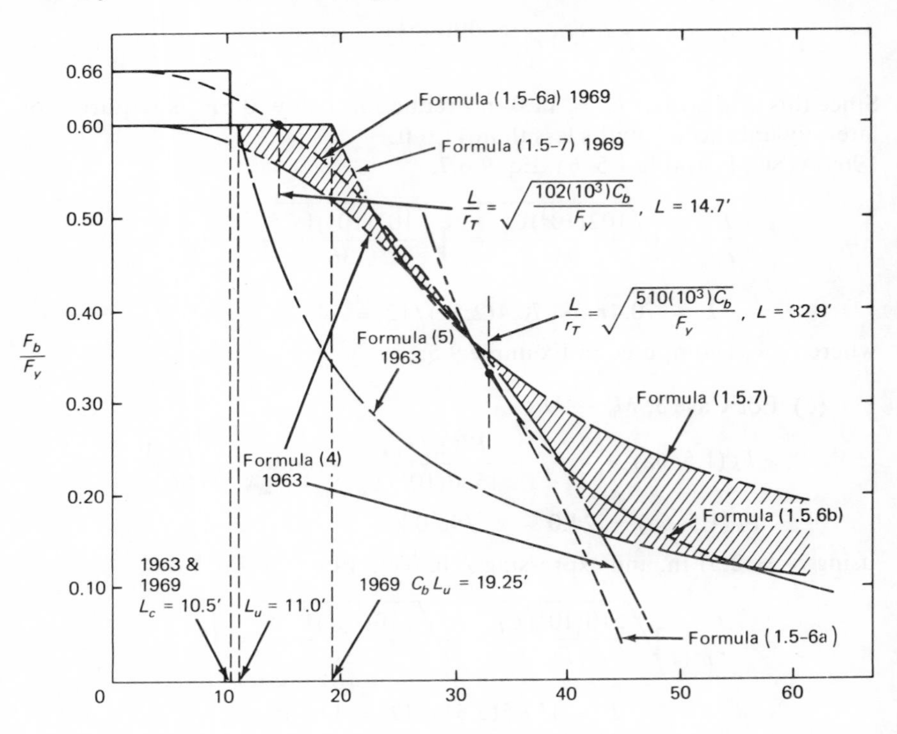

Fig. 9.6.1. Allowable stresses for W27 × 84, C_b = 1.75, A36 steel, for Example 9.6.1. (Cross-hatch shows increase in allowable stress of 1969 AISC over that of 1963 AISC.)

higher allowable stresses are permitted under 1969 AISC, a *proper* increase based on factors already discussed.

The CRC "basic" design method, as illustrated in Example 9.5.1b, is also acceptable under AISC–1969.

9.7. AISC–1969 PLASTIC DESIGN CRITERIA

In plastic design the analysis is made using plastic behavior, or limit analysis or collapse analysis as some may call it. In general it is an ultimate-strength approach as opposed to a working-stress approach. Basic plastic strength was discussed in Chapter 7 and limit analysis considering redistribution of moments is treated in Chapter 10.

Since the basic approach of plastic design is to develop the ultimate strength of the structure, generally lateral-torsional buckling must not be allowed to govern the failure mode. In other words, a reduced strength of section based on laterally unbraced length is not acceptable. Lateral bracing must be provided. AISC–2.9 states "Members shall be adequately braced to resist lateral and torsional displacements at the plastic hinge locations associated with the failure mechanism."

In accordance with AISC–2.9 unbraced length slenderness ratios must not exceed

$$\frac{L_{cr}}{r_y} \leq \frac{1375}{F_y} + 25 \tag{9.7.1}$$

when $+1.0 > M/M_p > -0.5$ (i.e., large moment gradient, including double curvature), and

$$\frac{L_{cr}}{r_y} \leq \frac{1375}{F_y} \tag{9.7.2}$$

when $-0.5 > M/M_p > -1.0$ (i.e., small moment gradient approaching uniform moment).

In Eqs. 9.7.1 and 9.7.2, M/M_p is the end moment ratio over a laterally unbraced length where a plastic hinge associated with the failure mechanism occurs at one end. The ratio is negative for single-curvature bending, and positive for double curvature.

Since the 1963 AISC permitted plastic design only for A36 steel, and 1969 AISC permits several steels, formulas were changed to become functions of F_y. Also, the change from using *bending* moment ratios in AISC 1963 to using *end* moment ratios in AISC 1969 must be kept in mind.

A comparison of the above provisions with the equations developed theoretically in Sec. 9.3, and with the 1963 provisions, appears in Fig. 9.7.1. It may be noted that Eq. 9.3.50, which is used as the basis for satisfying lateral buckling control for a "compact" section in the working stress method, compares favorably (as it should) with Eq. 9.7.1 for plastic design.

For unbraced segments where a plastic hinge is not expected to occur in the failure mechanism, the working stress equations, Formulas 1.5–6 and 1.5–7 are to be applied. To get the unit stresses into the working-stress range, the stresses P/A or M/S, computed using factored loads

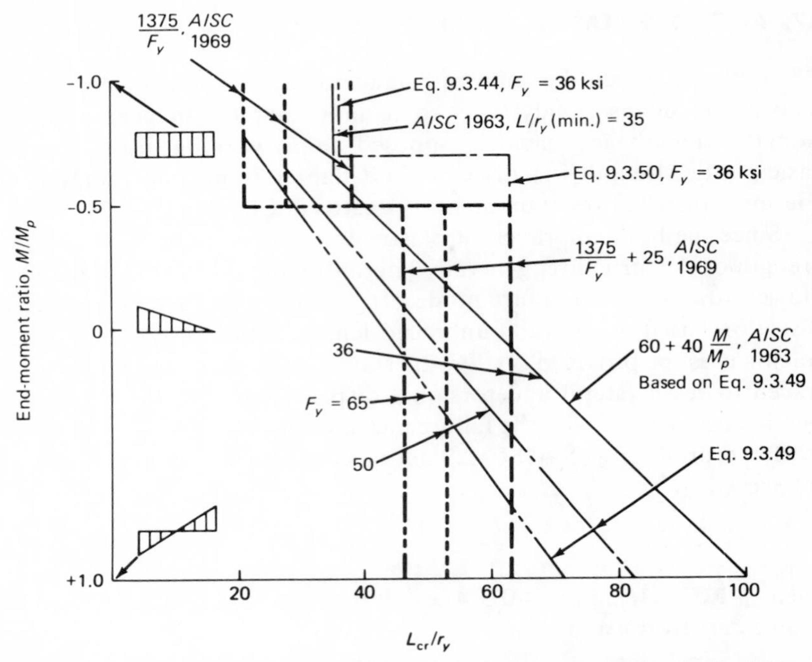

Fig. 9.7.1. Comparison of 1969 AISC critical slenderness ratio for plastic design with other recommendations and with 1963 AISC provision.

(i.e. design ultimate loads) are to be divided by the applicable load factor. Furthermore, if the last plastic hinge of the collapse mechanism is the only plastic hinge occurring over the unbraced length, the working-stress provisions apply just as when no plastic hinge is expected to form.

For weak-axis bending, lateral stability is not a factor and the slenderness ratio limitations are not applicable.

AISC–2.4, regarding beam-columns, provides an equation, Formula 2.4–4, for computing the reduced ultimate bending strength of a member subject to moment alone and having a slenderness ratio exceeding the limits of Eqs. 9.7.1 and 9.7.2. The maximum ultimate moment is given by

$$M_m = \left[1.07 - \frac{\sqrt{F_y}(L/r_y)}{3160}\right] M_p \leq M_p \tag{9.7.3}$$

It was felt that for the beam-column interaction one might desire to exceed the slenderness limitations ordinarily used for continuous beam design. However, since a beam-column in a plastically designed structure will be an integral part of a frame, continuity tends to provide restraint against lateral-torsional buckling not accounted for in AISC Formulas 1.5–6 and 1.5–7.

The use of the larger of Formulas 1.5–6 or 1.5–7 has already been shown to be conservative, and the CRC "basic" design method of computing an equivalent KL/r is overly complex for ordinary use and does not accurately consider inelastic behavior. The more accurate analysis of Galambos[16] shown in Fig. 9.7.2, shows Eq. 9.7.3 to be a realistic expres-

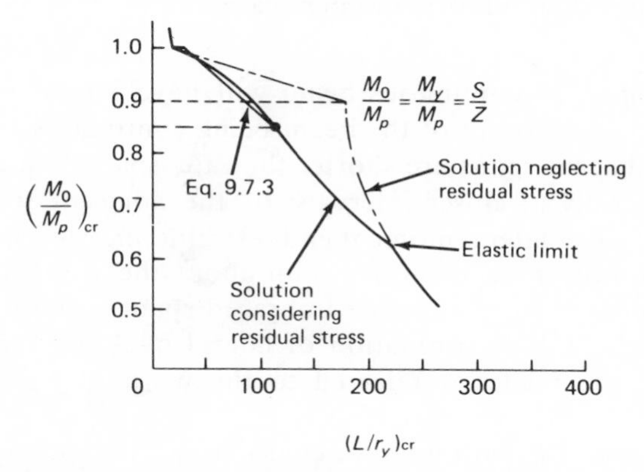

Fig. 9.7.2. Effects of residual stress and lateral-torsional buckling on the ability to develop M_p under pure bending. (From Ref. 16.)

sion, and its design use is recommended in Ref. 22.

Illustration of these provisions for plastic design appears in Chapter 10 for continuous beams and in Chapter 12 for beam-columns.

9.8. EFFECTIVE LENGTH

Design equations, such as those of AISC–1.5.1.4.6a are based on the most critical situation of uniform bending moment over the unbraced segment with assumed torsional simple support at its ends. This would mean that when considering torsional behavior only, the effective unbraced length kL, equals the actual unbraced length L. As is seen from Fig. 9.4.1, the effective length factor k varies from 0.5 for torsional fixity to 1.0 for torsional simple support.

Nearly any type of lateral brace or end connection will prevent rotation about the longitudinal axis of the beam ($\phi = 0$ in Fig. 9.3.1c). However, only rarely will there be much restraint against rotation of the flanges about a vertical axis as discussed in regard to Fig. 8.5.7, (i.e., warping restraint), so that $k = 1.0$ is reasonable for most cases.

Nonuniform Moment. To correct from the pure bending situation to that of the real loading condition, the CRC suggests[8] (permitted under AISC–

1.5.1.4.6a) computing an equivalent slenderness ratio, as discussed in Art. 9.4, rather than directly determining an effective length. Figures 9.4.1 and 9.4.2 may be used for determining the buckling stress according to Eq. 9.4.1 and then use Eq. 9.4.5 to get the equivalent slenderness ratio.

For linearly varying moment gradient, the C_b factor already discussed produces the equivalent uniform moment case.

Continuous Beams. A continuous beam will have lateral end restraint moments develop as a result of the beam being continuous over adjacent spans. If the adjacent spans are shorter than the span in question, or at least braced laterally at closer intervals; or the adjacent spans are less severely loaded, some lateral moment end restraint may develop.

Typically, however, such end restraint about the y–y axis cannot be assumed to be present since alternate unbraced spans could buckle in opposite directions. For determination of lateral buckling loads on continuous beams, the reader is referred to the work of Salvadori[21] and Hartmann.[15]

In past practice, the authors have considered it acceptable to treat an inflection point as a braced point, particularly when applying AISC Formula 5, since that formula had no provision for moment gradient. The corresponding 1969 AISC Formula 1.5–7 does include a C_b term which can account for an unbraced segment containing an inflection point. When the C_b term is used the inflection point should not be treated as braced.

For establishing ability to perform as a "compact" section, the authors believe it is proper to utilize an inflection point as a braced point since no provision is otherwise provided to account for the compression force in the critical flange decreasing to zero much sooner than the adjacent externally braced point. When the distance from the support to the inflection point is less than L_c, satisfying Eqs. 9.5.2 and 9.5.6, the sections may be treated as "compact" regarding lateral-torsional buckling.

Cantilever Beams. Unlike the flagpole type of column where the effective pin-end length is twice the actual length, the lateral buckling of a cantilever beam is not even as severe as the unbraced segment under uniform moment. If one considers the analogy to a column as discussed in Sec. 9.1, such a result will be logical. Since the moment at the end of the cantilever is zero, the compression force in the flange decreases from a maximum at one end to zero at the free end.

Theoretical analyses[8,14] show that it is conservative to use the full length as the effective length for lateral buckling of cantilevers. The authors suggest that the actual cantilever length be used for the AISC formulas; but, as implied in the CRC Guide[8] for cantilevers having complete fixity about both axes at the support end, that C_b be taken as 1.3 for a concentrated load on the free end, and 2.05 for uniform loading.

9.9.* LATERAL BUCKLING OF CHANNELS, ZEES, AND UNSYMMETRICAL I-SHAPED SECTIONS

The basic development of lateral buckling criteria has assumed that loads are applied vertically through the shear center. Furthermore, the resistance to lateral buckling considered that the shear forces which developed in the flanges were equal and the center of twist was located at mid-height.

Channels. Unless loaded through the shear center, a channel is subjected to combined bending and torsion. Since the shear center is not in the plane of the web (see Fig. 8.5.1), usual loadings through the centroid or in the plane of the web give rise to such combined stress. For loads in a plane parallel to the web, lateral buckling must be considered, even if the torsional moment may properly be neglected. The CRC Guide states "if an otherwise laterally unsupported channel has concentrated loads brought in by other members that frame into it, such loads can be considered as being applied at the shear center, *provided that the span of the framing member is measured from the channel shear center and the framing connections are designed for the moment and shear at the connection.*"

For design purposes, Hill[17] indicates that the lateral buckling equations for symmetrical I-shaped sections may be applied for channels (AISC–1.5.1.4.6a–1969 accepts use of Formula 1.5–7 only). Such a procedure is stated to err on the unsafe side by about 6 percent in extreme cases. For more exact determination of lateral buckling strength, Equation 9.3.32 or 9.3.33, may be used for theoretical elastic buckling stress along with the Eq. 9.4.5 approach of using an equivalent slenderness ratio. The torsion-bending constant, I_w, is different for channels than for I-shapes and can be obtained from Table 8.11.1.

Zees. The zee-section lateral buckling strength is complicated by the fact that loading in the plane of the web causes unsymmetrical bending, resulting because a principal axis does not lie in that plane. The general treatment of buckling under biaxial bending is found in Sec. 9.11. The effect of biaxial bending on zee sections was found[17] to reduce the critical moment, M_x, to 90–95 percent of the value given by Eq. 9.3.32 or 9.3.33. In addition, the torsion-bending constant I_w, is different than for channels or I-shapes.

For design purposes, in view of the fact that unbraced zees are relatively rare, AISC–1969 does not provide for them. The authors suggest use of the "rational method" using Eq. 9.3.32 or 9.3.33, or as an alternative it is suggested to use one-half the values obtained from AISC Formulas 1.5–6.

Unsymmetrical I-Shapes. I-shaped sections symmetrical about the y axis, but unsymmetrical about the x axis, are summarized in Refs. 8 and 14. The additional variable involved is e, the distance from the centroid of the girder cross section to the shear center (positive if the shear center lies between the centroid and compression flange, otherwise negative).

An approximate expression, for uniform moment, as derived by Hill[18] is

$$F_{cr} = \frac{\pi^2 E I_y}{S_c L^2}\left[e + \sqrt{e^2 + \frac{I_w}{I_y}\left[1 + \left(\frac{\lambda L}{\pi}\right)^2\right]}\right] \tag{9.9.1}$$

where L = effective length for torsional support and where $\lambda = \sqrt{GK/EI_w}$. Equation 9.9.1 corresponds to Eq. 9.3.35 for symmetrical sections.

Under AISC–1.5.1.4.6a such sections with an axis of symmetry in the plane of loading may be dealt with using Formulas 1.5–6 and 1.5–7. Such an approach will be conservative as long as the compression flange is larger than the tension flange. Equation 9.9.1 is also conservative under the same conditions; but is considered unsafe when the compression flange is smaller than the tension flange.

When the tension flange has the greater area, an equation derived by Winter[27] may be used in the "rational method,"

$$F_{cr} = \frac{\pi^2 E d}{2 S_c L^2}\left[I_c - I_t + I_y \sqrt{1 + \frac{4GK}{Ed^2 I_y}\left(\frac{L}{\pi}\right)^2}\right] \tag{9.9.2}$$

where I_c and I_t are the moments of inertia of the compression and tension flanges, respectively, computed about the y axis, and S_c is the section modulus referred to the compression flange.

EXAMPLE 9.9.1

Determine the allowable bending stress for a channel, C12 × 20.7, on a span of 24 ft with concentrated loads at the ⅓ points as shown in Fig. 9.9.1. Use steel with F_y = 50 ksi.

(a) Use AISC procedure.

(b) Use "basic" design procedure (i.e., equivalent slenderness ratio method) suggested by the Column Research Council.

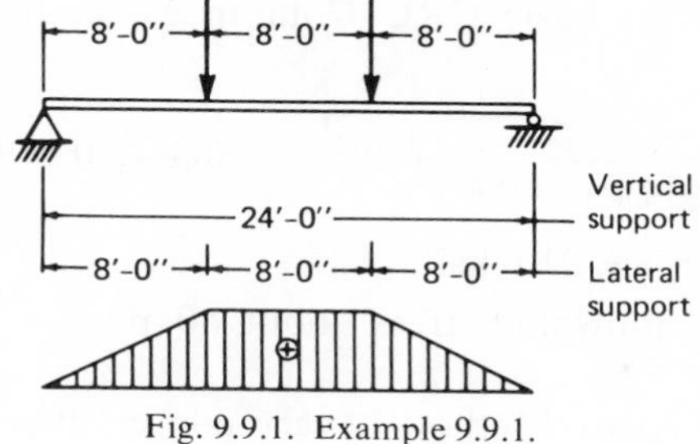

Fig. 9.9.1. Example 9.9.1.

Solution

(a) AISC Method. Assume loading to be through the shear center. For C12 × 20.7, $r_y = 0.80$ in., $d/A_f = 8.13$, $S_x = 21.5$ in.^{3} Since concentrated load points in this case provide lateral bracing, the center unbraced segment controls; thus $C_b = 1.0$.

Since only Formula 1.5–7 may be applied,

$$F_b(1.5\text{–}7) = \frac{12(10^3)}{Ld/A_f} = \frac{12(10^3)}{8(12)8.13} = 15.35 \text{ ksi}$$

(b) CRC "basic" design procedure. For this procedure the torsional properties are required. From Table 8.11.1,

$$K = \Sigma\tfrac{1}{3}\bar{b}t^3 = \tfrac{1}{3}[2(2.94)(0.501)^3 + 12.0(0.28)^3] = 0.33 \text{ in.}^4$$

$$I_w = \frac{t_f b^3 h^2}{12}\left[\frac{3bt_f + 2ht_w}{6bt_f + ht_w}\right]$$

$$= \frac{0.501(2.94)^3(11.5)^2}{12}\left[\frac{3(2.94)(0.501) + 2(11.5)(0.28)}{6(2.94)(0.501) + (11.5)(0.28)}\right]$$

$$= 126 \text{ in.}^6$$

$$\lambda = \frac{1}{a} = \sqrt{\frac{GK}{EI_w}} = \sqrt{\frac{0.33}{2.6(126)}} = \sqrt{\frac{1}{994}} = \frac{1}{31.5}$$

Using Eq. 9.3.33,

$$F_{cr} = \frac{\pi^2 E\sqrt{I_w I_y}}{L^3 S_x}\sqrt{1 + \frac{L^2 GK}{\pi^2 EI_w}}$$

$$= \frac{\pi^2(29{,}000)\sqrt{(126)(3.9)}}{(96)^2 21.5}\sqrt{1 + \left(\frac{96}{\pi(31.5)}\right)^2}$$

$$= 32.1\sqrt{1 + 0.94} = 44.8 \text{ ksi} > F_y/2 = 25 \text{ ksi}$$

Using Eq. 9.4.5, the equivalent slenderness ratio is found:

$$\left.\frac{L}{r}\right)_{\text{equiv.}} = \pi\sqrt{\frac{E}{44.8}} = \pi\sqrt{\frac{29{,}000}{44.8}} = 80$$

Applying this with the basic CRC Column Formula, Eq. 6.7.3, which gives

$$F_{cr} = 50\left[1 - \frac{50}{4\pi^2(29{,}000)}(80)^2\right] = 50(1 - 0.279) = 36.1 \text{ ksi}$$

$$F_b = 36.1/1.67 = 21.6 \text{ ksi}$$

which indicates an allowable stress about 40 percent higher than 1969 AISC permits.

Thus the 1969 AISC Method is very conservative for this channel, probably justifiably to offset the neglect of torsional moment when the load does not pass through shear center. Examination of the magnitude of the second term under the radical of Eq. 9.3.33 shows that instead of the second term being negligible, it is nearly equal to the first term. Such results will similarly occur on symmetrical I-shaped sections, so that the procedure for channels is probably in line with the two formula procedure used for I-shaped beams.

EXAMPLE 9.9.2

Repeat Example 9.9.1 using a girder (Fig. 9.9.2) composed of flange plates, $\frac{3}{4} \times 10$ for compression, and $\frac{3}{4} \times 5$ in tension, with a $\frac{3}{8} \times 13\frac{1}{2}$ web, and consider that the span is 42 ft, braced at the one-third points.

Solution

(a) AISC Method $C_b = 1.0$

$$r_T = \sqrt{\frac{0.75(10)^3/12}{10(0.75) + 5.115(0.375)/3}} = 2.77 \text{ in.}$$

$$\frac{L}{r_T} = \frac{14(12)}{2.77} = 60.6$$

which exceeds $L/r_T = 45$ (Table 9.6.1, $F_y = 50$ ksi), indicating the parabolic equation, Formula 1.5–6a is to be used.

$$F_b(1.5\text{–}6a) = 33.3 - \frac{(60.6)^2}{612} = 27.3 \text{ ksi}$$

$$F_b(1.5\text{–}7) = \frac{12(10^3)\,C_b}{Ld/A_f} = \frac{12(10^3)(1.0)}{14(12)(15.0)/7.5} = 35.7 \text{ ksi} > 0.6F_y\text{; use 30 ksi}$$

indicating stability is not controlling.

(b) CRC "basic" method using equivalent slenderness ratio. Use Eq. 9.9.1:

$$F_{cr} = \frac{\pi^2 EI_y}{S_c L^2}\left[e + \sqrt{e^2 + \frac{I_w}{I_y}\left[1 + \left(\frac{\lambda L}{\pi}\right)^2\right]}\right]$$

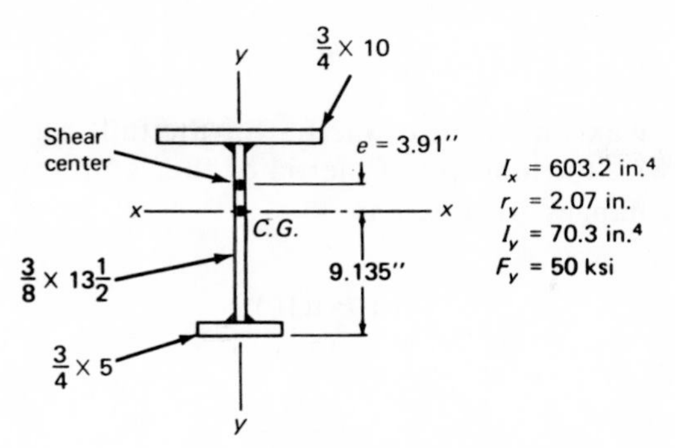

Fig. 9.9.2. Example 9.9.2.

Obtain torsional constants from Table 8.11.1,

$$\Sigma K = \Sigma \tfrac{1}{3} bt^3 = \tfrac{1}{3}[(10 + 5)(0.75)^3 + 15(0.375)^3] = 2.37 \text{ in.}^4$$

$$I_w = \frac{t_f h^2}{12}\left[\frac{b_1^3 b_2^3}{b_1^3 + b_2^3}\right] = \frac{0.75(14.25)^2}{12}\left[\frac{(10)^3(5)^3}{1000 + 125}\right] = 1{,}410 \text{ in.}^6$$

$$\lambda = \frac{1}{a} = \sqrt{\frac{GK}{EI_w}} = \sqrt{\frac{2.37}{2.6(1410)}} = \frac{1}{39.3}$$

Using $E/G = 2.6$,

$$e = \text{distance from centroid to shear center} = +3.91 \text{ in.}$$

$$S_c = I_x/\text{distance to compression flange} = 603.2/5.865 = 102.8 \text{ in.}^3$$

$$F_{cr} = \frac{\pi^2(29{,}000)(70.3)}{102.8(168)^2}\left[3.91 + \sqrt{(3.91)^2 + \frac{1410}{70.3}\left[1 + \left(\frac{168}{\pi(39.3)}\right)^2\right]}\right]$$

$$= 6.92\left[3.91 + \sqrt{15.28 + 20.05(1 + 2.12)}\right] = 88 \text{ ksi} > F_y/2$$

Even though F_{cr} exceeds F_y, inelastic buckling may still control.

The equivalent slenderness ratio

$$\left.\frac{KL}{r}\right)_{\text{equiv.}} = \pi\sqrt{\frac{E}{88}} = \pi\sqrt{\frac{29{,}000}{88}} = 57$$

which gives from the CRC basic formula with a 1.67 factor of safety, Eq. 6.7.3,

$$F_b = 50\left[1 - \frac{50}{4\pi^2(29{,}000)}(57)^2\right]\frac{1}{1.67} = 25.7 \text{ ksi}$$

which is less than obtained by AISC Formula 1.5–7; indicating that the AISC Formula is probably not conservative.

An alternate application of AISC Formula 1.5–7 was proposed by deVries[3] where

$$A_f = bt = \frac{5I_y}{b^2}$$

where I_y is the y axis moment of inertia for the full section if the flanges are equal; or twice the moment of inertia of the compression flange about the y axis if the flanges are different.

In this case,

$$I_y = 2\left[\frac{(0.75)(10)^3}{12}\right] = 125$$

$$A_{f\,\text{equiv.}} = \frac{5(125)}{(10)^2} = 6.25$$

which if used in Formula 1.5–7 gives

$$F_b(1.5\text{–}7) = \frac{12(10^3)}{168(15)/(6.25)} = 29.8 \text{ ksi} < 0.6F_y$$

It would appear that de Vries' suggestion gives the more correct result. Finally, one should conclude the CRC equivalent slenderness ratio gives a conservative result because the $\lambda L/\pi$ term in Eq. 9.9.1 is considerably larger than unity, indicating torsional stiffness as the predominant effect. Since such stiffness is not greatly affected by residual stress, the basic column formula overestimates the reduction in stress.

9.10.* LATERAL BRACING DESIGN

The designer is constantly faced with the question of how to design bracing, either cross bracing or sway bracing to take lateral loads, or point bracing of columns and beams sufficient to prevent buckling. Little is found in specifications but some information is to be found in the works of Zuk,[20] Winter,[21] Driscoll et al.,[22] and Lay and Galambos.[23] The following elastic buckling development follows that of Winter[21] wherein bracing is treated as a linear elastic spring. It is necessary to provide stiffness for the spring sufficient only to prevent displacement of the compressed element at that point. It is also necessary for the brace to have adequate strength to carry the spring force.

Bracing Forces and Stiffness for Elastic Stability. As is known for pin-end columns, such as in Fig. 9.10.1a, the elastic buckling load may be expressed as

$$P_{cr} = \frac{\pi^2 EA}{(L'/r)^2} \tag{9.10.1}$$

Consider that a spring is placed at mid-height such that a reaction can be developed equal to the spring constant k times the deflection d of the

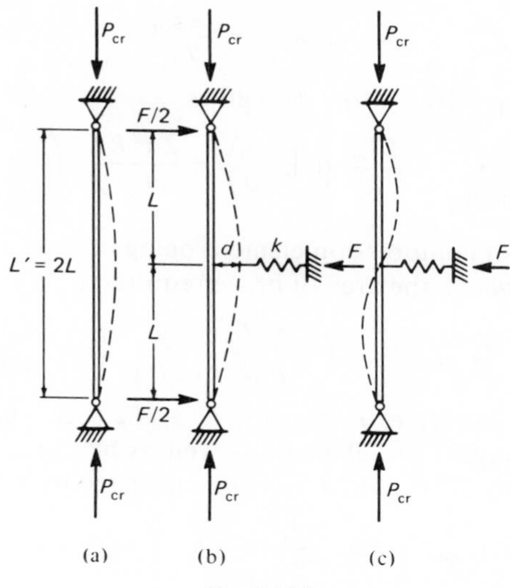

Fig. 9.10.1.

member. Thus, as in Fig. 9.10.1b,

$$F = kd \tag{9.10.2}$$

If no displacement occurs at mid-height, i.e., full bracing is provided, the column will buckle at a load nearly equal to

$$P_{cr} = \frac{\pi^2 EA}{(L/r)^2} \tag{9.10.3}$$

In other words, one may imagine that a hinge exists at mid-height. Failure then occurs when the column snaps into the two half-wave mode of Fig. 9.10.1c.

Taking moments about the imaginary hinge location with the column deflected by an amount d, as in Fig. 9.10.1b, gives

$$P_{cr}d = (F/2)\,L \tag{9.10.4}$$

and if the member was not initially straight, instead having an initial deflection, d_0, at mid-height, d in Eq. 9.10.4 may be replaced by $d_0 + d$:

$$P_{cr}(d + d_0) = FL/2 \tag{9.10.5}$$

Substituting Eq. 9.10.2 into 9.10.5 gives for the required spring constant

$$k_{reqd} = \frac{2P_{cr}(d + d_0)}{Ld} = \frac{2P_{cr}}{L}\left(1 + \frac{d_0}{d}\right) \tag{9.10.6}$$

This is the necessary stiffness required to make an elastic support as effective as a rigid unyielding one.

The bracing is subject to axial deformation; hence its spring constant equals the force per unit deformation, F/d:

$$k_{\text{bracing}} = \frac{F}{d} = \left.\frac{AE}{L}\right)_{\text{bracing}} \tag{9.10.7}$$

Substitution of Eq. 9.10.7 into 9.10.6 gives

$$\left.\frac{AE}{L}\right._{\substack{\text{bracing}\\ \text{reqd}}} = \frac{2P_{cr}}{L}\left(1 + \frac{d_0}{d}\right) = \frac{2\pi^2 EA_c}{(L/r)^2}\left(1 + \frac{d_0}{d}\right) \tag{9.10.8}$$

where A_c = area of compression element being braced.

To satisfy *stiffness*, the area of brace required is

$$A_{b\,\text{reqd}} = 2\pi^2\left(\frac{E_c}{E_b}\right)\left(\frac{L_b}{L_c}\right)\frac{A_c}{(L/r)^2}\left(1 + \frac{d_0}{d}\right) \tag{9.10.9}$$

Suppose, for example, $L/r = 75$, $E_c = E_b$ (i.e., elastic modulus applies for main compression element as well as brace), $L_b = L_c$ (conservative since usually $L_b < L_c$), $d_0/d = 1$ (as suggested by Winter[21]).

$$A_{b\,\text{reqd}} = 2\pi^2 \frac{A_c}{(75)^2}(1 + 1) = 0.007A_c$$

This would give $k_{act} = k_{reqd}$. For adequate safety, k_{act} should probably be twice k_{reqd}. While the area required for adequate stiffness can vary widely, it will still be only a modest requirement—one that is easily satisfied.

Adequate *strength* is required to carry the force F:

$$F_{\text{reqd}} = k_{\text{act}}d \tag{9.10.10}$$

Letting k_{ideal} be the stiffness for perfectly straight members; i.e., with $d_0 = 0$, the relationship between k_{act} and k_{ideal} is, from Eq. 9.10.6,

$$k_{\text{act}} = k_{\text{ideal}}\left(1 + \frac{d_0}{d}\right) \tag{9.10.11}$$

which when solved for d becomes

$$d = d_0\left(\frac{k_{\text{ideal}}}{k_{\text{act}} - k_{\text{ideal}}}\right) \tag{9.10.12}$$

Substitution of Eq. 9.10.12 into Eq. 9.10.10 gives a usable design equation in terms of initial imperfection d_0 rather than d,

$$F_{\text{reqd}} = d_0 \frac{k_{\text{ideal}}}{(1 - k_{\text{ideal}}/k_{\text{act}})} \tag{9.10.13}$$

Normal tolerances on crookedness of compression members would vary from 1/500 to 1/1000 of the length.[21] Considering accidental eccentricity of loading, Winter suggests taking d_0 from 1/250 to 1/500 of the length. With regard to k_{ideal}, for the two equal span systems of Figs. 9.10.1b and c, Eq. 9.10.6 gives

$$k_{\text{ideal}} = \frac{2P_{\text{cr}}}{L}$$

when $d_0 = 0$.

For situations with more than two equal spans, the same procedure may be used to obtain k_{ideal}. Examination of Fig. 9.10.2 for three equal

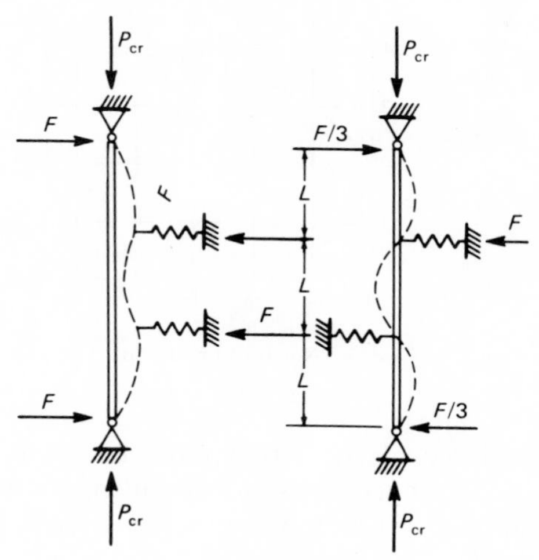

Fig. 9.10.2.

spans will show that the spring forces F can act either in the same or in opposite directions. Assuming they act in the *same* direction (Fig. 9.10.2a), using imaginary hinges at one-third points, and taking moments when slightly deflected at the brace points, gives

$$FL = kdL = P_{\text{cr}}d; \qquad k_{\text{ideal}} = \frac{P_{\text{cr}}}{L}$$

Assuming the forces F acting in *opposite* directions (Fig. 9.10.2b) gives

$$FL/3 = kdL = P_{\text{cr}}d; \qquad k_{\text{ideal}} = \frac{3P_{\text{cr}}}{L}$$

The configuration requiring the highest spring constant is the correct one, that which will permit the highest critical load. If a lesser stiffness is used, an alternate buckling mode will occur at a lower load, accompanied by displacement at the springs.

By the same process the k_{ideal} may be determined for any number of equal spans. In general,

$$k_{\text{ideal}} = \frac{\beta P_{\text{cr}}}{L} \tag{9.10.14}$$

where β varies from 2 for the two equal spans to 4 for infinite equal spans. The variation is given in Fig. 9.10.3.

For the working-stress method, where it is necessary to achieve a yield moment, M_y, or less, at ultimate condition the foregoing method

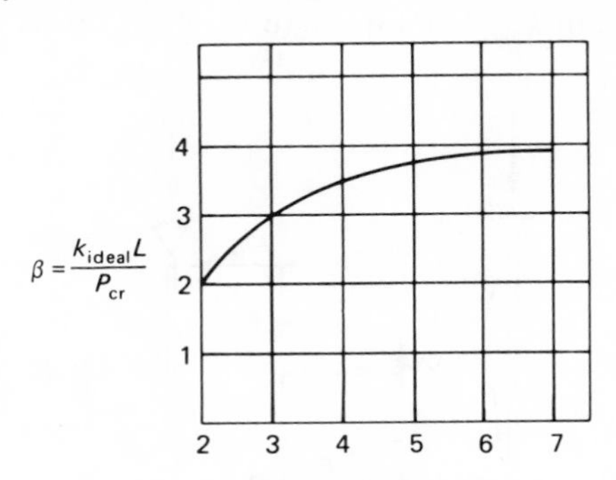

Fig. 9.10.3.

may be applied. If large plastic strain must be accommodated at the bracing points the design suggestions of Lay and Galambos,[23] as discussed below should be applied. For working-stress design where maximum allowable stress is $0.6F_y$,

1. Establish bracing locations and compute $P_{cr} = \pi^2 EI/L^2$ for the compression element (either entire column, or compression flange of beam).

2. Estimate d_0, say $L/250$ to $L/500$.

3. Estimate β from Fig. 9.10.3 based on number of equal unbraced lengths.

4. Determine k_{ideal}.

5. Select bracing area for stiffness equal to $2k_{ideal}$

$$A_{b\,reqd} \approx 2\beta\pi^2 \frac{A_c}{(L/r)^2} \tag{9.10.15}$$

obtained from Eq. 9.10.9 using β instead of 2, $E_c = E_b$, $L_b = L_c$, and $d_0 = d$. In Eq. 9.10.15, A_c and L/r are the area and slenderness ratio, respectively, of the *compression element being braced.*

6. Verify that the required force can be carried by the bracing element, using Eq. 9.10.13. Note that the ultimate strength of brace must exceed F_{reqd}, without any additional safety factor.

Bracing Requirements for Inelastic Steel Beams. When ability to accommodate large plastic strain is desired at bracing points, such as when plastic design or the compact section provision in working stress design is used, the procedure in the previous section may not be adequate. Lay and Galambos[23] have developed a set of rules for design where such high plastic strain (rotation capacity) is required to be accommodated.

In effect, bracing requirements are based on a rotation capacity R consistent with the beam unbraced length slenderness ratio given by Eq. 9.3.43. It has been found that within the laterally unbraced length "local buckling causes a curtailment of the load capacity of the member and therefore defines the rotation capacity of the beam.[23]

The derivation of Lay and Galambos[23] has determined the maximum lateral moment which can develop in the compression flange under a *uniform* moment, $M_x = M_p$, by using the strain distributions on the compression flange due to (a) compression due to planar bending; and (b) the lateral bending strains when local buckling occurs on the 'compression' side.

The design recommendations are:

1. For axial strength, the required cross-sectional area is

$$A_{b\,\text{reqd}} = \left[\frac{s - 1}{h - \sqrt{h}}\right]\left[\frac{2}{3}\right]\frac{A_c}{(L_{\text{avg}}/b)} \tag{9.10.16}$$

where $L_{\text{avg}} = \dfrac{2L_L L_R}{L_L + L_R}$

L_L = unbraced length to left of braced point

L_R = unbraced length to right of braced point

b = width of compression flange

s = strain at strain hardening divided by yield strain

h = elastic modulus divided by strain hardening modulus of elasticity

2. The axial stiffness requirement is satisfied when

$$\frac{L_b}{L_{\text{avg}}} \leq 0.57\left[\frac{s - 1}{h - \sqrt{h}}\right]\left[\frac{A_{b\,\text{act}}}{A_{b\,\text{reqd}}}\right]\left[\frac{L_a}{b}\right] \tag{9.10.17}$$

where L_a = longer of the two adjacent unbraced lengths.

In addition to the axial strength and stiffness requirements, Lay and Galambos indicated[23] that when only the compression flange is braced there are additional flexural strength and stiffness requirements which must be satisfied. These flexural requirements (not given here) give overly large and deep bracing members. It is now (1971) accepted that flexural requirements are unnecessary for lateral bracing locations away from beam vertical reactions. When the compression flange is braced, point restraint giving the necessary axial strength and stiffness is sufficient. Lateral bracing at vertical supports undoubtedly does need some flexural strength and stiffness to prevent a beam from tipping, but ordinary framing at such locations generally provides adequate flexural strength and stiffness.

EXAMPLE 9.10.1

Estimate what size bracing would be required for a W27 × 84 beam of A36 steel positioned as in Fig. 9.10.4. Assume the braces will be at-

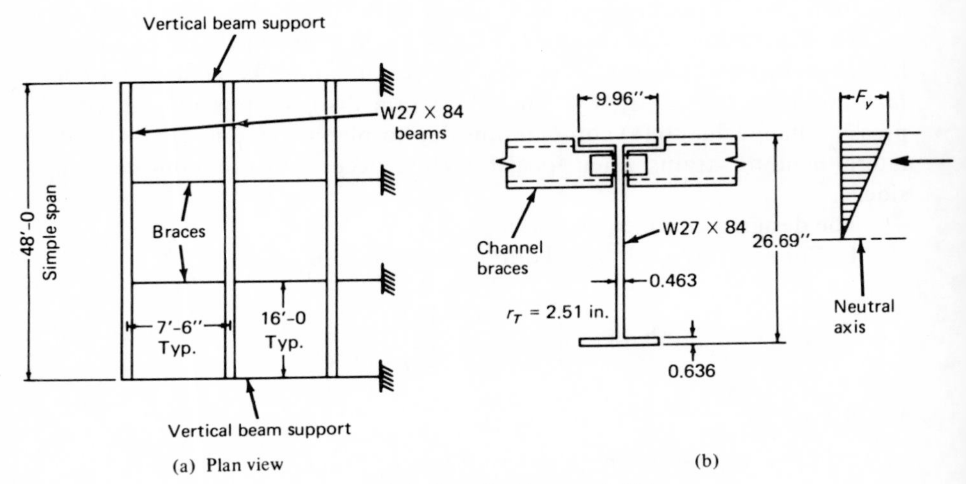

Fig. 9.10.4. Data for Example 9.10.1.

tached only to the compression flange, and that they will be located at the one-third points of a 48-ft span. Assume braces to be attached to both sides of beam and that they are 7.5 ft long.

Solution

(a) Consider working stress design requirements. Since bracing locations are given, establish P_{cr} for the compression zone. Assume maximum flange stress equals yield stress; i.e., allowable stress is $0.6F_y$. Total ultimate compressive load, P_u, when the applied moment is M_y, is

$$\begin{aligned} P_u &= 36(9.96)(0.636) + \tfrac{1}{2}(36)(13.35 - 0.64)(0.46) \\ &= 228 + 105 = 333 \text{ kips} \end{aligned}$$

To obtain the critical load, the slenderness ratio for lateral movement of the compression flange is

$$\frac{L}{r_T} = \frac{16(12)}{2.51} = 76.5$$

To account for inelastic buckling, the basic Column Research Council parabolic equation should be used; 1969 AISC parabolic Formula 1.5–6a times $\frac{3}{2}$ is about the same:

$$F_{cr} = F_y \left[1.0 - \frac{3}{2}\,\frac{F_y(L/r_T)^2}{1530(10^3)\,C_b}\right]$$

$$= 36.0\left[1.0 - \frac{(L/r_T)^2}{28{,}300}\right] = 36.0\left[1.0 - \frac{(76.5)^2}{28{,}300}\right] = 28.6 \text{ ksi}$$

$$P_{cr} \text{ for compression zone } = \frac{F_{cr}A}{2} = \frac{28.6(24.8)}{2} = 354 \text{ kips.}$$

There is no need for designing bracing for a compressive force exceeding P_u; use P_u since it is less than P_{cr}.

For three equal unbraced lengths, estimate $\beta = 3$ from Fig. 9.10.3. From Eq. 9.10.15,

$$A_{b\,\text{reqd}} \approx 2\beta\pi^2 \frac{A_c}{(L/r)^2}$$

$$= 2(3)\,\pi^2\,\frac{(24.8/2)}{(76.5)^2} = 0.125 \text{ sq in.}$$

While any nominal size will satisfy this area requirement, it must be remembered that as a compression element it must satisfy AISC–1.8.5.

$$r_{y\,\min} = \frac{L}{200} = \frac{7.5(12)}{200} = 0.45 \text{ in.}$$

Select C4 × 5.4 as the lightest section with $r_y \geq 0.45$ in. ($r_y = 0.45$ in.)

Check stiffness:

$$k_{\text{ideal}} = \frac{\beta P_{cr}}{L} = \frac{3(333)}{16(12)} = 5.2 \text{ kips/in.}$$

$$k_{\text{act}} = \frac{AE}{L_b} = \frac{1.59(29{,}000)}{7.5(12)} = 503 \text{ kips/in.}$$

which far exceeds the minimum of 2 times k_{ideal} to account for initial crookedness.

Check strength:

$$\left.\frac{L}{r_y}\right)_{\text{brace}} = \frac{7.5(12)}{0.45} = 200$$

$$F_{cr} = \frac{\pi^2 E}{(L/r_y)^2} = \frac{\pi^2(29{,}000)}{(200)^2} = 7.15 \text{ ksi}$$

$$P_{cr} = F_{cr}A_b = 7.15(1.59) = 11.2 \text{ kips}$$

If d_0 is assumed $L/500 = 48(12)/500 = 1.15$ in., using Eq. 9.10.13 gives

$$F_{\text{reqd}} = d_0\left(\frac{k_{\text{ideal}}}{1 - k_{\text{ideal}}/k_{\text{act}}}\right)$$

$$= 1.15\left(\frac{5.2}{1 - 0.01}\right) = 6.0 \text{ kips} < 11.2 \text{ kips}$$

Use C4 × 5.4.

It is noted that the minimum slenderness ratio of 200 required by the AISC specification assured proper bracing size in this case.

For short bracing members the preceding requirements are more likely to control, and will be found to be in rough agreement with the long-standing rule of thumb that bracing should provide a strength equal to 2 percent of the compressive strength of the member being braced.

(b) Consider plastic strength requirements. Assume normal plastic hinge rotation is required at the bracing points.

For axial strength, using Eq. 9.10.16:

$$A_{\text{reqd}} = \left[\frac{s-1}{h-\sqrt{h}}\right]\left(\frac{2}{3}\right)\frac{A_c}{(L_{\text{avg}}/b)}$$

where $s = \epsilon_{\text{strain hardening}}/\epsilon_y$ which may be taken as 12 for $F_y = 36$ ksi and can probably also be used for steels to about $F_y = 60$ ksi. For $h = E/E_{\text{strain hardening}}$; this may be approximated as $F_y/1000$. Certainly the use of such values is accurate enough for design purposes. For A36 steel, $s = 12, h = 36$.

$$\frac{s-1}{h-\sqrt{h}} = \frac{12-1}{36-\sqrt{36}} = 0.367, \text{ say } 0.37$$

$$L_{\text{avg}} = \frac{2L_L L_R}{L_L + L_R} = \frac{2(16)(16)(12)}{16+16} = 192 \text{ in.}$$

$$A_{b\,\text{reqd}} = 0.37\left(\frac{2}{3}\right)\frac{A_c}{192/9.96} = 0.013A_c$$

$$= 0.013(24.8/2) = 1.6 \text{ sq in.}$$

Try C4 × 5.4, $A = 1.59 \approx 1.6$ sq in. as used in part (a).

For axial stiffness, using Eq. 9.10.17:

$$\frac{L_b}{L_{\text{avg}}} \leq 0.57\left[\frac{s-1}{h-\sqrt{h}}\right]\left[\frac{A_{b\,\text{act}}}{A_{b\,\text{reqd}}}\right]\left[\frac{L_a}{b}\right]$$

$$\leq 0.57(0.37)(1.0)(192/9.96) = 4.06$$

$$L_b \leq 4.06L_{\text{avg}} = 4.06(16) = 65 \text{ ft} > 7.5 \text{ ft for brace} \qquad \text{OK}$$

For plastic design, the same brace used for working stress design is adequate but the area requirement is much greater than it was in working stress design.

Use C12 × 20.7.

Whenever it is anticipated that the main member will develop a plastic moment, M_p, bracing at that location should be provided meeting the requirements outlined above.

9.11. DESIGN OF LATERALLY UNSUPPORTED BEAMS—WORKING STRESS METHOD

Several examples are presented to illustrate working stress design procedure to include consideration of lateral-torsional buckling. Other considerations, such as deflection, shear and web crippling were treated and illustrated in Chapter 7.

EXAMPLE 9.11.1

A simply supported beam is loaded as shown in Fig. 9.11.1. The beam has transverse lateral support at the ends and every 7′-6″ along the span. Select the lightest W section of A36 steel, using AISC specification.

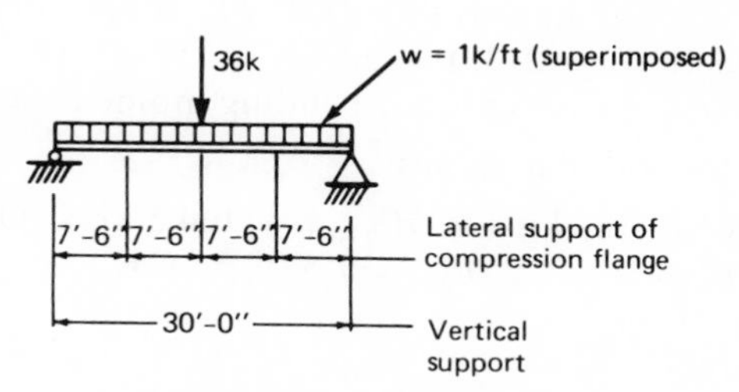

Fig. 9.11.1. Example 9.11.1.

Solution

(a) Estimate whether or not lateral supports are close enough to qualify beam as "compact section" under AISC–1.5.1.4.1. For "compact section," Eq. 9.5.2,

$$L_c = \text{max unbraced length} = \frac{76.0 b_f}{\sqrt{F_y}} \le \frac{20{,}000}{(d/A_f)F_y}$$

$$= 12.65 b_f (\text{for } F_y = 36 \text{ ksi})$$

$$b_{f\,\min} = \frac{L_c}{12.65} = \frac{7.5(12)}{12.65} = 7.1 \text{ in. if } L_c = 7.5 \text{ ft}$$

(b) Assume compact section; the allowable bending stress

$$F_b = 0.66F_y = 0.66(36) = 24 \text{ ksi}$$

$M = (1.0)(30)^2/8 + 36(30)/4 = 112.5 + 270 = 382.5$ ft-kips without beam weight.

$$S_{\text{reqd}} = \frac{M}{F_b} = \frac{382.5(12)}{24} = 191.25 \text{ in.}^3$$

(c) Selection from "Allowable Stress Design Selection Table," AISC Manual. This procedure is efficient if designer is certain "compact section" requirements can be satisfied. Select W24 × 84, $S_x = 197$ in.^{3} (One may also use Appendix Table A3.)

$$M_{DL} = (0.084)(30)^2/8 = 9.45 \text{ ft-kips}$$

$$f_b = \frac{M}{S_x} = \frac{(382.5 + 9.45)(12)}{197} = 23.9 < 24 \text{ ksi} \qquad \text{OK}$$

$$b_f = 9.01 > 7.1 \text{ in.} \qquad \text{OK}$$

Since other local buckling requirements for compact section are satisfied, use W24 × 84.

(d) Selection from Beam Tables, AISC Manual, for moderate unbraced lengths and loading distributions that are uniform or easily converted into uniform loading.

$$\begin{aligned} W_E &= \text{total load for equivalent uniformly loaded beam} \\ &= wL \text{ (actual uniform load)} + 2P \text{ (equiv. for conc. load)} \\ &= 1.084(30) + 2(36) = 32.5 + 72.0 = 104.5 \text{ kips} \\ L &= 30 \text{ ft} = \text{beam span for bending moment, } M_x \end{aligned}$$

For W24 × 84, find tabular values:

$$W = 105 \text{ kips} > W_{E\,\text{reqd}} = 104.5 \text{ kips} \quad \text{OK}$$
$$L_c = 9.5 \text{ ft} > L_{\text{act}} = 7.5 \text{ ft} \qquad \text{OK}$$

EXAMPLE 9.11.2

Select the lightest W section for the simply supported beam of Fig. 9.11.2. Lateral support is provided at the ends and at midspan. Assume

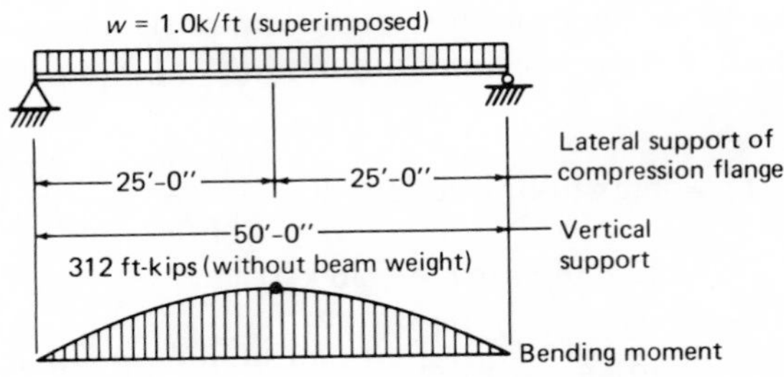

Fig. 9.11.2. Data for Examples 9.11.2 and 9.11.4.

deflection limitations need not be considered. Use A36 steel and AISC specifications.

Solution

(a) Approach directly with Beam Tables, since it is unlikely that L_c can be greater than 25 ft for any economical selection of beams. $C_b = 1.75$.

$$W_{E\,\text{reqd}} = wL = 1.0(50) = 50 \text{ kips (plus dead weight)}$$
$$L = 50 \text{ ft}$$

Not only is it likely that $L_c < 25$ ft, but also that $C_b L_u < 25$ ft; therefore, the allowable stress is in all probability below $0.6F_y$. Since AISC Formula 1.5–6 does not usually control for W sections, assume Formula 1.5–7 governs; in which case the allowable stress reduces in the direct ratio that L is greater than L_u, and is proportional to C_b. The beam tables have tabular loads based on maximum beam capacity when stability does not control, i.e., $F_b = 0.66F_y$ for beams which satisfy local-buckling requirements for "compact" sections. To illustrate tabular load adjustments to obtain estimated safe load; try W24 × 84:

$$W_{\text{safe capacity}} = (\text{Tabular Load})\left(\frac{L_u}{L}\right)\left(\frac{0.6F_y}{0.66F_y}\right)(C_b)$$

$$= 63\left(\frac{13.4}{25}\right)\left(\frac{22}{24}\right)(1.75) = 53.2 \text{ kips}$$

$$W_{E\,\text{reqd}} = 50 + (0.084)(50) = 54.2 \text{ kips} \approx W_{\text{safe}} \qquad \text{OK}$$

It is certain this beam will work; however, in cases where Formula 1.5–6 actually governs with a higher allowable stress, the above approach may give an excessively conservative result.

(b) Approach using AISC Manual Beam Curves. In these curves for A36 steel, the full range of allowable stress vs laterally unbraced length has been used to obtain total moment capacity. For all cases C_b has been taken conservatively as 1.0.

Assuming beam weight about 0.09 kip/ft,

$$M_{\text{reqd capacity}} = 1.09(50)^2/8 = 341 \text{ ft-kips}$$
$$L = \text{lateral unbraced length} = 25 \text{ ft}$$

In using the curves, the solid line *above* the intersection of M and L indicates the lightest section satisfying these requirements. Find, from AISC Manual, using $L/C_b = 14.3$ ft and $M = 341$ ft-kips,

W27 × 84, $M > 341$
W24 × 84, Might work

(c) Check AISC allowable stresses.

Try W24 × 84: $S_x = 197$ in.^{3}

Using Formula 1.5–7,

$$F_b = \frac{12{,}000\,C_b}{Ld/A_f} = \frac{12{,}000(1.75)}{25(12)\,3.47} = 20.2 < 0.6F_y$$

Though it is unlikely to govern here, examine Formula 1.5–6,

$$\frac{L}{r_T} = \frac{25(12)}{2.33} = 128.5; \qquad C_b = 1.75$$

$$\left.\frac{L}{r_T}\right)_{\substack{\text{lower}\\ \text{limit}}} = \sqrt{\frac{102(10^3)\,C_b}{F_y}} = 70$$

$$\left.\frac{L}{r_T}\right)_{\substack{\text{upper}\\ \text{limit}}} = \sqrt{\frac{510(10^3)\,C_b}{F_y}} = 158$$

Since 70 < 128.5 < 158, parabolic Formula 1.5–6a applies, Eq. 9.6.2,

$$F_b = \frac{2F_y}{3} - \frac{F_y^2(L/r_T)^2}{1530(10^3)\,C_b}$$

which for A36 steel becomes

$$F_b = 24.0 - \frac{(L/r_T)^2}{2070} = 24.0 - \frac{(128.5)^2}{2070} = 16.0 \text{ ksi}$$

which is less than given by Formula 1.5–7 and, therefore, does not control.

Bending stress under applied load:

$$f_b = \frac{M}{S_x} = \frac{341(12)}{197} = 20.8 \text{ ksi} > 20.2 \text{ ksi}$$

Exceeds allowable by only about 3 percent: use W24 × 84.

EXAMPLE 9.11.3

Select an economical W section for the beam of Fig. 9.11.3. Lateral support is provided at the vertical supports, concentrated load points, and at the end of the cantilever. Use A36 steel and the AISC specifications.

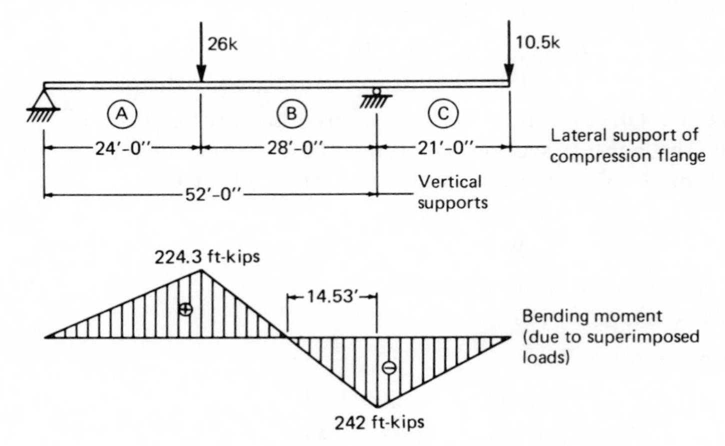

Fig. 9.11.3. Data for Example 9.11.3.

Solution

Three cases must be considered since each of the three laterally unbraced lengths is different and is subject to a different maximum bending moment. Assume the segment containing the largest moment governs.

(a) Segment Ⓒ:

$$M = 242 \text{ ft-kips}, \quad C_b = 1.75, \quad L = 21 \text{ ft}$$

From curves, AISC Manual using $L/C_b = 12$ ft, select

$$W21 \times 68, \quad M = 257 \text{ ft-kips}, \quad L_u = 12.2 \text{ ft}$$

Since $L > L_c$ but $L/C_b < L_u$, $F_b = 0.6F_y$.

(b) Segment Ⓐ:

$$M = 224.3 \text{ ft-kips}, \quad C_b = 1.75, \quad L = 24 \text{ ft}$$

From curves, using $L/C_b = 13.7$ ft, try

$$W21 \times 68, \quad M = 232 \text{ ft-kips}$$

(c) Segment Ⓑ:

$$M = 242 \text{ ft-kips}, \quad L = 28 \text{ ft}, \quad \text{and} \quad M_1/M_2 = +0.925,$$

$$C_b = 1.75 + 1.05(M_1/M_2) + 0.3(M_1/M_2)^2 = 2.98 > 2.3$$

Use $C_b(\max) = 2.3$.

From curves, using $L/C_b = 12.2$ ft

W21 × 68, same as segment Ⓒ.

(d) Check stresses for W21 × 68. Appears as if Segment Ⓐ may govern.

$$F_b(1.5\text{–}7) = \frac{12{,}000C_b}{Ld/A_f} = \frac{12{,}000(1.75)}{24(12)\,3.73} = 19.55 \text{ ksi} < 0.6F_y$$

$$\frac{L}{r_T} = \frac{24(12)}{2.14} = 134.5$$

Since $70 < L/r_T < 158$ (see Example 9.11.2), parabolic Formula 1.5–6a is to be used, which for A36 steel is

$$F_b\,(1.5\text{–}6a) = 24.0 - \frac{(L/r_T)^2}{2070} = 24.0 - \frac{(134.5)^2}{2070} = 15.2 \text{ ksi}$$

which is less than 19.55 ksi and therefore does not govern.

$$f_b = \frac{[224.3 + 15.9(DL)]\,12}{140} = 20.7 \text{ ksi} > 19.55 \text{ ksi} \qquad \text{N.G.}$$

(e) Try W21 × 73: $S_x = 151$ in.^{3}

$$F_b\,(1.5\text{–}7) = \frac{12{,}000(1.75)}{24(12)\,3.46} = 21 \text{ ksi}$$

$$f_b = \frac{(224.3 + 17.1)\,12}{151} = 19.2 \text{ ksi} < 21 \text{ ksi} \qquad \text{(Segment Ⓐ is OK)}$$

For segment Ⓒ, $M_{\max} = 242 + 16.1\,(DL) = 258.1$ ft-kips

$$F_b = 21.0 \text{ ksi for } C_b = 1.75$$

$$f_b = \frac{258.1(12)}{151} = 20.6 \text{ ksi} < 21 \text{ ksi} \qquad \text{OK}$$

For segment Ⓑ, $M_{\max} = 258.1$ ft-kips, $C_b = 2.3 > 1.75$ for segment Ⓒ; therefore, acceptable by inspection.

Use W21 × 73. (W24 × 68 is also acceptable and may be preferred.)

EXAMPLE 9.11.4

Repeat Example 9.11.2 (see Fig. 9.11.2) except use steel with F_y = 60 ksi.

Solution

Since no Beam Tables or Beam Curves are available for direct use, a more general approach must be made.

1. If Formula 1.5–7 controls, beam curves can give correct capacity without correction.

2. If Formula 1.5–6 controls with L/r_T not exceeding about 85, capacity will be roughly proportional to yield point.

(a) Use beam curves: convert applied moment to equivalent moment on a beam of Grade 50 steel,

$$M_{\text{reqd cap}} = 340 \text{ ft-kips} \quad \text{(Example 9.11.2) (incl. beam wt.)}$$

$$M_E = 340(50/60) = 283 \text{ ft-kips}; \quad L/C_b = 25/1.75 = 14.3 \text{ ft}$$

for entering beam curves assuming yielding controls. If stability controls, Formula (1.5–7) capacities can be read directly.

$$F_b\,(1.5\text{–}7) = 0.6F_y = \frac{12(10)^3}{L_u d/A_f}\,C_b$$

$$L_u\,(\text{for } F_y = 60 \text{ ksi}) = \frac{50}{60}\,L_u\,(\text{for } F_y = 50 \text{ ksi})$$

$$= 0.83L_u\,(\text{for } F_y = 50 \text{ ksi})$$

Select from AISC Manual Beam Curves (using L_u values from Beam Tables for F_y = 50 ksi):

M = 283 ft kips and equiv. L = 14.3 ft

W21 × 73 Formula (1.5–6) or (1.5–7)

W24 × 68 Formula (1.5–6)

To get closer try F_y = 50 ksi Beam Tables, W_{reqd} = 50(1.07) = 53.5 kips for L = 50 ft; or W_E = 107 kips for 25 ft.

$$\text{W14} \times 74, \quad W = \frac{99}{2}\left(\frac{60}{50}\right)\left(\frac{0.6}{0.66}\right)\frac{0.83(18.5)(1.75)}{25}$$

$$= \text{Tabular load } (L_u)\ 0.0636$$

$$= 49.5(18.5)\ 0.0636 = 58 \text{ kips} > 53.5 \text{ kips} \qquad \text{OK}$$

$$\text{W27} \times 84, \quad W = 93(9.5)\ 0.0636 = 56.2 \text{ kips} > 53.5 \qquad \text{OK}$$

On the deeper sections Formula 1.5–6 will usually control. From the above investigation, the lightest section seems to be about 74 to 84 pounds/ft.

(b) Examine W27 × 84 according to Formula (1.5–6).

$$\frac{L}{r_T} = \frac{25(12)}{2.51} = 119.5 > \sqrt{\frac{102(10^3)\,1.75}{60}} = 54 \text{ and } <122$$

$$F_b\,(1.5\text{–}6) = \frac{2F_y}{3} - \frac{F_y^2(L/r_T)^2}{1530(10^3)\,C_b}$$

which for $F_y = 60$ ksi becomes

$$F_b\,(1.5\text{–}6) = 40.0 - \frac{(L/r_T)^2}{744} = 40.0 - \frac{(119.5)^2}{744} = 20.8 \text{ ksi}$$

$$f_b = \frac{M}{S_x} = \frac{1.084(50)^2(12)/8}{212} = \frac{339(12)}{212} = 19.2 \text{ ksi} \qquad \text{OK}$$

Summary

Beam	$(Ld/A_f)/C_b$	F_b (1.5–7)	L/r_T	F_b (1.5–6)	$f_b = M/S_x$	Comment
W14 × 74	309	(36.0)	109	—	35.8	OK. Deflection?
W16 × 78	372	(32.2)	131	17.3	31.7	OK
W18 × 77	427	(28.1)	128	18.1	28.4	within 1% OK
W21 × 82	502	(23.9)	128	18.1	24.1	within 1% OK
W24 × 84	595	(20.1)	128.5	18.0	20.6	2½% high OK
W27 × 84	722	16.6	119.5	(20.8)	19.2	OK

The reader should study carefully the results in the above summary. Essentially, strength of the material governs for the shallow beams, while stability controls for the deep ones.

For lightest section; use W14 × 74 if deflection can be tolerated; or W27 × 84 if minimum deflection is desired.

EXAMPLE 9.11.5

What W section can be used for the beam of Example 9.11.4 if lateral support is increased to occur every 5 ft?

Solution

In this case, since the stability has been improved, the deeper sections can safely carry greater load. The deeper sections will also be the lightest ones.

Assume $L < L_c$, in which case the allowable stress will be $0.66F_y$ if the section can satisfy local buckling requirements for a "compact section."

$$M = 1.06(50)^2/8 = 331 \text{ ft-kips}$$

$$S_{\text{reqd}} = \frac{331(12)}{39.5} = 104 \text{ in.}^3$$

Try W21 × 55: $S_x = 110$ in.^{3}, $b_f = 8.215$ in.

Check whether "compact section" provisions, AISC–1.5.1.4.1, are satisfied.

Lateral support:

$$L_u = \frac{20{,}000}{F_y(d/A_f)} = \frac{20{,}000}{60(4.85)\,12} = 5.7 \text{ ft}$$

$$L_c = \frac{76.0b_f}{\sqrt{F_y}} = \frac{76.0b_f}{\sqrt{60}} = 9.8b_f = 9.8\left(\frac{8.21}{12}\right) = 6.7 \text{ ft} > 5.7 \text{ ft}$$

$$L_c = L_u = 5.7 \text{ ft}$$

The lateral support is adequate, since $L = 5 \text{ ft} < L_c$

Local buckling:

$$\frac{d}{t_w} = \frac{20.8}{0.375} = 55.5 < \frac{412}{\sqrt{F_y}} = 53 \qquad \text{N.G.}$$

$$\frac{b_f}{2t_f} = \frac{8.215}{2(0.522)} = 7.86 > \frac{52.2}{\sqrt{F_y}} = 6.75 \qquad \text{N.G.}$$

Under the 1963 AISC, if local buckling requirements were not satisfied, the maximum allowable bending stress was $0.6F_y$. Under 1969 AISC, a linear transition is provided if only the *flange* local buckling limitation of $52.2/\sqrt{F_y}$ is exceeded, but it is not greater than $95/\sqrt{F_y}$:

$$F_b = F_y\left[0.733 - 0.0014\sqrt{F_y}\,\frac{b_f}{2t_f}\right]$$

In this case the d/t_w is also too high; thus $F_b = 0.6F_y = 36.0$ ksi.

Computed stress:

$$f_b = \frac{M}{S_x} = \frac{330(12)}{110} = 36.0 \text{ ksi} = F_b \qquad \text{OK}$$

EXAMPLE 9.11.6

Using the "more precise analysis" permitted by AISC–1.5.1.4.6a—i.e., the CRC "basic" design procedure as discussed in Sec. 9.4, determine whether a W24 × 76 would be acceptable for the beam of Example 9.11.4; $F_y = 60$ ksi.

Solution

$$M = 1.076(50)^2/8 = 336 \text{ ft-kips}$$

Assume the loading acts at the centroid of the beam, and use Eqs. 9.4.1 and 9.4.3:

$$F_{\text{cr}} = \frac{C_4\sqrt{EI_yGK}}{S_xL} \qquad [9.4.1]$$

where

$$C_4 = C_1 \pi \sqrt{1 + \left(\frac{\pi}{\lambda L}\right)^2} \qquad [9.4.3]$$

Torsional properties of beams may be obtained from Ref. 8, Chap. 8.† For W24 × 76,

$$K = 2.84 \text{ in.}^4$$
$$\lambda = 1/96.9$$

The actual loading and bracing most closely corresponds to Case 3 of Fig. 9.4.1; for which case $C_1 = C_b = 1.75$, according to AISC–1969:

$$C_4 = 1.75\pi \sqrt{1 + (96.9\pi/300)^2} = 7.80$$

$$F_{cr} = \frac{7.80 E \sqrt{I_y K/2.6}}{S_x L}$$

$$= \frac{7.80(29{,}000)\sqrt{76.5(2.84)/2.6}}{175.4(300)} = 39.2 \text{ ksi}$$

Since 39.2 ksi > $F_y/2$, use Eq. 9.5.4 to find an equivalent L/r:

$$\left.\frac{L}{r}\right)_{\text{equiv}} = \pi \sqrt{\frac{29{,}000}{39.2}} = 86$$

The basic CRC parabola is then used for the inelastic range of buckling,

$$F_{cr} = F_y \left[1 - \frac{F_y}{4\pi^2 E}\left(\frac{L}{r}\right)^2\right]$$

$$= 60.0 \left[1 - \frac{60}{4\pi^2(29{,}000)}(86)^2\right] = 36.7 \text{ ksi}$$

$$F_b = \frac{F_{cr}}{1.67} = \frac{36.7}{1.67} = 22.0 \text{ ksi}$$

$$f_b = \frac{336(12)}{176} = 23.0 \text{ ksi}$$

A W24 × 76 would be overstressed about 4½ percent according to the "rational method" indicating only slight conservatism of AISC formulas. Whenever both terms under the square root sign in Eq. 9.4.3 are significant, use of the AISC Formulas is conservative.

9.12.* BIAXIAL BENDING OF DOUBLY SYMMETRICAL SECTIONS

A designer frequently encounters the situation where a lateral bending moment M_y is applied in combination with the vertical moment M_x. The

†Approximately the same values are obtainable from torsional properties in the 1970 AISC Manual.

design of roof purlins and crane girders are two of the most common situations. How does the moment M_y affect lateral stability? When all moment is applied in the weak direction, stability is not a problem and if "compact section" limitations on local buckling are satisfied, an allowable stress of $0.75F_y$ is permitted by AISC specifications.

The subject of lateral buckling in biaxial bending has been discussed little in the literature. The design suggestion which follows is that presented by Gaylord and Gaylord.[25]

From Eq. 9.3.31 for pure bending M_x with respect to the strong axis,

$$M_{cr}^2 = M_x^2 = \frac{\pi^2}{L^2}\left[\frac{\pi^2 E^2 I_w I_y{}'}{L^2} + EI_y GK\right] \tag{9.12.1}$$

or

$$\frac{M_x^2}{EI_y GK} = \frac{\pi^2}{L^2}\left[1 + \left(\frac{\pi}{\lambda L}\right)^2\right] \tag{9.12.2}$$

where $\lambda = 1/a = \sqrt{GK/EI_w}$.

When simultaneously a moment M_y is applied, Eq. 9.12.2 becomes[26]

$$\frac{M_x^2}{EI_y GK} + \frac{M_y^2}{EI_x GK} = \frac{\pi^2}{L^2}\left[1 + \left(\frac{\pi}{\lambda L}\right)^2\right] \tag{9.12.3}$$

which is applicable to I-shaped sections with two axes of symmetry. Furthermore, it is applicable for sections with point symmetry (such as the zee), and is approximately valid for channels when M_y does not exceed $0.25M_x$.[25]

Combinations of M_x and M_y which satisfy Eq. 9.12.3 will plot as an ellipse, as shown in Fig. 9.12.1 for a W14 × 74. In biaxial bending it is

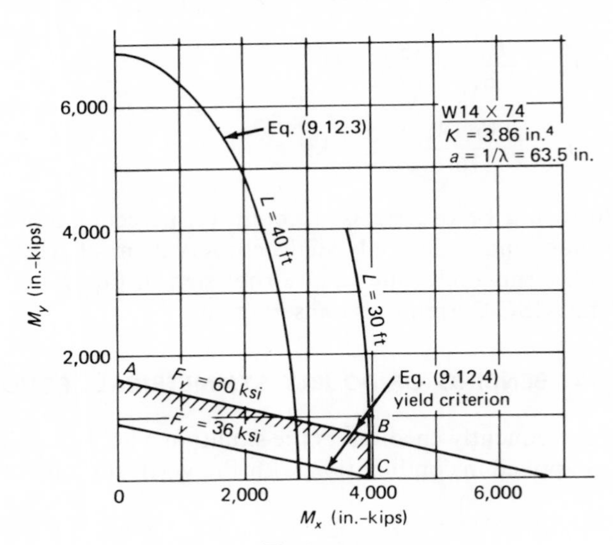

Fig. 9.12.1.

assumed for design that ultimate capacity is reached when the extreme fiber stress reaches F_y. Therefore, no matter how stable the beam, the combination of moments must satisfy:

$$\frac{M_x}{S_x} + \frac{M_y}{S_y} = F_y \tag{9.12.4}$$

assuming ideal elastic-plastic stress-strain conditions.

Since the relationships shown in Fig. 9.12.1 for W14 × 74 are typical, certain conclusions may be drawn. Consider a laterally unbraced length of 30 ft and steel with $F_y = 60$ ksi. Ideally, the ultimate condition is defined by the lines AB (yielding controls) and BC (buckling controls).

The most important observation is that for ordinary laterally unbraced lengths (say 25 ft or less) the line BC is nearly vertical; therefore, *simultaneous application of M_y does not appreciably affect the critical moment, M_x.*

Based on the aforementioned conclusion, the recommended design procedure[25] is as follows:

(a) For yielding controlling (Line AB of Fig. 9.12.1):

$$\frac{M_x}{S_x} + \frac{M_y}{S_y} \leq 0.6F_y \tag{9.12.5}$$

(b) For stability controlling (Line BC of Fig. 9.12.1):

$$\frac{M_x}{S_x} \leq \begin{array}{l}\text{Allowable stresses based on}\\ \text{AISC Formulas 1.5–6 or 1.5–7}\end{array} \tag{9.12.6}$$

The 1969 AISC specification has no specific provision for the above situation; it will always be conservative to require that the combined stress be within the allowable based on lateral buckling.

EXAMPLE 9.12.1

Design a W section to serve as a crane girder to carry a moment $M_x = 301$ ft-kips (without impact) and a top flange moment, $M_f = 30$ ft-kips. The moment M_f is based on a lateral force acting on the top flange equal to 10 percent of the lifted load and crane trolley weight in accordance with AISC–1.3.4. The moment M_f about the y axis is assumed to be resisted by one flange; in effect, this accounts for the torsional effect by using the flexure analogy (see Sec. 8.6). The approximation of equivalent systems is shown in Fig. 9.12.2. Assume the simple span of 24 ft is laterally braced only at the ends. Use $F_y = 50$ ksi.

Solution

(a) AISC–1969, using Eqs. 9.12.5 and 9.12.6 as criteria.

$M_x = 301(1.25) = 376$ ft-kips (using 25 percent impact)
$M_y = 30$ ft-kips

Obtain approximate section modulus, S_x, assuming that yielding controls; thus, $F_b = 0.6F_y = 30$ ksi. Solving Eq. 9.12.5 for S_x,

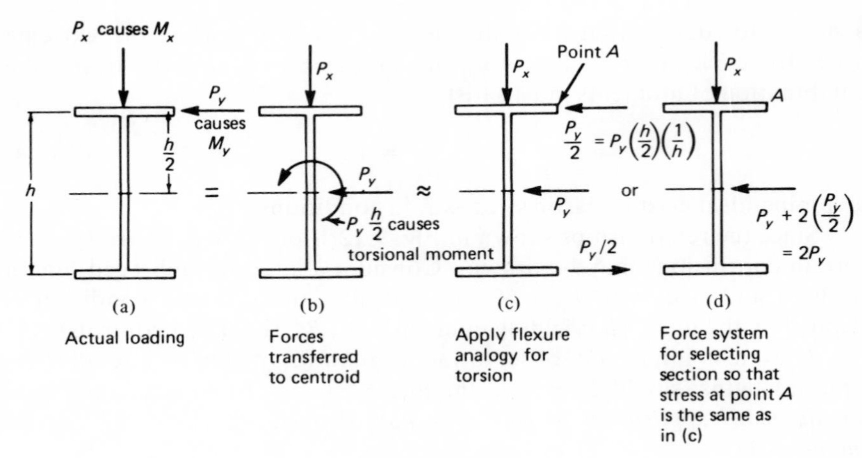

Fig. 9.12.2.

$$S_{x\,\text{reqd}} = \frac{M_x}{30} + \frac{M_y}{30}\left(\frac{S_x}{S_y}\right)$$

$$= \frac{(376 + 10)(12)}{30} + \frac{2(30)\,12}{30}(\approx 6) = 154 + 144 = 299 \text{ in.}^3$$

where a dead-load moment of 10 ft-kips is assumed using a 24-ft simple span; further, two times the moment carried by one flange is taken to be resisted by the entire section, in accordance with Fig. 9.12.2d. Try W24 × 130: $S_x = 332$ in.^{3}

$$S_{x\,\text{reqd}} = \frac{385.3(12)}{30} + \frac{60(12)(6.16)}{30} = 302 \text{ in.}^3$$

Try W21 × 127: $S_x = 284$ in.^{3}

$$S_{x\,\text{reqd}} = \frac{385(12)}{30} + \frac{60(12)(5.48)}{30} = 285.5 \text{ in.}^3$$

Check yield controlling criteria, Eq. 9.12.5:

$$f_{bx} = \frac{M_x}{S_x} = \frac{385(12)}{284} = 16.3 \text{ ksi}$$

$$f_{by} = \frac{2M_y}{S_y} = \frac{60(12)}{56.1} = \underline{12.9 \text{ ksi}}$$

$$f_b = 29.2 \text{ ksi} \approx 0.6F_y \qquad \text{OK}$$

Check stability controlling, Eq. 9.12.6, conservatively assuming $C_b = 1.0$:

$$\frac{Ld}{A_f} = 24(12)(1.65) = 475$$

$$F_b\,(1.5\text{–}7) = \frac{12{,}000}{475} = 25.3 \text{ ksi (controls)}$$

$$\frac{L}{r_T} = \frac{24(12)}{3.52} = 81.7$$

Using Table 9.6.1, $C_b = 1.0$ and

$$F_b\,(1.5\text{–}6a) = 33.3 - \frac{(L/r)^2}{612} = 33.3 - \frac{(81.7)^2}{612} = 22.5 \text{ ksi}$$

Since $f_{bx} = 16.3 \text{ ksi} < 25.3 \text{ ksi}$ OK

Use W21 × 127.

(b) AISC—using conservative approach of requiring combined stress to satisfy stability allowable.

Try W21 × 142:

$$f_{bx} + f_{by} = 14.6 + 12.2 = 26.8 \text{ ksi}$$

$$F_b\,(1.5\text{–}7) = \frac{12{,}000}{24(12)(1.49)} = 28.0 \text{ ksi} > 26.2 \text{ ksi} \qquad \text{OK}$$

No need of checking F_b (1.5–6).

Try W24 × 130:

$$f_{bx} + f_{by} = 14.0 + 13.4 = 27.4 \text{ ksi}$$

$$F_b\,(1.5\text{–}7) = \frac{12{,}000}{24(12)(1.93)} = 21.6 \text{ ksi}$$

$$\frac{L}{r_T} = \frac{24(12)}{3.74} = 77$$

$$F_b\,(1.5\text{–}6) = 33.3 - \frac{(77)^2}{612} = 23.6 \text{ ksi (controls)} < 27.4 \qquad \text{N.G.}$$

Use W21 × 142. (The authors believe this to be an excessively conservative approach.)

(c) AISC—but use the corrected flexure analogy for the torsional component of the loading, in accordance with Sec. 8.6.

Try W24 × 120: $a = 1/\lambda = 102$ (Ref. 8 of Chapter 8.)†

Of the 60 ft-kips used for M_y in the preceding parts, one-half is the actual lateral moment, M_y (Fig. 9.12.2b), while the other half is to approximate the torsional effect so that identical stress acts at point A in Figs. 9.12.2c and d.

Using Table 8.6.1, assuming torsional simple support, the true torsional effective M_y is obtained as follows:

$$\lambda L = 24(12)/102 = 2.82$$

find $\beta \approx 0.63$ for $a = 0.5$.

$$M_y\,(\text{for torsion}) \approx 0.63(30) = 18.9 \text{ ft-kips}$$

†Approximately the same value may be computed from torsional properties in the 1970 AISC Manual.

Check yield condition:

$$f_{bx} + f_{by} = \frac{384(12)}{300} + \frac{(30 + 18.9)(12)}{45.4}$$

$$= 15.4 + 13.0 = 28.4 \text{ ksi} < 0.6F_y \qquad \text{OK}$$

Check stability condition:

$$f_{bx} = 15.4 \text{ ksi} < F_b(1.5\text{–}7) = \frac{12{,}000}{24(12)(2.16)} = 19.3 \text{ ksi} \qquad \text{OK}$$

No need to check F_b (1.5–6), though in this case it gives an even higher allowable value; F_b (1.5–6) = 20.1 ksi.

Once again it is demonstrated that a better understanding of behavior may result in a saving in weight.

Use W24 × 120.

PROBLEMS

All problems are to be done in accordance with the latest AISC specification working stress provisions, unless otherwise indicated. Assume lateral support consists of translational restraint but not moment (rotational) restraint, unless otherwise indicated.

9.1. Plot the allowable bending stress vs. laterally unsupported length, (using C_b = 1.0 and 2.3), for the following beams:
1. W14 × 30
2. W14 × 142
3. Welded girder with flange plates, ⅝ × 20, and web plate, 5/16 × 60.

Use F_y = 36 ksi, and show both Formulas 1.5–6 and 1.5–7 for comparison.

9.2. On each of the plots for the three beams of Prob. 9.1, superimpose the allowable stress using the "Basic Design Procedure" of Sec. 9.4. (Use Eq. 9.4.1 for pure bending, Eq. 9.4.5, and the CRC column formula with a factor of safety of 1.67.) Discuss the comparison of AISC with "Basic Design Procedure."

9.3. Compute the allowable stress using the CRC "Single Formula Procedure," (either of Eqs. 9.4.9 or 9.4.10 with Eq. 9.4.5 and the CRC formula with a safety factor of 1.67) for the W14 × 30 and W14 × 142 of Prob. 9.1. Compare with results of Prob. 9.1, as well as Prob. 9.2, if that problem was solved.

9.4. Select the lightest W sections of A36 steel for the following conditions:

Span	Live Load	Lateral Support
(a) 20 ft	2 kips/ft	Continuous
(b) 20 ft	2 kips/ft	Ends and midspan
(c) 20 ft	2 kips/ft	Ends only

9.5. Design the lightest W section to carry a uniform live load of 0.8 kip/ft on a simply supported span of 35 ft. Select beams for two cases of lateral support, (1) continuously braced, and (2) braced at ends and midspan only.

Compare for the following steels:

(a) A36 steel
(b) A529 (F_y = 42 ksi)
(c) A572 Grade 50 (F_y = 50 ksi)
(d) A572 Grade 65 (F_y = 65 ksi)
(e) A514 (F_y = 100 ksi)
Assume no deflection restrictions.

9.6. A floor beam, laterally supported at the ends only and supporting vibration inducing heavy machinery, is subject to the loads shown in the accompanying figure. Select the lightest W section of A36 steel. (Consider AISC–1.33.)

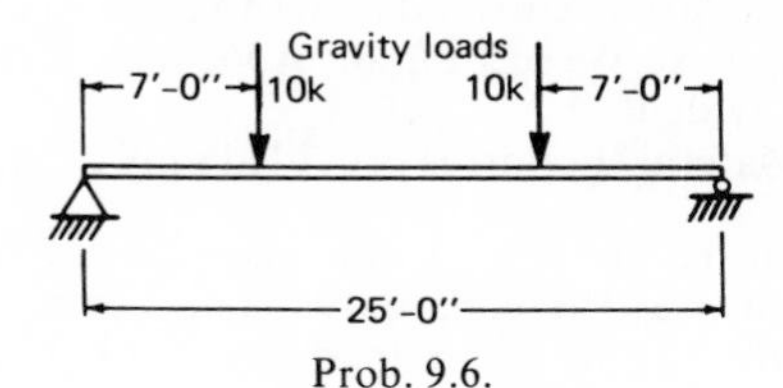

Prob. 9.6.

9.7. Select the lighest W section to serve as a uniformly loaded library floor beam carrying a live-load moment of 240 ft-kips on a simply supported span of 48 ft. Lateral support is at 12-ft intervals. Assume C_b = 1.0. Use (a) A36 steel, and (b) A572 Grade 65 (F_y = 65 ksi) steel. How much saving in weight, if any, could be obtained in each of these cases by using the CRC "basic" procedure?

9.8. Select the lightest W section for a simply supported beam on a span of 48 ft, carrying 2.5 kips/ft, with lateral support at 16-ft intervals. Compare results for A36 steel and A572 Grade 60 (F_y = 60 ksi) steel.

9.9. Determine the capacity P (neglect dead weight of the beam) which the W21 × 62 may be permitted to carry acting in the plane of the web. Lateral support is provided only at the ends and at midspan. Use A36 steel.

(a) Use AISC specs, 1963
(b) Use AISC specs, 1969
(c) Use CRC Basic Design Procedure with FS = 1.67
(d) Use CRC Single Formula Simplified Procedure, with FS = 1.67
(e) Use CRC Double Formula Procedure with FS = 1.67.

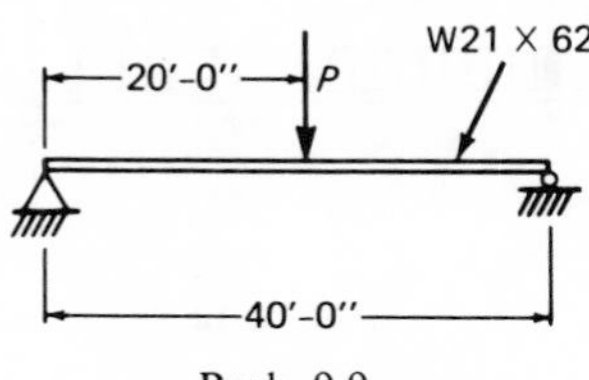

Prob. 9.9.

9.10. Select the lightest W section for the conditions shown. Assume there is no deflection limitation. Use material with F_y = 60 ksi.

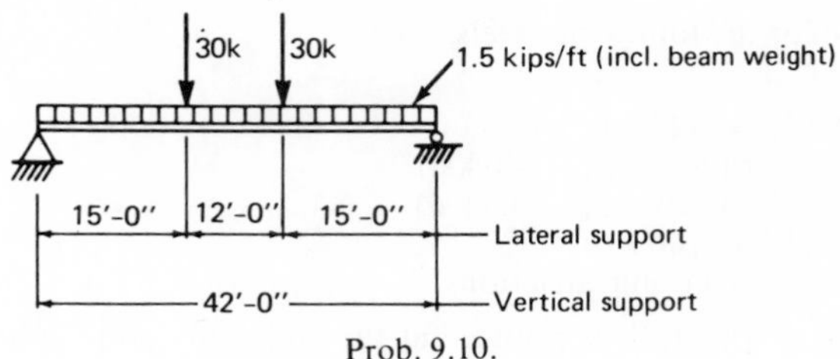

Prob. 9.10.

9.11. A beam is to serve as a floor beam on a simple span of 20 ft. The loading consists of a movable concentrated load (no impact) of 50 kips.
(a) Select the lightest W section of A36 steel when continuous lateral support is provided.
(b) Repeat (a) if lateral support is provided only at the ends.

9.12. A W10 × 33 is to be used as a simply supported beam on a span of 25 ft with lateral support at the ends only. The beam is required to support a plastered ceiling. If the dead load is 0.15 kip/ft (incl. beam weight), what uniform live load can be safely carried on the beam, using A36 steel? What percentage increase in live load can be gained if the beam is A572 Grade 60 steel? Comment.

9.13. Select the lightest W section to carry a library floor. The superimposed loading is 1.5 kips/ft. A plastered ceiling is to be under this beam and the architect requires that nothing deeper than a nominal W12 be used. Use A529 steel (F_y = 42 ksi).

9.14. Determine the lightest W sections to carry the loads shown under the following conditions:
(a) A36 steel; continuous lateral support
(b) A36 steel; lateral support at ends only
(c) A36 steel; lateral support at ends and at point *A*
(d) A529 steel (F_y = 42 ksi); lateral support at ends and point *A*
(e) A572 Grade 60 steel; lateral support at ends and point *A*

What are the maximum price differentials between the higher-strength steels and A36 steel which would justify using the higher-strength steels?

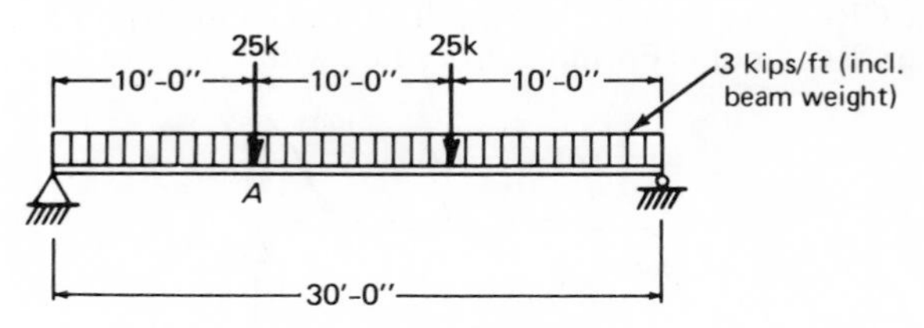

Prob. 9.14.

9.15. The beam shown of A572 Grade 50 (F_y = 50 ksi) steel is to be investigated for bending and shear stresses. Compare with allowable AISC values. External lateral support for the beam is provided only at the vertical supports.

w = 1.4 kips/ft
(incl. beam weight)
10′-0″
12k
W14 X 43
20′-0″
8′-0″

Prob. 9.15.

9.16. Select an economical W section for given loads using A572 Grade 50 steel, and considering lateral support at the ends and at point A only.

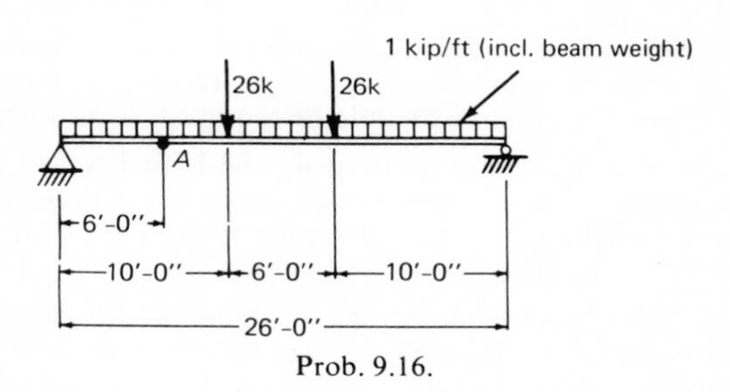

Prob. 9.16.

9.17. Select the lightest W section for a cantilever beam 16 ft long which must support a concentrated load of 10 kips on its free end. No lateral support is provided except at its supported end, and vertical deflection may be disregarded. Use A36 steel.

9.18. Determine the lightest W section for both of the situations shown. Assume lateral support is provided at the reactions and at the concentrated loads. (a) Use A36 steel, and (b) use A572 Grade 50 steel.

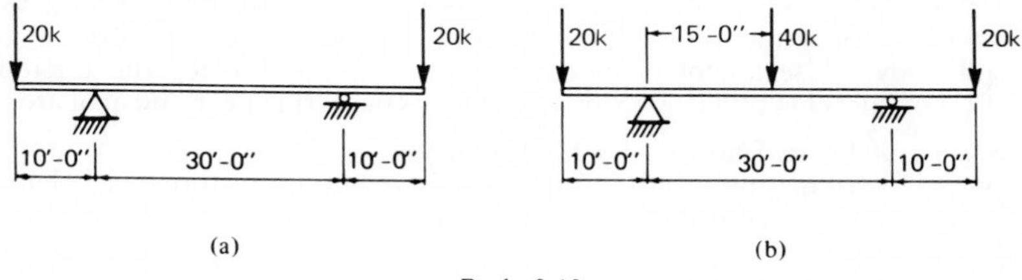

Prob. 9.18.

9.19. Select the lightest W section for the beam shown. Lateral support is provided at concentrated load points, reactions, and at end of cantilever. Use A36 steel.

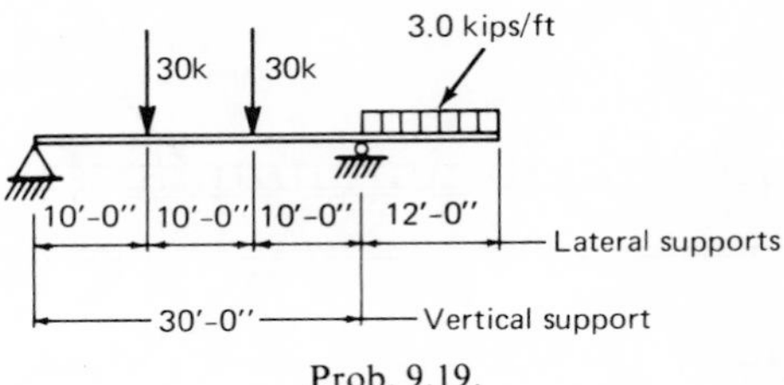

Prob. 9.19.

9.20. Redesign the beam of Prob. 9.6, using the CRC Basic Design Procedure, utilizing Figs. 9.4.1 and 9.4.2. Use FS = 1.67. If Prob. 9.6 has been solved, compare results and discuss.

9.21. Suppose the beam of Probs. 9.6 and 9.20 were torsionally fixed at the ends. Using the CRC Basic Design Procedure, redesign the beam.

9.22. Determine the allowable bending stress for the channel section and loading in Prob. 8.20 if lateral support exists only at the ends. *Neglecting torsion*, how much larger, if any, section would be required. Use A36 steel.

9.23. Assuming lateral support only at the ends, and further assuming the combined stress due to bending and torsion should not exceed the allowable based on lateral buckling, select a channel for the conditions of Prob. 8.21. Consider torsion in accordance with Chapter 8.

9.24. Design a built-up I-shaped beam with different sized flanges for the conditions of Prob. 9.14, part (b). What percent weight can be saved, if any, by using different sized flanges? Use beam depth and web thickness approximately the same as for the lightest rolled W shape which satisfies loading conditions. A36 steel.

9.25. For the beam selected for Prob. 9.4b estimate the size bracing (i.e., select a section) required. The bracing frames in on both sides and is attached to the compression flange. Length of bracing is 6 ft.

9.26. For the beams selected in Prob. 9.7, estimate the size bracing required. Assume bracing is 12 ft long, frames in on both sides of beam, and is attached to the compression flange. Preferably select channels.

9.27. Determine the adequacy of a W24 × 84 (with rail, 20 lb/ft) serving as a crane support girder. The simple span is 20 ft with lateral support at the ends only. Use accepted good practice in accounting for the torsional effect of lateral loading. Maximum moments occurring near midspan are

M_x = 100 ft-kips (incl. live load plus impact)
M_f = 10 ft-kips (assumed resisted by one flange using top flange lateral loading in accordance with AISC–1.3.4)

Use A36 steel.

9.28. Select the lightest W8 section to be used in an inclined position such that the plane of the web makes an angle of 30 degrees with the plane of loading. The beam is to be of A572 Grade 50 steel (F_y = 50 ksi), and it has lateral support only at the ends of the 22-ft simple span. The uniform gravity load is 0.4 kips/ft, in addition to the beam weight.

9.29. Design the lightest W section to serve as a crane support girder with the following requirements: crane capacity = 10 tons; maximum gravity reaction per end-truck wheel ≐ 18 kips; weight of trolley = 6 kips. Assume lateral support at the ends only and that deflection need not be restricted. Note AISC–1.3.3 and 1.3.4. Use A36 steel.

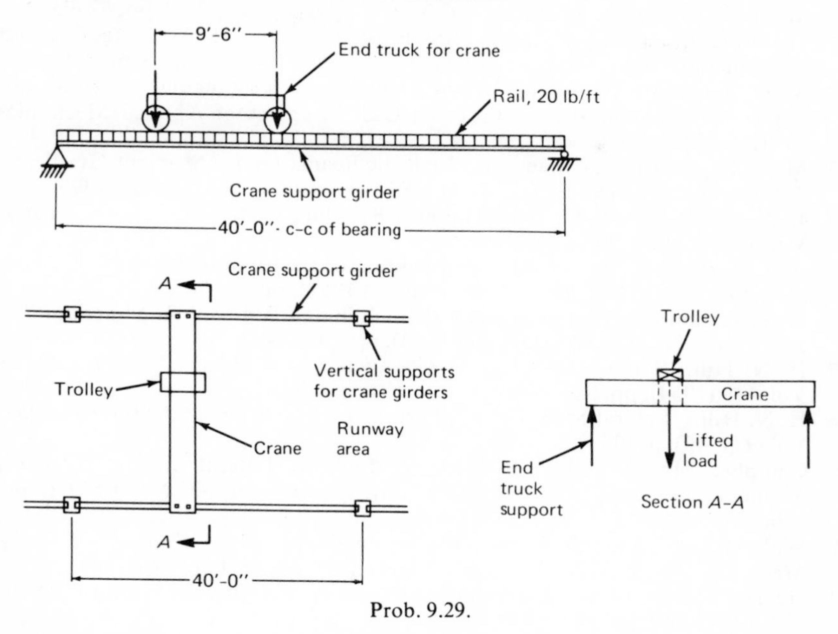

Prob. 9.29.

9.30. Redesign the crane support girder of Prob. 9.29, using a combination channel and W section.

SELECTED REFERENCES

1. S. P. Timoshenko and James M. Gere, *Theory of Elastic Stability*, 2nd ed., McGraw-Hill Book Company, 1961, Chap. 6.
2. Friederich Bleich, *Buckling Strength of Metal Structures*, McGraw-Hill Book Company, 1952, Chap. 4.
3. Karl de Vries, "Strength of Beams As Determined by Lateral Buckling," *Trans. ASCE*, Vol. 112 (1947), pp. 1245–1320.
4. R. A. Hechtman, J. S. Hattrup, E. F. Styer, and J. L. Tiedemann, "Lateral Buckling of Rolled Steel Beams," *Trans. ASCE*, Vol. 122 (1957), pp. 823–843.
5. W. J. Austin, S. Yegian, and T. P. Tung, "Lateral Buckling of Elastically End-Restrained I-Beams," *Trans. ASCE*, Vol. 122 (1957), pp. 374–388.
6. J. W. Clark and J. R. Jombock, "Lateral Buckling of I-Beams Subjected to Unequal End Moments," *J. Engg. Mech. Div. ASCE*, Vol. 83, No. EM3 (July 1957).
7. Mario G. Salvadori, "Lateral Buckling of I-Beams," *Trans. ASCE*, Vol. 120 (1955), pp. 1165–1182.
8. B. G. Johnston, *Column Research Council Guide to Design Criteria for Metal Compression Members*, 2nd ed., John Wiley & Sons, Inc., New York, 1966, Chap. 4.

9. Mario G. Salvadori, "Lateral Buckling of Eccentrically Loaded I-Columns," *Trans. ASCE*, Vol. 121 (1956), pp. 1163–1178.
10. M. G. Lay and T. V. Galambos, "Inelastic Steel Beams Under Uniform Moment," *J. Structural Div. ASCE*, Vol. 91, (December 1965), pp. 67–93.
11. *Commentary on Plastic Design*, ASCE Manual No. 41, Welding Research Council and American Society of Civil Engineers, 1961, Chap. 6, pp. 50–63.
12. M. G. Salvadori, "Lateral Buckling of Beams of Rectangular Cross Section Under Bending and Shear," Proc. 1st U.S. Congress of Applied Mechanics, 1951, pp. 403–406.
13. M. G. Lay and T. V. Galambos, "Inelastic Beams Under Moment Gradient," *J. Structural Div. ASCE*, Vol. 93, No. ST1 (February 1967), pp. 381–399.
14. J. W. Clark and H. N. Hill, "Lateral Buckling of Beams," *Trans. ASCE*, Vol. 127, Part II (1962), pp. 180–201.
15. A. J. Hartmann, "Elastic Lateral Buckling of Continuous Beams," *J. Structural Div. ASCE*, Vol. 93, No. ST4 (August 1967), pp. 11–26.
16. T. V. Galambos, "Inelastic Lateral Buckling of Beams," *J. Structural Div. ASCE*, Vol. 89, No. ST5 (October 1963), pp. 217–242.
17. H. N. Hill, "Lateral Buckling of Channels and Z-Beams," *Trans. ASCE*, Vol. 119 (1954), pp. 829–841.
18. H. N. Hill, "Lateral Stability of Unsymmetrical I-Beams," *J. Aeronaut. Sci.*, Vol. 9 (1942), p. 175.
19. Campbell Massey and F. S. Pitman, "Inelastic Lateral Stability Under a Moment Gradient," *J. Engg. Mech. Div. ASCE*, Vol. 92, No. EM2 (April 1966), pp. 101–111.
20. William Zuk, "Lateral Bracing Forces on Beams and Columns," *J. Engg. Mech. Div. ASCE*, Vol. 82, No. EM3 (July 1956), Proc. Paper No. 1032.
21. George Winter, "Lateral Bracing of Columns and Beams," *Trans. ASCE*, Vol. 125 (1960), pp. 807–845.
22. G. C. Driscoll et al., "Plastic Design of Multi-Story Frames, Lecture Notes and Design Aids," Fritz Engineering Laboratory Reports, Nos. 273.20 and 273.24, Lehigh University, Bethelem, Pa., 1965.
23. Maxwell G. Lay and T. V. Galambos, "Bracing Requirements for Inelastic Steel Beams," *J. Structural Div. ASCE*, Vol. 92, No. ST2 (April 1966), pp. 207–228.
24. G. Haaijer, "Plate Buckling in the Strain Hardening Range," *J. Engg. Mech. Div. ASCE*, Vol. 83, No. EM2 (April 1957).
25. E. H. Gaylord, Jr. and C. N. Gaylord, *Design of Steel Structures*, McGraw-Hill Book Company, New York, 1957, pp. 169–170.
26. C. O. Dohrenwend, "Action of Deep Beams Under Combined Vertical, Lateral and Torsional Loads," *J. Appl. Mech.*, Vol. 8 (1941), pp. A–130.
27. G. Winter, "Lateral Stability of Unsymmetrical I-Beams and Trusses in Bending," *Trans. ASCE*, Vol. 108 (1943), pp. 247–268.

10

Continuous Beam Design

10.1. INTRODUCTION

This chapter brings together theoretical concepts relating to both the working stress and the plastic design methods which have been presented in Chapters 7 and 9. In addition, the plastic design section presents the analysis procedure for the method as applied to continuous structures. It is assumed the reader is familiar with elastic methods of statically indeterminate analysis as required for the working stress method.

Before proceeding into this chapter, the reader should review Secs. 7.3 and 7.5 where the strength of the cross section in flexure and shear is treated, and where the working stress and plastic design approaches are contrasted. Also in Sec. 7.3 are examples of the two methods as applied to simple span beams.

The reader is also expected to be familiar with the lateral stability criteria considered in Chapter 9, particularly Secs. 9.1, 9.2, 9.6–9.8, and 9.11.

For the reader who wishes a compact study of plastic design, Secs. 7.3 and 7.5 give plastic moment and shear strength, including the concepts of plastic moment, plastic hinge, and shape factor. Section 7.3 contains the basic approach of using factored loads to obtain the required ultimate strength, as well as a design example for a simple beam. The plastic analysis and design sections of this chapter can then form the body of the study. When lateral stability considerations are applicable the plastic strength sections of Secs. 9.3, 9.7, and 9.10 may be studied.

10.2. PLASTIC STRENGTH OF A STATICALLY INDETERMINATE SYSTEM

In Sec. 7.3 it was demonstrated that on a simply supported beam the maximum moment at a single location increased under increasing load until a collapse condition was reached, at which point the maximum moment was the plastic moment, M_p. The plastic condition was assumed to exist at one point only with the remainder of the beam elastic.

Consider Fig. 10.2.1, where a simple beam is loaded with its ultimate concentrated load. As the load P is increased from zero such that the bending moment exceeds M_y a portion of the length is inelastic, until the

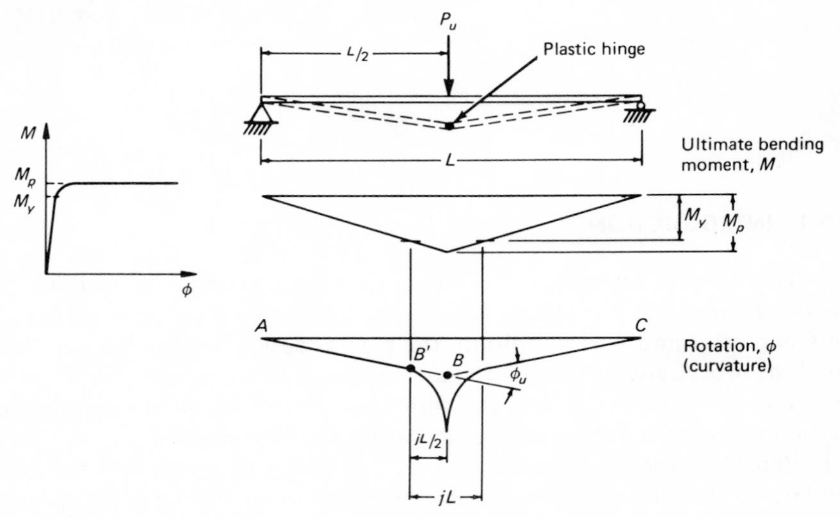

Fig. 10.2.1. Moment-curvature relationships for a plastic hinge.

moment reaches M_p, at which point the inelastic length is jL. The curvature at midspan is very large. The variation is as shown in Fig. 10.2.1; however, for practical purposes the beam may be treated as two rigid portions (AB and BC) connected by a hinge at B, known as a *plastic hinge*, and having a concentrated angle change, or hinge rotation, ϕ_u.

The actual length of a plastic hinge is dependent on the shape of the cross section and can vary from about one-tenth to as much as one-third of the span. As an example, consider the relationship between the ultimate moments and the beam segments AB and AB' in Fig. 10.2.1:

$$\frac{AB}{M_p} = \frac{AB'}{M_y} \quad \text{or} \quad \frac{L/2}{M_p} = \frac{L/2 - jL/2}{M_y} \tag{10.2.1}$$

Simplifying,

$$\frac{L}{M_p} = \frac{L(1 - j)}{M_y} \tag{10.2.2}$$

or

$$j = 1 - \frac{M_y}{M_p} = 1 - \frac{1}{f} \tag{10.2.3}$$

where f is the shape factor as discussed in Sec. 7.3.

EXAMPLE 10.2.1

Compute the length of the plastic hinge for the beam shown in Fig. 10.2.1 for (a) a W16 × 40 and (b) a rectangular beam with a width b and a depth d.

Solution

(a) For W16 × 40,

$$f = \frac{M_p}{M_y} = \frac{Z}{S} = \frac{72.8}{64.6} = 1.13$$

$$jL = L\left(1 - \frac{1}{f}\right) = L\left(1 - \frac{1}{1.13}\right) = 0.115L$$

(b) For rectangular section,

$$f = \frac{Z}{S} = \frac{\dfrac{bd^2}{4}}{\dfrac{bd^2}{6}} = 1.5$$

$$jL = L\left(1 - \frac{1}{1.5}\right) = 0.333L$$

Even though the distance jL may be as much as one-third of the span, as shown in Example 10.2.1, the simple assumption of a plastic hinge at a point has been amply demonstrated by tests. Beedle[1] and Massonnet and Save[2] have extensive discussions of theoretical and experimental verification of plastic-design procedures so individual research references are not included here.

Plastic Limit Load—Equilibrium Method. At the ultimate condition with the plastic limit load P_u acting, the requirements of equilibrium are still applicable. Consider first a statically determinate simply supported beam, as shown in Fig. 10.2.2. A collapse condition is achieved when the load P_u is large enough to cause the plastic moment M_p to occur at one location (in this case under the load). When the sufficient number of plastic hinges have been developed to allow instantaneous hinge rotations without developing increased resistance, a mechanism is said to have occurred.

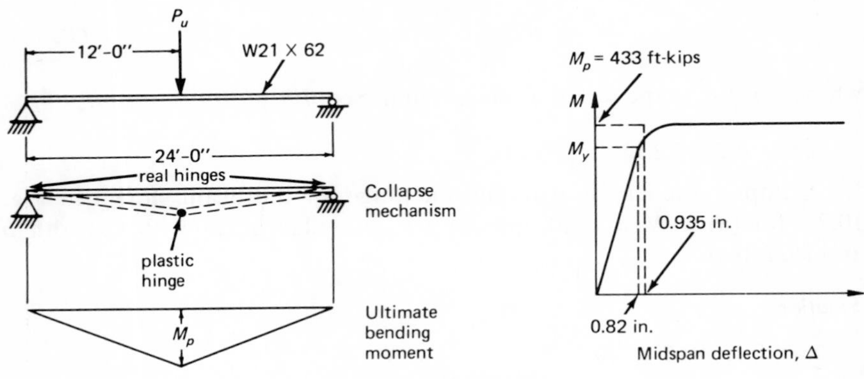

Fig. 10.2.2. Example 10.2.2.

EXAMPLE 10.2.2

Determine the ultimate load P_u for the W21 × 62 beam of Fig. 10.2.2. Assume the material has $F_y = 36$ ksi.

Solution

Using equilibrium,

$$M_p = \frac{P_u L}{4}$$

$$P_u = \frac{4M_p}{L} = \frac{4(F_y Z)}{L} = \frac{4(36)(144.1)}{24(12)} = 72.05 \text{ kips}$$

Using the simplified procedure of considering the behavior as ideally elastic-plastic, the deflection occurring when M_p is reached is based on the elastic equation, which is strictly valid only until M_y is reached. At a maximum moment of M_y,

$$\Delta y = \frac{PL^3}{48EI} = \frac{M_y L^2}{12EI} = \frac{F_y I L^2}{c(12)EI} = \frac{F_y L^2}{12cE}$$

$$= \frac{36(24)^2(144)}{12(10.5)(29{,}000)} = 0.82 \text{ in.}$$

Assuming a linear extension until $M = M_p$,

$$\Delta p = 0.82\frac{Z}{S} = 0.82\frac{144.1}{126.4} = 0.82(1.14) = 0.935 \text{ in.}$$

The ultimate deflection when M_p is achieved will be higher than this. However, it is the service-load deflection which is normally of concern, and this is correctly computed by the usual elastic procedures.

Note that the ultimate moment diagram for this statically determinate problem is the same shape as that which occurs if the maximum moment is down in the working stress range. From an infinitesimal load to ultimate load the moments at every point vary in the same ratio.

EXAMPLE 10.2.3

Determine the ultimate capacity of the fixed end W16 × 40 beam of Fig. 10.2.3. Assume the yield point to be $F_y = 36$ ksi.

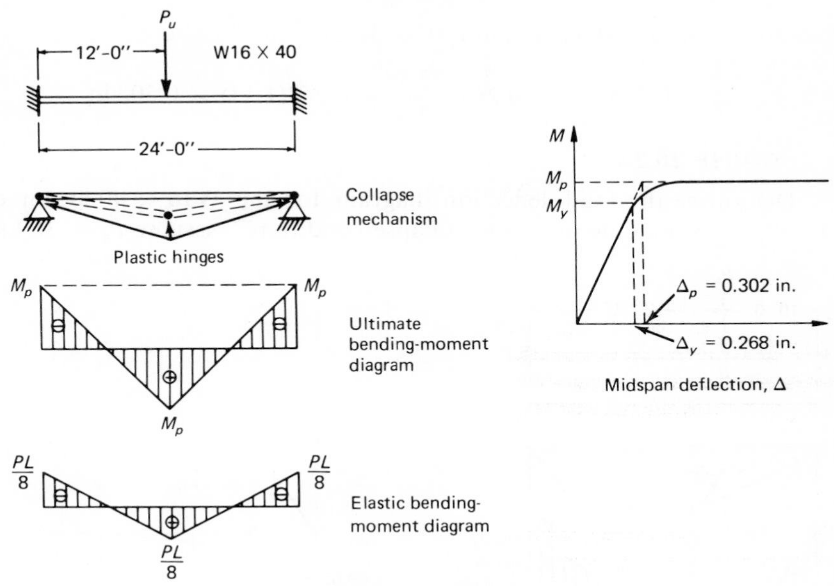

Fig. 10.2.3. Example 10.2.3.

Solution

Since the beam is statically indeterminate, three plastic hinges are required to form a mechanism. Using equilibrium,

$$2M_p = \frac{P_u L}{4}$$

$$M_p = F_y Z = 36(72.8)/12 = 218 \text{ ft-kips}$$

$$P_u = 8M_p/L = 8(218)/24 = 72.7 \text{ kips}$$

In order to determine the load-deflection diagram, it is necessary to know the loading history up to the collapse condition. In this case the elastic bending moments, as shown in Fig. 10.2.3, give equal positive and negative bending moment; therefore, the three plastic hinges form simultaneously. Moments will increase simultaneously in direct proportion until the collapse condition is reached. Again, as for the statically determinate case, when load is applied all moments vary in the same ratio.

This special statically indeterminate case offers no difference from a statically determinate one.

$$\Delta_y = \frac{PL^3}{192EI} = \frac{M_y L^2}{48EI} = \frac{F_y L^2}{48Ec}$$

$$= \frac{36(24)^2(144)}{48(8)(29{,}000)} = 0.268 \text{ in.}$$

Assuming a linear extension until $M = M_p$,

$$\Delta_p = 0.268\,\frac{Z}{S} = 0.268\,\frac{72.8}{64.6} = 0.268(1.13) = 0.302 \text{ in.}$$

EXAMPLE 10.2.4

Determine the load-deflection diagram for the W16 × 40 beam of Fig. 10.2.4 for loading up to the collapse condition. Assume $F_y = 36$ ksi.

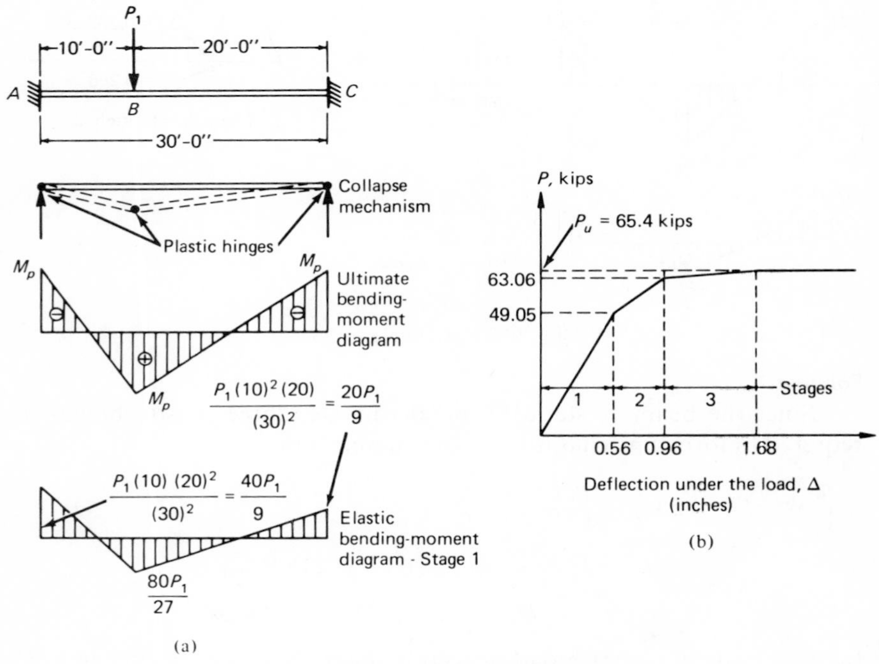

Fig. 10.2.4. Example 10.2.4.

Solution

The ultimate capacity P_u can be found directly without knowing the sequence of plastic-hinge formation. In this case, since the elastic moments are different at the three locations where plastic hinges will form, the plastic hinges will not form simultaneously.

At ultimate, from the equilibrium requirement,

$$M_p = \frac{P_u(10)(20)}{30} - M_p$$

$$P_u = 2M_p\frac{3}{20} = 0.3M_p$$

$$M_p = 218 \text{ ft kips (from Example 10.2.3)}$$

$$P_u = 0.3(218) = 65.4 \text{ kips}$$

The load-deflection diagram, however, requires examination of the loading stages.

Stage 1. From the elastic bending moments, the first plastic hinge will form at point A.

$$\frac{40P_1}{9} = M_p = 218 \text{ ft-kips}$$

$$P_1 = 9(218)/40 = 49.05 \text{ kips}$$

$$\Delta_1 = \frac{P(10)^3(20)^3}{3(30)^3EI} = \frac{49.05(8000)(1728)}{3(27)(29{,}000)(517)} = 0.56 \text{ in.}$$

Stage 2. With $P = 49.05$ kips applied one can consider that part of the available moment capacity at points B and C has been used up. The available capacity remaining is

$$M_B = M_p - 80(49.05)/27 = 218 - 145.3 = 72.7 \text{ ft-kips}$$
$$M_C = M_p - 20(49.05)/9 = 218 - 109.0 = 109.0 \text{ ft-kips}$$

As the load is increased above $P = 49.05$ kips, the added load acts on a different elastic system. The moments caused by the additional load are not distributed over the span in the same manner as moments caused by the first 49.05 kips; thus, the term "redistribution of moments" is applied. Figure 10.2.5a shows the elastic system and moments for stage 2. It is

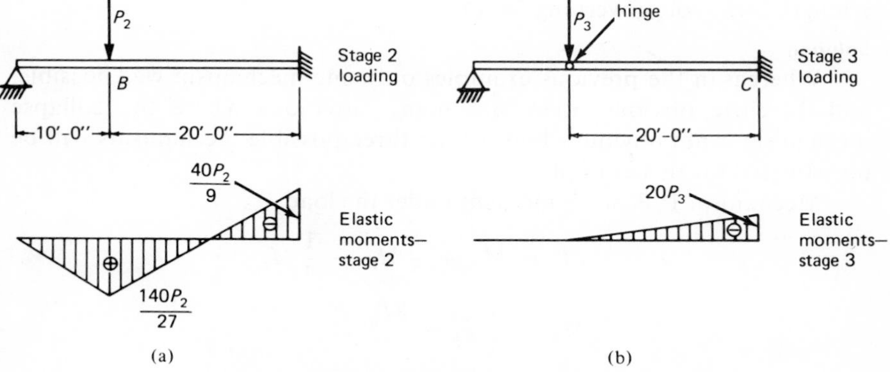

Fig. 10.2.5. Stages 2 and 3 for Example 10.2.4.

apparent that the next plastic hinge will form under the load

$$\frac{140P_2}{27} = 72.7$$

$$P_2 = 27(72.8)/140 = 14.01 \text{ kips}$$

$$\Delta_2 = \frac{Pa^2b^3(3L + a)}{12L^3EI} = \frac{14.01(10)^2(20)^3(90 + 10)(1728)}{12(30)^3(29{,}000)(517)} = 0.40 \text{ in.}$$

Stage 3. With a total of $P_1 + P_2 = 49.05 + 14.01 = 63.06$ kips applied, the moment capacity available at C is

$$M_C = M_p - 109.0 \text{ (Stage 1)} - 40(14.05)/9 \text{(Stage 2)} = 46.71 \text{ ft-kips}$$

$$20P_3 = 46.71; \qquad P_3 = 2.34 \text{ kips}$$

$$\Delta_3 = \frac{PL^3}{3EI} = \frac{2.34(20)^3(1728)}{3(29{,}000)(517)} = 0.72 \text{ in.}$$

The complete load-deflection diagram appears in Fig. 10.2.4b. The total ultimate load equals

$$P_u = P_1 + P_2 + P_3 = 49.05 + 14.01 + 2.34 = 65.4 \text{ kips}$$

Note that this statically indeterminate case, unlike the previous special case, exhibits the multistage load-deflection diagram typical of statically indeterminate systems. As load is increased moments vary in a different ratio for each stage of loading. Thus, the ultimate moment diagram is *not* equal to a constant times the elastic moments. Only for statically determinate cases, and a few special statically indeterminate ones can the ultimate diagram be obtained by multiplying the elastic moments by a constant.

EXAMPLE 10.2.5

Using the equilibrium method for the continuous beam of Fig. 10.2.6, determine the ultimate capacity P_u. Assume $F_y = 36$ ksi and that plastic strength is the sole governing factor.

Solution

Whereas in the previous examples only one mechanism was possible, and therefore obvious, there are many situations where the collapse mechanism is not obvious. Each of the three possible mechanisms will be investigated for this example.

Mechanism 1: Positive moment under the load P_u,

$$M_p = M_{s1} + \frac{1}{2}M_{s2} - \frac{1}{3}M_p$$

$$\frac{4}{3}M_p = \frac{16}{3}P_u + \frac{8P_u}{2}$$

$$M_p = 7P_u$$

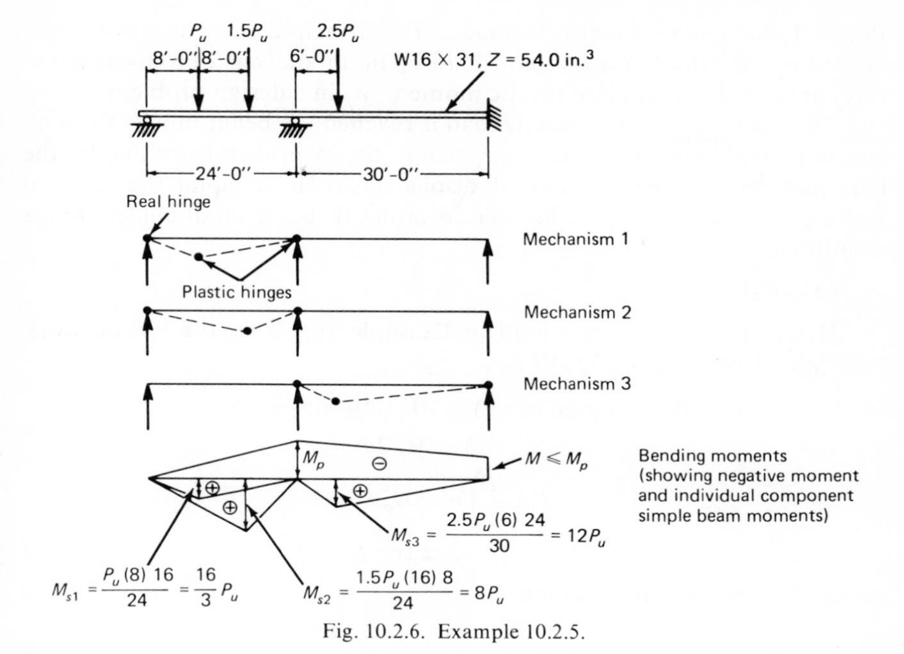

Fig. 10.2.6. Example 10.2.5.

Mechanism 2: Positive moment plastic hinge at the load $1.5P_u$,

$$M_p = \frac{M_{s1}}{2} + M_{s2} - \frac{2}{3} M_p$$

$$\frac{5}{3} M_p = \frac{8}{3} P_u + 8P_u$$

$$M_p = \frac{32}{5} P_u = 6.4P_u$$

Mechanism 3: Positive-moment plastic hinge at the load $2.5P_u$ with negative-moment plastic hinges at the ends of the 30-ft span,

$$M_p = M_{s3} - M_p$$

$$2M_p = 12P_u$$

$$M_p = 6P_u$$

Since the plastic moment capacity required is largest for Mechanism 1, it controls:

$$P_u = \frac{M_p}{7} = \frac{36(54.0)}{7(12)} = 23 \text{ kips}$$

In this case the collapse mechanism occurs in the 24-ft span while the 30-ft span remains stable and elastic. The use of cover plates as discussed in Example 10.3.3 may be a more economical solution.

Plastic Limit Load—Energy Method. The principle of virtual work may also be applied to obtain the limit load P_u in an analysis of a given structure, or to find the required plastic moment M_p in a design problem.

Consider that as the ultimate load is reached the beam moves through a virtual displacement δ. For equilibrium, the external work done by the load moving through the virtual displacement must equal the internal strain energy due the plastic moment rotating through small angles (hinge rotations).

EXAMPLE 10.2.6

Determine the ultimate load in Example 10.2.2 by the virtual-work principle. Referring to Fig. 10.2.7,

$$W_e \text{ (external work)} = W_i \text{ (internal work)}$$

$$P_u \delta = M_p \, 2\theta$$

$$P_u\left(\frac{\theta L}{2}\right) = 2M_p\theta$$

$$P_u = 4M_p/L$$

the same as previously obtained.

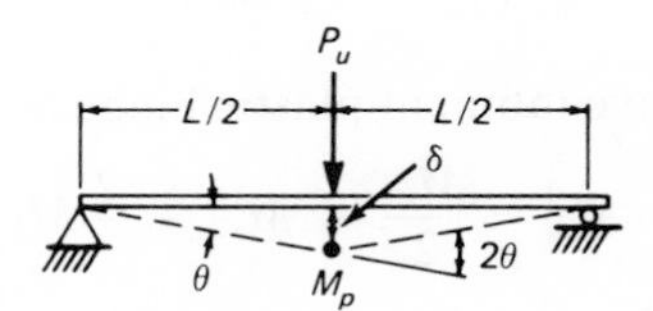

Fig. 10.2.7. Example 10.2.6.

EXAMPLE 10.2.7

Determine the ultimate load in Example 10.2.3 by the virtual-work principle. Referring to Fig. 10.2.8,

$$W_e = W_i$$

$$P_u\frac{\theta L}{2} = 2M_p\theta + M_p(2\theta)$$

$$P_u = 8M_p/L$$

which agrees with previous solution.

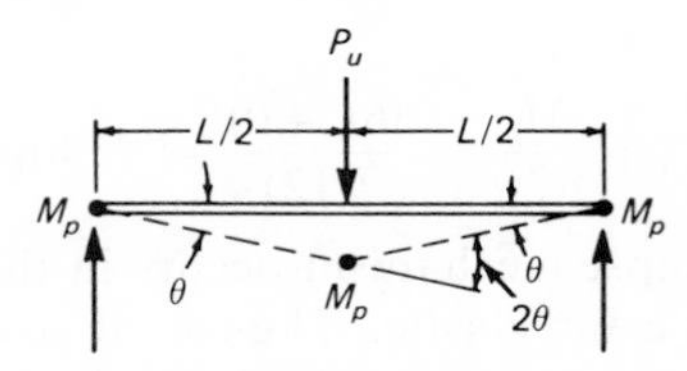

Fig. 10.2.8. Example 10.2.7.

EXAMPLE 10.2.8

Determine the ultimate load for Example 10.2.4 by the virtual-work method. Referring to Fig. 10.2.9,

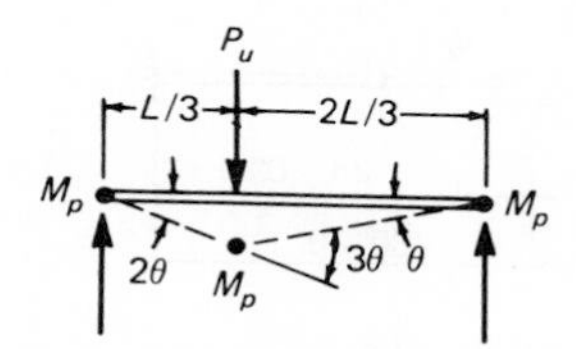

Fig. 10.2.9. Example 10.2.8.

$$W_e = W_i$$

$$P_u \frac{2\theta L}{3} = M_p(2\theta + \theta + 3\theta)$$

$$P_u = \frac{3}{2L}(6)M_p = 9M_p/L$$

For $L = 30$ ft, $P_u = 0.3M_p$ as before.

EXAMPLE 10.2.9

Determine the ultimate load P_u for the continuous beam of Example 10.2.5. Referring to Fig. 10.2.10,

Mechanism 1:

$$P_u(2\theta)(8) + 1.5P_u\theta(8) = M_p(3\theta + \theta)$$

$$28P_u = 4M_p$$

$$M_p = 7P_u$$

Mechanism 2:

$$P_u(\theta)(8) + 1.5P_u 2\theta(8) = M_p(3\theta + 2\theta)$$

$$32P_u = 5M_p$$

$$M_p = 6.4P_u$$

Mechanism 3:

$$2.5P_u(4\theta')(6) = M_p(4\theta' + 5\theta' + \theta')$$

$$60P_u = 10M_p$$

$$M_p = 6P_u$$

The equilibrium and energy procedures are equally applicable for frames where degrees of freedom in sidesway must be considered. For single-story frames, the reader is referred to Chapter 15. In multistory braced frames, the design procedure is to force the plastic hinges to form in the girders rather than in the columns; thus, the girders are treated as continuous beams. Where bracing contributes an axial force to the girder, the beam-column principles of Chapter 12 apply.

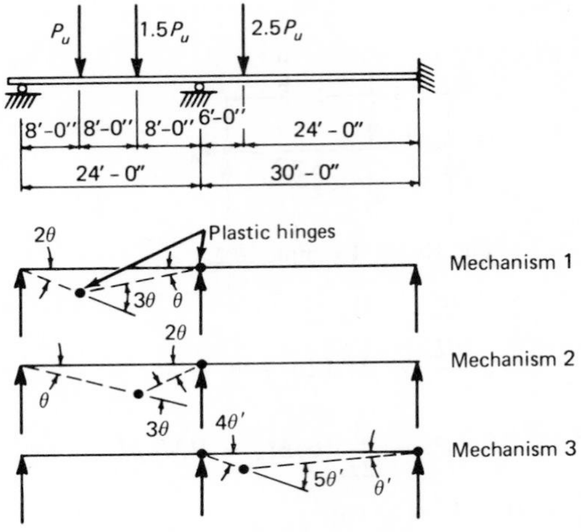

Fig. 10.2.10. Example 10.2.9.

10.3. DESIGN EXAMPLES—PLASTIC DESIGN

An elementary example of the design of a simple beam by plastic design was presented in Sec. 7.3. For additional elementary material on the philosophy, procedure, and experimental verification of plastic design, References 3 through 10 are suggested. Plastic-design examples are presented first because the ultimate-strength approach is the most realistic approach and is the authors' recommended procedure for most continuous-beam situations. The working-stress method might be the method of choice for the situations in which long distances between lateral supports exist. Furthermore, the working-stress method involves procedures which attempt to bring the result of that method close to the plastic-design result.

Plastic-design procedures make strength the basis for design rather than stress. However, in working-stress design, factors other than stress may control; so also in plastic design—factors other than strength may govern. Factors such as shear, instability of various types, and deflection, must also be considered. These factors are treated in Chapters 7 and 9, so that only incidental treatment arises in the following examples, which concentrate principally on flexure and its relationship to lateral-torsional buckling.

EXAMPLE 10.3.1

Select the lightest W section for the 2-span continuous beam of Fig. 10.3.1, using the plastic-design method. Use A36 steel. External lateral support is provided at the ends and midpoint of each span.

Solution

Temporarily neglecting the weight of the beam, apply the load factor to the service loads:

$$w_u = 3(1.7) = 5.1 \text{ kips/ft.}; \quad P_u = 40(1.7) = 68 \text{ kips}$$

Next, consider the various possible mechanisms for failure. In this case there are two, as shown in Fig. 10.3.1b.

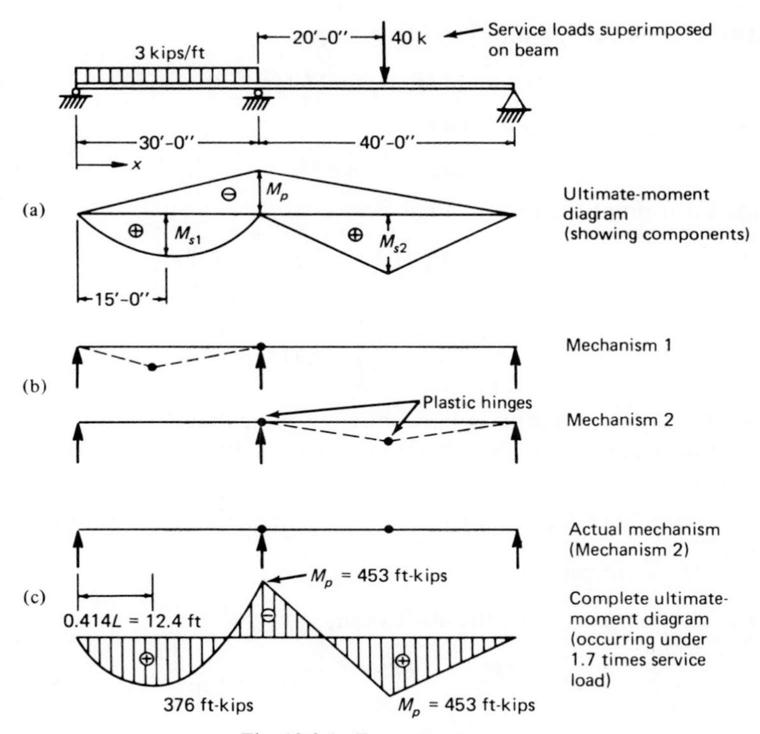

Fig. 10.3.1. Example 10.3.1.

(a) *Mechanism 1.* For a span with uniform load, simply supported at one end, and expected to develop a plastic hinge at the other, the location of maximum positive moment is not self-evident. At an unknown distance x from the discontinuous end, the net positive moment is

$$M_x = \frac{w}{2} x(L - x) - \frac{x}{L} M_p \tag{a}$$

Taking $dM_x/dx = 0$ and solving for x, the location of maximum moment

$$x = \frac{L}{2} - \frac{M_p}{wL} \tag{b}$$

Substituting Eq. (b) into Eq. (a) gives

$$M_x(\max) = \frac{w}{2}\left(\frac{L}{2} - \frac{M_p}{wL}\right)\left(L - \frac{L}{2} + \frac{M_p}{wL}\right) - \frac{M_p}{L}\left(\frac{L}{2} - \frac{M_p}{wL}\right) \qquad \text{(c)}$$

When a plastic hinge develops, M_x (max) $= M_p$; thus Eq. (c) becomes

$$M_p = \frac{wL^2}{8} - \frac{M_p^2}{2wL^2} - \frac{M_p}{2} + \frac{M_p^2}{wL^2} \qquad \text{(d)}$$

Substitute $M_s = wL^2/8$ into Eq. (d), giving

$$M_p^2 - 24M_pM_s + 16M_s^2 = 0 \qquad \text{(e)}$$

Solving the quadratic for M_p gives

$$M_p = 0.686M_s \qquad \text{(10.3.1)}$$

and the location of the plastic hinge is at

$$x = \frac{L}{2} - \frac{M_p}{wL}\left(\frac{L}{L}\right)\frac{8}{8} = \frac{L}{2} - \frac{M_p}{M_s}\left(\frac{L}{8}\right)$$

$$= L\left(\frac{1}{2} - \frac{0.686}{8}\right) = 0.414L$$

The results of this development are shown in Fig. 10.3.2.

For the specific example, Mechanism 1 (Fig. 10.3.1b) requires

$$M_p = 0.686M_{s1}; \quad M_{s1} = \frac{w_uL^2}{8} = \frac{5.1(30)^2}{8} = 573 \text{ ft-kips}$$

$$M_p = 0.686(573) = 394 \text{ ft-kips}$$

(b) *Mechanism 2:* For the 40-ft span,

$$M_{s2} = \frac{P_uL}{4} = \frac{68(40)}{4} = 680 \text{ ft-kips}$$

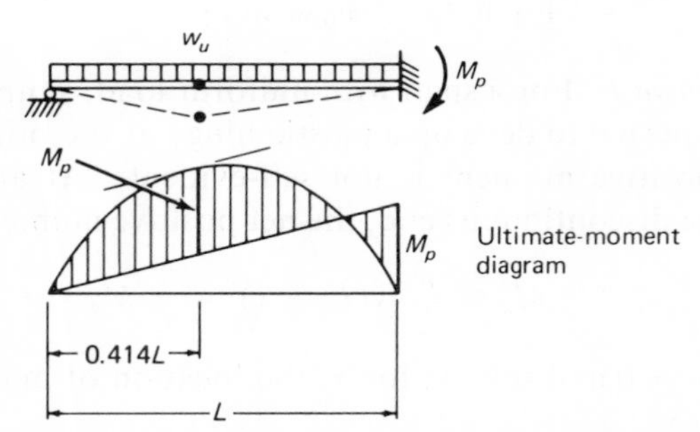

Fig. 10.3.2. Collapse condition for uniformly loaded span continuous at one end and simply supported at the other.

Equilibrium requires

$$M_p = M_{s2} - \frac{M_p}{2} = 680 - \frac{M_p}{2}$$

$$M_p = 2(680)/3 = 453 \text{ ft-kips}$$

The largest M_p requirement obtained from the possible mechanisms determines the controlling condition. Thus Mechanism 2 governs. A check by determining the complete ultimate moment diagram should be made to insure that M_p is the maximum moment. If a location is found where the computed ultimate moment exceeds M_p, the mechanism used is not correct. In this case, Fig. 10.3.1c shows $M_p = 453$ ft-kips is the maximum, so that Mechanism 2 is the correct one.

The maximum positive moment on the 30-ft span using Eq. (a) is

$$M = \frac{5.1}{2}(12.4)(17.6) - 0.414M_p$$

$$= 557 - 0.4(453) = 376 \text{ ft-kips} < 453 \qquad \text{OK}$$

(c) Select section:

$$\text{Reqd } Z = \frac{M_p}{F_y} = \frac{453(12)}{36} = 151 \text{ in.}^3$$

If beam weight is to be added, W14 × 84 ($Z = 145$) is overstressed, while W14 × 95 and W14 × 100 are not "compact"; try W14 × 111 with $Z = 196$ in.^{3} Because of the long laterally unbraced length, only sections with large values of r_y are considered.

(d) Check lateral support. According to AISC–2.9, the lateral support requirements of that article need not be applied in the region of the last hinge to form. However, even though the last hinge forms under the 40-kip load, the authors believe it to be appropriate to apply Formula 2.9–1a to the 20-ft segment between the concentrated load and the interior support where significant rotation must occur to develop the last hinge:

$$L_{cr} = \left[\frac{1375}{F_y} + 25\right] r_y = \left[\frac{1375}{36} + 25\right] r_y = 63.2r_y$$

for $M/M_p = +1.0$ (double curvature). For the W14 × 111, $r_y = 3.73$ in.

$$L_{cr} = \frac{63.2(3.73)}{12} = 19.6 \text{ ft.} \approx 20 \text{ ft} \qquad \text{OK}$$

For the exterior 20-ft segment, where at one end of the segment the final plastic hinge occurs under the 40-k load, AISC–1.5.1.4.6a must be satisfied:

$$F_b(1.5\text{–}7) = \frac{12{,}000C_b}{Ld/A_f} = \frac{12{,}000(1.75)}{240(1.13)} > 0.6F_y$$

$$f_b = \frac{M_p}{(LF)S_x} = \frac{453(12)}{1.7(176.3)} = 18.1 \text{ ksi} < 22 \text{ ksi} \qquad \text{OK}$$

For the shorter unbraced lengths on the 30-ft span, the flexural stresses f_b are equal to or less than the above value and therefore acceptable.

(e) Check shear. At interior end of 40-ft span,

$$V_u = 20 + \frac{453}{40} = 20 + 11.3 = 31.3 \text{ kips}$$

$$v_u = \frac{V_u}{td} = \frac{31.3}{0.54(14.37)} = 4.04 \text{ ksi} < 0.55F_y = 19.8 \text{ ksi} \qquad \text{OK}$$

As expected, except on short spans with heavy loading, shear does not govern.

(f) Deflection. If deflection appears to be a problem, service load moments must be determined and the deflection computed. Deflection at the collapse condition is not easily determined, nor is it generally of interest. In this case,

$$\frac{L}{d} = \frac{40(12)}{14} = 34$$

which is somewhat high, so that if there is a deflection limitation, service load deflection should be computed.

Use W14 × 111 – A36 Steel.

EXAMPLE 10.3.2

Select the lightest uniform section required for the continuous beam of Fig. 10.3.3 using the AISC plastic design method. Use $F_y = 60$ ksi, and assume lateral support is provided at the vertical supports and at each concentrated load.

Solution

The equilibrium method is illustrated, although the energy method is equally applicable. Apply the 1.7 load factor and compute the simple beam moments under the factored loads.

$$M_{s1} = \frac{15(1.7)(10)15}{25} = 153 \text{ ft-kips}$$

$$M_{s2} = \frac{25(1.7)(20)5}{25} = 170 \text{ ft-kips}$$

$$M_{s3} = \frac{50(1.7)10(20)}{30} = 567 \text{ ft-kips}$$

(a) *Mechanism 1:* Assume moment under 15-kip load equals M_p,

$$M_p = M_{s1} + \frac{1}{2}M_{s2} - \frac{10}{25}M_p$$

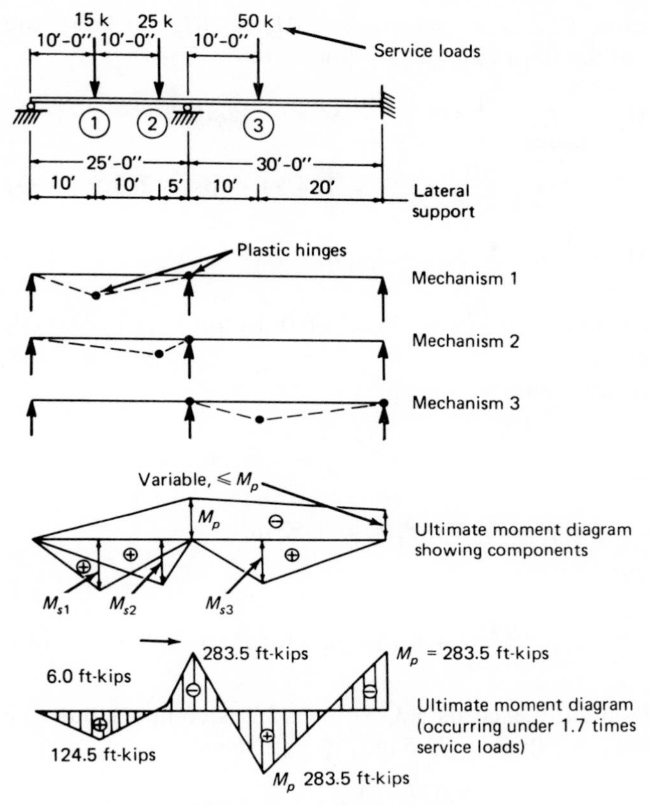

Fig. 10.3.3. Example 10.3.2.

$$\frac{35}{25} M_p = 153 + \frac{1}{2}(170) = 238$$

$$M_p = \frac{25}{35}(238) = 170 \text{ ft-kips}$$

(b) *Mechanism 2:* Assume moment under 25-kip load equals M_p,

$$M_p = \frac{1}{3} M_{s1} + M_{s2} - \frac{20}{25} M_p$$

$$\frac{45}{25} M_p = \frac{1}{3}(153) + 170 = 221$$

$$M_p = \frac{25}{45}(221) = 123 \text{ ft-kips}$$

(c) *Mechanism 3:* Assume moment under 50-kip load equals M_p,

$$M_p = M_{s3} - M_p$$

$$2M_p = 567; \quad M_p = 283.5 \text{ ft-kips} \qquad \text{Controls.}$$

If a section is selected which has $M_p = 283.5$ ft-kips, the resulting moments under the 15 and 25 kip loads are

$$M_1 = M_{s1} + \frac{1}{2}M_{s2} - \frac{10}{25}M_p$$

$$= 238 - \frac{10}{25}(283.5) = 124.5 \text{ ft-kips} < 283.5 \qquad \text{OK}$$

$$M_2 = \frac{1}{3}M_{s1} + M_{s2} - \frac{20}{25}M_p$$

$$= 221 - \frac{20}{25}(283.5) = -6.0 \text{ ft-kips} < 283.5 \qquad \text{OK}$$

(d) Select beam:

$$\text{Reqd } Z = \frac{283.5(12)}{60} = 56.7 \text{ in.}^3$$

Try W16 × 36: $Z = 64.0$ in.^{3}, $r_y = 1.52$ in.

Check AISC–2.7:

$$\frac{b_f}{2t_f} = \frac{6.99}{2(0.428)} = 8.15 > 6.3 \text{ for } F_y = 60 \text{ ksi} \qquad \text{N.G.}$$

W14 × 38 slightly exceeds 6.3, S15 × 42.9 is compact but r_y is too low. Try W12 × 50: $Z = 72.5$ in.^{3}, $r_y = 1.96$ in.

$$\frac{b_f}{2t_f} = \frac{8.077}{2(0.641)} = 6.3 = 6.3 \text{ limit} \qquad \text{OK}$$

$$\frac{d}{t_w} = \frac{12.19}{0.371} = 32.8 < 53.2 \qquad \text{OK}$$

Although it is likely the plastic hinge under the 50 kip load is the last to form, the designer may not be sure. The authors suggest that it is prudent to apply the critical length criteria of AISC–2.9 to all segments having an M_p at either end, eliminating need for knowing the sequence of plastic hinge formation. The 20-ft segment will obviously control,

$$L_{cr} = \left[\frac{1375}{F_y} + 25\right]r_y$$

$$= \left[\frac{1375}{60} + 25\right]r_y = 48\, r_y$$

$$= \frac{48(1.96)}{12} = 7.85 \text{ ft} < 20 \text{ ft} \qquad \text{N.G.}$$

$$r_y \text{ reqd} = \frac{240}{48} = 5.0 \text{ in.}$$

Lateral stability controls. Since an $r_y = 5.0$ cannot be achieved by any realistic design, additional lateral bracing must be provided.

Assume lateral bracing at 5-ft intervals in the 30-ft span. Then, at the most critical location,

$$\frac{M}{M_p} = \frac{-141.75}{283.5} = -0.5$$

which is the dividing condition between using AISC Formulas 2.9–1a or 2.9–1b.

$$L_{cr}\,(2.9\text{–}1a) = 48r_y = 7.85 \text{ ft} > 5 \text{ ft} \qquad \text{OK}$$

as previously computed.

$$L_{cr}\,(2.9\text{–}1b) = \frac{1375}{F_y}r_y = \frac{1375}{60}r_y = 23r_y$$

$$= 23(1.96)/12 = 3.8 \text{ ft} < 5 \text{ ft} \qquad \text{N.G.}$$

Based on the development in Sec. 9.3, Formula 2.9–1a is more nearly correct.

Use 5 ft spacing of lateral bracing on the 30 ft. span.

On the 25-ft span, check the 10-ft unbraced segment; for W12 × 50,

$$L_c = 9.8b_f = 9.8(8.077)/12 = 6.6 \text{ ft} < 10 \text{ ft} \qquad \text{N.G.}$$

Not "compact" for working-stress method.

$$F_b\,(1.5\text{–}7) = \frac{12{,}000\,C_b}{Ld/A_f} = \frac{12{,}000(1.75)}{120(2.35)} > 0.6F_y = 36 \text{ ksi}$$

$$f_b = \frac{M_u}{(LF)S_x} = \frac{124.5(12)}{1.7(64.7)} = 13.6 \text{ ksi} < 36 \text{ ksi} \qquad \text{OK}$$

Shear and deflection will not be critical for this problem.

Use W12 × 50, $F_y = 60$ ksi.

EXAMPLE 10.3.3

Redesign the beam of Example 10.3.2, using either cover plates or two different sections butt-spliced together. $F_y = 60$ ksi.

Solution

Assume that the change of section occurs near the inflection point to the right of the central support, giving rise to different plastic moments M_{p1} and M_{p2} at the two restrained supports (see Fig. 10.3.4).

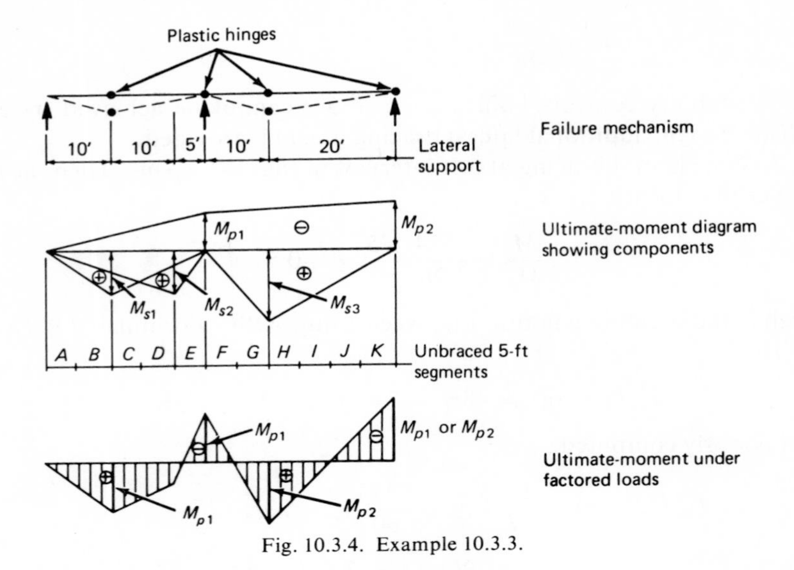

Fig. 10.3.4. Example 10.3.3.

(a) Consider first the smaller section supporting the 15- and 25-kip loads. Assume the positive movement under the 15-kip load, reaches M_{p1},

$$
\begin{aligned}
M_{p1} &= M_{s1} + \frac{1}{2} M_{s2} - \frac{10}{25} M_{p1} \\
&= 153 + 85 - 0.4\, M_{p1} \\
M_{p1} &= 170 \text{ ft-kips}
\end{aligned}
$$

Under the 25-kip load the moment is

$$
\begin{aligned}
M &= \frac{1}{3} M_{s1} + M_{s2} - \frac{20}{25} M_{p1} \\
&= 51 + 170 - 0.8(170) = 85 \text{ ft kips} < 170 \qquad \text{OK}
\end{aligned}
$$

(b) Consider lateral support requirements. For the 10-ft unbraced length between the 15- and 25-kip loads, the end-moment ratio under factored loads is

$$
\frac{M}{M_p} = \frac{-85}{170} = -0.5
$$

which is the boundary between use of AISC Formulas 2.9–1a and 2.9–1b.

$$
L_{cr}\ (2.9\text{–}1\text{b}) = \frac{1375}{F_y} r_y = 23 r_y
$$

$$
L_{cr}\ (2.9\text{–}1\text{a}) = \left[\frac{1375}{F_y} + 25\right] r_y = 48 r_y
$$

For $L = 10$ ft, $r_{min} = 120/23 = 5.2$ in., which cannot be achieved from any realistic design. Increase the lateral support to every 5 ft. This will be necessary in the 30-ft span also, as shown in Example 10.3.2.

Assuming the plastic hinge $(+M)$ between segments B and C (see Fig. 10.3.4) is the last hinge to form, only the working stress compact section requirements of AISC–1.5.1.4.1 need to be satisfied; i.e., $L < L_c$.

$$\frac{L_c}{b_f} = \frac{76.0}{\sqrt{F_y}} = \frac{76.0}{\sqrt{60}} = 9.8$$

$$\text{Min } b_f = \frac{60}{9.8} = 6.12 \text{ in. for 5-ft unbraced length}$$

This restriction applies for segments B and C adjacent to the last plastic hinge.

For segments E and F, rotation capacity must be assured by satisfying AISC Formulas 2.9–1.

$$\frac{M}{M_p} = \frac{+85}{170} = +0.5 \text{ (reverse curvature)}$$

$$L_{cr} \text{ (2.9–1a)} = 48r_y \text{ (previously computed)}$$

$$\text{Min } r_y = \frac{60}{48} = 1.25 \text{ in.}$$

(c) Select section for 25-ft span:

$$\text{Reqd } Z = \frac{170(12)}{60} = 34 \text{ in.}^3$$

$$\text{Min } b_f = 6.12 \text{ in.}$$

$$\text{Min } r_y = 1.25$$

From plastic-modulus table, try

W12 × 31, $Z = 44.1$, $r_y = 1.54$
W10 × 29, $Z = 34.7$, $r_y = 1.38$

Check flange width:

W10 × 29, $b_f = 5.80$ in. < 6.12 NG
W12 × 31, $b_f = 6.53$ in. > 6.12 OK

Check local buckling AISC–2.7:

For W12 × 31,

$$\frac{b_f}{2t_f} = \frac{6.525}{2(0.465)} > 6.3 \qquad \text{N.G.}$$

For W12 × 36

$$\frac{b_f}{2t_f} = \frac{6.565}{2(0.54)} = 6.07 < 6.3 \qquad \text{OK}$$

$$\frac{d}{t_w} = \frac{12.24}{0.305} = 40.2 < \frac{412}{\sqrt{F_y}} = \frac{412}{\sqrt{60}} = 53.1 \qquad \text{OK}$$

USE W12 × 36, F_y = 60 ksi, for the 25-ft span. For the 30-ft span, top and bottom cover plates will be added, since the 36-lb W12 is the heaviest within its group having the same proportions. Sections within a group having the same proportions can be economically butt welded together, but not sections having different proportions.

Since the section used has a greater plastic moment capacity than necessary, the ultimate moment diagram under factored loads is as shown in Fig. 10.3.5b. The diagram assumes the first plastic hinge is at the

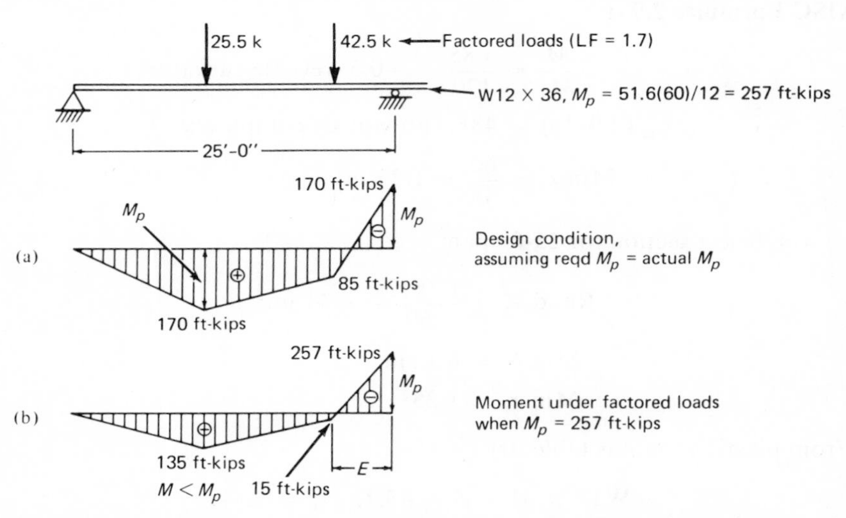

Fig. 10.3.5. Comparison of design condition and actual condition for 25-ft span of Example 10.3.3.

support, and that it does form under the application of the factored loads. The M_{p1} = 257 kips as provided is then used in the analysis of the 30-ft span.

In the actual condition, unbraced segment E is more severely loaded than in the design condition. Recheck $M/M_p = +15/257$, which is still > -0.5, so that AISC Formula 2.9–1a still applies and is satisfied.

(d) Determine plastic moment requirement on the 30 ft. span. When M_{p2} forms under the 50-kip load, equilibrium requires

$$M_{p2} = M_{s3} - \frac{1}{3}(M_{p2} - M_{p1}) - M_{p1} \quad \text{(long cover plates)}$$

if M_{p2} is developed at fixed end. This would require cover plates to extend into the support. If the cover plates are not extended to the support, equilibrium requires

$$M_{p2} = M_{s3} - M_{p1} \qquad \text{(short cover plates)}$$

If long cover plates are used,

$$M_{p2} = 567 - \frac{M_{p2}}{3} + \frac{257}{3} - 257; \ \frac{4}{3}M_p = 395.67, \quad M_{p2} = 296 \text{ ft-kips}$$

If shorter cover plates are used,

$$M_{p2} = 567 - 257 = 310 \text{ ft-kips}$$

Since there is little difference in requirements, the cover plates will be made as short as possible. The cover plates, top and bottom, must provide

$$M_p \text{ (cover)} = 310 - 257 = 53 \text{ ft-kips}$$

(e) Select plate size. For an area A_p representing the cover plate at one flange, and a distance d center-to-center of cover plates, the plastic modulus Z becomes

$$Z = 2A_p\left(\frac{d}{2}\right) = A_p d$$

$$\text{Reqd } A_p = \frac{Z}{d} = \frac{53(12)}{60d} = \frac{10.6}{d} = \frac{10.6}{12.5} = 0.85 \text{ sq in.}$$

assuming $d \approx 12.24 + 0.25$ ($^1/_4$-in. plates). If the plate width is somewhat less than the flange width of W12 × 36, say 6 in., the area A_p provided will be more than needed. One might reduce the width of plate to 5 in.

Check local buckling on plate as a stiffened element welded along two sides, according to AISC–2.7:

$$\frac{b}{t} = \frac{5}{0.25} = 20 < \frac{190}{\sqrt{F_y}} = \frac{190}{\sqrt{60}} = 24.5 \qquad \text{OK}$$

Use plates $^1/_4$ × 5 top and bottom. These would probably be welded continuously along the sides. Discussion of weld sizes and other requirements are contained in Chapter 5.

(f) Determine plate length. Referring to Fig. 10.3.6 showing the design condition for the 30-ft span, the reader may observe that cover plates will be required for only a short distance, L_1, which might easily be scaled from the diagram. For this straight line moment diagram, it is easily computed:

$$L_1 = \left(1 - \frac{257 + 257}{257 + 310}\right)(10 + 20) = 0.095(30) = 2.75 \text{ ft}$$

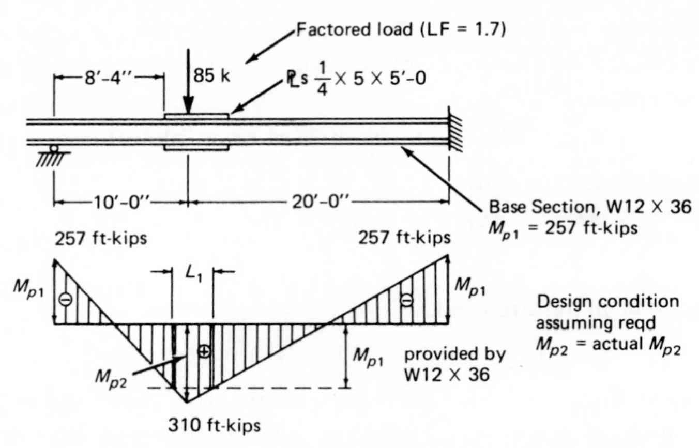

Fig. 10.3.6. Design condition for 30-ft span of Example 10.3.3.

Extension of the plates in both directions must satisfy AISC–1.10.4 regarding flange development; thus the welded connection must develop the cover plate's proportion of the capacity. The tension or compression capacity required from a plate is computed:

$$\text{Total reqd } Z = \frac{310(12)}{60} = 62 \text{ in.}^3$$

$$\% \ Z \text{ from plates} = \frac{A_p}{A_f}(100) = \frac{5(0.25)}{6.565(0.54)}(100) = 35.2\%$$

neglecting the web effect.

$$\begin{pmatrix}\text{Force in plate} \\ \text{at max moment}\end{pmatrix} \approx \frac{0.352 Z F_y}{d} = \frac{0.352(62)(60)}{12.5} = 105 \text{ kips}$$

AISC–1.10.4 requires that the necessary force in the plate be developed in the distance a' from the end of plate. Assuming no weld across the end of the plate and continuous weld along the sides, the distance a' = 2(width) = 2(5) = 10 in. At 10 in. from termination of the plate, the welds must have developed the proportionate part of the total flange force needed. For $^3\!/_{16}$-in. fillet welds placed by the submerged-arc process (AISC–1.14.7), and using E 70 electrodes, the capacity per inch is

$$R_w = \frac{3}{16}(21)(1.7) = 6.7 \text{ kips/in.}$$

using allowable stresses 1.7 times stated values according to AISC–2.8. In order to satisfy AISC–1.10.4, the maximum force which can be developed over 10 in. by two lines of fillet weld is

$$\text{Force} = 10(6.7)(2) = 134 \text{ kips}$$

Since this exceeds the maximum required force of 105 kips, the L_1 length previously computed governs.

Use PLs 5 ft long, beginning 8′-4″ from support as shown in Fig. 10.3.6.

In this example no splicing of member was used; however, splicing will frequently be desired on continuous beams. Splices are discussed in Sec. 10.5.

10.4. DESIGN EXAMPLES—WORKING-STRESS DESIGN

For ordinary continuous beam design, plastic design is the preferred method. Working-stress design employs some arbitrary adjustments to indirectly account for plastic behavior and redistribution of moments. Such arbitrary adjustments cannot reflect true behavior; consequently, a uniform safety factor is not obtained. Occasionally, for designs involving long laterally unsupported lengths where stability instead of strength controls, the working-stress method becomes the practical approach.

EXAMPLE 10.4.1.

Redesign the beam of Example 10.3.1 (Fig. 10.3.1) by the working-stress method. A36 steel.

Solution

The elastic bending moment diagram (see Fig. 10.4.1) under service loads must first be computed using a method of statically indeterminate analysis.

The four separate unbraced segments with their C_b values and design moments are shown in Fig. 10.4.1. If lateral bracing is sufficient to permit a plastic moment to form at a given maximum moment location, an allowable stress of $0.66F_y$ is permitted. Further, at the negative moment location if adequate rotation capacity (plastic strain after M_p has been reached) is available, redistribution of moments is accounted for by a 10 percent reduction in maximum negative moment, with the corresponding adjustment of positive moments to maintain equilibrium.

(a) Determine required section assuming reduction in $(-M)$ and using $0.66\,F_y$.

$$\text{Reqd } S = \frac{0.9(316)12}{24} = 142 \text{ in.}^3$$

which is the minimum section modulus required assuming the most favorable conditions. Since the distance to the point of inflection is 11.33 ft for segment C, try for $L_c > 11.33$ ft. Appendix Table A3 indicates that selection of a minimum weight beam with $L_c > 10$ ft is not practical.

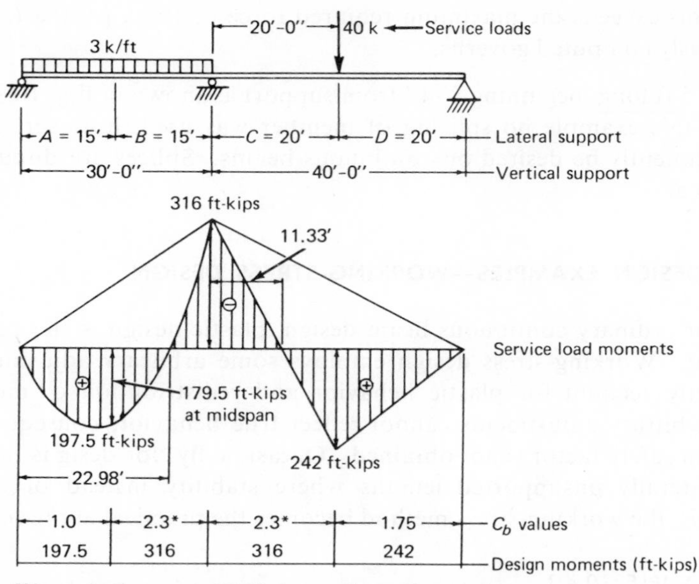

Fig. 10.4.1. Example 10.4.1.

(b) Assume no reduction in negative moment and an allowable of $0.6F_y$.

$$\text{Reqd } S = \frac{316(12)}{22} = 172.5 \text{ in.}^3$$

Premature lateral-torsional buckling must still be prevented, so the selected section must have $L_u C_b \geq 20$ ft (assuming AISC Formula 1.5–7 governs). With large unbraced lengths, the use of AISC Manual Beam Curves offers the most suitable approach. Enter with

$$M = 316 \text{ ft-kips}$$

$$L = \frac{\text{actual unbraced length}}{C_b} = \frac{20}{2.3} = 8.7 \text{ ft}$$

Try W24 × 76:

$$M = 322.5 \text{ ft-kips @ } L = 8.7 \text{ ft}$$

assuming $0.6F_y$ governs allowable stress.

(c) Check segment A (Fig. 10.4.1); W24 × 76:

$$f_b = \frac{197.5(12)}{175.4} = 13.5 \text{ ksi}$$

$$\frac{L}{C_b} = \frac{15}{1.0} = 15 \text{ ft} > L_u = 11.7 \text{ ft} \qquad \text{N.G.}$$

$$F_b(1.5\text{–}7) = \frac{12{,}000\, C_b}{Ld/a_f} = \frac{12{,}000(1.0)}{15(12)(3.90)} = 17.1 \text{ ksi} > 13.5 \text{ ksi}$$

which indicates beam is OK and no need to check AISC Formula 1.5–6.

(d) Check segment D; W24 × 76.

$$\frac{L}{C_b} = \frac{20}{1.75} = 11.4 < L_u = 11.7 \text{ ft} \qquad \text{OK}$$

$$F_b = 22 \text{ ksi}$$

$$f_b = \frac{242(12)}{175.4} = 16.6 \text{ ksi} < 22 \text{ ksi} \qquad \text{OK}$$

(e) Show final check for segments B and C; W24 × 76:

$$F_b\ (1.5\text{–}7) = \frac{12{,}000(2.3)}{20(12)(3.90)} = 29.5 \text{ ksi} > 0.6F_y$$

$$F_b = 0.6F_y = 22 \text{ ksi}$$

$$f_b = \frac{316(12)}{176} = 21.6 \text{ ksi} < 22 \text{ ksi} \qquad \text{OK}$$

Use W24 × 76, F_y = 36 ksi.

A comparison with Example 10.3.1 shows that the working-stress method produced the lighter section (W24 × 76 vs. W14 × 111). The reason is that plastic design requires development of the collapse mechanism; which because of the 20-ft laterally unsupported length required a larger r_y to satisfy the severe lateral bracing requirement. The working-stress method in effect regarded first yield in the negative zone as the limiting criterion; thus, lateral bracing did not have to insure rotation capacity in the plastic range.

EXAMPLE 10.4.2.

Redesign the continuous beam of Example 10.3.2 (Fig. 10.3.3) by the working-stress method. F_y = 60 ksi.

Solution

The elastic-bending-moment diagram under service load is shown in Fig. 10.4.2. To get a more appropriate comparison with plastic design, lateral support is provided every 5 ft on the 30-ft span and at the load and vertical support points on the 25-ft span.

(a) Determine section required using $0.9M_{max}$ with $0.66F_y$. L_c must exceed 5 ft.

$$\text{Reqd } S = \frac{0.9(185.8)(12)}{39.6} = 50.7 \text{ in.}^3$$

Try W14 × 38:

$$S = 54.7 \text{ in.}^3$$

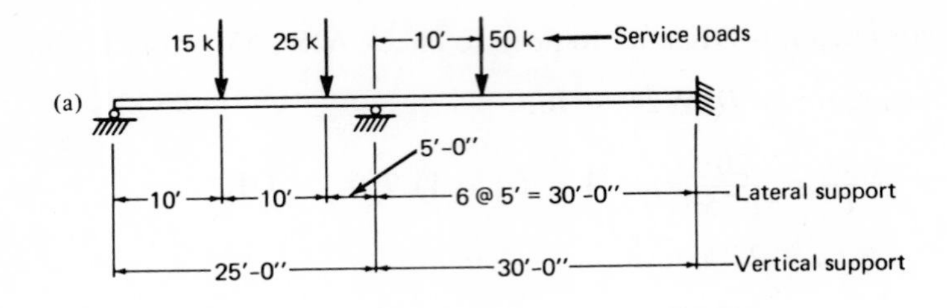

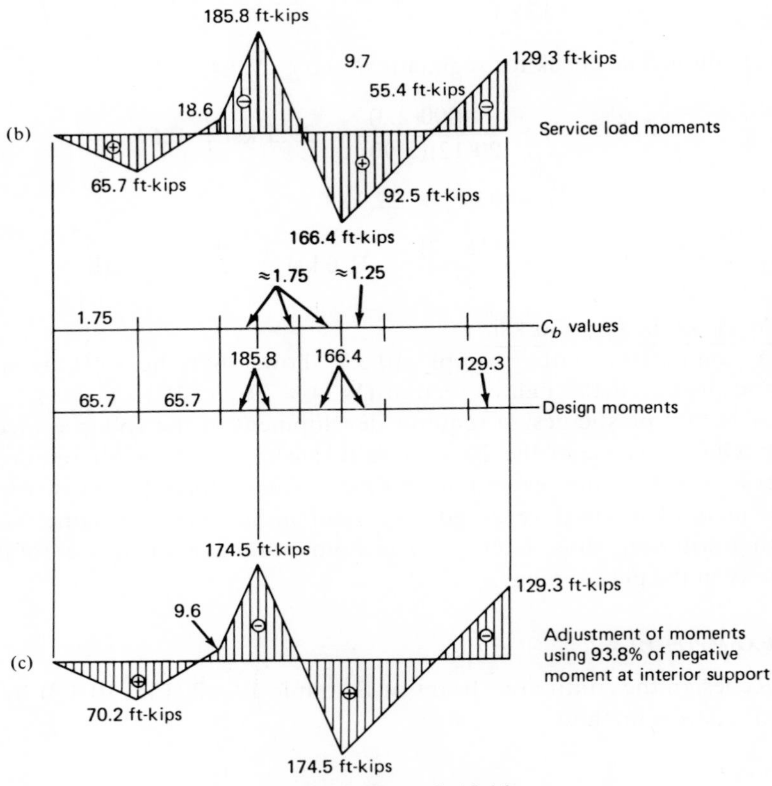

Fig. 10.4.2. Example 10.4.2.

$$\frac{L}{b_f} = \frac{5(12)}{6.776} = 8.85 < \frac{76.0}{\sqrt{F_y}} = 9.8 \qquad \text{OK}$$

$$\frac{b_f}{2t_f}\ \frac{6.776}{2(0.513)} = 6.62 < 6.7 \qquad \text{OK}$$

Note is made that this local buckling limitation for plastic design was 6.3 and prevented use of this section.

$$\frac{d}{t_w} = \frac{14.12}{0.313} = 45.2 < 53.2 \qquad \text{OK}$$

(b) Adjustment of the moment diagram to utilize the reduction in negative moment permitted under AISC–1.5.1.4.1 would make the negative moment 0.9(185.8) = 167 ft-kips. The resulting increase in positive moment will make the positive moment exceed the adjusted negative moment. Adjust only until the moments in question are equal:

$$x\,185.8 = \frac{50(10)20}{30} - \frac{129.3}{3} - \frac{2}{3}(185.8\,x)$$

which indicates that using 93.8 percent ($x = 0.938$) of the negative moment will equalize the positive and negative values at

$$-M = 0.938(185.8) = 174.5 \text{ ft-kips}$$

$$\text{Reqd } S = \frac{174.5(12)}{39.6} = 52.8 \text{ in.}^3$$

W14 × 38, $S = 54.7$ in.^{3} The adjusted moments are shown in Fig. 10.4.2c. The decrease in negative moment is 11.3 ft-kips. For equilibrium, the moments in span 1 become

$$-18.6 - 0.8(-11.3) = -9.6 \text{ ft-kips @ 25-kip load}$$
$$+65.7 - 0.4(-11.3) = +70.2 \text{ ft-kips @ 15-kip load}$$

These are only slightly different than the AISC indication of using the average of the change in negative moments [in this case 0.5(11.3)] to correct positive moments. Further, the authors believe the intention of the AISC method is followed by adjusting less than the full 10 percent and maintaining equilibrium.

(c) Check 10-ft unbraced lengths on 25-ft span for W14 × 38:

$$Ld/A_f = 120(4.06) = 388$$

$$F_b\,(1.5\text{–}7) = \frac{12{,}000C_b}{Ld/A_f} = \frac{12{,}000(1.75)}{388} > 0.6F_y$$

$$F_b = 0.6F_y = 36 \text{ ksi}$$

$$f_b = \frac{70.2(12)}{54.7} = 15.4 \text{ ksi} < 36 \text{ ksi} \qquad \text{OK}$$

Use W14 × 38, $F_y = 60$ ksi.

In plastic design, Example 10.3.2 shows that because of more severe local buckling and lateral bracing requirements a W14 × 38 was not acceptable, even though providing adequate strength. Note is made that if strength had controlled; i.e., plastic modulus Z for plastic-design method, and section modulus S for working-stress method, then W16 × 36 would have been the choice in both methods.

EXAMPLE 10.4.3.

Redesign the continuous beam of Example 10.3.3 by the working-stress method (Fig. 10.4.2). $F_y = 60$ ksi.

Solution

The requirement here is to use a cover plate on a basic section. A study of Fig. 10.4.2c shows that any cover plate would have to extend over the interior support and out into the region under the 50-kip load. The moments under service load conditions (neglecting the change caused by variable moment of inertia) show that the plastic-design solution of using cover plates only in the region of the 50-kip load would not be acceptable. It would be inadequate at the support.

(a) Select section for $M = 129.3$ ft-kips

$$\text{Reqd } S = \frac{129.3(12)}{39.6} = 39.2 \text{ in.}^3$$

Assuming "compact" section, $F_b = 0.66F_y = 39.6$ ksi

Try W12 × 31, W16 × 31, or W14 × 34:

Check W12 × 31:

$$\frac{b_f}{2t_f} = \frac{6.525}{2(0.465)} = 7.0 < 6.7 \qquad \text{N.G.}$$

$$\frac{L}{b_f} = \frac{60}{6.525} = 9.2 < 9.8 \qquad \text{OK}$$

$$\frac{d}{t_w} = \frac{12.09}{0.265} = 45.6 < 53.2 \qquad \text{OK}$$

$$F_b = F_y\left[0.733 - 0.0014\left(\frac{b_f}{2t_f}\right)\sqrt{F_y}\right] = 39.4 \text{ ksi}$$

$$f_b = \frac{129.3(1.03)12}{39.5} = 40.5 \text{ ksi} \approx 3\% \text{ overstress—say OK}$$

Dead-load effect estimated at 3 percent.

(b) Check W12 × 31 for 10 ft. unbraced length on 25-ft span. $L/b_f >$ 9.8—not compact.

$$Ld/A_f = 10(12)(3.98) = 477$$

$$F_b\ (1.5\text{–}7) = \frac{12{,}000C_b}{Ld/A_f} = \frac{12{,}000(1.75)}{477} = 44 \text{ ksi} > 0.6F_y$$

$$F_b = 0.6F_y = 36 \text{ ksi}$$

$$f_b = \frac{70.2(12)}{39.5} = 21.4 \text{ ksi} < 36 \text{ ksi} \qquad \text{OK}$$

Use W12 × 31, F_y = 60, as the base section.

(c) Determine plate size required where moment exceeds capacity of W12 × 31. For lateral bracing at 5-ft intervals in this region,

$$F_b = 39.4 \text{ ksi} \quad [\text{from part (a)}]$$

$$\text{Moment capacity} = \frac{39.4(39.5)}{12} = 129.2 \text{ ft-kips}$$

$$\text{Reqd from plates: } M \approx 174.5 - 129.2 = 45.3 \text{ ft-kips}$$

$$A_{\text{reqd}} = \frac{M}{df_{\text{avg}}} = \frac{45.4(12)}{12.34(38.5)} = 1.15 \text{ sq in.}$$

estimating $d \approx 12.09 + 0.25 = 12.34$ and $f_{\text{avg}} \approx 39.4(12.34/12.59) =$ 38.5 ksi. For a plate 6 in. wide,

$$\text{Thickness } t = \frac{1.15}{6} = 0.19 \text{ in.}$$

Try plates, top and bottom, $\frac{1}{4} \times 5\frac{3}{4}$, A = 1.44 sq in.; narrower than 6 in. to permit easier welding along the edges.

Check local buckling:

$$\frac{b}{t} = \frac{5.75}{0.25} = 23 < 24.5 \qquad \text{OK}$$

Therefore this cover plate continuously welded along its edges, a stiffened element, may be treated as "compact"; $F_b = 0.66 F_y = 39.6$ ksi.

(d) Recheck flexural capacity:

$$I = 239 \text{ (for W12} \times 31) + 1.44(76.4) = 348.2 \text{ in.}^4$$

$$S = \frac{\text{I}}{\text{d}/2} = \frac{348.2}{6.295} = 55.4 \text{ in.}^3$$

$$f_b = \frac{174.5(1.03)\,12}{55.4} = 38.9 \text{ ksi} < 39.6 \text{ ksi} \qquad \text{OK}$$

Use plates, top and bottom, $\frac{1}{4} \times 5\frac{3}{4}$, F_y = 60 ksi.

(e) Determine length for plates (see Fig. 10.4.3):

To right of 50-kip load,

$$L_1 = \frac{174.5 - 129.3}{174.4 + 129.3}(20) = \frac{45.2}{303.8}(20) = 2.98 \text{ ft.}$$

To left of interior support,

$$L_2 = \frac{174.5 - 129.3}{174.5 - 9.6}(5) = \frac{45.2}{164.9}(5) = 1.37 \text{ ft}$$

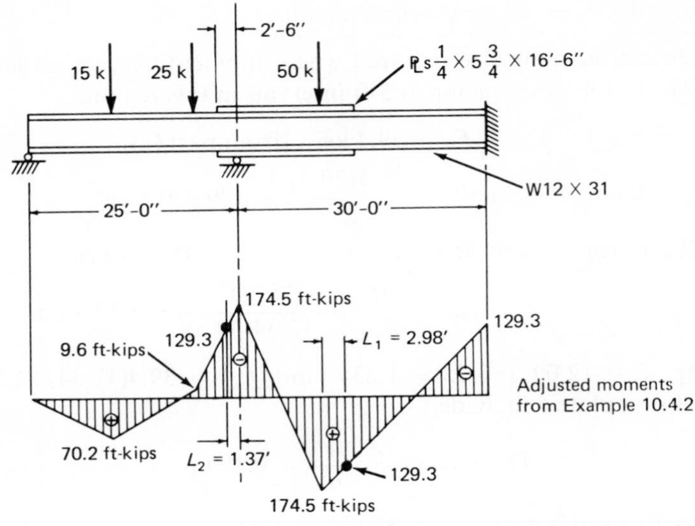

Fig. 10.4.3. Example 10.4.3.

$$\text{Force in plate} = \frac{A_{p1}}{A_{f1}}\left(\frac{M}{d}\right) = \frac{1.44}{3.03}\left[\frac{174.5(12)}{(11.9)}\right] = 83.5 \text{ kips}$$

For $\frac{3}{16}$-in. fillet weld placed by the *submerged-arc process* using E70 electrodes, the capacity R_w per inch is

$$R_w = \frac{3}{16}(21) = 3.94 \text{ kips/in.}$$

The maximum force which can be developed in the distance a' (2 times width = 2(5.75) = 11.5 in.) is

$$\text{Force} = 11.5(3.94)\,2 = 90.5 \text{ kips} > 83.5 \text{ kips} \qquad \text{OK}$$

$$\text{Plate length} = L_1 + 10.0 + L_2 + 2a'$$
$$= 2.98 + 10.0 + 1.37 + 2(11.5/12) = 16.27 \text{ ft.}$$

Use Pls. – 16'-6" long, beginning 2'-6" to left of interior support, as shown in Fig. 10.4.3.

10.5. SPLICES

While the design of connections is outside the scope of this chapter, the location of and strength requirements for beam splices are appropriately discussed here. It is obvious that if a splice is designed for the moment and shear capacity for the member spliced, full continuity is maintained and no special precautions are necessary. Some designers prefer to use shear splices at points of contraflexure and thus introduce a real hinge at a point of zero moment. There are two reasons why this should be avoided: (1) the point of contraflexure under service load is not at the same location that it occurs under factored loads (i.e., at its mechanism condition); (2) moments obtained assuming continuity are invalid if real hinges are inserted. Hart and Milek[11] have provided a good discussion of this problem.

AISC–1969 in Secs. 1.10.8 and 2.8 require connection design for the moments, shears, and axial loads to which they are to be subjected. AISC–1.10.8 refers to service loads, while AISC–2.8 refers to factored loads.

If full continuity is assumed when determining moments, either under service loads or under factored loads, then splices should provide that continuity. A reduced stiffness at a splice may prevent or reduce transmission of moment across that section and significantly change the resulting moment diagram.

EXAMPLE 10.5.1

Examine the effect of a shear splice at the point of contraflexure in the 40-ft span of the two-span continuous beam of Example 10.3.1. A36 steel. Assume 70 percent of the given load is live load.

Solution

(a) Full dead load plus live load. If full continuity is maintained, partial loading in some spans to account for live load in various locations is unnecessary. Note that the moment M_p = 453 ft-kips (see Fig. 10.5.1a) can develop even when the adjacent span has a reduced load. In other words, each span may be treated separately, as long as continuity exists so that the negative moment may be assumed to reach M_p.

(b) Effect of hinge at inflection point. Use of a shear splice in effect creates a hinge for all stages of loading (i.e., transforms system into statically determinate one). The maximum negative moment which can develop is limited to the shear at the hinge times the distance to the support. When load on the span with the hinge is reduced, the negative moment which can develop is reduced. As shown in Fig. 10.5.1b, when only 30 percent of the factored load [0.3(40) (1.7) = 20.4 kips] is on the 40-ft span, the maximum negative moment is

Shear at splice = 20.4(2/3) = 13.6 kips

Moment at support = 13.6(10) = 136 ft-kips

With the full factored load in the adjacent span, the maximum positive moment in the 30 ft. span is

$$V_{AB} = \frac{3(30)\,(1.7)}{2} - \frac{136}{30} = 76.5 - 4.53 = 71.97 \text{ kips}$$

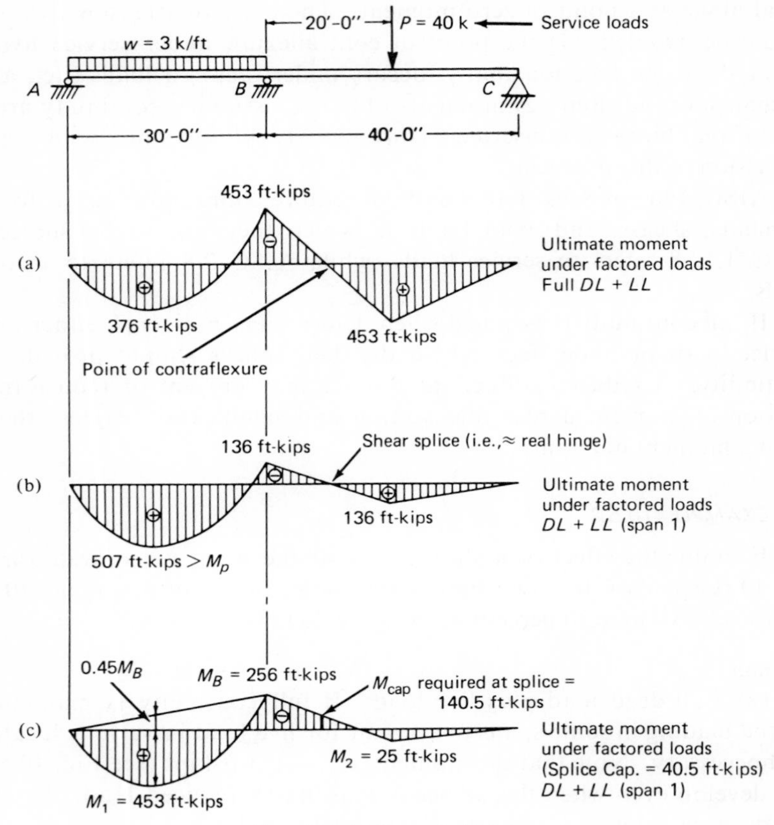

Fig. 10.5.1. Example 10.5.1.

$$M = \frac{(71.97)^2}{2(3)(1.7)} = 507 \text{ ft-kips} > 453 \text{ ft-kips} \qquad \text{N.G.}$$

Since the original design required $M_p = 453$ ft-kips, the 30-ft span is now inadequate as a result of the inability of the negative moment to exceed 136 ft-kips with the reduced load on the 40-ft span.

(c) Minimum strength required for splice. If the splice design is to provide less than full continuity, Fig. 10.5.1c illustrates the minimum capacity needed. If the section being spliced provides $M_p = 453$ ft-kips, the positive moment in the 30-ft span may not exceed that value. Thus, referring to Fig. 10.5.1c, and assuming the maximum occurs at approximately 0.45 L_1,

$$M_1 = \frac{3(1.7)(0.45)(0.55)}{2}(30)^2 - 0.45M_B = 453 \text{ ft-kips}$$

$$M_B = \frac{568 - 453}{0.45} = \frac{115}{0.45} = 256 \text{ ft-kips}$$

In order to have this moment develop, a moment capacity must be provided at the splice point:

$$M_2 = \frac{40(1.7)(0.30)(30)}{4} - \frac{M_B}{2} = 153 - 128 = 25 \text{ ft-kips}$$

$$M \text{ required at splice} = \frac{256 + 25}{2} = 140.5 \text{ ft-kips}$$

If the splice in this plastic design problem is designed for the shear as determined from Fig. 10.5.1a, plus the 140.5 ft-kip moment, the reduced loading of span *BC* will not cause overstress in span *AB*.

Finally, it is noted that reduced load in span *AB* has no detrimental effect on span *BC*.

The authors prefer that continuity be maintained by the design of splices for 100 percent of the moment capacity of members spliced; however, reduced capacities providing more economical designs may be used as long as the resulting designs are checked under possible partial loadings. Other examples and a more detailed procedure are given in Ref. 11.

PROBLEMS

All problems are to be done in accordance with latest AISC specifications, unless otherwise indicated.

10.1. Determine the maximum value for service load P. Assume adequate lateral support to satisfy any "compact section" requirements if necessary. Solve by (a) working-stress method, (b) plastic design.

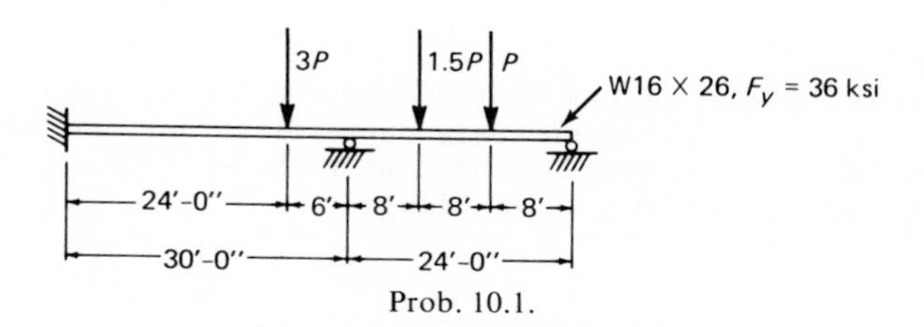

Prob. 10.1.

10.2. Determine the uniformly distributed service load a W24 × 100, F_y = 50 ksi, may be permitted to carry as a two-span continuous beam, having equal spans of 40 ft. Assume deflections do not control, and that external lateral support is provided at the reactions only. Solve by (a) working-stress method, (b) plastic design.

10.3. Select the lightest W section for the two-span continuous beam shown. Assume the given moment diagram is for fixed position service loads, with beam weight not included. F_y = 50 ksi. Use (a) working-stress method, (b) plastic design.

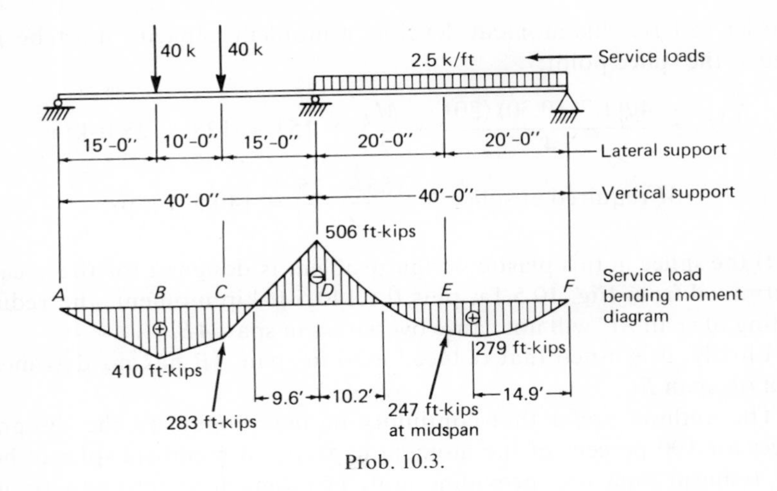

Prob. 10.3.

10.4. Determine the lightest W section which is suitable to serve as the two-span continuous beam shown. $F_y = 60$ ksi. Lateral supports exists at the vertical supports and under the concentrated load. Use (a) working-stress method, (b) plastic design.

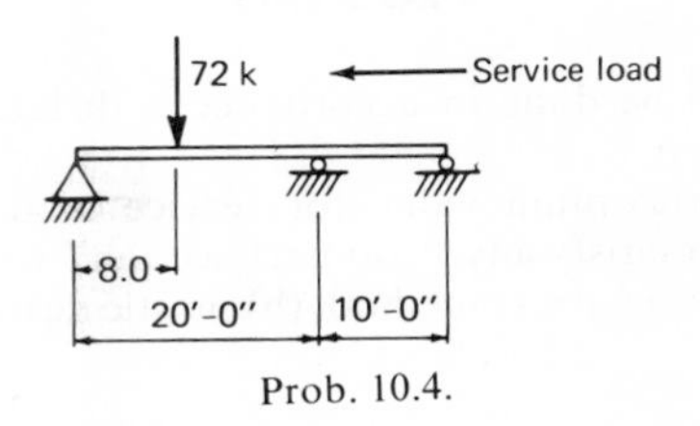

Prob. 10.4.

10.5. Select the lightest W section for the two-span continuous beam shown. The service load of 3 kips/ft may be assumed to include the beam weight. Lateral support is provided at ends and midpoint of each span. Use (a) working-stress method, (b) plastic design.

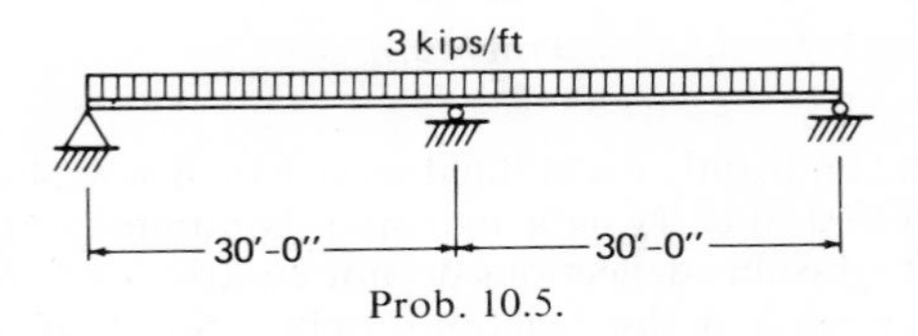

Prob. 10.5.

10.6. Select the lightest W section for the two-span continuous beam shown. Assume lateral support is provided every 5 ft. $F_y = 65$ ksi. Use (a) working-stress method, (b) plastic design.

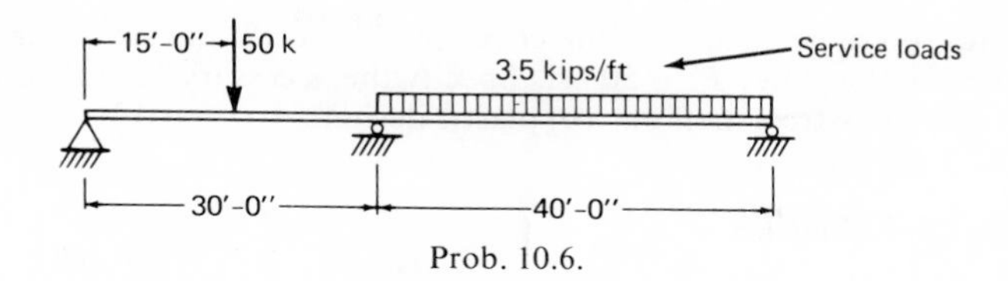

Prob. 10.6.

10.7. Select the lightest W section for the three-span continuous beam shown with fixed position service loads. Beam weight is not included. $F_y = 36$ ksi. Use (a) working-stress method, (b) plastic design.

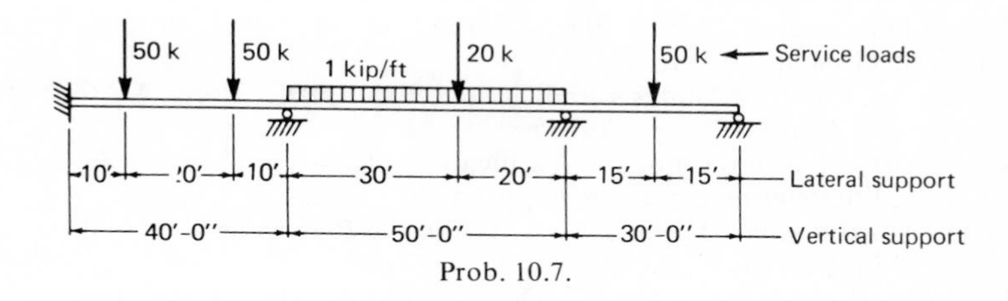

Prob. 10.7.

10.8. Select the lightest W section for a three-span (50 ft–65 ft–50 ft) continuous beam to carry a uniform dead load of 2 kips/ft in addition to the beam weight, and a uniform live load of 1.5 kips/ft. Live load deflection (maximum) may not exceed L/360. Lateral support is provided every 5 ft. $F_y = 60$ ksi. Use (a) working-stress method, (b) plastic design.

10.9. Repeat Prob. 10.8 using steel with $F_y = 50$ ksi and specify maximum lateral bracing for an economical design, instead of using the 5-ft distance of Prob. 10.8. Use (a) working-stress method, (b) plastic design.

10.10. Assume one splice is required for the beam selected for Prob. 10.3. Specify its location and the shear and moment for which it should be designed. Assume any loads may be reduced to 30 percent of their value while other loads remain at their maximum values. Use (a) working-stress method, (b) plastic design.

10.11. Same as Prob. 10.10. but splicing the beam of Prob. 10.7. Use (a) working-stress method, (b) plastic design.

10.12. For the beam shown with service loads, not including beam weight, (a) select the lightest W section assuming the same section extending over all three spans, (b) redesign, using a smaller base section with welded cover

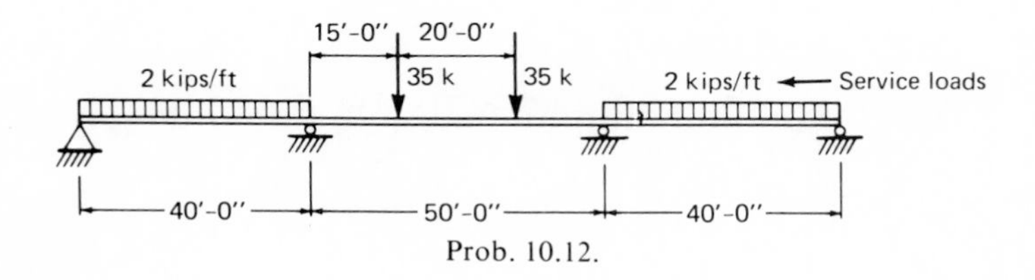

Prob. 10.12.

plates where needed. For the cover plate, specify the size, length, and location of the plate. $F_y = 50$ ksi. Specify the necessary lateral support. Use (a) working-stress method. (b) plastic design.

SELECTED REFERENCES

1. Lynn S. Beedle, *Plastic Design of Steel Frames*, John Wiley & Sons, Inc., New York, 1958.
2. C. E. Massonnet and M. A. Save, *Plastic Analysis and Design*, Vol. I, Beams and Frames, Blaisdell Publishing Company, Waltham, Mass., 1965.
3. Bruce G. Johnston, "Strength as a Basis for Structural Design," Proc., AISC National Engineering Conf., 1956, pp. 7–13.
4. Bruno Thürlimann, "Simple Plastic Theory," Proc. AISC National Engineering Conf., 1956, pp. 13–18.
5. Robert L. Ketter, "Analysis and Design Examples," Proc. AISC National Engineering Conf., 1956, pp. 19–35.
6. Lynn S. Beedle, "Experimental Verification of Plastic Theory," Proc., AISC National Engineering Conf., 1956, pp. 35–49.
7. Bruno Thürlimann, "Modifications to 'Simple Plastic Theory,'" Proc. AISC National Engineering Conf., 1956, pp. 50–57.
8. Frederick S. Merritt, "How to Design Steel by the Plastic Theory," *Engineering News-Record*, Apr. 4, 1957, pp. 38–43.
9. Bruce G. Johnston, C. H. Yang, and Lynn S. Beedle, "An Evaluation of Plastic Analysis as Applied to Structural Design," *Welding Journal Research Suppl.*, May 1953, pp. 1–16.
10. British Constructional Steelwork Association, "The Collapse Method of Design," Publication No. 5, 1952.
11. Willard H. Hart and William A. Milek, "Splices in Plastically Designed Continuous Structures," *Engineering Journal*, American Institute of Steel Construction, Vol. 2, No. 2 (April 1965), pp. 33–37.

11

Plate Girders

11.1. INTRODUCTION AND HISTORICAL DEVELOPMENT

A plate girder is a beam built up from plate elements to achieve a more efficient arrangement of material than is possible with rolled beams. Plate girders are economical where spans are long enough to permit saving in cost by proportioning for the particular requirements. Plate girders may be of riveted, bolted, or welded construction. Beginning with early railroad bridges during the period 1870–1900, riveted plate girders

Plate girders—Madison Square Garden Sports and Entertainment Center. Courtesy of Bethlehem Steel Corporation.

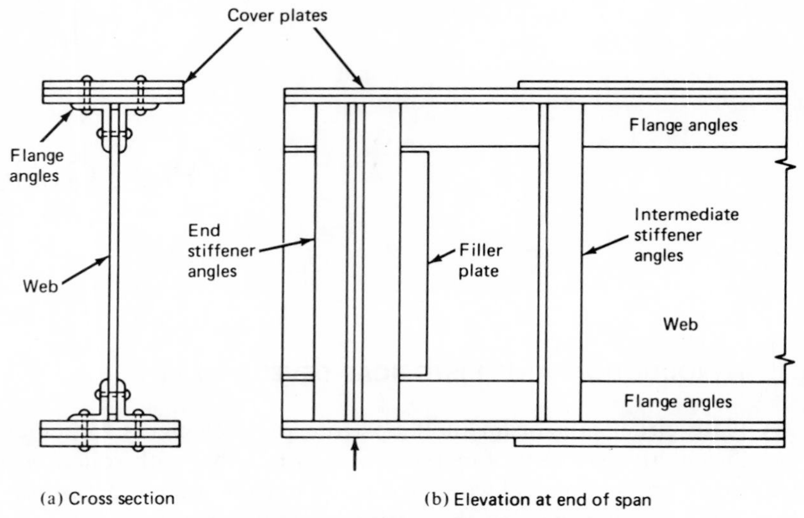

Fig. 11.1.1. Typical components of riveted plate girder.

(Fig. 11.1.1) composed of angles connected to a web plate, with or without cover plates, were extensively used in the United States on spans from about 50 to 150 ft.

Beginning in the 1950's when welding became more widely used (because of improved quality of welding and shop-fabricating economies resulting from increased use of automatic equipment) shop-welded plate girders composed of three plates (Fig. 11.1.2) gradually replaced riveted

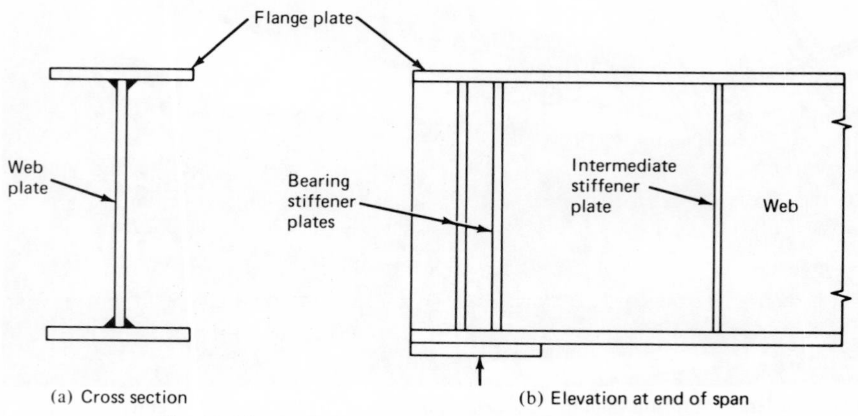

Fig. 11.1.2. Typical components of a welded plate girder.

girders. During this period also, high-tensile-strength bolts were displacing rivets in field construction. In the 1970's plate girders are nearly always shop welded using two flange plates and one web plate to make an I-shaped cross section.

Where practically all *riveted* girders were composed of plate and angle components having the same material yield point, the tendency now with *welded* girders is to combine materials of different strength. By changing materials at various locations *along the span* so that higher strength materials are available at locations of high moment and/or shear, or by using different strength material for flanges than for web (hybrid girders), more efficient and economical girders can be obtained.

Because few railroad bridges are being built today, discussion of economical spans and other dimensioning comments in this chapter will be limited to highway bridges, where most are continuous over two or more spans; or to buildings where some spans may be assumed as simply supported but more frequently are part of a rigid frame system.

Better understanding of plate-girder behavior, higher-strength steels, and improved welding techniques have combined to make plate girders economical in many situations formerly thought to be ideal for the truss. Generally, simple spans of 70 ft to 150 ft have traditionally been the domain for the plate girder. For bridges, continuous spans frequently using haunches (variable-depth sections) are now the rule for spans 90 ft or more. There are several three-span-continuous plate girders in the USA with center spans exceeding 400 ft, and longer spans are likely to be feasible in the future. The longest plate girder in the world is a three-span continuous structure over the Save River at Belgrade, Yugoslavia, with spans 246–856–246 ft. It is a double box girder in cross section varying in depth from 14 ft 9 in. at midspan to 31 ft 6 in. at the pier. The structure replaces a suspension bridge destroyed in World War II.

Three types of plate girders whose design is outside the scope of this chapter are shown in Fig. 11.1.3: (a) the box girder, providing improved torsional stiffness for long-span bridges; (b) The hybrid girder, providing variable material strength in accordance with stresses; and (c) the delta girder, providing improved lateral rigidity for long lengths of lateral unsupport.

Prior to studying the theoretical development in this chapter the reader is advised to review Chapter 6, Part II, where the basic elastic stability of plates is treated. If the theoretical development is to be omitted, the reader may proceed directly to Secs. 11.8 through 11.13 which treat the design procedure.

Since the design of riveted girders has been extensively treated in older

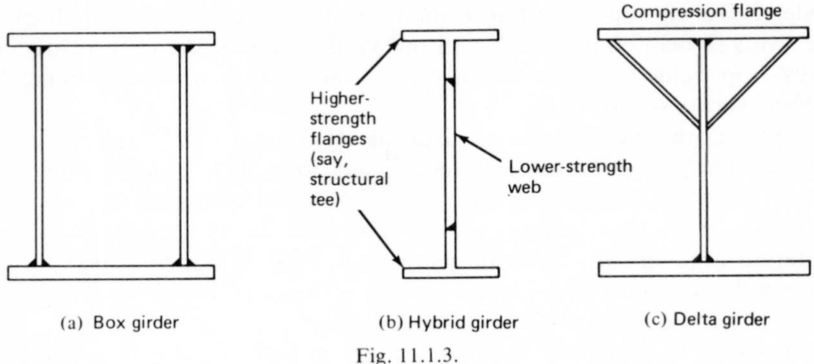

Fig. 11.1.3.

texts,[1,2] and such girders are rarely used at present, emphasis is placed on welded girders. No example of bolted or riveted girder design is given; however, high-strength bolted splices, commonly found in field connections, are treated in Chapter 13.

General Design Concepts. As with most other aspects of steel design, procedures are increasingly becoming based on ultimate strength. Until adoption of the 1961 AISC specification the basis for design rules was that elastic buckling should be prevented in plate elements. It was thus assumed that either yielding or elastic instability constituted failure.

The work of Basler and others at Lehigh University has resulted in present AISC provisions which consider post-buckling strength. A plate girder with properly designed regularly spaced stiffeners has the capacity, after instability of the web plate has occurred, to behave much like a truss with its web carrying diagonally the tension forces and the stiffeners carrying the compression forces. This trusslike behavior is referred to as *tension-field action.* Even classical buckling theory recognized that reserve capacity was available by using lower factors of safety against web buckling than for the overall strength of the member.

The classical theory and design procedure are still used by the AREA and AASHO (railroad bridge and highway bridge) specifications. The ultimate strength concept, including "tension-field" action, forms the basis for the 1969 AISC specification.

11.2. STABILITY RELATING TO LOADS CARRIED BY THE WEB PLATE

When a designer is free to arrange material to most efficiently carry load, it is apparent that for bending moment which is carried mostly by the flanges, a deep section is desired. A web is needed to make the flanges act as a unit and to carry the shear, but extra web thickness unnecessarily increases weight. So far as material is concerned, a thin web with stiffeners will give the lightest weight girder. Stability of the thin-web plate then is of primary concern.

Consider a segment of web plate, as in Fig. 11.2.1, where a represents

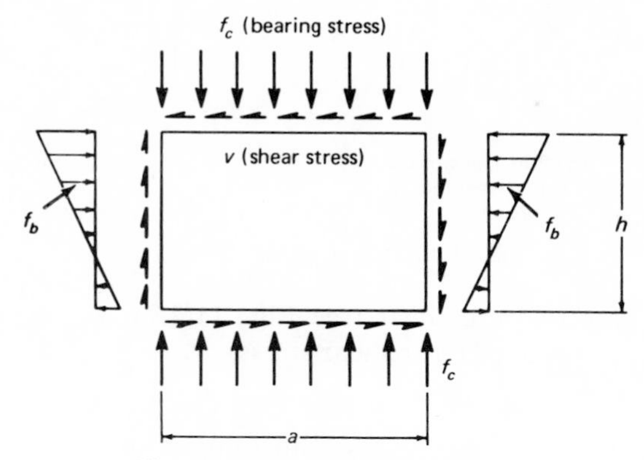

Fig. 11.2.1. Stresses on web plate.

the length between stiffeners and h is the clear height between longitudinal plate supports (i.e., between flanges, flange and longitudinal stiffener, or between longitudinal stiffeners). In general, such a plate segment is subjected to shear stresses along its edges, linearly varying normal stresses over the height h, and compressive stress along the length a due to loads bearing directly on the girder. The analysis of such a combined stress situation is complex and not suitable for use in design.

The procedure of this article is to consider separately the three types of stress (shear, v; uniform compression, f_c; and linearly varying compression, such as bending, f_b) on the plate edges. After separate treatment, a combined stress criterion may be established.

The stresses are assumed distributed according to classical beam theory where bending stress is computed $f = Mc/I$ and shear stress is computed as $v = VQ/It$.

In each of the separate treatments the basic approach is the same as illustrated in Secs. 6.14 and 6.15. The problems of inelastic buckling due to residual stress, accidental eccentricity, and initial crookedness apply here to give the strength vs. slenderness function relationship similar to that shown in Fig. 6.15.4.

Elastic Buckling Under Pure Shear. The elastic buckling stress for any plate is given by Eq. 6.14.25 as

$$F_{cr} = k\frac{\pi^2 E}{12(1 - \mu^2)(b/t)^2} \qquad [6.14.25]$$

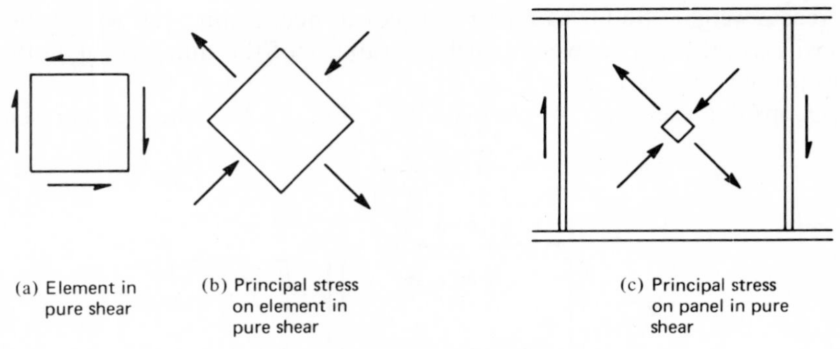

(a) Element in pure shear
(b) Principal stress on element in pure shear
(c) Principal stress on panel in pure shear

Fig. 11.2.2. Classical shear theory.

where for the case of pure shear (see Fig. 11.2.2), Eq. 6.14.25 may be written (using τ in place of F for shear stress)

$$\tau_{cr} = k \frac{\pi^2 E}{12(1 - \mu^2)\left(\dfrac{\text{short dimension}}{t}\right)^2} \tag{11.2.1}$$

where for the case of edges simply supported (i.e., displacement prevented but rotation about edges unrestrained)

$$k = 5.34 + 4.0\left(\frac{\text{short dimension}}{\text{long dimension}}\right)^2 \tag{11.2.2}$$

the development of which may be found in Ref. 18 (pp. 379–385) of Chapter 6.

For design purposes it may be desirable to put Eqs. 11.2.1 and 11.2.2 in terms of h, the unsupported web height, and a, the stiffener spacing. When this is done two cases must be considered.

(a) If $a/h \leq 1$ (see Fig. 11.2.3a), Eq. 11.2.1 becomes

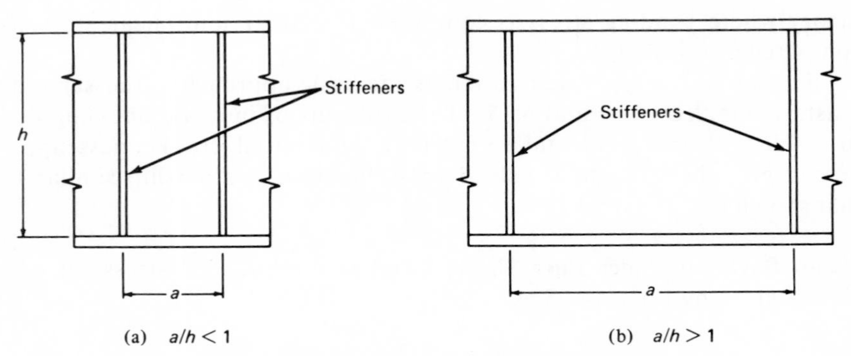

Fig. 11.2.3. Two cases of intermediate spacing.

$$\tau_{cr} = \frac{\pi^2 E\,[5.34 + 4.0(a/h)^2]}{12(1 - \mu^2)\quad (a/t)^2}\,\frac{(h/a)^2}{(h/a)^2} \tag{11.2.3}$$

(b) If $a/h \geq 1$ (see Fig. 11.2.3b), Eq. 11.2.1 becomes

$$\tau_{cr} = \frac{\pi^2 E\,[5.34 + 4.0(h/a)^2]}{12(1 - \mu^2)(h/t)^2} \tag{11.2.4}$$

It is apparent from Eqs. 11.2.3 and 11.2.4 that if one desires to use h/t as the stability ratio in the denominator, then two expressions for k are necessary. For all ranges of a/h, Eqs. 11.2.3 and 11.2.4 may be written

$$\tau_{cr} = \frac{\pi^2 E k}{12(1 - \mu^2)(h/t)^2} \tag{11.2.5}$$

where

$$k = 4.0 + 5.34/(a/h)^2 \quad \text{for} \quad a/h \leq 1 \tag{11.2.6}$$

$$k = 4.0/(a/h)^2 + 5.34 \quad \text{for} \quad a/h \geq 1 \tag{11.2.7}$$

For use in AISC–1.10.5, Eq. 11.2.5 has been put in nondimensional form, defining coefficient C_v as the ratio of shear stress at buckling to shear yield stress,

$$C_v = \frac{\tau_{cr}}{\tau_y} \quad \frac{\pi^2 E k}{\tau_y (12)(1 - \mu^2)(h/t)^2} \tag{11.2.8}$$

which is the C_v for *elastic* stability. Substitution of $E = 29{,}000$ ksi, $\mu = 0.3$, and $\tau_y = F_y/\sqrt{3}$ (Eq. 7.5.9),

$$C_v = \frac{\pi^2(29{,}000)\sqrt{3}}{12(1 - 0.09)}\,\frac{k}{F_y(h/t)^2}$$

$$C_v = \frac{45{,}000k}{F_y(h/t)^2} \tag{11.2.9}$$

valid for τ_{cr} below the elastic proportional limit, as shown in Fig. 11.2.4.

Inelastic Buckling Under Pure Shear. As in all stability situations, residual stresses and imperfections cause inelastic buckling as critical stresses approach yield stress. A transition curve for inelastic buckling was given by Basler[3] based on curve fitting and using test results from Lyse and Godfrey.[4] In the transition zone between elastic buckling and yielding,

$$\tau_{cr} = \sqrt{\tau_{\substack{\text{prop.}\\ \text{limit}}}\ \tau_{\substack{\text{cr ideal}\\ \text{elastic}}}} \tag{11.2.10}$$

The proportional limit is taken as $0.8\tau_y$, higher than for compression in

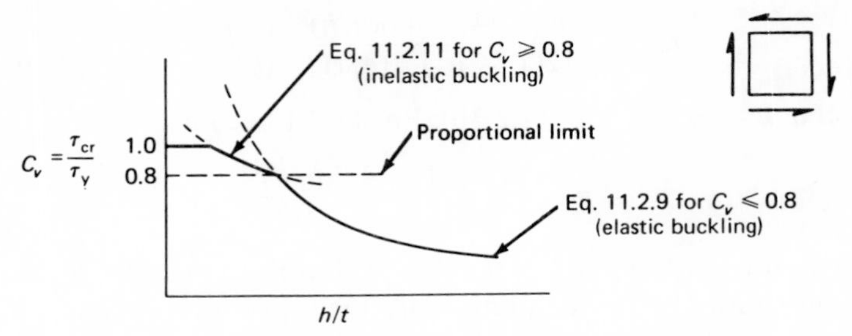

Fig. 11.2.4. Buckling of plates under pure shear.

flanges, because the effect of residual stress is less. Dividing Eq. 11.2.10 by τ_y to obtain C_v and using Eq. 11.2.9 gives

$$C_v = \frac{\tau_{cr}}{\tau_y} = \sqrt{(0.8)\frac{45{,}000k}{F_y(h/t)^2}}$$

$$= \frac{190}{h/t}\sqrt{\frac{k}{F_y}} \tag{11.2.11}$$

which is shown in Fig. 11.2.4.

Bending in the Plane of the Web. As for any plate stability problem, the elastic buckling stress is represented by Eq. 6.14.25,

$$F_{cr} = k\frac{\pi^2 E}{12(1-\mu^2)(b/t)^2} \qquad [6.14.25]$$

where for this case $b = h$.

The theoretical development of the k values for pure bending in the plane of the plate (Fig. 11.2.5), is to be found in Ref. 18 (pp 373–379)

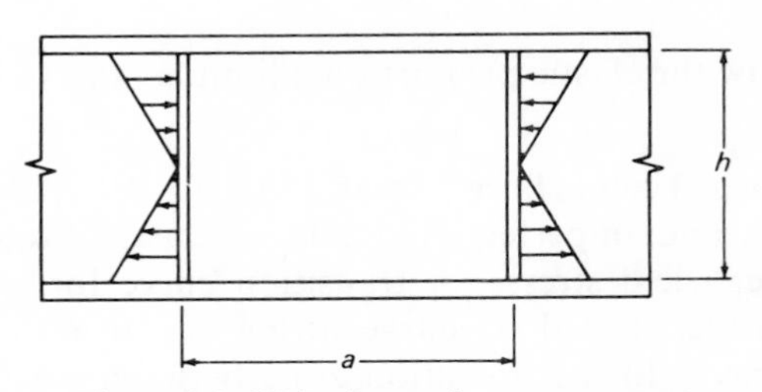

Fig. 11.2.5. Web plate under pure moment.

of Chapter 6. For any given type of loading, k varies with the aspect ratio, a/h (see Fig. 11.2.5), and with the support conditions along the edges. If the plate can be considered to have full fixity (full moment resistance against edge rotation) along the edges parallel to the direction of loading (i.e., at the edges joined to the flanges), the minimum value of k is 39.6 for any a/h ratio. If the flanges are assumed to offer no resistance to edge rotation, the minimum k value is 23.9. The variation of k with a/h ratio is given in Fig. 11.2.6.

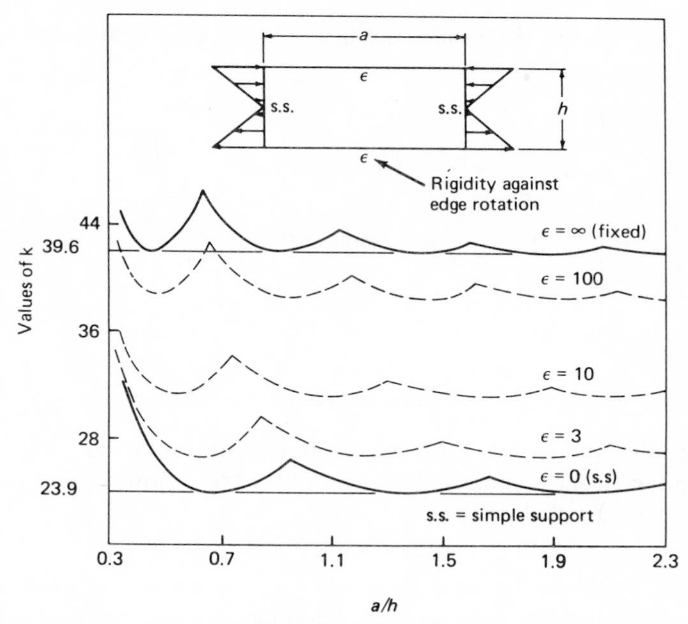

Fig. 11.2.6. Buckling coefficients for plates in pure bending. (From Ref. 5, p. 92.)

Thus the critical stress (using E = 29,000 ksi) may be said to lie between

$$F_{cr} = \frac{627\,(10)^3)}{(h/t)^2} \qquad \text{for } k = 23.9 \text{ (simple support at flanges)}$$

and

$$F_{cr} = \frac{1038\,(10^3)}{(h/t)^2} \qquad \text{for } k = 39.6 \text{ (full fixity at flanges)}$$

While each particular girder will have a different degree of flange restraint, fully welded flange to web connections will surely approach the full fixity case. It will be reasonable then to arbitrarily select a k value closer to 39.6, say 80 percent of the difference toward the higher value. One might say that

$$F_{cr} = \frac{954\,(10^3)}{(h/t)^2} \text{ ksi} \tag{11.2.12}$$

is representative of the buckling stress due to bending in the plane of the web. Bend-buckling of the web cannot occur if

$$\frac{h}{t} \leq \sqrt{\frac{954{,}000}{F_{cr}}} = \frac{975}{\sqrt{F_{cr}}} \tag{11.2.13}$$

Figure 11.2.7 shows the elastic stability relationship as rationalized by the foregoing logic.

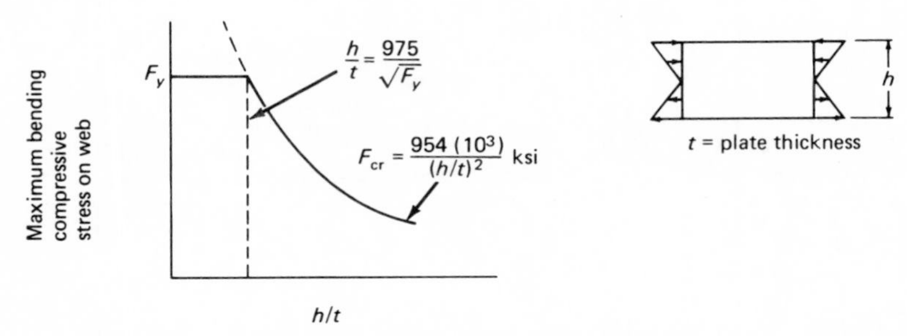

Fig. 11.2.7. Buckling of plates under bending in the plane of the web.

Since the web carries only a small part of the total bending moment to which the girder is subjected, neglect of the transition zone arising from inelastic buckling will not be significant.

Combined Shear and Bending. While there often are regions along a girder span where the case of pure bending or pure shear may be considered separately, usually combined shear and bending stresses are present. Timoshenko[6] has shown that when $\tau/\tau_{cr} < 0.4$ the effect of shearing stress on the critical value of bending stress is small. It is suggested by the CRC Guide[7] that the following approximate interaction formula agrees with theoretical computations for elastic stability,

$$\left[\frac{f_b}{F_{cr}\ (\text{pure bending})}\right]^2 + \left[\frac{v}{\tau_{cr}\ (\text{pure shear})}\right]^2 = 1 \tag{11.2.14}$$

where f_b and v are the bending and shear stresses, respectively, which in combination will cause elastic instability.

Elastic Buckling Under Transverse Compression. Heavy transverse loads are normally carried by bearing stiffeners so that compressive stresses f_c, as shown in Fig. 11.2.1 are usually of low magnitude. However, as the web plate is made thinner, even uniform loading may cause compressive stress high enough to buckle the web vertically. The localized yielding, i.e., web crippling, that may occur at the toe of fillets on beams has already been treated in Sec. 7.6. While the local effect certainly requires checking, thin-plate girder webs may also be subject to overall collapse, i.e., vertical buckling due to transverse compression.

Reexamine the situation treated in Sec. 6.14, the uniformly compressed plate, shown in Fig. 11.2.8. The solution obtained in Chap. 6,

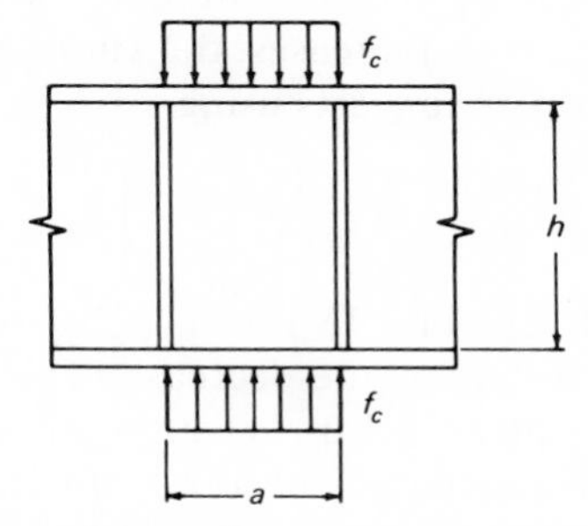

Fig. 11.2.8. Uniformly compressed plate.

Eq. 6.14.25, may be rewritten in terms of the plate girder. The dimension b, the width of loaded edge, becomes a, the stiffener spacing or width over which the load may be considered uniformly distributed. The length a in Fig. 6.14.6 becomes the unsupported web height h in the plate girder. Equations 6.14.25 and 6.14.26 become

$$F_{cr} = k\frac{\pi^2 E}{12(1 - \mu^2)(a/t)^2} \tag{11.2.15}$$

where

$$k = \left[\frac{1}{m(a/h)} + m(a/h)\right]^2 \tag{11.2.16}$$

with m an integer indicating the number of half waves which occur over the height h at buckling.

If one desires the stability ratio in the denominator of Eq. 11.2.15 to be h/t, multiply numerator and denominator by $(a/h)^2$, giving

$$F_{cr} = k_c\frac{\pi^2 E}{12(1 - \mu^2)(h/t)^2} \tag{11.2.17}$$

where

$$k_c = \frac{k}{(a/h)^2} = \left[\frac{1}{m(a/h)^2} + m\right]^2 \tag{11.2.18}$$

Since a single expression for k is desired for all a/h values, consider the actual variation over the practical range. When a/h becomes large, i.e., wide stiffener spacing compared to depth, $k_c \approx m^2$. The minimum value is with $m = 1$, essentially the Euler pinned column. Basler[8] states that since the compression stress actually varies from a maximum at the top of the web plate to near zero at the bottom, Eq. 11.2.18 is overly conservative. Therefore, k_c is suggested to have a minimum value of 2 instead of 1.

At the other extreme with closely spaced stiffeners, strength greatly increases. As shown in Fig. 6.15.2, Curve C, when the ratio of plate length to width of loaded edge increases, the value of k approaches 4. To satisfy both extremes Basler[8] suggests using

$$k_c = \left[\frac{4}{(a/h)^2} + 2\right] \tag{11.2.19}$$

which is compared with Eq. 11.2.18 for $m = 1$ in Fig. 11.2.9. For small h/a, i.e., widely spaced stiffeners, the decrease of intensity of compression from top to bottom means Eq. 11.2.19 should correctly exceed Eq. 11.2.18. As the stiffener spacing becomes closer, the variation in com-

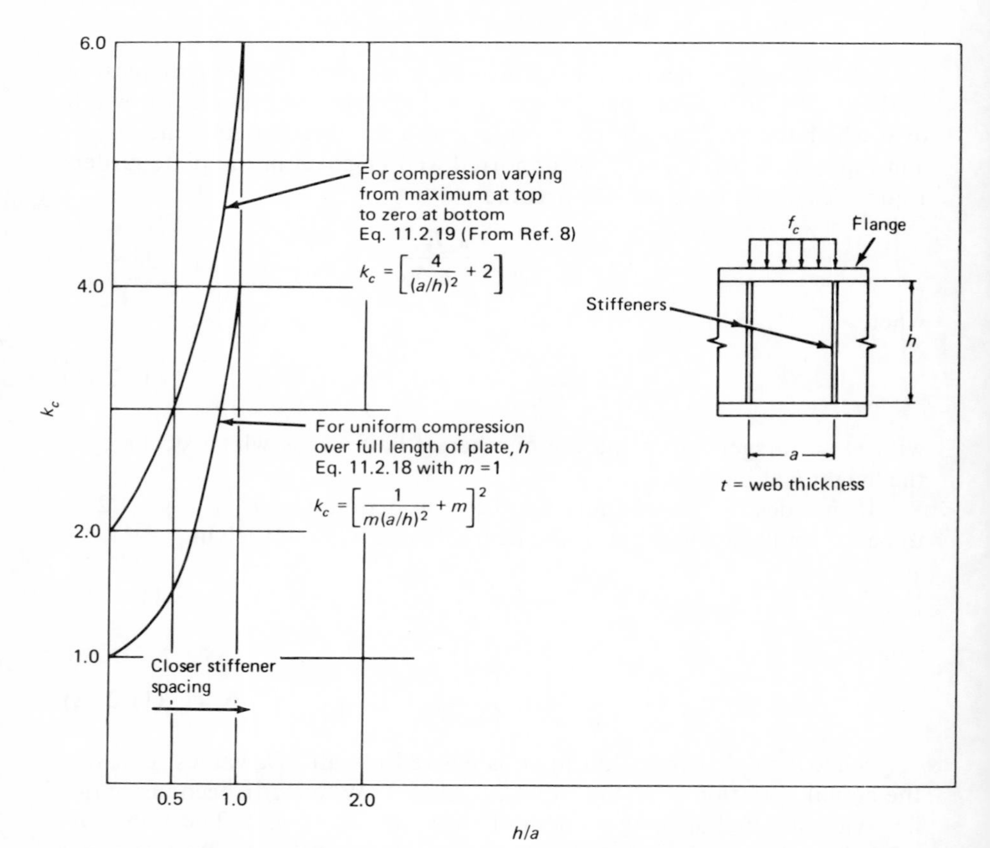

Fig. 11.2.9. Buckling coefficient for vertical buckling under transverse compression, $F_{cr} = k_c\pi^2E/[12(1 - \mu^2)(h/t)^2]$.

pressive stress over the depth h has less effect. For h/a greater than about 2, Eqs. 11.2.18 and 11.2.19 become essentially coincident. Thus, while the specific values of Eq. 11.2.19 have not been verified, the result is reasonable.

The preceding discussion and Eq. 11.2.19 apply to situations in which the flange offers negligible rotational restraint at the compressed edge of the web plate. When the loaded edges are fixed, or fully restrained against rotation, the k value is higher as shown in Fig. 6.15.2, curve C (dotted). For wide stiffener spacing (a/b of Fig. 6.15.2 small) the effect of end restraint increases stability, so that k varies from 5.5 upward as the ratio, girder height/stiffener spacing, decreases below 1.5. On the other hand, for closely spaced stiffeners, the minimum k value is about 4, the same as for no rotational restraint at the loaded edge. Basler, therefore, recommends[8] using

$$k_c = \left[\frac{4}{(a/h)^2} + 5.5\right] \tag{11.2.20}$$

which increases k_c for wide spacing of stiffeners and approaches the same value as Eq. 11.2.19 as a/h becomes small. Again, although the coefficient k_c cannot be rigorously derived, it seems logical.

For uniform loading on the flange of a girder, the unit stress, f_c, is obtained directly by dividing the uniform loading w per inch by the thickness t (Fig. 11.2.10a). Basler[8] recommends distributing concentrated loads over the panel width (Fig. 11.2.10b), or over the depth of the web (Fig. 11.2.10c), whichever gives the greater stress f_c.

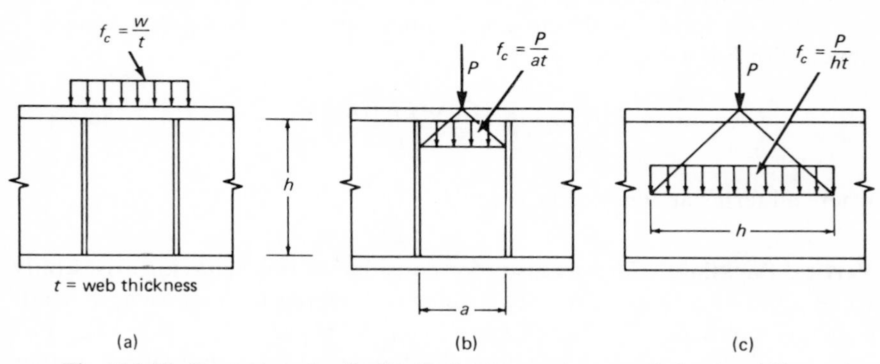

Fig. 11.2.10. Suggestions for distributing loads to evaluate vertical web stability.

11.3. STABILITY RELATING TO THE COMPRESSION FLANGE

Flange related stability, *lateral-torsional buckling*, was the subject of Chapter 9. The flange plates in rolled beams are attached together by a relatively thick web which brings both flanges into action (high torsional stiffness) when lateral instability is imminent. As the h/t for the web increases, the effect of the tension flange decreases (lateral bending stiffness predominates). Once h/t exceeds the value at which bend-buckling in the plane of the web occurs, the cross section behaves, for carrying bending stress, as if part of the web were not there. When bend-buckling has occurred, the vertical support the web is able to provide to the compression flange is significantly reduced and the possibility of *vertical buckling of the flange* must be considered. Furthermore, once this flange support of the web is reduced, depending upon how thick is the web and how much of it will perform integrally with the flange plate, the possibility of *torsional buckling* of the tee-shaped flange (flange with segment of web combined) arises. The typical buckling modes of the compression flange column are shown in Fig. 11.3.1.

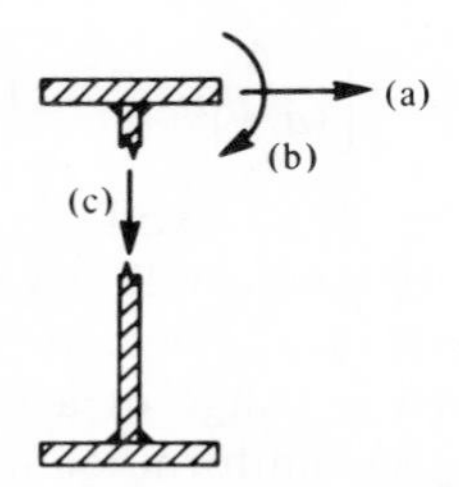

Fig. 11.3.1. (a) Lateral buckling, (b) torsional buckling, (c) vertical buckling.

Lateral-Torsional Buckling. This type of behavior has been theoretically treated in Chapter 9 with the elastic buckling stress computed in accordance with Eq. 9.4.1,

$$F_{cr} = \frac{C_4\sqrt{EI_y GK}}{S_x L} \qquad [9.4.1]$$

where all terms are defined in Sec. 9.4.

Vertical Buckling. The necessity for considering this behavior arises only when ultimate moment capacity is being considered for cases where bend-buckling can occur, i.e., when h/t exceeds $975/\sqrt{F_{cr}}$ (Eq. 11.2.13). For such cases, one may imagine that the flange is a compression member independent of the rest of the girder (see Fig. 11.3.2).

Neglecting higher-order terms, the curvature of the girder gives rise to flange force components which cause uniform compressive stresses on the edges of the web, as in Fig. 11.3.3. So long as the web is stable under the compressive stresses produced by the transverse components of the flange forces, the flange cannot buckle vertically. In the following derivation the flange is assumed to have *zero* stiffness to resist vertical buckling.

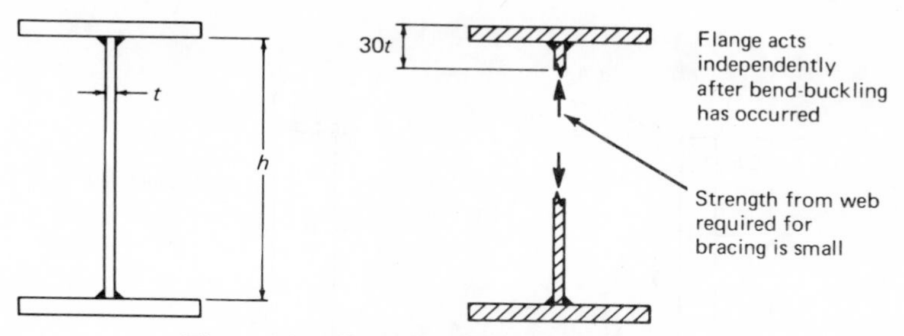

Fig. 11.3.2. Effect of bend-buckling of the web.

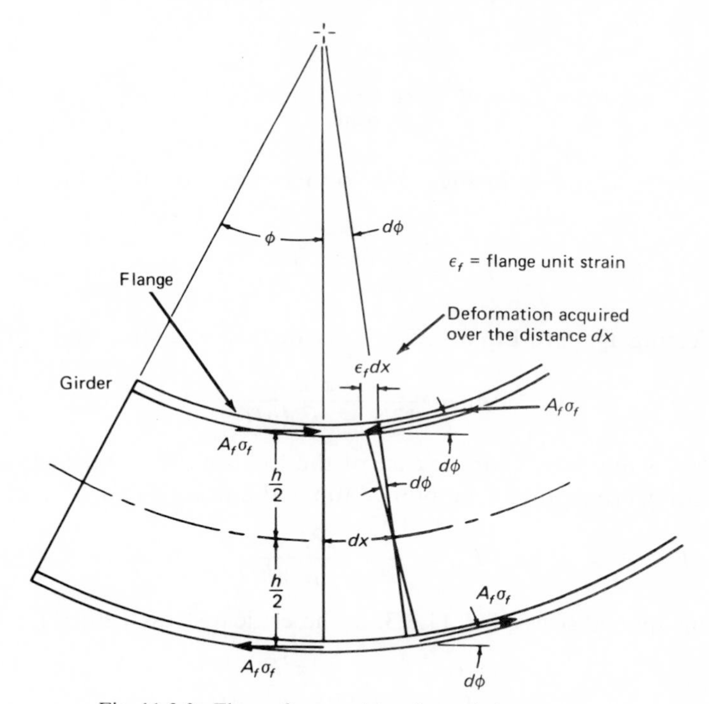

Fig. 11.3.3. Flange forces arising from girder curvature.

Referring to Fig. 11.3.3, the deformation acquired over the distance dx is

$$\epsilon_f \, dx = d\phi \frac{h}{2} \tag{11.3.1}$$

$$d\phi = \frac{2\epsilon_f}{h} dx \tag{11.3.2}$$

As shown in Fig. 11.3.4a, the vertical component causing compressive stress is $\sigma_f A_f \, d\phi$. After dividing by the area $t dx$ to obtain the compressive

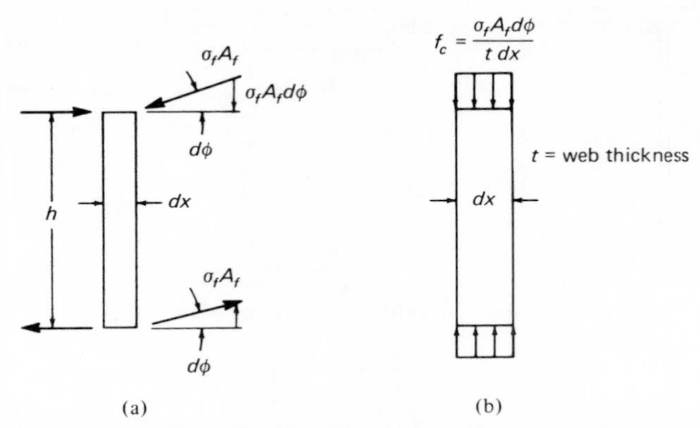

Fig. 11.3.4. Effect of flange force component normal to flange plate.

unit stress, f_c, as shown in Fig. 11.3.4b, one may substitute Eq. 11.3.2 for $d\phi$,

$$f_c = \frac{\sigma_f A_f \, d\phi}{t \, dx} = \frac{2\sigma_f A_f \epsilon_f}{th} \tag{11.3.3}$$

Referring again to Eq. 6.4.25, the elastic buckling stress for a plate,

$$F_{cr} = \frac{k\pi^2 E}{12(1 - \mu^2)(b/t)^2} \qquad [6.4.25]$$

where $b = h$ and $k = 1$ for the case of the Euler plate assumed free along edges parallel to loading, and pinned top and bottom. Thus

$$F_{cr} = \frac{\pi^2 E}{12(1 - \mu^2)(h/t)^2} \tag{11.3.4}$$

Equating applied stress, Eq. 11.3.3, to the critical stresss, Eq. 11.3.4, gives

$$\frac{2\sigma_f A_f \epsilon_f}{th} = \frac{\pi^2 E}{12(1 - \mu^2)(h/t)^2} \tag{11.3.5}$$

which, letting $th = A_w$, gives

$$\frac{h}{t} = \sqrt{\frac{\pi^2 E}{24(1 - \mu^2)}\left(\frac{A_w}{A_f}\right)\left(\frac{1}{\sigma_f \epsilon_f}\right)} \tag{11.3.6}$$

Conservatively assume that σ_f must reach the yield stress, F_y, for the flange to achieve its ultimate capacity. Furthermore, if residual stress F_r exists in the flange distributed as shown in Fig. 11.3.5, then the total flange strain will be that due to the sum of the residual stress plus the yield stress; therefore

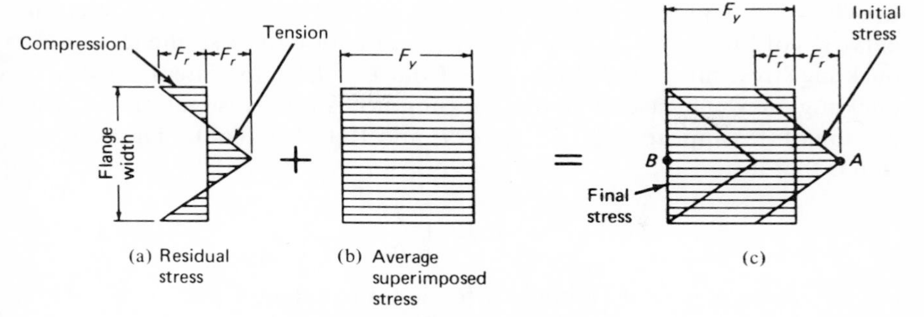

Fig. 11.3.5. Effect of residual stress.

$$\epsilon_f = (F_r + F_y)/E \tag{11.3.7}$$

It is the strain adjacent to the web that is of concern; in Fig. 11.3.5c the change is from F_r in tension (point A) to F_y in compression (point B).

Substitution of $\sigma_f = F_y$, $\epsilon_f =$ Eq. 11.3.7, $E = 29{,}000$ ksi, and $\mu = 0.3$, into Eq. 11.3.6 gives

$$\frac{h}{t} = \frac{19{,}500\sqrt{A_w/A_f}}{\sqrt{F_y(F_y + F_r)}} \tag{11.3.8}$$

a conservative estimate of the maximum h/t to prevent vertical buckling. Basler[9] suggests A_w/A_f will rarely be below 0.5 and that $F_r = 16.5$ ksi is realistic. If these substitutions are made,

$$\frac{h}{t} = \frac{13{,}800}{\sqrt{F_y(F_y + 16.5)}} \tag{11.3.9}$$

Equation 11.3.9 has been developed without regard to placement of stiffeners. The effect of stiffeners would certainly be to increase the buckling capacity above the value given by Eq. 11.3.4. Recent tests reported[12] on hybrid girders with A514 flanges ($F_y = 100$ ksi) indicate that h/t can conservatively be accepted in design as 250 if $a/h \leq 1.0$, and 200 for a/h between 1.0 and 1.5. The limitation of 200 may be expressed for other yield strengths as $2000/\sqrt{F_y}$ for $a/h \leq 1.5$.

Torsional Buckling. If the girder flange is made as wide as feasible to provide good lateral-torsional buckling properties, it is possible that torsional buckling of the flange plate will occur at a critical stress below the value given by Eq. 9.4.1. Essentially, the torsional buckling is the buckling of a uniformly compressed plate free along one edge and hinged at the other (Case E, Fig. 6.15.2). If strain hardening and inelastic buckling are considered, the critical stress variation is as shown in curve (b) of Fig. 6.15.4. In order for a safe situation to exist, the torsional-buckling stress must exceed the value from Eq. 9.4.1 for lateral-torsional buckling, in which case torsional buckling needs no investigation. A discussion of this appears in Ref. 7 (p. 129) and Ref. 9 (p. 668). The conclusion is that if

$$\frac{b_f}{t_f} \leq 12 + \frac{L}{b_f} \tag{11.3.10}$$

the possibility of primary failure by torsional buckling is prevented. This is for the most severe case of pure bending in the plane of the web. In Eq. 11.3.10, L is the distance between lateral bracing points.

11.4. POST-BUCKLING CONDITIONS IN THE WEB PLATE

The various types of web plate stability discussed in Art. 11.2 all were based on small-deflection theory wherein the buckled position is an unstable equilibrium one, close to the flat unbuckled position. The efforts of many researchers over the years to experimentally check such plate buckling theories have been relatively unsuccessful.[7] Such a small change in behavior occurs in the web plate of a girder at buckling that it is difficult to determine experimentally the critical load. After the critical stress is reached in a plate and a small out-of-plane displacement occurs, tensile membrane stresses arise in the middle plane which tend to stabilize the plate. Any given element of a plate is constrained by all adjacent elements. Because these membrane effects arise gradually during the loading of a plate girder, the sudden buckling as for a compressed bar does not appear. Figure 11.4.1 shows a comparison of load-deflection curves for an isolated bar and a plate such as a plate girder web.

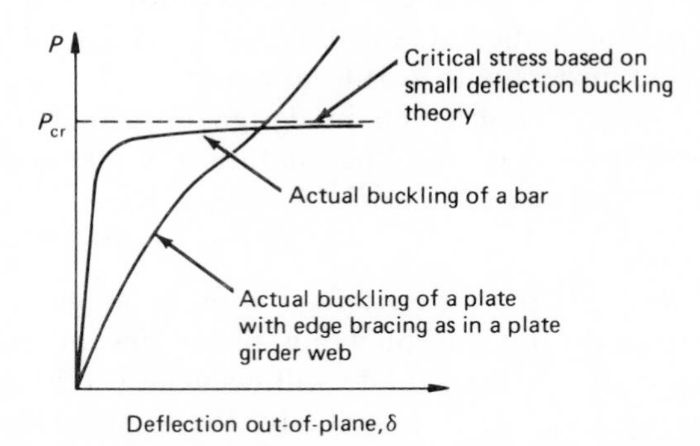

Fig. 11.4.1. Plate buckling compared with buckling of a slender bar.

The web of a plate girder cannot actually collapse without the flanges and stiffeners which surround it also collapsing. The result of web buckling is to cause a redistribution of stresses. As long as the flanges and stiffeners are *capable* of resisting an increased share of the load, the girder cannot fail.

Aircraft have long been designed considering post-buckling plate strength. If they were not, they would be unable to life their weight off the ground.

11.5. ULTIMATE BENDING STRENGTH; CONSIDERING POST-BUCKLED WEB

As discussed in Sec. 11.2, the web may buckle due to bending stress when, according to Eq. 11.2.13,

$$\frac{h}{t} \geq \frac{975}{\sqrt{F_{cr}}} \qquad [11.2.13]$$

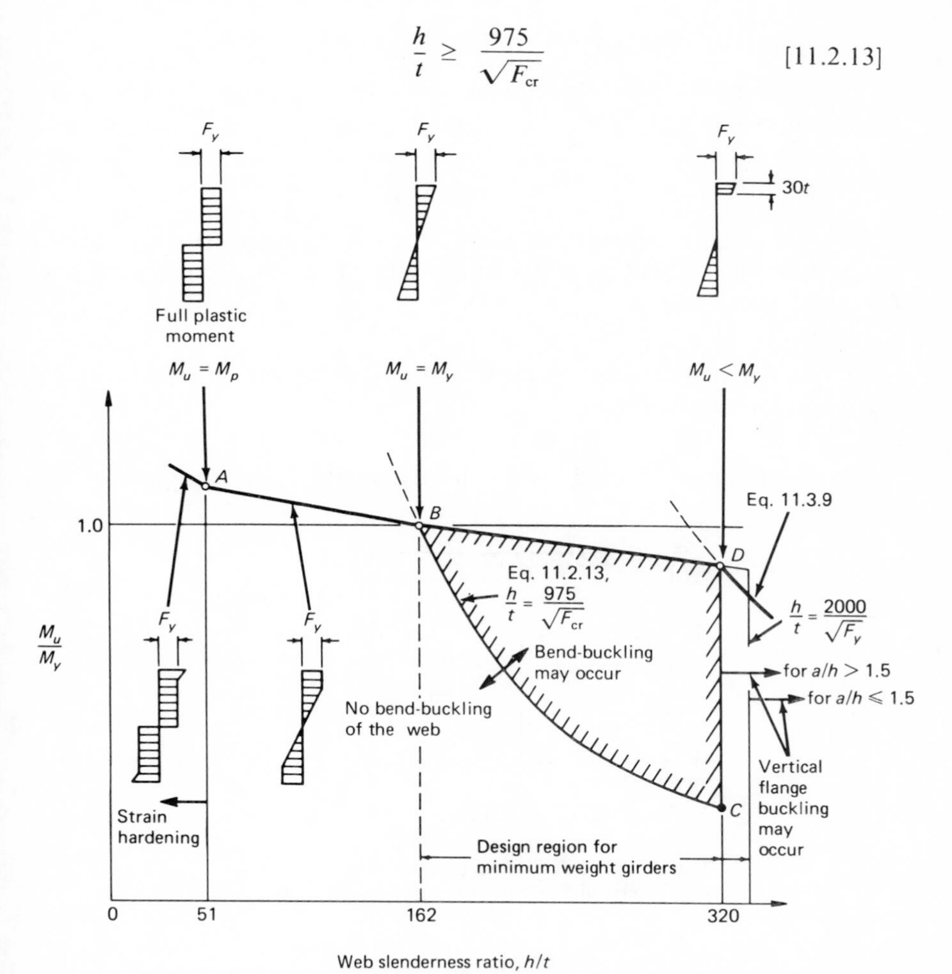

Fig. 11.5.1. Bending strength of girders as affected by bending stress in the web plate: A36 steel.

As described in Sec. 11.4, such buckling does not mean the usefulness of the girder has ended. Figure 11.5.1 shows the ultimate moment capacity of a girder as affected by h/t. It should be remembered that ultimate moment capacity of a deep girder is a function of

$$M_u = f\,\frac{h}{t},\left(\frac{L}{r_y},\ \frac{b}{t_f},\ \frac{A_w}{A_f}\right) \tag{11.5.1}$$

where h/t governs web stability (bend-buckling)

$\dfrac{L}{r_y}$ governs lateral flange stability (lateral-torsional buckling)

$\dfrac{b}{t_f}$ governs local buckling (or torsional buckling) of flange

$\dfrac{A_w}{A_f}$ determines the post-web-buckling effect on the flange

Assuming that lateral-torsional buckling and local buckling are prevented as assumed for Fig. 11.5.1, the variables become

$$M_u = f\left(\frac{h}{t},\ \frac{A_w}{A_f}\right) \tag{11.5.2}$$

When post-buckling strength of the girder is considered, the capacity is raised from line BC of Fig. 11.5.1 to line BD. The actual position of line BD varies with A_w/A_f.

EXAMPLE 11.5.1

Using $h/t = 320$, determine the expression for M_u/M_y (point D of Fig. 11.5.1) in terms of A_w/A_f.

Solution

Refer to Fig. 11.5.2, and use an effective depth of girder equal to $30t$,

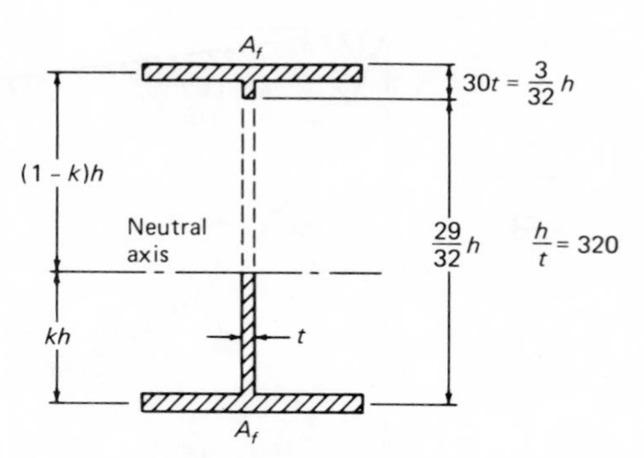

Fig. 11.5.2. Effective section in bending when vertical flange buckling is imminent.

as per Ref. 9. The remainder of the web in the compression zone is assumed to be nonparticipating.

(a) Determine location of neutral axis; equating static moments about the neutral axis:

$$A_f(kh) + \frac{t(kh)^2}{2} = A_f(1 - k)h + \frac{3}{32}h\left(\frac{29}{32}h - kh\right)t$$

Divide by $A_f h$:

$$k + k^2\frac{th}{2A_f} = (1 - k) + \frac{3}{32}\left(\frac{29}{32} - k\right)\frac{th}{A_f}$$

Noting that $th = A_w$ and letting $\rho = A_w/A_f$ give

$$k^2 + k\left(\frac{4}{\rho} + \frac{3}{16}\right) = \frac{2}{\rho} + \frac{87}{512}$$

$$k = \sqrt{\frac{183}{1024} + \frac{38}{16\rho} + \frac{4}{\rho^2}} - \left(\frac{3}{32} + \frac{2}{\rho}\right) \qquad \text{(a)}$$

(b) Determine moment of inertia:

$$I = A_f(\text{kh})^2 + \frac{1}{3}t(kh)^3 + A_f(1 - k)^2h^2 + \frac{3th}{32}\left(\frac{29}{32}h - kh\right)^2$$

Using $th = A_w$ and $\rho = A_w/A_f$,

$$I = A_f h^2\left[\frac{\rho}{3}k^3 + k^2 + (1 - k)^2 + \frac{3\rho}{32}\left(\frac{29}{32} - k\right)^2\right] \qquad \text{(b)}$$

Assuming the extreme fiber stressed to yield stress F_y,

$$M_u = F_y I/[(1 - k)h] \qquad \text{(c)}$$

To obtain M_y, the entire girder is considered effective. In developing the expression the flange-area concept will be used as shown in Fig. 11.5.3. The moment capacity of the web is approximately (Fig. 11.5.3a)

$$M_{\text{web}} = fth^2 16 \qquad \text{(d)}$$

which assumes web depth, distance between flange centroids, and overall depth are the same. The moment capacity of the equivalent flange area system (Fig. 11.5.3b) is

$$M_{\text{equiv}} = fA_f'h \qquad \text{(e)}$$

Equating Eqs. (d) and (e) gives for the equivalent flange area, A_f',

$$A_f' = th/6 = A_w/6 \qquad \text{(f)}$$

The total moment capacity of a girder where the stress $f = F_y$ then becomes

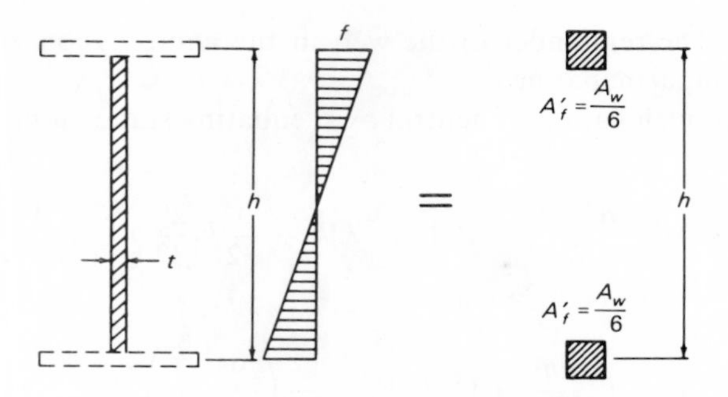

Fig. 11.5.3. Equivalent flange area to replace web.

$$M_y = F_y\left[A_f + \frac{A_w}{6}\right]h \tag{g}$$

$$= F_y A_f h\left(1 + \frac{\rho}{6}\right) \tag{h}$$

The vertical ordinate of point D in Fig. 11.5.1 is obtained by dividing Eq. (h) by Eq. (c):

$$\frac{M_u}{M_y} = \frac{\frac{\rho}{3}k^3 + k^2 + (1-k)^2 + \frac{3\rho}{32}\left(\frac{29}{32} - k\right)^2}{(1-k)(1+\rho/6)} \tag{i}$$

which is plotted in Fig. 11.5.4.

From Fig. 11.5.4, the variation in M_u/M_y might be approximated by a straight line for A_w/A_f from zero to three with a slope of $-(1.00 - 0.73)/3.0 = -0.09$.

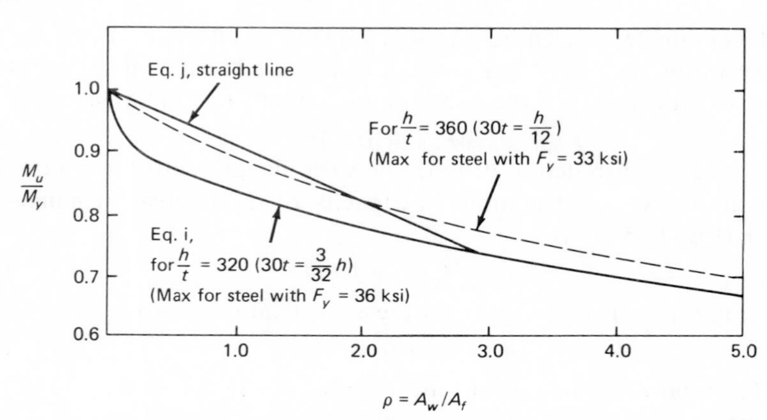

Fig. 11.5.4. Reduction in moment capacity considering post buckling strength at maximum h/t for A36 steel.

Thus, at $h/t = 320$,

$$\frac{M_u}{M_y} = 1.0 - 0.09\frac{A_w}{A_f} \tag{j}$$

It may be observed that the straight line agrees better for $h/t = 360$, the situation for which this linear equation was originally developed,[9] than it does for $h/t = 320$. For higher strength steels, for which the maximum h/t to prevent vertical buckling of the flange is less than 360, more of such a stiffer web participates with the compression flange, causing a greater reduction in M_u/M_y.

The linear reduction based on Eq. (j) does not seem conservative, but is within several percent of the more accurate curve using $30t$ as the effective depth of web participating with the compression flange.

Tests[9] have verified the correctness of this linear reduction method using $h/t = 360$ as its basis.

Reduced Nominal Flange Stress for $M_v < M_y$. By reference to Fig. 11.5.1, it may be reasonably assumed that M_u/M_y varies linearly from point B to D. Thus, the reduction in M_u/M_y per A_w/A_f per h/t greater than that at point B is

$$\frac{\text{Slope per } A_w/A_f}{320 - 162} = \frac{0.09}{158} = 0.00057 \text{ (say 0.0005)}$$

Thus M_u/M_y for the region from point B to D (Fig. 11.5.1), assuming linear variation, is

$$\frac{M_u}{M_y} = 1.0 - 0.0005\frac{A_w}{A_f}\left(\frac{h}{t} - \frac{975}{\sqrt{F_y}}\right) \tag{11.5.3}$$

If stress calculations are to be made using gross properties, then nominal ultimate stress is $M_u/S = F_{cr}$. Since $F_y = M_y/S$, the $M_u/M_y = F_{ult}/F_y$; thus

$$F_{ult} = F_y\left[1.0 - 0.0005\frac{A_w}{A_f}\left(\frac{h}{t} - \frac{975}{\sqrt{F_y}}\right)\right] \tag{11.5.4}$$

Equation 11.5.4 assumes no influence of stability relative to the compression flange. If, however, lateral-torsional buckling of the girder or torsional buckling of the compression flange gives $F_{cr} < F_y$, then F_{cr} should replace F_y in Eq. 11.5.4. In general, then,

$$F_{ult} = F_{cr}\left[1.0 - 0.0005\frac{A_w}{A_f}\left(\frac{h}{t} - \frac{975}{\sqrt{F_{cr}}}\right)\right] \tag{11.5.5}$$

In summary, the reader should keep in mind that if $F_{cr} \geq F_y$ and h/t

exceeds $\dfrac{975}{\sqrt{F_y}}$, the extreme fiber stress when $M = M_u$ is actually F_y. The section properties used are, however, those of a reduced section, as in Fig. 11.5.2. For the case of stable flanges

$$M_u = F_y S_{\text{reduced}} = F_{\text{ult}} S_{\text{full}} \tag{11.5.6}$$

Using a reduced stress on gross section provides the same capacity as if the real conditions are used. The foregoing treatment for the post-buckled web is the same concept used for stiffened plate elements in Chapter 6, Part II.

11.6. ULTIMATE SHEAR STRENGTH: POST-BUCKLED WEB

As discussed in Sec. 11.2, the buckling of plates under pure shear, both elastically and inelastically, gives critical shear stress as illustrated by line *ABCD* in Fig. 11.6.1. A plate stiffened by flanges and stiffeners has considerable post-buckling strength. For efficient use of web plate material in plate girders thin webs are necessary, which results in buckling at low shear stresses.

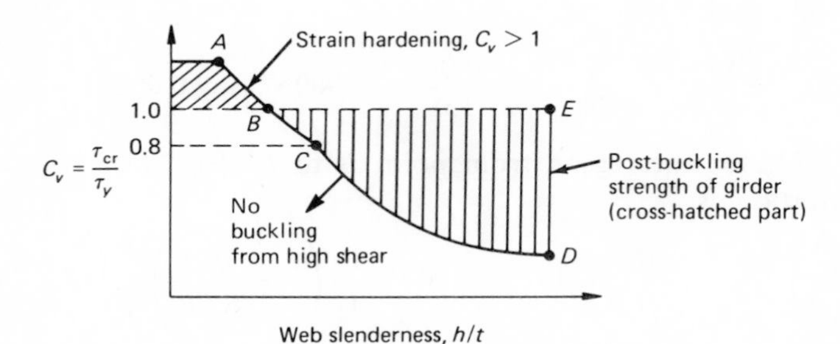

Fig. 11.6.1. Shear capacity available, considering post-buckling strength.

According to Basler,[3] the ability of a plate girder to perform in a manner similar to a truss was recognized as early as 1898. The tension forces are carried by membrane action of the web while the compression forces are carried by stiffeners. The recent work of Basler[3] has led to a theory which agreed with tests and provides criteria to insure achievement of truss action. The shear strength may then be raised from that based on buckling (*ABCD* on Fig. 11.6.1) to near or at a condition corresponding to shear yield in classical beam theory (*ABE* of Fig. 11.6.1).

If one prefers to compute stresses according to classical beam theory (using either VQ/It or V/A_g), strength under post buckling conditions will require using a fictitious stress equal to V_{ult}/A_g. The alternative is to consider the classical theory only until buckling has occurred, then use the tension-field membrane stresses along with compression forces in stiffeners to work with the post-buckling condition. The former is simpler when design procedures are involved.

The ultimate shear strength, in general, may be expressed as

$$V_{u\,\text{ultimate}} = V_{\text{cr buckling}} + V_{tf\,\text{tension-field action}} \tag{11.6.1}$$

As discussed in Sec. 11.2, the buckling strength, both elastic and inelastic, may be expressed as

$$V_{\text{cr}} = \tau_{\text{cr}} A_w = \tau_{\text{cr}} ht = \tau_y ht \left(\frac{\tau_{\text{cr}}}{\tau_y}\right) \tag{11.6.2}$$

Since $C_v = \tau_{\text{cr}}/\tau_y$, Eq. 11.6.2 becomes

$$V_{\text{cr}} = \tau_y ht\ C_v \tag{11.6.3}$$

where C_v is given by Eq (11.2.9) or Eq. (11.2.11) for elastic or inelastic buckling, respectively.

The next step is to develop the expression for the shear force V_{tf} which arises from post-buckling strength (i.e., tension-field action in the web). This trusslike behavior is illustrated by Fig. 11.6.2. The tension-field action, or membrane action, in the web develops a band of tensile

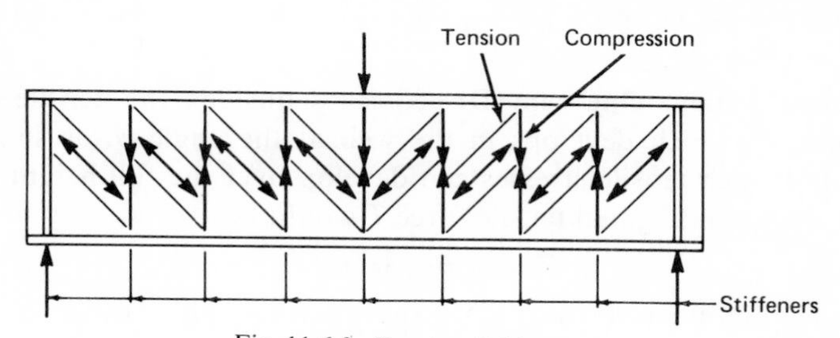

Fig. 11.6.2. Tension-field action.

forces which occur after the web has buckled under principal compression (ordinary beam theory). Equilibrium is maintained by the transfer of stress to the vertical stiffeners. As the load increases, the angle of the tension field changes to accommodate the greatest carrying capacity. Figure 11.6.3. shows a 50 × 50-in. panel with a ¼-in. web which has buckled under diagonal compression when subjected to pure shear. It also illustrates the anchorage requirement wherein the longitudinal component of the tension field must be transmitted to the flange in the adjacent panel (as shown by the vertical breaks in the whitewash at the flange in the corner of the adjacent panel where the tension field intersects the stiffener and flange).

Fig. 11.6.3. Tension-field in test plate girder. (From Ref. 3, Courtesy of Lehigh University.)

Tension-Field Action: Optimum Direction. Consider the tensile membrane stress, σ_t, which develops in the web at the angle ϕ, as shown in Fig. 11.6.4. If such tensile stresses can develop over the full height of the web, then the total diagonal tensile force T would be

$$T = \sigma_t th \cos \phi \tag{11.6.4}$$

the vertical component of which is the shear force V, given by

$$V = T \sin \phi = \sigma_t \, th \cos \phi \sin \phi \tag{11.6.5}$$

If such diagonal tensile stresses could develop *along* the flanges, vertical stiffness of the flanges would be required. Since the flanges have

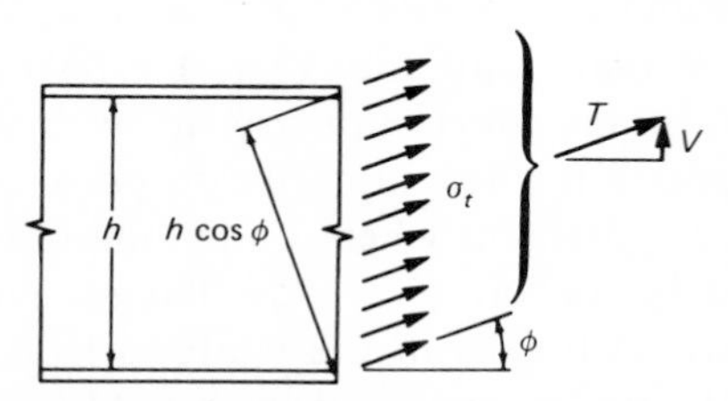

Fig. 11.6.4. Membrane stresses in tension-field action.

little vertical stiffness and are acting to their capacity in resisting flexure on the girder, the tension field actually can develop only over a band width such that the vertical component can be transferred at the vertical stiffeners. The stiffeners can be designed to carry the necessary compressive force. It will be assumed that the tension field (or partial tension field as some may prefer to call it) may develop over the band width s, shown in Fig. 11.6.5a.

The membrane tensile force tributory to one stiffener is $\sigma_t st$, and the partial shear force, ΔV_{tf}, developed by compression in the stiffener is

$$\Delta V_{tf} = \sigma_t st \sin \phi \tag{11.6.6}$$

and the angle ϕ may be expected to be the angle providing the maximum shear component from the partial tension field.

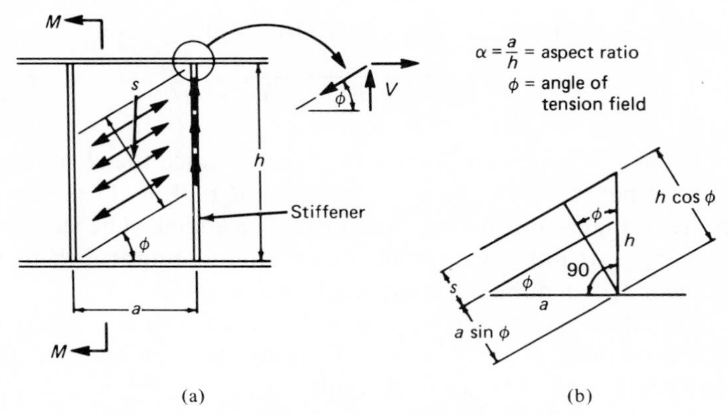

Fig. 11.6.5. Forces arising from tension field.

From the geometry shown in Fig. 11.6.5b,

$$s = h \cos \phi - a \sin \phi \tag{11.6.7}$$

where a = stiffener spacing. Substitution of Eq. 11.6.7 into Eq. 11.6.6 gives

$$\Delta V_{tf} = \sigma_t t(h \cos \phi - a \sin \phi) \sin \phi$$

$$= \sigma_t t\left(\frac{h}{2} \sin 2\phi - a \sin^2 \phi\right) \tag{11.6.8}$$

For maximum ΔV_{tf}, it is required that $d(\Delta V_{tf})/d\phi = 0$. Thus,

$$\frac{d(\Delta V_{tf})}{d\phi} = \sigma_t t\left(\frac{h}{2} (2) \cos 2\phi - 2a \sin \phi \cos \phi\right) = 0 \tag{11.6.9}$$

$$0 = h \cos 2\phi - a \sin 2\phi \tag{11.6.10}$$

or

$$\tan 2\phi = \frac{h}{a} = \frac{1}{a/h} \tag{11.6.11}$$

In order to calculate the maximum shear force available from the tension field component, Eq. 11.6.8 indicates that $\sin 2\phi$ and $\sin^2 \phi$ are needed. From the trigonometry of Eq. 11.6.11,

$$\sin 2\phi = \frac{1}{\sqrt{1 + (a/h)^2}} \tag{11.6.12}$$

also

$$\sin^2\phi = \frac{1 - \cos 2\phi}{2} = \frac{1}{2}\left[1 - \frac{a/h}{\sqrt{1 + (a/h)^2}}\right] \tag{11.6.13}$$

Substituting Eqs. 11.6.12 and 11.6.13 into Eq. 11.6.8 gives

$$\Delta V_{tf} = \sigma_t \frac{ht}{2}\left[\sqrt{1 + (a/h)^2} - a/h\right] \tag{11.6.14}$$

It is not practical to use Eq. 11.6.14 directly, since the shear contribution from the part of the section (such as M–M of Fig. 11.6.5) that cuts through the triangles outside the band s must be added. The state of stress in these triangles is unknown, requiring an alternate approach to finding the total shear V_{tf} when the optimum angle ϕ is reached.

An alternate way, as used by Basler,[3] is to cut a free body as in Fig. 11.6.6. The section is taken vertically midway between two adjacent stiffeners and horizontally at mid-depth. The mid-depth cut provides access to the tension field where the state of stress is known, and the shear resultant on each vertical face equals $V_{tf}/2$ from symmetry.

Shear Force Resulting Under Tension-Field Action. Using the free body of Fig. 11.6.6, horizontal force equilibrium requires

$$\begin{aligned} \Delta F_f &= (\sigma_t\, ta \sin\phi)\cos\phi \\ &= \sigma_t \frac{ta}{2}\sin 2\phi \end{aligned} \tag{11.6.15}$$

Rotational equilibrium, taken about point O, requires

$$\Delta F_f \frac{h}{2} - \frac{V_{tf}\, a}{2} = 0 \tag{11.6.16}$$

Solving Eq. 11.6.16 for ΔF_f and substituting into Eq. 11.6.15 give

$$\frac{V_{tf}\, a}{h} = \sigma_t \frac{ta}{2}\sin 2\phi \tag{11.6.17}$$

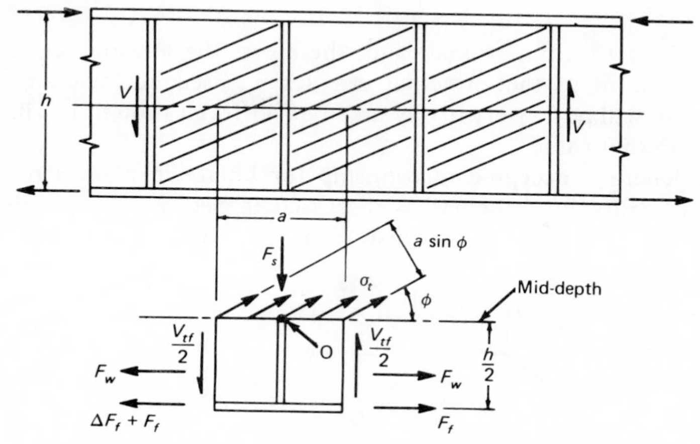

Fig. 11.6.6. Force in stiffener resulting from tension-field action.

and using Eq. 11.6.12 for sin 2ϕ results in

$$V_{tf\,\text{tension field action}} = \sigma_t \frac{ht}{2}\left[\frac{1}{\sqrt{1 + (a/h)^2}}\right] \qquad (11.6.18)$$

Failure Condition. The actual state of stress in the web involves both τ and σ_t, and the failure condition involving them needs to be established. In effect, failure of an element subjected to pure shear in combination with an inclined tension must be considered, as shown in Fig. 11.6.7. Two basic assumptions are involved: first, τ_{cr} remains at constant value from

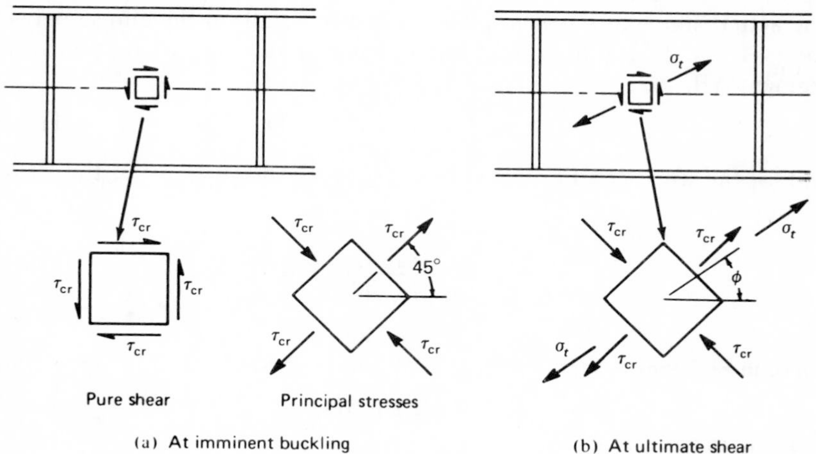

Fig. 11.6.7. State of stress.

buckling load to ultimate load and, therefore, the tension-field stress σ_t acts in addition to the principal stresses τ_{cr}; second, the angle ϕ in Fig. 11.6.7b will be conservatively taken as 45° even though it will always be less than that value.

The generally accepted relationship for failure in plane stress is the "energy of distortion" theory, as discussed in Sec. 2.7 and as shown in Fig. 11.6.8. The "energy of distortion" or Hencky-von Mises yield

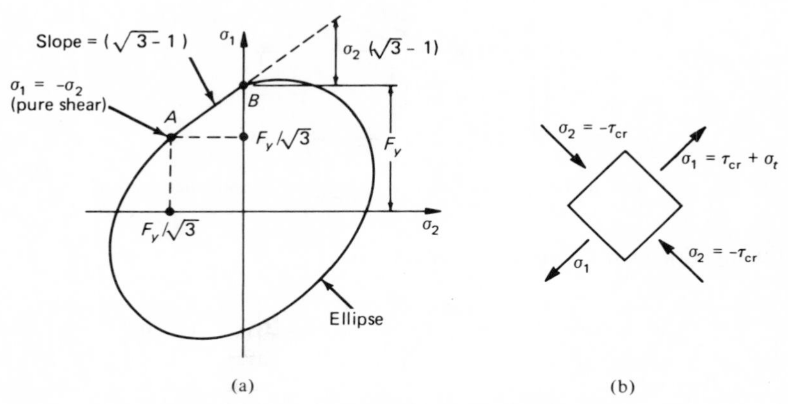

Fig. 11.6.8. Energy of distortion failure criterion.

criterion may be written

$$\sigma_1^2 + \sigma_2^2 - \sigma_1\sigma_2 = F_y^2 \tag{11.6.19}$$

which is the ellipse of Fig. 11.6.8. σ_1 and σ_2 are principal stresses. Point A represents the case of pure shear and Point B represents pure tension. The actual states of stress in plate girder webs fall on the ellipse between points A and B, and a straight line is a reasonable approximation of the segment AB:

$$\sigma_1 = F_y + \sigma_2(\sqrt{3} - 1) \tag{11.6.20}$$

and for the stress condition that $\sigma_1 = \tau_{cr} + \sigma_t$ and $\sigma_2 = -\tau_{cr}$, Eq. 11.6.20 becomes

$$\frac{\sigma_t}{F_y} = 1 - \frac{\tau_{cr}}{F_y/\sqrt{3}} = 1 - C_v \tag{11.6.21}$$

Force in Stiffener. Using Fig. 11.6.6b, vertical force equilibrium requires

$$F_s = (\sigma_t ta \sin\phi)\sin\phi \tag{11.6.22}$$

and substitution of Eq. 11.6.13 for $\sin^2\phi$ gives

$$F_s = \sigma_t \left(\frac{at}{2}\right) \left[1 - \frac{a/h}{\sqrt{1 + (a/h)^2}}\right] \tag{11.6.23}$$

Substituting Eq. 11.6.21 into Eq. 11.6.23 gives

$$F_s = \frac{F_y(1 - C_v)at}{2} \left[1 - \frac{a/h}{\sqrt{1 + (a/h)^2}}\right] \tag{11.6.24}$$

which is the force achieved under ultimate shear to accommodate tension-field action.

Ultimate Shear Capacity: Combined Buckling and Post-Buckling Strength. Since thin-web plate girders exhibit some strength in shear before diagonal buckling occurs (V_{cr} from Sec. 11.2) and further strength in the post-buckling range (V_{tf} from Eq. 11.6.18), their actual capacity is the sum of both components. Substituting Eqs. 11.6.3 and 11.6.18 into 11.6.1 gives

$$V_u = ht \left[\tau_y C_v + \frac{\sigma_t}{2\sqrt{1 + (a/h)^2}}\right] \tag{11.6.25}$$

Substituting Eq. 11.6.21 and using $\tau_y = F_y/\sqrt{3}$ give

$$V_u = F_y ht \left[\frac{C_v}{\sqrt{3}} + \frac{1 - C_v}{2\sqrt{1 + (a/h)^2}}\right] \tag{11.6.26}$$

11.7. COMBINED SHEAR AND TENSION STRESS

In the vast majority of cases the ultimate capacity in bending is not influenced by shear, nor is the ultimate shear capacity influenced by moment. Particularly, in very slender webs where bend-buckling may occur, the bending stress is redistributed as discussed in Sec. 11.5. so that the flanges carry an increased share. The shear capacity of the web, however, is not reduced as a result of bend-buckling because most of the shear capacity is from tension-field action with only a small contribution from the web that is adjacent to the flange. In stockier webs no bend-buckling may occur, but high web shear in combination with bending may cause yielding of the web adjacent to the flange; again resulting in a transfer of part of the web's share of the bending moment to the flange.

Since stability is not involved, a plastic analysis may provide rational treatment wherein the web adjacent to the flange which yields under bending moment is not available to carry shear, and the central region of the web which yields in shear is not available to resist bending. Referring to Fig. 11.7.1, the ultimate shear capacity may be expressed as

$$V = \tau_y y_0 t \tag{11.7.1}$$

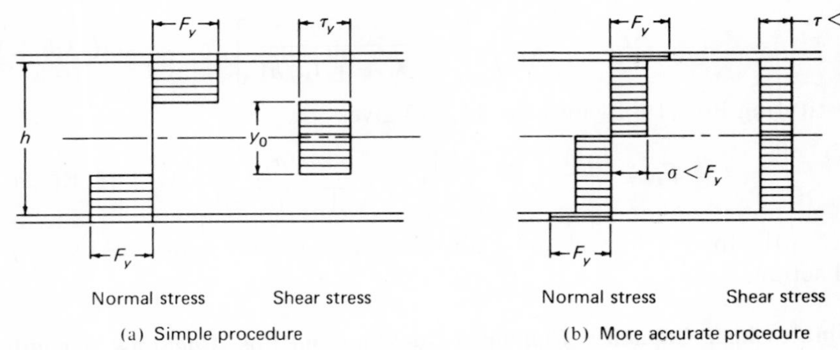

Fig. 11.7.1. Ultimate conditions under combined bending and shear.

which if no bending moment were present, $y_0 = h$, giving the maximum shear capacity

$$V_y = \tau_y th \tag{11.7.2}$$

Eliminating τ_y from Eqs. 11.7.1 and 11.7.2 gives

$$y_0 = \frac{V}{V_y} h \tag{11.7.3}$$

The moment capacity from Fig. 11.7.1 is

$$M_u = A_f F_y h + F_y t\left(\frac{h}{2}\right)\left(\frac{h}{2}\right) - F_y t\left(\frac{y_0}{2}\right)\left(\frac{y_0}{2}\right) \tag{11.7.4}$$

$$= F_y\left[A_f h + A_w \frac{h}{4} - A_w \frac{h}{4}\left(\frac{V}{V_y}\right)^2\right]$$

$$= F_y A_f h \left\{1 + \frac{1}{4}\left(\frac{A_w}{A_f}\right)\left[1 - \left(\frac{V}{V_y}\right)^2\right]\right\} \tag{11.7.5}$$

The ordinary beam-theory moment capacity at first yield with the web fully participating is (see Eq. g, p. 554)

$$M_y = F_y h A_f\left(1 + \frac{1}{6}\frac{A_w}{A_f}\right) \tag{11.7.6}$$

As the percentage of the maximum shear capacity utilized increases, the available ultimate moment capacity decreases.

In the absence of instability, but in the presence of high shear, the ultimate bending-moment capacity may be expressed as

$$M_u = M_y\left[\frac{1 + \frac{1}{4}\rho\,[1 - (V/V_y)^2]}{1 + \frac{1}{6}\,\rho}\right] \tag{11.7.7}$$

When $M_u = M_y$, $V/V_y = 0.577$, or approximately 0.6. When more than 60 percent of the maximum shear capacity is used, the available ultimate-moment capacity is reduced. Table 11.7.1 gives some values for M_u/M_y for various values of ρ in the practical range, with graphical illustration in Fig. 11.7.2.

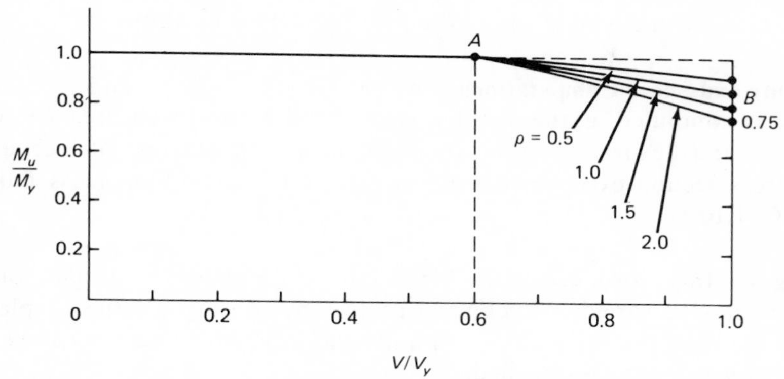

Fig. 11.7.2. Moment-capacity reduction under combined bending and shear.

TABLE 11.7.1.
Values of M_u/M_y in accordance with Eq. 11.7.7 for $V/V_y \geq 0.6$

$\rho = \dfrac{A_w}{A_f}$	For $\dfrac{V}{V_y} = 0.8$	For $\dfrac{V}{V_y} = 1.0$
0	1.0	1.0
0.5	0.964	0.923
1.0	0.935	0.856
1.5	0.908	0.800
2.0	0.885	0.750

Certainly it would be more accurate to consider both normal stress σ and shear stress τ acting on the web over the entire depth as in Fig. 11.7.1b. Using the Hencky-von Mises failure criterion, Eq. 11.6.19 with $\sigma_1, \sigma_2 = \dfrac{\sigma}{2} \pm \sqrt{\left(\dfrac{\sigma}{2}\right)^2 + \tau^2}$ (i.e., principal stresses), the failure criterion becomes

$$\sigma^2 + 3\tau^2 = F_y^2 \tag{11.7.8}$$

which would provide a more complex relationship for M_u/M_y. Considering that the thin plate girder web typically is subject to a combination of beam action and tension-field action at ultimate condition, the conservative simple approach of Eq. 11.7.7 seems justified.

11.8. AISC PROVISIONS FOR WEB AND FLANGE SELECTION—1969.

Specification requirements and limitations are summarized, including a brief statement of the concept involved. The reader is reminded that most provisions relate to ultimate strength behavior, the theory of which has been developed in Secs. 11.2 through 11.7. The present approach was introduced with the adoption of the 1961 AISC Specifications.

Flexure and Shear Computations. As discussed in Secs. 7.5 and 7.7, shear stress is computed as the average stress on the gross web area (AISC–1.5.1.2), and flexural stress is computed using the moment of inertia of the gross section except where holes exceed 15 percent of gross flange (AISC–1.10.1).

Flanges. In accordance with AISC–1.10.3, "Flanges of welded plate girders may be varied in thickness or width by splicing a series of plates or by the use of cover plates." In addition, local buckling provisions of AISC–1.9.1.2. must be satisfied.

Provisions for bolted or riveted girders are not discussed in this section, since their use is essentially obsolete.

Web Slenderness Ratio Relating to Flexure. A graphical description of behavior was provided by Fig. 11.5.1, showing ultimate moment capacity vs. web slenderness ratio.

Recall from Sec. 7.3 that maximum beam strength is obtained when the full plastic moment develops, and this occurs when the stability conditions of AISC–1.5.1.4.1 are satisfied. In terms of the web-slenderness ratio h/t, the fully plastic condition is the region at or to the left of point A, Fig. 11.5.1. Economical plate girders have much higher h/t values, in the region from B to D on Fig. 11.5.1.

As h/t increases, buckling of the web due to bending stress may result. Equation 11.2.13 gives a rational upper limit for h/t to prevent such instability,

$$\frac{h}{t} \leq \frac{975}{\sqrt{F_{cr}}} \qquad [11.2.13]$$

Using the basic AISC factor of safety, the maximum allowable stress in flexure, F_b, would be $0.6F_{cr}$. Substituting this into Eq. 11.2.13 gives

$$\frac{h}{t} \leq \frac{975\sqrt{0.6}}{\sqrt{F_b}} \leq \frac{755}{\sqrt{F_b}}$$

which AISC–1.10.6 prescribes as

$$\frac{h}{t} \leq \frac{760}{\sqrt{F_b}} \qquad (11.8.1)$$

preventing bend-buckling of the web.

When higher h/t values are used the web will carry less than its share of the moment computed by ordinary flexure theory. To permit the flange to carry part of the moment ordinarily carried by the web, the allowable flange stress must be reduced. This reduction is assumed to be linear, as shown from B to D in Fig. 11.5.1, and may be approximately expressed by Eq. 11.5.5,

$$F_{\text{ult}} = F_{\text{cr}}\left[1.0 - 0.0005\,\frac{A_w}{A_f}\left(\frac{h}{t} - \frac{975}{\sqrt{F_{\text{cr}}}}\right)\right] \qquad [11.5.5]$$

Again, reducing the expression into the working stress range by letting $F_b' = F_{\text{ult}}/1.67$ and $F_b = F_{\text{cr}}/1.67$, which gives

$$F_b' = F_b\left[1.0 - 0.0005\,\frac{A_w}{A_f}\left(\frac{h}{t} - \frac{760}{\sqrt{F_b}}\right)\right] \qquad (11.8.2)$$

which is AISC–1.10.6, Formula 1.10–5, F_b is the allowable stress (in ksi) after considering lateral-torsional buckling. Equation 11.8.2 is shown graphically in Fig. 11.8.1.

Web Slenderness Ratio Relating to Vertical Buckling of Flange. For very large values of h/t the web may be insufficiently stiff to prevent vertical buckling of the flange. Equation 11.3.9 provides a reasonable approximation for the maximum h/t to prevent such instability in the absence of intermediate transverse stiffeners,

$$\frac{h}{t} = \frac{13{,}800}{\sqrt{F_y(F_y + 16.5)}} \qquad [11.3.9]$$

which when rounded becomes

$$\frac{h}{t} = \frac{14{,}000}{\sqrt{F_y(F_y + 16.5)}} \qquad (11.8.3)$$

which is the AISC–1.10.2 general limitation.

In the presence of transverse stiffeners, recent studies[12] indicate higher values could be permitted. Thus 1969 AISC in accordance with test results and recommendations of the ASCE–AASHO Joint Committee, Subcommittee 1 on Hybrid Girder Design,[12] provides a higher value for maximum h/t,

$$\frac{h}{t} \le \frac{2000}{\sqrt{F_y}} \qquad (11.8.4)$$

when transverse stiffeners are provided such that stiffener spacing/web depth, $a/h \le 1.5$. This is developed using the recommended $h/t = 200$ for $F_y = 100$ ksi* and using the typical stability variation with yield point for the equation.

*Ref. 12, p. 1412.

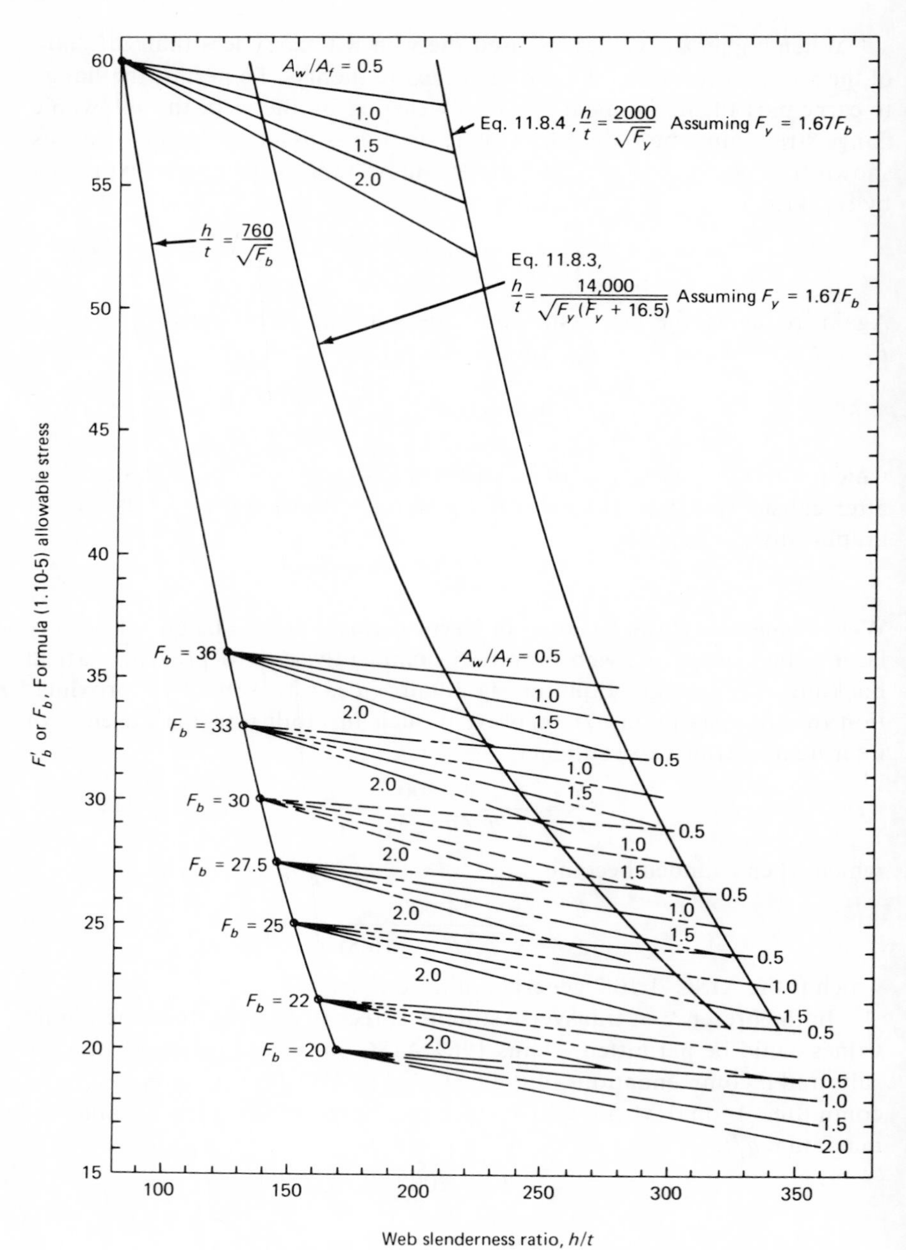

Fig. 11.8.1. Web slenderness effects and limitations.

Values for Eq. 11.8.3 and 11.8.4 are given in Table 11.8.1.

TABLE 11.8.1
Maximum h/t Limitations–1969 AISC

F_y (ksi)	h/t for Eq. 11.8.3 For $a/h > 1.5$	h/t for Eq. 11.8.4 For $a/h \leq 1.5$
36	322	333
42	282	309
45	266	298
50	243	283
55	223	270
60	207	258
65	192	248
100	130	200

Shear and Combined Shear and Tension as Affecting Web Selection. In addition to the stability limitations on h/t, the web area must be sufficient so that shear stress is not excessive. Maximum allowable shear stress is $0.4F_y$, and assuming stiffeners are to be used, the allowable value based on stiffener placement, as discussed in the next section, will usually be no greater than about 0.3 to $0.35F_y$.

Furthermore, locations of combined high moment and high shear may limit the shear stress to between $0.25F_y$ to $0.3F_y$.

As discussed in Sec. 11.7, Eq. 11.7.7 and Fig. 11.7.2 show the interaction between shear and bending moment. If one uses a conservative value of $\rho = A_w/A_f = 2.0$ and considers the strength reduction from points A to B of Fig. 11.7.2 as a straight line, the slope of AB would be

$$\text{Slope of } AB \text{ (Fig. 11.7.2)} = \frac{-0.25}{0.40} = -\frac{5}{8}$$

The reduction equation then becomes

$$\frac{M_u}{M_y} = 1 - \frac{5}{8}\left(\frac{V}{V_y} - 0.6\right) \leq 1 \tag{11.8.5}$$

$$M_u = M_y\left(\frac{11}{8} - \frac{5}{8}\frac{V}{V_y}\right) \leq M_y \tag{11.8.6}$$

Dividing by a 1.67 factor of safety and converting moments and shears to unit stresses gives

$$f_b = 0.6F_y\left(\frac{11}{8} - \frac{5}{8}\frac{f_v}{F_v}\right) \leq 0.6F_y \tag{11.8.7}$$

which gives

$$f_b \leq \left(0.825 - 0.375\frac{f_v}{F_v}\right)F_y \leq 0.6F_y \tag{11.8.8}$$

which is AISC–1.10.7, Formula 1.10–7. Note that f_b and f_v are the maximum flexural and shear stresses in the *web*. The adjacent relatively stiff flange prevents stability from influencing the strength of the web under combined stress.

11.9. AISC–1969; INTERMEDIATE TRANSVERSE STIFFENERS

Prior to the 1961 AISC specification, intermediate stiffeners were designed to prevent all web buckling by limiting the shearing stresses, and post-buckling strength was not considered. Since 1961, ultimate strength considering both buckling and post-buckling conditions forms the basis for design.

Placement Criteria Including Tension-Field Action. When the web plate is stiff enough so that diagonal buckling from shear— either elastic or inelastic—cannot occur, intermediate transverse stiffeners are not required. The critical buckling stresses, τ_{cr}, for elastic and inelastic conditions are given by Eqs. 11.2.9 and 11.2.11, respectively, and the relationship is shown in Fig. 11.2.4.

The allowable shear stress (using a factor of safety = 1.67) if stiffeners are to be avoided is then

$$F_v = \frac{\tau_{cr}}{1.67} \tag{11.9.1}$$

or in terms of $C_v = \tau_{cr}/\tau_y = \tau_{cr}\sqrt{3}/F_y$, Eq. 11.9.1 becomes

$$F_v = \frac{F_y C_v}{\sqrt{3}\,(1.67)} = \frac{F_y C_v}{2.89} \tag{11.9.2}$$

which is AISC–1.10.5, Formula 1.10–1, with an upper limit of $0.4F_y$ (see Sec. 7.5). In addition to situations where stiffeners are omitted, it is also to be applied when strain hardening governs capacity (i.e., $C_v > 1.0$) when stiffeners are used.

For practical purposes, the logic of Eq. 11.9.2 has been that traditionally used for designing webs for shear until the 1961 AISC specification. Equation 11.9.2 logically applies both *with* and *without* stiffeners if the objective is to prevent buckling, as it was in earlier design practice.

Under 1969–AISC, buckling prior to achieving ultimate shear capacity is recognized as shown in Fig. 11.6.1. The post-buckling behavior is similar to truss action as shown in Figs. 11.6.2 and 11.6.3, and is referred to as "tension-field" action. The ultimate shear is the sum of the buckling strength plus the "tension-field" strength, as given by Eq. 11.6.26,

$$V_u = F_y\, ht \left[\frac{C_v}{\sqrt{3}} + \frac{1 - C_v}{2\sqrt{1 + a/h)^2}}\right] \qquad [11.6.26]$$

Conversion into unit stress on gross web, V_u/ht, and dividing by the factor of safety 1.67, gives for the allowable stress

$$F_v = \frac{F_y}{2.89}\left[C_v + \frac{1 - C_v}{1.15\sqrt{1 + (a/h)^2}}\right] \tag{11.9.3}$$

which is AISC–1.10.5, Formula 1.10–2. This equation is to be applied when "tension-field" action can be expected; i.e., when stiffeners are used and buckling occurs before shear yielding occurs ($C_v < 1.0$).

While no theoretical upper limits exist for h/t except those of Eqs. 11.8.3 and 11.8.4, there are practical considerations which require a limit. It is considered[8] that fabrication, handling, and erection are facilitated when the smaller panel dimension, either a or h, does not exceed $260t$. If stiffeners are omitted and a/t becomes large, h/t, the smaller dimension ratio, is limited to 260. When stiffeners are used, the area of the stiffened panel is prevented from being greater than if both a/t and h/t equal 260; thus the arbitrary restriction may be stated

$$\frac{a}{h} \leq \left(\frac{260}{h/t}\right)^2 \leq 3.0 \tag{11.9.4}$$

with the arbitrary limit on a/h equal to 3.

The AISC allowable stresses and limitations, using Eqs. 11.9.2, 11.9.3, and 11.9.4, are shown in Fig. 11.9.1 for $F_y = 50$ ksi.

End Panels. Figure 11.6.5 shows that at the junction of stiffener and flange, equilibrium requires an axial tension to develop in the flange of the adjacent panel. If no such flange is available, as in an end panel, the tension-field cannot adequately develop. AISC–1.10.5.3, therefore, considers that only buckling strength is available. Stiffeners are provided with assurance that buckling will be prevented. Using Eq. 11.9.2,

$$F_v = \frac{F_y C_v}{2.89} \qquad [11.9.2]$$

and conservatively using elastic conditions (i.e., $C_v < 0.8$) and taking $a/h = 1$, gives

$$k = 4.0 + \frac{5.34}{(1.0)^2} = 9.34$$

$$C_v = \frac{45{,}000\,(9.34)}{F_y(h/t)^2}$$

$$F_v = \frac{45{,}000\,(9.34)}{2.89\,(h/t)^2} = \frac{145{,}000}{(h/t)^2}\ \text{ksi}$$

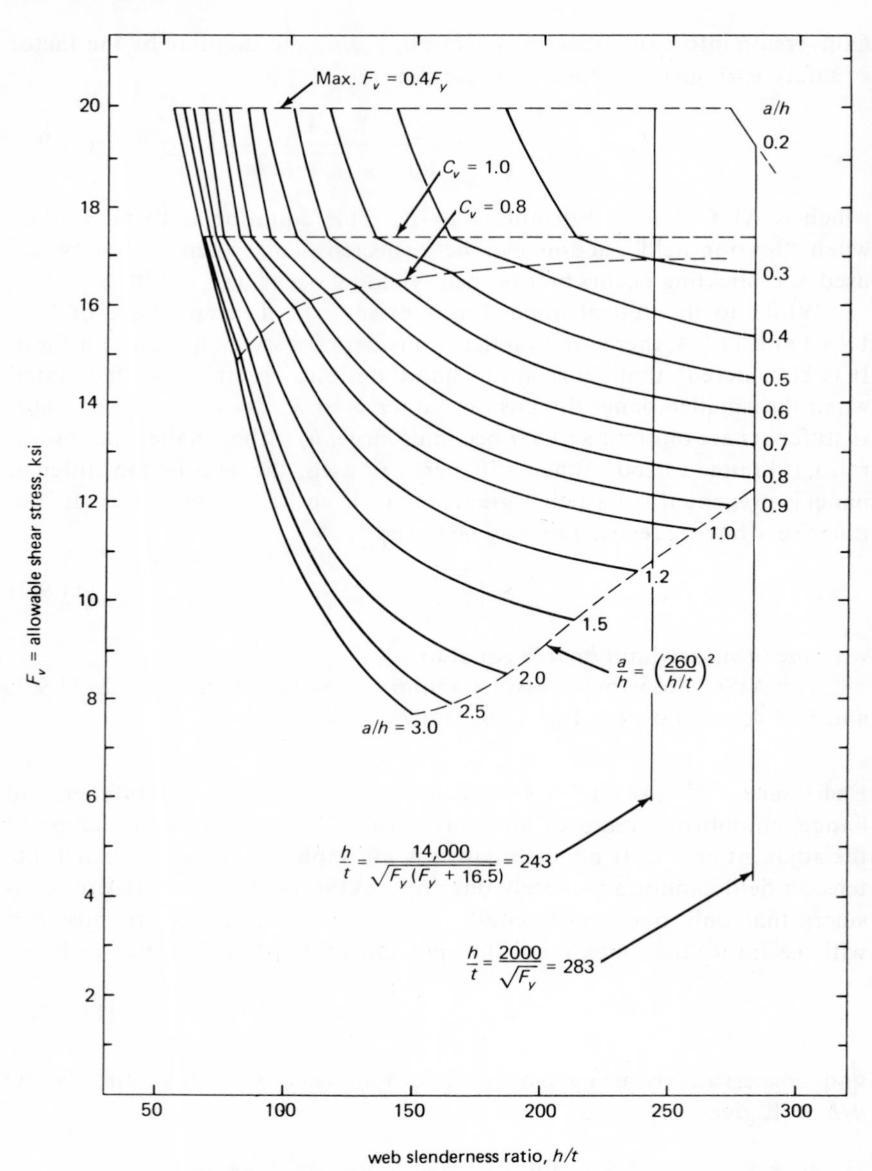

Fig. 11.9.1. Allowable shear stress on plate girders with intermediate transverse stiffeners for $F_y = 50$ ksi.

If f_v, the actual stress V/ht in the panel, is substituted for F_v, one obtains

$$\frac{h}{t} = \frac{380}{\sqrt{f_v}}$$

AISC–1.10.5.3 uses a slightly more restrictive expression and recognizes that the slenderness ratio relating to shear strength always involves the smaller of h or a. Thus

$$\text{Smaller of } a \text{ or } h \leq \frac{348t}{\sqrt{f_v}} \tag{11.9.6}$$

which applies for end panels; i.e., panels having *no adjacent panel*. The coefficient 348 is obtained by dividing the quantity 11,000 (the traditional value when f_v was in psi) by $\sqrt{1000}$; Thus

$$\frac{11{,}000}{31.6} = 348$$

Stiffness Requirement. Intermediate stiffeners must be sufficiently rigid to keep the web *at the stiffener* from deflecting when buckling of the web has occured. The stiffener must then have a rigidity EI_s which is related to the web plate rigidity, $Et^3a/[12(1 - \mu^2)]$.

AISC–1.10.5.4 provides for intermediate stiffeners that

$$I_s \geq \left(\frac{h}{50}\right)^4 \tag{11.9.7}$$

where I_s is the moment of inertia of the stiffener with respect to the center of the web.

Equation 11.9.7 is obviously an oversimplification, since it does not depend upon stiffener spacing or web thickness. Various theoretical relationships have been developed for the ratio of the stiffener rigidity to one panel of web-plate rigidity, which may be expressed as

$$\gamma_0 = \frac{EI_s}{Da} = \frac{EI_s[12(1 - \mu^2)]}{Et^3a} \tag{11.9.8}$$

where I_s = optimum stiffener moment of inertia
$D = Et^3/[12(1 - \mu^2)]$ = flexural rigidity per unit length of web plate

The following three expressions for γ_0 are representative of the results of investigations into the stiffness requirement for transverse stiffeners:

$$\gamma_0 = \frac{14}{(a/h)^3} \tag{11.9.9}$$

as proposed by Moore and reported in Ref. 7 (p. 136),

$$\gamma_0 = 4\left[7\left(\frac{h}{a}\right)^2 - 5\right] \tag{11.9.10}$$

as developed by Bleich (Ref. 13, p. 417), and

$$\gamma_0 = \frac{3.75}{(a/h)^4} \tag{11.9.11}$$

as suggested by McGuire (Ref. 14, p. 742) to give a stiffener flexural rigidity about twice the theoretical value, in accord with the recommendation of Timoshenko.[6]

According to Eq. 11.9.8 with $\mu = 0.3$,

$$I_s\text{reqd} = \frac{\gamma_0 t^3 a}{12(1 - \mu^2)} = \frac{\gamma_0 t^3 a}{10.92} \tag{11.9.12}$$

Equations 11.9.9 through 11.9.11 when substituted into Eq. 11.9.12 give

$$I_s = \frac{1.28h^4}{(a/t)^2\,(h/t)} \tag{11.9.13}$$

$$I_s = \frac{0.366h^4}{(h/t)^3}\,[7\,(h/a) - 5\,(a/h)] \tag{11.9.14}$$

$$I_s = \frac{0.343h^4}{(a/t)^3} \tag{11.9.15}$$

In order to compare with the AISC–1969 requirement, Eq. 11.9.7, let $h/t = 200$ and a/t be 170; consequently $a/h = 0.85$. Equations 11.9.13 through 11.9.15 become $I_s = \left(\frac{h}{46}\right)^4, \left(\frac{h}{48.4}\right)^4$, and $\left(\frac{h}{61.5}\right)^4$ respectively. The AISC provision, seems in the proper order of magnitude. The stiffness requirement will generally govern only when h/t or a/t does not exceed about 160–200.

Strength Requirement. Intermediate stiffeners carry their computed compression load only after buckling of the web has occurred. As the post-buckling truss like "tension-field" action increases, the stiffener force increases. The force in the stiffener under the ultimate capacity based on "tension-field" action was obtained as Eq. 11.6.24,

$$F_s = \frac{F_y(1 - C_v)\,at}{2}\left[1 - \frac{a/h}{\sqrt{1 + (a/h)^2}}\right] \tag*{[11.6.24]}$$

If this force is developed as the stiffener yields, the ultimate capacity of the girder has been achieved. Thus, the stiffener area required is

$$A_{st} = \frac{F_s}{F_{y,st}} = \frac{F_{y,w}(1 - C_v)\,at}{F_{y,st}\,2}\left[1 - \frac{a/h}{\sqrt{1 + (a/h)^2}}\right] \tag{11.9.16}$$

where $F_{y,w}$ = yield stress of web material

$F_{y,st}$ = yield stress of stiffener material

Equation 11.9.16 may be rewritten, letting $F_{y,w}/F_{y,st} = Y$ and multiplying and dividing by h, giving

$$A_{st} = \frac{1 - C_v}{2}\left[\frac{a}{h} - \frac{(a/h)^2}{\sqrt{1 + (a/h)^2}}\right]Yht \tag{11.9.17}$$

which assumes the compressive force from "tension-field" action is axially applied to the stiffener, i.e., the stiffeners are placed in pairs.

Sometimes stiffeners are alternated on each side of the web to gain better economy or they are placed all on one side to improve esthetics. Whenever the stiffeners are not placed in pairs a greater cross-sectional area must be provided to account for eccentric loading. Referring to Fig. 11.9.2a, the symmetrical pair of stiffeners reaches its plastic condition and

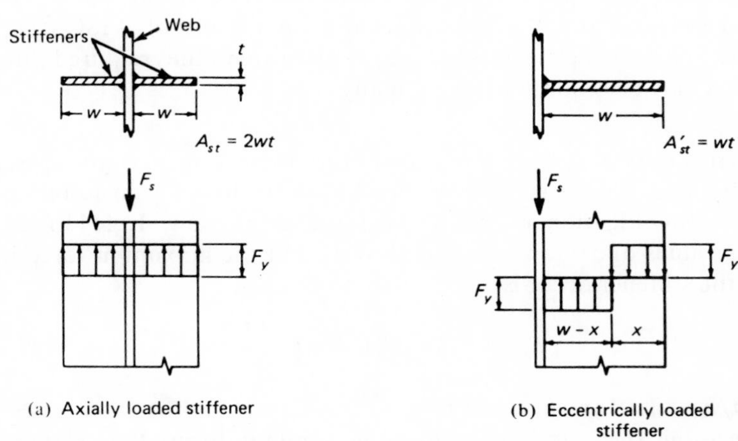

Fig. 11.9.2. Intermediate stiffeners at ultimate shear condition considering tension-field action.

the ultimate force F_s is

$$F_s = 2wtF_y = A_{st}F_y \qquad \text{(for concentric load)} \qquad (11.9.18)$$

On the other hand, an eccentrically loaded stiffener becomes plastic with a stress distribution as shown in Fig. 11.9.2b. For this case, force equilibrium requires

$$F_s = (w - x)tF_y - xtF_y \qquad (a)$$

and moment equilibrium requires

$$xtF_y\left(w - \frac{x}{2}\right) = F_y(w - x)t\left(\frac{w - x}{2}\right) \qquad (b)$$

Solving the quadratic gives $x = 0.293w$. Substitution for x in Eq. (a) gives

$$\begin{aligned} F_s &= [(w - 0.293w)t - 0.293wt]\,F_y \\ &= 0.414wtF_y = 0.414\,A'_{st}F_y \qquad \text{(for eccentric load)} \qquad (11.9.19) \end{aligned}$$

If single plate stiffeners are used on one side only, equating Eqs. 11.9.18 and 11.9.19 shows

$$A'_{st} = \frac{A_{st}}{0.414} = 2.42A_{st} \qquad (11.9.20)$$

AISC–1.10.5.4 uses a multiplier of 2.4 for single plate stiffeners. For a single angle whose center of gravity is closer to the web, the multiplier reduces to 1.8.

The stiffener force used in the foregoing development is that which occurs when the girder is loaded to its ultimate shear capacity. Because unrestricted yielding of the stiffener is assumed to occur prior to stiffener plate buckling, the yield stress is used to determine area required. In other words, Eq. 11.9.17 gives the area required if the panel is **fully stressed** in shear and the local buckling limitations of AISC–1.9.1 are satisfied. For panels stressed below allowable, the required area of stiffeners may be reduced proportionally.

Connection to Web. Determination of the connection strength presents a difficulty since the exact distribution of shear to be transferred is unknown. Investigation of the shear force expression, Eq. 11.6.24, for various values of a/h and h/t has shown[3] that the maximum force which can in the stiffener occur is

$$F_s(\text{max}) = 0.015h^2 \sqrt{\frac{F_y^3}{E}} \tag{11.9.21}$$

when $a/h = 1.18$ and $h/t = 1060/\sqrt{F_y}$.

Since the shear in the connection is nonuniform, Basler[3] has suggested that a safe procedure would be to consider the force F_s to be transferred over one-third of the girder depth. Thus, as an average for the full height, the shear per inch f_{vs} to be considered at ultimate load is

$$f_{vs} = \frac{3F_s}{h} \tag{11.9.22}$$

which upon substituting Eq. 11.9.21 and using $E = 29{,}000$ ksi with a 1.65 safety factor gives

$$f_{vs} = 0.045h \sqrt{\frac{F_y^3}{E}} \frac{1}{1.65} \tag{11.9.23}$$

$$= h \sqrt{\left(\frac{F_y}{340}\right)^3} \tag{11.9.24}$$

which is AISC Formula 1.10–4. The reader is reminded that F_y is the yield stress for the *web* material.

As with the area requirement when adjacent panels are not fully stressed, the design shear flow f_{vs} for the connection may also be reduced proportionally.

Connection to Flanges. Intermediate stiffeners are provided to assist the web; to stiffen and create nodal lines during buckling of the web and to accept compression forces transmitted directly from the web. At the compression flange, welding of the stiffener across the flange as shown in Fig. 11.9.3 provides stability to the stiffener and holds it perpendicular to the web; in addition, such welding provides restraint against torsional buckling of the compression flange.

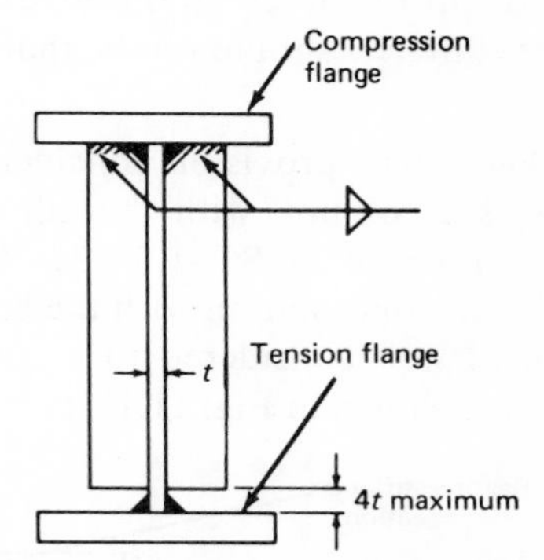

Fig. 11.9.3. Intermediate stiffener connection to flange.

On the tension flange, the effects of stress concentration increase the fatigue or brittle fracture possibilities, i.e., welding in no way helps the tension flange. Since the work of Basler[15] has shown that welding of stiffeners to the tension flange is unnecessary for proper functioning of stiffeners, AISC–1.10.5.4 permits stopping stiffeners "short of the tension flange a distance not to exceed 4 times the web thickness, provided bearing is not needed to transmit a concentrated load or reaction."

For situations where the stiffener serves as the attachment for lateral bracing, the welding to the compression flange should be designed to transmit 1 percent of the compressive force in the flange[15]. For important lateral bracing design in situations involving long unsupported lengths, the strength of lateral bracing connections should be designed using the principles of Sec. 9.10.

11.10. AISC–1969; BEARING STIFFENERS

Concentrated loads, such as at unframed end reactions, must be carried by stiffeners placed in pairs. Whenever the compressive stress in the vicinity of concentrated loads, as discussed in Sec. 7.6, exceeds the allowable stresses of AISC–1.10.10.1 bearing stiffeners must be provided. Furthermore, interior concentrated loads bearing on or through a flange plate for which web stability as governed by Eqs. 11.2.17, 11.2.19, and 11.2.20 (i.e., AISC–1.10.10.2) is not satisfied must be transmitted through bearing stiffeners.

Bearing stiffeners, unlike intermediate stiffeners, should be close fitting and connected to both tension and compression flanges; furthermore, they should extend approximately to the edge of the flange, whereas economical intermediate stiffeners should not be that wide.

Column Stability Criterion. This provision considers the overall stability of the bearing stiffeners as a column, with the allowable stress given by the ordinary column provisions of AISC–1.5.1.3. A portion of the web is logically assumed to act in combination with the bearing stiffener plates, or angles. The portion of web considered to act with the stiffener according to AISC–1.10.5.1 is shown in Fig. 11.10.1.

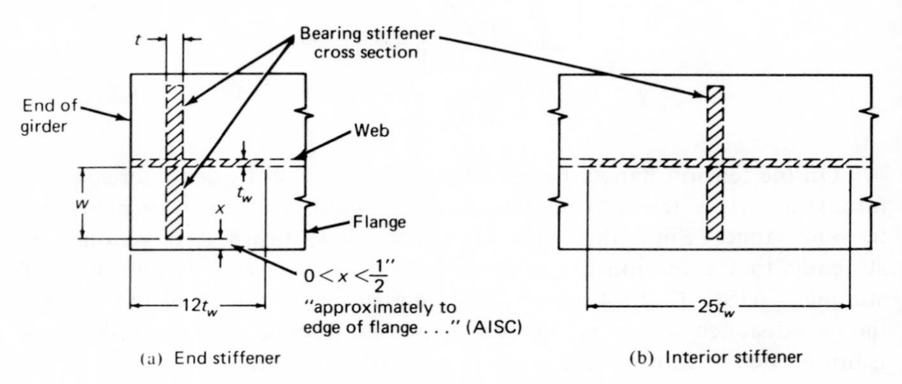

Fig. 11.10.1. Bearing stiffener effective cross sections.

The end restraint against column buckling provided by the flanges affords a reduced effective pin-ended length. The AISC specification states that the effective length may not be taken as less than three-quarters of the actual length; i.e.,

$$\left(\frac{KL}{r}\right)_{\text{eff}} \geq 0.75\,\frac{h}{r} \qquad (11.10.1)$$

where h = web plate depth

r = radius of gyration of the shaded portion shown in Fig. 11.10.1

Thus,

$$\text{Reqd } A_e = \frac{\text{concentrated load}}{F_a \text{ from AISC–1.5.1.3}} \qquad (11.10.2)$$

where F_a is computed using Eq. 11.10.1 and A_e is the shaded area of Fig. 11.10.1 which includes the stiffener plates plus the tributary web area.

Local-Buckling Criterion. Since the width w of the stiffener plates is governed by the flange width (see Fig. 11.10.1), the minimum thickness to prevent local buckling is obtained by applying AISC–1.9.1:

$$t_{\min} = \frac{w}{95.0/\sqrt{F_y}} \tag{11.10.3}$$

unless the special provisions of AISC–Appendix C are applied for cases where lesser thicknesses are desired to be used.

Compression Yield Criterion. Assuming a compression situation in which stability is assured, the stiffener plates are assumed capable of achieving yield stress under the concentrated load. Using a 1.67 factor of safety, the required area based on the yield condition AISC–1.5.1.3.4 is

$$\text{Reqd } A_g = \frac{\text{concentrated load}}{0.60F_y} \tag{11.10.4}$$

where A_g = area of only the stiffener plates.

Bearing Criterion. In order to bring bearing stiffener plates tight against the flanges, some part of the stiffener must be cut off so as to clear the flange-to-web fillet weld. The area of direct bearing is less than the gross area. The allowable stress in direct bearing where this lateral confinement is available is properly greater than $0.60F_y$. AISC–1.5.1.5.1, therefore, requires

$$\text{Reqd contact area} = \frac{\text{concentrated load}}{0.90F_y} \tag{11.10.5}$$

11.11. LONGITUDINAL WEB STIFFENERS

While longitudinal stiffeners are not as effective as transverse stiffeners, they are frequently desired on highway bridge girders for esthetic reasons. Summaries of theoretical studies of longitudinal stiffener effectiveness, as related to stiffener size and location, are to be found in Ref. 7 (pp. 137–141) and Ref. 13 (pp. 418–423).

Since the principal use of longitudinal stiffeners is in highway bridge design wherein the buckling strength is assumed to be the maximum strength available, most studies have directly related to stiffness requirements which improve buckling strength. Longitudinal stiffeners effect on, or relationship to, "tension-field" ultimate strength behavior has received little attention.

As discussed in Sec. 11.2, the buckling strength of the web plate in bending (Fig. 11.2.5) may be written

$$F_{cr} = \frac{\pi^2 E_t k}{12(1 - \mu^2)(h/t)^2}$$

If the plate is stiffened by a longitudinal stiffener, as shown in Fig. 11.11.1, the value of k will be significantly greater than for the unstiffened case. The stiffener used should be stiff enough so that when buckling occurs a nodal line will be formed along the line of the stiffener.

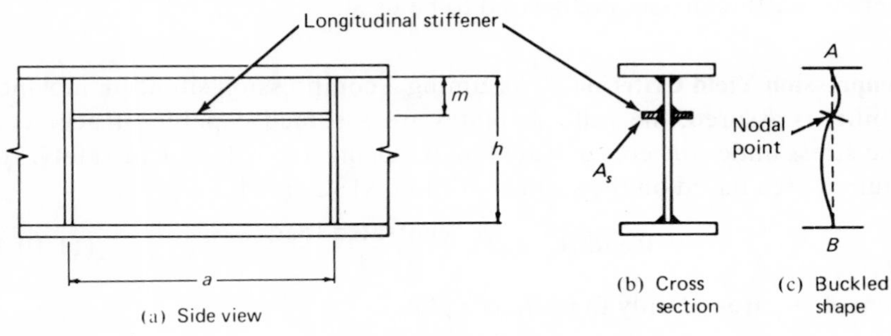

Fig. 11.11.1.

Under pure bending, the value of the buckling coefficient k has been found to be as high as 142.6 for the case where the flanges are assumed to provide full restraint to rotation at points A and B of Fig. 11.11.1c and $m = h/5$. For the case where the flanges provide no moment restraint at A and B (simply supported) the stiffener located at $m = h/5$ is also the optimum location. Such stiffener placement in the compression zone serves the purpose of maintaining the full effectiveness of the web in resisting bending stress, which is really the stiffener's principal function.

For webs subjected to pure shear, the longitudinal stiffener should be located at mid-height. For combined shear and bending the stiffener should be located so that $h/5 < m < h/2$; because of its principal function, however, it should preferably be closer to $h/5$.

For design there are two requirements: (1) a moment of inertia to insure adequate stiffness to create a nodal line along the stiffener, and (2) an area adequate to carry axial compression stress while acting integrally with the web.

The design requirement for stiffness can be expressed as a function of the rigidity of the web, using the same approach as discussed for transverse stiffeners. Substituting the web height h for the transverse stiffener spacing a in Eq. 11.9.12 gives

$$I_{s\,\text{reqd}} = \frac{\gamma_0 t^3 h}{10.92} \tag{11.11.1}$$

The results of theoretical studies[7] to determine γ_0 are shown in Fig. 11.11.2. It becomes apparent from a study of these curves that the selecting of the correct value to be used for γ_0 when the web is subjected to combined bending and shear is not a simple task. The AISC specification gives no information regarding longitudinal stiffeners. The AASHO–1969

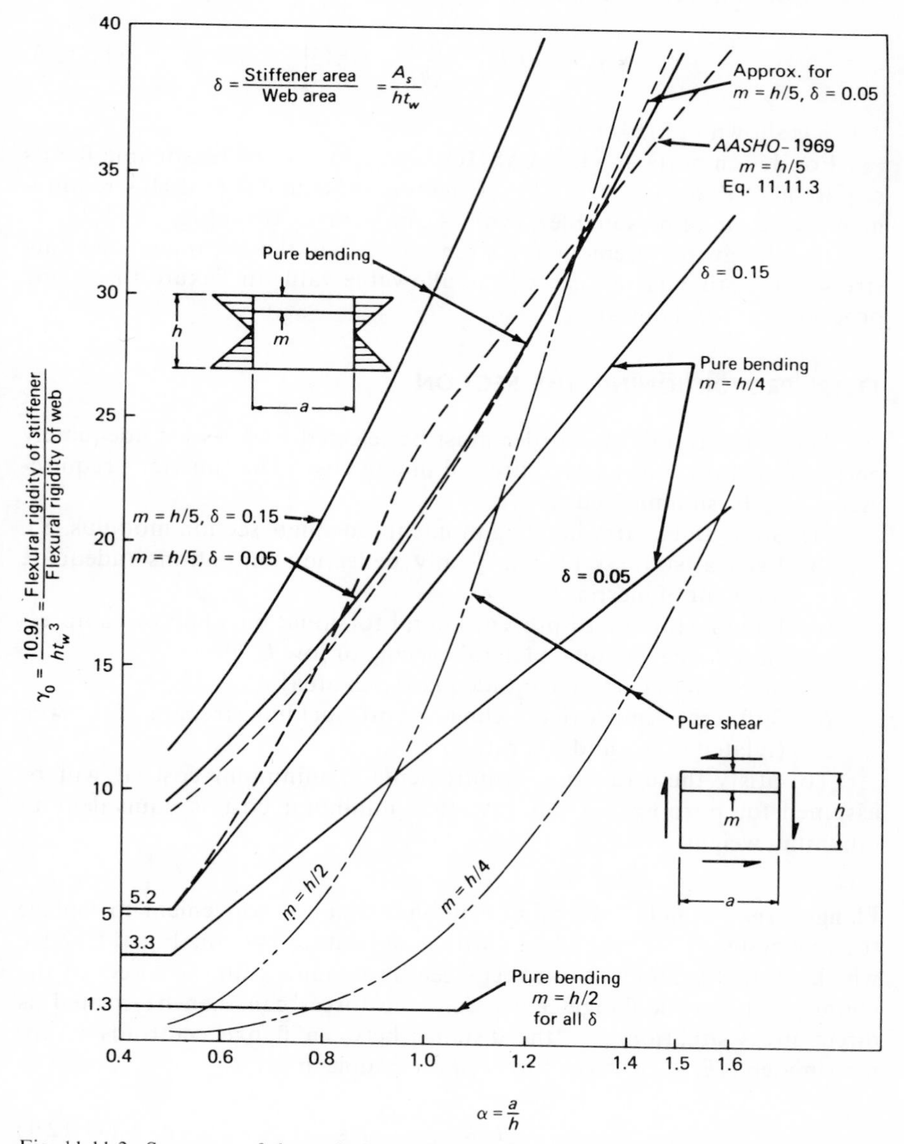

Fig. 11.11.2. Summary of theoretical studies of optimum stiffener relative rigidity, γ_0, for various web conditions. (From equations given by Ref. 7, p. 140.)

(1.7.73) gives the following expression:

$$I_{s\ \mathrm{reqd}} = t^3 h \left[2.4\left(\frac{a}{h}\right)^2 - 0.13\right] \tag{11.11.2}$$

which from Eq. 11.11.1 indicates

$$\gamma_0 = 10.92\left[2.4\left(\frac{a}{h}\right)^2 - 0.13\right] \tag{11.11.3}$$

which is shown in Fig. 11.11.2.

For design purposes, the AASHO expression seems reasonable for its requirement that $m = h/5$. For situations not under AASHO requirement, selection of reasonable γ_0 values can be made from Fig. 11.11.2.

The design requirement for strength is that the computed bending stress in the stiffener not exceed the allowable value in flexure (use same procedure as for flanges).

11.12. PROPORTIONING THE SECTION

The cross section of a girder must be selected such that it adequately performs its functions and requires minimum cost. The functions requirements may be summarized as:

(a) Strength to carry bending moment (adequate section modulus, S).
(b) Vertical stiffness to satisfy any deflection limitations (adequate moment of inertia, I).
(c) Lateral stiffness to prevent lateral-torsional buckling of compression flange (adequate lateral bracing or low L/b).
(d) Strength to carry shear (adequate web area).
(e) Stiffness to improve buckling or post-buckling strength of the web (related to h/t and a/h ratios).

To satisfy these function requirements at minimum cost, it will be assumed for purposes of this text that minimum cost is equivalent to minimum weight.

Flange-Area Formula. For simplicity in design it is convenient to replace the real system of Fig. 11.12.1a with a substitute system, Fig. 11.12.1b, which allows the moment to be replaced by a couple with the forces of the couple acting at the flange centroids. The forces can then be treated as direct stress situations. If the distance between flange centroids is approximately, $(h + d)/2$, the forces of the couple are

$$C = T = \frac{M}{(h + d)/2} \tag{11.12.1}$$

The effective area on which these forces act is equal to the flange plate plus additional area to represent the effectiveness of the web in resisting moment.

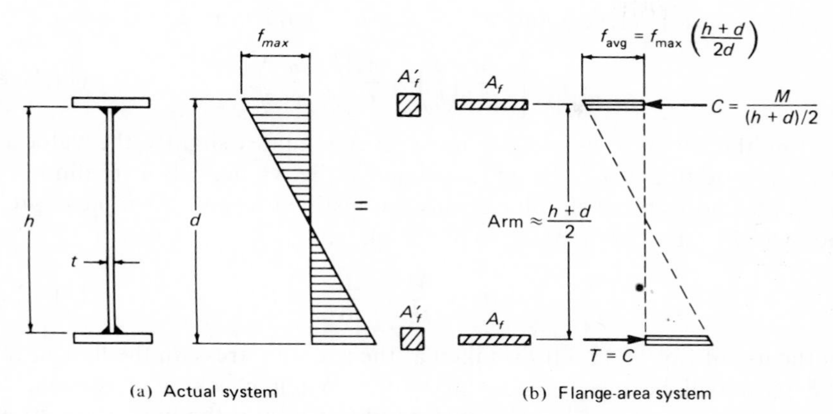

(a) Actual system (b) Flange-area system

Fig. 11.12.1. Flange-area formula development.

The average stress on the total effective area is

$$f_{avg} = \frac{\text{Force}}{\text{Area}} = \frac{M}{(h + d)/2}\left(\frac{1}{A_f + A'_f}\right) \tag{11.12.2}$$

The area A'_f must be taken such that the bending moment carried by the web is the same for both the real and substitute systems:

$$M_{\text{real system}} = f_{max}\left(\frac{h}{d}\right)\frac{th^2}{6} \tag{11.12.3}$$

$$M_{\text{substitute system}} = f_{max}\left(\frac{h + d}{2d}\right)A'_f\left(\frac{h + d}{2}\right) \tag{11.12.4}$$

Equating Eqs. 11.2.3 and 11.2.4 gives

$$A'_f = \frac{h}{d}\left(\frac{th^2}{6}\right)\left(\frac{2d}{h + d}\right)\left(\frac{2}{h + d}\right) = \frac{th}{6}\left(\frac{2h}{h + d}\right)^2 \tag{11.12.5}$$

which if $A_w = th$, and the squared term is neglected, becomes

$$A'_f = \frac{A_w}{6} \tag{11.12.6}$$

Next, solving Eq. 11.12.2 for A_f gives

$$A_f = \frac{M}{[(h + d)/2]f_{avg}} - A'_f \tag{11.12.7}$$

which, using Eq. 11.12.5 and $f_{avg} = f_{max}(h + d)/2d$, gives

$$A_f = \left[\frac{M}{f_{max}h}\left(\frac{d}{h}\right) - \frac{A_w}{6}\right]\left(\frac{2h}{h + d}\right)^2 \tag{11.12.8}$$

Letting the squared term equal unity overestimates slightly the value of A_f, while letting $d/h = 1$ underestimates the value. For preliminary design purposes these simplifications are justified to give a simple expression for the required area of one flange plate,

$$A_f = \frac{M}{fh} - \frac{A_w}{6} \tag{11.12.9}$$

In the use of Eq. 11.12.9 if f is taken as the average stress on the flange, the d/h term will be nearly accounted for. When checking a section, of course, the moment of inertia must be obtained and the maximum stress correctly computed.

Optimum Girder Depth. The variation in girder cross-sectional area is to be examined as a function of web depth to determine the depth which will give minimum area. Extended treatment of this subject has been given by Shedd,[1] Bresler and Lin,[17] and Blodgett.[18] The average gross area A_g of the girder for the entire span may be expressed

$$A_g = 2C_1A_f + C_2ht \tag{11.12.10}$$

where C_1 = factor to account for reducing flange size at regions of lower than maximum moment

C_2 = factor to account for reducing web thickness at regions of reduced shear

Substituting Eq. 11.12.9 into Eq. 11.12.10 gives

$$A_g = 2C_1\left(\frac{M}{fh} - \frac{ht}{6}\right) + C_2ht \tag{11.12.11}$$

To find minimum average gross area,

$$\frac{\partial A_g}{\partial h} = 0 \tag{11.12.12}$$

(a) Case 1. No depth restriction; desire length h/t. Assume K = constant $= h/t$; $t = h/K$. Equation 11.12.11 becomes

$$A_g = 2C_1\left(\frac{M}{fh} - \frac{h^2}{6K}\right) + C_2\frac{h^2}{K} \tag{11.12.13}$$

$$\frac{\partial A_g}{\partial h} = 0 = \frac{-2C_1 Mf}{f^2h^2} - \frac{4C_1 h}{6K} + \frac{2C_2 h}{K} \tag{11.12.14}$$

$$-6C_1 MK - 2C_1 h^3 f + 6C_2 h^3 f = 0 \tag{11.12.15}$$

from which

$$h = \sqrt[3]{\frac{3MC_1 K}{f(3C_2 - C_1)}} \tag{11.12.16}$$

and if one neglects the section reduction in regions of lower stress, $C_1 = C_2 = 1$, Eq. 11.12.16 becomes

$$h = \sqrt[3]{\frac{3MK}{2f}} \tag{11.12.17}$$

Using Eq. 11.12.13 with $C_1 = C_2 = 1$ and substituting for M/f from Eq. 11.12.17 gives

$$A_g = \frac{4h^2}{3K} - \frac{h^2}{3K} + \frac{h^2}{K} = \frac{2h^2}{K} \tag{11.12.18}$$

from which the girder weight per foot can be estimated using the fact that steel weight is 3.4 lb/sq in./linear ft.

$$Wt/ft = 3.4A_g = \frac{6.8h^2}{K} = 8.9\sqrt[3]{\frac{M^2}{f^2K}} \tag{11.12.19}$$

using inch units for the variables. Note that stiffeners will generally increase this value by 5 to 10 percent.

(b) Case 2. Minimum web thickness; t = const. Differentiating Eq. 11.12.11, $\partial A_g/\partial h = 0$, gives

$$\frac{-2C_1 Mf}{f^2h^2} - \frac{C_1 t}{3} + C_2 t = 0 \tag{11.12.20}$$

$$h = \sqrt{\frac{6C_1 M}{ft(3C_2 - C_1)}} \tag{11.12.21}$$

If $C_1 = C_2 = 1$,

$$h = \sqrt{\frac{3M}{ft}} \tag{11.12.22}$$

and

$$Wt/ft = 3.4A_g = 4.53ht = 7.85\sqrt{\frac{Mt}{f}} \qquad (11.12.23)$$

using inch units for the variables. Estimation for the weight of stiffeners should be added to the above value.

(c) Case 3. Heavy shear which governs web area; Z = constant = web area, ht. Equation 11.12.11 becomes

$$A_g = 2C_1\left(\frac{M}{fh} - \frac{Z}{6}\right) + C_2 Z \qquad (11.12.24)$$

from which it is apparent that minimum A_g results from maximum depth, h. This case usually does not govern.

If the same kind of steel is used throughout, the value of C_1 may vary from 0.7 to 0.9 when used with the maximum positive moment: 0.85 to 0.90 is the usual range. The value of C_2 is not as likely to vary except on continuous structures where it might be 1.05 when used with maximum positive moment, or 0.95 when used with maximum negative moment. Because of the complexity of evaluating C_1 and C_2 for continuous structures, it might be well to take them as unity.

Flange Plate Changes in Size. It is usually economical to reduce the size of flange plates in the region of low moment. While no specific rules can be made to help the designer to determine when it is desirable to change flange plate size, certain simple relationships are possible if only one change in flange size is desired.

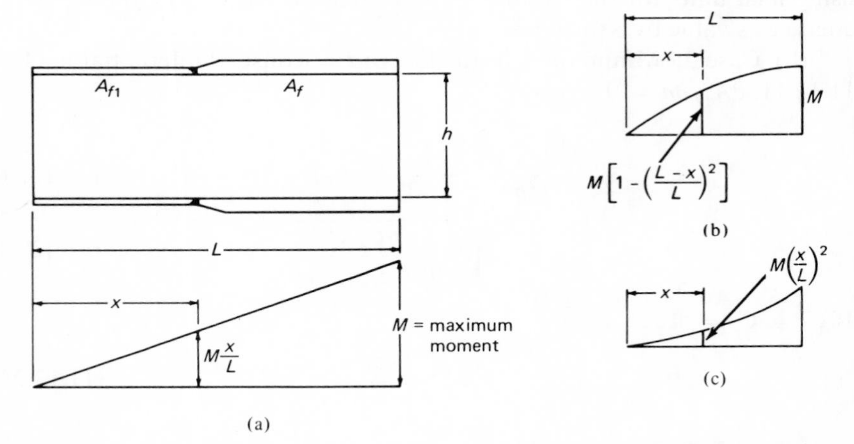

Fig. 11.12.2. Common moment variations for determining changes in flange plate size.

(a) Case 1. Linear variation in moment—two flange plate sizes. Consider the situation of Fig. 11.12.2a, and assuming both plates are stressed to their allowable values, the flange-area formula can be used for each plate.

$$A_f = \frac{M}{hf} - \frac{A_w}{6} \tag{11.12.25}$$

$$A_{f1} = \frac{M(x/L)}{hf} - \frac{A_w}{6} \tag{11.12.26}$$

The total flange volume in the length L is

$$\begin{aligned} V &= A_f(L - x) + A_{f1}x \\ &= \frac{M(L - x)}{hf} - \frac{A_w}{6}(L - x) + \frac{M}{hf}\left(\frac{x^2}{L}\right) - \frac{A_w}{6}x \\ &= \frac{M}{hf}\left[\frac{L^2 - xL + x^2}{L}\right] - \frac{A_w L}{6} \end{aligned} \tag{11.12.27}$$

For minimum volume,

$$\frac{\partial V}{\partial x} = 0 = 2x - L; \qquad x = \frac{L}{2}$$

which means

$$\frac{A_{f1}}{A_f} = \frac{\dfrac{M}{2h} - \dfrac{A_w}{6}}{\dfrac{M}{hf} - \dfrac{A_w}{6}} \approx \frac{1}{2} \tag{11.12.28}$$

(b) Case 2. Parabolic variation as for uniformly loaded simple beam (see Fig. 11.12.2b). The total volume in the length L is

$$V = \frac{M}{hf}\left[\frac{L^3 - L^2x + 2Lx^2 - x^3}{L^2}\right] - \frac{A_w L}{6}$$

$$\frac{\partial V}{\partial x} = 0 = x^2 - \frac{4}{3}Lx + \frac{L^2}{3}; \qquad x = \frac{L}{3}$$

and

$$A_{f1}/A_f \approx \frac{5}{9}. \tag{11.12.29}$$

(c) Case 3. Parabolic variation as for uniformly loaded cantilever (see Fig. 11.12.2c). The total volume in the length L is

$$V = \frac{M}{hf}\left[\frac{L^2 - L^2x + x^3}{L^2}\right] - \frac{A_w L}{6}$$

$$\frac{\partial V}{\partial x} = 0 = 3x^2 - L^2; \qquad x = \frac{L}{\sqrt{3}}$$

and

$$A_{f1}/A_f \approx \frac{1}{3}. \qquad (11.12.30)$$

The foregoing developments can provide a guide for change of plate sizes. Since making a change involves a butt-welded splice of the flange plate, enough material must be saved to more than offset the welding cost.

As a rule of thumb, unless 200 to 300 pounds of material are saved in a flange plate per added splice the added cost of the butt splice (considering a plate about 2 ft wide and 2 in. thick) is not justified.

Flange-Plate Proportions. According to most of the theory developed earlier in this chapter, the flange plate can be any width and thickness as long as it contributes to the girder the properties necessary to satisfy the functional requirements. However, nearly all testing has used flange plate dimensions which were considered reasonable.

In order to assist the engineer who is unsure of what constitutes such reasonable dimensions the following guidelines are suggested.

(a) Typically the ratio of girder flange width to girder depth, b/d, varies from about 0.3 for shallow girders to about 0.2 for deep girders.

(b) Plate widths should be in 2-in. increments.

(c) Plate-thickness increments should be as follows:

$\frac{1}{16}$ in.	$t \leq \frac{9}{16}$ in.
$\frac{1}{8}$ in.	$\frac{5}{8} = t \leq 1\frac{1}{2}$ in.
$\frac{1}{4}$ in.	over $1\frac{1}{2}$ in.

(d) Where lateral stability is of concern for the girder, plate width-to-thickness ratios, b/t, should be kept below the limits of AISC–1.9 at the point of maximum moment so that reduced thicknesses can be used to reduce the flange-plate size in regions of low moment. For such cases, flange-plate area reduction should be made by reducing the thickness.

(e) For laterally stable girders, the flange-plate area reduction in regions of lower moment may be accomplished by reducing the thickness; reducing the width; or reducing both thickness and width. A slight advantage in fatigue strength accrues by reducing the width rather than the thickness.[18] The transition slope should not exceed 1 in $2\frac{1}{2}$ for either width or thickness, and is usually 1 in 4 to 1 in 12 for the transition in width.[18]

(f) Excessive thickness of flange plate may require an arbitrarily larger weld size than required for strength in making the flange-to-web connection. AISC–1.17.6 should be considered when deciding on thickness.

11.13. PLATE GIRDER DESIGN EXAMPLE—AISC

Partially design a two-span continuous welded plate girder to support a uniform load of 4 kips per foot plus two fixed concentrated loads of 75 kips in each span as shown in Fig. 11.13.1. Lateral support is provided every 25 ft.

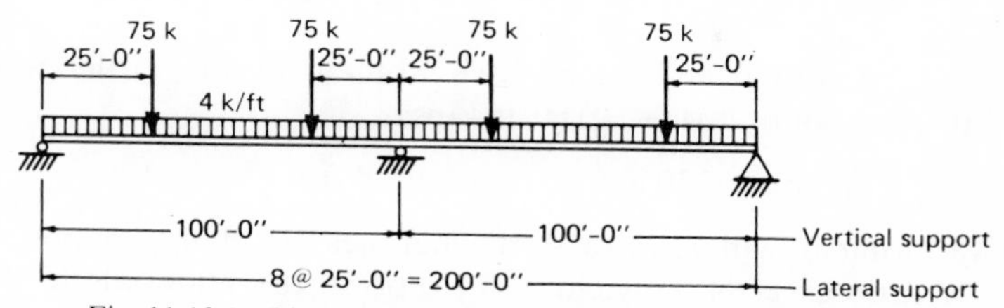

Fig. 11.13.1. Girder loading and support for design example.

Additional specifications and general comments:

(a) Assume all loading has fixed position (no moving or partial span loading).

(b) Girder should have constant depth of web plate; use A36 steel in positive moment zone and A572 Grade 50 for the negative moment zone.

(c) Assume $\frac{5}{16}$ in. is minimum web plate thickness available.

(d) Assume no depth restriction; also, that any thin-web deep girder that is selected can be feasibly transported to the construction site without excessive difficulty or cost.

(e) Note should be made that laterally unsupported distances as great as 25 ft are uncommon and are used here only for the purpose of illustrating procedure.

Solution

(*a*) *Preliminary Estimate of Weight.* The maximum moments due to the superimposed loads of Fig. 11.13.1 for the positive and negative regions are computed.

$$+M = 3950 \text{ ft-kips}$$
$$-M = 7100 \text{ ft-kips}$$

Since there is no depth limitation given in the statement, AISC limitations for h/t must be considered. Referring to Table 11.8.1 (based on AISC–1.10.2),

$$\text{Max } h/t = 322 \ (333 \text{ if } a/h \leq 1.5) \qquad \text{A36 steel}$$
$$\text{Max } h/t = 243 \ (283 \text{ if } a/h \leq 1.5) \qquad \text{A572 Grade 50}$$

Flange stress reduction will be required according to Fig. 11.8.1 (AISC–1.10.6) when

$h/t > 162$ A36 when $F_b = 0.6F_y$
$h/t > 139$ A572 Grade 50 when $F_b = 0.6F_y$

For weight estimate, try $K = h/t = 250$ and use Eq. 11.12.19, $M =$ 3,950 ft-kips, and $f \approx 21$ ksi:

$$\text{Wt/ft} = 8.9\sqrt[3]{\frac{M^2}{f^2K}} = 8.9\sqrt[3]{\frac{[3950(12)]^2}{(21)^2\,250}} = 245\text{ lb/ft}$$

Assume $w = 250$ lb/ft and the positive moment due to girder weight $\approx$ 175 ft kips; say total $+M = 4{,}150$ ft-kips:

$$\text{Wt/ft} \approx 250\text{ lb/ft (recomputed from formula)}$$

Negative moment will require somewhat heavier flange plates because $-M/+M$ ratio exceeds the yield point ratio of the materials to be used. Thus the average weight will be somewhat higher than 250 lb/ft. In addition, allowance of about 10 percent should be made for stiffeners. Use a value somewhat larger than 275 lb/ft. Try $w = 290$ lb/ft. The total moments and shears are then computed as shown in Fig. 11.13.2.

(*b*) *Determine Web Plate Sizes.* For $+M$ with A36 steel assume $C_1 = C_2 = 1$ and use Eq. 11.12.17. Evaluate optimum value of h for various h/t values,

$$h = \sqrt[3]{\frac{3MK}{2f}} = \sqrt[3]{\frac{3(4153)(12)320}{2(22)}} = 95.5$$

Assume $0.6F_y$ for f_b (arrow pointing to 22)

$K = h/t$	Formula h	h	t	A_w	$f_v = V/A_w$	Actual h/t
320	95.5	96	$\frac{5}{16}$	30.0	7.17 ksi	307
340*	97.4	98	$\frac{5}{16}$	30.6	7.03	314

*This value which slightly exceeds the upper limit for h/t is used only for the purpose of establishing economical depth.

Based on $+M$ the economical depth for bending moment requirements appears to be 96 to 98 for near maximum web slenderness ratios. The allowable shear stress for these h/t ratios for $F_y = 36$ ksi is about 10 ksi (see AISC Table 3-36 or compute from Eq. 11.9.3), indicating that the $\frac{5}{16}$ by 96 or 98 inch plate provides more web area than necessary.

Evaluate optimum h for $(-M)$ requirement,

$$h = \sqrt[3]{\frac{3(7471)(12)240}{2(30)}} = 102.5$$

assume $0.6F_y$ for f_b (arrow pointing to 30)

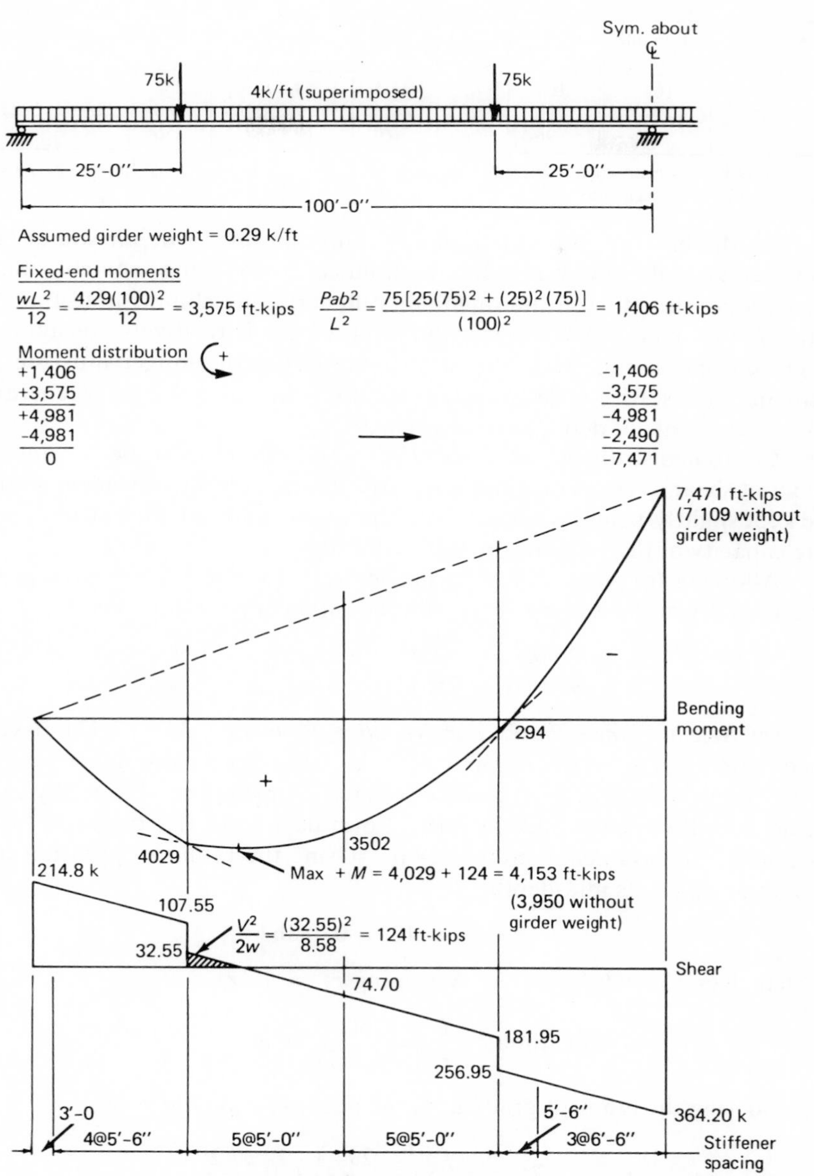

Fig. 11.13.2. Moments and shears for two-span continuous beam of illustrative example.

$K = h/t$	Formula h	h	t	A_w	$f_v = V/A_w$	Actual h/t
240	102.5	104	$^7/_{16}$	45.5	8.0 ksi	238
290*	107.4	108	$^3/_8$	40.5	9.0 ksi	288 > 283 N.G.
		96	$^3/_8$	36.0	10.1 ksi	256

*This value which slightly exceeds the upper limit for h/t is used only for the purpose of establishing economical depth.

On the basis of these preliminary computations, a depth somewhere between 96 and 108 inches seems economical for bending moment. Since it is economical to use intermediate stiffeners, one should note that for large h/t ratios the stiffener spacing requirement is relatively constant for a given shear stress. Thus, the deeper the girder the longer must be the stiffener plates, with little advantage to offset this extra weight. Generally, 2-in. increments of depth should be used.

Combined stress on the web (AISC–1.10.7) should also be considered; the unit shear stress at maximum negative moment in this problem should be kept slightly below what would normally be allowed so that the bending capacity of the web may be fully utilized.

Allowing for some flexibility in the design, use $h = 96$ in. which will mean that web slenderness is near but not at its upper limit. Try

$$^5/_{16} \times 96\ (h/t = 307) \text{ for } +M, \quad F_y = 36 \text{ ksi}$$
$$^3/_8 \times 96\ (h/t = 256) \text{ for } -M, \quad F_y = 50 \text{ ksi}$$

(*c*) *Select Flange Plates For Negative Moment.* $M = 7{,}471$ ft-kips, web plate $= {}^3/_8 \times 96$ ($A_w = 36.0$ sq in.). Allowable stress must be estimated considering the 25-ft distance of lateral unsupport. Normally plate girder stability is controlled by AISC Formula 1.5–6. If a typical ratio of $b/d = 0.25$ is assumed, then $b = 24$ in. Using the radius of gyration of a rectangle about its mid-depth,

$$r = b/\sqrt{12} = 0.288\text{b}$$

which gives $r \approx 6.9''$, say $7''$. Assuming $r_T \approx 1.2r_y$,

$$\text{Estimated } \frac{L}{r_T} = \frac{25(12)}{1.2(7)} = 35.7$$

Considering the moment gradient in the negative-moment zone,

$$C_b = 1.75 + 1.05\left(+\frac{294}{7471}\right) + 0.3\left(\frac{294}{7471}\right)^2$$
$$= 1.79 + \text{or} \approx 1.75 \qquad \text{(AISC–1.5.1.4.6a)}$$

Referring to Table 9.6.1 (AISC–1.5.1.4.6a), the allowable stress is $F_b = 0.6F_y$, since $L/r_T < 60$. To obtain the estimated reduction due to high web slenderness, see Fig. 11.8.1 (AISC–1.10.6). For $h/t = 256$,

$$F_b' \approx 28 \text{ ksi}$$

Using the flange-area formula, Eq. 11.12.9, gives the requirement for one flange as

$$A_f = \frac{M}{fh} - \frac{A_w}{6} = \frac{7471(12)}{28(96)} - \frac{36.0}{6} = 27.4 \text{ sq in.}$$

Try PL – $1\frac{1}{4} \times 22$: $A_f = 27.5$ sq in.

Check ordinary allowable stress:

$$r_T = \sqrt{\frac{\frac{1}{12}(22)^3(1.25)}{27.5 + 36.0/6}} = 5.75 \text{ in.}$$

$$\frac{L}{r_T} = \frac{300}{5.75} = 52.2 < 60, \quad \therefore F_b = 0.6F_y = 30 \text{ ksi}$$

Reduction according to AISC–1.10.6:

$$\frac{A_w}{A_f} = \frac{36.0}{27.5} = 1.31$$

From Fig. 11.8.1, using $h/t = 256$, find $F_b' = 27.7$ ksi. Try plate size $1\frac{1}{4} \times 24$, since assumed F_b' is slightly higher than correct value.

$$\frac{L}{r_T} < 60, \quad \therefore F_b = 30 \text{ ksi}$$

$$\frac{A_w}{A_f} = \frac{36.0}{30.0} = 1.2, \quad \frac{h}{t} = 256, \qquad F_b' = 27.8 \text{ ksi}$$

Check stress using flexure formula. Using AISC Manual (see p. 2–127 and 130) tables for computing properties,

$$\begin{aligned}
&1\tfrac{1}{4} \times 24: && 30(4729) = 141{,}870 \\
&2y = 97.25 \text{ for properties} && \\
&\tfrac{3}{8} \times 96: && = \underline{\ 27{,}648\ } \\
& && I = 169{,}518 \text{ in.}^4
\end{aligned}$$

$$f = \frac{7471(12)(48 + 1.25)}{169{,}518} = 26.1 \text{ ksi}$$

Verifying the allowable value, AISC–1.10.6,

$$F_b' = 30\left[1.0 - 0.0005\frac{A_w}{A_f}\left(\frac{h}{t} - \frac{760}{\sqrt{F_b}}\right)\right]$$

$$F'_b = 30\,[1.0 - 0.0005\,(1.2)(256 - 139)] = 30(1 - 0.070)$$
$$= 27.9 \text{ ksi}$$

The resulting understress is about $6\frac{1}{2}$ percent, but no other practical choice seems available which would narrow the gap. ($1\frac{1}{8} \times 24$ is overstressed about 2 percent). Stiffener spacing will probably be helped by leaving the moderate understress, rather than beginning with overstress.

Accept plates–$1\frac{1}{4} \times 24$.

(*d*) *Select Flange Plates for Positive Moment.* $M = 4153$ ft-kips; web plate $= \frac{5}{16} \times 96$ ($A_w = 30.0$ sq in.) Estimating flange width to be about 24 in., $L/r_T < 52.2$, as computed in part (c) above. In this region $C_b = 1$, in which case from Table 9.6.1 for $F_y = 36$ ksi,

$$F_b = 0.6F_y = 22 \text{ ksi}, \quad \text{since} \frac{L}{r_T} < 53$$

Referring to Fig. 11.8.1, the allowable bending stress resulting from flange stress reduction (AISC–1.10.6) is estimated as

$$F'_b \approx 19.6 \text{ ksi for } \frac{h}{t} = 307$$

$$\text{Reqd } A_f = \frac{M}{fh} - \frac{A_w}{6} = \frac{4153(12)}{19.6(96)} - \frac{30.0}{6} = 20.5 \text{ sq in.}$$

Try PL $\frac{7}{8} \times 24$: $\quad A_f = 21.0$ sq in.

Check stress:

$\frac{7}{8} \times 24$,	$21.0(4692) =$	98,530
$\frac{5}{16} \times 96$	$=$	23,000
	$I =$	121,530 in.^{4}

$$f = \frac{4153(12)(48 + 0.875)}{121{,}530} = 20.0 \text{ ksi}$$

The allowable stress F'_b, with $A_w/A_f = 1.43$, is

$$F'_b = 22.0[1.0 - 0.0005(1.43)(307 - 162)] = 19.7 \text{ ksi} < 20.0 \text{ ksi} \quad \text{OK}$$

Accept overstress as combined stress will cause no difficulty in the positive moment zone.

Accept plates, $\frac{7}{8} \times 24$.

(e) *Intermediate Stiffeners—Placement in Positive Moment Zone.* Web plate $= \frac{5}{16} \times 96$, $A_w = 30.0$ sq in., $F_y = 36$ ksi.

Exterior end, $V = 214.8$ kips. Use Eq. 11.9.6, AISC–1.10.5.3:

$$a \text{ or } h \leq \frac{348t}{\sqrt{f_v}} = \frac{348(5/16)}{\sqrt{7.16}} = 38.4 \text{ in.}$$

where

$$f_v = \frac{214.8}{30} = 7.16 \text{ ksi}$$

Use 3′ = 0″.

Panel 2:

$$V = 214.8 - 3(4.29) = 201.9 \text{ kips}$$

Equations 11.9.2 and 11.9.3, AISC Formulas (1.10–1) and (1.10–2), considering tension-field action are to be applied here:

$$f_v = \frac{201.9}{30} = 6.73 \text{ ksi}, \quad \frac{h}{t} = \frac{96}{0.3125} = 307$$

Using AISC Table 3-36, find $a/h \approx 0.7$, computed more exactly as

$$\left.\frac{a}{h}\right)_{\max} = \left(\frac{260}{h/t}\right)^2 \leq 3.0$$

$$= \left(\frac{260}{307}\right)^2 = 0.717$$

$$a = 0.717(96) = 68.8''(5.74')$$

Since lateral support occurs every 25 ft and a bearing stiffener is required at concentrated loads, the stiffener spacing is usually arranged to fit these limitations. Considering the first stiffener is at 3′-0″, 22′-0″ remains. Use 4 spaces @ 5′-6″.

For the region from 25 to 50 ft from the end, the a/h limitation of 0.717 still governs. Use 5 spaces @ 5′-0″.

For the region from 50 to 75 ft the maximum shear stress is still below the value in panel 2, so that a/h maximum still governs. Use 5 spaces @ 5′-0″.

(f) *Intermediate Stiffeners—Placement in Negative Moment Zone.* Web plate = 3/8 × 96, A_w = 36.0 sq in., F_y = 50 ksi,

Interior end:

$$V = 364.2 \text{ kips}$$

Use Eqs. 11.9.2 and 11.9.3 for this interior panel (there is an adjacent panel in the other span):

$$f_v = \frac{364.2}{36} = 10.1 \text{ ksi}, \quad \frac{h}{t} = \frac{96}{0.375} = 256$$

Extrapolating from AISC Table 3–50, $\frac{a}{h} \approx 1.0$ for $F_v \approx 11.5$ ksi. Computations give

$$\frac{h}{t} = 250, \quad F_v = 11.5 \text{ ksi}, \quad \frac{a}{h} = 1.0$$

$$\frac{h}{t} = 260, \qquad F_v = 11.5 \text{ ksi}, \qquad \frac{a}{h} = 1.0$$

$$\left.\frac{a}{h}\right)_{max} = \left(\frac{260}{256}\right)^2 = 1.03$$

Combined stress on web at interior end:

$$f_b \text{ (maximum on web)} = 26.1 \frac{48}{49.25} = 25.4 \text{ ksi}$$

from part (c)

$$\frac{f_b}{0.6F_y} = \frac{25.4}{30} = 0.846 > 0.75$$

In accordance with AISC-Formula 1.10–7, Eq. 11.8.8, the usable shear capacity must be reduced from its ordinary allowable value:

$$\left.\frac{f_v}{F_v}\right)_{\substack{max \\ permitted}} = \frac{0.825 - f_b/F_y}{0.375} = \frac{0.825 - 25.4/50}{0.375} = 0.845$$

The required allowable for the panel as computed by AISC Formula 1.10–2 is

$$\text{Reqd } F_v = \frac{f_v}{0.845} = \frac{10.1}{0.845} = 11.95 \text{ ksi}$$

Therefore the required $a/h = 0.9$, which gives $F_v = 12.1$ ksi. Try 7′-0″. ($a/h_1 = 0.875$) Revised below.

Panel 2:

$$V = 364.2 - 7(4.29) = 334.2 \text{ kips}$$

$$f_v = \frac{334.2}{36} = 9.3 \text{ ksi}$$

At this point,

$$\frac{f_b \text{ (on web)}}{0.6F_y} = \frac{25.4}{30}\left(\frac{5150}{7471}\right) = 0.58 < 0.75$$

Full shear capacity may be utilized. Therefore maximum $a/h = 1.03$ controls:

$$a = 1.03(96) = 99''$$

Since two spaces will not work, three spaces must be used. To prevent using a larger spacing in the region of higher shear, reduce the spacing in the first panel. ~~Try 4 spaces @ 6′-3″ for the entire 25 ft. section.~~ Revised below. The arrangement of intermediate stiffeners is shown in the design sketch, Fig. 11.13.7.

(g) *Location of Flange and Web Splices.* The location of splices is partially dependent upon the type of splices used; for a bolted field splice joining the A36 with the A572 sections both flanges and web usually would be spliced at the same location; for a field welded splice it might be preferable to splice the flanges at a location offset from the web splice by as much as ten feet. Such an offset reduces stress concentration and probably assists in getting proper alignment at the splice. For this example assume splices for flanges and web are to occur at the same location.

From previous calculations for stiffener spacing it is apparent that splicing could be done at, say, 26 ft from the interior support, just beyond the bearing stiffener. However, perhaps it can be made closer to the interior support.

Try 5′-9″ from 75-kip load (19′-3″ from interior support) which would be acceptable based on maximum $a/h = 0.717$ for the $^5/_{16}$-in. web. See part (e).

$$M = 1250 \text{ ft-kips}; \qquad V = 280 \text{ kips}$$

(scaled from Fig. 11.13.2).

$$f_b \text{ (on web)} = \frac{1250(12)48}{121{,}530} = 5.9 \text{ ksi} < 0.45F_y$$

Combined stress is acceptable.

$$f_v = \frac{280}{30} = 9.3 \text{ ksi}; \qquad \frac{h}{t} = 307$$

From AISC Table 3–36, find $a/h \approx 0.7$. Therefore, it is apparent that no higher shear stress can be accepted on the $^5/_{16}$-in. web. Splice at 20 ft from interior support. Revise stiffener spacing in end 25-ft section to 3 spaces at 6′-6″ and one space at 5′-6″, starting from interior support. A desirable 6-in. offset will exist between the splice and the nearest stiffener.

A summary showing complete stiffener spacing appears in Fig. 11.13.7.

(h) *Size of Intermediate Stiffeners.* Frequently, A36 stiffeners are suitable, with higher yield point material offering little if any saving. Try A36 steel for all stiffeners.

Panel 2 from exterior support: This is the first panel in which tension-field action is presumed to occur.

$$V = 201.9 \text{ kips}; \qquad f_v = 6.73 \text{ ksi}$$

$$\frac{a}{h} = \frac{66}{96} = 0.69, \qquad \frac{h}{t} = 307; \qquad F_v \approx 9.7 \text{ ksi}$$

Using AISC Table 3–36, find $A_{st}/A_w \approx 0.118$. (Same as using AISC Formula 1.10–3, or Eq. 11.9.17.)

$$\frac{f_v}{F_v} = \frac{6.73}{9.7} = 0.695$$

$$\text{Reqd } A_{st} = 0.118A_w(0.695) = 0.118(30.0)(0.695) = 2.46 \text{ sq in.}$$

This area requirement assumes stiffeners to be used in pairs. Furthermore, local buckling requirements of AISC–1.9.1 must be satisfied; i.e., $w/t = 95/\sqrt{F_y} = 15.8$.

The requirement for stiffness, Eq. 11.9.7, gives

$$\text{Reqd } I_s = \left(\frac{h}{50}\right)^4 = \left(\frac{96}{50}\right)^4 = 13.55 \text{ in.}^4$$

To find minimum acceptable stiffener width (Fig. 11.13.3),

$$\text{Reqd } r^2 = \frac{I}{A} = \frac{13.55}{2.46} = 5.5 \text{ in.}^2$$

$$\text{Provd } r^2 = \frac{tW^3}{12tW} = \frac{W^2}{12}$$

$$\text{Reqd } W = \sqrt{12(5.5)} = 8.1 \text{ in.}$$

On this basis, plates 4 inches wide could be used, which, considering the area requirement, would have to be $^3/_8 \times 4$ in. in pairs. A designer at this

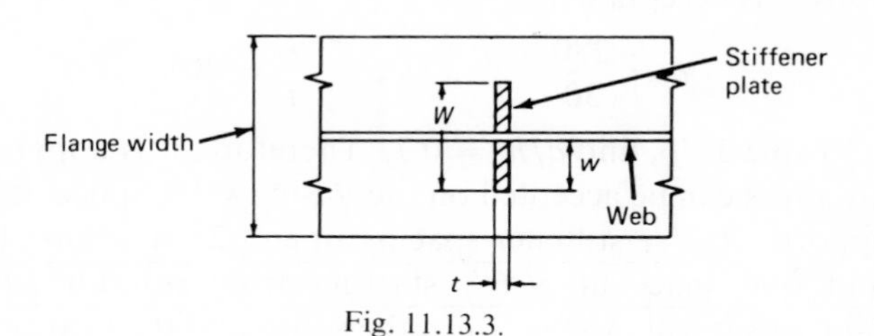

Fig. 11.13.3.

point must decide what stiffener width he considers as minimum. The authors prefer in this case 2 Ⓟs – $^5/_{16} \times 5$.

Since the stiffeners for the panel are about the minimum size based on local-buckling provisions, the reduced area required for the more interior panels on the $^5/_{16}$-in. web is of little consequence. Therefore, use 2 Ⓟs – $^5/_{16} \times 5$ for all intermediate stiffeners attached to the $^5/_{16}$-in. web.

Examine the end panel adjacent to the interior support:

$$V = 364.2 \text{ kips}; \qquad f_v = 10.1 \text{ ksi}$$

Based on combined stress, however, the required allowable value was 11.95 ksi. Applying AISC Formula 1.10-2,

$$\frac{a}{h} = \frac{78}{96} = 0.81, \qquad \frac{h}{t} = 256; \quad F_v \approx 12.6 \text{ ksi}$$

Using AISC Table 3–50, estimate $A_{st}/A_w \approx .125$, for stiffeners with $F_y = 50$ ksi.

$$\frac{f_v}{F_v} = \frac{\text{Reqd allowable}}{\text{Actual allowable}} = \frac{11.95}{12.6} = 0.95$$

$$\text{Reqd } A_{st} = 0.125\ A_w(0.95) = 0.125(36)(0.95) = 4.27 \text{ sq in.}$$

for A572 Grade 50 stiffeners. A36 stiffeners would require A_{st} = 4.27(50/36) = 5.92 sq in. A572 stiffeners appear to be economical:

$$2\text{PLs} - {}^{7}/_{16} \times 5, \qquad A_{st} = 4.38 \text{ sq in.}$$
$$2\text{PLs} - {}^{3}/_{8} \times 6, \qquad A_{st} = 4.50 \text{ sq in.}$$

The authors consider the ³/₈ × 6 ($w/t = 16$) better because they provide increased rigidity to the entire girder and are only slightly larger. Use <u>2PLs – ³/₈ × 6 of A572 Grade 50 for all intermediate stiffeners connected to the ³/₈-in. web.</u>

(i) Connection of Intermediate Stiffeners to Web. For the ⁵/₁₆-in. A36 web:

$$f_{vs} = h\sqrt{\left(\frac{F_y}{340}\right)^3} = 96\sqrt{\left(\frac{36}{340}\right)^3} = 3.3 \text{ kips/in.}$$

in accordance with AISC–1.10.5.4. Since panels are not fully stressed, this value can be reduced in direct proportion:

$$\text{Reqd } f_{vs} = 3.3(0.695) = 2.29 \text{ kips/in.}$$
$$\text{Min weld size } a = {}^{3}/_{16} \text{ (AISC–1.17.5)}$$

Determine maximum effective weld size (AISC–1.17.6 using shear stress on throat of fillet from AISC–Table 1.5.3),

$$a_{\text{max eff.}} = \frac{0.4F_y t_w}{2(0.707)21.0} = \frac{14.4(0.3125)}{2(0.707)21.0} = 0.151 \text{ in.}$$

Use ³/₁₆-in. weld of E70 electrodes (basic welding on A36 steel). The safe weld capacity per inch R_w, is

$$R_w = 0.151\ (0.707)21.0 = 2.24 \text{ kips/in.}$$

Since four lines of fillet welds (Fig. 11.13.4) are to provide 2.29 kips/in., then 2.29/4 = 0.57 kips/in. are required along each line.

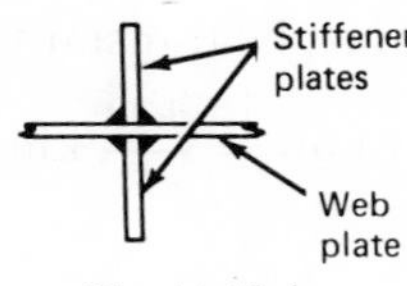

Fig. 11.13.4.

$$\text{Percent of continuous } {}^{3}/_{16} \text{ weld reqd} = \frac{0.57}{2.24}(100) = 25.4\%$$

For intermittent welding, minimum segment is $1\frac{1}{2}$ in., according to AISC–1.17.8; the capacity of this segment is

$$L_w R_w = 1.5(2.24) = 3.36 \text{ kips}$$

$$\text{Reqd pitch } p = \frac{3.36}{0.57} = 5.9 \text{ in.}$$

Use $\frac{3}{16}$ in.–$1\frac{1}{2}$ in. segments @ $5\frac{1}{2}''$ pitch, E70 electrodes, for connecting $\frac{5}{16} \times 5$ plates to $\frac{5}{16}$-in. web.

For the $\frac{3}{8}$-in. A572 web:

$$f_{vs} = h\sqrt{\left(\frac{50}{340}\right)^3} = 0.0564h = 0.0564(96) = 5.41 \text{ kips/in.}$$

Based on 95 percent stressed in shear for the most highly stressed panel,

$$\text{Reqd } f_{vs} = 0.95(5.41) = 5.14 \text{ kips/in.}$$

$$\text{Min weld size } a = \tfrac{3}{16} \text{ in.}$$

$$a_{\text{max eff}} = \frac{20.0(0.375)}{2(0.707)21.0} = 0.252 \text{ in.} > \tfrac{3}{16}, \qquad \text{try } \tfrac{3}{16} \text{ in.}$$

Using E70 electrodes (AISC–Table 1.17.2), along with the corresponding allowable stress (AISC–Table 1.5.3),

$$R_w \text{ (for } \tfrac{3}{16}\text{-in. weld)} = \tfrac{3}{16}(0.707)\,21.0 = 2.79 \text{ kips/in.}$$

For $1\frac{1}{2}$ in. segments, the required pitch is

$$p = \frac{1.5(2.79)}{(5.14/4)} = 3.26 \text{ in.}$$

which means nearly continuous $\frac{3}{16}$-in. weld. Try $\frac{1}{4}$-in. weld (maximum effective = 0.252 in.):

$$R_w = 0.25(0.707)21.0 = 3.71 \text{ kips/in.}$$

$$p = \frac{1.5(3.71)}{(5.14/4)} = 4.33 \text{ in.} \qquad \text{Not sufficient savings.}$$

Use $\frac{3}{16}$-in. continuous weld, E70 electrodes, for connecting $\frac{3}{8} \times 6$ plates to $\frac{3}{8}$-in. web.

(j) Flange to Web Connection—A36 Steel. The welding of the flanges to the web must provide for the horizontal shear developed at the joint. The shear flow which must be transmitted may be expressed as

$$\text{Shear flow} = \frac{VQ}{I} \text{ kips/in.}$$

where V = shear at section
Q = statical moment of flange area about neutral axis
I = moment of inertia of section

Welding along both sides of the web provides a shear flow capacity, which, if it exceeds the VQ/I requirement, may be reduced by the use of intermittent welding. Normally flange to web welding is made continuous, primarily because automatic fabricating procedures usually make it more economical to do so. However, for design purposes the minimum percent of continuous weld acceptable in each panel between stiffeners will be prescribed. If it is more economical for the fabricator to use more weld, it is permissible to do so. The following calculations conservatively assume manual shielded-arc welding is to be used.

$$\text{Min weld size } a = \tfrac{5}{16} \text{ in. (AISC–1.17.5)}$$

$$\text{Max effective size } a = \frac{14.4(0.3125)}{2(0.707)21.0} = 0.151 \text{ in.}$$

for E70 electrodes. For strength, the weld size required is

$$\frac{VQ}{I} = \frac{214.8(21)(48.44)}{121{,}350} = 1.80 \text{ kips/in.}$$

$$2a(0.707)21.0 = 1.80$$

$$\text{Reqd } a = \frac{1.80}{2(0.707)21.0} = 0.06 \text{ in.} < 0.151 \text{ in.}$$

<u>Use $\tfrac{5}{16}$-in. weld, E70 electrodes</u> (effective size = 0.151). The safe capacity for continuous weld on both sides of the web is

$$R_w = 2(0.707)(0.151)21.0 = 4.5 \text{ kips/in.}$$

$$\text{Max percent continuous weld} = \frac{1.80}{4.5}(100) = 40.0\%$$

For each panel along the A36, $\tfrac{5}{16}$-in. web, the minimum percentages of $\tfrac{5}{16}$-in. continuous weld required are similarly computed and given in Fig. 11.13.7. AISC–1.10.5.4. requires at least 23.1 percent of continuous weld ($1\tfrac{1}{2}$-in. segment @$6\tfrac{1}{2}$ in. maximum pitch).

(k) *Flange to Web Connection*–A572 Steel.

$$\text{Min weld size } a = \tfrac{5}{16}\text{-in. (AISC–1.17.5)}$$

$$\text{Max effective size } a = \frac{20.0(0.375)}{2(0.707)21.0} = 0.252 \text{ in.}$$

The maximum strength requirement is

$$\frac{VQ}{I} = \frac{364.2(30.0)48.625}{169{,}518} = 3.13 \text{ kips/in.}$$

The safe capacity of continuous $\tfrac{5}{16}$-in. weld on both sides of the web is

$$R_w = 2(0.252)(0.707)21.0 = 7.50 \text{ kips/in.}$$

$$\text{Max percent continuous weld} = \frac{3.13}{7.50}(100) = 41.7\%$$

Use $^5/_{16}$-in. weld, E70 electrodes. The minimum percent of continuous weld required for each panel along the A572 $^3/_8$-in. web is summarized in Fig. 11.13.7. AISC–1.10.5.4 requires at least 20 percent of continuous weld ($1\frac{1}{2}$-in. segment @ $7\frac{1}{2}$-in. maximum pitch).

(l) *Design of Bearing Stiffener—Interior Support.* As discussed in Sec. 11.10, bearing stiffeners are required at concentrated loads. At the interior support,

$$\text{Reaction} = 728.4 \text{ kips}$$

Since bearing stiffeners must extend "approximately to the edge of the flange plates" . . . , the width must be

$$w = \frac{24 - 0.375}{2} = 11.8 \text{ in.} \qquad \text{(say 11 in.)}$$

Column stability criterion (Fig 11.13.5):

$$r \approx .25(22.375) = 5.6 \text{ in.}$$

$$\frac{L}{r} \approx \frac{0.75(96)}{5.6} = 12.9$$

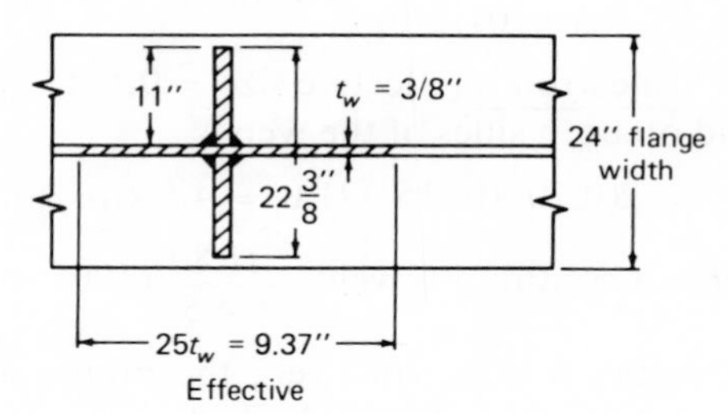

Fig. 11.13.5.

For A36 steel, $F_a = 21$ ksi

$$\text{Reqd area} = \frac{728.4}{21} = 34.7 \text{ sq in.}$$

$$\text{Reqd } t = \frac{34.7 - 9.37(0.375)}{2(11)} = 1.42 \text{ in.}$$

Local-buckling criterion:

$$\frac{w}{t} = 15.8 \qquad \text{for } F_y = 36 \text{ ksi}$$

$$\text{Reqd } t = \frac{11}{15.8} = 0.70 \text{ in.}$$

Compression yield criterion:

$$\text{Reqd } A = \frac{728.4}{0.6F_y} = \frac{728.4}{22} = 33.1 \text{ sq in.}$$

$$\text{Reqd } t = \frac{33.1}{2(11)} = 1.5 \text{ in.}$$

Bearing criterion:

$$\text{Reqd contact area} = \frac{728.4}{0.9F_y} = \frac{728.4}{32.4} = 22.5 \text{ sq in.}$$

$$A_c = 2(11 - 0.5)t$$

(0.5 — est. to allow for fillet weld)

$$\text{Reqd } t = \frac{22.5}{2(10.5)} = 1.07 \text{ in.}$$

Compression yield criterion governs. So as to properly distribute this reaction, try 4 PLs – 3/4 × 11 to serve as bearing stiffeners. (Alternative, 2 PLs – 1½ × 11.) Recheck r for column stability:

$$r = \sqrt{\frac{(1/12)(1.50)(22.375)^3}{22(1.50) + 9(0.375)}} = 6.2 \text{ in.}$$

By inspection, column stability still does not govern.
Use 4PLs – 3/4 × 11 for R = 728.4 kips. (A36 steel).

(m) *Connection for Bearing Stiffeners to Web.*

$$\text{Reqd} f_{vs} = \frac{728.4}{8(96)} = 0.95 \text{ kips/in.}$$

for eight lines of fillet weld (Fig 11.13.6). Use 5/16-in., with maximum

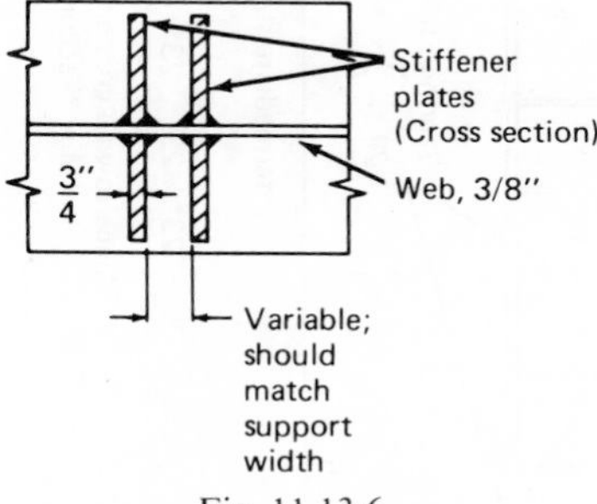

Fig. 11.13.6.

effective size = 0.252, and E70 electrodes, as determined in part (k). The capacity R_w of one line of continuous weld is

$$R_w = 0.252(0.707)21.0 = 3.75 \text{ kips/in.}$$

$$\text{Percent of continuous weld required} = \frac{0.95}{3.75}(100) = 25.3\%$$

Use intermittent 5/16-in. fillet welds, 1½ in. segments @ 5½" pitch (27 percent of continuous weld).

(n) *Compression Stress Directly on the Web (AISC–1.10.10.2).* This

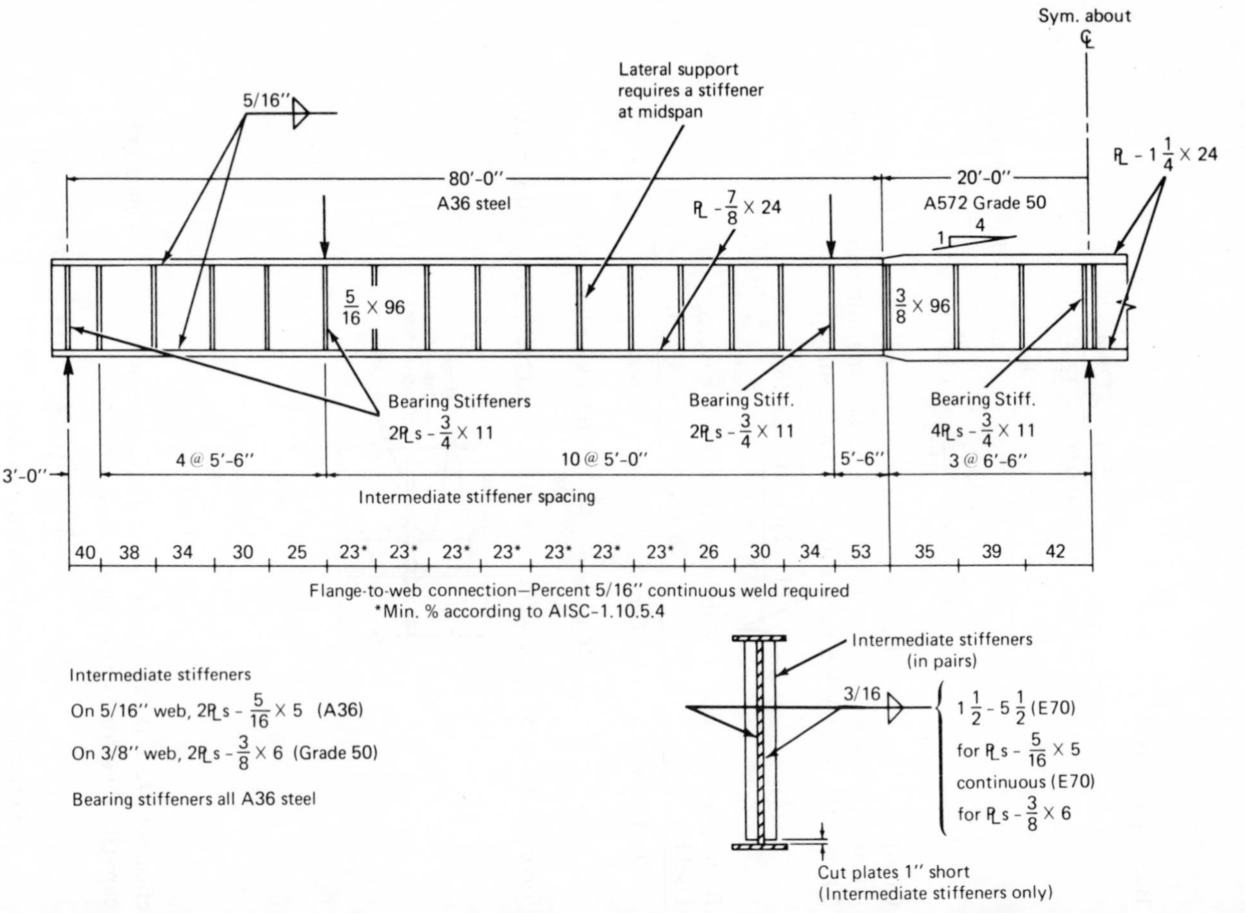

Fig. 11.13.7. Design sketch.

check assures adequate stability under direct compression, as discussed in Sec. 11.2, using Eqs. 11.2.17 with 11.2.19 (AISC–Formula 1.10–11).

The critical region will be the largest stiffener spacing on the thinnest web; therefore the 5′-6″ panel on the $\frac{5}{16}$-in. A36 web will be investigated.

$$\frac{a}{h} = \frac{66}{96} = 0.687; \qquad \frac{h}{t} = 307$$

$$\text{Allowable } f_c = \left[2 + \frac{4}{(a/h)^2}\right]\frac{10{,}000}{(h/t)^2} \text{ ksi}$$

$$= \left[2 + \frac{4}{(0.687)^2}\right]\frac{10{,}000}{(307)^2} = 1.11 \text{ ksi}$$

$$\text{Actual } f_c = \left[\frac{4}{12(5/16)}\right] = 1.07 \text{ ksi} < 1.11 \text{ ksi} \qquad \text{OK}$$

(o) *Design Sketch*. Every design must have all decisions summarized in a design sketch such as Fig. 11.13.7. The design of the bearing stiffeners for the exterior support and under the 75-kip load has been omitted because the procedure has been adequately illustrated in items (l) and (m):

Girder weight: (For one span)

A572 PLs	$1\frac{1}{4} \times 24 \times 20'$	$= 102.0(40.0)$	=	4,080
	$\frac{3}{8} \times 96 \times 20'$	$= 122(20.0)$	=	2,440
A36 PLs	$\frac{7}{8} \times 24 \times 80'$	$= 71.4(160)$	=	11,410
	$\frac{5}{16} \times 96 \times 80'$	$= 102(80)$	=	8,160
Stiffeners	$\frac{3}{8} \times 6 \times 8.0 \times 2 \times 3$	$= 7.65(8)6$	=	370
	$\frac{5}{16} \times 5 \times 8.0 \times 2 \times 13$	$= 5.31(8)26$	=	1,100
	$\frac{3}{4} \times 11 \times 8.0 \times 2 \times 4$	$= 28.1(8)8$	=	1,800
				29,360 lb

Average weight = <u>294 lb/ft</u>

PROBLEMS

Analysis

11.1. For the section shown, the shear force acting is 150 kips and the bending moment is 1,085 ft-kips.

(a) Compute the bending and shear unit stress distribution from top to bottom of the section.

(b) Compute total shear carried by the web and that carried by the flanges. What percent of the total is carried by each?

(c) Compute the total bending moment carried by the web and that carried by the flanges. What percent of the total is carried by each?

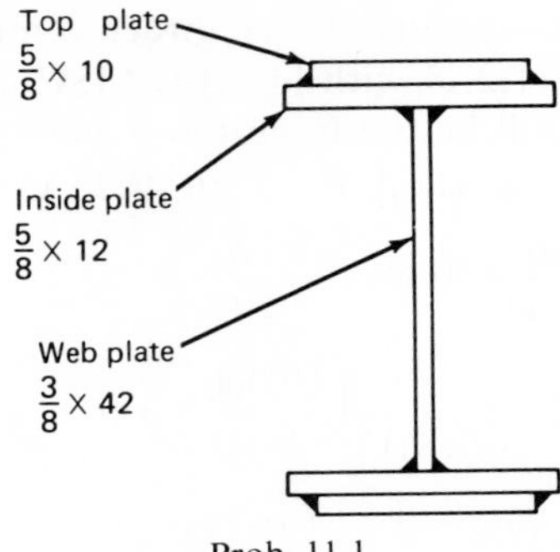

Prob. 11.1.

11.2. (This problem forms basis of Probs. 11.3 to 11.7.) A plate girder supported as shown must carry a live load of 10 kips/ft, not including girder weight, plus the concentrated dead load shown. Compute and draw to scale for later use the moment and shear envelopes for this girder.

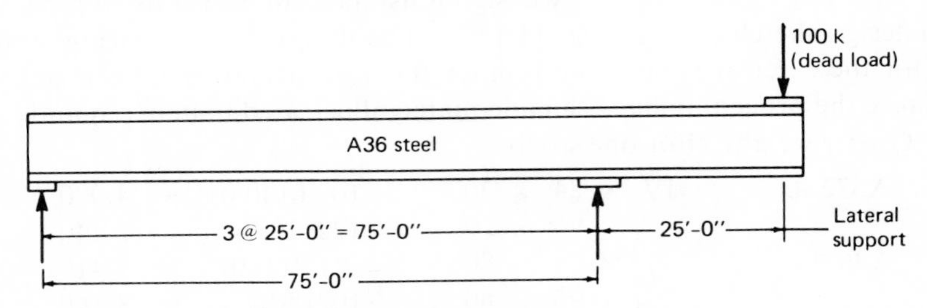

Assume girder weight is 390 lb/ft

Prob. 11.2.

11.3. (Uses results of Prob. 11.2.) For the given plate girder designed for the conditions of Prob. 11.2, compute and draw the moment capacity diagram according to AISC specs. Neglect any reduction which might result from combined shear and moment under AISC–1.10.7. Compare capacity with requirements from Prob. 11.2.

11.4. (Use diagrams of Prob. 11.2.) For the given plate girder, compute and draw shear capacity diagram based on location of intermediate stiffeners. Neglect any limitations of AISC–1.10.7 or 1.10.10. Compare shear capacity with requirements from Prob. 11.2.

11.5. (Use diagrams of Prob. 11.2.) Investigate combined shear and tension stress (AISC–1.10.7) and direct compression on web (AISC–1.10.10.2) by comparing the computed stress with the allowable value at all critical locations.

11.6. (Use diagrams of Prob. 11.2.) Compute and draw the shear capacity diagram for the girder based on flange to web connection. Compare by showing the capacity diagram superimposed on the requirement diagram from Prob. 11.2.

11.7. (Use information from Prob. 11.2.) Check the adequacy of each of the bearing stiffeners, including connection to the web.

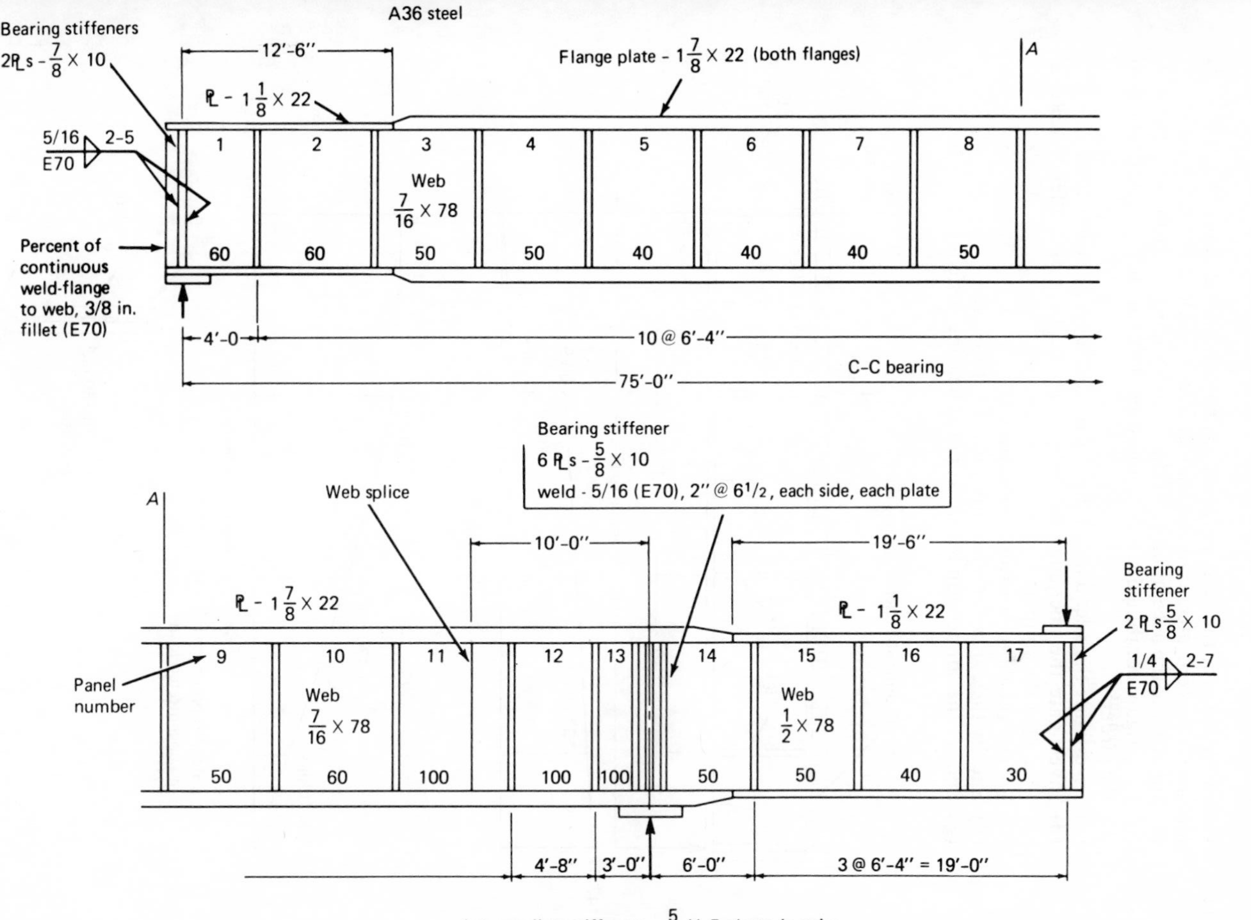

Probs. 11.3 through 11.7.

11.8. Given the data for the 50-ft simple span, having lateral support at the ends and at the concentrated loads located 18 ft from each end; AISC specs.

(a) Investigate the acceptability of the 84-in. spacing for stiffener panel 4.

(b) Investigate combined stress in the web at its most critical location in the girder.

(c) Investigate the adequacy of the size of intermediate stiffeners.

(d) Investigate the adequacy of the size of the bearing stiffener (2PLs = $\frac{1}{2} \times 8$) at the support.

(e) Specify the flange-to-web connection.

(f) Specify the connection for intermediate stiffeners.

(g) Specify the connection for the support bearing stiffener.

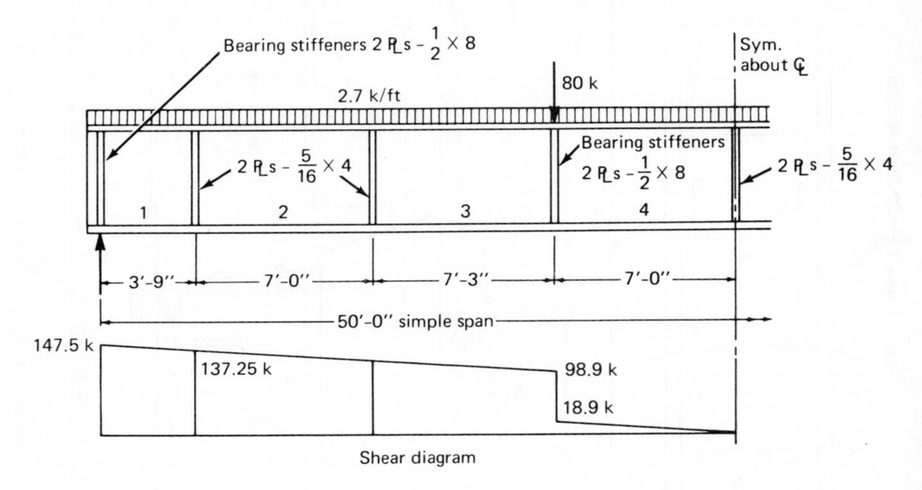

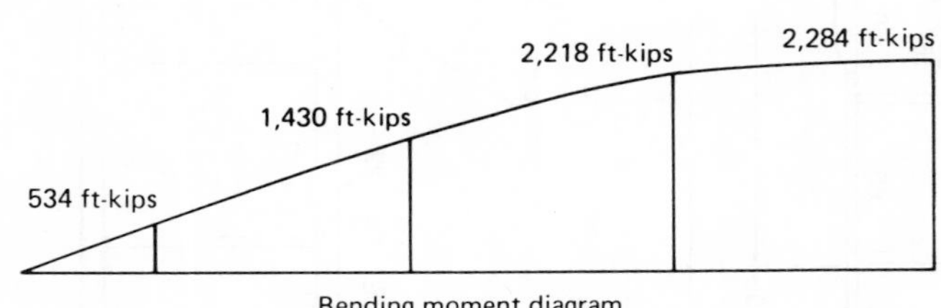

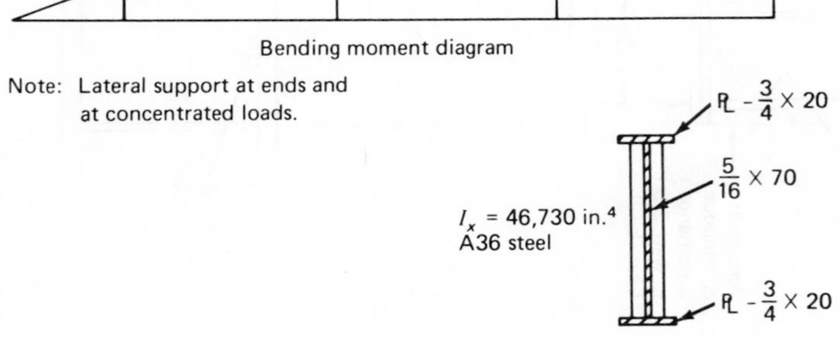

Prob. 11.8.

11.9. Using the given information concerning the portion of the plate girder, determine how close stiffener *B* must be to stiffener *A* in order for the design to be acceptable according to the AISC specification.

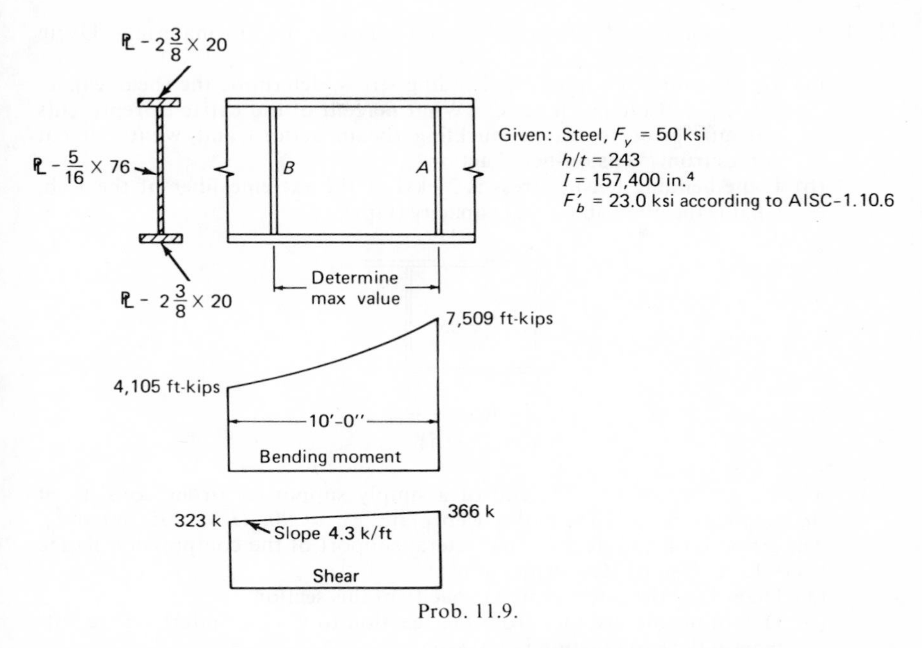

Prob. 11.9.

11.10. Using the given information concerning the portion of the plate girder, determine how close stiffener *B* must be to stiffener *A* in order for the design to be acceptable according to AISC specifications.

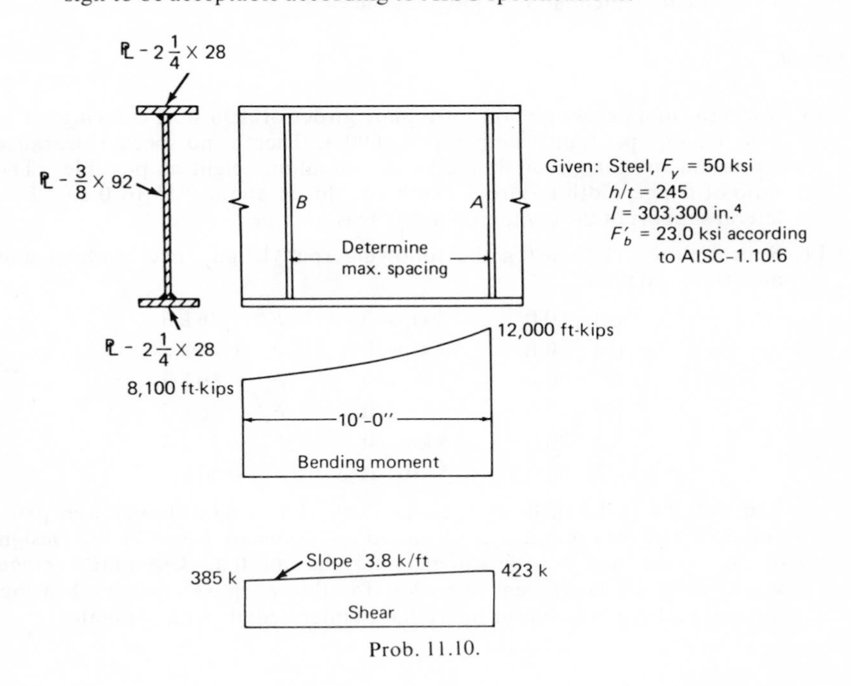

Prob. 11.10.

11.11. Given the plate girder interior panel as shown, of A36 material. Using AISC specs,

(a) Disregarding simultaneous bending stress, determine the shear capacity (kips) of the given panel. What percent of the capacity represents strength prior to elastic buckling (beam action) and what percent comes from "tension-field" action?

(b) If the bending tensile stress is 21 ksi at the extreme fiber of the web, what is the allowable shear capacity (kips)?

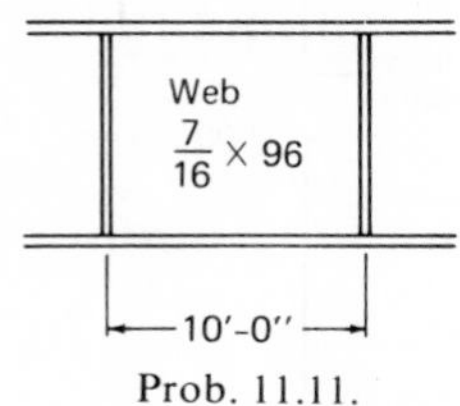

Prob. 11.11.

11.12. The cross section near the end of a simply supported girder consists of flange plates, $1\frac{1}{2} \times 18$, and a web plate, $\frac{3}{8} \times 90$, (I = 135,780 in.^{4}). The girder is of A36 steel and has lateral support of the compression flange every 12 ft. Use AISC specifications.

(a) Determine the safe moment capacity of this section.

(b) Determine the distance from the reaction to the first intermediate stiffener if the end reaction is 320 kips.

(c) Using E70 electrodes, determine the necessary flange to web welding in the end panel. If intermittent welding can be used, indicate the length of weld segment and the pitch distance.

Design

11.13. Select the main cross section for a plate girder of A36 steel carrying a live load of 3 kips per ft on a simple span of 90 ft. There is no specific clearance limitation and the girder should be as minimum weight as possible. The ratio of flange width to girder depth should be about 0.15 to 0.20. Full lateral support of the compression flange is provided.

11.14. Repeat Prob. 11.13 using the following span length, live loading, and material yield point:

(a)	60 ft	3 kips/ft	F_y = 36 ksi
(b)	70 ft	3 kips/ft	F_y = 50 ksi
(c)	80 ft	3 kips/ft	F_y = 60 ksi
(d)	60 ft	5 kips/ft	F_y = 36 ksi
(e)	70 ft	5 kips/ft	F_y = 50 ksi
(f)	80 ft	5 kips/ft	F_y = 60 ksi

11.15. Redesign the girder of Probs. 11.2 through 11.7 using a deeper web plate as dictated by the optimum depth equations discussed in Sec. 11.12. Design of connections and bearing stiffeners may be omitted. Compute average weight per foot for the entire girder; for this purpose consider bearing stiffener plates to weigh four times that of intermediate stiffener plates.

11.16. Design a two-span 140 ft–140 ft continuous welded plate girder to support uniform dead load of 1.25 kips/ft in addition to the girder weight, and a uniform live load of 2.25 kips/ft plus two concentrated loads of 60 kips (40 percent live load) located in each span 35 ft from the supports. Consider lateral support every 35 ft (supports, concentrated loads, and midspan.) Use AISC specifications.

(a) Consider live load to be applied as necessary to give maximum range of stresses.

(b) Use F_y = 50 ksi steel for negative moment zone and A36 steel for positive moment zone.

(c) Consider $^5/_{16}$ in. as the minimum web thickness available.

(d) Use A36 steel for stiffeners.

(e) Specify intermittent welding for connections if any material saving results, even though a cost analysis may later dictate continuous welding.

(f) Submit design sketch to approximate scale on $8^1/_2 \times 11$ paper showing all final decisions.

(g) Compute the total average weight per foot of the girder, including all stiffeners. Assume one splice is required in each span because of the lengths involved. An extra butt welded flange splice (two flanges) should be treated as adding 10 lb/ft to the average weight. (This may approximate the added cost effect of such splices.) Web splices in excess of one required per span should be considered as adding 6 lb/ft to the average girder weight. Economical designs should have an adjusted average girder weight under 430 lb/ft.

11.17. Redesign the girder of Prob. 11.16, except consider the spans 125 ft and the concentrated loads located 25 ft from the supports. Lateral support exists every 25 ft. AISC specifications. Other problem requirements are retained, except as indicated below:

(d) Stiffeners may be either A36 or F_y = 50 ksi

(g) Average girder minimum weight suggestion of 430 lb/ft should be disregarded.

11.18. Redesign the girder of Prob. 11.16, except consider the spans 200 ft without any concentrated loads. Lateral support is continuously provided by the floor system (i.e., it is not provided by intermediate stiffener attachments.) Use AISC specifications. Other problem requirements are retained, except as indicated below:

(b) Use material with F_y = 100 ksi for highest moment region, F_y = 50 ksi for medium moment, and F_y = 36 ksi for lowest moment regions.

(d) Omit restrictions on stiffener material.

(g) Consider that 85 ft maximum length sections will be spliced with field bolted splices (i.e., minimum of 5 pieces will be required.) Extra splices should be considered for comparison purposes by adjusting the average weight as per Prob. 11.16. Disregard the reference to 430 lb/ft.

Review of Theory

11.19. Explain the physical significance of the following C_v values for shear in plate girders. Describe the ultimate behavior based on shear strength in each case.

(a) $C_v \leq 0.8$
(b) $0.8 < C_v \leq 1.0$
(c) $C_v > 1.0$

11.20. What is the specific type of behavior that is prevented when h/t is kept below $760/\sqrt{F_b}$?

11.21. Show by a diagram of forces in equilibrium why stiffener spacing requirements are different for end panels and panels containing large holes than they are for interior panels of a plate girder.

11.22. In deriving AISC–Formula 1.10–3, the force in the stiffener was obtained and then divided by the yield stress to obtain the required area. Why was the yield point used instead of an allowable stress in the working range, such as $0.6F_y$?

11.23. Why is a reduced allowable flexural stress used for $h/t > 760/\sqrt{F_b}$ in AISC-Formula 1.10-5, when actually at ultimate moment the extreme fiber is assumed stressed to F_y for h/t both greater and less than $760/\sqrt{F_b}$. In other words, why does the AISC formula reduce the allowable stress when actually the stress at ultimate does reach F_y?

11.24. Consider the AISC-1.10.2 limitation on the clear distance between flanges:
(a) Show the stress condition on the web plate which gives rise to the limitation equation.
(b) State explicitly what the number 16.5 in the equation means.
(c) If h/t equals the limit value, show the effective girder cross section which might be used to compute moment capacity.

11.25. On a plate girder with $F_y = 50$ ksi, if the web $h/t = 185$, what will happen to the girder before ultimate moment capacity is reached? Particularly describe what happens in a panel between intermediate stiffeners where the moment is high and the shear is low. Be specific and use a sketch.

11.26. Refer to AISC-1.10.5.2.
(a) Explain by diagram the meaning of C_v and state why two different expressions are used for it. Quotations or formulas from the AISC Manual are an inadequate answer.
(b) Why is AISC-Formula 1.10-2 restricted to values of C_v less than 1.0?
(c) What does the second term of AISC-Formula 1.10-2 represent?

SELECTED REFERENCES

1. Thomas C. Shedd, *Structural Design in Steel*, John Wiley & Sons, Inc., New York, 1934, Chap. 3.
2. Edwin H. Gaylord, Jr. and Charles N. Gaylord, *Design of Steel Structures*, McGraw-Hill Book Company, 1957, Chap. 8.
3. Konrad Basler, "Strength of Plate Girders in Shear," *Trans. ASCE*, Vol. 128, Part II (1963), pp. 683–719. (Also as Paper No. 2967, J. Structural Div. *ASCE*, October 1961.)
4. I. Lyse and H. J. Godfrey, "Investigation of Web Buckling in Steel Beams," *Trans. ASCE*, Vol. 100 (1935), pp. 675.
5. George Gerard and Herbert C. Becker, *Handbook of Structural Stability*: Vol. I, *Buckling of Flat Plates*, National Advisory Committee for Aeronautics, NACA Tech. Note 3781, July 1957.

6. Stephen P. Timoshenko and James M. Gere, *Theory of Elastic Stability*, 2nd Ed., McGraw-Hill Book Company, 1961, Chap. 8.
7. Bruce G. Johnston, *The Column Research Council Guide to Design Criteria for Metal Compression Members*, 2nd Ed., John Wiley & Sons, Inc., New York, 1966, Chap. 5.
8. Konrad Basler, "New Provisions for Plate Girder Design," Proc., 1961 National Engineering Conf., AISC, New York.
9. Konrad Basler "Strength of Plate Girders in Bending," *Trans. ASCE*, Vol. 128, Part II (1963), pp. 655–686. (Also as Paper No. 2913, J. Structural Div., *ASCE*, August 1961.)
10. Charles E. Massonnet, "Stability Considerations in the Design of Steel Plate Girders," *Trans. ASCE*, Vol. 127, Part II (1962), pp. 420–447.
11. Geerhard Haaijer and Bruno Thürlimann, "Inelastic Buckling in Steel," *Trans. ASCE*, Vol. 125 (1960), pp. 308–344.
12. Joint ASCE–AASHO Committee on Flexural Members, Subcommittee 1, "Design of Hybrid Steel Beams," *J. Structural Div. ASCE*, Vol 94, No. ST6 (June 1968), pp. 1397–1426.
13. Friederich Bleich, *Buckling Strength of Metal Structures*, McGraw-Hill Book Company, 1952, Chap. 11.
14. William McGuire, *Steel Structures*, Prentice-Hall, Inc., Englewood Cliffs, N. J., 1968, pp. 734–779.
15. Konrad Basler and Bruno Thürlimann, "Plate Girder Research," Proc. National Engineering Conf., AISC, New York, 1959.
16. M. A. D'Apice, D. J. Fielding, and P. B. Cooper, "Static Tests on Longitudinally Stiffened Plate Girders," Welding Research Council Bulletin No. 117, October 1966. (Includes historical survey and bibliography on longitudinally stiffened plates.)
17. Boris Bresler, T. Y. Lin, and John Scalzi, *Design of Steel Structures*, 2nd ed., John Wiley & Sons, Inc., New York, 1968, pp. 497–554.
18. Omer W. Blodgett, *Design of Welded Structures*, James F. Lincoln Arc Welding Foundation, Cleveland, Ohio, 1966.

12

Combined Bending and Axial Load

12.1. INTRODUCTION

Nearly all members in a structure are subjected to both bending moment and axial load—either tension or compression. When the magnitude of one or the other is relatively small its effect is usually neglected and the member is designed either as a beam, or as an axially loaded column. For many situations neither effect can properly be neglected and the behavior under the combined loading must be considered in design. The general subject of strength and stability considerations and design procedures has been extensively treated by Massonnet[1] and Austin.[2]

Since bending is involved, all of the factors considered in Chapters 7 and 9 apply here also; particularly, the stability related factors, such as lateral-torsional buckling and local buckling of compression elements. When bending is combined with axial tension, the chance of instability is reduced and yielding usually governs the design. For bending in combination with axial compression, the possibility of instability is increased with all of the considerations of Chapter 6 applying. Furthermore, when axial compression is present, a secondary bending moment arises equal to the axial compression force times the deflection. Prior to 1961, most design specifications did not explicitly account for this secondary bending moment.

A number of categories of combined bending and axial load along with the likely mode of failure may be summarized as follows:

(a) Axial tension and bending; failure usually by yielding.

(b) Axial compression and bending about one axis; failure by instability in the plane of bending, without twisting. (Transversely loaded beam-columns which are stable with regard to lateral buckling are an example of this category.)

(c) Axial compression and bending about the strong axis: failure by lateral-torsional buckling.
(d) Axial compression and biaxial bending—torsionally stiff sections; failure by instability in one of the principal directions; (W shapes are usually in this category.)
(e) Axial compression and biaxial bending—thin-walled open sections; failure by combined twisting and bending on these torsionally weak sections.
(f) Axial compression, biaxial bending, and torsion; failure by combined twisting and bending when plane of bending does not contain the shear center.

It may be apparent from this summary that no single design procedure is likely to properly account for such varied behavior. Current design procedures generally are in one of three categories: (a) limitation on combined stress, (b) empirical interaction formulas, essentially related to working stress procedures, and (c) semi-empirical interaction procedures based on ultimate strength. Limitations on combined stress ordinarily cannot provide a proper criterion unless instability is prevented, or large safety factors are used. Interaction equations more accurately describe the true behavior by accounting for most of the stability situations. The AISC equations for beam-columns are of the interaction type.

12.2. DIFFERENTIAL EQUATION FOR AXIAL COMPRESSION AND BENDING

In order to understand the behavior, the basic situation of case (b), Sec. 12.1, will be treated. Failure by instability in the plane of bending is assumed. Consider the general case of Fig. 12.2.1, where the lateral loading $w(z)$ in combination with any end moments, M_1 or M_2, constitute the primary bending moment M_i a function of z. The primary moment causes the member to deflect, y, giving rise to a secondary

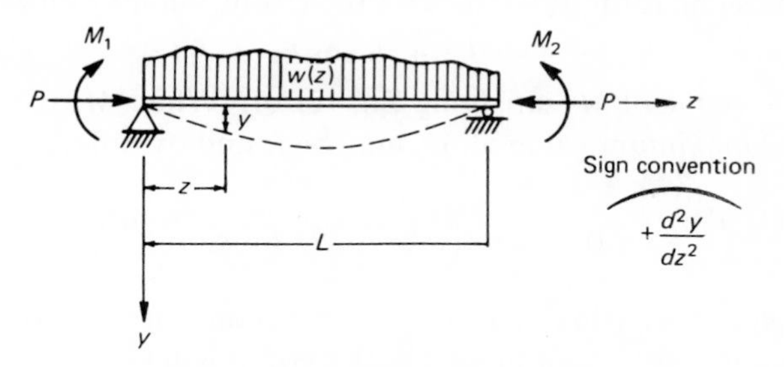

Fig. 12.2.1. General loading of beam-column.

moment Py. Stating the moment M_z at the section z of Fig. 12.2.1, gives

$$M_z = M_i + Py = -EI\frac{d^2y}{dz^2} \tag{12.2.1}$$

for sections with constant EI, and dividing by EI gives

$$\frac{d^2y}{dz^2} + \frac{P}{EI}y = -\frac{M_i}{EI} \tag{12.2.2}$$

For design purposes, the general expression for moment M_z is of greater importance than the deflection y. Differentiating Eq. 12.2.2 twice gives

$$\frac{d^4y}{dz^4} + \frac{P}{EI}\frac{d^2y}{dz^2} = -\frac{1}{EI}\frac{d^2M_i}{dz^2} \tag{12.2.3}$$

From Eq. 12.2.1,

$$\frac{d^2y}{dz^2} = -\frac{M_z}{EI} \quad \text{and} \quad \frac{d^4y}{dz^4} = -\frac{1}{EI}\frac{d^2M_z}{dz^2}$$

Substitution in Eq. 12.2.3 gives

$$-\frac{1}{EI}\frac{d^2M_z}{dz^2} + \frac{P}{EI}\left(\frac{-M_z}{EI}\right) = -\frac{1}{EI}\frac{d^2M_i}{dz^2}$$

or, simplifying and letting $k^2 = P/EI$,

$$\frac{d^2M_z}{dz^2} + k^2M_z = \frac{d^2M_i}{dz^2} \tag{12.2.4}$$

which is of the same form as the deflection differential equation, Eq. 12.2.2.

The homogeneous solution for Eq. 12.2.4 is

$$M_z = A\sin kz + B\cos kz$$

as first discussed in Sec. 6.2. To this must be added the particular solution which will satisfy the right-hand side of the differential equation. Since $M_i = f(z)$, where $f(z)$ is usually a polynomial in z, the particular solution will be of the same form; thus, the complete solution may be written

$$M_z = A\sin kz + B\cos kz + f_1(z) \tag{12.2.5}$$

where $f_1(z)$ = value of M_z satisfying Eq. 12.2.4. When M_z is a continuous function, the maximum value of M_z may be found by differentiation:

$$\frac{dM_z}{dz} = 0 = Ak\cos kz - Bk\sin kz + \frac{df_1(z)}{dz} \tag{12.2.6}$$

For most ordinary loading cases, such as concentrated loads, uniform loads, end moments, or combinations thereof, it will be shown that

$$\frac{df_1(z)}{dz} = 0$$

in which case a general expression for maximum M_z can be established; from Eq. 12.2.6,

$$Ak \cos kz = Bk \sin kz$$

$$\tan kz = \frac{A}{B} \tag{12.2.7}$$

At maximum M_z,

$$\sin kz = \frac{A}{\sqrt{A^2 + B^2}}, \quad \cos kz = \frac{B}{\sqrt{A^2 + B^2}} \tag{12.2.8}$$

and substitution of Eqs. 12.2.8 into Eq. 12.2.5 gives

$$M_{z\max} = \frac{A^2}{\sqrt{A^2 + B^2}} + \frac{B^2}{\sqrt{A^2 + B^2}} + f_1(z)$$

$$= \sqrt{A^2 + B^2} + f_1(z) \tag{12.2.9}$$

It is noted that whenever $df_1(z)/dz \neq 0$, Eq. 12.2.6 must be solved for kz and the result substituted into Eq. 12.2.5.

Case 1—Unequal End Moments Without Transverse Loading. Referring to Fig. 12.2.2, the primary moment, M_i, may be expressed

$$M_i = M_1 + \frac{M_2 - M_1}{L} z \tag{12.2.10}$$

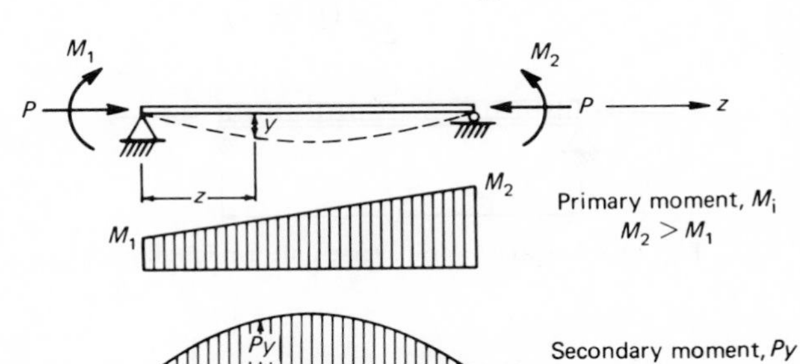

Fig. 12.2.2. Case 1—end moments without transverse loading.

Since

$$\frac{d^2 M_i}{dz^2} = 0$$

Eq. 12.2.4 becomes a homogeneous equation, in which case $f_1(z) = 0$ for Eq. 12.2.5. The maximum moment, Eq. 12.2.9 is

$$M_{z\max} = \sqrt{A^2 + B^2} \tag{12.2.11}$$

The constants A and B are evaluated by applying the boundary conditions to Eq. 12.2.5. The general equation is

$$M_z = A \sin kz + B \cos kz$$

and the conditions are

(1) at $z = 0, M_z = M_1$

$$\therefore B = M_1$$

(2) at $z = L, M_z = M_2$

$$M_2 = A \sin kL + M_1 \cos kL$$

$$\therefore A = \frac{M_2 - M_1 \cos kL}{\sin kL}$$

so that

$$M_z = \left(\frac{M_2 - M_1 \cos kL}{\sin kL}\right) \sin kz + M_1 \cos kz \tag{12.2.12}$$

and

$$M_{z\max} = \sqrt{\left(\frac{M_2 - M_1 \cos kL}{\sin kL}\right)^2 + M_1^2}$$

$$= M_2 \sqrt{\frac{1 - 2(M_1/M_2)\cos kL + (M_1/M_2)^2}{\sin^2 kL}} \tag{12.2.13}$$

Case 2—Transverse Uniform Loading. Referring to Fig. 12.2.3, the primary moment M_i may be expressed as

$$M_i = \frac{w}{2} z(L - z) \tag{12.2.14}$$

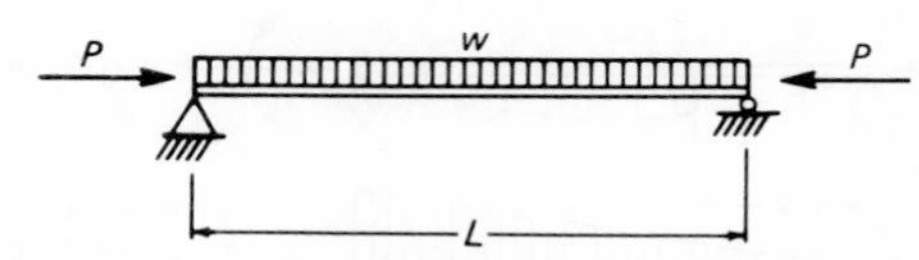

Fig. 12.2.3. Case 2—transverse uniform loading.

Since

$$\frac{d^2 M_i}{dz^2} = -w$$

$f_1(z) \neq 0$; the particular solution for the differential equation is required. Let $f_1(z) = C_1 + C_2 z$; i.e., any polynomial. Substitute the particular solution into Eq. 12.2.4,

$$\frac{d^2[f_1(z)]}{dz^2} = 0$$

$$0 + k^2(C_1 + C_2 z) = -w$$

Thus

$$C_1 = -w/k^2$$

$$C_2 = 0$$

Equation 12.2.5 then becomes

$$M_z = A \sin kz + B \cos kz - w/k^2 \tag{12.2.15}$$

Applying the boundary conditions,

(1) at $z = 0, M_z = 0$

$$0 = B - w/k^2; \quad \therefore \mathrm{B} = w/k^2$$

(2) at $z = L, M_z = 0$

$$0 = A \sin kL + \frac{w}{k^2} \cos kL - \frac{w}{k^2}$$

$$\therefore A = \frac{w}{k^2}\left(\frac{1 - \cos kL}{\sin kL}\right)$$

Since

$$\frac{df_1(z)}{dz} = 0,$$

Eq. 12.2.9 gives maximum moment,

$$M_{z\,\max} = \frac{w}{k^2}\sqrt{\left(\frac{1 - \cos kL}{\sin kL}\right)^2 + 1} - \frac{w}{k^2}$$

$$= \frac{w}{k^2}\left(\sec \frac{kL}{2} - 1\right)$$

$$= \frac{wL^2}{8}\underbrace{\left(\frac{8}{(kL)^2}\right)\left(\sec \frac{kL}{2} - 1\right)}_{\text{magnification factor due to axial load}} \tag{12.2.16}$$

Case 3—Equal End Moments Without Transverse Loading (Secant Formula). Consider that $M_1 = M_2 = M$ in Fig. 12.2.2, in which case Eq. 12.2.13 becomes

$$M_{z\,\max} = M\sqrt{\frac{2(1 - \cos kL)}{\sin^2 kL}} \tag{12.2.17}$$

$$= M\sqrt{\frac{2(1 - \cos kL)}{1 - \cos^2 kL}}$$

$$= M\left(\frac{1}{\cos kL/2}\right)$$

$$= M \sec \frac{kL}{2} \tag{12.2.18}$$

which was used in Eq. 6.11.3. It is noted that loading a beam-column with constant moment over the length causes a uniform compression due to bending, and constitutes the most severe loading to which such a member can be subjected. In view of this it would always be conservative to multiply the maximum primary moment due to *any* loading by sec $kL/2$, which is excessively conservative in most cases.

12.3. MOMENT MAGNIFICATION— SIMPLIFIED TREATMENT FOR MEMBERS IN SINGLE CURVATURE WITHOUT END TRANSLATION.

As an alternate to the differential equation approach, a simple approximate procedure is satisfactory for many common situations.

Assume a beam-column is subject to lateral loading q which causes a deflection δ_0 at midspan, as shown in Fig. 12.3.1. The secondary bend-

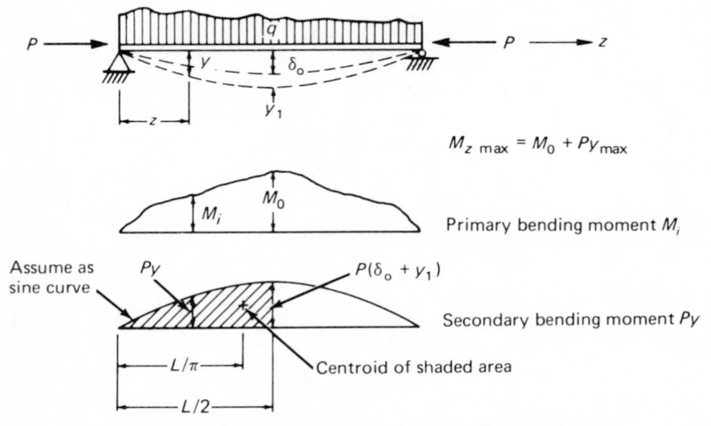

Fig. 12.3.1. Primary and secondary bending moment.

ing moment may be assumed to vary as a sine curve, which is nearly correct for members with no end restraint whose primary bending moment and deflection are maximum at midspan.

The portion of the midspan deflection y_1, due to the secondary bending moment, equals the moment of the M/EI diagram between the support and midspan (shaded portion of Fig. 12.3.1) taken about the support, according to the moment-area principle:

$$y_1 = \frac{P}{EI}(y_1 + \delta_0)\left(\frac{L}{2}\right)\frac{2}{\pi}\left(\frac{L}{\pi}\right) = (y_1 + \delta_0)\frac{PL^2}{\pi^2 EI} \tag{12.3.1}$$

or

$$y_1 = (y_1 + \delta_0)\frac{P}{P_e} \tag{12.3.2}$$

where $P_e = \pi^2 EI/L^2$. Solving for y_1:

$$y_1 = \delta_0 \left[\frac{P/P_e}{1 - \dfrac{P}{P_e}} \right] = \delta_0 \left(\frac{\alpha}{1 - \alpha} \right) \tag{12.3.3}$$

where $\alpha = P/P_e$. Since y_{max} is the sum of δ_0 and y_1,

$$y_{max} = \delta_0 + y_1 = \delta_0 + \delta_0 \left(\frac{\alpha}{1 - \alpha} \right) = \frac{\delta_0}{1 - \alpha} \tag{12.3.4}$$

The maximum bending moment including the axial effect becomes

$$M_{z\,max} = M_0 + P y_{max} \tag{12.3.5}$$

Substituting the expression for y_{max} in Eq. 12.3.5 and setting $P = \alpha P_e = \alpha \pi^2 EI/L^2$, Eq. 12.3.5 becomes

$$M_{z\,max} = M_0 \left(\frac{C_m}{1 - \alpha} \right) \tag{12.3.6}$$

where

$$C_m = 1 + \left(\frac{\pi^2 EI\, \delta_0}{M_0 L^2} - 1 \right) \alpha = 1 + \psi \alpha \tag{12.3.7}$$

The term $C_m/(1 - \alpha)$ may be considered the magnification factor.

EXAMPLE 12.3.1

Compare the differential equation amplification factor for the loading of Fig. 12.2.3, Eq. 12.2.16, with the approximate value, Eq. 12.3.6.

Solution

For the differential equation,

$$A_m = \text{magnification factor} = \frac{2}{(kL/2)^2} \left(\sec \frac{kL}{2} - 1 \right) \tag{a}$$

where

$$\frac{kL}{2} = \frac{L}{2} \sqrt{\frac{P}{EI}} = \frac{\pi}{2} \sqrt{\alpha}$$

For the approximate solution,

$$A_m = \frac{C_m}{1 - \alpha}$$

$$\delta_0 = \frac{5wL^4}{384EI}; \quad M_0 = \frac{wL^2}{8}$$

$$\frac{\delta_0}{M_0} = \frac{5L^2}{48EI}$$

$$C_m = 1 + \left(\frac{\pi^2 EI}{L^2}\,\frac{5L^2}{48EI} - 1\right)\alpha$$

$$= 1 + 0.028\alpha$$

$$A_m = \frac{1 + 0.028\alpha}{1 - \alpha} \tag{b}$$

α	sec $kL/2$	Eq. (a)	Eq. (b)
0.1	1.137	1.114	1.114
0.2	1.310	1.257	1.257
0.3	1.533	1.441	1.441
0.4	1.832	1.686	1.685
0.5	2.252	2.030	2.028
0.6	2.884	2.546	2.542
0.7	3.941	3.405	3.399
0.8	6.058	5.125	5.112
0.9	12.419	10.284	10.253

Obviously, there is no significant difference between the differential equation solution and the approximate solution for this case.

12.4. MOMENT MAGNIFICATION— MEMBERS SUBJECTED TO END MOMENTS ONLY; NO JOINT TRANSLATION

For the situation shown in Fig. 12.2.2, which has no transverse loading, the theoretical maximum moment is given by Eq. 12.2.13,

$$M_{z\,\max} = M_2\sqrt{\frac{(M_1/M_2)^2 - 2(M_1/M_2)\cos kL + 1}{\sin^2 kL}} \tag{12.2.13}$$

For this situation the maximum moment may be either (1) the larger end moment M_2 at the braced location (Fig. 12.4.1a), or (2) the amplified moment given by Eq. 12.2.13 which occurs at various locations out along the span (Fig. 12.4.1b), depending upon the ratio M_1/M_2 and the value of α, since $kL = \pi\sqrt{\alpha}$. In order to make a stress analysis, one needs to know whether the maximum moment occurs at a location away from the support, and if so, the correct *distance*. To eliminate the need for such information, the concept of equivalent uniform moment (Fig. 12.4.1c) is used. Thus when investigating stresses away from the supported point, use of the equivalent moment assumes $M_{z\,\max}$ to be at midspan.

To establish the equivalent moment, let the solution for uniform moment, Eq. 12.2.17 with $M = M_{\text{equiv}}$ be equated with Eq. 12.2.13. One obtains

$$M_{\text{equiv}} = M_2\sqrt{\frac{(M_1/M_2)^2 - 2(M_1/M_2)\cos kL + 1}{2(1 - \cos kL)}} \tag{12.4.1}$$

By the procedure used in Example 12.3.1 it may be shown that for

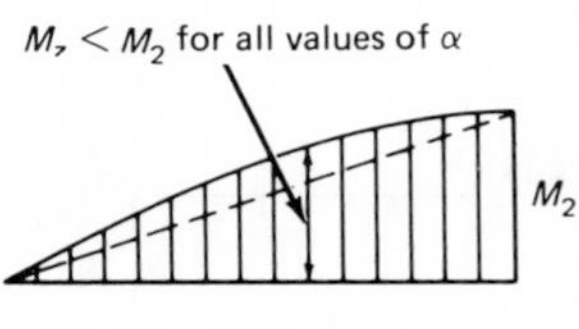

(a) Maximum moment at ends

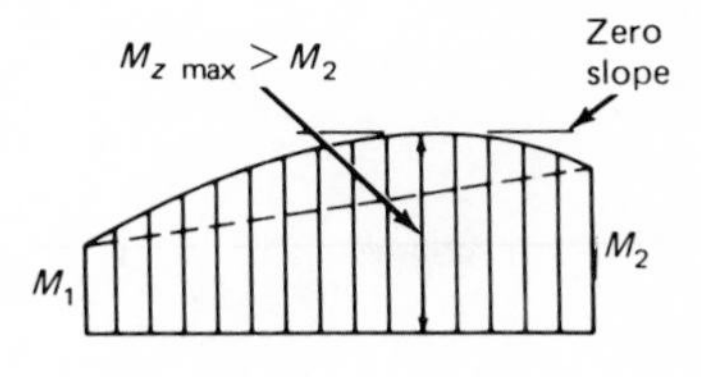

(b) Maximum moment *not* at ends

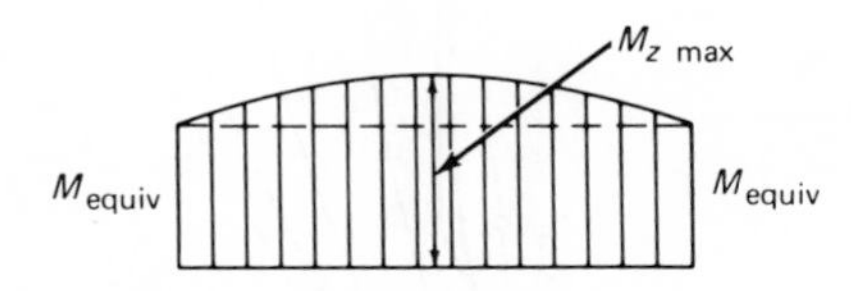

(c) Equivalent uniform moment with maximum magnified moment at midspan

Fig. 12.4.1.

uniform moment, the magnification factor is obtained from Eq. 12.2.18:

$$A_m = \sec\frac{kL}{2} = \frac{1}{1-\alpha} \tag{12.4.2}$$

and using the equivalent uniform moment, M_{equiv}, to replace M_1 and M_2, the full maximum moment may be expressed as

$$M_{z\,\max} = \frac{M_{\text{equiv}}}{1-\alpha} \tag{12.4.3}$$

which when compared with Eq. 12.3.6 may be written

$$M_{z\,\max} = M_2\frac{C_m}{1-\alpha} \tag{12.4.4}$$

where

$$C_m = M_{\text{equiv}}/M_2$$

$$= \sqrt{\frac{(M_1/M_2)^2 - 2(M_1/M_2)\cos kL + 1}{2(1-\cos kL)}} \tag{12.4.5}$$

Equation 12.4.5 does not consider lateral-torsional buckling, or fully cover the double-curvature cases where M_1/M_2 lies between -0.5 and -1.0. The actual failure of members bent in double curvature with bending-moment ratios -0.5 to -1.0 is generally one of "unwinding" through from double to single curvature in a sudden type of buckling, as discussed by Ketter,[4] among others.

Massonnet,[1] Horne,[18] and the AISC specification[19] have suggested alternate expressions for practical use. Figure 12.4.2 shows Eq. 12.4.5 compared with the recommendations of Massonnet and the AISC. One should note that for a given value of α, the curve shown terminates when the moment, M_2, at the end of the member exceeds the amplified moment.

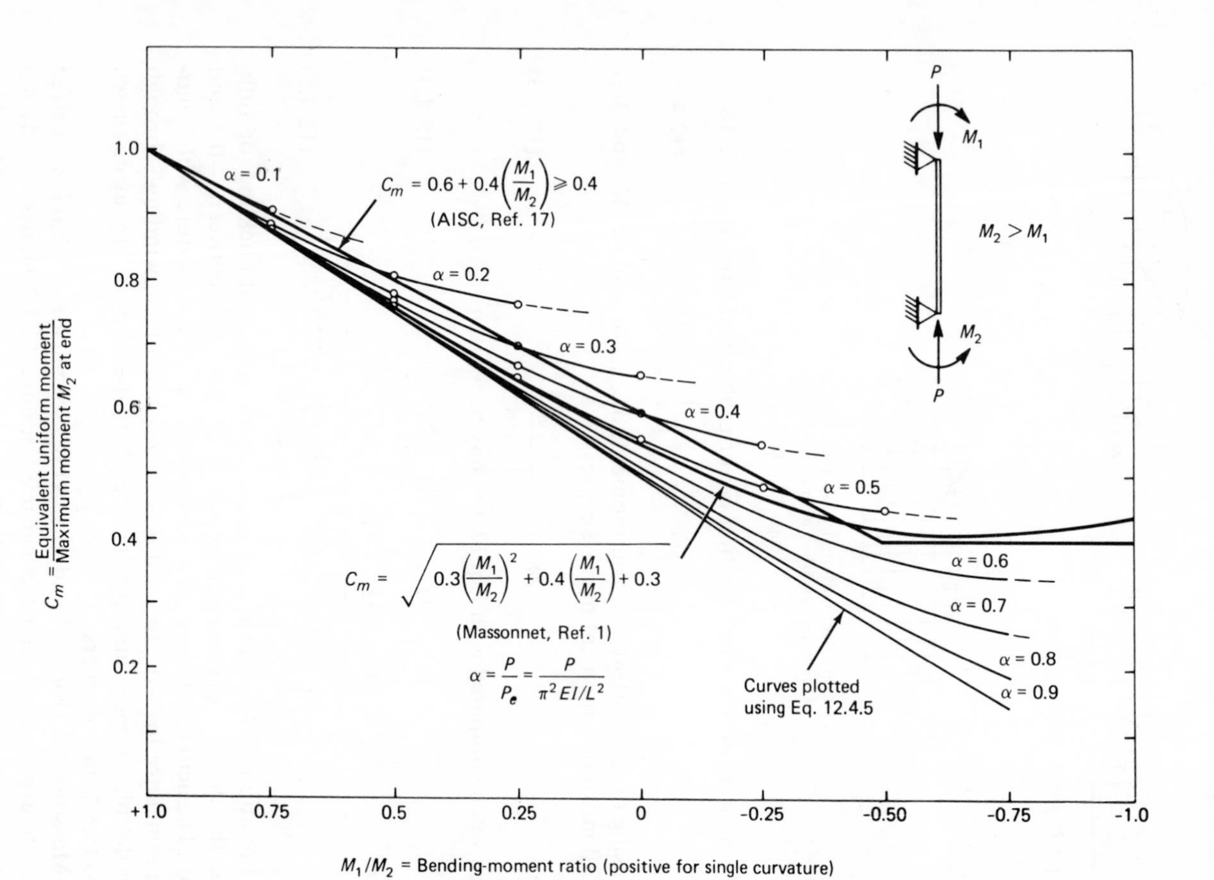

Fig. 12.4.2. Comparison of theoretical C_m with design recommendations for members subject to end moments only, without joint translation.

The most important situations are those where the amplified moment exceeds the moment at the end. The straight line recommended by AISC falls near the upper limit for C_m at any given bending-moment ratio, and thus seems to be a realistic and simple approximation. In design, a safety factor is used which overestimates α. By not taking C_m as a function of this increased α, a higher value of C_m will be used; thus it is a conservative procedure.

12.5. MOMENT MAGNIFICATION—MEMBERS SUBJECT TO SIDESWAY

The unbraced frame is dealt with in Chapter 14, where the development is presented to show that the maximum moment M_0 computed in a standard analysis of an elastic frame should be amplified to account for sidesway buckling. One may write

$$M_{\max} = M_0\left(\frac{C_m}{1 - \alpha}\right) \tag{12.5.1}$$

in the same manner as for the braced frame cases of Secs. 12.3 and 12.4.

Consider the sidesway situation of Fig. 12.5.1. Whatever the degree

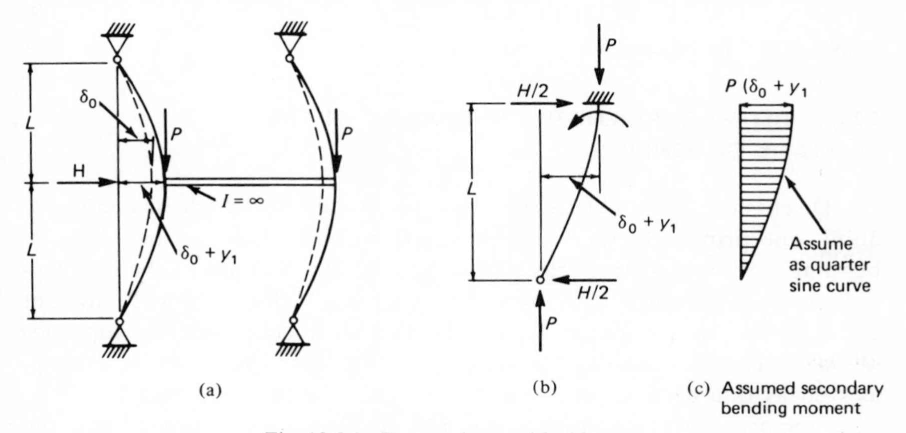

Fig. 12.5.1. Beam-column with sidesway.

of restraint at the top and bottom of the two story member, the deflection curve, and therefore the secondary bending moment (P times deflection) may be reasonably assumed as a sine curve. In which case the development used when no sidesway occurs (Fig. 12.3.1) is also valid here. Since $2L$ from Fig. 12.5.1 equals L for Fig. 12.3.1, Eq. 12.3.7 becomes

$$C_m = 1 + \left(\frac{\pi^2 EI\delta_0}{4L^2 M_0} - 1\right)\alpha \tag{12.5.2}$$

The larger effective length ($2L$ instead of L) is also used in the computation of α. Next, referring to Fig. 12.5.1,

$$\delta_0 = \frac{(H/2)\,L^3}{3EI} \tag{12.5.3}$$

$$M_0 = \frac{HL}{2} \tag{12.5.4}$$

Substitution of Eqs. 12.5.3 and 12.5.4 into Eq. 12.5.2 gives

$$C_m = 1 + \left[\frac{\pi^2 EI}{4L^2}\left(\frac{HL^3}{6EI}\right)\left(\frac{2}{HL}\right) - 1\right]\alpha$$

or

$$C_m = 1 + \left(\frac{\pi^2}{12} - 1\right)\alpha = 1 - 0.18\alpha \tag{12.5.5}$$

as suggested in the AISC Commentary–1.6.1.

A comparison of a theoretical analysis with Eq. 12.5.5 is presented in Chapter 14.

The AISC Commentary–1.6.1 recommends but the specification, AISC–1.6.1, *requires* use of

$$C_m = 0.85 \tag{12.5.6}$$

which seems to be *un*conservative.

12.6. * ULTIMATE STRENGTH—INSTABILITY IN THE PLANE OF BENDING

Discussion is limited to the basic beam-column which has moment about one principal axis and which fails by instability in the plane of bending, without twisting [case (b) of Sec. 12.1]. A study of the differential equation solution will show that the axial effect and the moment effect cannot be computed separately and then combined by superposition. It is a nonlinear relationship. Furthermore, because of bending, as well as due to residual stress, some fibers reach yield before others. Thus the problem is a nonlinear stability problem which must include the effects of inelastic action. The ultimate capacity is reached when the stiffness of the members has been reduced, because of yielding under combined stress, to the state where instability occurs. Similar to the treatment in Secs. 6.4 through 6.6, the ultimate axial load is the critical load on the partially yielded member. Since it is believed that an understanding of ultimate strength is necessary for knowledgable application of design procedures, the following example of developing the ultimate strength of a simple rectangular section is presented.

EXAMPLE 12.6.1

Determine moment-curvature-axial load relationships for a rectangular section. Assume no residual stress and that the steel exhibits ideal elastic-plastic behavior (Fig. 12.6.1d).

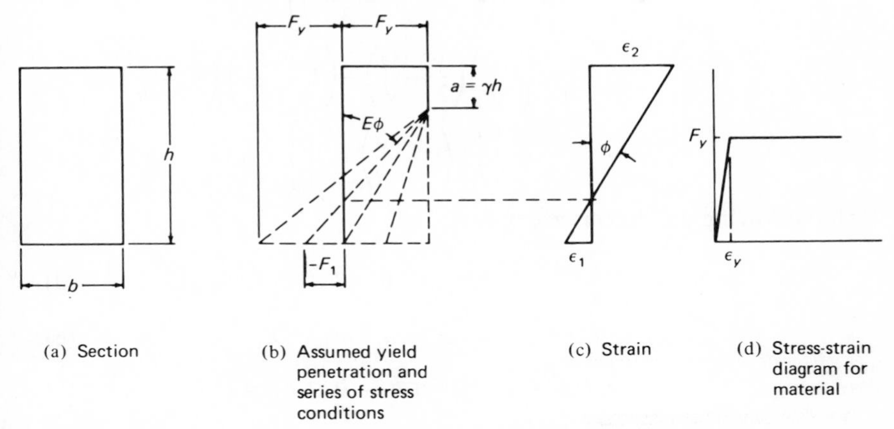

Fig. 12.6.1. Rectangular member under axial compression and bending, for $\epsilon_1 \leq \epsilon_y$.

Solution

The procedure used is that presented by Ketter, Kaminsky, and Beedle for wide-flange sections.[3] It is assumed that bending strain is linear even when some fibers are strained into the plastic range. For strains greater than ϵ_y, the stress is constant at the value F_y. The curvature at any section is then a function of the elastic portion of the cross-section.

The procedure is to establish a series of relationships between moment M, axial load P, and curvature ϕ for a given assumed depth of yield penetration a (see Fig. 12.6.1b). From similar triangles, the curvature equation is

$$\phi = \frac{F_y - F_1}{E(h - a)} = \frac{F_y(1 - F_1/F_y)}{h(1 - a/h)E} \tag{a}$$

The volume of the stress solid gives the axial load,

$$P = \left[F_y a + \frac{F + F_1}{2}(h - a)\, b\right]$$

$$= F_y bh\left(\frac{1}{2}\right)\left[1 + \frac{a}{h} + \frac{F_1}{F_y}\left(1 - \frac{a}{h}\right)\right] \tag{b}$$

and taking moments about the bottom edge gives

$$M = b\left[F_y a\left(h - \frac{a}{2}\right) + F_y\left(\frac{h-a}{2}\right)\left(\frac{2}{3}\right)(h-a) + F_1\left(\frac{h-a}{2}\right)\left(\frac{h-a}{3}\right)\right] - P\frac{h}{2}$$

$$= F_y\frac{bh^2}{6}\left\{3\left(\frac{a}{h}\right)\left(\frac{3}{2} - \frac{a}{h}\right) + \left(2 + \frac{F_1}{F_y}\right)\left(1 - \frac{a}{h}\right)^2 - \frac{3}{2}\left[\frac{F_1}{F_y}\left(1 - \frac{a}{h}\right) + 1\right]\right\} \quad \text{(c)}$$

If the following definitions are used:

$$\phi_y = \frac{2F_y}{hE} \qquad \text{for } a = 0, \quad F_1 = -F_y \qquad \text{(d)}$$

$$P_y = F_y bh \qquad \text{for } a = h, \quad F_1 = F_y \qquad \text{(e)}$$

$$M_y = F_y\frac{bh^2}{6} \qquad \text{for } a = 0, \quad F_1 = -F_y \qquad \text{(f)}$$

equations (a), (b), and (c) can be nondimensionalized by dividing by ϕ_y, P_y, and M_y respectively. Further, defining $R = F_1/F_y$ and $\gamma = a/h$, the dimensionless equations may be written

$$\frac{\phi}{\phi_y} = \frac{1 - R}{2(1-\gamma)} \qquad \text{(g)}$$

$$\frac{P}{P_y} = \frac{1}{2}[1 + \gamma + R(1-\gamma)] \qquad \text{(h)}$$

$$\frac{M}{M_y} = 3\gamma(1.5 - \gamma) + (2 + R)(1-\gamma)^2 - 1.5[R(1-\gamma) + 1] \qquad \text{(i)}$$

Equations (g), (h), and (i) are valid for elastic conditions ($\gamma = 0$), or for primary yielding (yielding in the compression zone.)

When $R = -1$, ϕ/ϕ_y may still increase until either $\beta = \gamma$ or ϕ becomes infinite, whichever occurs first. The additional contribution to Equations (g), (h), and (i) can be obtained by working with the shaded portion of Fig. 12.6.2.

As β increases from zero, the shaded area represents a decrease in the axial compression capacity:

$$\Delta P = \frac{1}{2}\beta h(2F_y)b = \beta bhF_y$$

or

$$\frac{\Delta P}{P_y} = \beta \qquad \text{(j)}$$

The additional moment capacity equals ΔP times the moment arm measured from middepth to the centroid of the shaded area:

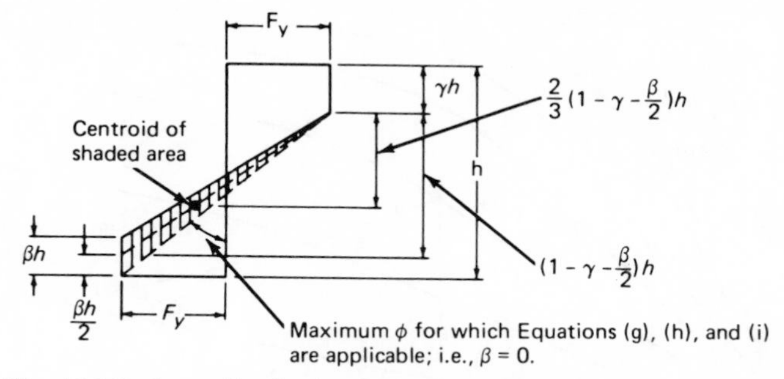

Fig. 12.6.2. Stress distribution for member under axial compression and bending: effect of yield in the tension zone.

$$\Delta M = \beta bhF_y\left[\gamma h + \frac{2}{3}(1 - \gamma - \beta/2)h - h/2\right]$$

or

$$\frac{\Delta M}{M_y} = \beta(2\gamma - 2\beta + 1) \tag{k}$$

When secondary yielding (yielding in the tension zone) occurs, i.e., when $\beta \geq 0$, Eqs. (g), (h), and (i) become

$$\frac{\phi}{\phi_y} = \frac{h}{2F_y}\left(\frac{2F_y}{h - \gamma h - \beta h}\right) = \left(\frac{1}{1 - \gamma - \beta}\right) \tag{l}$$

Setting $R = -1$ and subtracting Eq. (j) from Eq. (h) give

$$\frac{P}{P_y} = \gamma - \beta \tag{m}$$

Setting $R = -1$ and adding Eq. (k) to Eq. (i) give

$$\frac{M}{M_y} = \gamma - 2\gamma^2 + 1 + \beta(2\gamma - 2\beta + 1) \tag{n}$$

Figure 12.6.3 shows the interaction curves for $M - P - \phi$. The dotted line indicates the limit of applicability for Eqs. (g), (h) and (i) at $R = -1$ (first yield on the tensile fiber). The remainder of the curves are obtained using Eqs. (l), (m), and (n).

From the auxiliary curves of Fig. 12.6.3, a series of M–ϕ curves can be obtained for constant values of P/P_y. The graphical procedure is shown on Fig. 12.6.3 for $P/P_y = 0.3$. The results are shown in Fig. 12.6.4, which gives the beam-column strength for a short member where yielding governs.

EXAMPLE 12.6.2

Making use of the M–ϕ curves of Fig. 12.6.4, determine the P–M strength relationship for a 10 by 20 rectangular section for a slenderness ratio of 60. Assume $E = 29{,}000$ ksi, and $F_y = 50$ ksi.

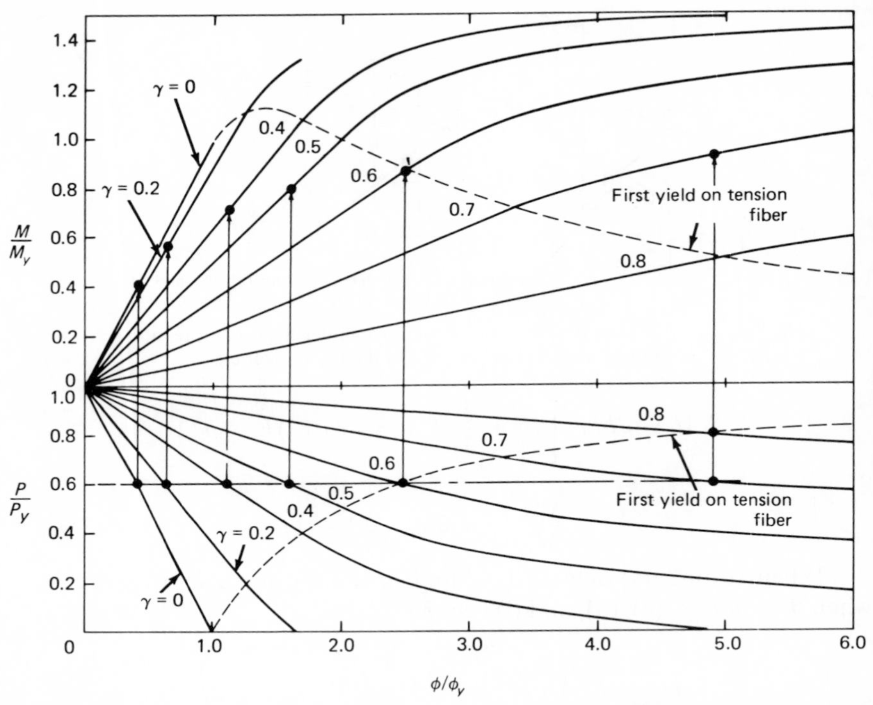

Fig. 12.6.3. Auxiliary curves of M—P—ϕ for constant values of yield penetration at compression face of rectangular beam. (Dashed lines indicate limits of applicability of Eqs. (g), (h), and (i) of Example 12.6.1)

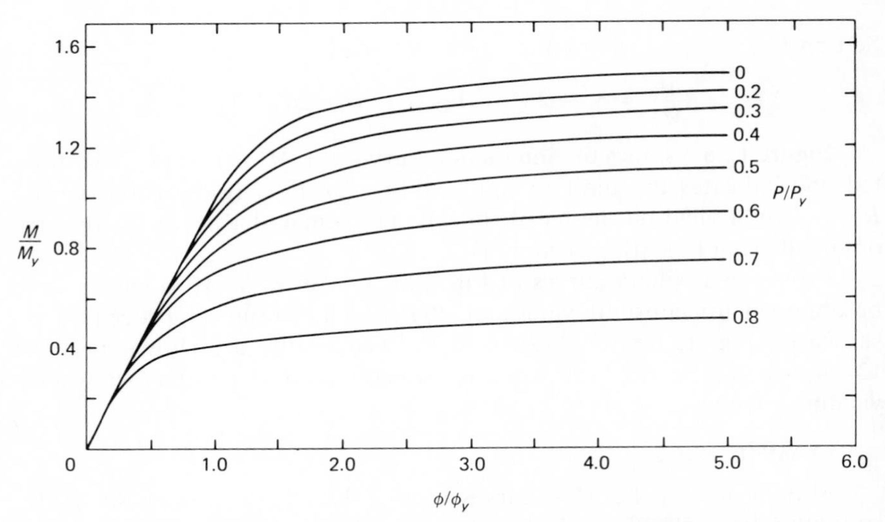

Fig. 12.6.4. Moment-curvature relationship for rectangular beam-column.

Solution

(a) Determine the values of P/P_y for $M/M_y = 0$.

$$P = \frac{\pi^2 EI}{L^2} = \frac{\pi^2 EA}{(L/r)^2}$$

$$\frac{P}{P_y} = \frac{\pi^2 EA}{(L/r)^2 F_y A}$$

which with $L/r = 60$, $E = 29{,}000$ ksi and $F_y = 50$ ksi becomes

$$\frac{P}{P_y} = \frac{\pi^2 (29{,}000)}{(60)^2\, 50} > 1 \qquad \therefore \frac{P}{P_y} = 1.0$$

since yielding governs rather than stability.

(b) For $P/P_y = 0.3$, determine maximum value of M/M_y. Begin by trying $M/M_y = 0.1$. In order to determine the beam-column strength of structural members, various numerical procedures are available. As an example of one such method, the procedure known as Newmark's method, will be illustrated. For a comprehensive development of this method see Refs. 34 and 35. For a given value of M/M_y, the following steps are to be used:

Step 1. Assume the deflection, Δ.

Step 2. Compute total moment (M/M_y) as the primary moment M_0 plus the secondary moment P times Δ.

Step 3. Using M/M_y, determine the curvature ϕ/ϕ_y from Fig. 12.6.4.

Step 4. Determine the concentrated angle changes assuming a parabolic distribution,

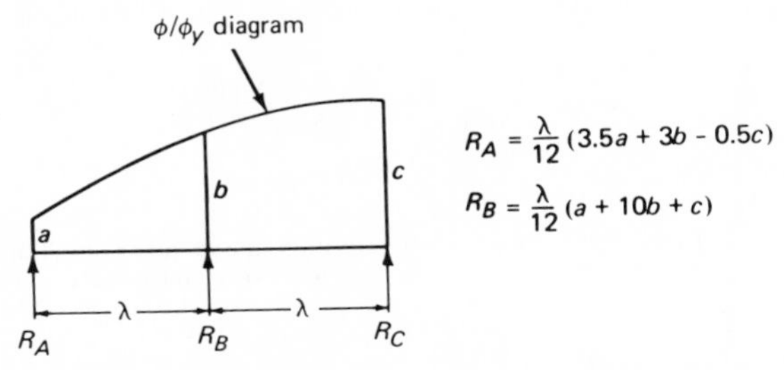

Concentrated angle changes (reactions to conjugate beam)

Step 5. Use numerical integration twice to obtain deflection.

Step 6. If the final deflection differs from the assumed value, then use the new assumption and repeat steps 2 through 5, repeating until the assumption and the result agree.

Step 7. After the deflection has converged, compute the end slope θ and plot M/M_y vs. θ, as shown for this case in Fig. 12.6.5. When the M/M_y value just reaches a high enough value that the deflection no longer converges to an equilibrium value, the member has achieved its ultimate strength. This is defined here as a failure by instability.

In order to illustrate the numerical procedure, the largest value for which convergence could be obtained, $M/M_y = 0.83$, is used to show the typical computation.

$M/M_y = 0.83$ $P/P_y = 0.3$ Sym. about ℄

Line	$\lambda = L/6$		λ		λ			$L/r = 60$ Multiplier
(1) Assume* deflection, Δ	0		2.09		3.51		4.01	in.
(2) Primary moment M_i	0.83		0.83		0.83		0.83	M_y
(3) Secondary Moment $P\Delta$	0		0.627		1.053		1.023	P_y
(4) do†	0		0.1881		0.3159		0.3609	M_y
(5) $M = M_i + P\Delta$	0.83		1.018		1.146		1.191	M_y
(6) ϕ/ϕ_y‡	0.85		1.17		1.60		1.80	
(7) Conc. ϕ	5.69		14.15		18.97		21.20	$\lambda\phi_y/12$
(8) Average slope		43.72		29.57		10.60		do
(9) Deflection, Δ	0		43.72		73.29		83.89	$\lambda^2\phi_y/12$
(10) do	0		2.09		3.51		4.01	in.

Where $\dfrac{\lambda^2\phi_y}{12} = \dfrac{\lambda^2 F_y}{12\,(h/2)\,E} = \dfrac{(L/r)^2 S_x F_y}{(6)^2 A\,(12)\,E} = \dfrac{(60)^2\,(50)}{36\,(0.3)\,(12)\,29000} = \dfrac{1}{20.90}$

*An initial estimate of the deflection may be made by considering only the primary bending moment:

$$\Delta = \frac{ML^2}{8EI} = \frac{0.83M_y(L/r)^2}{8EA} = \frac{0.83F_yS_x(L/r)^2}{8EA} = \frac{0.83\,(50)\,(60)^2}{8\,(29000)\,0.3} = 2.16\text{ in.}$$

which is amplified approximately by $\dfrac{\Delta}{1 - P/P_y} = \dfrac{2.16}{0.7} = 3.1$ in.

This can be used as a starting value. After several repetitions the values on line (1) are obtained.

†Since $P_y = 6M_y/h$ and $h = 20$, $P_y = 0.3M_y$, and line (3) is converted to have same multiplier as line (2).

‡From Fig. 12.6.4.

As a result of the trial process, the condition of instability for $P/P_y = 0.3$ and $KL/r = 60$ was found to be $M_i/M_y = 0.83$; one point on an ultimate strength interaction curve of P/P_y vs M/M_y for $KL/r = 60$.

To aid in understanding the concept, a rectangular section was used in the example; however, equations similar to those developed in this chapter have been found to agree well with an analysis of an I-shaped section having a linear distribution of residual stress across the flange

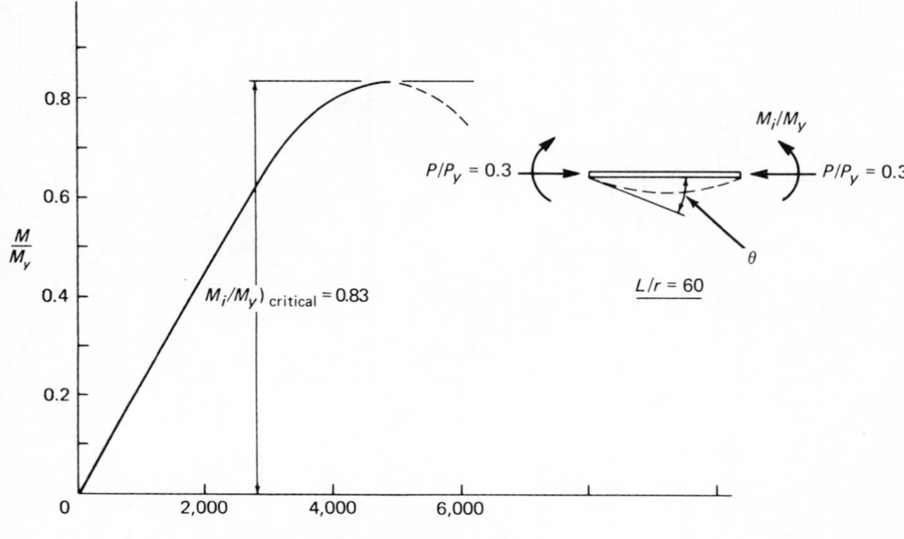

Fig. 12.6.5. Determining critical value for M_i/M_y for constant primary bending moment, with $P/P_y = 0.3$.

width (as per Fig. 6.6.6a), with a maximum residual stress of $0.3F_y$.[3] Using a W8 × 31 section which has reasonably typical cross-sectional properties, ultimate-strength interaction diagrams have been developed using the method illustrated herein.[4,5]

An alternate procedure assuming the flanges to be thin has been used[6-9] which reduces somewhat the graphical or numerical steps. However, residual stress has not been included; and the authors believe the method presented in this chapter offers a clearer understanding of the behavior.

For illustrative purposes, the ultimate-strength interaction curves for wide-flange sections with residual stress (as represented by the W8 × 31) loaded in compression combined with various ratios of end moments are presented in Fig. 12.6.6. Each point required to establish these curves involved the process illustrated in this article.

In studying the interaction curves of Fig. 12.6.6, the reader is reminded that the capacity P when $M = 0$ is based on the slenderness ratio, KL/r_x. As stated at the beginning of the article, the member was assumed to fail by instability in the plane of bending.

Other studies of instability in the plane of bending have included the effects of transverse loading.[4,10]

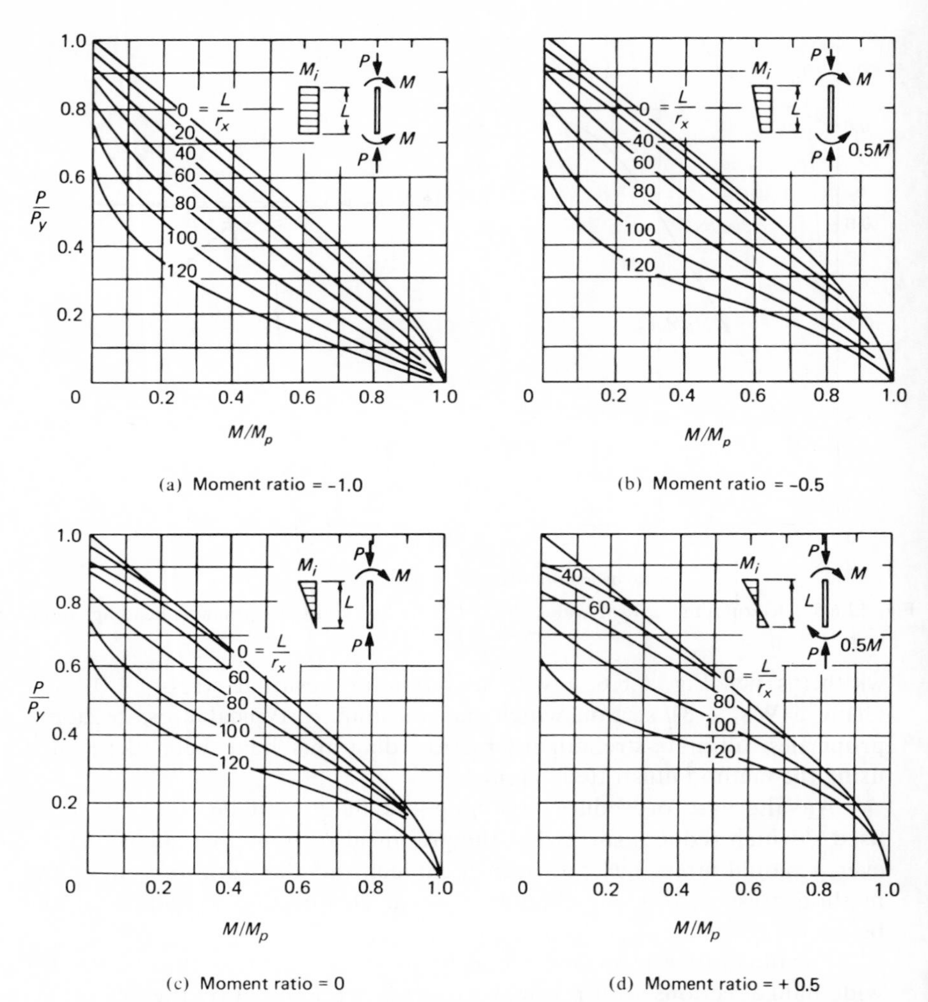

(a) Moment ratio = -1.0

(b) Moment ratio = -0.5

(c) Moment ratio = 0

(d) Moment ratio = + 0.5

Fig. 12.6.6. Ultimate-strength interaction curves (W8 × 31, F_y = 33 ksi, linear residual stress = F_r = $0.3F_y$, strong axis bending) for various end-moment ratios. NOTE: For F_y > 33 ksi, use adjusted L/r = actual L/r ($\sqrt{F_y/33}$). (Adapted from Figs. 7-10 of Ref. 4.)

12.7.* ULTIMATE-STRENGTH— FAILURE BY COMBINED BENDING AND TORSION

The ordinary beam-column unbraced over a finite length involves consideration of instability transverse or oblique to the plane of bending, involving torsional effects. This subject is an extension of lateral-torsional buckling in beams (Chapter 9), and involves both elastic and inelastic considerations. Interaction curves for a number of elastic-buckling situations have been developed, including (a) I-section columns with eccentric end loads in the plane of the web[11]; (b) I-columns with unequal end moments but without restraint to rotation about principal axes at the ends of the member[12]; and (c) I-columns with unequal end moments—hinged at ends for rotation about strong axis, but elastically restrained for rotation about the weak axis at the ends.[13] An excellent summary of the topic is presented by Massonnet[1] which includes discussion of plastic effects, and the more recent paper by Austin[2] which provides a brief summary.

A number of studies are available containing both analytical and experimental treatments of inelastic lateral-torsional buckling.[14–17]

In lieu of a rigorous theoretical development which may be found in the literature, including the previously cited references, the authors will attempt to make use of concepts already derived to give a rational basis for explaining failure by combined bending and torsion. A similar discussion is presented by Massonnet.[1]

Recall the Euler equation developed in Sec. 6.2:

$$P_{\text{cr}} = \frac{\pi^2 EI}{L^2} \qquad [6.2.7]$$

where in the development it was assumed that the member is in a slightly bent position. In other words, failure was assumed to occur by bending only. In Sec. 6.12 it was pointed out that in the slightly bent position the shear equals the buckling load P times the slope of the deflection curve. In Sec. 8.4, it was noted that only when flexural shears act in planes passing through the shear center is there flexure without torsion. Furthermore, in Sec. 7.8, it was shown that only when $I_{xy} = 0$, does the member bend in the same plane that the flexural shears act. Thus, the basic Euler expression, Eq. 6.2.7 implies buckling by bending about the weak axis of a member having at least one axis of symmetry ($I_{xy} = 0$); and furthermore, that there is no simultaneous twisting, which means the center of gravity position, G, and the shear center, O, coincide. Such a condition is true of a rectangle or an I-shaped member (see Fig. 12.7.1a).

In Sec. 8.11, torsional buckling of an axially loaded compression

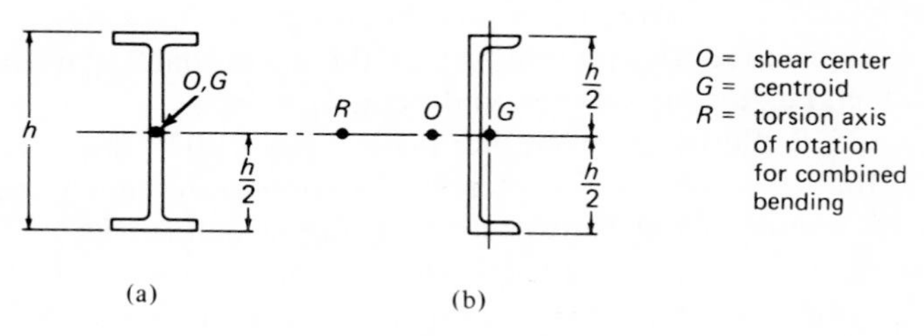

Fig. 12.7.1.

member was treated. When, as was assumed in that development, the shear center and the centroid coincide, torsional buckling is a possibility, but it is not coupled or linked to the buckling failure by bending. Upon comparison of the homogeneous differential equation for torsion, Eq. 8.5.12, when differentiated once with respect to z,

$$\frac{d^4\phi}{dz^4} - \lambda^2 \frac{d^2\phi}{dz^2} = 0$$

with Eq. 8.11.6 for torsional buckling,

$$\frac{d^4\phi}{dz^4} + p^2 \frac{d^2\phi}{dz^2} = 0 \qquad [8.11.6]$$

it may be noted they are identical if

$$p^2 = -\lambda^2$$

or

$$\left.\frac{\sigma_z J - GK}{EI_w}\right)_{\substack{\text{torsion}\\\text{buckling}}} = -\left.\frac{GK}{EI_w}\right)_{\text{torsion}} \qquad (12.7.1)$$

Thus one may observe from Eq. 12.7.1,

$$\left.GK\right)_{\substack{\text{torsion}\\\text{without}\\\text{compression}}} = \left.GK - \sigma_z J\right)_{\substack{\text{torsion with}\\\text{axial compression}}} \qquad (12.7.2)$$

which indicates that axial compression reduces the effective torsional rigidity. One may conclude torsional buckling is likely when the polar moment of inertia J is large compared to torsional rigidity GK.

When the shear center O and the center of gravity G do not coincide, as in Fig. 12.7.1b, buckling occurs by bending (with incremental flexural shears acting through G) in combination with torsion about O. The combined effect is a torsional failure about an axis at R, located more distant than O from G and in the same direction. The coupling effect to produce combined bending and torsion is proportional to the distance OG. Figure 12.7.1 gives two extremes, (a) with no coupling effect and (b) with large coupling effect. Locations of O and G for some other sections are given in Table 8.11.1.

It may now be observed that an applied moment combined with axial compression, or an axial compression eccentrically applied along a principal axis produces the same effect as the axial load alone produces on the channel, i.e., a coupled torsional-bending effect.

Figure 12.7.2 illustrates what happens. Note that the behavior is identical to that discussed in Chapter 9. The only difference is that here the torsional moment arises from shear, $P\,dx/dz$, which in turn arises from the axial load; whereas in Chapter 9, the torsional moment was a component, $M_0\,dx/dz$, of the moment applied in the yz plane. The loading involving the beam-column is the more severe one.

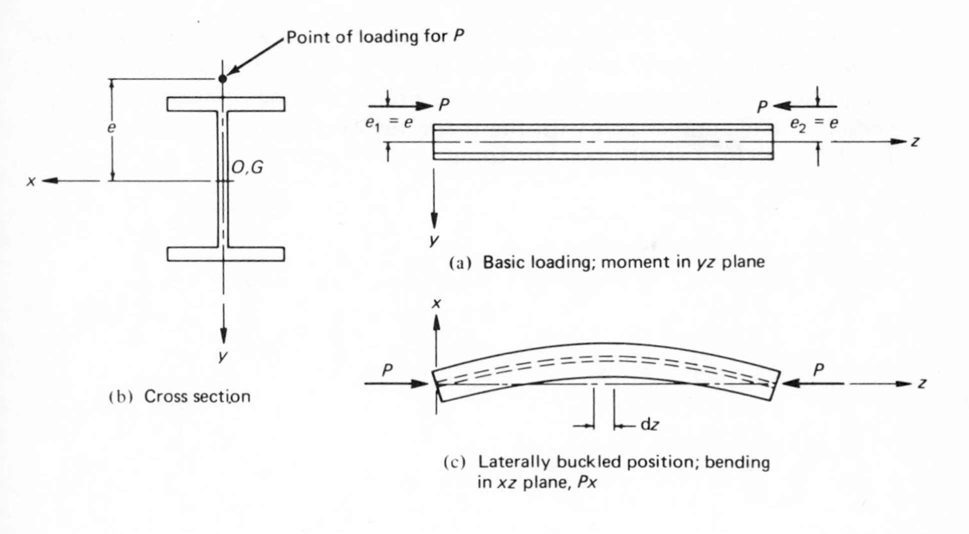

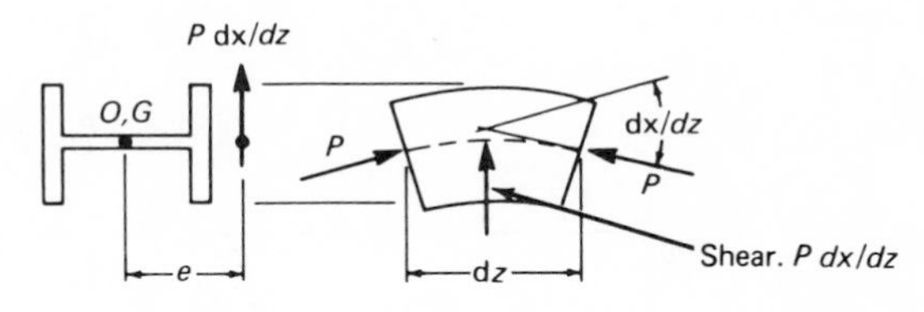

(d) Laterally buckled position; element dz giving torsional moment $(P\,dx/dz)\,e$

Fig. 12.7.2. Combined bending and torsion failure; lateral-torsional failure.

Tests by Massonnet[1] and computer studies[14,15] including inelastic effects indicate that interaction diagrams similar to those in Fig. 12.6.6 will result when including lateral-torsional buckling. The main difference is that P/P_y will be lower when based on KL/r_y (weak axis) rather than on KL/r_x (strong axis).

12.8. ULTIMATE-STRENGTH— INTERACTION EQUATIONS

The interaction curves discussed and illustrated in Sec. 12.6 may be approximated by simple interaction equations.

Case 1—No Instability. For the braced location where instability cannot occur (i.e., $Kl/r = 0$), the uppermost curves of Fig. 12.6.6 apply and may be approximated by

$$\frac{P}{P_y} + \frac{M}{1.18\,M_p} = 1.0 \qquad (12.8.1)$$

and $M/M_p \leq 1.0$. In the above equation, $P_y = A_g F_y$ and M_p = maximum moment capacity of the member in the absence of axial load (equals the plastic moment for all cases where premature local buckling is prevented). The comparison with the theoretical result in Fig. 12.8.1 shows Eq. 12.8.1 to be a good approximation.

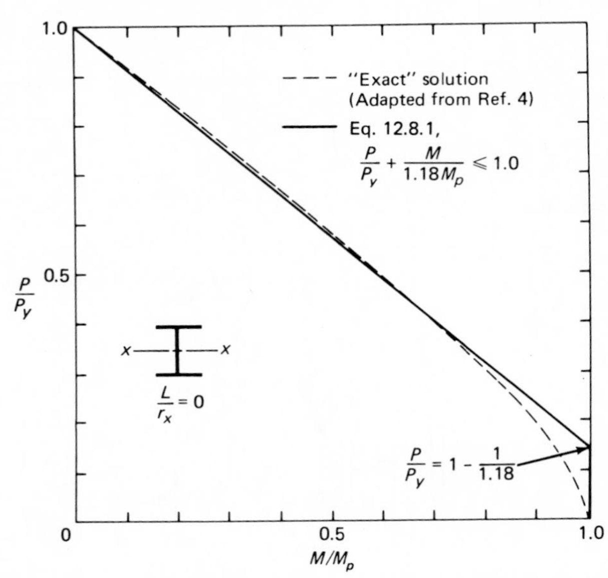

Fig. 12.8.1. "Exact" ultimate strength interaction curve for typical wide-flange sections (including residual stress) compared with interaction equation—Case 1, no instability.

Case 2—Instability in the Plane of Bending. The curves of Fig. 12.6.6 for various combinations of moments and values of L/r_x may be approximated by

$$\frac{P}{P_{cr}} + \frac{M_{equiv}}{M_p(1 - P/P_e)} = 1.0 \tag{12.8.2}$$

where

P_{cr} = buckling strength under axial load based on slenderness ratio for the axis of bending:

M_{equiv} = Eq. 12.4.4 (or its alternates from Fig. 12.4.2)

$P_e = \pi^2 EI/L^2$

Massonnet[1] has shown that Eq. 12.8.2 is a good approximation by comparing it with the curves of Galambos and Ketter.[5] A comparison in Fig. 12.8.2 of Eq. 12.8.2 with some curves from Fig. 12.6.6a shows the good correlation.

For primary bending moment from transverse loading, Lu and

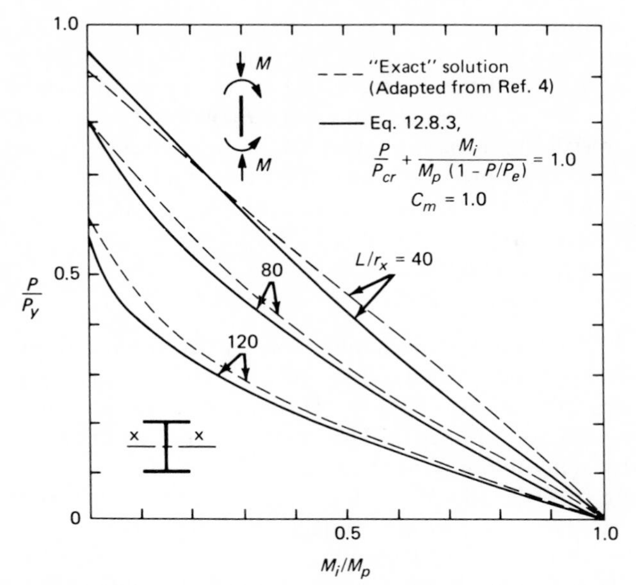

Fig. 12.8.2. "Exact" ultimate strength interaction curve for typical wide-flange sections (including residual stress) compared with interaction equation—Case 2, instability in the plane of bending ("exact" solution from Fig. 7 of Ref. 4). NOTE: For actual use, use adjusted L/r = Actual $L/r[\sqrt{F_y/33}]$.

Kamalvand[10] have shown that when M_{equiv} is replaced by $C_m M_i$, using C_m as given by Eq. 12.3.7, Eq. 12.8.2 is also a good representation for the "exact" solutions. Thus, in general, the interaction equation may be written

$$\frac{P}{P_{\text{cr}}} + \frac{C_m M_i}{M_p(1 - P/P_e)} = 1.0 \tag{12.8.3}$$

for all cases of instability in the plane of bending.

Case 3—Instability by Lateral-Torsional Buckling. Massonnet[1] has shown that with only slight error the form of Eq. 12.8.2 may also be used for this case. P_{cr} for this case may be governed by the slenderness ratio for the axis normal to the axis of bending. Further, since lateral-torsional buckling as a beam may occur for a moment less than M_p, use M_m instead of M_p in Eq. 12.8.3. Using the various definitions of C_m from Secs. 12.3 through 12.5, the general notation $C_m M_i$ should be used. Thus for instability under cases 2 or 3, the following interaction equation may be considered to apply:

$$\frac{P}{P_{\text{cr}}} + \frac{M_i C_m}{M_m(1 - P/P_e)} = 1.0 \tag{12.8.4}$$

where P = applied axial compression load
M_i = applied primary bending moment
$P_{cr} = A_g F_{cr}$ = ultimate strength for axially loaded compression member (Eq. 6.7.3 for F_{cr} would be a rational value)
M_m = maximum resisting moment in the absence of axial load. For adequately braced members of low slenderness ratios where local buckling is precluded, $M_m = M_p$.
C_m = factors discussed in Secs. 12.3–12.5

$$P_e = \frac{\pi^2 EI}{L^2} = \frac{\pi^2 EP_y}{F_y(L/r)^2}$$

The reduction in ultimate moment capacity M_m when lateral-torsional buckling influences the ultimate strength interaction, may be computed as $F_{cr}S_x$ using the CRC "Basic Design Procedure" of Sec. 9.4 with no safety factor applied. This will be consistent with working stress procedures involving AISC Formulas 1.5–6 and 1.5–7.

According to Driscoll, et al. (Ref. 27, p. 4.25), such a procedure is overly conservative. According to Ref. 27, "The behavior of beam-columns after lateral-torsional buckling has not yet been fully explored and available information is not sufficient to evaluate the effect of lateral-torsional buckling on rotation capacity. The limited results seem to indicate that for columns of low slenderness ratios (L/r = 40 or less) the end moment may not be reduced from the in-plane values significantly and drop off suddenly until a large rotation is reached." In other words, in real structures *without* pin-ended beam-columns the lateral-torsional buckling is restrained and the moment capacity does not drop off as drastically as indicated by the theoretical expressions developed in Chapter 9.

Figure 12.8.3 shows an empirical relationship for the reduction in ultimate moment.

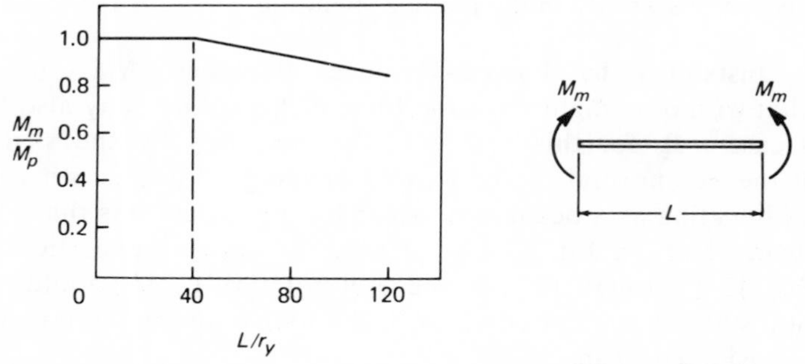

Fig. 12.8.3. Lateral buckling moment for uniform bending. (From Ref. 27; see also Ref. 17.)

12.9. BIAXIAL BENDING

The ultimate strength of members under axial compression and biaxial bending has received little attention until the last few years. Computer studies such as those of Birnstiel and Michalos,[21] Culver,[22,23] and Harstead, Birnstiel, and Leu,[25] illustrate that even with a number of simplifying assumptions, the analysis is one of considerable complexity. Some tests have been performed[24] which, though limited, have shown agreement with computer studies. The status of work on biaxial bending of compression members is summarized well by Chen and Santathadaporn.[26]

Simple plastic theory becomes inadequate when moments exist about two principal axes. When only one moment exists, plastic behavior is exhibited no matter what the value of axial compression. Figure 12.6.4 showed that under such loading the effective plastic moment reduces as axial compression increases, but plastic behavior does occur. It was further demonstrated how one might develop the interaction diagram for ultimate strength, P vs. M.

Upon application of an additional moment about the other principal axis, one might consider an interaction surface relating P, M_x, and M_y. Even for ideal elastic-plastic material, however, present plastic-analysis theorems neglect the influence of deformation on equilibrium. For zero length compression members, the concept of an interaction surface (see Fig. 12.9.1) may be thought of as a first step to the ultimate strength analysis under biaxial bending. Once this topic has been more fully

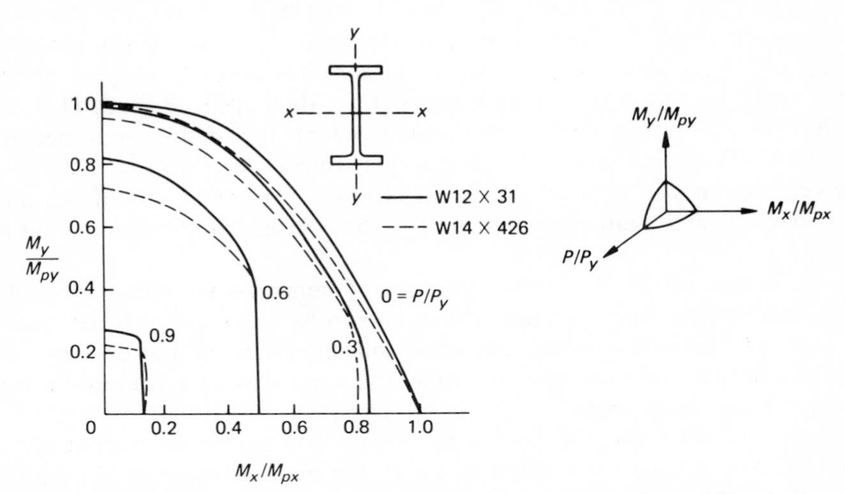

Fig. 12.9.1. Contours on interaction surface for short members where instability does not occur. (Adapted from Ref. 26.)

studied, ultimate strength analysis of space frames may become more realistic.

While few designers concern themselves greatly about the sequence of load application, nevertheless loading sequence affects ultimate strength. This is also true for uniaxial bending and compression, but it has less effect on that case than for the biaxial loading.

Figure 12.9.2 illustrates several loading sequences to obtain point A,

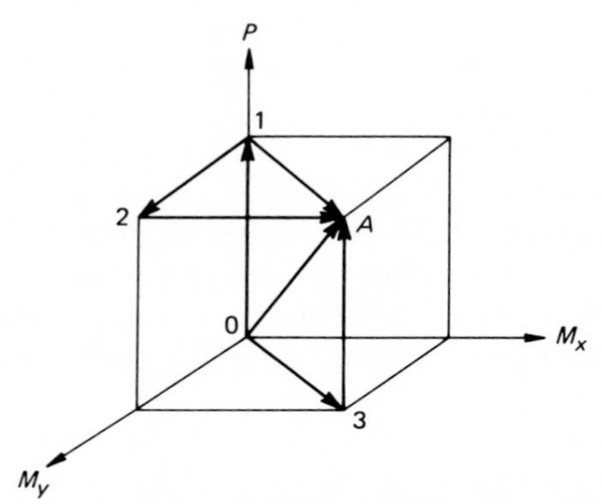

Fig. 12.9.2. Paths of loading for biaxial bending combined with axial force. (Adapted from Ref. 26.)

a particular value of P, M_x, and M_y. Point A may be reached by the following paths:

(a) Apply P first, then M_y, then M_x (path 0–1–2–A); (b) Apply P first, then apply M_x and M_y proportionally (path 0–1–A); (c) apply P, M_x, and M_y by increasing magnitude in constant proportion (path O–A); (d) apply M_x and M_y in constant proportion, then apply P (path O–3–A). Other combinations are possible and in general the loading may become applied via any path through space to get from O to A on Fig. 12.9.2. A given section will exhibit a different strength for each path of loading. Nearly all investigators to date (1971) have used proportional loading (path O–A).[26]

To conclude, the ultimate strength of compression members under biaxial bending is not sufficiently well known to make use of it for plastic design of rigid space frames; therefore, plastic design should be restricted to planar structures, or ones for which planar behavior represents a reasonable approximation.

For lack of adequate investigation, an interaction formula, such as Eq. 12.8.4, is usually assumed to apply for biaxial bending. Computer studies and some tests indicate that such a procedure is realistic for those cases investigated. Thus, the full interaction equation would be

$$\frac{P}{P_{cr}} + \frac{M_x C_{mx}}{M_{mx}(1 - P/P_{ex})} + \frac{M_y C_{my}}{M_{my}(1 - P/P_{ey})} \leq 1 \qquad (12.9.1)$$

where all terms are as defined following Eq. 12.8.4, except that now the quantities subscripted x and y must be evaluated for bending about the axis indicated by subscript.

Furthermore, torsional stresses frequently exist when biaxial bending occurs, and may require a more elaborate analysis.

12.10. AISC–1969 WORKING-STRESS DESIGN CRITERIA

As is usual with design specifications, current requirements reflect some historical developments. Until 1961, the AISC specification did not consider the amplification of bending moment due to deflection (For the simplified treatment applicable to braced systems, refer to Sec. 12.3). When axial compression is relatively small, neglect of the secondary effect makes little difference. AISC–1.6.1 *permits*:

(a) when $f_a/F_a \leq 0.15$

$$\frac{f_a}{F_a} + \frac{f_{bx}}{F_{bx}} + \frac{f_{by}}{F_{by}} \leq 1.0 \qquad (12.10.1)$$

which is AISC Formula 1.6–2; where

$f_a = P/A_g =$ nominal axial compression stress

$f_{bx}, f_{by} =$ flexural stresses based on primary bending moment about the x and y axes, respectively (see Sec. 12.3 if the full theoretical development has been omitted)

$F_a =$ allowable compression stress considering the member as loaded by axial compression only

$F_{bx}, F_{by} =$ allowable flexural stresses for the x and y axes, respectively, considering the member loaded in bending only

(b) When $f_a/F_a > 0.15$, the secondary moment magnification effects *must* be considered for combined stress at locations away from a support. Thus by using the ultimate strength interaction equation, Eq. 12.8.4, converting to unit stresses, and applying a factor of safety (FS) gives

$$\frac{\dfrac{P/A_g}{P_{cr}}}{(A_g\,\text{FS})} + \frac{(M_i/S)\,C_m}{\dfrac{M_m}{(S\text{FS})}\left[1 - \dfrac{P/A_g}{P_e/(A_g\,\text{FS})}\right]} \leq 1.0 \qquad (12.10.2)$$

which gives

$$\frac{f_a}{F_a} + \frac{f_b C_m}{F_b(1 - f_a/F'_e)} \leq 1.0 \qquad (12.10.3)$$

By analogy, for bending about both x and y axes, Eq. 12.10.3 would become

$$\frac{f_a}{F_a} + \frac{f_{bx} C_{mx}}{F_{bx}(1 - f_a/F'_{ex})} + \frac{f_{by} C_{my}}{F_{by}(1 - f_a/F'_{ey})} \leq 1.0 \qquad (12.10.4)$$

which is AISC Formula 1.6–1a. For F_{bx} and F_{by} in Eq. 12.10.4, C_b is to be taken as unity. Subscripts x and y refer to the axes of bending. In Eq. 12.10.4, C_m = factors discussed in Secs. 12.3–12.5, to be taken as follows:

(1) For *braced frames* without transverse loading between supports, $C_m = 0.6 - 0.4M_1/M_2 \geq 0.4$, where the moments carry a sign in accordance with end rotational direction; i.e., positive moment ratio for reverse curvature, negative moment ratio for single curvature. This case is discussed in Sec. 12.4.

(2) For *braced frames* with transverse lateral loading between supports, C_m = rational value as determined by the procedures of Sec. 12.3. Such rational values may be obtained from Eq. 12.3.7 after dividing α by a safety factor (1.67), from Table 12.10.1, or by a direct analysis. In

TABLE 12.10.1
C_m Values for Braced Frames (Ref. 20), Eq. 12.3.7

	Case	ψ	$C_m = 1 + \psi\alpha$
1	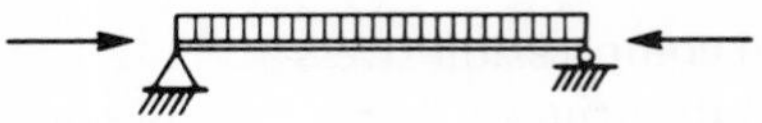	0	1.0
2	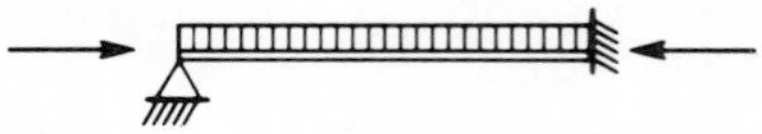	−0.3	$1 - 0.3f_a/F'_e$
3	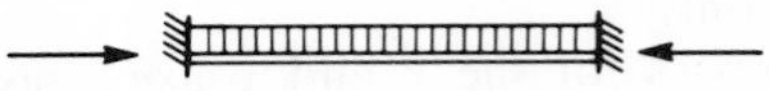	−0.4	$1 - 0.4f_a/F'_e$
4	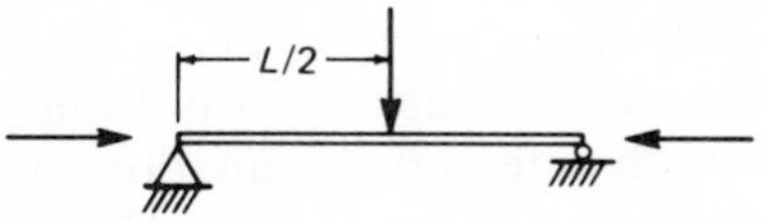 L/2	−0.2	$1 - 0.2f_a/F'_e$
5	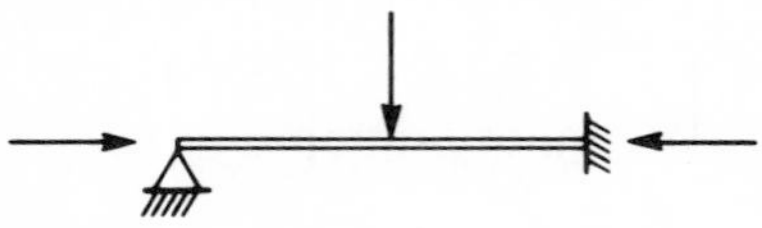	−0.4	$1 - 0.4f_a/F'_e$
6	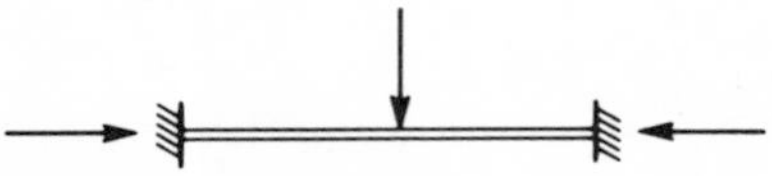	−0.6	$1 - 0.6f_a/F'_e$

lieu of such an analysis, however, AISC 1.6.1 permits using $C_m = 0.85$ for members with moment restraint at the ends, and $C_m = 1.0$ for members with simply supported ends.

(3) For *unbraced frames*, $C_m = 0.85$. The AISC Commentary[20] suggests that C_m may be taken conservatively as $C_m = 1 - 0.18 f_a/F'_e$. A thorough study of Chapter 14 on frames is recommended for advanced students.

(4) When $f_a/F_a > 0.15$, combined stress must be checked against a yield criterion,

$$\frac{f_a}{0.6F_y} + \frac{f_{bx}}{F_{bx}} + \frac{f_{by}}{F_{by}} \leq 1.0 \tag{12.10.5}$$

which is AISC Formula 1.6–1b.

Regarding bending moment adjustments to account for plastic-behavior moment redistribution, as discussed in Chapter 10, AISC–1.5.1.4.1 allows the 10 percent reduction in negative moment on a beam to be used in proportioning the column if the *beam or girder* satisfies "compact" section requirements and the unit compressive stress on the column does not exceed $0.15F_a$. The authors further believe the effective laterally unbraced length on the column, measured from the end where the adjustment occurs, should also satisfy "compact" section limitations; i.e., it should not exceed L_c.

In the application of Eq. 12.10.4 (AISC Formula 1.6–1a), the term F'_e refers to the *effective pin-end length* in the *plane of bending*:

$$F'_e = \frac{P_e}{A_g \text{ FS}} = \frac{\pi^2 EI}{A_g L^2 \text{ FS}}$$

$$= \frac{12\pi^2 E}{23(KL_b/r_b)^2} = \frac{149{,}000}{(KL_b/r_b)^2} \text{ ksi} \tag{12.10.6}$$

where K = effective length factor (see Sec. 6.9 and Chapter 14)
L_b = actual unbraced length in the plane of bending
r_b = radius of gyration in the plane of bending

It is noted that a nominal safety factor of $23/12 = 1.92$ is used; this is the maximum factor used for long axially loaded members (see Sec. 6.8) and is therefore conservative in the amplification term.

In the application of AISC Formula 1.6–1a which contains C_m, the allowable stress for the member in bending alone, F_b, must use $C_b = 1.0$. *Since C_m accounts for moment gradient, it should not again be accounted for in the F_b term.*

EXAMPLE 12.10.1

Investigate the acceptability of a W16 × 78 used as a beam-column under the loading shown in Fig. 12.10.1. Steel is A572 Grade 60.

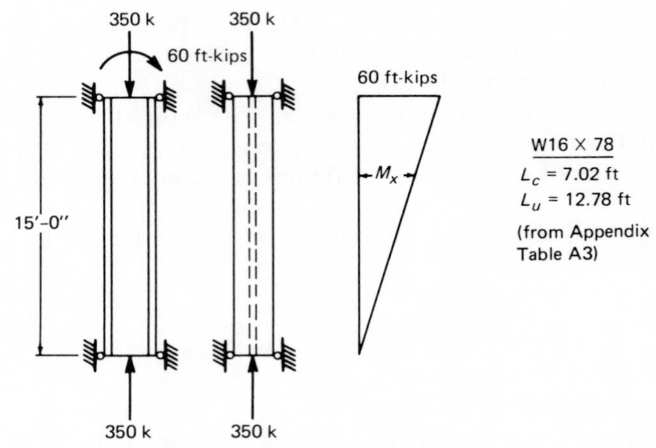

Fig. 12.10.1. Beam-column for Example 12.10.1.

Solution

(a) Column effect:

$$\frac{KL}{r_y} = \frac{15(12)}{2.01} = 90; \qquad F_a = 18.0 \text{ ksi} \quad \text{(AISC Table 1–60)}$$

$$f_a = \frac{P}{A} = \frac{350}{23.0} = 15.2 \text{ ksi}$$

$$\frac{f_a}{F_a} = \frac{15.2}{18.0} = 0.85 > 0.15; \qquad \text{use Formula 1.6–1a}$$

(b) Beam effect:

$$F_b\,(1.5\text{–}7) = \frac{12{,}000}{Ld/A_f} = \frac{12{,}000}{180(2.17)} = 30.7 \text{ ksi} < 0.60F_y$$

$$\frac{L}{r_T} = \frac{180}{2.29} = 78.5$$

$$F_b\,(1.5\text{–}6a) = 40.0 - \frac{(L/r_T)^2}{425} = 40.0 = \frac{(78.5)^2}{425}$$

$$= 40.0 - 14.5 = 25.5 \text{ ksi}$$

Since 30.7 > 25.5, F_b (1.5–7) controls. In both formulas C_b is taken as 1.0 (i.e., not used), because C_m will serve the same purpose.

$$f_b = \frac{60\,(12)}{128} = 5.63 \text{ ksi}$$

$$\frac{f_b}{F_b} = \frac{5.63}{30.7} = 0.184$$

(c) Moment magnification:

$$\frac{KL}{r_x} = \frac{15(12)}{6.75} = 26.75; \qquad F'_e = 208 \text{ ksi} \quad \text{(AISC Table 2)}$$

where the x axis is the axis of bending.

$$C_m = 0.6 - 0.4(M_1/M_2) = 0.60$$

$$\frac{C_m}{1 - f_a/F'_e} = \frac{0.60}{1 - 15.2/208} = \frac{0.60}{1 - 0.0735} = 0.646$$

(d) Check of AISC formulas:

For stability, Formula 1.6–1a,

$$\frac{f_a}{F_a} + \frac{f_b}{F_b}\left(\frac{C_m}{1 - f_a/F'_e}\right) = 0.85 + 0.184(0.646) = 0.95 < 1.0$$

For yielding, Formula 1.6–1b, at the braced point,

$$\frac{f_a}{0.60F_y} + \frac{f_b}{F_b} = \frac{15.2}{36} + \frac{5.63}{36} = 0.58 < 1.0$$

For the above equation which does not involve C_m, F_b(1.5–7) should use $C_b = 1.75$ for this problem. In which case $F_b > 0.60F_y$; use $0.60F_y$ = 36 ksi. <u>The W16 × 78 is acceptable for the given loading.</u>

12.11. DESIGN PROCEDURES AND EXAMPLES—WORKING STRESS METHOD

To aid in selection of a beam-column section, it is usually advantageous to convert, in an approximate way, the resulting bending moment into an equivalent axial compression load and then to make use of column tables. Occasionally, conversion of the axial load into equivalent moment will be helpful.

The interaction equation when stability controls, Eq. 12.10.3, may be written

$$\frac{P}{AF_a} + \frac{M}{F_bS}\left(\frac{C_m}{1 - f_a/F'_e}\right) = 1.0 \qquad (12.11.1)$$

Multiplying by AF_a gives

$$P + M\left(\frac{A}{S}\right)\left(\frac{F_a}{F_b}\right)\left(\frac{C_m}{1 - f_a/F'_e}\right) = F_aA = \text{Equiv. } P \qquad (12.11.2)$$

Next, examine the amplification term, which may be changed in form, using Eq. 12.10.6 for F'_e,

$$\frac{1}{1 - f_a/F'_e} = \frac{F'_e}{F'_e - f_a} = \frac{149(10^3)\,r^2}{(KL)^2\left(\dfrac{149(10^3)\,r^2}{(KL)^2} - \dfrac{P}{A}\right)} \qquad (12.11.3)$$

$$= \frac{149(10^3)\,Ar^2}{149(10^3)\,Ar^2 - P(KL)^2} \qquad (12.11.4)$$

Thus the equivalent column load may be expressed in the manner shown by Eq. 12.11.2, and that corresponding to AISC Formula 1.6–1a is

$$\text{Equiv. } P = P + MB\left(\frac{F_a}{F_b}\right)\left(\frac{C_m a}{a - P(KL)^2}\right) \qquad (12.11.5)$$

where B = bending factor = A/S

a = $149(10^3)\,Ar^2$ for axis of bending

Note is made that the allowable stress ratio (F_a/F_b) reduces Equiv. P while the amplification term $[C_m a/(a - P(KL)^2)]$ usually increases Equiv. P.

When the yield criterion controls, the equivalent column load may be expressed,

$$\text{Equiv. } P = P + MB\left(\frac{0.6F_y}{F_b}\right) \qquad (12.11.6)$$

corresponding to AISC Formula 1.6–1b when a second moment term is included for the possibility of bending about both axes.

When $f_a/F_a < 0.15$, the equivalent P may be computed similarly from Eq. 12.10.1 as

$$\text{Equiv. } P = P + MB\left(\frac{F_a}{F_b}\right) \qquad (12.11.7)$$

corresponding to AISC Formula 1.6–2 when a moment term for both axes of bending is included.

Several examples follow which demonstrate application of the interaction formulas, using principles established in Chapters 6, 7, and 9. The following examples use the approaches outlined above; however, alternative approaches and uses of various design aids have been proposed,[29-32] including nonconventional cases such as stepped columns.

EXAMPLE 12.11.1

Select the lightest W14 section to carry an axial compression force of 150 kips in combination with a moment of 500 ft-kips. The member is part of a *braced* system, with support provided in each direction at top and bottom of a 14-ft length. Conservatively assume the moment causes single curvature and varies as shown in Fig. 12.11.1. Use A36 steel and the AISC specification.

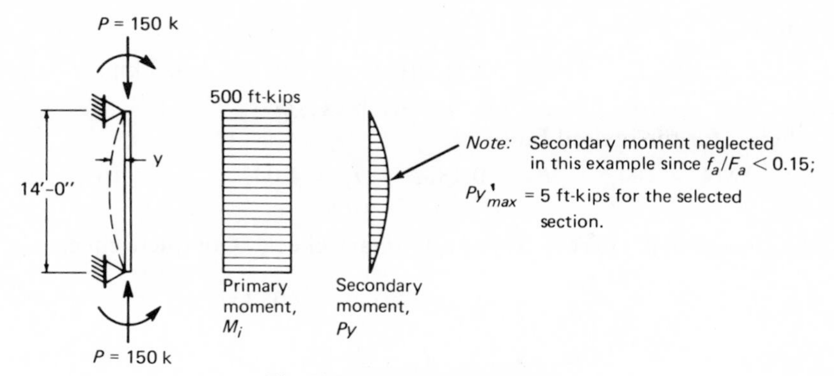

Fig. 12.11.1. Example 12.11.1.

Solution

Conservatively assume the equivalent pinned length factor $K = 1.0$. If adjacent member stiffness is known, the alignment chart which appears as Fig. 14.3.1a in Chapter 14, may be used to determine $K < 1.0$ for this braced frame.

Since at the start one may have no idea whether $f_a/F_a < 0.15$, use the simplest Equiv. P expression, Eq. 12.11.7,

$$\text{Equiv. } P = P + MB_x\left(\frac{F_a}{F_b}\right)$$

Referring to AISC Manual Column Tables: for W14 sections, find the average bending factor with respect to the strong axis, $B_x \approx 0.19$. Neglect temporarily F_a/F_b; i.e., assume $F_a \approx F_b$.

$$\text{Equiv. } P \approx 150 + 500(12)(0.19) = 150 + 1140 = 1{,}290 \text{ kips}$$

At this point, since the column part represents less than 15 percent (150/1,290) of the total, assume that AISC Formula 1.6–2 will apply.

Obtain trial sections from AISC Manual using $P = 1{,}290$ kips and $L = 14$ ft,

$$\text{W14} \times 219, \qquad B_x = 0.183, \qquad r_y = 4.08, \qquad L_c = 16.8$$
$$P = 1{,}230 \text{ kips}$$

Check:

$$\frac{KL}{r_y} = \frac{14(12)}{4.08} = 41.2; \qquad F_a = 19.1 \text{ ksi}$$

$$L = 14 \text{ ft} < L_c \text{ (Sec. 9.5)}; \qquad F_b = 0.66\,F_y = 24 \text{ ksi}$$

$$\frac{F_a}{F_b} = \frac{19.1}{24} = 0.795$$

Revise:

$$\text{Equiv. } P = 150 + 500(12)(0.183)(0.795) = 1{,}023 \text{ kips}$$

using closer approximations for the variables.

Select for the revised Equiv. P,

$$W14 \times 184, \qquad B_x = 0.183, \qquad r_y = 4.04, \qquad L_c = 16.6$$
$$P = 1{,}031 \text{ kips}$$

Little change in properties from preliminary check. Complete check:

$$\frac{KL}{r_y} = \frac{14(12)}{4.04} = 41.6; \qquad F_a = 19.05 \text{ ksi}$$

$$L = 14 \text{ ft} < L_c; \qquad F_b = 24 \text{ ksi}$$

$$f_a = \frac{P}{A} = \frac{150}{54.1} = 2.77 \text{ ksi}$$

$$\frac{f_a}{F_a} = \frac{2.77}{19.05} = 0.145 < 0.15; \qquad \text{Use Formula 1.6–2}$$

$$f_b = \frac{M}{S_x} = \frac{500(12)}{296} = 20.3 \text{ ksi}$$

$$\frac{f_a}{F_a} + \frac{f_b}{F_b} = 0.145 + 0.845 = 0.99 < 1.0 \qquad \text{OK}$$

<u>Use W14 × 184.</u>

Although the AISC specification does not require it, check Formula 1.6–1a to see the effect of the neglected secondary bending moment.

$$C_m = 0.6 - 0.4(M_1/M_2) = 0.6 - 0.4(-1.0) = 1.0$$

$$\frac{KL}{r_x} = \frac{14(12)}{6.49} = 25.9; \qquad F'_e = 222.4 \text{ ksi}$$

based on the strong axis of bending.

$$\text{Magnification Factor} = \frac{C_m}{1 - f_a/F'_e}$$
$$= \frac{1}{1 - 2.77/222.4} = 1.01$$

Obviously, using Formula 1.6–2 instead of Formula 1.6–1a involves insignificant error.

EXAMPLE 12.11.2

Select the lightest W section to carry an axial compression $P = 120$ kips applied with an eccentricity $e = 5$ in. as shown in Fig. 12.11.2. The member is part of a *braced* frame, and is conservatively assumed loaded in single curvature with constant e. Use A36 steel and the AISC specification.

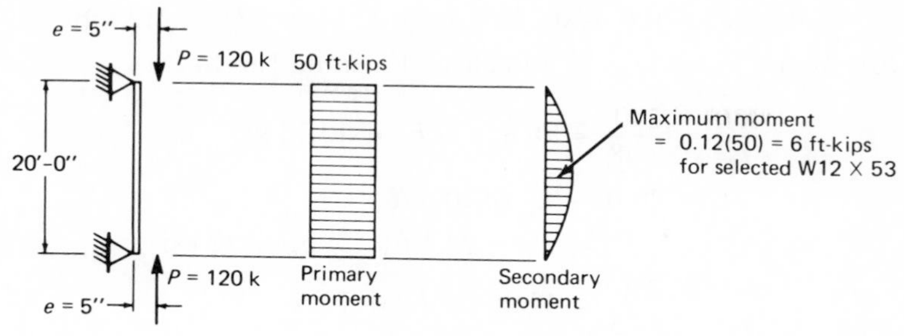

Fig. 12.11.2. Example 12.11.2.

Solution

(a) Compute the equivalent column load. Estimate bending factor and assume other factors to be unity.

$$\begin{aligned} &\text{W10,} \quad \text{Equiv. } P \approx 120 + 0.26(50)(12) = 276 \text{ kips} \\ &\text{W12,} \quad \text{Equiv. } P \approx 120 + 0.22(50)(12) = 252 \text{ kips} \\ &\text{W14,} \quad \text{Equiv. } P \approx 120 + 0.20(50)(12) = 240 \text{ kips} \end{aligned}$$

It is apparent that the axial compression effect exceeds 15 percent of the total equivalent load. Since maximum moment occurs at midheight when stability controls, only Formula 1.6–1a needs to be checked.

(b) Select trial sections. With a view toward satisfying Eq. 12.11.5, compute

$$P(KL)^2 = 120(240)^2 = 6.92 \times 10^6$$

$$C_m = 1.0$$

$$\left.\begin{aligned} a_x &\approx 60 \text{ for W10} \\ &\approx 71 \text{ for W12} \\ &\approx 96 \text{ for W14} \end{aligned}\right\} \text{ estimates based on initial values of Equiv. } P$$

$$\text{Magnification} = \frac{a_x}{a_x - P(KL)^2} \approx \frac{60}{60 - 7} = 1.13$$

The reduction from F_a/F_b probably exceeds magnification. Select for $L = 20$ ft,

W10 × 60	$P = 244$ k
W12 × 58	$P = 231$ k
W14 × 61	$P = 237$ k

(c) Check W12 × 58.

$$\frac{KL}{r_y} = \frac{20(12)}{2.51} = 95.5; \qquad F_a = 13.54 \text{ ksi}$$

$$L_c < L < L_u; \text{ i.e., } 10.6 \text{ ft} < 20 < 24.4 \text{ ft.}; \ F_b = 22 \text{ ksi}$$

$$\frac{F_a}{F_b} = \frac{13.54}{22} = 0.616$$

$$\text{Equiv. } P = 120 + 0.22(50)(12)(0.616)(1.13) = 212 \text{ kips}$$

(d) Revise to W12 × 53 and check Formula 1.6–1a.

$$\frac{KL}{r_y} = \frac{240}{2.48} = 96.8; \qquad F_a = 13.35 \text{ ksi}$$

$$L = 20 \text{ ft} > L_c = 10.6 \text{ ft}$$

$$< L_u = 22.1 \text{ ft}; \qquad F_b = 22 \text{ ksi}$$

$$\frac{KL}{r_x} = \frac{240}{5.23} = 45.9; \qquad F'_e = 70.5 \text{ ksi}$$

$$f_a = \frac{P}{A} = \frac{120}{15.6} = 7.70 \text{ ksi}$$

$$f_b = \frac{M}{S_x} = \frac{50(12)}{70.7} = 8.49 \text{ ksi}$$

Formula 1.6–1a, Eq. 13.10.3:

$$\frac{7.70}{13.35} + \frac{8.49}{22}\left(\frac{1.0}{1 - 7.70/70.5}\right) = 0.577 + 0.386(1.12)$$

$$= 1.009 \approx 1 \qquad \text{OK}$$

Use W12 × 53. A check of W10 and W14 sections will show that no lighter weight section is acceptable.

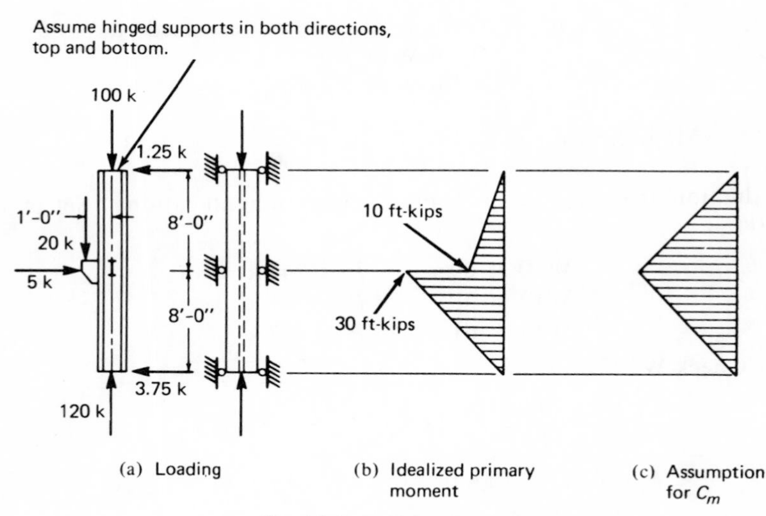

Fig. 12.11.3. Example 12.11.3.

EXAMPLE 12.11.3

Design a beam-column for the service loading conditions shown in Fig. 12.11.3. The member is part of a *braced* system, has support in the weak direction at mid-height, but only at top and bottom for the strong direction. Use AISC specification and A36 steel.

Solution

The particular features of this problem are (a) that bracing is not at the same points for both directions; and (b) lateral transverse loading causes the primary bending moment.

(a) Establish effective lengths. The member must be viewed separately as a column, then as a beam, as in Fig. 12.11.4.

For column action:

$$KL_x = 16 \text{ ft}; \qquad KL_y = 8 \text{ ft}$$

$$\text{Reqd } r_x/r_y \geq 2 \text{ if } L = 8 \text{ ft is valid for tables.}$$

For beam action:

$$\text{Laterally unbraced length} = 8 \text{ ft}$$

$$L \text{ for moment magnification} = 16 \text{ ft}$$

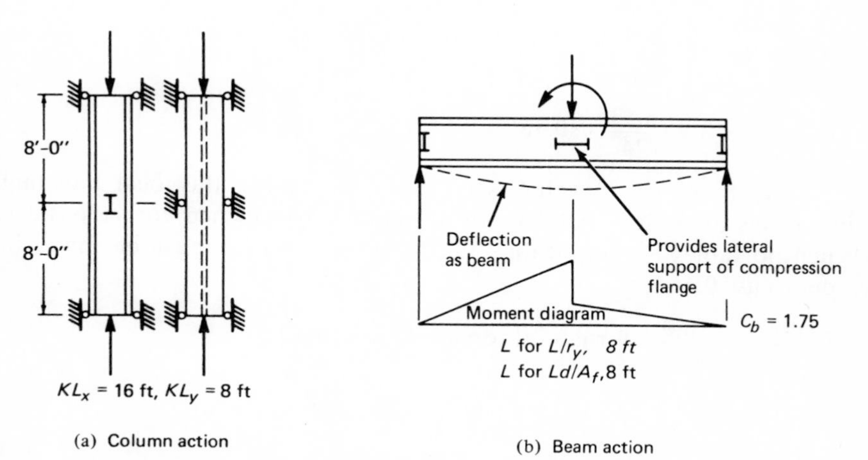

Fig. 12.11.4. Separate beam action and column action from Example 12.11.3.

(b) Estimate equivalent column load. The critical location for checking the interaction formula is at mid-height. Stability controls and AISC Formula 1.6–1a must be satisfied. Even though bracing in the weak direction is provided at mid-height, the point deflects as a result of bending, and failure would probably result by instability in the plane of bending.

Using Eq. 12.11.5 for several depths of section, and KL = 8 ft,

$$\text{W10} \quad \text{Equiv. } P \approx 120 + 0.27(30)(12) = 217 \text{ k}$$
$$\text{W12} \quad \text{Equiv. } P \approx 120 + 0.23(30)(12) = 203 \text{ k}$$

Using these values in AISC Column Tables suggests W10 × 39 or W12 × 40.

(c) Check W10 × 39:

Column action:

$$f_a = \frac{120}{11.48} = 10.45 \text{ ksi}$$

$$\frac{KL}{r_y} = \frac{8(12)}{1.98} = 48.5, \qquad \frac{r_x}{r_y} = 2.16 > 2.0$$

$$F_a = 18.49 \text{ ksi}$$

$$\frac{f_a}{F_a} = \frac{10.45}{18.49} = 0.57 > 0.15 \qquad \text{Use Formula 1.6–1a}$$

Beam action:

$$f_b = \frac{360}{42.2} = 8.53 \text{ ksi}$$

$$L_c = 8.4 \text{ ft} > \text{Actual unbraced length} = 8 \text{ ft}$$

$$F_b = 0.66F_y = 24 \text{ ksi}$$

$$\frac{f_b}{F_b} = \frac{8.53}{24} = 0.36$$

Note: Since an L_c value is given in the tables, the local buckling limitations of AISC–1.5.1.4.1 are automatically satisfied for A36 steel. When no tables are applicable, the following checks must be made in order to use $0.66F_y$:

$$\frac{L}{b_f} \le \frac{76.0}{\sqrt{F_y}} \quad \text{for lateral-torsional buckling}$$

$$\frac{b_f}{2t_f} \le \frac{52.2}{\sqrt{F_y}} \quad \text{for local buckling on flange}$$

$$\frac{d}{t_w} \le \frac{412}{\sqrt{F_y}}\left(1 - 2.33\frac{f_a}{F_y}\right) \ge \frac{257}{\sqrt{F_y}} \quad \text{for local buckling of web}$$

This has been discussed in Secs. 6.17 and 9.5.

Beam-column moment magnification:

$$\frac{KL}{r_x} = \frac{16(12)}{4.27} = 45 \quad \text{(for axis of bending)}$$

$$F'_e = \frac{149{,}000}{(45)^2} = 73.6 \text{ ksi}$$

$$\frac{f_a}{F'_e} = \frac{10.45}{73.6} = 0.142, \qquad 1 - \frac{f_a}{F'_e} = 0.858$$

$$C_m \approx 1 - 0.2\frac{f_a}{F'_e} = 0.972$$

(assumes moment variation of Fig. 12.11.3c
may approximate that of Fig. 12.11.3b)

$$\text{Magnification factor} = \frac{1}{1 - f_a/F'_e} = \frac{0.972}{0.858} = 1.13$$

Formula 1.6–1a:

$$\frac{f_a}{F_a} + \frac{f_b}{F_b}\frac{C_m}{(1 - f_a/F'_e)} = 0.57 + 0.36(1.13) = 0.97 < 1.0 \qquad \text{OK}$$

Use W10 × 39.

EXAMPLE 12.11.4

Select a W14 section for the service loading conditions of Fig. 12.11.5. The member is part of a *braced* system, is to be of A36 steel, and is to

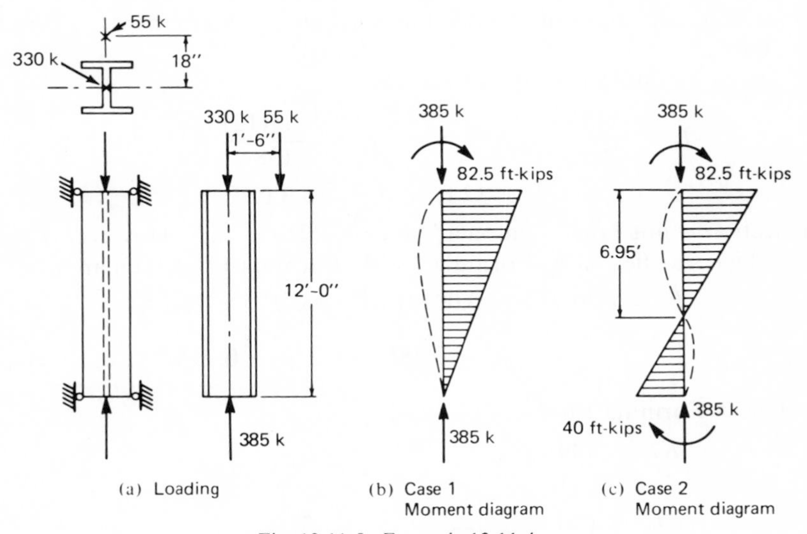

Fig. 12.11.5. Example 12.11.4.

be selected according to the AISC specification. Consider the following two cases of moment variation along the member:

(a) Case 1. Assume hinge at bottom, and maximum moment at the top.

(b) Case 2. Assume partial fixity at bottom so that an opposite sign bending moment of 40 ft-kips is developed at the bottom.

Solution

For this problem the design may be governed either by the stability criterion, Formula 1.6–1a, based on stresses out in the span (in the vicinity of a point about 0.4 of the span from the top); or by the yield criterion, Formula 1.6–1b, based on stresses at the braced point (the top).

(a) Case 1. Assuming the yield criterion governs,

$$\text{Equiv. } P = P + B_x M = 385 + 0.185(82.5)(12) = 568 \text{ kips}$$

$$\text{Reqd } A = \frac{P}{0.6F_y} = \frac{568}{22} = 25.8 \text{ sq in.}$$

Try W14 × 87: $A = 25.56$ sq in. For this section estimate Equiv. P for stability criterion:

$$C_m = 0.6 - 0.4M_1/M_2 = 0.6$$

$$a_x = 144.0 \times 10^6$$

$$P(KL)^2 = 385(144)^2 = 7.98 \times 10^6$$

$$\text{Magnification factor} = \frac{C_m a_x}{a_x - P(KL)^2} = \frac{0.6(144.0)}{(144.0 - 7.98)} = 0.635$$

Because the magnification factor is less than unity, it is unlikely Formula 1.6–1a will control, even though the allowable column stress, F_a, is less than $0.6F_y$.

Check Formula 1.6–1b: W14 × 87

$$f_a = \frac{385}{25.6} = 15.1 \text{ ksi}; \qquad f_b = \frac{82.5(12)}{138} = 7.2 \text{ ksi}$$

$$L = 12 \text{ ft} < L_c = 15.3 \text{ ft}$$

but section is not fully "compact" since $b_f/(2t_f) > 8.7$. However, $d/t_w = 33.3$ which satisfies AISC Formula 1.5–4. Formula 1.5–5 then applies,

$$F_b = F_y[0.733 - 0.0014(b_f/2t_f)\sqrt{F_y}] = 23.2 \text{ ksi}$$

$$\frac{f_a}{0.6F_y} + \frac{f_b}{F_b} = \frac{15.1}{22} + \frac{7.2}{23.2} = 0.686 + 0.310 = 0.996 < 1.0 \qquad \text{OK}$$

Check Formula 1.6–1a:

$$\frac{KL}{r_y} = \frac{144}{3.70} = 39; \qquad F_a = 19.3 \text{ ksi}$$

$$\frac{f_a}{F_a} = \frac{15.1}{19.3} = 0.782; \qquad \frac{f_b}{F_b} = \frac{7.2}{23.2} = 0.310$$

Magnification factor = 0.635, as computed above.

$$\frac{f_a}{F_a} + \frac{f_b}{F_b}\left(\frac{C_m}{1 - f_a/F'_e}\right) = 0.782 + 0.310(0.635) = 0.98 < 1 \qquad \text{OK}$$

In this case both Formulas 1.6–1a and 1.6–1b were close to the limit.

Use W14 × 87.

(b) Case 2. Since the stability of the member is increased when moments cause double curvature: obviously, the yield criterion will again control. The improved stability is utilized by $C_m = 0.6 - 0.4(-40/82.5) = 0.52$, reduced from 0.6 as used for Case 1. In Case 2 the distance from maximum moment to the point of inflection is reduced from 12 ft to about 7 ft, which would permit F_b to reach $0.66F_y$ if $L_c > 7$ ft.

Try W14 × 84, which is "compact" for local buckling; and $L_c = 12.7 > L = 6.95$ ft.

$$F_b = 0.66F_y$$

$$f_a = \frac{385}{24.7} = 15.6 \text{ ksi}; \qquad f_b = \frac{82.5(12)}{131} = 7.56 \text{ ksi}$$

Formula 1.6–1b:

$$\frac{f_a}{0.6F_y} + \frac{f_b}{F_b} = \frac{15.6}{22} + \frac{7.56}{24} = 0.71 + 0.32 = 1.03 \approx 1.0 \qquad \text{OK}$$

Since C_m is 0.52 for this case, by inspection the stability criterion, AISC Formula 1.6–1a, need not be checked. Accept the 3 percent overstress.

Use W14 × 84—A36.

EXAMPLE 12.11.5

The W8 × 24 section, as shown in Fig. 12.11.6, is laterally loaded to cause bending about its weak axis. The material is A572 Grade 50.

(a) Check the adequacy according to the AISC specification.
(b) Calculate the maximum deflection and amplified bending moment directly using $M_i + Py_{max}$ (see Sec. 12.3).
(c) Using the result of part (b), show what part of the AISC interaction equation to which it corresponds.

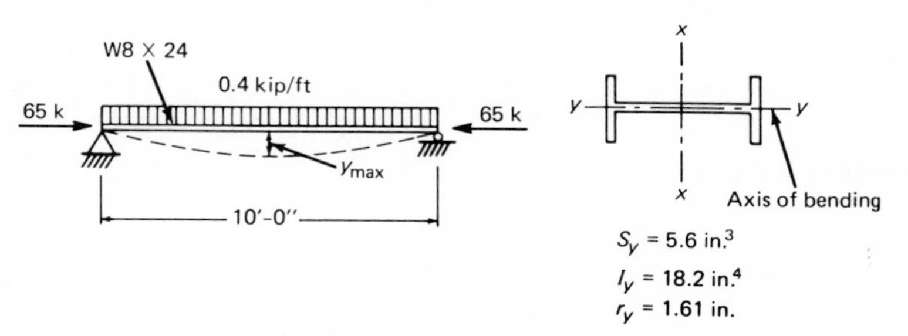

Fig. 12.11.6. Example 12.11.5.

Solution

(a) AISC specification check; since maximum moment occurs away from the supports, the stability criterion will govern. Check Formula 1.6–1a.

Note that for this problem buckling as a column occurs in the plane of bending, whereas in the previous four examples column buckling occurred about the axis which was not the axis of bending.

Column action:

$$\frac{KL}{r_y} = \frac{120}{1.61} = 74.5; \qquad F_a = 20.1 \text{ ksi}$$

$$f_a = \frac{P}{A} = \frac{65}{7.06} = 9.2 \text{ ksi}$$

$$\frac{f_a}{F_a} = \frac{9.2}{20.1} = 0.458$$

Beam action: Since bending is about the weak axis, lateral-torsional buckling is not a problem. Assuming the section is "compact" with regard to local buckling, it would be able to develop its plastic moment. Checking AISC–1.5.1.4.1,

$$\frac{b_f}{2t_f} = \frac{6.50}{0.796} = 8.14 > \frac{52.2}{\sqrt{50}} = 7.4 \qquad \text{N.G.}$$

which is a very conservative requirement for weak axis bending. Also, d/t_w does not have to be limited for weak axis bending.

Local buckling limitation precludes using $0.75F_y$; however, since $b_f/2t_f < 95.0/\sqrt{F_y} = 13.4$, a transition value between $0.6F_y$ and $0.75F_y$ may be used, (see AISC Supplement No. 1, Nov. 1970, Formula 1.5–5b)

$$\begin{aligned} F_b &= F_y[0.933 - 0.0035\sqrt{F_y}(b_f/2t_f)] \\ &= 0.731\,F_y = 36.6 \text{ ksi} \end{aligned}$$

$$M_y = \frac{0.4(10)^2}{8} = 5 \text{ ft-kips}$$

$$f_b = \frac{M_y}{S_y} = \frac{60}{5.6} = 10.7 \text{ ksi}$$

$$\frac{f_b}{F_b} = \frac{10.7}{36.6} = 0.292$$

Moment magnification:

$$F'_e = 26.9 \text{ ksi, using } \frac{KL}{r_y} = 74.5$$

$$\frac{f_a}{F'_e} = \frac{9.2}{26.9} = 0.342; \qquad 1 - \frac{f_a}{F'_e} = 0.658$$

$$C_m = 1.0 \quad \text{(from Table 12.10.1)}$$

$$\text{Magnification Factor} = \frac{C_m}{1 - f_a/F'_e} = \frac{1}{0.658} = 1.52$$

AISC Formula 1.6–1a:

$$\frac{f_a}{F_a} + \frac{f_b}{F_b}\left(\frac{C_m}{1 - f_a/F'_e}\right) = 0.458 + 0.292(1.52) = 0.90 < 1 \qquad \text{OK}$$

(b) Compute maximum deflection y_{max} and amplified stress directly using $M_i + Py_{max}$.

$$y_{max} = \frac{\delta_0}{1 - \alpha} \qquad [12.3.4]$$

$$\delta_0 = \text{primary deflection} = \frac{5wL^4}{384EI}$$

$$= \frac{5(0.4)(10)^4 1728}{384(29.5)(10^3)(18.2)} = 0.168 \text{ in.}$$

$$\alpha = \frac{P}{P_e} = \frac{f_a}{1.92F'_e} = \frac{0.342}{1.92} = 0.178$$

$$y_{max} = \frac{0.168}{1 - 0.178} = 0.204 \text{ in.}$$

$$M_{max} = M_i + Py_{max} = 5.0 + \frac{65(0.204)}{12}$$

$$= 5.0 + 1.1 = 6.1 \text{ ft-kips}$$

which is the *actual* magnified moment for the given loading on an W8 × 24.

(c) Show how $M_i + Py_{max}$ relates to the AISC Formula 1.6–1a. To compare part (b) with the design criterion, the factor of safety 1.92 must be applied to Eqs. 12.3.5 and 12.3.6, giving

$$M_i + P\frac{1.92\delta_0}{1 - 1.92\alpha} = \frac{M_i C_m}{1 - f_a/F'_e}$$

$$5 + \frac{65}{12}\left(\frac{1.92(0.168)}{1 - 0.342}\right) = \frac{5(1.0)}{1 - 0.342} = 5 + 2.6 = 7.6 \text{ ft-kips}$$

which compares with 6.1 ft-kips (without factor of safety) in part (b). If one divides both sides by the section modulus S, the identity is established,

$$\frac{M_i + P\left(\dfrac{1.92\delta_0}{1 - 1.92\alpha}\right)}{S} = \frac{f_b C_m}{1 - f_a/F'_e}$$

That the factor of safety (1.92) should appear in two places in the above equation may be verified by assuming that both P and M service loads are multiplied by a factor of 1.92, giving

$$\text{Overload moment} = 1.92M_i + 1.92P\left(\frac{1.92\delta_0}{1 - 1.92\alpha}\right)$$

The safe service load is obtained by dividing the overload moment by 1.92, which gives

$$\text{Service moment} = M_i + P\left(\frac{1.92\delta_0}{1 - 1.92\alpha}\right)$$

EXAMPLE 12.11.6

Select the lightest W section for the column member of the frame shown in Fig. 12.11.7. The joints are rigid to give frame action in the

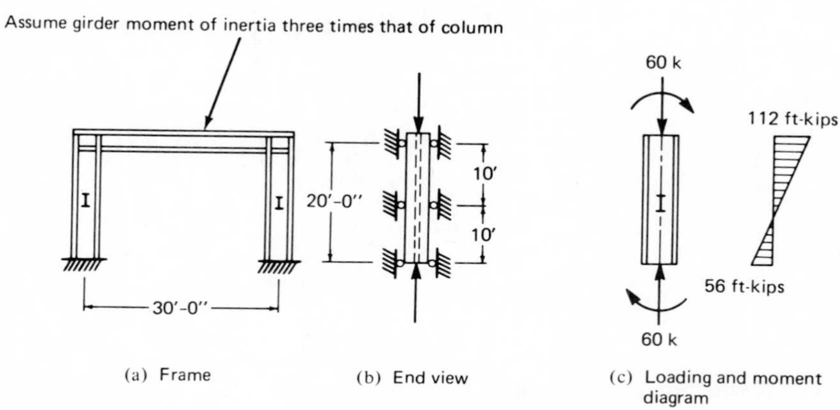

Fig. 12.11.7. Example 12.11.6.

plane of bending, but in the transverse direction sway bracing is provided and the attachments may be considered hinged. Use A36 steel and AISC specification. Solve for the following two cases:

(a) *Braced* frame in the plane of bending.
(b) *Unbraced* frame in the plane of bending.

Solution

(a) Frame *braced* in the plane of bending. Estimate equivalent column load:

$$K_x \approx 0.7,$$

or can be obtained from Alignment Chart, Fig. 14.3.1a, as

$$G_A = 1.0 \text{ (fixed)}, \qquad G_B = \frac{I/20}{3I/30} = 0.5, \qquad K_x = 0.73$$

$$P(K_x L)^2 = 60(0.7)^2(240)^2 = 16.9 \times 10^6$$

$$\text{Equiv. } P \approx P + B_x M$$

$$\text{W10,} \quad P = 60 + 0.26(112)(12) = 60 + 350 = 410 \text{ k}$$
$$\text{W12,} \quad P = 60 + 0.22(112)(12) = 60 + 296 = 356 \text{ k}$$
$$\text{W14,} \quad P = 60 + 0.20(112)(12) = 60 + 269 = 329 \text{ k}$$

Using the AISC column tables, with $K_y L_y = 10$ ft, preliminary sections are selected as

W10 × 72	$P = 396$ k @ $L = 10$ ft	$r_x/r_y = 1.72$
W12 × 65	$P = 367$ k	$r_x/r_y = 1.75$
W14 × 61	$P = 330$ k	$r_x/r_y = 2.44$

For values in the AISC tables to give correct column allowable loads, r_x/r_y of section must exceed $K_x L_x / K_y L_y = 0.7(20)/10 = 1.40$.

Before proceeding, the reader is reminded that due to plastic behavior, moment redistribution may occur (if local buckling and lateral-torsional buckling are prevented in the girder), which according to AISC–1.5.1.4.1 may be accounted for by a 10 percent reduction in negative moment on the girder. Such a reduced moment may also be used in proportioning the column for combined stress, if f_a/F_a does not exceed 0.15.

Under the conditions when moment redistribution is considered; i.e., plastic moment is assumed to occur at the top of the column, the designer should be wary of using an effective length factor K_x less than 1.0, since end restraint becomes a constant value (M_p), instead of increasing proportionally to the end rotation as it does when the system is elastic.

However, if one is certain the moment gradient is one causing double curvature, as in this case, then it seems rational to do as illustrated in this example.

For this example, the axial compression effect is about $60/329 = 0.18$ from Equiv. P calculation; therefore the reduction in negative moment at the top of the member may not be used.

Because of the moment gradient indicating double curvature, with $C_m = 0.6 - 0.4(112/56) = 0.4$ min, it is likely that the yield criterion at the braced point will control. Thus Formula 1.6–1b may be used to obtain required area for a more accurate choice of section. For a W14,

$$\text{Reqd } A = \frac{P_{\text{equiv.}}}{0.6F_y} = \frac{329}{22} = 14.9 \text{ sq in.}$$

Try W14 × 53, or W16 × 50. The W16 section does not appear in column tables but has the required area and according to the trend in calculating Equiv. P, may well be the lightest section. Of course, changing depth will affect the relative moments of inertia. For the *braced* frame such changes will have less effect than on the *unbraced* frame.

Check W16 × 50:

Formula 1.6–1b at the braced point:

$$f_a = \frac{60}{14.7} = 4.08 \text{ ksi}; \qquad f_b = \frac{112(12)}{80.8} = 16.62 \text{ ksi}$$

$$C_b = 1.75 + 1.05(-28/112) + 0.3(28/112)^2 = 1.41$$
$$L_c < L < C_b L_u; \qquad 7.5 < 10.0 < (1.41)(12.4)$$
$$F_b = 0.6F_y = 22 \text{ ksi}$$
$$f_a + f_b = 4.08 + 16.62 = 20.7 \text{ ksi} < 22 \text{ ksi} \qquad \text{OK}$$

Formula 1.6–1a for instability out in the span:

$$\frac{K_y L_y}{r_y} = \frac{1.0(10)(12)}{1.59} = 78; \qquad F_a = 15.6 \text{ ksi}$$
$$\frac{Ld}{A_f} = 10(12)(3.66) = 438$$

C_b is not used for computing F_b in the formula where C_m is used.

$$F_b = \frac{12{,}000}{438} = 27.4 \text{ ksi} > 0.6F_y = 22 \text{ ksi}$$

Since AISC Formula 1.5–7 shows $0.6F_y$ to be allowable, no check of Formula 1.5–6 is needed:

$$\frac{K_x L_x}{r_x} = \frac{0.7(20)(12)}{6.68} = 25.2; \qquad F_e' = 235.0 \text{ ksi}$$
$$\frac{f_a}{F_e'} = \frac{4.08}{235.0} = 0.017; \qquad 1 - \frac{f_a}{F_e'} = 0.983$$
$$\frac{f_a}{F_a} + \frac{f_b}{F_b}\left(\frac{C_m}{1 - f_a/F_e'}\right) = \frac{4.08}{15.6} + \frac{16.62}{22}\left(\frac{0.4}{0.983}\right)$$
$$= 0.262 \qquad + 0.755(0.41) = 0.57 < 1$$

Use W16 × 50.

(b) Frame *unbraced* in the plane of bending. The significant difference between this case and previous examples is that here the K_x value in the plane of bending exceeds one. Using the alignment chart from Chapter 14 (Fig. 14.3.1b) for *unbraced* frame (sidesway not prevented) for $G_A = 1.0$ (fixed at bottom) and $G_B = 0.5$ (see case a), find $K_x \approx 1.24$.

Equivalent column-load estimates are the same as for case (a), but in using column tables, correct allowable loads are obtained for $K_y L_y = 10$ ft only when r_x/r_y exceeds

$$K_x L_x / K_y L_y = 1.24(20)/10 = 2.48$$

Selection of W10 or W12 sections, if the stability criterion governs, will be controlled by column buckling in the plane of bending, while deeper sections will be controlled by weak axis column buckling.

From column tables,

W10, for $L = 1.24(20)/1.72 = 14.4$ ft, Equiv. P $\approx$ 395 kips, find W10 × 77 with $P \approx 380$ kips.

W12, for $L = 1.24(20)/1.75 = 14.2$ ft, Equiv. $P \approx 367$ kips, find W12 × 72 with $P \approx 376$ kips.

W14, for $L = 1.24(20)/2.44 = 10.2$ ft, Equiv. $P \approx 331$ kips, find W14 × 61 with $P \approx 330$ kips.

These preliminary sections are likely to be too heavy because F_a/F_b is usually a significant reduction effect, while amplification is normally not more than 10 to 20 percent. If the braced point controls, the section and procedure are as in case (a) where W16 × 50 was acceptable.

Check W16 × 50 by Formula 1.6–1a for stability:

$$K_yL_y/r_y = 78 \text{ as from part (a)}$$
$$K_xL_x/r_x = 1.24(240)/6.68 = 44.5$$
$$F_a = 15.6 \text{ ksi; weak axis governs}$$
$$F_b = 22 \text{ ksi; calculation same as for part (a)}$$
$$F'_e = 75.3 \text{ ksi; based on } KL/r = 44.5$$
$$C_m = 0.85 \text{ for unbraced frame}$$

$$\text{Magnification factor} = \frac{C_m}{1 - f_a/F'_e} = \frac{0.85}{1 - 4.08/75.3} = 0.90$$

$$\frac{f_a}{F_a} + \frac{f_b}{F_b}\left(\frac{C_m}{1 - f_a/F'_e}\right) = 0.262 + 0.755(0.90) = 0.94 < 1 \qquad \text{OK}$$

For the unbraced frame the stability requirement does not govern but it is considerably closer to governing than for the braced frame. Thus, in this frame example, a W16 × 50 may be used whether the frame is braced or unbraced.

Use W16 × 50.

EXAMPLE 12.11.7

Select an economical structural tee for use as the top chord in a roof truss, as shown in Fig. 12.11.8. The chord member is to be designed

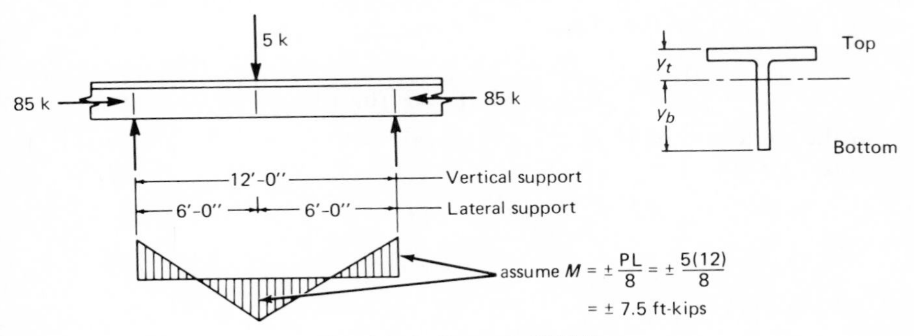

Fig. 12.11.8. Example 12.11.7.

as continuous over 12-ft spans, with transverse lateral support at 6-ft intervals. Use $F_y = 50$ ksi and the AISC specification.

Solution

This problem requires consideration of Formula 1.6.1a at midspan and Formula 1.6.1b at the supports. Furthermore, because of an unsymmetrical section both top and bottom extreme fibers must be examined.

(a) Estimate equivalent column load. Bending factors are not tabulated; therefore, Table 12.11.1 may serve as a guide.

TABLE 12.11.1
Approximate Bending Factors for Structural Tees

WT	$B_{xb} = \frac{A}{S_{x\,\text{stem}}}$	$B_{xt} = \frac{A}{S_{x\,\text{flange}}}$
WT9	1–1.2	0.3
WT8	1–1.5	0.3
WT7	1.2–2.0	0.3–0.5
WT6	1.5–2.0	0.5
WT5	2.0–2.5	0.6
WT4	2.5–3.5	0.7

If $(+M)$ governs, combined compression will be in the flange,

$$\text{Equiv. } P \approx P + B_{xt}M = 85 + 0.4(7.5)12 = 121 \text{ kips}$$

If $(-M)$ governs, combined compression will be in the stem,

$$\text{Equiv. } P \approx P + B_{xb}M = 85 + 1.5(90) = 85 + 135 = 220 \text{ kips}$$

(b) Select trial sections.

It seems likely that the $(-M)$ at the braced point will control;

$$f_a + f_b \leq 0.6F_y = 30 \text{ ksi}$$

$$\text{Reqd } A = \frac{220}{30} = 7.3 \text{ sq in.}$$

Also, bending is about 60 percent (135/220) of the total effect; i.e., $f_b/0.6F_y \approx 0.6$, which means

$$\text{Reqd } S_{xb} \approx \frac{M}{30(0.6)} = \frac{90}{30(0.6)} = 5.0 \text{ in.}^3$$

For the section to be fully effective, local buckling must be precluded. If this is desired

$$\text{Reqd } \frac{d}{t_w} \leq \frac{127}{\sqrt{F_y}} = 17.9$$

Trial sections:

			Computed	
WT6 × 36	$A = 10.6$	$S_{xb} = 4.54$	$B_{xb} = 2.34$	
WT7 × 26.5	$A = 7.79$	$S_{xb} = 4.96$	$B_{xb} = 1.58$,	d/t_w high
WT8 × 25	$A = 7.36$	$S_{xb} = 6.77$	$B_{xb} = 1.09$,	d/t_w high

For WT6

$$\text{Equiv. } P = 85 + 2.34(90) = 296 \text{ kips}$$

$$\text{Reqd } A = \frac{296}{30} = 9.85 \text{ sq in.}; \quad \text{Reqd } S_{xb} = 4.2 \text{ in.}^3$$

For WT7

$$\text{Equiv. } P = 85 + 1.58(90) = 227 \text{ kips}$$

$$\text{Reqd } A = \frac{227}{30} = 7.6 \text{ sq in.}; \quad \text{Reqd } S_{xb} = 4.8 \text{ in.}^3$$

Best choices appear to be:

WT6 × 32.5	$A = 9.55$	$S_{xb} = 4.2$
WT7 × 26.5	$A = 7.79$	$S_{xb} = 4.96$

(c) Check WT7 × 26.5 even though $d/t_w = 18.8$ exceeds the 17.9 basic limit of AISC–1.9.1.

Check AISC Formula 1.6–1b, yielding at braced points:

$$f_a = \frac{85}{7.79} = 10.9 \text{ ksi}$$

$$f_{bt} = \frac{90(1.38)}{27.7} = 4.5 \text{ ksi} \quad \text{(tension at top)}$$

$$f_{bb} = \frac{90}{4.96} = 18.1 \text{ ksi} \quad \text{(compression at bottom)}$$

$$F_b = Q_s(0.6F_y) = 0.958(30) = 28.8 \text{ ksi*}$$

$$\text{(bottom)} \quad f_a + f_b = 10.9 + 18.1 = 29.0 \text{ ksi} \approx 28.8 \text{ ksi} \quad \text{OK}$$

$$\text{(top)} \quad f_a + f_b = 10.9 - 4.5 = 6.4 \text{ ksi} \quad \text{OK}$$

Check AISC Formula 1.6–1a, stability out in $(+M)$ region:

$f_{bt} = 4.5$ ksi (compression at top)
$f_{bb} = 21.4$ ksi (tension at bottom)
$K = 1.0$ for truss joints as recommended by the Column Research Council (see Sec. 6.9).

*For discussion of the Q factor, see Sec. 6.18 and AISC Appendix Secs. C5 and C6.

$$\frac{KL}{r_y} = \frac{72}{1.92} = 37.5 \qquad \frac{KL}{r_x} = \frac{144}{1.88} = 76.5$$

$$\text{Eff. } KL/r = 76.5\,(C_c/C_c') = 76.5(107/109) = 75$$

where $C_c' = \sqrt{2\pi^2 E/(Q_a Q_s F_y)}$
$F_a = 19.99$ ksi (Table 1-50)
$F_b = Q_s(0.6F_y) = 28.8$ ksi

A structural tee, having r_x and r_y of comparable magnitude has no tendency to buckle in a plane other than the plane of bending; thus, $0.60F_y$ is the basic value to be used (AISC–1.5.1.4.6b) for F_b, modified by Q_s when d/t_w exceeds $127/\sqrt{F_y}$. Some ability probably exists to carry a moment exceeding the moment at first yield, but criteria to evaluate this extra capacity have not been established.

$$F_e' = 25.5 \text{ ksi} \quad \text{(based on axis of bending)}$$

$$C_m = 1.0 - 0.6\frac{f_a}{F_e'} \text{ (see Table 12.10.1)}$$

$$= 1.0 - 0.6(10.9)/25.5 = 1.0 - 0.6(0.428) = 0.744$$

$$\text{Magnification factor} = \frac{C_m}{1 - f_a/F_e'} = \frac{0.744}{0.572} = 1.30$$

$$\frac{f_a}{F_a} + \frac{f_b}{F_b}\left(\frac{C_m}{1 - f_a/F_e'}\right) = \frac{10.9}{19.99} + \frac{4.5}{28.8}(1.30) = 0.75 < 1.0$$

based on the compression fiber (the top) under flexure.

Use WT7 × 26.5. (Note: WT8 × 25 also is adequate and may be preferred.)

EXAMPLE 12.11.8.

Select the lightest W12 section to carry an axial compression in addition to biaxial bending, as shown in Fig. 12.11.9. Use $F_y = 50$ ksi.

Solution

Formulas 1.6–1a and 1.6–1b, Eqs. 12.10.8 and 12.10.9, are used in the three term form for biaxial bending.

(a) Determine equivalent column load. Assume that yielding at braced point governs:

$$\text{Equiv. } P \approx P + B_x M_x + B_y M_y$$
$$= 250 + 0.22(25)(12) + 0.8\,(9)(12)$$
$$= 250 + 66 + 86.5 = 402.5 \text{ kips}$$

(b) Select trial section. Using $KL_y = 7.5$ ft, select W12 × 53 (AISC Manual Column Tables):

$$r_x/r_y = 2.11 > KL_x/KL_y = 2.0 \qquad \text{OK}$$

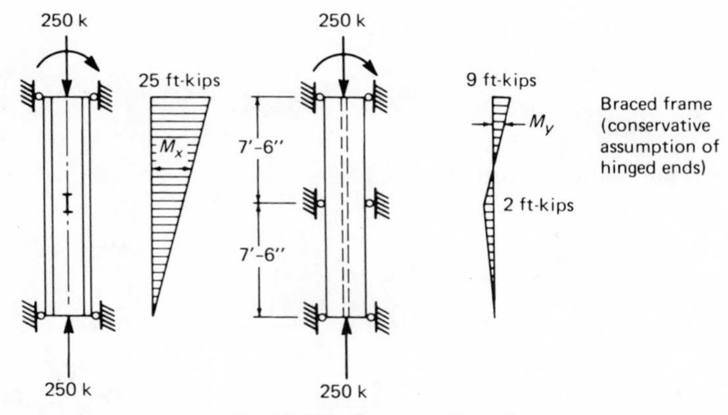

Fig. 12.11.9. Example 12.11.8.

(c) Investigate magnification factors:

$$P(KL_y)^2 = 250(1)^2(90)^2 = 2.02 \times 10^6$$
$$P(KL_x)^2 = 250(1)^2(180)^2 = 8.1 \times 10^6$$

$$\frac{C_{mx}\,a_x}{a_x - P(KL_x)^2} = \frac{0.6\,(63.6)}{63.6 - 8.1} = 0.687$$

$$\frac{C_{my}\,a_y}{a_y - P(KL_y)^2} = \frac{[0.6 - 0.4\,(2/9)]\,14.3}{14.3 - 2.02} = \frac{0.51(14.3)}{12.27} = 0.593$$

Since F_a/F_b is also a reduction effect, the computed equivalent column loads are too high if stability governs.

(d) Check Formula 1.6–1b, yielding governs:

$$\text{Reqd } A = \frac{402.5}{0.6\,F_y} = \frac{402.5}{30} = 13.4 \text{ sq in.}$$

W12 × 50 might satisfy. Check:

$$f_a = \frac{250}{14.7} = 17.0 \text{ ksi}$$

$$L = 7.5 \text{ ft.} > L_c = 7.3 \text{ ft}$$

$$F_{bx} = 0.6F_y = 30 \text{ ksi}$$

Section "compact" based on local buckling:

$$F_{by} = 0.75F_y = 37.5 \text{ ksi}$$

$$f_{bx} = \frac{25(12)}{64.7} = 4.6 \text{ ksi}; \qquad f_{by} = \frac{9(12)}{14.0} = 7.7 \text{ ksi}$$

$$\frac{f_a}{0.6F_y} + \frac{f_{bx}}{F_{bx}} + \frac{f_{by}}{F_{by}} = \frac{17.0}{30} + \frac{4.6}{30} + \frac{7.7}{37.5} = 0.93 < 1.0 \qquad \text{OK}$$

(e) Check Formula 1.6–1a, stability governs:

$$\frac{KL}{r_y} = \frac{90}{1.96} = 46; \qquad F_a = 24.96 \text{ ksi}$$

$$\text{Magnification factor (for } M_x) = \frac{C_{mx}\,a_x}{a_x - P(KL_x)^2}$$

$$= \frac{0.6\,(58.8)}{58.8 - 8.1} = 0.695$$

$$\text{Magnification factor (for } M_y) = \frac{0.51\,(8.4)}{8.4 - 2.02} = 0.671$$

$$\frac{f_a}{F_a} + \frac{f_{bx}}{F_{bx}}\left(\frac{C_{mx}}{1 - f_a/F'_e}\right) + \frac{f_{by}}{F_{by}}\left(\frac{C_{my}}{1 - f_a/F'_e}\right)$$

$$= \frac{17.0}{24.96} + \frac{4.6}{30}(0.695) + \frac{7.7}{37.5}\,(0.671)$$

$$= 0.68 + 0.11 + 0.14 = 0.93 < 1 \qquad \text{OK}$$

Use W12 × 50. (W12 × 45 gives about 3+% overstress.)

12.12. AISC–1969—PLASTIC DESIGN CRITERIA

The 1969 provisions for beam-columns differ considerably from those of the 1963 specification. The new provisions are more rational for all cases of loading and are essentially the equations discussed in Sec. 12.8. Furthermore, there is a fairly good correlation between the plastic design provisions and working stress provisions.

(a) For conditions at a braced location where instability is prevented; AISC–2.4 provides, as per Eq. 12.8.1,

$$\frac{P}{P_y} + \frac{M}{1.18M_p} \leq 1.0 \qquad [12.8.1]$$

and $M \leq M_p$. This is AISC Formula 2.4–3. In this equation, P and M are the ultimate loads (service load times load factor) to be designed for; $P_y = F_y A_g$; and M_p is the plastic moment capacity of the section. Eq. 12.8.1 is illustrated in Fig. 12.8.1.

(b) For locations where instability may govern, AISC–2.4 provides, as per Eq. 12.8.4,

$$\frac{P}{P_{cr}} + \frac{M_i}{M_m}\left(\frac{C_m}{1 - P/P_e}\right) \leq 1.0 \qquad [12.8.4]$$

which is AISC Formula 2.4–2. In the above equation,

P = factored service axial compression load

M_i = factored service primary bending moment

P_{cr} = ultimate strength of an axially loaded compression member, taken as $1.70\, F_a A_g$

M_m = maximum resisting moment in the absence of axial load, to be taken as

(*aa*) if braced to prevent lateral-torsional buckling,

$$M_m = M_p$$

(*ab*) If unbraced over the length L

$$M_m = \left[1.07 - \frac{\sqrt{F_y}(L/r_y)}{3160}\right] M_p \leq M_p$$

P_e = $1.92\, F_e' A_g$; i.e., $\pi^2 EI/L^2$

C_m = same as for working-stress method; see Sec. 12.10 for details

The reduction equation for M_m corresponds to the straight line reduction shown in Fig. 12.8.3 and is less conservative than using Formulas 1.5–6 and 1.5–7 as the basis for ultimate strength. There are two reasons which may explain this. (a) Formulas 1.5–6 and 1.5–7 were conservative approximations for the correct stability expression, Eq. 9.4.1, so those formulas indicated excessive reduction in moment capacity; and (b) plastic design uses the reduced M_m only for beam-columns, where end restraint provided by the integral rigid frame action increases the buckling and post-buckling strength over that which might be obtained for transversely loaded beams with simple framing. Typical values for M_m are given in Table 12.12.1.

TABLE 12.12.1
Moment Capacity Under Pure Bending,
As Used for Plastic Design of
Beam-Columns

L/r_y	M_m as a percent of M_p: $F_y = 36$	$F_y = 50$	$F_y = 60$
28.6			100
31.4		100	
36.9	100		
40	99	98	97
60	96	93	92
80	92	89	87
100	88	85	82
120	84	80	78

12.13. DESIGN EXAMPLES—PLASTIC-DESIGN METHOD

The general procedure for selecting sections for combined axial compression and bending is the same as that for working stress; i.e., compute an equivalent axial load and use column tables. The designer may also find that the manual, "Plastic Design of Braced Multistory Steel Frames"[28] provides useful design aids.

Expressed in terms of ultimate loads (factored service loads) the stability criterion, Eq. 12.8.4, converted into equivalent axial compression becomes

$$\text{Equiv. } P_{cr} = P + M_i \frac{P_{cr}}{M_m}\left(\frac{C_m}{1 - P/P_e}\right) \tag{12.13.1}$$

Since $P_{cr} = 1.70\, F_a A$ and M_m is normally $M_p = ZF_y$,

$$\text{Equiv. } P_{cr} = P + M_i \frac{1.70 F_a A}{F_y Z}\left(\frac{C_m}{1 - P/P_e}\right) \tag{12.13.2}$$

The magnification term may be written

$$\frac{C_m}{1 - P/P_e} = \frac{C_m}{1 - f_a/(1.92\, F_e')} = \frac{C_m}{1 - 0.52 f_a/F_e'} = \frac{C_m a}{a - 0.52P(KL)^2} \tag{12.13.3}$$

Also, $Z \approx 1.12S$ and $F_a \approx 0.5F_y$, which gives

$$\frac{1.70F_a A}{F_y Z} \approx \frac{1.7(0.5)F_y A}{1.12F_y Z} \approx 0.75B$$

where B = bending factor, A/S.

Thus the stability criterion, Eq. 12.13.1, (AISC Formula 2.4–2), becomes approximately

$$\text{Equiv. } P_{cr} = P + M_i(0.75)B\left(\frac{C_m a}{a - 0.52P(KL)^2}\right) \tag{12.13.4}$$

If stability governs,

$$\text{Equiv. } P_{cr} = 1.70\, F_a A$$

$$\text{Reqd } A = \frac{\text{Equiv. } P_{cr}}{1.70F_a} \tag{12.13.5}$$

On the other hand, if strength (yielding) governs at braced points, Eq. 12.8.1 must form the basis of the Equiv. P; thus

$$\text{Equiv. } P_y = P + M\frac{P_y}{1.18M_p} \tag{12.13.6}$$

Since P_y and AF_y and $M_p = ZF_y \approx 1.12SF_y$, Eq. 12.13.6 becomes

$$\text{Equiv. } P_y = P + M(0.76)B \tag{12.13.7}$$

where B = bending factor, A/S. Because the bending factor involves the working stress concept of section modulus, a function of the basic dimensions may be preferred. In which case, for strong-axis bending,

$$S = Ar_x^2/(d/2) = 2Ad(r_x/d)^2 \tag{12.13.8}$$

$$\frac{P_y}{1.18M_p} \approx \frac{AF_y}{1.18(1.12)SF_y} \approx \frac{1}{1.32(2)d(r_x/d)^2} = \frac{0.37}{d(r_x/d)^2} \tag{12.13.9}$$

Using $r_x/d \approx 0.43$ (see Appendix Table A1 or Ref. 28),

$$\text{Equiv. } P_y = P + 2M/d \tag{12.13.10}$$

Reference 28 suggests 2.1 instead of 2. This shows that for strong axis bending, $B_x \approx 2.6/d$.

Thus for selection purposes Eqs. 12.13.4 and 12.13.10 may be used. In addition, for the selected section, the plastic moment capacity must exceed the factored applied moment.

EXAMPLE 12.13.1

Repeat Example 12.11.1, except use plastic design (see Fig. 12.11.1). $F_y = 36$ ksi.

Solution

Assuming the plastic analysis of the structure has been made for preliminary sections using a gravity load factor of 1.7,

$$P_u = 255 \text{ kips}; \qquad M_u = 850 \text{ kips}$$

For this braced system, the equivalent-length factors are $K_x = K_y = 1.0$. Assuming that stability controls, use Eq. 12.13.4 for Equiv. P and as a first approximation assume

$$\begin{aligned} \text{Equiv. } P_{cr} &= P + M_i(0.75)B_x \\ &= 255 + 850(12)(0.75)(0.19) = 255 + 1454 = 1709 \text{ kips} \end{aligned}$$

When AISC column tables are available, Eq. 12.13.5 gives

$$\text{Reqd } AF_a = \frac{\text{Equiv. } P_{cr}}{1.70} = 1005 \text{ kips}; \qquad KL = 14 \text{ ft}$$

A comparison with Example 12.11.1 will show the approach to selecting a section when stability governs is identical for working stress and plastic design. From AISC Manual select as first trial, W14 × 184, with $B_x = 0.183$, and $a_x = 339.3 \times 10^6$.

Check W14 × 184: Use AISC Formula 2.4–2, Eq. 12.8.4, for stability:

$$\frac{KL}{r_y} = \frac{168}{4.04} = 41.6; \qquad F_a = 19.06 \text{ ksi}$$

$$P_{cr} = 1.70\, F_a A = 1.70(19.06)(54.1) = 1752 \text{ kips}$$

$$M_m = \left[1.07 - \frac{\sqrt{F_y}(L/r_y)}{3160}\right] M_p$$

$$= [1.07 - \sqrt{36}\,(41.6)/3160]M_p = 0.991 M_p$$

$$= 0.991\, F_y Z = 0.991\,(36)\,(338)/12 = 1003 \text{ ft-kips}$$

$$\frac{KL}{r_x} = \frac{168}{6.49} = 25.9; \qquad F_e' = 223 \text{ ksi}$$

$$P_e = (23/12)\, F_e' A = 1.92\,(223)\, 54.1 = 23{,}150 \text{ kips}$$

$$\frac{P}{P_{cr}} + \frac{M_i}{M_m}\left(\frac{C_m}{1 - P/P_e}\right) = \frac{255}{1752} + \frac{850}{1003}\left(\frac{1.0}{1 - 255/23{,}150}\right)$$

$$= 0.146 + 0.857 = 1.003 \approx 1 \qquad \text{OK}$$

Check the strength criterion, AISC Formula 2.4–3:

$$P_y = F_y A = 36\,(54.1) = 1947 \text{ kips}$$

$$M_p = F_y Z = 36\,(338)/12 = 1013 \text{ ft-kips}$$

$$\frac{P}{P_y} + \frac{M}{1.18 M_p} = \frac{255}{1947} + \frac{850}{1.18(1013)} = 0.13 + 0.71 = 0.84 < 1.0 \qquad \text{OK}$$

Use W14 × 184.

This is the same as obtained by working stress method, as would be expected for this statically determinate system. There would ordinarily be a difference, however, for a real structure where the analysis to obtain the loads would be performed differently; plastic analysis of a rigid frame structure vs. elastic analysis with perhaps some arbitrary adjustment of negative moments. For plastic analysis and design of frames, see Chapter 15.

EXAMPLE 12.13.2

Select the lightest W section to resist the factored ($F = 1.70$) gravity loading on the braced frame of Fig. 12.13.1 using plastic design according to AISC specification. Use A36 steel.

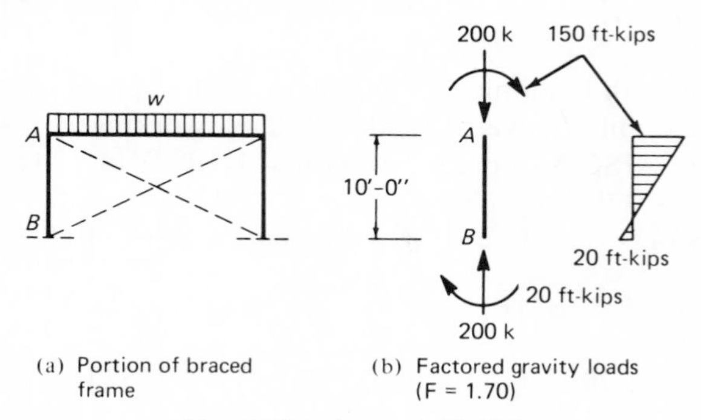

Fig. 12.13.1. Example 12.13.2.

Solution

Use the unbraced length as the equivalent pin-end length. The alignment chart for $K < 1.0$ (Fig. 14.3.1) should *not* be used for this case since the point of contraflexure is close to the bottom and a plastic condition is assumed to develop at the top. The failure mode for buckling would probably be one of single curvature, with the moment at the bottom close to zero and the restraint at the top reduced when the plastic condition is reached.

Assume that the strength criterion, AISC Formula 2.4-3, governs. Use Eq. 12.13.10 to obtain Equiv. P_y:

$$\begin{aligned}\text{Equiv. } P_y &= P + 2M/d \\ &= 200 + 2(150)(12)/d \\ &= 200 + 3600/d\end{aligned}$$

d	Equiv. P_y	$A = P_y/F_y$	Section
W12	500	13.9	W12 × 50, $A = 14.7$
W14	457	12.7	W14 × 48, $A = 14.1$

Check W14 × 48 according to AISC Formula 2.4–3:

$$P_y = AF_y = 14.1\,(36) = 508 \text{ kips}$$

$$M_p = ZF_y = 78.4\,(36)/12 = 236 \text{ ft-kips}$$

$$\frac{P}{P_y} + \frac{M}{1.18M_p} = \frac{200}{508} + \frac{150}{1.18(236)} = 0.393 + 0.539$$

$$= 0.932 < 1 \qquad \text{OK}$$

Check stability according to AISC Formula 2.4–2:

$$\frac{KL}{r_y} = \frac{120}{1.91} = 62.8; \qquad F_a = 17.16 \text{ ksi}$$

$$P_{cr} = 1.70\,F_aA = 1.70\,(17.16)\,14.1 = 411 \text{ kips}$$

$$M_m = \left[1.07 - \frac{\sqrt{F_y}(L/r_y)}{3160}\right]M_p = 0.95M_p = 224 \text{ ft-kips}$$

$$\frac{KL}{r_x} = \frac{120}{5.86} = 20.5; \qquad F'_e = 354 \text{ ksi}$$

$$P_e = (^{23}\!/_{12})F'_e = 1.92\,(354)\,14.1 = 9590 \text{ kips}$$

$$C_m = 0.6 - 0.4\,(20/150) = 0.55$$

$$\frac{P}{P_{cr}} + \frac{M_i}{M_m}\left(\frac{C_m}{1 - P/P_e}\right) = \frac{200}{411} + \frac{150}{224}\left(\frac{0.55}{1 - 200/9590}\right)$$

$$= 0.49 + 0.67(0.56) = 0.87 < 1 \qquad \text{OK}$$

The W14 × 48 satisfies the local buckling limitations of AISC–2.7:

$$\frac{b}{2t_f} = \frac{8.031}{2(0.593)} = 6.8 < 8.5 \qquad \text{OK}$$

For $P/P_y > 0.27$,

$$\frac{d}{w} = \frac{13.81}{0.339} = 40.7 < \frac{257}{\sqrt{F_y}} = 42.8 \qquad \text{OK}$$

Use W14 × 48.

EXAMPLE 12.13.3

Select a W shape to serve as the beam-column of a gabled frame as shown in Fig. 12.13.2. Use A36 steel. The structure is braced transverse to the plane of the frame, with bracing at top, bottom, and at mid-height on member AB.

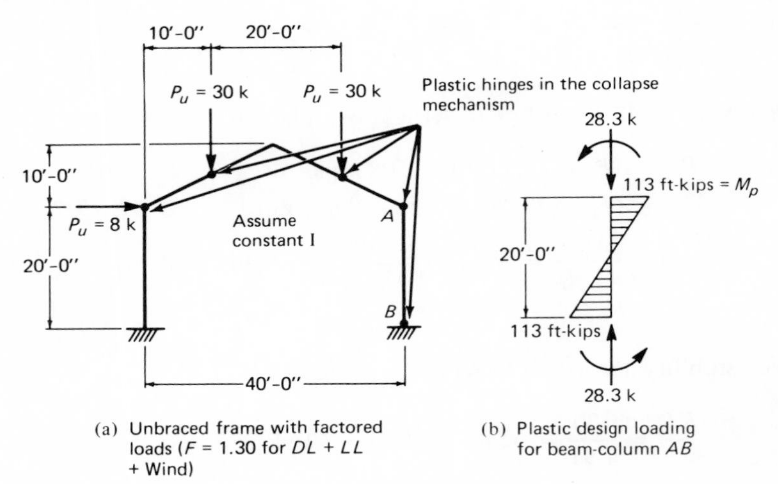

(a) Unbraced frame with factored loads (F = 1.30 for $DL + LL$ + Wind)

(b) Plastic design loading for beam-column AB

Fig. 12.13.2. Example 12.13.3.

Solution

Assume the plastic analysis has been made for the frame using constant moment of inertia.

(a) Estimate Equiv. P based on the strength criterion, AISC Formula 2.4–3. Using Eq. 12.13.10,

$$\text{Equiv. } P_y = P + 2M/d$$

$$\text{Equiv. } P_y = 28.3 + 2(113)(12)/8 \text{ (for W8)}$$

$$\text{Reqd } A = \frac{\text{Equiv. } P_y}{F_y} = \frac{367}{36} = 10.2 \text{ sq in.}$$

Try W8 × 35, A = 10.3 sq in.

(b) Check strength criterion, AISC Formula 2.4–3:

$$P_y = AF_y = 10.3(36) = 371 \text{ kips}$$
$$M_p = ZF_y = 34.7(36)/12 = 104 \text{ ft-kips}$$

It is immediately noted that Reqd M exceeds the plastic moment capacity of the section; since this cannot be, the section must be increased so that its $M_p \geq 113$ ft-kips. It appears W8 × 40 or W10 × 39 will satisfy the strength criterion.

(c) Investigate lateral bracing requirement. Since plastic hinges are expected to occur at the top and bottom of the member, the provisions of AISC–2.9 apply. With lateral support at top, bottom, and midheight, the moment gradient is $M/M_p = 0$ which is more than (-0.5):

$$L_{cr} = \left[\frac{1375}{F_y} + 25\right] r_y = \left[\frac{1375}{36} + 25\right] r_y = 63.2 r_y$$

If unbraced over a length of 10 ft,

$$\text{Reqd } r_y = \frac{120}{63.2} = 1.90$$

Both the W8 × 40 and W10 × 39 will satisfy this requirement.

(d) Complete check of strength requirement for W10 × 39:

$$P_y = AF_y = 11.5(36) = 413 \text{ kips}$$
$$M_p = ZF_y = 46.9(36)/12 = 141 \text{ ft-kips} > 113 \text{ ft-kips} \qquad \text{OK}$$

$$\frac{P}{P_y} + \frac{M}{1.18M_p} = \frac{28.3}{413} + \frac{113}{1.18(141)} = 0.07 + 0.68 = 0.75 < 1 \qquad \text{OK}$$

(e) Check stability requirement, AISC Formula 2.4–2.

Note: With large moment restraint at the base of the frame, say $G_B = 1.0$, and with I/L approximately the same for beam and column, $G_A = 1.0$. From Alignment Chart, Fig. 14.3.1b, find $K = 1.3$.

$$\frac{KL}{r_x} = \frac{1.3(240)}{4.27} = 73; \qquad F'_e = 30.0 \text{ ksi}$$

$$P_e = (23/12)F'_e A = 1.92(30.0)(11.5) = 662 \text{ kips}$$

$$C_m = 0.85 \quad \text{for an unbraced frame}$$

$$\frac{KL}{r_y} = \frac{120}{1.98} = 60.5 < 73; \quad F_a = 16.12 \text{ ksi}$$

$$P_{cr} = 1.70F_a A = 1.70(16.12)(11.5) = 315 \text{ kips}$$

$$M_m = \left[1.07 - \frac{\sqrt{F_y}(L/r_y)}{3160}\right] M_p = 0.955M_p = 135 \text{ ft-kips}$$

$$\frac{P}{P_{cr}} + \frac{M_i}{M_m}\left(\frac{C_m}{1 - P/P_e}\right) = \frac{28.3}{315} + \frac{113}{135}\left(\frac{0.85}{(1 - 28.3/662)}\right)$$
$$= 0.09 + 0.75 = 0.84 < 1 \qquad \text{OK}$$

The W10 × 39 has been verified to be a "compact" section.

Use W 10 × 39.

For practical design of beam-columns, there are a number of design aids and charts available.[27,28] Most of these use the concept of a plastic hinge developing at a moment M_{pc} which is less than M_p, when axial compression exists. This is really the strength criterion, AISC Formula 2.4–3, Eq 12.8.1, where solving for M/M_p gives

$$\frac{M}{M_p} = 1.18\,(1 - P/P_y) \qquad (12.13.5)$$

where M may be called M_{pc}, the reduced plastic-moment capacity. If $P/P_y \leq 0.15$, $M_{pc} = M_p$. To visualize this concept, refer to Fig. 12.6.4, where for different ratios of P/P_y elastic-plastic behavior occurs, but with large plastic hinge rotation occuring at lower levels of moment for large P/P_y.

EXAMPLE 12.13.4

Recheck the W14 × 48 of Example 12.13.2 using the Design Aids of "Plastic Design of Braced Multistory Steel Frames."[28] Use of such charts gives a better visual understanding of what is going on.

Solution

(a) Since the ratio of end moments is 20/150 = 0.133, use the charts for $q = 0$ (Must choose between +1, 0 and −1). Compute,

$$\frac{P}{P_y} = \frac{200}{508} = 0.39$$

(b) Check lateral-torsional buckling using Design Aid III–2a, as shown in Fig. 12.13.3.

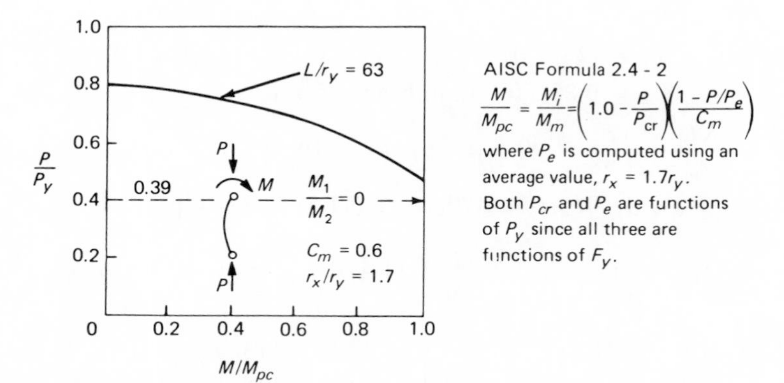

Fig. 12.13.3. Lateral-torsional buckling, A36 steel. (From Ref. 28.)

$$\frac{L}{r_y} = 62.8; \quad \text{find } M/M_{pc} = 1.0$$

(c) Check in-plane bending using Design Aid III–2b, as shown in Fig. 12.13.4, find

$$M/M_{pc} = 1.0$$

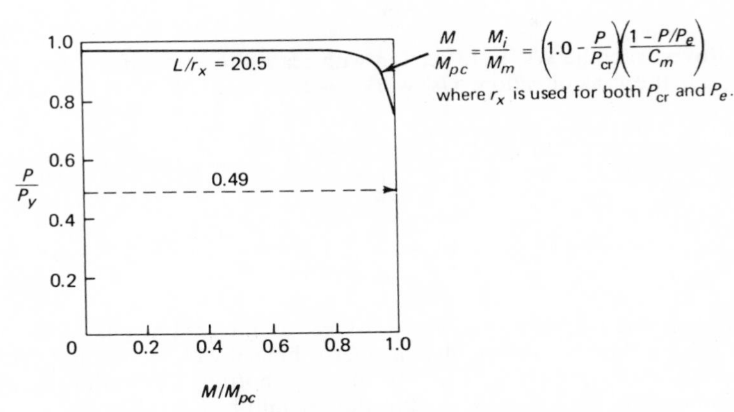

Fig. 12.13.4. In-plane bending, A36 steel. (From Ref. 28.)

Thus, from these two charts, stability does not control:

$$\frac{M}{M_{pc}} \approx 1$$

Since for W14 × 48, $M_p = 236$ ft-kips

$$M_{pc} = 1.18(1 - P/P_y)M_p$$
$$= 1.18(1 - 0.39)236 = 170 \text{ ft-kips} > 150 \text{ ft-kips} \qquad \text{OK}$$

If Fig. 12.13.3 had indicated a reduction for M/M_{pc}, this would have been conservative since actual $r_x/r_y = 3.07$, while the charts use 1.7. Charts are useful for designers, but are generally outside the scope of a textbook.

PROBLEMS

For all problems involving design or safety analysis, use the latest AISC specifications unless otherwise indicated. The requirement of W section is intended to include W, S, and M sections.

Problems 12.1 through 12.27 relate to design considerations. Problems 12.28 through 12.38 relate to other theoretical considerations.

12.1. Investigate the adequacy of the section according to the working-stress method. No translation of joints can occur, and external lateral support is provided at the ends only.

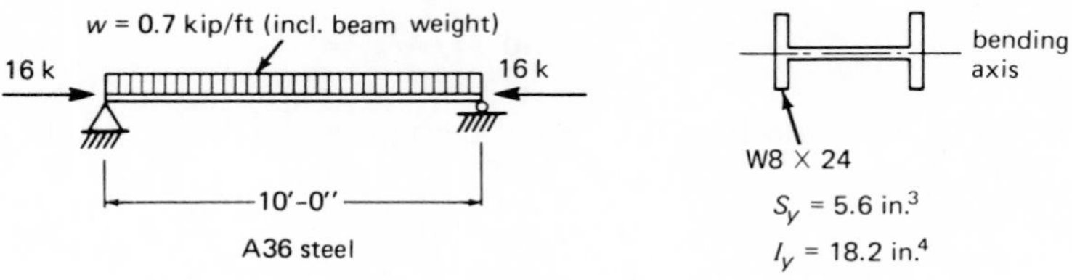

Prob. 12.1.

12.2. Investigate the W4 × 13 section with regard to safety if primary bending is in the weak direction. Use working-stress method. A572 Grade 50.

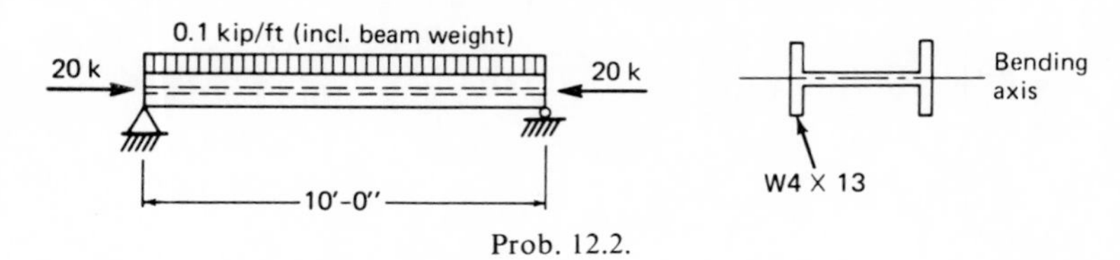

Prob. 12.2.

12.3. Determine the allowable load Q (kips) at the mid-height of the beam-column shown. Assume the member is hinged with respect to bending in both x and y directions at the top and bottom. Additionally, lateral support occurs in the weak direction at mid-height. Use working-stress method.

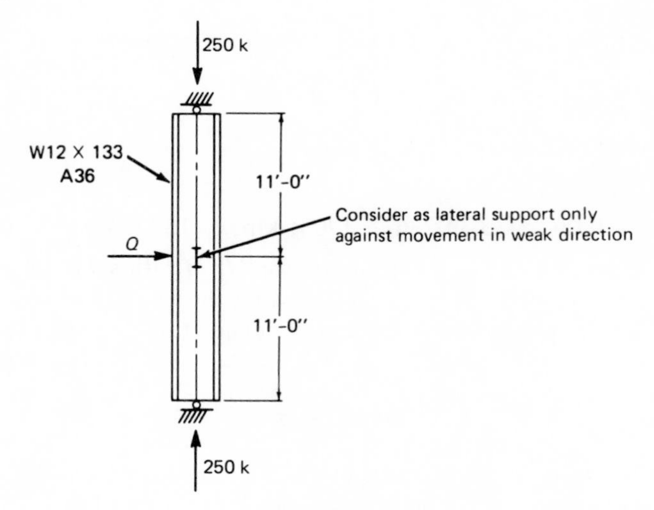

Prob. 12.3.

12.4. Investigate the adequacy of the given section, according to the working-stress method. No joint translation can occur and external lateral support is provided at the ends only.

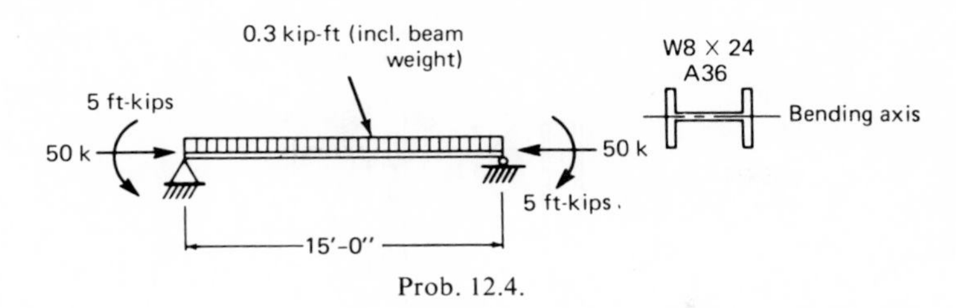

Prob. 12.4.

12.5. Determine the safe service load P which is permitted according to the working-stress method. Braced frame.

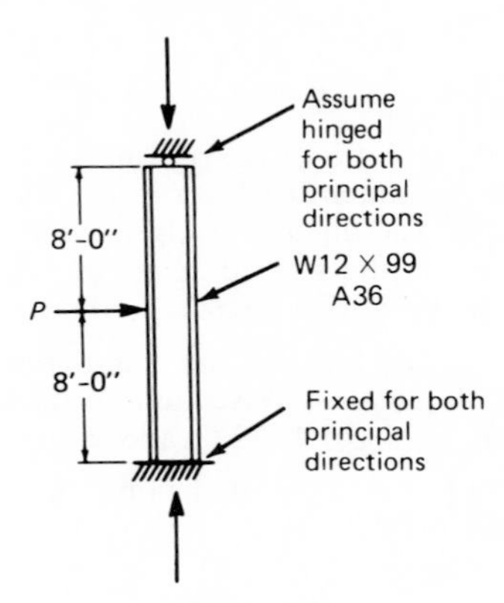

Prob. 12.5.

12.6. Determine the allowable axial load P which the W16 × 96 may be permitted to carry according to the working-stress method. Lateral support is provided at ends and at midspan. Compare for A36 and A572 Grade 45 steels.

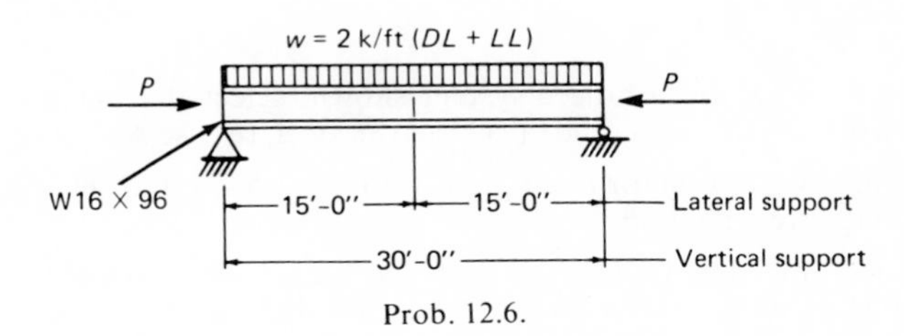

Prob. 12.6.

12.7. Select the lightest W14 section to carry a service load P = 250 kips with an eccentricity e = 24 in. with respect to the strong axis. Assume the member is part of a braced system, and conservatively assume the equivalent length equals the unbraced height. Use (a) A36 steel; (b) A572 Grade 60 steel. Use working-stress method.

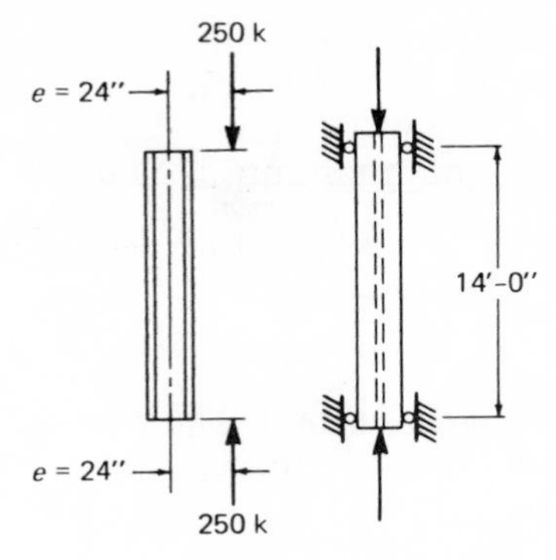

Prob. 12.7.

12.8. Select the lightest W14 section to carry an axial compression of 240 kips along with a bending moment of 450 ft-kips, which, for conservative simplicity, is assumed to be acting uniformly along the 15-ft equivalent pin end length of the member in the braced structure. (a) use A36 steel; (b) use A572 Grade 50. Use the working-stress method.

12.9. If the service load P is 125 kips for the beam-column loading and support of Prob. 12.6, select the lightest W section according to the working-stress method. Use (a) A36 steel; (b) A572 Grade 60.

12.10. Select the lightest W14 section for the beam-column shown. Use working-stress method and assume lateral buckling is adequately prevented and that $L < L_c$. Use A36 steel.

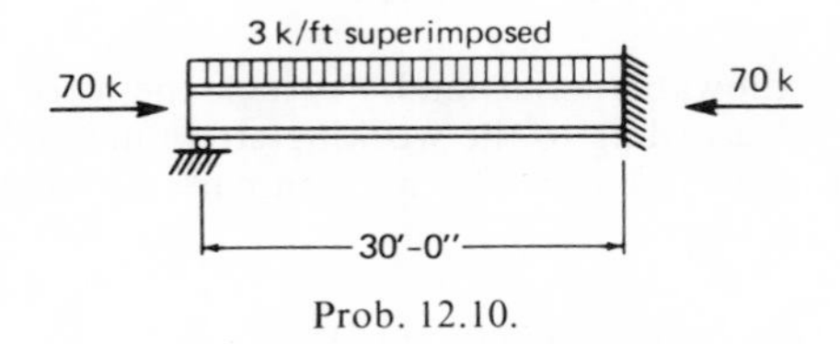

Prob. 12.10.

12.11. For the member of a braced system shown, select the lightest W section. Use working-stress method. (a) Use A36 steel; (b) use A572 Grade 60 steel.

12.12. Redesign the section for Example 12.11.6 for A572 Grade 60 steel. Use the working-stress method.

12.13. A frame braced against sidesway has a beam-column loaded as shown as a result of an elastic analysis. Select the lightest W section acceptable. Use A36 steel.

12.14. Redesign the member of Prob. 12.13 as part of an unbraced frame. Assume moment of inertia for beam and column to be the same. Use working-stress method.

12.15. Design the lightest W12 for column A of the unbraced frame shown. Preliminary design has selected W27 × 94 for all adjacent beams, and it has been decided column A will be approximately the same stiffness as those above and below it. Use working-stress method with A572 Grade 50 steel.

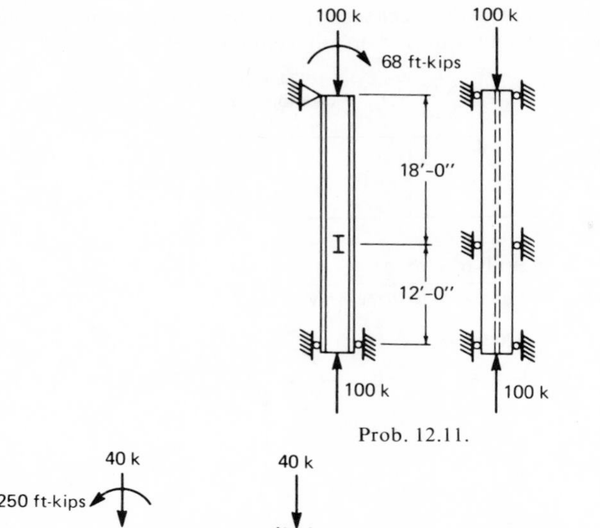

Prob. 12.11.

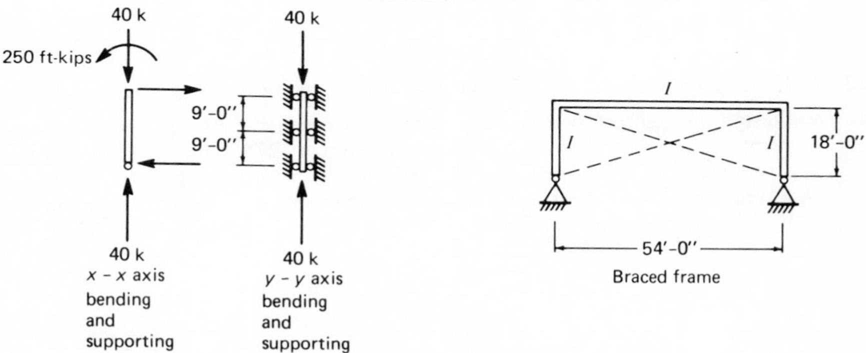

Prob. 12.13 and 12.14.

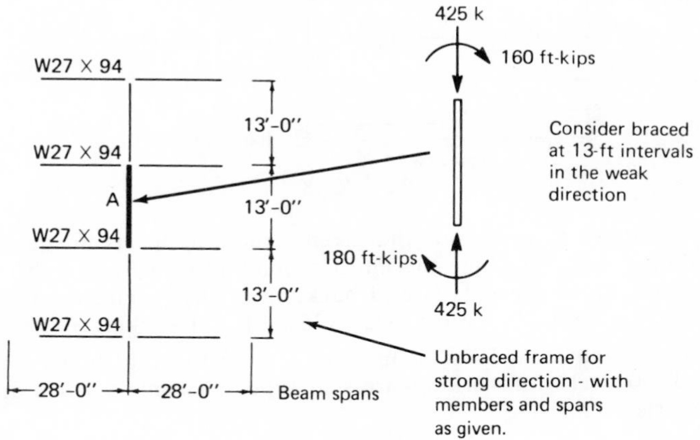

Prob. 12.15.

12.16. For the vierendeel truss (rigid frame) shown, investigate the safety of members *A* and *B*, according to the working-stress method. The uniform loading includes the weight of the steel section. A36 steel. Assume simple cross-bracing between given frame and an adjacent parallel one.

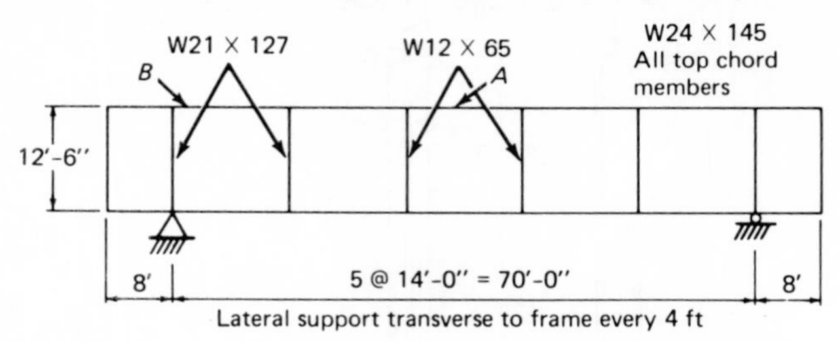

220 ft-kips
1.3 k/ft
215 k
220 ft-kips
215 k
Member *A*

90 k
545 ft-kips
1.3 k/ft
450 ft-kips
90 k
Member *B*

Probs. 12.16 and 12.17.

12.17. Redesign member *A* for A572 Grade 60 steel, using working-stress method.

12.18. Design the lightest W section to carry two eccentric 20-kip loads causing primary bending plus an axial compression of 275 kips. Use working-stress method. Assume torsional fixity at the vertical supports. Assume lateral bracing at midspan is adequate to prevent lateral-torsional buckling, but does permit enough rotation for torsional moments to develop.

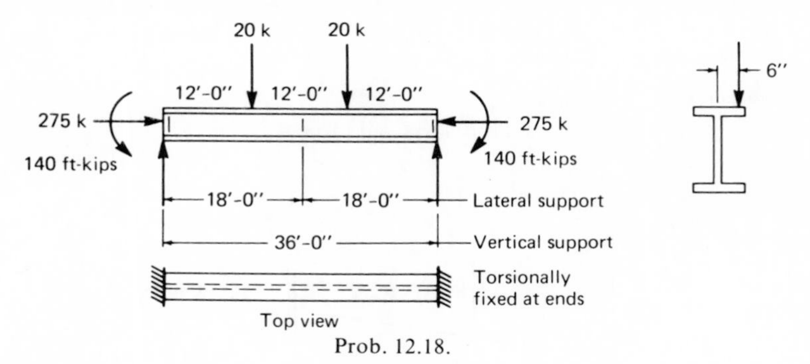

Prob. 12.18.

12.19. Determine whether or not the given structural tee is adequate for the loading shown. Assume the uniform loading is delivered through construction which prevents lateral buckling of the member. Assume continuity over the end supports. Use A36 steel, and working-stress method.

12.20. Select the lightest WT7 structural tee for the loading shown. Use A572 Grade 50 steel with working-stress method. Assume vertical and lateral supports at ends only.

12.21. Select an economical structural tee to serve as a continuous compression chord member of a truss to carry service loads as shown. For design purposes assume fixed ends on the member. Use working-stress method.

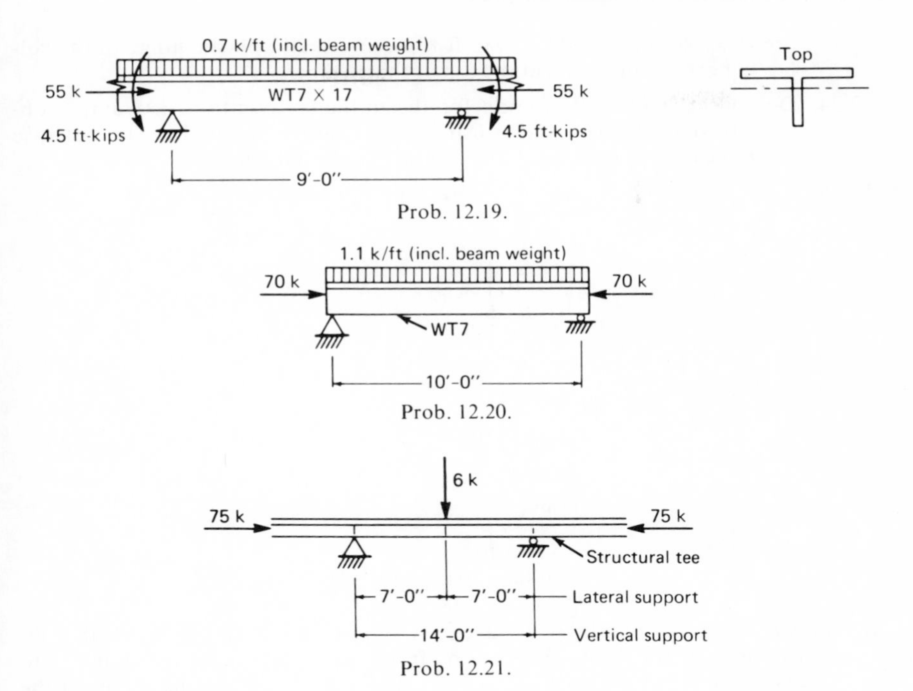

Prob. 12.19.

Prob. 12.20.

Prob. 12.21.

12.22. Select the lightest W section for the member of the unbraced frame using plastic-design procedure. Assume simple lateral bracing at 8-ft intervals transverse to the plane of the frame.

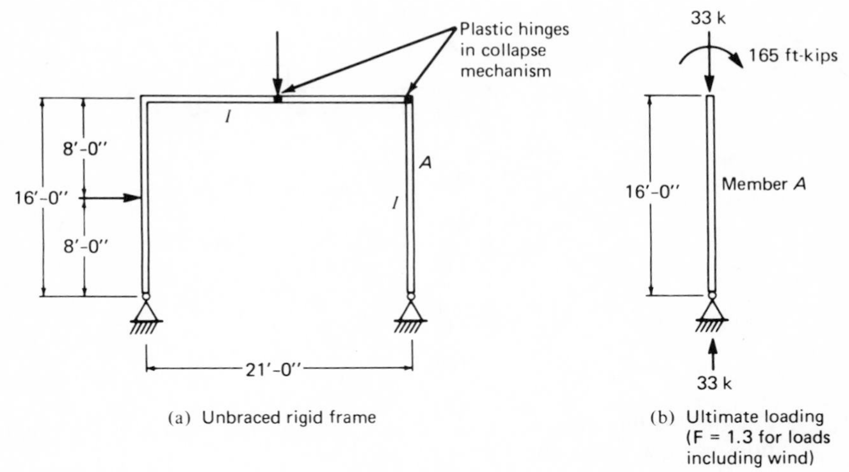

Prob. 12.22.

12.23. Redesign the member of Prob. 12.13 using the plastic design method. Assume ultimate axial force and moment (i.e., obtained using factored loads) are $P = 66$ kips and $M = 380$ ft-kips, and that a plastic hinge in the collapse mechanism occurs at the top of the member.

12.24. Design the lightest W section for the ultimate conditions (M and P obtained using factored loads) using plastic-design method. The collapse mechanism does *not* have a plastic hinge in this member.

12.25. Repeat Prob. 12.24 using $P_u = 500$ kips and $M_u = 60$ kips.

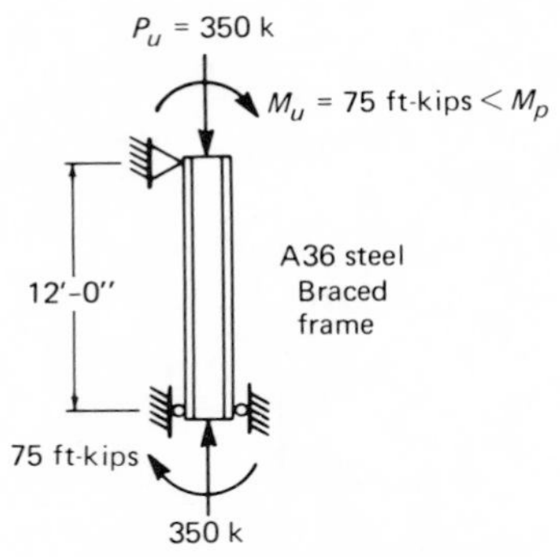

Prob. 12.24 and 12.25

12.26. A column in a building has reactions from beams framing in which contribute moments to it. The beams framing into the web are assumed to rest on seats, where reactions are assumed to be 2-in. from the center of the web. The reaction from the other beam is assumed to be acting at the face of the flange. Use A36 steel and select lightest W section. Use working-stress method.

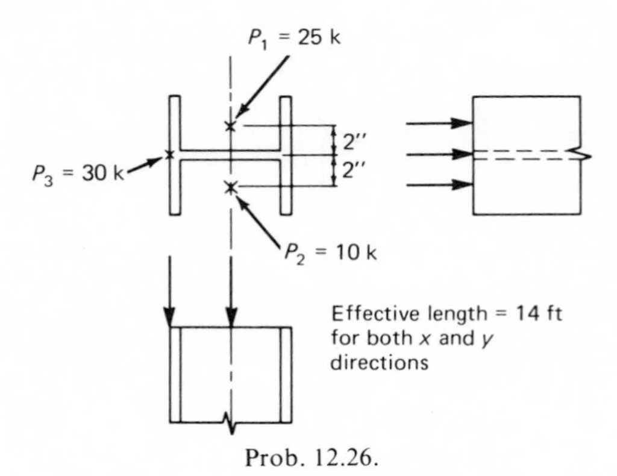

Prob. 12.26.

12.27. For the loading shown, select lightest sections for the following conditions using working-stress method:
(a) W14 using A36 steel
(b) W in any depth using $F_y = 50$ ksi
(c) W14 using $F_y = 60$ ksi
(d) W14 using $F_y = 70$ ksi

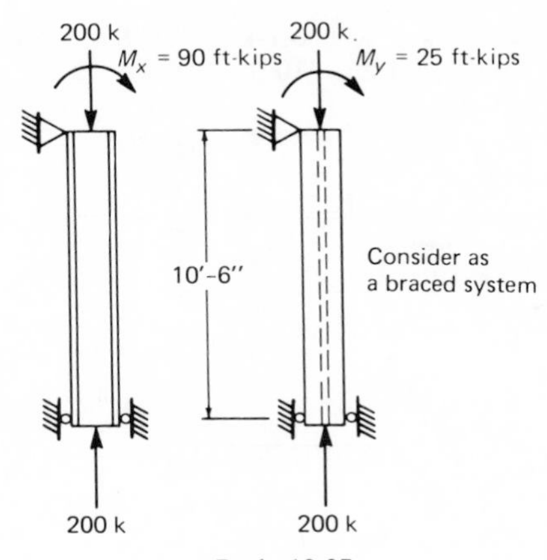

Prob. 12.27.

Problems Relating to Theoretical Considerations

12.28 through 12.32. For the given loading and support conditions, develop the differential equation for M_z, the moment in the plane of bending, and determine the maximum value for M_z.

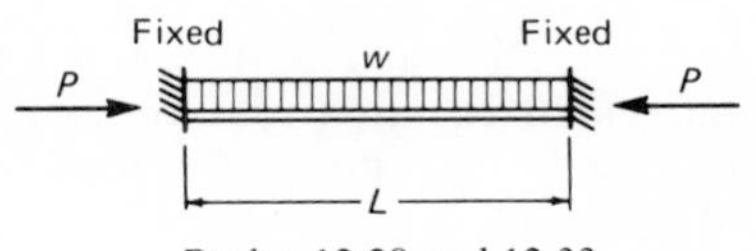

Probs. 12.28 and 12.33

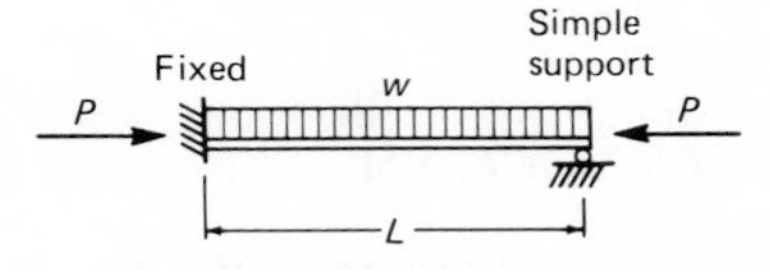

Probs. 12.29 and 12.34

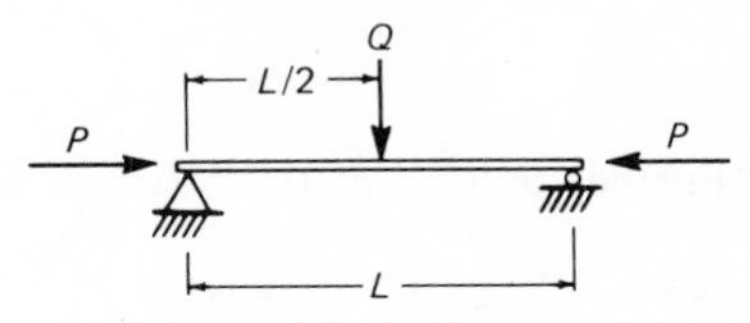

Probs. 12.30 and 12.35

12.33 through 12.37. For each loading condition used in Probs. 12.28 through 12.32, compute the maximum total compressive stress on the extreme fiber of a W4 × 13 bent in the weak direction using (a) the differential equation solution and (b) the approximate procedure of

$$M_{max} = M_{imax} + Py_{max}$$

For unknowns use $P = 20$ kips, $w = 0.1$ kip/ft, $Q = 0.5$ kips, $a = 3$ ft, $M_1 = 0.6$ ft-kips, $M_2 = 1.0$ ft-kips, and $L = 10$ ft.

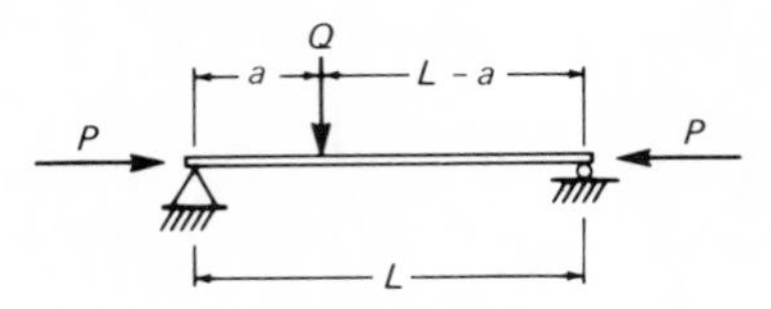

Probs. 12.31 and 12.36

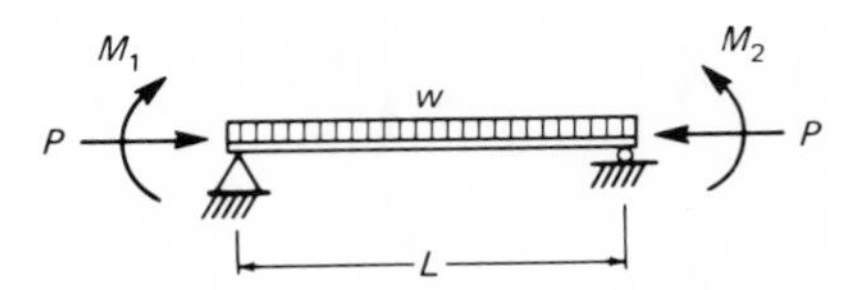

Probs. 12.32 and 12.37.

12.38. (a) Develop the differential equation solution for the loading of Prob. 12.10, and determine the expression for maximum bending moment. Assume the fixed-end moment may be approximated as $wL^2/8$.

(b) For the W14 selected in Prob. 12.10 compute the maximum bending moment using both the differential equation and the approximate method,

$$M_{\max} = M_{i\max} + Py_{\max}$$

(c) Develop the differential equation solution assuming zero slope is required at the fixed end. How much effect does moment magnification have on the end moment?

SELECTED REFERENCES

1. Charles Massonnet, "Stability Considerations in the Design of Steel Columns," *J. Structural Div. ASCE*, Vol. 85, No. ST7 (September 1959), pp. 75–111.
2. Walter J. Austin, "Strength and Design of Metal Beam-Columns," *J. Structural Div. ASCE*, Vol. 87, No. ST4 (April 1961), pp. 1–32.
3. Robert L. Ketter, Edmund L. Kaminsky, and Lynn S. Beedle, "Plastic Deformation of Wide-Flange Beam-Columns," *Trans. ASCE*, Vol. 120 (1955), pp. 1028–1069.
4. Robert L. Ketter, "Further Studies of the Strength of Beam-Columns," *J. Structural Div. ASCE*, Vol. 87, No. ST6 (August 1961), pp. 135–152.
5. Theodore V. Galambos and Robert L. Ketter, "Columns Under Combined Bending and Thrust," *J. Engg. Mech. Div. ASCE*, Vol. 85, No. EM2 (April 1959), pp. 1–30.
6. George F. Hauck and Seng-Lip Lee, "Stability of Elasto-Plastic Wide-Flange Columns," *J. Structural Div. ASCE*, Vol. 89, No. ST6 (December 1963), pp. 297–324.
7. S. L. Lee and G. F. Hauck, "Buckling of Steel Columns Under Arbitrary End Loads," *J. Structural Div. ASCE*, Vol. 90, No. ST2 (April 1964), pp. 179–200.

8. S. L. Lee and S. C. Anand, "Buckling of Eccentrically Loaded Steel Columns," *J. Structural Div. ASCE*, Vol. 92, No. ST2 (April 1966), pp. 351–370.
9. Edwin C. Rossow, George B. Barney, and Seng-Lip Lee, "Eccentrically Loaded Steel Columns with Initial Curvature," *J. Structural Div. ASCE*, Vol. 93, No. ST2 (April 1967), pp. 339–358.
10. Le-Wu Lu and Hassan Kamalvand, "Ultimate Strength of Laterally Loaded Columns," *J. Structural Div. ASCE*, Vol. 94, No. ST6 (June 1968), pp. 1505–1524.
11. B. G. Johnston, "Lateral Buckling of I-Section Columns with Eccentric End Loads in the Plane of the Web," *Trans. ASME*, Vol. 62 (1941), p. A–176.
12. Mario G. Salvadori, "Lateral Buckling of I-beams," *Trans. ASCE*, Vol. 120 (1955), pp. 1165–1182.
13. M. Salvadori, "Lateral Buckling of Eccentrically Loaded I-Columns," *Trans. ASCE*, Vol. 121 (1956), pp. 1163–1178.
14. Constancio Miranda and Morris Ojalvo, "Inelastic Lateral-Torsional Buckling of Beam Columns," *J. Engg. Mech. Div. ASCE*, Vol. 91, No. EM6 (December 1965), pp. 21–37.
15. Yushi Fukumoto and T. V. Galambos, "Inelastic Lateral-Torsional Buckling of Beam Columns," *J. Structural Div. ASCE*, Vol. 92, No. ST2 (April 1966), pp. 41–61.
16. T. V. Galambos, P. F. Adams, and Y. Fukumoto, "Further Studies on the Lateral-Torsional Buckling of Steel Beam-Columns," Bulletin No. 115, Welding Research Council, July 1966.
17. Ralph C. Van Kuren and T. V. Galambos, "Beam Column Experiments," *J. Structural Div. ASCE*, Vol. 90, No. ST2 (April 1964), pp. 223–256.
18. M. R. Horne, "The Stanchion Problem in Frame Structures Designed According to Ultimate Carrying Capacity," *Proc. Inst. Civil Engrs.*, Vol. 5, No. 1, Part III (April 1956), pp. 105–146.
19. "Specification for the Design, Fabrication and Erection of Structural Steel for Buildings," American Institute of Steel Construction, 1969.
20. "Commentary on the Specification for the Design, Fabrication, and Erection of Structural Steel for Buildings," American Institute of Steel Construction, 1969.
21. Charles Birnstiel and James Michalos, "Ultimate Load of H-Columns Under Biaxial Bending," *J. Structural Div. ASCE*, Vol. 89, No. ST2 (April 1963), pp. 161–197.
22. Charles G. Culver, "Exact Solution of Biaxial Bending Equations," *J. Structural Div. ASCE*, Vol. 92, No. ST2 (April 1966), pp. 63–83.
23. Charles G. Culver, "Initial Imperfections in Biaxial Bending," *J. Structural Div. ASCE*, Vol. 92, No. ST3 (June 1966), pp. 119–135.
24. Charles Birnstiel, "Experiments on H-Columns Under Biaxial Bending," *J. Structural Div. ASCE*, Vol. 94, No. ST10 (October 1968), pp. 2429–2449.
25. Gunnar A. Harstead, Charles Birnstiel, and Keh-Chun Leu, "Inelastic Behavior of H-Columns Under Biaxial Bending," *J. Structural Div. ASCE*, Vol. 94, No. ST10 (October 1968), pp. 2371–2398.
26. Wai F. Chan and Sakda Santathadaporn, "Review of Column Behavior Under Biaxial Loading." *J. Structural Div. ASCE*, Vol. 94, No. ST12 (December 1968), pp. 2999–3021.
27. George C. Driscoll, Jr., et al., "Plastic Design of Multistory Frames," Lecture Notes, Fritz Engineering Laboratory Report No. 273.20, 1965.

28. *Plastic Design of Braced Multistory Steel Frames*, American Iron and Steel Institute, New York, 1968, pp. 8–12, 98–111.
29. Ira Hooper, "Design of Beam-Columns," *Engineering Journal*, AISC, Vol. 4, No. 2 (April 1967), pp. 41–61.
30. Moe A. Rubinsky, "Rapid Selection of Beam-Columns," *Engineering Journal*, AISC, Vol. 5, No. 3 (July 1968), pp. 100–122.
31. William Y. Liu, "Steel Column Bending Amplification Factor," *Engineering Journal*, AISC, Vol. 2, No. 2 (April 1965), pp. 50–51.
32. Benjamin Koo, "Amplification Factors for Beam-Columns," *Engineering Journal*, AISC, Vol. 5, No. 2 (April 1968), pp. 66–71.
33. Suresh T. Dalal, "Some Nonconventional Cases of Column Design," *Engineering Journal*, AISC, Vol. 6, No. 1 (January 1969), pp. 28–39.
34. N. M. Newmark, "Numerical Procedure for Computing Deflections, Moments, and Buckling Loads," *Trans. ASCE*, Vol. 108 (1943), pp. 1161–1234.
35. Ping-Chun Wang, *Numerical and Matrix Methods in Structural Mechanics*, John Wiley & Sons, Inc., New York, 1966, pp. 96–109.

13

Connections

13.1. TYPES OF CONNECTIONS

Under AISC–1.2 for working stress design and AISC–2.1 for plastic design, steel construction is defined in three categories according to the type of connections used. The following three types are indicated.

(a) AISC Type 1. *Rigid frame*, where full continuity is provided at the connection so that the original angles between intersecting members are held virtually constant; i.e., with rotational restraint on the order of 90 percent or more of that necessary to prevent any angle change. Such connections are used under both the working-stress and plastic-design methods.

(b) AISC Type 2. *Simple framing*, where rotational restraint at the ends of members is as little as practicable. For beams, simple framing provides only shear transfer at the ends. One may consider the framing simple if the original angle between intersecting pieces may change up to 80 percent of the amount it would theoretically change if frictionless hinged connections could be used. Design of simply supported beams under working-stress method uses Type 2 connections. Simple framing is not used in plastic design, except for connections of members transverse to the plane of the frame in which the plastic strength is to develop. Two or more plastically designed planar systems may be linked together by means of simple framing combined with cross-bracing.

(c) AISC Type 3. *Semirigid framing*, where rotational restraint is between 20 and 90 percent of that necessary to prevent any relative angle change. Alternatively, one may consider that with semirigid framing the moment transmitted across the joint is neither zero (or a small amount) as in simple framing, nor is it the full continuity moment as assumed in elastic rigid-frame analysis. AISC–1.2 states that design of construction using Type 3 connections may be used when "the connections of beams and girders possess a *dependable and known* moment capacity intermediate in degree between the rigidity of Type 1 and the flexibility of Type 2."

Semirigid connections are not used in plastic design and only rarely used in working-stress design, primarily because of the difficulty of evaluating the degree of restraint.

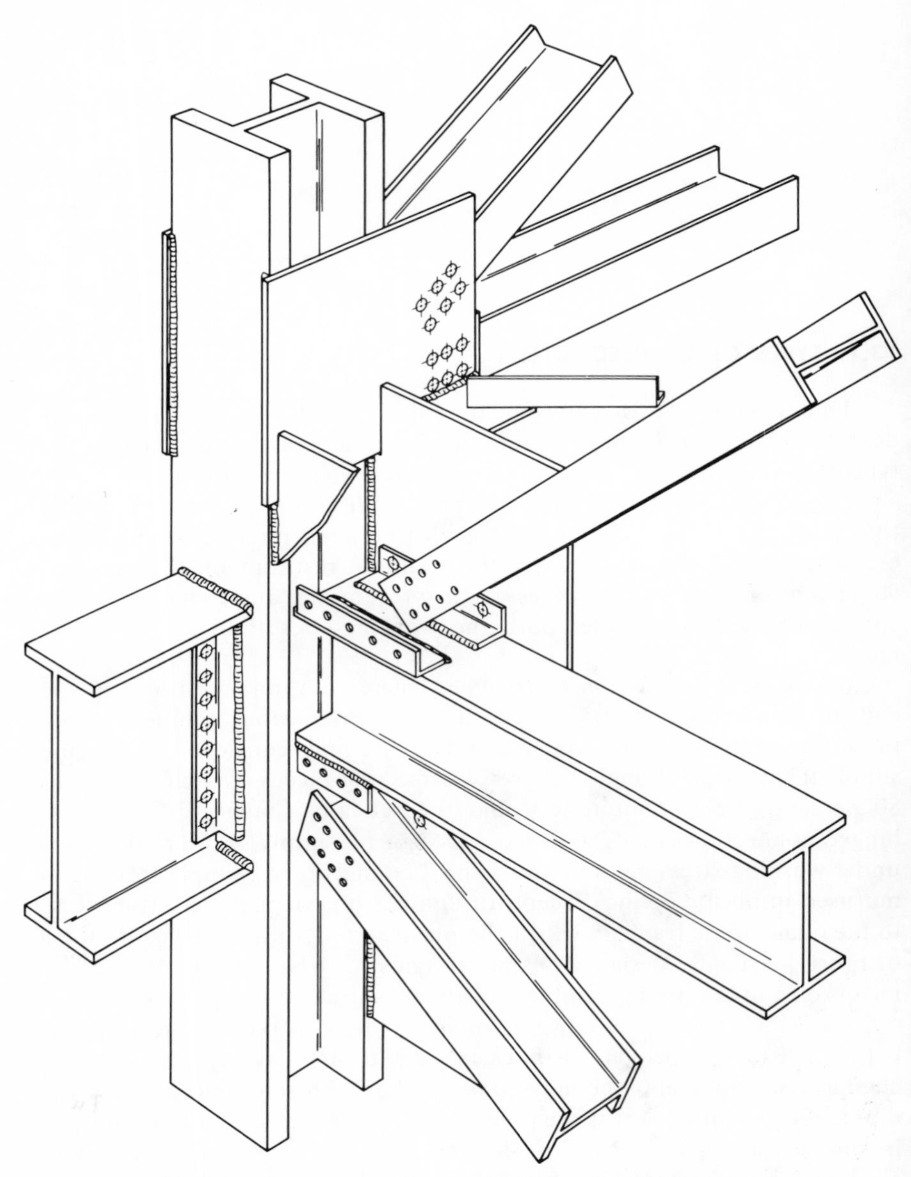

Fig. 13.1. Connections. Courtesy Bethlehem Steel Corporation.

Beam Line. In order to better understand the practical distinction between the AISC framing types, the beam line developed by Batho & Rowan[1] and used by Sourochnikoff[2] is a useful graphical device.

Consider as shown in Fig. 13.1.1 a beam AB loaded in any manner

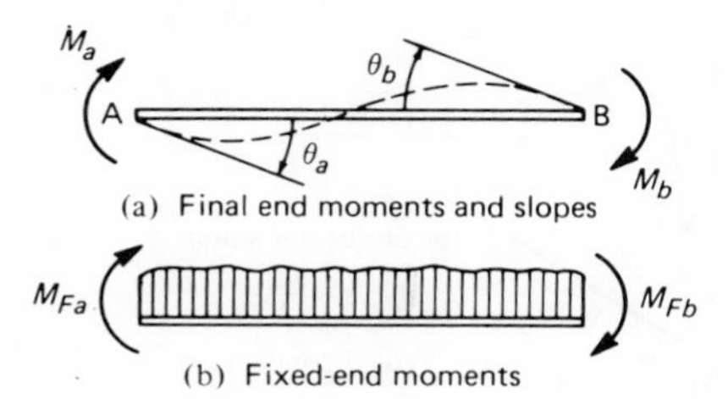

Fig. 13.1.1. Moments and rotations for slope-deflection equations (shown with positive signs).

and subject to end moments M_a and M_b, and with end slopes θ_a and θ_b. The moments necessary to have $\theta_a = \theta_b = 0$ are designated M_{Fa} and M_{Fb}, the fixed-end moments. Writing the slope deflection equations,

$$\left.\begin{aligned} M_a &= M_{Fa} + \frac{4EI}{L}\,\theta_a + \frac{2EI}{L}\,\theta_b \\ M_b &= M_{FB} + \frac{2EI}{L}\,\theta_a + \frac{4EI}{L}\,\theta_b \end{aligned}\right\} \tag{13.1.1}$$

Solving Eqs. 13.1.1 for θ_a and θ_b gives

$$\left.\begin{aligned} \frac{6EI}{L}\,\theta_a &= 2(M_a - M_{Fa}) - (M_b - M_{Fb}) \\ \frac{6EI}{L}\,\theta_b &= -(M_a - M_{Fa}) + 2(M_b - M_{Fb}) \end{aligned}\right\} \tag{13.1.2}$$

Subtracting the second equation from the first gives

$$\frac{6EI}{L}(\theta_a - \theta_b) = 3(M_a - M_b) - 3(M_{Fa} - M_{Fb}) \tag{13.1.3}$$

If symmetrical loading is considered, then

$$M_b = -M_a, \quad \theta_b = -\theta_a, \quad M_{Fb} = -M_{Fa} \tag{13.1.4}$$

in which case Eq. 13.1.3 becomes

$$\frac{2EI}{L}\,\theta_a = M_a - M_{Fa}$$

or

$$M_a = M_{Fa} + \frac{2EI}{L}\,\theta_a \tag{13.1.5}$$

which may be called the *beam-line equation.* When $\theta_a = 0$ (a full fixity condition), $M_a = M_{Fa}$; and for a hinged end where $M_a = 0$, the slope becomes $\theta_a = -M_{Fa}/(2EI/L)$.

Figure 13.1.2 shows a diagram of the beam-line equation and also the moment-rotation behavior of typical connections of Types 1, 2, and 3.

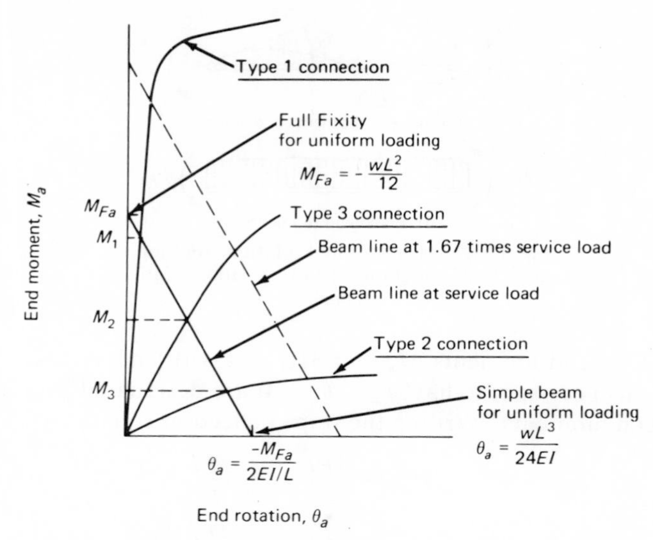

Fig. 13.1.2. Moment-rotation characteristics of AISC connection types.

The typical rigid connection would have to carry an end moment M_1, about 90 percent or more of M_{Fa}; hence, its degree of restraint may be said to be 90 percent. The simple connection (Type 2) may have to resist only 20 percent or less of the moment M_{Fa}, as indicated by the moment M_2, while the semirigid connection would be expected to resist some intermediate value M_3, at perhaps 50 percent of the fixed-end moment M_{Fa}.

If one is able to establish the moment-rotation characteristics of a particular connection, then the strength can be designed so that the resulting end rotation θ is compatible with that caused by the loads. A summary and discussion of the moment-rotation characteristics of various connection arrangements is to be found in Ref. 3.

13.2. FRAMED BEAM CONNECTIONS

These simple framing connections, AISC Type 2, are used to connect beams to other beams or to column flanges. For the most part these connections are standardized and given in the AISC Manual under the headings, "Framed Beam Connections" and "Heavy Framed Beam Connections." Typical bolted and welded framed connections are shown in Fig. 13.2.1. It is intended in such connections that the angles be as flexible as possible. The connection to the column (2 rows of 5 fasteners shown in Fig. 13.2.1a) is usually made in the field while the connection

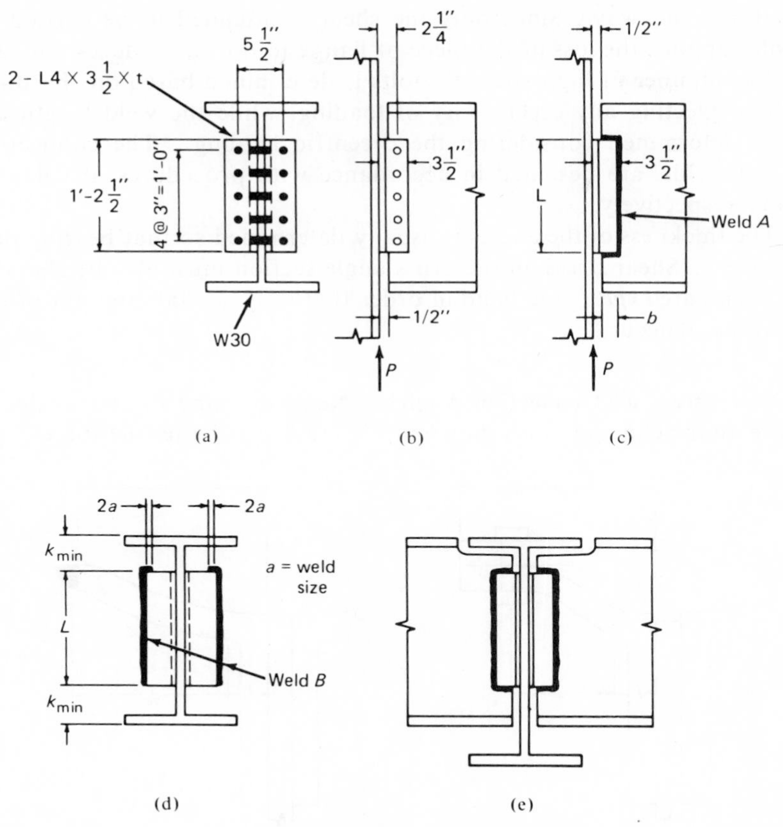

Fig. 13.2.1. Framed beam connections.

to the beam web (one row of 5 fasteners shown in Fig. 13.2.1b) is usually made in the shop. Generally on plans, shop fastener holes are shown as in Fig. 13.2.1b, while field fastener holes are shown as solid black dots.

In today's fabrication practice, the shop connection is usually welded, while the field connection may be either bolted or welded; thus any combination in Fig. 13.2.1 of (a) with (b) or (c); or (d) with (b) or (c), may be used.

When these angles, sometimes known as *clip angles*, are used to attach a beam to a column there is a clearance setback of about ½ in. so that if the beam is too long, within acceptable tolerances, the angles may be relocated without cutting off a piece of the beam. When beams intersect and are attached to other beams so that the flanges of both are at the same elevation, as in Fig. 13.2.1e, the beams framing in have their flanges coped, or cut away. Since only the shear is intended to be carried for simple support, the loss of the piece of flange in no way reduces strength.

The number of high-strength bolts is determined based on the direct shear, neglecting any eccentricity of loading, while the weld length and size is determined considering the eccentric loading. The connectors, bolts or welds, are designed in accordance with procedures of Chapters 4 and 5 respectively.

The thickness of the angles is usually determined so that bearing does not govern. Shear stress on the gross angle section must also be checked. The angles are *expected* to bend in order that only a small amount of end restraint will occur.

Flexural Stress on Connection Angles. Next, examine the stress due to flexure on the clip angles, as shown in Fig. 13.2.2. The tensile force T per

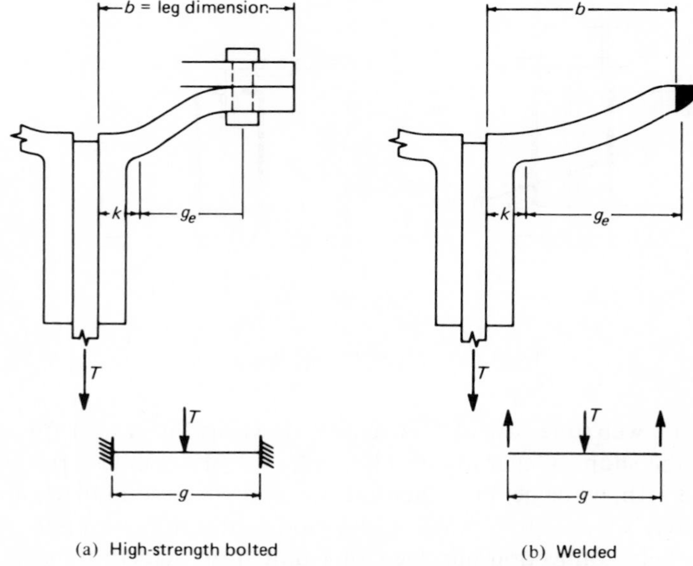

Fig. 13.2.2. Behavior at tension edge of clip angles.

inch at the top of the angles is obtained by taking the applied moment (reaction times eccentricity of applied load) and computing the flexural stress on the projecting angle legs.

If the angle thickness is t and the angle length is L, the stress at the top of the angles is

$$f = \frac{Mc}{I} = \frac{Pe(L/2)}{\frac{1}{12}(2t)(L)^3} = \frac{3Pe}{tL^2} \tag{13.2.1}$$

and

$$T = (f)(2)(\text{angle thickness } t) = \frac{6Pe}{L^2} \tag{13.2.2}$$

where T is the load on the top one inch length of two connection angles.

The load T bends the connection angles as shown in Fig. 13.2.2. For the high-strength bolted connection the clamping force provides fixity at the bolt and the critical section in flexure is probably at the base of the fillet of the angle, a distance k from the back of angle (see AISC Manual properties).

If the bolted connection is considered a fixed-end beam (Fig. 13.2.2a) because of the clamping due to initial tension in the bolts, and the welded connection is considered a simply supported beam (Fig. 13.2.2b) because of the lack of initial tension, the bending stress in the angles may be approximated as:

For bolted angles,

$$f = \frac{M}{S} = \frac{M}{\frac{1}{6}(1)(t^2)} = \frac{6M}{t^2} \tag{13.2.3}$$

$$M \approx \frac{Tg}{8} \tag{13.2.4}$$

$$f = \frac{0.75Tg}{t^2} = \frac{4.5Peg}{t^2L^2} = \frac{9.0Peg_e}{t^2L^2} \tag{13.2.5}$$

For welded angles,

$$M \approx \frac{Tg}{4} \tag{13.2.6}$$

$$f = \frac{1.5Tg}{t^2} = \frac{9.0Peg}{t^2L^2} = \frac{18.0Peg_e}{t^2L^2} \tag{13.2.7}$$

From Eq. 13.2.5 or 13.2.7 one may observe that thick angles will reduce the stress; however, the deformation will also reduce, offsetting the desired flexibility. For some brackets treated later in Secs. 13.4 and 13.5, the rigidity may be desired. For framed connections, adequate deformation of the angles is desired to get close to simple support conditions. The deflections may be approximated as follows:

For bolted angles,

$$\Delta \approx \frac{T(2g_e)^3}{192EI} = \frac{T(2g_e)^3}{192E(1/12)t^3} \tag{13.2.8}$$

from Eq. 13.2.5,

$$T = \frac{ft^2}{1.5g_e}$$

$$\Delta = \frac{fg_e^2}{3.0Et} \tag{13.2.9}$$

For welded angles,

$$\Delta \approx \frac{T(2g_e)^3}{48EI} = \frac{T(2g_e)^3}{48E(1/12)t^3} \tag{13.2.10}$$

and using T from Eq. 13.2.7 gives

$$\Delta = \frac{fg_e^2}{1.5Et} \tag{13.2.11}$$

Equations 13.2.9 and 13.2.11 show that for constant deformation Δ the thicker the angle the greater must be the stress. Further, for a constant stress f the thinner the angle the greater the deflection.

EXAMPLE 13.2.1

Determine the capacity for the 5 row framed beam connection of Fig. 13.2.1 for connecting a W30 × 99 to a column with a ¾-in. flange. Use ¾ in. diam. A325 bolts and compute for (a) friction-type connection, and (b) bearing-type connection with no threads in the shear plane. A36 steel.

Solution

(a) Connection to web of W30 × 99, $t_w = 0.522$ in.

$$R_{DS} = 0.4418(15)2 = 13.25 \text{ kips} \quad \text{(friction-type)}$$

$$R_{DS} = 0.4418(22)2 = 19.44 \text{ kips} \quad \text{(bearing-type)}$$

$$R_B = 1.35(36)(3/4)(0.522) = 19.0 \text{ kips} \quad \text{(bearing-type only)}$$

Neglecting reaction eccentricity with respect to the connector line,

$$\text{Total shear} = 5(13.25) = 66.3 \text{ kips} \quad \text{(friction-type)}$$

$$= 5(19.44) = 97.2 \text{ kips} \quad \text{(bearing-type)}$$

$$\text{Total bearing} = 5(19.0) = 95.0 \text{ kips}$$

Considering eccentricity with respect to connector line,

$$e = 2\tfrac{1}{4} \text{ in.}$$

assuming reaction at face of column. If the reduced effective eccentricity is used as discussed in Sec. 4.7,

$$e_{\text{eff}} = e - \frac{1 + 2n}{4} = 2.25 - \frac{1 + 2(5)}{4} = 2.25 - \frac{11}{4} < 0$$

Or one may conclude that as long as the gage distance to the connector line does not exceed $^{11}/_4 = 2.75$ in., eccentric shear need not be considered.

(b) Connection to the ¾-in. flange. The shear capacity for fasteners in single shear is

$$R_{SS} = 0.4418(15) = 6.63 \text{ kips} \quad \text{(friction-type)}$$

$$R_{SS} = 0.4418(22) = 9.72 \text{ kips} \quad \text{(bearing-type)}$$

Bearing on $\frac{3}{4}$-in. flange will obviously not govern on this bearing-type connection.

Neglecting reaction eccentricity with respect to face of column,

$$\begin{aligned} \text{Total shear} &= 10(6.63) = 66.3 \text{ kips} \quad \text{(friction-type)} \\ &= 10(9.72) = 97.2 \text{ kips} \quad \text{(bearing-type)} \end{aligned}$$

Considering eccentricity with respect to connector line,

$$e = 2\tfrac{1}{4} \text{ in.}$$

assuming the reaction to be along the line of the 5 connectors fastened to the beam web. The reduced eccentricity from Sec. 4.7 is only used for eccentric shear and is not applicable for shear and tension. The nominal forces on the most highly stressed connectors are

$$Af_v = \frac{P}{10} = 0.10P \qquad \text{(direct shear)}$$

$$Af_t = \frac{Pe(6)}{4(3)^2 + 4(6)^2} = \frac{P(2.25)(6)}{180} = 0.075P \qquad \text{(tension)}$$

Friction-type:

$$F'_v = 15.0(1 - f_t A_b / T_i) = 15.0(1 - 0.075P/28)$$

For connectors fully stressed in shear, $f_v = 0.10 \text{ P}/A$:

$$\frac{0.10P}{0.4418} = 15(1 - 0.75P/28)$$

$$P = \frac{15(0.4418)28}{2.8 + 0.50} = 56.2 \text{ kips} < 66.3 \text{ kips}$$

Combined stress would reduce capacity about 10 percent for the friction-type connection.

Bearing type:

$$F'_t = 50 - 1.6f_v = 50 - 1.6\left(\frac{0.10P}{0.4418}\right)$$

and maximum tension stress is $f_t = 0.075P/A$. Thus

$$\frac{0.075P}{0.4418} = 50 - \frac{0.16P}{0.4418}$$

$$P = \frac{50(0.4418)}{0.085} = 260 \text{ kips} > 97.2 \text{ kips}$$

Combined stress would not affect the bearing-type connection.

The AISC tables for high-strength bolted "Framed Beam Connections" do not consider eccentricity of load. The capacity would be

$$P = 66.3 \text{ kips (shear on friction-type)}$$

or

$$P = 182(0.522) = 95.0 \text{ kips}$$

(bearing on beam web for bearing-type)

(c) Angle thickness. For the friction-type connection there is no check of bearing stress so there is no related thickness requirement. For the bearing-type connection, the two angles must have a total thickness exceeding 0.522 in. (beam web) if the maximum capacity is to be carried; thus, $^5/_{16}$ for each angle would be satisfactory.

For $^5/_{16}$-in.-thick angles, examine the flexural stress in accordance with Eq. 13.2.5:

$$f = \frac{9.0\,Pe g_e}{t^2 L^2} = \frac{9.0(95.0)(2.25)(2.25 - 11/16)}{(^5/_{16})^2(14.5)^2} = 146 \text{ ksi}$$

using angles $4 \times 3^1/_2 \times {}^5/_{16}$. This indicates that the yield stress is reached and permanent deformation of the angles occurs under service load. Examine the deformation in accordance with Eq. 13.2.9:

$$\Delta = \frac{f(g_e)^2}{3.0Et} = \frac{146(1.5625)^2}{3.0(29{,}000)(0.3125)} = 0.0131 \text{ in.}$$

The longest beam span uniformly loaded of a W30 × 99 having a reaction of 95 kips is

$$M = F_b S = 24(270) = 540 \text{ ft-kips} \qquad 8(540)$$

$$M = \frac{WL}{8} = \frac{95(2)L}{8}$$

$$L = \frac{8(540)}{95(2)} = 22.6 \text{ ft}$$

The end slope is

$$\theta = \frac{WL^2}{24EI} = \frac{ML}{3EI} = \frac{540(12)(22.6)(12)}{3(29{,}000)4000} = 0.00505 \quad \text{radian}$$

Assuming rotation about the bottom of the angle, the deformation at the top of the angles would be

$$\Delta = 14.5\,\theta = 14.5(0.00505) = 0.073 \text{ in.} > 0.0131$$

Based on these approximate calculations, the angles must yield to accommodate rotation at the end of the beam, and even if such rotation occurs, the full simple beam end slope is not likely to occur. Some end moment will exist but it is generally not large enough to significantly reduce the midspan positive moment. The conclusion is that simple framed connections should use the thinnest angles consistent with bearing and other practical limitations.

EXAMPLE 13.3.2

Design the connection for a W10 × 29 having a 30-kip reaction and a W24 × 100 with a 120-kip reaction to frame into opposite sides of a plate girder having a ³/₈-in. web as shown in Fig. 13.2.3. The connection

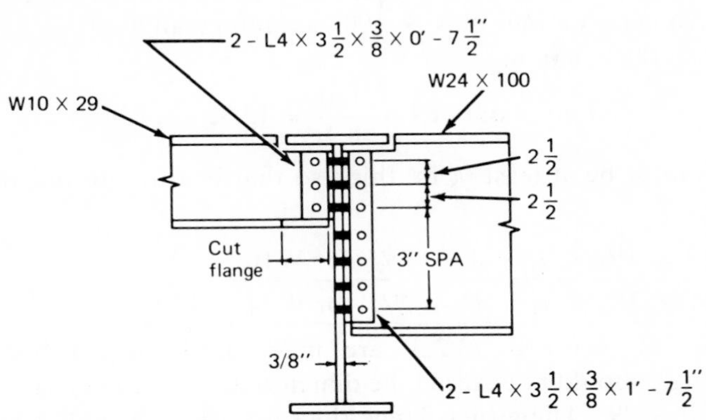

Fig. 13.2.3. Framed connection having unequal reactions.

is to be of ³/₄-in. diam. A325 bolts in a bearing-type connection, and threads are excluded from the shear planes. Base material is A36 steel.

Solution

(a) Connections to webs of W10 × 29 and W24 × 100.

$$R_B = 1.35(36)(0.289)(0.75) = 10.53 \text{ kips (W10 web)}$$

$$R_B = 1.35(36)(0.468)(0.75) = 17.07 \text{ kips (W24 web)}$$

$$R_{DS} = 2(22)(0.4418) = 19.45 \text{ kips (double shear)}$$

$$\text{Number of bolts} = \frac{30}{10.53} = 2.85, \quad \text{say 3} \quad \text{(W10)}$$

$$\text{Number of bolts} = \frac{120}{17.07} = 7.04, \quad \text{say 7} \quad \text{(W24)}$$

(b) Connection to plate-girder web. For this connection, the top bolts common to both sides will be governed by double shear or bearing on the ³/₈-in. plate, while the remainder are governed by single shear or bearing on the ³/₈-in. plate.

$$R_B = 1.35(36)(0.375)(0.75) = 13.7 \text{ kips} \quad (^3/_8\text{-in. web})$$

$$R_{DS} = 2(22)(0.4418) = 19.45 \text{ kips}$$

$$R_{SS} = 22(0.4418) = 9.72 \text{ kips}$$

For the bolts common to both sides, bearing governs:

$$R = 13.7 \text{ kips/bolt}$$

Six bolts in common (two vertical rows of three) are assumed because it makes a convenient pattern with the three needed for the web of the W10 × 29. Each of the top six bolts then carries 30/6 = 5 kips from the W10 × 29. The remainder is available for W24 × 100 reaction; i.e., 13.7 − 5 = 8.7 kips.

Since 8.7 < 9.72, it will be conservative to assume 8.7 as the capacity per bolt available to the W24 × 100, assuming all bolts will carry an equal part of the reaction:

$$\text{Number of bolts} = \frac{120}{8.7} = 13.8, \quad \text{say } 14$$

The angles must be at least $^3/_8$ in. thick so that bearing on the angles will not control.

Use 2—L 4 × $3^1/_2$ × $^3/_8$ × 0′ − $7^1/_2$″ for W10 × 29.

Use 2—L 4 × $3^1/_2$ × $^3/_8$ × 1′ − $7^1/_2$″ for W24 × 100.

The angles (see Fig. 13.2.3) are made nonstandard because the length of angle should not exceed the dimension T which is exactly $8^1/_2$ in. for the W10 × 29. The girder flange thickness is such that it requires a cope which encroaches on the T dimension; thus the $2^1/_2$-in. spacing is prescribed so that adequate edge distance will be available on the web of the W10 × 29.

Weld Capacity in Eccentric Shear on Angles Connections. Since no initial tension is involved with welded connections, the eccentricity of loading, even though small, is considered. The principles of Chap. 5 (Sec. 5.16) are used working with the welds as lines.

EXAMPLE 13.2.2

Determine the capacity for weld A on the angle connection shown in Fig. 13.2.1. The beam is a W30 × 99 and the weld size is $^1/_4$ in. The angles are 4 × $3^1/_2$ × $^5/_{16}$, 1′-$2^1/_2$″ in length. Use E70 electrodes on A36 steel.

Solution

Using I_p from Table 5.16.1 and referring to Fig. 13.2.1c,

$$I_p = \frac{2(3)^3 + 6(3)(14.5)^2 + (14.5)^3}{12} + \frac{(3)^3(14.5)}{2[2(3) + 14.5]}$$

$$= 574 + 9.6 = 583.6 \text{ in.}^3$$

When properties of lines are used, one may either consider that stresses are in ksi as computed with a 1-in. effective throat, or one may consider the unit stress has been multiplied by the 1-in. effective throat to give the stresses in kips per inch.

$$f'_y = \frac{P}{2(20.5)} = 0.0244P \quad \text{(direct shear component)}$$

$$\overline{x} = \frac{(3)^2}{2(3) + 14.5} = 0.44 \text{ in.}$$

The x and y components of torsional stress are

$$f_x'' = \frac{P(3.50 - 0.44)(3.50 - 0.44 - 0.50)}{2(583.6)} = 0.00672P$$

$$f_y'' = \frac{P(3.50 - 0.44)(7.25)}{2(583.6)} = 0.0190P$$

$$f_r = P\sqrt{(0.0244 + 0.0067)^2 + (0.190)^2} = 0.364P$$

The capacity per inch of weld is

$$R_w = (5/16)(0.707)21.0 = 4.64 \text{ kips/in.}$$

$$P = \frac{4.64}{0.0364} = 127 \text{ kips}$$

It is in the foregoing manner that the capacities are computed based on "weld A" for the AISC Manual table for welded "Framed Beam Connections."

Tests of welded angle connections[4,5] have demonstrated that performance of web angles agrees generally with assumptions.

Weld Capacity in Tension and Shear on Angle Connections. This is the field-welded connection shown in Fig. 13.2.1d. There is not agreement regarding the strength analysis for this situation. Reference 6 considers the strength as an eccentric shear situation in the plane of the welds. With the eccentric load as in Fig. 13.2.4b, the angles bear against themselves for a distance of $L/6$ from the top, and the torsional stress over the remaining $5/6$ of the length is resisted by the weld. Neglecting the effects

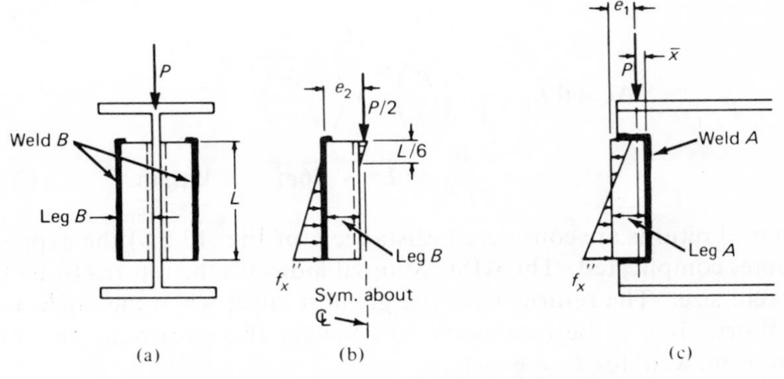

Fig. 13.2.4. Field-welded connection for web framing angles.

of the returns at the top, moment equilibrium requires

$$\underbrace{\frac{1}{2} f_r \left(\frac{5}{6} L\right)}_{\text{force/inch}} \underbrace{\frac{2}{3} L}_{\text{arm}} = \frac{P}{2} e_2 \tag{13.2.12}$$

$$f_x'' = \frac{9 P e_2}{5 L^2} \text{ kips/in.} \tag{13.2.13}$$

The direct-shear component is

$$f_y' = \frac{P}{2L} \text{ kips/in.} \tag{13.2.14}$$

$$\text{Actual } f_r = \sqrt{\left(\frac{P}{2L}\right)^2 + \left(\frac{9}{5}\frac{P e_2}{L^2}\right)^2}$$

$$= \frac{P}{2L^2}\sqrt{L^2 + 12.9 e_2^2} \text{ kips/in.} \tag{13.2.15}$$

which is the equation of Ref. 6 for which a nomograph is available (Nomograph No. 6). This procedure neglects the eccentricity e_1, which tends to cause tension at the top of the weld lines. The authors do not believe the tendency is great for the angle legs to spread at the bottom as indicated by the stress distribution of Fig. 13.2.4b.

Instead of the procedure of Ref. 6, the authors consider the flexural stress distribution of Fig. 13.2.4c to be more appropriate. The flexural component is

$$f_x'' = \frac{Mc}{I} = \frac{P e_1 (L/2)}{2L^3/12} = \frac{3 P e_1}{L^2} \tag{13.2.16}$$

where the returns at the top of angles are neglected. The direct shear component is

$$f_y' = \frac{P}{2L} \tag{13.2.17}$$

$$\text{Actual } f_r = \sqrt{\left(\frac{P}{2L}\right)^2 + \left(\frac{3 P e_1}{L^2}\right)^2}$$

$$= \frac{P}{2L^2}\sqrt{L^2 + 36 e_1^2} \quad \text{kip/in.} \tag{13.2.18}$$

Or, if returns are considered (distance b of Fig. 13.2.5) the expression becomes complicated. The AISC Manual indicates the returns to be twice the weld size. The returns have the greatest effect when the angle length L is short. It may be reasonable to consider the returns to be $L/12$ (2 times $\frac{1}{4}$-in. weld for $L = 6$ in.).

Using, from Table 5.16.1, $S = I/\bar{y}$ referred to the top of the configuration which is in tension,

$$S = 2\left[\frac{4bd + d^2}{6}\right] \tag{13.2.19}$$

which for $d = L$ and $b = L/12$ becomes

$$S = \frac{4L^2}{9} \tag{13.2.20}$$

The flexural-stress component, as shown in Fig. 13.2.5, is

$$f_x'' = \frac{M}{S} = \frac{Pe_1}{S} = \frac{Pe_1}{4L^2/9} = \frac{9Pe_1}{4L^2} \tag{13.2.21}$$

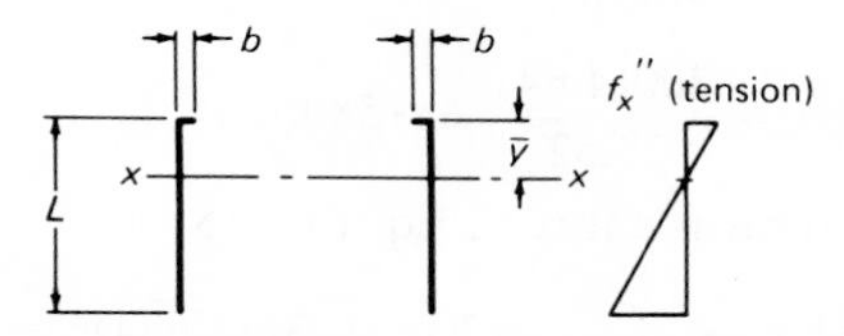

Fig. 13.2.5. Weld configuration for web angles and beam seats.

The direct-shear component is

$$f_y' = \frac{P}{2L + 2L/12} = \frac{6P}{13L} \tag{13.2.22}$$

$$\text{Actual } f_r = \sqrt{\left(\frac{6P}{13L}\right)^2 + \left(\frac{9Pe_1}{4L^2}\right)^2}$$

$$= \frac{P}{2L^2}\sqrt{1.08L^2 + 20.25e_1^2} \text{ kips/in.} \tag{13.2.23}$$

Since the direct shear is carried by the two lines of weld and little is carried by the returns, one should use $f_y' = P/(2L)$ and the 1.08 coefficient of Eq. 13.2.23 becomes 1.0. For practical usage, the equation for capacity of weld B on legs B of Fig. 13.2.4, using the configuration of Fig. 13.2.5, becomes

$$\text{Actual } f_r = \frac{P}{2L^2}\sqrt{L^2 + 20.25e_1^2} \text{ kips/in.} \tag{13.2.24}$$

EXAMPLE 13.2.3

Determine the capacity of weld B on Fig. 13.2.4 if $\frac{5}{16}$-in. weld is used and $L = 20$ in. E70 electrodes are used in manual shielded arc-welding. $4 \times 3 \times \frac{3}{8}$ angles are used.

Solution

(a) Best procedure, Eq. 13.2.24

$$R_w = 0.707(5/16)21 = 4.64 \text{ kips/in.}$$

$$\text{Actual } f_r = \frac{P}{2L^2}\sqrt{L^2 + 20.25e_1^2}$$

$$e_1 = 3.00 - \bar{x} = 3.00 - 0.25 = 2.75 \text{ in.}$$

$$\bar{x} = \frac{2(2.5)(1.25)}{2(2.5) + 20} = \frac{6.25}{25} = 0.25 \text{ in.}$$

$$\text{Actual } f_r = \frac{P}{2(20)^2}\sqrt{(20)^2 + 20.25(2.75)^2} = \frac{23.52P}{800}$$

$$P\,(\text{capacity}) = \frac{800(4.64)}{23.52} = 158 \text{ kips}$$

(b) Neglecting returns entirely, Eq. 13.2.18,

$$\text{Actual } f_r = \frac{P}{800}\sqrt{(20)^2 + 36(2.654)^2} = \frac{25.55P}{800}$$

$$P\,(\text{capacity}) = \frac{800(4.64)}{25.55} = 145 \text{ kips}$$

(c) Using Ref. 6 equation, Eq. 13.2.15,

$$\text{Actual } f_r = \frac{P}{800}\sqrt{(20)^2 + 12.9e_2^2}$$

$$e_2 = \text{4-in. leg}$$

$$\text{Actual } f_r = \frac{P}{800}(24.65)$$

$$P\,(\text{capacity}) = \frac{800(4.64)}{24.65} = 150 \text{ kips}$$

The authors believe method (a) to be appropriate, $P = 159$ kips. The most conservative low estimate is 145 kips. Reference 6 equation gives reasonable results but the rationale of its development seems unrealistic. Equation 13.2.24 forms the basis for capacities for "weld B" of the AISC Manual tables for welded "Framed Beam Connections." For the AISC tables the full leg dimension (3 in. for this case) is used for e_1 rather than the arm to the centroid of weld A.

13.3. SEATED BEAM CONNECTIONS—UNSTIFFENED

As an alternative to framed-beam connections using web angles, or other attachments to the beam web, a beam may be supported on a seat, either unstiffened or stiffened. In this section the unstiffened seat is treated, as shown in Fig. 13.3.1. The seat ang is designed to carry the entire reaction but it must always be used with a top, or clip, angle. The sole function of the top angle is to provide lateral support of the compression flange.

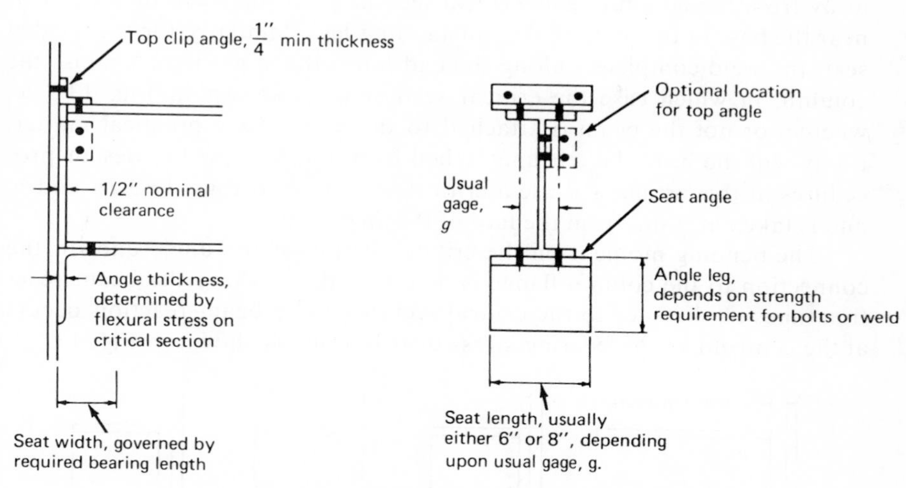

Fig. 13.3.1. Seated beam connection—unstiffened.

As with the case of the framed-beam connection, the seated connection is intended to transfer only the vertical reaction and should not give significant restraining moment on the end of the beam; thus the seat and the top angle should be relatively flexible.

The thickness of seat angle is determined by the flexural stress on a critical section of the angle, as shown in Fig. 13.3.2. If a bolted connec-

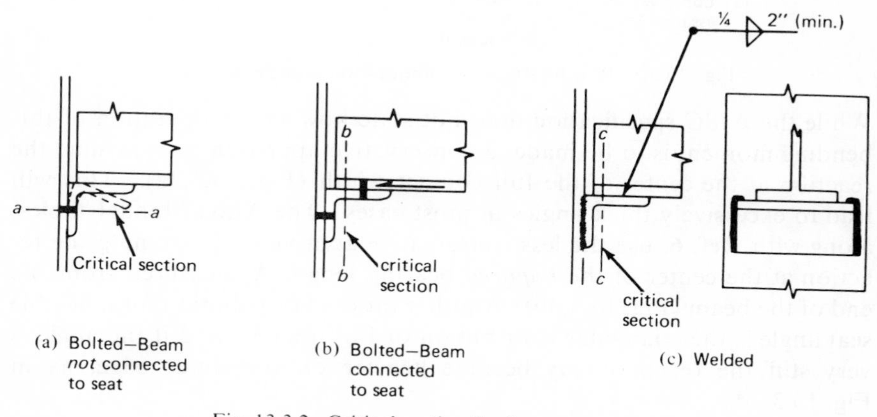

Fig. 13.3.2. Critical section for flexure on seats.

tion is used without attachment to the beam (Fig. 13.3.2a), the critical section should probably be taken as the net section through the upper bolt line. When the beam is attached to the seat as in Fig. 13.3.2b, the rotation of the beam at the end creates a force which tends to restrain the pull away from the column. The critical section for flexure will then be at or near the base of the fillet on the outstanding leg. Similarly for the welded seat, the weld completely along the end holds the angle tight against the column, in which case the critical section is as shown in Fig. 13.3.2c, whether or not the beam is attached to the seat. As a practical matter, rarely will the beam be left unattached from the seat, so the design procedures of this article will use a critical section as in Figs. 13.3.2b and c, and is taken at $^3/_8$-in. from the face of the angle.

The bending moment on the critical section of the angle and on the connection to the column flange is determined by taking the beam reaction times its distance to the critical sections. The beam reaction occurs at the centroid of the bearing stress distribution, as shown in Fig. 13.3.3.

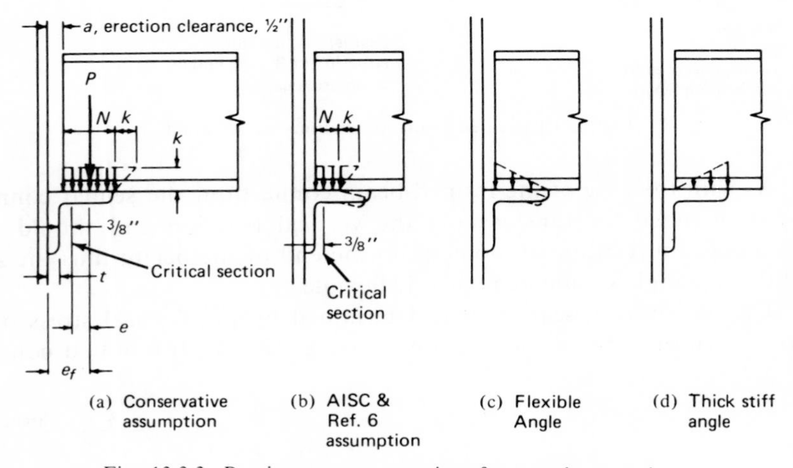

Fig. 13.3.3. Bearing stress assumptions for seated connections.

While the AISC specification does not state how the computation of this bending moment is to be made, a conservative approach is to assume the reaction at the center of the full contact width (Fig. 13.3.3a). This will lead to excessively thick angles in most cases. The AISC Manual tables, along with Ref. 6, use the less conservative approach of assuming the reaction at the center of the *required* bearing length N measured from the end of the beam (Fig. 13.3.3b). Another rational distribution for a flexible seat angle is the triangular distribution of Fig. 13.3.3c, and if the angle is very stiff the reaction may become heavier on the outer edge, as in Fig. 13.3.3d.

The design of unstiffened seats involves the following steps:

(a) Determine the seat width.

(b) Determine the moment arms e and e_f.

(c) Determine the length and thickness of the angle.

(d) Determine the supporting angle leg dimension, and the weld size; or the number and placement of bolts.

The seat width is based on the bearing length N required in accordance with AISC–1.10.10.1:

$$N = \frac{P}{0.75F_y t_w} - k \geq k \tag{13.3.1}$$

where t_w = thickness of web

k = distance from outer face of flange to web toe of fillet

Generally the seat width should not be less than 3 in., with the AISC tables indicating a standard 4 in. seat width.

The moment arms e and e_f are obtained as follows, referring to Fig. 13.3.3a:

$$e_f = a + \frac{N}{2} \tag{13.3.2}$$

$$e = e_f - t - {}^3\!/_8 \tag{13.3.3}$$

The bending moment on the critical section of the angle is

$$M = Pe$$

$$f_b = \frac{M}{S} = \frac{Pe}{\frac{1}{6}bt^2} = \frac{6Pe}{bt^2} \tag{13.3.4}$$

Using the allowable stress on solid rectangular sections bent about their weak axis, AISC–1.5.1.4.3,

$$F_b = 0.75F_y$$

in which case Eq. 13.3.4 becomes

$$t^2 = \frac{6Pe}{0.75F_y b} = \frac{8Pe}{F_y b} \tag{13.3.5}$$

This length of the seat angle is generally taken as either 6 in. or 8 in. for a beam gage g of $3\frac{1}{2}$ in. and $5\frac{1}{2}$ in. respectively.

The number of bolts, which are in combined shear and tension is determined in accordance with the principles of Sec. 4.9.

The weld size and length are obtained using the principles of Sec. 5.16. with Eq. 13.2.24 applicable to this case; direct shear and bending about the x–x axis for the configuration of Fig. 13.2.4 with the returns $b \approx L/12$.

EXAMPLE 13.3.1

Design the seat angle to support a W12 × 40 beam on a 25-ft span, assuming the beam has adequate lateral support. Use A36 steel.

Solution

In many cases it will be wise practice to design the seat for the maximum reaction when the beam is fully stressed in flexure.

(a) Determine seat width, length, and thickness:

$$M = 0.66F_yS_x = 24(51.9)/12 = 103.8 \text{ ft-kips}$$

$$P = \frac{wL}{2} = \frac{8M}{2L} = \frac{4(103.8)}{25} = 16.6 \text{ kips}$$

Bearing length required is

$$N = \frac{P}{0.75F_yt_w} - k = \frac{16.6}{27(0.294)} - 1.25 = 0.84 \text{ in.} < k$$

Use $N = k = 1.25$ in. Following AISC Manual, "Seated Beam Connections" recommendation, use 4-in. seat width.

$$e_f = \frac{1.25}{2} + \frac{3}{4} = 1.375 \text{ in.}$$

using $a = 0.75$ in. to allow for possible mill underrun.

Trying $t = \frac{1}{2}$ in.,

$$e = e_f - t - \tfrac{3}{8} = 1.375 - 0.50 - 0.375 = 0.50 \text{ in.}$$

Since $g = 5\frac{1}{2}$ in. for W12 × 40, use angle length of 8 in. The angle thickness required is then, by Eq. 13.3.5,

$$t^2 = \frac{8Pe}{F_yb} = \frac{8(16.6)(0.50)}{36(8)} = 0.23; \qquad t = 0.48 \text{ in.}$$

Use seat angle, $\frac{1}{2}$ in. thick and 8 in. long.

(b) Determine bolted connection to column, using $\frac{3}{4}$ in. diam. A325 bolts in a bearing-type connection with no threads in the shear plane.

$$R_{SS} = 9.72 \text{ kips}; \qquad R_B = 18.19 \text{ kips}; \qquad R_t = 17.67 \text{ kips}$$

$$n \approx \sqrt{\frac{6M}{Rp}} = \sqrt{\frac{6(16.6)(1.38)}{9.7(3)(2)}} = 1.5$$

assuming two vertical rows of connectors.

Try 2 bolts:

$$Af_v = \frac{P}{n} = \frac{16.6}{2} = 8.3 \text{ kips} < 9.72 \text{ kips} \qquad \text{OK}$$

Since the bolts lie on the center of gravity, no moment of inertia can be computed using Σy^2. However, since initial tension exists, the initial compression, according to Eq. 4.9.22 is

$$f_{ti} = \frac{\Sigma T_i}{bd} = \frac{2(28)}{8(3)} = 2.33 \text{ ksi}$$

The change in stress due to moment, Eq. 4.9.23, is

$$f_{tb} = \frac{6M}{bd^2} = \frac{6Re}{bd^2} = \frac{6(16.6)1.38}{8(3)^2} = 1.91 \text{ ksi}$$

Since $1.91 < 2.33$, the initial precompression is not eliminated and the connection may be considered safe.

Use 2 bolts, with seat angle, L $4 \times 3 \times \frac{1}{2} \times 0' - 8''$

(c) Determine welded connection to column, Using Eq. 13.2.24, with allowable weld value for E70 electrodes with shielded metal-arc welding:

Max weld size $= \frac{1}{2} - \frac{1}{16} = \frac{7}{16}$ in.

Min weld size = AISC–Table 1.17.5 based on thickest material being joined.

Try L = 4-in. supported leg:

$$f_r = \frac{P}{2L^2}\sqrt{L^2 + 20.25e_1^2} \qquad [13.2.24]$$

where $e_1 = e_f$

$$f_r = \frac{16.6}{2(4)^2}\sqrt{(4)^2 + 20.25(1.38)^2} = 3.8 \text{ kips/in.}$$

$$R_w = a(0.707)21.0 = 14.85a$$

$$\text{Weld size reqd } a = \frac{3.8}{14.85} = 0.257 \text{ in.,} \quad \text{say } \tfrac{1}{4} \text{ in.}$$

It is noted that weld capacities given in the AISC Manual tables, "Seated Beam Connections," will be found in agreement with Eq. 13.2.24 if e_f is measured to the center of the contact bearing width of the seat (conservative assumption of Fig. 13.3.3a). For those tables,

$$e_f = \frac{N}{2} + \frac{3}{4} = \frac{3.5 - 0.75}{2} + 0.75 = 2.13 \text{ in.}$$

which upon substitution into Eq. 13.2.24 with $R_w = 14.85a = 14.85(0.25) = 3.71$ kips/in. (for $\frac{1}{4}$ in. weld), gives a capacity of

$$P = \frac{R_w(2L^2)}{\sqrt{L^2 + 20.25e_f^2}} = \frac{3.71(2)(4)^2}{\sqrt{(4)^2 + 20.25(2.13)^2}} = 11.4 \text{ kips}$$

In other words, the AISC Manual table indicates a larger weld size is necessary:

$$\text{Reqd } a = 0.25\left(\frac{16.6}{11.4}\right) = 0.364 \text{ in.,} \quad \text{say } \tfrac{3}{8} \text{ in.}$$

Use L 4 × 3½ × ½ × 0′ − 8″ with ⅜ in. weld. Since the W12 × 4 flange width is 8 in., the beam must be "blocked" or "cut" to have a reduced flange width over the seat so that the necessary welding can be done and will not be interfered with by the flange. The final designs are shown in Fig. 13.3.4.

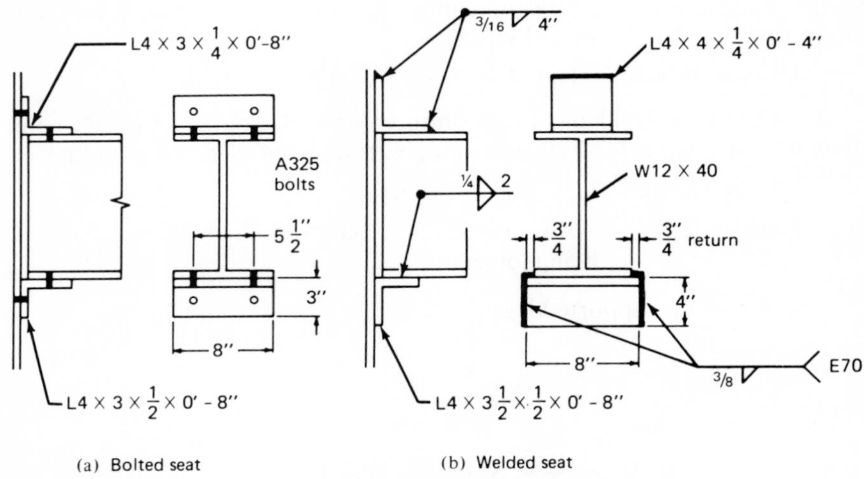

Fig. 13.3.4. Designs for Example 13.3.1.

13.4. STIFFENED SEAT CONNECTIONS

When reactions become heavier than desirable for unstiffened seats, stiffeners may be used with the seat angle in bolted construction, or a T-shaped stiffened seat may be used in welded construction. The unstiffened seat may become excessively thick when the beam reaction exceeds about 40 kips. There are no AISC restrictions, however, to the maximum load which may be carried by unstiffened seats.

The stiffened seat as discussed in this article is not intended to be part of a moment resisting connection, but rather it is only to support vertical loads. Here the stiffened seat is treated as AISC Type 2 construction; i.e., "simple framing."

There are two basic types of loading used on stiffened seats: the common one where the reaction is carried with the beam web directly in line with the stiffener, as shown in Fig. 13.4.1; the other is with a beam oriented so the plane of the web is at 90 degrees to the plane of the stiffener, as in Fig. 13.4.2. Furthermore, a difference in behavior arises depending on the angle at which the stiffener is cut, as shown in Fig. 13.4.3. If the angle θ is approximately 90 degrees, the stiffener behaves similarly to an unstiffened element under uniform compression, and local buckling

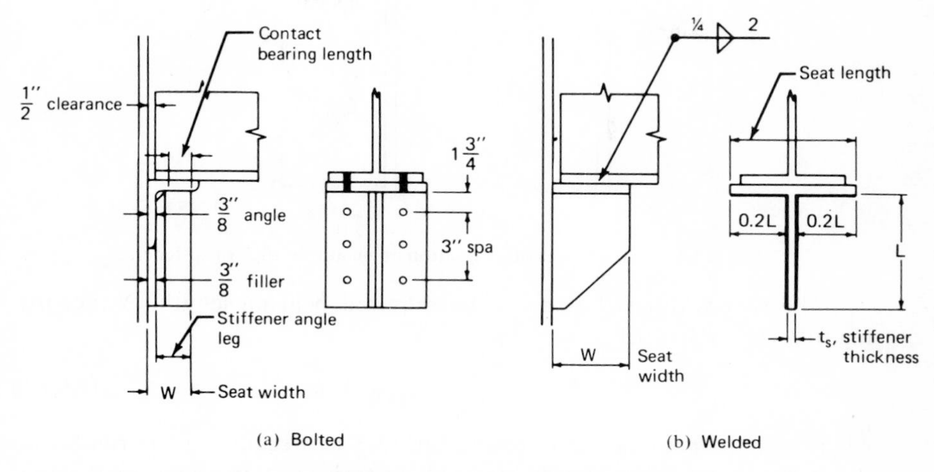

Fig. 13.4.1. Stiffened seat—beam web in line with stiffener.

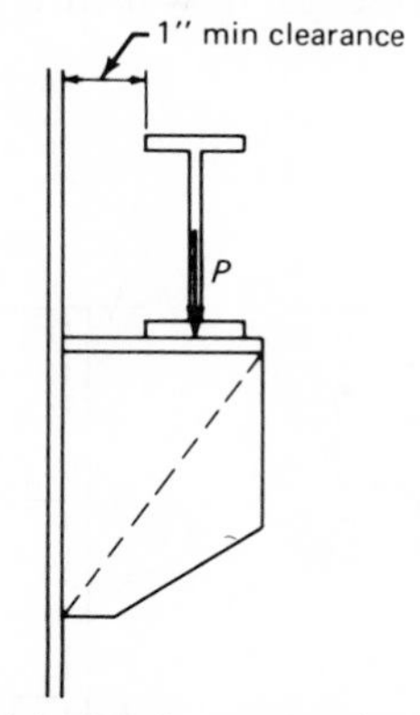

Fig. 13.4.2. Bracket supporting concentrated load.

may be prevented by satisfying AISC–1.9.1.2. When the supporting plate is cut to create a triangular bracket plate a different behavior results, and this case is discussed in Sec. 13.5.

The steps in the design of stiffened seats are as follows:

(a) Determine the seat width.
(b) Determine the eccentricity of load, e_s.
(c) Determine the stiffener thickness, t_s.
(d) Determine the angle sizes and arrangement of bolts; or the weld size and length.

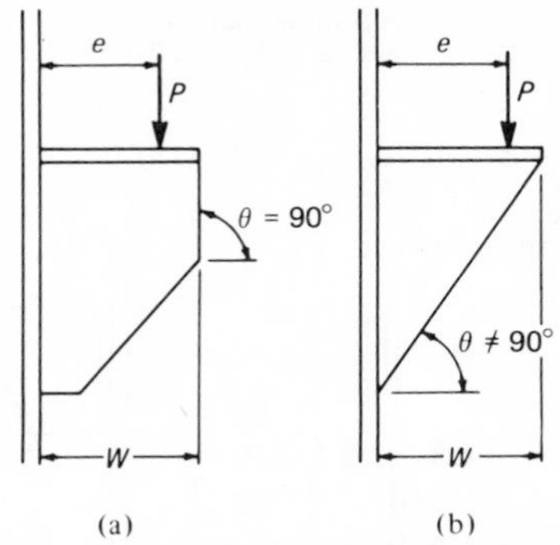

(a) (b)

Fig. 13.4.3. Two cases of inclination angle at free edge of stiffener.

The seat width is based on the required bearing length, N, according to AISC–1.10.10.1:

$$N = \frac{P}{0.75 F_y t_w} - k \geq k \tag{13.4.1}$$

where t_w and k are defined following Eq. 13.3.1. Because of the rigidity of the stiffener, the most highly stressed portion is at the edge of the seat rather than at the interior side as it was for the unstiffened seat (see Fig. 13.4.4).

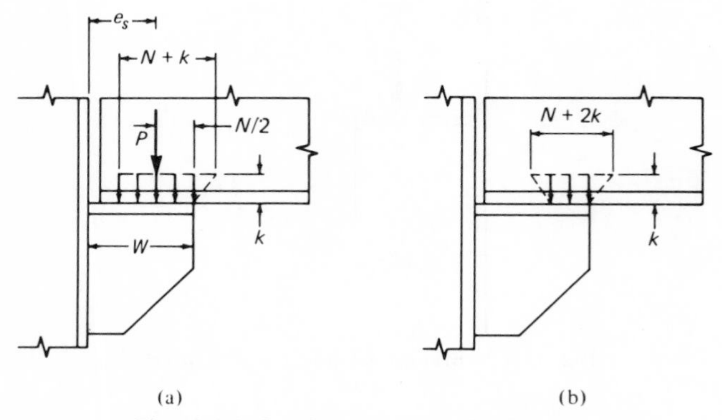

(a) (b)

Fig. 13.4.4. Bearing stress on stiffened seats.

The AISC Manual indirectly suggests seat widths of 4, 5, and 6 in. which will accommodate beam bearing lengths of 3½, 4¼, and 5¼ in., respectively, for bolted stiffened seats. For welded stiffened seats the AISC tables give seat widths ranging from 4 to 9 in. The thickness of the seat plate should be comparable to the flange of the supported beam.

Assuming the beam reaction P is located at $N/2$ from the edge of the seat the *thickness of the stiffener* t_s may be established to satisfy several criteria:

$$(1) \quad t_s \geq t_w \tag{13.4.2}$$

in order to be at least as thick as the beam web:

$$(2) \quad t_s \geq \frac{W}{95/\sqrt{F_y}} \tag{13.4.3}$$

according to AISC–1.9.1.2 so as to prevent local buckling:

$$(3) \quad t_s \geq \frac{P}{0.90F_y(W - 0.5)2} \qquad \text{(for stiffener angles)} \tag{13.4.4}$$

according to AISC-1.5.1.5, bearing on contact area. It is assumed that ½ in. of the stiffener angle is clipped off so as to get close bearing under the seat angle. Eq. 13.4.4 assumes no eccentricity of load with respect to center of bearing contact length.

$$(4) \quad f_b = \frac{P}{A} + \frac{M}{S}$$

$$= \frac{P}{Wt_s} + \frac{P(e_s - W/2)}{t_s W^2/6} = \frac{P}{t_s W^2}(6e_s - 2W)$$

$$t_s \geq \frac{P(6e_s - 2W)}{0.90F_y W^2} \qquad \text{(for welded stiffener)} \tag{13.4.5}$$

where the combined bending and direct stress on the rectangular stiffener plate is taken as a bearing stress. The neutral axis for flexure is taken at $W/2$. While one may argue that the bearing allowable stress should not be used, it seems that $0.60F_y$ is overly conservative. Many designers will prefer to use that value, however.

$$(5) \quad t_s \geq 74.2a/F_y \tag{13.4.6}$$

for welding of weld size a with E70 electrodes. Assuming two lines of fillet weld such that the weld may be fully effective and not overstress the stiffener plate in shear, it is required that

$$2(0.707a)(21.0) = 0.4F_y t_s$$

which gives Eq. 13.4.6. For some of the common types of steel this means

$$t_s \geq 2.06a \qquad \text{(for A36 steel)}$$

$$t_s \geq 1.48a \qquad (\text{for } F_y = 50 \text{ ksi})$$

Once the stiffener dimensions have been established, the connection must be designed to transmit the reaction at the moment arm, e_s. For the bolted connection, AISC tables, "Stiffened Seated Beam Connections," consider only direct shear in determining fastener group capacities. One may reason, as in Example 13.3.1, that as long as initial compression between the pieces in contact is not reduced to zero due to flexure, the moment component need not be considered.

For the welded connection suggested by the AISC Manual, as shown in Fig. 13.4.1b, the weld configuration is subject to direct shear and flexure using the combined stress at the top of the weld as critical. Thus the configuration is identical to that used for web framing angles (see Fig. 13.2.1d) except the return is longer. Using $d = L$ and $b = 0.2L$ in the S values from Table 5.16.1 gives

$$\bar{y} = \frac{L^2}{2(L + b)} = \frac{L^2}{2(1.2L)} = \frac{L}{2.4}$$

$$S_x = \frac{L(L + 4b)}{3} = \frac{L(L + 0.8L)}{3} = 0.6L^2$$

Then,

$$f_x'' = \frac{M}{S_x} = \frac{Pe_s}{0.6L^2} \text{ kips/in.}$$

$$f_y' = \frac{P}{2(L + 0.2L)} = \frac{P}{2.4L} \text{ kips/in.}$$

$$f_r = \sqrt{\left(\frac{Pe_s}{0.6L^2}\right)^2 + \left(\frac{P}{2.4L}\right)^2}$$

$$= \frac{P}{2.4L^2}\sqrt{16e_s^2 + L^2} \text{ kips/in.} \tag{13.4.7}$$

Equation 13.4.7 is used for obtaining capacities given in the AISC Manual tables, "Stiffened Seated Beam Connections," when e_s is taken as $0.8W$.

EXAMPLE 13.4.1

Design a welded stiffened seat to support a W30 × 99 with a reaction of 150 kips. The steel is A572 Grade 50.

Solution

The bearing length required is

$$N = \frac{P}{0.75F_y t_w} - k = \frac{150}{37.5(0.522)} - 1.438 = 6.21 \text{ in.}$$

$$\text{Reqd } W = 6.21 + 0.5\,(\text{setback}) = 6.71 \text{ in.} \qquad \text{Use 7 in.}$$

Since the W30 × 99 flange thickness is 0.67 in., use $\frac{5}{8}$-in. seat plate. Minimum weld size for welding on $\frac{5}{8}$-in. seat and 0.67-in. flange is $\frac{1}{4}$ inch.

The stiffener thickness is next to be established:

$$t_s \geq t_w = 0.522 \text{ in.} \qquad [13.4.2]$$

$$t_s \geq \frac{W}{95/\sqrt{F_y}} = \frac{7}{13.4} = 0.52 \text{ in.} \qquad [13.4.3]$$

$$e_s = W - \frac{N}{2} = 7.0 - 3.13 = 3.87 \text{ in.}$$

$$t_s \geq \frac{P(6e_s - 2W)}{0.90 F_y W^2} = \frac{150(23.2 - 14)}{45(7)^2} = 0.06 \text{ in.} \qquad [13.4.5]$$

Even if the allowable stress is reduced to $0.60F_y$, this stress requirement is not controlling. If a $^9/_{16}$-in. stiffener plate is used, Eq. 13.4.6 would permit a maximum effective weld size of

$$a_{\text{max eff}} = \frac{F_y t_s}{74.2} = \frac{50(9/16)}{74.2} = 0.379 \text{ in., say } \frac{3}{8} \text{ in.}$$

For weld size and length, use $e_s = 0.8W$ as is used for AISC Manual tables. Using Eq. 13.4.7 and assuming $0.8W$ (5.6 in.) is about $L/4$:

$$f_r = \frac{P}{2.4L^2}\sqrt{16\left(\frac{L^2}{16}\right) + L^2} = \frac{1.41P}{2.4L} = 0.59\frac{P}{L}$$

if $^5/_{16}$-in. weld is used,

$$R_w = 0.707(5/16)21 = 4.64 \text{ kips/in.}$$

$$\text{Reqd } L \approx \frac{0.59(150)}{4.64} = 19.1 \text{ in.}$$

Try 20 in. with $^5/_{16}$-in. weld:

$$f_r = \frac{150}{2.4(20)^2}\sqrt{16(5.6)^2 + (20)^2} = 4.7 \text{ kips/in.} \approx 4.64 \text{ kips/in.} \qquad \text{OK}$$

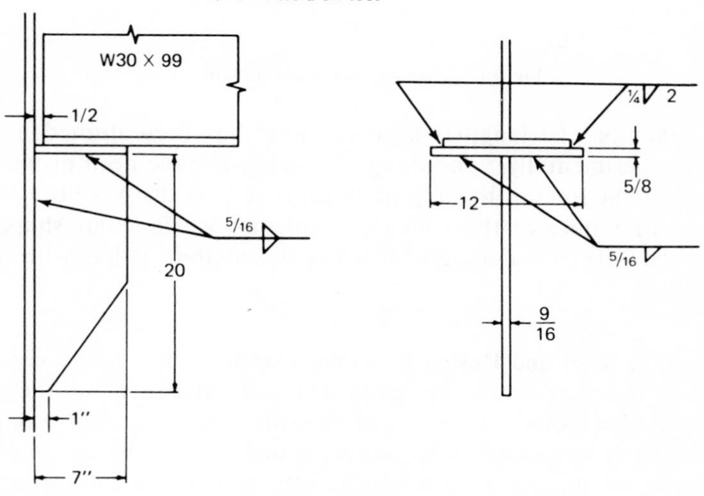

Fig. 13.4.5. Design for Example 13.4.1.

Accept the overstress in view of the fact that using $e_s = 0.8W$ is believed to overestimate the moment arm.

Use $\frac{5}{16}$ in. weld with $L = 20$ in. Use stiffener plate, $\frac{9}{16} \times 7 \times 1'$-$8''$; and seat plate $\frac{5}{8} \times 7 \times 1'$-$10''$. The seat plate width equals the flange width (10.46 in.) plus enough to easily make the welds (approx. 4 times the weld size is often used). The final design is shown in Fig. 13.4.5.

13.5. TRIANGULAR BRACKET PLATES

When the stiffener for a bracket is cut into a triangular shape, as in Fig. 13.4.3b, the plate behaves in a different manner than when the free edge is parallel to the direction of applied load in the region where the greatest stress occurs, as in Fig. 13.4.5. The triangular bracket plate arrangement and notation is shown in Fig. 13.5.1.

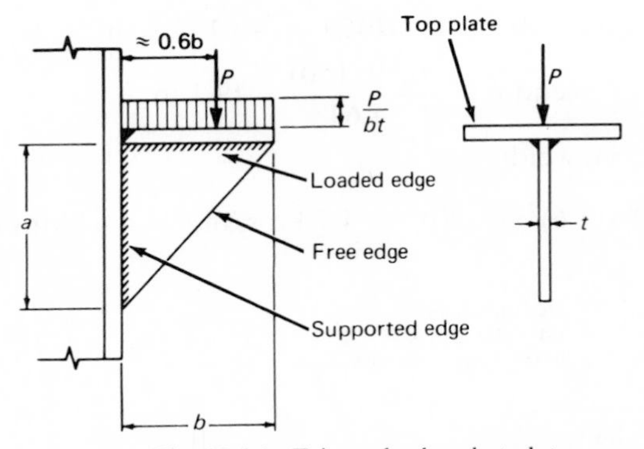

Fig. 13.5.1. Triangular bracket plate.

The behavior of triangular bracket plates has been studied analytically[9] and experimentally[10] and design suggestions have been proposed.[11] For small stiffened plates to support beam reactions there is little danger of buckling or failure of the stiffener if cut into a triangular shape. In general, it provides a stiffer support when so cut than if left with a rectangular shape.

Most Exact Analysis and Design Recommendations. For many years design of such brackets was either empirical without benefit of theory or tests, or when in doubt, angle or plate stiffeners were used along the diagonal edge. The recommendations presented here are based on certain assumptions: (a) the top plate is solidly attached to the supporting column; (b) the load P is distributed (though not necessarily uniformly) and has its centroid at approximately $0.6b$ from the support; and (c) the ratio b/a, loaded edge to supported edge, lies between 0.50 and 2.0.

The original theoretical analysis was concerned with elastic buckling; however, the experimental work showed that triangular bracket plates have considerable post-buckling strength. Yielding along the free edge frequently occurs prior to buckling, at which point redistribution of stresses occurs. A considerable margin of safety against collapse was observed indicating the ultimate capacity may be expected to be at least 1.6 times the buckling load.

The maximum stress was found to occur at the free edge; however, because of the complex nature of the stress distribution, the stress on the free edge is not obtainable by any simple process. Because of this difficulty, a ratio z was established between the average stress, P/bt, on the loaded edge to the maximum stress $f_{\max}$ on the free edge. The original theoretical expression for z[11] was revised as a result of the tests which conformed closely to what one could realistically expect in practice. The relationship is given[10] as

$$z = \frac{P/bt}{f_{\max}} = 1.39 - 2.2\left(\frac{b}{a}\right) + 1.27\left(\frac{b}{a}\right)^2 - 0.25\left(\frac{b}{a}\right)^3 \tag{13.5.1}$$

which for practical purposes may be obtained from Fig. 13.5.2.

If yielding controls strength, and $0.60F_y$ is considered a safe allowable

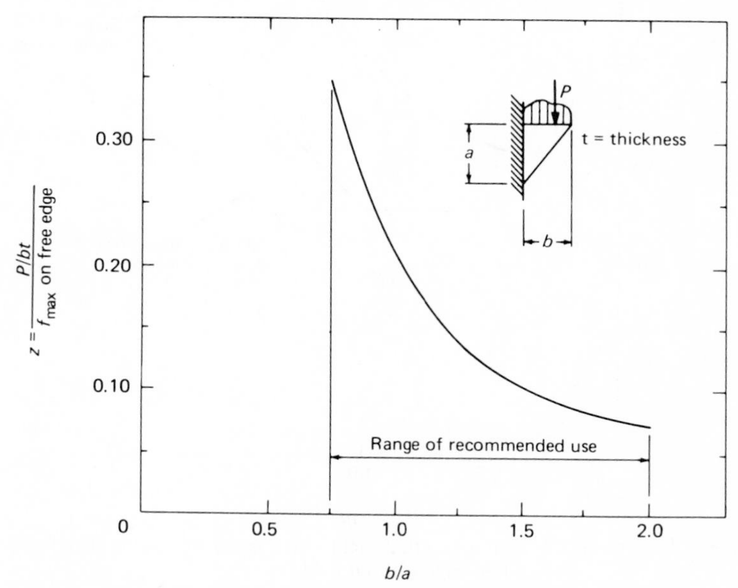

Fig. 13.5.2. Coefficient used to obtain maximum stress on free edge.

stress, the safety criterion becomes

$$f_{max} = \frac{P/bt}{z} \leq 0.60F_y \tag{13.5.2}$$

For stability controlling, the plate buckling equation, as discussed in Chapter 6, is

$$f_{cr} = \frac{k\pi^2 E}{12(1 - \mu^2)(b/t)^2} \tag{13.5.3}$$

where f_{cr} is the principal stress along the diagonal free edge. In order to guarantee that yielding is achieved without buckling, let $f_{cr} = F_y$ and

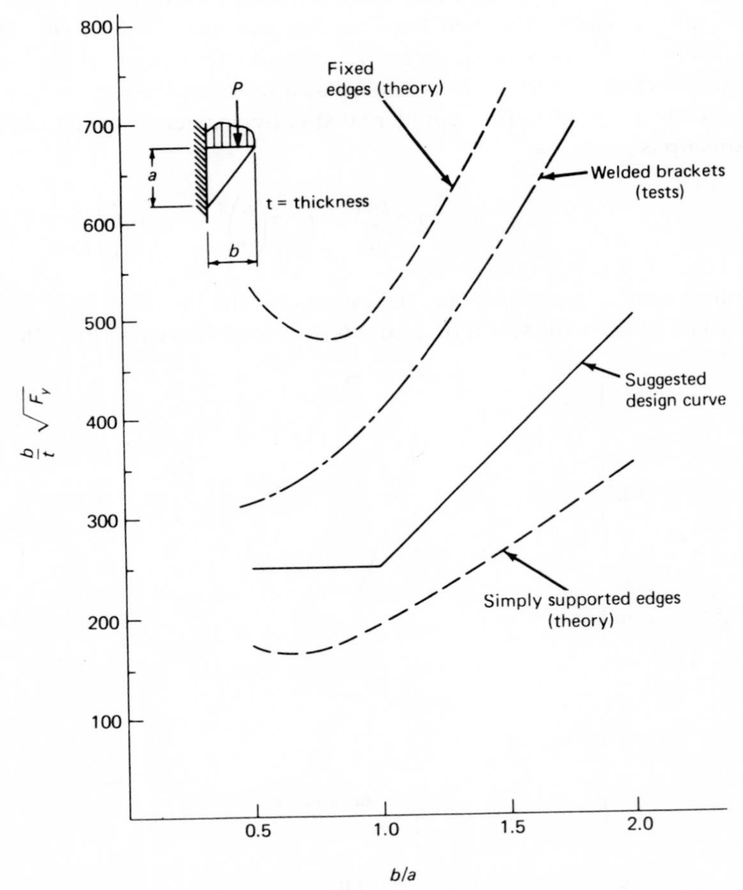

Fig. 13.5.3. Critical b/t values so that yield stress is reached along diagonal free-edge without buckling.

solve Eq. 13.5.3 for maximum b/t,

$$\frac{b}{t} \leq \sqrt{\frac{k\pi^2 E}{12(1-\mu^2)F_y}} = 162\sqrt{\frac{k}{F_y}} \tag{13.5.4}$$

where $E = 29{,}000$ ksi and $\mu = 0.3$.

Fig. 13.5.3 gives the variation in $(b/t)\sqrt{F_y}$ with b/a for the theoretical studies[9] (fixed and simply supported), the welded bracket tests result,[10] and the authors' suggested design curve. The design requirement may be expressed as

$$\text{For } 0.5 \leq b/a \leq 1.0; \qquad \frac{b}{t} \leq \frac{250}{\sqrt{F_y}} \tag{13.5.5}$$

$$\text{For } 1.0 \leq b/a \leq 2.0; \qquad \frac{b}{t} \leq \frac{250(b/a)}{\sqrt{F_y}} \tag{13.5.6}$$

Satisfying the above limits means that yielding along the *diagonal free edge* will occur prior to buckling.

It is noted that the b/t limits suggested here are higher than those of Ref. 11 (p. 552), which were based solely on the theoretical studies.[9] Reference 11 suggests a coefficient of 180 instead of 250 in Eq. 13.5.5 and reaches a maximum of 300 instead of 500 as indicated by Eq. 13.5.6 for $b/a = 2.0$. The reason for the higher values is found in the test results which showed the principal stress along the diagonal free edge to be lower relative to the stress on the loaded edge than had been established by the theoretical study. In other words, the z value, Eq. 13.5.1, as determined by tests is substantially smaller than assumed for the design suggestion of Ref. 11.

Approximate Beam Analysis for Stress. For many years stress on triangular, as well as other shaped, bracket plates has been based on a beam analysis (see Ref. 8) using the relationships from Fig. 13.5.4.

The combined stress at the free edge on the critical section is computed as follows:

$$f_{max} = \frac{P}{A} + \frac{Mc}{I} = \frac{P/\sin\phi}{bt\sin\phi} + \frac{(P/\sin\phi)(e\sin\phi)(b\sin\phi/2)}{(b\sin\phi)^3 t/12}$$

$$= \frac{P}{bt\sin^2\phi}\left[1 + \frac{6e}{b}\right] \tag{13.5.7}$$

or, in terms of e_s, Eq. 13.5.7 becomes

$$f_{max} = \frac{P}{bt\sin^2\phi}\left(\frac{6e_s}{b} - 2\right) \tag{13.5.8}$$

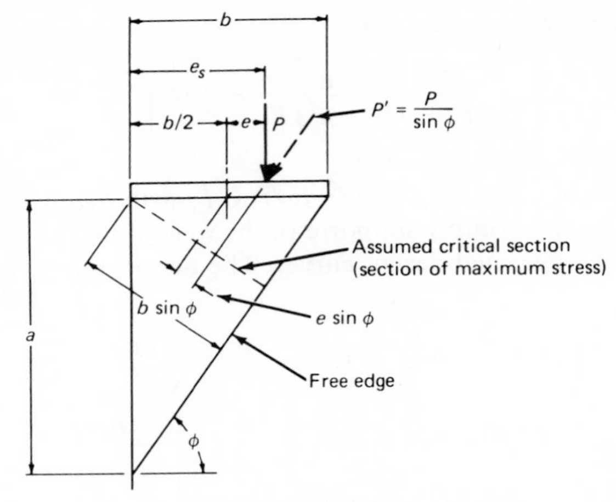

Fig. 13.5.4. Beam analysis for bracket plates.

If the maximum stress is limited to $0.6F_y$, the required thickness for stress is

$$t \geq \frac{P}{b\,(0.60F_y)\sin^2\phi}\left(\frac{6e_s}{b} - 2\right) \tag{13.5.9}$$

When the beam-analysis approach is used, one may imagine as suggested in Ref. 11, that a strip of width $(b \sin\phi)/4$ acts as a compression element. The problem then arises of what slenderness ratio limitation, if any, should be used. Ref. 11 gives no suggestion. The authors suggest that one may imagine a strip of plate acting as a column and use Eq. 13.5.4 with $k = 1.0$ for the pin-end plate; letting $b = L_d$, the length of the diagonal,

$$\frac{L_d}{t} \leq \frac{162}{\sqrt{F_y}} \tag{13.5.10}$$

Plastic Strength of Bracket Plates. Reference 11 suggests that to develop the full plastic strength of brackets the b/t ratios should be restricted to about $\frac{1}{3}$ of those limitations for achieving first yield on the free edge. The test results[10] indicated that ultimate strengths of at least 1.6 times buckling strengths could be achieved due to post-buckling strength. Probably to be certain of developing the plastic capacity of the bracket, it would be realistic to use half of the limitations of Eqs. 13.5.5 and 13.5.6.

To establish the plastic strength of a bracket plate used in rigid frame structures, one may follow the approach of Ref. 11, as shown in Fig. 13.5.5. This method assumes that plastic strength develops on the critical section. Equating the sum of the stresses normal to the critical section

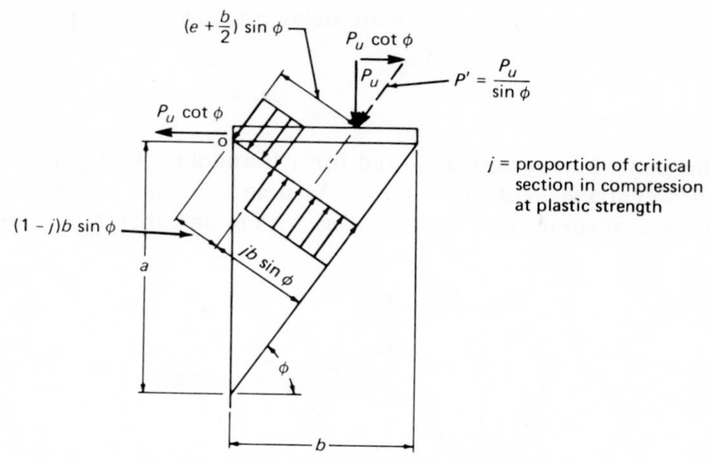

Fig. 13.5.5. Plastic-strength analysis.

to P' gives for $\Sigma F = 0$,

$$\frac{P_u}{\sin\phi} = jb\sin\phi\, F_y t - (1 - j)b\sin\phi\, F_y t \tag{13.5.11}$$

which upon solving for j gives

$$j = \frac{P_u}{2\sin^2\phi\, btF_y} + \frac{1}{2} \tag{13.5.12}$$

Moment equilibrium about point O gives

$$P_u\left(\frac{b + 2e}{2}\right) = \frac{b^2 t\sin^2\phi\, F_y}{2}[-2j^2 + 4j - 1] \tag{13.5.13}$$

Letting $A = P_u/(\sin^2\phi\, btF_y)$, Eq. 13.5.12 becomes

$$j = \tfrac{1}{2}(A + 1) \tag{13.5.14}$$

and Eq. 13.5.13 becomes

$$A\left(\frac{b + 2e}{2}\right) = -2j^2 + 4j - 1 \tag{13.5.15}$$

Substitution of Eq. 13.5.14 into Eq. 13.5.15 gives

$$A^2 + A\left(\frac{2e}{b}\right) - 1 = 0 \tag{13.5.16}$$

wherein

$$A = \frac{1}{b}[\sqrt{4e^2 + b^2} - 2e] \tag{13.5.17}$$

and using the definition of A gives for the ultimate load

$$P_u = F_y t\sin^2\phi\left(\sqrt{4e^2 + b^2 - 2e}\right) \tag{13.5.18}$$

In addition to the triangular plate being adequate, the top plate must carry the ultimate force $P_u \cot \phi$.

EXAMPLE 13.5.1

Determine the thickness required for a triangular bracket plate 25 in. by 20 in. to carry a load of 40 kips. Assume the load is located 15 in. from the face of support as shown in Fig. 13.5.6, and that A36 material is used.

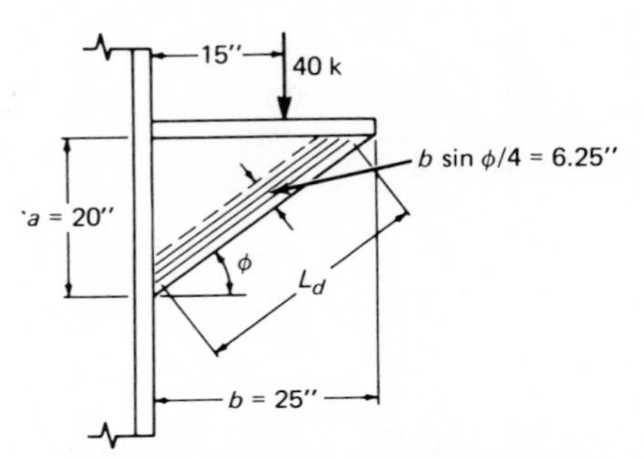

Fig. 13.5.6. Bracket for Example 13.5.1.

Solution

(a) Use the more exact method; working stress design. Since the load is approximately at the 0.6 point along the loaded edge, the bracket fits the assumption of this method. Using Eq. 13.5.2,

$$f_{avg} = \frac{P}{btz} = 0.60F_y$$

where from Fig. 13.5.2, for $b/a = 25/20 = 1.25$, find $z = 0.135$; for the yield criterion

$$t \geq \frac{P}{bz\,(0.60F_y)} = \frac{40}{25\,(0.135)\,22} = 0.54 \text{ in.}$$

For stability, using Eq. 13.5.6,

$$t \geq \frac{b\sqrt{F_y}}{250\,(b/a)} = \frac{25\sqrt{36}}{250\,(1.25)} = 0.48 \text{ in.}$$

<u>Use $^9/_{16}$-in. plate.</u>

(b) Use the approximate beam analysis; working stress method. Using Eq. 13.5.9,

$$t \geq \frac{P}{b\,(0.60F_y)\sin^2\phi}\left(\frac{6e_s}{b} - 2\right)$$

where $\sin^2\phi = (0.625)^2 = 0.39$

$$t \geq \frac{40}{25\,(22)\,0.39}\left(\frac{6\,(15)}{25} - 2\right) = 0.30 \text{ in.}$$

which by comparison with the previous method is not conservative.

Considering stability according to Eq. 13.5.10 gives

$$t \geq \frac{L_d\sqrt{F_y}}{162} = \frac{(\approx 27)\,(6)}{162} = 1 \text{ in.}$$

where L_d was taken along the center of the edge strip. The authors conclude that the beam analysis does not provide realistic answers; the stress requirement is low and the stability requirement is high, though past practice may well have used either a 1-in. plate, or an edge stiffener on a thinner plate.

Use 1 in. plate.

(c) Plastic-strength method. Use a load factor of 1.7. Using Eq. 13.5.18 for the stress requirement,

$$t \geq \frac{P_u}{F_y \sin^2\phi\,[\sqrt{4e^2 + b^2} - 2e]}$$

Using $e = 15 - 25/2 = 2.5$ in.

$$t \geq \frac{1.7\,(40)}{36\,(0.39)\,[\sqrt{4\,(2.5)^2 + (25)^2} - 2\,(25)]}$$

$$= \frac{1.7\,(40)}{36(0.39)(20.5)} = 0.24 \text{ in.}$$

For stability, using *one-half* of Eq. 13.5.6

$$t \geq \frac{b\sqrt{F_y}}{125\,(b/a)} = \frac{25\sqrt{36}}{125\,(1.25)} = 0.96 \text{ in.}$$

Use 1-in. plate as a conservative practice to assure deformation well beyond first yield along the free edge.

The authors note that if a 1-in. plate is adequate to develop the full plastic strength of the bracket, the resulting plastic strength would be about 4 times (1.0/0.24) the required P_u.

13.6. CONTINUOUS BEAM-TO-COLUMN CONNECTIONS

The connections are AISC Type 1—rigid frame connections. It is the design intent to have full transfer of moment and little or no relative rotation of members within the joint. Since the flanges of a beam carry most of the moment as a couple consisting of tension and compression flange forces acting at a moment arm approximately equal to the beam depth, it is the transfer of these essentially axial forces for which provision must be made. Since the shear is carried primarily by the web of a beam, full continuity requires that it be transferred directly from the web.

Columns being rigidly framed by beams may have attachments to both flanges, as in Figs. 13.6.1a, b, and c, or only to one flange, as in Fig.

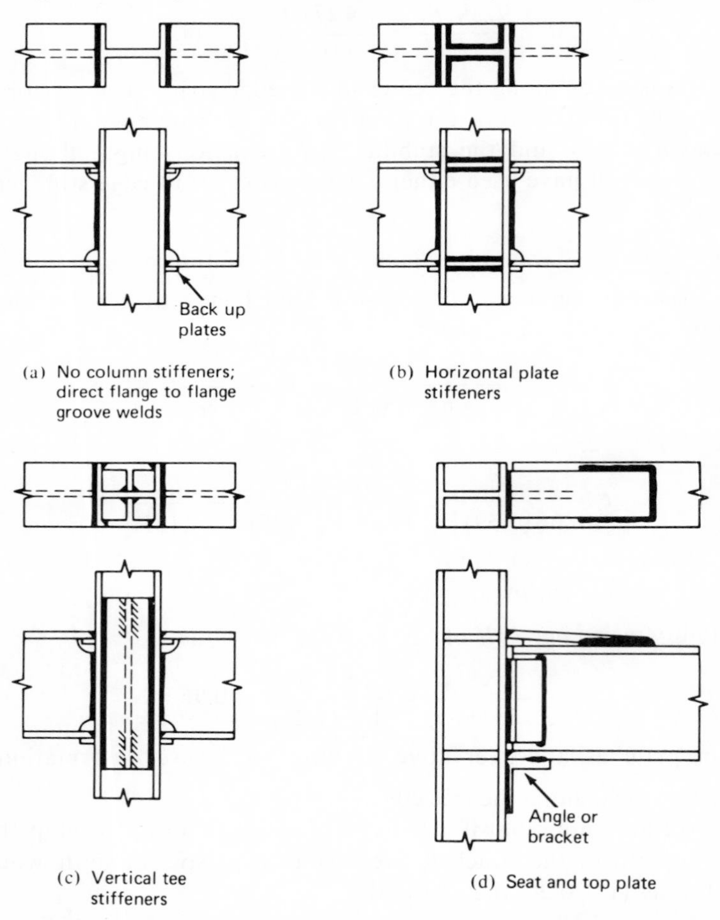

Fig. 13.6.1. Continuous beam-to-column connections: welded attachment to column flange.

13.6.1d, and in Fig. 13.6.2. Alternatively, the rigid attachment of beams may be to the web, from either or both sides, as in Fig. 13.6.3. When the rigid system has rigid attachments *either* to the flanges or the web (but not to both) the system is said to be a two-way, or planar, rigid frame. When the rigid frame system consists of continuous connections to both flange (or flanges) and web (either or both sides), the system becomes a four-way system, or space frame.

The variety of arrangements for a continuous beam-to-column connection is so great as to preclude any complete listing or illustration; however, those shown in Figs. 13.6.1, 13.6.2, and 13.6.3 are believed to be the most common in current (1971) design use. Most connections are partly shop welded and then completed in the field by either welding or fastening with high-strength bolts.

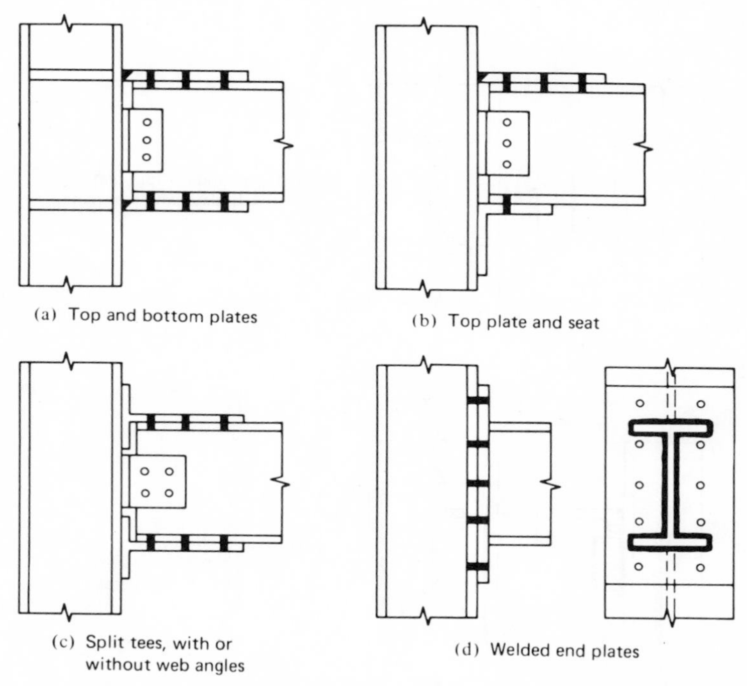

Fig. 13.6.2. Continuous beam-to-column connections: bolted attachments.

The principal design concern is with transmission of loads through the joint, and the localized deformations relating thereto. Rigid framing may be used to greatest advantage either (a) in plastically designed structures; or (b) in working-stress design when the beams framing to columns are "compact" and the 10 percent reduction permitted under AISC–1.5.1.4.1 may be used. In either case, it is the objective of the connection to be able to develop the full plastic-moment capacity at the joint and in addition be able to undergo plastic-hinge rotation.

Tests have adequately demonstrated the ability of beam-to-column rigid-frame connections to develop the plastic hinge moment and exhibit adequate rotation capacity (ductility).[12–15]

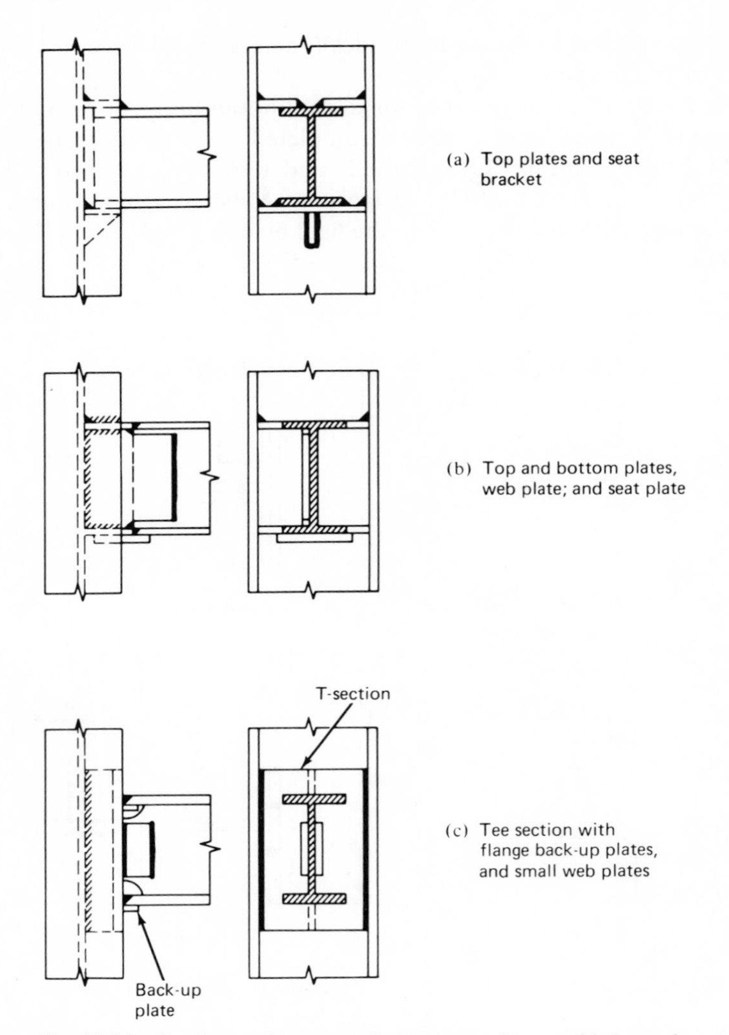

Fig. 13.6.3. Continuous beam-to-column connections: welded attachment to column web.

Horizontal Stiffener in Compression Region of Connection. When the forces in beam flanges are transmitted as compression or tension forces to column flanges, horizontal stiffeners, as in Figs. 13.6.1a and d and Fig. 13.6.2a, may be required to prevent web crippling in a region where the beam flange causes compression. The phenomenon to be protected against is similar to that related to beams as treated in Sec. 7.6, and to plate girders as treated in Sec. 11.2.

The following development, and resulting AISC provisions are from research conducted at Lehigh University.[12]

(a) Conservative Approach—1969 AISC–1.15.5. Consider a beam compression flange bearing against a column as in Fig. 13.6.4a. By the

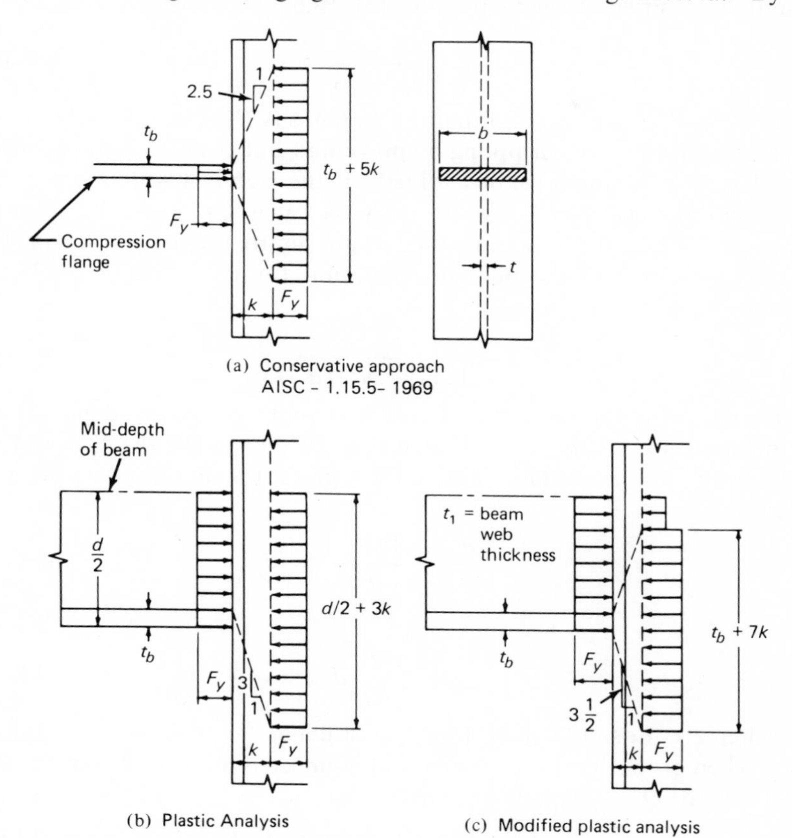

Fig. 13.6.4. Strength of column web in compression region of connection.

time yielding occurs the load has been distributed along the base of the fillet (k from the face of flange) on a 1:2.5 slope such that for equilibrium

$$Q_c = F_y(t_b + 5k)t \tag{13.6.1}$$

$$Q_b = F_y t_b b = F_y A_f \tag{13.6.2}$$

where Q_c and Q_b represent the compression capacities in the column web and the beam flange, respectively. Equating $Q_c = Q_b$ and solving for the minimum t to prevent web crippling,

$$t \geq \frac{A_f}{t_b + 5k} \tag{13.6.3}$$

where A_f = area of flange delivering concentrated load. If beam and column have different yield strengths, Eq. 13.6.3 becomes

$$t \geq \frac{C_1 A_f}{t_b + 5k} \tag{13.6.4}$$

where $C_1 = F_{y\,\text{beam}}/F_{y\,\text{column}}$. Equation 13.6.4 is AISC Formula 1.15–1.

Not only must web crippling be prevented but overall buckling of the web plate as a column must be avoided. If the web is a wide column uniformly compressed over the width $(t_b + 5k)$, it is similar to the situation of Fig. 11.2.8 except that when stiffeners are omitted $a/h \rightarrow \infty$. From Fig. 11.2.9, $k_c \rightarrow 1.0$ and the buckling equation for the plate, Eq. 11.2.17, becomes

$$F_{\text{cr}} = \frac{\pi^2 E}{12(1 - \mu^2)(d_c/t)^2} \tag{13.6.5}$$

where the unsupported depth of web h is replaced by the depth of the column section, d_c. Since it is desired that yield stress be reached without buckling, $F_{\text{cr}} \geq F_y$. Thus Eq. 13.6.5 gives for minimum t without stiffeners

$$\frac{d_c}{t} \leq \frac{162}{\sqrt{F_y}} \tag{13.6.6}$$

or

$$t \geq \frac{\sqrt{F_y}}{162} d_c \tag{13.6.7}$$

When $t <$ Eqs. 13.6.4 or 13.6.7, a stiffener (area $= A_{st}$) is required which when properly proportioned and connected adds $A_{st} F_y$ capacity to Q_c. Thus, for $Q_c = Q_b$, one obtains

$$F_y A_f = F_y (t_b + 5k) t + A_{st} F_y \tag{13.6.8}$$

or, solving for A_{st} gives for the stiffener area required,

$$A_{st} \geq A_f - (t_b + 5k) t \tag{13.6.9}$$

or, in general, if yield strengths are different for beam flange, column web, and stiffener, Eq. 13.6.8 gives

$$A_{st} \geq [A_f C_1 - t(t_b + 5k)] C_2 \tag{13.6.10}$$

where $C_2 = F_{y\,\text{column}}/F_{y\,\text{stiffener}}$. Equation 13.6.10 is AISC Formula 1.15–4.

(b) Plastic Analysis. This approach assumes the beam develops a plastic-hinge condition as shown in Fig. 13.6.4b. The column compression in the web is assumed to exist only on the compression side of the beam's neutral axis and distributes below the flange on a 1:3 slope, in which case

$$Q_c = F_y\left[\frac{d}{2} + 3k\right] t \tag{13.6.11}$$

$$Q_b = F_y A/2 \tag{13.6.12}$$

where A = full area of attached *beam*. By equating $Q_c = Q_b$ the minimum column web thickness t to prevent web crippling is

$$t \geq \frac{A}{d + 6k} \tag{13.6.13}$$

To be completely valid, this analysis assumes a beam-web thickness great enough to create the pressure and cause the column web to yield right up to the beam neutral axis. Usually the beam web is thinner than the column web.

(c) Modified Plastic Analysis. Here the beam-flange compression force is assumed to distribute along the column web on a 1:3.5 slope as shown in Fig. 13.6.4c. Outside the length $t + 7k$ the stress on the column web is less than F_y since it is assumed the portion of the load in the beam web is transmitted directly to the adjacent column web. If F_y acts on the beam web, a lower stress will act on the thicker column web:

$$Q_c = F_y(t_b + 7k)t \tag{13.6.14}$$

$$Q_b = A_f F_y + F_y(3.5k)t_1 \tag{13.6.15}$$

where t_1 = beam web thickness, and A_f = beam flange area. Equating Eqs (13.6.14) and (13.6.15) and solving for minimum t to prevent web crippling gives

$$t \geq \frac{A_f + 3.5kt_1}{t_b + 7k} \tag{13.6.16}$$

It is noted that $t + 5k$ was established for actual beams welded as in Fig. 13.6.1a, while $t + 7k$ was obtained for a concentrated load on a wide-flange section in an isolated test arrangement. Both were stated to be conservative.[12]

Horizontal Stiffener in Tension Region of Connection. At the beam tension flange it is the deformation caused to the column flange which is the main concern, as shown in Fig. 13.6.5. A yield-line analysis[12] was performed on the portion of the column flange of width q and length p, as in Fig. 13.6.5. Placing a line load on the system, the ultimate capacity was developed as

$$P_u = F_y t_c^2\left[\frac{4/\beta + \beta/\beta_1}{2 - \beta_1/\alpha}\right] \tag{13.6.17}$$

where t_c = column flange thickness

$$\beta = p/q$$
$$\alpha = x_1/q$$
$$\beta_1 = \frac{\beta}{4}[\sqrt{\beta^2 + 8\alpha} - \beta]$$

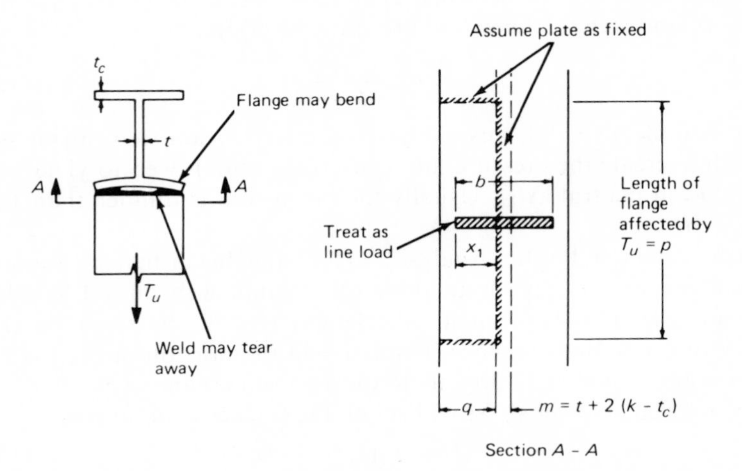

Fig. 13.6.5. Strength of column flange in tension region of connection.

It is stated[12] that a conservative approximation for Eq. 13.6.17 is

$$P_u = 3.5F_y t_c^2 \tag{13.6.18}$$

Thus the two portions of the column flange of width x_1 may carry $2P_u$. The central portion of width m may be assumed to be stressed to the yield stress:

$$Q_c = 7F_y t_c^2 + F_y tm \tag{13.6.19}$$
$$Q_b = F_y A_f \tag{13.6.20}$$

where Q_c and Q_b are the tension capacities in the column web and beam flange, respectively. Since the distribution length, $t + 5k$ or $t + 7k$ is conservative for the compression loading, similar conservatism may be obtained for the tension case by using $0.8Q_c$ instead of Q_c.[12] Equating $Q_b = 0.8Q_c$ and solving for t_c gives

$$t_c = \sqrt{\frac{1.25F_y A_f - F_y tm}{7F_y}} = \sqrt{\frac{A_f}{7}\left(1.25 - \frac{tm}{A_f}\right)} \tag{13.6.21}$$

Since tm/A_f varies from 0.15 to 0.20,[12] one may safely use the lower value of 0.15. Equation 13.6.21 then becomes

$$t_c = 0.397\sqrt{A_f} \tag{13.6.22}$$

or when column yield stress and beam yield stress differ, the minimum flange thickness without stiffeners is

$$t_c \geq 0.4\sqrt{C_1 A_f} \tag{13.6.23}$$

which is AISC Formula 1.15–3.

EXAMPLE 13.6.1

Design the connection for the rigid framing of two W16 × 40 beams to the flanges of a W12 × 65 column using A572 Grade 50 steel, as shown in Fig. 13.6.6. Use A36 steel for stiffeners.

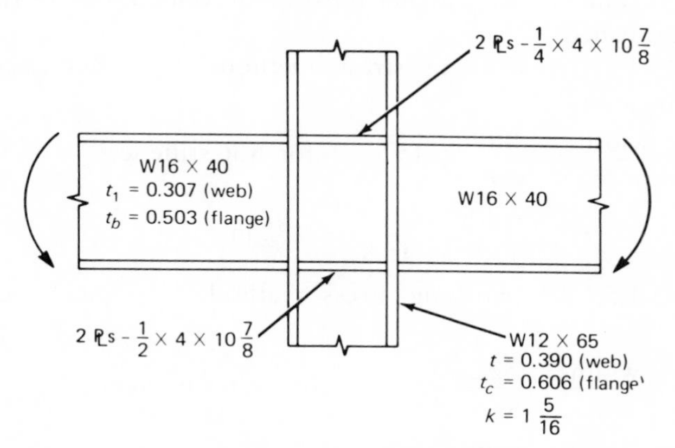

Fig. 13.6.6. Connection with horizontal stiffeners for Example 13.6.1.

Solution

(a) Compression region. According to AISC–1.15.5, or Eqs. 13.6.4 and 13.6.7,

$$t \geq \frac{C_1 A_f}{t_b + 5k} = \frac{1.0(7.00)(0.503)}{0.503 + 5(1^5/_{16})} = \frac{3.52}{5.503} = 0.50 \text{ in.}$$

or

$$t \geq \frac{d_c}{162/\sqrt{F_y}} = \frac{d_c}{22.9} = \frac{12.12}{22.9} = 0.53 \text{ in.}$$

Thus stiffeners are required since the column web (W12 × 65, t = 0.390 in.) has less thickness than necessary for preventing web crippling and overall web buckling.

As a comparison, compute the thickness required using the plastic analysis and the modified plastic analysis. For the plastic analysis, Eq. 13.6.13 requires

$$t \geq \frac{A}{d + 6k} = \frac{11.8}{16 + 6(1^5/_{16})} = 0.50 \text{ in.}$$

For the modified plastic analysis, Eq. 13.6.16 requires

$$t \geq \frac{A_f + 3.5kt_1}{t_b + 7k} = \frac{7.0(0.503) + 3.5(1\tfrac{5}{16})(0.307)}{0.503 + 7(1\tfrac{5}{16})} = 0.51 \text{ in.}$$

All methods indicate a compression stiffener is required.

The area of stiffener A_{st} required according to AISC–1.15.5, Eq. 13.6.10, is

$$A_{st} \geq [A_f C_1 - t(t_b + 5k)]\, C_2$$

$$= [3.52(1.0) - 0.390(7.06)]\, \frac{50}{36} = 1.07 \text{ sq in.}$$

Since local buckling of the stiffener must be prevented, one of the following must be satisfied:

1. AISC–1.9.1.2, working stress method, "compact" section *not* utilized at the joint.

$$\frac{\text{width}}{t_s} \leq \frac{95}{\sqrt{F_y}} \leq 15.8 \qquad \text{for A36 stiffeners}$$

$$t_s \geq \frac{\text{width}}{15.8} = \frac{5}{15.8} = 0.32 \text{ in.}$$

2. AISC–1.5.1.4.1, working stress method, "compact" section requirement.

$$\frac{\text{width}}{t_s} \leq \frac{52.2}{\sqrt{F_y}} = 8.7 \qquad \text{for A36 stiffeners}$$

$$t_s \geq \frac{\text{width}}{8.7} = \frac{5}{8.7} = 0.58 \text{ in.}$$

3. AISC–2.7, plastic-design method.

$$\frac{\text{width}}{t_s} \leq 8.5$$

$$t_s \geq \frac{5}{8.5} = 0.59 \text{ in.}$$

In the above three calculations the plate width of 5 in. was used; obtained from column flange width less web thickness divided by 2 and rounded to the lower even inch. Since the beam flange is only 8 in. wide, a 4-in.-wide plate would be adequate and should be used if plastic hinging is assumed at this connection. Assuming compact section is needed,

$$t_s \geq \frac{4}{8.7} = 0.46 \text{ in.,} \quad \text{say } \tfrac{1}{2} \text{ in.}$$

$$A_{st}\,\text{provd} = 2(4)(0.5) = 4.0 \text{ sq in.} > 1.07 \qquad \text{OK}$$

2 plates, $\frac{1}{2} \times 4$ are acceptable for the compression zone.

(b) Tension region. Check Eq. 13.6.21:

$$t_c = \sqrt{\frac{A_f}{7}\left(1.25 - \frac{tm}{A_f}\right)}$$

$$m = t + 2(k - t_c) = 0.390 + 2(1.31 - 0.606) = 1.80 \text{ in.}$$

$$\frac{tm}{A_f} = \frac{0.390(1.60)}{3.52} = 0.20$$

$$t_c = \sqrt{\frac{3.52}{7}(1.25 - 0.20)} = 0.73 \text{ in.}$$

or, using AISC Formula 1.15–3, Eq. 13.6.23,

$$t_c \geq 0.4\sqrt{C_1 A_f} = 0.4\sqrt{1.0(3.52)} = 0.75 \text{ in.}$$

Since the flange thickness of the W12 × 65 (actual $t_c = 0.606$) is less than 0.73 in., a stiffener is required. The equilibrium requirement for the tension region may be assumed the same as for the compression region; thus the reqd $A_{st} = 1.87$ still applies. The local-buckling requirements do not apply to the tension zone; therefore, somewhat thinner tension plates could be used.

Use 2 PLs—$\frac{1}{2}$ × 4 × $10\frac{7}{8}$, A36, for compression side.

Use 2 PLs—$\frac{1}{4}$ × 4, $10\frac{7}{8}$, A36, for tension side.

(c) Connection of plates to column. The forces to be considered in the design of welding are shown in Fig. 13.6.7. When beams frame from

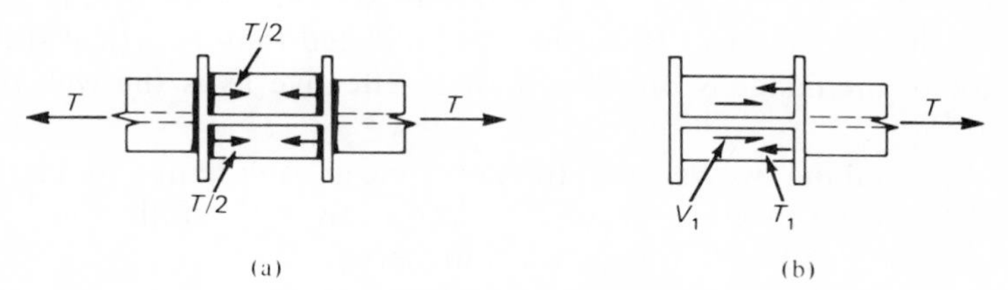

Fig. 13.6.7. Weld requirements for horizontal plates.

both sides and contribute equal flange forces T, the weld on the ends of the plates must be designed to carry all of the force T which is not directly taken by the web. For this example, Fig. 13.6.7a applies, and one may consider that the tributary length of column web is $t + 5k$, as from Fig. 13.6.4a.

$$\text{Force to stiffeners} = \frac{A_{st}}{t(t_b + 5k) + A_f}(T)$$

$$= \frac{4.0}{(0.39)(7.06) + 4.0}(24)(3.52) = 50 \text{ kips}$$

where the flange force, $T = 0.66F_y A_f$ for the working-stress method. For plastic design, the flange force is $T = F_y A_f$, and for designing the connection the allowable weld stresses must be multiplied by the load factor.

Maximum effective weld size, using shielded metal-arc welding,

$$a_{max} = \frac{0.6F_y}{2(0.707)21} = \frac{30}{29.7} = 1.01 \text{ in.}$$

$$\text{Min. weld size } a_{min} = \tfrac{1}{4} \text{ in.}$$

For fillets top and bottom of plates, capacity required per inch is

$$\frac{50}{2(8)} = 3.12 \text{ kips/in.}$$

$$\text{Reqd } a = \frac{3.12}{0.707(21)} = 0.21 \text{ in.,} \quad \text{say } \frac{1}{4} \text{ in.}$$

Use $\frac{1}{4}$ in. fillet weld, top and bottom, on both tension and compression plates where they bear against the column flanges. Along the column web, fillet weld is required on only one side of plate.

When a beam frames in from only one side, as in Fig. 13.6.7b, the weld forces T_1 are designed as in the symmetrical case. However, in addition the shear forces V_1 must also be developed to take the proportion of the unbalanced flange force which comes to the stiffener plates; in this case $V_1 = T_1$.

Vertical Plate and Tee Stiffeners. Sometimes it may be desirable to use vertical plates, or structural tee sections as shown in Fig. 13.6.1c. Particularly, they may be useful in a four-way system where beams are attached to the tee sections. Research has indicated that a vertical stiffener at the toe of the flange is only one-half as effective as is the web of the column.[12]

Thus, assuming two vertical stiffeners (one at each flange toe) stressed to one-half the web capacity of Eq. 13.6.1 gives the following as the counterpart to Eq. 13.6.8, for vertical stiffeners

$$F_y A_f = F_y(t_b + 5k)t + 2\left(\frac{F_y}{2}\right)(t_b + 5k)t_s \tag{13.6.24}$$

which upon solving for the stiffener thickness t_s gives

$$t_s \geq \frac{A_f}{t_b + 5k} - t \tag{13.6.25}$$

or, if yield strengths are different for beam flange, column web, and stiffener, Eq. 13.6.24 gives

$$t_s \geq \left[\frac{C_1 A_f}{t_b + 5k} - t\right] C_2 \tag{13.6.26}$$

where $C_1 = F_{y\,\text{beam}}/F_{y\,\text{column}}$ and $C_2 = F_{y\,\text{column}}/F_{y\,\text{stiffener}}$. Similar reasoning may be used to obtain the vertical stiffener thickness required based on the plastic and modified plastic analyses assuming each stiffener accommodates one-half the load the web carries.

If the vertical stiffener is a plate, then overall buckling must be prevented by satisfying Eq. 13.6.7. When structural tees are used, the web attachment precludes the overall buckling.

For the design of the tee stiffener and connection when a beam is attached to it, as in Fig. 13.6.3c, some special considerations are necessary. The beam flange, if equal in width to the tee, acts on the tee as shown in Fig. 13.6.8a. For analysis, one might consider uniform loading on a two-span beam as shown in Fig. 13.6.8b, wherein the load would be transmitted $^5/_8$ into the tee web and $^3/_{16}$ each to the column flanges. Ref. 6 suggests that when the beam flange extends full width of the tee one may assume the effective flange width b_E (Fig. 13.6.8a) tributary to the tee web to be $^3/_4$ of the beam flange width.

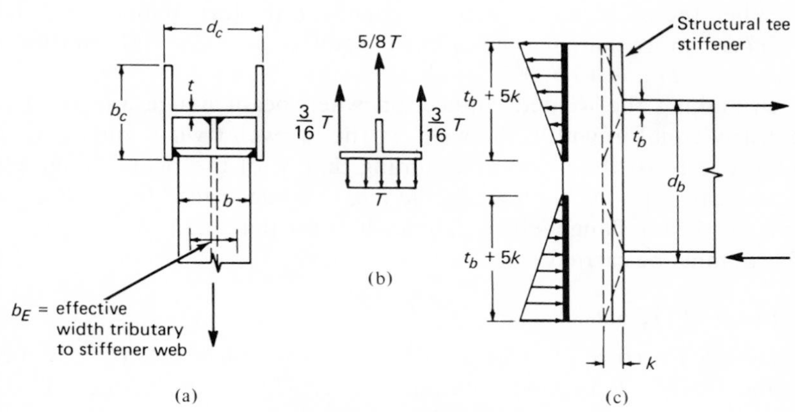

Fig. 13.6.8. Structural tee stiffener.

One may summarize the design requirements when the *beam flange width is approximately the same as the tee flange width:*

(a) The tee web thickness t_w must satisfy Eq. 13.6.4:

$$t_w \geq \frac{0.75 C_1 A_f}{t_s + 5k} \tag{13.6.27}$$

where $0.75A_f$ = tributary part of beam flange area A_f
k = distance to root of fillet; tee section
t_s = flange thickness of tee

(b) The structural tee flange thickness t_s must be able to carry the beam tensile flange force without excessive deformation; hence Eq. 13.6.23 should be satisfied. This will be conservative since the equation was derived for the free-edge condition at the flange toes, whereas here there is a welded connection. Using $0.75A_f$ as in (a),

$$t_s \geq 0.4\sqrt{C_1(0.75)A_f} = 0.35\sqrt{C_1 A_f} \tag{13.6.28}$$

(c) The structural tee flange width b_s must extend fully between the column flanges:

$$b_s = d_c - 2t_c \tag{13.6.29}$$

where d_c = column overall depth
t_c = flange thickness of column

(d) The structural tee depth d_s must be adequate to be nearly flush with the outer edges of the column flanges:

$$d_s = \frac{b_c - t}{2} \tag{13.6.30}$$

where b_c = column flange width
t = column web width

When the beam flange width is significantly less than the tee flange width (say, more than an inch or two), A_f instead of $0.75A_f$ should be used in Eqs. 13.6.27 and 13.6.28.

In making the welded connection when beam flange and tee flange are nearly equal in width, the weld on the tee web (two segments of two fillets) is to resist the moment assuming that $^3/_4$ of the beam flange effect must be carried.[6] At the toes of the tee, it is suggested[6] to consider that $^1/_3$ of the beam flange effect (somewhat greater than the $^3/_{16}$ of Fig. 13.6.8b) must be carried.

EXAMPLE 13.6.2

Design a vertical tee stiffener connection to frame a W14 × 61 beam into the web of a W12 × 65 column. Use A 572 Grade 50 steel. Use the type of connection shown in Fig. 13.6.8.

Solution

Since the beam flange width (10.00 in.) is approximately the same as the clear distance between column flanges (12.12 − 1.21 = 10.91 in.), Eqs. 13.6.27 and 13.6.28 may be applied.

(a) $t_s \geq 0.35\sqrt{C_1 A_f}$ [13.6.28]

$= 0.35\sqrt{1.0(10.00)(0.643)} = 0.89$ in.

(b) Estimating tee flange thickness as $^7/_8$ in., and $k \approx 1$ in.

$$t_w \geq \frac{0.75C_1A_f}{t_s + 5k} = \frac{0.75(1.0)(10.00)(0.643)}{0.875 + 5(1.0)} = 0.82 \text{ in.}$$

(c) $b_s = d_c - 2t_c = 12.12 = 2(0.606) = 10.91$ in.

(d) $d_s = (12.00 - 0.39)/2 = 5.81$ in.

Try W12 × 99 cut into tees:

$$t_s = 0.921, \qquad t_w = 0.582, \qquad k = 1^5/_8 \text{ in.}$$

$$\text{Reqd } t_w \geq \frac{0.75(1.0)(10.00)(0.643)}{0.921 + 5(1.63)} = 0.54 \text{ in} < 0.58 \qquad \text{OK}$$

Try W12 × 99 cut as shown in Fig. 13.6.9a.

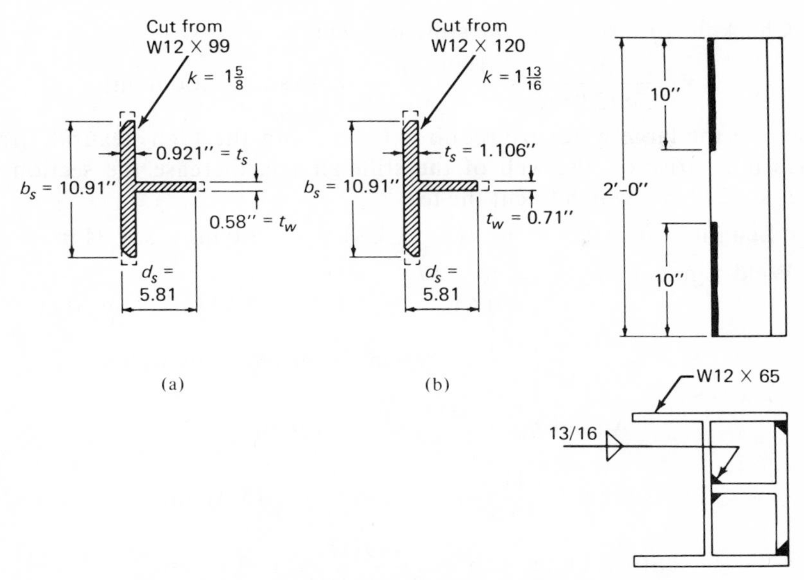

Fig. 13.6.9. Example 13.6.2. Tee selection and web welding.

(e) Welding on stiffener web. Referring to Fig. 13.6.8c, the length of tee stiffener required is

$$\text{Length} = d_b + 5k = 13.91 + 5(1.63) = 22.06 \text{ in.}$$

Try length of tee = 1′-10″.

Length of weld, upper, and lower ends:

$$t_b + 5k = 0.643 + 5(1.63) = 8.79 \text{ in.}$$

Try 9 in. weld at each end, connecting web of tee to web of column.

Assume the moment contributed by the center $^3/_4$ of the flange is tributary to the tee web:

$$0.75M = 0.75F_bS = 0.75\,(32.8)\,92.2/12 = 189 \text{ ft-kips}$$

where F_b = 32.8 ksi for partly "compact," AISC Formula 1.5–5.
The resisting section modulus of the two 9 in. weld segments treated as lines is

$$\left[\frac{(22)^3 - (22 - 18)^3}{12}\right] \frac{2}{22} = 80 \text{ in.}^2$$

$$\text{Actual } R_w = \frac{M}{2S} = \frac{189(12)}{2(80)} = 14.2 \text{ kips/in.}$$

$$\text{Allowable } R_w = a(0.707)\,21 = 14.85a$$

One-inch fillet welds on each side of web would be required.

Check flexural stress on the stiffener web:

$$f = \frac{M}{0.58(22)^2/6} = \frac{189(12)}{46.8} = 48.4 \text{ ksi}; \qquad \text{too high!}$$

Based on the large weld size required (1 in.) and the high flexural stress which may arise on the web of the stiffener tee, increase the section to W12 × 120 from which to cut the tee.

$$\text{Length} = d_b + 5k = 13.91 + 5(1.812) = 23.0 \text{ in.,} \quad \text{say 24 in.}$$

$$\text{Weld segment lengths} = t_b + 5k$$
$$= 0.643 + 5(1.812) = 9.73 \text{ in.,} \quad \text{say 10 in.}$$

$$S = \frac{[(24)^3 - (4)^3]}{12(12)} = 95.5 \text{ in.}^3 \quad \text{(provided by weld)}$$

$$\text{Actual } R_w = \frac{189(12)}{2(95.5)} = 11.9 \text{ kips/in.}$$

$$\text{Reqd } a = \frac{11.9}{14.85} = 0.80 \text{ in.,} \quad \text{say 13/16 in.}$$

$$\text{Stress on tee web} = \frac{189(12)}{0.71(24)^2/6} = 33.2 \text{ ksi}$$

Use W12 × 120 as shown in Fig. 13.6.9b.

(f) Welding on stiffener flanges. Assume conservatively that the tee connection to the column flange (Figs. 13.6.8a and b) may have to carry as much as ⅓ of the beam flange force. The concentrated forces from the beam flange may be considered as distributed along the column flange over the distance $t_b + 5t_s$, as shown in Fig. 13.6.10.

Since the W14 × 61 flange may carry an allowable stress, $F_b = 32.8$ ksi, the tensile force assumed tributary to the groove weld on one flange is

$$T = f\,t_b\left(\frac{b}{3}\right) = 32.8(0.643)(3.33) = 70.2 \text{ kips}$$

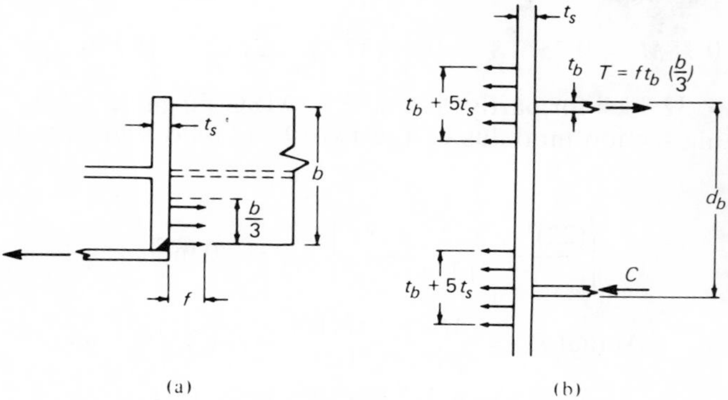

Fig. 13.6.10. Forces carried by weld along tee stiffener flanges.

$$t_b + 5t_s = 0.643 + 5(1.106) = 6.17 \text{ in.}$$

$$\text{Reqd weld capacity} = \frac{70.2}{6.17} = 11.4 \text{ kips/in.}$$

Using a partial penetration groove weld, E70 electrode material, the required effective throat dimension is

$$\text{Eff throat} = \frac{11.4}{21} = 0.541 \text{ in.}$$

$$\text{Reqd groove depth} = 0.541 + 0.125 = 0.67 \text{ in.}$$

if manual shielded metal-arc welding is used.

$$\text{Min eff throat} = \sqrt{\frac{t_t}{6}} = \sqrt{\frac{1.106}{6}} = 0.43 \text{ in.} < 0.541 \text{ in.} \qquad \text{OK}$$

Use $^{11}/_{16}$ in. single bevel groove weld along the edges of the flange of the tee.

(g) Effect of beam shear force. Ordinarily the length of weld used is so large that the additional capacity required to carry end shear is negligible. A W14 × 61 might be expected to carry something on the order of 50 kips shear. In which case it is resisted by

$$\text{Total length of weld} = \underset{\text{web}}{4(10)} + \underset{\text{flanges}}{2(24)} = 88 \text{ in.}$$

$$\text{Reqd extra weld capacity} = \frac{50}{88} = 0.57 \text{ kips/in.}$$

$$\text{Added size for web fillets} = \frac{0.57}{0.707(21)} = 0.04 \text{ in.}$$

For conservatism one could increase the fillet-weld size from $^{13}/_{16}$ to $^{3}/_{4}$ in.

$$\text{Added size for flange groove welds} = \frac{0.57}{21} = 0.03 \text{ in.}$$

One may conservatively increase the partial penetration groove depth to $^{3}/_{4}$ in.

Actually the shear component and the flexure component act at 90 degrees to one another so that adding the requirements algebraically is overly safe. The authors would consider the welds safe without an increase for direct shear.

Top Tension Plates. When the beam is connected to the column flange and the column is stiffened by vertical or horizontal plate stiffeners, or when the beam is connected to the column web through a vertical tee stiffener, a simple means of transmitting the moment from the beams is with a tension plate at the top of the beam combined with (a) a bottom compression plate and web plates for shear; (b) a bottom seat angle; or

(c) a bottom bracket (stiffened seat). See Figs. 13.6.1d and 13.6.2a and b. The behavior of such top plate connections has been verified by research studies.[16,17]

The design of seat angles and brackets has been treated in Secs. 13.3 and 13.4. Transmission of tension and compression forces into the column has been treated earlier in this section. Emphasis here is on the top plate, a tension member whose design is illustrated in the following example.

EXAMPLE 13.6.3

Design the top tension plate and its connection by welding or by A325 high-strength bolts for transferring the full end moment from a W14 × 61 to a column. Use A572 Grade 50 steel. Assume the connection to be of the type shown in Figs. 13.6.1d or 13.6.2b.

Solution

Since the W14 × 61 is partly "compact" for Grade 50 steel, the maximum allowable flexural stress is $0.655F_y$. Design for full capacity. The flange force T is approximately

$$T = \frac{M}{d_b} = \frac{0.655F_yS_x}{d_b} = \frac{32.8(92.2)}{13.91} = 215 \text{ kips}$$

(a) Determine plate size.

$$\text{Reqd plate area} = \frac{215}{32.8} = 6.55 \text{ sq in.}$$

The W14 × 61 flange width is 10.0 in., so plate width must be less than 10.0 in., say 9 in.

$$\text{Reqd plate thickness} = \frac{6.55}{9} = 0.73 \text{ in.}$$

Use PL—¾ × 9, A_p = 6.75 sq in., for welded construction.

(b) For welding plate to beam flange, try ⅜-in. fillet weld, E70 electrodes, assuming shielded metal-arc process,

$$R_w = a(0.707)\,21 = (0.375)(0.707)\,21 = 5.56 \text{ kips/in.}$$

$$\text{Reqd } L_w = \frac{215}{5.56} = 38.6 \text{ in.}$$

To reduce the required weld length, try ½-in. weld:

$$\text{Reqd } L_w = \frac{215}{7.42} = 29.0 \text{ in.}$$

Use ½ in. weld, 9 in. on end and 10 in. on each side for a total of 29 in.

The design is summarized in Fig. 13.6.11a.

The plate is connected to the column flange by a full penetration

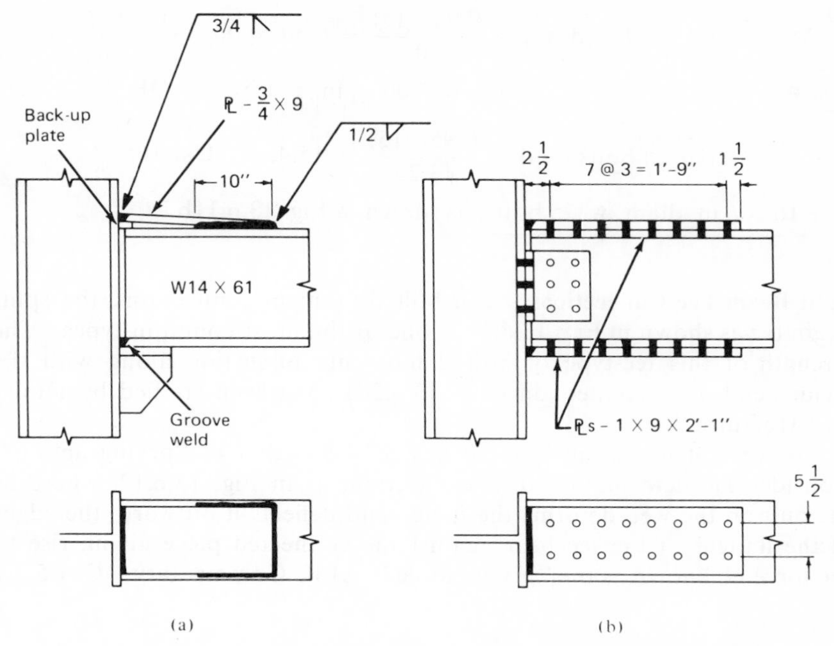

Fig. 13.6.11. Example 13.6.3. Moment connection using top plate.

single bevel groove weld, using a backup bar. When large moment is to be transmitted by the compression flange, it should be connected directly to the column flange by a groove weld. This will permit the seat angle or bracket to be designed to carry shear, without flexure. When a seat is used it will serve as backup to make the flange groove weld. Otherwise, a backup plate is needed below the compression flange.

(c) Using ⅞ in. diam. A325 bolts in a bearing-type joint with threads excluded from the shear plane,

$$\begin{aligned} R_{SS} &= 0.601(22) = 13.25 \text{ kips (single shear)} \\ R_B &= \text{does not control} \end{aligned}$$

When bolts are used, the holes in the W14 × 61 flange exceed 15 percent of the gross flange, so that according to AISC–1.10.1 the excess must be deducted. In this case

$$\text{Percent } \frac{A_{\text{net}}}{A_{\text{gross}}} = \frac{10.00(0.640) - 2(1)(0.64)}{10.00(0.64)} = \frac{1.28}{6.40} = 20\%$$

Thus 5 percent of the gross flange effect must be deducted and the tensile capacity of the flange becomes 95 percent of 215 kips.

For the plate as a tension member, the effect of all holes must be deducted:

$$\text{Reqd } A_{net} = \frac{0.95(215)}{30} = 6.80 \text{ sq in.}$$

Use ℞ 1 × 9, $A_{net} = 9.00 - 2(1) = 7.00$ sq in. > 6.80 OK

$$\text{No. of bolts reqd} = \frac{0.95(215)}{13.25} = 15.4, \qquad \text{Use 16.}$$

Use 16 ⅞ in. diam. A325 bolts, as shown in Fig. 13.6.11b.

Split-Beam Tee Connections. For bolted moment connections, the split-beam tee as shown in Fig. 13.6.2c is one of the most common types. The strength of this tee-type of bolted moment connection along with the welded end-plate connection (Fig. 13.6.2d) have been studied by Douty and McGuire.[15]

A concept merely mentioned in Sec. 4.8 is that of "prying action." Consider the deformation of a tee section, as in Fig. 13.6.12, where as the pull on the web deforms the flange and deflects it outward, the edges of the flanges of the tee bear against the connected piece giving rise to the force Q, known as the "prying force." AISC Commentary—C-1.5.2.1

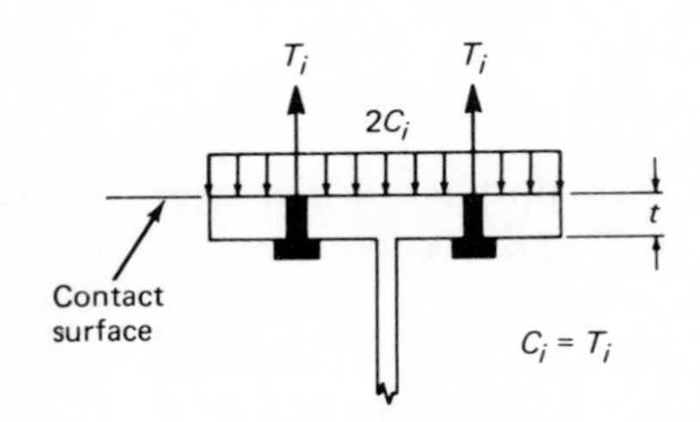

(a) No external load.

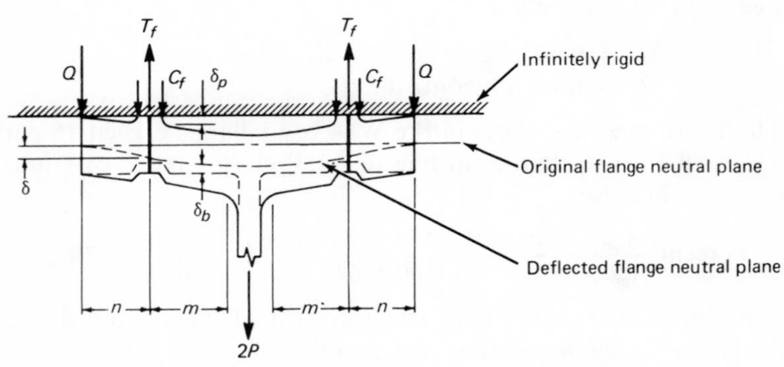

(b) External load and prying action

Fig. 13.6.12. Prying action.

states, "Any additional fastener tension resulting from prying action due to distortion of the connection details should be added to the stress calculated directly from the applied tension . . . ."

When bolted connections are not subject to distortion such as evident in Fig. 13.6.12b, the treatment of Sec. 4.8 for bolts in tension is valid. However, when a thick flanged tee distorts, the flange tips tend to dig in, giving rise to the force Q.

Using the same reasoning as for Eqs. 4.8.2 through 4.8.9, one may obtain the expression for the prying force Q, as presented in Ref. 15. Before external load is applied,

$$C_i = T_i \tag{13.6.31}$$

After external load is applied, prying action occurs giving for the equilibrium requirement

$$P + C_f + Q = T_f \tag{13.6.32}$$

The increased bolt length is

$$\delta_b = \left(\frac{T_f - T_i}{A_b E}\right) t = \left[\frac{P + Q - (T_i - C_f)}{A_b E}\right] t \tag{13.6.33}$$

The increased plate thickness is

$$\delta_p = \left(\frac{C_i - C_f}{A_p E}\right) t = \left(\frac{T_i - C_f}{A_p E}\right) t \tag{13.6.34}$$

Assume the forces are uniformly distributed along the tee section length w which is tributary to one bolt. The tee flange thickens an amount δ_p and remains in contact with the support, the bolt length increases an amount δ_b, and the neutral plane of the tee flange deflects an amount δ as shown in Fig. 13.6.12. All of these deformations are equal in magnitude. The neutral plane deflection may be computed by ordinary elastic beam deflection procedures, as described in detail in Ref. 15.

For design recommendations, Ref. 15 provides the following equations to compute the prying force.

(a) For the *working stress* method, where contact between surfaces is maintained because the bolt elongation is not great enough to entirely relieve the initial precompression, the prying force is

$$Q = \left[\frac{\dfrac{1}{2} - \dfrac{wt^4}{30nm^2 A_b}}{\dfrac{3n}{4m}\left(\dfrac{n}{4m} + 1\right) + \dfrac{wt^4}{30nm^2 A_b}}\right] P \tag{13.6.35}$$

It is recommended[15] that

$$P + Q \leq (0.75)(\text{proof load}) \tag{13.6.36}$$

and assume the prying force acts at $3m/4$ from the bolt line. Note is made that in using Eq. 13.6.35, Q is never taken as negative. When the numerator is negative, $Q = 0$.

(b) For the *plastic-design* method where loads are considered at ultimate capacity when initial precompression is entirely relieved, the prying force is

$$Q = \left[\frac{\dfrac{1}{2} - \dfrac{wt^4}{30nm^2 A_b}}{\dfrac{n}{m}\left(\dfrac{n}{3m} + 1\right) + \dfrac{wt^4}{6nm^2 A_b}}\right] P_u \tag{13.6.37}$$

It is recommended[15] that

1. For a connection at the last plastic hinge to form, and at other connections where $Q = 0$,

$$P_u + Q \leq (1.33)(\text{proof load}) \tag{13.6.38}$$

2. At all other connections,

$$P_u + Q \leq (1.15)(\text{proof load}) \tag{13.6.39}$$

and assume the prying force acts at the flange tip.

EXAMPLE 13.6.4

Design a split-beam tee connection, such as in Fig. 13.6.2c, to enable a plastic hinge to develop in a W14 × 53 beam framing to the flange of a W14 × 158 column. Use A572 Grade 50 steel and ¾ in. diam. A325 bolts in a bearing-type joint. Use (a) working-stress design method, and (b) plastic-design method.

Solution

(a) Working-Stress Method. Since the W14 × 53 satisfies the local buckling requirements for a "compact" section, and if the maximum laterally unbraced length is $L_c = 7.22$ ft, the allowable stress is $0.66F_y$.

$$M = 0.66F_y S_x = 33(77.8)/12 = 214 \text{ ft-kips}$$

If all moment is transmitted by the tees, the force of the couple is

$$\text{Force} = \frac{M}{d_b} = \frac{214(12)}{13.94} = 184.5 \text{ kips}$$

1. Check whether the tensile force can be accommodated by the bolts in tension:

$$\text{Allowable bolt capacity } R_t = 0.4418(40) = 17.6 \text{ kips}$$

Only 8 bolts will fit, as shown in Fig. 13.6.13; therefore, the maximum tensile force which may be carried is

$$T = 8(17.6) = 141 \text{ kips} < 184.5 \text{ kips} \qquad \text{N.G.}$$

When this difficulty arises one may use a stub beam or a tee stub attached to the bottom of the main beam (Fig. 13.6.13) to increase the

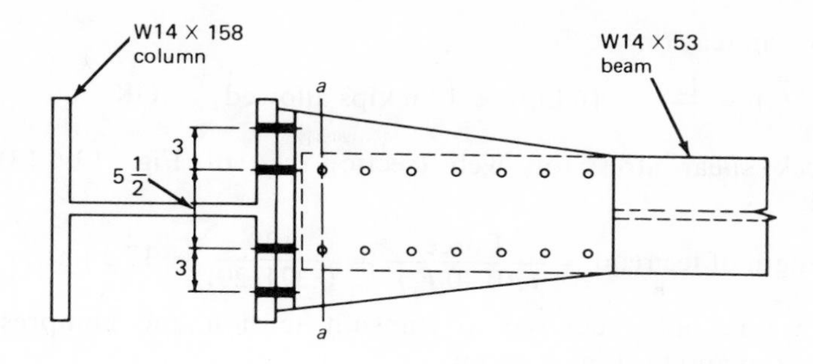

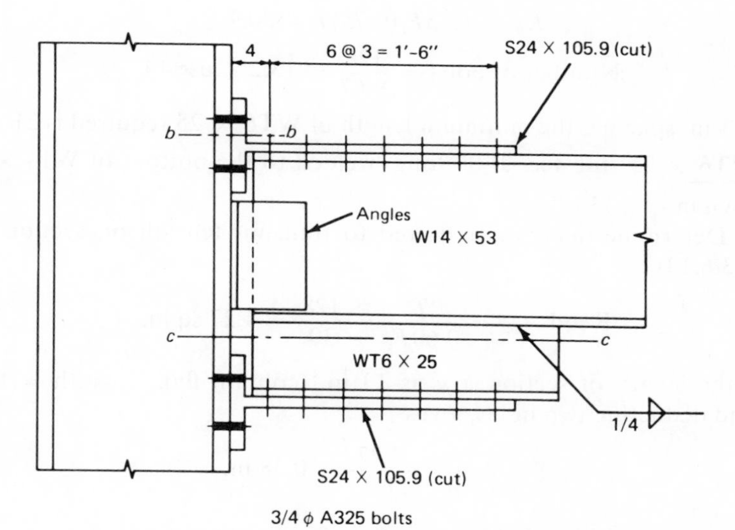

Fig. 13.6.13. Example 13.6.4. Split-beam tee connection with tee stub.

moment arm of the couple. Actually, when designing for the support moment, one might have used a beam size required for the midspan moment and then used the stub beam to gain the increased capacity required at the support. The necessary moment arm is

$$\text{Arm} = \frac{214(12)}{141} = 18.2 \text{ in.}$$

$$\text{Extra depth reqd} = 18.2 - 13.94 = 4.26 \text{ in.}$$

Try as stub beam a WT6 × 25, $t_w = 0.371$ in., $t_f = 0.641$ in., $b_f = 8.077$ in., whose dimensions are comparable to the main W14 × 53 beam.

$$\text{Force of couple} = \frac{214(12)}{13.94 + 6.10} = 128 \text{ kips}$$

Using 8 bolts in tension,

$$f_t A = \frac{128}{8} = 16 \text{ kips} < 17.6 \text{ kips allowed} \qquad \text{OK}$$

2. Check shear stress on web (section c–c of Fig. 13.6.13) of WT6 × 25:

$$\text{Length of tee reqd} = \frac{\text{Force}}{t_w(0.40\,F_y)} = \frac{128}{0.371(20)} = 17.3 \text{ in.}$$

3. Determine bolts required to transmit tension and compression forces at the top and bottom of beam:

$$R_{SS} = (0.4418)\,22 = 9.67 \text{ kips} \qquad \text{Controls}$$

$$R_B = 1.35F_y(0.75)\,t = 50.7t$$

$$\text{Number of bolts} = \frac{128}{9.67} = 13.2; \quad \text{use 14}$$

Using 3-in. spacing, the minimum length of WT6 × 25 required is 21 in.: Use WT6 × 25 stub tee, 2′-0″ long, welded to the bottom of W14 × 53, as shown in Fig. 13.6.13.

4. Determine thickness required to transmit tension on section a–a (Fig. 13.6.13):

$$\text{Reqd } A_{\text{net}} = \frac{T}{0.60F_y} = \frac{128}{30} = 4.27 \text{ sq in.}$$

Using the length of section a–a as 13 in. (column flange width = 15.55 in.), and deducting two holes, gives

$$t \geq \frac{4.27}{13 - 2(0.875)} = \frac{4.27}{11.25} = 0.38 \text{ in.}$$

$$t \geq \frac{9.67}{50.7} = 0.19 \text{ in.} \qquad \text{(bearing does not control)}$$

5. Determine flange thickness required for flexure on section b–b, Fig. 13.6.13:

$$\text{Moment on flange} = T_f m - Q(m + n)$$

using notation of Fig. 13.6.12b, and assuming no effect from C_f. The distance m may be taken as $(g - 2k_1)/2$ for W shapes and $(g - 2c)/2$ for S shapes, in terms of AISC Manual dimensions for detailing.

Estimating $Q = 0$, $g = 4$, and k_1 or $c = 1.0$,

$$M = T_f m \approx \frac{128}{2}\left(\frac{4 - 2(1.0)}{2}\right) = 64 \text{ in.-kips}$$

$$f_b = \frac{M}{S} = \frac{64}{14t_f^2/6} \leq 0.75F_y \text{ (AISC–1.5.1.4.3)}$$

$$\text{Reqd } t_f = \sqrt{\frac{6(64)}{14(37.5)}} = 0.86 \text{ in.}$$

Try tee cut from S24 × 79.9, $t_f = 0.871$, $t_w = 0.50$. Check prying force. Use Eq. 13.6.35:

$$n = (7.00 - 4.00)/2 = 1.50 \text{ in.}$$

$$m = (4.00 - 0.625)/2 = 1.69 \text{ in.}$$

$$\frac{wt^4}{30nm^2 A_b} = \frac{14(0.871)^4}{30(1.50)(1.69)^2(0.4418)} = 0.141$$

$$\frac{3n}{4m}\left(\frac{n}{4m} + 1\right) = \frac{3(1.50)}{4(1.69)}\left(\frac{1.50}{4(1.69)} + 1\right) = 0.815$$

$$Q = \left(\frac{0.500 - 0.141}{0.815 + 0.141}\right) P = \left(\frac{0.359}{0.956}\right) P = 0.375P$$

In accordance with Eq. 13.6.36, using the computed force per bolt of 16 kips,

$$P + Q = P + 0.375P = 1.375(16) = 22 \text{ kips}$$

$$0.75 \text{ (proof load)} = 0.75(28) = 21 \text{ ksi} < 22 \text{ kips} \qquad \text{N.G.}$$

It is probably wise to revise tee section to obtain a thicker flange and reduce prying action.

Try S24 × 105.9, $t_f = 1.102$, $t_w = 0.625$ in.

$$n = (7.875 - 4.00)/2 = 1.94 \text{ in.}$$

$$m = (4.00 - 0.75)/2 = 1.62 \text{ in.}$$

$$\frac{wt^4}{30nm^2 A_b} = 0.306; \qquad \frac{3n}{4m}\left(\frac{n}{4m} + 1\right) = 1.165$$

$$Q = \left(\frac{0.500 - 0.306}{1.165 + 0.306}\right) P = \left(\frac{0.194}{1.471}\right) P = 0.132P$$

$$P + Q = 1.132P = 1.132(16) = 18.1 \text{ kips} < 21 \text{ kips} \qquad \text{OK}$$

Use tees cut from S24 × 105.9 to carry tensile and compressive forces, according to working stress method.

(b) Plastic-Design Method. Check adequacy of design obtained in part (a):

$$M_p = F_y Z = 50(87.1)/12 = 364 \text{ ft-kips}$$

$$\text{Force} = \frac{M_p}{d_b + \text{depth of tee}} = \frac{364(12)}{13.94 + 6.10} = 218 \text{ kips}$$

$$\text{Net area reqd} = \frac{\text{Force}}{F_y} = \frac{218}{50} = 4.35 \text{ sq in.}$$

$$\text{Net area provd} = 0.625[14 - 2(7/8)] = 7.65 \text{ sq in.} > 4.35 \qquad \text{OK}$$

1. Check connection to beam flange:

Bolt capacity (single shear) = 1.7(9.67) = 16.4 kips (AISC–2.8)

$$\text{Load per bolt} = \frac{218}{14} = 15.6 \text{ kips} < 16.4 \text{ kips} \qquad \text{OK}$$

Shear stress on web of 2-ft length of WT6 × 25:

$$= \frac{218}{24(0.371)} = 24.5 \text{ ksi} < 27.5 \text{ ksi} \qquad \text{OK}$$

$$F_v = 0.55F_y = 27.5 \text{ ksi} \quad \text{(AISC-2.5)}$$

2. Check connection to column flange:

$$\text{Bolt capacity (tension)} = 1.7(17.6) = 30 \text{ kips}$$

$$\text{Load per bolt} = \frac{218}{8} = 27.2 \text{ kips} < 30 \text{ kips} \qquad \text{OK}$$

Prying force, using Eq. 13.6.37, and checking Eq. 13.6.39,

$$\frac{wt^4}{6nm^2A_b} = \frac{14(1.102)^4}{6(1.94)(1.62)^2(0.4418)} = 1.528$$

$$\frac{n}{m}\left(\frac{n}{3m} + 1\right) = \frac{1.62}{1.94}\left(\frac{1.62}{3(1.94)} + 1\right) = 1.067$$

$$Q = \left(\frac{0.500 - 0.306}{1.067 + 1.528}\right)P_u = \left(\frac{0.194}{2.595}\right)P_u = 0.075P_u$$

$$P_u + Q = P_u + 0.075P_u = 1.075P_u = 1.075(27.2) = 29.2 \text{ kips}$$

$$(1.15)(\text{proof load}) = 1.15(28) = 32.2 \text{ kips} > 29.2 \qquad \text{OK}$$

3. Check bending on tee flange at the column connection:
   Moment along bolt line (Fig. 13.6.12b):

$$Qn = 2.04(1.94) = 3.96 \text{ in.-kips/bolt}$$

Moment at base of fillet (section *b–b* of Fig. 13.6.13):

$$T_f m - Q(n + m) = 29.2(1.62) - 2.04(3.46) = 40.25 \text{ in.-kips/bolt}$$

$$\text{Reqd } Z = \frac{M}{F_y} = \frac{4(40.25)}{50} = 3.22 \text{ in.}^3$$

$$\text{Provd } Z = \tfrac{1}{4}wt_f^2 = \tfrac{1}{4}(14)(1.102)^2 = 4.25 \text{ in.}^3 > 3.22 \qquad \text{OK}$$

Not discussed in this example is the development of the shear capacity of the W14 × 53. A pair of angles may be attached to the beam web for the purpose of providing whatever shear is required. The final design is shown in Fig. 13.6.13.

End-Plate Connections. This simple type of moment resisting connection consists of a plate welded to the end of the beam and then bolted to the column, as shown in Fig. 13.6.2d.

Behavior of this type of connection has been studied by several researchers,[13–15] and the following discussion is based on their work.

In order to determine the adequacy of the bolts used, which are subject to combined bending (tension) and shear, the methods used are those described in Sec. 4.9.

The methods are summarized below:

(a) Neglect Initial Tension (Fig. 4.9.6). This method uses a neutral axis for flexure based on individual bolt areas acting in tension and the contact area of the joined pieces acting in compression. This is recommended for bolted joints without significant pretension; not for high-strength bolted joints.

(b) Consider Initial Tension (Fig. 4.9.8). This method is based on the assumption that because of initial tension the entire contact area of the pieces being joined is effective so that the neutral axis is taken at the centroid of the bolt group. This is recommended for working stress design of high-strength bolted joints.

(c) For plastic design, an ultimate strength approach has been illustrated[18] which assumes the bolts in tension are acting at their initial pretension value (proof load for A325 bolts) and the contact surfaces in the compression region are stressed to yield stress. As shown in Sec. 4.8, when the applied load equals the pretension load there will be little if any compression remaining between the contact surfaces.

For situations where end plates and bolts extend beyond the flanges, Ref. 15 indicates the bolts chiefly effective are those immediately adjacent to the tension flange. For this case, "the bolts and the part of the end plate symmetrical about the tension flange of the beam may be treated as an equivalent T-stub connection and proportioned to develop the force in the beam flange," using the methods discussed earlier in this section.

When end plate connections do not extend outside the tension flange, it is proposed[15] that the end plate thickness be determined to "resist a moment equal to the beam flange force times the distance from the center of the beam flange to the nearest row of bolts." It is further suggested that the stress computed in this fashion could be permitted to reach the yield stress when the beam flange reaches yield stress.

EXAMPLE 13.6.5

Design an end plate connection to develop the full moment and shear capacity of a W14 × 53 of A36 steel. Use ⅞ in. diam. A325 bolts in a bearing-type joint.

Solution

(a) Working-stress design:

$$M = 0.66F_yS_x = 24(77.8)/12 = 155.5 \text{ ft-kips}$$

If a plate is used which does not extend beyond the beam, estimate the moment arm for the connectors to be about 8 in.

$$\text{Force} = \frac{155.5(12)}{8} = 233 \text{ kips}$$

$$R_t = 0.6013(40) = 24.05 \text{ kips/bolt (allowable)}$$

$$\text{Number of bolts for tension} \approx 10$$

Try 12 bolts in tension region and 4 at compression side as shown in Fig. 13.6.14.

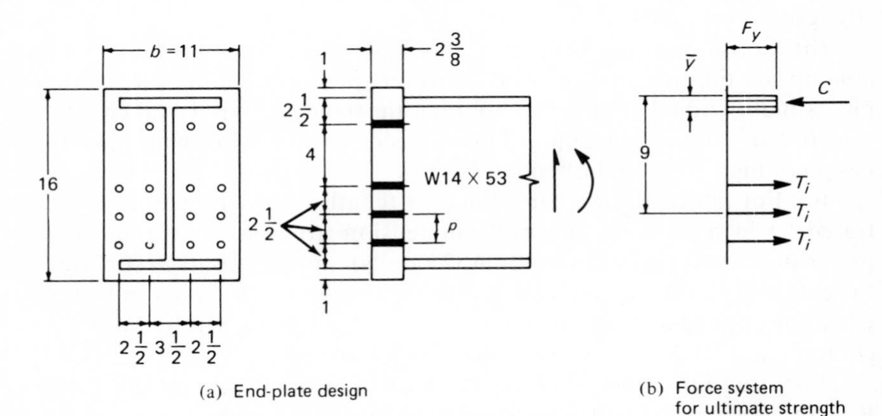

Fig. 13.6.14. Example 13.6.5. End plate connection.

In accordance with the concept of considering initial tension, the stress due to external moment on the plate at the extreme bolt, using Eq. 4.9.23, is

$$f_t = \frac{My}{I} = \frac{155.5(12)4.5}{(11)(16)^3/12} = 2.23 \text{ ksi}$$

The nominal tensile force in a bolt equals

$$f_t bp/4 = 2.23(11)(2.5)/4 = 15.4 \text{ kips} < 24.05 \qquad \text{OK}$$

Plate thickness:

$$\text{Flange force} = 8.062(0.658)(24)\left(\frac{13.28}{13.94}\right) = 121 \text{ kips}$$

$$M = 121(2.17) = 263 \text{ in.-kips}$$

$$S = \tfrac{1}{6}bt^2 = 11t^2/6$$

$$F_b = 0.75F_y = 27 \text{ ksi}$$

$$t = \sqrt{\frac{263(6)}{27(11)}} = 2.31 \text{ in.,} \quad \text{say } 2\tfrac{3}{8} \text{ in.}$$

$2\frac{1}{2} - \frac{0.658}{2} = 2.17$

$A_f\,(0.66F_y)$

M

Flange

First bolt line

Use 16 ⅞ in. diam. A325 bolts, as shown in Fig. 13.6.14, with a $2\frac{3}{8} \times 11 \times 1'\text{-}4''$ end plate.

(b) Plastic Design. Using an ultimate strength approach with the compression region stressed to F_y, and taking the tensile bolts acting at their pretension value (proof load for A325), the neutral axis is located so that $C = T$,

$$F_y b\bar{y} = 12T_i$$

$$36(11)\bar{y} = 12(39); \qquad \bar{y} = 1.18 \text{ in.}$$

$$M_u = \left[36(11)\frac{(1.18)^2}{2} + 4(39)(6.5 + 9 + 11.5)\right]\frac{1}{12} = 374 \text{ ft-kips}$$

The bolted-connection capacity exceeds the plastic-moment capacity of the beam:

$$M_p = 87.1(36)/12 = 261 \text{ ft-kips} < 374 \text{ ft-kips} \qquad \text{OK}$$

AISC–2.8 allows a tensile capacity under factored load of 1.7 times the working-stress value, i.e., 1.7(24.05) = 40.9 kips. This is slightly greater than the proof load used so the computed M_u is a reasonable estimate of the capacity based on the bolts.

Plate thickness:

$$\text{Flange force} = A_f F_y = 8.062(0.658)\,36 = 191 \text{ kips}$$

$$M = 191(2.17) = 415 \text{ in.-kips}$$

$$Z = \text{plastic modulus} = \frac{bt^2}{4} = \frac{11t^2}{4}$$

$$\text{Reqd } t = \sqrt{\frac{415(4)}{36(11)}} = 2.05 \text{ in.} < 2\tfrac{3}{8} \text{ used} \qquad \text{OK}$$

Shear: If the full shear capacity must simultaneously be carried,

$$V_u = 0.55F_y d t_w = 0.55(36)(13.94)(0.370) = 102 \text{ kips}$$

$$f_v = \frac{V_u}{(\text{No. of bolts})(\text{area})} = \frac{102}{16(0.6013)} = 10.6 \text{ ksi}$$

$$F_t' = 1.7(50 - 1.6f_v) = 1.7[50 - 1.6(10.6)] = 56.2 \text{ ksi}$$

The allowable tensile force per bolt in the presence of shear is

$$R_t' = 0.6013(56.2) = 33.8 \text{ kips} < 39 \text{ kips used}$$

Reducing the tensile force in bolts used for computing M_u gives

$$36(11)\bar{y} = 12(33.8); \qquad \bar{y} = 1.03 \text{ in.}$$

$$M_u = 322 \text{ ft-kips}$$

which still exceeds M_p and is satisfactory.

13.7. CONTINUOUS BEAM-TO-BEAM CONNECTIONS

When beams frame transversely to other beams or girders they may be attached to either or both sides of the girder web using simple framed beam connections (Sec. 13.2) or using simple seats in combination with a

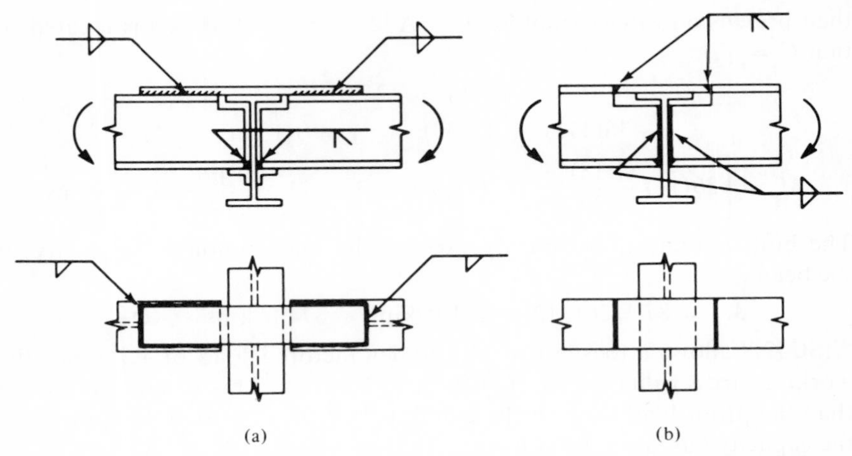

Fig. 13.7.1. Intersecting beam connections: tension flanges *not attached* to each other.

framed beam connection (Sec. 13.3). When full continuity of the beam is desired, the connection must develop a higher degree of rigidity than provided by framed connections. For beam-to-beam connections the principal objective is to provide a means of allowing the tensile force developed in one beam flange to be carried across to the adjacent beam framing opposite the girder web. These connections may be divided into two categories; (a) those with intersecting tension flanges not rigidly attached to one another, as in Fig. 13.7.1; and (b) those with rigidly attached intersecting tension flanges, as in Fig. 13.7.2.

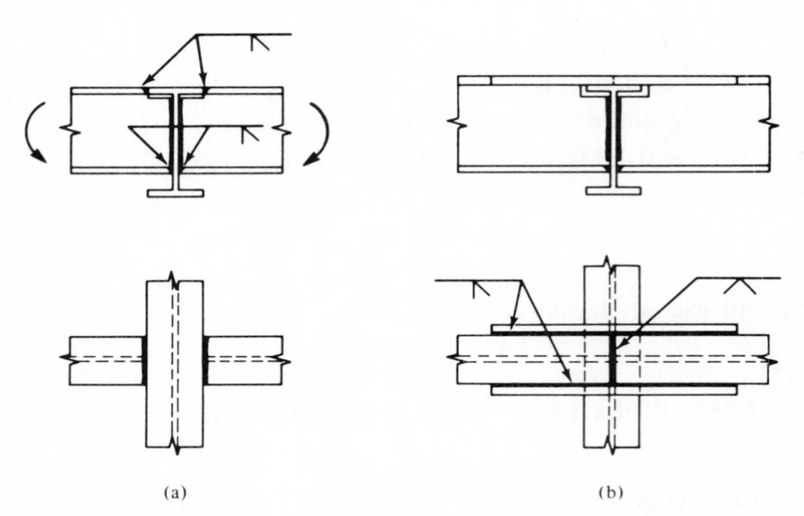

Fig. 13.7.2. Intersecting beam connections: tension flanges *attached* to each other.

When the intersecting tension flanges are not rigidly attached (Fig. 13.7.1) the connection design is essentially a tension member design at the tension flange along with a shear connection.

On the other hand, the designer should be cautious when designing intersecting beams with tension flanges attached, since this case becomes one of biaxial instead of uniaxial stress. With biaxial stress the possibility of brittle fracture increases (see Sec. 2.10). In addition, a biaxial stress yield criterion must be used, such as Eq. 2.7.2, to establish safety:

$$\sigma_y^2 = \sigma_1^2 + \sigma_2^2 + \sigma_1\sigma_2 \qquad [2.7.2]$$

where σ_1, σ_2 are the principal stresses acting. When the beams frame at 90° to one another the axial stresses in the flanges are principal stresses.

The designer should make certain the σ_y of Eq. 2.7.2 does not exceed $0.60F_y$ for working stress design, nor F_y for plastic design.

13.8. RIGID-FRAME KNEES

In the design of rigid frames according to either the working-stress or the plastic-design method, the safe transmission of stress at the junction of beam and column is of great importance. When members join with their webs lying in the plane of the frame, the junction is frequently referred to as a knee joint. Typical knee joints are (a) the square knee, with or without a diagonal or other stiffener (Fig. 13.8.1a and b); (b) the square knee with a bracket (Fig. 13.8.1c); (c) the straight haunched knee (Fig. 13.8.1d); and (d) the curved haunched knee (Fig. 13.8.1e).

Working-stress design ordinarily assumes the span of the members to extend from center to center of adjacent knees, and that the moment of inertia of the member varies in accordance with the moment of inertia of a cross section taken at right angles to lines extending center-to-center of knees. Moments and shears may then be determined using statically indeterminate analysis of a frame (involving variable moment of inertia if knees are haunched).

In plastic design using square knees without brackets or haunches, the plastic hinge forms in the member and the knee is proportioned to prevent failure occurring in that region. When haunches are used it is possible for the plastic hinge to occur either within or outside of the haunch region.

Considerable research has been conducted on rigid frame knees and the design concepts have been summarized in ASCE Manual No. 41,[18] which forms the basis for what follows.

To be adequately designed, a knee connection must (a) transfer the end moment between the beam and the column, (b) transfer the beam-end shear into the column, and (c) transfer the shear at the top of the column into the beam. Furthermore, in performing the three functions relating to

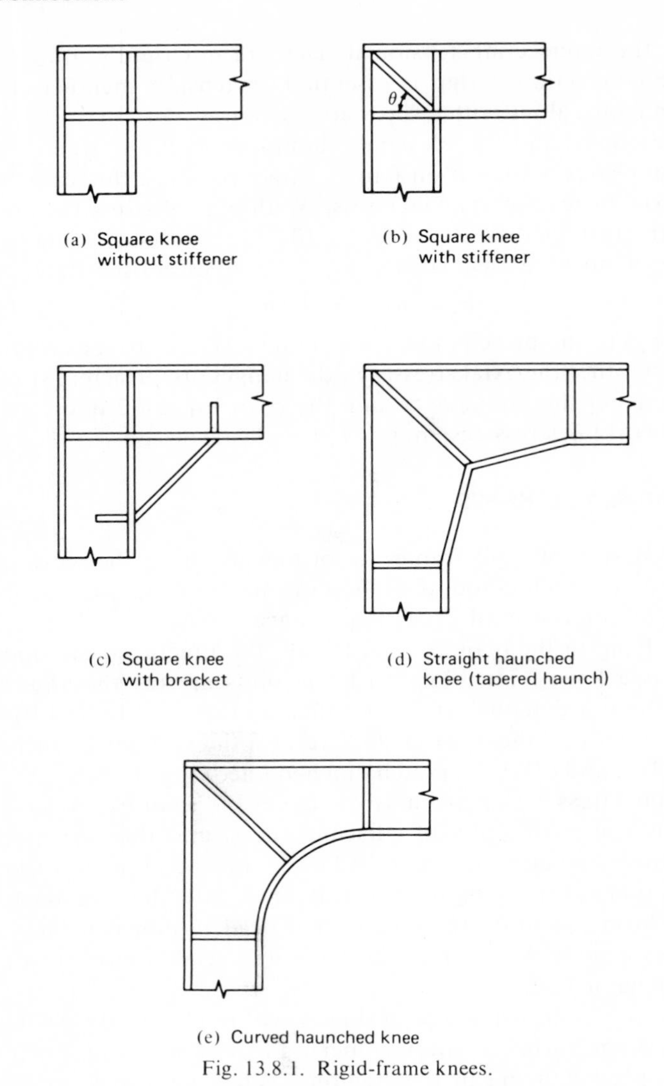

Fig. 13.8.1. Rigid-frame knees.

strength, the knee must deform in a manner consistent with the analysis by which moments and shears were determined.

If a plastic hinge associated with the failure mechanism is expected to form at or near the knee, adequate rotation capacity must be built into the connection. Square knees have the greatest rotation capacity but are also the most flexible (i.e., deform elastically the most under service load conditions). Curved knees are the stiffest but have the least rotation capacity. Since straight tapered knees provide reasonable stiffness along with adequate rotation capacity, in addition to the fact that they are cheaper than curved haunches to fabricate, the straight haunched knees are commonly used.

Shear Transfer in Square Knees Without Brackets or Haunches. In the design of a rigid frame having square knees, two rolled sections may come together at right angles as shown in Fig. 13.8.1a. A frame analysis, either elastic or plastic, will have established what moments and shears act on the boundaries of the square knee region, as shown in Fig. 13.8.2a. The forces carried by the flanges must be transmitted by shear into the web, as shown in Fig. 13.8.2b.

Assuming that all bending moment is carried by the flanges, and approximating the distance between flange centroids to be $0.95d_b$, the flange force is

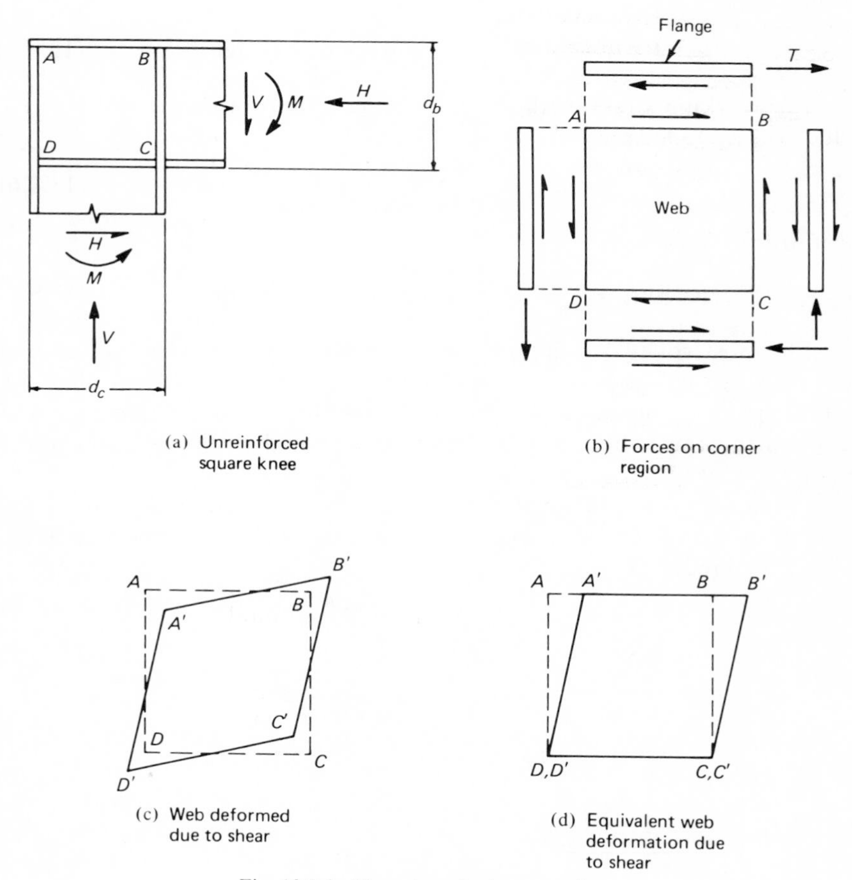

Fig. 13.8.2. Shear transfer in square knee.

$$T = \frac{M}{0.95d_b} \qquad (13.8.1)$$

The shear capacity of the web across AB is

$$V_{ab} = F_v t_w d_c \qquad (13.8.2)$$

Equating Eqs. 13.8.1 and 13.8.2 and solving for t_w gives

$$t_w = \frac{M}{F_v(0.95d_b)(d_c)} \qquad (13.8.3)$$

For plastic design, the value for F_v is the shear yield stress, taken as $0.55F_y$ by AISC–2.5. Eq. 13.8.3 then becomes

$$\text{Reqd } t_w = \frac{1.053M}{0.55F_y d_b d_c} = \frac{1.92M}{F_y d_b d_c} \qquad (13.8.4)$$

or, letting $A_{bc} = d_b d_c$ = area within the corner, and adjusting for M to be in ft-kips,

$$\text{Reqd } t_w = \frac{23M}{A_{bc}F_y} \quad \text{(for plastic design)} \qquad (13.8.5)$$

For working stress method, the maximum allowable shear is $0.40F_y$, which in Eq. 13.8.3 gives

$$\text{Reqd } t_w = \frac{1.053M}{0.40F_y d_b d_c} = \frac{2.64M}{F_y d_b d_c} \qquad (13.8.6)$$

or if M is in ft-kips,

$$\text{Reqd } t_w = \frac{32M}{A_{bc}F_y} \quad \text{(for working-stress design)} \qquad (13.8.7)$$

For most rolled W sections the required web thickness will exceed that provided by the section so that reinforcement is usually needed. A doubler plate is sometimes used to lap over the web to obtain the total required thickness, but usually a pair of diagonal stiffeners is the most practical solution.

When diagonal stiffeners are used, the horizontal component of the stiffener force, $T_s \cos\theta$, participates with the web. Equilibrium (Fig. 13.8.3) then requires

$$T = V_{ab} + T_s \cos\theta \qquad (13.8.8)$$

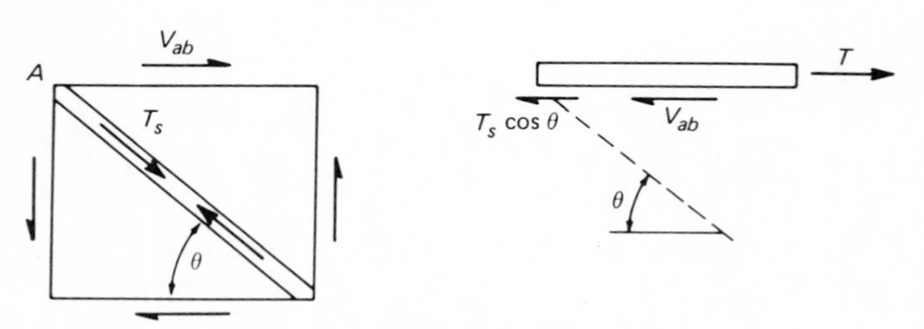

Fig. 13.8.3. Diagonal stiffener effect.

or

$$\frac{M}{0.95d_b} = F_v t_w d_c + A_{st} F_a \cos\theta \tag{13.8.9}$$

Solving for the stiffener area A_{st} gives

$$\text{Reqd } A_{st} = \frac{1}{\cos\theta}\left(\frac{M}{0.95d_b F_a} - \frac{F_v t_w d_c}{F_a}\right) \tag{13.8.10}$$

For plastic design, $F_a = F_y$ and $F_v = 0.55F_y$, which gives, for Eq. 13.8.10,

$$\text{Reqd } A_{st} = \frac{1}{\cos\theta}\left[\frac{1.053M}{d_b F_y} - 0.55t_w d_c\right] \tag{13.8.11}$$

For working stress method, $F_a = 0.60F_y$ assuming no instability, and $F_v = 0.40F_y$, which gives

$$\text{Reqd } A_{st} = \frac{1}{\cos\theta}\left[\frac{1.76M}{d_b F_y} - 0.67t_w d_c\right] \tag{13.8.12}$$

The moment M has inch units in Eqs. 13.8.11 and 13.8.12.

EXAMPLE 13.8.1

Design the square-knee connection to join a W27 × 94 girder to a W14 × 74 column. Use plastic design with a corner moment of 376 ft-kips under factored load (LF = 1.7). Use A36 steel and E70 electrodes with shielded metal-arc welding.

Solution

Refer to Fig. 13.8.4.

(a) Check whether or not diagonal stiffener is required. Using Eq. 13.8.5,

$$\text{Reqd } t_w = \frac{23M}{A_{bc}F_y} = \frac{23(376)}{26.91(14.19)\,36} = 0.63 \text{ in.}$$

$$\text{Actual } t_w = 0.49 \text{ in. for W27} \times 94 < 0.63 \text{ in.}$$

A diagonal stiffener is required.

(b) Determine stiffener size.

$$\tan\theta = \frac{26.91}{14.19} = 1.895; \quad \cos\theta = 0.467$$

Using Eq. 13.8.11,

$$\text{Reqd } A_{st} = \frac{1}{0.467}\left[\frac{1.053(376)(12)}{26.91(36)} - 0.55(0.49)\,14.19\right]$$

$$= \frac{1}{0.467}(4.91 - 3.82) = 2.34 \text{ sq in.}$$

<u>Use 2 PLs ½ × 2½.</u> The local-buckling requirement of AISC–2.7 is satisfied.

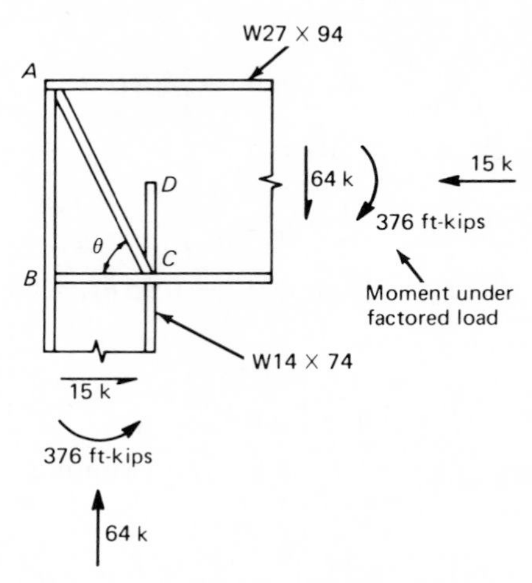

Fig. 13.8.4. Square knee of Example 13.8.1.

(c) Determine fillet weld size along length AB. This weld must be adequate to transmit the plastic moment flange force into the beam web.

$$\text{Flange force} = F_y A_f = 36(10.072)(0.783) = 302 \text{ kips}$$

$$\text{Fillet weld capacity} = a(0.707)(21)(2)(1.7) = 50.5a$$

Note that for plastic design the allowable stress is multiplied by the load factor 1.7.

$$\text{Available length for weld} = 26.91 - 2(0.747) = 25.41 \text{ in.}$$

$$\text{Reqd } a = \frac{302}{25.41(50.5)} = 0.24 \text{ in.}$$

Use $\frac{1}{4}$-in. E70 fillet weld along length AB (both sides of girder web).

(d) Determine fillet weld size along length BC. The connection of the column web to the beam flange must carry the stress due to flexure and axial load combined with shear. At the most highly stressed location, assuming a stress equal to yield stress resisted by fillets on each side of web,

$$\text{Direct component} = F_y t_w = 36(0.450) = 16.2 \text{ kips/in.}$$

$$\text{Shear component} = \frac{V}{d_c - 2t_f} = \frac{15}{14.19 - 2(0.783)} = 2.4 \text{ kips/in.}$$

$$\text{Resultant loading} = \sqrt{(16.2)^2 + (2.4)^2} = 16.4 \text{ kips/in.}$$

$$\text{Reqd } a = \frac{16.4}{2(0.707)(21)(1.7)} = 0.32 \text{ in.}$$

Use $\frac{3}{8}$-in. E70 fillet weld along length BC (both sides of girder web).

(e) Determine weld required along stiffener. This weld must develop the stiffener capacity, which in plastic design may be computed using the yield stress:

$$T_s = F_y A_{st} = 36(2.5)(0.5)\,2 = 90 \text{ kips}$$

$$\text{Reqd } a = \frac{90}{2(0.707)(21)\,1.7} = 0.18 \text{ in.}$$

Use $\frac{1}{4}$-in. E70 fillet weld along diagonal AC (both sides of girder web).

(f) Determine extent of stiffener required from point C vertically into girder. The capacity of the W27 × 94 web to carry the concentrated load from the W14 × 74 flange at C is, using AISC Formula 1.15–1,

$$\text{Web capacity} = F_y t_w(t_b + 5k)$$
$$= 36(0.49)[0.783 + 5(1.50)] = 146 \text{ kips}$$

Since the diagonal stiffener is already performing the task of transferring the shear resulting from moment on the W27 × 94, it may not be included in the resistance of the beam web against the column flange compression force. Thus a vertical stiffener CD must carry

$$\text{Stiffener } CD \text{ force} = 302 - 146 = 156 \text{ kips}$$

If $\frac{5}{16}$-in. weld is used, the stiffener length CD must be

$$\text{Length} = \frac{156}{2(\frac{5}{16})(0.707)(21)\,1.7} = 9.9 \text{ in.}$$

Use 10 in. for distance CD. Use 2 PLs —$\frac{7}{16}$ × $4\frac{1}{2}$ × 0′-10″, which taper from full width at point C to zero width at point D.

Straight Haunched Knees. Straight haunched knees, sometimes called *tapered haunches*, may in many cases extend over a significant portion of the span; in which case they are not really connections but rather they are an integral part of a variable depth frame. The design of tapered haunches involves consideration of three factors; (a) bending moment along the tapered region; (b) shear stresses and flange force transmission within and adjacent to the haunched section; and (c) local and lateral-torsional buckling resistance.

Examine the straight haunched knee of Fig. 13.8.5 showing a moment diagram over the haunched portion. Use of a haunch complicates a plastic-strength frame analysis because it increases the number of possible plastic hinge locations. For working stress method no special disadvantage arises. Even though the moment variation may not be linear along the haunch, a linear assumption satisfying the moment requirements for M_1 and M_2 will be satisfactory in flexure at intermediate points.

The following development relates to plastic design requirements as suggested in ASCE Manual No. 41,[19] and the 1963 AISC Specification

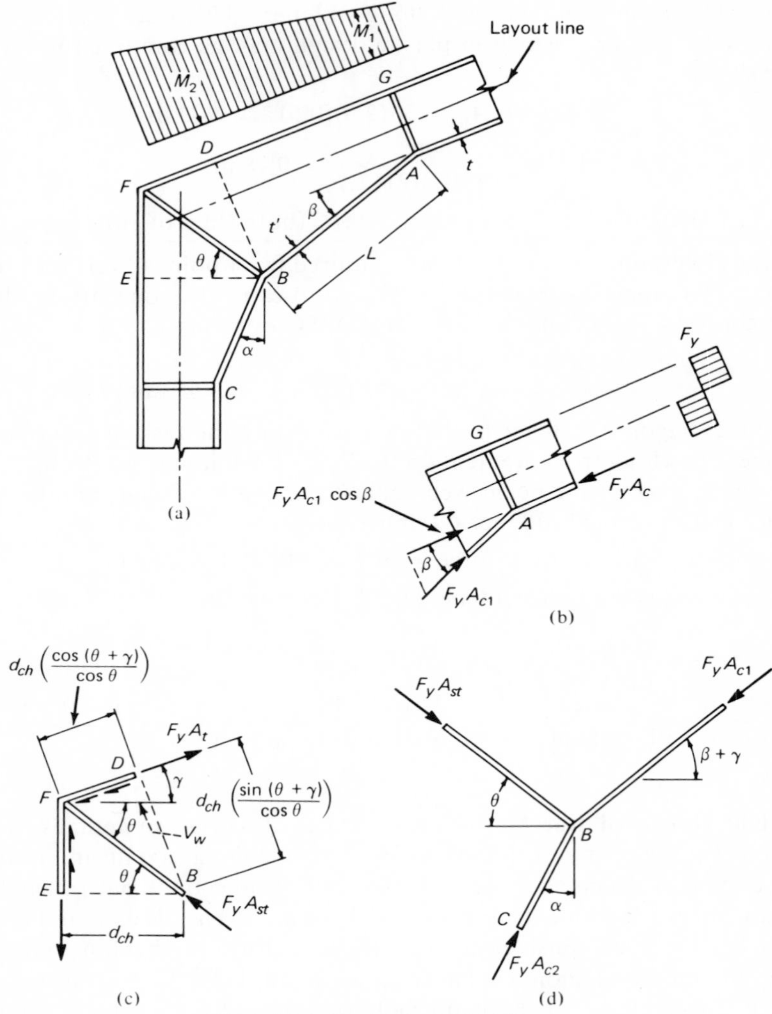

Fig. 13.8.5. Straight haunched knee.

Commentary, as well as Refs. 6 and 18. For working stress design, the reader is referred to Ref. 20; however, the general philosophy is similar to that of plastic design.

The design for flexure requires that the sections at moments M_1 and M_2 have a plastic moment capacity exceeding those moments. For computation, the resisting sections are taken perpendicular to the layout line; i.e., AG for resisting M_1 and BD for resisting M_2. Since the compressive force in the flange of the haunch acts in the direction AB the moment provided at any section is proportional to the component in the direction parallel to the layout line; thus the effective area is $A_{c1}\cos\beta$. It is usually desirable to increase the flange area A_{c1} in the haunched region to give equal flange forces so that the resisting section may be considered symmetrical. Thus the required area of the flange in the haunch region AB is

$$A_{c1} = \frac{A_c}{\cos\beta} \tag{13.8.6}$$

where A_c = compression flange area outside of haunch region. If the plate width is kept the same (as it usually should be) within and outside the haunch region; then

$$t' = \frac{t}{\cos\beta} \tag{13.8.7}$$

When the angle β is less than 20° the increase in area required is less than 6 percent and most practical design would neglect such a difference.[19]

The design for shear in the corner is similar to that for a square knee without haunch, except the region to be considered is $EBDF$ and the trigonometry is more complicated. The flange force at D is partly carried as shear on the web over the distance FD. A stiffener should be used along BF, and its size may be determined considering force equilibrium at F (tension in flanges) and at B (compression in flanges). The larger required stiffener force satisfying equilibrium will determine stiffener size.

Consider first the equilibrium of forces at the tension corner (corner at F), as shown in Fig. 13.8.5c. Summation of horizontal forces gives

$$F_y A_{st}\cos\theta + V_w\cos\gamma = F_y A_t\cos\gamma \tag{13.8.8}$$

$$A_{st} = \frac{\cos\gamma}{\cos\theta}\left(A_t - \frac{V_w}{F_y}\right) \tag{13.8.9}$$

Since plastic strength in shear is taken as $0.55F_y$, the shear force V_w carried by the web over the distance FD is

$$V_w = 0.55F_y t_w d_{ch}\left[\frac{\cos(\theta+\gamma)}{\cos\theta}\right] \tag{13.8.10}$$

which upon substitution into Eq. 13.8.9 gives

$$\text{Reqd } A_{st} = \frac{\cos\gamma}{\cos\theta}\left[A_t - 0.55t_w d_{ch}\left(\frac{\cos(\theta+\gamma)}{\cos\theta}\right)\right] \tag{13.8.11}$$

based on the tension corner.

Consider next the equilibrium of the forces at the compression corner (corner at B), as shown in Fig. 13.8.5d. Summation of horizontal forces gives

$$F_y A_{st}\cos\theta + F_y A_{c2}\sin\alpha = F_y A_{c1}\cos(\beta+\gamma) \tag{13.8.12}$$

$$\text{Reqd } A_{st} = \frac{A_{c1}\cos(\beta+\gamma) - A_{c2}\sin\alpha}{\cos\theta} \tag{13.8.13}$$

In compression there is no significant amount of shear transfer into the web, so that the small shear forces along CB and BA are neglected. The stiffener size is based on the larger requirement of Eqs. 13.8.11 and 13.8.13. Because of the web participation near F, and its lack of participation near B, equilibrium at B usually controls—i.e., Eq. 13.8.13.

Regarding lateral bracing, AISC–2.9, indicates that for severe cases where bending moment is nearly constant along an unbraced length and equal in magnitude to the plastic moment (i.e., rotational moments give $-0.5 > M/M_p > -1.0$):

$$L_{cr} = \frac{1375}{F_y} r_y \tag{13.8.14}$$

For $F_y = 36$ ksi, and taking $r_y \approx b_f/\sqrt{12}$ for rectangular flanges, gives

$$L_{cr} = \frac{38.2}{\sqrt{12}} b_f = 11 b_f \tag{13.8.15}$$

It would seem AISC–2.9 applies for this situation even though the 1963 AISC Commentary indicated that for A36 steel L_{cr} should not exceed $6b_f$. The 1969 AISC Commentary does not specifically refer to tapered haunches and indicates AISC–2.9 should be applied to all lateral bracing situations in connections near plastic hinges.

Curved Haunched Knees. For better appearance, curved haunches are sometimes used. Detailed treatment of curved haunches may be found in Refs. 6 and 19.

13.9. COLUMN BASES

The design of column bases involves two main considerations. The compressive forces in the column flanges must be distributed by means of a base plate to the supporting medium so that bearing stresses are within permissible values allowed by specifications. The second concern is with the connection, or anchorage, of the base plate and column to the concrete foundation. For frame analysis it may be of importance to evaluate the moment-rotation characteristics of the entire anchorage, including the base plate, anchor bolts, and concrete footing, to determine its stiffness and degree of fixity.

Requirements for Bases Under Axial Load. The method presented and terminology used is that from the AISC Manual, "Column Base Plates." This procedure is not required by the AISC specification but may be considered as accepted good practice. The dimensions and loads are given in Fig. 13.9.1.

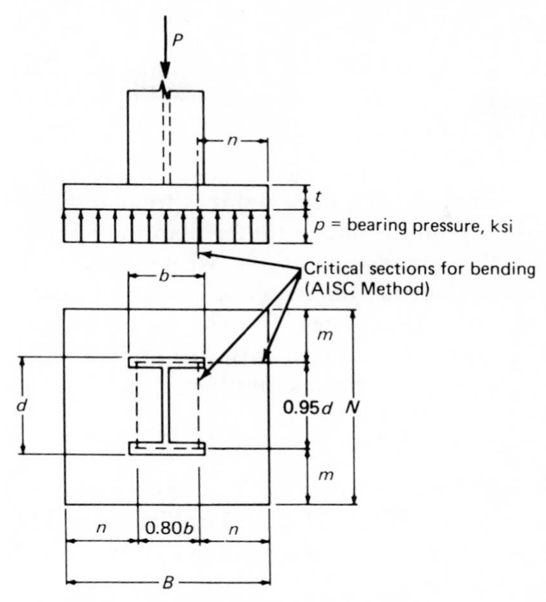

Fig. 13.9.1. Column base plate dimensions.

It is assumed the stress distribution under the base plate is uniform and that the plate projections beyond the critical sections act as cantilever beams.

The bending moments on cantilevers of spans m and n are respectively

$$M = \frac{pNn^2}{2} \quad \text{(on section parallel to column web)} \tag{13.9.1}$$

$$= \frac{pBm^2}{2} \quad \text{(on section parallel to column flanges)} \tag{13.9.2}$$

The unit stress is

$$f = \frac{M}{S} = \frac{pNn^2/2}{Nt^2/6} = \frac{3pn^2}{t^2} \tag{13.9.3}$$

or

$$f = \frac{3pm^2}{t^2} \tag{13.9.4}$$

The larger of Eqs. 13.9.3 or 13.9.4 governs.

AISC–1.5.1.4.3 allows a stress of $0.75F_y$ for bending on solid rectangular sections. Thus the required plate thickness can be found:

$$f = \frac{3pm^2}{t^2} \le 0.75F_y \tag{13.9.5}$$

$$\text{Reqd } t = \sqrt{\frac{3pm^2}{0.75F_y}} \quad \text{or} \quad \sqrt{\frac{3pn^2}{0.75F_y}} \tag{13.9.6}$$

EXAMPLE 13.9.1

Design a base plate for a W14 × 142 column of A36 steel to carry 775 kips axial compression resulting from dead load, live load, and wind. Assume allowable unit pressure under plate is 1.0 ksi.

Solution

Once the column has been designed it is frequently best to design the base plate for the column capacity rather than the applied load, in order to avoid having the connection as the weakest element. The column capacity, of course, depends on the effective (equivalent pinned) length. AISC tables, "Column Base Plates," provide base plate sizes assuming the shortest practical effective length, in most cases 6 ft.

(a) Determine plan dimensions for plate:

$$\text{Reqd } A = \frac{775}{1.33(1.0)} = 582 \text{ sq in.}$$

Note the allowable stress in increased 33⅓ percent for loading which includes wind in accordance with AISC–1.5.6. The plate should be approximately square, with slight differences in dimensions B and N to give nearly equal values for m and n.

$$0.80b = 0.80(15.50) = 12.40 \text{ in.}$$

$$0.95d = 0.95(14.75) = 14.00 \text{ in.}$$

Try $B = 24$ in. and $N = 25$ in.

Provd $A = 25(24) = 600$ sq in. > 582 OK

(b) Determine plate thickness:

$$n = 0.5(24 - 12.40) = 5.8 \text{ in.} \qquad \text{Controls}$$

$$m = 0.5(25 - 14.00) = 5.5 \text{ in.}$$

$$p = \frac{775}{600} = 1.29 \text{ ksi}$$

$$\text{Reqd } t = \sqrt{\frac{3pn^2}{0.75F_y}} = \sqrt{\frac{3(1.29)(5.8)^2}{27(1.33)}} = 1.91 \text{ in.}$$

Use PL 2 × 24 × 2′-1″. Note AISC–1.21.3 which states that plates over 2 in. but not over 4 in. in thickness may be straightened by rolling or pressing. When over 4 in. thick, extra thickness must be specified so that the bearing surface can be planed.

Column Bases to Resist Moment. Column bases frequently must resist moment in addition to axial compression. The situation has some similarities to the behavior of bolted connections discussed in Chapter 4, and in many respects is analogous to the situation of reinforcing bars in concrete construction. The axial force causes a precompression between the base plate and the contact surface (frequently a concrete wall or footing).

When the moment is applied the precompression on the tension side in flexure is reduced, often to zero, leaving only the anchor bolt to provide the tensile force resistance. On the compression side, the contact area remains in compression. The anchorage will have an ability to undergo rotational deformation depending primarily on the length of anchor bolt available to deform elastically. Also, the behavior is influenced by whether or not the anchor bolts are given an initial pretension (similar to the installation of high-strength bolts as discussed in Chapter 4). The moment-rotation characteristics of column anchorages are treated in detail in Ref. 21.

A number of elaborate methods are available for designing moment resisting bases, with variations depending on the magnitude of the eccentricity of loading and the specific details of the anchorage. Some simple details are shown in Fig. 13.9.2.

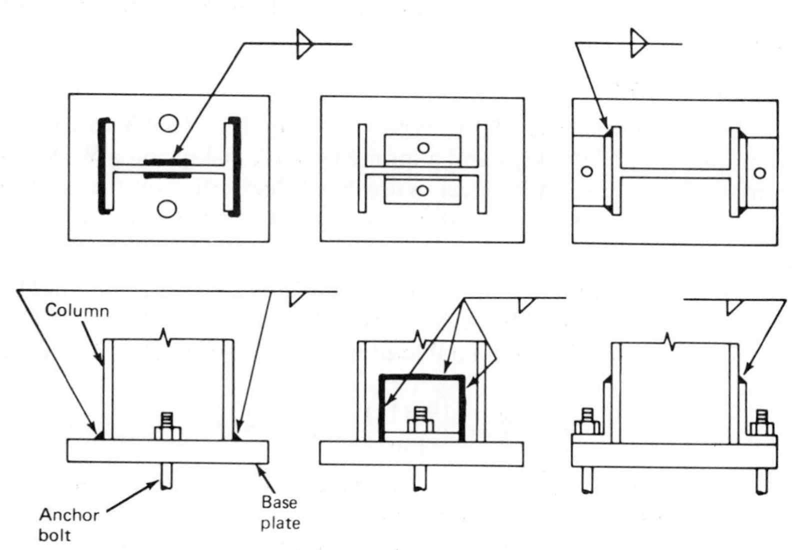

Fig. 13.9.2. Column bases.

When the eccentricity of loading, $e = M/P$, is small such that it does not exceed $\frac{1}{6}$ of the plate dimension N in the direction of bending (i.e., within the middle third of the plate dimension), the ordinary combined stress formula applies. Thus for small e

$$f = \frac{P}{A} \pm \frac{M}{S} \tag{13.9.7}$$

where $M = Pe$

$$S = Ar^2/(N/2) = AN/6$$
$$r^2 = N^2/12$$

$$f = \frac{P}{A} \pm \frac{6Pe}{AN} = \frac{P}{A}\left[1 \pm \frac{6e}{N}\right] \tag{13.9.8}$$

where N = plate dimension in the direction of bending. Eq. 13.9.8 is correct for $e \leq N/6$ when there is no bolt pretension, and is considered satisfactory for practical purposes at least up to $e = N/2$ without serious error.

When the eccentricity e exceeds $N/6$, part of the base plate at the tension face becomes inactive and the stress distribution becomes as shown in Fig. 13.9.3. A simple practical assumption to use for such a situation is that the resultant of the triangular distribution is located directly under the compression flange of the column. When large moment is designed for, generally the attachment arrangement to the base of the column becomes more complex than those of Fig. 13.9.2. For a more detailed presentation on designing column bases for moment, the reader is referred to Ref. 6 (Sec. 3.3, pp. 1–32).

EXAMPLE 13.9.2

Design a column base of the type shown in Fig. 13.9.3 to carry an axial compression of 150 kips and a moment of 190 ft-kips on a W14 × 84 column section. Use A36 steel with E70 electrodes in the shielded metal-arc process for welding, and the working-stress method.

Solution

(a) Determine number and size of anchor bolts needed. Try a 24-in. by 28-in. plate as suggested by AISC tables, "Column Base Plates." These are intended for much heavier axial load but without moment. The eccentricity of load is

$$e = \frac{M}{P} = \frac{190(12)}{150} = 15.2 \text{ in.}$$

Since $e > N/6$, or 15.2 in. > 28/6 in. there is some tendency for uplift and the compressive stress distribution will be as shown in Fig. 13.9.3. Assuming the compressive force resultant lies directly beneath the compression flange for bending, the internal couple moment arm is 15.7 in. if the anchor bolts are centered 2 in. from column face. Taking moments about the compression flange gives for the required tensile force

$$T = \frac{150(8.5)}{15.7} = 81.2 \text{ kips}$$

$$\text{Reqd } A_{\text{net}} = \frac{T}{F_a} = \frac{T}{0.60F_y} = \frac{81.2}{22} = 3.69 \text{ sq in.}$$

From AISC table, "Threaded Fasteners," select for tension three $1\frac{1}{2}$ in. diam. anchor bolts. Provd $A_{\text{net}} = 3(1.41) = 4.23$ sq in.

<u>Use 6 $1\frac{1}{2}$ in. diam. A36 threaded rods for anchor bolts.</u>

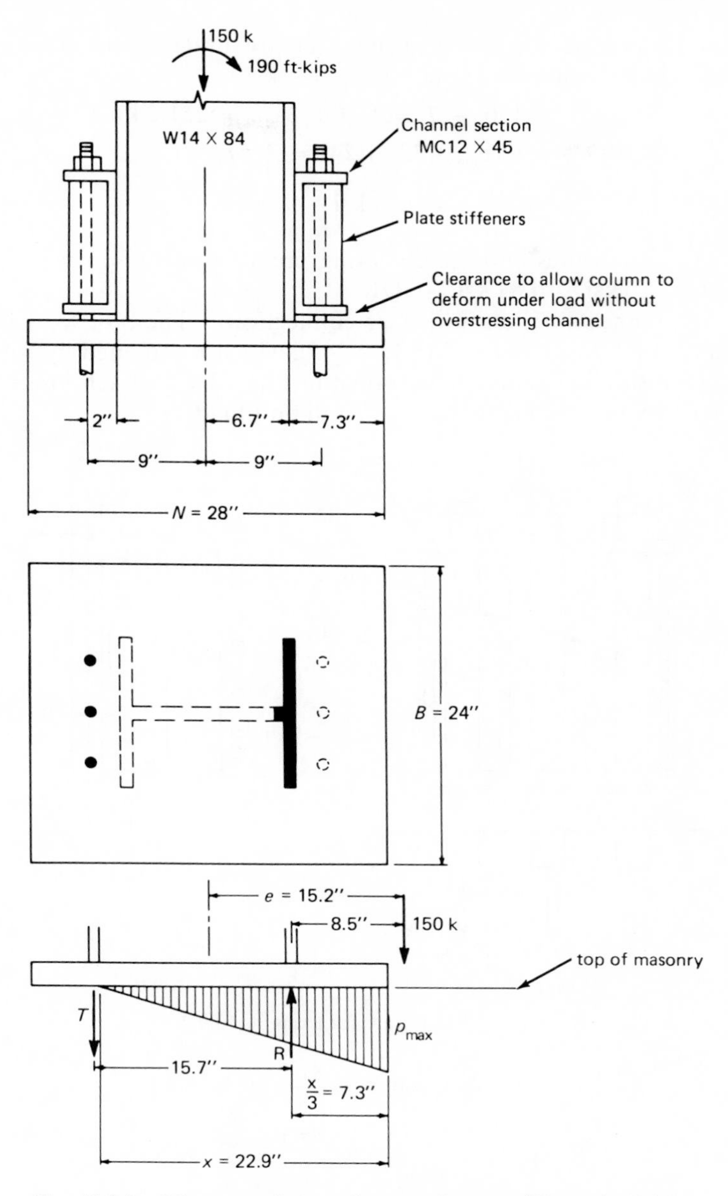

Fig. 13.9.3. Moment-resisting column anchorage of Example 13.9.2.

(b) Determine whether bearing pressure under plate is acceptable. Assume 28-day concrete strength is $f'_c = 4$ ksi.

$$R = 150 + T = 150 + 81.2 = 231.2 \text{ kips}$$

$$R = \tfrac{1}{2}(p_{max})(22.9)(24) = 275 p_{max}$$

$$p_{max} = \frac{231.2}{275} = 0.84 \text{ ksi}$$

AISC–1.5.5 indicates allowable bearing pressure on masonry is $0.25 f'_c = 0.25(4) = 1.0 \text{ ksi} > 0.84 \text{ ksi}$ OK

(c) Channel stiffeners and their connections. The bolt load is transmitted through the channel (MC12 × 45), which is stiffened by four plate stiffeners inside the channel, as shown in Fig. 13.9.4. Each interior plate stiffener may be assumed to carry the full bolt load.

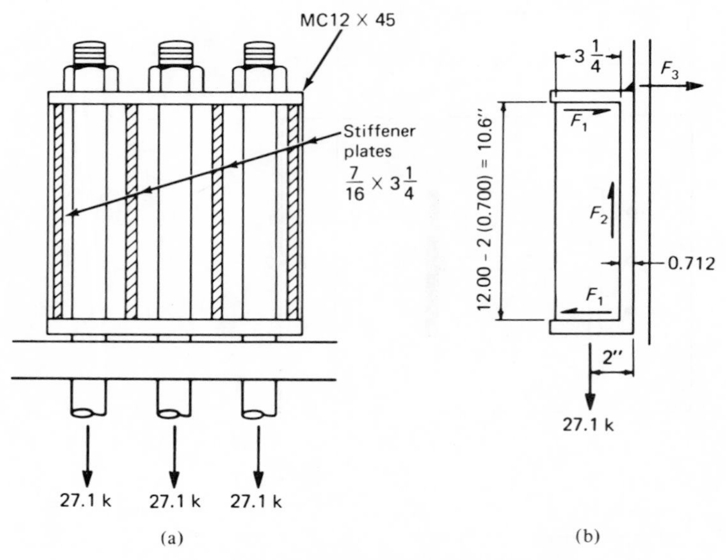

Fig. 13.9.4. Details for Example 13.9.2.

$$\text{Load to stiffener} = \frac{81.2}{3} = 27.1 \text{ kips}$$

$$\text{Reqd } A_{st} = \frac{27.1}{0.60 F_y} = \frac{27.1}{22} = 1.23 \text{ sq in.}$$

<u>Use PLs $\frac{7}{16} \times 3\frac{1}{4}$ as stiffeners.</u>

As shown in Fig. 13.9.4b, horizontal weld forces F_1 and a vertical weld force F_2 combine to resist the 27.1 kips acting with an eccentricity of 2 in. from channel web:

$$f'_y = \frac{P}{L} = \frac{27.1}{2(3.25) + 10.60} = 1.585 \text{ kips/in.}$$

$$\bar{x} \text{ from channel web } = \frac{(3.25)^2}{2(3.25) + 10.60} = 0.618 \text{ in.}$$

Using Table 5.16.1, case 5,

$$I_p = \frac{8(3.25)^3 + 6(3.25)(10.60)^2 + (10.60)^3}{12} - \frac{(3.25)^4}{2(3.25) + 10.625}$$

$$= 304.5 - 6.5 = 298 \text{ in.}^3$$

$$f_y'' = \frac{Mx}{I_p} = \frac{27.1(2.0 - 0.712)(3.25 - 0.618)}{298} = 0.308 \text{ kips/in.}$$

$$f_x'' = \frac{My}{I_p} = \frac{27.1(2.0 - 0.712)(5.3)}{298} = 0.620 \text{ kips/in.}$$

$$f_r = \sqrt{(1.585 + 0.308)^2 + (0.620)^2} = 1.99 \text{ kips/in.}$$

Using fillets on both sides of stiffener plates the required size is

$$\text{Reqd } a = \frac{f_r}{R_w} = \frac{1.99}{2(0.707)21} = \frac{1.99}{29.6} = 0.067 \text{ in.}$$

<u>Use minimum fillet size ¼ in.</u> in accordance with AISC–1.17.5.

As an alternative one may assume the forces F_1 carry the moment and the force F_2 carries the shear. Such an approach will give

$$F_2 = 27.1 \text{ kips}; \qquad F_1 = \frac{27.1(2)}{10.6} = 5.12 \text{ kips}$$

$$f_y' = \frac{27.1}{10.6} = 2.55 \text{ kips/in.}$$

$$f_x'' = \frac{5.12}{3.25} = 1.58 \text{ kips/in.}$$

Using ¼ in. fillet on both sides of stiffener,

$$R_w = \tfrac{1}{4}(0.707)(21)2 = 7.4 \text{ kips/in.}$$

Either approach is considered acceptable. The strength requirement is low so that intermittant weld (say several 1½ in. segments) would be acceptable. Weld should be made on both sides of stiffener so as to avoid eccentricity with respect to the plane of the stiffener.

(d) Connection to the column flange. The sum of the forces F_1 must be carried by a fillet weld connecting the channel flange to the column web, indicated in Fig. 13.9.4 by the force F_3.

$$F_3 = \frac{81.2(2)}{12} = 13.53 \text{ kips}$$

The length available for weld is the flange width of the W14 × 84; i.e., about 12 in. Using ¼ in. weld, the length required to be welded is

$$L_w = \frac{13.53}{\tfrac{1}{4}(0.707)21} = \frac{13.53}{3.71} = 3.65 \text{ in.}$$

<u>Use 4 in. of ¼ in. fillet weld to carry force F_3.</u>

The shear to be carried by weld along the channel web is 81.2 kips. Using ¼-in. fillet, the length of weld required is

$$L_w = \frac{81.2}{2(3.71)} = 10.95 \text{ in. (12 in. available)}$$

Use ¼-in. continuous fillet weld along the channel web to connect to the column flange.

This example has been presented to illustrate some of the reasoning which may be used to design the welds for a column base. The stiffened channel arrangement represents but one of many possible and commonly used means of transmitting the bolt forces into the column. When only one or two anchor bolts are required on each side, often angles are used as anchor bolt sleeves with their toes welded directly to the column flange. Frequently wing plates are used in various positions to better distribute the loads over the base plate.

13.10. BEAM SPLICES

There are several reasons why a rolled beam or plate girder may be spliced, such as: (a) the full length may not be available from the mill; (b) the fabricator may find it economical to splice even when the full length could be obtained; (c) the designer may desire to use splice points as an aid to cambering; and (d) the designer may desire a change in section to fit the variation in strength required along the span.

On each side of a joint, one must provide for the transverse shear and bending moment. For welded plate girders, and frequently for rolled beams, the splice may be accomplished by a full-penetration groove weld. Splices of material made entirely in the shop are nearly always groove welded. Field splices are becoming more frequently all welded, though usually using lapping of pieces and fillet-welded connections, instead of groove welding, where dimensional control is a critical factor. Since full-penetration-groove welded splices are as strong as the base material, they require no further comment here.

Many field welded splices for beams and welded plate girders use lapping splice plates and high-strength bolts as connectors. In this article the four- and eight-plate beam splices are treated.

Prior to the almost universal use of welding to fabricate plate girders, riveted or bolted girders had splices made of many plates through which a complex load distribution had to be considered. Treatment of such splices is adequately presented in older texts* and because of their rare current use are omitted here.

*See Thomas C. Shedd, *Structural Design in Steel* (New York: John Wiley & Son, Inc., 1934), pp. 142–166.

Splices are designed for the moment M and the shear V which occur at the spliced point, or they are designed for some arbitrary or specification prescribed higher values. AISC–1.10.8 requires groove welded splices to "develop the full strength of the smaller spliced section." "Other types of splices in cross sections of plate girders and in beams shall develop the strength required by the stresses, at the point of splice." Earlier AISC specifications required splices to develop not less than 50 percent of the effective strength of the material spliced. AASHO–1969 requires designing for "not less than 75% of the strength of the member."

For beam splices, each element of the splice is designed to do the work which the material underlying the splice plates could do if uncut. Plates on the flange do the work of the flange, while plates over the web do the work of the web. One typical arrangement of a four-plate splice is shown in Fig. 13.10.1. When the beam flange is thick and there are

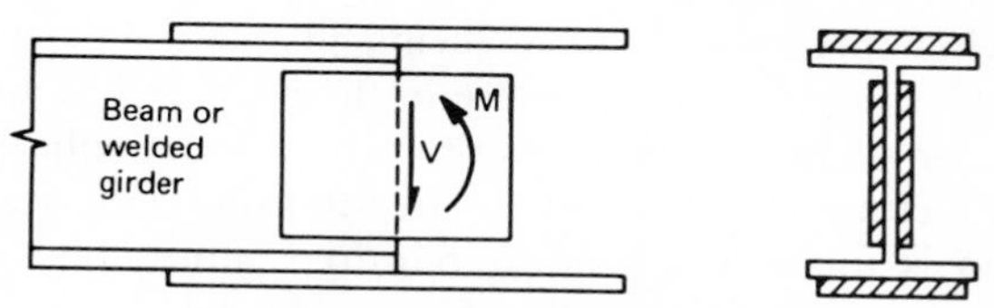

Fig. 13.10.1. Four-plate beam splice.

large forces to be transmitted, plates on both inside and outside of the flanges may be desired, so the eight-plate splice as shown in Fig. 13.10.2 may be used. In the eight-plate splice the flange splices can be designed so that the centroid of the splice area coincides with the flange centroid.

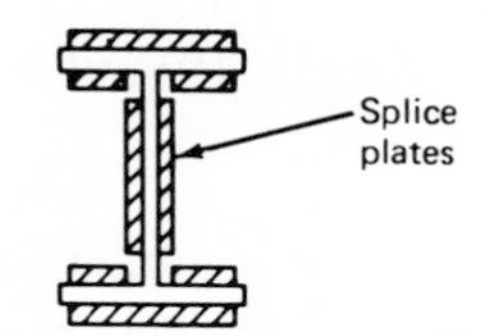

Fig. 13.10.2. Cross section of eight-plate splice.

When determining the forces to be designed for, one recognizes that the splice actually covers a finite distance along the span (perhaps 1 or 2 ft). Along this distance the moment and shear vary. In accordance with the principles used for designing bolted connections, the forces designed for should be those acting on the center of gravity of the bolt group. From Fig. 13.10.3 it may be noted that theoretically the moment at the centroid of the connectors on one side of the splice is different from that at the centroid on the other. Some designers, therefore, propor-

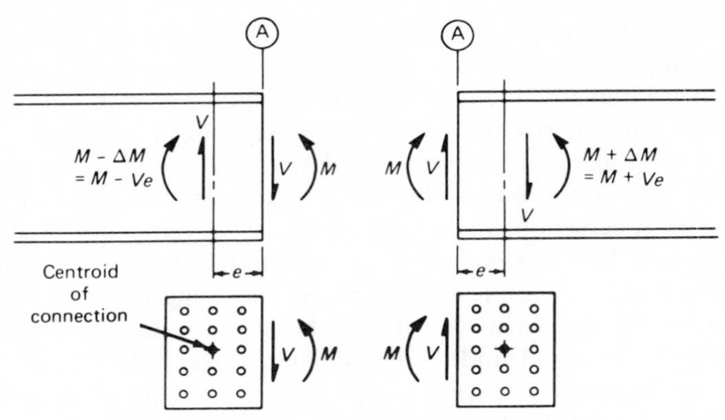

Fig. 13.10.3. Forces acting on web splice plates.

tion the connection on each side of the splice for a moment; $M_1 = M + Ve$. Such a procedure rarely seems justified. In those unusual situations when a splice is located where both shear and bending moment are high, such a procedure might seem desirable. Most splices are located in regions where either the shear or the bending moment are low so that often the design is for a specification required minimum strength. As discussed in Sec. 10.6 for plastic design, one must be wary of designing a splice for a low moment just because the splice is near an inflection point. *If moments for the structure were computed using theory of statically indeterminate structures without a hinge within the span, one should not later design a splice which has low stiffness to act as a hinge.*

For most situations the authors recommend designing for the actual or specification required forces *at the splice* and neglecting any eccentricity effect. Furthermore, if one considers the reduced effective eccentricity as given by Eq. 4.7.14, the effective value will be nearly zero for most situations.

EXAMPLE 13.10.1

Design a rolled-beam splice (four-plate type of Fig. 13.10.1) for a W24 × 84 beam of A36 steel to be located where the bending moment is 270 ft-kips and the shear is 70 kips. Use A36 splice material with ¾ in. diam. A325 bolts in a bearing-type joint. In addition to AISC requirements, assume the splice must have not less than 50 percent of the capacity of the material spliced.

(a) Total capacity of W24 × 84.

$$M = F_b S_x = 0.66 F_y S_x = 24(197)/12 = 394 \text{ ft-kips}$$

$$V = F_v d t_w = 0.40 F_y d t_w = 14.5(24.09)(0.47) = 164 \text{ kips}$$

(b) Design conditions.

Actual $M = 270$ ft-kips $>$ 50 percent of 394 ft-kips

Design M = 270 ft-kips
Actual V = 70 kips < 50 percent of 164 kips
Design V = 82 kips

Even though the design is to be made for the combination of 270 ft-kips moment and 82 kips vertical shear acting simultaneously, it is probable that these two values (or even 270 ft-kips and 70 kips) do not occur under the same loading condition. In an actual structure the designer will or can evaluate the chances of the loads acting simultaneously and consider them accordingly.

(c) Web plates. The web plates are selected to carry the shear on the gross section (AISC–1.10.1).

$$\text{Reqd } A_g = \frac{V}{0.4F_y} = \frac{82}{14.5} = 5.65 \text{ sq in.}$$

Available depth = T dimension = 21 in.

$$\text{Reqd } t = \frac{A_g}{2(\text{depth})} = \frac{5.65}{2(21)} = 0.134 \text{ in.}$$

A thickness of ¼-in. should be considered a practical minimum.

Use 2 PLs ¼ × 21.

(d) Flange plates (Method 1). In this approach, flange splice area is provided which will do the work the unspliced flange was required to do.

$$A_f(\text{W24} \times 84) = 9.015(0.772) = 6.96 \text{ sq in.}$$

$$\text{Percent of full capacity reqd} = \frac{270}{392.6}(100) = 68.8\%$$

$$\text{Reqd } A_{\text{splice plate}} = 6.96(0.688) = 4.79 \text{ sq in.}$$

AISC–1.10.1 requires deduction for area of holes exceeding 15 percent of gross flange.

Try PL $\frac{9}{16}$ × 9, A_g = 5.06 sq in. Use 2 lines of fasteners.

Ordinary deduction for 2 holes = $2(\frac{3}{4} + \frac{1}{8})(\frac{9}{16})$ (principles of Chapter 3)	=	0.985 sq in.
15 percent of gross flange splice plate = 0.15(5.06)	=	0.76
Excess that must be deducted	=	0.225 sq in.

Effective area proved = 5.06 − 0.23 = 4.83 sq in. > 4.79 reqd OK

Use 2 PLs $\frac{9}{16}$ × 9.

It would also be acceptable to consider the increased effectiveness resulting from the splice plate being centered farther from the neutral axis than the flange. In such a treatment,

$$\text{Reqd effective area} = 4.79\left(\frac{23.32}{24.56}\right) = 4.55 \text{ sq in.}$$

assuming a $\frac{9}{16}$-in. plate to get the value 24.56 in. It does not appear that this would justify use of a ½ × 9 plate.

(e) Flange plates (Method 2). In this method, one works with the entire splice. The moment capacity provided by the $\frac{1}{4} \times 21$ web splice plates is

$$M = fS = \frac{20(386)}{10.5(12)} = 61.3 \text{ ft-kips}$$

where $f = 24\left(\frac{10.5}{12.04 + 0.5625}\right) = 20.0$ ksi at edge of plate.

$$I = 2(\tfrac{1}{4})(21)^3/12 = 386 \text{ in.}^4$$

The "flange area formula" developed in Sec. 11.12 as Eq. 11.12.9 may be used:

$$A_f = \frac{M}{fh} - \frac{A_w}{6} \qquad [11.12.9]$$

The first term is the total effect and from it is subtracted the web effect. In this case the moment to be carried by the flange is

$$M_{\text{flange}} = M - M_{\text{web}} = 270.0 - 61.3 = 208.7 \text{ ft-kips}$$

$$\text{Reqd } A_{\text{splice plate}} = \frac{208.7(12)}{23.5(24.56)} = 4.35 \text{ sq in.}$$

Essentially the same as previously computed.

(f) Check PL $\frac{9}{16} \times 9$ using $f = Mc/I$.

I (web plates) .... = 386
I (flat plates) $2(4.83)(12.326)^2$ .... = 1,462
I = 1,848 in.^{4}

$$f = \frac{270(12)(12.61)}{1848} = 22.2 \text{ ksi} < 24 \text{ ksi} \quad \text{Understressed.}$$

Use PLs $\frac{9}{16} \times 9$, since $\frac{1}{2} \times 9$ plates will be overstressed.

(g) Flange bolts. A325 bearing-type joint.

$$R_{SS} = 9.72 \text{ kips;} \qquad R_B > 9.72 \text{ kips}$$

In order to carry the full capacity of the splice material,

$$P = f_{\text{avg}} A_{\text{net}} = 24\left(\frac{12.33}{12.61}\right)(4.83) = 113.3 \text{ kips}$$

$$\text{Number of bolts} = \frac{P}{R_{ss}} = \frac{113.3}{9.72} = 11.7, \quad \text{say } 12.$$

Use 2 rows of 6 bolts, each side of splice.

(h) Web Bolts. The web plates must carry $V = 82$ kips and $M = 61.3$ ft-kips. Using Eq. 4.7.22 to gain an estimate of the connectors required, try 2 rows of 6 bolts each; 5 spaces at approximately 3.6 in. vertically:

$$f_s A = \frac{82}{12} = 6.84 \text{ kips} \downarrow$$

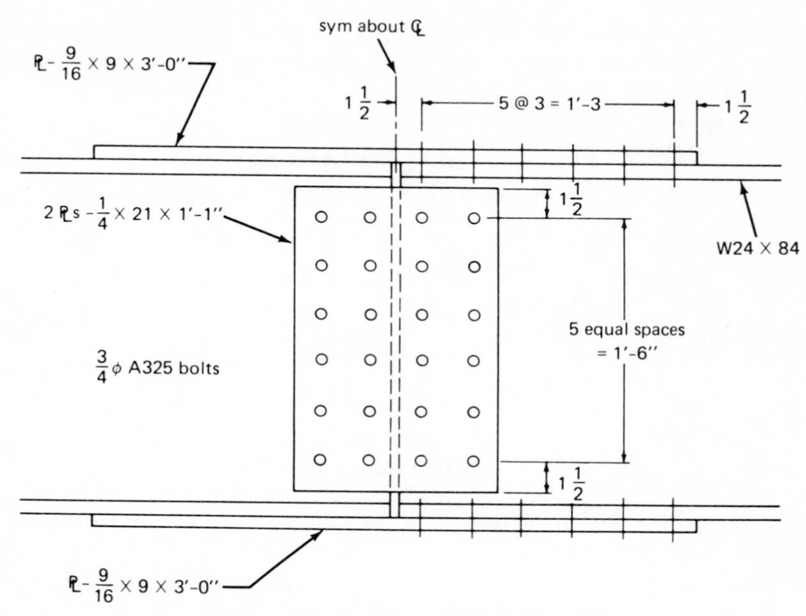

Fig. 13.10.4. Design for Example 13.10.1.

$$f_x A = \frac{My}{\Sigma x^2 + \Sigma y^2} = \frac{61.3(12)9}{481} = 13.75 \text{ kips} \rightarrow$$

where

$$\Sigma x^2 + \Sigma y^2 = 12(1.5)^2 + 4(9)^2 + 4(5.4)^2 + 4(1.8)^2 = 481 \text{ in.}^2$$

$$f_y A = \frac{Mx}{\Sigma x^2 + \Sigma y^2} = \frac{61.3(12)(1.5)}{481} = 2.29 \text{ kips} \downarrow$$

$$\text{Actual } R = \sqrt{(6.84 + 2.29)^2 + (13.75)^2} = 16.5 \text{ kips}$$

For allowable bolt values,

$$R_{DS} = 19.42 \text{ kips}$$

$$R_B = 1.35(36)(0.47)(0.75) = 17.15 \text{ kips} > 16.5 \text{ kips} \qquad \text{OK}$$

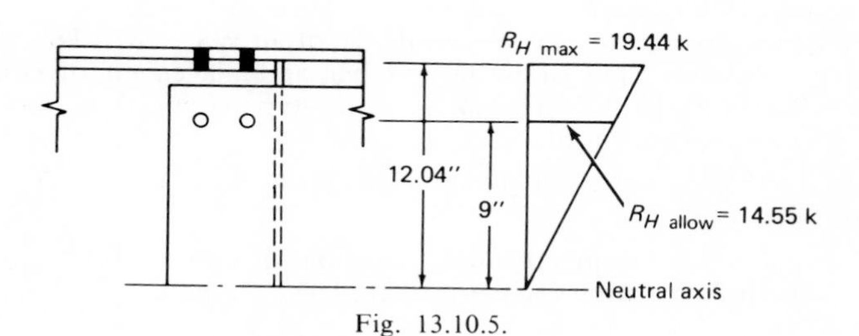

Fig. 13.10.5.

However, since there are bolts in the flange plates at this same section, the allowable horizontal component in the web bolts must be restricted so as not to overstress the flange bolts. The two flange connectors at a section acting in single shear can carry a total load of

$$R_{H\,\max} = 2(9.72) = 19.44 \text{ kips}$$

The allowable value on the web plate fastener most distant from the neutral axis is (see Fig. 13.10.5)

$$R_{H\,\text{allow}} = 19.44\left(\frac{9}{12.04}\right) = 14.55 \text{ kips} > f_x A = 13.75 \qquad \text{OK}$$

<u>Use 2 vertical rows of 6 bolts each, each side of joint.</u>

The final design is shown in Fig. 13.10.4.

PROBLEMS

All problems are to be done in accordance with the latest AISC specification unless otherwise indicated.

13.1. Determine the capacity for the framed beam connection shown. Show all calculations and compare with tabulated values from the AISC Manual.
(a) Friction-type connection.
(b) Bearing-type connection, with threads excluded from shear plane.

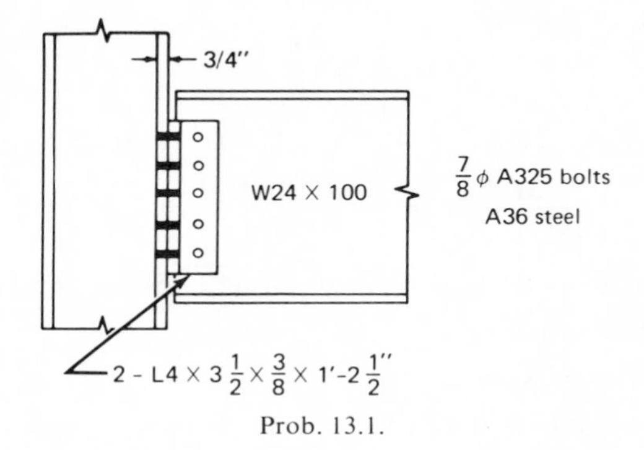

Prob. 13.1.

13.2. Design the framed beam connection for a W16 × 50 beam with a reaction of 50 kips to connect to the column flange of an W8 × 67. Determine the angles to be used, and show the number and placement of connectors. Use A36 steel and A325 bolts in a bearing-type connection, with threads excluded from the shear plane.

13.3. Redesign the framed beam connection of Prob. 13.1 as a friction-type connection.

13.4. Design the welded framed beam connection for a W33 × 118 connecting to a column flange which is $\frac{3}{4}$ in. thick. The beam reaction is 125 kips.

E70 electrodes are to be used with the shielded metal-arc process, and the base steel is A572 Grade 50. Design using basic calculations and then check with applicable AISC Manual tables.

13.5. What is the maximum allowable reaction the W14 × 30 may safely transmit to the seat angle *B*? What size should be used for angle *A* and what is its function? Assume the connection is bearing-type with threads excluded from the shearing plane.

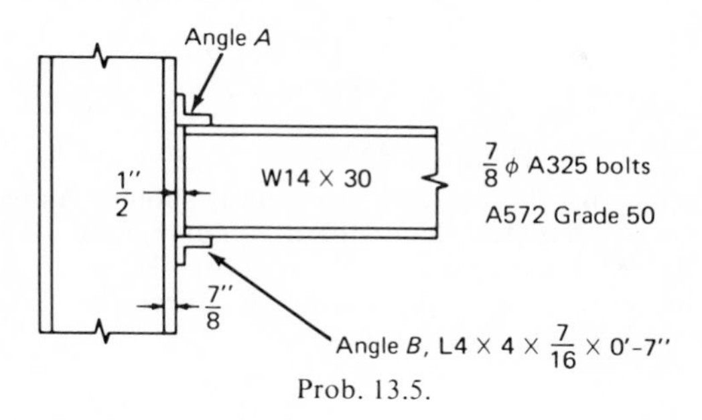

Prob. 13.5.

13.6. For the unstiffened seat of Prob. 13.5, if E70 electrodes are used in the shielded metal-arc process for welding, what is the maximum capacity which could be achieved? Would the 7-in. length of seat angle be satisfactory?

13.7. Assume a W16 × 58 is to have the maximum allowable reaction for a 15-ft simple span, uniformly loaded. Design the unstiffened seat to be used and specify its welded connection. Show all calculations and compare result with that given by the AISC Manual tables. Use A36 steel, and E70 electrodes with the shielded metal-arc process.

13.8. Redesign the unstiffened seat of Prob. 13.7 using ¾ in. diam. A325 bolts in a friction-type connection.

13.9. Design an unstiffened seat angle to support a W10 × 21 beam on the web of a W12 × 65 column. The beam reaction is 22 kips. There is no beam on the opposite side of the web of the column. Use ¾ in. diam. A325 bolts (bearing-type connection with threads excluded from shear planes) and A36 steel. Design by calculations, and do not merely select from AISC Manual. Specify a size for the top clip angle.

13.10. Design a welded stiffened seat similar to that of Fig. 13.4.4, to support a W21 × 82 with a reaction of 130 kips. Use E70 electrodes with the shielded metal-arc process and A572 Grade 60 for base material. The stiffened seat is attached to a ¾-in. flange.

13.11. Design a welded stiffened seat to support an W18 × 64 with a reaction of 90 kips. The beam lies at right angles to the stiffened seat *instead* of parallel to it as in Fig. 13.4.4. Use E70 electrodes, shield metal-arc process, and A572 Grade 55 steel.

13.12. Specify the weld size and the bracket plate thickness for the given situation. Use P = 33 kips, e = 6 in., and L_w = 10 in. Use E70 electrodes in the shielded metal-arc process, and A36 steel for base material.

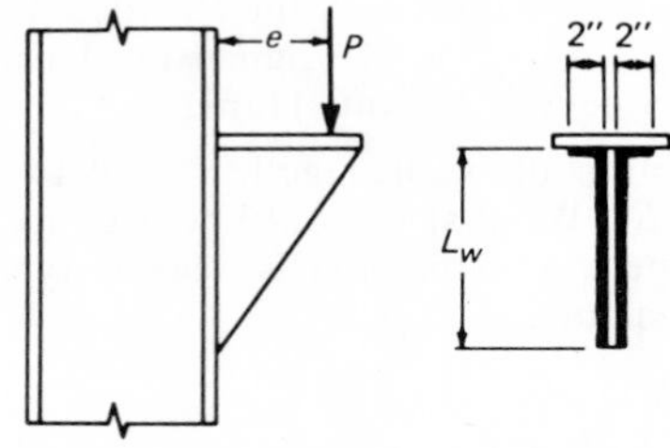

Probs. 13.12 and 13.13.

13.13. Repeat Prob. 13.12 using $P = 45$ kips, $e = 3$ in., and $L_w = 11$ in.

13.14. Design a bolted bracket plate for the loading shown. As part of the design, determine (a) size and length of angles; (b) number and location of fasteners in angles; (c) thickness and dimensions of bracket plate. Neglect concern about dimensions of the seat plate. Connection is bearing-type with threads excluded from the shear plane.

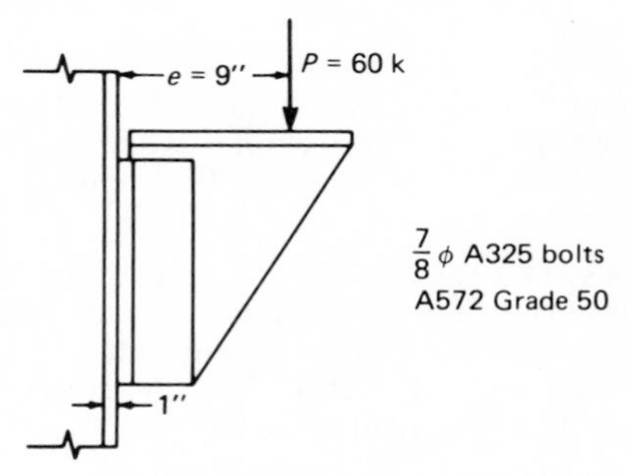

Prob. 13.14.

13.15. Repeat the design of Prob. 13.14 using a welded connection with E70 electrodes and the shielded metal-arc process.

13.16. Check the connector number and determine the bracket thickness to support a load of 20 kips to be applied 12½ in. from centerline of an W8 × 31 column. Use ⅞ in. diam. A325 bolts (friction-type connection) and A36 steel. Use a rational analysis for plate thickness.

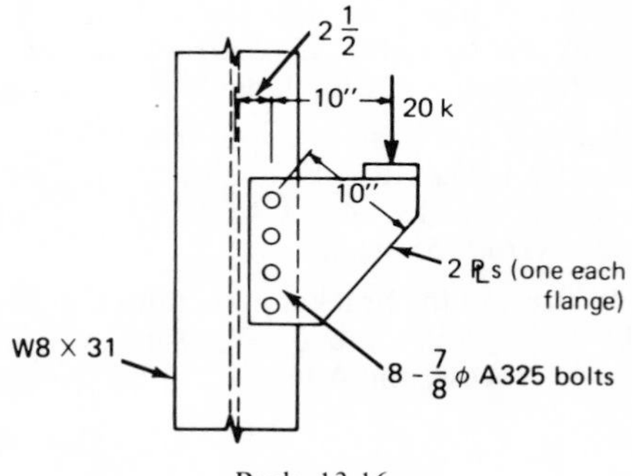

Prob. 13.16.

13.17. Design the continuous beam-to-column connection for W21 × 55 beams to connect to both flanges of a W14 × 84 column. Use the type of connection shown in Fig. 13.6.1b if stiffeners are required. Use the plastic design method assuming a plastic hinge must be able to form in the beams at this connection. Use A36 steel, and E70 electrodes with shielded metal-arc welding. The shear under factored (LF = 1.7) loads is 50 kips.

13.18. Design for the conditions of Prob. 13.17 except use vertical tee stiffeners as shown in Fig. 13.6.1c if stiffeners are required.

13.19. Design for the conditions of Prob. 13.17 except consider the beam to frame in from one side only, and use the type of connection shown in Fig. 13.6.1d.

13.20. The split-beam connection shown is subject to a moment of 80 ft-kips and an end reaction of 30 kips. (a) Determine whether the structural tees are adequate, and if not increase their size. (b) Determine the number of ⅞ in. diam. A325 bolts to connect the tees to the column flanges, the tees to the beam flanges, the clip angles to the column flange, and the clip angles to the beam web. Assume connections are bearing-type and threads are excluded from shear planes. Use working-stress method with A36 steel.

13.21. For the split-beam tee connection with stud beam shown, determine the allowable moment and shear capacity at the face of the column when A325 bolts are used in a bearing-type connection (no threads in shear planes) along with A36 steel. Use working-stress method.

13.22. Design a split-tee connection as shown in Fig. 13.6.2c to connect a W16 × 50 beam to a W14 × 127 column. Use A572 Grade 50 steel and assume the beam's full flexural capacity is to be developed. The shear is 20 kips under service load. Use ¾ in. diam. A325 bolts, or if it seems advantageous use A490 bolts. Use a bearing-type connection with no threads in the shear planes. Use working-stress method.

13.23. Redesign the connection of Prob. 13.22 using an end plate type of connection as shown in Fig. 13.6.2d.

13.24. Design a vertical tee stiffener (as in Figs. 13.6.3 and 13.6.8) connection to frame a W16 × 64 beam into the web of a W12 × 92 column. Use A572

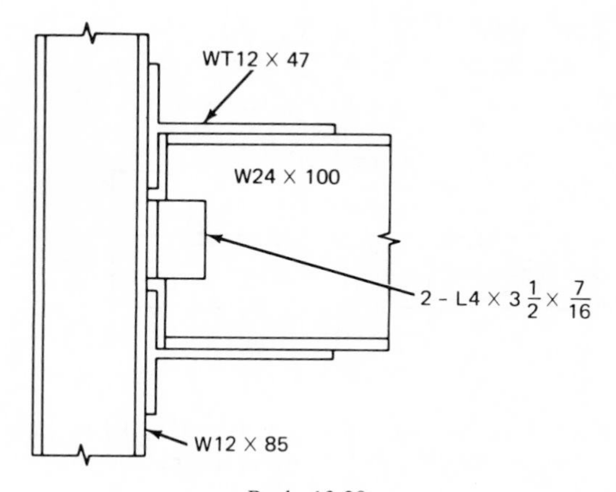

Prob. 13.20.

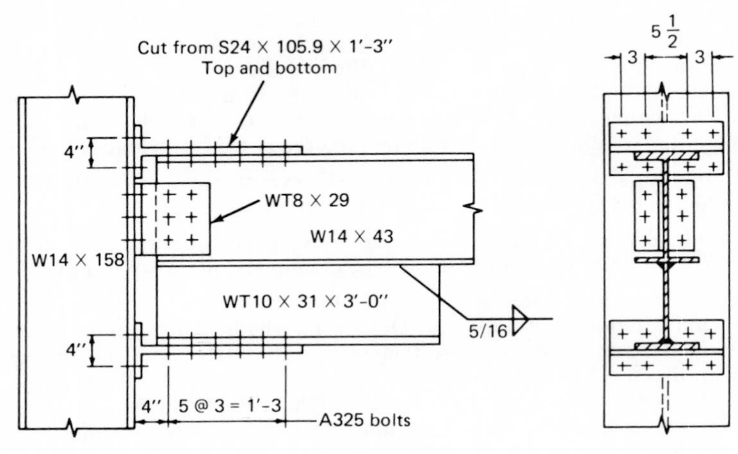

Prob. 13.21.

Grade 60 steel, and shielded metal-arc welding with E80 electrodes. The beam is required to develop maximum flexural capacity and a service load shear of 30 kips. Use working-stress method.

13.25. Redesign the connection of Prob. 13.22 using (a) top plate and seat connection similar to that of Fig. 13.6.11a, and (b) top and bottom plate connection similar to that of Fig. 13.6.11b. For welding use shielded metal-arc process with E70 electrodes.

13.26. Specify the plate size, weld size using E70 electrodes, and length of weld to develop the full continuity of the W10 × 39 beam. Consider only plate *A* and not other aspects of the connection. A36 steel and shielded metal-arc process are used.

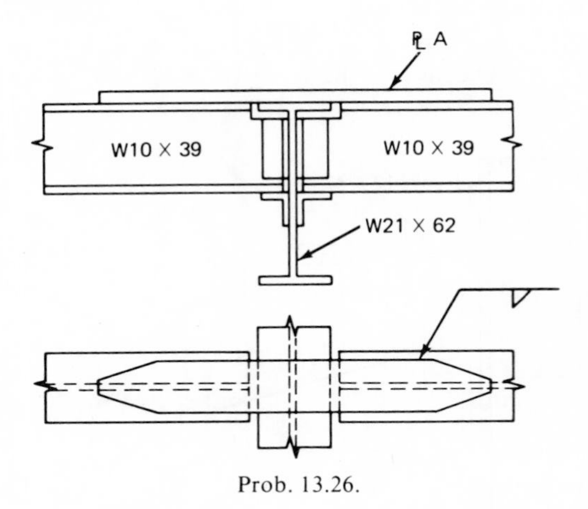

Prob. 13.26.

13.27. Design a square knee connection (Fig. 13.8.1a or b) to join a W24 × 84 girder to a W30 × 108 column. Assume the plastic moment develops in the girder at collapse condition. Use A36 steel and E70 electrodes (shielded metal-arc process). Use plastic design.

13.28. A W14 × 202 of A572 Grade 50 steel has a concentric reaction of 850 kips resulting from dead load plus wind load. Using the AISC Manual design procedure, select a column base plate. Allowable pressure on masonry is 750 psi.

13.29. Design the smallest size base plate for a W12 × 65 column to carry an axial compression load of 220 kips. The load is due to dead load plus live load. Allowable bearing pressure on masonry is 750 psi.

13.30. A W14 × 119 column base connection of the type in Fig. 13.9.3 is subject to a moment of 320 ft-kips and a direct compression of 210 kips. The base plate is $3\frac{1}{2}$ in. by 28 in. by 30 in., with $N = 30$ in. Determine the total force to be resisted by the anchor bolts, the maximum stress in the concrete, and the maximum bending stress in the base plate.

13.31. Design a four-plate beam splice for a W14 × 53 beam of A36 steel. Design for full capacity using working-stress method, with $\frac{3}{4}$ in. diam. A325 bolts in a bearing-type connection with no threads in shear planes.

13.32. Design an eight-plate beam splice as shown in Fig. 13.10.2 for a W27 × 94 section of A36 steel. Use $\frac{3}{4}$ in. diam. A325 bolts (friction-type connection), and design for full shear capacity and 80 percent of full moment capacity. Use working-stress method.

13.33. Design an eight-plate beam splice for a welded girder consisting of $1\frac{7}{8}$ × 22 flanges and a $\frac{7}{16}$ × 78 web. Use a design moment of 4000 ft-kips and design shear of 300 kips. Use working-stress method with A36 steel, and $\frac{7}{8}$ in. diam. A325 bolts in a friction-type connection.

13.34. Design an eight-plate splice for a welded girder consisting of $\frac{7}{8}$ × 24 flanges and a $\frac{5}{16}$ × 96 web of A36 steel. Design for 270 kips shear and 3,000 ft-kips moment. Use working stress method, with $\frac{3}{4}$ in. diam. A490 bolts in a bearing-type connection with no threads in shear plane.

SELECTED REFERENCES

1. C. Batho and H. C. Rowan, "Investigations of Beam and Stanchion Connections," 2nd Report, Steel Structures Research Committee, Dept. of Scientific and Industrial Research of Great Britain, His Majesty's Stationery Office, London, 1934.
2. Basil Sourochnikoff, "Wind Stresses in Semi-Rigid Connections of Steel Framework," *Trans. ASCE*, Vol. 115 (1950), pp. 382–402.
3. Leo Schenker, Charles G. Salmon, and Bruce G. Johnston, *Structural Steel Connections*, Armed Forces Special Weapons Project, Report No. 352, Engineering Research Institute, University of Michigan, June 1954.
4. Bruce Johnston and Lloyd F. Green, "Flexible Welded Angle Connections," *Welding Journal*, Vol. 19, No. 10 (October 1940).
5. Bruce G. Johnston, and G. R. Diets, "Tests of Miscellaneous Welded Building Connections," *Welding Journal*, Vol. 21, No. 1 (January 1942).

6. Omer W. Blodgett, *Design of Welded Structures*, James F. Lincoln Arc Welding Foundation, Cleveland, Ohio, 1966.
7. I. Lyse and N. G. Schreiner, "An Investigation of Welded Seat Angle Connections," *Welding Journal*, Vol. 14, No. 2 (February 1935).
8. Cyril D. Jensen, "Welded Structural Brackets," *Welding Journal*, Vol. 15, No. 10 (October 1936), p. 9-s.
9. Charles G. Salmon, "Analysis of Triangular Bracket-Type Plates," *J. Engg. Mech. Div. ASCE*, Vol. 88, No. EM6 (December 1962), pp. 41–87.
10. Charles G. Salmon, Donald R. Buettner, and Thomas C. O'Sheridan, "Laboratory Investigation of Unstiffened Triangular Bracket Plates," *J. Structural Div. ASCE*, Vol. 90, No. ST2 (April 1964), pp. 257–278.
11. Lynn S. Beedle et al., *Structural Steel Design*, The Ronald Press Company, New York, 1964, pp. 550–555.
12. J. D. Graham, A. N. Sherbourne, R. N. Khabbaz, and C. D. Jensen, "Welded Interior Beam-to-Column Connections," American Institute of Steel Construction, Inc., New York, 1959.
13. Lynn S. Beedle and Richard Christopher, "Tests of Steel Moment Connections," *Engineering Journal*, AISC, Vol. 1, No. 4 (October 1964), pp. 116–125.
14. A. B. Onderdonk, R. P. Lathrop, and Joseph Coel, "End Plate Connections in Plastically Designed Structures," *Engineering Journal*, AISC, Vol. 1, No. 1 (January 1964), pp. 24–27.
15. Richard T. Douty and William McGuire, "High Strength Bolted Moment Connections," *J. Structural Div. ASCE*, Vol. 91, No. ST2 (April 1965), pp. 101–128.
16. R. F. Pray and C. D. Jensen, "Welded Top Plate Beam-Column Connections," *Welding Journal*, Vol. 35, No. 7 (July 1956), p. 338-s.
17. J. L. Brandes and R. M. Mains, "Report of Tests of Welded Top-plate and Seat Connections," *Welding Journal*, Vol. 24, No. 3 (March 1944), p. 146-s.
18. *Plastic Design in Steel*, American Institute of Steel Construction, New York, 1959.
19. *Commentary on Plastic Design in Steel*, Welding Research Council and American Society of Civil Engineers, ASCE Manual of Engineering Practice No. 41, 1961.
20. John D. Griffiths, "Single Span Rigid Frames in Steel," American Institute of Steel Construction, Inc., New York, 1948.
21. Charles G. Salmon, Leo Schenker, and Bruce G. Johnston, "Moment-Rotation Characteristics of Column Anchorages," *Trans. ASCE*, Vol. 122 (1957), pp. 132–154.

14

Frames—Braced and Unbraced

14.1. GENERAL

As discussed in Sec. 6.9, the effective length of the column members in frames is greatly dependent on whether the frame is braced or unbraced. For the braced frame the equivalent length KL is equal to or less than the actual length. For the unbraced frame, the equivalent length KL always is greater than the actual length.

In order to understand frame behavior and be able to extend the

Vierendeel truss (unbraced frame). Courtesy Bethlehem Steel Corporation.

principles to multistory frame design in Chapter 15, consider in Fig. 14.1.1 the forces which arise in a column member of a frame as a result of lateral deflection due to a force such as wind. The moments M_Δ and shears Q_Δ are those portions of the moments and shears required to balance the moment $P\Delta$. In addition, there will be moments and shears due to gravity loads at the particular floor level. Equilibrium in Fig. 14.1.1a requires

$$P\Delta = Q_\Delta h + 2M_\Delta \tag{14.1.1}$$

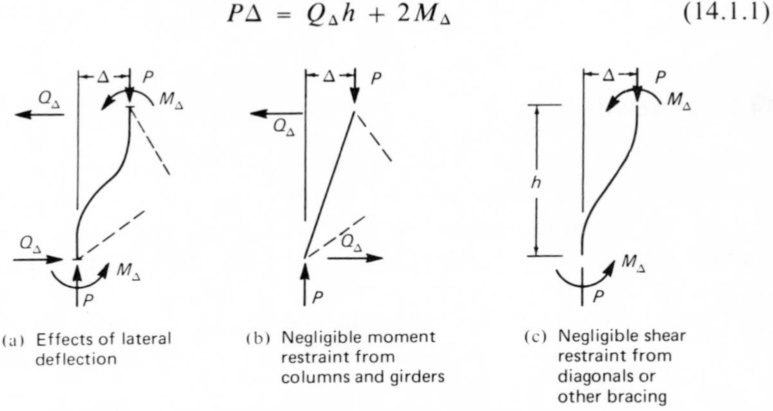

Fig. 14.1.1. Secondary bending moment due to $P\Delta$ in frames.

The lateral deflection Δ is commonly called *drift*[1] when it results from wind loading in multistory frames, as shown in Fig. 14.1.2. Drift consists of two parts; that resulting directly from horizontal load, and that arising from vertical load times the drift.

A frame will deflect under lateral loading such as wind regardless of the pattern of its component members. However, the manner in which equilibrium is maintained against the moment $P\Delta$ differs depending on the restraint conditions. If the building were a vertical pin-jointed struss under lateral loading there would be no continuity at the joints to allow the moment M_Δ to develop. In which case, as in Fig. 14.1.1b,

$$Q_\Delta = \frac{P\Delta}{h} \tag{14.1.2}$$

Diagonal and horizontal members (web members of the truss) would have to carry the entire shear Q_Δ.

On the other hand, if the members are rigidly joined together but without diagonal members, there would be little shear resistance. Neglecting the shear resistance entirely would result in

$$M_\Delta = \frac{P\Delta}{2} \tag{14.1.3}$$

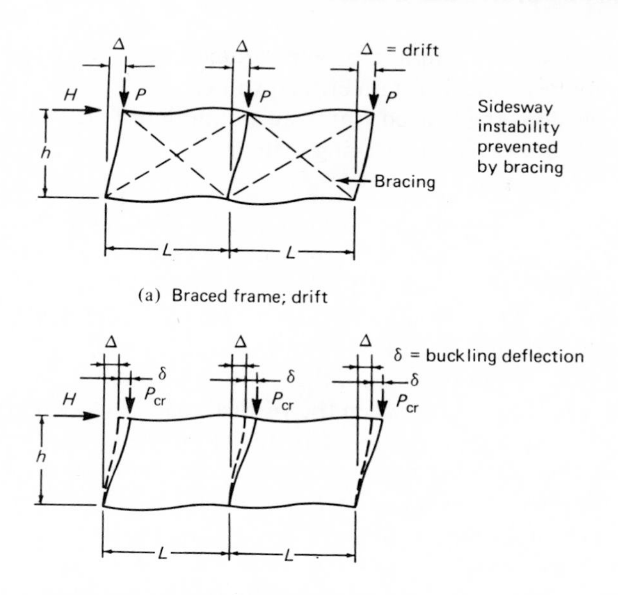

Fig. 14.1.2. Comparison of braced and unbraced frames.

as in Fig. 14.1.1c. In this case, the girders and columns would have to accommodate the moment M_{Δ}.

Braced Frame. A braced frame has negligible moment resistance from columns and girders to counterbalance $P\Delta$. In other words, Eq. 14.1.2 is assumed to represent the braced frame when it is dealt with in simplified design procedures. As is shown later in this chapter. there is both flexural resistance and shear resistance developed in the braced and the unbraced frame. It is the relative magnitudes of these resistances which make the difference between the braced and unbraced frame.

Basically, a braced frame is more appropriately defined as one in which *sidesway buckling is prevented* by bracing elements of the structure other than the structural frame itself. As will be seen in the next section, the theoretical elastic stability analysis of a braced frame assumes no relative joint displacements, which obviously could occur only with infinitely stiff bracing. However, it is practical for design and reasonably correct to assume negligible moment resistance as implied by Eq. 14.1.2, and to assume that for stability purposes the sidesway mode is thereby also prevented.

The term "sidesway" is used to refer to stable elastic lateral movement of a frame, usually due to lateral loads, such as wind. Sidesway buckling is the sudden lateral movement, such as δ of Fig. 14.1.2b, caused by axial loads reaching a certain critical value.

In conclusion, the braced frame accommodates the $P\Delta$ moments by developing shears Q_Δ in the bracing system.

Unbraced Frame. In the unbraced frame, as Fig. 14.1.2b, if the horizontal load H is maintained constant and the compressive loads P are increased sufficiently to cause failure, such failure will occur with a side lurch known as sidesway buckling. The lateral deflection will *suddenly* become greater than the drift as shown in Fig. 14.1.2b. For cases where there is no lateral loading H and, therefore, no initial deflection; the sudden sidesway will still occur when the vertical load reaches a critical value.

The practical design treatment of the unbraced frame assumes that, referring to Fig. 14.1.1c, no shears Q_Δ are capable of developing and Eq. 14.1.3 applies. Any $P\Delta$ effects are balanced by column and girder moments in the unbraced frame.

Elastic buckling of frames has received considerable attention of researchers so that classical analysis methods are widely available.[2] Detailed study of one-story frames has been presented by Goldberg,[3] and Zweig and Kahn.[4] Galambos[5] has studied the effect of partial base fixity on one and two story frames. Multistory frames have been treated by Switzky and Wang.[6]

Since the use of digital computers has become widespread, the matrix formulation for the solution of elastic buckling loads using flexibility and/or stiffness coefficients seems to be the most efficient approach.[7,8] The stiffness and flexibility coefficients are developed in Sec. 14.2 and used with the well-known slope deflection method in explaining frame behavior in the remainder of this chapter.

Since some fibers of a cross section usually reach yield prior to buckling; inelastic buckling probably most often governs the strength of a frame. Many studies of inelastic buckling have been made: including for braced frames the work of Ojalvo and Levi;[9] Levi, Driscoll, and Lu;[10] and for unbraced frames the work of Merchant;[11] Yura and Galambos,[12] Lu,[13] Levi, Driscoll, and Lu,[14] and Korn and Galambos.[15] Inelastic buckling studies have been extended to the hybrid frame[16] (different yield strength steels used in combination) and to space frames.[17]

14.2. ELASTIC BUCKLING OF FRAMES

The distinction between the braced and unbraced frame has been made in Sec. 14.1. In addition, there are two kinds of loading which may contribute to instability. For the *braced* frame of Fig. 14.2.1a, there are no primary bending moments; the only loading is axial compression. The

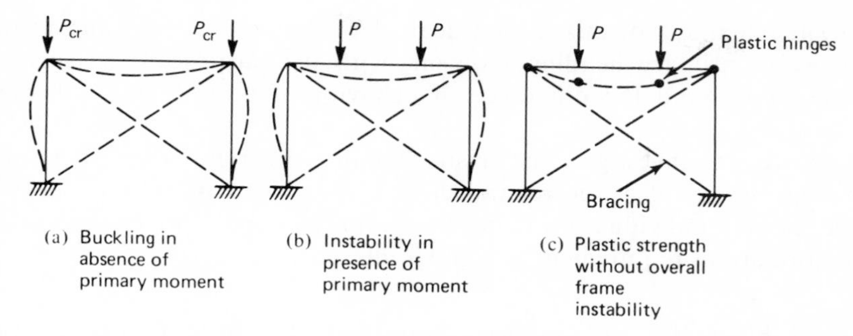

Fig. 14.2.1. Strength of braced frames.

critical load for such a situation is usually defined as a buckling load. No bending of members occurs until the buckling load is reached.

For Fig. 14.1.2b, bending moments exist even when the structure is entirely stable. Due to the rigid joints, moments are transmitted into column members. In addition to the primary bending moments in the column members, the axial compression induces secondary moments equal to P times the deflection. This was discussed in Chapter 12. Under certain combinations of axial compression and moment, the lateral deflection of the column increases without achieving equilibrium; this is usually referred to as instability, or instability in the presence of primary bending moment.

When primary bending moments are present, enough plastic hinges may develop prior to achieving frame instability so that a mechanism forms, in which case for the braced frame the ultimate strength is the plastic strength (Fig. 14.2.1c).

The strength of unbraced frames, as shown in Fig. 14.2.2, also may be separated into three categories; buckling in the absence of primary moment (Fig. 14.2.2a), instability in the presence of primary moment (Fig. 14.2.b), and plastic strength (Fig. 14.2.2c). For the unbraced frame, achieving plastic strength frequently (though not always) means achieving a mechanism associated with overall geometric instability.

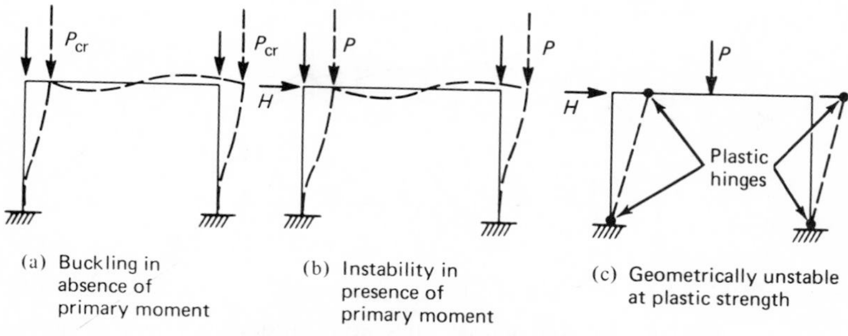

Fig. 14.2.2. Strength of unbraced frames.

The remainder of this section deals with the case of buckling in the absence of primary bending moment, with the purpose of having the reader understand the difference in behavior of braced and unbraced frames.

In order to investigate the elastic stability of a rigid frame, it is first necessary to establish the relationships between end moments and end slopes for the individual frame member and then apply the compatibility of deformations requirement for rigid joints.

General Flexibility and Stiffness Coefficients for Beam-Columns. The reader is presumed to have some familiarity with the slope deflection equations used in frame analysis when axial effect is not considered. For a prismatic section without axial load and without transverse load, as in Fig. 14.2.3a,

$$M_a = \theta_a\left(\frac{4EI}{L}\right) + \theta_b\left(\frac{2EI}{L}\right) \tag{14.2.1a}$$

$$M_b = \theta_a\left(\frac{2EI}{L}\right) + \theta_b\left(\frac{4EI}{L}\right) \tag{14.2.1b}$$

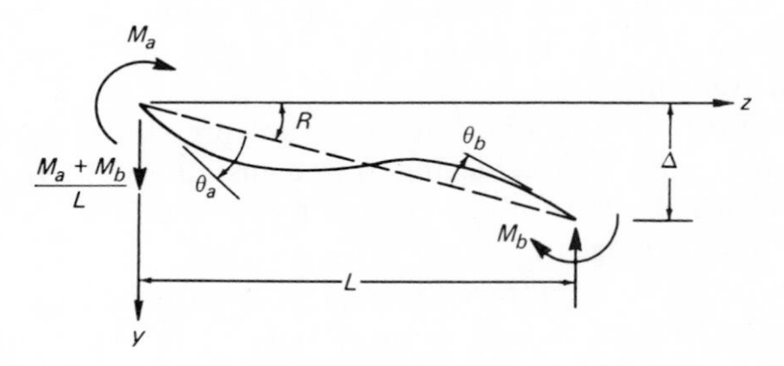

(a) Without axial load

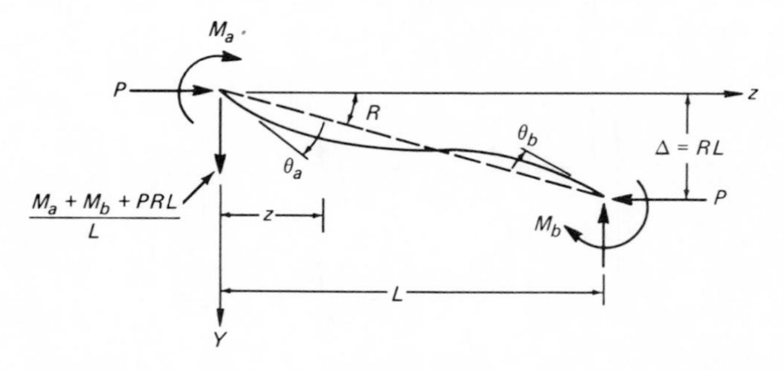

(b) Including axial compression

Fig. 14.2.3. Definition of terms and sign convention for slope-deflection equations.

Using Fig. 14.2.3b, the equations for M_a and M_b will be developed very similarly to the treatment in Sec. 12.2. The following more general approach begins by expressing the moment at any section z of Fig. 14.2.3b:

$$-EI\frac{d^2y}{dz^2} = M_z = M_a + Py - \left(\frac{M_a + M_b + PRL}{L}\right)z \tag{14.2.2}$$

$$\frac{d^2y}{dz^2} + \frac{P}{EI}y = -\frac{M_a}{EI} + \frac{M_a + M_b}{EIL}z + \frac{P}{EI}Rz \tag{14.2.3}$$

Letting $k^2 = P/EI$, the solution for Eq. 14.2.3 is

$$y = A\sin kz + B\cos kz - \frac{M_a}{P} + \frac{M_a + M_b}{PL}z + Rz \tag{14.2.4}$$

Applying the boundary conditions of zero deflection at $z = 0$ and $y = RL$ at $Z = L$ gives

$$B = \frac{M_a}{P}$$

and

$$A = \frac{1}{P\sin kL}(M_a\cos kL + M_b)$$

Then

$$y = \frac{M_a}{P}\left[\frac{\sin k(L - z)}{\sin kL} - \frac{(L - z)}{L}\right] - \frac{M_b}{P}\left(\frac{\sin kz}{\sin kL} - \frac{z}{L}\right) + Rz \tag{14.2.5}$$

Differentiating once to obtain the slope,

$$\frac{dy}{dz} = \frac{M_a}{P}\left[\frac{-k\cos k(L - z)}{\sin kL} + \frac{1}{L}\right] - \frac{M_b}{P}\left(\frac{k\cos kz}{\sin kL} - \frac{1}{L}\right) + R \tag{14.2.6}$$

when $z = 0, \dfrac{dy}{dz} = \theta_a + R$ and when $z = L, \dfrac{dy}{dz} = \theta_b + R.$

Letting $\phi^2/L^2 = k^2 = P/EI$, θ_a and θ_b after some manipulation of terms may be expressed as

$$\theta_a = \frac{M_aL}{EI}\left(\frac{\sin\phi - \phi\cos\phi}{\phi^2\sin\phi}\right) + \frac{M_bL}{EI}\left(\frac{\sin\phi - \phi}{\phi^2\sin\phi}\right) \tag{14.2.7a}$$

$$\theta_b = \frac{M_aL}{EI}\left(\frac{\sin\phi - \phi}{\phi^2\sin\phi}\right) + \frac{M_bL}{EI}\left(\frac{\sin\phi - \phi\cos\phi}{\phi^2\sin\phi}\right) \tag{14.2.7b}$$

where the bracketed terms are known as *flexibility coefficients*, f_{ii}, f_{ij}, f_{ji}, and f_{jj}. To obtain the beam-column counterparts to Eq. 14.2.1, solve Eqs. 14.2.7 (i.e., invert the matrix of coefficients) to obtain

$$M_a = \theta_a \frac{EI}{L}\left(\frac{\phi \sin \phi - \phi^2 \cos \phi}{2 - 2\cos \phi - \phi \sin \phi}\right) + \theta_b \frac{EI}{L}\left(\frac{\phi^2 - \phi \sin \phi}{2 - 2\cos \phi - \phi \sin \phi}\right) \tag{14.2.8a}$$

$$M_b = \theta_a \frac{EI}{L}\left(\frac{\phi^2 - \phi \sin \phi}{2 - 2\cos \phi - \phi \sin \phi}\right) + \theta_b \frac{EI}{L}\left(\frac{\phi \sin \phi - \phi^2 \cos \phi}{2 - 2\cos \phi - \phi \sin \phi}\right) \tag{14.2.8b}$$

where the bracketed terms are known as *stiffness coefficients*. The θ values are the end slopes measured with reference to the axis of the member.

Note that since $\phi^2 = PL^2/EI$, $\phi = 0$ means no axial compression and Eq. 14.2.8 should become Eq. 14.2.1. To verify the coefficient 4 when $\phi = 0$ in Eq. 14.2.8a, the numerator and denominator of the bracketed term must be differentiated four times in accordance with L'Hospital's Rule and then apply the $\phi = 0$ limit.

In order to simplify the use of Eqs. 14.2.8 for slope deflection solution of frame buckling problems, let the bracketed expressions for stiffness coefficients be referred to as S_{ii}, S_{ij}, S_{ji}, and S_{jj}. Because they are symmetrical $S_{ji} = S_{ij}$ and $S_{jj} = S_{ii}$. Thus Eqs. 14.2.8 becomes

$$M_a = \theta_a \frac{EI}{L} S_{ii} + \theta_b \frac{EI}{L} S_{ij} \tag{14.2.9a}$$

$$M_b = \theta_a \frac{EI}{L} S_{ij} + \theta_b \frac{EI}{L} S_{ii} \tag{14.2.9b}$$

Braced Frame—Slope-Deflection Method. The analysis of the rigid frame of Fig. 14.2.4 is presented using the slope deflection method. Using clockwise rotations and rotational end moments as positive, the slope-deflection equations are as follows, using Eqs. 14.2.9,

$$M_{12} = \theta_1 \frac{EI_c}{h} S_{ii} + \theta_2 \frac{EI_c}{h} S_{ij} \tag{14.2.10}$$

$$M_{21} = \theta_1 \frac{EI_c}{h} S_{ij} + \theta_2 \frac{EI_c}{h} S_{ii} \tag{14.2.11}$$

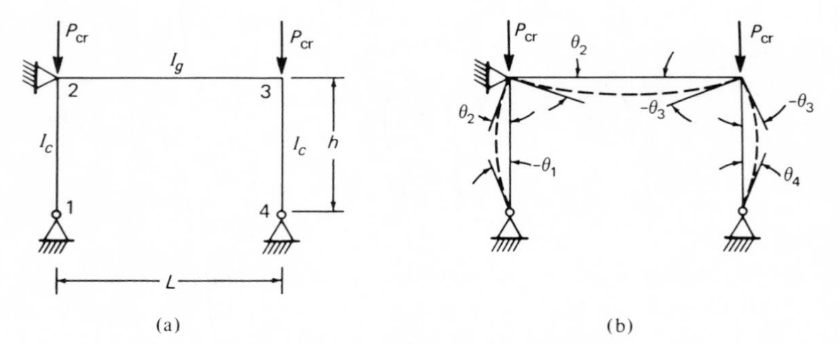

Fig. 14.2.4. Braced frame—hinged base.

$$M_{23} = \theta_2 \frac{EI_g}{L} S_{ii} + \theta_3 \frac{EI_g}{L} S_{ij} = \frac{2EI_g}{L} \theta_2 \qquad (14.2.12)$$

Because no axial compression acts on member 2–3, $S_{ii} = 4$ and $S_{ij} = 2$ for Eq. 14.2.12. Use of symmetry reduces the number of moment equations from six to three, and $\theta_3 = -\theta_2$. The compatibility equations for the joints are

$$M_{12} = 0 \qquad (14.2.13)$$

$$M_{21} + M_{23} = 0 \qquad (14.2.14)$$

From Eq. 14.2.10, with $M_{12} = 0$, one obtains

$$\theta_1 = -\frac{S_{ij}}{S_{ii}} \theta_2 \qquad (14.2.15)$$

Substituting Eq. 14.2.15 into Eq. 14.2.11 gives

$$M_{21} = \theta_2 \frac{EI_c}{h} \left[S_{ii} - \frac{S_{ij}^2}{S_{ii}} \right] \qquad (14.2.16)$$

also

$$M_{23} = \theta_2 \frac{2EI_g}{L} \qquad [14.2.12]$$

Using Eq. 14.2.14 gives

$$\theta_2 \left\{ \frac{EI_c}{h} \left[S_{ii} - \frac{S_{ij}^2}{S_{ii}} \right] + \frac{2EI_g}{L} \right\} = 0 \qquad (14.2.17)$$

which is the stability equation. Noting that θ_2 cannot be zero if buckling occurs, the bracketed term must be zero to satisfy the equation. Thus

$$S_{ii} - \frac{S_{ij}^2}{S_{ii}} = -\frac{2I_g h}{I_c L} \qquad (14.2.18)$$

which in terms of ϕ becomes

$$\frac{\phi^2 \sin \phi}{\sin \phi - \phi \cos \phi} = -\frac{2I_g h}{I_c L} \qquad (14.2.19)$$

EXAMPLE 14.2.1

Determine the buckling load P_{cr} and equivalent pinned length KL for a braced rigid frame, as in Fig. 14.2.4, which has $I_g = 2100$ in.^{4} (W24 × 76); $I_c = 797$ in.^{4} (W14 × 74); $L = 36$ ft; and $h = 14$ ft.

Solution

The stability equation to be satisfied in Eq. 14.2.19,

$$\frac{\sin \phi - \phi \cos \phi}{\phi^2 \sin \phi} = -\frac{I_c L}{2I_g h} = \frac{-797(36)}{2(2100)(14)} = -0.487 \qquad [14.2.19]$$

The smallest value of ϕ which satisfies the buckling equation is the critical value; i.e., the one that governs. One should recognize that if I_g approaches zero, the column becomes an isolated pinned column with $\phi^2 = \pi^2$. Then, as girder stiffness increases ϕ should become somewhat greater than π. For an infinitely stiff girder, from Fig. 6.9.1 which indicates $K = 0.7$ for such a case, one would expect $\phi^2 = (\pi/0.7)^2 = (4.49)^2$.

The solution for ϕ is obtained by trial using values greater than π but less than 4.49, as shown in Fig. 14.2.5. Thus $\phi = 3.60$ to a close approximation.

Comparing with the isolated pinned column,

$$\frac{\phi^2 EI}{h^2} = \frac{\pi^2 EI}{(Kh)^2} \tag{14.2.20}$$

it may be noted that the equivalent pinned length factor K may be expressed

$$K = \frac{\pi}{\phi} \tag{14.2.21}$$

which for this problem means $K = \pi/3.60 = 0.87$. In other words, to account for frame buckling the column member could be designed using $0.87h$ as the pinned length of the isolated member. The K factor axis is also shown on Fig. 14.2.5 so that for various frame properties one could obtain the K factor directly. Note that a large increase in girder stiffness achieves only a small reduction in K.

The buckling load is

$$P_{cr} = \frac{(3.60)^2 EI_c}{h^2} = \frac{(3.60)^2 29{,}000(797)}{(14)^2(12)^2} = 9{,}000 \text{ kips}$$

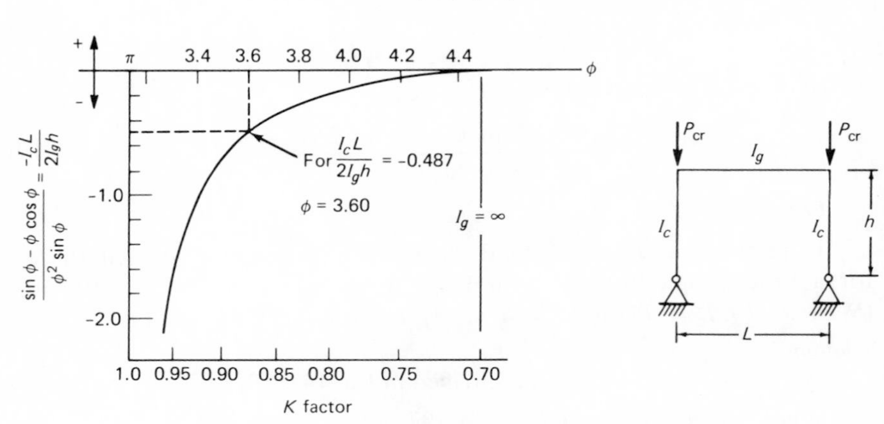

Fig. 14.2.5. Braced frame—hinged base.

Unbraced Frame—Slope-Deflection Method. Next, the same frame which was treated as braced against sidesway at joint 2 is now to be analyzed as unbraced; i.e., with the horizontal support at joint 2 removed, as in Fig. 14.2.6. Even for the unbraced frame it is conceivable to consider a symmetrical buckling mode (Fig. 14.2.1a) exactly as when it was braced. It will be demonstrated that the sidesway mode (Fig. 14.2.6b) will occur under a smaller buckling load than the value obtained for the symmetrical case.

Equations 14.2.9 are applied with the added factor that the axes of some members are tilted due to sidesway, so that letting θ represent *total* rotation, Δ/h must be subtracted from it to get the end slope measured with respect to the member axis.

$$M_{12} = \left(\theta_1 - \frac{\Delta}{h}\right)\frac{EI_c}{h}S_{ii} + \left(\theta_2 - \frac{\Delta}{h}\right)\frac{EI_c}{h}S_{ij} \tag{14.2.22}$$

$$M_{21} = \left(\theta_1 - \frac{\Delta}{h}\right)\frac{EI_c}{h}S_{ij} + \left(\theta_2 - \frac{\Delta}{h}\right)\frac{EI_c}{h}S_{ii} \tag{14.2.23}$$

$$M_{23} = \theta_2\frac{EI_g}{L}S_{ii} + \theta_3\frac{EI_g}{L}S_{ij} \tag{14.2.24}$$

With no axial compression considered on member 2–3, $S_{ii} = 4$ and $S_{ij} = 2$ for Eq. 14.2.24. This time if the structure is symmetrical, the sidesway buckling gives an antisymmetrical deflected curve; thus, $\theta_3 = \theta_2$ and only three moment equations are needed instead of six. The compatibility equations are

$$M_{12} = 0 \tag{14.2.25}$$

$$M_{21} + M_{23} = 0 \tag{14.2.26}$$

which are the same as for the braced case. In addition, the sum of the

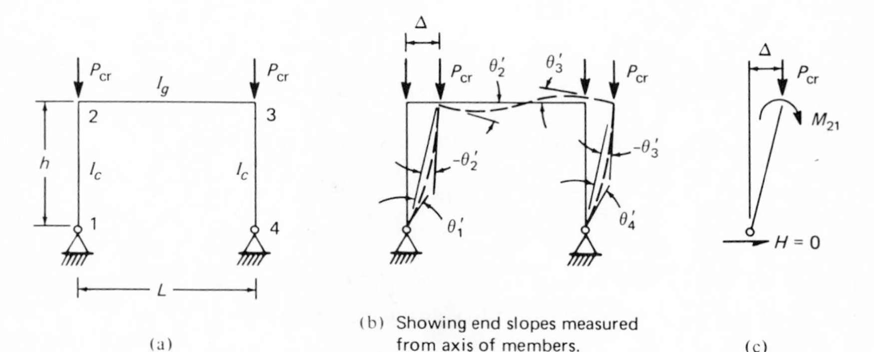

Fig. 14.2.6. Unbraced frame—hinged base. ($\theta' = \theta$ from Eq. 14.2.9 and Fig. 14.2.3.)

base shears must be zero since no external horizontal force is acting. Because of antisymmetry $H_1 = -H_4$ so that only one column member needs to be considered. Referring to Fig. 14.2.6c,

$$H = \frac{M_{21} + P_{cr}\Delta}{h} = 0 \tag{14.2.27}$$

Equations 14.2.25, 14.2.26, and 14.2.27 then become

$$\left.\begin{aligned} &\theta_1 \frac{EI_c}{h} S_{ii} + \theta_2 \frac{EI_c}{h} S_{ij} + \Delta \frac{EI_c}{h^2}(-S_{ii} - S_{ij}) = 0 \\ &\theta_1 \frac{EI_c}{h} S_{ij} + \theta_2 \left(\frac{EI_c}{h} S_{ii} + \frac{6EI_g}{L}\right) + \Delta \frac{EI_c}{h^2}(-S_{ii} - S_{ij}) = 0 \\ &\theta_1 \frac{EI_c}{h} S_{ij} + \theta_2 \frac{EI_c}{h} S_{ii} + \Delta \frac{EI_c}{h^2}(\phi^2 - S_{ii} - S_{ij}) = 0 \end{aligned}\right\} \tag{14.2.28}$$

Multiplying the first of Eqs. 14.2.28 by S_{ij} and the second and third by S_{ii}; then subtracting the second from the first and the third from the second, obtaining

$$\left.\begin{aligned} &\theta_2\left[\frac{EI_c}{h}(S_{ij}^2 - S_{ii}^2) - \frac{6EI_g}{L} S_{ii}\right] + \Delta \frac{EI_c}{h^2}\left(S_{ii}^2 - S_{ij}^2\right) = 0 \\ &\theta_2\left(\frac{6EI_g}{L}\right) + \Delta \frac{EI_c}{h^2}\left(-\phi^2\right) = 0 \end{aligned}\right\} \tag{14.2.29}$$

Since θ_2 and Δ cannot be zero (if buckling occurs), the determinate of the coefficients must be zero. Thus

$$\frac{EI_c}{h^2}(S_{ii}^2 - S_{ij}^2)\left[\frac{6EI_g}{L}\phi^2 \frac{S_{ii}}{(S_{ii}^2 - S_{ij}^2)} + \frac{\phi^2 EI_c}{h} - \frac{6EI_g}{L}\right] = 0 \tag{14.2.30}$$

Since $S_{ii} \neq S_{ij}$ the bracketed term must be zero,

$$\frac{1}{\phi^2} - \frac{S_{ii}}{(S_{ii}^2 - S_{ij}^2)} = \frac{I_c L}{6 I_g h} \tag{14.2.31}$$

or if one solves Eq. 14.2.9 in terms of θ, it will be found that

$$\frac{S_{ii}}{(S_{ii}^2 - S_{ij}^2)} = \frac{\sin\phi - \phi\cos\phi}{\phi^2 \sin\phi} \tag{14.2.32}$$

which is the bracketed flexibility coefficient from Eq. 14.2.7a. Using the identity of Eq. 14.2.32 in the stability equation, Eq. 14.2.31 gives

$$\phi \tan\phi = \frac{6 I_g h}{I_c L} \tag{14.2.33}$$

EXAMPLE 14.2.2

Determine the buckling load P_{cr} and the equivalent pinned length KL for the unbraced frame consisting of the same members and span as in Example 14.2.1.

Solution

$$\frac{6I_g h}{I_c L} = \frac{6(2000)(14)}{797(36)} = +6.13$$

$$\phi \tan \phi = +6.13$$

which may be solved by trial for the smallest ϕ which satisfies the equation. From Fig. 14.2.7 the value of ϕ is 1.355. Since from Eq. 14.2.21,

$$K = \frac{\pi}{\phi} = \frac{\pi}{1.355} = 2.32$$

Thus if the frame is unbraced it may be designed as an isolated member of length $2.32h$, whereas if it were braced the equivalent length would be $0.87h$. The range of K values may be studied for this frame in Fig. 14.2.7. With hinged bases, even with an infinitely rigid beam the equivalent pinned-length factor is never less than 2 for an unbraced frame.

The buckling load is

$$P_{cr} = \frac{(1.355)^2 EI_c}{h^2} = \frac{(1.355)^2\, 29{,}000\, (797)}{(14)^2\,(12)^2} = 1{,}500 \text{ kips}$$

or about ⅙ of what it was for the same frame when braced.

14.3. GENERAL EQUATION FOR EFFECTIVE LENGTH

For ordinary design, it is entirely impractical to analyze an entire frame to determine its buckling strength and its effective length (equiv-

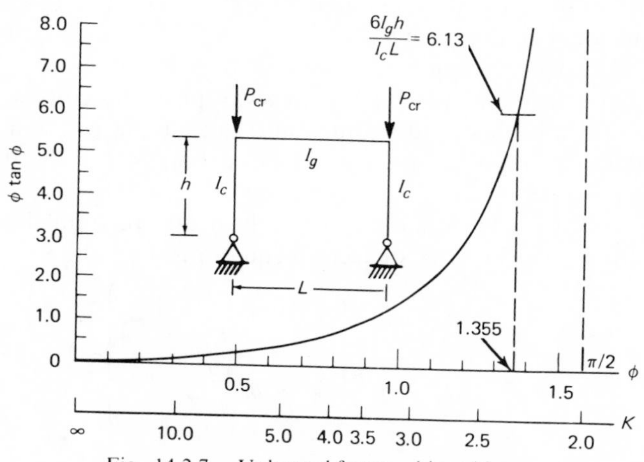

Fig. 14.2.7. Unbraced frame—hinged base.

alent pin-end length). Thus it is desirable to have some general way of obtaining the K factor without the full analysis.

A number of investigators have provided charts to permit easy determination of frame buckling loads and effective lengths for commonly encountered situations. Effective length factors are available for one-story, one-bay frames, with vertical loads applied to the columns at an intermediate point in addition to the load at the top;[18] one and two-story, one bay wide frames;[5] and the general case of an elastic rotationally restrained column (both with and without sidesway elastic restraint).[19] Buckling load data from which effective lengths can be obtained has been summarized for frames one bay wide and up to five stories high.[6]

The most commonly used procedure for obtaining effective length is to use the alignment charts from the CRC Guide,[20] originally developed by O. J. Julian and L. S. Lawrence, and presented in detail by T. C. Kavanagh.[21] The alignment chart method using Fig. 14.3.1 is also suggested by the AISC Commentary as satisfying the "rational method" requirement of AISC–1.8.3. Chu and Chow[22] have presented some modifications to use the alignment chart method for unsymmetrical frames.

Alignment Chart Equation—Braced Frame (No Sidesway Buckling). The following assumptions are used in the development of the elastic stability equation:

(a) All columns reach their respective critical loads simultaneously.
(b) The structure is assumed to consist of symmetrical rectangular frames.
(c) At a joint, the restraining moment provided by the girders is distributed among the columns in proportion to their stiffnesses.
(d) The girders are elastically restrained at their ends by the columns, and at the onset of buckling the rotations of the girder at its ends are equal and opposite.
(e) The girders carry no axial loads.

Consider a framework with no sidesway possible, consisting of two columns and two beams rigidly joined at a but having rotation elastically restrained at their opposite ends, as shown in Fig. 14.3.2. No translation of joints is possible.

Next, consider a beam-column with elastically restrained ends, as in Fig. 14.3.2b. Using the slope deflection equations, Eq. 14.2.9,

$$\left.\begin{aligned} M_a &= \theta_a \frac{EI}{L} S_{ii} + \theta_b \frac{EI}{L} S_{ij} \\ M_b &= \theta_a \frac{EI}{L} S_{ij} + \theta_b \frac{EI}{L} S_{ii} \end{aligned}\right\} \tag{14.2.9}$$

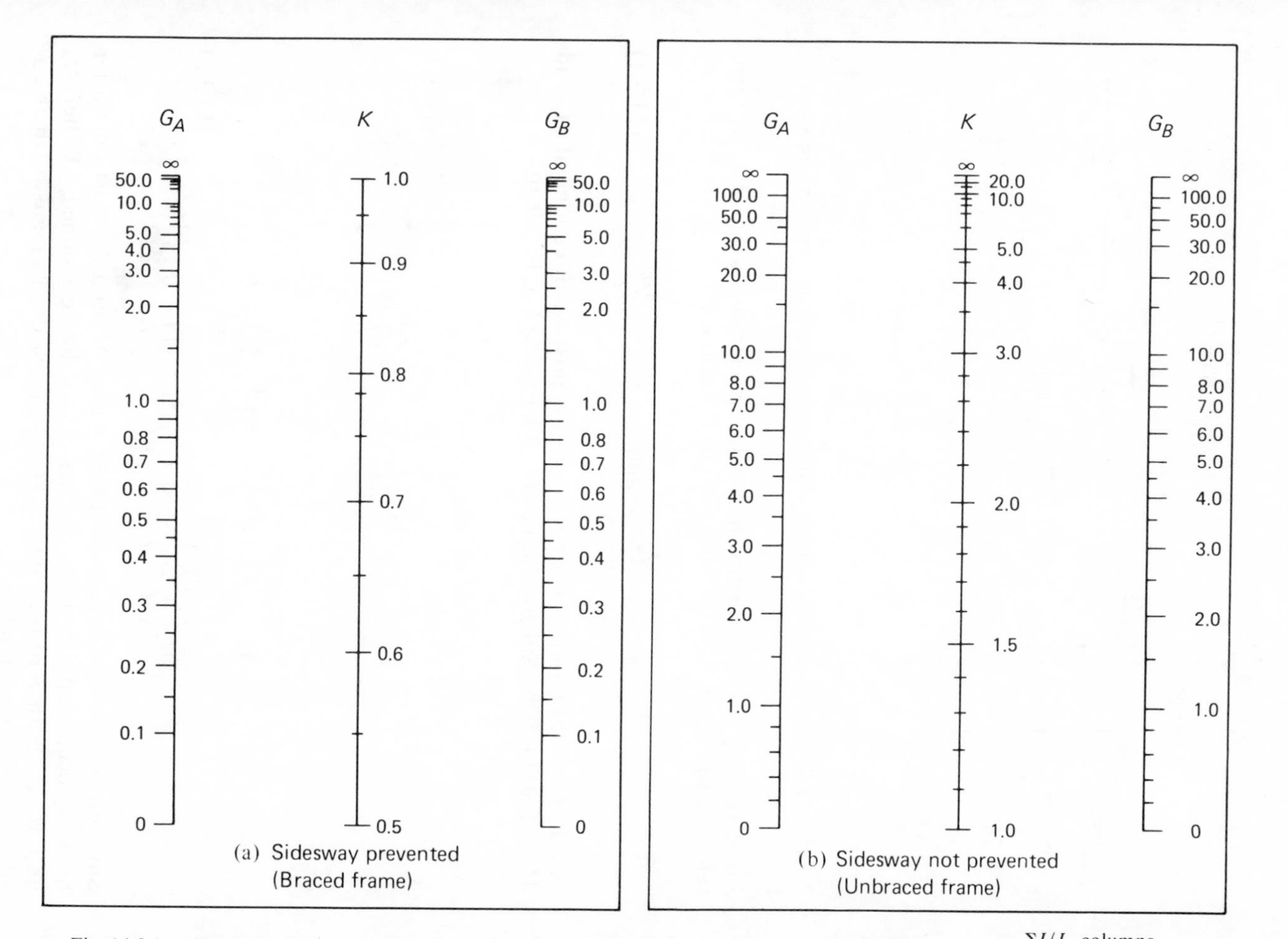

Fig. 14.3.1. Alignment Charts for effective length of columns in continuous frames (Ref. 21), where $G = \dfrac{\Sigma I/L, \text{ columns}}{\Sigma I/L, \text{ girders}}$.

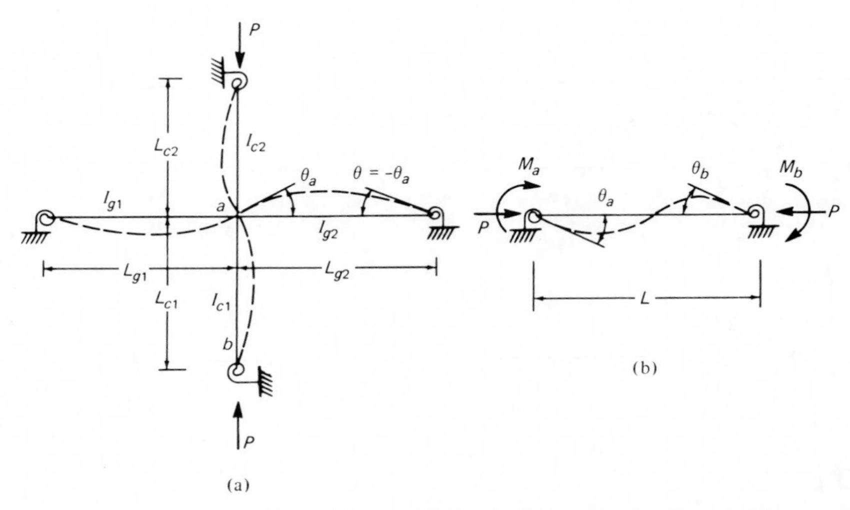

Fig. 14.3.2. Position of braced frame with elastically restrained ends—no sidesway.

For elastically restrained ends, if the elastic restraints are α and β; then, referring to Fig. 14.2.3b for positive sign convention,

$$\theta_a = -\frac{M_a}{\alpha} \quad \text{and} \quad \theta_b = -\frac{M_b}{\beta} \tag{14.3.1}$$

i.e., the restraint moment is opposite to the positive directions of M_a and M_b of Fig. 14.2.3b. Substitution of Eq. 14.3.1 into Eq. 14.2.9 gives

$$\left.\begin{aligned} M_a &= -M_a \frac{EI}{\alpha L} S_{ii} - M_b \frac{EI}{\beta L} S_{ij} \\ M_b &= -M_a \frac{EI}{\alpha L} S_{ij} - M_b \frac{EI}{\beta L} S_{ii} \end{aligned}\right\} \tag{14.3.2}$$

or

$$\left.\begin{aligned} M_a\left(\frac{EI}{\alpha L} S_{ii} + 1\right) + M_b\left(\frac{EI}{\beta L} S_{ij}\right) &= 0 \\ M_a\left(\frac{EI}{\alpha L} S_{ij}\right) + M_b\left(\frac{EI}{\beta L} S_{ii} + 1\right) &= 0 \end{aligned}\right\} \tag{14.3.3}$$

Since no applied moments are assumed acting, M_a and M_b exist only after buckling occurs and their magnitudes cannot be determined. Either M_a and M_b are both zero (i.e., no buckling) or the determinant of their coefficients must be equal to zero.

Thus it is required that

$$\left(\frac{EI}{\alpha L}S_{ii} + 1\right)\left(\frac{EI}{\beta L}S_{ii} + 1\right) - \left(\frac{EI}{\alpha L}\frac{EI}{\beta L}S_{ij}^2\right) = 0 \tag{14.3.4}$$

The same approach could have been used with Eqs. 14.2.7 involving the flexibility coefficients, with Eqs. 14.3.1 for θ_a and θ_b. The determinant of coefficients would then become

$$\left(\frac{L}{EI}f_{ii} + \frac{1}{\alpha}\right)\left(\frac{L}{EI}f_{ii} + \frac{1}{\beta}\right) - \left(\frac{L}{EI}f_{ij}\right)^2 = 0 \tag{14.3.5}$$

Equations 14.3.4 and 14.3.5 are identical; however because the flexibility coefficients have a single term denominator, Eq. 14.3.5 is easier to solve.

Expanding Eq. 14.3.5 gives

$$\frac{1}{\alpha}\left(\frac{1}{\beta}\right) + \left(\frac{L}{EI}\right)\left(\frac{1}{\alpha} + \frac{1}{\beta}\right)f_{ii} + \left(\frac{L}{EI}\right)^2(f_{ii}^2 - f_{ij}^2) = 0 \tag{14.3.6}$$

Substitution of the flexibility coefficients, f_{ii} and f_{ij} from Eq. 14.2.7a and multiplying by $(EI/L)^2$ give

$$\frac{1}{\alpha\beta}\left(\frac{EI}{L}\right)^2 + \left(\frac{EI}{L}\right)\left(\frac{1}{\alpha} + \frac{1}{\beta}\right)\left(\frac{\sin\phi - \phi\cos\phi}{\phi^2\sin\phi}\right) + \left[\frac{\sin\phi - \phi\cos\phi}{\phi^2\sin\phi}\right]^2 - \left[\frac{\sin\phi - \phi}{\phi^2\sin\phi}\right]^2 = 0 \tag{14.3.7}$$

or

$$\frac{\phi^2}{\alpha\beta}\left(\frac{EI}{L}\right)^2 + \left(\frac{EI}{L}\right)\left(\frac{1}{\alpha} + \frac{1}{\beta}\right)\left(1 - \frac{\phi}{\tan\phi}\right) + \frac{1}{\phi^2}\left[1 - \frac{\phi}{\tan\phi}\right]^2 - \frac{1}{\phi^2}\left[1 - \frac{\phi}{\sin\phi}\right]^2 = 0 \tag{14.3.8}$$

Expanding the last two terms of Eq. 14.3.8 gives

$$\begin{aligned}
\frac{1}{\phi^2}&\left[1 - 2\phi\left(\frac{1}{\tan\phi} - \frac{1}{\sin\phi}\right) + \phi^2\left(\frac{1}{\tan^2\phi} - \frac{1}{\sin^2\phi}\right) - 1\right] \\
&= \frac{1}{\phi}\left[-2\left(\frac{\cos\phi - 1}{\sin\phi}\right) + \phi\left(\frac{\cos^2 - 1}{\sin^2\phi}\right)\right] \\
&= \frac{1}{\phi}\left[-2\left(\frac{-2\sin^2\phi/2}{2\sin\frac{\phi}{2}\cos\frac{\phi}{2}}\right) + \phi\left(\frac{-\sin^2\phi}{\sin^2\phi}\right)\right] \\
&= \frac{2}{\phi}\tan\frac{\phi}{2} - 1
\end{aligned}$$

Substitution of the above expression into Eq. 14.3.8 gives

$$\frac{\phi^2}{\alpha\beta}\left(\frac{EI}{L}\right)^2 + \left(\frac{1}{\alpha} + \frac{1}{\beta}\right)\left(\frac{EI}{L}\right)\left(1 - \frac{\phi}{\tan\phi}\right) + \frac{2}{\phi}\tan\frac{\phi}{2} - 1 = 0 \qquad (14.3.9)$$

Next, the elastic restraint factors α and β must be established. Consider the frame of Fig. 14.3.2a; if on the girders which have no axial compression the θ at the far end from joint a equals $-\theta_a$, the slope deflection equations for the moments on the girders at end a become, using Eqs. 14.1.1,

$$\left.\begin{aligned} M_a \text{ (for girder 1)} &= \theta_a\left(\frac{4EI_{g1}}{L_{g1}}\right) - \theta_a\left(\frac{2EI_{g1}}{L_{g1}}\right) = \frac{2EI_{g1}}{L_{g1}}\theta_a \\ M_a \text{ (for girder 2)} &= \theta_a\left(\frac{4EI_{g2}}{L_{g2}}\right) - \theta_a\left(\frac{2EI_{g2}}{L_{g2}}\right) = \frac{2EI_{g2}}{L_{g2}}\theta_a \end{aligned}\right\} \qquad (14.3.10)$$

or the sum of the reactive moments developed due to girder stiffness M_{ag} may be expressed

$$M_{ag} = \sum \frac{2EI_g}{L_g}\theta_a \qquad (14.3.11)$$

The moments at a on the column members may be expressed, using Eq. 14.2.9a,

$$M_a \text{ (for col. 1)} = \theta_a\left(\frac{EI_{c1}}{L_{c1}}\right)S_{ii} - \theta_a\left(\frac{EI_{c1}}{L_{c1}}\right)S_{ij}$$

$$M_a \text{ (for col. 2)} = \theta_a\left(\frac{EI_{c2}}{L_{c2}}\right)S_{ii} - \theta_a\left(\frac{EI_{c2}}{L_{c2}}\right)S_{ij}$$

or

$$\left.\begin{aligned} M_a \text{ (col. 1)} &= \left(\frac{EI_{c1}}{L_{c1}}\right)(S_{ii} - S_{ij})\,\theta_a \\ M_a \text{ (col. 2)} &= \left(\frac{EI_{c2}}{L_{c2}}\right)(S_{ii} - S_{ij})\,\theta_a \end{aligned}\right\} \qquad (14.3.12)$$

The sum of the moments at joint a on the column members M_{ac} may be written

$$M_{ac} = \sum \frac{EI_c}{L_c}(S_{ii} - S_{ij})\,\theta_a \qquad (14.3.13)$$

where it is assumed that $(S_{ii} - S_{ij})$ is the same for all column members framing at joint a.

Solving for θ_a from Eq. 14.3.13 and substituting into Eq. 14.3.12a give

$$M_a\,(\text{col. 1}) = \frac{(EI_c/L_c)(S_{ii} - S_{ij})M_{ac}}{\sum \frac{EI_c}{L_c}(S_{ii} - S_{ij})} \tag{14.3.14}$$

Since $(S_{ii} - S_{ij})$ is assumed identical for all column members, Eq. 14.3.14 becomes

$$M_a\,(\text{col. 1}) = \frac{\frac{EI_c}{L_c}}{\sum \frac{EI_c}{L_c}} M_{ac} \tag{14.3.15}$$

With no external joint moment acting, $M_{ac} = -M_{ag}$; thus

$$M_a\,(\text{col. 1}) = -\frac{2EI_c}{L_c} \frac{\sum \frac{EI_g}{L_g}}{\sum \frac{EI_c}{L_c}} \theta_a \tag{14.3.16}$$

From Eqs. 14.3.1, M_a (col. 1) $= -\alpha\theta_a$; then, from Eq. 14.3.16,

$$\alpha = \frac{2EI_c}{L_c} \frac{\sum \frac{EI_g}{L_g}}{\sum \frac{EI_c}{L_c}} \quad \text{(for joint } a\text{)} \tag{14.3.17}$$

and similarly,

$$\beta = \frac{2EI_c}{L_c} \frac{\sum \frac{EI_g}{L_g}}{\sum \frac{EI_c}{L_c}} \quad \text{(for joint } b\text{)} \tag{14.3.18}$$

Defining, as in the AISC Commentary Fig. C1.8.2,

$$G = \frac{\sum \frac{I_c}{L_c}}{\sum \frac{I_g}{L_g}} \tag{14.3.19}$$

in which case the elastic restraint factors become

$$\alpha = \frac{2EI}{L}\left(\frac{1}{G_A}\right)$$
$$\beta = \frac{2EI}{L}\left(\frac{1}{G_B}\right) \tag{14.3.20}$$

with the subscripts A and B referring to the two ends of the column member ab. The subscripts on EI/L have been dropped; the EI/L in Eqs. 14.3.20 and well as in the stability equation, Eq. 14.3.9 refer to the column ab.

Substitution of Eqs. 14.3.20 into Eq. 14.3.9 and using $\phi = \pi/K$, where K = equivalent pinned length factor, gives

$$\frac{G_A G_B}{4}\left(\frac{\pi^2}{K^2}\right) + \left(\frac{G_A + G_B}{2}\right)\left(1 - \frac{\pi/K}{\tan \pi/K}\right) + \frac{2}{\pi/K}\tan\frac{\pi}{2K} = 1 \tag{14.3.21}$$

which is the equation used for Fig. 14.3.1a, the alignment chart for the braced frame.

If the far ends of the girders do not have rotations equal and opposite to θ_a (i.e., symmetrical single curvature) but are instead either fixed or hinged, adjustments on G may be made.

For girder far ends fixed, θ_b in Eq. 14.2.1 becomes zero, and Eq. 14.3.11 becomes

$$M_{ag} = \sum \frac{4EI_g}{L_g}\theta_a \tag{14.3.22}$$

and Eq. 14.3.20 will be

$$\left.\begin{aligned} \alpha &= \frac{4EI_c}{L_c}\left(\frac{1}{G_A}\right); & \beta &= \frac{4EI_c}{L_c}\left(\frac{1}{G_B}\right) \\ \alpha &= \frac{2EI_c}{L_c}\left(\frac{2}{G_A}\right); & \beta &= \frac{2EI_c}{L_c}\left(\frac{2}{G_B}\right) \end{aligned}\right\} \tag{14.3.23}$$

For girder far ends hinged, $M_b = 0$ and $\theta_b = -\theta_a/2$; thus Eqs. 14.3.20 are

$$\left.\begin{aligned} \alpha &= \frac{3EI_c}{L_c}\left(\frac{1}{G_A}\right); & \beta &= \frac{3EI_c}{L_c}\left(\frac{1}{G_B}\right) \\ \alpha &= \frac{2EI_c}{L_c}\left(\frac{1.5}{G_A}\right); & \beta &= \frac{2EI_c}{L_c}\left(\frac{1.5}{G_B}\right) \end{aligned}\right\} \tag{14.3.24}$$

In other words, to adjust for the far ends of girders fixed, divide G by 2.0; and for the far ends hinged, divide G by 1.5.

Alignment Chart Equation—Unbraced Frame (Sidesway Buckling Possible. The unbraced member ab, as shown in Fig. 14.3.3, having elastically restrained ends, has the following slope-deflection expressions:

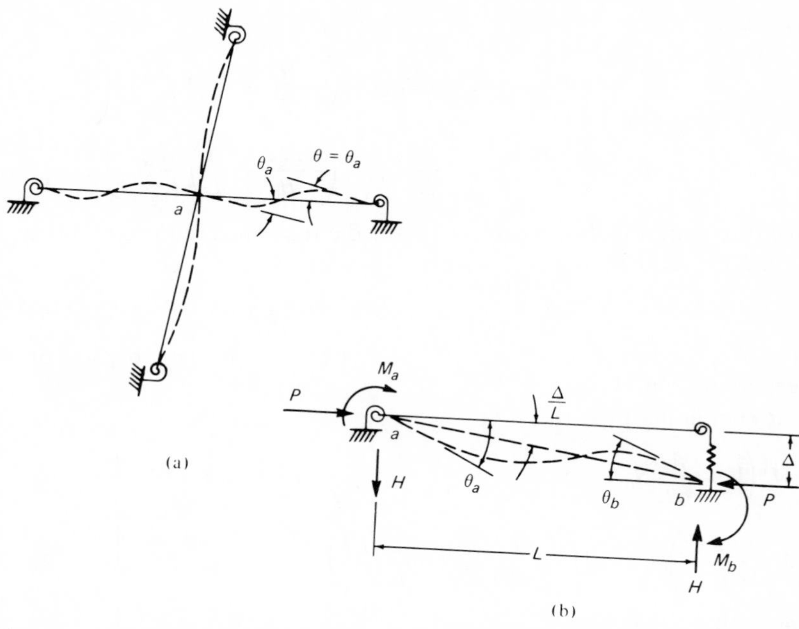

Fig. 14.3.3. Portion of unbraced frame with elastically restrained ends—with sidesway.

$$\left.\begin{aligned} M_a &= \left(\theta_a - \frac{\Delta}{L}\right)\frac{EI}{L}S_{ii} + \left(\theta_b - \frac{\Delta}{L}\right)\frac{EI}{L}S_{ij} \\ M_b &= \left(\theta_a - \frac{\Delta}{L}\right)\frac{EI}{L}S_{ij} + \left(\theta_b - \frac{\Delta}{L}\right)\frac{EI}{L}S_{ii} \end{aligned}\right\} \qquad (14.3.25)$$

where the original θ of Eqs. 14.2.9, which was measured from the axis connecting the ends of the member is replaced by θ which represents the total rotation.

The assumptions for the unbraced frame are the same as for the braced frame, except assumption (d) which for the braced frame assumed the girders to be in single curvature. For the unbraced frame, the girder is assumed to be in double curvature, (Fig. 14.3.3a) with the rotation at both ends equal in magnitude and direction.

To begin the development, assume elastic rotational restraints such that

$$\theta_a = -\frac{M_a}{\alpha}; \qquad \theta_b = -\frac{M_b}{\beta} \qquad (14.3.26)$$

the same as for the braced case. Substitution of Eqs. 14.3.26 into

Eqs. 14.3.25 gives

$$\left.\begin{aligned} M_a &= \left(-\frac{M_a}{\alpha} - \frac{\Delta}{L}\right)\frac{EI}{L}S_{ii} + \left(-\frac{M_b}{\beta} - \frac{\Delta}{L}\right)\frac{EI}{L}S_{ij} \\ M_b &= \left(-\frac{M_a}{\alpha} - \frac{\Delta}{L}\right)\frac{EI}{L}S_{ij} + \left(-\frac{M_b}{\beta} - \frac{\Delta}{L}\right)\frac{EI}{L}S_{ii} \end{aligned}\right\} \quad (14.3.27)$$

Since three unknowns are involved a third equation is required—to satisfy rotational equilibrium,

$$M_a + M_b - HL + P\Delta = 0 \quad (14.3.28)$$

where the net horizontal force H must be zero in the absence of any external horizontal force.

Rearranging Eqs. 14.3.27 gives

$$M_a\left(1 + \frac{EI}{\alpha L}S_{ii}\right) + M_b\left(\frac{EI}{\beta L}S_{ij}\right) + \frac{\Delta}{L}\left(\frac{EI}{L}\right)(S_{ii} + S_{ij}) = 0$$

$$M_a\left(\frac{EI}{\alpha L}S_{ij}\right) + M_b\left(1 + \frac{EI}{\beta L}S_{ii}\right) + \frac{\Delta}{L}\left(\frac{EI}{L}\right)(S_{ii} + S_{ij}) = 0 \quad (14.3.29)$$

along with Eq. 14.3.28,

$$M_a\left(\frac{1}{L}\right) + M_b\left(\frac{1}{L}\right) + \frac{\Delta}{L}(P) = 0$$

Because there are no applied moments; M_a, M_b, and Δ can exist only after buckling occurs. Equation 14.3.29 may be satisfied when M_a, M_b, and Δ are zero (i.e., no buckling) or the determinant of the coefficients must be equal to zero.

Using $P = \phi^2 EI/L^2$, the determinant of the coefficients becomes

$$\phi^2\left[1 + \left(\frac{1}{\alpha} + \frac{1}{\beta}\right)\left(\frac{EI}{L}\right)S_{ii} + \left(\frac{1}{\alpha\beta}\right)\left(\frac{EI}{L}\right)^2(S_{ii}^2 - S_{ij}^2)\right]$$
$$- 2(S_{ii} - S_{ij}) - \left(\frac{EI}{L}\right)\left(\frac{1}{\alpha} + \frac{1}{\beta}\right)(S_{ii}^2 - S_{ij}^2) = 0 \quad (14.3.30)$$

The alternate method would be to use Eq. 14.2.7 involving the flexibility coefficients, f_{ii} and f_{ij}; in which case the resulting determinant is

$$\phi^2\left[f_{ii}^2 - f_{ij}^2 + \left(\frac{1}{\alpha} + \frac{1}{\beta}\right)\left(\frac{EI}{L}\right)f_{ii} + \frac{1}{\alpha\beta}\left(\frac{EI}{L}\right)^2\right]$$
$$- 2(f_{ii} - f_{ij}) - \left(\frac{EI}{L}\right)\left(\frac{1}{\alpha} + \frac{1}{\beta}\right) = 0 \quad (14.3.31)$$

Equations 14.3.30 and 14.3.31 are different forms of the identical stability equation. Equation 14.3.31 is easier to solve since the coefficients have a single denominator term. Substitution of the flexibility coefficient function of ϕ from Eqs. 14.2.7 into Eq. 14.3.31 gives

$$\phi^2\left[\frac{\cancel{\sin^2\phi} - 2\phi\cos\phi\sin\phi + \phi^2\cos^2\phi - \cancel{\sin^2\phi} + 2\phi\sin\phi - \phi^2}{\phi^4\sin^2\phi}\right]$$

$$+\left(\frac{1}{\alpha}+\frac{1}{\beta}\right)\left(\frac{EI}{L}\right)\left(\frac{\phi^2(\cancel{\sin\phi} - \phi\cos\phi) - \phi^2\cancel{\sin\phi}}{\phi^2\sin\phi}\right)$$

$$+\frac{\phi^2}{\alpha\beta}\left(\frac{EI}{L}\right)^2 - 2\left(\frac{\phi - \phi\cos\phi}{\phi^2\sin\phi}\right) = 0$$

or combining first and last term and further reducing give

$$\frac{-\cancel{2\cos\phi}\sin\phi + \phi\cos^2\phi + \cancel{2\sin\phi} - \phi - \cancel{2\sin\phi} + \cancel{2\cos\phi}\sin\phi}{\phi\sin^2\phi}$$

$$+\left(\frac{1}{\alpha}+\frac{1}{\beta}\right)\left(\frac{EI}{L}\right)\left(\frac{-\phi\cos\phi}{\sin\phi}\right) + \frac{\phi^2}{\alpha\beta}\left(\frac{EI}{L}\right)^2 = 0$$

which, after multiplying by $\tan\phi$ finally becomes

$$\left[\frac{\phi^2}{\alpha\beta}\left(\frac{EI}{L}\right)^2 - 1\right]\tan\phi - \left(\frac{1}{\alpha}+\frac{1}{\beta}\right)\left(\frac{EI}{L}\right)^2 = 0 \qquad (14.3.32)$$

the elastic stability equation for the unbraced frame.

The development of the expression for the elastic restraint factors α and β is similar to that illustrated for the braced frame. The only difference is that the assumed deformation for the girders differs; in this case θ at the far end of the girder from joint a equals θ_a as shown in Fig. 14.3.3a.

Using the slope deflection equations for no axial load, Eq. 14.1.1, the moment at end a for each girder is

$$M_a\,(\text{one girder}) = \theta_a\left(\frac{4EI_g}{L_g}\right) + \theta_a\left(\frac{2EI_g}{L_g}\right) = \frac{6EI_g}{L_g}\theta_a \qquad (14.3.33)$$

or the sum of reactive moments developed due to girder stiffness M_{ag}, may be expressed

$$M_{ag} = \sum\frac{6EI_g}{L_g}\theta_a \qquad (14.3.34)$$

For the column, the assumptions behind Eq. 14.3.15 are still valid;

$$M_a\,(\text{col. 1}) = \frac{\dfrac{EI_c}{L_c}}{\sum\dfrac{EI_c}{L_c}}M_{ac} \qquad [14.3.15]$$

where M_{ac} is the total moment acting on the column members; and in the absence of any external joint moment, $M_{ac} = -M_{ag}$. Thus

$$M_a \text{ (col. 1)} = -\frac{6EI_c}{L_c} \frac{\sum \frac{EI_g}{L_g}}{\sum \frac{EI_c}{L_c}} \theta_a \tag{14.3.35}$$

and by comparison with Eqs. 14.3.17 through 14.3.20, α and β may be written

$$\left.\begin{aligned} \alpha &= \frac{6EI}{L}\left(\frac{1}{G_A}\right) \\ \beta &= \frac{6EI}{L}\left(\frac{1}{G_B}\right) \end{aligned}\right\} \tag{14.3.36}$$

Substitution of Eqs. 14.3.36 into Eq. 14.3.32 and using $\phi = \pi/K$ give

$$\left[\frac{(\pi/K)^2 G_A G_B}{36} - 1\right] \tan\frac{\pi}{K} - \left(\frac{G_A + G_B}{6}\right)\frac{\pi}{K} = 0$$

or

$$\frac{G_A G_B (\pi/K)^2 - 36}{6(G_A + G_B)} = \frac{\pi/K}{\tan(\pi/K)} \tag{14.3.37}$$

which is the equation used for Fig. 14.3.1b, the alignment chart for the unbraced frame.

For the case of the far ends of the girder hinged and fixed, the girder stiffnesses are $3EI_g/L_g$ and $4EI_g/L_g$ respectively. In which case, to adjust for far ends of girders hinged, multiply G by 2.0; and for the far ends of girders fixed, multiply G by 1.5.

14.4. STABILITY OF FRAMES UNDER PRIMARY BENDING MOMENTS

Referring to Figs. 14.2.1 and 14.2.2, the distinction should be noted between (a) buckling in the absence of primary bending moments; i.e., *no* moments exist until buckling occurs; and (b) magnification of primary bending moments (which exist even without the presence of compressive loads) due to axial compression P, times deflection Δ. The buckling of frames, which involved solving for the compressive loads P, which make a determinant equal to zero, was treated in the preceding section.

This article considers frame behavior when primary moments also exist. Consider a simple rectangular fixed base frame subject to primary moments as in Fig. 14.4.1a. If simple bending theory is used, and the stiffness of the girder is *not* considered reduced due to axial compression, any ordinary procedure of statically indeterminate analysis method will give the moments.

If compressive loads P are applied, there will be additional moments $P\Delta$, as in Fig. 14.4.1. The total effect may be treated as a magnification

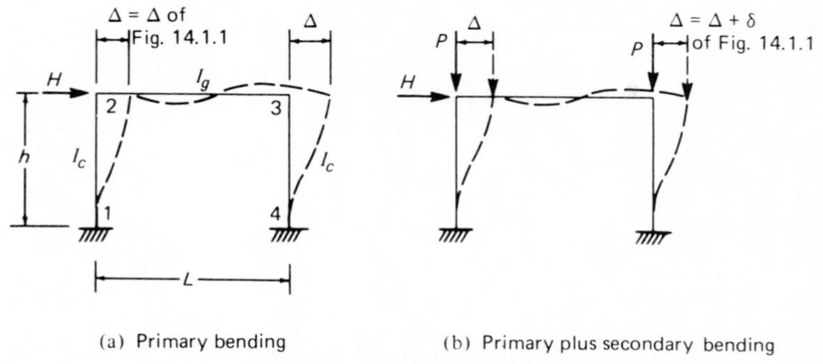

(a) Primary bending (b) Primary plus secondary bending

Fig. 14.4.1. Frame with fixed bases.

factor times the primary bending moments. This has been discussed in detail in Chapter 12 for members with no translation of joints (braced frames such as in Fig. 14.2.1b), and has been approximately presented for the unbraced frame beam-column in Sec. 12.5.

The following treatment is intended to provide an understanding of the mathematics and compare the actual magnification factor with the simple expression of Sec. 12.5 which is suggested in the AISC Commentary.

Primary Bending of Fixed-Base Frame. In order to determine magnification of primary moments, the primary moments must first be determined. While design practice would probably use the moment distribution method, the slope deflection method is used here so that the reader may compare what follows with the modifications required to include the $P\Delta$ effect.

For the frame of Fig. 14.4.1a, use Eqs. 14.2.1 noting that θ is the angle measured from the axis of the member, and that for θ to represent the total rotation, Δ/h must be subtracted from it. The moment equations are

$$\left.\begin{aligned} M_{12} &= \left(\theta_1 - \frac{\Delta}{h}\right)\frac{4EI_c}{h} + \left(\theta_2 - \frac{\Delta}{h}\right)\frac{2EI_c}{h} \\ M_{21} &= \left(\theta_1 - \frac{\Delta}{h}\right)\frac{2EI_c}{h} + \left(\theta_2 - \frac{\Delta}{h}\right)\frac{EI_c}{h} \\ M_{23} &= \theta_2\frac{4EI_g}{L} + \theta_3\frac{2EI_g}{L} \end{aligned}\right\} \qquad (14.4.1)$$

Because of symmetry in the structure,

$$\left.\begin{aligned} M_{43} &= M_{12} \\ M_{34} &= M_{21} \\ \theta_2 &= \theta_3 \end{aligned}\right\} \qquad (14.4.2)$$

Also, because of the fixed base, $\theta_1 = 0$. Rewriting Eqs. 14.4.1 gives

$$\left.\begin{aligned} M_{12} &= \theta_2\left(\frac{2EI_c}{h}\right) - \frac{\Delta}{h}\left(\frac{6EI_c}{h}\right) \\ M_{21} &= \theta_2\left(\frac{4EI_c}{h}\right) - \frac{\Delta}{h}\left(\frac{6EI_c}{h}\right) \\ M_{23} &= \theta_2\left(\frac{6EI_g}{L}\right) \end{aligned}\right\} \qquad (14.4.3)$$

For equilibrium, it is required that

$$M_{21} + M_{23} = 0 \qquad (14.4.4)$$

and also the shears on the columns must equal H:

$$\frac{M_{12} + M_{21}}{h} + \frac{M_{43} + M_{34}}{h} + H = 0 \qquad (14.4.5)$$

Substitution of Eqs. 14.4.3 into Eqs. 14.4.4 and 14.4.5 gives

$$\left.\begin{aligned} \theta_2\left(\frac{4EI_c}{h} + \frac{6EI_g}{L}\right) - \frac{\Delta}{h}\left(\frac{6EI_c}{h}\right) &= 0 \\ \theta_2\left(\frac{12EI_c}{h}\right) - \frac{\Delta}{h}\left(\frac{24EI_c}{h}\right) &= -Hh \end{aligned}\right\} \qquad (14.4.6)$$

Solving for θ_2 and Δ/h gives

$$\theta_2 = \frac{Hh^2}{4EI_c}\left(\frac{1}{1 + 6I_g h/I_c L}\right)$$

or

$$\theta_2 = \frac{Hh^2}{4EI_c}\left(\frac{I_c L/I_g h}{I_c L/I_g h + 6}\right) \qquad (14.4.7)$$

and

$$\frac{\Delta}{h} = \frac{Hh^2}{12EI_c}\left(\frac{2I_c L/I_g h + 3}{I_c L/I_g h + 6}\right) \qquad (14.4.8)$$

Substitution into the M_{12} and M_{21}, Eqs. 14.4.3, gives

$$M_{12} = \frac{-Hh}{2}\left(\frac{I_c L/I_g h + 3}{I_c L/I_g h + 6}\right) \qquad (14.4.9)$$

and

$$M_{21} = \frac{-Hh}{2}\left(\frac{3}{I_cL/I_gh + 6}\right) \tag{14.4.10}$$

M_{12} and M_{21} are the primary moments which are magnified when the axial loads P are applied.

Magnification Factor for Fixed-Base Frame. The frame of Fig. 14.4.1b is investigated next to determine the magnified moments M_{12} and M_{21}. Use the slope-deflection equations, Eqs. 14.2.9, which include effect of axial compression on stiffness. Again the rotation angle Δ/h must be subtracted from the full angle to obtain the value measured from the member axis:

$$\left.\begin{aligned} M_{12} &= \left(\theta_1 - \frac{\Delta}{h}\right)\frac{EI_c}{h}S_{ii} + \left(\theta_2 - \frac{\Delta}{h}\right)\frac{EI_c}{h}S_{ij} \\ M_{21} &= \left(\theta_1 - \frac{\Delta}{h}\right)\frac{EI_c}{h}S_{ij} + \left(\theta_2 - \frac{\Delta}{h}\right)\frac{EI_c}{h}S_{ii} \\ M_{23} &= \theta_2\frac{4EI_g}{L} + \theta_3\frac{2EI_g}{L} \end{aligned}\right\} \tag{14.4.11}$$

whereas for primary moment determination, the compression effect on the girder stiffness is not considered. The same symmetry and fixed-base condition ($\theta_1 = 0$) as in Eqs. 14.4.2 apply:

$$\left.\begin{aligned} M_{12} &= \theta_2\frac{EI_c}{h}S_{ij} - \frac{\Delta}{h}\left(\frac{EI_c}{h}\right)(S_{ii} + S_{ij}) \\ M_{21} &= \theta_2\frac{EI_c}{h}S_{ii} - \frac{\Delta}{h}\left(\frac{EI_c}{h}\right)(S_{ii} + S_{ij}) \\ M_{23} &- \theta_2\left(\frac{6EI_g}{L}\right) \end{aligned}\right\} \tag{14.4.12}$$

The equilibrium conditions are

$$M_{21} + M_{23} = 0 \tag{14.4.13}$$

the same as when the axial force P was not present; and

$$\frac{M_{12} + M_{21}}{h} + \frac{M_{43} + M_{34}}{h} + \frac{2P\Delta}{h} + H = 0 \tag{14.4.14}$$

where the term $P\Delta/h$ is added to the previous condition of Eq. 14.4.5. Let $P = \phi^2EI_c/h^2$.

Substitution of Eqs. 14.4.12 into Eqs. 14.4.13 and 14.4.14, and recalling $M_{43} = M_{12}$ and $M_{34} = M_{21}$, gives

$$\theta_2\left(S_{ii} + \frac{6I_g h}{I_c L}\right) - \frac{\Delta}{h}(S_{ii} + S_{ij}) = 0$$

$$\theta_2[2(S_{ii} + S_{ij})] - \frac{\Delta}{h}[4(S_{ii} + S_{ij}) - 2\phi^2] = \frac{-Hh^2}{EI_c} \qquad (14.4.15)$$

Equations 14.4.15 compare with Eqs. 14.4.6 when Eqs. 14.4.6 are divided by EI_c/h.

Solving Eqs. 14.4.15 for θ_2 and Δ/h gives

$$\frac{\Delta}{h} = \left(\frac{S_{ii} + 6I_g h/I_c L}{S_{ii} + S_{ij}}\right)\theta_2 \qquad (14.4.16)$$

and

$$\theta_2 = \frac{Hh^2}{2EI_c}\left\{\frac{(S_{ii} + S_{ij})(I_c L/I_g h)}{(S_{ii}^2 - S_{ij}^2 - \phi^2 S_{ii})(I_c L/I_g h) + 6[2(S_{ii} + S_{ij}) - \phi^2]}\right\} \qquad (14.4.17)$$

Note that Eq. 14.4.17 becomes Eq. 14.4.7 when $P = 0$. When $P = 0$, $\phi^2 = Ph^2/EI_c = 0$, $S_{ii} = 4$, and $S_{ij} = 2$.

Substituting Eqs. 14.4.16 and 14.4.17 into Eqs. 14.4.12 gives

$$M_{12} = \theta_2 \frac{EI_c}{h}(S_{ij} - S_{ii} - 6I_g h/I_c L)$$

$$= \frac{-Hh}{2}\left\{\frac{(S_{ii} + S_{ij})(S_{ii} - S_{ij} + 6I_g h/I_c L)(I_c L/I_g h)}{(S_{ii}^2 - S_{ij}^2 - \phi^2 S_{ii})I_c L/I_g h + 6[2(S_{ii} + S_{ij}) - \phi^2]}\right\}$$

$$M_{12} = \frac{-Hh}{2}\left\{\frac{(S_{ii}^2 - S_{ij}^2)I_c L/I_g h + 6(S_{ii} + S_{ij})}{(S_{ii}^2 - S_{ij}^2 - \phi^2 S_{ii})I_c L/I_g h + 6[2(S_{ii} + S_{ij}) - \phi^2]}\right\} \qquad (14.4.18)$$

and

$$M_{21} = \frac{-Hh}{2}\left\{\frac{6(S_{ii} + S_{ij})}{(S_{ii}^2 - S_{ij}^2 - \phi^2 S_{ii})I_c L/I_g h + 6[2(S_{ii} + S_{ij}) - \phi^2]}\right\} \qquad (14.4.19)$$

which become the same as Eqs. 14.4.9 and 14.4.10 when $\phi = 0$; i.e., no axial compression.

The magnification factor is the ratio of the moment including the $P\Delta$ effect to the primary moment without $P\Delta$. Thus, dividing Eq. 14.4.18 by Eq. 14.4.9 and Eq. 14.4.19 by Eq. 14.4.10 gives for the moments at the bottom of the column

$$A_{m12} = \frac{\text{Eq. 14.4.18}}{\text{Eq. 14.4.9}}$$

$$= \frac{(I_cL/I_gh + 6)[(S_{ii}^2 - S_{ij}^2)I_cL/I_gh + 6(S_{ii} + S_{ij})]}{(I_cL/I_gh + 3)\{(S_{ii}^2 - S_{ij}^2 - \phi^2 S_{ii})I_cL/I_gh + 6[2(S_{ii} + S_{ij}) - \phi^2]\}} \tag{14.4.20}$$

and for moments at the top of the column,

$$A_{m21} = \frac{\text{Eq. 14.4.19}}{\text{Eq. 14.4.10}}$$

$$= \frac{6(I_cL/I_gh + 6)(S_{ii} + S_{ij})}{3\{(S_{ii}^2 - S_{ij}^2 - \phi^2 S_{ii})I_cL/I_gh + 6[2(S_{ii} + S_{ij}) - \phi^2]\}} \tag{14.4.21}$$

The determination of the actual magnification factor is excessively complicated to be used in design practice. The AISC Commentary suggests that the amplification factor may be expressed by using Eq. 12.5.5 for C_m as follows:

$$A_m = \frac{C_m}{1 - \alpha} = \frac{1 - 0.18\alpha}{1 - \alpha} \tag{14.4.22}$$

where $\alpha = P/P_{cr}$.

The term P_{cr} is the buckling load occurring in the absence of primary moment. From the discussion in Sec. 14.3, P_{cr} may be determined for the present problem by letting $H = 0$, and setting the determinant of the coefficients in Eqs. 14.4.15 equal to zero. This is equivalent to setting the denominator of Eqs. 14.4.17 through 14.4.21 equal to zero. Setting the determinant equal to zero and substituting the ϕ expressions of Eqs. 14.2.8 for S_{ii} and S_{ij}, one obtains after considerable manipulation

$$\phi \sin \phi \cos \phi \left(\frac{\cot \phi}{\phi} - \frac{I_cL}{6I_gh} \right) = 1 \tag{14.4.23}$$

which for example, if $I_cL/I_gh = 1$ one obtains $\phi_{cr} = 0.865\pi$. For the fixed-base frame, the buckling load would be

$$P_{cr} = \frac{(0.865\pi)^2 EI_c}{h^2}$$

and

$$\alpha = \frac{P}{P_{cr}} = \left(\frac{\phi}{0.865\pi} \right)^2, \qquad \text{for } \frac{I_cL}{I_gh} = 1.0$$

The effective-length factor,

$$K = \frac{\pi}{\phi_{cr}} = \frac{1.0}{0.865} = 1.16$$

Table 14.4.1 provides a comparison of the theoretical magnification factor with $(1 - 0.18\alpha)/(1 - \alpha)$, the suggested practical design equation. In design, when computing α the effective length KL is used and also a safety factor is applied. The K in design would usually be determined using the alignment chart, Fig. 14.3.1, instead of from a theoretical elastic-buckling solution.

TABLE 14.4.1
Comparison of Theoretical Magnification Factor A_{m21}
With $A_m = \dfrac{1 - 0.18\alpha}{1 - \alpha}$, and $A_m = \dfrac{0.85}{1 - \alpha}$

	$\frac{I_c L}{I_g h} = 0.5$			1.0			5.0		
ϕ	Exact*	$\frac{1 - 0.18\alpha}{1 - \alpha}$	$\frac{0.85}{1 - \alpha}$	Exact*	$\frac{1 - 0.18\alpha}{1 - \alpha}$	$\frac{0.85}{1 - \alpha}$	Exact*	$\frac{1 - 0.18\alpha}{1 - \alpha}$	$\frac{0.85}{1 - \alpha}$
0.60	1.041	1.037	0.888	1.050	1.042	0.894	1.092	1.073	0.926
1.00	1.124	1.110	0.964	1.151	1.128	0.983	1.305	1.243	1.102
1.40	1.278	1.249	1.108	1.350	1.296	1.157	1.836	1.655	1.539
1.80	1.572	1.512	1.381	1.756	1.641	1.514	3.949	3.335	3.271
2.20	2.233	2.107	1.997	2.842	2.560	2.467	—	—	—
2.60	4.691	4.326	4.298	11.558	9.874	10.048	—	—	—

*Eq. 14.4.21.

The magnification factor using C_m suggested by the AISC Commentary, Eq. 14.4.22, seems to be in line with the theoretical value, though somewhat lower. The use of $0.85/(1 - \alpha)$ as required by AISC–1.6.1 seems even less conservative. However, when actual structures are built, attachments which are designed as nonload supporting do actually contribute some bracing effect. In other words, a real building can never be as flexible as the skeleton elastic frame. In addition, exterior walls, partitions, stairways, etc., all tend to add to the overall stiffness. The amount of additional stiffness will depend upon the actual structures and in many cases can be large.

For further study of the instability of frames in the presence of primary moment, the reader is referred to the work of Bleich,[2] Lu,[23] McGuire,[24] and Galambos.[25]

14.5. BRACING REQUIREMENTS—BRACED FRAME.

One of the decisions faced by the designer is the determination of whether a frame is braced or unbraced. Most efficient use of material in compression members is obtained when the frame is braced so that sidesway buckling or instability cannot occur. Some design guidelines are provided by Galambos.[26] AISC–1.8.2 indicates that lateral stability of frames is provided by attachment "to diagonal bracing, shear walls, and adjacent structures having adequate lateral stability, or to floor slabs or roof decks secured horizontally by walls or bracing systems parallel to the plane of the frame."

No indication is provided by the AISC specification of the amount of stiffness required to prevent sidesway buckling. It has been suggested[26] that "in many instances nonstructural building elements, curtain walls for example, can also provide the necessary stiffness against sidesway buckling."

The following presentation from Galambos[26] may provide assistance in making the necessary engineering judgment regarding the strength required to create a braced frame.

Stiffness Required from Bracing. It is the objective to use bracing to convert Fig. 14.2.2a into Fig. 14.2.1a. A simple and conservative procedure is to idealize the braced frame, as in Fig. 14.5.1. The following

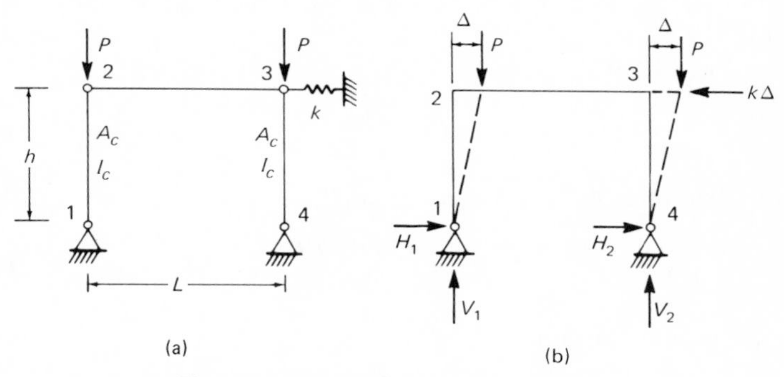

Fig. 14.5.1. Idealized bracing arrangement.

assumptions are made:

(a) Column does not participate in resisting sidesway.
(b) Columns are hinged at ends.
(c) Bracing acts independently as a spring at the top of columns.

Equilibrium using summation of moments about point 1 of Fig. 14.5.1 gives

$$P\Delta + P(\Delta + L) - k\Delta h - V_2 L = 0$$

$$V_2 = P + 2P\left(\frac{\Delta}{L}\right) - k\Delta\frac{h}{L} \tag{14.5.1}$$

Similarly, moments about point 2 gives

$$V_1 = P - 2P\left(\frac{\Delta}{L}\right) + k\Delta\frac{h}{L} \tag{14.5.2}$$

Moments about the hinges at 2 and 3 give

$$H_1 = V_1 \frac{\Delta}{h} \tag{14.5.3}$$

$$H_2 = V_2 \frac{\Delta}{h} \tag{14.5.4}$$

Summation of horizontal forces gives

$$H_1 + H_2 - k\Delta = 0 \tag{14.5.5}$$

Substitution of Eqs. 14.5.1 and 14.5.2 into Eqs. 14.5.3 and 14.5.4, and then into Eq. 14.5.5, gives

$$\frac{\Delta}{h}(2P - kh) = 0 \tag{14.5.6}$$

where Eq. 14.5.6 is the approximate buckling equation. For a simple rectangular frame, as in Fig. 14.2.4, the exact buckling equation is given by Eq. 14.2.19.

From Eq. 14.5.6,

$$P_{cr} = \frac{kh}{2} \tag{14.5.7}$$

where for Example 14.2.1 the theoretical P_{cr} is

$$P_{cr} = \frac{(3.60)^2 EI}{h^2} \tag{14.5.8}$$

and for the pin-end column,

$$P_{cr} = \frac{\pi^2 EI}{h^2} \tag{14.5.9}$$

Equating Eqs. 14.5.7 and 14.5.9 gives

$$k = \frac{2\pi^2 EI}{h^3} \tag{14.5.10}$$

In design, if the buckling load P_{cr} is assumed to be twice the service load P (i.e., factor of safety of 2.0 against buckling), then Eq. 14.5.7 gives the required k for a given P:

$$k_{\text{reqd}} = \frac{2P_{cr}}{h} = \frac{2(2P)}{h} = \frac{4P}{h} \tag{14.5.11}$$

For a frame of several bays, Eq. 14.5.6 becomes

$$\frac{\Delta}{h}(\Sigma P - kh) = 0$$

where ΣP = summation of loads causing buckling. When the factor of safety (2.0) is applied and service loads P are used, Eq. 14.5.11 becomes

$$k_{\text{reqd}} = \frac{2\Sigma P}{h} \tag{14.5.12}$$

Equation 14.5.12 assumes that only one of the bays in a multibay structure is braced. The ΣP includes the design loads in all columns for which the particular brace is to provide support.

Diagonal Bracing. When cross-bracing is used it is generally assumed that it can only act in tension; i.e., the diagonal which would be in compression buckles slightly and becomes inactive. Under a horizontal

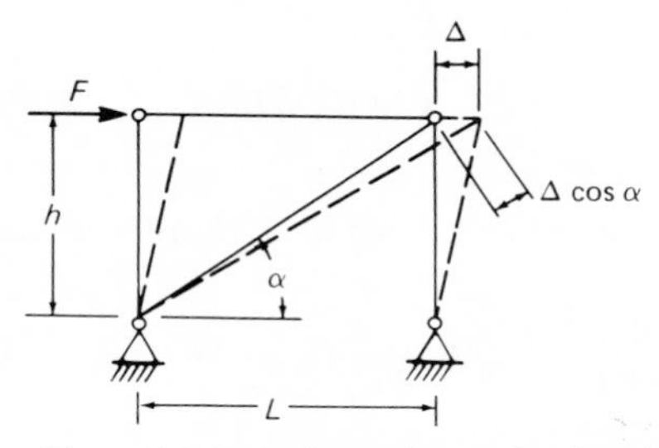

Fig. 14.5.2. Deformation of the bracing member.

force F the diagonal brace in Fig. 14.5.2 must carry the force

$$\text{Brace force} = \frac{F}{\cos\alpha} \tag{14.5.13}$$

and the elongation of the brace, $\Delta \cos\alpha$, is

$$\text{Elongation} = \frac{(\text{brace force})(\text{brace length})}{(\text{area of brace})\, E} \tag{14.5.14}$$

or

$$\Delta\cos\alpha = \frac{(F/\cos\alpha)\sqrt{h^2 + L^2}}{A_b E} \tag{14.5.15}$$

Solving for F, using $\cos\alpha = L/\sqrt{h^2 + L^2}$, gives

$$F = \frac{A_b E L^2}{(h^2 + L^2)^{3/2}}\Delta \tag{14.5.16}$$

Since $F = k\Delta$ according to Fig. 14.5.1b,

$$k = \frac{A_b E L^2}{(h^2 + L^2)^{3/2}} \tag{14.5.17}$$

using the required k in terms of the service loads, Eq. 14.5.12, the required brace area is determined:

$$A_b = \frac{2\left[1 + \left(\frac{L}{h}\right)^2\right]^{3/2} \Sigma P}{\left(\frac{L}{h}\right)^2 E} \tag{14.5.18}$$

EXAMPLE 14.5.1

Determine the area required to convert the unbraced frame of Example 14.2.2, Fig. 14.2.6a, into a braced frame.

Solution

For a frame with $I_g = 2100$, $I_c = 797$, $L = 36$ ft, and $h = 14$ ft, the *braced* frame buckling load from Example 14.2.1 is

$$P_{cr} = \frac{(3.60)^2 EI_c}{h^2}$$

Assuming $P_{cr}/2$ equals the service load and there are two loads, one on each column,

$$\Sigma P = 2\left(\frac{P_{cr}}{2}\right) = \frac{(3.60)^2 EI_c}{h^2}$$

The required area of the bracing, using Eq. 14.5.18, is

$$A_b = \frac{2\left[1 + \left(\frac{36}{14}\right)^2\right]^{3/2} (3.60)^2 E(797)}{\left(\frac{36}{14}\right)^2 E\,(14)^2(144)}$$

$$= \frac{2(7.6)^{3/2}(12.95)(797)}{6.6(196)(144)} = 2.32 \text{ sq-in.}$$

The brace area required is about 10 percent of the column cross-sectional area (21.8 sq in. for W14 × 74). It is noted that the example unbraced frame was very flexible with a higher than usual effective length factor K of 2.32; thus, a somewhat heavier than usual brace was required.

Usually bracing is designed to carry about 5 percent of the vertical column loads since horizontal forces applied at panel points will provide adequate stiffness to create a braced frame.

14.6. OVERALL STABILITY WHEN PLASTIC HINGES FORM

Braced Frame. For plastic design there are no special instability problems associated with the braced frame. Such frames are usually designed to cause the plastic hinges associated with the collapse mechanism to form in the girders. The columns are then designed as beam-columns in accordance with concepts treated in Chapter 12. Design procedures for the one- and two-story braced frames are presented in Chapter 15.

Unbraced Frame—1963 Criterion. When plastic hinges form in an unbraced frame, the overall frame buckling strength may be reduced. The subject of inelastic buckling of unbraced frames has received considerable attention.[13,15,27]

Under the 1963 specification, AISC permitted design of unbraced frames of one and two stories by plastic-design procedures provided that the following equation was satisfied,

$$\frac{2P}{P_y} + \frac{L}{70r} \leq 1.0 \tag{14.6.1}$$

where P = ultimate load, taken as the factored service load

$P_y = F_y A$

L/r = slenderness ratio based on actual unbraced length of the column for bending in the plane of the frame

Equation 14.6.1 is a conservative limitation obtained from the analysis of a frame shown in Fig. 14.6.1. Under the loads of Fig. 14.6.1a the

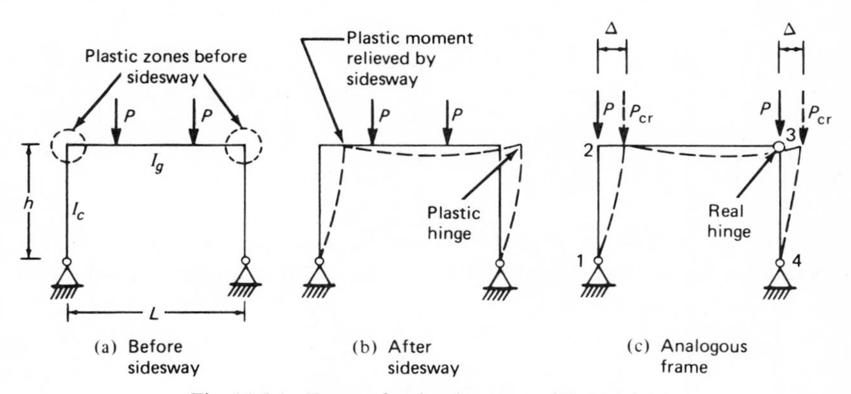

Fig. 14.6.1. Frames for development of Eq. 14.6.1.

upper corners of the frame may become plastic prior to sidesway instability. As the sidesway occurs, however, one plastic hinge is relieved while the other undergoes plastic hinge rotation. Thus Fig. 14.6.1c could represent the analogous frame to be analyzed for overall stability. Using a real hinge, instead of a plastic hinge, underestimates the actual frame strength.

Using slope-deflection equations, the moments in the frame of Fig.

14.6.1c may be expressed

$$\left.\begin{aligned}
M_{12} &= \frac{EI_c}{h}\left[S_{ii}\left(\theta_1 - \frac{\Delta}{h}\right) + S_{ij}\left(\theta_2 - \frac{\Delta}{h}\right)\right]\\
M_{21} &= \frac{EI_c}{h}\left[S_{ij}\left(\theta_1 - \frac{\Delta}{b}\right) + S_{ii}\left(\theta_2 - \frac{\Delta}{h}\right)\right]\\
M_{23} &= \frac{EI_g}{L}[4\theta_2 + 2\theta_3]\\
M_{32} &= \frac{EI_g}{L}[2\theta_2 + 4\theta_3]
\end{aligned}\right\} \quad (14.6.2)$$

From the fact that $M_{32} = 0$, $\theta_3 = -\theta_2/2$. The three displacements θ_1, θ_2, and Δ require the following three equilibrium conditions:

$$\begin{aligned}
M_{12} &= 0\\
M_{21} + M_{23} &= 0\\
H &= \frac{M_{21} + 2P_{cr}\Delta}{h}
\end{aligned} \quad (14.6.3)$$

Substituting Eqs. 14.6.2 into 14.6.3, using $P_{cr} = \phi^2 EI_c/h^2$,

$$\left.\begin{aligned}
&\theta_1\left(\frac{EI_c}{h}S_{ii}\right) + \theta_2\left(\frac{EI_c}{h}S_{ij}\right) + \Delta\frac{EI_c}{h^2}(-S_{ii} - S_{ij}) = 0\\
&\theta_1\left(\frac{EI_c}{h}S_{ij}\right) + \theta_2\left(\frac{EI_c}{h}S_{ii} + \frac{3EI_g}{L}\right) + \Delta\frac{EI_c}{h^2}(-S_{ii} - S_{ij}) = 0\\
&\theta_1\left(\frac{EI_c}{h}S_{ij}\right) + \theta_2\left(\frac{EI_c}{h}S_{ii}\right) + \Delta\frac{EI_c}{h^2}(2\phi^2 - S_{ii} - S_{ij}) = 0
\end{aligned}\right\} \quad (14.6.4)$$

Multiplying the first of Eqs. 14.6.4 by S_{ij} and the second by S_{ii}; then subtracting the second from the first and the third from the second, gives

$$\left.\begin{aligned}
&\theta_2\left[\frac{EI_c}{h}(S_{ij}^2 - S_{ii}^2) - 3\frac{EI_g}{L}S_{ii}\right] + \Delta\frac{EI_c}{h^2}(S_{ii}^2 - S_{ij}^2) = 0\\
&\theta_2\left(\frac{3EI_g}{L}\right) + \Delta\frac{EI_c}{h^2}(-2\phi^2) = 0
\end{aligned}\right\} \quad (14.6.5)$$

Setting the determinant of the coefficients of θ_2 and Δ in Eqs. 14.6.5 equal to zero gives

$$\frac{S_{ii}}{S_{ii}^2 - S_{ij}^2} + \frac{EI_cL}{3EI_gh} - \frac{1}{2\phi^2} = 0 \quad (14.6.6)$$

Replacing $S_{ii}/(S_{ii}^2 - S_{ij}^2)$ by its identity as the flexibility coefficient f_{ii}

from Eq. 14.2.7a gives

$$\frac{1}{2\phi^2} - \frac{\sin\phi - \phi\cos\phi}{\phi^2\sin\phi} = \frac{I_c L}{3 I_g h} \tag{14.6.7}$$

which is the stability equation as shown plotted in Fig. 14.6.2.

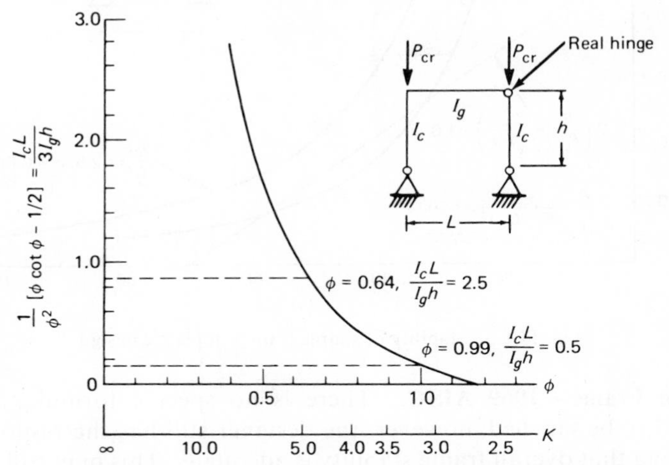

Fig. 14.6.2. Frame stability when approaching plastic-collapse mechanism.

In order to provide a comparison with Eq. 14.6.1, determine $\phi = 0.64$ for $I_c L / I_g h = 2.5$ (a value about as high as may be reasonably expected). Since at buckling

$$P = \frac{\phi^2 E I}{L^2}$$

dividing by $F_y A$ and using $I = A r^2$ give

$$\frac{P}{F_y A} = \frac{\phi^2 E}{F_y (L/r)^2} \tag{14.6.8}$$

or

$$\frac{P}{P_y} = \frac{\phi^2 E}{F_y (L/r)^2} \tag{14.6.9}$$

which is shown in Fig. 14.6.3 compared with Eq. 14.6.1. Eq. 14.6.1 was developed when plastic design was limited to A36 steel and from Fig. 14.6.3 is still probably acceptable as a conservative approximation for simple frames. As a more realistic alternative, ϕ may be determined from Fig. 14.6.2 and used in Eq. 14.6.9.

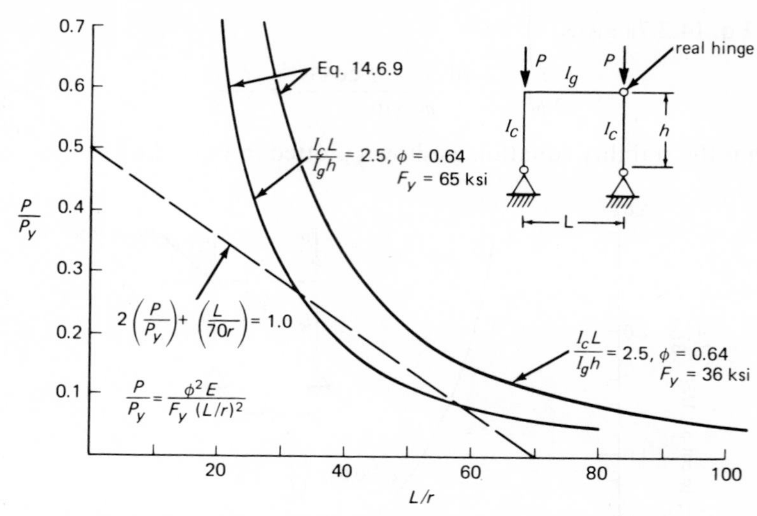

Fig. 14.6.3. Stability of simple frames in plastic design.

Unbraced Frame—1969 AISC. There is no specific formula, such as Eq. 14.6.1 to be satisfied; however, the designer still has the responsibility for insuring that overall frame stability is adequate. This may still be done using Eq. 14.6.1 and being guided by Fig. 14.6.3.

Use of AISC Formula 2.4–2, the stability equation for the beam-column as developed in Sec. 12.8, provides safety against overall frame stability in the process of providing safety for the individual member. Note that the effective length factors K, for bending in the plane of the frame, are to be determined and used in AISC Formula 2.4–2. The establishment of these values (which exceed 1.0) using the alignment chart (Fig. 14.3.1) in nearly all cases effectively accounts for overall frame stability and therefore Eq. 14.6.1 is no longer required.

Unbraced frames are permitted under AISC 1969 plastic design for only one- and two-story frames. Frames of more than two stories utilizing plastic strength must be braced.

Two- and Three-Story Unbraced Frames. The following procedure for frame buckling results from work at Lehigh University.[28] Even though it has been recommended[13,28] for up to three stories, AISC 1969 permits only two stories of unbraced frame. The following steps are suggested after the frame has been designed plastically assuming it to be braced:

(a) Determine the capacity P_u when the plastic collapse mechanism occurs.

(b) Determine the capacity $(P_{cr})_e$ when elastic buckling occurs.

(c) Determine the inelastic buckling load $(P_{cr})_i$ from the following empirical relationship.[28]

$$P_{cr} = \frac{3.4(P_{cr})_e}{1 + 3(P_{cr})_e/P_u} \tag{14.6.10}$$

(d) If $(P_{cr})_i/(P_{cr})_e \leq 0.4$, the carrying capacity of the frame is not affected by overall buckling and will be expected to reach the plastic load based on a braced frame.

Merchant[11] has also provided an empirical method for obtaining the inelastic buckling load as follows:

$$\frac{1}{(P_{cr})_i} = \frac{1}{(P_{cr})_e} + \frac{1}{P_u} \tag{14.6.11}$$

The accuracy of Merchants' formula has also been studied by Korn.[29]

Throughout this chapter, overall stability of frames has been treated. The design of frames incorporating some of these overall stability concepts is presented in Chapter 15.

SELECTED REFERENCES

1. T. R. Higgins, "Effective Column Length—Tier Buildings." *Engineering Journal*, AISC, Vol. 1, No. 1 (January 1964), pp. 12–15.
2. Frederich Bleich, *Buckling Strength of Metal Structures*, McGraw-Hill Book Company, 1952, Chaps. 6–7.
3. John E. Goldberg, "Buckling of One-Story Frames and Buildings," *J. Structural Div. ACSE*, Vol. 86, No. ST10 (October 1960), pp. 53–85.
4. Alfred Zweig and Albert Kahn, "Buckling Analysis of One-Story Frames," *J. Structural Div. ASCE*, Vol. 94, No. ST9 (September 1968), pp. 2107–2134.
5. Theodore V. Galambos, "Influence of Partial Base Fixity on Frame Stability," *J. Structural Div. ASCE*, Vol. 86, No. ST5 (May 1960), pp. 85–108.
6. Harold Switsky and Ping Chun Wang, "Design and Analysis of Frames for Stability," *J. Structural Div. ASCE*, Vol. 95, No. ST4 (April 1969), pp. 695–713.
7. Chu-Kia Wang, "Stability of Rigid Frames With Nonuniform Members," *J. Structural Div. ASCE*, Vol. 93, No. ST1 (February 1967), pp. 275–294.
8. Ottar P. Halldorsson and Chu-Kia Wang, "Stability Analysis of Frameworks by Matrix Methods," *J. Structural Div. ASCE*, Vol. 94, No. ST7 (July 1968), pp. 1745–1760.
9. M. Ojalvo and V. Levi, "Columns in Planar Continuous Structures," *J. Structural Div. ASCE*, Vol. 89, No. ST1 (February 1963), pp. 1–23.
10. Victor Levi, George C. Driscoll, Jr., and Le-Wu Lu, "Structural Subassemblages Prevented From Sway," *J. Structural Div. ASCE*, Vol. 91, No. ST5 (October 1965), pp. 103–127.
11. W. Merchant, "The Failure Load of Rigid Jointed Frameworks as Influenced by Stability," *Structural Engineer*, Vol. 32 (July 1954), pp. 185–190.

12. Joseph A. Yura and Theodore V. Galambos, "Strength of Single Story Steel Frames," *J. Structural Div. ASCE*, Vol. 91, No. ST5 (October 1965), pp. 81–101.
13. Le-Wu Lu, "Inelastic Buckling of Steel Frames," *J. Structural Div. ASCE*, Vol. 91, No. ST6 (December 1965), pp. 185–214.
14. Victor Levi, George C. Driscoll, Jr., and Le-Wu Lu, "Analysis of Restrained Columns Permitted to Sway," *J. Structural Div. ASCE*, Vol. 93, No. ST1 (February 1967), pp. 87–107.
15. Alfred Korn and Theodore V. Galambos, "Behavior of Elastic-Plastic Frames," *J. Structural Div. ASCE*, Vol. 94, No. ST5 (May 1968), pp. 1119–1142.
16. Peter Arnold, Peter F. Adams, and Le-Wu Lu, "Strength and Behavior of an Inelastic Hybrid Frame," *J. Structural Div. ASCE*, Vol. 94, No. ST1 (January 1968), pp. 243–266.
17. William W. McVinnie and Edwin H. Gaylord, "Inelastic Buckling of Unbraced Space Frames," *J. Structural Div. ASCE*, Vol. 94, No. ST8 (August 1968), pp. 1863–1885.
18. Kamal Hassan, "On the Determination of Buckling Length of Frame Columns," Publications, International Association for Bridge and Structural Engineering, Vol. 28-I, 1968, pp. 91–101 (in German).
19. German Gurfinkel and Arthur R. Robinson, "Buckling of Elastically Restrained Columns," *J. Structural Div. ASCE*, Vol. 91, No. ST6 (December 1965), pp. 159–183.
20. B. G. Johnston, *Column Research Council Guide to Design for Metal Compression Members*, 2nd ed., John Wiley & Sons, Inc., New York, 1966, pp. 40–51, 165–167.
21. Thomas C. Kavanagh, "Effective Length of Framed Columns," *Trans. ASCE*, Vol. 127, Part II (1962), pp. 81–101.
22. Kuang-Han Chu and Hsueh-Lien Chow, "Effective Column Length in Unsymmetrical Frames," Publications, International Association for Bridge and Structural Engineering, Vol. 29-I, 1969, pp. 1–15.
23. Le-Wu Lu, "Stability of Frames Under Primary Bending Moments," *J. Structural Div. ASCE*, Vol. 89, No. ST3 (June 1963), pp. 35–62.
24. William McGuire, *Steel Structures*, Prentice-Hall, Inc., Englewood Cliffs, N.J., 1968, pp. 448–474.
25. Theodore V. Galambos, *Structural Members and Frames*, Prentice-Hall, Inc., Englewood Cliffs, N.J., 1968, pp. 176–189.
26. Theodore V. Galambos, "Lateral Support for Tier Building Frames," *Engineering Journal*, American Institute of Steel Construction, Vol. 1, No. 1 (January 1964), pp. 16–19, Disc. by Ira Hooper, Vol. 1, No. 4 (October 1964), p. 141.
27. Fred Moses, "Inelastic Frame Buckling," *J. Structural Div. ASCE*, Vol. 90, No. ST6 (December 1964), pp. 105–122.
28. George C. Driscoll, Jr., et al., "Plastic Design of Multi-Story Frames," Lecture Notes, Fritz Engineering Lab., Rep. No. 273.20, Lehigh University, Bethlehem, Pa., 1965, Chaps. 13–14.
29. Alfred Korn, "The Approximation of Stability Effects in Frames," Publications, International Association for Bridge and Structural Engineering, Vol. 28-II, 1968, pp. 101–112.

15

Design of Rigid Frames

15.1. INTRODUCTION

Many concepts and procedures developed in previous chapters are combined in this chapter by the use of illustrative examples. Plastic analysis and design concepts are extended from the continuous beam treatment in Chapter 10 and the working stress method is compared with plastic design. However, no development is included regarding methods of statically indeterminate analysis of frames. The reader is assumed to be familiar with elastic methods such as moment distribution and slope deflection.

Primarily one-story frames, flat and gabled, are treated; both by plastic design and working stress methods. A brief discussion of multi-story frame design is presented, including factors relating to the plastic design of multi-story braced frames.

15.2. PLASTIC STRENGTH ANALYSIS OF ONE-STORY FRAMES

As discussed in Chapter 10 for continuous beams, the plastic strength of a structure may be obtained either by using the equilibrium method, or the energy method. In braced frames, where joints cannot displace (i.e., no sidesway degree of freedom), the plastic strength may be obtained exactly as for continuous beams. For unbraced frames the sidesway mechanism creates a somewhat more complicated analysis than that for continuous beams. The following examples are intended to illustrate principles of plastic analysis for one story unbraced frames. For a more extended treatment than is included here, the reader is referred to the AISC plastic design manual,[1] textbooks devoted entirely to plastic design[2,3] and published papers on strength and analysis of frames.[4,5]

Welded multi-story rigid frame. Courtesy Bethlehem Steel Corporation.

Equilibrium Method. As discussed in Sec. 10.2, equilibrium must be satisfied at every stage of loading from a small load until the collapse mechanism has been achieved. When a sufficient number of plastic hinges have been developed to allow instantaneous hinge rotations without developing increased resistance, a mechanism is said to have occurred.

EXAMPLE 15.2.1

Determine the plastic strength for the frame of Fig. 15.2.1, using the equilibrium method.

Solution

The elastic moment diagram shape is shown in Fig. 15.2.1b. For ease in solution it is often desirable to show the diagram with a horizontal base line as shown in Fig. 15.2.1d. Assuming overall frame stability is adequate, the collapse mechanism is that shown in Fig. 15.2.1c.

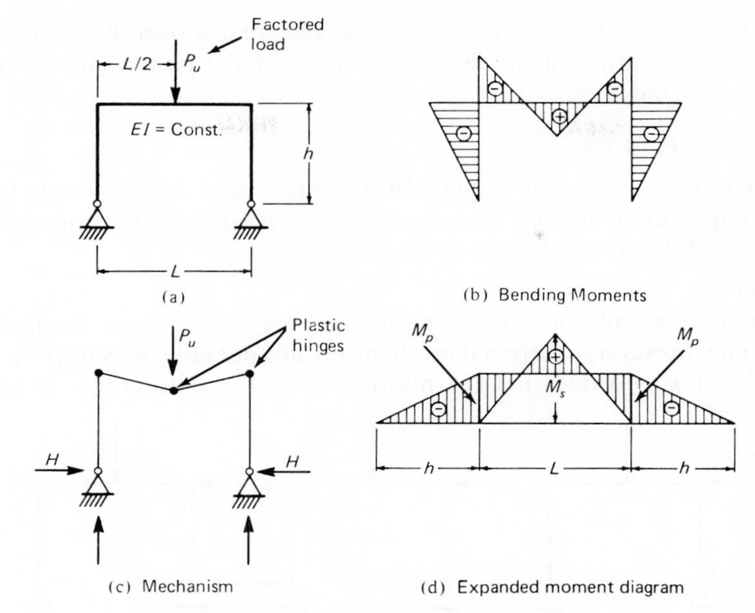

Fig. 15.2.1. Example 15.2.1.

The simple beam moment for the girder is

$$M_s = \frac{P_u L}{4} \tag{a}$$

which when superimposed on the moments at the ends of the girder ($H \times h$), gives for equilibrium when the mechanism occurs,

$$M_p = M_s - M_p = \frac{P_u L}{4} - M_p \tag{b}$$

$$P_u = \frac{8 M_p}{L} \tag{c}$$

From the concepts of plastic design it might have been expected that only two plastic hinges should be required for a collapse mechanism, since the structure is statically indeterminate to the first degree. In this case both corner hinges form simultaneously, since the horizontal base reactions are equal and opposite. Depending on the ratio of h to L either the midspan plastic hinge, or the two corner plastic hinges, will occur first. For small values of h/L the positive moment plastic hinge forms first and the structure remains stable until the corner plastic hinges form. If, however, h/L is large, the corner plastic hinges occur first and the structure has reached its collapse condition. If the beam and columns are of different cross section the structure can be designed so the positive moment plastic hinge occurs first. This should be the objective. The occurrence first of the corner plastic hinges creates overall frame instability prior to utilizing the flexural strength of the girder; such a result should be avoided.

EXAMPLE 15.2.2

Determine the plastic strength of the same frame as in Example 15.2.1 except in addition apply a horizontal load $0.5P_u$ at the top of the column (see Fig. 15.2.2a). Use the equilibrium method.

Solution

Again, two plastic hinges should provide the collapse mechanism. This time, however, additional mechanisms are possible, as shown in Fig. 15.2.2. It is possible that (a) plastic hinges form at points 2 and 4

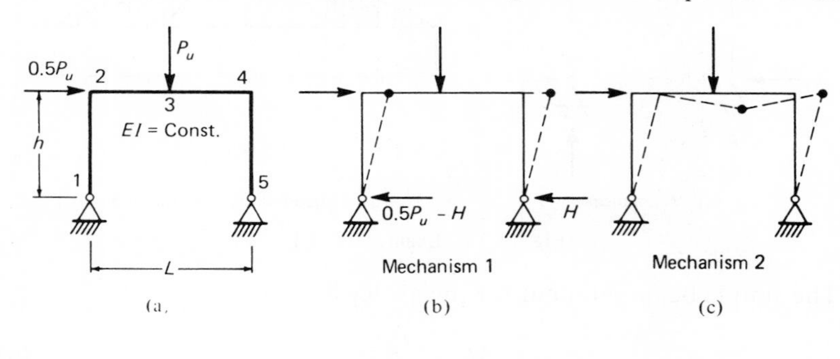

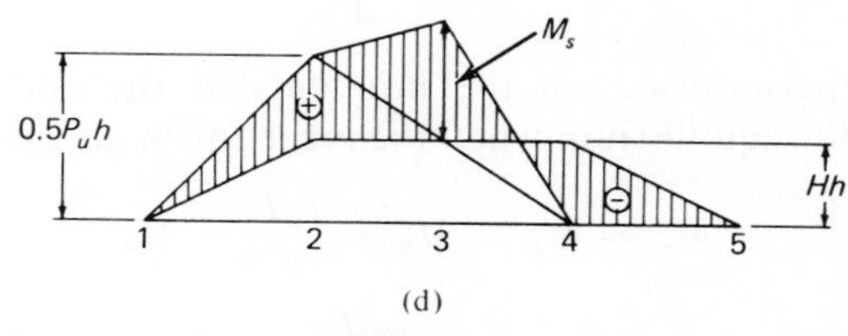

Fig. 15.2.2. Example 15.2.2.

(Fig. 15.2.2b); (b) plastic hinges form at points 3 and 4 (Fig. 15.2.2c); and also (c) plastic hinges form at points 2, 3, and 4. The last situation would occur only if two of the hinges formed simultaneously.

(a) Plastic hinges at points 2 and 4,

$$M_s = \frac{P_u L}{4} \tag{a}$$

Equilibrium requires the positive moment M_p at point 2 $(0.5P_u h - Hh)$ to equal the negative moment M_p at point 4(Hh):

$$M_p = 0.5P_u h - M_p \tag{b}$$

$$M_p = \frac{P_u h}{4}; \qquad P_u = \frac{4M_p}{h} \tag{c}$$

Further, in order that Eq. (c) is valid, the resulting moment at point 3 cannot exceed M_p:

$$M_3 = M_s + \frac{1}{2}(0.5P_u h) - M_p \tag{d}$$

$$= \frac{P_u L}{4} + \frac{P_u h}{4} - \frac{P_u h}{4} = \frac{P_u L}{4} \leq M_p \tag{e}$$

Equation (e) requires $P_u L/4 < P_u h/4$, which means if $L < h$ plastic hinges form at points 2 and 4. If $L > h$ the plastic hinges will form at points 3 and 4.

(b) Plastic hinges at points 3 and 4. For this, Eq. (d) is to be used letting $M_3 = M_p$,

$$M_p = M_s + \frac{1}{2}(0.5P_u h) - M_p \tag{f}$$

$$M_p = \frac{P_u}{8}(L + h) \tag{g}$$

$$P_u = \frac{8M_p}{L + h} \tag{h}$$

The above analysis has assumed constant moment of inertia (constant M_p) for both girder and column. If the girder and column are different a combined mechanism with plastic hinges at points 2, 3, and 4 could be achieved.

EXAMPLE 15.2.3

Determine the plastic capacity P_u for the gabled frame of Fig. 15.2.3 using the equilibrium method. All elements of the frame are identical, having a plastic moment capacity of 200 ft-kips.

Solution

Equilibrium requires a compatibility between the moment diagrams of the component loadings as shown in Fig. 15.2.3b. The moment due to H may be thought of as causing negative bending, while the other two components are causing positive bending. The moment diagrams for the components are shown on a horizontal baseline in Fig. 15.2.3c.

The maximum so-called positive moments, M_{s1} and M_{s2} are

$$M_{s1} = 7.5P_u$$

$$M_{s2} = 0.5P_u h = 7.5P_u$$

Several possible collapse mechanisms must be considered.

(a) Consider a sidesway mechanism (Fig. 15.2.3d) with plastic hinges at points 2 and 6. Equilibrium requires at point 2:

$$M_2 = M_p = M_{s2} - Hh \tag{a}$$

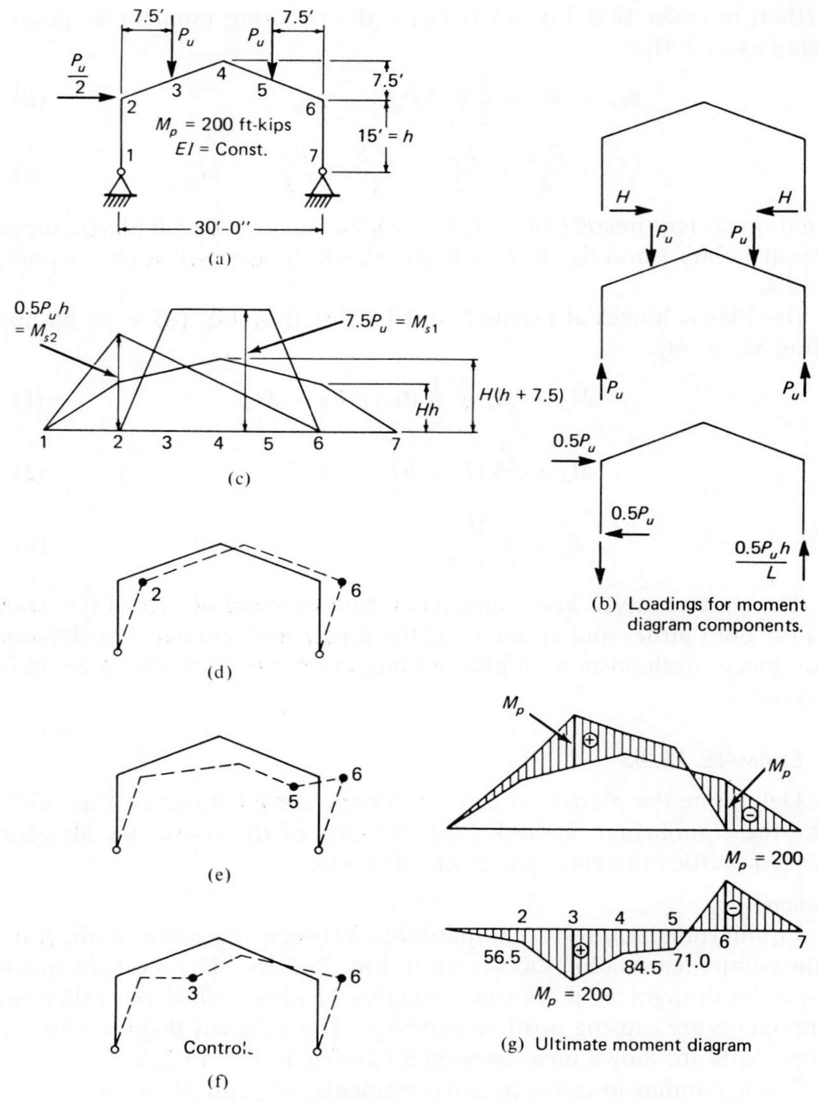

Fig. 15.2.3. Gabled frame analysis—Example 15.2.3.

and at point 6:

$$M_6 = M_p = Hh = 15H \tag{b}$$

Thus,

$$M_p = M_{s2} - M_p = 7.5P_u - M_p$$
$$M_p = 3.75P_u \tag{c}$$

(b) Consider a combination mechanism with plastic hinges at points 5 and 6 (Fig. 15.2.3e). At point 5,

$$M_5 = M_p = M_{s1} + \frac{1}{4} M_{s2} - H(h + 3.75)$$

$$M_p = 7.5 P_u + \frac{7.5}{4} P_u - \frac{M_p}{15}(15 + 3.75)$$

$$M_p = 4.17 P_u$$

Actually, the mechanism consisting of plastic hinges at points 5 and 6 could have been eliminated as a possibility by comparing the total positive moment at points 3 and 5 (Fig. 15.2.3c) which will indicate that point 3 will achieve a plastic hinge before point 5 can attain it.

(c) Consider the mechanism with plastic hinges at points 3 and 6:

$$M_3 = M_p = M_{s1} + \frac{3}{4} M_{s2} - H(h + 3.75)$$

$$M_p = 7.5 P_u + \frac{3}{4}(7.5 P_u) - \frac{M_p}{15}(15 + 3.75)$$

$$M_p = 5.83 P_u \qquad \text{Governs}$$

The largest M_p, or the smallest P_u, for a mechanism to occur indicates the governing one. A check may be made by determining the ultimate moment diagram assuming $M_p = 200$ ft-kips. Thus

$$P_u = \frac{200}{5.83} = 34.2 \text{ kips}$$

$$H = \frac{M_p}{15} = \frac{200}{15} = 13.33 \text{ kips}$$

$$M_{s1} = 7.5 P_u = 7.5(34.2) = 256.5 \text{ ft-kips}$$

$$M_{s2} = 256.2 \text{ ft-kips}$$

At the critical locations, the ultimate moment is

$$M_2 = 256.5 - 200 = 56.5 \text{ ft-kips} < M_p \qquad \text{OK}$$

$$M_3 = 256.5 + \tfrac{3}{4}(256.5) - 13.33(18.75) = 200 \text{ ft-kips} = M_p$$

$$M_4 = 256.5 + \tfrac{1}{2}(256.5) - 13.33(22.5) = 84.5 \text{ ft-kips} < M_p$$

$$M_5 = 256.5 + \tfrac{1}{4}(256.5) - 13.33(18.75) = 71.0 \text{ ft-kips} < M_p$$

$$M_6 = 13.33(15) = 200 \text{ ft-kips} = M_p$$

The final ultimate-moment diagram is shown in Fig. 15.2.3g, both as it would be from graphical superposition and as it would be if the net moment is plotted on a horizontal baseline.

EXAMPLE 15.2.4

Determine the plastic moment required for the frame of Fig. 15.2.4. Consider (a) uniform vertical loading w alone; and (b) uniform vertical loading in combination with uniform lateral loading.

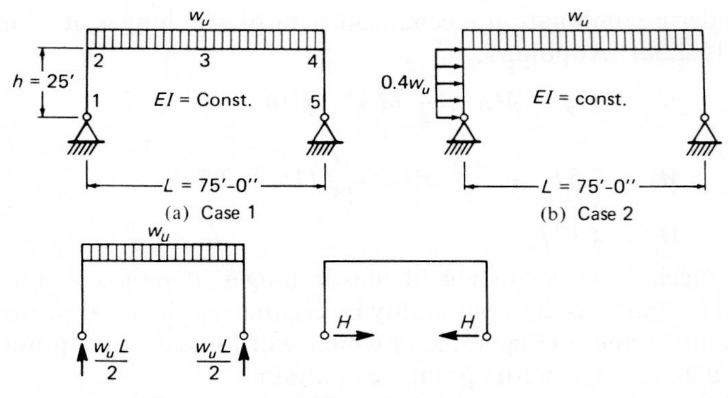

(a) Case 1 (b) Case 2

(c) Components of loading - Case 1

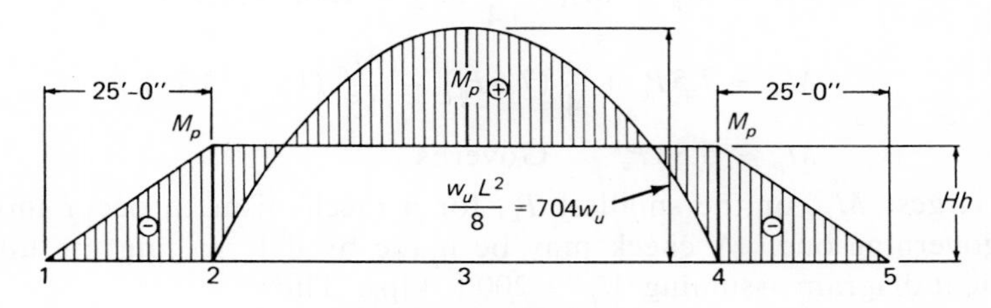

(d) Moment diagram - Case 1

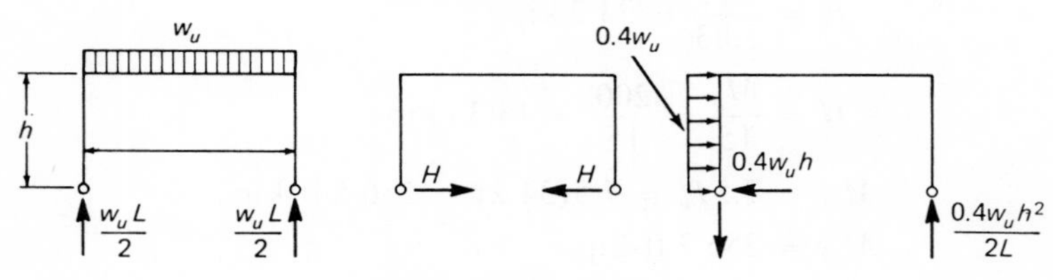

(e) Components of loading - Case 2

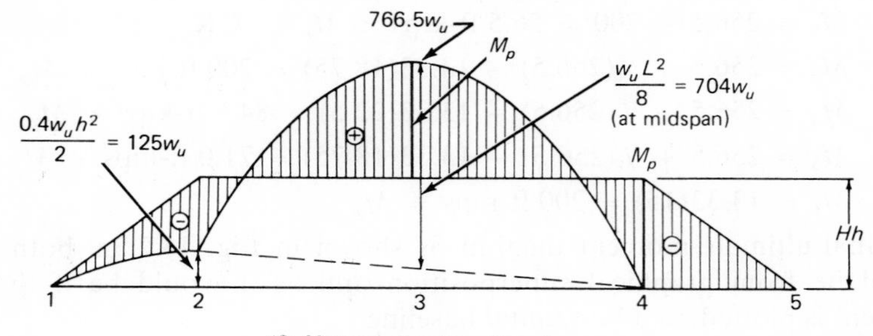

(f) Moment diagram - Case 2

Fig. 15.2.4. Example 15.2.4.

Solution

(a) Vertical loading only. The simple beam moment is

$$M_s = \frac{w_u L^2}{8} = \frac{w_u (75)^2}{8} = 704 w_u \quad \text{(a)}$$

For a properly designed frame, the plastic hinges will occur at the ends and midspan of the girder:

$$M_p = M_s - M_p$$

$$M_p = \frac{1}{2} M_s = \frac{w_u L^2}{16} = 352 w_u \quad \text{(b)}$$

and the superposition of the bending moments due to the loading components is shown in Fig. 15.2.4d.

(b) Combined vertical and lateral loading. There is an additional component of loading as seen in Fig. 15.2.4e which contributes an unsymmetrical effect of the same bending moment sign as caused by the vertical uniform loading. It may be observed from Fig. 15.2.4f that the maximum moments will occur slightly to the left of point 3 and at point 4. One could mathematically solve for the exact location and magnitude of the plastic moment. However, for design purposes one may use a graphical construction and divide the maximum value of the parabolic curve by two to establish the horizontal line *Hh*. In other words, equalize the positive and negative moments. The moment at point 2 is obviously less than that at point 4 so that no plastic hinge will form at point 2.

By scaling, the maximum ordinate of the parabola near point 3 is $768 w_u$. In which case

$$M_p = \frac{768 w_u}{2} = 384 w_u \quad \text{(c)}$$

The uniform lateral loading causes about 9½ percent reduction in vertical carrying capacity, or alternatively it would require a section having about 9½ percent greater moment capacity.

EXAMPLE 15.2.5

Re-solve Example 15.2.2 if the bases are fixed instead of hinged, as shown in Fig. 15.2.5. Use $L = 80$ ft and $h = 20$ ft.

Solution

Using the equilibrium method, separate the loading components (Fig. 15.2.5b) and then superimpose the moment diagrams for each component of load, as shown in Fig. 15.2.5c. Since the structure is statically indeterminate to the third degree, four plastic hinges are required for the collapse mechanism.

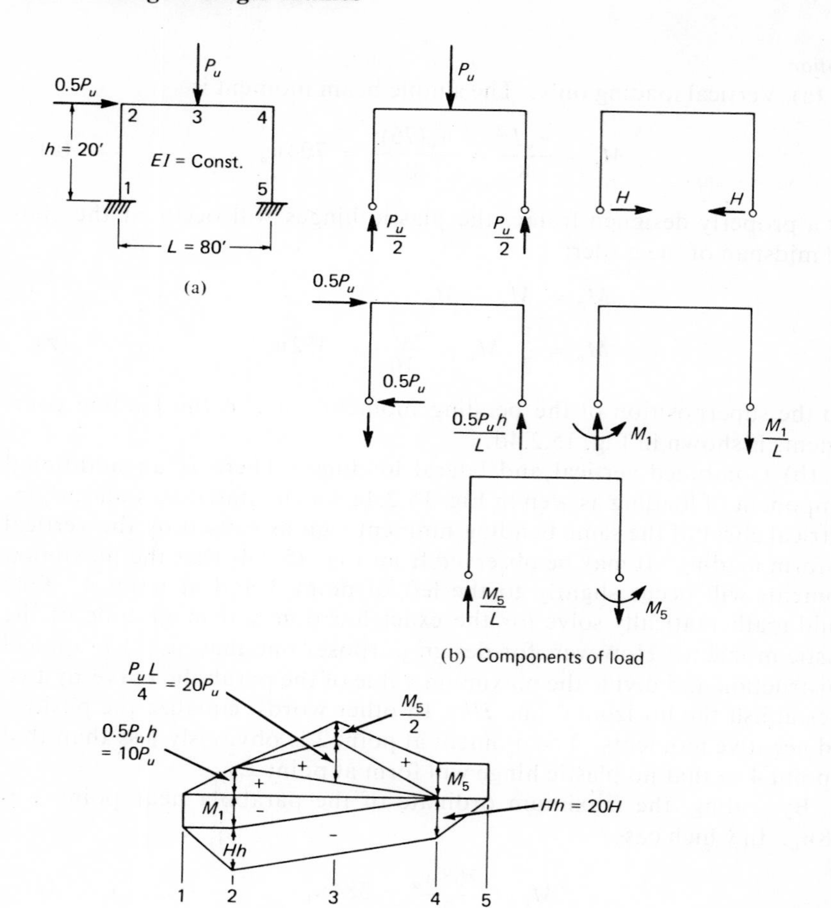

(a)

(b) Components of load

(c) Components of moment diagram

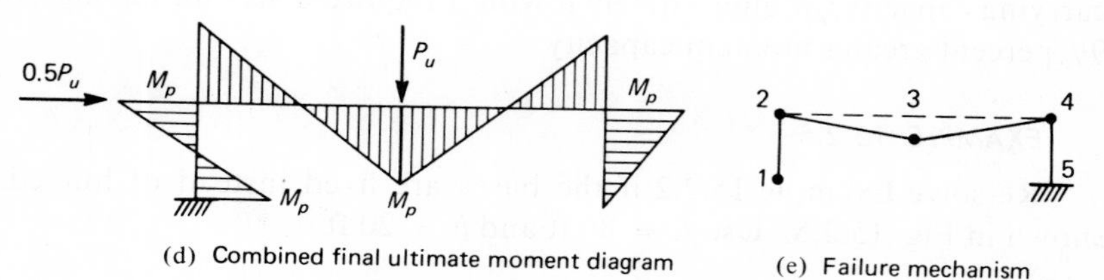

(d) Combined final ultimate moment diagram

(e) Failure mechanism

Fig. 15.2.5. Example 15.2.5.

A study of the moment-diagram components indicates maximum moments at points 1 and 5 and probable maximums at points 3 and 4. Assume this combination and check the resulting moment at point 2.

At point 1

$$M_1 = M_p \tag{a}$$

At point 5

$$M_5 = M_p \tag{b}$$

At point 4

$$\begin{aligned} M_4 &= 20H - M_5 = 20H - M_p = M_p \\ 20H &= 2M_p \\ H &= M_p/10 \end{aligned} \tag{c}$$

At point 3

$$M_3 = \frac{M_5}{2} + 20P_u + 5P_u - \frac{M_1}{2} - 20H \tag{d}$$

$$\begin{aligned} M_p &= \frac{M_p}{2} + 25P_u - \frac{M_p}{2} - 2M_p \\ M_p &= 8.33P_u \end{aligned} \tag{e}$$

Check point 2

$$\begin{aligned} M_2 &= 10P_u - M_1 - 20H \\ &= 10P_u - M_p - 2M_p \\ &= 10P_u - 25P_u = -15P_u > M_p \qquad \text{N.G.} \end{aligned}$$

Thus a plastic hinge occurs at point 2. In which case a beam mechanism is a probability. Furthermore, if a negative moment plastic hinge occurs at point 2, a plastic moment must also exist at point 1. The equilibrium conditions with plastic hinges at 1, 2, 3, and 4 are

$$M_1 = M_p \tag{g}$$

At point 2

$$\begin{aligned} M_2 &= 20H + M_1 - 10P_u \\ M_p &= 20H + M_p - 10P_u \\ 0 &= 20H - 10P_u \end{aligned} \tag{h}$$

At point 3

$$\begin{aligned} M_3 &= 0.5M_5 + 20P_u + 5P_u - 0.5M_1 - 20H \\ M_p &= 0.5M_5 + 25P_u - 0.5M_p - 10P_u \\ 1.5M_p &= 0.5M_5 + 15P_u \end{aligned} \tag{i}$$

At point 4

$$\begin{aligned} M_4 &= 20H - M_5 \\ M_p &= 10P_u - M_5 \end{aligned} \tag{j}$$

Multiply Eq. (j) by 0.5 and add to Eq. (i), which gives

$$2M_p = 20P_u$$

$$M_p = 10P_u \qquad \text{(k)}$$

The resulting moment at point 5, using Eq. (j), is

$$M_5 = M_p - 10P_u = 10P_u - 10P_u = 0$$

which is less than M_p, as it should be. The correct mechanism, containing plastic hinges at points 1, 2, 3, and 4 and the ultimate moment diagram appear in Fig. 15.2.5.

Energy Method. Just as discussed for beams in Chapter 10, the energy method may also be used for the plastic analysis of frames. For certain frame arrangements, the energy method will be found easier. The following examples show how the energy method may be applied to the same structures previously analyzed by the equilibrium method.

EXAMPLE 15.2.6

Repeat Example 15.2.1 except use the energy method.

Solution

The mechanism is shown in Fig. 15.2.6. The plastic hinge locations

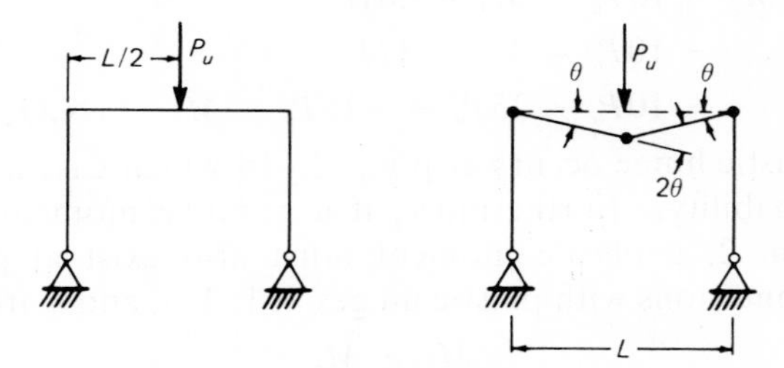

Fig. 15.2.6. Example 15.2.6.

are assumed and from the geometry the angles θ are established. The external work done by the load equals the internal strain energy of the plastic moments moving through their angles of rotation:

$$W_e = W_i$$

$$P_u \frac{\theta L}{2} = M_p(\theta + 2\theta + \theta)$$

$$P_u = \frac{8M_p}{L}$$

exactly as obtained in Example 15.2.1.

EXAMPLE 15.2.7

Repeat Example 15.2.2 using the energy method.

Solution

The possible mechanisms are shown in Fig. 15.2.7.

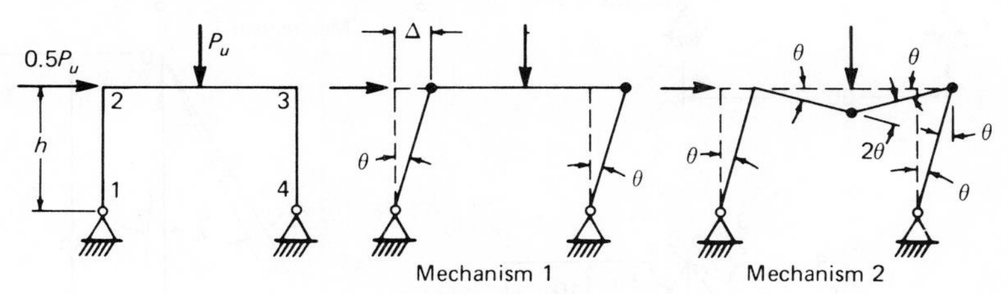

Fig. 15.2.7. Example 15.2.7.

(a) Mechanism 1

$$0.5P_u\theta h = M_p(\theta + \theta)$$

$$P_u = \frac{4M_p}{h}$$

(b) Mechanism 2

$$0.5P_u\theta h + P_u\theta\frac{L}{2} = M_p(2\theta + 2\theta)$$

$$P_u = \frac{8M_p}{L + h}$$

The results are identical to those of Example 15.2.2.

EXAMPLE 15.2.8

Repeat Example 15.2.3 using the energy method.

Solution

The possible mechanisms are shown in Fig. 15.2.8.

(a) Mechanism 1

$$0.5P_u(15\theta) = M_p(2\theta)$$

$$M_p = 3.75P_u$$

(b) Mechanism 2. In order to treat this more complex mechanism, the concept of *instantaneous center* is used. When plastic hinges form at points 5 and 6, three rigid bodies remain which rotate as the structure moves. Segment 1–2–3–4–5 rotates about point 1; segment 6–7 rotates about point 7; segment 5–6 rotates and translates an amount which is controlled by the movement of points 5 and 6 on the adjacent rigid segments. If the body is rigid, point 5′ is perpendicular to line 1–5, and point 6′ is perpendicular to line 6–7. Thus, the points 5 and 6 may be

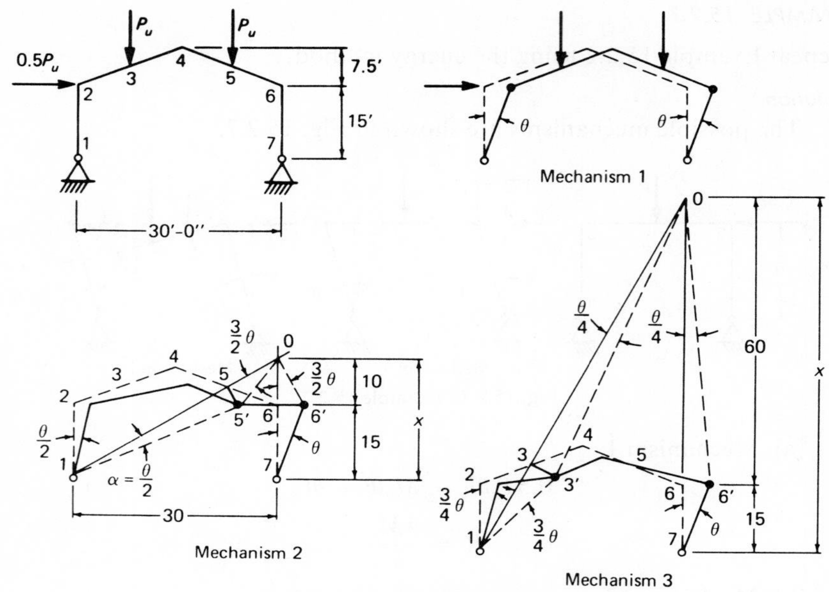

Fig. 15.2.8. Example 15.2.8.

thought of as rotating about point 0, the intersection of line 1–5 and line 6–7; i.e., the instantaneous center.

The first step in the energy method using the instantaneous center is to determine its location; since point 5 is 22.5 ft horizontally and 18.75 ft vertically from point 1, the vertical distance to point 0 from point 7 is

$$\frac{x}{30} = \frac{18.75}{22.5}, \qquad x = 25 \text{ ft}$$

Next, a reference angle θ is established arbitrarily as shown in Fig. 15.2.8. By proportion, the angle of rotation with respect to point 0 is $3\theta/2$. The rigid-body segment 5–6 rotates through this angle $3\theta/2$. By inverse proportion as distance 0–5 is to 1–5, the rotation of rigid body 1–2–3–4–5 about point 1 is

$$\frac{1/4}{3/4} = \frac{\alpha}{3\theta/2}, \qquad \alpha = \frac{\theta}{2}$$

The relative plastic-hinge rotation at point 5 is

$$\frac{\theta}{2} + \frac{3\theta}{2} = 2\theta$$

The relative plastic-hinge rotation at point 6 is

$$\theta + \frac{3\theta}{2} = 2.5\theta$$

In order to compute external work done by the applied loads, the vertical distances moved due to rotation of points 3 and 5 and the horizontal distance moved at point 2 are required.

The vertical displacement of point 3 equals the angle of rotation times the horizontal projection from points 2 to 3. The load at point 3 moves vertically through a distance

$$\frac{\theta}{2}(7.5) = 3.75\theta$$

The load at point 5 moves vertically through the distance

$$\frac{\theta}{2}(22.5) = 11.25\theta$$

The load at point 2 moves horizontally through the distance

$$\frac{\theta}{2}(15) = 7.5\theta$$

The complete energy equation then becomes

$$W_e = W_i$$

$$0.5P_u(7.5\theta) + P_u(3.75\theta) + P_u(11.25\theta) = M_p(2\theta + 2.5\theta)$$

$$M_p = \frac{18.75}{4.5}P_u = 4.17P_u$$

the same as from the equilibrium method.

(c) Mechanism 3. The instantaneous center is found by intersecting the line 1–3 with line 6–7.

$$\frac{x}{30} = \frac{18.75}{7.5}, \qquad x = 75 \text{ ft}$$

If θ is defined by Fig. 15.2.8, then the angle of rotation with respect to 0 is $\theta/4$, since the distance 0–6 is four times the distance 6–7. Since the distance 0–3 is three times the distance 3–1, the angle 3–1–3′ is $3\theta/4$ (3 times the rotation angle about 0).

The external work done by the various loads is

Load at 2, $$0.5P_u\left(\frac{3\theta}{4}\right)(15) = \frac{22.5}{4}P_u\theta$$

Load at 3, $$P_u\left(\frac{3\theta}{4}\right)(7.5) = \frac{22.5}{4}P_u\theta$$

Load at 5, $$P_u\left(\frac{\theta}{4}\right)(7.5) = \frac{7.5}{4}P_u\theta$$

The internal strain energy is

Moment at 3, $$M_p\left(\frac{3\theta}{4} + \frac{\theta}{4}\right) = M_p\theta$$

Moment at 6, $$M_p\left(\theta + \frac{\theta}{4}\right) = M_p \frac{5\theta}{4}$$

$$W_e = W_i$$

$$P_u\theta\left(\frac{22.5}{4} + \frac{22.5}{4} + \frac{7.5}{4}\right) = M_p\theta\left(1 + \frac{5}{4}\right)$$

$$P_u\left(\frac{52.5}{4}\right) = M_p\left(\frac{9}{4}\right)$$

$$M_p = \frac{52.5}{9} P_u = 5.83 P_u \qquad \text{Governs}$$

The result is identical with Example 15.2.3.

EXAMPLE 15.2.9

Repeat Example 15.2.4 using the energy method.

Solution

(a) For case 1 with no lateral loading, the mechanism is shown in Fig. 15.2.9a. The external work done by uniformly distributed load equals

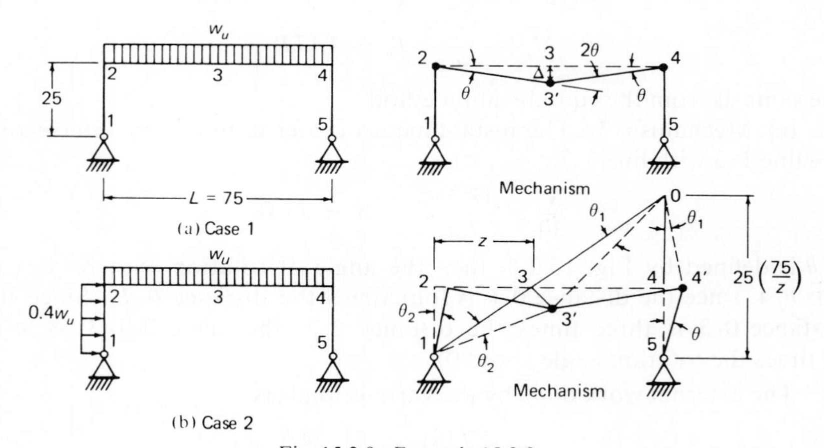

Fig. 15.2.9. Example 15.2.9.

the intensity of load w_u times the area displaced as the mechanism forms. In Fig. 15.2.9a the area displaced is the triangular area 2–3–4–3′. Thus,

$$W_e = w_u \frac{1}{2}(\Delta) L = w_u \frac{1}{2}\left(\theta \frac{L}{2}\right) L = w_u\theta \frac{L^2}{4}$$

Finally, equating

$$W_e = W_i$$

$$w_u\theta \frac{L^2}{4} = M_p(\theta + 2\theta + \theta) = 4M_p\theta$$

$$M_p = \frac{w_u L^2}{16} = \frac{w_u(75)^2}{16} = 352 w_u$$

(b) Case 2—combined vertical and lateral loading. For this situation, the position of the plastic hinge at point 3 is an unknown distance z from point 2, as shown in Fig. 15.2.9b. One could assume several positions of point 3 and determine the maximum M_p (or the minimum w_u), or the problem can be set up mathematically to determine minimum w_u.

Try $z = L/2 = 37.5$ ft:

$$\theta_1 = \theta$$

$$\theta_2 = \theta$$

$$W_e \text{ (vertical load)} = w_u\theta\left(\frac{L^2}{4}\right) = 1408\, w_u\theta$$

$$W_e \text{ (horizontal load)} = 0.4 w_u\left(\frac{1}{2}\right)\theta h^2 = 125 w_u\theta$$

$$W_e = W_i$$

$$w_u\theta(1408 + 125) = M_p(2\theta + 2\theta)$$

$$M_p = 382.8\, w_u$$

Try $z = 35$ ft:

$$\theta_1 = \theta\left(\frac{z}{75 - z}\right) = \frac{35}{40}\theta = 0.875\theta$$

$$\theta_2 = \theta_1\left(\frac{75 - z}{z}\right) = \frac{40}{35}\theta_1 = \theta$$

Vertical Δ at point 3,

$$\Delta = z\theta$$

$$W_e = W_i$$

$$w_u\left(\frac{1}{2}\right)35\theta(75) + 125 w_u\theta = M_p(1.875\theta + 1.875\theta)$$

$$w_u(1312.5 + 125) = 3.75 M_p$$

$$M_p = 383.0 w_u$$

This result is little different from that obtained for $z = L/2$. The answer in other words is not too sensitive to the exact position of the plastic hinge, and the approximate procedure illustrated is considered adequate for design purposes.

If one wishes, the "exact" solution may be obtained, as follows:

$$W_e \text{ (vertical load)} = w_u\left(\frac{1}{2}\right)z\theta(75) = 37.5w_u\theta z$$

$$W_e \text{ (horizontal load)} = 125w_u\theta$$

$$W_i \text{ (at point 3)} = M_p(\theta + \theta_1)$$

$$= M_p\left(\theta + \theta\frac{z}{75 - z}\right)$$

$$= M_p\theta\left(\frac{75}{75 - z}\right)$$

$$W_i \text{ (at point 4)} = M_p\theta\left(\frac{75}{75 - z}\right)$$

$$W_e = W_i$$

$$w_u(125 + 37.5z) = M_p\left(\frac{150}{75 - z}\right)$$

$$M_p = w_u\left[\frac{(125 + 37.5z)(75 - z)}{150}\right] \qquad \text{(a)}$$

$$M_p = w_u(-z^2 + 71.67z + 150)$$

$$\frac{dM_p}{dz} = 0 = -2z + 71.67$$

$$z = 35.83 \text{ ft}$$

Substituting z into Eq. (a) above gives

$$M_p = w_u\left[\frac{125 + 37.5(35.83)}{150}\right](75 - 35.83) = 383.5w_u$$

This result is obviously little different than the approximate values.

EXAMPLE 15.2.10

Repeat Example 15.2.5 using the energy method.

Solution

The possible mechanisms are shown in Fig. 15.2.10.

(a) Mechanism 1: The beam mechanism shown is always a good possibility when the vertical load is larger than the horizontal load.

$$W_e = W_i$$

$$P_u(40\theta) = M_p(\theta + 2\theta + \theta)$$

$$M_p = 10P_u$$

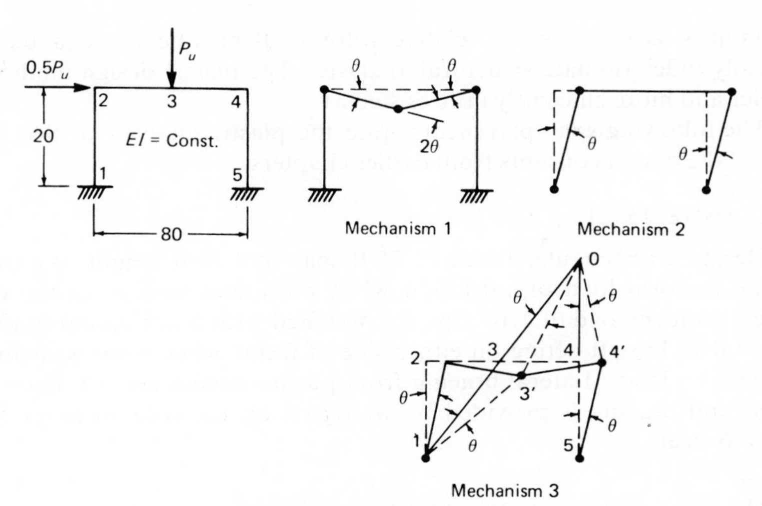

Fig. 15.2.10. Example 15.2.10.

(b) Mechanism 2: The sidesway mechanism is not expected to control since the larger load is the vertical one:

$$0.5P_u(20\theta) + P_u(0) = M_p(\theta + \theta + \theta + \theta)$$
$$M_p = 2.5P_u$$

(c) Mechanism 3: This mechanism is a combination of mechanisms 1 and 2. Note that the angle change at point 2 is equal and opposite for mechanisms 1 and 2; thus the combination would not expect a plastic hinge at point 2:

$$W_e = W_i$$
$$0.5P_u(20\theta) + P_u(40\theta) = M_p(\theta + 2\theta + 2\theta + \theta)$$
$$50P_u = 6M_p$$
$$M_p = 8.33P_u$$

The largest value of M_p is $10P_u$, so Mechanism 1, the beam mechanism, governs. It is noted that in the energy method it was not established that a plastic hinge develops at point 1 as determined in the equilibrium method; unnecessary here since no rotation occurs at point 1 when the mechanism is achieved.

In conclusion, either the equilibrium method or the energy method may be used for the plastic analysis of frames. It is suggested that one method be used to check the other.

15.3. PLASTIC-DESIGN EXAMPLES—ONE-STORY FRAMES

One-story frames are most practicably designed by the plastic-design method. Working stress design, as illustrated in the next article, requires preliminary assumptions of relative stiffness of members and a tedious statically indeterminate structural analysis. The plastic design method is simpler and more efficiently utilizes steel.

The following examples incorporate the plastic analysis of Sec. 15.2 with all the design concepts from earlier chapters.

EXAMPLE 15.3.1

Design a rectangular frame of 75-ft span and 25-ft height to carry a vertical uniform load of 1.0 kip/ft when no lateral load is acting; or a vertical uniform load of 1.10 kips/ft combined with a horizontal uniform load of 0.44 kips/ft acting on either side of frame when wind is included with gravity load. Lateral bracing from purlins occurs every 6 ft on the girder, and bracing is provided by wall girts on the column every 5 ft. Use A36 steel.

Solution

The frame is shown in Fig. 15.3.1.

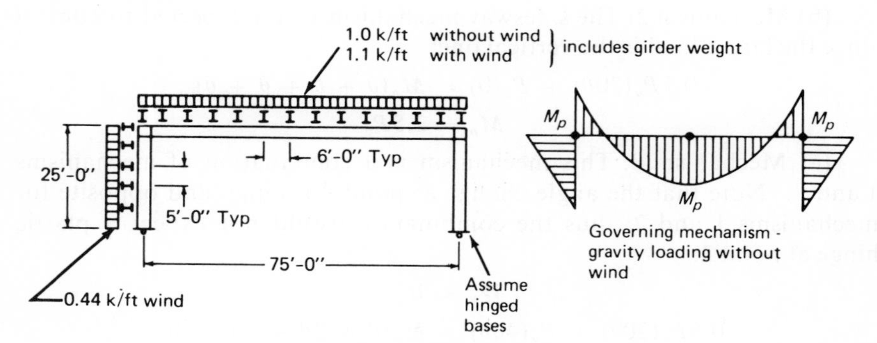

Fig. 15.3.1. Example 15.3.1.

(a) Factored Loads. Apply load factors to the given service loads. Two cases must be considered, loading with and without wind. AISC–2.1 prescribes factors of 1.7 and 1.3 for loads without wind and with wind respectively. The factored loads are therefore

Case 1

$$w_u = 1.0(1.7) = 1.7 \text{ kips/ft}$$

Case 2

$$w_u = 1.0(1.3) = 1.43 \text{ kips/ft (vertical)}$$

$$0.4w_u = 0.572 \text{ kip/ft (horizontal)}$$

(b) Consider vertical loading without wind; Case 1. Using the plastic analysis as illustrated in Example 15.2.4,

$$\text{Reqd } M_p = \frac{w_u L^2}{16} = \frac{1.7(75)^2}{16} = 598 \text{ ft-kips}$$

$$\text{Reqd } Z = \frac{M_p}{F_y} = \frac{598(12)}{36} = 199 \text{ in.}^3$$

Try W24 × 76, $M_p = 36(201.1)/12 = 603$ ft-kips, or a deeper heavier section if deflection is a critical factor. Check local buckling requirements of AISC–2.7.

$$\frac{b}{2t_f} = \frac{8.985}{2(0.682)} = 6.6 < 8.5 \qquad \text{OK}$$

$$\frac{d}{t} = \frac{23.91}{0.44} = 54.3 < 68.7 \qquad \text{OK}$$

(c) Check whether case 2 including wind may govern. Using plastic analysis from Example 15.2.4,

$$\text{Reqd } M_p = 384 w_u = 384(1.43) = 549 \text{ ft-kips}$$

Since this is less than required for vertical load only, Case 1 controls. The reader is reminded that use of the 1.3 load factor for wind instead of 1.7 is the same as allowing a 30 percent overstress for wind, because of its temporary nature. This is essentially the same as AISC–1.5.6 for working stress method.

(d) Check W24 × 76 for action as a beam-column. Determine the effective-length factor K,

$$G_A \approx 10.0 \text{ (Estimated for hinge at bottom)}$$

$$G_B = \frac{I/L \text{ (column)}}{I/L \text{ (girder)}} = \frac{I/25}{I/75} = 3.0$$

From Fig. 14.3.1b, unbraced frame, find $K_x = 2.25$.

$$\frac{K_x L}{r_x} = \frac{2.25(25)(12)}{9.69} = 69.7; \qquad F_e' = 30.7 \text{ ksi (AISC Appendix Table 2)}$$

$$\frac{K_y L}{r_y} = \frac{1.0(5)(12)}{1.92} = 31.2$$

$$P_e = (23/12)AF_e' = (23/12)(22.4)(30.7) = 1320 \text{ kips}$$

$$C_m = 0.85, \text{ for an unbraced frame.}$$

$$F_a = 16.46 \text{ ksi (for } KL/r = 69.7)$$

$$P_{cr} = 1.70F_aA = 1.70(16.46)(22.4) = 625 \text{ kips}$$

$$M_m = \left[1.07 - \frac{\sqrt{F_y}(L/r_y)}{3160}\right] M_p > M_p, \qquad \text{use } M_m = M_p$$

Check column stability using AISC Formula 2.4–2:

$$\frac{P}{P_{cr}} + \frac{M_i}{M_m}\left(\frac{C_m}{1 - P/P_e}\right) = \frac{63.8}{625} + \frac{598}{603}\left(\frac{0.85}{1 - 63.8/1320}\right)$$

$$= 0.102 + 0.885 = 0.987 < 1.0 \qquad \text{OK}$$

Check strength criterion, AISC Formula 2.4.3:

$$P_y = F_y A = 36(22.4) = 805 \text{ kips}$$

$$\frac{P}{P_y} + \frac{M}{1.18M_p} = \frac{63.8}{805} + \frac{598}{1.18(603)} = 0.08 + 0.84 = 0.92 < 1.0 \qquad \text{OK}$$

Since $P/P_y < 0.15$, the strength criterion is automatically satisfied and the full plastic moment may be utilized. Check local buckling on beam-column, AISC Formula 2.7–1a,

$$\frac{d}{t_w} = \frac{412}{\sqrt{F_y}}\left(1 - 1.4\frac{P}{P_y}\right) = 68.7 - 96.1\frac{P}{P_y}$$

$$= 68.7 - 96.1(0.08) = 61.0 > 54.3 \text{ actual} \qquad \text{OK}$$

Use W24 × 76 for beam and column members.

(e) Check shear. Outside the corner connection,

$$V_u = 1.7(75)/2 = 63.8 \text{ kips}$$

$$v_u = \frac{V_u}{t_w d} = \frac{63.8}{0.44(23.91)} = 6.06 \text{ ksi} < 0.55F_y \qquad \text{OK}$$

Inside the corner connection, referring to Fig. 15.3.2, the shear produced

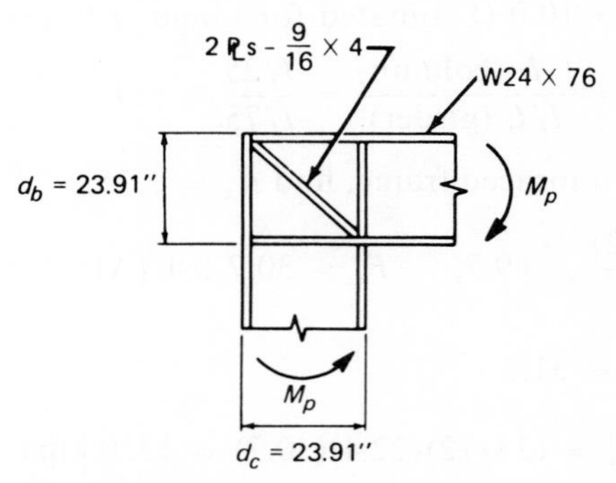

Fig. 15.3.2. Corner connection for Example 15.3.1.

may be expressed, according to Eq. 13.8.5 as

$$\text{Reqd } t_w = \frac{23M}{A_{bc}F_y} = \frac{23(603)}{(23.91)^2 36} = 0.67 \text{ in.}$$

Since the web thickness is only 0.44 in., a diagonal web stiffener is required.

(f) Determine size of diagonal stiffener plate. Force to be carried by diagonal stiffeners,

$$\text{Shear force} = \frac{M}{d_b}\left(\frac{0.67 - 0.44}{0.67}\right) = \frac{603(12)}{23.91}\left(\frac{0.23}{0.67}\right) = 103.5 \text{ kips}$$

$$\text{Stiffener force} = (\text{shear force})\sqrt{2}$$
$$= 103.5(1.414) = 146.4 \text{ kips}$$

$$\text{Reqd } A_{st} = \frac{\text{stiffener force}}{F_y} = \frac{146.4}{36} = 4.07 \text{ sq in.}$$

Since the flange width of W24 × 76 is 8.985 in. and the web thickness is 0.44 in., the width available for a stiffener plate is (8.985 − 0.44)/2 = 4.25 in.

Try 4-in. wide plate. For local buckling,

$$\frac{b}{t} \le 8.5; \qquad t_{\min} = \frac{4}{8.5} = 0.47 \text{ in.}$$

Use 2 PLs, $\frac{9}{16}$ × 4, $A = 4.50$ sq in.

(g) Check lateral support, AISC–2.9. Little moment variation occurs over the 6-ft laterally unbraced length. Thus, the end moment ratio on the segment ≈ −1, requiring use of AISC Formula 2.9–1b,

$$\frac{L_{cr}}{r_y} = \frac{1375}{F_y} = 38.2$$

$$L_{cr} = 38.2r_y = 38.2(1.92)/12 = 6.1 \text{ ft}, \quad > 6 \text{ ft} \qquad \text{OK}$$

Consider lateral bracing acceptable at 6 ft on the girder and 5 ft on the column.

(h) Overall frame stability. Use of AISC Formula 2.4–2 using $K > 1.0$ is considered to automatically satisfy the overall stability necessary to achieve plastic strength. As a further check one may apply Eq. 14.6.1 as used under 1963 AISC specification,

$$\frac{2P}{P_y} + \frac{L/r}{70} = 2(0.08) + \frac{31}{70} = 0.60 < 1.0$$

where $L/r = 25(12)/9.69 = 31$, based on the unbraced length rather than the effective length. The 1969 AISC procedure seems preferable and sufficient, but the designer may feel a greater degree of confidence if the above equation is also satisfied.

EXAMPLE 15.3.2

Redesign the frame of Example 15.3.1 (Fig. 15.3.1) if the column cannot be deeper than 14.50 in. actual depth.

Solution

For this situation the column and the beam will have different plastic-moment capacities.

(a) Select section for plastic strength. There are many possible solutions using various sizes of column. Try a W14 × 84 section with z = 145 in.^{3} Assuming vertical loading governs, the positive plastic moment required at girder midspan is

$$M_p = \frac{w_u L^2}{8} - M_p\,(\text{column}) = \frac{1.7(75)^2}{8} - \frac{145(36)}{12}$$
$$= 1195 - 435 = 760 \text{ ft-kips}$$

$$\text{Reqd } Z = \frac{760(12)}{36} = 253 \text{ in.}^3$$

A W27 × 94 might be selected to satisfy strength. For practicality of making a corner connection the two members should have approximately the same flange width. Revise sections.

Try W14 × 74 column, b_f = 10.072 in.

$$M_p\,(\text{column}) = 126(36)/12 = 378 \text{ ft-kips}$$
$$\text{Reqd } M_p\,(\text{girder}) = 1195 - 378 = 817 \text{ ft-kips}$$
$$\text{Reqd } Z = \frac{817(12)}{36} = 273 \text{ in.}^3$$

Try W27 × 94 girder, b_f = 9.99 in.

(b) Check whether loading which includes wind may govern. Refer to Fig. 15.2.4 for plastic-strength analysis using the same section throughout and Fig. 15.3.3 for the result using different column and girder sections,

$$M_s\,(\text{at knee}) = 0.2 w_u h^2 = 0.2(0.572)(25)^2 = 71.5 \text{ ft-kips}$$
$$M_s\,(\text{at girder midspan}) = \frac{w_u L^2}{8} = \frac{1.43(75)^2}{8} = 1{,}005 \text{ ft-kips}$$

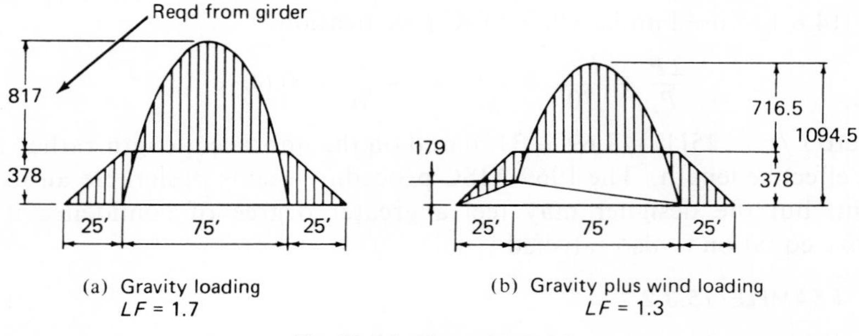

Fig. 15.3.3. Example 15.3.2.

The required plastic moment from the girder assuming the plastic hinge to occur approximately at midspan is

$$M_p = 1005 + \tfrac{1}{2}(71.5) - M_p\,(\text{column})$$
$$= 1005 + 35.75 - 376 = 664.75 \text{ ft-kips}$$

which is less than the 819 ft-kips required from gravity load alone. Wind loading does not control.

(c) Check beam-column capacity. If axial compression were large, plastic moment might not govern selection of column. In which case, using Eq. 12.13.10 one should compute equivalent P_y,

$$\text{Equiv. } P_y = P + \frac{2M}{d}$$

$$= 63.8 + \frac{2(378)12}{14} = 63.8 + 644 \approx 708 \text{ kips}$$

$$\text{Reqd } A = \frac{\text{Equiv. } P_y}{F_y} = \frac{708}{36} = 19.7 \text{ sq in.}$$

The selected W14 × 74 has $A = 21.8$ sq in., and so is likely to be acceptable or closely so. With the axial effect 15 percent or less, the girder may be selected solely for M_p; then checked for stability as follows:

$$G_A \approx 10.0 \text{ (hinged at bottom)}$$

$$G_B = \frac{I/L\text{ (column)}}{I/L\text{ (girder)}} = \frac{797/25}{3270/75} = \frac{31.8}{43.5} = 0.73$$

From Fig. 14.3.1b, find $K_x = 1.8$.

$$\frac{K_x L_x}{r_x} = \frac{1.8(25)(12)}{6.05} = 89.2; \qquad F'_e = 18.8 \text{ ksi}$$

$$\frac{K_y L_y}{r_y} = \frac{1.0(5)\,12}{2.48} = 24.2$$

$$F_a = 14.3 \text{ ksi (based on } KL/r = 89.2)$$

$$P_e = (23/12)AF'_e = (23/12)(21.8)(18.8) = 785 \text{ kips}$$

$$P_{cr} = 1.70F_aA = 1.70(14.3)(21.8) = 528 \text{ kips}$$

$$M_m = M_p = 378 \text{ ft-kips}$$

AISC Formulas 2.4–2,

$$\frac{P}{P_{cr}} + \frac{M_i}{M_m}\left(\frac{C_m}{1 - P/P_e}\right) = \frac{63.8}{528} + \frac{378}{378}\left(\frac{0.85}{1 - 63.8/785}\right)$$

$$= 0.121 + 0.925 = 1.046$$

Accept the section even though the criterion indicates an overstress of about 4½ percent, since the hinged-base assumption ($G_A = 10.0$) underestimates the moment restraint. AISC Formula 2.4–3 does not control, since $P/P_y < 0.15$,

$$P_y = F_yA = 36(21.8) = 785 \text{ kips}$$

$$\frac{P}{P_y} = \frac{63.8}{785} = 0.081 < 0.15 \qquad \text{OK}$$

Check local buckling:

$$\frac{d}{t} = \frac{412}{\sqrt{F_y}}\left(1 - 1.4\,\frac{P}{P_y}\right) = 68.7 - 96.1(0.081) = 60.9$$

$$\text{Actual } \frac{d}{t} = \frac{14.19}{0.45} = 31.5 < 60.9 \qquad \text{OK}$$

Use W14 × 74 for the column.

(d) Check the W27 × 94 beam for local buckling under AISC–2.7:

$$\frac{b_t}{2t_f} = \frac{9.990}{2(0.747)} = 6.7 < 8.5 \qquad \text{OK}$$

$$\frac{d}{t} = \frac{26.91}{0.49} = 54.9 < 68.7 \qquad \text{OK}$$

Use W27 × 94 for the girder.

(e) Check shear in the corner region. This is illustrated in Example 13.8.1.

(f) Check lateral support, AISC 2–9. For the girder, W27 × 94,

$$L_{cr} = 38.2\,r_y = 38.2\left(\frac{2.12}{12}\right) = 6.8 \text{ ft} > 6.0 \text{ ft} \qquad \text{OK}$$

For the column, W14 × 74,

$$L_{cr} = 38.2 r_y = 38.2\left(\frac{2.48}{12}\right) > 5.0 \text{ ft} \qquad \text{OK}$$

Final decision: use W14 × 74 column, W27 × 94 girder with a corner joint containing diagonal stiffeners, similar to Fig. 15.3.2.

EXAMPLE 15.3.3

Design a gabled frame as shown in Fig. 15.3.4a to carry a uniform vertical gravity of 1.1 kips/ft, and a horizontal wind load of 0.6 kips/ft. Use A572 Grade 50 steel.

Solution

(a) Establish plastic strength requirement for gravity loading alone. Apply the 1.7 load factor and compute simple beam moment at point 3,

$$M_{s3} = \frac{w_u L^2}{8} = \frac{1.7(1.1)(55)^2}{8} = 707 \text{ ft-kips}$$

plotted in Fig. 15.3.4b. Then superimpose the effect of the equal horizontal reactions H at points 1 and 5 until the positive and negative moments are equal. Referring back to Example 15.2.3 for the general procedure, the moment due to H is

$$M\,(\text{top of gable}) = 28H$$

$$M\,(\text{top of column}) = 20H$$

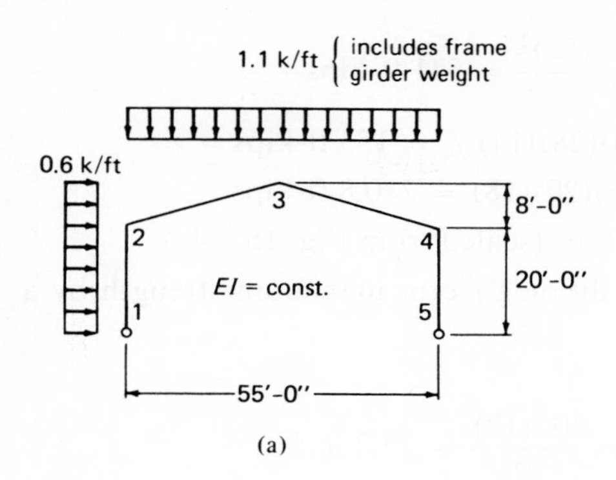

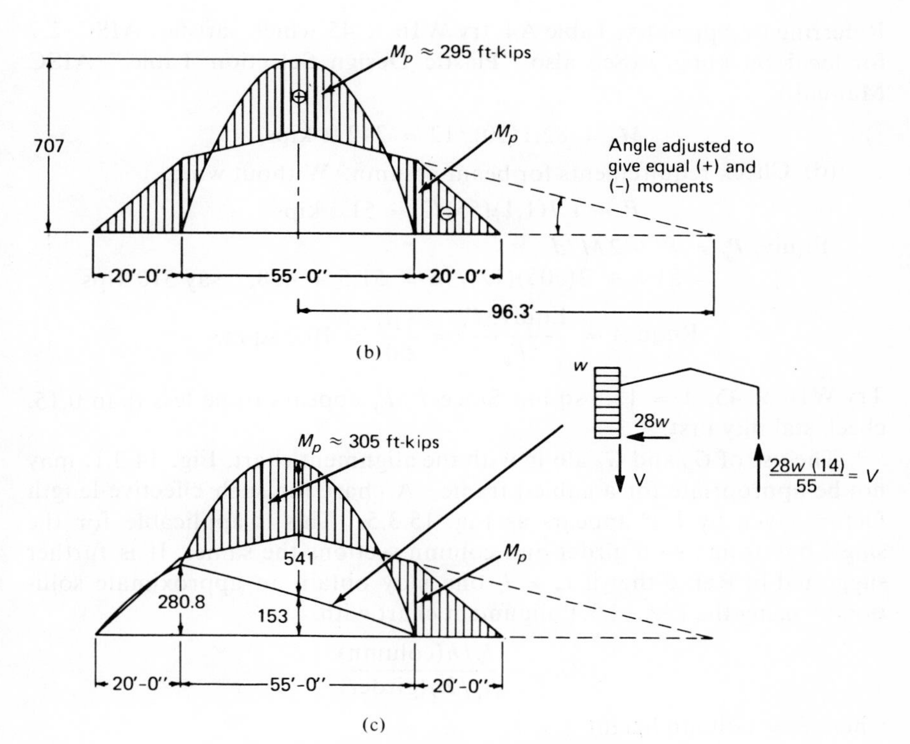

Fig. 15.3.4. Gabled frame of Example 15.3.3.

$$\text{Distance to rotation point} = 27.5\left(\frac{28H}{8H}\right) = 96.3 \text{ ft}$$

from midspan of the frame

$$M_p = 295 \text{ ft-kips (scaled from Fig. 15.3.4b)}$$

(b) Establish plastic-strength requirement for loading with wind. Apply 1.3 load factor.

$$M_{s3} = \frac{1.3(1.1)(55)^2}{8} = 541 \text{ ft-kips}$$

$$M_{s3}(\text{wind}) = 1.3(0.6)(28)(14)\,{}^1\!/\!_2 = 153 \text{ ft-kips}$$

$$M_{s2}(\text{wind}) = 1.3(0.6)(20)(18) = 280.8 \text{ ft-kips}$$

$$M_p = 305 \text{ ft-kips (scaled from Fig. 15.3.4c)}$$

Thus for this example the wind condition governs maximum strength by a slim margin.

(c) Select section:

$$\text{Reqd } Z = \frac{M_p}{F_y} = \frac{305(12)}{50} = 73.2 \text{ in.}^3$$

Referring to Appendix, Table A4, try W16 × 45 which satisfies AISC–2.7 for local buckling. (See also "Plastic Design Selection Table," AISC Manual.)

$$M_p = 82.1(50)/12 = 342 \text{ ft-kips}$$

(d) Check requirements for beam-column. Without wind,

$$P = 1.7(1.1)(55)/2 = 51.5 \text{ kips}$$

$$\text{Equiv. } P_y = P + 2M/d$$

$$= 51.5 + 2(305)(12)/16 = 51.5 + 458, \quad \text{say } 510 \text{ kips}$$

$$\text{Reqd } A = \frac{\text{Equiv. } P_y}{F_y} = \frac{510}{50} = 10.2 \text{ sq in.}$$

Try W16 × 45, $A = 13.3$ sq in. Since P/P_y appears to be less than 0.15, check stability first.

The use of G_A and G_B along with the alignment chart, Fig. 14.3.1, may not be appropriate for a gabled frame. A chart for such effective-length factors given by Lu[6] appears as Fig. 15.3.5. This is applicable for the single bay frame with girder and column sections the same. It is further suggested in Ref. 6 that if $I_g \neq I_c$ one may obtain an approximate solution by using the Fig. 14.3.1 alignment chart with

$$G_{\text{top}} = \frac{I_c/h(\text{column})}{I_g/2q(\text{girder})}$$

where h = column height
q = length of girder to ridge

In this example (see Fig. 15.3.5),

$$\frac{f}{h} = \frac{8}{20} = 0.4 \quad \text{and} \quad \frac{L}{h} = \frac{55}{20} = 2.75$$

find $K_x \approx 2.25$ for hinged base ($G = 10.0$).

$$\frac{K_x L}{r_x} = \frac{2.25(20)\,12}{6.64} = 81.3; \quad F'_e = 22.65 \text{ ksi}; \quad F_a = 18.77 \text{ ksi}$$

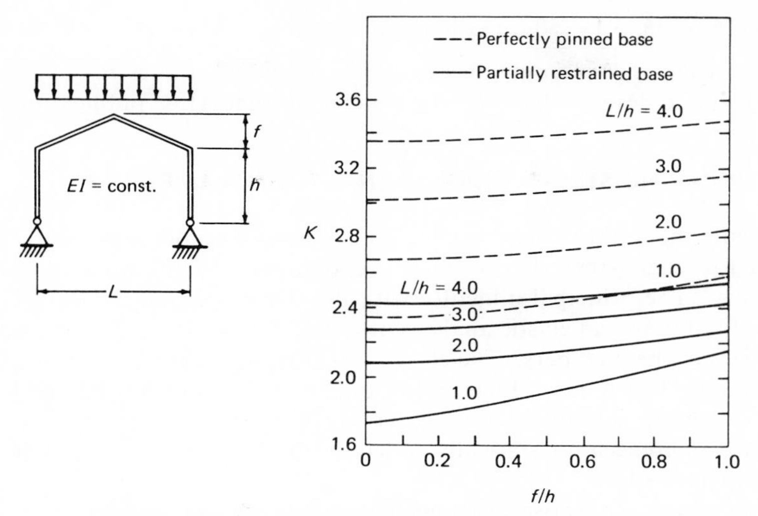

Fig. 15.3.5. Effective length factors for gabled frames (From Ref. 6.)

Assume lateral bracing will be adequate so that KL/r_y will not control.

$$P_e = (23/12)AF'_e = (23/12)(13.3)(22.65) = 575 \text{ kips}$$

$$P_{cr} = 1.70F_aA = 1.70(18.77)(13.3) = 423 \text{ kips}$$

$$M_m = M_p = 342 \text{ ft-kips}$$

AISC Formula 2.4–2 for stability,

$$\frac{P}{P_{cr}} + \frac{M_i}{M_m}\left(\frac{C_m}{1 - P/P_e}\right) = \frac{51.5}{423} + \frac{295}{342}\left(\frac{0.85}{1 - 51.5/575}\right)$$

$$= 0.122 + 0.0805 = 0.927 < 1.0 \qquad \text{OK}$$

AISC Formula 2.4–3 for strength:

$$P_y = 50(13.3) = 666 \text{ kips}$$

$$\frac{P}{P_y} = \frac{51.5}{666} = 0.078 < 0.15 \qquad \text{OK}$$

(e) Check shear in corner region. Use M_p for wind condition since it is the largest,

$$\text{Reqd } t_w = \frac{23M}{A_{bc}F_y} = \frac{23(305)}{(16.12)^2(50)} = 0.54 \text{ in.} > t_w = 0.346$$

Diagonal stiffeners required.

(f) Lateral-support requirement:

$$L_{cr} = \left(\frac{1375}{F_y}\right) r_y = \left(\frac{1375}{50}\right) r_y = 27.5r_y = \frac{27.5(1.57)}{12} = 3.6 \text{ ft}$$

Lateral support from purlins or joists is required every $3\frac{1}{2}$ ft.

Use W16 × 45, F_y = 50 for column and girder.

More detail concerning plastic design of gabled frames may be found in Ref. 1, and design tables are available in another AISC publication.[7]

15.4. WORKING-STRESS DESIGN—ONE-STORY FRAMES

While the authors believe the plastic design method is the method of choice for single-story rigid frames, a comparison of the methods seems desirable. The principal difficulty with the working stress method is the necessity of doing an elastic analysis of a rigid frame without knowing the final sizes of the members. The elastic analysis, using the matrix displacement method, ordinary slope deflection, moment distribution, or some other method, is outside the scope of this text; the many available standard references on statically indeterminate structures should be referred to as needed.*

For practical purposes, various design aids are available giving elastic solutions for standard frame dimensions and loadings.[8]

For design, an elastic analysis is needed to select member sizes; however, one cannot determine moments, shears and axial forces without knowing the relative moments of inertia. This problem does not arise in plastic design. However, there is one advantage in the working-stress method; if lateral bracing is not satisfactory to develop the plastic-strength mechanism, the working-stress method allows for design using reduced strengths based on lateral stability. Plastic design requires the mechanism to develop.

EXAMPLE 15.4.1

Design the frame of Example 15.3.1 using the working-stress method. A36 steel.

Solution

In the working-stress method the elastic moments under service load are required. In determining the elastic moments for this design, the girder and column are assumed to have constant moment of inertia. Figure 15.4.1 shows the two loading cases and the corresponding bending-moment diagrams. The relatively lengthy process of doing the statically indeterminate analysis has not been illustrated here. When wind loading is included, AISC–1.5.6 allows a 33⅓ percent overstress. The use of 0.75 of the applied loading has the same effect as using the full loading and 1.33 of the allowable stresses.

*See, for example, C. K. Wang, *Statically Indeterminate Structures* (New York: McGraw-Hill Book Company, 1953); or C. K. Wang, *Matrix Methods of Structural Analysis*, 2d ed. (Scranton, Pa.: International Textbook Company, 1970).

(a) Select the W section. For the loading without wind, the maximum moment occurring at the corner (381 ft-kips) may be reduced according to AISC–1.5.1.4.1.

Assume $F_b = 0.66F_y = 24$ ksi

$$\text{Reduced } -M = 0.9(381) = 343 \text{ ft-kips}$$

which would require for equilibrium an increase in the positive moment,

$$\text{Increased } +M = 322 + (381 - 343) = 360 \text{ ft-kips}$$

Thus the full reduction in negative moment should not be made. Adjust until positive and negative moments are equal:

$$+M = -M = 351.5 \text{ ft-kips}$$

$$\text{Reqd } S = \frac{351.5(12)}{24} = 175.7 \text{ in.}^3$$

Try W24 × 84, $S = 197$ in.^{3} (The W24 × 76 will be slightly overstressed on the girder and also may not satisfy beam-column requirements.)

$$f = \frac{351.5(12)}{197} = 21.5 \text{ ksi}$$

Note that "compact" section requirements must be satisfied even though stress is less than $0.60F_y$ because if the section is noncompact, the unadjusted moment of 381 ft-kips would have to be used.

Check AISC–1.5.1.4.1 for "compact":

$$\frac{b_f}{2t_f} = \frac{9.015}{2(0.772)} = 5.84 < 8.7 \qquad \text{OK}$$

$$\frac{d}{t} = \frac{24.09}{0.47} = 51.2 < 68.7 \qquad \text{OK}$$

(b) Check beam-column. When axial compression is low as in this case, selection as a beam is the correct approach. Exactly as for plastic design, determine G_A and G_B and find K_x from Fig. 14.3.1b. $K_x = 2.25$ as in Example 15.3.1,

$$\frac{K_xL}{r_x} = \frac{2.25(25)12}{9.79} = 69; \qquad \frac{K_yL}{r_y} = \frac{1.0(5)12}{1.95} = 30.8$$

$$F_a = 16.53 \text{ ksi}; \qquad F'_e = 31.37 \text{ ksi}$$

$$\frac{f_a}{F_a} = \frac{37.5/24.7}{16.53} = 0.092 < 0.15$$

AISC Formula 1.6.2 may be used:

$$\frac{f_a}{F_a} + \frac{f_b}{F_b} = 0.092 + \frac{21.5}{24} = 0.092 + 0.895 = 0.99 < 1.0 \qquad \text{OK}$$

Local buckling, AISC–1.5.1.4.1d:

$$\frac{d}{t} = \frac{412}{\sqrt{F_y}}(1 - 2.33f_a/F_y) = 68.7 - 4.4(37.5/24.7) = 62 > 51.2 \qquad \text{OK}$$

(c) Lateral bracing, W24 × 84:

$$L_c = \frac{76.0b_f}{\sqrt{F_y}} = 12.7b_f = \frac{12.7(9.015)}{12} = 9.5 \text{ ft}$$

or

$$\frac{20{,}000}{(d/A_f)F_y} = \frac{556}{d/A_f} = \frac{556}{3.47(12)} > 9.5 \text{ ft}; \quad L_c = 9.5 \text{ ft}.$$

Since lateral support is provided at 5-ft intervals on the column and at 6-ft intervals on the girder, lateral torsional buckling is adequately prevented.

(d) Check shear at the corner connection. Using Eq. 13.8.7, as discussed in AISC Commentary–1.5.1.2,

$$\text{Reqd } t_w = \frac{32M}{A_{bc}F_y} = \frac{32(351.5)}{(24.09)^2(36)} = 0.54 \text{ in.} > t_w = 0.47 \text{ in.}$$

Diagonal stiffeners are required.

(e) Since $f_a/F_a < 0.15$, there is no reason to question the overall stability of the frame. Use of AISC Formula 1.6–1a involving the magnification factor $C_m/(1 - f_a/F'_e)$ would be the appropriate check if one is desired.

(f) Conclusions. Use W24 × 84 section which compares with W24 × 76 used in plastic-design method. The greater designer time required in doing an elastic-frame analyses (with and without wind), and the somewhat arbitrary provision for reducing negative moment 10 percent, leave the conclusion that plastic design is the preferred method for designing one story (and probably two-story) frames.

EXAMPLE 15.4.2

Redesign the rectangular frame of Example 15.4.1 by the working-stress method using a column no deeper than 14.5 in. overall (compares with plastic design Example 15.3.2). A36 steel.

Solution

The frame dimensions and loadings are shown in Fig. 15.4.1a and b.

(a) Determination of moments. There is no direct process for this except to estimate from experience the relative moments of inertia of the column and girder. For this problem one might examine a W14 section which has a depth not exceeding 14.5 in., say the W14 × 119 (I_c = 1370), and compare it to the section used when $I_c = I_g$, W24 × 76 (I_g = 2100). Of course, one would not have known that value without doing an elastic

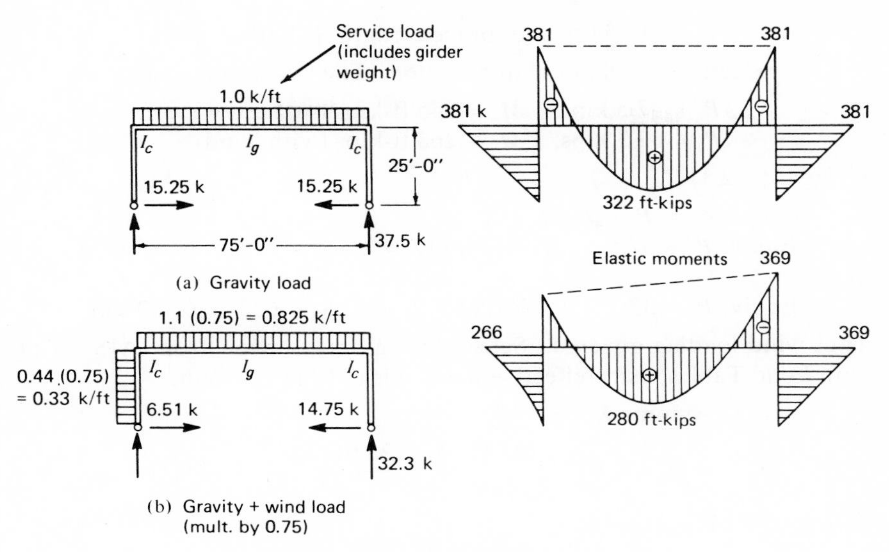

Fig. 15.4.1. Frame of Example 15.4.1: $I_g = I_c$.

analysis with $I_c = I_g$, as in Example 15.4.1. From the above, it seems the ratio I_g/I_c will be at least 2.0.

Assume $I_g/I_c = 4$ so as to compare results with plastic design; the elastic moments are given in Fig. 15.4.2.

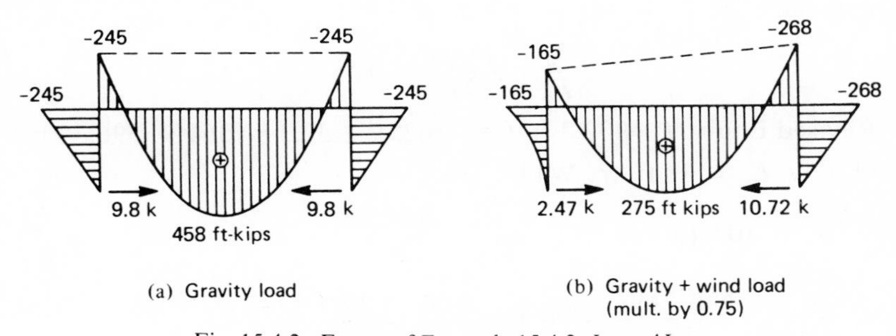

Fig. 15.4.2. Frame of Example 15.4.2: $I_g = 4I_c$.

(b) Selection of section for girder. Clearly, gravity load without wind governs. No negative moment reduction may be made even if section is "compact," because the positive moment is already the larger one.

$$\text{Assume } F_b = 0.66F_y = 24 \text{ ksi}$$

$$\text{Reqd } S = \frac{458(12)}{24} = 229.0 \text{ in.}^3$$

Try W27 × 94, $S = 243$ in.^{3} From Appendix, Table A3, this section is compact for local buckling; also

$$L_c = 10.54 \text{ ft} > \text{unbraced length} = 6 \text{ ft} \qquad \text{OK}$$

(c) Selection of beam-column section. The two loading cases are

$$P = 37.5 \text{ kips}, \quad M = 245 \text{ ft-kips (without wind)}$$
$$P = 32.3 \text{ kips}, \quad M = 268 \text{ ft-kips (with wind)}$$

Using Eq. 12.11.7,

$$\text{Equiv. } P \approx P + B_x M$$
$$\text{Equiv. } P = 37.5 + 0.194(245)\,12 = 37.5 + 570 = 607.5 \text{ kips}$$

or

$$\text{Equiv. } P = 32.3 + 0.194(268)\,12 = 32.8 + 622 = 654.3 \text{ kips}$$

Thus wind loading governs. Select preliminary section from AISC Column Load Tables, using effective weak axis unbraced length,

$$\text{Assume } K_x \approx 1.8^*$$

$$\text{Equiv. } L_y = \frac{25.0(1.8)}{r_x/r_y} = \frac{25.0(1.8)}{2.45} = 18.4 \text{ ft}$$

Estimate W14 × 111:

$$G_A = 10.0 \text{ (hinged)}; \qquad G_B = \frac{1270/25}{3270/75} = \frac{50.5}{43.5} = 1.15$$

Find $K_x = 1.9$

$$\frac{K_x L}{r_x} = \frac{1.9(25)\,12}{6.23} = 91.5; \qquad F_a = 14.03 \text{ ksi}$$

if "compact," $F_b = 0.66F_y = 24$ ksi

$$\frac{F_a}{F_b} = \frac{14.03}{24} = 0.585$$

$$\text{Revised Equiv. } P = 32.3 + 0.585(622) = 32.3 + 364 = 396.3 \text{ kips}$$

For Equiv. $L_y \approx 18$ ft, try W14 × 84:

$$G_A = 10.0 \text{ (hinged)}; \qquad G_B = \frac{928/25}{43.5} = 0.85; \qquad K_x = 1.85$$

$$\frac{K_x L}{r_x} = \frac{1.85(25)\,12}{6.13} = 90.5 > \frac{K_y L}{r_y} = \frac{1.0(5)\,12}{r_y}$$

$$F_a = 14.14 \text{ ksi}; \qquad F_e' = 18.20 \text{ ksi}$$

$$f_a = \frac{P}{A} = \frac{32.3}{24.7} = 1.31 \text{ ksi}$$

$$\frac{f_a}{F_a} = \frac{1.31}{14.14} = 0.092 < 0.15$$ AISC Formula (1.6–2) may be used

$$f_b = \frac{268(12)}{131} = 24.6 \text{ ksi} > 0.66F_y$$

*Theoretically should be ≥ 1.0 if frictionless hinge exists at base ($G = \alpha$). Using $G = 10.0$ assumes some restraint, in which case K may be less than 2.0.

For flexure alone a heavier section is required. Based on the strength concepts used in plastic design the above section is probably acceptable with the 2½ percent flexural overstress, since the axial effect is less than 15 percent of the total. The usual working stress solution would dictate used of a heavier section.

The next "compact" section is the preliminary W14 × 111,

$$f_b = \frac{268(12)}{176} = 18.3 \text{ ksi} \qquad \text{Too low}$$

Try W14 × 95, even though not fully compact (satisfies requirements to use AISC Formula 1.5–5 as partially compact):

$$f_b = \frac{268(12)}{151} = 21.3 \text{ ksi}$$

$$F_b = 0.65F_y = 23.4 \text{ ksi} \qquad \text{(see Appendix, Table A3)}$$

The values for $K_x = 1.85$ and $KL/r_x = 90.5$ as computed above are negligibly changed:

$$f_a = \frac{P}{A} = \frac{32.3}{27.9} = 1.16 \text{ ksi}$$

AISC Formula 1.6–2:

$$\frac{f_a}{F_a} + \frac{f_b}{F_b} = \frac{1.16}{14.1} + \frac{21.3}{23.4} = 0.08 + 0.91 = 0.99 < 1$$

Use W14 × 95 for column member. This compares with W14 × 74 using plastic design (Example 15.3.2).

The check for required diagonal stiffeners should be made as in other examples. The lateral support spacing requirement for elastic design is much less severe than required in plastic design. Maximum unbraced length $= L_c = 10.5$ ft on the girder and $= L_c = 15.4$ ft on the column. In plastic design, lateral bracing was necessary at 6.5 ft on the girder and at 7.9 ft on the column.

EXAMPLE 15.4.3

Redesign the gabled frame of Example 15.3.3 by the working-stress method. A572 Grade 50 steel.

Solution

Assuming the moment of inertia of all frame elements to be the same, the elastic analysis is performed and is shown in Fig. 15.4.3. It is seen that the wind loading case governs.

(a) Select the section to satisfy beam-column requirements.

$$\text{Equiv. } P \approx P + B_x M$$
$$= 25.9 + 0.2(227)12 = 25.9 + 545 = 570 \text{ kips}$$

Loading is about 95 percent flexure; select as a beam. If the section is "compact," the negative moment may be reduced to 90 percent of its

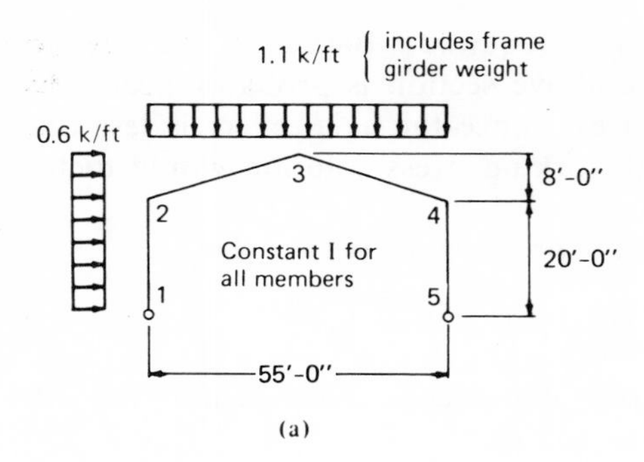

(a)

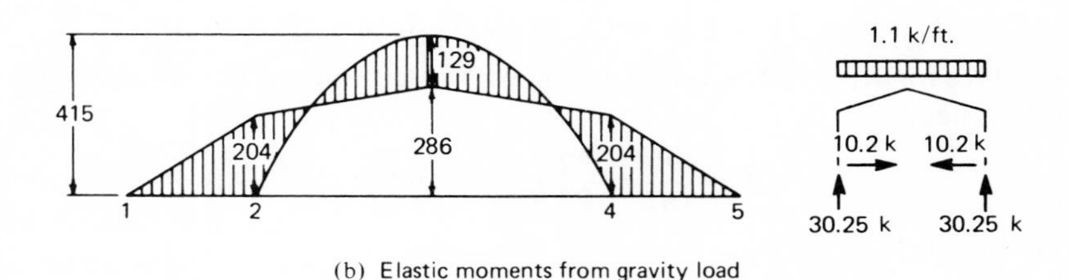

(b) Elastic moments from gravity load

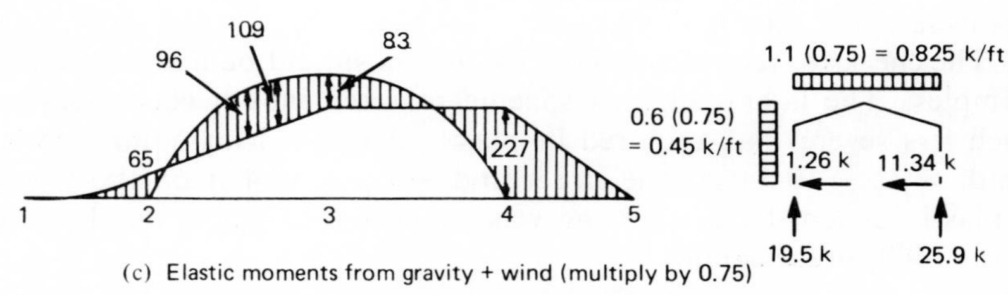

(c) Elastic moments from gravity + wind (multiply by 0.75)

Fig. 15.4.3. Gabled frame of Example 15.4.3.

elastic value:

$$-M = 0.9(227) = 204 \text{ ft-kips}$$

$$\text{Reqd } S = \frac{(204)\,12}{33(0.95)} = 78 \text{ in.}^3$$

where $F_b = 0.66F_y = 33$ ksi and 0.95 accounts for the percentage of capacity required for flexure. The increased positive moment is about 130, well below the negative moment of 204 used for design, and thus may be ignored.

Try W16 × 50, which is "compact" for $F_y = 50$ ksi according to Appendix, Table A3, and $L_c = 6.33$ ft.

$$f_b = \frac{204(12)}{80.8} = 30.3 \text{ ksi} < 33 \text{ ksi} \qquad \text{OK}$$

Using Fig. 15.3.5 with $f/h = 0.4$ and $L/h = 2.75$, find $K_x \approx 2.25$ for partially restrained bases.

$$\frac{K_x L}{r_x} = \frac{2.25(20)(12)}{6.68} = 81; \qquad F_a = 18.81 \text{ ksi}$$

$$f_a = \frac{25.9}{14.7} = 1.76 \text{ ksi}$$

$$\frac{f_a}{F_a} = \frac{1.76}{18.81} = 0.094 < 0.15 \qquad \text{May use AISC Formula 1.6–2}$$

$$\frac{f_a}{F_a} + \frac{f_b}{F_b} = 0.094 + \frac{30.3}{33} = 0.09 + 0.92 = 1.01 \approx 1 \qquad \text{OK}$$

Again, shear must be checked in corner regions to ascertain whether diagonal stiffeners are needed. Lateral support is necessary at the top of the column and at intervals of 6′ 4″ ($L_c = 6.33$).

Use W16 × 50, $F_y = 50$ ksi.

Plastic design was able to use a W16 × 45, but lateral bracing was required every 3½ ft.

15.5. MULTISTORY BRACED FRAMES—PLASTIC DESIGN

The design of multistory frames using the working stress method offers no particular unique problems in concept or in the selection of members; but the analysis does involve additional complexities. However, the analysis techniques are outside the scope of this text. The plastic design of braced multistory frames, first accepted in the 1969 AISC specification, involves some design concepts which are unique and which the authors believe justify some attention in this article.

Plastic design of *unbraced* multistory frames has also received attention by researchers,[9,10] and methods are recommended[9,11,12] but are outside the scope of this text.

The theoretical and experimental background for plastically designing bracing multistory frames was first presented to the engineering profession in the summer of 1965.[9] For additional experimental and theoretical studies the reader is referred to Refs. 10 and 14 through 17.

Since 1965, designers' acceptance has broadened; and with the publication of the design manual[13] in 1968 along with the 1969 AISC specification acceptance, plastic design of braced multistory rigid frames is expected to rapidly become a standard method.

Examine the multistory rigid frames of Fig. 15.5.1. A building is assumed to consist of a series of planar frames; i.e., all members lie in

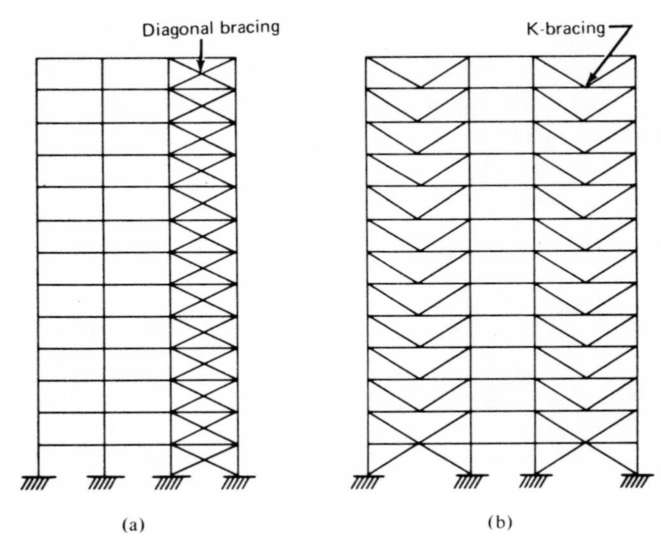

Fig. 15.5.1. Braced multistory rigid frames.

a single plane and all loads are applied in that plane. All connections between beams and columns in the frame are assumed and designed to be rigid. In the series of planar frames, or bents, one or more of the bents must be braced for the full height of the building, which may be accomplished by using diagonal or K-bracing as shown in Fig. 15.5.1. Such bracing creates a vertical cantilever frame capable of carrying the horizontal wind loads. When some bents are braced and some are unbraced, but supported by diaphragm action of the floor system between braced bents, all of the wind loads must be assumed transmitted to the braced bents, while the supported bents may be designed for gravity load only. A plan of a braced multistory structure showing braced and supported bents is shown in Fig. 15.5.2. Note that floor framing, such as with open web

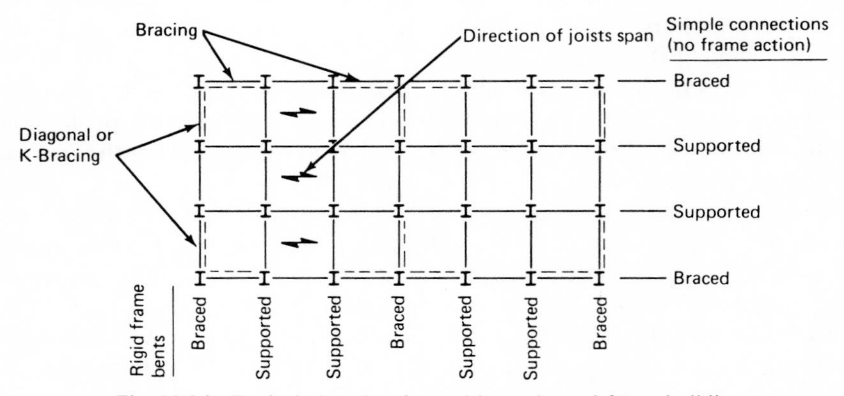

Fig. 15.5.2. Typical plan view for multistory braced-frame building.

joists, is a one-way system transmitting loads to the planar frames. In the direction transverse to the frame bents having rigid connections (AISC Type 1), the framing system frequently uses simple connections (AISC Type 2), but with adequate diagonal or other bracing to carry wind loads. Since nearly all of the load is transmitted to the frame bents, it is not too critical whether the transverse system is rigid or not, so the usual procedure is to make it simply supported. In other words, there are frames in one direction and simple beams in the other.

Supported Bent: Girder Design. Since a supported bent is designed to carry only gravity loads, the factored floor load (load factor = 1.7) is used along with the American Standard Building Code percent live load reduction, as described in Sec. 1.4. The girders are designed so the plastic hinges will form in the girders, giving a 3-hinged beam mechanism as shown in Fig. 15.5.3. With the end hinges forming in the girder, the

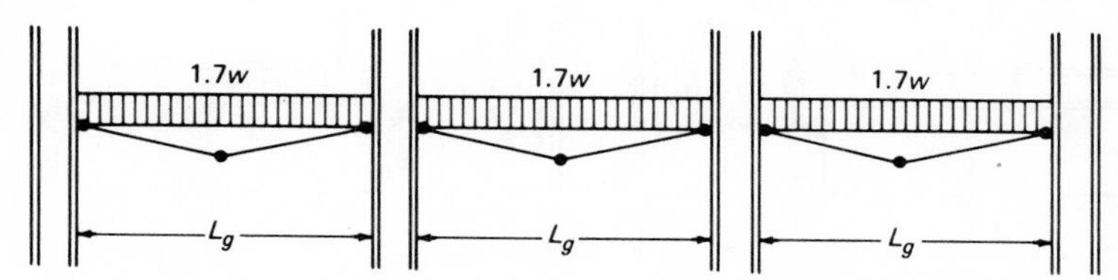

Fig. 15.5.3. Typical floor level for supported bent.

clear span L_g is used for determining the required plastic moment:

$$\text{Reqd } M_p = \frac{w_u L_g^2}{16} \tag{15.5.1}$$

Supported Bent: Column Design. The column axial loads are tabulated by accumulating the sum of the shears from the roof to each floor level. Referring to the typical interior joint in Fig. 15.5.4, equilibrium requires

$$M_{jU} + M_{jL} = -(M_{jA} + M_{jB}) \tag{15.5.2}$$

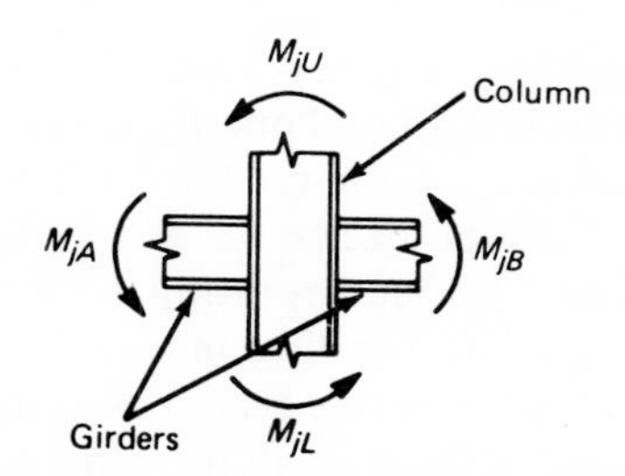

Fig. 15.5.4. Interior column joint for supported bent.

where M_{jU} and M_{jL} = moments about the center of the joint, on the column above and below the girder respectively (defining positive as counterclockwise rotation on the joint)

M_{jA} = moment about the center of the joint contributed from the girder on the left side of the column

M_{jB} = moment about the center of the joint contributed from the girder on the right side of the column

Additional torsional moment on the joint contributed from members framing in transversely to the bent should also be included if significant.

The terminology, including that above, used throughout this article is in accordance with Ref. 13 wherever feasible.

Next, the girder moments M_{jA} and M_{jB} at the *center* of the joint must be related to the plastic hinge moments at the face of the column. Referring to Fig. 15.5.5, the moments at the center of the column from the

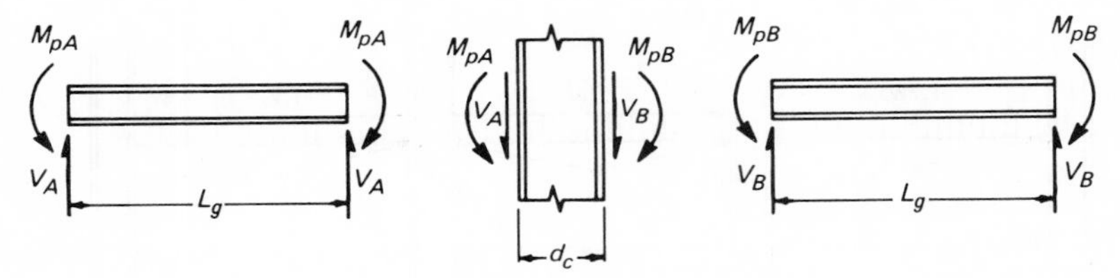

Fig. 15.5.5. Forces on column from girders.

girders are

$$\left.\begin{aligned} M_{jA} &= M_{pA} + V_A \frac{d_c}{2} = M_{pA} + \frac{w_u L_g}{2}\left(\frac{d_c}{2}\right) \\ -M_{jB} &= M_{pB} + V_B \frac{d_c}{2} = M_{pB} + \frac{w_u L_g}{2}\left(\frac{d_c}{2}\right) \end{aligned}\right\} \tag{15.5.3}$$

where M_{pA} and M_{pB} refer to the plastic hinge moments on the girders. Substituting Eq. 15.5.1 into Eqs. 15.5.3 gives

$$\begin{aligned} M_{jA} &= M_{pA}(1 + 4d_c/L_g) \\ -M_{jB} &= M_{pB}(1 + 4d_c/L_g) \end{aligned} \tag{15.5.4}$$

At the roof level, M_{jU} in Eq. 15.5.2 is zero, but at all other floor levels the total column moment is distributed one-half above and one-half below the joint. This is acceptable unless the column size or height is drastically different above and below a floor level.

At each floor level, with the end-rotational moments, M_{jU} and M_{jL}, equal and of the same sign, the bending moments are of opposite sign; thus, full factored gravity load will give rise to double curvature bending.

Selection of the girder section is done using the principles of the plastic design sections of Chapters 7, 9, and 10, while the columns are selected in accordance with procedures of Chapter 12.

EXAMPLE 15.5.1

Design the typical floor girders and adjacent exterior columns shown in Fig. 15.5.6 for a supported bent of a multistory braced frame using plastic design. A36 steel.

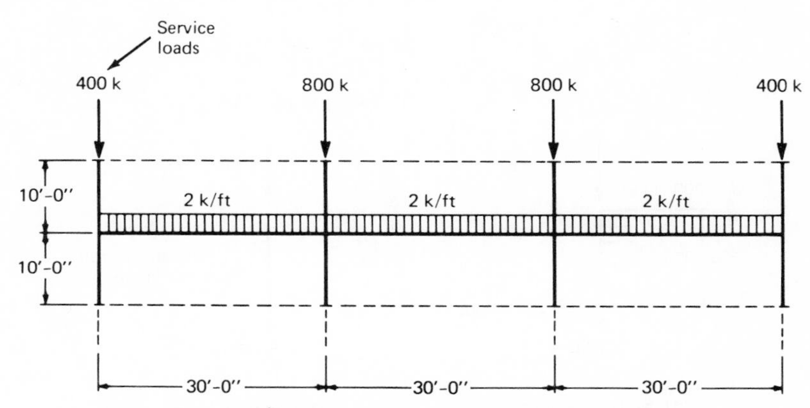

Fig. 15.5.6. Typical floor for supported bent of Example 15.5.1.

Solution

(a) Girder design. For gravity load, apply 1.7 load factor and use Eq. 15.5.1:

$$\text{Reqd } M_p = \frac{w_u L_g^2}{16} = \frac{1.7(2)(29)^2}{16} = 179 \text{ ft-kips}$$

$$\text{Reqd } Z = \frac{M_p}{F_y} = \frac{179(12)}{36} = 59.7 \text{ in.}^3$$

Use W18 × 35. $Z = 66.8$ in.^{3} Local buckling requirements of AISC–2.7 are satisfied and lateral support is provided in accordance with AISC–2.9.

(b) Column design. The moment at the center of the column due to the girder, according to Eq. 15.5.4 is

$$M_j = M_p(1 + 4d_c/L_g)$$

for both ends of each girder. Thus, assuming a W14 section,

$$M_j = 179[1 + 4(1.17)/29] = 208 \text{ ft-kips}$$

and the axial load to the column below the floor level is

$$P_{\text{exterior}} = [400 + 2(15)]\,1.7 = 731 \text{ kips}$$

$$P_{\text{interior}} = [800 + 2(30)]\,1.7 = 1{,}460 \text{ kips}$$

Assume lateral support to the columns occurs only at the top and bottom of the story height ($KL_y = 10$ ft). The loadings for which the columns are to be designed are shown in Fig. 15.5.7. Assume that the strength criterion controls and use Eq. 12.13.10 as the design approach,

$$\text{Equiv. } P_y = P + 2M/d$$
$$= 731 + 2(208)\,12/d = 731 + 5000/d$$
$$\text{For W12,} \quad \text{Equiv. } P_y = 1149$$
$$\text{W14,} \quad \text{Equiv. } P_y = 1089$$
$$\text{Reqd } A = \frac{P_y}{F_y} = 32 \text{ sq in. for W12}$$
$$= 30.3 \text{ sq in. for W14}$$

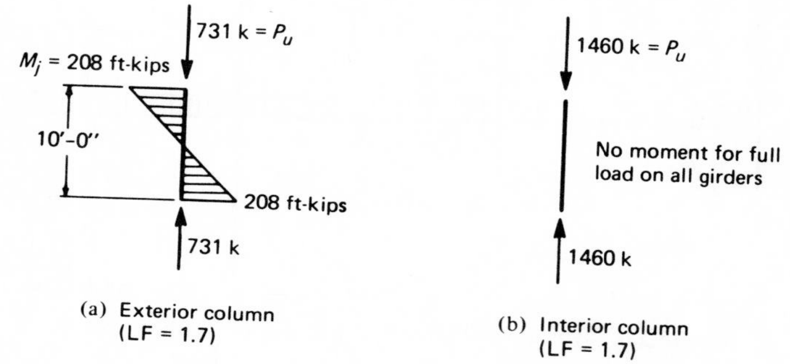

Fig. 15.5.7. Factored loads for design of columns.

Try W12 × 106: $A = 31.2$ sq in.

W14 × 111: $A = 32.7$ sq in. (W14 × 103 does not satisfy local buckling requirements of AISC–2.7).

Check W12 × 106: AISC Formula 2.4–3 for strength

$$P_y = AF_y = 31.2(36) = 1123 \text{ kips}$$
$$M_p = ZF_y = 164(36)/12 = 492 \text{ ft-kips} > M \qquad \text{OK}$$
$$\frac{P}{P_y} + \frac{M}{1.18M_p} = \frac{731}{1123} + \frac{208}{1.18(492)} = 0.65 + 0.36 = 1.01 \approx 1 \qquad \text{OK}$$

AISC Formula 2.4–2 for stability:

$$\frac{L}{r_y} = \frac{10(12)}{3.11} = 38.6; \qquad F_a = 19.31 \text{ ksi}$$
$$P_{cr} = 1.70F_aA = 1.70(19.31)\,31.2 = 1024 \text{ kips}$$
$$M_m = \left[1.07 - \frac{\sqrt{F_y}(L/r_y)}{3160}\right] M_p = M_p$$
$$\frac{KL}{r_x} = \frac{120}{5.46} = 22; \qquad F_e' = 308.5 \text{ ksi}$$
$$P_e = (23/12)\,F_e'A = 1.92(308.5)(31.2) = 18{,}500$$
$$C_m = 0.6 - 0.4(208/208) < 0.4; \qquad \text{use } 0.4$$

$$\frac{P}{P_{cr}} + \frac{M_i}{M_m} \frac{C_m}{1 - P/P_e} = \frac{731}{1024} + \frac{208}{492} \frac{0.4}{1 - 731/18{,}500}$$

$$= 0.715 + 0.177 = 0.892 < 1 \qquad \text{OK}$$

Use W12 × 106 for exterior column, and W18 × 35 for the girder.

Braced Bent: Girder Design for Gravity Load. If diagonal bracing is used as in Fig. 15.5.1a, the girder's plastic hinge mechanism will be the same as in Fig. 15.5.3. When K-bracing is used as in Fig. 15.5.1b an additional vertical support for the girder is provided. For either diagonal or K-bracing, the girder located in the bay with the bracing must resist an axial force in addition to bending.

The bracing is assumed to remain elastic as the plastic-hinge mechanism forms in the girder. The K-bracing causes four effects (see Fig. 15.5.8) not found in the supported bent or in bays of the braced bent without bracing: (1) provides a vertical support thus reducing the bending-moment span; (2) reduces girder deflection; (3) contributes part of a floor load P_{bV} to the column one-story above the location of the girder reaction to the column in unbraced bays (see Fig. 15.5.9); and (4) causes axial load in the girder.

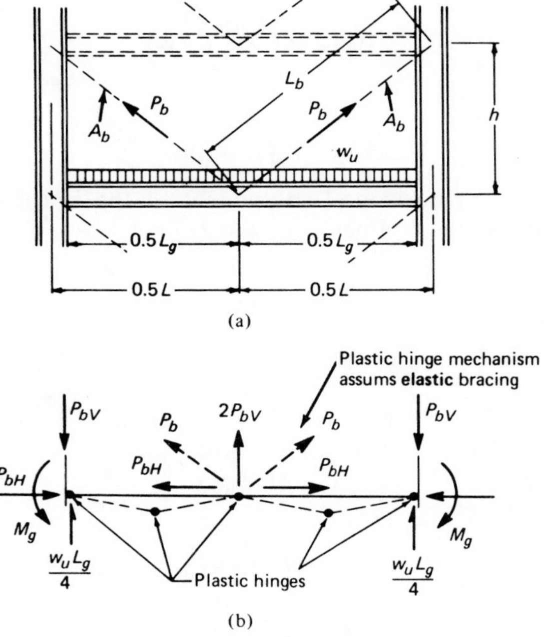

Fig. 15.5.8. Forces and plastic-hinge mechanism arising from gravity load for girders with K-bracing.

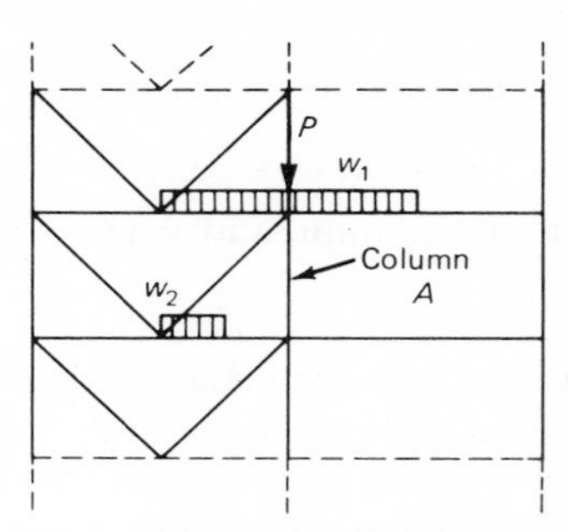

Fig. 15.5.9. Tributary uniform loads to top of column A when K-bracing is used.

It is considered satisfactory for design[13] to simplify the analysis and omit the full interaction analysis between K-bracing and floor girders. Thus it is suggested[13] to use the following procedure:

1. Assume the bracing provides a rigid support, rather than an elastic support, for establishing the girder plastic hinge mechanism.

2. Design the girders as laterally loaded girders in a braced frame system (i.e., assume no relative displacement of the girder supports).

Using the above assumptions, equilibrium at the brace support requires the vertical components of the force in the brace to equal the sum of the end shears on the two girder segments of length, L_g:

$$2P_{bV} = 2\left[\frac{w_u(0.5L_g)}{2}\right]$$

or

$$P_{bV} = \frac{w_u L_g}{4} \tag{15.5.5}$$

From Eq. 15.5.5 the horizontal component of the brace force and the full brace force may be obtained from the geometry as

$$P_{bH} = P_{bV}\left(\frac{0.5L}{h}\right) \tag{15.5.6}$$

$$P_b = P_{bV}\left(\frac{L_b}{h}\right) \tag{15.5.7}$$

With the brace force and its components known, the plastic moment M_g, axial force P_g, and shear V_g, on the girder are determined as

$$M_g = \frac{w_u(0.5L_g)^2}{16} = \frac{w_u L_g^2}{32} \tag{15.5.8}$$

$$P_g = P_{bH} = \frac{w_u L_g}{4}\left(\frac{0.5L}{h}\right) = \frac{w_u L_g L}{8h} \tag{15.5.9}$$

$$V_g = P_{bV} = \frac{w_u L_g}{4} \tag{15.5.10}$$

In order to prevent yielding of the brace, the area required is

$$A_b = \frac{P_b}{F_y} = \frac{w_u L_g L_b}{4F_y h} \tag{15.5.11}$$

EXAMPLE 15.5.2

Redesign the typical floor girder of Example 15.5.1 (see Fig. 15.5.6 for the dimensions and loads) but with K-bracing. Use A36 steel.

Solution

(b) Brace forces and components. Applying the 1.7 load factor for gravity load, and using Eqs. 15.5.5, 15.5.6, and 15.5.7, one obtains

$$P_{bV} = \frac{w_u L_g}{4} = \frac{2(1.7)(30 - 1.0)}{4} = 24.7 \text{ kips}$$

$$P_{bH} = P_{bV}\left(\frac{0.5L_g}{h}\right) = 24.7\left(\frac{0.5(29)}{10}\right) = 35.8 \text{ kips}$$

$$P_b = P_{bV}\frac{L_b}{h} = 24.7\left(\frac{\sqrt{(15)^2 + (10)^2}}{10}\right) = 24.7\left(\frac{18.03}{10}\right) = 44.5 \text{ kips}$$

The minimum brace area required to prevent yielding under factored gravity loads is given by Eq. 15.5.11 but AISC–2.3 requires that the axial force in any member of the bracing system not exceed $0.85P_y$. Thus the minimum area required for the brace is

$$\text{Min } A_b = \frac{P_b}{0.85F_y} = \frac{44.5}{0.85(36)} = 1.45 \text{ sq in.}$$

The reason for this arbitrary limitation is to insure that the brace does not yield prior to the girder plastic strength being developed. Premature yielding in the bracing might occur due to residual stress and/or secondary moments in the bracing system. Under the gravity loading of this example, the bracing force is tension; however, under lateral wind loading, the bracing force is compression and the $0.85F_y$ stress limitation would offset instability effects from residual stress, secondary bending, and lateral-torsional buckling. For both compression and tension braces, the $0.85F_y$ limitation may reduce lateral deformation and help assure adequate stiffness to prevent sidesway buckling of the frame.

(b) Girder design. The plastic moment and axial force required to be resisted are, according to Eqs. 15.5.8 and 15.5.9,

$$M_g = \frac{w_u L_g^2}{32} = \frac{2(1.7)(29)^2}{32} = 89.4 \text{ ft-kips}$$

$$P_g = \frac{w_u L_g L}{8h} = \frac{2(1.7)(29)(30)}{8(10)} = 37.0 \text{ kips}$$

Since the unbraced length of 15 ft for strong-axis bending is relatively short, it is possible strength rather than stability may control; thus, using design Eq. 12.13.10,

$$\text{Equiv. } P_y \approx P + 2M/d$$

$$\text{Equiv. } P_y = 37.0 + 2(89.4)(12)/12 = 216 \text{ kips (for W12)}$$

$$\text{Reqd } A = \frac{\text{Equiv. } P_y}{F_y} = \frac{216}{36} = 6.00 \text{ sq in.}$$

$$\text{Min } Z = \frac{M_g}{F_y} = \frac{89.4(12)}{36} = 29.8 \text{ in.}^3$$

Try W12 × 27, which according to Appendix, Table A4, satisfies local buckling requirements for plastic design.

Check strength, AISC Formula 2.4–3:

$$P_y = F_y A = 36(7.95) = 287 \text{ kips}$$

$$M_p = F_y Z = 36(38.0)/12 = 114 \text{ ft-kips}$$

$$\frac{P}{P_y} = \frac{37.0}{287} = 0.13 < 0.15 \qquad \text{Formula 2.4–3 is satisfied}$$

Check stability, AISC Formula 2.4–2:

$$\frac{KL}{r_x} = \frac{15(12)}{5.07} = 35.6; \qquad F_a = 19.54 \text{ ksi}; \qquad F'_e = 118.5 \text{ ksi}$$

The floor system provides weak-direction bracing:

$$P_{cr} = 1.7AF_a = 1.70(7.95)(19.54) = 265 \text{ kips}$$

$$P_e = (23/12)\,AF'_e = 1.92(7.95)(118.5) = 1{,}813 \text{ kips}$$

$$M_m = M_p, \quad \text{since lateral support is adequate}$$

$$C_m = 1 - 0.4\frac{f_a}{F'_e} = 1 - 0.4\frac{P}{P_e} = 1 - 0.008 = 0.99$$

which is obtained from the AISC Commentary, or from Table 12.10.1.

$$\frac{P}{P_{cr}} + \frac{M_i}{M_p}\left(\frac{C_m}{1 - P/P_e}\right) = \frac{37}{265} + \frac{89.4}{114}\left(\frac{0.99}{1 - 0.02}\right) = 0.14 + 0.79$$

$$= 0.93 < 1 \qquad \text{OK}$$

Use W12 × 27 for the gravity loading case. W14 × 26 is also satisfactory.

Braced Bent—Drift. Drift, which was first defined and discussed in Sec. 14.1, is the lateral deflection of a multistory building resulting from wind loading. It is not to be confused with the sidesway deflection resulting from overall frame instability, but drift occurring under *factored* combined wind and gravity loads may influence overall frame stability.

Common practice[13] has been to limit the drift under service loads (elastic deflection) to about 0.0025 h_t, where h_t is the total building height.

In plastic design, the maximum effects of wind combined with gravity loading are computed according to AISC–1969 using a 1.3 load factor. The lateral wind-loading tributary to a braced bent must be used (see Fig. 15.5.2). The wind ordinarily tributary to a supported bent must be assumed transferred to the nearest braced bent mainly through diaphragm action of the floor system.

The transfer of the factored gravity and wind loads throughout the system may be done using any analysis procedure appropriate to the structural arrangement; however, for the braced frame it is suggested[13] that analysis as vertical pin-connected cantilever trusses is an acceptable procedure.

The $P\Delta$ moments must also be included in the analysis. For simple analysis, one may assume that under loads factored by 1.3, the drift index (Δ/h) may also be increased by that factor. Furthermore, reasonable values of $P\Delta$ moments will be obtained if the same drift index used for the overall structure is used for each story. Since actual $P\Delta$ cannot be established until the entire structure has been designed, the foregoing seems to be a reasonable approach.

The factored drift index suggested[13] is 0.004 which is said to be 0.0025 (1.3) plus an allowance for additional drift due to $P\Delta$. For a given story, the total shear is the horizontal factored wind force, ΣH, above a given floor level, plus the Q_Δ shear according to Eq. 14.1.2:

$$\text{Story shear} = \Sigma H + Q_\Delta \tag{15.5.12}$$

$$Q_\Delta = \frac{\Sigma P\Delta}{h} \qquad [14.1.2]$$

(Review Sec. 14.1 and Fig. 14.1.1 for braced-frame assumptions.) Using the drift index 0.004,

$$Q_\Delta = 0.004\Sigma P \tag{15.5.13}$$

where ΣP is the sum of all P loads from the stories above the one in question. Finally,

$$\text{Story shear} = \Sigma H + 0.004\Sigma P \tag{15.5.14}$$

The shear in Eq. 15.5.14 must be resisted below the given floor level.

The analysis of the vertical truss may be done separately for factored gravity load (LF = 1.3) and for the shears indicated by Eq. 15.5.14, and then combined.

The foregoing brief presentation of some of the general concepts relating to multistory plastic design was intended to familiarize the reader with certain factors of design not treated elsewhere in the text. A more detailed presentation and a complete design example of a braced multistory frame is to be found in the publication *Plastic Design of Braced Multistory Steel Frames*.[13] Other brief treatments are to be found in Refs. 18 and 19.

Connections and joints for single and multistory frames are extremely important, and are dealt with in Chapter 13.

SELECTED REFERENCES

1. *Plastic Design in Steel*, American Institute of Steel Construction, Inc., New York, 1959.
2. Lynn S. Beedle, *Plastic Design of Steel Frames*, John Wiley & Sons, Inc., New York, 1958.
3. C. E. Massonnet and M. A. Save, *Plastic Analysis and Design, Vol. I, Beams and Frames*, Blaisdell Publishing Company, Waltham, Mass., 1965.
4. Lynn S. Beedle, "Plastic Strength of Steel Frames," Proc. *ASCE*, Vol. 81, No. 764 (August 1955).
5. Edward R. Estes, Jr., "Design of Multi-span Rigid Frames by Plastic Analysis," Proc. of National Engineering Conf., AISC, 1955, pp. 27–39.
6. Le-Wu Lu, "Effective Length of Columns in Gabled Frames," *Engineering Journal*, AISC, Vol. 2, No. 1 (January 1965), pp. 6–7.
7. *Steel Gables and Arches*, American Institute of Steel Construction, Inc., New York, 1963.
8. John D. Griffiths, *Single Span Rigid Frames in Steel*, American Institute of Steel Construction, Inc., New York, 1948.
9. George C. Driscoll, Jr., et al., *Plastic Design of Multi-Story Frames*, Lecture Notes, Fritz Engineering Laboratory Report No. 273.20, Lehigh University, Bethlehem, Pa., 1965.
10. George C. Driscoll and Lynn S. Beedle, "Research in Plastic Design of Multistory Frames," *Engineering Journal*, AISC, Vol. 1, No. 3 (July 1964), pp. 92–100.
11. Oscar DeBuen, "Plastic Design of a Three-Story Steel Frame," *Engineering Journal*, AISC, Vol. 2, No. 3 (July 1965), pp. 71–75.
12. J. Hartley Daniels, "A Plastic Method For Unbraced Frame Design," *Engineering Journal*, AISC, Vol. 3, No. 4 (October 1966), pp. 141–149.
13. *Plastic Design of Braced Multistory Steel Frames*, American Iron and Steel Institute, New York, 1968.
14. Joseph A. Yura and George C. Driscoll, Jr., "Plastic Design of Multi-Story Buildings—A Progress Report," *Engineering Journal*, AISC, Vol. 2, No. 3 (July 1965), pp. 76–84.
15. M. Ojalvo and V. Levi, "Columns in Planar Continuous Structures," *J. Structural Div. ASCE*, Vol. 89, No. ST1 (February 1963), pp. 1–23.
16. John E. Goldberg, "Lateral Buckling of Braced Multistory Frames," *J. Structural Div. ASCE*, Vol. 94, No. ST12 (December 1968), pp. 2963–2983.
17. Joseph A. Yura and Le-Wu Lu, "Ultimate Load Tests of Braced Multistory Frames," *J. Structural Div. ASCE*, Vol. 95, No. ST10 (October 1969), pp. 2243–2263.
18. Le-Wu Lu, "Design of Braced Multi-Story Frames by the Plastic Method," *Engineering Journal*, AISC, Vol. 4, No. 1 (January 1967), pp. 1–9.
19. James B. Williams and Theodore V. Galambos, "Economic Study of a Braced Multi-Story Steel Frame," *Engineering Journal*, AISC, Vol. 5, No. 1 (January 1968), pp. 2–11.

16

Composite Steel-Concrete Construction

16.1. HISTORICAL BACKGROUND

Steel framing supporting cast-in-place reinforced concrete slab construction was historically designed on the assumption that the concrete slab acts independently of the steel in resisting loads. No consideration was given to the composite effect of the steel and concrete acting together. This neglect was justified on the basis that the bond between concrete floor or deck and the top of the steel beam could not be depended upon. However, with the advent of welding, it became practical to provide me-

Stud shear connectors. Courtesy Gregory Industries, Inc., Lorain, Ohio.

chanical shear connectors to resist the horizontal shear which develops during bending.

Steel beams encased in concrete were widely used from the early 1900's until the development of lightweight materials for fire protection in the past 20 years. Some such beams were designed compositely and some were not. In the early 1930's bridge construction began to use composite sections. Not until the early 1960's was it economical to use composite construction for buildings. However, current practice (1971) utilizes composite action in nearly all situations where concrete and steel are in contact, both on bridges and buildings.

The study of composite structural action began with the 1927 work of MacKay et al.[1] which dealt with encasement of steel beams in concrete. MacKay noted that beams which were fully encased and also were supporting a reinforced concrete slab cast monolithically with them indicated good interaction between the steel beams and the slab. MacKay concluded that such beams could, within practical limits, be considered as acting in a composite manner. In these early studies of encased beams, bond was depended upon to provide the interaction between steel and concrete. Composite beams, including encased beams as well as slabs on I-shaped beams, exhibited sufficient reserve strength so that Caughey[2] in 1929 recommended that they be designed on the basis of a homogeneous section wherein the concrete area is transformed into an equivalent area of steel.

Since stresses in a wide slab supported on a steel beam are not uniform across the slab width, the ordinary flexure formula, $f = Mc/I$, does not apply. Similarly to T-sections entirely of reinforced concrete, an equivalent width of the wide slab is used such that the flexure formula may be applied to obtain the correct moment capacity.

Theoretical studies to determine the correct effective width have been carried out by von Kármán,[3] Reissner,[4] and Lee.[5] The effective widths determined from elastic theory provide rational values to be used in computing the strength of tee-sections and composite beams.

Viest,[6] in his 1960 review of research, notes that the important factor in composite action is that the bond between concrete and steel remain unbroken. As designers began to place slabs on top of supporting steel beams, investigators began to study the behavior of mechanical shear connectors. The shear connectors provided the interaction necessary for the composite action between slab and steel I-shaped beams which had previously been supplied by bond for the fully encased beams.

Early studies of mechanical shear connectors began in 1923 with the work of Voellmy reported by Ros.[7] Since that time studies have been carried out by many investigators, including Viest,[8] Newmark, et al.,[9] Culver and Coston,[10] and Thurliman.[11] Newmark was concerned with incomplete interaction and showed that slip between two interconnected elements such as a beam and slab will be so small that a design on the basis of full composite action will give results satisfactory for all practical purposes.

16.2. COMPOSITE ACTION

Composite behavior is developed when two load-carrying structural members such as a concrete floor system and the supporting steel beams are integrally connected and deflect as a single unit. Typical examples of composite cross sections are shown in Fig. 16.2.1. The extent to which

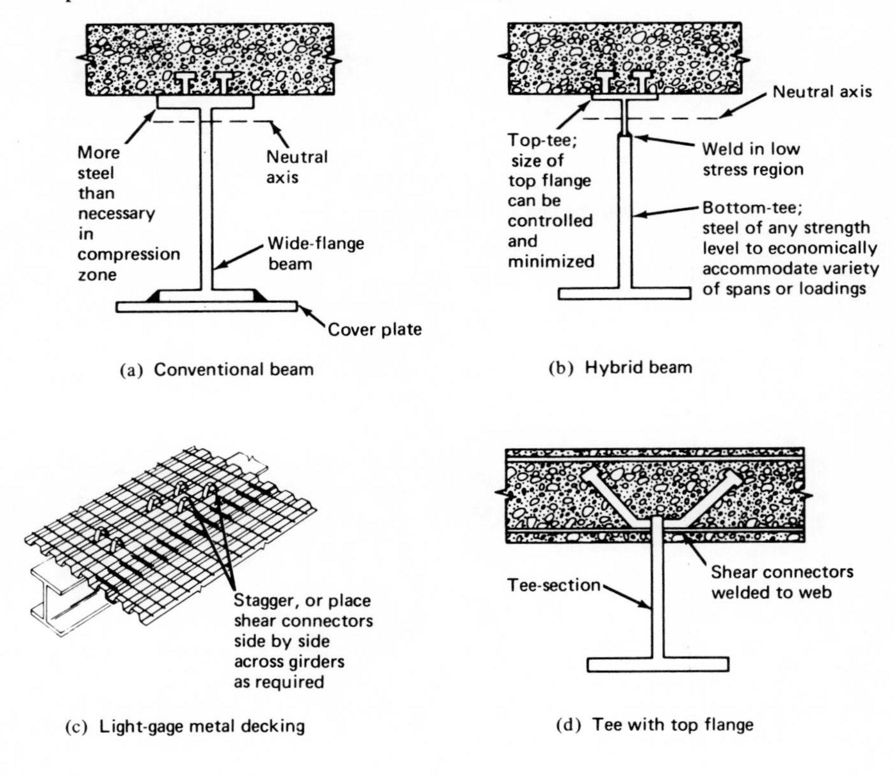

(e) Encasement (rarely used, 1971)

Fig. 16.2.1. Various types of composite steel-concrete sections.

composite action is developed depends on the provisions made to insure stress and strain continuity over the entire cross section.

In developing the concept of composite behavior, consider first the noncomposite beam of Fig. 16.2.2a, wherein if friction between the slab

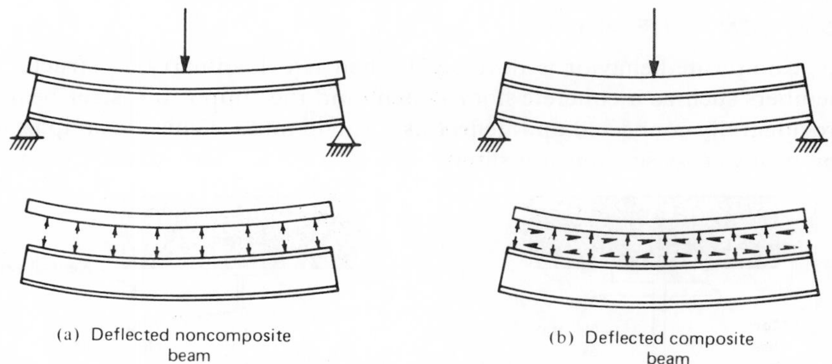

Fig. 16.2.2. Comparison of deflected beams with and without composite action.

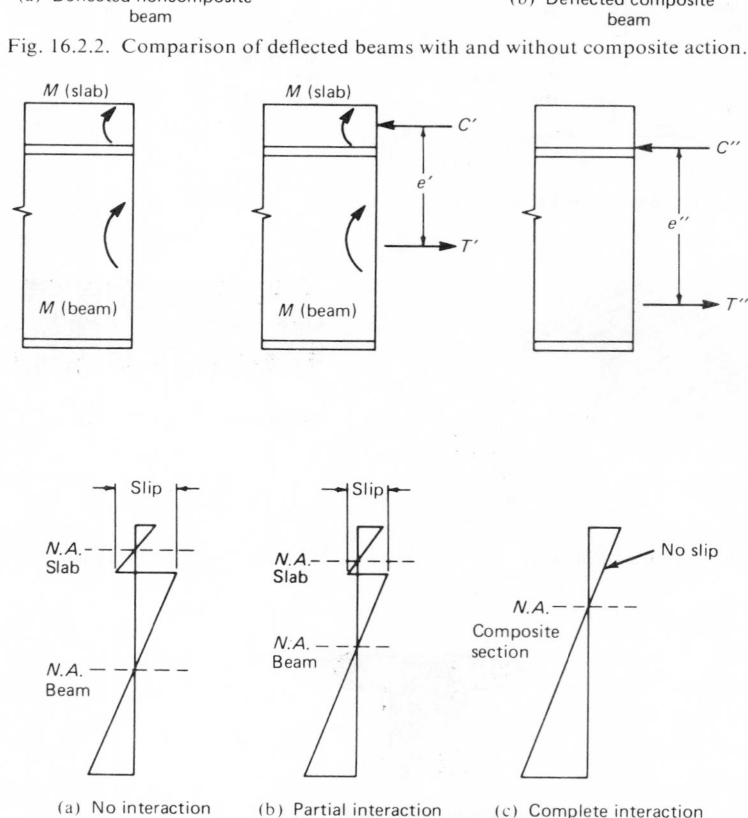

Fig. 16.2.3. Strain distribution in composite beams.

and beam is neglected the beam and slab each carry separately a part of the load. This is further shown in Fig. 16.2.3a. When the slab deforms under vertical load, its lower surface is in tension and elongates; while the upper surface of the beam is in compression and shortens. Thus a discontinuity will occur at the plane of contact. Since friction is neglected, only vertical internal forces act between the slab and beam.

When a system acts compositely (Fig. 16.2.2b and 16.2.3c) no relative slippage occurs between the slab and beam. Horizontal forces (shears) are developed which when acting on the lower surface of the slab compress and shorten it, while those acting on the upper surface of the beams elongate it.

By an examination of the strain distribution that occurs when there is no interaction between the concrete slab and the steel beam (Fig. 16.2.3a), it is seen that the total resisting moment is equal to

$$\Sigma M = M_{\text{slab}} + M_{\text{beam}} \tag{16.2.1}$$

It is noted that for this case there are two neutral axes; one at the center of the slab and the other at the center of the beam. The horizontal slippage resulting from the bottom of the slab in tension and the top of the beam in compression is also indicated.

Consider next the case where only partial interaction is present, Fig. 16.2.3b. The neutral axis of the slab is closer to the beam and that of the beam closer to the slab. Due to the partial interaction, the horizontal slippage has now decreased. The result of the partial interaction is the partial development of the compressive and tensile forces C' and T', the net forces in the concrete slab and steel beam, respectively. The resisting moment of the section is now increased by the amount $T'e'$ or $C'e'$.

When complete interaction between the slab and the beam is developed, no slippage occurs and the resulting strain diagram is shown in Fig. 16.2.3c. Under this condition a single neutral axis occurs which lies below that of the slab and above that of the beam. In addition, the compressive and tensile forces C'' and T'' are larger than the C' and T' existing with partial interaction. The resisting moment of the fully developed composite section then becomes

$$\Sigma M = T''e'' \quad \text{or} \quad C''e'' \tag{16.2.2}$$

16.3. ADVANTAGES AND DISADVANTAGES

The basic advantages resulting from composite design are

1. Reduction in the weight of steel
2. Shallower steel beams
3. Increased floor stiffness
4. Increased span length for a given member
5. Increased overload capacity

A weight savings in steel of 20 to 30 percent is often possible by taking full advantage of a composite system. Such a weight reduction in the supporting steel beams usually permits the use of a shallower as well as a lighter member. This advantage may reduce the height of a multistoried building significantly so as to provide savings in other building materials such as outside walls and stairways.

The stiffness of a composite floor is substantially greater than that of a concrete floor with its supporting beams acting independently. Normally the concrete slab acts as a one-way plate spanning between the supporting beams. In composite design, an additional use is made of the slab by its action in a direction parallel to and in combination with the supporting steel beams. The net effect is to greatly increase the moment of inertia of the floor system in the direction of the steel beams. The increased stiffness considerably reduces the live-load deflections and, if shoring is provided during construction, also reduces dead-load deflections. Assuming full composite action, the ultimate strength of the section greatly exceeds the sum of the strengths of the slab and the beam considered separately, providing high overload capacity.

The overall economy of using composite construction when considering total building costs appears to be good and is steadily improving. The development of new combinations of floor systems is continually expanding and the use of high-strength steels and hybrid members (see Fig. 6.2.1b) can be expected to produce further economies. In addition, composite wall systems and even composite columns are beginning to find their way into building construction.

While there are no major disadvantages in composite construction, there are some limitations which should be recognized. These are:

1. Effect of continuity
2. Long-term deflections

At the present time (1971), only the portion of the concrete slab acting in compression is considered to be effective. In the case of continuous beams, the advantage of composite behavior is reduced in the area of negative beam moments. This restriction seems overly conservative since floor slabs usually carry two-way steel reinforcement or at least, temperature steel in the direction parallel to the supporting steel beams. At columns in buildings, of course, there is no bar reinforcement to provide continuity of the composite action. On bridges, bar reinforcement is frequently provided to carry the tensile capacity required of the concrete in zones of negative moment; thus using composite action throughout both positive and negative moment zones.

The matter of long-term deflections might be important if the composite section is resisting a substantial portion of the dead loads or if the live loads are of a long duration. This problem, however, can be minimized by assuming a reduced effective width of the slab or by assuming an increased modulus of elasticity ratio n. This is discussed more fully in Sec. 16.10.

16.4. EFFECTIVE WIDTH

In order to compute the section properties of a composite section it is desirable to utilize the concept of effective width. Referring to Fig. 16.4.1, consider the composite section under stress in which the slab is infinitely wide. The intensity of the stress, σ_x, maximum over the steel

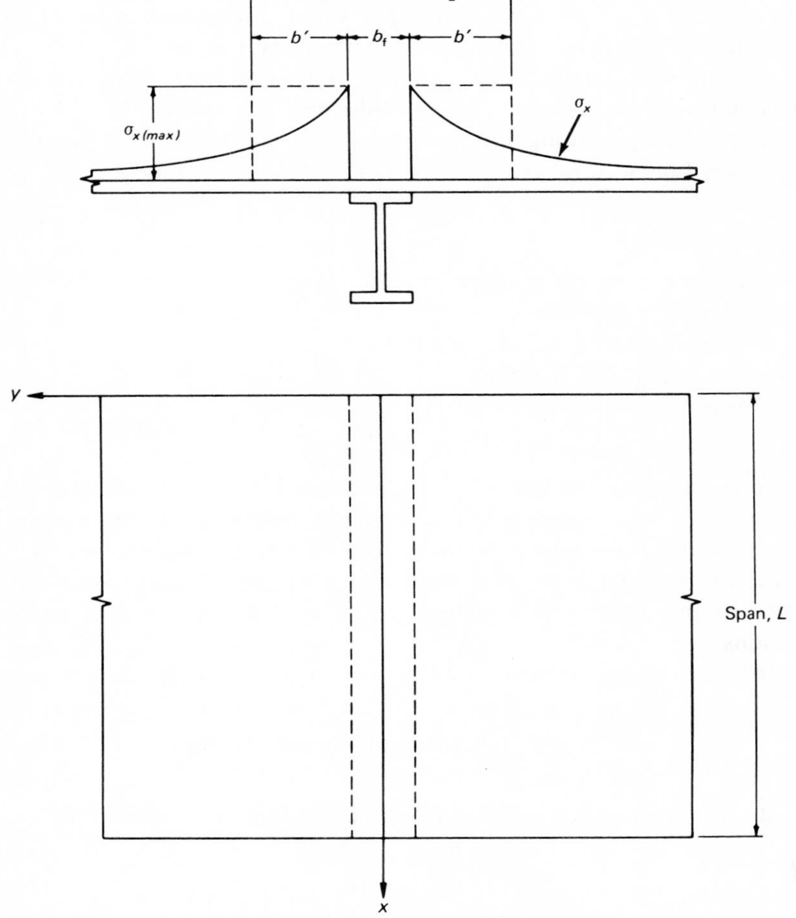

Fig. 16.4.1. Nonuniform distribution of compressive stress σ_x, and effective width b_E.

beam, decreases nonlinearly as the distance from the supporting beam increases.

The effective width of a composite member can be taken as

$$b_E = b_f + 2b' \tag{16.4.1}$$

where $2b'$ times the maximum stress, $\sigma_{x\max}$ is equal to the area under the curves for σ_x. Various investigators, including Timoshenko[12] and von Kármán[3] have derived expressions for the effective width of homogeneous beams with wide flanges; and Johnson[13] has shown the expressions to be valid also for beams in which the flange and stem are of different materials.

The analysis for effective width involves theory of elasticity applied to plates, using an infinitely long continuous beam on equidistant supports, with an infinitely wide flange having a small thickness compared to the beam depth. The total compression carried by the equivalent system is the same as that carried by the actual system. The value of b' depends on the span length and type of loading. Johnson[13] has shown that for loading which produces bending moment having a half-sine wave shape, the effective width is

$$b_E = b_f + \frac{2L}{\pi(3 + 2\mu - \mu^2)} \tag{16.4.2}$$

where L = span length of beam
b_f = steel beam flange width
μ = Poisson's ratio for the slab

Assuming $\mu = 0.2$ for concrete,

$$b_E = b_f + \frac{2L}{\pi(3 + 2(0.2) - (0.2)^2)} = b_f + 0.196L \tag{16.4.3}$$

As a simplification for design purposes, AISC–1.11.1 has adopted the same method of computing effective flange widths as the American Concrete Institute Building Code[14] does for reinforced concrete beams. Referring to Fig. 16.4.2, the maximum value of the effective width b_E permitted by the AISC–1.11.1 is the least value computed by the following relations:

(a) For an interior girder with slab extending on both sides of girder:

(1) $b_E \leq L/4$ (16.4.4a)
(2) $b_E \leq b_0$ (for equal beam spacing) (16.4.4b)
(3) $b_E \leq b_f + 16t_s$ (16.4.4c)

(b) For an exterior girder with slab extending only on one side:

(1) $b_E \leq L/12 + b_f$ (16.4.5a)
(2) $b_E \leq \frac{1}{2}(b_0 + b_f)$ (16.4.5b)
(3) $b_E \leq b_f + 6t_s$ (16.4.5c)

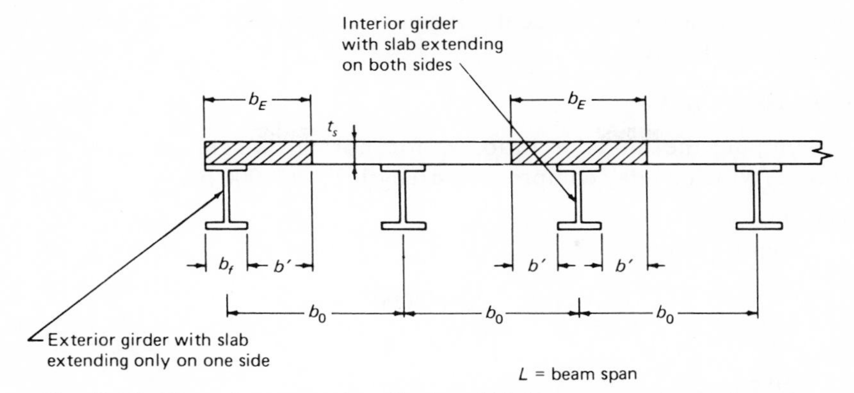

Fig. 16.4.2. Dimensions governing effective width b_E on composite steel-concrete beams.

Similarly, for highway bridge design,[15] the effective width according to AASHO–1969–1.7.99 is identical with AISC except Eq. 16.4.4c for an interior girder is replaced by

$$(3)\quad b_E \leq 12t_s \tag{16.4.6}$$

and for an exterior girder, Eqs. 16.4.5a and c are replaced by

$$(1)\quad b_E \leq L/12 \tag{16.4.7a}$$

$$(3)\quad b_E \leq 6t_s \tag{16.4.7b}$$

16.5. COMPUTATION OF SECTION PROPERTIES

The section properties of a composite section can be computed by the transformed area method. In contrast to reinforced concrete design, where the steel area is transformed into an equivalent concrete area, the concrete is transformed into equivalent steel. As a result, the concrete area is reduced by using a slab width equal to b_E/n, where n is the ratio of steel modulus of elasticity, E_s, to concrete modulus of elasticity, E_c.

Modular Ratio, n. The modulus of elasticity of concrete in psi may be taken[14] as

$$E_c = w^{1.5} 33 \sqrt{f'_c} \tag{16.5.1}$$

where w is weight of concrete in pounds per cubic feet and f'_c is taken with the units of pounds per square inch. For ordinary weight concrete, weighing approximately 145 lb cu ft, the value may be taken as

$$E_c = 57{,}800 \sqrt{f'_c} \tag{16.5.2}$$

The modular ratio n is frequently taken as the nearest whole number to that given by Eq. 16.5.1.

EXAMPLE 16.5.1

Compute the modular ratio, n, for normal weight concrete, (145 lb/cu ft) with a 28-day compressive strength f'_c of 3,000 psi.

Solution

From Eq. 16.5.1,

$$E_c = \frac{(145)^{1.5} 33 \sqrt{3000}}{1000} = 3{,}170 \text{ ksi}$$

which gives

$$n = \frac{E_s}{E_c} = \frac{29{,}000}{3170} = 9.15 \approx 9$$

The minimum value for n permitted by the ACI Code and the AASHO Specifications is 6. For practical design purposes, the value of n from Table 16.5.1 may be used.

TABLE 16.5.1.
Practical Design Values for
Modular Ratio n

Concrete strength f'_c psi	Modular ratio $n = E_s/E_c$
3,000	9
3,500	$8\frac{1}{2}$
4,000	8
4,500	$7\frac{1}{2}$
5,000	7
6,000	$6\frac{1}{2}$

Effective Section Modulus. A complete beam may be considered as a steel member to which has been added a cover plate on the top flange. This "cover plate" being concrete is considered to be effective only when the top flange is in compression. In continuous beams, the concrete slab is usually ignored in regions of negative moment. If the neutral axis falls within the concrete slab, present practice is to consider only that portion of the concrete slab which is in compression.

EXAMPLE 16.5.2

Compute the section properties of the composite section shown in Fig. 16.5.1 according to AISC specifications assuming $f'_c = 3{,}000$ psi and $n = 9$.

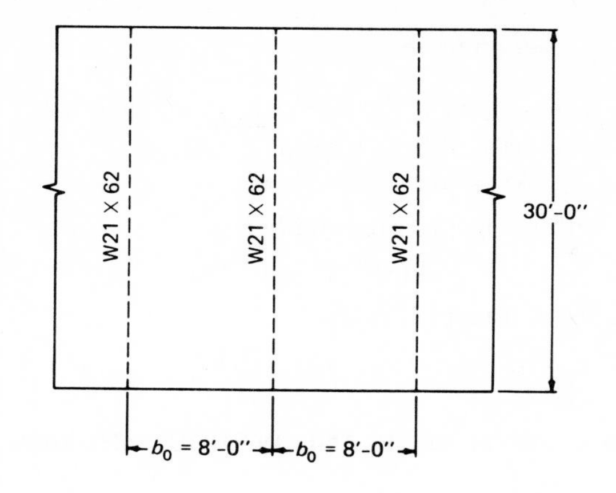

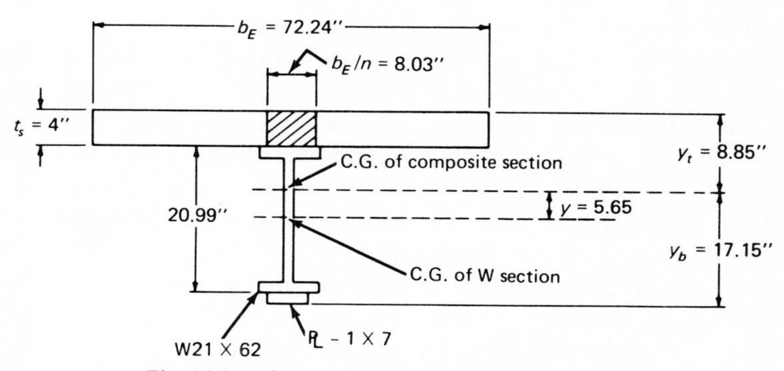

Fig. 16.5.1. Composite section for Example 16.5.2.

Solution

First, determine effective width:

$$b_E = (0.25)(\text{span length}) = 0.25(30)12 = 90 \text{ in.}$$
$$b_E = b_o = 8(12) = 96 \text{ in.}$$
$$b_E = b_f + 2(8)t_s = 8.24 + 64 = 72.24 \text{ in.} \quad \leftarrow \text{Controls}$$

The width of equivalent steel is $b_E/n = 8.03$ in. The computation of centroid and moment of inertia is shown in Table 16.5.2.

TABLE 16.5.2

Element	Transformed Area, A	Moment Arm from Centroid, y	Ay	Ay^2	I_0
Slab	32.12	+12.495	+401.34	5016.8	42.8
W21 × 62	18.23	0	0	0	1326.8
Cover Pl.	7.0	−10.995	−76.97	846.7	
Totals	57.35		+324.37	5863.5	1369.6
$Ay^2 + I_0 = I_x$					7233.1 in.^{4}

$$y = \frac{+324.37}{57.35} = +5.65 \text{ in.}$$

$$I = I_x - Ay^2 = 7233.1 - 57.35(5.65)^2 = 5403 \text{ in.}^4$$
$$y_t = 10.50 - 5.65 + 4.0 = 8.85 \text{ in.}$$
$$y_b = 10.50 + 5.65 + 1.0 + 17.15 \text{ in.}$$

The section modulus with respect to the top fiber, S_t, is

$$S_t = I/y_t = 5403/8.85 = 610 \text{ in.}^3$$

and the section modulus with respect to the bottom fiber, S_b, is

$$S_b = I/y_b = 5403/17.15 = 315 \text{ in.}^3$$

16.6. SERVICE LOAD STRESSES WITH AND WITHOUT SHORING

The actual stresses that result due to a given loading on a composite member are dependent upon the manner of construction.

The simplest construction occurs when the steel beams are placed first and used to support the concrete slab formwork. In this case the steel beam acting noncompositely (i.e., by itself) supports the weight of the forms, the wet concrete, and its own weight. Once forms are removed and concrete has cured, the section will act compositely to resist all dead and live loads placed after the curing of concrete. Such construction is said to be *without temporary shoring* (i.e., unshored).

Alternatively, to reduce the service load stresses, the steel beams may be supported on temporary shoring. In which case, the steel beam, forms, and wet concrete, are carried by the shores. After curing of the concrete, the shores are removed and the section acts compositely to resist all loads. This system is called *shored* construction.

The following example illustrates the difference in service load stresses under the two systems of construction.

EXAMPLE 16.6.1

For the steel W21 × 62 with the 1 by 7-in. plate of Fig. 16.5.1, determine the service load stresses considering that (a) construction is without temporary shoring, and (b) construction uses temporary shores. The dead- and live-load moment to be superimposed on the system after the concrete is cured is 560 ft-kips.

Solution

The composite section properties as computed in Example 16.5.2 are

$$S_{top} = 610 \text{ in.}^3 \text{ (top of concrete)}$$
$$S_{bottom} = S_{tr}\text{ (AISC–1969)} = 315 \text{ in.}^3 \text{ (bottom of steel)}$$

From AISC Manual tables, "Composite Design," the noncomposite

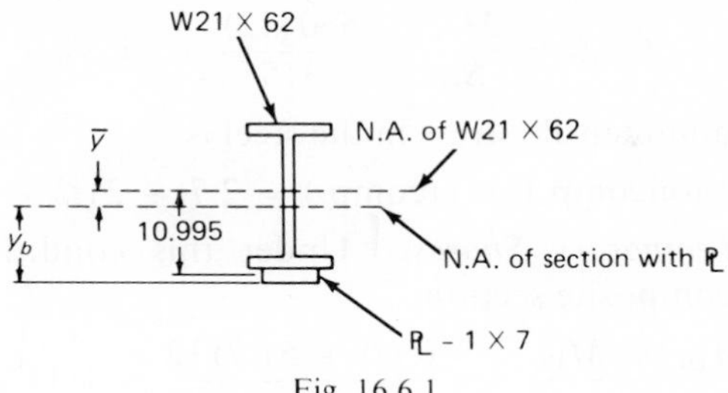

Fig. 16.6.1.

properties may be obtained or computed as follows: For the steel section alone, refer to Fig. 16.6.1.

$$y = \frac{7.0\,(10.995)}{7.0 + 18.23} = 3.05 \text{ in.}$$

$$y_b = 10.495 - 3.05 + 1.00 = 8.45 \text{ in.}$$

$$\begin{aligned} I &= I_0(\text{W21} \times 62) + A_p y^2 - A y^2 \\ &= 1326.8 + 7.0\,(10.995)^2 - 25.23\,(3.05)^2 \\ &= 1326.8 + 846 - 235 = 1938 \text{ in.}^4 \end{aligned}$$

$$S_{st} = \frac{1938}{13.55} = 143 \text{ in.}^3 \text{ (top)}$$

$$S_{sb} = \frac{1938}{8.45} = 229 \text{ in.}^3 \text{ (bottom)}$$

(a) *Without Temporary Shores.* Weight due to the concrete slab and steel beam,

$$w\text{(concrete slab)} = \frac{4}{12}(8)0.15 = 0.40$$

$$w\text{(steel beam)} = \underline{0.06}$$

$$0.46 \text{ kips/ft}$$

$$M\text{(DL on noncomposite)} = \frac{0.46\,(30)^2}{8} = 51.7 \text{ ft-kips}$$

$$f_{\text{top}} = \frac{M_{\text{DL}}}{S_{st}\text{(steel section)}} = \frac{51.7\,(12)}{143} = 4.3 \text{ ksi}$$

$$f_{\text{bottom}} = \frac{M_{\text{DL}}}{S_{\text{sb}}\text{(steel section)}} = \frac{51.7\,(12)}{229} = 2.7 \text{ ksi}$$

The additional stresses after concrete has cured are

$$f_{\text{top}} = \frac{M_{\text{LL}}}{S_{\text{top}}\text{(composite)}} = \frac{560\,(12)}{610\,(9)} = 1.22 \text{ ksi (concrete stress)}$$

where the stress in the concrete is $1/n$ times the stress on equivalent steel (transformed section).

$$f_{\text{bottom}} = \frac{M_{\text{LL}}}{S_{\text{tr}}} = \frac{560\,(12)}{315} = 21.3 \text{ ksi}$$

The total maximum tensile stress in the steel is

$$f = f(\text{noncomp.}) + f(\text{comp.}) = 2.7 + 21.3 = 24.0 \text{ ksi}$$

(b) *With Temporary Shores.* Under this condition all loads are resisted by the composite section.

$$f_{\text{top}} = \frac{M_{\text{DL}} + M_{\text{LL}}}{S_{\text{top}}\,(\text{composite})} = \frac{(560 + 51.7)\,12}{610\,(9)} = 1.34 \text{ ksi on concrete}$$

$$f_{\text{bottom}} = \frac{M_{\text{DL}} + M_{\text{LL}}}{S_{\text{tr}}} = \frac{(560 + 51.7)\,12}{315} = 23.3 \text{ ksi}$$

Stress distributions for both with and without shores are given in Fig. 16.6.2. Since the dead load was small in this example use of shores gave

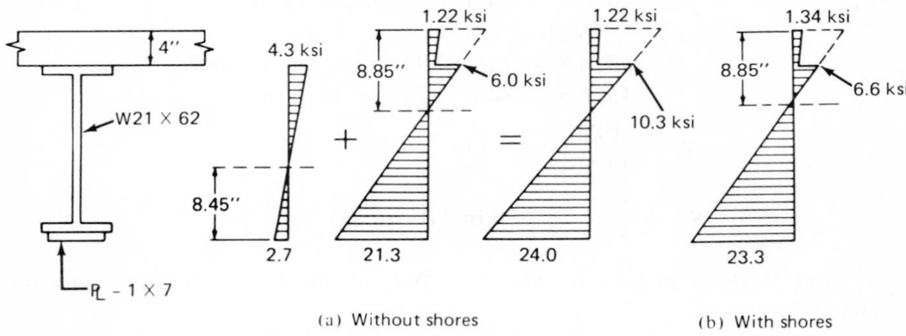

Fig. 16.6.2. Service load stresses.

insignificant reduction in service load stress. Where thicker slabs are used the dead load stresses may become as high as 30 percent, in which case using or not using shores will make a significant difference.

16.7. ULTIMATE STRENGTH OF FULLY COMPOSITE SECTIONS

The ultimate strength of a composite section is dependent upon the yield strength and section properties of the steel beam, the concrete slab strength and the interaction capacity of the shear connectors joining the slab to the beam.

The provisions of AISC–1.11 are nearly all based on ultimate strength behavior, even though all relationships are adjusted to be in the service load range. These ultimate strength concepts were applied to design practice as recommended by the ASCE-ACI Joint Committee on Composite Construction,[16] and further modified as a result of research at Lehigh University.[17]

The ultimate strength in terms of an ultimate moment capacity gives a clearer understanding of composite behavior as well as provides a more accurate measure of the true factor of safety. The true factor of safety is the ratio of the ultimate moment capacity to the actually applied moment. In both cases, whether the *slab* is termed "adequate" or "inadequate" as compared to the tensile yield capacity of the beam, the *connection* between the slab and beam is considered adequate in the following development. Complete shear transfer at the steel-concrete interface is assumed.

In determining the ultimate moment capacity, the concrete is assumed to take only compressive stress. Although concrete is able to sustain a limited amount of tensile stress, the tensile strength at the strains occuring during the development of the ultimate moment capacity is negligible.

The procedure for determining the ultimate moment capacity depends on whether the neutral axis falls within the concrete slab or within the steel beam. If the neutral axis falls within the slab, the slab is said to be adequate; i.e., the slab is capable of resisting the total compressive force. If the neutral axis falls within the steel beam, the slab is considered inadequate, i.e., the slab is able to resist only a portion of the compressive force, the remainder being taken by the steel beam. Figure 16.7.1 shows the stress distribution for these two cases.

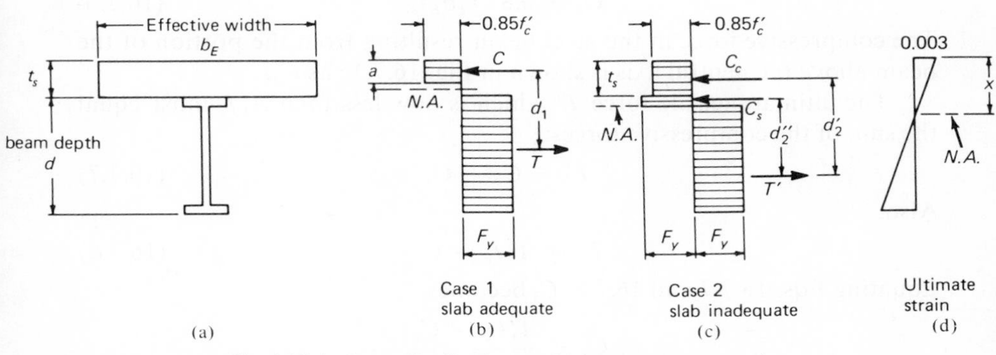

Fig. 16.7.1. Stress distribution at ultimate moment capacity.

Case 1—Slab Adequate. Referring to Fig. 16.7.1b and assuming the Whitney rectangular stress block (uniform stress of $0.85f'_c$ acting over a depth a), the ultimate compressive force C is

$$C = 0.85 f'_c a b_E \tag{16.7.1}$$

The ultimate tensile force T is the yield strength of the beam times its area:

$$T = A_s F_y \tag{16.7.2}$$

Equating the ultimate compressive force C to the ultimate tensile force T gives

$$a = \frac{A_s F_y}{0.85 f'_c b_E} \tag{16.7.3}$$

According to the rectangular stress block approach,[14] the neutral axis x, as shown in Fig. 16.7.1d, equals $a/0.85$ for $f'_c = 4{,}000$ psi. The ultimate moment capacity M_u becomes

$$M_u = Cd_1 \quad \text{or} \quad Td_1 \tag{16.7.4}$$

Since the slab is assumed adequate, it is capable of developing a compressive force equal to the full yield capacity of the steel beam. Expressing the ultimate moment in terms of the steel force gives

$$M_u = A_s F_y \left(\frac{d}{2} + t_s - \frac{a}{2} \right) \tag{16.7.5}$$

The usual procedure is to determine the depth of the stress block a by Eq. 16.7.3, and if a is less than the slab thickness t_s, determine the ultimate moment capacity by Eq. 16.7.5.

Case 2—Slab Inadequate. If the depth a of the stress block as determined in Eq. 16.7.3 exceeds the slab thickness, the stress distribution will be as shown in Fig. 16.7.1c. The ultimate compressive force C_c in the slab is

$$C_c = 0.85 f'_c b_E t_s \tag{16.7.6}$$

The compressive force in the steel beam resulting from the portion of the beam above the neutral axis is shown in Fig. 16.7.1c as C_s.

The ultimate tensile force T' which is now less than $A_s f_y$ must equal the sum of the compressive forces:

$$T' = C_c + C_s \tag{16.7.7}$$

Also,

$$T' = A_s f_y - C_s \tag{16.7.8}$$

Equating Eqs. 16.7.7 and 16.7.8, C_s becomes

$$C_s = \frac{A_s F_y - C_c}{2}$$

or

$$C_s = \frac{A_s F_y - 0.85 f'_c b_E t_s}{2} \tag{16.7.9}$$

Considering the compressive forces C_c and C_s, the ultimate moment capacity M_u for Case 2 is

$$M_u = C_c d'_2 + C_s d''_2 \tag{16.7.10}$$

the moment arms d'_2 and d''_2 are as shown in Fig. 16.7.1c.

Whenever the Case 2 situation occurs, the steel beam is assumed to accommodate plastic strain in both tension and compression at ultimate strength. Certainly, it is implied that such a steel section satisfy the requirements of "compact" sections; that is, it should have proportions which insure its ability to develop its plastic moment capacity. Little research has been performed on Case 2 situations because they rarely occur in practice.

EXAMPLE 16.7.1

Determine the ultimate moment capacity of the composite section shown in Fig. 16.7.2. Assume A36 steel, $f'_c = 3{,}000$ psi and $n = 9$.

Solution

Referring to Fig. 16.7.2, determine the adequacy of the concrete slab. Assuming the slab is fully adequate; i.e., Case 1,

$$a = \frac{A_s F_y}{0.85 f'_c b_E} = \frac{10.6(36)}{0.85\,(3)\,60} = 2.49 \text{ in.} < t_s \qquad \text{OK}$$

$$C = 0.85 f'_c a b_E = 0.85\,(3)\,(2.49)\,(60) = 381 \text{ kips}$$

$$T = A_s F_y = 10.6\,(36) = 381 \text{ kips (checks)}$$

$$\text{Arm } d_1 = \frac{d}{2} + t - \frac{a}{2} = 7.925 + 4.0 - 1.245 = 10.68 \text{ in.}$$

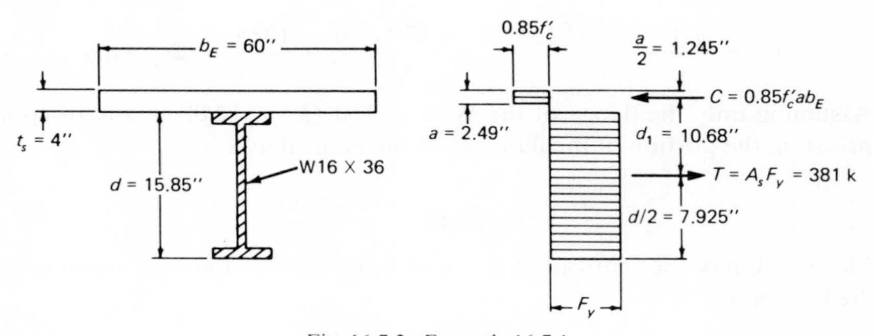

Fig. 16.7.2. Example 16.7.1.

Ultimate composite moment capacity,

$$M_u = Cd_1 = Td_1$$
$$= 381(10.68)/12 = 340 \text{ ft-kips}$$

EXAMPLE 16.7.2

Determine the ultimate moment capacity of the composite section shown in Fig. 16.7.3. Assume A36 Steel, $f'_c = 3{,}000$ psi, and $n = 9$.

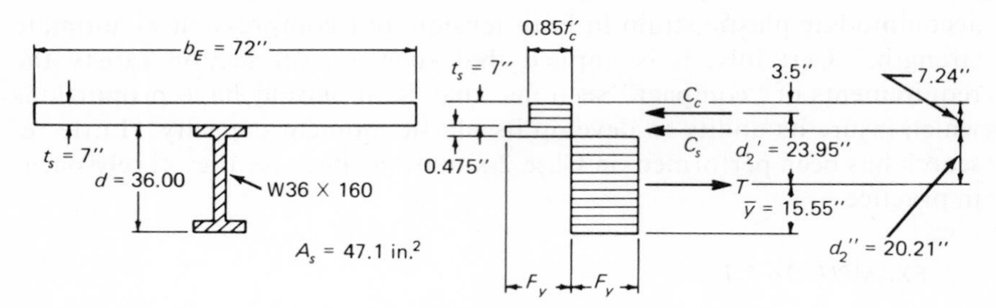

Fig. 16.7.3. Example 16.7.2.

Solution

Referring to Fig. 16.7.3 determine whether or not the slab is adequate. Assuming the slab is adequate to balance tensile capacity of the steel section (i.e., Case 1),

$$a = \frac{A_s F_y}{0.85 f'_c b_E} = \frac{47.1(36)}{0.85(3)(72)} = 9.22 \text{ in.} \qquad t_s = 7 \text{ in.} \qquad \text{N.G.}$$

Since the concrete slab is only 7 in. thick, the slab is inadequate to carry a compressive force equal to the tensile force that can be developed by the W36 × 160. Thus, Case 2 applies. Using Eq. 16.7.6,

$$C_c = 0.85 f'_c b_E t_s = 0.85(3)72(7) = 1285 \text{ kips}$$

Using Eq. 16.7.9,

$$C_s = \frac{A_s F_y - 0.85 f'_c b_E t_s}{2} = \frac{47.1(36) - 1285}{2} = 205 \text{ kips}$$

Assuming only the flange of the W36 × 160 (b_f = 12.00 in.) is in compression, the portion of the flange d_f to the neutral axis is

$$d_f = \frac{205}{36(12.00)} = 0.475 \text{ in.}$$

The location of the centroid of the tension portion of the steel beam from the bottom is

$$\bar{y} = \frac{47.1(18) - 0.475(12)35.76}{47.1 - 0.475(12)} = 15.55 \text{ in.}$$

Referring to Fig. 16.7.3, the ultimate composite-moment capacity from Eq. 16.7.10 is

$$\begin{aligned} M_u &= C_c d_2' + C_s d_2'' \\ &= [1285(23.95) + 205(20.21)]/12 = 2{,}910 \text{ ft-kips} \end{aligned}$$

In the development of the ultimate strength of composite sections, it has been implicitly assumed that sufficient interaction between the concrete slab and the steel beam existed. Such interaction is usually assured by providing a sufficient number of "shear connectors." This matter is treated in Sec. 16.8.

The foregoing development and examples have emphasized the computation of ultimate moment capacity. Tests have verified that such capacities are achieved. Furthermore, whether during construction the steel beam is *shored* or *unshored*, the ultimate strength is identical. Even though service load stresses, as discussed in Sec. 16.6, are lower when temporary shoring is used to support the beam during construction than when such shoring is omitted, the same ultimate moment capacity is achieved. Current (1971) AISC design practice, as discussed in Sec. 16.8, uses this ultimate strength concept as the basis for its working stress method.

16.8. SHEAR CONNECTORS

The horizontal shear that develops between the concrete slab and the steel beam during loading must be resisted so the composite section acts monolithically. Although the bond developed between the slab and the steel beam may be significantly high, it cannot be depended upon to provide the required interaction. Neither can the frictional force developed between the slab and the steel beam.

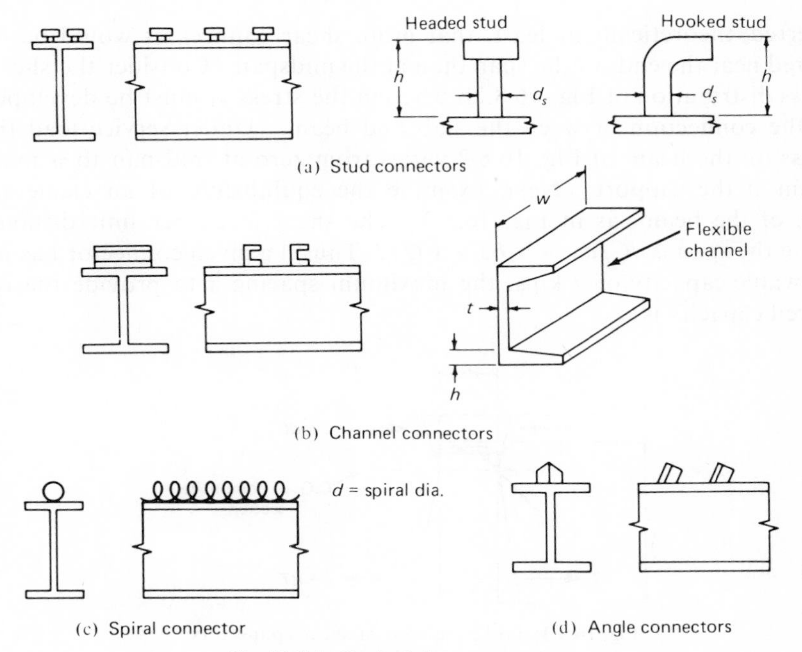

Fig. 16.8.1. Typical shear connectors.

Instead, mechanical shear connectors attached to the top of the beam must be provided. Typical shear connectors are shown in Fig. 16.8.1.

Ideally, the shear connectors should be stiff enough to provide the complete interaction shown in Fig. 16.2.3c. This, however, would require that the stiffeners be infinitely rigid. Also, by referring to the shear diagram of a uniformly loaded beam as shown in Fig. 16.8.2, it would be

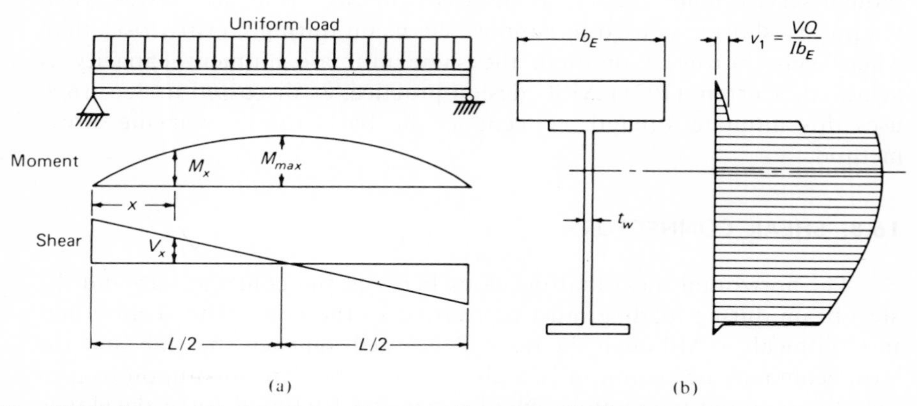

Fig. 16.8.2. Shear variation for uniform loading and distribution across a steel-concrete composite section.

inferred, theoretically at least, that more shear connectors would be required near the ends of the span than at the midspan. Consider the shear-stress distribution of Fig. 16.8.2b wherein the stress v_1 must be developed by the connection between the slab and beam. Under service load the stress on the beam of Fig. 16.8.2 varies from zero at midspan to a maximum at the support. Next, examine the equilibrium of an elemental slice of the beam, as in Fig. 16.8.3. The shear force per unit distance along the span is $dC/dx = v_1 b_E = VQ/I$. Thus if a given connector has an allowable capacity of q kips, the maximum spacing s to provide the required capacity is

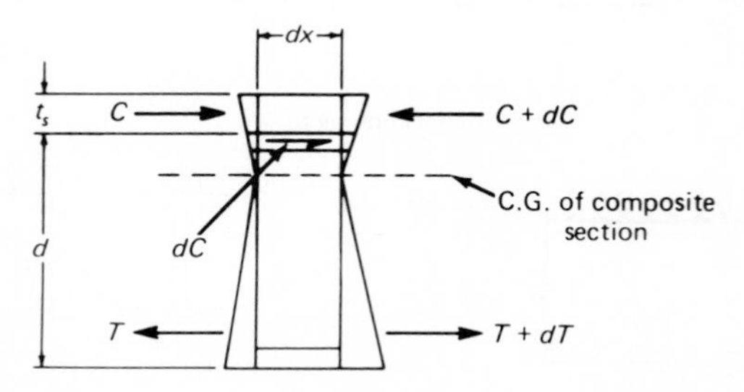

Fig. 16.8.3. Force required of shear connectors at service loads.

$$s = \frac{q}{VQ/I} \tag{16.8.1}$$

Composite design until recent years has used Eq. 16.8.1 to space connectors. AASHO-1.7.101-1969[15] requires using Eq. 16.8.1 to design for fatigue; then a check is required for ultimate strength.

If one uses an ultimate-strength concept, the shear connectors under ultimate bending moment share equally in carrying the total maximum compressive force developed in the concrete slab. This would mean, referring to Fig. 16.8.2a, that shear connection is required to transfer the compressive force developed in the slab at midspan to the steel beam in the distance $L/2$, since no compressive force can exist in the slab at the end of the span where zero moment exists. The ultimate compressive force to be accommodated could not exceed that which the concrete can carry:

$$C_{\max} = 0.85 f'_c b_E t_s \tag{16.8.2}$$

or if the ultimate tensile force below the bottom of the slab is less than $C_{\max}$,

$$T_{\max} = A_s f_y \tag{16.8.3}$$

Thus if a given connector has an ultimate capacity q_{ult} the total number of connectors N required between the points of maximum and zero bending moment is

$$N = \frac{C_{\max}}{q_{\text{ult}}} \quad \text{or} \quad \frac{T_{\max}}{q_{\text{ult}}} \tag{16.8.4}$$

whichever is smaller. Under the ultimate-strength approach, the total required number of shear connectors are distributed uniformly over the region of the beam between maximum and zero bending moment.

The determination of the connector capacity analytically is complex, since the shear connector deforms under load and the concrete which surrounds it is also a deformable material. Moreover, the amount of deformation a shear connector undergoes is dependent upon factors such as its own shape and size, its location along the beam, the location of the maximum moment, and the manner in which it is attached to the top flange of the steel beam. In addition, any particular shear connector may yield sufficiently to cause a slipping between the beam and the slab. In the latter case the adjacent shear connectors pick up the additional shear.

As a result of the extremely complex behavior of shear connectors, their capacities are not based solely on a theoretical analysis. In order to develop a rational approach, a number of research programs[6,18] were undertaken to develop the push-out strengths of the various types of shear connectors. Fig. 16.8.4 shows a typical push-out test. Observations of experiments simulating both static and dynamic loadings indicate that the

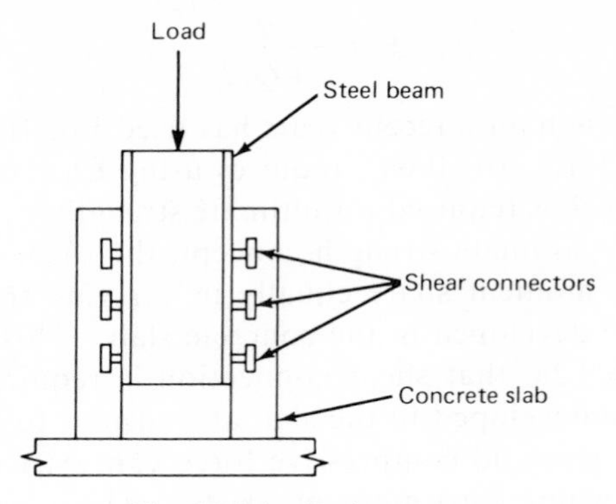

Fig. 16.8.4. Push-out test.

shear connectors will not fail if the average load per connector is kept below a "limiting load," defined as that load which causes 0.003-in. slip. The slippage is also a function of the strength of the concrete which surrounds the shear connector. Relating ultimate capacity to a specified slip is realistic for bridge design where fatigue strength is important. When connectors are to be positioned in accordance with shear distribution at service loads, use of capacities based on slip limitation is consistent with such elastic analysis. So called "ultimate" capacities of connectors used prior to 1965[17] were based on slip limitation, giving values about one-third of the ultimate strengths obtained if actual failure of a connector is the criterion.

When ultimate flexural capacity of the composite section is the basis for design, the connectors must be adequate to satisfy equilibrium of the concrete slab between the points of maximum and zero moment, as discussed in developing Eqs. 16.8.2, 16.8.3, and 16.8.4. Slip is not the major criterion for this equilibrium requirement. As stated by the Lehigh investigators[17] "the magnitude of slip will not reduce the ultimate moment provided that (1) the equilibrium condition is satisfied, and (2) the magnitude of slip is no greater than the lowest value of slip at which an individual connector might fail.

Expressions for the ultimate connector capacity, without regard to the magnitude of slip, are as follows:[17]

(a) Hooked or headed shear stud connectors (Fig. 16.8.1a). The ultimate capacity is proportional to the square of the stud diameter d_s and to the square root of the compressive strength of concrete, f'_c[10,11]. The expression is

$$q_{\text{ult}} = 930\, d_s^2 \sqrt{f'_c} \qquad \text{for } d_s \leq {}^7\!/_8 \text{ in.} \tag{16.8.5}$$

where d_s = stud diameter, in. and q_{ult} = capacity for one stud, lb.

(b) Channel connectors (Fig. 16.8.1b).

$$q_{\text{ult}} = 550\,(h + 0.5t)\,w\sqrt{f'_c} \qquad (16.8.6)$$

where h = maximum thickness of channel flange, in., t = thickness of channel web, in.; and w = length of channel, in.

(c) Spiral connectors (Fig. 16.8.1c)

$$q_{\text{ult}} = 8{,}000d\sqrt[4]{f'_c} \qquad (16.8.7)$$

where d = diameter of spiral bar, in.

AASHO–1.7.101 (1969) revises the so-called "ultimate capacities from earlier specification editions (which were based on slip) into allowable capacities related to fatigue under service loads. In addition, the ultimate capacities based on failure are given and are identical with Eqs. 16.8.5 and 16.8.6.

(d) Angle connectors (Fig. 16.8.1d). While these connectors are used, there is no generally recognized ultimate strength expression available.

Most design specifications permit other connectors if the capacities are developed from test data.

Connector Design—Ultimate-Strength Concept. It may be noted that the connection and the beam must resist the same ultimate load. However, under service loads the *beam* resists dead and live loads, but unless shores are used the *connectors* resist essentially only the live load. Working-stress method might design the connection only for live load; however, an increased factor of safety should be used, since the ultimate capacity would otherwise be inadequate.

1969 AISC–1.11 uses ultimate-strength concept but converts both the forces to be designed for and the connector capacities into the service load range by dividing them by a factor. The loads to be carried, either Eq. 16.8.2 or Eq. 16.8.3 are divided by a nominal factor of 2. Thus, for design under service loads,

$$V_h = \frac{C_{\max}}{2} = \frac{0.85f'_cA_c}{2} \qquad (16.8.8)$$

which is AISC Formula 1.11–3, and where $A_c = b_E t_s$, the effective concrete area. Equation 16.8.3 divided by 2 becomes

$$V_h = \frac{T_{\max}}{2} = \frac{A_sF_y}{2} \qquad (16.8.9)$$

which is AISC Formula 1.11–4. In Eqs. 16.8.8 and 16.8.9,

V_h = horizontal shear to be resisted between the points of maximum positive moment and points of zero moment, the smaller of Eq. 16.8.8 or Eq. 16.8.9 to be used

f'_c = 28-day compressive strength of concrete

$A_c = b_E t_s$ = effective concrete area
A_s = area of steel beam
F_y = yield point stress for steel beam

The connector ultimate capacities must also be divided by factors to give "allowable values" for working-stress method. AISC–1969 allowable values are obtained by dividing ultimate capacities from Eqs. 16.8.5 and 16.8.6 by a factor of safety of 2.5. AISC allowable values are given in Table 16.8.1 for hooked or headed studs and channels.

Working-stress method will actually give a higher safety factor than 2.5 because the ultimate load to be carried is divided only by 2 while the capacity per connector is divided by 2.5.

The number of connectors required is obtained by dividing the smaller value of V_h by the allowable shear per connector:

$$N = \frac{\text{smaller } V_h}{q} \tag{16.8.10}$$

where q = allowable load from Table 16.8.1.

TABLE 16.8.1 (from AISC–1.11.4)
AISC Design Capacities for Shear Connectors

Connector	Allowable Horizontal Shear Load q, kips (Applicable only to concrete made with ASTM C33 aggregates)		
	f'_c = 3,000	f'_c = 3,500	f'_c = 4,000
½″ diam., 2″ hooked or headed stud	5.1	5.5	5.9
⅝″ diam., 2½″ hooked or headed stud	8.0	8.6	9.2
¾″ diam., 3″ hooked or headed stud	11.5	12.5	13.3
⅞″ diam., 3½″ hooked or headed stud	15.6	16.8	18.0
3″ channel, 4.1 lb	4.3w*	4.7w	5.0w
4″ channel, 5.4 lb	4.6w	5.0w	5.3w
5″ channel, 6.7 lb	4.9w	5.3w	5.6w

*w = length of channel, in.

The smaller value of V_h, determined by Eqs. 16.8.8 or 16.8.9, is used since it represents the maximum force to give equilibrium at ultimate strength, as discussed in developing Eqs. 16.8.2 and 16.8.3. To provide more shear resistance than either the concrete slab or the steel beam could develop would be needless. Also, it is doubtful that an excessive number of shear connectors would perceptibly reduce deflection.

EXAMPLE 16.8.1

Determine the number of ¾-in. diam. 3-in. shear stud connectors required for the composite section of Example 16.7.2, shown in Fig. 16.8.5 according to AISC Specifications. Assume a uniform loading and simple beam supports. F_y = 36 ksi.

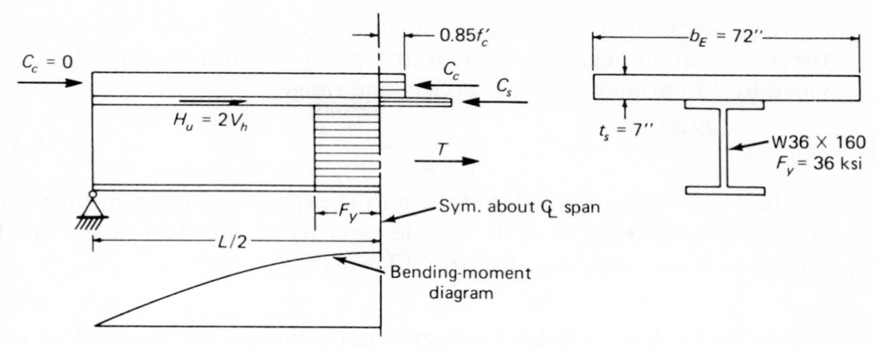

Fig. 16.8.5. Example 16.8.1.

Solution

Using Eqs. 16.8.8 and 16.8.9,

$$V_h = \frac{0.85 f'_c A_c}{2} = \frac{0.85(3.0)72(7)}{2} = 642.6 \text{ kips}$$

or

$$V_h = \frac{A_s F_y}{2} = \frac{47.1(36)}{2} = 847.6 \text{ kips}$$

From Table 16.8.1, the allowable shear per connector is 11.5 kips. Taking the smaller value of V_h, the number of shear connectors, N, required for each half of the span is therefore

$$N = \frac{642.6}{11.5} = 55.8$$

Use 56 $^3/_4 \times$ 3-in. studs per half span.

Ordinarily, AISC design would not involve an ultimate strength analysis which would permit direct computation of H_u (Fig. 16.8.5). In lieu of such computation the two-formula procedure is necessary.

In the case of continuous beams, the longitudinal reinforcing steel within the effective width of the concrete slab may be assumed to act compositely with the steel beam in the areas of negative moment. The total horizontal shear to be resisted by the shear connectors between the interior support and each adjacent point of contraflexure equals the maximum tensile force which can be developed in the reinforced concrete slab; i.e., neglecting tensile capacity of the concrete.

$$T_{\text{slab}} = A_{sr} F_{yr} \tag{16.8.11}$$

where A_{sr} = total area of longitudinal reinforcing steel at the interior support located within the effective flange width

F_{yr} = specified minimum yield point of the longitudinal reinforcing steel

For working-stress design, the ultimate shear force developed between maximum negative moment and point of contraflexure, T_{slab}, is divided by 2 to bring it into the service load range,

$$V_h = \frac{T_{slab}}{2} = \frac{A_{sr}F_{yr}}{2} \tag{16.8.12}$$

It is logically presumed that the tensile capacity of the reinforced concrete slab will be less than the tensile capacity of the steel beam, so that for negative moment, only Eq. 16.8.12 is used.

Connector Design—Elastic Concept. Even though AISC–1969 is nominally a working-stress method, the ultimate strength concept was the basis for the design equations and methods. Design according to AASHO–1969[15] for bridges actually uses a working stress concept for fatigue based on a slip limitation as previously discussed. The so-called "limiting load" capacities for shear connectors are divided by 3 to get allowable values. The superimposed service load elastic shears acting on the composite section are used as design loads.

EXAMPLE 16.8.2

Redesign the shear connectors for the beam of Example 16.8.1 using the working stress concept of AASHO 1969 with $^3/_4 \times$ 3-in. stud connectors. Assume 500,000 cycles of loading is required. Assume shoring is to be used so that all dead load and live load shear acts on the composite section. Spacing of beams = 7′ −0, F_y = 36 ksi. Use uniform live load of 3.5 kips/ft and a beam span of 45 ft.

Solution

(a) Loads and shears:

Beam weight (W36 × 160) = 0.16
Slab weight 7/12(6.0)0.15 = 0.53
Additional live load = 3.5

w = 4.19 k/ft

$$V_{max}(\text{support}) = \frac{4.19(45)}{2} = 94 \text{ kips}$$

Considering partial span loading of additional live load,

$$V_{1/4\,Pt} = \frac{0.69}{2}\left(\frac{45}{2}\right) + 3.5(45)(0.75)(0.375) = 52.1 \text{ kips}$$

$$V_{mid\ span} = \frac{1}{8}(3.5)45 = 19.7 \text{ kips}$$

The envelope of maximum shear carried by the composite section is given in Fig. 16.8.6.

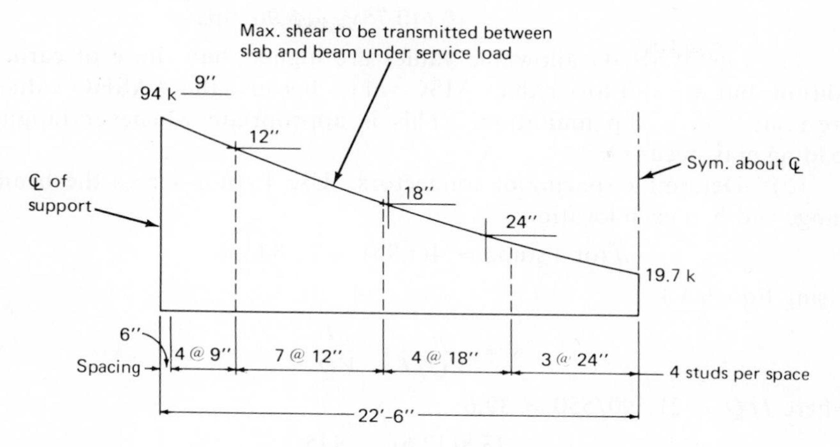

Fig. 16.8.6. Shear diagram and stud spacing according to elastic theory.

(b) Compute composite section properties ($n = 9$)

	Effective Area, A	Arm from C.G. of steel beam, y	Ay	Ay^2	I_0
Slab, 72(7)/9 W36 × 160	56.0 47.1	21.5 —	1204.0 —	25,886 —	227 9760
	103.1		1204.0	25,886	9987

$$I_x = Ay^2 + I_0 = 25{,}886 + 9987 = 35{,}873 \text{ in.}^4$$

$$\bar{y} = \frac{1204.0}{103.1} = 11.68 \text{ in.}$$

$$I = 35{,}873 - 103.1(11.68)^2 = 21{,}790 \text{ in.}^4$$

$$y_t = 18.0 + 7.0 - 11.68 = 13.32 \text{ in.}$$

$$y_b = 18.0 + 11.68 = 29.68 \text{ in.}$$

$$S_t = \frac{21{,}790}{13.32} = 1636 \text{ in.}^3 \qquad S_b = \frac{21{,}790}{29.68} = 734 \text{ in.}^3$$

Determine the static moment of the effective concrete area about the centroid of the composite section,

$$Q = 56.0(y_t - 3.5) = 56.0(9.82) = 550 \text{ in.}^3$$

(c) Determine the allowable load for $^3/_4 \times 3$-in. stud connectors. AASHO–1.7.101 gives an allowable service load capacity based on fatigue for 500,000 cycles of loading as

$$\text{Allowable } q = 10.6\, d_s^2$$
$$= 10.6(0.75)^2 = 5.96 \text{ kips}$$

The 1969 AASHO allowable values are higher than those of earlier editions but are still lower than AISC values because the AASHO values are related to a slip limitation. This is appropriate whenever fatigue loading may occur.

(d) Determine spacing of connectors. Use 4 studs across the beam flange width at each location:

$$q \text{ for 4 studs} = 4(5.95) = 23.8 \text{ kips}$$

Using Eq. 16.8.1,

$$s = \frac{q}{VQ/I} = \frac{qI}{VQ}$$

where $I/Q = 21{,}790/550 = 39.6$ in.

$$s = \frac{23.8(39.6)}{V} \quad \frac{945}{V\,(\text{kips})}$$

The values are computed in the table below and the spacing is determined graphically on the shear diagram of Fig. 16.8.6.

s, in.	V, kips
8	118
9	105
12	79
18	53
24	39.4

The elastic method requires nearly 30 percent more connectors than the procedure based on ultimate-strength concept. If slip is to be the limiting criterion, however, the difference is not unreasonable.

16.9. AISC DESIGN FOR FLEXURE

As discussed in Sec. 16.6, the actual stresses that occur in a given composite member depend on the manner of construction. The slab formwork must be supported by either the steel beam acting alone or by temporary shoring which also would support the beam. When temporary shoring is used service load stresses will be lower than when such shoring is not used, since *all* loads will be supported by the composite section. If the system is built without temporary shoring, the steel beam alone must support itself and the slab without benefit of composite action.

For economical construction, it is desired to avoid use of shoring wherever possible. In Sec. 16.7 it was shown that no matter which construction system is used, the ultimate moment capacity is identical. It is a simple procedure, therefore, to design as if the entire load is to be carried compositely (i.e. assume shores are used) *even when shores are not to be used.* Strength is assured; however, it is necessary to insure that the steel beam does not too closely approach yield under service load conditions.

In order to resist loads compositely, the concrete strength must be adequately developed. AISC–1.11.2.2 requires that 75 percent of the compressive strength f'_c of the concrete must be developed before composite action may be assumed.

The AISC design procedure for flexure may be summarized by the following steps:

(a) Select section as if shores are to be used; the required composite section modulus, S_{tr}, with reference to the tension fiber is

$$S_{tr}\,(\text{reqd}) = \frac{M_D + M_L}{F_b} \tag{16.9.1}$$

where $F_b = 0.66F_y$ or $0.6F_y$ depending on whether or not "compact section" local buckling limitations are satisfied. Lateral support is assumed adequately provided by the concrete slab and its shear connector attachments.

(b) Check AISC Formula 1.11–2. When shores are actually not to be used, service load stress on the steel section must be assured of being less than yield stress. AISC–1.11.2.2 uses an indirect procedure for checking this. The section modulus of the composite section, S_{tr}, may not exceed (or be considered more effective than) the following:

$$S_{tr}\,(\text{effective}) \le \left(1.35 + 0.35\frac{M_L}{M_D}\right) S_s \tag{16.9.2}$$

which is AISC Formula 1.11–2. In this formula M_L refers to the moment caused by loads which are to be carried compositely, and M_D refers to loads carried by the steel beam.

To understand the development of Eq. 16.9.2, the reader is referred back to Sec. 16.6 where service load stresses are computed for construction with and without shores. Service load tension stresses on the steel beam may be expressed in general as

$$f_b = \frac{M_D}{S_s} + \frac{M_L}{S_{tr}} \le k_1 F_y \qquad \text{without shores} \tag{a}$$

$$f_b = \frac{M_D + M_L}{S_{tr}} \le k_2 F_y \qquad \text{with shores} \tag{b}$$

where S_s = section modulus of the steel beam referred to its bottom flange (tension flange)

S_{tr} = section modulus of composite section referred to its bottom flange (tension flange)

k_1, k_2 = constants to obtain the allowable stresses in tension without shores and with shores, respectively

Divide Eq. (a) by Eq. (b), letting $kS_s = S_{tr}$:

$$\frac{k_1}{k_2} \geq \frac{\dfrac{M_D}{S_s} + \dfrac{M_L}{kS_s}}{\dfrac{M_D + M_L}{kS_s}} = \frac{kM_D + M_L}{M_D + M_L} \tag{c}$$

$$\frac{k_1}{k_2}(M_D + M_L) - M_L \geq kM_D \tag{d}$$

Divide by M_D,

$$k \leq \frac{k_1}{k_2}\left(1 + \frac{M_L}{M_D}\right) - \frac{M_L}{M_D} \tag{e}$$

Replacing k by S_{tr}/S_s gives the AISC formula in general terms.

$$S_{tr} \leq \left[\frac{k_1}{k_2} + \frac{M_L}{M_D}\left(\frac{k_1}{k_2} - 1\right)\right] S_s \tag{f}$$

The AISC value of $k_1/k_2 = 1.35$ is obtained if a compact section ($F_b = 0.66F_y$) is allowed to reach a service load stress of $0.89F_y$ ($0.89/0.66 = 1.35$). The same 1.35 value is obtained for noncompact sections ($F_b = 0.60F_y$) if the service load stress is allowed to reach $0.81F_y$ ($0.81/0.60 = 1.35$). As is seen from Eq. (f), this limitation of stress is valid no matter what ratio of M_L to M_D is used.

(c) Check stress on steel beam supporting the loads acting before concrete has hardened.

$$\text{Reqd } S_s = \frac{M_D}{F_b} \tag{16.9.3}$$

where F_b may be $0.66F_y$, $0.60F_y$, or some lower value if adequate lateral support is not provided. It is to be noted that Eq. 16.9.3 is frequently controlling on the compression fiber (top in positive moment zone), particularly if a steel cover plate is used on the bottom.

(d) Incomplete composite action. When fewer connectors are used than necessary to develop full composite action, an effective section modulus may be obtained by linear interpolation. AISC-1.11.2.2 allows

$$S_{\text{eff}} = S_s + \frac{V'_h}{V_h}(S_{tr} - S_s) \tag{16.9.4}$$

where V_h = design horizontal shear for full composite action

V'_h = actual capacity of connectors used; less than V_h

S_s and S_{tr} as defined previously in this section

For this case, S_{eff} is used in design calculations in place of that computed from beam dimensions, and is the quantity that may not exceed the value given by Eq. 16.9.2.

EXAMPLE 16.9.1

Design an interior member of the floor shown in Fig. 16.9.1 assuming it is constructed without temporary shoring. Assume $F_y = 36$ ksi, $n = 9$, $f'_c = 3{,}000$ psi, $f_c = 1{,}350$ psi and a 4-in. slab.

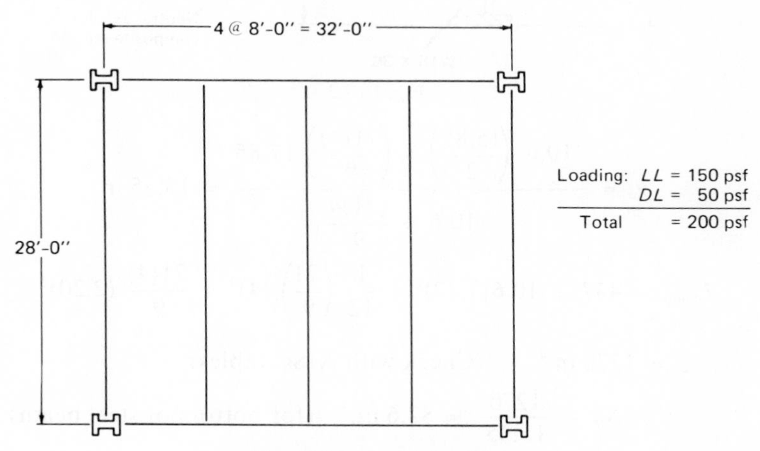

Fig. 16.9.1. Example 16.9.1.

Solution

Estimate load per interior beam .... 8(0.2) = 1.6
Estimate weight of beam .... = 0.04

Total .... 1.64 kips/ft

(a) Select beam as if shores were to be used:

$$M = \frac{1.64(28)^2}{8} = 161 \text{ ft-kips}$$

$$S_{tr}\,(\text{reqd}) = \frac{161(12)}{24} = 80.5 \text{ in.}^3$$

assuming compact section, using $0.66F_y$.

Try a W16 × 36:

$A = 10.6$ in.^{2}, $I_x = 447$ in.^{4}; $S = 56.5$ in.^{3}; $b_f = 6.99$ in.

(Selected from AISC Manual, "Composite Beam Selection Table")

Determine effective width (see Fig. 16.9.2):

$$b_E = \tfrac{1}{4} \text{ of span} = \frac{28(12)}{4} = 84 \text{ in.}$$

$$b_E = \text{beam spacing} = 96 \text{ in.}$$

$$b_E = 16\,(\text{thickness of slab}) + b_f$$
$$= 16(4) + 6.99 = 70.99 = 71 \text{ in.} \qquad \text{Controls}$$

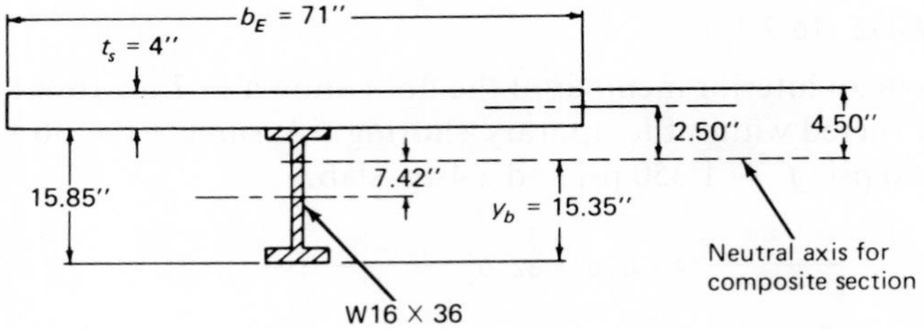

$b_E = 71''$
$t_s = 4''$
2.50″
4.50″
7.42″
15.85″
$y_b = 15.35''$
Neutral axis for composite section
W16 × 36

Fig. 16.9.2.

$$y_b = \frac{10.6\left(\dfrac{15.85}{2}\right) + \left(\dfrac{71(4)}{9}\right)17.85}{10.6 + \dfrac{(71)4}{9}} = 15.35 \text{ in.}$$

$$I_{\text{comp}} = 447 + 10.6(7.42)^2 + \frac{1}{12}\left(\frac{71}{9}\right)(4)^3 + \frac{71(4)}{9}(2.20)^2$$

$$= 1270 \text{ in.}^4 \qquad \text{(Check with AISC tables)}$$

$$S_{tr} = \frac{1270}{15.35} = 82.6 \text{ in.}^3 \quad \text{(for bottom of steel beam)}$$

$$S_{\text{top}} = \frac{1269.4}{4.50} = 282.4 \text{ in.}^3 \quad \text{(for top of concrete)}$$

Recomputing moments, $w = 0.05(8) + 0.036 = 0.436$ kips/ft

$$M_D = \frac{0.436\,(28)^2}{8} = 42.8 \text{ ft-kips}$$

$$M_L = \frac{0.150\,(8)\,(28)^2}{8} = 117.7 \text{ ft-kips}$$

$$\text{Total} \qquad 160.5 \text{ ft-kips}$$

Check stresses:

At top of concrete slab; allowable $f_c = 0.45 f'_c = 1.35$ ksi

$$f_c = \frac{160.5\,(12)}{9\,(282.4)} = 0.758 \text{ ksi} < 1.35 \text{ ksi} \qquad \text{OK}$$

At bottom of steel beam; allowable $F_b = 0.66 F_y = 24$ ksi

$$f_b = \frac{160.5\,(12)}{82.6} = 23.3 \text{ ksi} < 24.0 \text{ ksi} \qquad \text{OK}$$

(b) Check Eq. 16.9.2 to determine maximum transformed section modulus S_{tr} that can be used.

$$S_{tr} = \left(1.35 + 0.35\,\frac{117.7}{42.8}\right) 56.5 = 130.0 \text{ in.}^3 > 82.6 \text{ in.}^3 \qquad \text{OK}$$

Thus, shores need not be used.

(c) Check steel stress for loads carried noncompositely, using Eq. 16.9.3,

$$f_b = \frac{M_D}{S_s} = \frac{42.8\,(12)}{56.5} = 9.1 \text{ ksi} < 0.60F_y$$

The above assumes adequate lateral support during construction so that the laterally unbraced length is less than $L_u = 20{,}000/(d/A_f)F_y$. Stress on the steel section resulting from noncomposite loads is more likely to govern when a cover plate is used on the bottom than when such a plate is not used. The beam section selected is therefore satisfactory. Use W16 × 36.

(d) Design shear connectors:

$$\text{From Eq. 16.8.8,} \quad V_h = \frac{0.85\,(3)\,71\,(4)}{2} = 362 \text{ kips}$$

$$\text{From Eq. 16.8.9,} \quad V_h = \frac{10.6(36)}{2} = 190.5 \text{ kips}$$

From Table 16.8.1, $\frac{5}{8} \times 2\frac{1}{2}$-in. headed stud, $q = 8.0$ k/stud

$$N = \frac{V_h}{q} = \frac{190.5}{8.0} = 23.8, \text{ say } 24.$$

Use 24 shear connectors on each side of the centerline at midspan. Use a uniform spacing with 2 studs at a section across the beam width:

$$s = \frac{L/2}{N/2} = \frac{28\,(12)}{23.8} = 14.1 \text{ in.}$$

Use a 14-in. spacing for the pairs of stud connectors, starting at the support.

EXAMPLE 16.9.2

Design a composite section, without shores, for use as an interior floor beam of an office building. $f'_c = 3{,}000$ psi; $n = 9$; $F_y = 36$ ksi

Span	= 30 ft	Live load	= 150 psf
Beam spacing	= 8 ft	Partitions	= 25 psf
Slab thickness	= 5 in.	Ceiling	= 7 psf

Solution

(a) Determine moments:

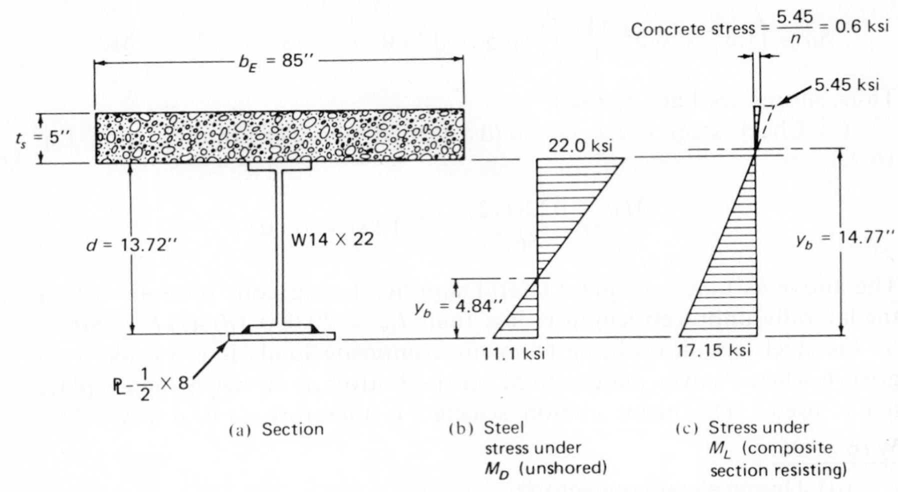

Fig. 16.9.3. Solution for Example 16.9.2, showing stresses under service loads.

$$
\begin{aligned}
\text{5-in. slab, } \frac{5}{12}\,(8)\,0.15 &= 0.50 \text{ kips/ft}\\
\text{Steel beam (assumed)} &= \underline{0.03}\\
&\; 0.53 \text{ kips/ft}
\end{aligned}
$$

$$M_D = \frac{1}{8}\,(0.53)\,(30)^2 = 62 \text{ ft-kips}$$

$$
\begin{aligned}
\text{Live load } 0.15\,(8) &= 1.2 \text{ kips/ft}\\
\text{Partitions } 0.025\,(8) &= 0.2\\
\text{Ceiling } 0.007\,(8) &= \underline{0.05}\\
&\; 1.45 \text{ kips/ft}
\end{aligned}
$$

$$M_L = \frac{1}{8}\,(1.45)\,(30)^2 = 163 \text{ ft kips}$$

(b) Select section

$$S_{tr}\,(\text{reqd}) = \frac{M_D + M_L}{0.66F_y} = \frac{225\,(12)}{24} = 112.5 \text{ in.}^3$$

If shores are not used, stress on the steel section prior to developing composite action must not be excessive. Assuming adequate lateral bracing such that unbraced length $L < L_u$,

$$S_s\,(\text{reqd}) = \frac{M_D}{0.60F_y} = \frac{62\,(12)}{22} = 33.8 \text{ in.}^3$$

If the unbraced length $L < L_c$ and the section is "compact" for local buckling, $0.66F_y$ could be used for the allowable stress. Use AISC Manual "Composite Beam Selection Table" to select this member. Find

W14 × 22 with 4 sq in. cover plate, whose properties are given in AISC Manual, "Properties of Composite Beams,"

$$\text{Composite section properties:} \quad S_{top} = 362.0 \text{ in.}^3 \qquad S_{bottom} = 113.0 \text{ in.}^3$$

$$\text{Steel section alone:} \quad S_{top} = 34.4 \text{ in.}^3 \qquad S_{bottom} = 65.8 \text{ in.}^3$$

Further checking is required, even for the properties, since the effective width of concrete slab b_E may not be the same as used for computing the properties in the AISC Manual.

(c) Check effective width and compute properties.
Effective width:

$$b_E \leq L/4 = 90 \text{ in.}$$
$$\leq 16t_s + b_f = 16(5) + 5.0 = \underline{85.0 \text{ in.} \quad \text{(governs)}}$$
$$\leq \text{spacing of beams} = 96 \text{ in.}$$

Since $b_E = 85$ in. was used to obtain the AISC Manual properties, the values are correct and are not rechecked here.

(d) Check stress under $M_D + M_L$ (ultimate-strength concept; nominal working-stress procedure). Since girder weight is 32 lb/ft $\approx$ 30 as assumed, original moments are used

$$f_b = \frac{M_D + M_L}{S_{tr}} = \frac{225(12)}{113} = 23.9 \text{ ksi} < 24 \text{ ksi} \qquad \text{OK}$$

W14 × 22 is "compact" in A36 steel; thus $F_b = 0.66F_y = 24$ ksi.

(e) Check AISC Formula 1.11–2:

$$S_{tr} = \left(1.35 + 0.35\frac{M_L}{M_D}\right)S_S$$
$$= \left(1.35 + 0.35\frac{163}{62}\right)65.8$$
$$= 2.27(65.8) = 149 \text{ in.}^3$$

Since actual $S_{tr} = S_{bottom} = 113$ does not exceed the upper limit of 149, shores are not required to prevent service load stresses from becoming too close to the yield stress. As discussed earlier, the same conclusion could be reached if actual service-load stresses were computed, as follows, for no shoring,

$$f_b = \frac{M_D}{S_s} + \frac{M_L}{S_{tr}} = \frac{62(12)}{65.8} + \frac{163(12)}{113}$$
$$= \frac{62(12)}{65.8} + \frac{163(12)}{113} = 11.3 + 17.3 = 28.6 \text{ ksi}$$

which is acceptable, since it does not exceed $0.89F_y = 32$ ksi at service load. AISC requires the formula check, instead of the stress check, because without the understanding of ultimate strength, the $0.89F_y$ might appear as an unsafe value.

(f) Check stress on steel beam prior to development of composite action for the unshored system. The maximum stress occurs in compression at the top of the beam:

$$f_b = \frac{M_D}{S_s} = \frac{62\,(12)}{34.4} = 21.6 \text{ ksi} \approx F_b = 0.60F_y$$

Maximum laterally unsupported length during construction is $L = L_u = 5.7$ ft.

<u>Use W14 × 22 with ½ × 8 cover plate.</u> The section with service load stresses is given in Fig. 16.9.3. Design of connectors is not illustrated since no new principles are involved.

16.10. DEFLECTIONS

In order to accurately determine the deflections of composite members, a number of factors must be taken into account which are not normally considered. These are: the method of construction, the separation of the live-load and dead-load moments, and the effect of creep in the concrete slab.

The method of construction determines the manner in which the composite cross section carries the dead-load stresses. If the steel is shored from below during the hardening of the concrete slab, the composite section will assume both the dead-load *and* the live-load stresses. However, if the steel beam is *not* shored, the steel beam will assume the dead load stresses and the composite section will take only the live-load stresses.

If the construction is *without* shoring, the total deflection will be the sum of the dead load deflection of the steel beam and the live-load deflection of the composite section.

EXAMPLE 16.10.1

Determine the total deflection of the composite section in Example 16.9.1, and check against maximum deflection permitted by AISC if no shoring is used (see Fig. 16.10.1.).

Solution

Dead-load deflection:

$$\Delta_{DL} = \frac{5wL^4}{384EI} = \frac{5\left[\dfrac{8(50) + 36}{12}\right](28)^4(12)^4}{384(29)\,10^6(447)} = 0.46 \text{ in.}$$

Live load deflection,

$$\Delta_{LL} = \frac{5wL^4}{384EI_{comp}} = \frac{5\,\dfrac{8(150)}{12}\,(28)^4(12)^4}{384\,(29)\,10^6\,(1270)} = 0.45\text{ in.}$$

Total Deflection,

$$\Delta = \Delta_{DL} + \Delta_{LL} = 46 + 0.45 = 0.91\text{ in., say }\frac{15}{16}\text{ in.}$$

Check maximum deflection permitted

$$\Delta_{max} = \frac{L}{360} = \frac{28\,(12)}{360} = 0.93\text{ in.}$$

Since $\Delta_{actual} < 0.933$ in., deflection criteria is satisfied.

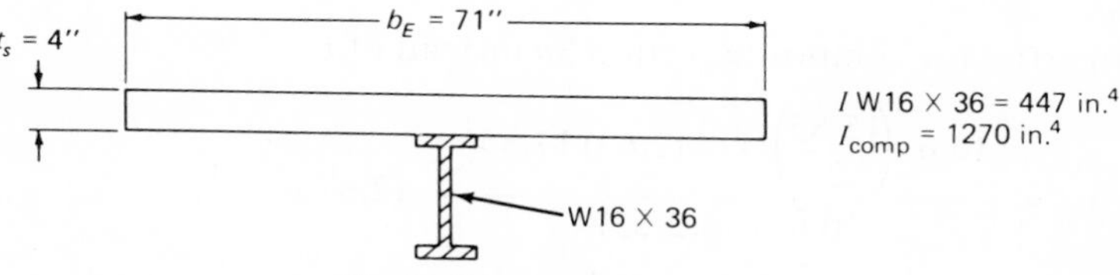

Fig. 16.10.1. Example 16.10.1.

If shoring provides support during the hardening of the concrete, the total deflection will be a function of the total composite section. Account must be taken of the fact that concrete is subject to creep under long-term loadings. This inelastic behavior may be approximated by multiplying the modular ratio n by a factor to reduce the net effective width. The result is a reduced moment of inertia for the composite section which is used in computing the dead-load deflection. The live-load deflection is then usually computed on the basis of the elastic composite moment of inertia. Occasionally, the conservative approach is to use the reduced composite moment inertia when the live loads are expected to remain for extended periods of time.

Because the concrete slab in building construction is normally not too thick (say $t_s \leq 5$ in.) creep deflection is not considered to be a problem. AISC gives no indication that one need be concerned with anything but live-load short-time deflection. The ACI–ASCE Joint Committee[16] recommends using one-half the concrete modulus of elasticity, $E_c/2$ instead of E_c when computing sustained load creep deflection. AASHO[15] uses $E_c/3$ instead of E_c. Such arbitrary procedures can at best give an estimate of creep effects, probably not better than ± 30 percent. The steel section, exhibiting no creep, and representing the principal carrying element, insures that creep problems will usually be minimal.

EXAMPLE 16.10.2

Determine the total deflection of the composite section in Example 16.9.1 if temporary shoring is provided. Assume a value of $3n$ for the modular ratio, i.e., $E_c' = E_c/3$. (See Fig. 16.10.2.)

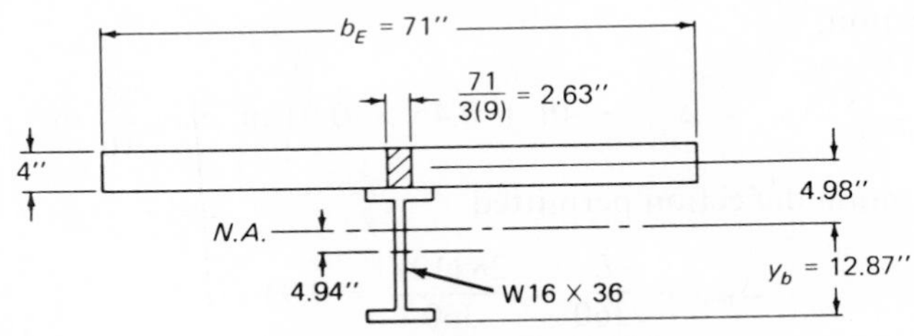

Fig. 16.10.2. Example 16.10.2.

Solution

Properties for composite section $3n$ instead of n:

$$y_b = \frac{10.6\left(\frac{15.85}{2}\right) + 4(2.63)\,17.85}{10.6 + 4(2.63)} = 12.87 \text{ in.}$$

$$I_{comp} = 447 + 10.6(4.94)^2 + \frac{1}{12}(2.63)(4)^3 + 2.63(4)(4.98)^2$$

$$= 1106.5 \text{ in.}^4$$

Dead-load deflection:

$$\Delta_{DL} = \frac{5\left[\frac{8(50) + 36}{12}\right](28)^4(12)^4}{384(29)10^6(1106.5)} = 0.19 \text{ in.}$$

Live-load deflection, using elastic I from Example 16.10.1:

$$\Delta_{LL} = \frac{5\,\frac{8(150)}{12}(28)^4(12)^4}{384(29)10^6(1270)} = 0.45 \text{ in.}$$

Total deflection, $\Delta = \Delta_{DL} + \Delta_{LL}$:

$$\Delta = 0.19 + 0.45 = 0.64 \text{ in.,} \quad \text{say } 5/8 \text{ in.}$$

It should be noted that the difference in the computed deflection of the two previous examples was only about 5/16 in. The authors consider it doubtful that this difference would be worth the extra expense of shoring. In general, the advantage of providing shoring during the hardening of the concrete is minimal and therefore is not recommended unless deflection criteria govern.

PROBLEMS

16.1. Determine the composite section properties for the given section, using AISC procedures.

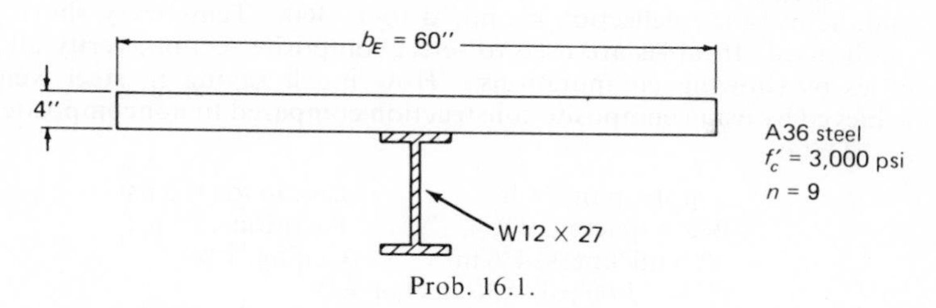

Prob. 16.1.

16.2. Determine the composite section properties for the member of Prob. 16.1 according to AISC procedures if a W18 × 50 section of A572 Grade 50 steel is used.

16.3. Determine the ultimate moment capacity of the composite section in Prob. 16.1 using ultimate-strength concepts.

16.4. Determine the ultimate-moment capacity of the composite section in Prob. 16.2 using ultimate-strength concepts.

16.5. Determine for the beam of Prob. 16.1 the number of shear connectors required by AISC if $^1/_2$ × 2 in. studs are used.

16.6. Repeat Prob. 16.5 using A572 Grade 60 steel instead of A36.

16.7. Repeat Prob. 16.5 using a W18 × 50 and A572 Grade 50 steel instead of a W12 × 27 and A36 steel.

16.8. Using AISC procedures, select a W section for span BD without using a bottom cover plate on the basis of composite design, and design the required shear connectors. Use A36 steel, a 4-in.-thick slab with f'_c = 3,000 psi and n = 9. Assume a live load of 100 psf and no temporary shoring is used. Limit the live load deflection to $L/360$.

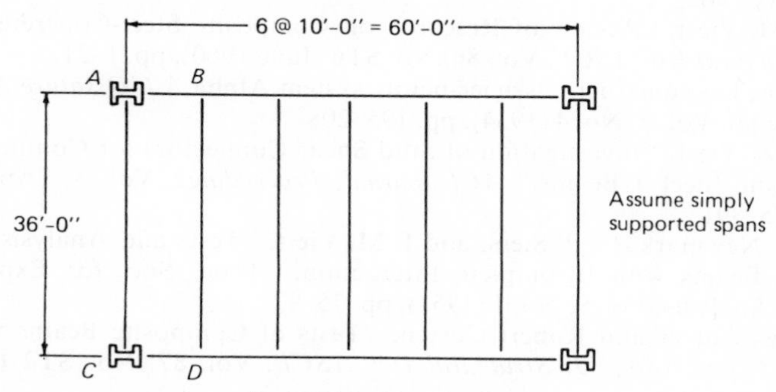

Prob. 16.8. Framing plan.

16.9. Repeat Prob. 16.8 using a cover plate on the bottom flange.

16.10. Repeat Prob. 16.8 using temporary shoring.

16.11. Repeat Prob. 16.8 selecting a W section for the exterior span *AB*.

16.12. Repeat Prob. 16.8 using A572 Grade 60 steel and a W12 section.

16.13. Design the lightest-weight composite section for use as an interior floor beam of an office building. The dead-load deflection is not to exceed $^7/_8$ in. and the live-load deflection is limited to $L/360$. Temporary shoring will not be used. If tables are used to select composite section, verify all properties by showing computations. How much saving in steel weight is achieved by using composite construction compared to noncomposite?

Data:

Simple span, 28 ft	Live load, 125 psf
Beam spacing, 9 ft	Partitions, 25 psf
Slab thickness, $4\,^1/_2$ in.	Ceiling, 7 psf

$f'_c = 3{,}000$ psi, A36 steel, $n = 9$

16.14. Repeat Prob. 16.13 using minimum feasible depth and shores.

16.15. Repeat Prob. 16.13 using A572 Grade 50 steel.

16.16. Repeat Prob. 16.14 using A572 Grade 50 steel.

16.17. Repeat Prob. 16.13 using ultimate-strength method (not AISC method) and a factor of safety of 2.5.

SELECTED REFERENCES

1. H. M. MacKay, P. Gillespie, and C. Leluau, "Report on the Strength of Steel I-Beams Haunched with Concrete." *Engineering Journal*, Engineering Institute of Canada, Vol. 6, No. 8 (1923), pp. 365–369.
2. R. A. Caughey, "Composite Beams of Concrete and Structural Steel," Proceedings, 41st Annual Meeting, Iowa Engineering Society, 1929.
3. Theodore von Kármán, "Die Mittragende Breitte," Collected Works of Theodore von Kármán, Volume II, p. 176.
4. Eric Reissner, "Über die Berechnung von Plattenbalkan," *Der Stahlbau*, December 1954.
5. J. A. N. Lee, "Effective Widths of Tee-Beams," *Structural Engineer* (London), January 1962.
6. Ivan M. Viest, "Review of Research on Composite Steel-Concrete Beams," *J. Structural Div. ASCE*, Vol. 86, No. ST6 (June 1960), pp. 1–21.
7. M. Ros, Les constructions acier-béton, system Alpha," *L'Ossature Metallique* (Bruxelle), Vol. 3, No. 4 (1934), pp. 195–208.
8. Ivan M. Viest, "Investigation of Stud Shear Connectors for Composite Concrete and Steel T-Beams," *ACI Journal, Proceedings*, Vol. 53 (April 1956), pp. 875–891.
9. N. M. Newmark, C. P. Siess, and I. M. Viest, "Tests and Analysis of Composite Beams with Incomplete Interaction," Proc. Soc. for Experimental Stress Analysis, Vol. 9, No. 1 (1951), pp. 75–92.
10. Charles Culver and Robert Costen, "Tests of Composite Beams with Stud Shear Connectors," *J. Structural Div. ASCE*, Vol. 87, No. ST2 (February 1961), pp. 1–17.
11. Bruno Thurliman, "Fatigue and Static Strength of Stud Shear Connectors," *ACI Journal, Proceedings*, Vol. 55 (June 1959), pp. 1287–1302.

12. S. Timoshenko and J. Goodier, *Theory of Elasticity*, McGraw-Hill Book Company, 1959, Chap. 6.
13. John E. Johnson and Albert D. M. Lewis, "Structural Behavior in a Gypsum Roof-Deck System," *J. Structural Div. ASCE*, Vol. ST2 (April 1966), pp. 283–296.
14. ACI Committee 318, "Building Code Requirements for Reinforced Concrete," American Concrete Institute, Detroit, Mich., 1963.
15. *Standard Specifications for Highway Bridges*, 10th Edition, American Association of State Highway Officials, Washington, D.C., 1969.
16. Joint ASCE-ACI Committee on Composite Construction, "Tentative Recommendations for the Design and Construction of Composite Beams and Girders for Buildings," *J. Structural Div. ASCE*, Vol. 86, No. ST12 (December 1960), pp. 73–92.
17. Roger G. Slutter and George C. Driscoll, "Flexural Strength of Steel-Concrete Composite Beams," *J. Structural Div. ASCE*, Vol. 91, No. ST2 (April 1965), pp. 71–99.
18. I. M. Viest, R. S. Fountain, and C. P. Siess, "Development of the New AASHO specification for Composite Steel and Concrete Bridges." Highway Research Board Bulletin No. 174, Washington, D.C., 1959.

Appendix

Table A1
Approximate Radius of Gyration

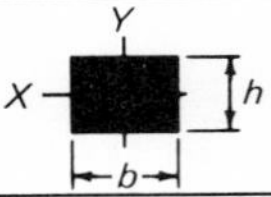 $r_x = 0.29h$ $r_y = 0.29b$	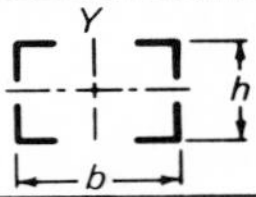 $r_x = 0.42h$ $r_y = 0.42b$	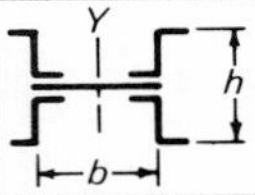 $r_x = 0.31h$ $r_y = 0.48b$
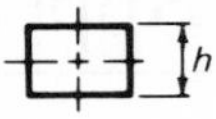 $r_x = 0.40h$ h = mean h	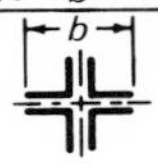 r_y = same as for 2 L	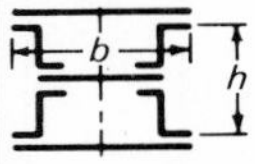 $r_x = 0.37h$ $r_y = 0.28b$
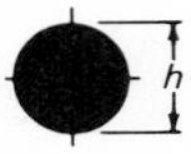 $r_x = 0.25h$	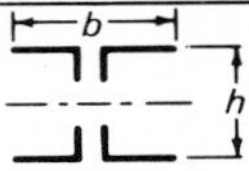 $r_x = 0.42h$ r_y = same as for 2 L	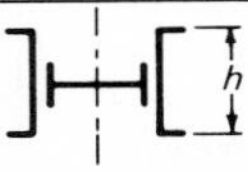 $r_x = 0.31h$
 $r = \sqrt{\frac{H^2 + h^2}{16}}$ $r = 0.35H_m$	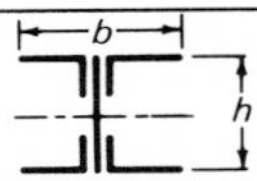 $r_x = 0.39h$ $r_y = 0.21b$	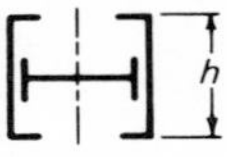 $r_x = 0.31h$
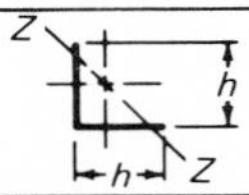 $r_x = 0.31h$ $r_y = 0.31h$ $r_z = 0.197h$	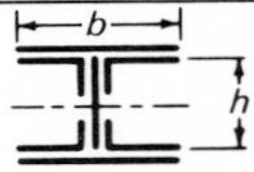 $r_x = 0.45h$ $r_y = 0.235b$	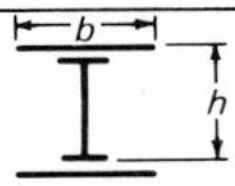 $r_x = 0.40h$ $r_y = 0.21b$
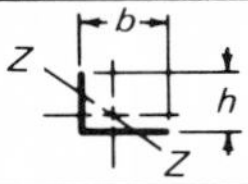 $r_x = 0.29h$ $r_y = 0.32b$ $r_z = 0.18\frac{h + b}{2}$	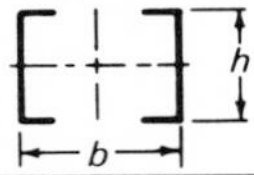 $r_x = 0.36h$ $r_y = 0.45b$	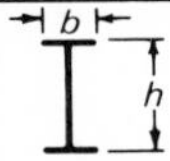 $r_x = 0.38h$ $r_y = 0.22b$
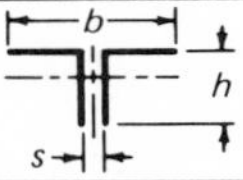 $r_x = 0.31h$ $r_y = 0.215b$ $= b(0.21+0.02s)$	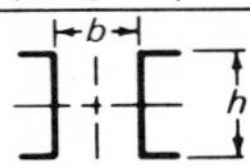 $r_x = 0.36h$ $r_y = 0.60b$	 $r_x = 0.39h$
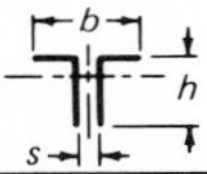 $r_x = 0.32h$ $r_y = 0.21b$ $= b(0.19+0.02s)$	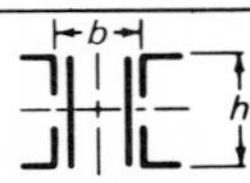 $r_x = 0.36h$ $r_y = 0.53b$	 $r_x = 0.35h$
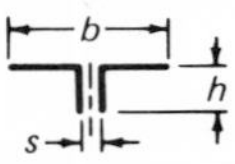 $r_x = 0.29h$ $r_y = 0.24b$ $= b(0.23+0.02s)$	 $r_x = 0.39h$ $r_y = 0.55b$	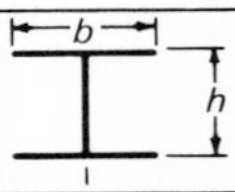 $r_x = 0.435h$ $r_y = 0.25b$
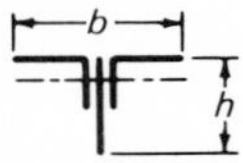 $r_x = 0.30h$ $r_y = 0.17b$	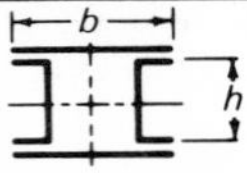 $r_x = 0.42h$ $r_y = 0.32b$	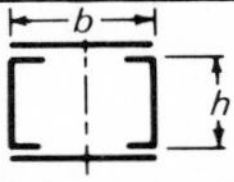 $r_x = 0.42h$
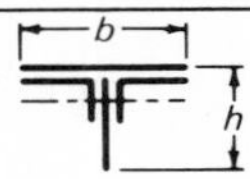 $r_x = 0.25h$ $r_y = 0.21b$	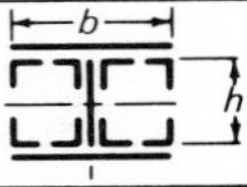 $r_x = 0.44h$ $r_y = 0.28b$	 $r_x = 0.42h$
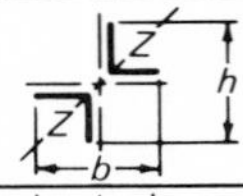 $r_x = 0.21h$ $r_y = 0.21b$ $r_z = 0.19h$	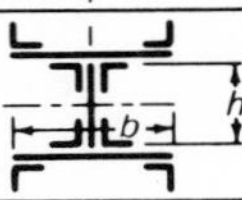 $r_x = 0.50h$ $r_y = 0.28b$	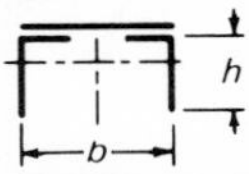 $r_x = 0.285h$ $r_y = 0.37b$
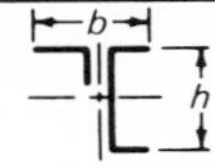 $r_x = 0.38h$ $r_y = 0.19b$	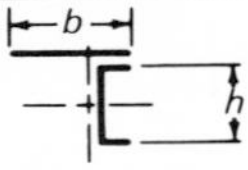 $r_x = 0.39h$ $r_y = 0.21b$	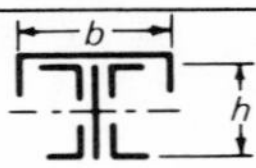 $r_x = 0.42h$ $r_y = 0.23b$

* J.A.L. Waddell, "Bridge Engineering," John Wiley & Sons, Inc., New York, 1916. Reproduced by permission.

TABLE A2
Radius of Gyration of Wide Flange Sections for Use in AISC Formula 1.5–6

$$r_T = \sqrt{I_f/[A_f + (A_w/6)]}$$

Beam	r_T	Beam	r_T	Beam	r_T	Beam	r_T
W36 × 300	4.41	W21 × 142	3.54	W14 × 193	4.36	W12 × 27	1.73
W36 × 280	4.40	W21 × 127	3.52	W14 × 184	4.35	W10 × 112	2.89
W36 × 260	4.37	W21 × 112	3.50	W14 × 176	4.34	W10 × 100	2.86
W36 × 245	4.35	W21 × 96	2.37	W14 × 167	4.33	W10 × 89	2.84
W36 × 230	4.33	W21 × 82	2.34	W14 × 158	4.31	W10 × 77	2.82
W36 × 194	3.09	W21 × 73	2.15	W14 × 150	4.30	W10 × 72	2.81
W36 × 182	3.07	W21 × 68	2.13	W14 × 142	4.29	W10 × 66	2.80
W36 × 170	3.06	W21 × 62	2.12	W14 × 320	4.57	W10 × 60	2.79
W36 × 160	3.04	W21 × 55	2.09	W14 × 136	4.08	W10 × 54	2.78
W36 × 150	3.02	W18 × 114	3.20	W14 × 127	4.07	W10 × 49	2.77
W36 × 135	2.96	W18 × 105	3.19	W14 × 119	4.06	W10 × 45	2.20
W33 × 240	4.20	W18 × 96	3.17	W14 × 111	4.04	W10 × 39	2.19
W33 × 220	4.17	W18 × 85	2.35	W14 × 103	4.03	W10 × 33	2.17
W33 × 200	4.15	W18 × 77	2.34	W14 × 95	4.02	W10 × 20	1.56
W33 × 152	2.96	W18 × 70	2.32	W14 × 87	4.01	W10 × 25	1.55
W33 × 141	2.94	W18 × 64	2.31	W14 × 84	3.31	W10 × 21	1.52
W33 × 130	2.91	W18 × 60	1.97	W14 × 78	3.30	W8 × 67	2.29
W33 × 118	2.86	W18 × 55	1.96	W14 × 74	2.75	W8 × 58	2.27
W30 × 210	4.01	W18 × 50	1.95	W14 × 68	2.74	W8 × 48	2.25
W30 × 190	3.99	W18 × 45	1.93	W14 × 61	2.72	W8 × 40	2.23
W30 × 172	3.97	W16 × 96	3.13	W14 × 53	2.17	W8 × 35	2.21
W30 × 132	2.70	W16 × 88	3.12	W14 × 48	2.16	W8 × 31	2.20
W30 × 124	2.68	W16 × 78	2.29	W14 × 43	2.15	W8 × 28	1.79
W30 × 116	2.66	W16 × 71	2.28	W14 × 38	1.79	W8 × 24	1.78
W30 × 108	2.63	W16 × 64	2.26	W14 × 34	1.78	W8 × 20	1.42
W30 × 99	2.60	W16 × 58	2.25	W14 × 30	1.75	W8 × 17	1.40
W27 × 177	3.75	W16 × 50	1.85	W12 × 190	3.51		
W27 × 160	3.73	W16 × 45	1.84	W12 × 161	3.46		
W27 × 145	3.71	W16 × 40	1.83	W12 × 133	3.42		
W27 × 114	2.59	W16 × 36	1.81	W12 × 120	3.40		
W27 × 102	2.57	W14 × 426	4.64	W12 × 106	3.38		
W27 × 94	2.55	W14 × 398	4.61	W12 × 99	3.37		
W27 × 84	2.51	W14 × 370	4.58	W12 × 92	3.35		
W24 × 160	3.79	W14 × 342	4.55	W12 × 85	3.34		
W24 × 145	3.77	W14 × 314	4.51	W12 × 79	3.33		
W24 × 130	3.74	W14 × 287	4.48	W12 × 72	3.32		
W24 × 120	3.21	W14 × 264	4.45	W12 × 65	3.31		
W24 × 110	3.19	W14 × 246	4.43	W12 × 58	2.75		
W24 × 100	3.18	W14 × 237	4.42	W12 × 53	2.74		
W24 × 94	2.35	W14 × 228	4.40	W12 × 50	2.19		
W24 × 84	2.33	W14 × 219	4.39	W12 × 45	2.18		
W24 × 76	2.30	W14 × 211	4.38	W12 × 40	2.17		
W24 × 68	2.27	W14 × 202	4.37	W12 × 36	1.76		
				W12 × 31	1.75		

TABLE A3
Elastic Section Modulus Table, including Limiting Values of Unbraced Length (L_c and L_u) and Maximum Allowable Flexural Stress*

SX	SHAPE	COMPACT SECTION-SATISFIES AISC 1.5.1.4.1 (YES OR NO)							
		FY=	36.00	42.00	45.00	50.00	55.00	60.00	65.00
1280.87	W14X730		YES	YES	YES	YES	YES	YES	YES
		LC=	18.88	17.48	16.88	16.02	15.27	14.62	14.05
		LU=	181.21	155.32	144.97	130.47	118.61	108.72	100.36
		FB=	23.76	27.72	29.70	33.00	36.30	39.60	42.90
1151.61	W14X665		YES	YES	YES	YES	YES	YES	YES
		LC=	18.62	17.24	16.65	15.80	15.06	14.42	13.86
		LU=	170.47	146.12	136.38	122.74	111.58	102.28	94.41
		FB=	23.76	27.72	29.70	33.00	36.30	39.60	42.90
1105.13	W36X300		YES	YES	YES	YES	YES	YES	YES
		LC=	17.58	16.27	15.72	14.91	14.22	13.61	13.08
		LU=	35.27	30.23	28.22	25.39	23.09	21.16	19.53
		FB=	23.76	27.72	29.70	33.00	36.30	39.60	42.90
1035.55	W14X605		YES	YES	YES	YES	YES	YES	YES
		LC=	18.38	17.02	16.44	15.60	14.87	14.24	13.68
		LU=	160.08	137.21	128.06	115.26	104.78	96.05	88.66
		FB=	23.76	27.72	29.70	33.00	36.30	39.60	42.90
1031.19	W36X280		YES	YES	YES	YES	YES	YES	YES
		LC=	17.51	16.21	15.66	14.86	14.17	13.56	13.03
		LU=	33.04	28.32	26.43	23.79	21.63	19.82	18.30
		FB=	23.76	27.72	29.70	33.00	36.30	39.60	42.90
951.09	W36X260		YES	YES	YES	YES	YES	YES	YES
		LC=	17.47	16.17	15.62	14.82	14.13	13.53	13.00
		LU=	30.45	26.10	24.36	21.92	19.93	18.27	16.86
		FB=	23.76	27.72	29.70	33.00	36.30	39.60	42.90
932.19	W14X550		YES	YES	YES	YES	YES	YES	YES
		LC=	18.16	16.81	16.24	15.41	14.69	14.06	13.51
		LU=	150.11	128.66	120.09	108.08	98.25	90.06	83.14
		FB=	23.76	27.72	29.70	33.00	36.30	39.60	42.90
892.52	W36X245		YES	YES	YES	YES	YES	YES	YES
		LC=	17.42	16.13	15.58	14.78	14.10	13.50	12.97
		LU=	28.61	24.53	22.89	20.60	18.73	17.17	15.85
		FB=	23.76	27.72	29.70	33.00	36.30	39.60	42.90
838.93	W14X500		YES	YES	YES	YES	YES	YES	YES
		LC=	17.95	16.62	16.05	15.23	14.52	13.90	13.36
		LU=	140.43	120.37	112.34	101.11	91.92	84.26	77.77
		FB=	23.76	27.72	29.70	33.00	36.30	39.60	42.90
835.47	W36X230		YES	YES	YES	YES	YES	YES	PTY
		LC=	17.39	16.10	15.55	14.75	14.06	13.47	12.94
		LU=	26.78	22.95	21.42	19.28	17.53	16.07	14.83
		FB=	23.76	27.72	29.70	33.00	36.30	39.60	42.84
811.05	W33X240		YES	YES	YES	YES	YES	YES	YES
		LC=	16.74	15.50	14.97	14.20	13.54	12.97	12.46
		LU=	30.69	26.31	24.55	22.10	20.09	18.41	17.00
		FB=	23.76	27.72	29.70	33.00	36.30	39.60	42.90
757.46	W14X455		YES	YES	YES	YES	YES	YES	YES
		LC=	17.76	16.44	15.88	15.07	14.37	13.75	13.21
		LU=	131.39	112.62	105.11	94.60	86.00	78.83	72.77
		FB=	23.76	27.72	29.70	33.00	36.30	39.60	42.90
740.57	W33X220		YES	YES	YES	YES	YES	YES	YES
		LC=	16.68	15.45	14.92	14.16	13.50	12.92	12.41
		LU=	28.06	24.05	22.45	20.20	18.37	16.84	15.54
		FB=	23.76	27.72	29.70	33.00	36.30	39.60	42.90
707.36	W14X426		YES	YES	YES	YES	YES	YES	YES
		LC=	17.62	16.31	15.76	14.95	14.25	13.65	13.11
		LU=	125.42	107.51	100.34	90.30	82.09	75.25	69.46
		FB=	23.76	27.72	29.70	33.00	36.30	39.60	42.90

*(Satisfying local buckling requirements of AISC–1.5.1.4.1 indicated by YES, NO, or PTY (partly). PTY indicates F_b computed from AISC Formula 1.5–5.)

Table A3 (*continued*)

SX	SHAPE	COMPACT SECTION–SATISFIES AISC 1.5.1.4.1 (YES OR NO)							
		FY=	36.00	42.00	45.00	50.00	55.00	60.00	65.00
669.58	W33X200		YES	YES	YES	YES	YES	PTY	PTY
		LC=	16.62	15.39	14.86	14.10	13.45	12.87	12.37
		LU=	25.41	21.78	20.32	18.29	16.63	15.24	14.07
		FB=	23.76	27.72	29.70	33.00	36.30	39.52	42.62
663.56	W36X194		YES	YES	YES	YES	YES	YES	YES
		LC=	12.79	11.84	11.43	10.85	10.34	9.90	9.51
		LU=	19.37	16.60	15.50	13.95	12.68	11.62	10.73
		FB=	23.76	27.72	29.70	33.00	36.30	39.60	42.90
656.87	W14X398		YES	YES	YES	YES	YES	YES	YES
		LC=	17.51	16.21	15.66	14.85	14.16	13.56	13.03
		LU=	119.25	102.21	95.40	85.86	78.05	71.55	66.04
		FB=	23.76	27.72	29.70	33.00	36.30	39.60	42.90
649.92	W30X210		YES	YES	YES	YES	YES	YES	YES
		LC=	15.94	14.76	14.26	13.52	12.89	12.35	11.86
		LU=	30.26	25.94	24.21	21.79	19.81	18.16	16.76
		FB=	23.76	27.72	29.70	33.00	36.30	39.60	42.90
621.22	W36X182		YES	YES	YES	YES	YES	YES	YES
		LC=	12.74	11.79	11.39	10.81	10.30	9.87	9.48
		LU=	18.15	15.56	14.52	13.07	11.88	10.89	10.05
		FB=	23.76	27.72	29.70	33.00	36.30	39.60	42.90
608.04	W14X370		YES	YES	YES	YES	YES	YES	YES
		LC=	17.39	16.10	15.55	14.75	14.06	13.47	12.94
		LU=	113.00	96.86	90.40	81.36	73.96	67.80	62.58
		FB=	23.76	27.72	29.70	33.00	36.30	39.60	42.90
586.04	W30X190		YES	YES	YES	YES	YES	YES	YES
		LC=	15.87	14.69	14.19	13.47	12.84	12.29	11.81
		LU=	27.39	23.48	21.91	19.72	17.93	16.43	15.17
		FB=	23.76	27.72	29.70	33.00	36.30	39.60	42.90
579.09	W36X170		YES	YES	YES	YES	YES	YES	NO
		LC=	12.69	11.75	11.35	10.77	10.27	9.83	0.00
		LU=	16.93	14.51	13.55	12.19	11.08	10.16	9.38
		FB=	23.76	27.72	29.70	33.00	36.30	39.60	39.00
559.39	W14X342		YES	YES	YES	YES	YES	YES	YES
		LC=	17.27	15.99	15.45	14.65	13.97	13.38	12.85
		LU=	106.48	91.27	85.18	76.66	69.69	63.89	58.97
		FB=	23.76	27.72	29.70	33.00	36.30	39.60	42.90
541.04	W36X160		YES	YES	YES	YES	YES	NO	NO
		LC=	12.66	11.72	11.32	10.74	10.24	0.00	0.00
		LU=	15.74	13.49	12.59	11.33	10.30	9.44	8.71
		FB=	23.76	27.72	29.70	33.00	36.30	36.00	39.00
528.21	W30X172		YES	YES	YES	YES	YES	PTY	PTY
		LC=	15.81	14.64	14.14	13.42	12.79	12.25	11.77
		LU=	24.72	21.19	19.78	17.80	16.18	14.83	13.69
		FB=	23.76	27.72	29.70	33.00	36.30	39.40	42.48
511.85	W14X314		YES	YES	YES	YES	YES	YES	YES
		LC=	17.13	15.86	15.32	14.54	13.86	13.27	12.75
		LU=	99.82	85.56	79.85	71.87	65.33	59.89	55.28
		FB=	23.76	27.72	29.70	33.00	36.30	39.60	42.90
502.90	W36X150		YES	YES	YES	YES	NO	NO	NO
		LC=	12.63	11.69	11.30	10.46	0.00	0.00	0.00
		LU=	14.53	12.46	11.62	10.46	9.51	8.72	8.05
		FB=	23.76	27.72	29.70	33.00	33.00	36.00	39.00
492.76	W14X320		YES	YES	YES	YES	YES	YES	YES
		LC=	17.63	16.32	15.77	14.96	14.27	13.66	13.12
		LU=	96.32	82.56	77.05	69.35	63.04	57.79	53.34
		FB=	23.76	27.72	29.70	33.00	36.30	39.60	42.90
492.75	W27X177		YES	YES	YES	YES	YES	YES	YES
		LC=	14.87	13.76	13.30	12.61	12.03	11.52	11.06
		LU=	28.42	24.36	22.73	20.46	18.60	17.05	15.74
		FB=	23.76	27.72	29.70	33.00	36.30	39.60	42.90

Table A3 (*continued*)

SX	SHAPE	COMPACT SECTION-SATISFIES AISC 1.5.1.4.1 (YES OR NO)							
		FY=	36.00	42.00	45.00	50.00	55.00	60.00	65.00
486.42	W33X152		YES	YES	YES	YES	YES	YES	NO
		LC=	12.20	11.30	10.91	10.35	9.87	9.45	0.00
		LU=	16.86	14.45	13.48	12.14	11.03	10.11	9.33
		FB=	23.76	27.72	29.70	33.00	36.30	39.60	39.00
465.44	W14X287		YES	YES	YES	YES	YES	YES	YES
		LC=	17.02	15.76	15.22	14.44	13.77	13.18	12.67
		LU=	92.97	79.69	74.38	66.94	60.85	55.78	51.49
		FB=	23.76	27.72	29.70	33.00	36.30	39.60	42.90
446.84	W33X141		YES	YES	YES	YES	YES	NO	NO
		LC=	12.17	11.27	10.89	10.33	9.85	0.00	0.00
		LU=	15.39	13.19	12.31	11.08	10.07	9.23	8.52
		FB=	23.76	27.72	29.70	33.00	36.30	36.00	39.00
444.50	W27X160		YES	YES	YES	YES	YES	YES	PTY
		LC=	14.80	13.70	13.23	12.55	11.97	11.46	11.01
		LU=	25.77	22.09	20.61	18.55	16.86	15.46	14.27
		FB=	23.76	27.72	29.70	33.00	36.30	39.60	42.85
438.59	W36X135		YES	YES	YES	NO	NO	NO	NO
		LC=	12.35	10.58	9.88	0.00	0.00	0.00	0.00
		LU=	12.35	10.58	9.88	8.89	8.08	7.41	6.84
		FB=	23.76	27.72	29.70	30.00	33.00	36.00	39.00
427.39	W14X264		YES	YES	YES	YES	YES	YES	YES
		LC=	16.91	15.66	15.12	14.35	13.68	13.10	12.58
		LU=	87.13	74.69	69.71	62.74	57.03	52.28	48.26
		FB=	23.76	27.72	29.70	33.00	36.30	39.60	42.90
413.45	W24X160		YES	YES	YES	YES	YES	YES	YES
		LC=	14.87	13.77	13.30	12.62	12.03	11.52	11.06
		LU=	29.95	25.67	23.96	21.56	19.60	17.97	16.58
		FB=	23.76	27.72	29.70	33.00	36.30	39.60	42.90
404.77	W33X130		YES	YES	YES	YES	NO	NO	NO
		LC=	12.14	11.24	10.86	9.91	0.00	0.00	0.00
		LU=	13.76	11.79	11.01	9.91	9.00	8.25	7.62
		FB=	23.76	27.72	29.70	33.00	33.00	36.00	39.00
402.84	W27X145		YES	YES	YES	YES	PTY	PTY	PTY
		LC=	14.74	13.64	13.18	12.50	11.92	11.41	10.97
		LU=	23.45	20.10	18.76	16.88	15.34	14.07	12.98
		FB=	23.76	27.72	29.70	33.00	36.22	39.32	42.39
397.40	W14X246		YES	YES	YES	YES	YES	YES	YES
		LC=	16.83	15.58	15.05	14.28	13.61	13.03	12.52
		LU=	82.35	70.59	65.88	59.29	53.90	49.41	45.61
		FB=	23.76	27.72	29.70	33.00	36.30	39.60	42.90
382.24	W14X237		YES	YES	YES	YES	YES	YES	YES
		LC=	16.79	15.54	15.02	14.25	13.58	13.00	12.49
		LU=	79.87	68.46	63.89	57.50	52.27	47.92	44.23
		FB=	23.76	27.72	29.70	33.00	36.30	39.60	42.90
379.74	W30X132		YES	YES	YES	YES	YES	YES	YES
		LC=	11.13	10.31	9.96	9.45	9.01	8.62	8.28
		LU=	16.12	13.81	12.89	11.60	10.55	9.67	8.92
		FB=	23.76	27.72	29.70	33.00	36.30	39.60	42.90
372.47	W24X145		YES	YES	YES	YES	YES	PTY	PTY
		LC=	14.82	13.72	13.25	12.57	11.99	11.48	11.03
		LU=	27.07	23.20	21.66	19.49	17.72	16.24	14.99
		FB=	23.76	27.72	29.70	33.00	36.30	39.50	42.59
367.80	W14X228		YES	YES	YES	YES	YES	YES	YES
		LC=	16.74	15.50	14.97	14.20	13.54	12.97	12.46
		LU=	77.48	66.41	61.99	55.79	50.71	46.49	42.91
		FB=	23.76	27.72	29.70	33.00	36.30	39.60	42.90
358.30	W33X118		YES	YES	YES	NO	NO	NO	NO
		LC=	11.94	10.23	9.55	0.00	0.00	0.00	0.00
		LU=	11.94	10.23	9.55	8.59	7.81	7.16	6.61
		FB=	23.76	27.72	29.70	30.00	33.00	36.00	39.00

Table A3 (*continued*)

SX	SHAPE		COMPACT SECTION-SATISFIES AISC 1.5.1.4.1 (YES OR NO)						
		FY=	36.00	42.00	45.00	50.00	55.00	60.00	65.00
354.58	W30X124		YES	YES	YES	YES	YES	YES	NO
		LC=	11.10	10.28	9.93	9.42	8.98	8.60	0.00
		LU=	15.01	12.87	12.01	10.81	9.83	9.01	8.31
		FB=	23.76	27.72	29.70	33.00	36.30	39.60	39.00
352.64	W14X219		YES	YES	YES	YES	YES	YES	YES
		LC=	16.70	15.46	14.94	14.17	13.51	12.93	12.43
		LU=	74.92	64.22	59.94	53.94	49.04	44.95	41.49
		FB=	23.76	27.72	29.70	33.00	36.30	39.60	42.90
339.22	W14X211		YES	YES	YES	YES	YES	YES	YES
		LC=	16.67	15.44	14.91	14.15	13.49	12.91	12.41
		LU=	72.59	62.22	58.07	52.26	47.51	43.55	40.20
		FB=	23.76	27.72	29.70	33.00	36.30	39.60	42.90
330.68	W24X130		YES	YES	YES	PTY	PTY	PTY	PTY
		LC=	14.77	13.68	13.21	12.53	11.95	11.44	10.99
		LU=	24.05	20.61	19.24	17.31	15.74	14.43	13.32
		FB=	23.76	27.72	29.70	32.80	35.87	38.91	41.93
327.94	W30X116		YES	YES	YES	YES	YES	NO	NO
		LC=	11.08	10.26	9.91	9.40	8.96	0.00	0.00
		LU=	13.77	11.80	11.01	9.91	9.01	8.26	7.62
		FB=	23.76	27.72	29.70	33.00	36.30	36.00	39.00
324.86	W14X202		YES	YES	YES	YES	YES	YES	YES
		LC=	16.62	15.39	14.86	14.10	13.45	12.87	12.37
		LU=	70.11	60.10	56.09	50.48	45.89	42.07	38.83
		FB=	23.76	27.72	29.70	33.00	36.30	39.60	42.90
317.15	W21X142		YES	YES	YES	YES	YES	YES	YES
		LC=	13.86	12.83	12.39	11.76	11.21	10.73	10.31
		LU=	31.02	26.58	24.81	22.33	20.30	18.61	17.18
		FB=	23.76	27.72	29.70	33.00	36.30	39.60	42.90
309.98	W14X193		YES	YES	YES	YES	YES	YES	YES
		LC=	16.58	15.35	14.83	14.07	13.41	12.84	12.34
		LU=	67.47	57.83	53.98	48.58	44.16	40.48	37.37
		FB=	23.76	27.72	29.70	33.00	36.30	39.60	42.90
299.19	W30X108		YES	YES	YES	YES	YES	NO	NO
		LC=	11.06	10.24	9.89	8.90	8.09	0.00	0.00
		LU=	12.37	10.60	9.89	8.90	8.09	7.42	6.85
		FB=	23.76	27.72	29.70	33.00	36.30	36.00	39.00
299.15	W27X114		YES	YES	YES	YES	YES	YES	YES
		LC=	10.62	9.84	9.50	9.01	8.59	8.23	7.91
		LU=	15.92	13.65	12.74	11.46	10.42	9.55	8.82
		FB=	23.76	27.72	29.70	33.00	36.30	39.60	42.90
299.07	W24X120		YES	YES	YES	YES	YES	YES	PTY
		LC=	12.75	11.81	11.41	10.82	10.32	9.88	9.49
		LU=	21.40	18.35	17.12	15.41	14.01	12.84	11.85
		FB=	23.76	27.72	29.70	33.00	36.30	39.60	42.87
295.81	W14X184		YES	YES	YES	YES	YES	YES	YES
		LC=	16.53	15.30	14.78	14.02	13.37	12.80	12.30
		LU=	64.95	55.67	51.96	46.76	42.51	38.97	35.97
		FB=	23.76	27.72	29.70	33.00	36.30	39.60	42.90
284.10	W21X127		YES	YES	YES	YES	YES	YES	PTY
		LC=	13.78	12.76	12.33	11.69	11.15	10.67	10.26
		LU=	28.04	24.03	22.43	20.19	18.35	16.82	15.53
		FB=	23.76	27.72	29.70	33.00	36.30	39.60	42.78
281.91	W14X176		YES	YES	YES	YES	YES	YES	YES
		LC=	16.50	15.28	14.76	14.00	13.35	12.78	12.28
		LU=	62.34	53.43	49.87	44.88	40.80	37.40	34.52
		FB=	23.76	27.72	29.70	33.00	36.30	39.60	42.90
274.42	W24X110		YES	YES	YES	YES	PTY	PTY	PTY
		LC=	12.71	11.76	11.36	10.78	10.28	9.84	9.45
		LU=	19.72	16.91	15.78	14.20	12.91	11.83	10.92
		FB=	23.76	27.72	29.70	33.00	36.29	39.39	42.47

Table A3 (*continued*)

SX	SHAPE	COMPACT SECTION-SATISFIES AISC 1.5.1.4.1 (YES OR NO)							
		FY=	36.00	42.00	45.00	50.00	55.00	60.00	65.00
269.13	W30X 99		YES	YES	PTY	PTY	NO	NO	NO
		LC=	10.94	9.38	8.75	7.87	0.00	0.00	0.00
		LU=	10.94	9.38	8.75	7.87	7.16	6.56	6.06
		FB=	23.76	27.72	29.68	32.78	33.00	36.00	39.00
267.30	W14X167		YES	YES	YES	YES	YES	YES	YES
		LC=	16.46	15.24	14.72	13.97	13.32	12.75	12.25
		LU=	59.61	51.09	47.68	42.92	39.01	35.76	33.01
		FB=	23.76	27.72	29.70	33.00	36.30	39.60	42.90
266.28	W27X102		YES	YES	YES	YES	YES	YES	NO
		LC=	10.57	9.79	9.45	8.97	8.55	8.19	0.00
		LU=	14.16	12.14	11.33	10.20	9.27	8.50	7.84
		FB=	23.76	27.72	29.70	33.00	36.30	39.60	39.00
263.21	W12X190		YES	YES	YES	YES	YES	YES	YES
		LC=	13.37	12.38	11.96	11.34	10.82	10.35	9.95
		LU=	70.81	60.69	56.65	50.98	46.35	42.48	39.21
		FB=	23.76	27.72	29.70	33.00	36.30	39.60	42.90
253.41	W14X158		YES	YES	YES	YES	YES	YES	PTY
		LC=	16.41	15.19	14.68	13.92	13.27	12.71	12.21
		LU=	57.01	48.87	45.61	41.05	37.32	34.21	31.57
		FB=	23.76	27.72	29.70	33.00	36.30	39.60	42.84
249.58	W21X112		YES	YES	YES	PTY	PTY	PTY	PTY
		LC=	13.72	12.70	12.27	11.64	11.10	10.62	10.21
		LU=	24.79	21.24	19.83	17.84	16.22	14.87	13.73
		FB=	23.76	27.72	29.70	32.93	36.02	39.09	42.13
248.94	W24X100		YES	YES	YES	PTY	PTY	PTY	NO
		LC=	12.66	11.72	11.32	10.74	10.24	9.81	0.00
		LU=	17.93	15.37	14.35	12.91	11.74	10.76	9.93
		FB=	23.76	27.72	29.70	32.81	35.89	38.94	39.00
242.78	W27X 94		YES	YES	YES	YES	YES	NO	NO
		LC=	10.54	9.76	9.43	8.94	8.40	0.00	0.00
		LU=	12.83	11.00	10.27	9.24	8.40	7.70	7.11
		FB=	23.76	27.72	29.70	33.00	36.30	36.00	39.00
240.17	W14X150		YES	YES	YES	YES	YES	PTY	PTY
		LC=	16.37	15.16	14.64	13.89	13.24	12.68	12.18
		LU=	54.45	46.67	43.56	39.20	35.64	32.67	30.15
		FB=	23.76	27.72	29.70	33.00	36.30	39.50	42.59
226.73	W14X142		YES	YES	YES	YES	PTY	PTY	PTY
		LC=	16.36	15.14	14.63	13.88	13.23	12.67	12.17
		LU=	51.71	44.32	41.37	37.23	33.85	31.02	28.64
		FB=	23.76	27.72	29.70	33.00	36.15	39.23	42.29
222.16	W12X161		YES	YES	YES	YES	YES	YES	YES
		LC=	13.21	12.23	11.81	11.20	10.68	10.23	9.83
		LU=	62.03	53.16	49.62	44.66	40.60	37.21	34.35
		FB=	23.76	27.72	29.70	33.00	36.30	39.60	42.90
220.91	W24X 94		YES	YES	YES	YES	YES	YES	YES
		LC=	9.56	8.85	8.55	8.11	7.73	7.40	7.11
		LU=	15.05	12.90	12.04	10.84	9.85	9.03	8.34
		FB=	23.76	27.72	29.70	33.00	36.30	39.60	42.90
220.10	W18X114		YES	YES	YES	YES	YES	YES	YES
		LC=	12.49	11.56	11.17	10.59	10.10	9.67	9.29
		LU=	29.37	25.18	23.50	21.15	19.22	17.62	16.27
		FB=	23.76	27.72	29.70	33.00	36.30	39.60	42.90
216.00	W14X136		YES	YES	YES	YES	YES	PTY	PTY
		LC=	15.55	14.40	13.91	13.20	12.58	12.05	11.57
		LU=	49.17	42.15	39.34	35.40	32.19	29.50	27.23
		FB=	23.76	27.72	29.70	33.00	36.30	39.46	42.55
211.67	W27X 84		YES	YES	PTY	PTY	NO	NO	NO
		LC=	10.51	9.42	8.79	7.91	0.00	0.00	0.00
		LU=	10.99	9.42	8.79	7.91	7.19	6.59	6.08
		FB=	23.76	27.72	29.67	32.77	33.00	36.00	39.00

Table A3 (*continued*)

COMPACT SECTION-SATISFIES AISC 1.5.1.4.1 (YES OR NO)

SX	SHAPE	FY=	36.00	42.00	45.00	50.00	55.00	60.00	65.00
202.23	W18X105		YES	YES	YES	YES	YES	YES	YES
		LC=	12.44	11.52	11.13	10.56	10.07	9.64	9.26
		LU=	27.14	23.26	21.71	19.54	17.76	16.28	15.03
		FB=	23.76	27.72	29.70	33.00	36.30	39.60	42.90
202.01	W14X127		YES	YES	YES	YES	PTY	PTY	PTY
		LC=	15.50	14.35	13.86	13.15	12.54	12.01	11.53
		LU=	46.42	39.79	37.13	33.42	30.38	27.85	25.71
		FB=	23.76	27.72	29.70	33.00	36.11	39.19	42.24
197.62	W21X 96		YES	YES	YES	YES	YES	YES	YES
		LC=	9.54	8.83	8.53	8.09	7.71	7.38	7.09
		LU=	18.50	15.86	14.80	13.32	12.11	11.10	10.24
		FB=	23.76	27.72	29.70	33.00	36.30	39.60	42.90
196.28	W24X 84		YES	YES	YES	YES	YES	YES	NO
		LC=	9.51	8.80	8.51	8.07	7.69	7.37	0.00
		LU=	13.37	11.46	10.69	9.62	8.75	8.02	7.40
		FB=	23.76	27.72	29.70	33.00	36.30	39.60	39.00
189.39	W14X119		YES	YES	PTY	PTY	PTY	PTY	PTY
		LC=	15.46	14.31	13.83	13.12	12.51	11.97	11.50
		LU=	43.87	37.60	35.10	31.59	28.71	26.32	24.30
		FB=	23.76	27.72	29.68	32.78	35.85	38.89	41.91
184.43	W18X 96		YES	YES	YES	YES	PTY	PTY	PTY
		LC=	12.40	11.48	11.09	10.52	10.03	9.60	9.23
		LU=	24.89	21.33	19.91	17.92	16.29	14.93	13.78
		FB=	23.76	27.72	29.70	33.00	36.27	39.37	42.45
182.54	W12X133		YES	YES	YES	YES	YES	YES	YES
		LC=	13.05	12.08	11.67	11.07	10.55	10.10	9.71
		LU=	52.88	45.32	42.30	38.07	34.61	31.72	29.28
		FB=	23.76	27.72	29.70	33.00	36.30	39.60	42.90
176.27	W14X111		YES	PTY	PTY	PTY	PTY	PTY	PTY
		LC=	15.43	14.28	13.80	13.09	12.48	11.95	11.48
		LU=	41.11	35.24	32.89	29.60	26.91	24.67	22.77
		FB=	23.76	27.59	29.44	32.50	35.53	38.53	41.50
175.35	W24X 76		YES	YES	YES	YES	YES	NO	NO
		LC=	9.48	8.78	8.48	8.04	7.67	0.00	0.00
		LU=	11.86	10.17	9.49	8.54	7.76	7.11	6.57
		FB=	23.76	27.72	29.70	33.00	36.30	36.00	39.00
168.01	W21X 82		YES	YES	YES	YES	YES	YES	YES
		LC=	9.45	8.75	8.46	8.02	7.65	7.32	7.04
		LU=	15.81	13.55	12.65	11.38	10.35	9.48	8.75
		FB=	23.76	27.72	29.70	33.00	36.30	39.60	42.90
166.06	W16X 96		YES	YES	YES	YES	YES	YES	PTY
		LC=	12.17	11.27	10.88	10.32	9.84	9.42	9.05
		LU=	28.62	24.53	22.90	20.61	18.73	17.17	15.85
		FB=	23.76	27.72	29.70	33.00	36.30	39.60	42.80
163.62	W14X103		PTY	PTY	PTY	PTY	PTY	PTY	PTY
		LC=	15.38	14.24	13.76	13.05	12.44	11.91	11.44
		LU=	38.49	32.99	30.79	27.71	25.19	23.09	21.32
		FB=	23.67	27.37	29.19	32.21	35.19	38.14	41.06
163.36	W12X120		YES	YES	YES	YES	YES	YES	YES
		LC=	13.00	12.03	11.63	11.03	10.52	10.07	9.67
		LU=	48.08	41.21	38.46	34.61	31.47	28.84	26.62
		FB=	23.76	27.72	29.70	33.00	36.30	39.60	42.90
156.10	W18X 85		YES	YES	YES	YES	YES	YES	YES
		LC=	9.32	8.63	8.34	7.91	7.54	7.22	6.94
		LU=	20.34	17.43	16.27	14.64	13.31	12.20	11.26
		FB=	23.76	27.72	29.70	33.00	36.30	39.60	42.90
153.05	W24X 68		YES	YES	YES	PTY	NO	NO	NO
		LC=	9.45	8.72	8.14	7.33	0.00	0.00	0.00
		LU=	10.18	8.72	8.14	7.33	6.66	6.11	5.64
		FB=	23.76	27.72	29.70	32.83	33.00	36.00	39.00

Table A3 (*continued*)

SX	SHAPE	COMPACT SECTION-SATISFIES AISC 1.5.1.4.1 (YES OR NO)							
		FY=	36.00	42.00	45.00	50.00	55.00	60.00	65.00
151.31	W16X 88		YES	YES	YES	YES	PTY	PTY	PTY
		LC=	12.14	11.24	10.85	10.30	9.82	9.40	9.03
		LU=	26.19	22.45	20.95	18.86	17.14	15.71	14.50
		FB=	23.76	27.72	29.70	33.00	36.18	39.27	42.33
150.68	W21X 73		YES	YES	YES	YES	YES	YES	YES
		LC=	8.75	8.10	7.83	7.42	7.08	6.78	6.51
		LU=	13.37	11.46	10.70	9.63	8.75	8.02	7.41
		FB=	23.76	27.72	29.70	33.00	36.30	39.60	42.90
150.63	W14X 95		PTY	PTY	PTY	PTY	PTY	PTY	PTY
		LC=	15.35	14.21	13.73	13.02	12.42	11.89	11.42
		LU=	35.67	30.57	28.53	25.68	23.34	21.40	19.75
		FB=	23.44	27.08	28.87	31.83	34.76	37.65	40.51
144.51	W12X106		YES	YES	YES	YES	YES	YES	YES
		LC=	12.90	11.95	11.54	10.95	10.44	9.99	9.60
		LU=	43.34	37.15	34.67	31.20	28.37	26.00	24.00
		FB=	23.76	27.72	29.70	33.00	36.30	39.60	42.90
141.71	W18X 77		YES	YES	YES	YES	YES	YES	YES
		LC=	9.27	8.58	8.29	7.87	7.50	7.18	6.90
		LU=	18.61	15.95	14.89	13.40	12.18	11.16	10.31
		FB=	23.76	27.72	29.70	33.00	36.30	39.60	42.90
139.92	W21X 68		YES	YES	YES	YES	YES	YES	YES
		LC=	8.72	8.08	7.80	7.40	7.06	6.76	6.49
		LU=	12.41	10.63	9.92	8.93	8.12	7.44	6.87
		FB=	23.76	27.72	29.70	33.00	36.30	39.60	42.90
138.12	W14X 87		PTY	PTY	PTY	PTY	PTY	PTY	PTY
		LC=	15.30	14.17	13.68	12.98	12.38	11.85	11.39
		LU=	32.98	28.27	26.39	23.75	21.59	19.79	18.27
		FB=	23.20	26.77	28.53	31.43	34.29	37.12	39.91
134.66	W12X 99		YES	YES	YES	YES	YES	YES	PTY
		LC=	12.86	11.91	11.50	10.91	10.41	9.96	9.57
		LU=	40.76	34.94	32.61	29.35	26.68	24.45	22.57
		FB=	23.76	27.72	29.70	33.00	36.30	39.60	42.78
130.94	W14X 84		YES	YES	YES	PTY	PTY	PTY	PTY
		LC=	12.69	11.74	11.35	10.76	10.26	9.83	9.44
		LU=	30.53	26.17	24.43	21.98	19.98	18.32	16.91
		FB=	23.76	27.72	29.70	32.82	35.90	38.95	41.97
129.49	W24X 61		YES	YES	YES	YES	NO	NO	NO
		LC=	7.41	6.86	6.48	5.83	0.00	0.00	0.00
		LU=	8.10	6.94	6.48	5.83	5.30	4.86	4.48
		FB=	23.76	27.72	29.70	33.00	33.00	36.00	39.00
128.21	W18X 70		YES	YES	YES	YES	YES	YES	YES
		LC=	9.23	8.55	8.26	7.83	7.47	7.15	6.87
		LU=	16.90	14.48	13.52	12.16	11.06	10.14	9.36
		FB=	23.76	27.72	29.70	33.00	36.30	39.60	42.90
127.76	W16X 78		YES	YES	YES	YES	YES	YES	YES
		LC=	9.06	8.39	8.10	7.69	7.33	7.02	6.74
		LU=	21.31	18.26	17.04	15.34	13.94	12.78	11.80
		FB=	23.76	27.72	29.70	33.00	36.30	39.60	42.90
126.42	W21X 62		YES	YES	YES	YES	YES	YES	NO
		LC=	8.69	8.05	7.77	7.38	7.03	6.70	0.00
		LU=	11.17	9.58	8.94	8.04	7.31	6.70	6.19
		FB=	23.76	27.72	29.70	33.00	36.30	39.60	39.00
126.30	W10X112		YES	YES	YES	YES	YES	YES	YES
		LC=	10.99	10.17	9.83	9.32	8.89	8.51	8.18
		LU=	52.87	45.32	42.30	38.07	34.61	31.72	29.28
		FB=	23.76	27.72	29.70	33.00	36.30	39.60	42.90
125.02	W12X 92		YES	YES	YES	YES	PTY	PTY	PTY
		LC=	12.83	11.87	11.47	10.88	10.38	9.93	9.54
		LU=	38.16	32.71	30.53	27.48	24.98	22.90	21.13
		FB=	23.76	27.72	29.70	33.00	36.26	39.36	42.43

Table A3 (*continued*)

SX	SHAPE	COMPACT SECTION-SATISFIES AISC 1.5.1.4.1 (YES OR NO)							
		FY=	36.00	42.00	45.00	50.00	55.00	60.00	65.00
121.08	W14X 78		YES	PTY	PTY	PTY	PTY	PTY	PTY
		LC=	12.66	11.72	11.32	10.74	10.24	9.81	9.42
		LU=	28.37	24.31	22.69	20.42	18.56	17.02	15.71
		FB=	23.76	27.60	29.45	32.51	35.54	38.54	41.51
117.04	W18X 64		YES	YES	YES	YES	YES	YES	YES
		LC=	9.19	8.51	8.22	7.80	7.44	7.12	6.84
		LU=	15.48	13.27	12.39	11.15	10.13	9.29	8.57
		FB=	23.76	27.72	29.70	33.00	36.30	39.60	42.90
115.95	W16X 71		YES	YES	YES	YES	YES	YES	YES
		LC=	9.01	8.34	8.06	7.65	7.29	6.98	6.71
		LU=	19.45	16.67	15.56	14.00	12.73	11.67	10.77
		FB=	23.76	27.72	29.70	33.00	36.30	39.60	42.90
115.72	W12X 85		YES	YES	YES	PTY	PTY	PTY	PTY
		LC=	12.77	11.82	11.42	10.84	10.33	9.89	9.50
		LU=	35.68	30.58	28.54	25.69	23.35	21.41	19.76
		FB=	23.76	27.72	29.70	32.88	35.97	39.03	42.06
113.64	W24X 55		YES	YES	YES	NO	NO	NO	NO
		LC=	6.92	5.93	5.53	0.00	0.00	0.00	0.00
		LU=	6.92	5.93	5.53	4.98	4.53	4.15	3.83
		FB=	23.76	27.72	29.70	30.00	33.00	36.00	39.00
112.41	W10X100		YES	YES	YES	YES	YES	YES	YES
		LC=	10.91	10.10	9.76	9.26	8.83	8.45	8.12
		LU=	48.15	41.27	38.52	34.66	31.51	28.89	26.66
		FB=	23.76	27.72	29.70	33.00	36.30	39.60	42.90
112.30	W14X 74		YES	YES	YES	YES	YES	YES	YES
		LC=	10.63	9.84	9.50	9.02	8.60	8.23	7.91
		LU=	25.73	22.05	20.58	18.52	16.84	15.43	14.25
		FB=	23.76	27.72	29.70	33.00	36.30	39.60	42.90
109.68	W21X 55		YES	YES	PTY	PTY	PTY	NO	NO
		LC=	8.67	8.02	7.63	6.87	6.24	0.00	0.00
		LU=	9.54	8.18	7.63	6.87	6.24	5.72	5.28
		FB=	23.76	27.72	29.65	32.75	35.82	36.00	39.00
107.83	W18X 60		YES	YES	YES	YES	YES	YES	YES
		LC=	7.97	7.38	7.13	6.76	6.45	6.17	5.93
		LU=	13.32	11.42	10.66	9.59	8.72	7.99	7.38
		FB=	23.76	27.72	29.70	33.00	36.30	39.60	42.90
107.10	W12X 79		YES	PTY	PTY	PTY	PTY	PTY	PTY
		LC=	12.75	11.80	11.40	10.81	10.31	9.87	9.48
		LU=	33.24	28.49	26.59	23.93	21.76	19.94	18.41
		FB=	23.76	27.65	29.51	32.58	35.62	38.64	41.62
104.22	W16X 64		YES	YES	YES	YES	YES	YES	YES
		LC=	8.97	8.30	8.02	7.61	7.25	6.94	6.67
		LU=	17.58	15.07	14.06	12.66	11.51	10.55	9.73
		FB=	23.76	27.72	29.70	33.00	36.30	39.60	42.90
103.00	W14X 68		YES	YES	YES	YES	YES	PTY	PTY
		LC=	10.59	9.81	9.47	8.99	8.57	8.20	7.88
		LU=	23.73	20.34	18.98	17.09	15.53	14.24	13.14
		FB=	23.76	27.72	29.70	33.00	36.30	39.43	42.51
99.70	W10X 89		YES	YES	YES	YES	YES	YES	YES
		LC=	10.84	10.04	9.70	9.20	8.77	8.40	8.07
		LU=	43.63	37.40	34.90	31.41	28.56	26.18	24.16
		FB=	23.76	27.72	29.70	33.00	36.30	39.60	42.90
98.22	W18X 55		YES	YES	YES	YES	YES	YES	YES
		LC=	7.95	7.36	7.11	6.74	6.43	6.15	5.91
		LU=	12.12	10.39	9.69	8.72	7.93	7.27	6.71
		FB=	23.76	27.72	29.70	33.00	36.30	39.60	42.90
97.53	W12X 72		PTY	PTY	PTY	PTY	PTY	PTY	PTY
		LC=	12.70	11.76	11.36	10.78	10.28	9.84	9.45
		LU=	30.53	26.17	24.42	21.98	19.98	18.31	16.91
		FB=	23.67	27.36	29.19	32.20	35.19	38.14	41.06

Table A3 (*continued*)

SX	SHAPE	COMPACT SECTION-SATISFIES AISC 1.5.1.4.1 (YES OR NO)							
		FY=	36.00	42.00	45.00	50.00	55.00	60.00	65.00
94.12	W16X 58		YES	YES	YES	YES	YES	YES	PTY
		LC=	8.93	8.27	7.99	7.58	7.22	6.92	6.64
		LU=	15.93	13.65	12.74	11.47	10.43	9.56	8.82
		FB=	23.76	27.72	29.70	33.00	36.30	39.60	42.83
93.20	W21X 49		YES	YES	YES	YES	NO	NO	NO
		LC=	6.88	6.37	6.15	5.55	0.00	0.00	0.00
		LU=	7.71	6.61	6.17	5.55	5.04	4.62	4.27
		FB=	23.76	27.72	29.70	33.00	33.00	36.00	39.00
92.23	W14X 61		YES	YES	YES	PTY	PTY	PTY	PTY
		LC=	10.55	9.77	9.44	8.95	8.53	8.17	7.85
		LU=	21.40	18.34	17.12	15.40	14.00	12.84	11.85
		FB=	23.76	27.72	29.70	32.80	35.87	38.92	41.93
88.95	W18X 50		YES	YES	YES	YES	YES	YES	PTY
		LC=	7.91	7.32	7.08	6.71	6.40	6.13	5.89
		LU=	10.99	9.42	8.79	7.91	7.19	6.59	6.08
		FB=	23.76	27.72	29.70	33.00	36.30	39.60	42.81
88.01	W12X 65		PTY	PTY	PTY	PTY	PTY	PTY	PTY
		LC=	12.66	11.72	11.32	10.74	10.24	9.81	9.42
		LU=	27.77	23.80	22.22	20.00	18.18	16.66	15.38
		FB=	23.39	27.01	28.80	31.74	34.66	37.53	40.38
86.10	W10X 77		YES	YES	YES	YES	YES	YES	YES
		LC=	10.76	9.96	9.62	9.13	8.70	8.33	8.00
		LU=	38.57	33.06	30.86	27.77	25.25	23.14	21.36
		FB=	23.76	27.72	29.70	33.00	36.30	39.60	42.90
81.51	W21X 44		YES	YES	YES	NO	NO	NO	NO
		LC=	6.56	5.63	5.25	0.00	0.00	0.00	0.00
		LU=	6.56	5.63	5.25	4.72	4.29	3.94	3.63
		FB=	23.76	27.72	29.70	30.00	33.00	36.00	39.00
80.66	W16X 50		YES	YES	YES	YES	YES	YES	YES
		LC=	7.46	6.91	6.67	6.33	6.04	5.78	5.55
		LU=	12.65	10.84	10.12	9.11	8.28	7.59	7.00
		FB=	23.76	27.72	29.70	33.00	36.30	39.60	42.90
80.13	W10X 72		YES	YES	YES	YES	YES	YES	YES
		LC=	10.73	9.93	9.60	9.10	8.68	8.31	7.98
		LU=	36.23	31.05	28.98	26.08	23.71	21.73	20.06
		FB=	23.76	27.72	29.70	33.00	36.30	39.60	42.90
78.89	W18X 45		YES	YES	YES	PTY	PTY	NO	NO
		LC=	7.89	7.30	7.05	6.69	6.33	0.00	0.00
		LU=	9.67	8.28	7.73	6.96	6.33	5.80	5.35
		FB=	23.76	27.72	29.70	32.94	36.03	36.00	39.00
78.11	W12X 58		YES	YES	PTY	PTY	PTY	PTY	PTY
		LC=	10.57	9.78	9.45	8.96	8.55	8.18	7.86
		LU=	24.37	20.89	19.50	17.55	15.95	14.62	13.50
		FB=	23.76	27.72	29.68	32.78	35.85	38.89	41.91
77.77	W14X 53		YES	YES	YES	YES	YES	YES	YES
		LC=	8.50	7.87	7.61	7.22	6.88	6.59	6.33
		LU=	17.61	15.10	14.09	12.68	11.53	10.57	9.75
		FB=	23.76	27.72	29.70	33.00	36.30	39.60	42.90
73.69	W10X 66		YES	YES	YES	YES	YES	PTY	PTY
		LC=	10.67	9.88	9.55	9.06	8.63	8.27	7.94
		LU=	33.75	28.93	27.00	24.30	22.09	20.25	18.69
		FB=	23.76	27.72	29.70	33.00	36.30	39.57	42.68
72.36	W16X 45		YES	YES	YES	YES	YES	YES	YES
		LC=	7.43	6.87	6.64	6.30	6.01	5.75	5.52
		LU=	11.38	9.75	9.10	8.19	7.44	6.82	6.30
		FB=	23.76	27.72	29.70	33.00	36.30	39.60	42.90
70.67	W12X 53		YES	PTY	PTY	PTY	PTY	PTY	PTY
		LC=	10.55	9.77	9.44	8.95	8.53	8.17	7.85
		LU=	22.11	18.95	17.68	15.92	14.47	13.26	12.24
		FB=	23.76	27.47	29.31	32.35	35.35	38.33	41.27

Table A3 (*continued*)

SX	SHAPE	COMPACT SECTION-SATISFIES AISC 1.5.1.4.1 (YES OR NO)							
		FY=	36.00	42.00	45.00	50.00	55.00	60.00	65.00
70.22	W14X 48		YES	YES	YES	YES	YES	PTY	PTY
		LC=	8.47	7.84	7.58	7.19	6.85	6.56	6.30
		LU=	15.96	13.68	12.77	11.49	10.45	9.57	8.84
		FB=	23.76	27.72	29.70	33.00	36.30	39.57	42.67
68.30	W18X 40		YES	YES	YES	YES	NO	NO	NO
		LC=	6.35	5.88	5.68	5.39	0.00	0.00	0.00
		LU=	8.15	6.99	6.52	5.87	5.33	4.89	4.51
		FB=	23.76	27.72	29.70	33.00	33.00	36.00	39.00
67.06	W10X 60		YES	YES	YES	YES	PTY	PTY	PTY
		LC=	10.63	9.84	9.51	9.02	8.60	8.23	7.91
		LU=	31.08	26.64	24.86	22.37	20.34	18.64	17.21
		FB=	23.76	27.72	29.70	33.00	36.10	39.18	42.23
64.72	W12X 50		YES	YES	YES	YES	YES	YES	YES
		LC=	8.52	7.89	7.62	7.23	6.89	6.60	6.34
		LU=	19.66	16.85	15.73	14.15	12.87	11.79	10.89
		FB=	23.76	27.72	29.70	33.00	36.30	39.60	42.90
64.43	W16X 40		YES	YES	YES	YES	YES	PTY	NO
		LC=	7.38	6.84	6.60	6.26	5.97	5.72	0.00
		LU=	10.18	8.73	8.15	7.33	6.66	6.11	5.64
		FB=	23.76	27.72	29.70	33.00	36.30	39.45	39.00
62.71	W14X 43		YES	YES	YES	PTY	PTY	PTY	PTY
		LC=	8.44	7.81	7.55	7.16	6.83	6.54	6.28
		LU=	14.29	12.25	11.43	10.29	9.35	8.57	7.91
		FB=	23.76	27.72	29.70	32.90	35.98	39.05	42.08
60.41	W10X 54		YES	PTY	PTY	PTY	PTY	PTY	PTY
		LC=	10.58	9.79	9.46	8.98	8.56	8.19	7.87
		LU=	28.35	24.30	22.68	20.41	18.55	17.01	15.70
		FB=	23.76	27.69	29.55	32.63	35.68	38.70	41.69
60.40	W 8X 67		YES	YES	YES	YES	YES	YES	YES
		LC=	8.74	8.09	7.82	7.42	7.07	6.77	6.50
		LU=	39.77	34.09	31.81	28.63	26.03	23.86	22.02
		FB=	23.76	27.72	29.70	33.00	36.30	39.60	42.90
58.17	W12X 45		YES	YES	YES	YES	YES	PTY	PTY
		LC=	8.48	7.85	7.59	7.20	6.86	6.57	6.31
		LU=	17.78	15.24	14.22	12.80	11.63	10.66	9.84
		FB=	23.76	27.72	29.70	33.00	36.30	39.43	42.52
57.85	W18X 35		YES	YES	YES	NO	NO	NO	NO
		LC=	6.33	5.76	5.38	0.00	0.00	0.00	0.00
		LU=	6.72	5.76	5.38	4.84	4.40	4.03	3.72
		FB=	23.76	27.72	29.70	30.00	33.00	36.00	39.00
56.31	W16X 36		YES	PTY	PTY	PTY	PTY	PTY	NO
		LC=	7.38	6.83	6.60	6.26	5.72	5.24	0.00
		LU=	8.74	7.49	6.99	6.29	5.72	5.24	4.84
		FB=	23.76	27.67	29.53	32.60	35.65	38.66	39.00
54.58	W10X 49		PTY	PTY	PTY	PTY	PTY	PTY	PTY
		LC=	10.55	9.77	9.44	8.95	8.53	8.17	7.85
		LU=	25.83	22.14	20.66	18.60	16.90	15.50	14.30
		FB=	23.67	27.37	29.19	32.21	35.19	38.14	41.07
54.57	W14X 38		YES	YES	YES	YES	YES	YES	PTY
		LC=	7.15	6.62	6.39	6.06	5.78	5.54	5.32
		LU=	11.39	9.76	9.11	8.20	7.46	6.83	6.31
		FB=	23.76	27.72	29.70	33.00	36.30	39.60	42.79
51.95	W 8X 58		YES	YES	YES	YES	YES	YES	YES
		LC=	8.67	8.03	7.76	7.36	7.02	6.72	6.45
		LU=	35.15	30.12	28.12	25.30	23.00	21.09	19.46
		FB=	23.76	27.72	29.70	33.00	36.30	39.60	42.90
51.94	W12X 40		YES	YES	YES	PTY	PTY	PTY	PTY
		LC=	8.44	7.81	7.55	7.16	6.83	6.54	6.28
		LU=	16.00	13.71	12.80	11.52	10.47	9.60	8.86
		FB=	23.76	27.72	29.70	32.81	35.88	38.93	41.95

Table A3 (*continued*)

SX	SHAPE	COMPACT SECTION-SATISFIES AISC 1.5.1.4.1 (YES OR NO)							
		FY=	36.00	42.00	45.00	50.00	55.00	60.00	65.00
49.13	W10X 45		YES	YES	YES	YES	YES	YES	PTY
		LC=	8.46	7.83	7.57	7.18	6.85	6.55	6.30
		LU=	22.67	19.43	18.14	16.32	14.84	13.60	12.56
		FB=	23.76	27.72	29.70	33.00	36.30	39.60	42.88
48.45	W14X 34		YES	YES	YES	PTY	PTY	PTY	PTY
		LC=	7.12	6.59	6.37	6.04	5.76	5.51	5.30
		LU=	10.11	8.66	8.08	7.28	6.61	6.06	5.60
		FB=	23.76	27.72	29.70	32.96	36.06	39.13	42.17
47.03	W16X 31		YES	YES	YES	YES	NO	NO	NO
		LC=	5.83	5.39	5.21	4.94	0.00	0.00	0.00
		LU=	7.13	6.11	5.70	5.13	4.67	4.28	3.95
		FB=	23.76	27.72	29.70	33.00	33.00	36.00	39.00
45.88	W12X 36		YES	YES	YES	YES	YES	YES	YES
		LC=	6.92	6.41	6.19	5.88	5.60	5.36	5.15
		LU=	13.40	11.49	10.72	9.65	8.77	8.04	7.42
		FB=	23.76	27.72	29.70	33.00	36.30	39.60	42.90
43.22	W 8X 48		YES	YES	YES	YES	YES	YES	YES
		LC=	8.56	7.93	7.66	7.27	6.93	6.63	6.37
		LU=	30.19	25.88	24.15	21.74	19.76	18.11	16.72
		FB=	23.76	27.72	29.70	33.00	36.30	39.60	42.90
42.19	W10X 39		YES	YES	YES	PTY	PTY	PTY	PTY
		LC=	8.43	7.80	7.54	7.15	6.82	6.53	6.27
		LU=	19.64	16.84	15.71	14.14	12.86	11.78	10.88
		FB=	23.76	27.72	29.70	32.90	35.99	39.05	42.09
41.78	W14X 30		PTY	PTY	PTY	PTY	PTY	PTY	NO
		LC=	7.10	6.57	6.35	6.03	5.63	5.16	0.00
		LU=	8.61	7.38	6.89	6.20	5.63	5.16	4.77
		FB=	23.72	27.43	29.27	32.29	35.29	38.26	39.00
39.43	W12X 31		YES	YES	YES	YES	YES	PTY	PTY
		LC=	6.88	6.37	6.16	5.84	5.57	5.33	5.12
		LU=	11.61	9.95	9.29	8.36	7.60	6.97	6.43
		FB=	23.76	27.72	29.70	33.00	36.30	39.41	42.49
38.09	W16X 26		YES	YES	NO	NO	NO	NO	NO
		LC=	5.61	4.81	0.00	0.00	0.00	0.00	0.00
		LU=	5.61	4.81	4.49	4.04	3.67	3.36	3.10
		FB=	23.76	27.72	27.00	30.00	33.00	36.00	39.00
35.46	W 8X 40		YES	YES	YES	YES	PTY	PTY	PTY
		LC=	8.52	7.89	7.62	7.23	6.89	6.60	6.34
		LU=	25.29	21.67	20.23	18.20	16.55	15.17	14.00
		FB=	23.76	27.72	29.70	33.00	36.18	39.27	42.33
35.05	W10X 33		PTY	PTY	PTY	PTY	PTY	PTY	PTY
		LC=	8.40	7.78	7.51	7.13	6.80	6.51	6.25
		LU=	16.37	14.03	13.09	11.78	10.71	9.82	9.06
		FB=	23.60	27.28	29.09	32.09	35.06	37.99	40.89
34.93	W14X 26		YES	YES	YES	YES	YES	NO	NO
		LC=	5.30	4.91	4.74	4.50	4.29	0.00	0.00
		LU=	7.00	6.00	5.60	5.04	4.58	4.20	3.87
		FB=	23.76	27.72	29.70	33.00	36.30	36.00	39.00
34.13	W12X 27		YES	PTY	PTY	PTY	PTY	PTY	PTY
		LC=	6.86	6.35	6.13	5.82	5.55	5.31	5.10
		LU=	10.06	8.62	8.05	7.24	6.58	6.03	5.57
		FB=	23.76	27.68	29.55	32.62	35.67	38.69	41.68
31.15	W 8X 35		YES	PTY	PTY	PTY	PTY	PTY	PTY
		LC=	8.47	7.84	7.57	7.18	6.85	6.56	6.30
		LU=	22.56	19.33	18.05	16.24	14.76	13.53	12.49
		FB=	23.76	27.68	29.54	32.62	35.66	38.68	41.67
30.78	W10X 29		YES	YES	YES	YES	YES	YES	YES
		LC=	6.12	5.66	5.47	5.19	4.95	4.74	4.55
		LU=	13.13	11.25	10.50	9.45	8.59	7.88	7.27
		FB=	23.76	27.72	29.70	33.00	36.30	39.60	42.90

Table A3 (*continued*)

SX	SHAPE	COMPACT SECTION-SATISFIES AISC 1.5.1.4.1 (YES OR NO)							
		FY=	36.00	42.00	45.00	50.00	55.00	60.00	65.00
28.77	W14X 22		YES	YES	YES	NO	NO	NO	NO
		LC=	5.27	4.84	4.52	0.00	0.00	0.00	0.00
		LU=	5.65	4.84	4.52	4.06	3.69	3.39	3.13
		FB=	23.76	27.72	29.70	30.00	33.00	36.00	39.00
27.42	W 8X 31		PTY	PTY	PTY	PTY	PTY	PTY	PTY
		LC=	8.44	7.81	7.55	7.16	6.83	6.54	6.28
		LU=	20.04	17.18	16.03	14.43	13.12	12.02	11.10
		FB=	23.59	27.26	29.08	32.07	35.03	37.96	40.86
26.42	W10X 25		YES	YES	YES	YES	YES	YES	PTY
		LC=	6.08	5.63	5.44	5.16	4.92	4.71	4.52
		LU=	11.37	9.75	9.10	8.19	7.44	6.82	6.30
		FB=	23.76	27.72	29.70	33.00	36.30	39.60	42.72
25.29	W12X 22		YES	YES	YES	YES	YES	YES	YES
		LC=	4.25	3.93	3.80	3.60	3.44	3.29	3.16
		LU=	6.42	5.50	5.14	4.62	4.20	3.85	3.55
		FB=	23.76	27.72	29.70	33.00	36.30	39.60	42.90
24.26	W 8X 28		YES	YES	YES	YES	PTY	PTY	PTY
		LC=	6.90	6.39	6.17	5.85	5.58	5.34	5.13
		LU=	17.39	14.90	13.91	12.52	11.38	10.43	9.63
		FB=	23.76	27.72	29.70	33.00	36.28	39.38	42.46
21.47	W10X 21		YES	PTY	PTY	PTY	PTY	PTY	PTY
		LC=	6.06	5.61	5.42	5.15	4.91	4.70	4.51
		LU=	9.14	7.83	7.31	6.58	5.98	5.48	5.06
		FB=	23.76	27.56	29.41	32.46	35.48	38.47	41.44
21.39	W12X 19		YES	YES	YES	YES	YES	YES	YES
		LC=	4.23	3.91	3.78	3.59	3.42	3.19	2.95
		LU=	5.32	4.56	4.26	3.83	3.48	3.19	2.95
		FB=	23.76	27.72	29.70	33.00	36.30	39.60	42.90
20.80	W 8X 24		YES	PTY	PTY	PTY	PTY	PTY	PTY
		LC=	6.86	6.35	6.13	5.82	5.55	5.31	5.10
		LU=	15.10	12.94	12.08	10.87	9.88	9.06	8.36
		FB=	23.76	27.67	29.53	32.60	35.65	38.66	41.65
18.77	W10X 19		YES	YES	YES	YES	YES	YES	YES
		LC=	4.24	3.92	3.79	3.60	3.43	3.28	3.15
		LU=	7.15	6.13	5.72	5.15	4.68	4.29	3.96
		FB=	23.76	27.72	29.70	33.00	36.30	39.60	42.90
17.55	W12X 16		YES	YES	YES	PTY	PTY	PTY	NO
		LC=	4.15	3.55	3.32	2.98	2.71	2.49	0.00
		LU=	4.15	3.55	3.32	2.98	2.71	2.49	2.29
		FB=	23.76	27.72	29.70	32.96	36.06	39.14	39.00
17.00	W 8X 20		YES	YES	YES	YES	YES	PTY	PTY
		LC=	5.56	5.14	4.97	4.71	4.49	4.30	4.13
		LU=	11.32	9.70	9.06	8.15	7.41	6.79	6.27
		FB=	23.76	27.72	29.70	33.00	36.30	39.44	42.53
16.79	W 6X 25		YES	YES	YES	YES	YES	YES	PTY
		LC=	6.41	5.94	5.74	5.44	5.19	4.97	4.77
		LU=	20.15	17.27	16.12	14.50	13.18	12.09	11.16
		FB=	23.76	27.72	29.70	33.00	36.30	39.60	42.75
16.16	W10X 17		YES	YES	YES	YES	YES	YES	YES
		LC=	4.23	3.91	3.78	3.59	3.42	3.27	3.15
		LU=	6.03	5.17	4.82	4.34	3.95	3.62	3.34
		FB=	23.76	27.72	29.70	33.00	36.30	39.60	42.90
14.81	W12X 14		PTY	PTY	PTY	NO	NO	NO	NO
		LC=	3.45	2.96	2.76	0.00	0.00	0.00	0.00
		LU=	3.45	2.96	2.76	2.48	2.26	2.07	1.91
		FB=	23.70	27.40	29.23	30.00	33.00	36.00	39.00
14.10	W 8X 17		YES	PTY	PTY	PTY	PTY	PTY	PTY
		LC=	5.54	5.13	4.95	4.70	4.48	4.29	4.12
		LU=	9.35	8.02	7.48	6.73	6.12	5.61	5.18
		FB=	23.76	27.53	29.38	32.43	35.44	38.43	41.39

Table A3 (*continued*)

SX	SHAPE	COMPACT SECTION-SATISFIES AISC 1.5.1.4.1 (YES OR NO)							
		FY=	36.00	42.00	45.00	50.00	55.00	60.00	65.00
13.76	W10X 15		YES	YES	YES	PTY	PTY	PTY	PTY
		LC=	4.22	3.90	3.77	3.58	3.26	2.98	2.75
		LU=	4.98	4.26	3.98	3.58	3.26	2.98	2.75
		FB=	23.76	27.72	29.70	32.96	36.06	39.14	42.19
13.45	W 6X 20		YES	PTY	PTY	PTY	PTY	PTY	PTY
		LC=	6.35	5.88	5.68	5.39	5.13	4.92	4.72
		LU=	16.49	14.13	13.19	11.87	10.79	9.89	9.13
		FB=	23.76	27.66	29.52	32.59	35.63	38.64	41.62
11.82	W 8X 15		YES	YES	YES	YES	YES	YES	YES
		LC=	4.23	3.92	3.79	3.59	3.42	3.28	3.15
		LU=	7.18	6.16	5.75	5.17	4.70	4.31	3.98
		FB=	23.76	27.72	29.70	33.00	36.30	39.60	42.90
10.51	W10X 11		PTY	PTY	PTY	PTY	PTY	NO	NO
		LC=	3.77	3.23	3.02	2.72	2.47	0.00	0.00
		LU=	3.77	3.23	3.02	2.72	2.47	2.26	2.09
		FB=	23.46	27.09	28.89	31.85	34.78	36.00	39.00
10.14	W 6X 16		YES	YES	YES	YES	YES	YES	YES
		LC=	4.25	3.93	3.80	3.60	3.44	3.29	3.16
		LU=	12.06	10.33	9.64	8.68	7.89	7.23	6.67
		FB=	23.76	27.72	29.70	33.00	36.30	39.60	42.90
10.10	W 6X 15		PTY	PTY	PTY	PTY	PTY	PTY	PTY
		LC=	6.33	5.86	5.66	5.37	5.12	4.90	4.71
		LU=	12.45	10.67	9.96	8.96	8.15	7.47	6.89
		FB=	23.01	26.53	28.27	31.12	33.94	36.72	39.46
9.92	W 5X 18		YES	YES	YES	YES	YES	YES	YES
		LC=	5.30	4.91	4.74	4.50	4.29	4.10	3.94
		LU=	19.08	16.35	15.26	13.74	12.49	11.45	10.56
		FB=	23.76	27.72	29.70	33.00	36.30	39.60	42.90
9.87	W 8X 13		YES	YES	PTY	PTY	PTY	PTY	PTY
		LC=	4.22	3.90	3.77	3.58	3.41	3.27	3.14
		LU=	5.87	5.03	4.70	4.23	3.84	3.52	3.25
		FB=	23.76	27.72	29.65	32.75	35.81	38.85	41.86
8.52	W 5X 16		YES	YES	YES	YES	YES	PTY	PTY
		LC=	5.27	4.88	4.72	4.47	4.26	4.08	3.92
		LU=	16.66	14.28	13.33	12.00	10.90	10.00	9.23
		FB=	23.76	27.72	29.70	33.00	36.30	39.46	42.55
7.79	W 8X 10		PTY	PTY	PTY	PTY	PTY	PTY	PTY
		LC=	4.15	3.85	3.71	3.39	3.08	2.82	2.60
		LU=	4.71	4.03	3.76	3.39	3.08	2.82	2.60
		FB=	23.46	27.10	28.90	31.87	34.80	37.69	40.56
7.23	W 6X 12		YES	YES	YES	YES	PTY	PTY	PTY
		LC=	4.22	3.90	3.77	3.58	3.41	3.27	3.14
		LU=	8.61	7.38	6.88	6.20	5.63	5.16	4.76
		FB=	23.76	27.72	29.70	33.00	36.22	39.31	42.38
5.43	W 4X 13		YES	YES	YES	YES	YES	YES	YES
		LC=	4.28	3.96	3.83	3.63	3.46	3.31	3.18
		LU=	15.58	13.36	12.47	11.22	10.20	9.35	8.63
		FB=	23.76	27.72	29.70	33.00	36.30	39.60	42.90
5.07	W 6X 8		PTY	PTY	PTY	PTY	PTY	PTY	PTY
		LC=	4.15	3.85	3.71	3.52	3.36	3.22	3.09
		LU=	6.06	5.20	4.85	4.37	3.97	3.64	3.36
		FB=	23.31	26.91	28.69	31.62	34.51	37.37	40.19

TABLE A4
Plastic Section Modulus Table

ZX	SHAPE	RY	SATISFIES AISC-2.7 (YES OR NO)						
		FY=	36.00	42.00	45.00	50.00	55.00	60.00	65.00
1664.12	W14X730	4.687	YES	YES	YES	YES	YES	YES	YES
1483.00	W14X665	4.616	YES	YES	YES	YES	YES	YES	YES
1320.70	W14X605	4.549	YES	YES	YES	YES	YES	YES	YES
1255.33	W36X300	3.727	YES	YES	YES	YES	YES	YES	YES
1177.18	W14X550	4.487	YES	YES	YES	YES	YES	YES	YES
1168.31	W36X280	3.700	YES	YES	YES	YES	YES	YES	YES
1076.74	W36X260	3.651	YES	YES	YES	YES	YES	YES	YES
1049.48	W14X500	4.429	YES	YES	YES	YES	YES	YES	YES
1008.88	W36X245	3.621	YES	YES	YES	YES	YES	YES	NO
943.48	W36X230	3.585	YES	YES	YES	YES	YES	NO	NO
938.21	W14X455	4.376	YES	YES	YES	YES	YES	YES	YES
918.44	W33X240	3.521	YES	YES	YES	YES	YES	YES	YES
869.92	W14X426	4.340	YES	YES	YES	YES	YES	YES	YES
837.09	W33X220	3.476	YES	YES	YES	YES	YES	YES	NO
802.38	W14X398	4.306	YES	YES	YES	YES	YES	YES	YES
767.22	W36X194	2.494	YES	YES	YES	YES	YES	YES	YES
755.13	W33X200	3.430	YES	YES	YES	YES	NO	NO	NO
737.48	W14X370	4.272	YES	YES	YES	YES	YES	YES	YES
734.47	W30X210	3.385	YES	YES	YES	YES	YES	YES	YES
717.13	W36X182	2.473	YES	YES	YES	YES	YES	YES	YES
673.42	W14X342	4.238	YES	YES	YES	YES	YES	YES	YES
667.39	W36X170	2.452	YES	YES	YES	YES	YES	YES	NO
660.37	W30X190	3.342	YES	YES	YES	YES	YES	NO	NO
624.02	W36X160	2.418	YES	YES	YES	YES	YES	NO	NO
611.19	W14X314	4.204	YES	YES	YES	YES	YES	YES	YES
593.96	W30X172	3.295	YES	YES	YES	NO	NO	NO	NO
592.31	W14X320	4.168	YES	YES	YES	YES	YES	YES	YES
580.41	W36X150	2.381	YES	YES	YES	YES	NO	NO	NO
558.37	W33X152	2.393	YES	YES	YES	YES	YES	YES	NO
557.02	W27X177	3.155	YES	YES	YES	YES	YES	YES	YES
551.33	W14X287	4.169	YES	YES	YES	YES	YES	YES	YES
513.27	W33X141	2.352	YES	YES	YES	YES	YES	NO	NO
509.40	W36X135	2.283	YES	YES	NO	NO	NO	NO	NO
502.47	W14X264	4.141	YES	YES	YES	YES	YES	YES	YES
500.58	W27X160	3.120	YES	YES	YES	YES	YES	NO	NO
466.19	W33X130	2.294	YES	YES	YES	YES	NO	NO	NO
464.41	W14X246	4.118	YES	YES	YES	YES	YES	YES	YES
463.70	W24X160	3.236	YES	YES	YES	YES	YES	YES	NO
452.29	W27X145	3.087	YES	YES	YES	NO	NO	NO	NO
445.33	W14X237	4.105	YES	YES	YES	YES	YES	YES	YES
437.02	W30X132	2.182	YES	YES	YES	YES	YES	YES	YES
427.11	W14X228	4.095	YES	YES	YES	YES	YES	YES	YES
416.57	W24X145	3.192	YES	YES	YES	YES	NO	NO	NO
414.64	W33X118	2.215	YES	YES	NO	NO	NO	NO	NO
408.13	W14X219	4.083	YES	YES	YES	YES	YES	YES	YES
407.88	W30X124	2.157	YES	YES	YES	YES	YES	YES	NO
391.58	W14X211	4.070	YES	YES	YES	YES	YES	YES	YES
377.73	W30X116	2.118	YES	YES	YES	YES	YES	NO	NO
373.62	W14X202	4.061	YES	YES	YES	YES	YES	YES	YES
369.20	W24X130	3.133	YES	YES	NO	NO	NO	NO	NO
357.09	W21X142	3.039	YES	YES	YES	YES	YES	YES	YES
355.33	W14X193	4.049	YES	YES	YES	YES	YES	YES	YES
346.00	W30X108	2.062	YES	YES	YES	YES	NO	NO	NO
342.92	W27X114	2.112	YES	YES	YES	YES	YES	YES	YES
337.77	W14X184	4.040	YES	YES	YES	YES	YES	YES	YES
336.67	W24X120	2.682	YES	YES	YES	YES	YES	NO	NO
321.01	W14X176	4.024	YES	YES	YES	YES	YES	YES	YES

Table A4 (*continued*)

ZX	SHAPE	RY			SATISFIES	AISC-2.7	(YES OR NO)		
		FY=	36.00	42.00	45.00	50.00	55.00	60.00	65.00
317.99	W21X127	3.011	YES	YES	YES	YES	NO	NO	NO
312.19	W30X 99	2.003	YES	YES	NO	NO	NO	NO	NO
311.57	W12X190	3.249	YES	YES	YES	YES	YES	YES	YES
307.91	W24X110	2.660	YES	YES	YES	NO	NO	NO	NO
304.67	W27X102	2.077	YES	YES	YES	YES	YES	YES	NO
303.26	W14X167	4.012	YES	YES	YES	YES	YES	YES	NO
286.36	W14X158	4.003	YES	YES	YES	YES	YES	NO	NO
278.71	W24X100	2.629	YES	YES	NO	NO	NO	NO	NO
278.20	W21X112	2.966	YES	YES	NO	NO	NO	NO	NO
277.92	W27X 94	2.040	YES	YES	YES	YES	NO	NO	NO
270.45	W14X150	3.992	YES	YES	YES	YES	NO	NO	NO
259.35	W12X161	3.203	YES	YES	YES	YES	YES	YES	YES
254.74	W14X142	3.971	YES	YES	YES	NO	NO	NO	NO
252.87	W24X 94	1.923	YES	YES	YES	YES	YES	YES	YES
247.67	W18X114	2.761	YES	YES	YES	YES	YES	YES	YES
243.37	W27X 84	1.967	YES	YES	NO	NO	NO	NO	NO
242.87	W14X136	3.768	YES	YES	YES	YES	NO	NO	NO
226.76	W18X105	2.735	YES	YES	YES	YES	YES	NO	NO
226.43	W21X 96	1.968	YES	YES	YES	YES	YES	YES	YES
226.10	W14X127	3.759	YES	YES	YES	NO	NO	NO	NO
224.27	W24X 84	1.890	YES	YES	YES	YES	YES	YES	NO
211.18	W14X119	3.749	YES	YES	NO	NO	NO	NO	NO
209.92	W12X133	3.157	YES	YES	YES	YES	YES	YES	YES
206.16	W18X 96	2.707	YES	YES	YES	NO	NO	NO	NO
200.50	W24X 76	1.849	YES	YES	YES	YES	YES	NO	NO
195.84	W14X111	3.732	YES	NO	NO	NO	NO	NO	NO
191.63	W21X 82	1.928	YES	YES	YES	YES	YES	YES	YES
186.70	W12X120	3.126	YES	YES	YES	YES	YES	YES	YES
186.12	W16X 96	2.709	YES	YES	YES	YES	YES	NO	NO
181.04	W14X103	3.724	NO	NO	NO	NO	NO	NO	NO
177.60	W18X 85	1.995	YES	YES	YES	YES	YES	YES	YES
175.68	W24X 68	1.786	YES	YES	NO	NO	NO	NO	NO
172.22	W21X 73	1.756	YES	YES	YES	YES	YES	YES	YES
169.07	W16X 88	2.675	YES	YES	YES	NO	NO	NO	NO
166.08	W14X 95	3.705	NO	NO	NO	NO	NO	NO	NO
163.71	W12X106	3.106	YES	YES	YES	YES	YES	YES	NO
160.49	W18X 77	1.978	YES	YES	YES	YES	YES	YES	YES
159.77	W21X 68	1.736	YES	YES	YES	YES	YES	YES	NO
151.88	W12X 99	3.092	YES	YES	YES	YES	NO	NO	NO
151.58	W24X 61	1.376	YES	YES	YES	YES	NO	NO	NO
151.57	W14X 87	3.698	NO	NO	NO	NO	NO	NO	NO
147.69	W10X112	2.674	YES	YES	YES	YES	YES	YES	YES
145.55	W16X 78	1.953	YES	YES	YES	YES	YES	YES	YES
145.39	W14X 84	3.020	YES	YES	NO	NO	NO	NO	NO
144.80	W18X 70	1.953	YES	YES	YES	YES	YES	YES	YES
144.24	W21X 62	1.706	YES	YES	YES	YES	NO	NO	NO
140.41	W12X 92	3.078	YES	YES	YES	NO	NO	NO	NO
134.04	W14X 78	3.003	YES	NO	NO	NO	NO	NO	NO
133.77	W24X 55	1.332	YES	YES	YES	NO	NO	NO	NO
131.82	W18X 64	1.933	YES	YES	YES	YES	YES	NO	NO
131.64	W16X 71	1.932	YES	YES	YES	YES	YES	YES	YES
130.26	W10X100	2.649	YES	YES	YES	YES	YES	YES	YES
129.30	W12X 85	3.070	YES	YES	NO	NO	NO	NO	NO
125.71	W14X 74	2.476	YES	YES	YES	YES	YES	NO	NO
125.52	W21X 55	1.649	YES	YES	NO	NO	NO	NO	NO
122.83	W18X 60	1.634	YES	YES	YES	YES	YES	YES	YES
119.31	W12X 79	3.052	YES	NO	NO	NO	NO	NO	NO
117.86	W16X 64	1.907	YES	YES	YES	YES	YES	YES	YES
114.87	W14X 68	2.461	YES	YES	YES	YES	NO	NO	NO
114.54	W10X 89	2.625	YES	YES	YES	YES	YES	YES	YES

Table A4 (*continued*)

ZX	SHAPE	RY	SATISFIES AISC-2.7 (YES OR NO)						
		FY=	36.00	42.00	45.00	50.00	55.00	60.00	65.00
111.79	W18X 55	1.610	YES	YES	YES	YES	YES	YES	YES
108.43	W21X 49	1.303	YES	YES	YES	YES	NO	NO	NO
108.14	W12X 72	3.038	NO	NO	NO	NO	NO	NO	NO
106.11	W16X 58	1.884	YES	YES	YES	YES	YES	NO	NO
102.41	W14X 61	2.445	YES	YES	NO	NO	NO	NO	NO
101.03	W18X 50	1.590	YES	YES	YES	YES	YES	NO	NO
97.91	W10X 77	2.601	YES	YES	YES	YES	YES	YES	YES
97.12	W12X 65	3.022	NO	NO	NO	NO	NO	NO	NO
95.33	W21X 44	1.261	YES	YES	YES	NO	NO	NO	NO
91.67	W16X 50	1.538	YES	YES	YES	YES	YES	YES	YES
90.74	W10X 72	2.587	YES	YES	YES	YES	YES	YES	NO
89.70	W18X 45	1.552	YES	YES	NO	NO	NO	NO	NO
87.25	W14X 53	1.920	YES	YES	YES	YES	YES	YES	NO
86.57	W12X 58	2.509	YES	YES	NO	NO	NO	NO	NO
82.94	W10X 66	2.579	YES	YES	YES	YES	NO	NO	NO
82.08	W16X 45	1.517	YES	YES	YES	YES	YES	YES	NO
78.47	W14X 48	1.906	YES	YES	YES	YES	NO	NO	NO
78.34	W18X 40	1.271	YES	YES	YES	YES	NO	NO	NO
78.17	W12X 53	2.482	NO	NO	NO	NO	NO	NO	NO
75.02	W10X 60	2.568	YES	YES	YES	NO	NO	NO	NO
72.78	W16X 40	1.500	YES	YES	YES	YES	NO	NO	NO
72.61	W12X 50	1.958	YES	YES	YES	YES	YES	NO	NO
70.23	W 8X 67	2.120	YES	YES	YES	YES	YES	YES	YES
69.88	W14X 43	1.888	YES	YES	NO	NO	NO	NO	NO
67.17	W10X 54	2.557	YES	NO	NO	NO	NO	NO	NO
66.76	W18X 35	1.222	YES	YES	YES	NO	NO	NO	NO
64.97	W12X 45	1.943	YES	YES	YES	YES	NO	NO	NO
63.89	W16X 36	1.444	YES	NO	NO	NO	NO	NO	NO
61.52	W14X 38	1.484	YES	YES	YES	YES	NO	NO	NO
60.43	W10X 49	2.541	NO	NO	NO	NO	NO	NO	NO
59.78	W 8X 58	2.095	YES	YES	YES	YES	YES	YES	YES
57.65	W12X 40	1.935	YES	YES	NO	NO	NO	NO	NO
55.03	W10X 45	2.004	YES	YES	YES	YES	YES	NO	NO
54.58	W14X 34	1.459	YES	YES	NO	NO	NO	NO	NO
53.92	W16X 31	1.126	YES	YES	YES	YES	NO	NO	NO
51.53	W12X 36	1.495	YES	YES	YES	YES	YES	YES	NO
49.00	W 8X 48	2.077	YES	YES	YES	YES	YES	YES	YES
47.11	W14X 30	1.409	NO	NO	NO	NO	NO	NO	NO
47.00	W10X 39	1.977	YES	YES	NO	NO	NO	NO	NO
44.07	W12X 31	1.473	YES	YES	YES	NO	NO	NO	NO
43.90	W16X 26	1.067	YES	YES	NO	NO	NO	NO	NO
39.97	W14X 26	1.039	YES	YES	YES	YES	YES	NO	NO
39.85	W 8X 40	2.041	YES	YES	YES	NO	NO	NO	NO
38.91	W10X 33	1.938	NO	NO	NO	NO	NO	NO	NO
38.05	W12X 27	1.443	YES	NO	NO	NO	NO	NO	NO
34.71	W 8X 35	2.031	YES	NO	NO	NO	NO	NO	NO
34.64	W10X 29	1.334	YES	YES	YES	YES	YES	YES	YES
33.00	W14X 22	.994	YES	YES	NO	NO	NO	NO	NO
30.39	W 8X 31	2.014	NO	NO	NO	NO	NO	NO	NO
29.60	W10X 25	1.314	YES	YES	YES	YES	NO	NO	NO
29.27	W12X 22	.838	YES	YES	YES	YES	YES	YES	YES
27.16	W 8X 28	1.620	YES	YES	YES	NO	NO	NO	NO
24.82	W12X 19	.808	YES	YES	YES	YES	YES	YES	YES
24.11	W10X 21	1.251	YES	NO	NO	NO	NO	NO	NO
23.12	W 8X 24	1.605	YES	NO	NO	NO	NO	NO	NO
21.58	W10X 19	.864	YES	YES	YES	YES	YES	YES	YES
20.59	W12X 16	.757	YES	YES	NO	NO	NO	NO	NO
19.09	W 8X 20	1.202	YES	YES	YES	YES	NO	NO	NO
19.01	W 6X 25	1.523	YES	YES	YES	YES	NO	NO	NO
18.63	W10X 17	.832	YES	YES	YES	YES	YES	YES	NO

Table A4 (*continued*)

ZX	SHAPE	RY	SATISFIES AISC-2.7 (YES OR NO)						
		FY=	36.00	42.00	45.00	50.00	55.00	60.00	65.00
17.36	W12X 14	.737	NO	NO	NO	NO	NO	NO	NO
15.96	W10X 15	.796	YES	YES	NO	NO	NO	NO	NO
15.83	W 8X 17	1.159	NO	NO	NO	NO	NO	NO	NO
15.02	W 6X 20	1.501	YES	NO	NO	NO	NO	NO	NO
13.56	W 8X 15	.863	YES	YES	YES	YES	YES	NO	NO
12.18	W10X 11	.770	NO	NO	NO	NO	NO	NO	NO
11.58	W 6X 16	.956	YES	YES	YES	YES	YES	YES	YES
11.38	W 8X 13	.827	YES	YES	NO	NO	NO	NO	NO
11.35	W 5X 18	1.277	YES	YES	YES	YES	YES	YES	YES
11.27	W 6X 15	1.448	NO	NO	NO	NO	NO	NO	NO
9.62	W 5X 16	1.264	YES	YES	YES	YES	NO	NO	NO
8.83	W 8X 10	.821	NO	NO	NO	NO	NO	NO	NO
8.22	W 6X 12	.904	YES	YES	YES	NO	NO	NO	NO
6.27	W 4X 13	.992	YES	YES	YES	YES	YES	YES	YES
5.6.9	W 6X 8	.869	NO	NO	NO	NO	NO	NO	NO

Index